Lecture Notes in Computer Science 2396
Edited by G. Goos, J. Hartmanis, and J. van Leeuwen

AF828258

Springer
Berlin
Heidelberg
New York
Barcelona
Hong Kong
London
Milan
Paris
Tokyo

Terry Caelli Adnan Amin
Robert P.W. Duin Mohamed Kamel
Dick de Ridder (Eds.)

Structural, Syntactic, and Statistical Pattern Recognition

Joint IAPR International Workshops
SSPR 2002 and SPR 2002
Windsor, Ontario, Canada, August 6-9, 2002
Proceedings

Springer

Volume Editors

Terry Caelli
University of Alberta, Dept. of Computing Science
Athabasca Hall, Room 409, Edmonton, Alberta, Canada T6G 2H1
E-mail: tcaelli@ualberta.ca

Adnan Amin
University of New South Wales, School of Computer Science and Engineering
Sydney 2052, NSW, Australia
E-mail: cse.unsw.edu.au

Robert P.W. Duin
Dick de Ridder
Delft University of Technology, Dept. of Applied Physics
Pattern Recognition Group, Lorentzweg 1, 2628 CJ Delft, The Netherlands
E-mail: {duin,dick}@ph.tn.tudelft.nl

Mohamed Kamel
University of Waterloo, Dept. of Systems Design Engineering
Waterloo, Ontario, Canada N2L 3G1
E-mail: mkamel@uwaterloo.ca

Cataloging-in-Publication Data applied for

Die Deutsche Bibliothek - CIP-Einheitsaufnahme

Advances in pattern recognition : joint IAPR international workshops ; proceedings / SSPR 2002 and SPR 2002, Windsor, Ontario, Canada, August 6 - 9, 2002. Terry Caelli ... (ed.). - Berlin ; Heidelberg ; New York ; Barcelona ; Hong Kong ; London ; Milan ; Paris ; Tokyo : Springer, 2002
(Lecture notes in computer science ; Vol. 2396)
ISBN 3-540-44011-9

CR Subject Classification (1998): I.5, I.4, I.2.10, I.2, G.3

ISSN 0302-9743
ISBN 3-540-44011-9 Springer-Verlag Berlin Heidelberg New York

This work is subject to copyright. All rights are reserved, whether the whole or part of the material is concerned, specifically the rights of translation, reprinting, re-use of illustrations, recitation, broadcasting, reproduction on microfilms or in any other way, and storage in data banks. Duplication of this publication or parts thereof is permitted only under the provisions of the German Copyright Law of September 9, 1965, in its current version, and permission for use must always be obtained from Springer-Verlag. Violations are liable for prosecution under the German Copyright Law.

Springer-Verlag Berlin Heidelberg New York
a member of BertelsmannSpringer Science+Business Media GmbH

http://www.springer.de

© Springer-Verlag Berlin Heidelberg 2002

Typesetting: Camera-ready by author, data conversion by DA-TeX Gerd Blumenstein
Printed on acid-free paper SPIN 10873552 06/3142 5 4 3 2 1 0

Preface

This volume contains all papers presented at SSPR 2002 and SPR 2002 hosted by the University of Windsor, Windsor, Ontario, Canada, August 6-9, 2002.

This was the third time these two workshops were held back-to-back. SSPR was the ninth International Workshop on Structural and Syntactic Pattern Recognition and the SPR was the fourth International Workshop on Statistical Techniques in Pattern Recognition. These workshops have traditionally been held in conjunction with ICPR (International Conference on Pattern Recognition), and are the major events for technical committees TC2 and TC1, respectively, of the International Association of Pattern Recognition (IAPR).

The workshops were held in parallel and closely coordinated. This was an attempt to resolve the dilemma of how to deal, in the light of the progressive specialization of pattern recognition, with the need for narrow-focus workshops without further fragmenting the field and introducing yet another conference that would compete for the time and resources of potential participants.

A total of 116 papers were received from many countries with the submission and reviewing processes being carried out separately for each workshop. A total of 45 papers were accepted for oral presentation and 35 for posters. In addition four invited speakers presented informative talks and overviews of their research. They were:

Tom Dietterich, Oregon State University, USA
Sven Dickinson, the University of Toronto, Canada
Edwin Hancock, University of York, UK
Anil Jain, Michigan State University, USA

SSPR 2002 and SPR 2002 were sponsored by the IAPR and the University of Windsor.

We would like to thank our sponsors and, in particular, the members of the program committees of both workshops for performing the hard work of reviewing the many submissions which led to a selection of high quality papers.

Special thanks to our host, Majid Ahmadi, and his colleagues, for running the event smoothly. Moreover, special thanks to Sue Wu for helping prepare the proceedings.

We also appreciate the help of the editorial staff at Springer-Verlag and, in particular, Alfred Hofmann, for supporting this publication in the LNCS series.

August 2002

Terry Caelli
Adnan Amin
Bob Duin
Mohamed Kamel
Dick de Ridder

SSPR and SPR 2002

General Chairman

Terry Caelli

Dept. of Computing Science
University of Alberta
Alberta, Canada
tcaelli@ualberta.ca

Local Chairman

Majid Ahmadi

Dept. of Electrical and Computer Engineering
University of Windsor, Canada
ahmadi@uwindsor.ca

Conference Information Technology Manager

Dick de Ridder

Faculty of Applied Sciences
Delft University of Technology, The Netherlands
dick@ph.tn.tudelft.nl

Supported by

International Association of Pattern Recognition

SSPR Committee

Co-chairmen

Adnan Amin

School of Computer Science and Engineering
University of New South Wales
Sydney, Australia
amin@cse.unsw.edu.au

Terry Caelli

Dept. of Computing Science
University of Alberta
Alberta, Canada
tcaelli@ualberta.ca

Program Committee

K. Abe (Japan)
W. Bischof (Canada)
K. Boyer (USA)
H. Bunke (Switzerland)
F. Casacuberta (Spain)
S. Dickinson (Canada)
I. Dinstein (Israel)
A. Fred (Portugal)
G. Gimel'farb (N.Zealand)
E. Hancock (UK)
R. Haralick (USA)
J. I nesta (Spain)
J. Jolion (France)
W. Kropatsch (Austria)
B. Lovell (Australia)
J. Oommen (Canada)
P. Perner (Germany)
A. Sanfeliu (Spain)
G. Sanniti di Baja (Italy)
K. Tombre (France)
S. Venkatesh (Australia)

SPR Committee

Co-chairmen

Robert P.W. Duin
Faculty of Applied Sciences
Delft University of Technology
Delft, The Netherlands
duin@ph.tn.tudelft.nl

Mohamed Kamel
Dept. of Systems Design Engineering
University of Waterloo
Waterloo, Ontario, Canada
mkamel@uwaterloo.ca

Program Committee

V. Brailovsky (Israel)
L. P. Cordella (Italy)
B. V. Dasarathy (USA)
F. J. Ferri (Spain)
J. Ghosh (USA)
M. Gilloux (France)
T. M. Ha (Switzerland)
J-P. Haton (France)
T. K. Ho (USA)
A. K. Jain (USA)
J. Kittler (UK)
M. Kudo (Japan)
L. Kuncheva (UK)
L. Lam (Hong Kong)
J. Novovicova (Czech Rep.)
E. Nyssen (Belgium)
P. Pudil (Czech Rep.)
S. Raudys (Lithuania)
P. Rocket (UK)
F. Roli (Italy)
S. Singh (UK)
C. Y. Suen (Canada)

Reviewers

The program committees for both SSPR and SPR were kindly assisted by the following reviewers:

R. Alquezar (Spain)
A. Belaid (France)
M. Berger (France)
G. Boccignone (Italy)
F. Corato (Spain)
K. Daoudi (France)
C. de la Higuera (Spain)
D. de Ridder (The Netherlands)
C. De Stefano (Italy)
D. Deugo (Spain)
P. Fieguth (Canada)
P. Foggia (Italy)
J. Grim (Czech Rep.)
P. Juszczak (The Netherlands)
B. Miner (Canada)
F. Palmieri (Italy)
E. Pekalska (The Netherlands)
S. Rajan (USA)
P. Rockett (UK)
L. Rueda (Canada)
F. Serratosa (Spain)
M. Skurichina (The Netherlands)
P. Somol (UK)
A. Strehl (USA)
F. Tortorella (Italy)
N. Wanas (Canada)
S. Wesolkowski (USA)
A. Whitehead (Canada)
S. Zhong (USA)

Table of Contents

Documents and OCR

Image Shape Analysis and Application

Poster Papers

Graphs, Strings, Grammars and Language

Image Analysis and Feature Extraction

Documents and OCR

SPR

Density Estimation and Distribution Models

Multi-classifiers and Fusion

Feature Extraction and Selection

General Methodology

Documents and OCR

Image Shape Analysis and Application

Poster Papers

Face Recognition

Speech and Multimedia

Data and Cluster Analysis

Classification

Image Analysis and Vision

Spectral Methods for View-Based 3-D Object Recognition Using Silhouettes

Diego Macrini[1], Ali Shokoufandeh[2], Sven Dickinson[3],
Kaleem Siddiqi[4], and Steven Zucker[5]

[1] Department of Computer Science, University of Toronto
[2] Department of Mathematics and Computer Science, Drexel University
[3] Department of Computer Science, University of Toronto
[4] Centre for Intelligent Machines
School of Computer Science, McGill University
[5] Center for Computational Vision and Control, Yale University

Abstract. The shock graph is an emerging shape representation for object recognition, in which a 2-D silhouette is decomposed into a set of qualitative parts, captured in a directed acyclic graph. Although a number of approaches have been proposed for shock graph matching, these approaches do not address the equally important indexing problem. We extend our previous work in both shock graph matching and hierarchical structure indexing to propose the first unified framework for view-based 3-D object recognition using shock graphs. The heart of the framework is an improved spectral characterization of shock graph structure that not only drives a powerful indexing mechanism (to retrieve similar candidates from a large database), but also drives a matching algorithm that can accommodate noise and occlusion. We describe the components of our system and evaluate its performance using both unoccluded and occluded queries. The large set of recognition trials (over 25,000) from a large database (over 1400 views) represents one of the most ambitious shock graph-based recognition experiments conducted to date. This paper represents an expanded version of [12].

1 Introduction

There are two approaches to 3-D object recognition. One assumes a 3-D object-centered model, and attempts to match 2-D image features to viewpoint-invariant 3-D model features, e.g., [2,11,7]. Over the last decade, this approach has given way to a viewer-centered approach, where the 3-D model is replaced by a collection of 2-D views. These views can be represented in terms of segmented features, such as lines or regions, e.g., [22], or in terms of the photometric "appearance" of the object, e.g., [21,13]. Although these latter, appearance-based recognition schemes have met with great success, it must be understood that they address the task of exemplar-based recognition. When faced with novel exemplars belonging to known classes, they simply do not scale up.

To achieve such categorical, or generic, object recognition requires a representation that is invariant to within-class shape deformations. One such powerful

T. Caelli et al. (Eds.): SSPR&SPR 2002, LNCS 2396, pp. 1–14, 2002.
© Springer-Verlag Berlin Heidelberg 2002

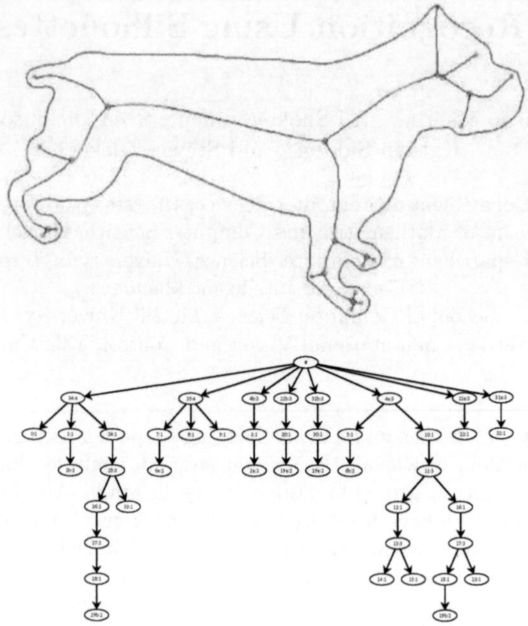

Fig. 1. A two-dimensional shape and its corresponding shock graph. The nodes represent groups of singularities (shocks) along with their geometric attributes. The edges are between adjacent shock groups and in a direction opposite to Blum's grassfire flow [19]

representation is offered by the shock graph [20], which represents the silhouette of an object in terms of a set of qualitatively defined parts, organized in a hierarchical, directed acyclic graph. Figure 1 illustrates an example of a two-dimensional shape, its shocks (singularities), and the resulting shock graph. In previous work, we introduced the first algorithm for matching two shock graphs, and showed that it could be used to recognize novel exemplars from known classes [19]. Since then, other approaches to shock graph matching have emerged, including [14] and [15]. However, earlier approaches, including our own, have not been extensively tested on noisy graphs, occluded scenes, or cluttered scenes.

A shock graph representation of shape suggests the use of graph matching techniques for shape recognition. However, matching (graph or otherwise) is only half the problem. Without an effective *indexing* mechanism with which to narrow a large database down to a small number of candidates, recognition degenerates to matching a query to each model in the database. In the case of view-based object recognition, in which a large number of objects map to an even larger number of views, such a linear search is intractable. Unfortunately, very few researchers, in either the computer vision or graph algorithms communities, have addressed the important problem of graph indexing. How, then, can we exploit the power of the shock graph to perform view-based object recognition?

In recent work, we introduced a novel indexing method which maps the structure of a directed acyclic graph to a point in low-dimensional space [18]. This same mapping, in fact, was used as the basis for our shock graph matching algorithm [19]. Using standard, nearest-neighbor search methods, this compact, structural signature was used to retrieve structurally similar candidates from a database. The highest scoring candidates, in turn, were compared to the query using our matching algorithm, with the "closest" candidate used to "recognize" the object. Our experiments showed that the target ranked highly among the candidates, even in the presence of noise and occlusion.

Armed with a unified approach to the indexing and matching of graphs, we now turn to the problem of view-based object recognition using shock graphs. In fact, we are not the first to apply shock graphs to this problem. In recent work, Cyr and Kimia [3,15] explore the important problem of how to partition the view sphere of a 3-D object using a collection of shock graphs. However, they do not address the shock graph indexing problem, resorting to a linear search of all views in the database in order to recognize an object. Even for small object databases, the number of views required per object renders this approach intractable. In this paper, we unify our shock graph indexing and matching techniques to yield a novel, effective method for view-based 3-D object recognition.

2 A Compact Encoding of Graph Structure

In [19], we introduced a transformation mapping the structure of a directed acyclic graph to a point in low-dimensional space. As mentioned earlier, this mapping was the heart of an algorithm for matching two *shock trees*, derivable from shock graphs in linear time. This same transformation later gave rise to an indexing mechanism, which used the low-dimensional, structural signature of a shock tree to select structurally similar candidates from a database of shock trees [18]. In this latter paper, we analyzed the stability of a tree's signature to certain restricted classes of perturbations.

In a recent paper on matching multi-scale image decompositions, we have strengthened this encoding from undirected, unique rooted trees to directed acyclic graphs, yielding a more powerful characterization of graph structure [17]. This new formulation has led to a broader stability analysis that accommodates *any* graph perturbation in terms of node addition and/or deletion. Furthermore, we extend our matching algorithm to deal with directed acyclic graphs rather than undirected, unique rooted trees. Due to space constraints, we will summarize our new encoding of graph structure; details of the new encoding, as well as an analysis of its stability can be found in [17].

To encode the structure of a DAG, we turn to the domain of eigenspaces of graphs, first noting that any graph can be represented as an antisymmetric $\{0, 1, -1\}$ adjacency matrix, with 1's (-1's) indicating a forward (backward) edge between adjacent nodes in the graph (and 0's on the diagonal). The eigenvalues of a graph's adjacency matrix encode important structural properties of the graph, and are stable under minor perturbations in structure. Our goal, therefore, is to map the eigenvalues of a DAG to a point in some low-dimensional space, providing a stable, compact encoding of structure.

Specifically, let T be a DAG whose maximum branching factor is $\Delta(T)$, and let the subgraphs of its root be $T_1, T_2, \ldots, T_{\delta(T)}$, as shown in Figure 2. For each subgraph, T_i, whose root degree is $\delta(T_i)$, we compute[1] the magnitudes of the eigenvalues of T_i's submatrix, sort them in decreasing order by absolute value, and let S_i be the sum of the $\delta(T_i) - 1$ largest absolute values. The sorted S_i's become the components of a $\Delta(T)$-dimensional vector assigned to the DAG's root. If the number of S_i's is less than $\Delta(T)$, then the vector is padded with zeroes. We can recursively repeat this procedure, assigning a vector to each nonterminal node in the DAG, computed over the subgraph rooted at that node. We call each such vector a *topological signature vector*, or TSV. The details of this transformation, the motivation for each step, and an evaluation of its properties is given in [17].

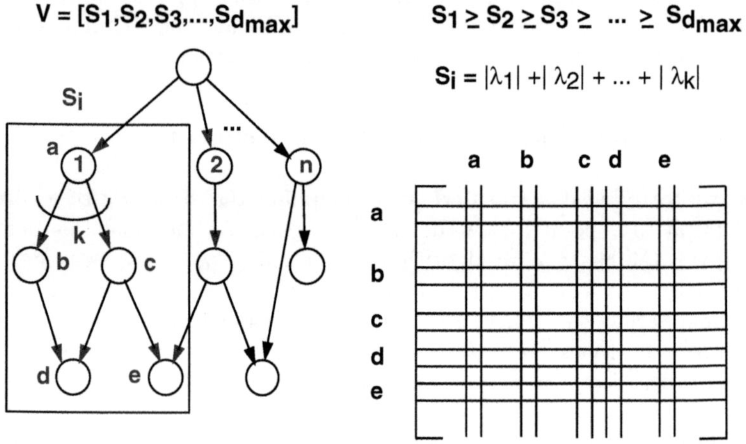

Fig. 2. Forming the Topological Signature Vector (TSV) for the root. For a given DAG rooted at a child (e.g., a) of the root, compute the sum of the magnitudes of the k largest eigenvalues (k is the out-degree of a) of the adjacency submatrix defining the DAG rooted at a. The sorted sums, one per child of the root, define the components of the TSV assigned to the root. The process can be repeated, defining a TSV for each non-leaf node in the DAG. The dimensionality of a shock graph's TSV's is equal to the maximum branching factor of the shock graph and not the size of the graph

[1] We use SVD to compute the magnitudes of the eigenvalues.

3 Shock Graph Indexing

Given a query shape, represented by a shock graph, the goal of indexing is to efficiently retrieve, from a large database, similar shock graphs that might account for the query or some portion thereof (in the case of an occluded query or a query representing a cluttered scene). These *candidate* model graphs will then be compared directly with the query, i.e., *verified*, to determine which candidate model best accounts for the query. We therefore seek an effective index for shock graph recognition that possesses a number of important properties, including:

1. low dimensionality
2. captures both local and global structural properties
3. low ambiguity
4. stable to minor perturbations of graph structure
5. efficiently computed

Our topological signature vector, in fact, satisfies these five criteria. Its dimensionality is bounded by the graph's maximum branching factor, *not* the size of the graph (criteria 1); for shock graphs, the branching factor is typically low (< 5). TSV's for nodes high in the graph capture global structure while lower nodes capture local structure (criteria 2). The components of a node's vector are based on summing the largest eigenvalues of its subgraph's adjacency submatrix. Although our dimensionality-reducing summing operation has cost us some uniqueness, our partial sums still have very low ambiguity (criteria 3).[2] From our improved sensitivity analysis, described in [17], we have shown our index to be stable to minor perturbations of the DAG's structure (criterion 4). Moreover, as shown in [19], these sums can be computed even more efficiently (criterion 5) than the eigenvalues themselves. The vector labeling of all DAGs isomorphic to T not only has the same vector labeling, but spans the same subspace in $R^{\Delta(T)-1}$. Moreover, this extends to any DAG which has a subgraph isomorphic to a subgraph of T.

3.1 A Database for Model DAGs

Our spectral characterization of a DAG's structure suggests that a model DAG's structure can be represented as a vector in δ-dimensional space, where δ is an upper bound on the degree of any vertex of any image or model DAG. If we could assume that an image DAG represents a properly segmented, unoccluded object, then the TSV computed at the query DAG's root, could be compared with those topological signature vectors representing the roots of the model DAGs. The vector distance between the image DAG's root TSV and a model

[2] Moreover, if p is the probability that a query graph and a model graph have different structure but are isospectral, then the probability that the k vectors corresponding to the query graph's nodes are identical to the k vectors corresponding to the model graph's nodes is p^k. This suggests the use of a collection of indexes rather than a single index, as will be discussed later.

DAG's root TSV would be inversely proportional to the structural similarity of their respective DAGs, as finding two subgraphs with "close" eigenvalue sums represents an approximation to finding the largest subgraph isomorphism.

Unfortunately, this simple framework cannot support either cluttered scenes or large occlusion, both of which result in the addition or deletion of significant structure. In either case, altering the structure of the DAG will affect the TSV's computed at its nodes. The signatures corresponding to the roots of those subgraphs (DAGs) that survive the occlusion will not change. However, the signature of the root of a subgraph that has undergone any perturbation will change which, in turn, will affect the signatures of any of its ancestor nodes, including the root of the entire DAG. We therefore cannot rely on indexing solely with the root's signature. Instead, we will exploit the local subgraphs that survive the occlusion.

We can accommodate such perturbations through a local indexing framework analogous to that used in a number of geometric hashing methods, e.g., [9,5]. Rather than storing a model DAG's root signature, we will store the signatures of *each* node in the model DAG, as shown in Figure 3. At each such point (node signature) in the database, we will associate a pointer to the object model containing that node as well as a pointer to the corresponding node in the model DAG (allowing access to node label information). Since a given model subgraph can be shared by other model DAGs, a given signature (or location in δ-dimensional space) will point to a list of (model object, model node) ordered pairs. At runtime, the signature at each node in the query DAG becomes a separate index, with each nearby candidate in the database "voting" for one or more (model object, model node) pairs. Nearby candidates are retrieved using a nearest neighbor retrieval method, described in [8].

3.2 Accumulating Local Evidence

Each node in the query DAG will generate a set of (model object, model node) votes. To collect these votes, we set up an accumulator with one bin per model object, as shown in Figure 4. Furthermore, we can weight the votes that we add to the accumulator according to two important factors. Given a query node and a model node (retrieved from the database),

1. we weight the vote according to the distance between their respective TSV's – the closer the signatures, the more weight the vote gets.
2. we weight the vote according to the complexity of its corresponding subgraph, allowing larger and more complex subgraphs (or "parts") to have higher weight. This can be easily accommodated within our eigenvalue framework, for the richer the structure, the larger its maximum eigenvalue:

   Theorem 1 (Lovász and Pelikán [10]) *Among the graphs with n vertices, the star graph ($K_{1,n-1}$), has the largest eigenvalue ($\sqrt{n-1}$), while the path on n nodes (P_n) has the smallest eigenvalue* ($2\cos\pi/(n+1)$).

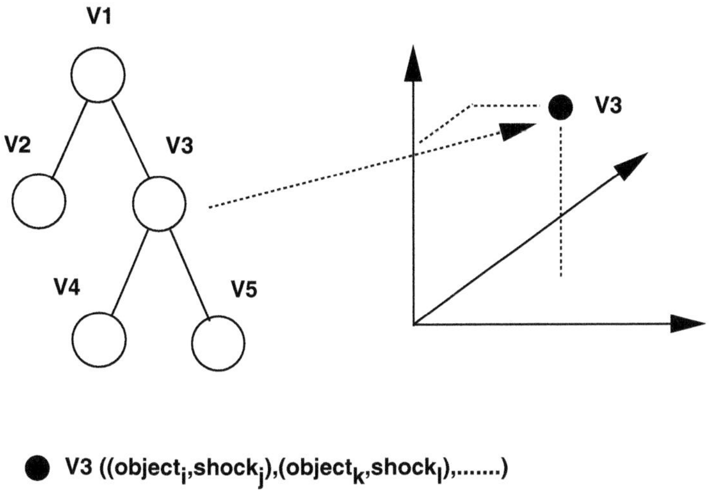

Fig. 3. Populating the Database. For every non-leaf node, n_i, in every model view shock graph, insert into a point database at the location defined by n_i's TSV the label of n_i and the object (model and view) that contains n_i. If multiple nodes collide at the same location, then maintain a list of objects (or "parts") that share the TSV

Since the size of the eigenvalues, and hence their sum, is proportional to both the branching factor (in some sense, richness or information content) as well as the number of nodes, the magnitude of the signature is used to weight the vote.

Before assembling the components of our weighting function, we must address one final issue. Namely, for a query point q and neighboring model point m, we would like to increase the weight of the vote for an object model M if m represents a larger proportion of M. Similarly, we would like to increase the weight of the vote for M if q represents a larger proportion of the query. Equivalently, we favor models that can cover a larger proportion of the image, while at the same time, we favor models which have a larger proportion covered by the query. These two goals are in direct competition, and their relative merit is a function of the task domain. Our weighting function can now be specified as:

$$W = \frac{(1-\omega)||q||}{T_q(1+||m-q||)} + \frac{\omega||m||}{T_m(1+||m-q||)} \qquad (1)$$

where q is the TSV of the query DAG node, m the TSV of the model DAG node (that is sufficiently close), T_q and T_m are the sums of the TSV norms of the entire query and model DAGs, respectively, and convexity parameter $\omega, 0 \leq \omega \leq 1$ is the weighting affecting the roles of the opposing goals described above. The first

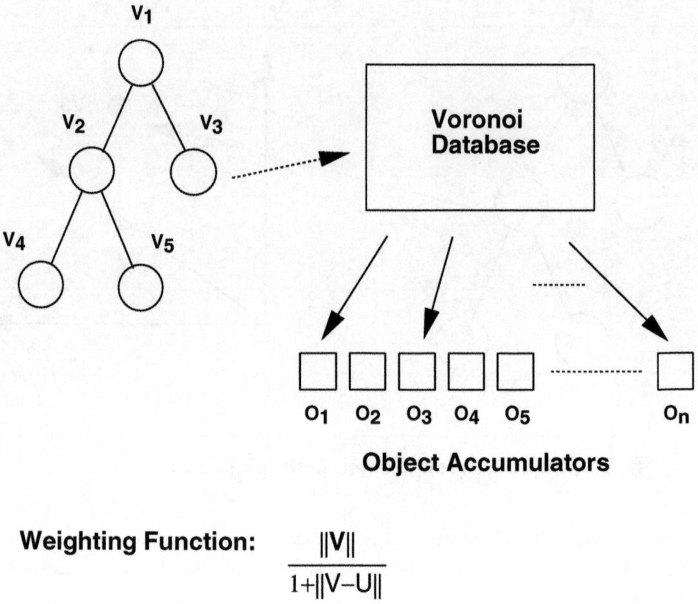

Fig. 4. Accumulating Evidence for Candidate Models. For each non-leaf node in the query DAG, find the nearest neighbors in the database. Each nearest neighbor defines a list of objects which contain that part (DAG). For each object whose nodel label (of the root of the DAG defining the TSV) matches that of the query, accumulate evidence for that model. In general, evidence is weighted proportionally to the size and complexity of the part and inversely proportionally to the distance between the query and neighbor

term favors models that cover a larger proportion of the image, while the second favors models with more nodes accounted for.

Once the evidence accumulation is complete, those models whose support is sufficiently high are selected as candidates for verification. The bins can, in effect, be organized in a heap, requiring a maximum of $O(\log k)$ operations to maintain the heap when evidence is added, where k is the number of non-zero object accumulators. Once the top-scoring models have been selected, they must be individually verified according to the matching algorithm described in the next section.

4 Shock Graph Matching

Our spectral characterization of graph structure forms the backbone of our indexing mechanism, as described in the previous section. Moreover, this *same* spectral characterization forms the backbone of our matching algorithm, thereby unifying the mechanisms of indexing and matching [16]. In previous work [19], we showed that a shock graph could be transformed into a unique rooted undirected *shock tree* in linear time. We introduced a novel algorithm for computing the distance between two shock trees (including correspondence) in the presence of noise and occlusion. As mentioned earlier, we have strengthened our indexing and matching framework to include directed acyclic graphs.

We will now briefly describe the algorithm; details can be found in [17]. Having already computed the TSV at each node in both the query graph as well as the model graph (at compile time), we can use this information to compute node correspondence. Specifically, we set up a bipartite graph spanning the nodes of the query graph and the nodes of the model graph. The edge weights in the bipartite graph will be a function of both the structural similarity of the directed acyclic subgraphs rooted at these nodes and the similarity of the nodes' contents. We then compute a maximum cardinality, maximum weight matching in the bipartite graph, leading to a selection of edges defining the final node correspondence.

This procedure, unfortunately, will not enforce the hierarchical constraints imposed by a shock graph, allowing inversions in the computed correspondence.[3] Instead, we take a greedy approach and select only the best edge in the bipartite solution to add to the correspondence set. We then recursively continue, computing the matching between the subgraphs rooted at these two nodes and adding its best edge to the solution set. The maximum weight maximum cardinality matching is based on an objective function that measures the quality of the correspondence between matched regions while penalizing for unmatched nodes in the image, the model, or both, depending on the a priori conditions on query generation.

Before stating the algorithm, let us define some of its components. Let $G = (V_1, E_1)$ and $H = (V_2, E_2)$ be the two DAGs to be matched, with $|V_1| = n_1$ and $|V_2| = n_2$. Define d to be the maximum degree of any vertex in G and H, i.e., $d = \max(\delta(G), \delta(H))$. For each vertex v, we define $\chi(v) \in R^{d-1}$ as the unique topological signature vector (TSV), introduced in Section 2.[4] The bipartite edge weighted graph $\mathcal{G}(V_1, V_2, E_{\mathcal{G}})$ is represented as a $n_1 \times n_2$ matrix $\Pi(G, H)$ whose (u, v)-th entry has the value:

$$\phi(u,v) = \mathcal{W}(u,v) \times e^{-(||\chi(u)-\chi(v)||)}, \tag{2}$$

where $\mathcal{W}(u, v)$ denotes the similarity between u and v, assuming that u and v are compatible in terms of their nodes (more on this later), and has the value zero otherwise. Using the scaling algorithm of Goemans, Gabow, and Williamson [6], we can compute the maximum cardinality, maximum weight matching in $\mathcal{G}$, resulting in a list of node correspondences between G and H, called $\mathcal{M}_1$, that can

[3] An inversion occurs when an ancestor/descendant in one graph is mapped to a descendant/ancestor in the other graph.

[4] Note that if the maximum out-degree of a node is d, then excluding the edge from the node's parent, the maximum number of children is $d - 1$. Also note that if $\delta(v) < d$, then then the last $d - \delta(v)$ entries of χ are set to zero to ensure that all χ vectors have the same dimension.

```
procedure isomorphism(G,H)
    Φ(G, H) ← ∅ ;solution set
    d ← max(δ(G), δ(H)) ;TSV degree
    for u ∈ V_G { ;compute TSV at each node and unmark all nodes in G
        compute χ(u) ∈ R^{d-1} (see Section 2)
        unmark u
    }
    for v ∈ V_H { ;compute TSV at each node and unmark all nodes in H
        compute χ(v) ∈ R^{d-1} (see Section 2)
        unmark v
    }
    call match(root(G),root(H))
    return(cost(Φ(G, H))
end

procedure match(u,v)
    do
        {
        let G_u ← rooted unmarked subgraph of G at u
        let H_v ← rooted subgraph of H at v
        compute |V_{G_u}| × |V_{H_v}| weight matrix Π(G_u, H_v)
        M ← max cardinality, max weight bipartite matching
                in G(V_{G_u}, V_{H_v}) with weights from Π(G_u, H_v) (see [6])
        (u', v') ← max weight pair in M
        Φ(G, H) ← Φ(G, H) ∪ {(u', v')}
        call match(u',v')
        mark G'_u
        mark G'_v
        }
    while (G_u ≠ ∅ and H_v ≠ ∅)
```

Fig. 5. Algorithm for matching two hierarchical structures

be ranked in decreasing order of similarity. The precise algorithm, whose complexity is $O(n^3)$, is given in Figure 5; additional details, analysis, and examples are given in [17].

5 Experiments

We have systematically tested our integrated framework using both occluded and unoccluded queries. With over 27,000 trials and a database of over 1400 graphs, this represents one of the most comprehensive set of shock graph experiments to date. Our database consists of views computed from 3-D graphics models obtained from the public domain. Using a graphics modeling tool (3D Studio Max), each model is centered in a uniformly tessellated view sphere, and a silhouette is generated for each vertex in the tessellation. A shock graph is computed for each silhouette [4], and each node of the resulting graph is added to the model database, as described in Section 3. A sampling of the object views is shown in Figure 6.

In the first set of experiments, we evaluate the performance of the system on a set of unoccluded queries to an object view database. The database contains 1408 views describing 11 objects (128 uniformly sampled views per object). We then remove each view from the database and use it as a query to the remaining

Fig. 6. Some example object views drawn from our database

views. For each node of the query DAG, the indexing module (see Section 3) will return all neighbors within a radius of 40% of the norm of (query) node. Evidence for models (containing a neighbor) is then accumulated, and the model bins are sorted. The indexer will return at most the highest scoring 50 candidates, but will return fewer if the sorted bins' contents drop suddenly. The candidates are matched to the query, using the matcher (see Section 4), and sorted according to similarity. If the query object (from which the query view was drawn) is the same as the model object from which the most similar candidate view is drawn, recognition is said to be successful, i.e., the object label is correct.[5]

Figure 7(a) plots recognition performance as a function of increasing number of objects (with 128 views per new object), while Figure 7(b) fixes the number of objects (11) and plots recognition performance as a function of sampling resolution. Recognition performance is very high, with better than 90% success until sampling resolution drops below 32 views (over the entire view sphere) per object. This demonstrates both the efficacy of the recognition framework and the viewpoint invariance of the shock graph, respectively. The most complex component of the algorithm is the matcher. However, with a fixed number (50) of verifications per query, independent of database size, complexity therefore varies as a function of nearest neighbor search and bin sorting, both of which are sublinear in the number of database views.

In the final experiment, shown in Figure 7(c), we plot recognition performance as a function of degree of occlusion (for the entire database) for occluded queries. To generate an occluded query, we randomly choose a node in the query DAG and delete the subgraph rooted at that node, provided that the node "mass" of the graph does not drop by more than 50%. As can be seen from the

[5] Note that if multiple views (perhaps from different objects) are tied for "most similar", then each can be considered to be "most similar."

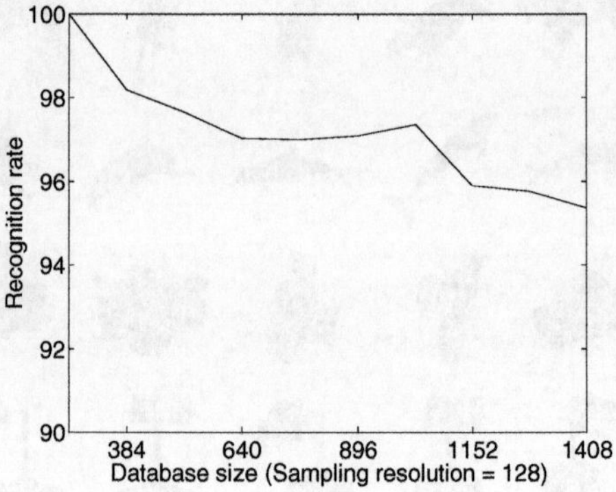

(a)

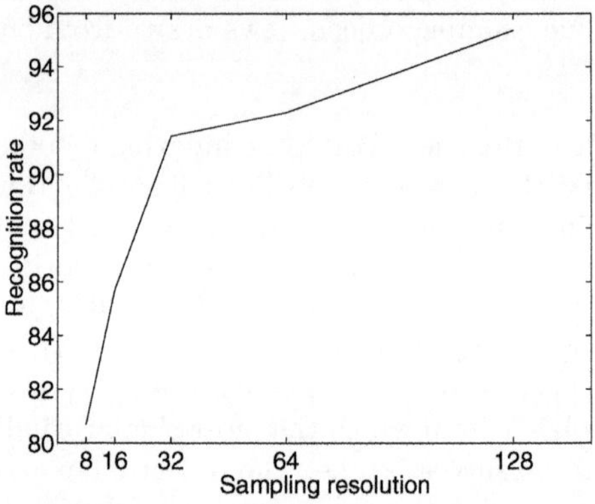

(b)

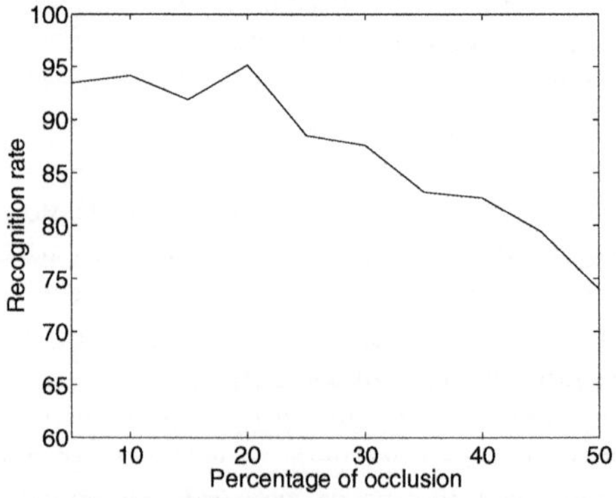

(c)

Fig. 7. Recognition Performance: (a) Recognition performance as a function of object database size; (b) Recognition performance as a function of sampling resolution; and (c) Recognition performance as a function of degree of occlusion plot, performance decreases gradually as a function of occluder size (or, more accurately, the amount of "missing data"), reflecting the framework's ability to recognize partially visible objects.

It should be noted that in the above experiments, erroneous matches may be due to either ambiguous views (views shared by different objects) or to queries representing "degenerate" views, in which the removed view acting as a query was the last view of its class and therefore not expected to match other views on the object.

6 Conclusions

We have presented a unified mechanism for shock graph indexing and matching, and have applied it to the problem of view-based 3-D object recognition. Our spectral-based indexing framework quickly and effectively selects a small number of candidates, *including the correct one*, from a large database of model views from which our spectral-based matcher computes an accurate distance measure. Our scaling experiments demonstrate the framework's ability to effectively deal with large numbers of views, while our occlusion experiments establish its robustness. Current work is focused on view-cell clustering and strengthening the indexer to include more geometric and node label information. In particular, it is known that nodes related to ligature are likely to be less stable and hence should be given less weight by both the indexer and the matcher [1].

Acknowledgements

The authors would like to acknowledge the programming support of Maxim Trokhimtchouk, Carlos Phillips, and Pavel Dimitrov. The authors would also like to express their thanks to Norio Katayama for the use of their SR-tree implementation. Finally, the authors would like to acknowledge the generous support of NSERC, FCAR, CFI, CITO, and NSF.

References

1. J. August, K. Siddiqi, and S. W. Zucker. Ligature instabilities in the perceptual organization of shape. *Computer Vision and Image Understanding*, 76(3):231–243, 1999. 13
2. R. Brooks. Model-based 3-D interpretations of 2-D images. *IEEE Transactions on Pattern Analysis and Machine Intelligence*, 5(2):140–150, 1983. 1
3. M. Cyr and B. Kimia. 3d object recognition using shape similarity-based aspect graph. In *Proceedings, ICCV*, Vancouver, B.C., 2001. 3
4. P. Dimitrov, C. Phillips, and K. Siddiqi. Robust and efficient skeletal graphs. In *IEEE Conference on Computer Vision and Pattern Recognition*, Hilton Head, SC, June 2000. 10
5. P. Flynn and A. Jain. 3D object recognition using invariant feature indexing of interpretation tables. *CVGIP:Image Understanding*, 55(2):119–129, March 1992. 6

6. H. Gabow, M. Goemans, and D. Williamson. An efficient approximate algorithm for survivable network design problems. *Proc. of the Third MPS Conference on Integer Programming and Combinatorial Optimization*, pages 57–74, 1993. 9, 10
7. D. Huttenlocher and S. Ullman. Recognizing solid objects by alignment with an image. *International Journal of Computer Vision*, 5(2):195–212, 1990. 1
8. Norio Katayama and Shin'ichi Satoh. The sr-tree: an index structure for high-dimensional nearest neighbor queries. In *Proceedings of the 1997 ACM SIGMOD international conference on Management of data*, pages 369–380. ACM Press, 1997. 6
9. Y. Lamdan, J. Schwartz, and H. Wolfson. Affine invariant model-based object recognition. *IEEE Transactions on Robotics and Automation*, 6(5):578–589, October 1990. 6
10. L. Lovász and J. Pelicán. On the eigenvalues of a tree. *Periodica Math. Hung.*, 3:1082–1096, 1970. 6
11. D. Lowe. *Perceptual Organization and Visual Recognition.* Kluwer Academic Publishers, Norwell, MA, 1985. 1
12. D. Macrini, A. Shokoufandeh, S. Dickinson, K. Siddiqi, and S. Zucker. View-based 3-D object recognition using shock graphs. In *Proceedings, Internal Conference on Pattern Recognition*, Quebec City, August 2002. 1
13. H. Murase and S. Nayar. Visual learning and recognition of 3-D objects from appearance. *International Journal of Computer Vision*, 14:5–24, 1995. 1
14. M. Pelillo, K. Siddiqi, and S. Zucker. Matching hierarchical structures using association graphs. *IEEE Transactions on Pattern Analysis and Machine Intelligence*, 21(11):1105–1120, November 1999. 2
15. T. Sebastian, P. Klein, and B. Kimia. Recognition of shapes by editing shock graphs. In *Proceedings, ICCV*, Vancouver, B.C., 2001. 2, 3
16. A. Shokoufandeh and S. Dickinson. A unified framework for indexing and matching hierarchical shape structures. In *Proceedings, 4th International Workshop on Visual Form*, Capri, Italy, May 28–30 2001. 9
17. A. Shokoufandeh, S. Dickinson, C. Jonsson, L. Bretzner, and T. Lindeberg. The representation and matching of qualitative shape at multiple scales. In *Proceedings, ECCV*, Copenhagen, May 2002. 3, 4, 5, 9, 10
18. A. Shokoufandeh, S. Dickinson, K. Siddiqi, and S. Zucker. Indexing using a spectral encoding of topological structure. In *Proceedings, IEEE CVPR*, pages 491–497, Fort Collins, CO, June 1999. 3
19. K. Siddiqi, A. Shokoufandeh, S. Dickinson, and S. Zucker. Shock graphs and shape matching. *International Journal of Computer Vision*, 30:1–24, 1999. 2, 3, 5, 9
20. Kaleem Siddiqi and Benjamin B. Kimia. A shock grammar for recognition. In *IEEE Computer Society Conference on Computer Vision and Pattern Recognition*, pages 507–513, 1996. 2
21. M. Turk and A. Pentland. Eigenfaces for recognition. *Journal of Cognitive Neuroscience*, 3(1):71–86, 1991. 1
22. S. Ullman and R. Basri. Recognition by linear combinations of models. *IEEE Transactions on Pattern Analysis and Machine Intelligence*, 13(10):992–1006, October 1991. 1

Machine Learning for Sequential Data: A Review

Thomas G. Dietterich

Oregon State University
Corvallis, Oregon, USA
tgd@cs.orst.edu
http://www.cs.orst.edu/~tgd

Abstract. Statistical learning problems in many fields involve sequential data. This paper formalizes the principal learning tasks and describes the methods that have been developed within the machine learning research community for addressing these problems. These methods include sliding window methods, recurrent sliding windows, hidden Markov models, conditional random fields, and graph transformer networks. The paper also discusses some open research issues.

1 Introduction

The classical supervised learning problem is to construct a classifier that can correctly predict the classes of new objects given training examples of old objects [19]. A typical application is in optical character recognition where the objects are images of hand-written characters and the classes are the 26 alphabetic letters. The classifier takes an image as input and produces a letter as output. This task is typically formalized as follows.

Let $\mathbf{x}$ denote an image of a hand-written character and $y \in \{A, \ldots, Z\}$ denote the corresponding letter class. A training example is a pair $(\mathbf{x}, y)$ consisting of an image and its associated class label. We assume that the training examples are drawn independently and identically from the joint distribution $P(\mathbf{x}, y)$, and we will refer to a set of N such examples as the training data.

A classifier is a function h that maps from images to classes. The goal of the learning process is to find an h that correctly predicts the class $y = h(\mathbf{x})$ of new images $\mathbf{x}$. This is accomplished by searching some space $\mathcal{H}$ of possible classifiers for a classifier that gives good results on the training data without overfitting.

Over the past 10 years, supervised learning has become a standard tool in many fields, and practitioners have learned how to take new application problems and view them as supervised learning problems. For example, in cellular telephone fraud detection, each $\mathbf{x}$ describes a telephone call, and y is 0 if the call is legitimate and 1 if the call originated from a stolen (or cloned) cell phone [8]. Another example involves computer intrusion detection where each x describes a request for a computer network connection and y indicates whether that request is part of an intrusion attempt. A third example is part-of-speech tagging in

T. Caelli et al. (Eds.): SSPR&SPR 2002, LNCS 2396, pp. 15–30, 2002.
© Springer-Verlag Berlin Heidelberg 2002

which each $\mathbf{x}$ describes a word and each y gives the part-of-speech of that word (noun, verb, adjective, etc.).

One thing that is apparent in these (and other) applications is that they do not quite fit the supervised learning framework. Rather than being drawn independently and identically (iid) from some joint distribution $P(\mathbf{x}, y)$, the training data actually consist of *sequences* of $(\mathbf{x}, y)$ pairs. These sequences exhibit significant sequential correlation. That is, nearby $\mathbf{x}$ and y values are likely to be related to each other.

For example, before a cell phone is stolen, all of the y values will be 0. Afterwards, all of the y values will be 1. Similarly, computer intrusions exhibit significant clustering—particularly denial of service attacks. Other kinds of attacks are deliberately spaced over time to avoid detection, which is a form of temporal anti-correlation. In part-of-speech tagging, sequences of parts of speech are constrained by the grammar of the language. Hence, in English, a sequence such as (verb verb adjective verb verb) would be very unlikely. Sequential patterns are present even in the original task of character recognition: Character sequences usually form words rather than random sequences of letters.

Sequential patterns are important because they can be exploited to improve the prediction accuracy of our classifiers. In English, for example, if the classifier determines that one letter is Q, then the next letter is almost certain to be U. In telephone fraud detection, it is only possible to detect fraud by looking at the distribution of typical (legitimate) phone calls and then to see that this distribution changes when the telephone is stolen. Any single phone call, viewed in isolation, appears to be perfectly legitimate.

The *sequential supervised learning* problem can be formulated as follows. Let $\{(\mathbf{x}_i, \mathbf{y}_i)\}_{i=1}^N$ be a set of N training examples. Each example is a pair of sequences $(\mathbf{x}_i, \mathbf{y}_i)$, where $\mathbf{x}_i = \langle \mathbf{x}_{i,1}, \mathbf{x}_{i,2}, \ldots, \mathbf{x}_{i,T_i} \rangle$ and $\mathbf{y}_i = \langle y_{i,1}, y_{i,2}, \ldots, y_{i,T_i} \rangle$. For example, in part-of-speech tagging, one $(\mathbf{x}_i, \mathbf{y}_i)$ pair might consist of $\mathbf{x}_i = \langle$do you want fries with that$\rangle$ and $\mathbf{y}_i = \langle$verb pronoun verb noun prep pronoun$\rangle$. The goal is to construct a classifier h that can correctly predict a new label sequence $\mathbf{y} = h(\mathbf{x})$ given an input sequence $\mathbf{x}$.

This task should be contrasted with two other, closely-related tasks. The first of these is the time-series prediction problem. Here the task is to predict the $t+1^{\text{st}}$ element of a sequence $\langle y_1, \ldots, y_t \rangle$. This can be extended in two ways. First, we can consider the case where each y_t is a vector $\mathbf{y}_t$. The time-series task becomes to predict simultaneously a whole collection of parallel time series: Predict $\mathbf{y}_{t+1}$ given $\langle \mathbf{y}_1, \ldots, \mathbf{y}_t \rangle$. Second, we can consider the case when there are other "features" or co-variates $\langle \mathbf{x}_1, \ldots, \mathbf{x}_t, \mathbf{x}_{t+1} \rangle$ available.

There are two key differences between time-series prediction and sequential supervised learning. First in sequential supervised learning, the entire sequence $\langle \mathbf{x}_1, \ldots, \mathbf{x}_T \rangle$ is available before we make any predictions of the y values, whereas in time-series prediction, we have only a prefix of the sequence up to the current time $t+1$. Second, in time-series analysis, we have the true observed y values up to time t, whereas in sequential supervised learning, we are not given any y values and we must predict them all.

The second closely-related task is sequence classification. In this task, the problem is to predict a single label y that applies to an entire input sequence $\langle \mathbf{x}_1, \mathbf{x}_2, \ldots, \mathbf{x}_T \rangle$. For example, given a sequence of images of hand-written characters, the task might be to determine the identity of the person who wrote those characters (hand-writing identification). In these kinds of problems, each training example consists of a pair $(\mathbf{x}_i, y_i)$, where $\mathbf{x}_i$ is a sequence $\langle \mathbf{x}_{i,1}, \ldots, \mathbf{x}_{i,T_i} \rangle$ and each y_i is a class label (such as a person's identification number). A similar problem arises in recognizing whole words on handwritten checks. The $\mathbf{x}_i$ could be a sequence of hand-written letters, and y_i could be a word such as "hundred".

All of these problems are closely related, and sometimes a solution to one can be converted into a solution for another. For example, one strategy for recognizing a handwritten word (e.g., "hundred") would be first to solve the sequential supervised learning problem of recognizing the individual letters ⟨H, U, N, D, R, E, D⟩, and then assembling them into the entire word. This works for cases where the class label y can be decomposed into sub-parts (in this case, individual letters). But no similar strategy would work for recognizing an individual's identity from their handwriting.

Similarly, some methods for sequential supervised learning make their predictions by scanning the sequence from left-to-right, and such methods can typically be applied to time-series problems as well. However, methods that analyze the entire sequence of $\mathbf{x}_t$ values before predicting the y_t labels typically can give better performance on the sequential supervised learning problem.

2 Research Issues in Sequential Supervised Learning

Now let us consider three fundamental issues in sequential supervised learning: (a) loss functions, (b) feature selection, and (c) computational efficiency.

2.1 Loss Functions

In classical supervised learning, the usual measure of success is the proportion of (new) test data points correctly classified. This is known as the 0/1 loss, since a loss of 1 is received for every misclassified test point and a loss of 0 for every correctly-classified test point. More recently, researchers have been studying non-uniform loss functions. These are usually represented by a cost matrix $C(i, j)$, which gives the cost of assigning label i to an example whose true label is j. In such cases, the goal is to find a classifier with minimum expected cost. One strategy for developing such a classifier is to learn a conditional density estimator $P(y|\mathbf{x})$ and then classify a new data point $\mathbf{x}$ according to the formula

$$y = \operatorname*{argmin}_i \sum_j P(j|\mathbf{x}) C(i, j).$$

This formula chooses the class whose expected cost is minimum.

In sequential supervised learning problems, many different kinds of loss functions are encountered. Statistical learning methods are needed that can minimize the expected loss for all of these different loss functions. First, we will consider some of the loss functions that have appeared in various applications. Second, we will discuss how these different loss functions might be incorporated into learning and prediction.

In some problems, the goal is to predict the entire output sequence of labels $\mathbf{y}_i$ correctly, and any error in this sequence counts as an error for the entire sequence. Other problems exhibit the opposite extreme: the goal is to predict correctly as many individual labels $y_{i,t}$ in the sequence as possible. One can imagine problems intermediate between these extremes.

In many applications, different kinds of errors have different costs. Consider cellular telephone fraud. The real goal here is to determine the time t^* at which the telephone was stolen (or cloned). As described above, we can view this as a sequential supervised learning problem in which $y_t = 0$ for $t < t^*$ and $y_t = 1$ for $t \geq t^*$. Consider the problem of making a prediction $\bar{t}$ for the value of t^*. One strategy would be to apply the learned classifier h to classify each element $\mathbf{x}_{i,t}$ and predict $\bar{t} = t$ for the earliest time t for which $h(\mathbf{x}_{i,t}) = 1$. A typical form for the loss function assesses a penalty of $c_1(t^* - \bar{t})$ if $\bar{t} < t^*$ and a penalty of $c_2(\bar{t} - t^*)$ if $\bar{t} > t^*$. In the telephone fraud case, the first penalty is the cost of lost business if we prematurely declare the telephone to be stolen. The second penalty is the cost of the fraudulent calls when we are late in declaring the telephone to be stolen. More complex loss functions can be imagined that take into account the cost of each individual telephone call. This argument applies to any form of monitoring of financial transactions. It also applies to systems that must determine when manufacturing equipment begins to malfunction.

Another kind of loss function applies to problems of event detection. Suppose that the input sequence $\mathbf{x}_i$ consists of infrequent events superimposed on "normal" signals. For example, in high-energy physics, these might be detections of rare particles. In astronomy, these might be sightings of events of interest (e.g., gamma ray bursts). The loss function should assign a cost to missed events, to extra events, and to events that are detected but not at the correct time.

Finally, a loss function closely related to event detection arises in the problem of hyphenation. Consider the problem of learning to hyphenate words so that a word processor can determine where to break words during typesetting (e.g., "porcupine" $\rightarrow$ "00101000"). In this case, the input sequence $\mathbf{x}_i$ is a string of letters, and the output sequence $\mathbf{y}_i$ is a sequence of 0's and 1's, such that $y_{i,t} = 1$ indicates that a hyphen can legally follow the letter $x_{i,t}$. Each opportunity for a hyphen can be viewed as an event. False positive hyphens are very expensive, because they lead to incorrectly-hyphenated words that distract the reader. False negative hyphens are less of a problem—provided that at least one hyphen is correctly identified. Furthermore, hyphens near the middle of long words are more helpful to the typesetting program than hyphens near the ends of the words. This is a case where the loss function involves a global analysis of the predicted sequence $\mathbf{y}_i$ but where not all of the individual y_t predictions need to be correct.

How can these kinds of loss functions be incorporated into sequential supervised learning? One approach is to view the learning problem as the task of predicting the (conditional) joint distribution of all of the labels in the output sequence: $P(\mathbf{y}_i|\mathbf{x}_i)$. If this joint distribution can be accurately predicted, then all of the various loss functions can be evaluated, and the optimal decisions can be chosen. There are two difficulties with this: First, predicting the entire joint distribution is typically very difficult. Second, computing the optimal decisions given the joint distribution may also be computationally infeasible.

Some loss functions only require particular marginal probabilities. For example, if the loss function is only concerned with the number of correct individual predictions $y_{i,t}$, then the goal of learning should be to predict the individual marginal probabilities $P(y_{i,t}|\mathbf{x}_i)$ correctly. If the loss function is only concerned with classifying the entire sequence correctly, then the goal should be to predict $\text{argmax}_{\mathbf{y}_i} P(\mathbf{y}_i|\mathbf{x}_i)$ correctly. We will see below that there are learning algorithms that directly optimize these quantities.

2.2 Feature Selection and Long-Distance Interactions

Any method for sequential supervised learning must employ some form of divide-and-conquer to break the overall problem of predicting $\mathbf{y}_i$ given $\mathbf{x}_i$ into subproblems of predicting individual output labels $y_{i,t}$ given some subset of information from $\mathbf{x}_i$ (and perhaps other predicted values $y_{i,u}$). One of the central problems of sequential supervised learning is to identify the relevant information subset for making accurate predictions.

In standard supervised learning, this is known as the *feature selection problem*, and there are four primary strategies for solving it. The first strategy, known as the wrapper approach [12], is to generate various subsets of features and evaluate them by running the learning algorithm and measuring the accuracy of the resulting classifier (e.g., via cross-validation or by applying the Akaiki Information Criterion). The feature subsets are typically generated by forward selection (starting with single features and progressively adding one feature at a time) or backward elimination (starting with all of the features and progressively removing one feature at a time). For some learning algorithms, such as linear regression, this can be implemented very efficiently.

The second strategy is to include all possible features in the model, but to place a penalty on the values of parameters in the fitted model. This causes the parameters associated with useless features to become very small (perhaps even zero). Examples of this approach include ridge regression [10], neural network weight elimination [24], and L_1-norm support vector machines (SVMs; [5]).

The third strategy is to compute some measure of feature relevance and remove low-scoring features. One of the simplest measures is the mutual information between a feature and the class. This (or similar measures) forms the basis of recursive-partioning algorithms for growing classification and regression trees. These methods incorporate the choice of relevant features into the tree-growing process [3,21]. Unfortunately, this measure does not capture interactions between features. Several methods have been developed that identify such interactions including RELIEFF [14], Markov blankets [13], and feature racing [17].

The fourth strategy is to first fit a simple model and then analyze the fitted model to identify the relevant features. For example, Chow and Liu [4] describe an efficient algorithm for fitting a tree-structured Bayesian network to a data set. This network can then be analyzed to remove features that have low influence on the class. Kristin Bennett (personal communication, 2001) fits L_1-norm SVMs to drug binding data to remove irrelevant features prior to fitting a more complex SVM regression model.

In sequential supervised learning, most authors have assumed that a fixed-sized neighborhood of features is relevant for predicting each output value. For example, suppose we assume a neighborhood of size 3. Then we will employ $\mathbf{x}_{i,t-1}, \mathbf{x}_{i,t}$, and $\mathbf{x}_{i,t+1}$ to predict $y_{i,t}$. However, this has two drawbacks. First, not all of the features in each feature vector $\{\mathbf{x}_{i,u}\}_{u=t-1}^{t+1}$ are necessarily relevant. Second, there may be longer-range interactions that are missed. For example, consider the problem of predicting the pronunciation of English words from their spelling. The only difference between the words "thought" and "though" is the final "t", yet this influences the pronunciation of the initial "th" (changing it from unvoiced to voiced). An even more extreme case is the pair "photograph" and "photography" in which the final "y" changes the pronunciation of every vowel in the word.

Of the four feature-selection strategies discussed above, it is unlikely that the first two are feasible for sequential supervised learning. There are so many potential features to consider in a long sequence, that a direct search of possible feature subsets becomes completely intractable (even with greedy algorithms). The third and fourth approaches are more promising, but with long sequences, they still raise the possibility of overfitting. Hence, any successful methodology for feature selection (and for handling long distance interactions) will probably need to combine human expertise with statistical techniques rather than applying statistical techniques alone.

2.3 Computational Efficiency

A third challenge for sequential supervised learning is to develop methods for learning and classification that are computationally efficient. We will see below that some of the learning algorithms that have been proposed for sequential supervised learning are computationally expensive.

Even after learning, it may be computationally expensive to apply a learned classifier to make minimum-cost predictions. Even relatively efficient methods such as the Viterbi algorithm can be slow for complex models.

These computational challenges are probably easier to solve than the statistical ones. As in many other computational problems, it is usually possible to identify a series of approximate methods that are progressively more expensive and more accurate. The cheapest methods can be applied first to generate a set of possible candidate solutions which can then be evaluated more carefully by the more expensive methods.

3 Machine Learning Methods for Sequential Supervised Learning

In this section, we will briefly describe six methods that have been applied to solve sequential supervised learning problems: (a) sliding-window methods, (b) recurrent sliding windows, (c) hidden Markov models, (d) maximum entropy Markov models, (e) input-output Markov models, (f) conditional random fields, and (g) graph transformer networks.

3.1 The Sliding Window Method

The sliding window method converts the sequential supervised learning problem into the classical supervised learning problem. It constructs a *window classifier* h_w that maps an input window of width w into an individual output value y. Specifically, let $d = (w - 1)/2$ be the "half-width" of the window. Then h_w predicts $y_{i,t}$ using the window $\langle x_{i,t-d}, x_{i,t-d+1}, \ldots, x_{i,t}, \ldots, x_{i,t+d-1}, x_{i,t+d} \rangle$. In effect, the input sequence $\mathbf{x}_i$ is padded on each end by d "null" values and then converted into N_i separate examples.

The window classifier h_w is trained by converting each sequential training example $(\mathbf{x}_i, \mathbf{y}_i)$ into windows and then applying a standard supervised learning algorithm. A new sequence $\mathbf{x}$ is classified by converting it to windows, applying h_w to predict each y_t and then concatenating the y_t's to form the predicted sequence $\mathbf{y}$.

The obvious advantage of this sliding window method is that permits any classical supervised learning algorithm to be applied. Sejnowski and Rosenberg [23] applied the backpropagation neural network algorithm with a 7-letter sliding window to the task of pronouncing English words. A similar approach (but with a 15-letter window) was employed by Qian and Sejnowski [20] to predict protein secondary structure from the protein's sequence of amino acid residues. Provost and Fawcett [8] addressed the problem of cellular telephone cloning by applying the RL rule learning system to day-long windows from telephone calling logs.

Although the sliding window method gives adequate performance in many applications, it does not take advantage of correlations between nearby y_t values. To be more precise, the only relationships between nearby y_t values that are captured are those that are predictable from nearby x_t values. If there are correlations among the y_t values that are independent of the x_t values, then these are not captured.

3.2 Recurrent Sliding Windows

One way that sliding window methods can be improved is to make them recurrent. In a recurrent sliding window method, the predicted value $\overline{y}_{i,t}$ is fed as an input to help make the prediction for $y_{i,t+1}$. Specifically, with a window of half-width d, the most recent d predictions, $\overline{y}_{i,t-d}, \overline{y}_{i,t-d+1}, \ldots, \overline{y}_{i,t-1}$, are used

Table 1. Left-to-right and right-to-left

Method	Direction of letter processing	% correct Level of Aggregation Word	Letter
Sliding Window		12.5	69.6
Recurrent Sliding Window	Left-to-Right	17.0	67.9
Recurrent Sliding Window	Right-to-Left	24.4	74.2

as inputs (along with the sliding window $\langle x_{i,t-d}, x_{i,t-d+1}, \ldots, x_{i,t}, \ldots, x_{i,t+d-1}, x_{i,t+d}\rangle$) to predict $y_{i,t}$.

Bakiri and Dietterich [1] applied this technique to the English pronunciation problem using a 7-letter window and a decision-tree algorithm. Table 1 summarizes the results they obtained when training on 1000 words and evaluating the performance on a separate 1000-word test data set. The baseline sliding window method correctly pronounces 12.5% of the words and 69.6% of the individual letters in the words. A recurrent sliding window moving left-to-right improves the word-level performance but worsens the pronunciations of individual letters. However, a right-to-left sliding window improves both the word-level and letter-level performance. Indeed, the percentage of correct word-level pronunciations has nearly doubled!

Clearly, the recurrent method captures predictive information that was not being captured by the simple 7-letter sliding window. But why is the right-to-left scan superior? It appears that in English, the right-to-left scan is able to capture long-distance effects such as those mentioned above for "thought" and "photography". For example, the right-most window can correctly pronounce the "y" of "photography". This information is then available when the system attempts to pronounce the "a". And this information in turn is available when the system is pronouncing the second "o", and so on. Because the stress patterns in English are determined by the number of syllables to the right of the current syllable, a right-to-left recurrent window is able to correctly predict these stresses, and hence, choose the correct pronunciations for the vowels in each syllable.

One issue arises during training: What values should be used for the $\overline{y}_{i,t}$ inputs when training the window classifier? One approach would be to first train a non-recurrent classifier, and then use its $\overline{y}_{i,t}$ predictions as the inputs. This process can be iterated, so that the predicted outputs from each iteration are employed as inputs in the next iteration. Another approach is to use the correct labels $y_{i,t}$ as the inputs. The advantage of using the correct labels is that training can be performed with the standard supervised learning algorithms, since each training example can be constructed independently. This was the choice made by Bakiri and Dietterich.

In addition to recurrent decision trees, many other classifiers can be made recurrent. Recurrent neural networks are of particular interest. Figure 1 shows two of the many architectures that have been explored. Part (a) shows a network in which the output units are fed as inputs to the hidden units at the next time step. This is essentially identical to the recurrent decision trees employed by

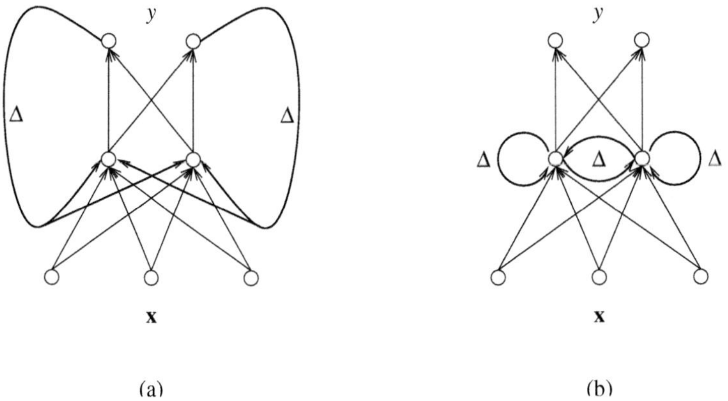

Fig. 1. Two recurrent network architectures: (a) outputs are fed back to hidden units; (b) hidden units are fed back to hidden units. The **Δ** symbol indicates a delay of one time step

Bakiri and Dietterich, except that during training, the predicted outputs $\overline{y}_{i,t-1}$ are used as the inputs at time t. Networks similar to this were first introduced by Jordan [11]. Part (b) shows a network in which the hidden unit activations at time $t-1$ are fed as additional inputs at time t. This allows the network to develop a representation for the recurrent information that is separate from the representation of the output y values. This architecture was introduced by Elman [7].

These networks are usually trained iteratively via a procedure known as backpropagation-through-time (BPTT) in which the network structure is "unrolled" for the length of the input and output sequences $\mathbf{x}_i$ and $\mathbf{y}_i$ [22]. Recurrent networks have been applied to a variety of sequence-learning problems [9].

3.3 Hidden Markov Models and Related Methods

The hidden Markov Model (HMM; see Figure 2(a)) is a probabilistic model of the way in which the $\mathbf{x}_i$ and $\mathbf{y}_i$ strings are generated—that is, it is a representation of the joint distribution $P(\mathbf{x}, \mathbf{y})$. It is defined by two probability distributions: the transition distribution $P(y_t|y_{t-1})$, which tells how adjacent y values are related, and the observation distribution $P(\mathbf{x}|y)$, which tells how the observed $\mathbf{x}$ values are related to the hidden y values. These distributions are assumed to be stationary (i.e., the same for all times t).

In most problems, $\mathbf{x}$ is a vector of features $(x_1, \ldots, x_n)$, which makes the observation distribution difficult to handle without further assumptions. A common assumption is that each feature is generated independently (conditioned on y). This means that $P(\mathbf{x}|y)$ can be replaced by the product of n separate distributions $P(x_j|y)$, $j = 1, \ldots, n$.

The HMM generates $\mathbf{x}_i$ and $\mathbf{y}_i$ as follows. Suppose there are K possible labels $1, \ldots, K$. Augment this set of labels with a start label 0 and a terminal

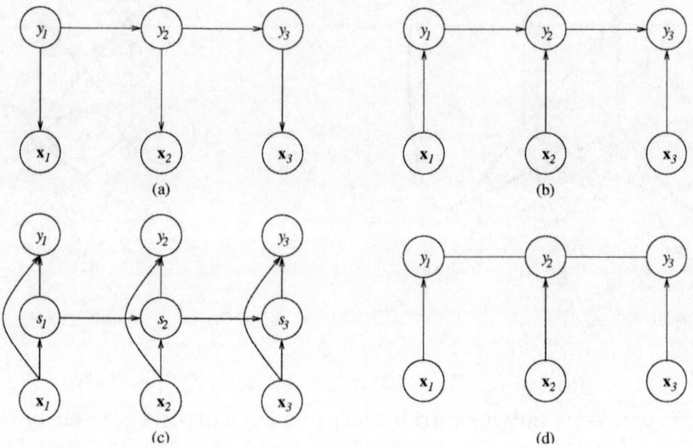

Fig. 2. Probabilistic models related to hidden Markov models: (a) HMM, (b) maximum entropy Markov model, (c) input-output HMM, and (d) conditional random field

label $K+1$. Let $y_{i,0} = 0$. Then, generate the sequence of y values according to $P(y_{i,t}|y_{i,t-1})$ until $y_{i,t} = K+1$. At this point, set $T_i := t$. Finally, for each $t = 1, \ldots, T_i$, generate $\mathbf{x}_{i,t}$ according to the observation probabilities $P(\mathbf{x}_{i,t}|y_{i,t})$.

In a sequential supervised learning problem, it is straightforward to determine the transition and observation distributions. $P(y_{i,t}|y_{i,t-1})$ can be computed by looking at all pairs of adjacent y labels (after prepending 0 at the start and appending $K+1$ to the end of each $\mathbf{y}_i$). Similarly, $P(x_j|y)$ can be computed by looking at all pairs of x_j and y.

The most complex computation is to predict a value $\overline{\mathbf{y}}$ given an observed sequence $\mathbf{x}$. This computation depends on the nature of the loss function. Because the HMM is a representation of the joint probability distribution $P(\mathbf{x}, \mathbf{y})$, it can be applied to compute the probability of any particular $\mathbf{y}$ given any particular $\mathbf{x}$: $P(\mathbf{y}|\mathbf{x})$. Hence, for an arbitrary loss function $L(\overline{\mathbf{y}}, \mathbf{y})$, the optimal prediction is

$$\overline{\mathbf{y}} = \operatorname*{argmin}_{\mathbf{z}} \sum_{\mathbf{y}} P(\mathbf{y}|\mathbf{x}) L(\mathbf{z}, \mathbf{y}).$$

However, if the sequences are of length L and there are K labels, then direct evaluation of this equation requires $O(K^L)$ probability evaluations, which is usually impractical.

There are two notable cases where this computation can be performed in $O(K^2 L)$ time. The first is where the loss function depends on the entire sequence. In this case, the goal is usually to find the $\mathbf{y}$ with the highest probability: $\overline{\mathbf{y}} = \operatorname{argmax}_{\mathbf{y}} P(\mathbf{y}|\mathbf{x})$. This can be computed via the Viterbi algorithm, which is a dynamic programming algorithm that computes, for each class label u and each time step t, the probability of the most likely path starting at time 0 end ending at time t with class u. When the algorithm reaches the end of the sequence, it has computed the most likely path from time 0 to time T_i and its probability.

The second interesting case is where the loss function decomposes into separate decisions for each y_t. In this case, the so-called Forward-Backward algorithm can be applied. It performs a left-to-right pass, which fills a table of $\alpha_t(y_t)$ values which represent $P(y_1, \ldots, y_t|\mathbf{x}_1, \ldots, \mathbf{x}_t)$, and a right-to-left pass, which fills a table of $\beta_t(y_t)$ values which represent $P(y_t, \ldots, y_{T_i}|\mathbf{x}_{t+1}, \ldots, \mathbf{x}_{T_i})$. Once these two passes are complete, the quantity

$$\gamma_t(u) = \frac{\alpha_t(u) \cdot \beta_t(u)}{\sum_v \alpha_t(v) \cdot \beta_t(v)}$$

gives the desired probability: $P(y_t = u|\mathbf{x})$. This probability can be applied to choose the predicted value $\overline{y}_t$ that minimizes the loss function.

Although HMMs provide an elegant and sound methodology, they suffer from one principal drawback: The structure of the HMM is often a poor model of the true process producing the data. Part of the problem stems from the Markov property. Any relationship between two separated y values (e.g., y_1 and y_4) must be communicated via the intervening y's. A first-order Markov model (i.e., where $P(y_t)$ only depends on y_{t-1}) cannot in general capture these kinds of relationships.

Sliding window methods avoid this difficulty by using a window of $\mathbf{x}_t$ values to predict a single y_t. However, the second problem with the HMM model is that it generates each $\mathbf{x}_t$ only from the corresponding y_t. This makes it difficult to use an input window. In theory, one could replace the output distribution $P(\mathbf{x}_t|y_t)$ by a more complex distribution $P(\mathbf{x}_t|y_{t-1}, y_t, y_{t+1})$ which would then allow an observed value $\mathbf{x}_t$ to influence the three y values. But it is not clear how to represent such a complex distribution compactly.

Several directions have been explored to try to overcome the limitations of the HMM: Maximum Entropy Markov models (MEMMs), Input-Output HMMs (IOHMMs), and conditional random fields (CRFs); see Figure 2. All of these are *conditional* models that represent $P(\mathbf{y}|\mathbf{x})$ rather than $P(\mathbf{x}, \mathbf{y})$. They do not try to explain how the $\mathbf{x}$'s are generated. Instead, they just try to predict the $\mathbf{y}$ values given the $\mathbf{x}$'s. This permits them to use arbitrary features of the $\mathbf{x}$'s including global features, features describing non-local interactions, and sliding windows.

The Maximum Entropy Markov Model learns $P(y_t|y_{t-1}, \mathbf{x}_t)$. It is trained via a maximum entropy method that attempts to maximize the conditional likelihood of the data: $\prod_{i=1}^{N} P(\mathbf{y}_i|\mathbf{x}_i)$. The maximum entropy approach represents $P(y_t|y_{t-1}, \mathbf{x}_t)$ as a log-linear model:

$$P(y_t|y_{t-1}, \mathbf{x}) = \frac{1}{Z(y_{t-1}, \mathbf{x})} \exp\left(\sum_\alpha \lambda_\alpha f_\alpha(\mathbf{x}, y_t)\right),$$

where $Z(y_{t-1}, \mathbf{x})$ is a normalizing factor to ensure that the probabilities sum to 1. Each f_α is a boolean feature that can depend on y_t and on *any* properties of the input sequence $\mathbf{x}$. For example, in their experiments with MEMMs, McCallum, et al. [18] employed features such as "$\mathbf{x}$ begins with a number", "$\mathbf{x}$ ends with a question mark", etc. Hence, MEMMs support long-distance interactions.

The IOHMM is similar to the MEMM except that it introduces hidden state variables s_t in addition to the output labels y_t. Sequential interactions are modeled by the s_t variables. To handle these hidden variables during training, the Expectation-Maximization (EM; [6]) algorithm is applied. Bengio and Frasconi [2] report promising results on various artificial sequential supervised learning and sequence classification problems.

Unfortunately, the MEMM and IOHMM models suffer from a problem known as the *label bias problem.* To understand the origins of the problem, consider the MEMM and note that

$$\begin{aligned}\sum_{y_t} P(y_t|y_{t-1}, \mathbf{x}_1, \ldots, \mathbf{x}_t) &= \sum_{y_t} P(y_t|y_{t-1}, \mathbf{x}_t) \cdot P(y_{t-1}|\mathbf{x}_1, \ldots, \mathbf{x}_{t-1}) \\ &= 1 \cdot P(y_{t-1}|\mathbf{x}_1, \ldots, \mathbf{x}_{t-1}) = P(y_{t-1}|\mathbf{x}_1, \ldots, \mathbf{x}_{t-1})\end{aligned}$$

This says that the total probability mass "received" by y_{t-1} (based on $\mathbf{x}_1, \ldots, \mathbf{x}_{t-1}$) must be "transmitted" to labels y_t at time t regardless of the value of $\mathbf{x}_t$. The only role of $\mathbf{x}_t$ is to influence *which* of the labels receive more of the probability at time t. In particular, *all* of the probability mass *must* be passed on to some y_t even if $\mathbf{x}_t$ is completely incompatible with y_t.

For example, suppose that there are two labels $\{1, 2\}$ and that the input string $\mathbf{x} =$"rob" is supposed to get the label string "111" and $\mathbf{x} =$"rib" is supposed to get the label string "222". Consider what happens with the input string $\mathbf{x} =$"rib". After observing the $\mathbf{x}_1 = r$, the probability of y_1 is evenly split between labels "1" and "2": $P(y_1 = 1|\mathbf{x}_1 = r) = P(y_1 = 2|\mathbf{x}_1 = r) = 0.5$. After observing $\mathbf{x}_2 = i$, the probability remains equally split, because the 0.5 probability for $P(y_1 = 1|\mathbf{x}_1 = r)$ must be passed on to $P(y_2 = 1|\mathbf{x}_1 = r, \mathbf{x}_2 = i)$, since the $y_1 = 1 \rightarrow y_2 = 2$ transition has probability 0. After observing $\mathbf{x}_3 = b$, the probability of $y_3 = 1$ and $y_3 = 2$ remains equally split. So the MEMM has completely ignored the "i"! The same problem occurs with the hidden states s_t of the IOHMM.

3.4 Conditional Random Fields

Lafferty, McCallum, and Pereira [15] introduced the conditional random field (CRF; Figure 2(d)) to try to overcome the label bias problem. In the CRF, the relationship among adjacent pairs y_{t-1} and y_t is modeled as an Markov Random Field *conditioned on* the $\mathbf{x}$ inputs. In other words, the way in which the adjacent y values influence each other is determined by the input features.

The CRF is represented by a set of *potentials* $M_t(y_{t-1}, y_t|\mathbf{x})$ defined as

$$M_t(y_{t-1}, y_t|\mathbf{x}) = \exp\left(\sum_\alpha \lambda_\alpha f_\alpha(y_{t-1}, y_t, \mathbf{x}) + \sum_\beta \lambda_\beta g_\beta(y_t, \mathbf{x})\right),$$

where the f_α are boolean features that encode some information about y_{t-1}, y_t, and arbitrary information about $\mathbf{x}$, and the g_β are boolean features that encode some information about y_t and $\mathbf{x}$. As with MEMM's and IOHMM's, arbitrarily long-distance information about $\mathbf{x}$ can be incorporated into these features.

As with HMM's, CRF's assume two special labels 0 and $K+1$ to indicate the start and end of the sequence. Let $M_t(\mathbf{x})$ be the $(K+2) \times (K+2)$ matrix of potentials for all possible pairs of labels for y_{t-1} and y_t.

The CRF computes the conditional probability $P(\mathbf{y}|\mathbf{x})$ according to

$$P(\mathbf{y}|\mathbf{x}) = \frac{\prod_{t=1}^{L} M_t(y_{t-1}, y_t|\mathbf{x})}{\left[\prod_{t=1}^{L} M_t(\mathbf{x})\right]_{0,K+1}},$$

where L is one more than the length of the strings, $y_0 = 0$, $y_L = K+1$, and the denominator is the $(0, K+1)$ entry in the matrix product of the M_t potential matrices. The normalizer in the denominator is needed because the potentials M_t are unnormalized "scores".

The training of CRFs is expensive, because it requires a global adjustment of the λ values. This global training is what allows the CRF to overcome the label bias problem by allowing the $\mathbf{x}_t$ values to modulate the relationships between adjacent y_{t-1} and y_t values. Algorithms based on iterative scaling and gradient descent have been developed both for optimizing $P(\mathbf{y}|\mathbf{x})$ and also for separately optimizing $P(y_t|\mathbf{x})$ for loss functions that depend only on the individual labels.

Lafferty, et al. compared the performance of the HMM, MEMM, and CRF models on a part-of-speech tagging problem. For a basic configuration, in which the MEMM and CRF features were defined to provide the same information as the HMM, the error rates of the three methods were HMM: 5.69%, MEMM: 6.37%, and CRF: 5.55%. This is consistent with the hypothesis that the MEMM suffers from the label bias problem but the HMM and the CRF do not. Lafferty et al. then experimented with providing a few simple spelling-related features to the MEMM and CRF models, something that is impossible to incorporate into the HMM. The resulting error rates where MEMM: 4.81% and CRF: 4.27%. Even more dramatic results are observed if we consider only "out of vocabulary" words (i.e., words that did not appear in any training sentence): HMM: 45.99%, MEMM: 26.99%, CRF: 23.76%. The spelling-related features provide powerful information for describing out of vocabulary words, whereas the HMM must rely on default observation probabilities for these words.

3.5 Graph Transformer Networks

In a landmark paper on handwritten character recognition, LeCun, Bottou, Bengio, and Haffner [16] describe a neural network methodology for solving complex sequential supervised learning problems. The architecture that they propose is shown in Figure 3. A graph transformer network is a neural network that transforms an input graph into an output graph. For example, the neural network in the figure transforms an input graph, consisting of the linear sequence of $\mathbf{x}_t$, into an output graph, consisting of a collection of u_t values. Each $\mathbf{x}_t$ is a feature vector attached to an edge of the graph; each u_t is a pair of a class label and a

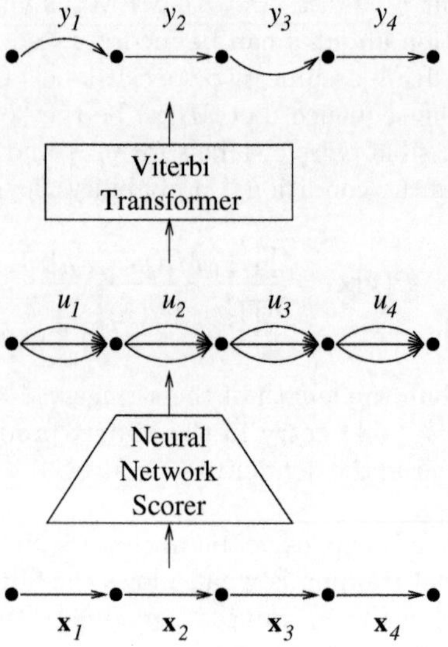

Fig. 3. The GTN architecture containing two graph transformers: a neural network and a Viterbi transformer

score. The Viterbi transformer analyzes the graph of u_t scores and finds the path through the graph with the lowest total score. It outputs a graph containing only this path, which gives the predicted y_t labels.

The architecture is trained globally by gradient descent. In order to do this, each graph transformer must be differentiable with respect to any internal tunable parameters. LeCun et al. describe a method called *discriminative forward training* that adjusts the parameters in the neural network to reduce the score along paths in the u graph corresponding to the correct label sequence $\mathbf{y}$ and to increase the scores of the other paths. An advantage of this approach is that arbitrary loss functions can be connected to the output of the Viterbi transformer, and the network can be trained to minimize the loss on the training data.

4 Concluding Remarks

Sequential supervised learning problems arise in many applications. This paper has attempted to describe the sequential supervised learning task, discuss the main research issues, and review some of the leading methods for solving it. The four central research issues are (a) how to capture and exploit sequential correlations, (b) how to represent and incorporate complex loss functions, (c) how to identify long-distance interactions, and (d) how to make the learning algorithms fast. Our long-term goal should be to develop a toolkit of off-the-shelf techniques for sequential supervised learning. Although we are still some distance from this goal, substantial progress has already been made, and we can look forward to more exciting work in the near future.

References

1. G. Bakiri and T. G. Dietterich. Achieving high-accuracy text-to-speech with machine learning. In R. I. Damper, editor, *Data Mining Techniques in Speech Synthesis*. Chapman and Hall, New York, NY, 2002. 22
2. Y. Bengio and P. Frasconi. Input-output HMM's for sequence processing. *IEEE Transactions on Neural Networks*, 7(5):1231–1249, September 1996. 26
3. L. Breiman, J. H. Friedman, R. A. Olshen, and C. J. Stone. *Classification and Regression Trees*. Wadsworth International Group, 1984. 19
4. C. Chow and C. Liu. Approximating discrete probability distributions with dependence trees. *IEEE Transactions on Information Theory*, 14:462–467, 1968. 20
5. N. Cristianini and J. Shawe-Taylor. *An Introduction to Support Vector Machines*. Cambridge University Press, 2000. 19
6. A. P. Dempster, N. M. Laird, and D. B Rubin. Maximum-likelihood from incomplete data via the EM algorithm. *J. Royal Stat. Soc.*, B39:1–38, 1977. 26
7. J. L. Elman. Finding structure in time. *Cognitive Science*, 14:179–211, 1990. 23
8. T. Fawcett and F. Provost. Adaptive fraud detection. *Knowledge Discovery and Data Mining*, 1:291–316, 1997. 15, 21
9. C. L. Giles, G. M. Kuhn, and R. J. Williams. Special issue on dynamic recurrent neural networks. *IEEE Transactions on Neural Networks*, 5(2), 1994. 23
10. A. E. Hoerl and R. W. Kennard. Ridge regression: biased estimation of nonorthogonal components. *Technometrics*, 12:55–67, 1970. 19
11. M. I. Jordan. Serial order: A parallel distributed processing approach. ICS Rep. 8604, Inst. for Cog. Sci., UC San Diego, 1986. 23
12. Ron Kohavi and George H. John. Wrappers for feature subset selection. *Artificial Intelligence*, 97(1–2):273–324, 1997. 19
13. Daphne Koller and Mehran Sahami. Toward optimal feature selection. In *Proc. 13th Int. Conf. Machine Learning*, pages 284–292. Morgan Kaufmann, 1996. 20
14. Igor Kononenko, Edvard Šimec, and Marko Robnik-Šikonja. Overcoming the myopic of inductive learning algorithms with RELIEFF. *Applied Intelligence, 7*(1): 39–55, 1997. 20
15. John Lafferty, Andrew McCallum, and Fernando Pereira. Conditional random fields: Probabilistic models for segmenting and labeling sequence data. In *Int. Conf. Machine Learning*, San Francisco, CA, 2001. Morgan Kaufmann. 26
16. Y. LeCun, L. Bottou, Y. Bengio, and P. Haffner. Gradient-based learning applied to document recognition. *Proceedings of the IEEE*, 86(11):2278–2324, 1998. 27
17. Oded Maron and Andrew W. Moore. Hoeffding races: Accelerating model selection search for classification and function approximation. In *Adv. Neural Inf. Proc. Sys. 6*, 59–66. Morgan Kaufmann, 1994. 20
18. Andrew McCallum, Dayne Freitag, and Fernando Pereira. Maximum entropy Markov models for information extraction and segmentation. In *Int. Conf. on Machine Learning*, 591–598. Morgan Kaufmann, 2000. 25
19. Thomas M. Mitchell. *Machine Learning*. McGraw-Hill, New York, 1997. 15
20. N. Qian and T. J. Sejnowski. Predicting the secondary structure of globular proteins using neural network models. *J. Molecular Biology*, 202:865–884, 1988. 21

21. J. R. Quinlan. *C4.5: Programs for machine learning.* Morgan Kaufmann, 1993. 19
22. D. E. Rumelhart, G. E. Hinton, and R. J. Williams. Learning internal representations by error propagation. In *Parallel Distributed Processing – Explorations in the Microstructure of Cognition*, chapter 8, pages 318–362. MIT Press, 1986. 23
23. T. J. Sejnowski and C. R. Rosenberg. Parallel networks that learn to pronounce english text. *Journal of Complex Systems*, 1(1):145–168, February 1987. 21
24. A. S. Weigend, D. E. Rumelhart, and B. A. Huberman. Generalization by weight-elimination with application to forecasting. *Adv. Neural Inf. Proc. Sys. 3*, 875–882, Morgan Kaufmann, 1991. 19

Graph-Based Methods for Vision: A Yorkist Manifesto

Edwin Hancock and Richard C. Wilson

Department of Computer Science, University of York
York Y01 5DD, UK

Abstract. This paper provides an overview of our joint work on graph-matching. We commence by reviewing the literature which has motivated this work. We then proceed to review our contributions under the headings of 1) the probabilistic framework, 2) search and optimisation, 3) matrix methods, 4) segmentation and grouping, 5) learning and 6) applications.

1 Introduction

For the past decade, we have been involved in a programme of activity aimed at developing a probabilistic framework for graph matching. This paper provides an overview of the main achievements of this programme of work. We commence, with a review of the literature to set our own work in context. The achievements are then summarised under the headings of 1) the probabilistic framework, 2) search and optimisation, 3) matrix methods, 4) segmentation and grouping, 5) learning and 6) applications. We conclude the paper by outlining our future plans.

The work summarised here has been undertaken with a number of present and former colleagues who have all made important contributions. These include Andrew Finch, Andrew Cross, Benoit Huet, Richard Myers, Mark Williams, Simon Moss, Bin Luo, Andrea Torsello, Marco Carcassoni and Antonio Robles-Kelly.

2 Literature Review

We set our work in context with a brief review of the related literature. Some of the pioneering work on graph matching was undertaken in the early 1970's by Barrow and Burstall [46] and by Fischler and Enschlager [55]. These two studies provided proof of concept for the use of relational structures in high-level pictorial object recognition. Over the intervening three decades, there has been a sustained research activity. Broadly speaking the work reported in the literature can be divided into three areas. The first of these is concerned with defining a measure of relational similarity. Much of the early work here was undertaken in the structural pattern recognition literature. For instance, Shapiro and Haralick [64] showed how inexact structural representations could be compared by

T. Caelli et al. (Eds.): SSPR&SPR 2002, LNCS 2396, pp. 31–46, 2002.
© Springer-Verlag Berlin Heidelberg 2002

counting consistent subgraphs. This similarity measure was refined by Eshera and Fu [53] and by Sanfeliu and Fu [61] who showed how the concept of string edit distance could be extended to graphical structures. The formal basis of graph edit distance has recently been extended by Bunke and his coworkers [48,49] who have shown, among other things, that the edit distance is related to the size of the maximum common subgraph. More recently Tirthapura, Sharvit, Klein and Kimia have shown how the classical Levenshtein distance can be used to match shock-graphs representing 2D skeletal shapes [69].

Much of the work described above adopts a heuristic or goal directed approach to measuring graph similarity. The second issue addressed in our literature survey is that of how to develop more principled statistical measures of similarity. This endeavour involves the modelling of the processes of structural error present in the graph-matching problem. Wong and You [72] made one of the first contributions here by defining an entropy measure for structural graph-matching. Boyer and Kak [47] also adopted an information theoretic approach, but worked instead with attribute relations. Using a probabilistic relaxation framework Christmas, Kittler and Petrou [50] have developed a statistical model for pairwise attribute relations.

The third issue is that of optimisation. Here there have been several attempts to use both continuous and discrete optimisation methods to locate optimal graph matches. Turning our attention first to discrete optimisation methods, there have been several attempts to apply techniques such as simulated annealing [56], genetic search [52] and tabu search [71] to the graph matching problem. However, continuous optimisation methods provide attractive alternatives since their fixed points and convergence properties are usually better understood than their discrete counterparts. However, the main difficulty associated with mapping a discretely defined search problem onto a continuous optimisation method is that of embedding. There are several ways in which this embedding can be effected for the problem of graph matching. The most straightforward of these is to pose the graph-matching problem as that of recovering a permutation matrix which preserves edge or adjacency structure. For instance, Kosowsky and Yuille have cast the problem into a statistical physics setting and have recovered a continuous representation of the permutation matrix using mean-field update equations [73]. Gold and Rangarajan [54] have exploited the stochastic properties of Sinkhorn matrices to recover the matches using a soft-assign update algorithm. Umeyama [70] takes a more conventional least-squares approach and shows how an eigendecomposition method can be used to recover the permutation matrix. An alternative representation has recently been developed by Pelillo [60] which involves embedding the association graph. Matches are located by using the replicator equations of evolutionary game-theory to locate the maximal clique of the association graph, i.e. the maximum common subgraph, of the two graphs being matched. This method has subsequently also been applied to shock-graph matching [74].

Closely related to this work on recovering permutation structure by continuous embedding is the literature on spectral graph theory. This is a term applied to a family of techniques that aim to characterise the global structural properties of graphs using the eigenvalues and eigenvectors of the adjacency matrix [51]. In the computer vision literature there have been a number of attempts to use spectral properties for graph-matching, object recognition and image segmentation. Umeyama has an eigendecomposition method that matches graphs of the same size [70]. Borrowing ideas from structural chemistry, Scott and Longuet-Higgins were among the first to use spectral methods for correspondence analysis [62]. They showed how to recover correspondences via singular value decomposition on the point association matrix between different images. In keeping more closely with the spirit of spectral graph theory, yet seemingly unaware of the related literature, Shapiro and Brady [65] developed an extension of the Scott and Longuet-Higgins method, in which point sets are matched by comparing the eigenvectors of the point proximity matrix. Here the proximity matrix is constructed by computing the Gaussian weighted distance between points. The eigen-vectors of the proximity matrices can be viewed as the basis vectors of an orthogonal transformation on the original point identities. In other words, the components of the eigenvectors represent mixing angles for the transformed points. Matching between different point-sets is effected by comparing the pattern of eigenvectors in different images. Shapiro and Brady's method can be viewed as operating in the attribute domain rather than the structural domain. Horaud and Sossa[57] have adopted a purely structural approach to the recognition of line-drawings. Their representation is based on the immanental polynomials for the Laplacian matrix of the line-connectivity graph. By comparing the coefficients of the polynomials, they are able to index into a large data-base of line-drawings. In another application involving indexing into large data-bases, Sengupta and Boyer[63] have used property matrix spectra to characterise line-patterns. Various attribute representations are suggested and compared. Shokoufandeh, Dickinson and Siddiqi [67] have shown how graphs can be encoded using local topological spectra for shape recognition from large data-bases.

3 Probability Distributions

We embarked on our programme of work motivated by the need for a probabilistic framework so that graph matching could be approached in a statistically robust manner. At the time (1992), there were three pieces of work that addressed this problem. Wong and You [72], in early work in the structural pattern recognition area had made one of the first attempts to cast the graph matching problem into an information theoretic setting by defining the relative entropy of two graphs. The second contribution came from Boyer and Kak [47], who cast the problem of structural stereopsis into a mutual information setting, and by choosing Gaussian distributions to measure the distribution of attributes, arrived at a least squares framework for relational matching. In a more ambitious piece of work, Christmas, Kittler and Petrou [75] showed how pairwise attribute relations could be modelled probabilistically and used to match graphs using an iterative probabilistic relaxation scheme. Taking a critical view of this existing work, we identified two areas where there appeared to be scope for improvement and further work. First, the existing work in the area relied only on pairwise relations. In other words, there had been little attempt to model constraints on the graph matching problem beyond the edge-level. Second, there appeared to be a move away from the use of purely structural constraints in favour of attribute relations. Hence, we embarked on a programme of work aimed at modelling the distribution of structural errors and using the distribution to develop robust graph matching algorithms.

3.1 Edge and Face Compatibilities

Our initial study followed Christmas, Kittler and Petrou and persued the graph-matching problem in the setting of probabilistic relaxation. Using the product-form support function developed by Kittler and Hancock [76], we set about the graph matching problem by attempting to model the compatibility co-efficients for graph matching [5,9]. In the probabilistic framework, these take the form of conditional probabilities. When the distribution of structural errors follows a uniform distribution, then we discovered that the compatibilities followed a particularly simple and intuitively appealing distribution rule. In the case of edges and triangular faces, the compatibilities were simply proportional to the edge and face densities in the graphs being studied. Moreover, the entire compatibility model was free of parameters.

3.2 Dictionaries

Encouraged by these findings, our next step was more ambitious. Here we aimed to cast the graph-matching problem into a discrete relaxation setting [2]. Rather than modelling structural corruption at the level of edge or face units, we aimed to extend the model to entire edge-connected neighbourhoods of nodes. To do this we adopted a dictionary model of structural errors [4,7]. This allowed for neighbourhoods to match to one-another under a cyclic permutation of nodes with possible dummy node insertions. Our structural model hence commenced from a dictionary of possible mappings between each data-graph neighbourhood and each permuted and padded neighbourhood of the model graph. With this structural model to hand, the next step was to develop probability distributions for the different types of error that can occur in graph matching by discrete relaxation. Discrete relaxation is an iterative process which relies on replacing symbolic label assignments so as to maximise a global consistency criterion [4]. Hence, at any iterative epoch there are two types of error that can occur. The first of these are label placement of assignment errors. The second are structural errors due to node and edge dropout or insertion. We commenced from a simple distribution model in which these two error processes were modelled by uniform and memoryless distributions. Based on these two assumptions, we arrived at a probability distribution for the label configurations contained in the structural dictionary. Again the model had a simple intuitive form. Each dictionary item had a probability that took an exponential form. The exponential was a function of two variables. The first of these was the Hamming distance between the current configuration of label placements and those demanded by the permutations in the dictionary. The second variable was the size difference or number of padding nodes needed. The distribution was controlled by two parameters. These are the probability of label misplacement and the probability of dummy node insertion. The former parameter was gradually reduced to zero (annealed) as the discrete relaxation process iterated. The latter parameter was estimated from the size difference of the graphs being matched.

3.3 Graph Editing

In addition to allowing graph-matching, the probabilistic framework also allows us to rectify the effects of structural corruption. The aim here is to use maximum a posteriori probability estimation to restore noise corrupted graphs. This is done by developing statistical tests for the removal ands re-insertion of nodes, together with their associated edges [11]. This method provides significant performance advantages over other methods for dealing with structural error such as null-labelling or locating maximal cliques of the association graph [4,7].

3.4 Edit Distance

One of the shortcomings of the method described above is that it requires the enumeration of a dictionary of structures so that it may be applied to the graph-matching problem. The number of dictionary items has a potentially exponential complexity with the degree of the nodes in the graph when padding with dummy insertions is allowed. To overcome this problem, we turned to string edit distance or Levenshtein distance [77] as a means of computing the similarity of labelled neighbourhoods. There had been previous attempts to extend the idea of edit distance to graphs. The idea of doing this had already been explored by Eshera and Fu [53], and by Sanfeliu and Fu [61] in their early work on structural matching. However, although effective the extension to graphs, lacks some of the formal neatness of the string treatment. Hence, we adopted an approach in which each neighbourhood is represented by a string, and the different permutations are implicitly traversed in the edit matrix. In this way we lift the exponential overhead on dictionary enumeration [21]. The location of the optimal string edit path is computed using Dijkstra's algorithm. Since, the Hamming distance of the original probability distribution is a special case of the Levenshtein distance, we again use a family of exponential distributions to model the probability of match between neighbourhoods.

Recently, we have taken this work one step further. We make use of the recent observation by Bunke [49] that the size of the maximum common subgraph and the edit distance are related to one another. Hence, by locating the maximum common subgraph, we may estimate edit distance. The recent work of Pelillo [60] provides a means of locating the max-clique using a quadratic programming method. We have made use of this result to efficiently compute tree edit distance [37].

4 Search and Optimisation

Once a measure of graph similarity is to hand, then a means of locating the matching configuration of maximum similarity is one of configurational optimisation. In our initial experiments with the methods described in the previous section we used simple hill-climbing methods. These are prone to convergence to local optima. Hence, an important theme in our work has been to explore the use of global optimisation methods for graph matching.

4.1 Genetic Search

The simplest way to overcome local convergence problems is to adopt a stochastic optimisation framework. There are a number of well known techniques here, including the extensively studied process of simulated annealing. However, genetic search offers a topical method which has furnished us with an effective route to graph matching.

Evolutionary or genetic search is a population-based stochastic optimisation process which draws is inspiration from population genetics. The process relies on three different operations to modify a population of possible solutions to discretely defined problems which have been encoded in a string of symbols which resemble the arrangement of genes on a chromosome. Mutation is a background operator which randomly switches individual symbols. Crossover or recombination, exchanges randomly selected subparts of pairs of strings residing in the population to produce new ones. Finally, selection uses a fitness measure for the solutions candidates which can be retained in the population. In a series of papers [8,12,22,24,26,28] we have investigated how genetic search can be tailored to our probabilistic framework for graph-matching. The reason for doing this is that there are several features of our formulation that have an natural assonance with genetic search. The first of these is that we have a probabilistic measure of graph similarity, which naturally plays the role of a fitness function [8]. Second, the development of this model relies on the specification of the assignment error probability. This probability can be used to control the re-assignment of the labels to nodes via mutation operator. However, despite these appealing conceptual isomorphisms, there are a number of more difficult problems to be solved in mapping the graph-matching problem onto a genetic search procedure. The most obvious of these is that GA's usually rely on a string encoding, and a graph is a relational structure. Although the encoding issue is not one of major importance, the main difficultly arises when considering how to implement crossover since we no longer have the convenience of strings. To overcome this problem we have developed a graph-based crossover mechanism. This involves performing a graph-cuts and recombining the two subgraphs so formed. Normally, this would be rather intractable since we would have to produce cuts in different graphs which traversed the same number of edges so that the subgraphs could be recombined. We circumvent this problem by working with the Delauney graphs of point-sets. Here we bisect planar point-sets with a straight line and recombine the point-sets at the cut line. The recombined point-sets can be re-triangulated to produce new Delaunay graphs. Structural errors may be corrected by deleting and re-inserting points in the recombination step. We have performed both theoretical and empirical analysis on the resulting graph-matching process. First, we have established the convergence properties of the resulting algorithm using a Gaussian model of the distribution of population fitness. Second, we have embarked on an exhaustive investigation of the best choice of mutation, crossover, and selection operators for use in conjunction with the graph-matching problem. Finally, we have exploited the population architecture of the optimisation process to develop a least commitment graph-matching method which allows multiple ambiguous solutions to be maintained.

4.2 Softened Representations

One way to escape local optima in discretely defined optimisation problems such as graph matching, consistent labelling and constraint satisfaction is to soften the representation. In other words, the aim is to replace the symbolic labels by real-valued assignment variables or by label probabilities. Probabilistic relaxation is such a scheme, but it only posses local convergence properties. A method which posses global convergence properties is mean field annealing. The have been several attempts to cast the graph-matching problem into a mean field setting. These include the early work of Simic [68] and Suganathan [78]. The idea here is to locate an energy function for the assignment or mean field variables through a process of local averaging over the configuration space of the matching problem. The mean field variables are updated by exponentiating the derivatives of the energy function. As the process proceeds, the constant of the exponentials is annealed, so as to increasingly penalise departures from consistency. These methods are also closely akin to the soft-assign method of Gold and Rangarajan [54].

However, these early mean-field methods are often referred to as "naive" since they commence from an ad hoc energy function rather than a principled probability distribution. More recently, the apparatus of variational inference have been deployed to develop mean field equations [58,79]. The idea here is to find the set of mean field variables which have minimum Kullback divergence to a prior distribution for the discrete entities for the problem in hand. We have adopted this approach to develop soft assign equations for graph matching [11,13]. Our approach is midway between the naive one and the full variational treatment. The reason for the compromise is, of course, that of tractability. Rather than using the variational approach to find minimum Kullback divergence update equations for the assignment variables, we use it to find an equivalent free energy for the dictionary-based probability distribution. This energy is couched in terms of assignment variables. We follow Gold and Rangarajan [54] and update the assignment variables by exponentiating the derivatives of the free energy. The result is a non-quadratic energy function which proves more robust than its quadratic counterpart.

4.3 Tabu Search

In addition to stochastic and mean-field methods for optimisation methods there are also deterministic methods which can yield good results for the graph-matching problem. Recently, tabu search has been used with great effect for a number of path-based planning and scheduling problems. The method uses an aspiration criterion to intensify and diversify the search procedure. We have used this search procedure with our probabilistic similarity measure to perform a deterministic brushfire search for graph matches [6,17]. We commence from the most confident matches and propagate the search for matches across the edges of the graph to adjacent nodes. This method provides a search procedure which is both efficient and effective.

5 Matrix Methods

One of the disappointments of the early matrix-based methods for graph-matching is that while they are extremely elegant, they can not cope with graphs of different size and are not robust to structural error. For this reason we have recently attempted to extend our probabilistic framework for graph-matching to the matrix domain. Our aim here has been to combine probabilistic methods with ideas from spectral graph theory [51].

5.1 Singular Value Decomposition

The method of Umeyama attempts to match graphs by comparing the singular vectors of the adjacency matrix. One of the problems which limits this method is that there is no clear way to compare the singular vectors of matrices of different size. To overcome this problem, we have recently presented a study which aims to cast the Umeyama [70] algorithm into the setting of the EM algorithm. Commencing from a simple model for the correspondence process in which assignment errors follow a Bernoulli distribution, we have developed a mixture model for graph-matching [27]. This leads to a utility measure for the graph-matching problem which is obtained from the trace of the weighted correlation of the adjacency matrices for the two graphs being matched. The weight matrix allows for both correspondence errors and difference in the size of the two adjacency matrices.

We find optimal matches in a two-step iterative EM algorithm. In the M or maximisation step, the optimal matches are found by performing singular value decomposition on the weighted adjacency correlation matrix. In the E-step, the weight matrix used to perform adjacency matrix correlation is updated.

5.2 Adjacency Matrix Markov Chains

Spectral methods hold out another promising route to graph-matching. It is well known that the leading eigenvector of the transition probability matrix for a Markov chain is the steady state random walk on the graphical model of the Markov chain. Viewed slightly differently, if we commence with the adjacency matrix for a graph and convert it into a transition probability matrix using row and column normalisation, then the leading eigenvector provides us with a string representation of the graph. By converting graphs into strings in this way, then graph-matching can be effected using string-matching methods. With this goal in mind, we have recently reported a method which aims to match graphs using the Levenshtein distance between steady state random walks [43].

6 Segmentation and Grouping

To apply the matching methods listed above, which are entirely generic in nature, we require graph structures to be segmented from image data. The ideas of error modelling and graph editing also provide theoretical tools which can be used to develop segmentation and grouping algorithms.

6.1 Regions and Arrangements

The idea of using a Bernoulli distribution to model structural errors also has applications to the pairwise grouping of image entities. To do this we have used the Bernoulli distribution to model the distribution of edge weights or grouping strength between image tokens [35,36,42]. These may be viewed as the elements of a weighted adjacency matrix. We have developed a simple EM algorithm based on this Bernoulli model, which encourages blocks to form in the permuted adjacency matrix. This algorithm uses the eigenvectors of the adjacency matrix to update the connectivity weights, and shares much in common with the normalised cuts method of Shi and Malik [66].

6.2 Meshes and Surfaces

The explicit graph edit operations used to correct structural errors in the graph-matching process can also be used to modify relational structures so as to improve their consistency with raw image data. One application in which this arises is the control of surface meshes. Here the aim is to have a fine mesh in the proximity of high curvature surface detail, and a coarse grained mesh in the proximity of relatively unstructured surface regions. We have developed an adaptive surface mesh which uses split and merge operations to control the placement of mesh facets so as to optimise a bias-variance criterion [23]. Here facets may be added or subtracted so that they minimise the sum of variance and squared bias. In this way we achieve good data-closeness while avoiding over fitting with too many mesh facets.

7 Learning

Our probabilistic framework has provided us with a means of measuring the similarity of graphs. With good similarity measures to hand, then we can address a number of problems that involve learning from structural representations.

7.1 Graph Clustering

The simplest task that arises in machine learning with structural representations is how to cluster similar graphs together. This problem is more complicated than that of using feature or pattern vectors, since it is notoriously difficult to define the mean and covariance of a set of graphs. Hence, central clustering methods can not easiliy be used to learn classes of graphs. However, there are several methods which allow data to be clustered using pairwise or relational attributes rather than ordinal ones. Two such methods are multidimensional scaling and pairwise clustering. We have recently applied both methods to the clustering of shock-tress using edit distances between pairs of graphs [38,44]. The results prove promising and indicate that shape classes can be learnt from structural representations.

7.2 Structural Variations

One of the shortcoming of simply using a pattern of distances, it that it does not capture the variations in detailed structure associated with a particular class of graph-structure. In a recent study we have turned to spectral graph theory to furnish us with a representation suitable for this purpose. One of the problems with computing the mean and variance of a set of graphs is that they may potentially contain different numbers of node and edges, and even if the structures are the same then correspondences between nodes are required. To overcome this problem we have use the leading eigenvectors of the adjacency matrices of sets of graphs to define fuzzy clusters of nodes. From the fuzzy clusters, we compute features including volume, perimeters and Cheeger constants. By ordering the clusters according to the magnitude of the associated eigenvalue, we construct feature-vectors of fixed length. ¿From these vectors it is possible to construct eigenspaces which trace out the main modes of structural variation in the graphs [45].

8 Applications

The research described in the previous sections is of a predominantly methodological and theoretical nature. However, we have applied the resulting techniques to a large variety of real world problems. In this section we summarise these.

8.1 Model Alignment

Graph-matching provides correspondences between image entities. The entities may be points, lines, regions or surfaces depending on the abstraction adopted. However, once correspondence information is to hand then it may be used for the purposes of detailed model alignment. We have explored a number of different approaches to using the correspondence information delivered by our graph-matching methods for model alignment.

Rigid Models The first class of alignment process are those which involve rigid transformations. The rigid geometries most frequently used in computer vision are the similarity, affine and perspective transformations. To exploit our probabilistic framework for measuring the consistency of graph correspondences, we have developed a novel dual-step EM algorithm [14,18]. Here we work with a mixture model over the space of correspondences. The aim is estimate the parameters of any the three rigid transformations that bring noisy point sets of potentially different size into correspondence with one another. If this is done using the EM algorithm then the aim is to find the set of parameters that optimise an expected log likelihood function. If the alignment errors between the different point sets follow a Gaussian distribution then the expected log-likelihood is a weighted squared error function, where the weights are the a posteriori alignment probabilities. Starting from first principles, we have shown how the correspondence probabilities delivered by our dictionary-based structural error model can be used to impose additional constraints on the alignment process. This takes the form of an additional weighting of the contributions to the expected log-likelihood function. We develop a dual step EM algorithm which interleaves the tasks of finding point correspondences and estimating alignment parameters. The method has been applied to recovering plane perspective pose. We have also shown how the method relates to the Procrustes alignment process for point-sets subject to similarity transformations [31].

Non-rigid Models For non-rigid alignment we have adopted a slightly different approach. Here we have posed the alignment and correspondence problems in a setting which involves maximising the cross entropy between the probability distributions for alignment and correspondence errors [59]. We have applied the resulting matching process to the problem of aligning unlabelled point distribution models to medical image sequences. In this way we are able to impose relational constraints on the arrangement of landmark points, and improve the robustness of the matching process to both measurement and structural errors [30,41].

8.2 Object Recognition

Another important application of the graph matching methods is to object recognition, and in particular that of recognising an object from a large number of image instances. This problem is one of practical significance since it arises in the retrieval of images from large data-bases.

Line-Patterns The main application vehicle for the object recognition has been the retrieval of trade marks abstracted in terms of line pattern arrangements. Here we have used a coarse to fine search procedure. We commence with a representation of the line patterns based on a so-called relational histogram. This is a histogram of affine invariant relative angle and length ratio attributes. However, only those pairs of lines that are connected by a line-centre adjacency graph can contribute to the histogram [20]. The next step is to refine the candidate matches using a set-based representation of the edge attributes. Here we use a robust variant of the Hausdorf distance to find the closest matching sets of attributes [29]. Finally, the pruned set of matches is searched for the best match using a full graph-based matching scheme [19].

Surface Topography The framework described above can be extended from line patterns to surface patches. One of our complementary interests is in the process of shape-from-shading [15]. This is concerned with recovering surface patches from local shading variations. We have shown how the robust Haussdorff distance can be used to recognise 3D objects from 2D views using the arrangement surface patches together with their shape-index attributes [25].

View-Based Object Recognition The final application studied has been view based object recognition. Here the aim is to a large number of views of a 3D object as the viewpoint is slowly varied.

9 The Future

One of the main challenges facing the structural pattern recognition community is that of developing methods for shape modelling that are comparable in performance and versatility to those developed by the computer vision community. Here eigenspace methods have proved powerful for both modelling variations in boundary shape and appearance attributes. Although the vision community have recently been pre-occupied with the use of geometric methods, the recent use of shock-graphs have increased their confidence and curiosity in the use of structural methods. However, while recent work has demonstrated the utility of graph-based methods for matching and recognition, it is fair to say that the available methodology for learning structural descriptions is limited. In many ways the graphical models community has much from which we can learn. They have developed principled information theoretic methods for learning the structure of both Bayes nets and other graphical structures such as decision trees. It must be stressed that these are inference structures rather than relational models. Hence, they are rather different from the relational structures encountered in image analysis and computer vision which are predominantly concerned with the representation of spatially organised data. What we do have to offer is a methodology for measuring the similarity of relational structures at a relatively fine-grained level. Moreover, if spectral techniques can be further tamed then there is scope for considerable convergence of methodology.

References

1. J. Kittler and E. R. Hancock, "Combining Evidence in Probabilistic Relaxation," *International Journal of Pattern Recognition and Artificial Intelligence*, **3**, pp.29–52, 1989.
2. E. R. Hancock and J. Kittler, "Discrete Relaxation," *Pattern Recognition*, **23**, pp.711–733, 1990. 34
3. E. R. Hancock and J. Kittler, "Edge-labelling using Dictionary-based Probabilistic Relaxation," *IEEE Transactions on Pattern Analysis and Machine Intelligence*,**12**, pp 165–181 1990.
4. R. C. Wilson, A. N. Evans and E. R. Hancock, "Relational Matching by Discrete Relaxation", *Image and Vision Computing*, **13** , pp. 411–422, 1995. 34, 35
5. R. C. Wilson and E. R. Hancock, "A Bayesian Compatibility Model for Graph Matching", *Pattern Recognition Letters*, **17**, pp. 263–276, 1996. 34
6. M. L. Williams, R. C. Wilson, and E. R. Hancock, "Multiple Graph Matching with Bayesian Inference", *Pattern Recognition Letters*, **18**, pp. 1275–1281, 1997. 38
7. R. C. Wilson and E. R. Hancock, "Structural Matching by Discrete Relaxation", *IEEE Transactions on Pattern Analysis and Machine Intelligence*, **19**, No.6, pp.634-648, 1997. 34, 35
8. A. D. J. Cross, R. C. Wilson and E. R. Hancock, "Genetic Search for Structural Matching", *Pattern Recognition*, **30**, pp.953-970, 1997. 36
9. A. M. Finch, R. C. Wilson and E. R. Hancock, "Matching Delaunay Graphs", *Pattern Recognition*, **30**, pp. 123–140, 1997. 34
10. A. M. Finch, R. C. Wilson and E. R. Hancock, "Symbolic Graph Matching with the EM Algorithm", *Pattern Recognition*, **31**, pp. 1777–1790, 1998.
11. R. C. Wilson, A. D. J. Cross and E. R. Hancock, "Structural Matching with Active Triangulations", *Computer Vision and Image Understanding*, **72**, pp. 21–38, 1998. 35, 37
12. A. D. J. Cross and E. R. Hancock, "Matching Buildings in Aerial Stereogramms using Genetic Search and Matched Filters", *ISPRS Journal of Photogrammetry and Remote Sensing*, **53**, pp. 95–107, 1998. 36
13. A. M. Finch, R. C. Wilson and E. R. Hancock, "An Energy Function and Continuous Edit Process for Graph Matching", *Neural Computation*, **10**, pp. 1873–1894, 1998. 37
14. A. D. J. Cross and E. R. Hancock, "Graph Matching with a Dual Step EM Algorithm", *IEEE Transactions on Pattern Analysis and Machine Intelligence*, **20**, pp. 1236–1253,, 1998. 41
15. P. L Worthington and Edwin R Hancock, "New Constraints on Data-Closeness and Needle Map Consistency for Shape-from-Shaping", *IEEE Transactions on Pattern Analysis and Machine Intelligence,* IEEE Computer Society Press, **vol. 21**, no. 12, pp 1250-1267, 1999. 42
16. R. C. Wilson and E. R. Hancock, "Graph matching with hierarchical discrete relaxation", *Pattern Recognition Letters*, **20**, pp 1041-1052, 1999.
17. M. L. Williams and R. C. Wilson and E. R. Hancock, "Deterministic Search For Relational Graph Matching", *Pattern Recognition*, **32**, pp. 1255-1271, 1999. 38
18. S. Moss and Richard C. Wilson and E. R. Hancock, "A mixture model of pose clustering", *Pattern Recognition Letters*, **20**, pp 1093-1101, 1999. 41
19. B. Huet and E. R. Hancock, "Shape recognition from large image libraries by inexact graph matching", *Pattern Recognition Letters*, **20**, pp. 1259-1269, 1999. 42

20. B. Huet and E. R. Hancock, "Line Pattern Retrieval Using Relational Histograms", *IEEE Transactions on Pattern Analysis and Machine Intelligence*, **21**, pp. 1363-1370, 1999. 42
21. R. Myers, R. C. Wilson and E. R. Hancock, "Bayesian Graph Edit Distance", *IEEE Transactions on Pattern Analysis and Machine Intelligence*, **22**, pp 628-635, 2000. 35
22. A. D. J. Cross and E. R. Hancock, "Convergence of a Hill Climbing Genetic Algorithm for Graph Matching", *Pattern Recognition*, **33**, pp 1863-1880, 2000. 36
23. R. C. Wilson and E. R. Hancock, "Bias Variance Analysis for Controlling Adaptive Surface Meshes", *Computer Vision and Image Understanding* **77**, pp 25-47, 2000. 39
24. R. Myers, E. R. Hancock, "Genetic Algorithms for Ambiguous Labelling Problems", *Pattern Recognition*, **33**, pp. 685–704, 2000. 36
25. P. L. Worthington and E. R. Hancock, "Object Recognition using Shape-from-shading", *IEEE Transactions on Pattern Analysis and Machine Intelligence*, **33**, pp. 535–562, 2001. 42
26. R. Myers and E. R. Hancock, "Least Commitment Graph Matching with Genetic Algorithms", *Pattern Recognition*, **34**, pp 375-394, 2001. 36
27. B. Luo and E. R. Hancock "Structural Matching using the EM algorithm and singular value decomposition", *IEEE Transactions on Pattern Analysis and Machine Intelligence*, **23**, pp. 1120—1136, 2001. 38
28. R. Myers and E. R. Hancock, "Empirical Modelling of Genetic Algorithms", *Evolutionary Computation*, **9**, pp. 461–493, 2001. 36
29. B. Huet and E. R. Hancock, "Relational Object Recognition from Large Structural Libraries", *Pattern Recognition*, 2002. 42
30. M. Carcassoni and E. R. Hancock, "Spectral Correspondence for Point Pattern Matching", *Pattern Recognition*, to appear 2002. 41
31. B. Luo and E. R. Hancock, "Iterative Procrustes Analysis using the EM Algorithm, *Image and Vision Computing*, to appear 2002. 41
32. B. Huet and E. R. Hancock, "Object Recognition from Large Structural Libraries", *Advances in Structural and Syntactic Pattern Recognition, Springer*, Lecture Notes in Computer Science, **1451**, pp. 190–199, 1998.
33. S. Moss and Edwin R. Hancock, "Structural Constraints for Pose Clustering", *Computer Analysis of Images and Patterns, Springer Lecture Notes in Computer Science*, **1689**, Franc Solina and Ales Leonardis eds.,pp. 632-640, 1999.
34. A. Torsello and E. R. Hancock, "A Skeletal Measure of 2D Shape Similarity", *Springer Lecture Notes in Computer Science*, **2059**, Edited by C. Arcelli, L. P. Cordella and G. Sannitii di Baja, pp. 260–271, 2001.
35. A. Robles-Kelly and E. R. Hancock, "An Expectation-Maximisation Framework for Perceptual Grouping", *Springer Lecture Notes in Computer Science*, **2059**, Edited by C. Arcelli, L. P. Cordella and G. Sannitii di Baja, pp. 594–605, 2001. 39
36. A. Kelly and E. R. Hancock, "A maximum likelihood framework for grouping and segmentation", *Springer Lecture notes in Computer Science*, **2134**, pp. 251–266, 2001. 39
37. A. Torsello and E. R. Hancock, "Efficiently computing weighted tree edit distance using relaxation labeling", *Springer Lecture notes in Computer Science*, **2134**, pp. 438–453, 2001. 35
38. B. Luo, A. Robles-Kelly, A. Torsello, R. C. Wilson, E. R. Hancock, " Discovering Shape Categories by Clustering Shock Trees", *Springer Lecture notes in Computer Science*, **2124**, pp. 151-160, 2001. 40

39. B. Huet and E. R. Hancock, "Fuzzy Relational Distance for Large-scale Object Recognition", *IEEE Computer Society Computer Vision and Pattern Recognition Conference, IEEE Computer Society Press*, pp. 138–143, 1998.
40. S. Moss and E. R. Hancock, "Pose Clustering with Density Estimation and Structural Constraints", *IEEE Computer Society Conference on Computer Vision and Pattern Recognition, IEEE Computer Society Press*, Fort Collins, vol II, pp. 85-91, 1999.
41. M. Carcassoni and E. R. Hancock, "Point Pattern Matching with Robust Spectral Correspondence", *IEEE Computer Society Conference on Computer Vision and Pattern Recognition, IEEE Computer Society Press*, **I**, pp. 649-655, 2000. 41
42. A. Robles-Kelly and E. R. Hancock, "A Maximum Likelihood Framework for Iterative Eigendecomposition", *Eighth International Conference on Computer Vision*, IEEE Computer Society Press, Vancouver, Canada, pp. 654–661, 1999. 39
43. A. Robles-Kelly and E. R. Hancock, " Graph Matching using Adjacency Matrix Markov Chains", *British Machine Vision Conference*, pp. 383–390, 2001. 39
44. A. Torsello, B. Luo, A. Robles-Kelly, R. C. Wilson and E. R. Hancock, "A Probabilistic Framework for Graph Clustering", *IEEE Computer Vision and Pattern Recognition Conference*, pp. 912-919, 2001. 40
45. A. Torsello and E. R. Hancock, "Matching and Embedding through Edit Union of Trees", ECCV 2002, to appear. 40
46. H. G. Barrow and R. J. Popplestone. Relational descriptions in picture processing. *Machine Intelligence*, VI:377–396, 1971. 31
47. K. Boyer and A. Kak. Structural Stereopsis for 3D Vision. *IEEE Transactions on Pattern Analysis and Machine Intelligence*, 10:144–166, 1988. 32, 33
48. H. Bunke and K. Shearer. A graph distance metric based on the maximal common subgraph. *Pattern Recognition Letters*, 19:255–259, 1998. 32
49. H. Bunke. Error correcting graph matching: On the influence of the underlying cost function. *IEEE Transactions on Pattern Analysis and Machine Intelligence*, 21:917–922, 1999. 32, 35
50. W. J. Christmas, J. Kittler, and M. Petrou. Structural matching in computer vision using probabilistic relaxation. *IEEE PAMI*, 17(8):749–764, 1995. 32
51. F. R. K. Chung. *Spectral Graph Theory.* American Mathmatical Society Ed., CBMS series 92, 1997. 33, 38
52. A. D. J. Cross, R. C. Wilson, and E. R. Hancock. Inexact graph matching with genetic search. *Pattern Recognition*, 30(6):953–970, 1997. 32
53. M. A. Eshera and K. S. Fu. An image understanding system using attributed symbolic representation and inexact graph-matching. *Journal of the Association for Computing Machinery*, 8(5):604–618, 1986. 32, 35
54. S. Gold and A. Rangarajan. A graduated assignment algorithm for graph matching. *IEEE PAMI*, 18(4):377–388, 1996. 32, 37
55. M. Fischler and R. Elschlager. The representation and matching of pictorical structures. *IEEE Transactions on Computers*, 22(1):67–92, 1973. 31
56. L. Herault, R. Horaud, F. Veillon, and J. J. Niez. Symbolic image matching by simulated annealing. In *Proceedings of British Machine Vision Conference*, pages 319–324, 1990. 32
57. R. Horaud and H. Sossa. Polyhedral object recognition by indexing. *Pattern Recognition*, 28(12):1855–1870, 1995. 33
58. M. I. Jordan and R. A. Jacobs. Hierarchical mixture of experts and the EM algorithm. *Neural Computation*, 6:181–214, 1994. 37
59. B. Luo and E. R. Hancock. Relational constraints for point distribution models. *Springer Lecture Notes in Computer Science*, 2124:646–656, 2001. 41

60. M. Pellilo. Replicator equations, maximal cliques, and graph isomorphism. *Neural Computation*, 11(8):1933–1955, 1999. 32, 35
61. A. Sanfeliu and K. S. Fu. A distance measure between attributed relational graphs for pattern recognition. *IEEE Trans. Systems, Man and Cybernetics*, 13(3):353–362, May 1983. 32, 35
62. G. L. Scott and H. C. Longuet-Higgins. An Algorithm for Associating the Features of 2 Images. *Proceedings of the Royal Society of London Series B-Biological*, 244(1309):21–26, 1991. 33
63. K. Sengupta and K. L. Boyer. Modelbase partitioning using property matrix spectra. *Computer Vision and Image Understanding*, 70(2):177–196, 1998. 33
64. L. G. Shapiro and R. M. Haralick. A metric for comparing relational descriptions. *IEEE Transactions on Pattern Analysis and Machine Intelligence*, 7(1):90–94, January 1985. 31
65. L. S. Shapiro and J. M. Brady. Feature-based Correspondence - An Eigenvector Approach. *Image and Vision Computing*, 10:283–288, 1992. 33
66. J. Shi and J. Malik. Normalized cuts and image segmentation. *PAMI*, 22(8):888–905, August 2000. 39
67. A. Shokoufandeh, S. J. Dickinson, K. Siddiqi, and S. W. Zucker. Indexing using a spectral encoding of topological structure. In *Proc. of the IEEE Conf. on Computer Vision and Pattern Recognition*, pages 491–497, 1999. 33
68. P. D. Simic. Constrained nets for graph matching and other quadratic assignment problems. *Neural Computation*, 3:268–281, 1991. 37
69. S. áTirthapura, D. Sharvit, P. Klein, and B. B. Kimia. Indexing based on edit-distance matching of shape graphs. *Multimedia Storage And Archiving Systems III*, 3527:25–36, 1998. 32
70. S. Umeyama. An eigen decomposition approach to weighted graph matching problems. *IEEE Transactions on Pattern Analysis and Machine Intelligence*, 10:695–703, 1988. 32, 33, 38
71. M. L. áWilliams, R. C. Wilson, and E. R. Hancock. Deterministic search for relational graph matching. *Pattern Recognition*, 32(7):1255–1271, 1999. 32
72. A. K. C. Wong and M. You. Entropy and distance of random graphs with application to structural pattern recognition. *IEEE Transactions on Pattern Analysis and Machine Intelligence*, 7:509–609, 1985. 32, 33
73. A. L. Yuille, P. Stolorz, and J. Utans. Statistical Physics, Mixtures of Distributions, and the EM Algorithm. *Neural Computation*, 6:334–340, 1994. 32
74. M. Pelillo, K. Siddiqi, and S. W. Zucker. Matching hierarchical structures using association graphs. *IEEE PAMI*, 21(11):1105–1120, 1999. 32
75. W. J. Christmas, J. Kittler, and M. Petrou. Structural matching in computer vision using probabilistic relaxation. *IEEE Transactions on Pattern Analysis and Machine Intelligence*, 17(8):749–764, 1995. 33
76. J. Kittler and E. R. Hancock. Combining evidence in probabilistic relaxation. *IEEE PRAI*, 3:29–51, 1989. 34
77. V. Levenshtein. Binary codes capable of correcting deletions, insertions, and reversals. *Soviet Physics-Doklady*, 10:707–710, 1966. 35
78. P. N. Suganthan, E. K. Teoh, and D. P. Mital. Pattern-recognition by graph matching using the potts mft neural networks. 37
79. T. Hofmann and J. M. Buhmann. Pairwise data clustering by deterministic annealing. *PAMI*, 19(2):192–192, February 1997. 37

Reducing the Computational Cost of Computing Approximated Median Strings*

Carlos D. Martínez-Hinarejos, Alfonso Juan,
Francisco Casacuberta, and Ramón Mollineda

Departament de Sistemes Informàtics i Computació
Institut Tecnològic d'Informàtica, Universitat Politècnica de València
Camí de Vera s/n, 46022, València, Spain

Abstract. The k-Nearest Neighbour (k-NN) rule is one of the most popular techniques in Pattern Recognition. This technique requires good prototypes in order to achieve good results with a reasonable computational cost. When objects are represented by strings, the Median String of a set of strings could be the best prototype for representing the whole set (i.e., the class of the objects). However, obtaining the Median String is an **NP**-Hard problem, and only approximations to the median string can be computed with a reasonable computational cost. Although proposed algorithms to obtain approximations to Median String are polynomial, their computational cost is quite high (cubic order), and obtaining the prototypes is very costly. In this work, we propose several techniques in order to reduce this computational cost without degrading the classification performance by the Nearest Neighbour rule.

1 Introduction

Many pattern classification techniques, such as k-Nearest Neighbour (k-NN) classification, require good prototypes to represent pattern classes. Sometimes, clustering techniques can be applied in order to obtain several subgroups of the class training data, where each subgroup has internal similarities [1]. Each cluster is usually represented by a prototype, and several prototypes can be obtained for each class (one per cluster). One important problem in Pattern Recognition is the selection of an appropiate prototype for a given cluster of data points.

Although the feature vector is the most common data point representation, there are many applications where strings (sequences of discrete symbols) are more appropriate as data representation (e.g., chromosomes, character contours, shape contours, etc.). The optimal prototype of a cluster of strings is the *(generalized) median string*. The median string of a given set of strings is defined as a string which minimizes the sum of distances to each string of the set. The problem of searching the median string is a **NP**-Hard problem [2]. Therefore, only approximations to median string can be achieved in a reasonable time.

* Work partially supported by Spanish Ministerio de Ciencia y Tecnología under projects TIC2000-1703-C03-01 and TIC2000-1599-C02-01 and by European Union under project IST-2001-32091.

T. Caelli et al. (Eds.): SSPR&SPR 2002, LNCS 2396, pp. 47–55, 2002.
© Springer-Verlag Berlin Heidelberg 2002

One of these approximations is the *set median string*. In this case, the search for the string is constrained to the given input set and is a polynomial problem [3,4]. In some cases, the set median string cannot be a good approximation to a median string (as in the extreme case of a set of two strings).

Other heuristic approaches were proposed in [5,6,7]. Based on the proposal presented in [5] (systematic perturbation of the set median), a new greedy simple algorithm was proposed in [8] to efficiently compute a good approximation to the median string of a set. This algorithm was improved by iterative refinement as described in [9]. Exhaustive experimentation with this algorithm and k-Nearest Neighbour classifiers is reported in [15]. The results presented in [15] show that the algorithm provides prototypes which give better classification results than set median.

In this work, we propose several methods to reduce the computational cost of the proposed algorithm. NN classifier results are presented to show the performance of the obtained prototypes.

2 Approximations to Median String

In this section, we present different methods to obtain median string approximations. Let Σ^* be the free monoid over the alphabet Σ. Given a finite set S of strings such that $S \subset \Sigma^*$, the median string of S is defined by:

$$m_S = \underset{t \in \Sigma^*}{\operatorname{argmin}} \sum_{r \in S} d(t, r) \tag{1}$$

where d is the distance used to compare two strings (usually, edit distance or normalized edit distance [10]). In other words, m_S is the string with the lowest accumulated distance to the set S. However, with this definition, no adequate median strings can be obtained (e.g., in a set with two strings, both of them achieve the minimum accumulated distance to the set). Therefore, considering the definition of the mean vector for Euclidean spaces, which uses square distances, an alternative definition (as proposed in [11]) could be:

$$m_S = \underset{t \in \Sigma^*}{\operatorname{argmin}} \sum_{r \in S} (d(t, r))^2 \tag{2}$$

As we pointed out above, computing m_S is a **NP**-Hard problem, and only approximations to m_S can be built in a reasonable time by using heuristics which attempt to optimize the accumulated distance for one of the previous definitions.

The set median string can be used as an alternative to median string. Given the set S, the set median string of S is defined as:

$$sm_S = \underset{t \in S}{\operatorname{argmin}} \sum_{r \in S} d(t, r) \tag{3}$$

that is, the search space is reduced to the set S, and not to all the free monoid Σ^*. The edit distance can be computed in a time $O(|t| \cdot |r|)$, where $|t|$ is the length of the string t. Therefore, obtaining the set median has a computational complexity of $O(|S|^2 \cdot l_S^2)$, where l_S is the maximum length of the strings from S.

Another approximation to median string can be obtained by using a refinement process. This process is based on applying the editing operations (insertion, deletion and substitution) over each position on the string, looking for a reduction of the accumulated distance defined in Equation 1 or Equation 2. This process is repeated until there is no improvement, and it needs an initial string which can be the set median string. Given a set of strings S, the specification of the process is:

For each position i in the current approximated median string M

1. Build alternatives
   Substitution: Make $M_{sub} = M$. For each symbol $a \in \Sigma$
   - Make M'_{sub} the result string of substituting the ith symbol of M by symbol a.
   - If the accumulated distance of M'_{sub} to S is lower than the accumulated distance from M_{sub} to S, then make $M_{sub} = M'_{sub}$.

   Deletion: Make M_{del} the result string of deleting the ith symbol of M.
   Insertion: Make $M_{ins} = M$. For each symbol $a \in \Sigma$
   - Make M'_{ins} the result of adding a at position i of M.
   - If the accumulated distance from M'_{ins} to S is lower than the accumulated distance from M_{ins} to S, then make $M_{ins} = M'_{ins}$.
2. Choose an alternative: from the set $\{M, M_{sub}, M_{del}, M_{ins}\}$, take the string M' with the least accumulated distance to S. Make $M = M'$.

M is the returned string at the end of the process. The time complexity of this optimization process is $O(|\Sigma| \cdot |S| \cdot l_S^3)$ for each iteration, which is a very high cost although the number iteration is low. Therefore, it will be convenient to use techniques which could reduce this cost.

3 Techniques for Reducing the Computational Cost

In this section, we introduce two techniques which allow us to reduce the computational complexity of approximated median string computations. These techniques are called division and local optimization.

3.1 The Division Technique

As we showed in Section 2, the time complexity of the process to obtain the approximated median string is cubic with the length of the strings. Therefore, it seems that reducing the length of the strings would be the most influential action on the complexity.

Following this idea, the division technique acts by dividing the strings of S into d substrings. Therefore, given a string s, this string is divided to give the strings s^1, s^2, ..., s^d such as $s = s^1 \cdot s^2 \cdots s^d$. From the set of strings S, this division provides d sets of strings S^1, S^2, ..., S^d as result. Then, an approximated median string can be obtained for each set S^1, S^2, ..., S^d, that is, M^1, M^2, ..., M^d and a final approximated median string of S will be $M = M^1 \cdot M^2 \cdots M^d$.

More formally, the process is:

1. $S^i = \emptyset$ for $i = 1, \ldots, d$
2. For each string $s \in S$
   (a) Divide s into s^1, s^2, ..., s^d
   (b) Make $S^i = S^i \bigcup \{s^i\}$ for $i = 1, \ldots, d$
3. Compute the approximated median string M^i of S^i for $i = 1, \ldots, d$
4. $M = M^1 \cdot M^2 \cdots M^d$

The main complexity of this procedure is due to the approximated median string computation, whose complexity is still $O(|\Sigma| \cdot |S| \cdot l_S^3)$ for each iteration. Nevertheless, with the previous division, the approximated median string is computed for each S^i, where $|S^i| = |S|$ and, therefore, a total number of $d \cdot |S|$ strings are involved in the process. Futhermore, the maximum length of the strings of each S^i is $\frac{l_S}{d}$.

The final time complexity is proportional to $|\Sigma| \cdot d \cdot |S| \cdot \frac{l_S}{d}^3$, and the complexity is then $O(|\Sigma| \cdot |S| \cdot l_S^3 \frac{1}{d^2})$; that is, the real complexity is reduced by a factor d^2.

3.2 The Local Optimization Technique

From the algorithm presented in Section 2, the substitution and insertion are the edit operations which involve the majority of the edit distance calculations. This is due to the use of all the symbols in the alphabet Σ; that is, we try to substitute the current symbol by all the symbols in Σ and the insertion on the current position of all the symbols in Σ.

However, in practice, the natural symbol sequence is usually correlated so that a symbol can preceed or follow another symbol with a certain probability. This fact leads to modifying the possible symbols to be inserted or substituted on the string, using only the most likely symbols in these operations. The only way we have to determine the chosen symbols without an exhaustive study of the corpus is by using the weight matrix which is used in the edit distance calculation. In our current approach, we propose:

- For substitution: to use only the two closest symbols (according to the weight matrix) to the current one.
- For insertion: to use only the previous position symbol and its two closest symbols.

Therefore, the algorithm is modified in the following way:

For each position i in the current approximated median string M

1. Build alternatives
   Substitution: Make $M_{sub} = M$. For each symbol $a \in \text{nearest}(M_i)$
   - Make M'_{sub} the result string of substituting the ith symbol of M by symbol a.

- If the accumulated distance of M'_{sub} to S is lower than the accumulated distance from M_{sub} to S, then make $M_{sub} = M'_{sub}$.

Deletion: Make M_{del} the result string of deleting the ith symbol of M.
Insertion: Make $M_{ins} = s$. For each symbol $a \in \{M_{i-1}\} \bigcup \text{nearest}(M_{i-1})$
- Make M'_{ins} the result of adding a at position i of M.
- If the accumulated distance from M'_{ins} to S is lower than the accumulated distance from M_{ins} to S, then make $M_{ins} = M'_{ins}$.

2. Choose an alternative: from the set $\{M, M_{sub}, M_{del}, M_{ins}\}$, take the string M' with the least accumulated distance to S. Make $M = M'$.

where nearest(M_i) gives the two nearest symbols to symbol M_i according to the weight matrix used and M is the returned string at the end of the process. The definition of nearest can be easily extended to more than two symbols.

With this modification, we avoid the factor $|\Sigma|$ in the complexity (because we only try with a low number of symbols). Therefore, the final time complexity of this approximation is $O(|S| \cdot l_S^3)$ for each iteration, which gives an assymptotic complexity reduction.

4 Experimental Framework

This section is devoted to describing the corpus we used and the experiments we carried out to compare the cost and performance of the different approximation methods described in Sections 2 and 3.

4.1 The Chromo Corpus

The data used in this paper was extracted from a database of approximately 7,000 chromosome images that were classified by cytogenetic experts [12]. Each digitized chromosome image was automatically transformed into a string through a procedure that starts with obtaining an idealized, one-dimensional density profile that emphasizes the band pattern along the chromosome. The idealized profile is then mapped nonlinearly into a string composed of symbols from the alphabet $\{1, 2, 3, 4, 5, 6\}$. Each symbol in this alphabet represents a different absolute density level. Then, the resulting string is difference coded to represent signed differences of successive symbols, using the alphabet $\Sigma = \{=, A, B, C, D, E, a, b, c, d, e\}$ ("=" for a difference of 0; "A" for +1; "a" for -1; etc.). For instance, the string "1221114444333" is difference coded as "AA=a==C====a==". A total of 4400 samples were collected, 200 samples of each of the 22 non-sex chromosome types. See [14,13] for details about this preprocessing.

The chromosome dataset comprises 200 string samples for each of the 22 non-sex chromosome types, i.e. a total of 4400 samples. An additional piece of information contained in this dataset is the location of the *centromere* in each chromosome string. However, this position is difficult to accurately determine in a fully automatic way and, thus, we have decided not to use it in this work.

4.2 Experiments and Results

The experiments to compare the different approximations were carried out with a 2-fold cross-validation of the chromo corpus. Each class of chromosomes was divided into several clusters in order to get several prototypes for each class. Different number of clusters were obtained for each class, from 1 (i.e., no clustering) up to 9 clusters and from 10 up to 90 clusters. The set median and different approximated median strings (for both definition in Equations 1 and 2) were obtained, using the normalized edit distance and the same weights as in [15]. The optimization techniques described in section 3 were also applied, using 2 and 3 divisions in the division technique.

In this prototype extraction process, the length of the compared strings was taken as the basic cost unit, i.e., when two strings s and t are compared, the total cost is incremented in $|s| \cdot |t|$. The comparison for the different approximations using the median string definition of Equation 1 is given in Figure 1. The results using the definition given in Equation 2 are very similar. You can see the large difference from set median to all the approximated median strings (at least one order of magnitude), and also the large difference between the original method (not optimized), the division method and the local optimization method (from 3 to 5 times lower cost).

After obtaining the prototypes, several classification experiments were performed using a classical NN classifier [1] in order to quantify the degradation of the prototypes by the use of the different techniques. The results obtained

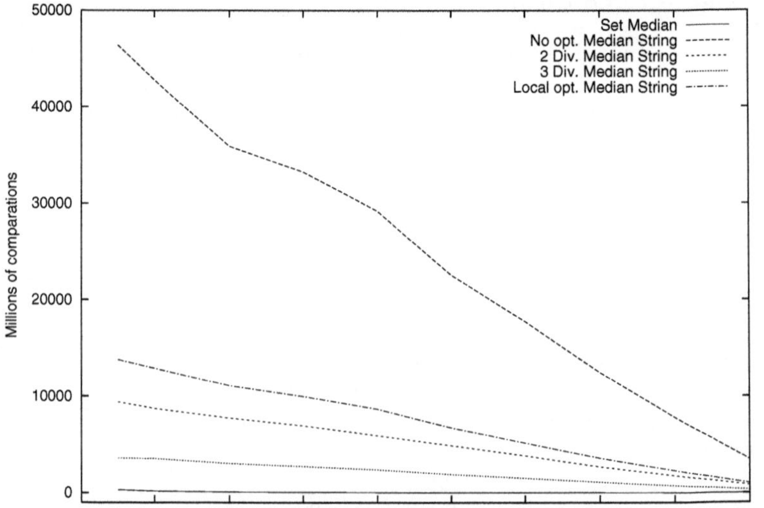

Fig. 1. Time cost in millions of comparisons performed for the different approximations to median string using the definition of Equation 1

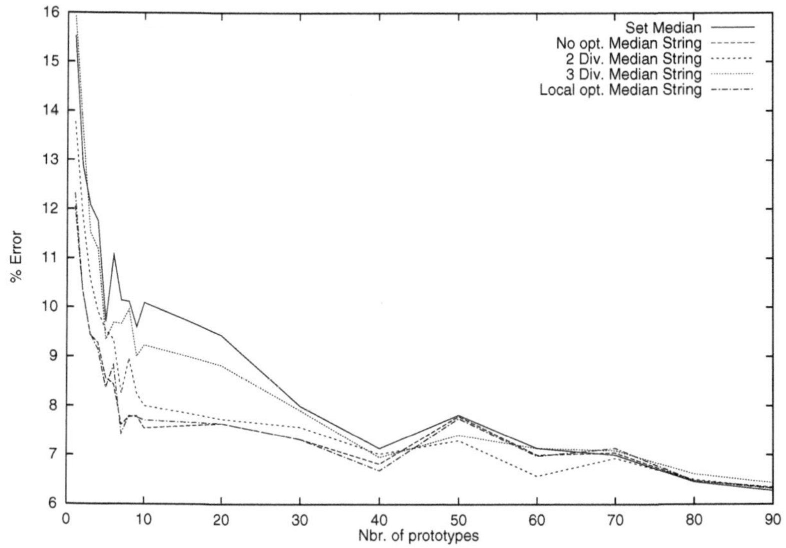

Fig. 2. Set median and approximated median string results using a NN classifier using the prototype given by the definition in Equation 1

are shown in Figure 2 (using the definition given in 1) and Figure 3 (using the definition given in 2). In both graphics, you can see that in general the median strings perform much better than set median and that the local optimization technique is the one that obtains the most similar results to the not optimized approximated median string.

5 Conclusions and Future Work

In this work, we have proposed two different techniques to reduce the computational cost to obtain an approximated median string. Even for a reduced number of divisions, the division technique provides a high reduction in the number of comparisons, although the classification results show us a clear degradation in the prototypes quality when increasing the number of divisions. The local optimization technique gives a lower reduction than the division technique, but the given prototypes are as good as those obtained by the not optimized process.

However, these conclusions are limited because the experiments were carried out with only the chromosome corpus and using only a NN classifier. Therefore, future work is directed to extending these conclusions using other corpora and more powerful classifiers (as k-NN classifiers), and to verify the effects of combining both techniques.

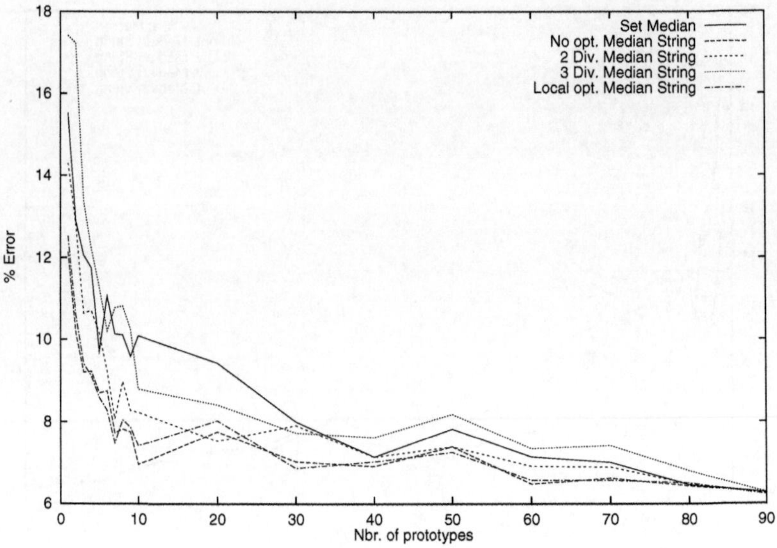

Fig. 3. Set median and approximated median string results using a NN classifier using the prototype given by the definition in Equation 2

Acknowledgements

The authors wish to thank Dr. Jens Gregor for providing the preprocessed chromosome data used in this work, and the anonymous reviewers for their criticism and suggestions.

References

1. Duda, R. O., Hart, P., Stork, D. G., 2001. Pattern Classification. John Wiley. 47, 52
2. de la Higuera, C., Casacuberta, F., 2000. The topology of strings: two np-complete problems. Theoretical Computer Science 230, 39–48. 47
3. Fu, K. S., 1982. Syntactic Pattern Recognition. Prentice-Hall. 48
4. Juan, A., Vidal, E., 1998. Fast Median Search in Metric Spaces. In: Proceedings of the 2nd International Workshop on Statistical Techniques in Pattern Recognition. Vol. 1451 of Lecture Notes in Computer Science. Springer-Verlag, Sydney, pp. 905–912. 48
5. Kohonen, T., 1985. Median strings. Pattern Recognition Letters 3, 309–313. 48
6. Kruzslicz, F., 1988. A greedy algorithm to look for median strings. In: Abstracts of the Conference on PhD Students in Computer Science. Institute of informatics of the József Attila University. 48
7. Fischer, I., Zell, A., 2000. String averages and self-organizing maps for strings. In: Proceeding of the Second ICSC Symposium on Neural Computation. pp. 208–215. 48

8. Casacuberta, F., de Antonio, M., 1997. A greedy algorithm for computing approximate median strings. In: Proceedings of the VII Simposium Nacional de Reconocimiento de Formas y Análisis de Imágenes. pp. 193–198. 48
9. Martínez, C. D., Juan, A., Casacuberta, F., 2000. Use of Median String for Classification. In: Proceedings of the 15th International Conference on Pattern Recognition. Vol. 2. Barcelona (Spain), pp. 907–910. 48
10. Vidal, E., Marzal, A., Aibar, P., 1995. Fast computation of normalized edit distances. IEEE Transactions on Pattern Analysis and Machine Intelligence 17 (9), 899–902. 48
11. Martínez, C., Juan, A., Casacuberta, F., 2001. Improving classification using median string and nn rules. In: Proceedings of IX Simposium Nacional de Reconocimiento de Formas y Análisis de Imágenes. pp. 391–394. 48
12. Lundsteen, C., Philip, J., Granum, E., 1980. Quantitative Analysis of 6895 Digitized Trypsin G-banded Human Metaphase Chromosomes. Clinical Genetics 18, 355–370. 51
13. Granum, E., Thomason, M., 1990. Automatically Inferred Markov Network Models for Classification of Chromosomal Band Pattern Structures. Cytometry 11, 26–39. 51
14. Granum, E., Thomason, M. J., Gregor, J. On the use of automatically inferred Markov networks for chromosome analysis. In C Lundsteen and J Piper, editors, *Automation of Cytogenetics*, pages 233–251. Springer-Verlag, Berlin, 1989. 51
15. Martínez-Hinarejos, C. D., Juan, A., Casacuberta, F., Median String for k-Nearest Neighbour classification, *Pattern Recognition Letters*, acepted for revision. 48, 52

Tree k-Grammar Models for Natural Language Modelling and Parsing

Jose L. Verdú-Mas, Mikel L. Forcada,
Rafael C. Carrasco, and Jorge Calera-Rubio*

Departament de Llenguatges i Sistemes Informàtics, Universitat d'Alacant
E-03071 Alacant, Spain
{verdu,mlf,carrasco,calera}@dlsi.ua.es

Abstract. In this paper, we compare three different approaches to build a probabilistic context-free grammar for natural language parsing from a tree bank corpus: (1) a model that simply extracts the rules contained in the corpus and counts the number of occurrences of each rule; (2) a model that also stores information about the parent node's category, and (3) a model that estimates the probabilities according to a generalized k-gram scheme for trees with $k = 3$. The last model allows for faster parsing and decreases considerably the perplexity of test samples.

1 Introduction

Context-free grammars are the customary way of representing syntactical structure in natural language sentences. In many natural-language processing applications, obtaining the correct syntactical structure for a sentence is an important intermediate step before assigning an interpretation to it. Choosing the correct parse for a given sentence is a crucial task if one wants to interpret the meaning of the sentence, due to the *principle of compositionality* [13, p. 358], which states, informally, that the interpretation of a sentence is obtained by *composing* the meaning of its constituents according to the groupings defined by the parse tree.

But ambiguous parses are *very* common in real natural-language sentences (e.g., those longer than 15 words). Some authors (e.g. [7]) propose that a great deal of syntactic disambiguation may actually occurs without the use of any semantic information; that is, just by selecting a preferred parse tree. It may be argued that the preference of a parse tree with respect to another is largely due to the relative frequencies with which those choices have led to a successful interpretation. This sets the ground for a family of techniques which use a probabilistic scoring of parses to the correct parse in each case.

Probabilistic scorings depend on parameters which are usually estimated from data, that is, from parsed text corpora such as the Penn Treebank [11]. The most straightforward approach is that of *tree bank grammars* [6]. Treebank

* The authors wish to thank the Spanish CICyT for supporting this work through project TIC2000-1599.

T. Caelli et al. (Eds.): SSPR&SPR 2002, LNCS 2396, pp. 56–63, 2002.
© Springer-Verlag Berlin Heidelberg 2002

grammars are probabilistic context-free grammars in which the probabilities that a particular nonterminal is expanded according to a given rule are estimated as the relative frequency of that expansion by simply counting the number of times it occurs in a manually-parsed corpus. This is the simplest probabilistic scoring scheme, and it is not without problems. Better results were obtained with *parent-annotated* labels [8] where each node stores contextual information in the form of the category of the node's parent. This fact is in agreement with the observation put forward by Charniak [6] that simple PCFGs, directly obtained from a corpus, largely overgeneralize. This property suggests that, in these models, a large probability mass is assigned to incorrect parses and, therefore, any procedure that concentrates the probability on the correct parses will increase the likelihood of the samples.

In this spirit, we introduce a generalization of the classic k-gram models, widely used for string processing [2], to the case of trees. The PCFGs obtained in this way consist of rules that include information about the context where the rule is applied. One might call these PCFGs *offspring-annotated* CFGs (by analogy to Johnson's [8] *parent-annotation* concept).

2 A Generalized k-Gram Model

Recall that k-gram models are stochastic models for the generation of sequences $s_1, s_2, ...$ based on conditional probabilities, that is:

1. the probability $P(s_1 s_2 \dots s_t | M)$ of a sequence in the model M is computed as a product $p_M(s_1) p_M(s_2|s_1) \cdots p_M(s_t|s_1 s_2 \dots s_{t-1})$, and
2. the dependence of the probabilities p_M on previous history is assumed to be restricted to the immediate preceding context, in particular, the last $k-1$ words: $p_M(s_t|s_1 \dots s_{t-1}) = p_M(s_t|s_{t-k+1} \dots s_{t-1})$.

Note that in this kind of models, the probability that the observation s_t is generated at time t is computed as a function of the subsequence of length $k-1$ that immediately precedes s_t (this is called a *state*). However, in the case of trees, it is not obvious what context should be taken in to account. Indeed, there is a natural preference when processing strings (the usual left-to-right order) but there are at least two standard ways of processing trees: ascending (or bottom-up) analysis and descending (or top-down) analysis. Ascending tree automata recognize a wider class of tree languages [12] and, therefore, they allow for richer descriptions.

Therefore, our model will compute the expansion probability for a given node as a function of the subtree of depth $k-2$ that the node generates[1], i.e., every *state* stores a subtree of depth $k-2$. In the particular case $k=2$, only the label of the node is taken into account (this is analogous to the standard bigram model for strings) and the model coincides with the simple rule-counting approach used

[1] Note that in our notation a single node tree has depth 0. This is in contrast to strings, where a single symbol has length 1.

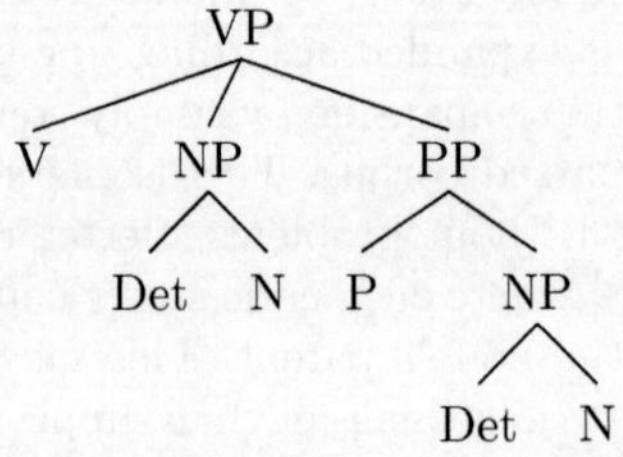

Fig. 1. A sample parse tree of depth 3

in treebank grammars. For instance, for the tree depicted in Fig. 1, the following rules are obtained:

$$\begin{aligned} \mathrm{VP} &\rightarrow \mathrm{V\ NP\ PP} \\ \mathrm{NP} &\rightarrow \mathrm{Det\ N} \\ \mathrm{PP} &\rightarrow \mathrm{P\ NP} \end{aligned}$$

However, in the case $k = 3$, which will be called *child-annotated* model, the expansion probabilities depend on states that are defined by the node label, the number of descendents the node and the sequence of labels in the descendents (if any). Therefore, for the same tree the following rules are obtained in this case:

$$\begin{aligned} \mathrm{VP}_{\mathrm{V,NP,PP}} &\rightarrow \mathrm{V\ NP}_{\mathrm{Det,N}}\ \mathrm{PP}_{\mathrm{P,NP}} \\ \mathrm{NP}_{\mathrm{Det,N}} &\rightarrow \mathrm{Det\ N} \\ \mathrm{PP}_{\mathrm{P,NP}} &\rightarrow \mathrm{P\ NP}_{\mathrm{Det,N}} \end{aligned}$$

where each state has the form $\mathrm{X}_{\mathrm{Z}_1,\ldots,\mathrm{Z}_m}$. This is equivalent to performing a relabelling of the parse tree before extracting the rules.

Finally, in the parent-annotated model (PA) described in [8] the states depend on both the node label and the node's parent label:

$$\begin{aligned} {}^{\mathrm{S}}\mathrm{VP} &\rightarrow \mathrm{V}\ {}^{\mathrm{VP}}\mathrm{NP}\ {}^{\mathrm{VP}}\mathrm{PP} \\ {}^{\mathrm{VP}}\mathrm{NP} &\rightarrow \mathrm{Det\ N} \\ {}^{\mathrm{VP}}\mathrm{PP} &\rightarrow \mathrm{P}\ {}^{\mathrm{PP}}\mathrm{NP} \\ {}^{\mathrm{PP}}\mathrm{NP} &\rightarrow \mathrm{Det\ N} \end{aligned}$$

It is obvious that the $k = 3$ and PA models incorporate contextual information that is not present in the case $k = 2$ and, then, a higher number of rules for a fixed number of categories is possible. In practice, due to the finite size of the training corpus, the number of rules is always moderate. However, as higher values of k lead to a huge number of possible rules, huge data sets would be necessary in order to have a reliable estimate of the probabilities for values above $k = 3$. A detailed mathematical description of *offspring-annotated* models can be found in [14].

3 Experimental Results

3.1 General Conditions

We have performed experiments to assess the structural disambiguation performance of k-gram models as compared to standard treebank grammars and Johnson's [8] parent-annotation scheme, that is, to compare their relative ability for selecting the best parse tree. We have also used the perplexity as an indication of the quality of each model. To build training corpora and test sets of parse trees, we have used English parse trees from the Penn Treebank, release 3, with small, basically structure-preserving modifications:

- insertion of a root node (ROOT) in all sentences, (as in Charniak [6]) to encompass the sentence and final periods, etc.;
- removal of nonsyntactic annotations (prefixes and suffixes) from constituent labels (for instance, NP-SBJ is reduced to NP);
- removal of empty constituents; and
- collapse of single-child nodes with the parent node when they have the same label (to avoid rules of the form A $\rightarrow$ A which would generate an infinite number of parse trees for some sentences).

In all experiments, the training corpus consisted of all of the trees (41,532) in sections 02 to 22 of the *Wall Street Journal* portion of Penn Treebank, modified as above. This gives a total number of more than 600,000 subtrees. The test set contained all sentences in section 23 having less than 40 words.

3.2 Structural Disambiguation Results

All grammar models were rewritten as standard context-free grammars, and Chappelier and Rajman's [5] probabilistic extended Cocke-Younger-Kasami parsing algorithm was used to obtain all possible parse trees for each sentence in the test sets and to compute their individual and total probabilitites; for each sentence, the most likely parse was compared to the corresponding tree in the test set using the customary PARSEVAL evaluation metric [1,10, p. 432] after eliminating any parent and child annotation of nodes in the most likely tree delivered by the parser. PARSEVAL gives partial credit to incorrect parses by establishing three measures:

- *labeled precision* (P) is the fraction of correctly-labeled nonterminal bracketings (constituents) in the most likely parse which match the parse in the treebank,
- *labeled recall* (R) is the fraction of brackets in the treebank parse which are found in the most likely parse with the same label, and
- *crossing brackets* (X) refers to the fraction of constituents in one parse cross over constituent boundaries in the other parse.

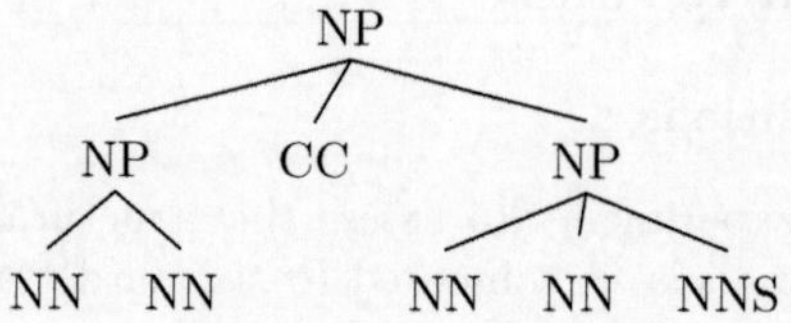

Fig. 2.

The crossing brackets measure does not take constituent labels into account and will not be shown here. Some authors (see, e.g. [4]) have questioned partial-credit evaluation metrics such as the PARSEVAL measures; in particular, if one wants to use a probability model to perform structural disambiguation before assigning some kind of interpretation ot the parsed sentence, it may well be argued that the exact match between the treebank tree and the most likely tree is the only possible relevant measure. It is however, very well known that the Penn Treebank, even in its release 3, still suffers from problems. One of the problems worth mentioning (discussed in detail by Krotov et al. [9]) is the presence of far too many partially bracketed constructs according to rules like NP → NN NN CC NN NN NNS, which lead to very flat trees, when one can, in the same treebank, find rules such as NP → NN NN, NP → NN NN NNS and NP → NP CC NP, which would lead to more structured parses such as the one in Fig. 2. Some of these flat parses may indeed be too flat to be useful for semantic purposes; therefore, if one gets a more refined parse, it may or may not be the one leading to the correct interpretation, but it may never be worse than the flat, unstructured one found in the treebank.

For this reason, we have chosen to give, in addition to the exact-match figure, the percentage of trees having 100% recall, because these are the trees in which the most likely parse is either exactly the treebank parse or a refinement thereof in the sense of the previous example.

Here is a list of the models which were evaluated:

- A standard treebank grammar, with no annotation of node labels (k=2), with probabilities for 15,140 rules.
- A child-annotated grammar (k=3), with probabilities for 92,830 rules.
- A parent-annotated grammar (PARENT), with probabilities for 23,020 rules.
- A both parent- and child-annotated grammar (BOTH), with probabilities for 112,610 rules.

As expected, the number of rules obtained increases as more information is conveyed by the node label, although this increase is not extreme. On the other hand, as the generalization power decreases, some sentences in the test set become unparsable, that is, they cannot be generated by the grammar. The results in table 1 show that:

- The parsing performance of parent-annotated and child-annotated PCFG is similar and better than those obtained with the standard treebank PCFG.

Table 1. Parsing results with different annotation schemes: labelled recall R, labelled precision P, fraction of sentences with total labelled recall $f_{R=100\%}$, fraction of exact matches, fraction of sentences parsed, and average time per sentence in seconds

MODEL	R	P	$f_{R=100\%}$	EXACT	PARSED	t
k=2	70.7%	76.1%	10.4%	10.0%	100%	57
k=3	79.6%	74.3%	19.9%	13.4%	94.6%	7
PARENT	80.0%	81.9%	18.5%	16.3%	100%	340
BOTH	80.5%	74.5%	22.7%	15.5%	79.6%	4

This performance is measured both with the customary PARSEVAL metrics and by counting the number of maximum-likelihood trees that (a) match their counterparts in the treebank exactly, and (b) contain all of the constituents in their counterpart (100% labeled recall, $f_{R=100\%}$). The fact that child-annotated grammars do not perform better than parent-annotated ones may be due to their larger number of parameters compared to parent-annotated PCFG, which may make them hard to estimate accurately from currently available treebanks (there are, on average, only about 6 subtrees per rule in the experiments).

- The average time to parse a sentence shows that child annotation leads to parsers that are much faster. This is not surprising because the number of possible parse trees considered is drastically reduced; this is, however, not the case with parent-annotated models.

It may be worth mentioning that an analysis of parse trees produced by child-annotated models tend to be more structured and refined than parent-annotated and unannotated parses which tend to use rules that lead to flat trees in the sense mentioned.

3.3 Perplexity Results

We have also used the perplexity of a test sample $S = \{w_1, ..., w_{|S|}\}$ as an indication of the quality of the model, $P = \frac{1}{|S|} \sum_{l=1}^{|S|} \log_2 p(w_l|M)$, where $p(w_l|M)$ is the sum of the probabilities of all of the parse trees of the sentence w_l. Since unparsable sentences would produce an infinite perplexity, we have studied the perplexity of the test set for linear combinations of two models M_i and M_j with $p(w_l|M_i, M_j) = \lambda p(w_l|M_i) + (1 - \lambda)p(w_l|M_j)$. The mixing parameter $\lambda \in [0, 1]$ was chosen, in steps of 0.05, in order to minimize the perplexity.

The best results were obtained with a mixture of the child-annotated ($k = 3$) and the parent-annotated models with a heavier component (65%) of the first one. When parsing, the recall and precision of that mixture were respectively 82.1% and 81.0% and the fraction of sentences with total labelled recall $f_{R=100\%}$ scored 22.2%, similar to using both annotation models at the same time but covering all the test set. The minimum perplexity $P_{\min}$ and the corresponding value of λ obtained are shown in the table 2.

Table 2. Mixture parameter $\lambda_{\min}$ that gives the minimum test set perplexity for each linear combination. The lowest perplexity was obtained with a combination of the k=3 and parent-annotation models. All mixture models covered all the set test

MIXTURE MODEL	$P_{\min}$	$\lambda_{\min}$
$k = 2$ and $k = 3$	90.8	0.25
$k = 2$ and PARENT	108.7	0.6
$k = 2$ and BOTH	94	0.3
$k = 3$ and PARENT	88	0.65

4 Conclusion

We have introduced a new probabilistic context-free grammar model, offspring-annotated PCFG, in which the grammar variables are specialized by annotating them with the subtree they generate up to a certain level. In particular, we have studied offspring-annotated models with $k = 3$, that is, child-annotated models, and have compared their parsing performance to that of unannotated PCFG and of parent-annotated PCFG [8]. Child-annotated models are related to probabilistic bottom-up tree automata [12] . The experiments show that:

- The parsing performance of parent-annotated and child-annotated PCFG is similar.
- Parsers using child-annotated grammars are much faster because the number of possible parse trees considered is drastically reduced; this is, however, not the case with parent-annotated models.
- Child-annotated grammars have a larger number of parameters than parent-annotated PCFG which makes it difficult to estimate them accurately from currently available treebanks.
- Child-annotated models tend to give very structured and refined parses instead of flat parses, a tendency not so strong for parent-annotated grammars.
- The perplexity of the test sample decreases when a combination of models with child-annotated and parent-annotated is used to predict string probabilities.

We plan to study the use of statistical confidence criteria as used in grammatical inference algorithms [3] to eliminate unnecessary annotations by merging states, therefore reducing the number of parameters to be estimated. Indeed, offspring-annotation schemes (for a value of $k \geq 3$) may be useful as starting points for those state-merging mechanisms, which so far have always started with the complete set of different subtrees found in the treebank (ranging in the hundreds of thousands). We also plan to study the smoothing of offspring-annotated PCFGs and to design parsers which can profit from these.

References

1. Ezra Black, Steven Abney, Dan Flickinger, Claudia Gdaniec, Ralph Grishman, Philip Harrison, Donald Hindle, Robert Ingria, Frederick Jelinek, Judith Klavans, Mark Liberman, Mitch Marcus, Salim Roukos, Beatrice Santorini, and Tomek Strzalkowski. A procedure for quantitatively comparing the syntatic coverage of english grammars. In *Proc. Speech and Natural Language Workshop 1991*, pages 306–311, San Mateo, CA, 1991. Morgan Kauffmann. 59
2. Peter F. Brown, Vincent J. Della Pietra, Peter V. deSouza, Jenifer C. Lai, and Robert L. Mercer. Class-based n-gram models of natural language. *Computational Linguistics*, 18(4):467–479, 1992. 57
3. Rafael C. Carrasco, Jose Oncina, and Jorge Calera-Rubio. Stochastic inference of regular tree languages. *Machine Learning*, 44(1/2):185–197, 2001. 62
4. John Carroll, Ted Briscoe, and Antonio Sanfilippo. Parser evaluation: A survey and a new proposal. In *Proceedings of the International Conference on Language REsources and Evaluation*, pages 447–454, Granada, Spain, 1998. 60
5. J.-C. Chappelier and M. Rajman. A generalized CYK algorithm for parsing stochastic CFG. In *Actes de TAPD'98*, pages 133–137, 1998. 59
6. Eugene Charniak. Treebank grammars. In *Proceedings of the Thirteenth National Conference on Artificial Intelligence*, pages 1031–1036. AAAI Press/MIT Press, 1996. 56, 57, 59
7. L. Frazier and K. Rayner. Making and correcting errors during sentence comprehension: Eye movements in the analysis of structurally ambiguous sentences. *Cognitive Psychology*, 14:178–210, 1982. 56
8. Mark Johnson. PCFG models of linguistic tree representations. *Computational Linguistics*, 24(4):613–632, 1998. 57, 58, 59, 62
9. Alexander Krotov, Robert Gaizauskas, Mark Hepple, and Yorick Wilks. Compacting the Penn Treebank grammar. In *Proceedings of COLING/ACL'98*, pages 699–703, 1998. 60
10. Christopher D. Manning and Hinrich Schütze. *Foundations of Statistical Natural Language Processing*. MIT Press, 1999. 59
11. Mitchell P. Marcus, Beatrice Santorini, and Mary Ann Marcinkiewicz. Building a large annotated corpus of english: the penn treebank. *Computational Linguistics*, 19:313–330, 1993. 56
12. Maurice Nivat and Andreas Podelski. Minimal ascending and descending tree automata. *SIAM Journal on Computing*, 26(1):39–58, 1997. 57, 62
13. A. Radford, M. Atkinson, D. Britain, H. Clahsen, and A. Spencer. *Linguistics: an introduction*. Cambridge Univ. Press, Cambridge, 1999. 56
14. J.R. Rico-Juan, J. Calera-Rubio, and R.C. Carrasco. Probabilistic k-testable tree-languages. In A.L. Oliveira, editor, *Proceedings of 5th International Colloquium, ICGI 2000, Lisbon (Portugal)*, volume 1891 of *Lecture Notes in Computer Science*, pages 221–228, Berlin, 2000. Springer. 58

Algorithms for Learning Function Distinguishable Regular Languages

Henning Fernau[1]* and Agnes Radl[2]

[1] School of Electrical Engineering and Computer Science, University of Newcastle
University Drive, NSW 2308 Callaghan, Australia
fernau@cs.newcastle.edu.au
[2] Wilhelm-Schickard-Institut für Informatik, Universität Tübingen
Sand 13, D-72076 Tübingen, Germany
agnes.radl@student.uni-tuebingen.de

Abstract. Function distinguishable languages were introduced as a new methodology of defining characterizable subclasses of the regular languages which are learnable from text. Here, we give details on the implementation and the analysis of the corresponding learning algorithms. We also discuss problems which might occur in practical applications.

1 Introduction

Identification in the limit from positive samples, also known as *exact learning from text* as proposed by Gold [10], is one of the oldest yet most important models of grammatical inference. Since not all regular languages can be learned exactly from text, the characterization of *identifiable* subclasses of regular languages is a useful line of research, because the regular languages are a very basic language family, see also the discussions in [12] regarding the importance of finding *characterizable* learnable language classes.

In [4], we introduced the so-called function-distinguishable languages as a rich source of examples of identifiable language families. Among the language families which turn out to be special cases of our approach are the k-reversible languages [1] and (reversals of) the terminal-distinguishable languages [13,14], which belong, according to Gregor [11], to the most popular identifiable regular language classes. Moreover, we have shown [4] how to transfer the ideas underlying the well-known identifiable language classes of k-testable languages, k-piecewise testable languages and threshold testable languages to our setting. In a nutshell, an identification algorithm for f-distinguishable languages assigns to every finite set of samples $I_+ \subseteq T^*$ the smallest f-distinguishable language containing I_+ by subsequently merging states which cause conflicts to the definition of f-distinguishable automata, starting with the simple prefix tree automaton accepting I_+.

* Work was done while the author was with Wilhelm-Schickard-Institut für Informatik, Universität Tübingen, Sand 13, D-72076 Tübingen, Germany

T. Caelli et al. (Eds.): SSPR&SPR 2002, LNCS 2396, pp. 64–73, 2002.
© Springer-Verlag Berlin Heidelberg 2002

Another interesting property of each class f-DL of function distinguishable languages is established in [6]: the approximability of the whole class of regular languages in the sense that, given any regular language L, a learning algorithm for f-DL infers, given L, the smallest language $L' \in f$-DL including L.

Applications of the learnability of function-distinguishable languages have been reported in [8] for the identifiability of parallel communicating grammar systems and in [7] for inferring document type definitions of XML documents.

Here, we aim at giving more details on the implementation and analysis of the learning algorithms for function-distinguishable languages. We also give a proof of a counterexample originally given by Radhakrishnan and Nagaraja.

2 General Definitions

Σ^* is the set of words over the alphabet Σ. Σ^k ($\Sigma^{<k}$) collects the words whose lengths are equal to (less than) k. λ denotes the empty word. Pref(L) is the set of prefixes of L and $u^{-1}L = \{ v \in \Sigma^* | uv \in L \}$ is the quotient of $L \subseteq \Sigma^*$ by u.

Regular languages can be characterized by (deterministic) finite automata $A = (Q, T, \delta, q_0, Q_F)$, where Q is the state set, $\delta \subseteq Q \times T \times Q$ is the transition relation, $q_0 \in Q$ is the initial state and $Q_F \subseteq Q$ is the set of final states. As usual, δ^* denotes the extension of the transition relation to arbitrarily long input words. The language defined by an automaton A is written $L(A)$. An automaton is called *stripped* iff all states are accessible from the initial state and all states lead to some final state. Observe that the transition function of a stripped deterministic finite automaton is not total in general.

We denote the minimal deterministic automaton of the regular language L by $A(L)$. Recall that $A(L) = (Q, T, \delta, q_0, Q_F)$ can be described as follows: $Q = \{u^{-1}L | u \in \text{Pref}(L)\}$, $q_0 = \lambda^{-1}L = L$; $Q_F = \{u^{-1}L | u \in L\}$; and $\delta(u^{-1}L, a) = (ua)^{-1}L$ with $u, ua \in \text{Pref}(L)$, $a \in T$. According to our definition, any minimal deterministic automaton is stripped.
The *product automaton* $A = A_1 \times A_2$ of two automata $A_i = (Q_i, T, \delta_i, q_{0,i}, Q_{F,i})$ for $i = 1, 2$ is defined as $A = (Q, T, \delta, q_0, Q_F)$ with $Q = Q_1 \times Q_2$, $q_0 = (q_{0,1}, q_{0,2})$, $Q_F = Q_{F,1} \times Q_{F,2}$, $((q_1, q_2), a, (q_1', q_2')) \in \delta$ iff $(q_1, a, q_1') \in \delta_1$ and $(q_2, a, q_2') \in \delta_2$.

3 Function Distinguishable Languages

In order to avoid cumbersome case discussions, let us fix now T as the input alphabet of the finite automata we are going to discuss.

Definition 1. *Let F be some finite set. A mapping $f : T^* \to F$ is called a* distinguishing function *if $f(w) = f(z)$ implies $f(wu) = f(zu)$ for all $u, w, z \in T^*$.*

In the literature, we can find the terminal function [14]

$$\text{Ter}(x) = \{ a \in T \mid \exists u, v \in T^* : uav = x \}$$

and, more generally, the k-terminal function [5]

$$\mathrm{Ter}_k(x) = (\pi_k(x), \mu_k(x), \sigma_k(x)), \quad \text{where}$$
$$\mu_k(x) = \{ a \in T^{k+1} \mid \exists u, v \in T^* : uav = x \}$$

and $\pi_k(x)$ [$\sigma_k(x)$] is the prefix [suffix] of length k of x if $x \notin T^{<k}$, and $\pi_k(x) = \sigma_k(x) = x$ if $x \in T^{<k}$. The example $f(x) = \sigma_k(x)$ leads to the k-reversible languages, confer [1,5]. In particular, the trivial distinguishing function, whose range is a singleton set, characterizes the 0-reversible languages.

To every distinguishing function f, a finite automaton $A_f = (F, T, \delta_f, f(\lambda), F)$ can be associated by setting $\delta_f(q, a) = f(wa)$, where $w \in f^{-1}(q)$ can be chosen arbitrarily, since f is a distinguishing function. Here, we will formally introduce function distinguishable languages and discuss some formal language properties.

Definition 2. *Let $A = (Q, T, \delta, q_0, Q_F)$ be a finite automaton. Let $f : T^* \to F$ be a distinguishing function. A is called* f-distinguishable *if:*

1. *A is deterministic.*
2. *For all states $q \in Q$ and all $x, y \in T^*$ with $\delta^*(q_0, x) = \delta^*(q_0, y) = q$, we have $f(x) = f(y)$.*
   (In other words, for $q \in Q$, $f(q) := f(x)$ for some x with $\delta^*(q_0, x) = q$ is well-defined.)
3. *For all $q_1, q_2 \in Q$, $q_1 \neq q_2$, with either (a) $q_1, q_2 \in Q_F$ or (b) there exist $q_3 \in Q$ and $a \in T$ with $\delta(q_1, a) = \delta(q_2, a) = q_3$, we have $f(q_1) \neq f(q_2)$.*

Informally speaking, an automaton is f-distinguishable if backward nondeterminism conflicts can be resolved with the help of the distinguishing function f.

A language is *f-distinguishable* iff it can be accepted by an f-distinguishable automaton. The class of f-distinguishable languages is denoted by f-DL.

We need a suitable notion of a canonical automaton in the following.

Definition 3. *Let $f : T^* \to F$ be a distinguishing function and let $L \subseteq T^*$ be a regular set. Let $A(L, f)$ be the stripped subautomaton of the product automaton $A(L) \times A_f$. $A(L, f)$ is called* f-canonical automaton *of L.*

Theorem 1. *Let $f : T^* \to F$ be some distinguishing function. Then, $L \subseteq T^*$ is f-distinguishable iff $A(L, f)$ is f-distinguishable.* □

This characterization was proved in [4] and used in order to establish the inferability of f-DL. $A(L, f)$ was employed to construct a characteristic sample for L (with respect to f), and moreover, the $A(L, f)$ (note that $A(L, f)$ is usually larger than $A(L)$) are the hypothesis space of the learning algorithm.

4 An Extended Example

Radhakrishnan showed [13, Example 3.4] that the language L described by $ba^*c + d(aa)^*c$ lies in Ter-DL but its reversal does not. Consider the deterministic (minimal) automaton $A(L)$ with transition function δ (see Table 1). Is $A(L)$

Table 1. The transition functions δ, δ_{Ter} and δ_{inferred}

	a	b	c	d
$\to 0$	–	1	–	2
1	1	–	3	–
2	4	–	3	–
$3 \to$	–	–	–	–
4	2	–	–	–

Ter		a	b	c	d
$\emptyset$	$\to 0$	–	1	–	2
$\{b\}$	1	$1'$	–	3	–
$\{a,b\}$	$1'$	$1'$	–	$3'$	–
$\{d\}$	2	4	–	$3''$	–
$\{a,d\}$	$2'$	4	–	$3'''$	–
$\{b,c\}$	$3 \to$	–	–	–	–
$\{a,b,c\}$	$3' \to$	–	–	–	–
$\{c,d\}$	$3'' \to$	–	–	–	–
$\{a,c,d\}$	$3''' \to$	–	–	–	–
$\{a,d\}$	4	$2'$	–	–	–

Ter		a	b	c
$\emptyset$	$\to 0$	–	1	–
$\{b\}$	1	$1'$	–	3
$\{a,b\}$	$1'$	$1'$	–	$3'$
$\{b,c\}$	$3 \to$	–	–	–
$\{a,b,c\}$	$3' \to$	–	–	–

Ter-distinguishable? We have still to check whether it is possible to resolve the backward nondeterminism conflicts (the state 3 occurs two times in the column labelled c). This resolution possibility is formalized in the second and third condition in Definition 2. As to the second condition, the question is whether it is possible to assign Ter-values to states of $A(L)$ in a well-defined manner: assigning $\text{Ter}(0) = \emptyset$ and $\text{Ter}(4) = \{a, d\}$ is possible, but should we set $\text{Ter}(1) = \{b\}$ (since $\delta^*(0, b) = 1$) or $\text{Ter}(1) = \{a, b\}$ (since $\delta^*(0, ba) = 1$)?; similar problems occur with states 2 and 3. Let us therefore try another automaton accepting L, whose transition function δ_{Ter} is given by Table 1, we indicate the Ter-values of the states in the first column of the table. As the reader may verify, δ_{Ter} basically is the transition table of the stripped subautomaton of the product automaton $A(L) \times A_{\text{Ter}}$. One source of backward nondeterminism may arise from multiple final states, see condition 3.(a) of Def. 2. Since the Ter-values of all four finite states are different, this sort of nondeterminism can be resolved. Let us consider possible violations of condition 3.(b) of Def. 2. In the column labelled a, we find multiple occurrences of the same state entry:

- $\delta_{\text{Ter}}(1, a) = \delta_{\text{Ter}}(1', a) = 1'$: since $\text{Ter}(1) = \{b\} \neq \text{Ter}(1') = \{a, b\}$, this conflict is resolvable.
- $\delta_{\text{Ter}}(2, a) = \delta_{\text{Ter}}(2', a) = 4$: since $\text{Ter}(2) = \{d\} \neq \text{Ter}(2') = \{a, d\}$, this conflict is resolvable.

Observe that the distinguishing function f can be also used to design efficient backward parsing algorithms for languages in f-DL. The only thing one has to know are the f-values of all prefixes of the word w to be parsed. Let us try to check that $daac$ belongs to the language L in a backward fashion. For the prefixes, we compute: $\text{Ter}(d) = \{d\}$, $\text{Ter}(da) = \text{Ter}(daa) = \{a, d\}$, and $\text{Ter}(daac) = \{a, c, d\}$. Since $\text{Ter}(w_1) = \{a, c, d\}$, we have to start our backward parse in state $3'''$. The column labelled c reveals that after reading the last letter c, we are in state $2'$. After reading the penultimate letter a, we are therefore in state 4. Reading the second letter a brings us into state 2, since the Ter-value

Table 2. The transition functions of A_{L^R} and of $A(L^R, \mathrm{Ter})$; X is one of $\{a,b,c\}$, $\{a,c,d\}$, $\{b,c\}$ and $\{c,d\}$

	a	b	c	d
$\to 0$	–	–	1	–
1	2	3	–	3
2	1	3	–	–
$3 \to$	–	–	–	–

	a	b	c	d
$\to (0,\{\lambda\})$	–	–	$(1,\{c\})$	–
$(1,\{c\})$	$(2,\{a,c\})$	$(3,\{b,c\})$	–	$(3,\{c,d\})$
$(1,\{a,c\})$	$(2,\{a,c\})$	$(3,\{a,b,c\})$	–	$(3,\{a,c,d\})$
$(2,\{a,c\})$	$(1,\{a,c\})$	$(3,\{a,b,c\})$	–	–
$(3,X) \to$	–	–	–	–

of the prefix left to be read is $\{d\} = \mathrm{Ter}(2)$. Finally, reading d brings us to the initial state 0; hence, *daac* is accepted by the automaton.

Let us discuss why L^R described by $ca^*b + c(aa)^*d$ is *not* in Ter-DL, as already Radhakrishnan claimed (without proof) [13, Example 3.4]. Table 2 shows the transition function of the minimal deterministic automaton A_{L^R} and the transition function of $A(L^R, \mathrm{Ter})$. As the reader may verify, $A(L^R, \mathrm{Ter})$ is *not* Ter-distinguishable. Our characterization theorem implies that L^R is not Ter-distinguishable either. A similar argument shows that L^R is not σ_1-distinguishable. On the contrary, L^R is σ_2-distinguishable. This can be seen by looking at $A(L^R, \sigma_2)$.

5 Inference Algorithm

We present an algorithm which receives an input sample set $I_+ = \{w_1, \dots, w_M\}$ (a finite subset of the language $L \in f$-DL to be identified) and finds the smallest language $L' \in f$-DL which contains I_+.

The *prefix tree acceptor* $PTA(I_+) = (Q, T, \delta, q_0, Q_F)$ of a finite sample set $I_+ = \{w_1, \dots, w_M\} \subset T^*$ is a deterministic finite automaton which is defined as follows: $Q = \mathrm{Pref}(I_+)$, $q_0 = \lambda$, $Q_F = I_+$ and $\delta(v, a) = va$ for $va \in \mathrm{Pref}(I_+)$.

A simple merging state inference algorithm `f-Ident` for f-DL now starts with the automaton $A_0 = PTA(I_+)$ and merges two arbitrarily chosen states q and q' which cause a conflict to the first or the third of the requirements for f-distinguishing automata.[1] This yields an automaton A_1. Again, choose two conflicting states p, p' and merge them to obtain an automaton A_2 and so forth, until one comes to an automaton A_t which is f-distinguishable. In this way, we get a chain of automata $A_0, A_1, \dots, A_t$. Observe that each A_i is stripped, since A_0 is stripped. In a fashion analogous to the algorithm ZR designed by Angluin for inferring 0-reversible languages, a description of the algorithm `f-Ident`, where $f : T^* \to F$, can be given as follows:

Algorithm 1 (`f-Ident`).
Input: a nonempty positive sample $I_+ \subseteq T^*$.

[1] One can show that the second requirement won't ever be violated when starting the merging process with A_0 which trivially satisfies that condition.

Output: $A(L,f)$, where L is the smallest f-distinguishable language containing I_+.
*** Initialization
Let $A_0 = (Q_0, T, \delta_0, q_{0,0}, Q_{F,0}) = PTA(I_+)$.
For each $q \in Q_0$, compute $f(q)$.
Let π_0 be the trivial partition of Q_0.
Initialize the successor function s by defining $s(\{q\}, a) := \delta_0(q, a)$ for $q \in Q_0$, $a \in T$.[2]
Initialize the predecessor function p by $p(\{q\}, a) := (q', f(q'))$, with $\delta_0(q', a) := q$.[3]
Let LIST contain all pairs $\{q, q'\} \subseteq Q_0$ with $q \neq q'$, $q, q' \in Q_{F,0}$ and $f(q) = f(q')$.
Let $i := 0$.
*** Merging
While LIST$\neq \emptyset$ do begin
    Remove some element $\{q_1, q_2\}$ from LIST.
    Consider the blocks $B_1 = B(q_1, \pi_i)$ and $B_2 = B(q_2, \pi_i)$.
    If $B_1 \neq B_2$, then begin
        Let π_{i+1} be π_i with B_1 and B_2 merged.
        For each $a \in T$, do begin
            If both $s(B_1, a)$ and $s(B_2, a)$ are defined and not equal,
                then place $\{s(B_1, a), s(B_2, a)\}$ on LIST.
            If $s(B_1, a)$ is defined, then set $s(B_1 \cup B_2, a) := s(B_1, a)$;
            otherwise, set $s(B_1 \cup B_2, a) := s(B_2, a)$.
            For each $z \in F$, do begin
                If there are $(p_i, z) \in p(B_i, a)$, $i = 1, 2$, then:
                    If $B(p_1, \pi_i) \neq B(p_2, \pi_i)$, then place $\{p_1, p_2\}$ on LIST.
                    Set $p(B_1 \cup B_2, a) := p(B_1, a)$.
                If $(p_2, z) \in p(B_2, a)$ then:
                    If there is no p_1 with $(p_1, z) \in p(B_1, a)$,
                        then add (p_2, z) to $p(B_1 \cup B_2, a)$.
            end *** for z
        end *** for a
        Increment i by one.
        If $i = |Q_0| - 1$, then LIST$= \emptyset$.
    end *** if
end *** while

An induction shows that any pair $\{q, q'\}$ ever placed on LIST obeys $f(q) = f(q')$.

Example 1. Let us illustrate the work of `f-Ident` by means of an example (with $f = \text{Ter}$): Consider $I_+ = \{bc, bac, baac\}$. Since $\text{Ter}(bac) = \text{Ter}(baac)$, initially (only) the state pair $\{bac, baac\}$ of the PTA is placed onto LIST. In the first pass through the while-loop, these two states are merged. Since both $s(\{bac\}, u)$ and $s(\{baac\}, u)$ are undefined for any letter u, $s(\{bac, baac\}, u)$ will be also undefined. Since $\text{Ter}(ba) = \text{Ter}(baa)$, the pair $\{ba, baa\}$ is placed on LIST when investigating the predecessors. In the next pass through the while-loop, $\{ba\}$ and $\{baa\}$ are merged, but no further mergeable pairs are created,

[2] According to the initialization, both s and p may be undefined for some arguments.
[3] Note that this state q' is uniquely defined in $PTA(I_+)$.

since in particular, the predecessors b and ba of ba and baa, respectively, have different Ter-values. Hence, the third transition function δ_{inferred} of Table 1 is inferred; for clarity, we indicate the Ter-values of the states in the first column of the table.

The somewhat peculiar names of the states were chosen in order to make the comparison with the Ter-distinguishable automaton presented in Section 4 easier for the reader. In terms of the block notion used in the inference algorithm, we have $0 = \{\lambda\}$, $1 = \{b\}$, $1' = \{ba, baa\}$, $3 = \{bc\}$, and $3' = \{bac, baac\}$. Observe that the resulting automaton is not the minimal automaton of the obtained language ba^*c, which is obtainable by merging state 1 with $1'$ and state 3 with $3'$.

Theorem 2 (Correctness, see [4]). *If $L \in f$-DL is enumerated as input to the algorithm* `f-Ident`*, it converges to the f-canonical automaton $A(L, f)$.* □

Here, we will give a detailed complexity analysis valid for the popular RAM computation model where arbitrarily large integers fit into one register or memory unit. This means that values of the s and p functions can be compared in unit time. The following analysis is based on an implementation which uses the operations UNION (of two disjoint subsets, i.e., classes, of a given n-element universe) and FIND (the class to which a given element belongs).[4]

Theorem 3 (Time complexity). *By using a standard union-find algorithm, the algorithm* `f-Ident` *can be implemented to run in time*

$$O(\alpha(2(|F| + 1)(|T| + 1)n, n)(|F| + 1)(|T| + 1)n),$$

where α is the inverse Ackermann function[5] and n is the total length of all words in I_+ from language L, when L is the language presented to the learner for f-DL.

Proof. In any case, $PTA(I_+)$ has basically n states; these states comprise the universe of the union-find algorithm. UNION will be applied no more than $n - 1$ times, since then the inferred automaton will be trivial,

How many FIND operation will be triggered? Two FIND operations will be needed to compute the blocks B_1 and B_2 to which a pair (q_1, q_2) taken from LIST belongs. Apart from the initialization, a certain number of new elements is put onto LIST each time a UNION operation is performed. More precisely, each letter $a \in T$ may cause $\{s(B_1, a), s(B_2, a)\}$, as well as $|F|$ "predecessor pairs" $\{p_1, p_2\}$, to be put onto LIST. In the initialization phase, no more than $\min\{|F|^2, n^2\}$ elements are put onto LIST. So, no more than

$$(|F| + 1)|T|(n - 1) + \min\{|F|^2, n^2\} \leq (|F| + 1)(|T| + 1)n$$

elements are ever put onto LIST. □

[4] A thorough analysis of various algorithms is contained in [15]. A simplified analysis can be found in [3]. Angluin's algorithm ZR for 0-reversible languages works similarly.

[5] For the exact definition of α, we refer to [15].

Observe that this basically leads to an $O(\alpha(|T|^{2k+1}n, n)|T|^{2k+1}n)$ algorithm for k-reversible languages; but note that we output a different type of canonical automata compared with Angluin. When k is small compared to n (as it would be in realistic applications, where k could be considered even as a fixed parameter), our algorithm for k-reversible language inference would run in nearly linear time, since the inverse Ackermann function is an extremely slowly growing function, while Angluin [1] proposed a cubic learning algorithm (always outputting the minimal deterministic automaton). Similar considerations are also true for the more involved case of regular tree languages [9].

Note that the performance of f-`Ident` depends on the size of A_f (since the characteristic sample $\chi(L, f)$ we defined above depends on this size) and is in this sense "scalable", since "larger" A_f permit larger language families to be identified. More precisely, we can show:

Theorem 4. *Let f and g be distinguishing functions. If A_f is a homomorphic image of A_g, then f-DL $\subseteq$ g-DL.* □

This scalability leads to the natural question which distinguishing function one has to choose in one's application.

Let us assume that the user knows several "typical" languages $L_1, \ldots, L_r$. Possibly, the choice of $f_{L_1} \times \cdots \times f_{L_r}$ as distinguishing function has a range which is too large for practical implementation. Recall that the identification algorithm proposed in [4] exponentially depends on the size of the range of the distinguishing function. Therefore, the following problem is of interest:
Problem: Given $L_1, \ldots, L_r$, find a distinguishing function f with minimal range such that $L_1, \ldots, L_r$ lie all within f-DL. Although we expect this problem to be NP-hard, we have yet no proof. We even suspect the problem is hard if $r = 1$.

6 A Retrievable Implementation

Under `www-fs.informatik.uni-tuebingen.de/~fernau/GI.htm`, the program **inference** written in C++ can be retrieved. Usage:
inference -a [**-k**|**-t**|**-k***n*|**-t***n*] [*fdist*] *sample*

You can specify your own distinguishing function in the file *fdist* as the transition relation of a finite automaton. The format of *fdist* is a table where the rows are terminated by line feeds or by carriage returns and line feeds. Columns are separated by space characters or tabs or both. *fdist* must contain all the symbols used in *sample*. The learner generalizes according to the specified algorithm and writes the inferred automaton to standard output.

If the **-k** or **-t** option is given, *fdist* will be ignored. The **-k** option invokes a k-reversible inference algorithm. If **k** is followed by a positive integer n, σ_n will be used as the distinguishing function, else σ_0 will be used. The **-t** option causes the learner to use Ter as the distinguishing function—basically corresponding to the terminal distinguishable languages. If **-t** is followed by a positive integer n, Ter_n will be used as the distinguishing function. As an example, consider invoking `inference -a -t sampleRN` with the samples *bc*, *bac*, *dc*, *daac*, *baac* and *daaaac* listed in the file `sampleRN`; this yields δ_{Ter} from Table 1.

Moreover, the program can be used to infer document type definitions (DTD) for XML documents base on the inference of function distinguishable languages, as explained in [7]. Usage:
inference -x [**-k**|**-t**|**-k***n*|**-t***n*] *xml-doc* [*dtd-file*]
Most of the options are used as explained before. Additionally, the created DTD rules will be written to the file *dtd-file*, if specified. An existing file *dtd-file* will be overwritten. Note that the DTD will most probably be incomplete because no attribute rules are inferred.

As a peculiarity, let us finally mention the 1-unambiguity requirement for DTDs: If some rule violates this requirement, the DTD-rule will be
`<!ELEMENT tag ANY >`.
The one-unambiguity is checked according to [2].

At the mentioned website you can also find a similar program for learning regular tree languages, also see [9].

References

1. D. Angluin. Inference of reversible languages. *J. of the ACM*, 29(3):741–765, 1982. 64, 66, 71
2. A. Brüggemann-Klein and D. Wood. One-unambiguous regular languages. *Information and Computation*, 142(2):182–206, 1998. 72
3. T. H. Cormen, C. E. Leiserson, R. L. Rivest, and C. Stein. *Introduction to Algorithms.* MIT Press, 2nd. edition, 2001. 70
4. H. Fernau. Identification of function distinguishable languages. In H. Arimura, S. Jain, and A. Sharma, editors, *Proceedings of the 11th International Conference Algorithmic Learning Theory ALT 2000*, volume 1968 of *LNCS/LNAI*, pages 116–130. Springer, 2000. 64, 66, 70, 71
5. H. Fernau. *k*-gram extensions of terminal distinguishable languages. In *International Conference on Pattern Recognition (ICPR 2000)*, volume 2, pages 125–128. IEEE/IAPR, IEEE Press, 2000. 66
6. H. Fernau. Approximative learning of regular languages. In L. Pacholski and P. Ružička, editors, *SOFSEM'01; Theory and Practice of Informatics*, volume 2234 of *LNCS*, pages 223–232. Springer, 2001. 65
7. H. Fernau. Learning XML grammars. In P. Perner, editor, *Machine Learning and Data Mining in Pattern Recognition MLDM'01*, volume 2123 of *LNCS/LNAI*, pages 73–87. Springer, 2001. 65, 72
8. H. Fernau. Parallel communicating grammar systems with terminal transmission. *Acta Informatica*, 37:511–540, 2001. 65
9. H. Fernau. Learning tree languages from text. In J. Kivinen, editor, *Conference on Learning Theory COLT'02*, to appear in the *LNCS/LNAI* series of Springer. 71, 72
10. E. M. Gold. Language identification in the limit. *Information and Control*, 10:447–474, 1967. 64
11. J. Gregor. Data-driven inductive inference of finite-state automata. *International Journal of Pattern Recognition and Artificial Intelligence*, 8(1):305–322, 1994. 64

12. C. de la Higuera. Current trends in grammatical inference. In F. J. Ferri et al., editors, *Advances in Pattern Recognition, Joint IAPR International Workshops SSPR+SPR'2000*, volume 1876 of *LNCS*, pages 28–31. Springer, 2000. 64
13. V. Radhakrishnan. *Grammatical Inference from Positive Data: An Effective Integrated Approach*. PhD thesis, Department of Computer Science and Engineering, Indian Institute of Technology, Bombay (India), 1987. 64, 66, 68
14. V. Radhakrishnan and G. Nagaraja. Inference of regular grammars via skeletons. *IEEE Transactions on Systems, Man and Cybernetics*, 17(6):982–992, 1987. 64, 65
15. R. E. Tarjan and J. van Leeuwen. Worst-case analysis of set union algorithms. *J. of the ACM*, 31:245–281, 1984. 70

Non-bayesian Graph Matching without Explicit Compatibility Calculations

Barend Jacobus van Wyk[1,2] and Michaël Antonie van Wyk[3]

[1] Kentron, a division of Denel, Centurion, South Africa
[2] University of the Witwatersrand, Johannesburg, South Africa
ben.van.wyk@kentron.co.za
[3] Rand Afrikaans University, Johannesburg, South Africa
mavw@ing.rau.ac.za

Abstract. This paper introduces a novel algorithm for performing Attributed Graph Matching (AGM). A salient characteristic of the Interpolator-Based Kronecker Product Graph Matching (IBKPGM) algorithm is that it does not require the explicit calculation of compatibility values between vertices and edges, either using compatibility functions or probability distributions. No assumption is made about the adjacency structure of the graphs to be matched. The IBKPGM algorithm uses Reproducing Kernel Hilbert Space (RKHS) interpolator theory to obtain an unconstrained estimate to the Kronecker Match Matrix (KMM) from which a permutation sub-matrix is inferred.

1 Introduction

An object can be described in terms of its parts, the properties of these parts and their mutual relationships. Representation of the structural descriptions of objects by attributed relational graphs reduces the problem of matching to an *Attributed Graph Matching* (AGM) problem.

According to [1], graph matching algorithms can be divided into two major approaches. In general, the first approach constructs a state-space, which is searched using heuristics to reduce complexity [2-7]. The second approach, which is the one adopted here, is based on function optimization techniques which include Bayesian, linear-programming, continuation, eigen-decomposition, polynomial transform, genetic, neural network and relaxation-based methods [1, 8–19].

Except for some earlier approaches not suited to sub-graph matching, such as [8–10], most optimization-based algorithms require the explicit calculation of compatibility values between vertices and edges, either using compatibility functions or probability distributions. In addition, some Bayesian-based approaches fail when graphs are fully connected [14–19]. The focus of this paper is on matching fully-connected, undirected attributed graphs, using a non-Bayesian optimization framework, without the explicit calculation of compatibility values or matrices.

The outline of the presentation is as follows: In section 2, we briefly review the concept of an attributed graph, formulate the *Attributed Graph Matching* (AGM)

T. Caelli et al. (Eds.): SSPR&SPR 2002, LNCS 2396, pp. 74–83, 2002.
© Springer-Verlag Berlin Heidelberg 2002

problem, and introduce the *Kronecker Product Graph Matching* KPGM formulation. The *Interpolator-Based Kronecker Product Graph Matching* (IBKPGM) algorithm is presented in section 3. Numerical results obtained during the evaluation of our algorithm are presented in section 4.

2 KPGM Formulation

The focus of this paper is on matching graphs where a duplicate graph, say

$$G = (V, E, \{\mathbf{A}_i\}_{i=1}^r, \{\mathbf{B}_j\}_{j=1}^s) \tag{1}$$

is matched to a reference graph, say

$$G' = (V', E', \{\mathbf{A}'_i\}_{i=1}^r, \{\mathbf{B}'_j\}_{j=1}^s) \tag{2}$$

where $\mathbf{A}_i \in \mathbb{R}^{n \times n}$, $\mathbf{B}_j \in \mathbb{R}^{n \times 1}$, $\mathbf{A}'_i \in \mathbb{R}^{n' \times n'}$and $\mathbf{B}'_j \in \mathbb{R}^{n' \times 1}$ represent the edge attribute adjacency matrices and vertex attribute vectors respectively. The reference and duplicate graphs each have r edge attributes and s vertex attributes. The number of vertices of G' (respectively, G) is $n' := |V'|$ (respectively, $n := |V|$). Here we consider the general case of sub-graph matching. *Full-graph Matching* (FGM) refers to matching two graphs having the same number of vertices (i.e. $n' = n$) while *Sub-graph Matching* (SGM) refers to matching two graphs having a different number of vertices (i.e. $n' > n$).

We say that G is *matched* to some sub-graph of G' if there exists an $n \times n'$ permutation sub-matrix $\mathbf{P}$ such that $\mathbf{A}_i = \mathbf{P}\mathbf{A}'_i\mathbf{P}^T$ and $\mathbf{B}_j = \mathbf{P}\mathbf{B}'_j$ where $i = 1, ..., r$ and $j = 1, ..., s$.

We now observe that vertex attribute vectors converted to diagonal matrices, using the diag(·) operation in linear algebra, satisfy the same expression as edge attribute matrices do, namely $\operatorname{diag}\mathbf{B}_j \approx \mathbf{P}\operatorname{diag}(\mathbf{B}'_j)\,\mathbf{P}^T$, with exact equality holding for the ideal case (i.e. when G is just a permuted sub-graph of G'). This means that these converted vertex attribute vectors may be considered as additional edge attribute matrices. Observing that

$$\operatorname{vec}(\mathbf{A}_i - \mathbf{P}\mathbf{A}'_i\mathbf{P}^T) \equiv \operatorname{vec}\mathbf{A}_i - \mathbf{\Phi}\operatorname{vec}\mathbf{A}'_i \tag{3}$$

where $\mathbf{\Phi} = \mathbf{P} \otimes \mathbf{P}$ and $\otimes$ denotes the *Kronecker Product* for matrices, the AGM problem can be expressed as

$$\min_{\mathbf{\Phi}} \left(\sum_{i=1}^{r+s} \|\operatorname{vec}\mathbf{A}_i - \mathbf{\Phi}\operatorname{vec}\mathbf{A}'_i\|^2 \right) \tag{4}$$

subject to $\mathbf{\Phi} = \mathbf{P} \otimes \mathbf{P}$, $\mathbf{P} \in \operatorname{Per}(n, n')$ where $\operatorname{Per}(n, n')$ is the set of all $n \times n'$ permutation sub-matrices. Here $\|\cdot\|$ denotes some matrix norm. Different norms will yield different solutions to the above minimization problem. The IBKPGM algorithm in essence obtains an approximation to $\mathbf{\Phi}$, from which $\mathbf{P} \in \operatorname{Per}(n, n')$ can be derived.

The following definitions and observation are used in the sequel:

Definition 1. *The matrix* $\mathbf{\Phi} = \mathbf{P} \otimes \mathbf{P}$ *minimizing Eq. (4), subject to* $\mathbf{P} \in Per(n, n')$, *is termed the* Constrained Kronecker Match Matrix.

Definition 2. *An unconstrained approximation to the Constrained Kronecker Match Matrix, say* $\widehat{\mathbf{\Phi}}$, *is termed the* Unconstrained Kronecker Match Matrix.

Definition 3. *An unconstrained approximation to the permutation matrix, say* $\widehat{\mathbf{P}}$, *is termed the* Unconstrained Permutation Sub-matrix.

Observation 1 *Given a Constrained Kronecker Match Matrix,* $\mathbf{\Phi} = \mathbf{P} \otimes \mathbf{P}$, *such that* $\mathbf{P} \in Per(n, n')$, *we can retrieve the unknown permutation sub-matrix* $\mathbf{P}$ *by*

$$P_{ij} := \frac{\sum_{k,l} \mathbf{\Phi}_{kl}}{n}, \tag{5}$$

where $i = 1, ..., n$, $j = 1, ..., n'$, $k = (i-1)n + 1, ..., (i-1)n + n$ *and* $l = (j-1)n' + 1, ..., (j-1)n' + n'$.

The space generated by all Unconstrained Kronecker Match Matrices contains the space generated by all Constrained Kronecker Match Matrices. The general approach we follow is to first derive an unconstrained Kronecker Match Matrix from which we then infer a Constrained Kronecker Match Matrix.

3 The Interpolator-Based Approach

The *Interpolator Based Kronecker Product Graph Matching* (IBKPGM) algorithm uses *Reproducing Kernel Hilbert Space* (RKHS) interpolator theory, the general framework of which is presented in [20], to obtain an unconstrained approximation to $\mathbf{\Phi}$. The RIGM algorithm [21] is based on the same theory, but the way in which the permutation sub-matrix $\mathbf{P}$ is obtained, is completely different.

3.1 The Interpolator Equation

The AGM process can be viewed to be comprising of the mapping

$$\mathbf{F} : \mathbb{R}^{n' \times n'} \to \mathbb{R}^{n \times n}, \qquad \mathbf{F}(\mathbf{X}) := \mathbf{P}\,\mathbf{X}\,\mathbf{P}^T, \tag{6}$$

with the fixed parameter matrix $\mathbf{P} \in \text{Per}(n, n')$ the unknown quantity to be determined. In doing so, the AGM problem is cast into a system identification problem which can be solved by selecting an appropriate RKHS-based interpolator to model the above mapping. This interpolator contains an approximation to $\mathbf{P}$ implicitly. The input-output pairs used for establishing the interpolative constraints associated with $\mathbf{F}$ is the set $\{(\mathbf{X}_k, \mathbf{Y}_k)\}_{k=1}^{r+s}$, where $\mathbf{X}_k := \mathbf{A}'_k$ and $\mathbf{Y}_k := \mathbf{A}_k$.

The framework presented in [20] will now be used to find a solution to the AGM problem. Consider the space of all functions of the form $F : \mathbb{R}^{(n')^2 \times 1} \to \mathbb{R}$, $F(\mathbf{X}) = \sum_{i,j=1}^{n'} \varphi_{ij} X_{ij}$. The component functions F_{ij} of the function $\mathbf{F}$

clearly belong to this space. We choose as basis for this space the functions $e_{ij}(\mathbf{X}) := X_{ij}$ where $i,j = 1,\ldots,n'$. Next, we endow this space with the inner product $(F,G) := \sum_{i,j=1}^{n'} \varphi_{ij}\gamma_{ij}$ to obtain an RKHS with reproducing kernel given by $K(\mathbf{Y},\mathbf{Z}) = \sum_{i,j=1}^{n'} e_{ij}(\mathbf{Y})\, e_{ij}(\mathbf{Z}) = (\mathrm{vec}\,\mathbf{Y})^T \mathrm{vec}\,\mathbf{Z}$. Here vec$(\cdot)$ denotes the vectorization operation of linear algebra. Given the training set $\{(\mathbf{A}'_k, \mathbf{A}_k)\}_{k=1}^{r+s}$ we therefore have the interpolative constraints, $(F_{ij}, K(\mathbf{A}'_k,\cdot)) = F_{ij}(\mathbf{A}'_k) = A_{k\,|\,ij}$ for each $k = 1,\ldots,(r+s)$ and $i,j = 1,\ldots,n$. As shown in [20], the minimum-norm interpolator has the form $\widetilde{F}_{ij}(\cdot) = \sum_{l=1}^{r+s} C_{l\,|\,ij}\, K(\mathbf{A}'_l,\cdot)$ where the coefficients $C_{l\,|\,ij}$ are the unknown parameters. Evaluation of the function $\widetilde{F}_{ij}$ at the points $\mathbf{X} = \mathbf{A}'_k$ yields the following system of simultaneous equations,

$$A_{k\,|\,ij} = \sum_{l=1}^{r+s} C_{l\,|\,ij}\, K(\mathbf{A}'_l, \mathbf{A}'_k)\,, \qquad k = 1,\ldots,r+s\,, \tag{7}$$

for $i,j = 1,\ldots,n$. Assembling these results into a matrix, we obtain

$$\mathbf{A}_k = \sum_{l=1}^{r+s} \mathbf{C}_l\, G_{l\,k} \tag{8}$$

where $\mathbf{C}_l := (C_{l\,|\,ij})$, $G_{l\,k} := K(\mathbf{A}'_l, \mathbf{A}'_k)$ and $k = 1,\ldots,r+s$.

By introducing the matrices $\mathbf{A} \in \mathbb{R}^{n^2\times(r+s)}$ and $\mathbf{G} \in \mathbb{R}^{(r+s)\times(r+s)}$ where $\mathbf{A} := (\mathrm{vec}\,\mathbf{A}_1,\ldots,\mathrm{vec}\,\mathbf{A}_{r+s})$ and $\mathbf{G} := (\mathbf{G}_1,\ldots,\mathbf{G}_{r+s})$, we can express the complete problem in the form of a single matrix equation, namely $\mathbf{A} = \mathbf{C}\,\mathbf{G}$, where the only unknown is the matrix $\mathbf{C} := (\mathrm{vec}\,\mathbf{C}_1,\ldots,\mathrm{vec}\,\mathbf{C}_{r+s})$. Conditions under which the Gram matrix $\mathbf{G}$ is invertible are stated in [23]. These conditions are assumed to be satisfied by the attribute matrices of an attributed graph, and hence the coefficients of the interpolator are described by the matrix expression,

$$\mathbf{C} = \mathbf{A}\,\mathbf{G}^{-1}\,. \tag{9}$$

Up to this point the development coincides with that of the RIGM algorithm [21]. The way in which the permutation sub-matrix is inferred is totally different and is presented next.

3.2 Inferring the Unconstrained Kronecker Match Matrix

Proposition 1. *An Unconstrained Kronecker Match Matrix is given by*

$$\widehat{\mathbf{\Phi}} = \begin{pmatrix} \widehat{\Phi}_1^T \\ \vdots \\ \widehat{\Phi}_{n^2}^T \end{pmatrix} \tag{10}$$

where

$$\widehat{\Phi}_i^T = \left(\sum_{l=1}^{r+s} C_{l|i} \mathrm{vec}\,\mathbf{A}'_l\right)^T, \tag{11}$$

$\mathbf{C} := (\mathrm{vec}\,\mathbf{C}_1, \ldots, \mathrm{vec}\,\mathbf{C}_{r+s})$, $\mathrm{vec}\,\mathbf{C}_l := (C_{l|i})$ *and* $i = 1, ..., n^2$.

Proof. Eq. (8) can be written in the form

$$\mathrm{vec}\,\mathbf{A}_k = \mathrm{vec}\left(\sum_{l=1}^{r+s} \mathbf{C}_l\, G_{l\,k}\right) = \sum_{l=1}^{r+s} \mathrm{vec}(\mathbf{C}_l) G_{l\,k} = \mathbf{C}\,\mathbf{G}_k \tag{12}$$

with $\mathbf{C} := (\mathrm{vec}\,\mathbf{C}_1, \ldots, \mathrm{vec}\,\mathbf{C}_{r+s})$ and $\mathbf{G}_k := (G_{1\,k}, \ldots, G_{r+s,\,k})^T$. By comparing Eq. (12) with Eq. (3) it is clear that

$$\mathrm{vec}\,\mathbf{A}_k = \mathbf{C}\,\mathbf{G}_k = \widehat{\mathbf{\Phi}}\,\mathrm{vec}\mathbf{A}'_k. \tag{13}$$

Expanding Eq. (13) and performing a row-wise decomposition we obtain

$$A_{k|i} = \left(\sum_{l=1}^{r+s} C_{l|i} \mathrm{vec}\mathbf{A}'_l\right)^T \mathrm{vec}\mathbf{A}'_k, \tag{14}$$

where $\mathrm{vec}\,\mathbf{A}_k := A_{k|i}$, $k = 1, ..., r+s$ and $i = 1, ..., n^2$. From Eq. (14) it follows that $\widehat{\Phi}_i^T = \left(\sum_{l=1}^{r+s} C_{l|i} \mathrm{vec}\mathbf{A}'_l\right)^T$ which concludes the proof. □

3.3 Inferring an Approximation to P from $\widehat{\mathbf{\Phi}}$

Observation 2 *If an Unconstrained Permutation Sub-matrix* $\widehat{\mathbf{P}}$ *is used as the weight matrix of an* optimal assignment problem, *the solution of which is a permutation sub-matrix* $\overline{\mathbf{P}}$ *representing the optimal assignment of* $\widehat{\mathbf{P}}$, *then the permutation sub-matrix* $\overline{\mathbf{P}}$ *is the closest to* $\widehat{\mathbf{P}}$ *in the mean-square-error sense, that is,* $\|\overline{\mathbf{P}} - \widehat{\mathbf{P}}\|_F$ *is a minimum where* $\|\cdot\|_F$ *denotes the Frobenius norm [22].*

Some optimal assignment algorithms might require $\widehat{\mathbf{P}}$ to be a positive matrix. If $\widehat{\mathbf{P}}$ is not a positive matrix, we can add a constant to $\widehat{\mathbf{P}}$ to ensure that all its elements are non-negative.

Proposition 2. *Let* $\overline{\mathbf{P}}$ *be the permutation sub-matrix closest to* $\widehat{\mathbf{P}}$ *in the mean-square-error sense. If the Unconstrained Kronecker Match Matrix is given as* $\widehat{\mathbf{\Phi}} = \widehat{\mathbf{P}} \otimes \widehat{\mathbf{P}}$, *then* $\overline{\mathbf{P}}$ *will also be the permutation sub-matrix closest to* $\widetilde{\mathbf{P}}$ *in the same sense where*

$$\widetilde{P}_{ij} := \frac{\sum_{k,l} \widehat{\Phi}_{kl}}{n}, \tag{15}$$

$\widehat{\mathbf{\Phi}} := \widehat{\Phi}_{kl}$, $i = 1, ..., n$, $j = 1, ..., n'$, $k = (i-1)n+1, ..., (i-1)n+n$, $l = (j-1)n'+1, ..., (j-1)n'+n'$ *and* $\sum_{i,j} \widehat{P}_{ij}$ *is positive.*

Proof. Observe that $\frac{\sum_{k,l} \widehat{\Phi}_{kl}}{n} = \widehat{P}_{ij} \sum_{p,q} \widehat{P}_{pq}$ where $k = (i-1)n+1, ..., (i-1)n+n$, $l = (j-1)n'+1, ..., (j-1)n'+n'$, $p = 1, ..., n$, $q = 1, ..., n'$, $i = 1, ..., n$ and $j = 1, ..., n'$. This leads to

$$\widetilde{\mathbf{P}} = \alpha \widehat{\mathbf{P}} \tag{16}$$

where α is a constant. Noting that the optimal assignment procedure yields the same answer when its weight matrix is multiplied by a positive constant concludes the proof. □

By assuming that $\widehat{\mathbf{\Phi}}$, as given by Proposition 1, is the Kronecker Product of an Unconstrained Permutation Sub-matrix $\widehat{\mathbf{P}}$ with itself, we can therefore use Proposition 2 to obtain a permutation sub-matrix $\overline{\mathbf{P}}$ which will serve as an approximation to $\mathbf{P}$. The resultant algorithm has a complexity of $O(n^4)$ when $n > (r+s)^2$ and the Kuhn-Munkres optimal assignment algorithm is used to approximate $\mathbf{P}$.

4 Numerical Experiments

In order to evaluate the performance of the IBKPGM algorithm, the following procedure was used: Firstly, the parameters n', n, r and s were fixed. For every iteration, a reference graph G' was generated randomly with all attributes distributed between 0 and 1. An $n \times n'$ permutation sub-matrix, $\mathbf{P}$, was also generated randomly, and then used to permute the rows and columns of the edge attribute adjacency matrices and the elements of the vertex attribute vectors of G'. Next, an independently generated noise matrix (vector, respectively) was added to each edge attribute adjacency matrix (vertex attribute vector, respectively) to obtain the duplicate graph G. The element of each noise matrix or vector was obtained by multiplying a random variable—uniformly distributed on the interval $[-1/2, 1/2]$—by the noise magnitude parameter ε. Different graph matching algorithms were then used to determine a permutation sub-matrix which approximates the original permutation sub-matrix $\mathbf{P}$. The proportion correct vertex-vertex assignments was calculated for a given value of ε after every 300 trials for each algorithm. From a probabilistic point of view, this approximates how well an algorithm performs for a given noise magnitude.

4.1 Numerical Results

The performance of the IBKPGM algorithm was compared to the performance of the GAGM [1], EGM [10], CGGM [24], PTGM [9] and RIGM [21] algorithms. A comparison is also made against the *Faugeras-Price Relaxation Labelling* (FPRL) algorithm [25]. The computational complexity of the IBKPGM algorithm is $O(n^4)$.

Figure 4.1 presents the estimated probability of correct vertex-vertex match as a function of noise magnitude ε for the case $(n, r, s) = (30, 3, 3)$. Globally, the performance curves for the GAGM and CGGM algorithms are closely spaced and well separated from the performance curves of the IBKPGM and FPRL algorithms for large values of ε, which in turn are well separated from the performance curves of the PTGM and EGM algorithms. In figure 2, the performance of the IBKPGM algorithm is shown for values of n = 100, 150, and 200 where $r = 3$ and $s = 3$. Figure 3 depicts the estimated probability of correct vertex-vertex

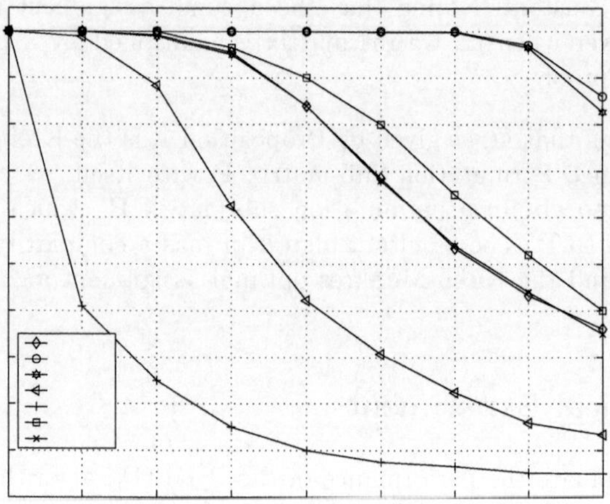

Fig. 1. Matching of (30,3,3) attributed graphs: Estimated probability of correct vertex-vertex matching versus ε

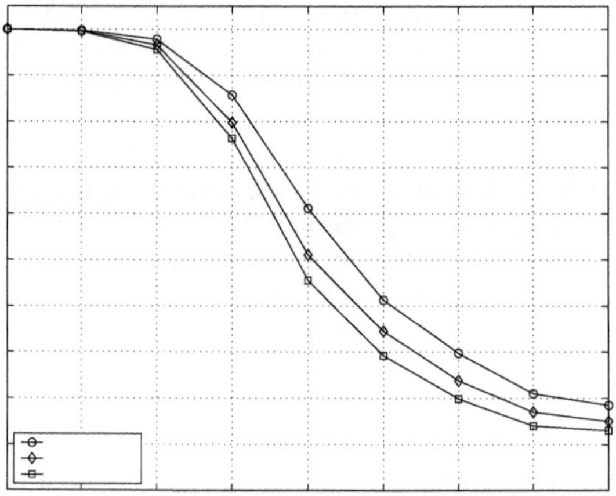

Fig. 2. Matching of (100,3,3), (150,3,3) and (200,3,3) attributed graphs: Estimated probability of correct vertex-vertex matching versus ε

matching for the case $(n'/n, r, s) = (15/5, 5, 5)$. The results of this experiment indicate that the estimated probability of a correct vertex-vertex match is higher than 0.8 for noise values up to nearly 0.3 when a third of the vertices are missing. When only two or three vertices are missing, the experiment is trivial, since the IBKPGM algorithm almost always finds the correct vertex-vertex match when no noise is present. The curve IBKPGM(Augmented) indicates the performance

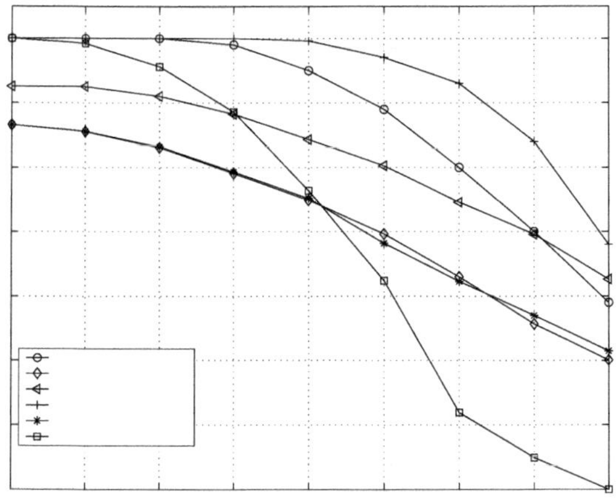

Fig. 3. Matching of (15/5,5,5) attributed graphs: Estimated probability of correct vertex-vertex matching versus ε

of the IBKPGM algorithm when five additional attributes were added to each edge and vertex of each graph. The additional attributes were derived from the existing attributes by squaring them. This procedure significantly improves the sub-graph matching performance of the IBKPGM algorithm for small values of ε.

5 Conclusion

The *Kronecker Product Graph Matching* (KPGM) formulation was presented and the *Interpolator-Based Kronecker Product Graph Matching* (IBKPGM) algorithm, based on this formulation, was introduced. The IBKPGM algorithm incorporates a general approach to a wide class of graph matching problems.

It was demonstrated that the performance of the IBKPGM algorithm is comparable to the performance of a typical gradient-based relaxation method such as the FPRL algorithm when performing full-graph matching. The performance curves of the IBKPGM are almost identical to those of the RIGM [21] algorithm. This phenomenon cannot be explained at present, and is a topic for further investigation.

References

1. Gold, S., Rangarajan, A.: A Graduated Assignment Algorithm for Graph Matching, IEEE Trans. Patt. Anal. Machine Intell., Vol. 18(4)(1996)377–388 74, 79
2. You, M., Wong, K. C.: An Algorithm for Graph Optimal Isomorphism, Proc. ICPR, (1984) 316–319
3. Tsai, W.-H., Fu, K.-S.: Error-Correcting Isomorphisms of Attributed Relation Graphs for Pattern Recognition, IEEE Trans. Syst. Man Cybern., Vol. 9(1979)757–768

4. Tsai, W.-H., Fu, K.-S.: Subgraph Error-Correcting Isomorphisms for Syntactic Pattern Recognition, IEEE Trans. Systems, Man, Cybernetics, Vol. 13(1983)48–62
5. Depiero, F., Trived, M., Serbin, S.: Graph Matching using a Direct Classification of Node Attendance, Pattern Recognition, Vol. 29(6)(1996)1031–1048
6. Eshera, M. A., Fu, K-S.: A Graph Distance Measure for Image Analysis, IEEE Trans. Syst. Man Cybern., Vol. 14(3)(1984)
7. Bunke, H., Messmer, B.: Recent Advances in Graph Matching, Int. J. Pattern Recognition Artificial Intell., Vol. 11(1)(1997)169–203
8. Almohamad, H. A. L., Duffuaa, S. O.: A Linear Programming Approach for the Weighted Graph Matching Problem, IEEE Trans. Patt. Anal. Machine Intell., Vol. 15(5)(1993)522–525
9. Almohamad, H. A. L.: Polynomial Transform for Matching Pairs of Weighted Graphs, Applied Mathematical Modelling, Vol. 15(4)(1991)216–222 79
10. Umeyama, S.: An Eigendecomposition Approach to Weighted Graph Matching Problems, IEEE Trans. Patt. Anal. Machine Intell., Vol. 10(5)(1988)695–703 79
11. Hummel, R. A., Zucker, S. W.: On the Foundations of Relaxation Labelling Processes, IEEE Trans. Patt. Anal. Machine Intell., Vol. 5(3)(1983)267–286
12. Peleg, S.: A New Probabilistic Relaxation Scheme, IEEE Trans. Patt. Anal. Machine Intell., 2(4)(1980)362–369
13. Christmas, W. J., Kittler, J., Petrou, M.: Structural Matching in Computer Vision using Probabilistic Relaxation, IEEE Trans. Patt. Anal. Machine Intell., Vol. 17(8)(1995)749–764
14. Finch, A. M., Wilson, R. C., Hancock, R.: Symbolic Matching with the EM Algorithm, Pattern Recognition, Vol. 31(11)(1998)1777–1790
15. Williams, M. L., Wilson, R. C., Hancock, E. R.: Multiple Graph Matching with Bayesian Inference, Pattern Recognition Letters, Vol. 18(1997)1275–1281
16. Cross, A. D. J., Hancock, E. R.: Graph Matching with a Dual Step EM Algorithm, IEEE Trans. Patt. Anal. Machine Intell., Vol. 20(11)(1998)1236–1253
17. Wilson, R. C., Hancock, E. R.: A Bayesian Compatibility Model for Graph Matching, Pattern Recognition Letters, Vol. 17(1996)263–276
18. Cross, A. D. J., Wilson, C., Hancock, E. R.: Inexact Matching Using Genetic Search, Pattern Recognition, Vol. 30(6)(1997)953–970
19. Wilson, R. C., Hancock, E. R.: Structural Matching by Discrete Relaxation, IEEE Trans. Patt. Anal. Machine Intell., Vol. 19(8)(1997)634–648
20. van Wyk, M. A., Durrani, T. S.: A Framework for Multi-Scale and Hybrid RKHS-Based Approximators, IEEE Trans. Signal Proc., Vol. 48(12)(2000)3559–3568 76, 77
21. van Wyk, M. A., Durrani, T. S., van Wyk, B. J.: A RKHS Interpolator-Based Graph Matching Algorithm, To appear in IEEE Trans. Patt. Anal. Machine Intell, July 2002 76, 77, 79, 81
22. Van Wyk, M. A., Clark, J.: An Algorithm for Approximate Least-Squares Attributed Graph Matching, in Problems in Applied Mathematics and Computational Intelligence, N. Mastorakis (ed.), World Science and Engineering Society Press, (2001) 67-72 78
23. Luenberger, D. G.: Optimization by Vector Space Methods, New York, NY: John Wiley & Sons, 1969 77
24. van Wyk, B. J., van Wyk, M. A., Virolleau, F.: The CGGM Algorithm and its DSP implementation, Proc. 3rd European DSP Conference on Education and Research, ESIEE-Paris, 20-21 September, 2000 79
25. Faugeras, O. D., Price, K. E.: Semantic Description of Aerial Images Using Stochastic Labeling, IEEE Trans. Patt. Anal. Machine Intell., Vol. 3(6)(1981)633–642 79

Spectral Feature Vectors for Graph Clustering

Bin Luo[1,2], Richard C. Wilson[1], and Edwin R. Hancock[1]

[1] Department of Computer Science, University of York
York YO1 5DD, UK
[2] Anhui University, P.R. China
{luo,wilson,erh}@cs.york.ac.uk

Abstract. This paper investigates whether vectors of graph-spectral features can be used for the purposes of graph-clustering. We commence from the eigenvalues and eigenvectors of the adjacency matrix. Each of the leading eigenmodes represents a cluster of nodes and is mapped to a component of a feature vector. The spectral features used as components of the vectors are the eigenvalues, the cluster volume, the cluster perimeter, the cluster Cheeger constant, the inter-cluster edge distance, and the shared perimeter length. We explore whether these vectors can be used for the purposes of graph-clustering. Here we investigate the use of both central and pairwise clustering methods. On a data-base of view-graphs, the vectors of eigenvalues and shared perimeter lengths provide the best clusters.

1 Introduction

Graph clustering is an important yet relatively under-researched topic in machine learning [9,10,4]. The importance of the topic stems from the fact that it is a key tool for learning the class-structure of data abstracted in terms of relational graphs. Problems of this sort are posed by a multitude of unsupervised learning tasks in knowledge engineering, pattern recognition and computer vision. The process can be used to structure large data-bases of relational models [11] or to learn equivalence classes. One of the reasons for limited progress in the area has been the lack of algorithms suitable for clustering relational structures. In particular, the problem has proved elusive to conventional central clustering techniques. The reason for this is that it has proved difficult to define what is meant by the mean or representative graph for each cluster. However, Munger, Bunke and Jiang [1] have recently taken some important steps in this direction by developing a genetic algorithm for searching for median graphs.

Generally speaking, there are two different approaches to graph clustering. The first of these is pairwise clustering[7] . This requires only that a set of pairwise distances between graphs be supplied. The clusters are located by identifying sets of graphs that have strong mutual pairwise affinities. There is therefore no need to explicitly identify an representative (mean, mode or median) graph for each cluster. Unfortunately, the literature on pairwise clustering is much less developed than that on central clustering. The second approach is to embed

T. Caelli et al. (Eds.): SSPR&SPR 2002, LNCS 2396, pp. 83–93, 2002.
© Springer-Verlag Berlin Heidelberg 2002

graphs in a pattern space[8]. Although the pattern spaces generated in this way are well organised, there are two obstacles to the practical implementation of the method. Firstly, it is difficult to deal with graphs with different numbers of nodes. Secondly, the node and edge correspondences must be known so that the nodes and edges can be mapped in a consistent way to a vector of fixed length.

In this paper, we attempt to overcome these two problems by using graph spectral methods to extract feature vectors from symbolic graphs [2]. Spectral graph theory is a branch of mathematics that aims to characterise the structural properties of graphs using the eigenvalues and eigenvectors of the adjacency matrix, or of the closely related Laplacian matrix. There are a number of well-known results. For instance, the degree of bijectivity of a graph is gauged by the difference between the first and second eigenvalues (this property has been widely exploited in the computer vision literature to develop grouping and segmentation algorithms). In routing theory, on the other hand, considerable use is made of the fact that the leading eigenvector of the adjacency matrix gives the steady-state random walk on the graph. Here we adopt a different approach. Our aim is to use the leading eigenvectors of the adjacency matrix to define clusters of nodes. From the clusters, we extract structural features and using the eigenvalue order to index the components we construct feature-vectors. The length of the vectors are determined by the number of leading eigenvalues. The graph spectral features explored include the eigenvalue spectrum, cluster volume, cluster perimeter, cluster Cheeger constant, shared perimeter and cluster distances. The specific technical goals in this paper are two-fold. First, we aim to investigate whether the independent or principal components of the spectral feature vectors can be used to embed graphs in a pattern space suitable for clustering. Second, we investigate which of the spectral features results in the best clusters.

2 Graph Spectra

In this paper we are concerned with the set of graphs $G_1, G_2, .., G_k, ..., G_N$. The kth graph is denoted by $G_k = (V_k, E_k)$, where V_k is the set of nodes and $E_k \subseteq V_k \times V_k$ is the edge-set. Our approach in this paper is a graph-spectral one. For each graph G_k we compute the adjacency matrix A_k. This is a $|V_k| \times |V_k|$ matrix whose element with row index i and column index j is

$$A_k(i,j) = \begin{cases} 1 & \text{if } (i,j) \in E_k \\ 0 & \text{otherwise} \end{cases} . \tag{1}$$

From the adjacency matrices $A_k, k = 1...N$, we can calculate the eigenvalues λ_k by solving the equation $|A_k - \lambda_k I| = 0$ and the associated eigenvectors ϕ_k^ω by solving the system of equations $A_k \phi_k^\omega = \lambda_k^\omega \phi_k^\omega$. We order the eigenvectors according to the decreasing magnitude of the eigenvalues, i.e. $|\lambda_k^1| > |\lambda_k^2| > \ldots |\lambda_k^{|V_k|}|$. The eigenvectors are stacked in order to construct the modal matrix $\Phi_k = (\phi_k^1 | \phi_k^2 | \ldots | \phi_k^{|V_k|})$.

We use only the first n eigenmodes of the modal matrix to define spectral clusters for each graph. The components of the eigenvectors are used to compute the probabilities that nodes belong to clusters. The probability that the node indexed $i \in V_k$ in graph k belongs to the cluster with eigenvalue order ω is

$$s_{i,\omega}^k = \frac{|\Phi_k(i,\omega)|}{\sum_{\omega\prime=1}^{n} |\Phi_k(i,\omega\prime)|}. \quad (2)$$

3 Spectral Features

Our aim is to use spectral features for the modal clusters of the graphs under study to construct feature-vectors. To overcome the correspondence problem, we use the order of the eigenvalues to establish the order of the components of the feature-vectors. We study a number of features suggested by spectral graph theory.

3.1 Unary Features

We commence by considering unary features for the arrangement of modal clusters. The features studied are listed below:

Leading Eigenvalues: Our first vector of spectral features is constructed from the ordered eigenvalues of the adjacency matrix. For the graph indexed k, the vector is $B_k = (\lambda_k^1, \lambda_k^2, ..., \lambda_k^n)^T$.

Cluster Volume: The volume $Vol(S)$ of a subgraph S of a graph G is defined to be the sum of the degrees of the nodes belonging to the subgraph, i.e $Vol(S) = \sum_{i \in S} deg(i)$, where $deg(i)$ is the degree of node i. By analogy, for the modal clusters, we define the volume of the cluster indexed ω in the graph-indexed k to be

$$Vol_k^\omega = \frac{\sum_{i \in V_k} s_{i\omega}^k deg(i)}{\sum_{\omega=1}^{n} \sum_{i \in V_k} s_{i\omega}^k deg(i)}. \quad (3)$$

The feature-vector for the graph-indexed k is $B_k = (Vol_k^1, Vol_k^2, ..., Vol_k^n)^T$.

Cluster Perimeter: For a subgraph S the set of perimeter nodes is $\Delta(S) = \{(u,v)|(u,v) \in E \wedge u \in S \wedge v \notin S\}$. The perimeter length of the subgraph is defined to be the number of edges in the perimeter set, i.e. $\Gamma(S) = |\Delta(S)|$. Again, by analogy, the perimeter length of the modal cluster indexed ω is

$$\Gamma_k^\omega = \frac{\sum_{i \in V_k} \sum_{j \in V_k} s_{i\omega}^k (1 - s_{j\omega}^k) A_k(i,j)}{\sum_{\omega=1}^{n} \sum_{i \in V_k} \sum_{j \in V_k} s_{i\omega}^k (1 - s_{j\omega}^k) A_k(i,j)}. \quad (4)$$

The perimeter values are ordered according to the modal index of the relevant cluster to form the graph feature vector $B_k = (\Gamma_k^1, \Gamma_k^2, ..., \Gamma_k^n)^T$.

Cheeger Constant: The Cheeger constant for the subgraph S is defined as follows. Suppose that $\hat{S} = V - S$ is the complement of the subgraph S. Further let $E(S, \hat{S}) = \{(u, v) | u \in S \wedge v \in \hat{S}\}$ be the set of edges that connect S to $\hat{S}$. The Cheeger constant for the subgraph S is

$$H(S) = \frac{|E(S, \hat{S})|}{min[vol(S), vol(\hat{S})]}. \quad (5)$$

The cluster analogue of the Cheeger constant is

$$H_k^\omega = \frac{\Gamma_k^\omega}{min[Vol_k^\omega, Vol_k^{\hat{\omega}}]}, \quad (6)$$

where

$$Vol_k^{\hat{\omega}} = \sum_{\omega=1}^{n} \sum_{i \in V_k} s_{i,\omega}^k deg(i) - Vol_k^\omega. \quad (7)$$

is the volume of the complement of the cluster indexed ω. Again, the cluster Cheeger numbers are ordered to form a spectral feature-vector $B_k = (H_k^1, H_k^2, ..., H_k^n)^T$.

3.2 Binary Features

In addition to the unary cluster features, we have studied pairwise cluster attributes.

Shared Perimeter: The first pairwise cluster attribute studied is the shared perimeter of each pair of clusters. For the pair subgraphs S and T the perimeter is the set of nodes belong to the set $P(S, T) = \{(u, v) | u \in S \wedge v \in T\}$. Hence, our cluster-based measure of shared perimeter for the clusters is

$$U_k(u, v) = \frac{\sum_{(i,j) \in E_k} s_{i,u}^k s_{j,v}^k A_k(i, j)}{\sum_{(i\prime,j\prime) \in E_k} s_{i\prime,u}^k s_{j\prime,v}^k}. \quad (8)$$

Each graph is represented by a shared perimeter matrix U_k. We convert these matrices into long vectors. This is obtained by stacking the columns of the matrix U_k in eigenvalue order. The resulting vector is $B_k = (U_k(1,1), U_k(1,2), ..., U_k(1,n), U_k(2,1)....., U_k(2,n,), ...U_k(n,n))^T$ Each entry in the long-vector corresponds to a different pair of spectral clusters.

Cluster Distances: The between cluster distance is defined as the path length, i.e. the minimum number of edges, between the most significant nodes in a pair of clusters. The most significant node in a cluster is the one having the largest co-efficient in the eigenvector associated with the cluster. For the cluster indexed u in the graph indexed k, the most significant node is $i_u^k = \arg\max_i s_{iu}^k$. To compute the distance, we note that if we multiply the adjacency matrix A_k by itself l times, then the matrix $(A_k)^l$ represents the distribution of paths of length l in the graph G_k. In particular, the element $(A_k)^l(i,j)$ is the number of paths of length l edges between the nodes i and j. Hence the minimum distance between the most significant nodes of the clusters u and v is $d_{u,v} = \arg\min_l (A_k)^l(i_u^k, i_v^k)$.

If we only use the first n leading eigenvectors to describe the graphs, the between cluster distances for each graph can be written as a n by n matrix which can be converted to a $n \times n$ long-vector $B_k = (d_{1,1}, d_{1,2}, ...d_{1,n}, d_{2,1}.....d_{n,n})^T$.

4 Embedding the Spectral Vectors in a Pattern Space

In this section we describe two methods for embedding graphs in eigenspaces. The first of these involves performing principal components analysis on the covariance matrices for the spectral pattern-vectors. The second method involves performing multidimensional scaling on a set of pairwise distance between vectors.

4.1 Eigendecomposition of the Graph Representation Matrices

Our first method makes use principal components analysis and follows the parametric eigenspace idea of Murase and Nayar [8]. The relational data for each graph is vectorised in the way outlined in Section 3. The N different graph vectors are arranged in view order as the columns of the matrix $S = [B_1|B_2|\dots|B_k|\dots|B_N]$.

Next, we compute the covariance matrix for the elements in the different rows of the matrix S. This is found by taking the matrix product $C = SS^T$. We extract the principal components directions for the relational data by performing an eigendecomposition on the covariance matrix C. The eigenvalues λ_i are found by solving the eigenvalue equation $|C - \lambda I| = 0$ and the corresponding eigenvectors $\boldsymbol{e}_i$ are found by solving the eigenvector equation $C\boldsymbol{e}_i = \lambda_i \boldsymbol{e}_i$.

We use the first 3 leading eigenvectors to represent the graphs extracted from the images. The co-ordinate system of the eigenspace is spanned by the three orthogonal vectors by $E = (\boldsymbol{e}_1, \boldsymbol{e}_2, \boldsymbol{e}_3)$. The individual graphs represented by the long vectors $B_k, k = 1, 2, \dots, N$ can be projected onto this eigenspace using the formula $\boldsymbol{x}_k = \boldsymbol{e}^T B_k$. Hence each graph G_k is represented by a 3-component vector $\boldsymbol{x}_k$ in the eigenspace.

4.2 Multidimensional Scaling

Multidimensional scaling(MDS)[3] is a procedure which allows data specified in terms of a matrix of pairwise distances to be embedded in a Euclidean space. The classical multidimensional scaling method was proposed by Torgenson[12] and Gower[6]. Shepard and Kruskal developed a different scaling technique called ordinal scaling[5]. Here we intend to use the method to embed the graphs extracted from different viewpoints in a low-dimensional space.

To commence we require pairwise distances between graphs. We do this by computing the L2 norms between the spectral pattern vectors for the graphs. For the graphs indexed $i1$ and $i2$, the distance is

$$d_{i1,i2} = \sum_{\alpha=1}^{K} \left[B_{i1}(\alpha) - B_{i2}(\alpha) \right]^2 . \tag{9}$$

The pairwise similarities $d_{i1,i2}$ are used as the elements of an $N \times N$ dissimilarity matrix D, whose elements are defined as follows

$$D_{i1,i2} = \begin{cases} d_{i1,i2} & \text{if } i_1 \neq i2 \\ 0 & \text{if } i1 = i2 \end{cases} . \tag{10}$$

In this paper, we use the classical multidimensional scaling method to embed our the view-graphs in a Euclidean space using the matrix of pairwise dissimilarities D. The first step of MDS is to calculate a matrix T whose element with row r and column c is given by $T_{rc} = -\frac{1}{2}[d^2_{rc} - \hat{d}^2_{r.} - \hat{d}^2_{.c} + \hat{d}^2_{..}]$, where $\hat{d}_{r.} = \frac{1}{N}\sum_{c=1}^{N} d_{rc}$ is the average dissimilarity value over the rth row, $\hat{d}_{.c}$ is the similarly defined average value over the cth column and $\hat{d}_{..} = \frac{1}{N^2}\sum_{r=1}^{N}\sum_{c=1}^{N} d_{r,c}$ is the average similarity value over all rows and columns of the similarity matrix T.

We subject the matrix T to an eigenvector analysis to obtain a matrix of embedding co-ordinates X. If the rank of T is $k, k \leq N$, then we will have k non-zero eigenvalues. We arrange these k non-zero eigenvalues in descending order, i.e. $\lambda_1 \geq \lambda_2 \geq \ldots \geq \lambda_k > 0$. The corresponding ordered eigenvectors are denoted by $\boldsymbol{e}_i$ where λ_i is the ith eigenvalue. The embedding co-ordinate system for the graphs obtained from different views is $X = [\boldsymbol{f}_1, \boldsymbol{f}_2, \ldots, \boldsymbol{f}_k]$, where $\boldsymbol{f}_i = \sqrt{\lambda_i}\boldsymbol{e}_i$ are the scaled eigenvectors. For the graph indexed i, the embedded vector of co-ordinates is $\boldsymbol{x}_i = (X_{i,1}, X_{i,2}, X_{i,3})^T$.

5 Experiments

Our experiments have been conducted with 2D image sequences for 3D objects which undergo slowly varying changes in viewer angle. The image sequences for three different model houses are shown in Figure 1. For each object in the view sequence, we extract corner features. From the extracted corner points we construct Delaunay graphs. The sequences of extracted graphs are shown in Figure 2. Hence for each object we have 10 different graphs. In table 1 we list the number of feature points in each of the views. From inspection of the graphs in Figure 2 and the number of feature points in Table 1 it is clear that the different graphs for the same object undergo significant changes in structure as the viewing direction changes. Hence, this data presents a challenging graph clustering problem.

Our aim is to investigate which combination of spectral feature-vector and embedding strategy gives the best set of graph-clusters. In other words, we aim to see which method gives the best definition of clusters for the different objects.

Table 1. Number of feature points extracted from the three image sequences

Image Number	1	2	3	4	5	6	7	8	9	10
CMU	30	32	32	30	30	32	30	30	30	31
MOVI	140	134	130	136	137	131	139	141	133	136
Chalet	40	57	92	78	90	64	113	100	67	59

In Figure 3 we compare the results obtained with the different spectral feature vectors. In the centre column of the figure, we show the matrix of pairwise Euclidean distances between the feature-vectors for the different graphs (this is best viewed in colour). The matrix has 30 rows and columns (i.e. one for each of the images in the three sequences with the three sequences concatenated), and the images are ordered according to the position in the sequence. From top-to-bottom, the different rows show the results obtained when the feature-vectors are constructed using the eigenvalues of the adjacency matrix, the cluster volumes, the cluster perimeters, the cluster Cheeger constants, the shared perimeter length and the inter-cluster edge distance. From the pattern of pairwise distances, it is clear that the eigenvalues and the shared perimeter length give the best block structure in the matrix. Hence these two attributes may be expected to result in the best clusters.

To test this assertion, in the left-most and right-most columns of Figure 3 we show the leading eigenvectors of the embedding spaces for the spectral feature-vectors. The left-hand column shows the results obtained with principal components analysis. The right-hand column shows the results obtained with multidimensional scaling. From the plots, it is clear that the best clusters are obtained when MDS is applied to the vectors of eigenvalues and shared perimeter length. Principal components analysis, on the other hand, does not give a space in which there is a clear cluster-structure.

We now embark on a more quantitative analysis of the different spectral representations. To do this we plot the normalised squared eigenvalues $\hat{\lambda}_i^2 =$

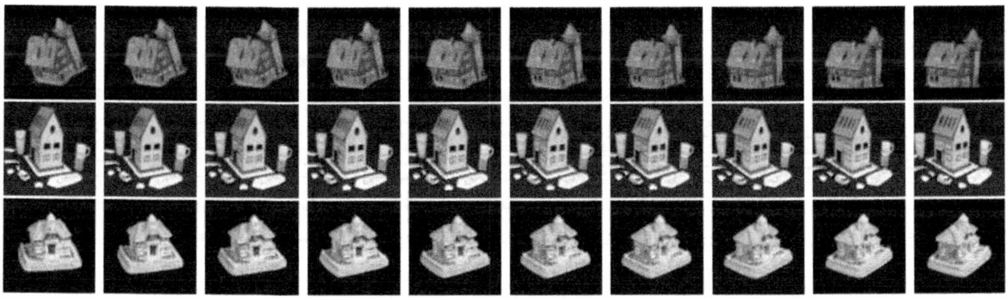

Fig. 1. Image sequences

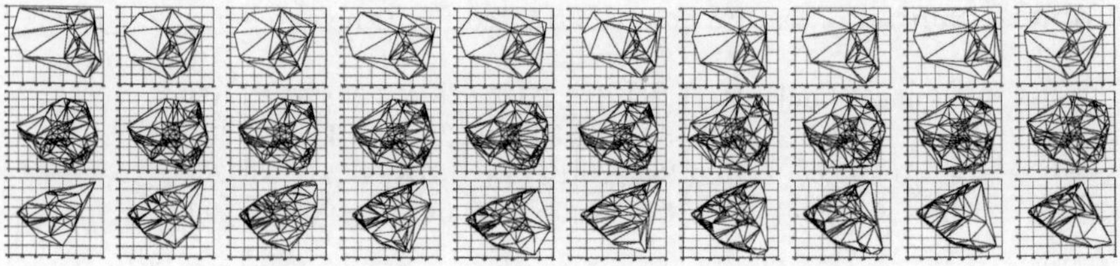

Fig. 2. Graph representation of the sequences

$\frac{\lambda_i^2}{\sum_{i=1}^{n} \lambda_i^2}$ against the eigenvalue magnitude order i. In the case of the parametric eigenspace, these represent the fraction of the total data variance residing in the direction of the relevant eigenvector. In the case of multidimensional scaling, the normalised squared eigenvalues represent the variance of the inter-graph distances in the directions of the eigenvectors of the similarity matrix. The first two plots are for the case of the parametric eigenspaces. The left-hand side plot of Figure 4 is for the unary attribute of eigenvalues, while the middle plot is for the pairwise attribute of shared perimeters. The main feature to note is that of the binary features the vector of adjacency matrix eigenvalues has the fastest rate of decay, i.e. the eigenspace has a lower latent dimensionality, while the vector of Cheeger constants has the slowest rate of decay, i.e. the eigenspace has greater dimensionality. In the case of the binary attributes, the shared perimeter results in the eigenspace of lower dimensionality.

In the last plot of Figure 4 we show the eigenvalues of the graph similarity matrix. We repeat the sequence of plots for the three house data-sets, but merge the curves for the unary and binary attributes into a single plot. Again the vector of adjacency matrix eigenvalues gives the space of lower dimensionality, while the vector of inter-cluster distances gives the space of greatest dimensionality.

Finaly, we compare the performances of the graph embedding methods using measures of their classification accuracy. Each of the six graph spectral features mentioned above are used. We have assigned the graphs to classes using the the K-means classifier. The classifier has been applied to the raw Euclidean distances, and to the distances in the reduced dimension feature-spaces obtained using PCA and MDS. In Table 2 we list the number of correctly classified graphs. From the table, it is clear that the eigevalues and the shared perimeters are the best features since they return higher correct classification rates. Cluster distance is the worst feature for clustering graphs. We note also that classification in the feature-space produced by PCA is better than in the original feature vector spaces. However, the best results come from the MDS embedded class spaces.

6 Conclusions

In this paper we have investigated how vectors of graph-spectral attributes can be used for the purposes of clustering graphs. The attributes studied are the

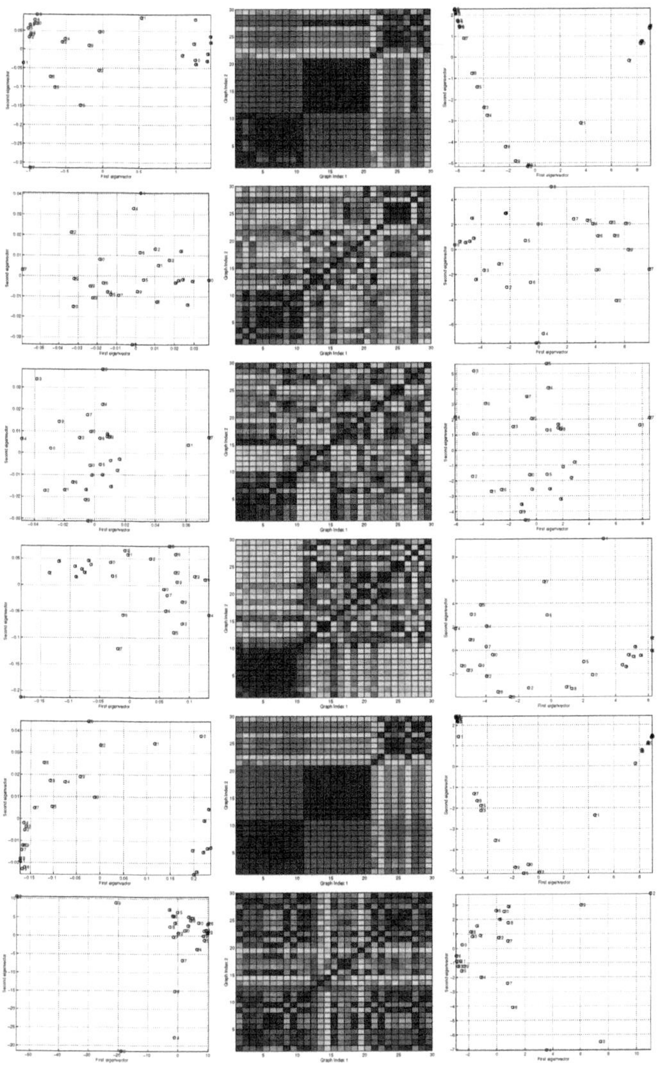

Fig. 3. Eigenspace and MDS space embedding using the spectral features of binary adjacency graph spectra, cluster volumes, cluster perimeters, cluster Cheeger constants, shared perimeters and cluster distances

leading eigenvalues, and, the volumes, perimeters, shared perimeters and Cheeger numbers for modal clusters. The best clusters emerge when we apply MDS to the vectors of leading eigenvalues. The best clusters result when we use cluster volume or shared perimeter.

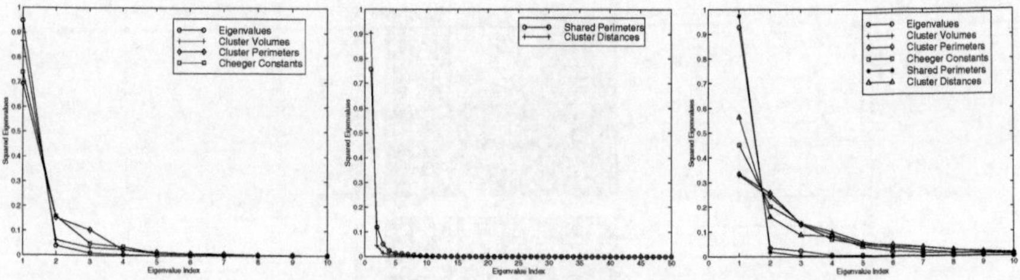

Fig. 4. Comparison of graph spectral features for eigenspaces. The left plot is for the unary features in eigenspace, the middle plot is for the binary features in eiegenspace and the right plot is for all the spectral features in MDS space

Table 2. Correct classifications

Features	Eigenvalues	Volumes	Perimeters	Cheeger constants	Shared Perimeters	Distances Distances
Raw vector	29	26	26	13	25	12
PCA	29	27	26	17	25	12
MDS	29	28	27	16	29	17

Hence, we have shown how to cluster purely symbolic graphs using simple spectral attributes. The graphs studied in our analysis are of different size, and we do not need to locate correspondences. Our future plans involve studying in more detail the structure of the pattern-spaces resulting from our spectral features. Here we intend to investigate the use of ICA as an alternative to PCA as a means of embedding the graphs in a pattern-space. We also intend to study how support vector machines and the EM algorithm can be used to learn the structure of the pattern spaces. Finally, we intend to investigate whether the spectral attributes studied here can be used for the purposes of organising large image data-bases.

References

1. H. Bunke. Error correcting graph matching: On the influence of the underlying cost function. *IEEE Transactions on Pattern Analysis and Machine Intelligence*, 21:917–922, 1999. 83
2. F. R. K. Chung. *Spectral Graph Theory.* American Mathmatical Society Ed., CBMS series 92, 1997. 84
3. Chatfield C. and Collins A. J. *Introduction to multivariate analysis.* Chapman & Hall, 1980. 87
4. R. Englert and R. Glantz. Towards the clustering of graphs. In *2nd IAPR-TC-15 Workshop on Graph-Based Representation*, 1999. 83
5. Kruskal J. B. Nonmetric multidimensional scaling: A numerical method. *Psychometrika*, 29:115–129, 1964. 87

6. Gower J. C. Some distance properties of latent root and vector methods used in multivariate analysis. *Biometrika*, 53:325–328, 1966. 87
7. B. Luo, A. Robles-Kelly, A. Torsello, R. C. Wilson, and E. R. Hancock. Clustering shock trees. In *Proceedings of GbR*, pages 217–228, 2001. 83
8. H. Murase and S. K. Nayar. Illumination planning for object recognition using parametric eigenspaces. *IEEE Transactions on Pattern Analysis and Machine Intelligence*, 16(12):1219–1227, 1994. 84, 87
9. S. Rizzi. Genetic operators for hierarchical graph clustering. *Pattern Recognition Letters*, 19:1293–1300, 1998. 83
10. J. Segen. Learning graph models of shape. In *Proceedings of the Fifth International Conference on Machine Learning*, pages 25–29, 1988. 83
11. K. Sengupta and K. L. Boyer. Organizing large structural modelbases. *PAMI*, 17(4):321–332, April 1995. 83
12. Torgerson W. S. Multidimensional scaling. i. theory and method. *Psychometrika*, 17:401–419, 1952. 87

Identification of Diatoms by Grid Graph Matching

Stefan Fischer, Kaspar Gilomen, and Horst Bunke

Institute of Computer Science and Applied Mathematics
University of Bern, Switzerland
{fischer,bunke}@iam.unibe.ch

Abstract. Diatoms are unicellular algae found in water and other places wherever there is humidity and enough light for photo synthesis. In this paper a graph matching based identification approach for the retrieval of diatoms from an image database is presented. The retrieval is based on the matching of labeled grid graphs carrying texture information of the underlying diatom. A grid graph is a regular, rectangular arrangement of nodes overlaid on an image. Each node of the graph is labeled with texture features describing a rectangular sub-region of the object. Properties of gray level co-occurrence matrices as well as of Gabor wavelets are used as texture features. The method has been evaluated on a diatom database holding images of 188 different diatoms belonging to 38 classes. For the identification of these diatoms recognition rates of more than 90 percent were obtained.

1 Introduction

In this paper an approach to the identification of diatoms based on the matching of labeled grid graphs is presented. The work has been done in the framework of the ADIAC project [1] which aims at the automatic identification and classification of diatoms. Diatoms are unicellular algae found in water and other places wherever there is humidity and enough light for photo synthesis. Diatom identification has a number of applications in areas such as environmental monitoring, climate research and forensic medicine [14]. One of the great challenges in diatom identification is the large number of classes involved[1]. Experts estimate the number of diatom species between 15000 and 20000, or even higher. Diatoms are characterized by an ornamented cell wall composed of silica, which is highly resistant and remains after chemical cleaning where all organic contents are removed. The cell wall consist of two valves that fit within each other like the pieces of a petri dish. The identification as well as the taxonomy of diatoms is based on the morphology of these silica valves. Example images of valves of four different classes of diatoms are shown in Figure 1. As can be seen diatoms

[1] In terms of biologists diatoms are hierarchically classified in *genus*, *species*, *subspecies* and so forth, but in this paper we'll use the term *class* in the pattern recognition sense.

T. Caelli et al. (Eds.): SSPR&SPR 2002, LNCS 2396, pp. 94–103, 2002.
© Springer-Verlag Berlin Heidelberg 2002

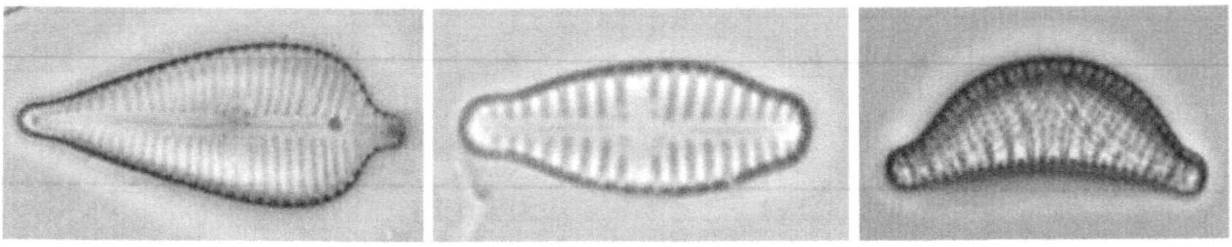

a.) b.) c.)

Fig. 1. Example images of diatom valves in face view (top-down view): a.) *Gomphonema augur var. augur* (Ehrenberg), b.) *Gomphonema olivaceum* (Möller), c.) *Epithemia sorex var. sorex* (Kützing)

are of different shape and morphological structure. Furthermore, for some diatoms the size and the shape of the valve changes during the live cycle, but the morphology of the valve keeps mostly the same. While in previous studies the variation of the shape and shape based identification have been studied ([4], [5], [10], [11], [13], [17]), the objective of the present paper is the identification of diatoms based on the morphology of the valve.

In the retrieval approach presented here, texture features of the morphological structure of diatom valves are extracted and taken as node labels of a grid graph. Grid graphs are a special class of graphs which correspond to a regular, rectangular arrangement of nodes overlaid on an image. In Figure 2 an example of a grid graph is shown. The inner nodes of the graph are connected with their four neighbors in horizontal and vertical direction. The outer nodes are connected with two or three neighbors, depending whether they correspond to a corner or not. For the diatom images stored in the reference image database, labeled grid-graphs are pre-computed and attached to each image. During the retrieval phase the similarity between the query image and images in the database is evaluated by matching the corresponding grid graphs. In this phase the query image is overlaid with a grid and texture properties are computed in image regions surrounding the nodes. To take the changing size of diatoms into regard, the position of the nodes is varied. A distance measure between the different graphs extracted from the query image and all graphs stored in the database is computed. As result of the query the most similar images found in the database are returned to the user.

The remainder of the paper is organized as follows. In Section 2 a short overview of grid graph matching techniques found in other pattern recognition applications will be given. The representation of objects, in our case diatoms, will be described in Section 3. The matching procedure will be outlined in Section 4. Experimental results will be reported in Section 5 and conclusions drawn in Section 6.

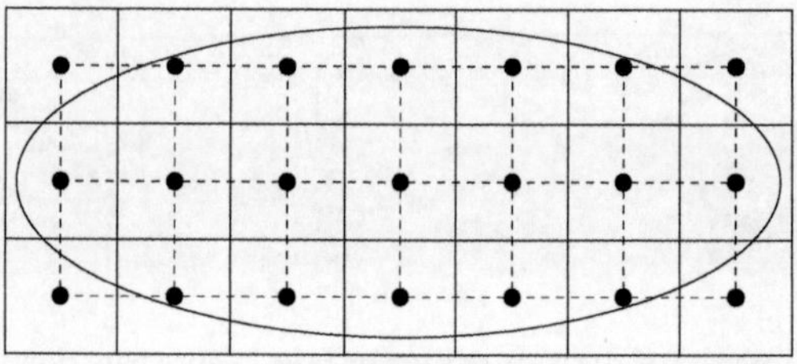

Fig. 2. Grid graph with 3×7 nodes representing the underlying object outlined by an ellipse

2 Related Work in Grid Graph Matching

Labeled graph matching has been successfully used in numerous object recognition tasks [3]. Grid graphs are a special subclass of graphs. Most applications of grid graph matching are focused on face recognition tasks such as the detection, tracking, or identification of persons ([9], [15], [18]). There are two main sources of information which are used for face recognition based on grid graph matching. One source is geometrical features, for example, the position of the nose, mouth and eyes. The second source is gray level and texture information of the skin. An approach that exploits both sources of information is the so-called dynamic link architecture [9]. This approach is divided into a training and a recall phase. In the training phase, a sparse grid is build for each person in the reference database. The grid is overlaid on the facial region of a person's image and the response of a set of Gabor filters is measured at the grid nodes. The Gabor filters are tuned to different orientations and scales. In the recall phase, the reference grid of each person is overlaid on the face image of a test person and is deformed so that a cost function is minimized. The cost function is based on the differences between the feature vectors stored at the nodes of the reference grids and the feature vectors computed at certain pixel coordinates in the test image. Additionally, costs for the distortion between the reference grid and the variable graph built on the image of the test person are taken into account. The cost function is a measure of similarity of the model grid graph to the test graph. Similar ideas are used in the grid graph matching technique proposed in this paper.

3 Object Representation by Means of Grid Graphs

An important characteristic used in the identification of diatoms is the morphology of the valve face. In this paper we propose a complementary approach based on texture measures computed in local image regions. The identification of texture has been extensively studied in the computer vision literature. There are statistical methods that measure variance, entropy or energy of the gray level distribution of an image. Moreover, perceptual techniques have been proposed, which

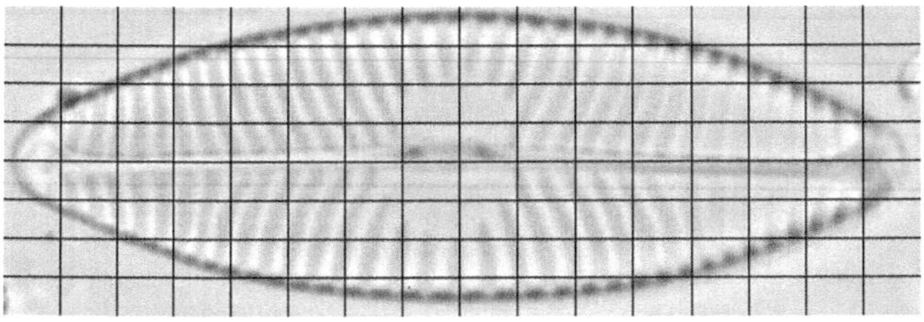

Fig. 3. Example image of windowing with a grid of dimension 16×8

are able to identify the direction, orientation and regularity of textures [12]. Among the most widely used texture measures are those derived from gray level co-occurrence matrices [7], and features based on Gabor wavelets [8,16]. These features have been adopted in the system described in this paper.

From the general point of view, almost any visual pattern can be represented via a graph containing nodes labeled with local features and links encoding the spatial relationship between the features [2]. It can be observed that many diatom valves consist of areas having relatively homogeneous texture. Thus, a diatom can be divided into separate areas and the texture can be measured in each area. The spatial relationships between such areas are preserved by overlaying a grid on the object. This is visualized in Figure 3. The example image is overlaid by a grid of 16×8 rectangular regions. The morphology of the valve face is then described by average values of characteristic properties inside these rectangular regions. As such properties, 13 features of the gray level co-occurrence matrix and the mean and standard deviation of 4 Gabor functions with different orientation are used. The features of each rectangular region are assigned as an attribute vector to the corresponding node of the grid graph as shown in Figure 2. For further details about the 13 textural features used as attributes of the grid graphs, the reader is referred to [6].

The textural features as well as the grid graph matching approach described in the next section are not invariant w.r.t object rotation. From the theoretical point of view, the development of an invariant recognition procedure would be very interesting. However the actual application doesn't require for rotational invariance because the images of both the diatoms in the database and the query diatoms are acquired by a human operator, who manually aligns the objects such that they appear in a standard pose. Diatom image acquisition is quite time consuming. Hence this manual alignment doesn't add any significant overhead to the overall retrieval process. (The issue of scale invariance will be addressed in Section 4.)

4 Grid Graph Matching

Based on the grid graph representation, the problem of diatom identification can be formulated as a labeled graph matching task, where the goal is to find an optimal one-to-one correspondence between the nodes of an input graph and the nodes of a graph stored in the database. A good correspondence is one that respects the spatial relationships between the nodes, and exhibits a high degree of similarity between the labels of the corresponding nodes [2,3].

In our implementation, the dissimilarity of two grid graphs is measured as the sum of distances of the nodes of two grid graphs. That is, the distance $\delta(G_1, G_2)$ between two graphs $G_1 = (V_1, E_1)$ and $G_2 = (V_2, E_2)$, where V_i and E_i denotes the set of nodes and edges of graph G_i, respectively, is defined as:

$$\delta(G_1, G_2) = \frac{1}{|M(V_1, V_2)|} \sum_{(v_i, v_j) \in M(V_1, V_2)} d(v_i, v_j). \tag{1}$$

The quantity $d(v_i, v_j)$ denotes the distance of feature vectors of a pair of nodes (see Eqn. (2)), and $M(V_1, V_2)$ is the set of all pairs of nodes $v_i \in V_1$ and $v_j \in V_2$ with similar spatial positions. This means if p_i is the position of the node v_i and p_j is the position of the node v_j, then the constraint $|p_i - p_j| < \epsilon$ is fulfilled for each pair (v_i, v_j) in M. To gain size invariance, the node positions are coded as distances relative to the center of the bounding box of the object.

In the simple version of the grid graph matching approach no distortion of the nodes' position is allowed. Thus, the above constraint becomes $p_i = p_j$. In this case the distance between two nodes $v_i \in V_1$ and $v_j \in V_2$ is defined as the distances between the feature vectors

$$d(v_i, v_j) = \frac{1}{N} \sum_{n=1}^{N} |\mathrm{f}_{i,n} - \mathrm{f}_{j,n}| \tag{2}$$

where $\mathrm{f}_{i,n}$ is the n-th feature of the node v_i and N is the number of features in the feature vectors. As the ranges of the various features are different they are normalized in a pre-processing step. In our approach the min-max normalization is used to normalize all feature values to the interval $[0, 1]$. For each feature f the minimum f_{min} and the maximum f_{max} over the training set are computed and the normalized feature value $\mathrm{f}^{'}$ is calculated as

$$\mathrm{f}^{'} = \frac{\mathrm{f} - \mathrm{f}_{min}}{\mathrm{f}_{max} - \mathrm{f}_{min}}. \tag{3}$$

Whenever a feature of a query diatom has a value smaller than f_{min}, or larger than f_{max}, the normalized value is set equal to 0, or 1, respectively.

The grid graph distance defined by Eqn. (1) is sensitive to geometric distortions. In order to improve its robustness, a second grid graph distance, called flexible grid graph distance, is introduced. In flexible grid graph matching, each node of the query graph can be translated, by a small degree, in the image plane. Note that the nodes of the database graph remain fixed.

In the flexible grid graph distance an additional cost is introduced for each node, indicating whether a translation is applied to the node or not. If no translation is applied this additional cost is equal to zero. Formally, let the translation cost be defined as

$$t(v_i, v_j) = t(v_j) \tag{4}$$

where v_i belongs to a graph from the database and v_j to a query graph. Moreover,

$$t(v) = \begin{cases} c & \text{if } v \text{ has been translated} \\ 0 & otherwise \end{cases} . \quad (5)$$

Using this additional term in the cost function we define the flexible node distance as (cf. Eqn. (2))

$$d'(v_i, v_j) = \min_T \left[\frac{1}{N} \sum_{n=1}^{N} (|\mathrm{f}_{i,l} - \mathrm{f}_{j,l}|) + t(v_i, v_j) \right] \quad (6)$$

In this equation, T is the set of all translations that can be applied to the nodes. The aim of each translation is to reduce the matching cost of the two feature vectors, i.e. to make $\sum (|\mathrm{f}_{i,l} - \mathrm{f}_{j,l}|)$ smaller. However, because there is a cost for each translation, a trade-off between the cost resulting from the feature vector difference and from the node translations arises. In the flexible distance $d'(v_i, v_j)$ this cost is minimized over all possible translations of the nodes. The parameter c is the same for all nodes and has to be chosen empirically. Theoretically, a more general model could be adopted, where $t(v)$ depends on the actual degree of translation. This generalization, however, results in many more free parameters in the cost function that need to be defined. To avoid this problem a constant cost of node translations has been used.

Given the flexible distance of a pair of nodes defined by Eqn. (6), the flexible grid graph distance δ' is defined as follows

$$\delta'(G_1, G_2) = \frac{1}{|M(V_1, V_2)|} \sum_{(v_i, v_j) \in M(V_1, V_2)} d'(v_i, v_j). \quad (7)$$

5 Experimental Results

The proposed grid graph matching based identification approach for the retrieval of diatoms is evaluated on a test database of 188 images of different diatoms. The diatoms belong to 38 different classes. In the Appendix a sample image for each class can be found.

In the test database, most of the classes are only represented by 3 to 6 images. Actually, the minimum and maximum number of representatives is 3 and 9 images, respectively. Because of this limited number of images the performance of the approach is validated using the leave-one-out technique. This means each sample in the databases is once used for testing and all other samples are used as prototypes. This procedure is repeated until each sample is used exactly once for testing.

In a first test the standard grid graph matching procedure is used, and in the second test the flexible matching is applied. The results of the first test are visualized in Figure 4. The recognition rate achieved using only features of the gray level co-occurrence matrix are drawn as a dashed line, the rate using features of Gabor wavelets as dotted line, and the combination of both sets of

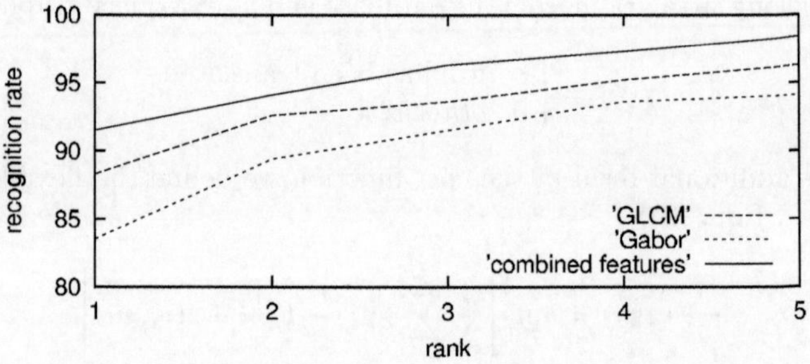

Fig. 4. Results for identification of diatoms using standard grid graph matching

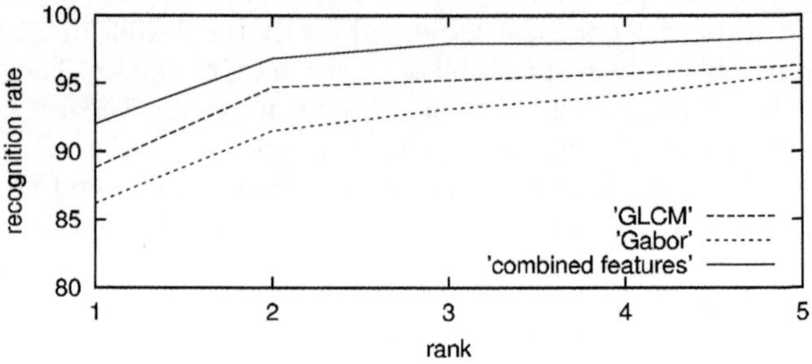

Fig. 5. Results for identification of diatoms using flexible grid graph matching

features as solid line. On the y-axis the recognition rate is given and on the x-axis the highest rank taken into regard. Thus, for example, rank 2 in the chart represents the accumulated recognition rate for all samples whose real class is detected as the most, or second most, similar image by the matching procedure. As can be seen in Figure 4 the recognition rate using only features of the gray level co-occurrence matrix is slightly higher than the recognition rate obtained by features of the Gabor wavelets. The highest recognition rate is obtained by combining both feature sets.

In the second test the flexible grid graph matching is used which allows distortions of the nodes. As can be seen in Figure 5 the recognition rates for all three feature sets are higher than in the previous test. Especially when including the second rank better results are obtained. Instead of 94% in Fig. 4 now nearly 97% are obtained. Thus, if a set of similar images is returned for a query image the probability that images of the same class are in the result set is higher for flexible grid graph matching then for the simple matching approach.

6 Conclusion

In this paper we have proposed a flexible grid graph matching based identification approach for the retrieval of diatoms from an image database. The application of graph matching methods for the identification of diatoms turned out to be especially useful for the description of morphological properties, which change during the live cycle of diatoms. As features texture properties of gray level co-occurrence matrices and Gabor wavelets of local image regions were used.

On a complex database holding images of 38 different classes of diatoms recognition rates of nearly 98% were achieved if the first three ranks are taken into regard. These rates are impressive regarding the difficulty of the considered task. As one can conclude from the images shown in the Appendix, some classes are quite similar in shape, and others have a similar valve structure. Another complication arises from the small size of the available database. There are individuals in some classes that are quite different from the individuals in the same class. These outliers are very prone to being misclassified by the selected classification procedure.

The system described in this paper is a small prototype that was built to study the feasibility of grid graph matching for automatic identification and retrieval. The recognition rate of 98%, considering the first three ranks, seems very promising for real application of the system. In future versions of the system it is planed to significantly increase the number of diatom classes. From such an increase, a drop of the recognition performance must be expected. However, in the context of ADIAC [1] several other methods for diatom identification are under development. They are based on different features, for example, shape and global texture features, and different recognition procedures, for example, decision tree based classification. These methods have characteristics that are quite complementary to the approaches described in this paper. Therefore, it can be expected that the combination of these methods with the grid graph matching approach proposed in this paper will further improve the recognition rate. Such a combination will be one of our future research topics.

Acknowledgment

The work has been done in the framework of the EU-sponsored Marine Science and Technology Program (MAST-III), under contract no. MAS3-CT97-0122. Additional funding came from the Swiss Federal Office for Education and Science (BBW 98.00.48). We thank our project partners Micha Bayer and Stephen Droop from Royal Botanic Garden Edinburgh and Steve Juggins and co-workers at Newcastle University for preparing the images in the ADIAC image database and for useful discussions and hints.

References

1. Automatic Diatom Identification And Classification. Project home page: http://www.ualg.pt/adiac/. 94, 101
2. E. Bienenstock and C. von der Malsburg. A neural network for invariant pattern recognition. *Europhysics Letters*, 4:121–126, 1987. 97, 98
3. H. Bunke. Recent developments in graph matching. In *Proceedings of the 15th International Conference on Pattern Recognition (ICPR '00)*, volume 2, pages 117–124, Barcelona, Spain, September 3–8 2000. 96, 98
4. S. Fischer, M. Binkert, and H. Bunke. Feature based retrieval of diatoms in an image database using decision trees. In *Proceedings of the 2nd International Conference on Advanced Concepts for Intelligent Vision Systems (ACIVS 2000)*, pages 67–72, Baden-Baden, Germany, August 2000. 95
5. S. Fischer, M. Binkert, and H. Bunke. Symmetry based indexing of diatoms in an image database. In *Proceedings of the 15th International Conference on Pattern Recognition (ICPR '00)*, volume 2, pages 899–902, Barcelona, Spain, September 3–8 2000. 95
6. K. Gilomen. Texture based identification of diatoms (in German). Master's thesis, University of Bern, 2001. 97
7. R. M. Haralick, K. Shanmugam, and I. Dinstein. Textural features for image classification. *IEEE Transactions on Systems, Man, and Cybernetics*, pages 610–621, 1973. 97
8. A. K. Jain and F. Farrokhnia. Unsupervised texture segmentation using Gabor filters. *Pattern Recognition*, 24(12):1167–1186, 1991. 97
9. M. Lades, J. Vorbrüggen, J. Buhmann, J. Lange, C. von der Malsburg, R. Würtz, and W. Konen. Distortion invariant object recognition in the dynamic link architecture. *IEEE Transaction on Computers*, 42(3):300–311, 1993. 96
10. D. Mou and E. F. Stoermer. Separating tabellaria (bacillariophyceae) shape groups: A large sample approach based on fourier descriptor analysis. *Journal of Phycology*, 28:386–395, 1992. 95
11. J. L. Pappas and E. F. Stoermer. Multidimensional analysis of diatom morphological phenotypic variation and relation to niche. *Ecoscience*, 2:357–367, 1995. 95
12. I. Pitas. *Digital image processing algorithms.* Prentice Hall, London, 1993. 97
13. E. F. Stoermer. A simple, but useful, application of image analysis. *Journal of Paleolimnology*, 15:111–113, 1996. 95
14. E. F. Stoermer and J. P. Smol, editors. *The Diatoms: Applications for the Environmental and Earth Science.* Cambridge University Press, 1999. 94
15. A. Tefas, C. Kotropoulos, and I. Pitas. Using support vector machines to enhance the performance of elastic graph matching for frontal face authentication. *IEEE Transactions on Pattern Analysis and Machine Intelligence*, 23:735–746, 2001. 96
16. M. Tuceryan and A. K. Jain. Texture analysis. In C. H. Chen, L. F. Pau, and P. S. P. Wang, editors, *The Handbook of Pattern Recognition and Computer Vision*, pages 207–248. World Scientific Publishing Co, 2 edition, 1998. 97
17. M. Wilkinson, J. Roerdink, S. Droop, and M. Bayer. Diatom contour analysis using morphological curvature scale spaces. In *Proceedings of the 15th International Conference on Pattern Recognition (ICPR '00)*, pages 656–659, Barcelona, Spain, September 3-7 2000. 95
18. L. Wiskott, J. Fellous, N. Krüger, and C. von der Malsburg. Face recognition by elastic bunch graph matching. *IEEE Transactions on Pattern Analysis and Machine Intelligence*, 19(7):775–779, 1997. 96

Appendix

For more images see: `http://www.ualg.pt/adiac/`.

Table 1. Example images of the classes 1 to 38 included in the test database

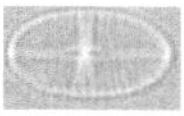 *Achnanthes saxonica*	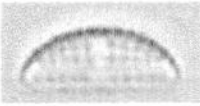 *Amphora pediculus*	 *Brachysira vitrea*	 *Caloneis silicula*
 *Cymatopleura solea*	 *Cymbella cornuta*	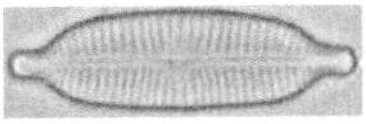 *Cymbella hybrida*	 *Cymbella subequalis*
 *Delphineis minutissima*	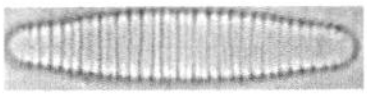 *Diatoma vulgaris*	 *Encyonema neogracile*	 *Epithemia sorex*
 *Eunotia bilunaris*	 *Eunotia exigua*	 *Fragilaria vaucheriae*	 *Fragilaria virescens*
 *Frustulia rhomboides*	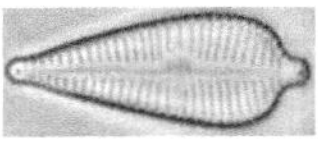 *Gomphonema augur*	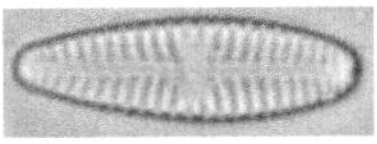 *Gomphonema olivaceum*	 *Gomphonema parvulum*
 *Gomphonema truncatum*	 *Luticola mutica*	 *Navicula lanceolata*	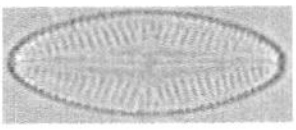 *Navicula palpebralis*
 *Navicula protracta*	 *Navicula radiosa*	 *Navicula rhynchocephala*	 *Nitzschia recta*
 *Peronia fibula*	 *Pinnularia silvatica*	 *Pinnularia subcapitata*	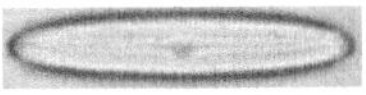 *Pinnularia viridi*
 *Reimeria sinuata*	 *Sellaphora pupula*	 *Stauroneis phoenicenteron*	 *Staurosirella pinnata*
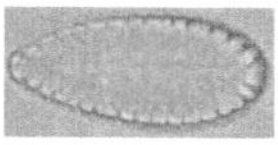 *Surirella brebissonii*	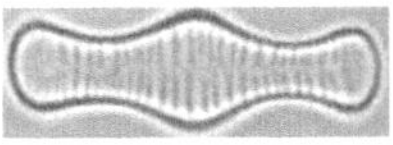 *Tabellaria flocculosa*		

String Edit Distance, Random Walks and Graph Matching

Antonio Robles-Kelly * and Edwin R. Hancock

Department of Computer Science, University of York
York, Y01 5DD, UK
{arobkell,erh}@cs.york.ac.uk

Abstract. This paper shows how the eigenstructure of the adjacency matrix can be used for the purposes of robust graph-matching. We commence from the observation that the leading eigenvector of a transition probability matrix is the steady state of the associated Markov chain. When the transition matrix is the normalised adjacency matrix of a graph, then the leading eigenvector gives the sequence of nodes of the steady state random walk on the graph. We use this property to convert the nodes in a graph into a string where the node-order is given by the sequence of nodes visited in the random walk. We match graphs represented in this way, by finding the sequence of string edit operations which minimise edit distance.

1 Introduction

Graph-matching is a task of pivotal importance in high-level vision since it provides a means by which abstract pictorial descriptions can be matched to one-another. Unfortunately, since the process of eliciting graph structures from raw image data is a task of some fragility due to noise and the limited effectiveness of the available segmentation algorithms, graph-matching is invariably approached by inexact means [15,13]. The search for a robust means of inexact graph-matching has been the focus of sustained activity over the last two decades. Early work drew heavily on ideas from structural pattern recognition and revolved around extending the concept of string edit distance to graphs [13,6,4]. More recent progress has centred around the use of powerful optimisation and probabilistic methods, with the aim of rendering the graph matching process robust to structural error.

Despite proving effective, these methods lack the elegance of the matrix representation first used by Ullman in his work on subgraph isomorphism [17]. The task of posing the inexact graph matching problem in a matrix setting has proved to be an elusive one. This is disappointing since a rich set of potential tools are available from the field of mathematics referred to as spectral graph theory. This is the term given to a family of techniques that aim to characterise the global structural properties of graphs using the eigenvalues and eigenvectors

* Supported by CONACYT, under grant No. 146475/151752.

T. Caelli et al. (Eds.): SSPR&SPR 2002, LNCS 2396, pp. 104–112, 2002.
© Springer-Verlag Berlin Heidelberg 2002

of the adjacency matrix [5]. In the computer vision literature there have been a number of attempts to use spectral properties for graph-matching, object recognition and image segmentation. Umeyama has an eigendecomposition method that matches graphs of the same size [18]. Borrowing ideas from structural chemistry, Scott and Longuet-Higgins were among the first to use spectral methods for correspondence analysis [14]. They showed how to recover correspondences via singular value decomposition on the point association matrix between different images. In keeping more closely with the spirit of spectral graph theory, yet seemingly unaware of the related literature, Shapiro and Brady [16] developed an extension of the Scott and Longuet-Higgins method, in which point sets are matched by comparing the eigenvectors of the point proximity matrix. Here the proximity matrix is constructed by computing the Gaussian weighted distance between points. The eigen-vectors of the proximity matrices can be viewed as the basis vectors of an orthogonal transformation on the original point identities. In other words, the components of the eigenvectors represent mixing angles for the transformed points. Matching between different point-sets is effected by comparing the pattern of eigenvectors in different images. Shapiro and Brady's method can be viewed as operating in the attribute domain rather than the structural domain. Horaud and Sossa[8] have adopted a purely structural approach to the recognition of line-drawings. Their representation is based on the immanental polynomials for the Laplacian matrix of the line-connectivity graph. By comparing the coefficients of the polynomials, they are able to index into a large data-base of line-drawings. Shokoufandeh, Dickinson and Siddiqi [2] have shown how graphs can be encoded using local topological spectra for shape recognition from large data-bases.

In a recent paper Luo and Hancock [11] have returned to the method of Umeyama and have shown how it can be rendered robust to differences and graph-size and structural errors. Commencing from a Bernoulli distribution for the correspondence errors, they develop an expectation maximisation algorithm for graph-matching. Correspondences are recovered in the M or maximisation step of the algorithm by performing singular value decomposition on the weighted product of the adjacency matrices for the graphs being matched. The correspondence weight matrix is updates in the E or expectation step. However, since it is iterative the method is relatively slow and is sensitive to initialisation.

The aim in this paper is to investigate whether the eigenstructure of the adjacency matrix can be used to match graphs using a search method rather than by iteration. To do this we draw on the theory of Markov chains. We consider a Markov chain whose transition probability matrix is the normalised edge-weight matrix for a graph. The steady random walk for the Markov chain on the graph is given by the leading eigenvector of the transition probability, i.e. edge weight, matrix. Hence, by considering the order of the nodes defined by the leading eigenvector, we are able to convert the graph into a string. This opens up the possibility of performing graph matching by performing string alignment by minimising the Levenshtein or edit distance [10,20]. We can follow Wagner and use dynamic programming to evaluate the edit distance between strings and hence recover correspondences [20]. It is worth stressing that although there been attempts to extend the string edit idea to trees and graphs [21,12,13,15], there is considerable current effort aimed at putting the underlying methodology on a rigourous footing. For instance, Bunke and his co-workers who have demonstrated the relationship between graph edit distance and the size of the maximum common subgraph.

2 Random Walks on Graphs

The relationship between the leading eigenvector of the adjacency matrix and the steady state random walk has been exploited in a number areas including routing theory and information retrieval. We are interested in the weighted graph $G = (V, E, P)$ with node index-set V and edge-set $E \subseteq V \times V$. The off-diagonal elements of the transition probability matrix P are the weights associated with the edges. In this paper, we exploit a graph-spectral property of the transition matrix P to develop a surface height recovery method. This requires that we have the eigenvalues and eigenvectors of the matrix P to hand. To find the eigenvectors of the transition probability matrix, P, we first solve polynomial equation

$$|P - \lambda I| = 0 \tag{1}$$

The unit eigenvector ϕ_i associated with the eigenvalue λ_i is found by solving the system of linear equations

$$P\phi_i = \lambda_i \phi_i \tag{2}$$

and satisfies the condition $\phi_i^T \phi = 1$.

Consider a random walk on the graph G. The walk commences at the node j_1 and proceeds via the sequence of edge-connected nodes $\Gamma = \{j_1, j_2, j_3, ...\}$ where $(j_i, j_{i-1}) \in E$. Suppose that the transition probability associated with the move between the nodes j_l and j_m is $P_{l,m}$. If the random walk can be represented by a Markov chain, then the probability of visiting the nodes in the sequence above is $P_\Gamma = P(j_1) \prod_{l=1}^{|V|} P_{j_{l+1}, j_l}$. This Markov chain can be represented using the transition probability matrix P whose element with row l and column m is $P_{l,m}$. Further, let $Q_t(i)$ be the probability of visiting the node indexed i after t-steps of the random walk and let $Q_t = (Q_t(1), Q_t(2), ...)^T$ be the vector of probabilities. After t time steps $Q_t = (P^T)^t Q_0$. If λ_i are the eigenvalues of P and ϕ_i are the corresponding eigenvectors of unit length, then

$$P = \sum_{i=1}^{|V|} \lambda_i \phi_i \phi_i^T$$

As a result, after t applications of the Markov transition probability matrix

$$P^t = \sum_{i=1}^{|V|} \lambda_i^t \phi_i \phi_i^T$$

If the row and columns of the matrix P sum to unity, then $\lambda_1 = 1$. Furthermore, from spectral graph theory [5] provided that the graph G is not a bipartite graph, then the smallest eigenvalue $\lambda_{|V|} > -1$. As a result, when the Markov chain approaches its steady state, i.e. $t \to \infty$, then all but the first term in the above series become negligible. Hence,

$$\lim_{t\to\infty} P^t = \phi_1 \phi_1^T$$

This establishes that the leading eigenvector of the transition probability matrix is the steady state of the Markov chain. For a more complete proof of this result see the book by Varga [19] or the review of Lovasz [9]. As a result, if we visit the nodes of the graph in the order defined by the magnitudes of the co-efficients of the leading eigenvector of the transition probability matrix, then the path is the steady state Markov chain. In this paper we aim to exploit this property to impose a string ordering on the nodes of a graph, and to use this string ordering property for matching the nodes in different graphs by minimising string edit distance.

Our goal is to match the nodes in a "data" graph $G_D = (V_D, E_D, P_D)$ to their counterparts in a "model" graph $G_M = (V_M, E_M, P_M)$. Suppose that the leading eigenvector for the data-graph transition probability matrix P_D matrix is denoted by $\phi_D^* = (\phi_D^*(1), ..., \phi_D^*(|V_D|))^T$ while that for the model graph transition probability matrix P_M is denoted by $\phi_M^* = (\phi_M^*(1), ...\phi_M^*(|V_M|))^T$. The associated eigenvalues are λ_D^* and λ_M^*. The designation of the two graphs as "data" and "model" is a matter of convention. Here we take the data graph to be the graph which possesses the largest leading eigenvalue, i.e. $\lambda_D^* > \lambda_M^*$.

Our aim is to use the sequence of nodes defined by the rank order of the magnitudes of the components of the leading eigenvector as a means of locating correspondences. The rank order of the nodes in the data graph is given by the string of sorted node-indices $X = (j_1, j_2, j_3, ..., j_{|V_D|})$ where $\phi_D^*(j_1) > \phi_D^*(j_2) > \phi_D^*(j_3) > ... > \phi_D^*(j_{|V_D|})$. The subscript n of the node-index $j_n \in V_D$ is hence the rank-order of the eigenvector component $\phi_D^*(j_n)$. The rank-ordered list of model-graph nodes is $Y = (k_1, k_2, k_3, ..., k_{|V_M|})$ where $\phi_M^*(k_1) > \phi_M^*(k_2) > \phi_M^*(k_3) > ... > \phi_M^*(k_{|V_M|})$.

We augment, the information provided by the leading eigenvectors, with morphological information conveyed by the degree of the nodes in the two graphs. Suppose that deg(i) is the degree of node i. We establish the morphological affinity $\beta_{i,j}$ of nodes $i \in V_D$ and $j \in V_M$ using their degree ratio. Specifically, the morphological affinity of the nodes is taken to be

$$\beta_{i,j} = \exp\left(-\frac{\max(\deg(i), \deg(j)) - \min(\deg(i), \deg(j))}{\max(\deg(i), \deg(j))}\right) \quad (3)$$

If the degree ratio is one then the affinity measure is maximum. If the ratio is small (i.e. $\beta_{i,j} << 1$) then the affinity is zero.

3 Edit Distance

By taking the leading eigenvectors of the model-graph and data-graph adjacency matrices, we have converted the two graphs into strings. Our aim in this paper is to explore whether we can use string edit distance to robustly match the graphs when they are represented in this way. Let X and Y be two strings of symbols drawn from an alphabet Σ. We wish to convert X to Y via an ordered sequence of operations such that the cost associated with the sequence is minimal. The original string to string correction algorithm defined *elementary edit operations*, $(a, b) \neq (\epsilon, \epsilon)$ where a and b are symbols from the two strings or the NULL symbol, ϵ. Thus, changing symbol x to y is denoted (x, y), inserting y is denoted (ϵ, y), and deleting x is denoted (x, ϵ). A sequence of such operations which transforms X into Y is known as an *edit transformation* and denoted $\Delta =< \delta_1, ..., \delta_{|\Delta|} >$. Elementary costs are assigned by an elementary weighting function $\gamma : \Sigma \cup \{\epsilon\} \times \Sigma \cup \{\epsilon\} \mapsto \Re$; the cost of an edit transformation, $W(\Delta)$, is the sum of its elementary costs. The edit distance between X and Y is defined as

$$\mathbf{d}(X, Y) = \min\{W(\Delta) | \Delta \text{ transforms } X \text{ to } Y\} \tag{4}$$

We aim to to locate correspondence matches by seeking the edit-path that minimises the edit distance between the strings representing the steady state random walks on the two graphs. To this end, suppose that $\delta_l = (a, b)$ and $\delta_{l+1} = (c, d)$ represent adjacent states in the edit path between the stready state random walks X and Y. The cost of the edit path is given by

$$W(\Delta) = \sum_{\delta_l \in \Delta} \gamma_{\delta_l \rightarrow \delta_{l+1}} \tag{5}$$

where $\gamma_{\delta_l \rightarrow \delta_{l+1}}$ is the cost of the transition between the states $\delta_l = (a, b)$ and $\delta_{l+1} = (c, d)$. Since, we commenced with a probabilistic characterisation of the matching problem using Markov chains, we define the elementary edit cost to be the negative logarithm of the transition probability for the edit operation. Hence,

$$\gamma_{(a,b)\rightarrow(c,d)} = -\ln P((a, b) \rightarrow (c, d)) \tag{6}$$

We adopt a simple model of the transition probability. The probability is a product of the node similarity weight, and the edge probabilities. Hence we write

$$P((a, b) \rightarrow (c, d)) = \beta_{a,b}\beta_{c,d}R_D(a, c)R_M(b, d) \tag{7}$$

where R_D and R_M are matrices of compatibility weights. The elements of the matrices are assigned according to the following distribution rule

$$R_D(a, c) = \begin{cases} P_D & if(a, c) \in E_D \\ P_\epsilon & \text{if } a = \epsilon \text{ or } c = \epsilon \\ 0 & \text{otherwise} \end{cases} \tag{8}$$

where E_D is the edge set of the data-graph, P_D is the associated normalised transition probability matrix and P_ϵ is the probability associated with a match to the null symbol ϵ. The compatibility weight is hence zero if either the symbol pair (a, c) is unconnected by an edge of the data-graph, or the symbol pair (b, d) is unconnected by a model graph edge. As a result, edit operations which violate edge consistency on adjacent nodes in the strings are discouraged.

The optimal set of correspondences between the two sequences of nodes is found by minimising the stringf edit distance. The optimal sequence of correspondence Δ^* satisfies the condition

$$\Delta^* = \arg\max_{\Delta} W(\Delta) \tag{9}$$

In practice, we find the optimal edit sequence using Dijkstra's algorithm. Since both the data-graph random walk X and the model-graph random walk are edge-connected, the edit path coils around neighbourhoods in the graphs, while "zippering" the strings together.

4 Experiments

We have conducted some experiments with the CMU house sequence. This sequence consists of a series of images of a model house which have been captured from different viewpoints. To construct graphs for the purposes of matching, we have first extracted corners from the images using the corner detector of Luo, Cross and Hancock [3]. The graphs used in our experiments are the Delaunay triangulations of these points. The Delaunay triangulations of the example images are shown in Figure 1a. We have matched pairs of graphs representing increasingly different views of the model house. To do this, we have matched the first image in the sequence, with each of the subsequent images. In Figure 1 b, c and d we show the sequence of correspondence matches. In each case the left-hand graph contains 34 nodes, while the right-hand graphs contain 30, 32 and 34 nodes. From the Delaunay graphs it is clear that there are significant structural differences in the graphs. The numbers of correctly matched nodes in the sequence are respectively 29, 24 and 20 nodes. By comparison, the more complicated iterative EM algorithm of Luo and Hancock [11] gives 29, 23 and 11 correct correspondences. As the difference in viewing direction increases, the fraction of correct correspondences decreases from 80% for the closest pair of images to 60% for the most distant pair of images.

We have conducted some comparison with a number of alternative algorithms. The first of these share with our method the feature of using matrix factorisation to locate correspondences and have been reported by Umeyama [18] and Shapiro and Brady [16]. Since these two algorithms can not operate with graphs of different size, we have taken pairs of graphs with identical numbers of nodes from the CMU sequence; these are the second and fourth images which both contain 32 nodes. Here the Umeyama method and the Shapiro and Brady method both give 6 correct correspondences, while both the Luo and Hancock [11] method and our own give 22 correct correspondences.

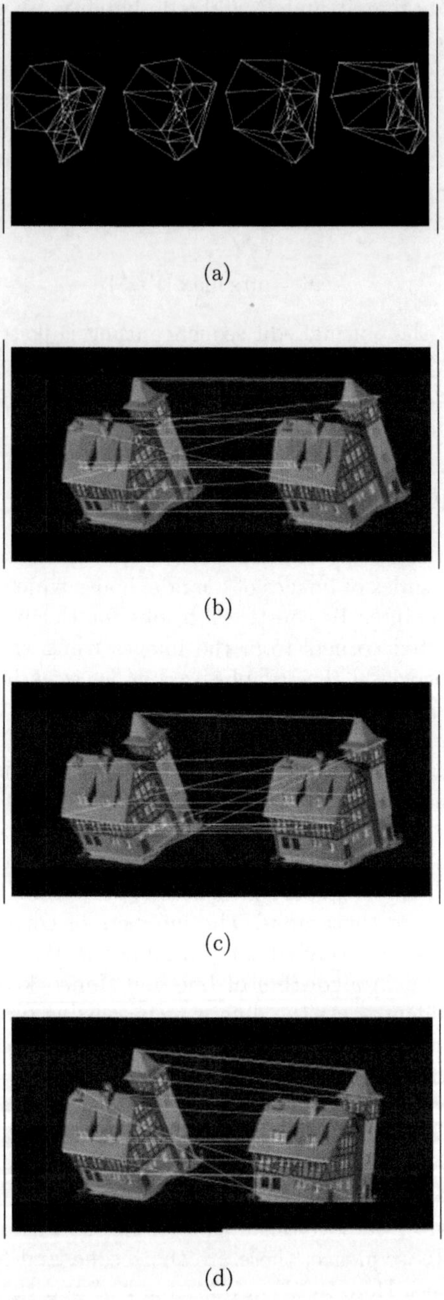

(a) (b) (c) (d)

Fig. 1. Delaunay triangulations and sequence of correspondences

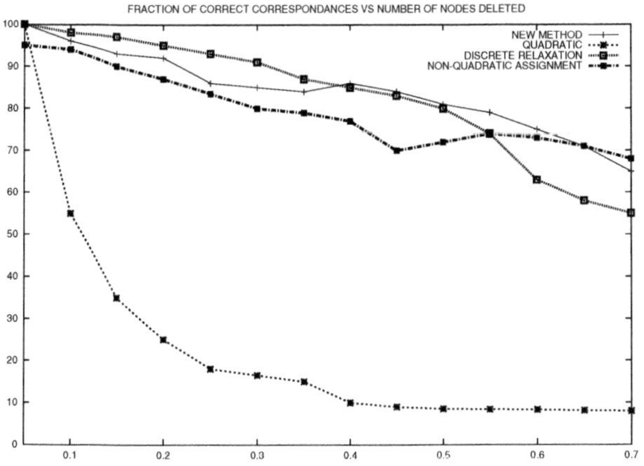

Fig. 2. Sensitivity study results

Finally, we have conducted some experiments with synthetic data to measure the sensitivity of our matching method to structural differences in the graphs and to provide comparison with alternatives. Here we have generated random point-sets and have constructed their Delaunay graphs. We have simulated the effects of structural errors by randomly deleting nodes and re-triangulating the remaining point-set. In Figure 2 we show the fraction of correct correspondences as a function of the fraction of nodes. The performance curve for our new method (marked as "Evidence combining" on the plot) is shown as the lightest of the curves. Also shown on the plot are performance curves for the Wilson and Hancock discrete relaxation scheme, [22], the Gold and Rangarajan [7] quadartic assignment method and the Finch, Wilson and Hancock [1] non-quadratic assignment method. In the case of random node deletion, our method gives performance that is significantly better than the Gold and Rangarajan method, and intermediate in perfomance between the discrete relaxation and non-quadratic assignment methods.

5 Conclusions

The work reported in this paper provides a synthesis of ideas from spectral graph-theory and structural pattern recognition. We use the result from spectral graph theory that the steady state random walk on a graph is given by the leading eigenvector of the adjacency matrix. This allows us to provide a string ordering of the nodes in different graphs. We match the resulting string representations by minimising edit distance. The edit costs needed are computed using a simple probailistic model of the edit transitions which is designed to preserve the edge order on the correspondences.

References

1. R. C. Wilson A. M. Finch and E. R. Hancock. An energy function and continuous edit process for graph matching. *Neural Computation*, 10(7):1873–1894, 1998. 111
2. K. Siddiqi A. Shokoufandeh, S. J. Dickinson and S. W. Zucker. Indexing using a spectral encoding of topological structure. In *Proceedings of the Computer Vision and Pattern Recognition*, 1998. 105
3. Luo Bin and E. R. Hancock. Procrustes alignment with the em algorithm. In *8th International Conference on Computer Analysis of Images and Image Patterns*, pages 623–631, 1999. 109
4. H. Buke. On a relation between graph edit distance and maximum common subgraph. *Pattern Recognition Letters*, 18, 1997. 104
5. Fan R. K. Chung. *Spectral Graph Theory.* American Mathematical Society, 1997. 105, 107
6. M. A. Eshera and K. S. Fu. A graph distance measure for image analysis. *SMC*, 14(3):398–408, May 1984. 104
7. S. Gold and A. Rangarajan. A graduated assignment algorithm for graph matching. *PAMI*, 18(4):377–388, April 1996. 111
8. R. Horaud and H. Sossa. Polyhedral object recognition by indexing. *Pattern Recognition*, 1995. 105
9. L. Lovász. Random walks on graphs: a survey. *Bolyai Society Mathematical Studies*, 2(2):1–46, 1993. 107
10. V. I. Levenshtein. Binary codes capable of correcting deletions, insertions and reversals. *Sov. Phys. Dokl.*, 6:707–710, 1966. 105
11. Bin Luo and E. R. Hancock. Structural graph matching using the EM algorithm and singular value decomposition. *To appear in IEEE Trans. on Pattern Analysis and Machine Intelligence*, 2001. 105, 109
12. B. J. Oommen and K. Zhang. The normalized string editing problem revisited. *PAMI*, 18(6):669–672, June 1996. 106
13. A. Sanfeliu and K. S. Fu. A distance measure between attributed relational graphs for pattern recognition. *IEEE Transactions on Systems, Man and Cybernetics*, 13:353–362, 1983. 104, 106
14. G. Scott and H. Longuet-Higgins. An algorithm for associating the features of two images. In *Proceedings of the Royal Society of London*, number 244 in B, 1991. 105
15. L. G. Shapiro and R. M. Haralick. Relational models for scene analysis. *IEEE Transactions on Pattern Analysis and Machine Intelligence*, 4:595–602, 82. 104, 106
16. L. S. Shapiro and J. M. Brady. A modal approach to feature-based correspondence. In *British Machine Vision Conference*, 1991. 105, 109
17. S. Ullman. Filling in the gaps. *Biological Cybernetics*, 25:1–6, 76. 104
18. S. Umeyama. An eigen decomposition approach to weighted graph matching problems. *PAMI*, 10(5):695–703, September 1988. 105, 109
19. R. S. Varga. *Matrix Iterative Analysis.* Springer, second edition, 2000. 107
20. R. A. Wagner. The string-to-string correction problem. *Journal of the ACM*, 21(1), 1974. 105, 106
21. J. T. L. Wang, B. A. Shapiro, D. Shasha, K. Zhang, and K. M. Currey. An algorithm for finding the largest approximatelycommon substructures of two trees. *PAMI*, 20(8):889–895, August 1998. 106
22. R. C. Wilson and E. R. Hancock. Structural matching by discrete relaxation. *PAMI*, 19(6):634–648, June 1997. 111

Learning Structural Variations in Shock Trees

Andrea Torsello and Edwin R. Hancock

Department of Computer Science, University of York
Heslington, York, YO10 5DD, UK
atorsell@cs.york.ac.uk

Abstract. In this paper we investigate how to construct a shape space for sets of shock trees. To do this we construct a super-tree to span the union of the set of shock trees. We learn this super-tree and the correspondences of the node in the sample trees using a maximizing likelihood approach. We show that the likelihood is maximized by the set of correspondences that minimizes the sum of the tree edit distance between pair of trees, subject to edge consistency constraints. Each node of the super-tree corresponds to a dimension of the pattern space. Individual such trees are mapped to vectors in this pattern space.

1 Introduction

Recently, there has been considerable interest in the structural abstraction of 2D shapes using shock-graphs [9]. The shock-graph is a characterization of the differential structure of the boundaries of 2D shapes. Although graph-matching allows the pairwise comparison of shock-graphs, it does not allow the shape-space of shock-graphs to be explored in detail. In this paper we take the view that although the comparison of shock-graphs, and other structural descriptions of shape, via graph matching or graph edit distance has proved effective, it is in some ways a brute-force approach which is at odds with the non-structural approaches to recognition which have concentrated on constructing shape-spaces which capture the main modes of variation in object shape. Hence, we aim to address the problem of how to organize shock-graphs into a shape-space in which similar shapes are close to one-another, and dissimilar shapes are far apart. In particularly, we aim to do this in a way such that the space is traversed in a relatively uniform manner as the structures under study are gradually modified. In other words, the aim is to embed the graphs in a vector-space where the dimensions correspond to principal modes in structural variation.

There are a number of ways in which this can be achieved. The first is to compute the edit-distance between shock-graphs and use multidimensional scaling to embed the individual graphs in a low-dimensional space [6]. However, as pointed out above, this approach does not necessarily result in a shape-space where the dimensions reflect the modes of structural variation of the shock-graphs. Furthermore, pairwise distance algorithms consistently underestimate the distance between shapes belonging to different clusters. When two shapes are similar, the node-correspondences can be estimated reliably, but as shapes move farther

T. Caelli et al. (Eds.): SSPR&SPR 2002, LNCS 2396, pp. 113–122, 2002.
© Springer-Verlag Berlin Heidelberg 2002

apart in shape space the estimation becomes less reliable. This is due to the fact that correspondences are chosen to minimize the distance between trees: as the shock-trees move further apart the advantage the "correct" correspondence has over alternative ones diminishes. Until, eventually, a match which yields a lower distance is selected. The result of this is a consistent underestimation of the distance as the shapes move further apart in shape space.

The second approach is to extract feature vectors from the graphs and use these as a shape-space representation. A shape-space can be constructed from such vectors by performing modal analysis on their covariance matrix. However, when graphs are of different size, then the problem of how to map the structure of a shock-graph to a vector of fixed length arises. It may be possible to circumvent the problem using graph spectral features.

In this paper we take a different approach to the problem. We aim to embed shock trees in a pattern space by mapping them to vectors of fixed length. We do this as follows. We commence from a set of shock-trees representing different shapes. From this set we learn a super-tree model of which each tree can be considered a noisy sample. In particular, we assume that each node feature is detected with a probability that depends on its weight, but that the hierarchical relation between two detected nodes is always correct. That is, our model has every possible node and the sampling error is in the existence of nodes in our samples, not in their relational structure. Hence, the structure of each sample tree can be obtained from the structure of the super-tree with node removal operations only. We learn this super-tree and the correspondences between the nodes in the sample trees using a maximizing likelihood approach. We show that the likelihood is maximized by the set of correspondences that minimizes the sum of the tree edit distance between pair of trees, subject to edge consistency constraints. To embed the individual shock-trees in a vector-space we allow each node of the super-tree to represent a dimension of the space. Each shock-tree is represented in this space by a vector which has non-zero components only in the directions corresponding to its constituent nodes. The non-zero components of the vectors are the weights of the nodes. In this space, the edit distance between trees is the L1 norm between their embedded vectors.

2 Tree Edit-Distance

The idea behind edit distance is that it is possible to identify a set of basic edit operations on nodes and edges of a structure, and to associate with these operations a cost. The edit-distance is found by searching for the sequence of edit operations that will make the two graphs isomorphic with one-another and which has minimum cost. By making the evaluation of structural modification explicit, edit distance provides a very effective way of measuring the similarity of relational structures. Moreover, the method has considerable potential for error tolerant object recognition and indexing problems. Transforming node insertions in one tree into node removals in the other allows us to use only structure reducing operations. This, in turn, means that the edit distance between two trees is completely determined by the subset of nodes left after the optimal removal sequence. In this section we show how to find the set correspondences that minimizes the edit distance between two trees. To find he edit distance we make use of results presented in [10]. We call $\mathcal{C}(t)$ the closure of tree t, $E_v(t)$ the edit operation that removes node v from t and $\mathcal{E}_v(\mathcal{C}(t))$ the equivalent edit operation that removes v from the closure. The first result is that edit and closure operations commute: $\mathcal{E}_v(\mathcal{C}(t)) = \mathcal{C}(E_v(t))$. For the second result we need some more definitions: We call a subtree s of $\mathcal{C}t$ *obtainable* if for each node v of s if there cannot be two children a and b so that (a, b) is in $\mathcal{C}t$. In other words, for s to be obtainable, there cannot be a path in t connecting two nodes that are siblings in s. We can, now, introduce the following:

Theorem 1. *A tree $\hat{t}$ can be generated from a tree t with a sequence of node removal operations if and only if $\hat{t}$ is an obtainable subtree of the directed acyclic graph $\mathcal{C}t$.*

By virtue of the theorem above, the node correspondences yielding the minimum edit distance between trees t and t' form an obtainable subtree of both $\mathcal{C}t$ and $\mathcal{C}t'$. Hence, we reduce the problem to the search for a common substructure: the maximum common obtainable subtree (MCOS).

We commence by transforming the problem from the search of the minimum edit cost linked to the removal of some nodes, to the maximum of a utility function linked to the nodes that are retained. To do this we assume that we have a weight w_i assigned to each node i, that the cost of matching a node i to a node j is $|w_i - w_j|$, and that the cost of removing a node is equivalent to matching it to a node with weight 0. We define the set $M \subset \mathcal{N}^t \times \mathcal{N}^{t'}$ the set of pair of nodes in t and t' that match, the set $L^t = \{i \in \mathcal{N}^t | \forall x, < i, x > \notin M\}$ composed of nodes in the first tree that are not matched to any node in the second, and the set $R^{t'} = \{j \in \mathcal{N}^{t'} | \forall x, < x, j > \notin M\}$, which contains the unmatched nodes of the second tree. With these definitions the edit distance becomes:

$$d(t, t') = \sum_{i \in L^t} w_i + \sum_{j \in R^{t'}} w_j + \sum_{<i,j> \in M} |w_i - w_j| =$$
$$= \sum_{i \in \mathcal{N}^t} w_i + \sum_{j \in \mathcal{N}^{t'}} w_j - 2 \sum_{<i,j> \in M} \min(w_i, w_j). \quad (1)$$

We call he *utility* of the match M. the quantity

$$\mathcal{U}(M) = \sum_{<i,j> \in M} \min(w_i, w_j).$$

Clearly the match that maximizes the utility minimizes the edit distance. That is, Let $O \subset \mathcal{P}(\mathcal{N}t \times \mathcal{N}t')$ be the set of matches that satisfy the obtainability constraint, the node correspondence $M^* = (N_t^*, N_{t'}^*)$ is

$$M^* = \underset{M \in O}{\operatorname{argmax}} \mathcal{U}(M),$$

and the closure of the MCOS is the restriction to N_t^* of $\mathcal{C}t$.

Let us assume that we know the utility of the best match rooted at every descendent of v and w. We aim to find the set of siblings with greatest total utility. To do this we make use of a derived structure similar to the association graph introduced by Barrow in [1]. The nodes of this structure are pairs drawn from the Cartesian product of the descendents of v and w and each pair correspond to a particular association between a node in one tree to a node in the other. We connect two such associations if and only if there is no inconsistency between the two associations, that is the corresponding subtree is obtainable. Furthermore, we assign to each association node (a, b) a weight equal to the utility of the best match rooted at a and b. The maximum weight clique of this graph is the set of consistent siblings with maximum total utility, hence the set of children of v and w that guarantee the optimal isomorphism. Given a method to obtain a maximum weight clique, we can use it to obtain the solution to our isomorphism problem. We refer again to [10] for heuristics for the weighted clique problem.

3 Edit-Intersection and Edit-Union

The edit distance between two trees is completely determined by the set of nodes that do not get removed by edit operations, that is, in a sense, the *intersection* of the sets of nodes. Furthermore, the distance, and hence, the intersection, determines the probability of a match. We would like to extend the concept to more than two trees so that we can compare a shape tree to a whole set T of trees. Moreover, this allows us to determine how a new sample relates to a previous distribution of trees.

Formally, we assume that the set T of tree samples is drawn from an unknown distribution of trees τ that we want to learn. We assume that we have no sampling error in the detection of the hierarchical relation between two nodes in a sample, that is if we detect two nodes, we detect them with the correct ancestor descendent relation. On the other hand, we assume an exponential distribution for the node weight for a node i of tree t:

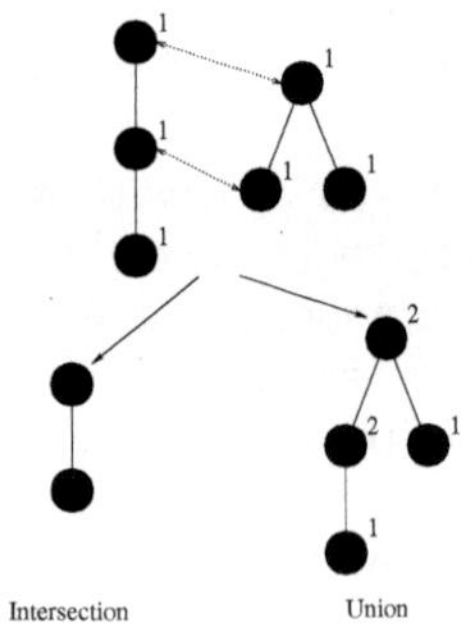

Fig. 1. Union and intersection of trees

$$p_i^t(x) = k \exp\left[-|x - \theta_i^t|\right],$$

where θ_i is a parameter of node i's weight distribution we want to estimate, and k is a normalizing constant.

The log-likelihood function based on the samples T and the set of nodes parameters $\Omega = \{\omega_i\}$ is

$$\mathcal{L} = \sum_{t \in T} \sum_{i \in \mathcal{N}(t)} \log p_i^t = \sum_{t \in T} \sum_{i \in \mathcal{N}(t)} \log k - \sum_{t \in T} \sum_{i \in \mathcal{N}(t)} |w_i^t - \theta_i^t|.$$

We can estimate θ assuming we know the correspondences $C(t, s) \subseteq \mathcal{N}(t) \times \mathcal{N}(s)$ between two trees $t, s \in T$. Fixing this correspondences and estimating θ, we can write the variable part of the log-likelihood function as:

$$\hat{\mathcal{L}} = -\sum_{t\in T}\sum_{s\in T}\sum_{<i,j>\in C(t,s)} |w_i^t - w_j^s|.$$

The structure we want to learn must maximize this function, subject to a consistency constraint on the correspondences. That is if node a in tree t_1 is matched to node b in tree t_2 and to node c in tree t_3, then b must be matched to c, i.e.

$$< a, b >\in C(t_1, t_2)\wedge < a, c >\in C(t_1, t_3) \Rightarrow< b, c >\in C(t_2, t_3).$$

To find the match we calculate a *union* of the nodes: a structure from which we can obtain any tree in our set removing appropriate nodes, as opposed to the intersection of nodes, which is a structure that can obtained removing nodes from the original trees (see Figure 1). Any such structure has the added advantage of implicitly creating an embedding space for our trees: assigning to each node a coordinate in a vector space V, we can associate each tree t to a vector $v \in V$ so that $v_i = w_i^t$, where w_i^t is the weight of the node of t associated with node i of the union, $w_i^t = 0$ if no node in t is associated with i.

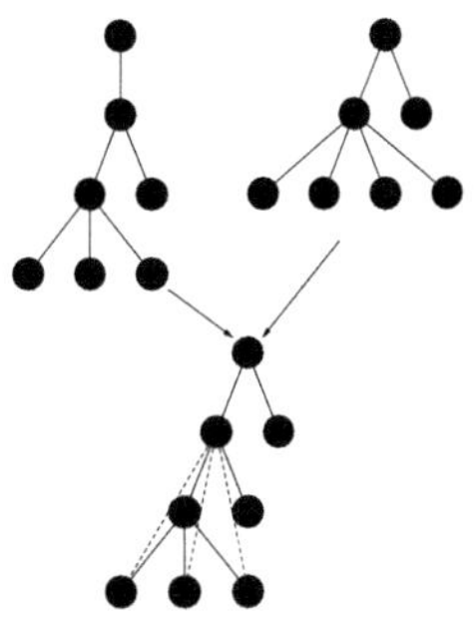

Fig. 2. Edit-union of two trees

The structure of the union of two trees is completely determined by the set of matched nodes: it can be obtained by iteratively merging the nodes of the two trees that are matched. The result will be a directed acyclical graph with multiple paths connecting various nodes (see Figure 2). This structure, thus, has more links than necessary and cannot be obtain from the first tree by node removal operations alone. Removing the superfluous edges. we obtain a tree starting from which we can obtain either one of the original trees by node removal operations alone. Furthermore, this reduced structure maintains the same transitive closure, hence the same hierarchical relation between nodes.

Since the node weights are positive, we can rewrite the variable component of the log-likelihood function as:

$$\hat{\mathcal{L}} = -\sum_{t\in T}\sum_{s\in T}\sum_{i\in\mathcal{N}(t)} w_i^t - \sum_{t\in T}\sum_{s\in T}\sum_{j\in\mathcal{N}(s)} w_j^s + 2\sum_{t\in T}\sum_{s\in T}\sum_{<i,j>\in M(t,s)} \min(w_i^t, w_j^s),$$

where $M(t,s)$ is the set of matches between the nodes of the trees t and s. From this we can see that the set of matches M that maximizes the log likelihood maximizes the sum of the utility functions $\sum_{t\in T}\sum_{s\in T}\mathcal{U}(M(t,s))$ and, hence, minimizes the sum of the edit distances between each pair of samples.

3.1 Joining Multiple Trees

Learning the super-structure, or equivalently finding the structure that minimizes the total distance between trees in the set is computationally infeasible, but we propose a suboptimal iterative approach that iteratively extends the union adding a new tree to it. We want to find the match between the union and the nodes to be added consistent with the obtainability constraint that minimizes the sum of the edit distance between the new tree and each tree in the set. Unfortunately the union operation is not closed in the set of trees, that is the union is not necessarily a tree since it is not always possible to find a tree such that we can edit it to obtain the original trees. For an example where the union of two trees in not a tree see Figure 3. In this figure α and β are subtrees. Because of the constraints posed by matching trees α and trees β, nodes b and b' cannot be matched and neither b can be a child of b' nor b' a child of b. The only option is to keep the two paths as separate alternatives: this way we can obtain the first tree removing the node b' and the second removing b.

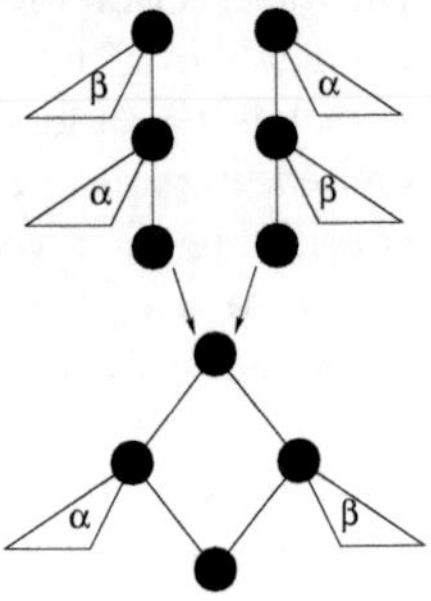

Fig. 3. Edit-union is not a tree

For this reason we cannot use our tree edit distance algorithm unchanged to find the matches between the union and the new tree because it would fail on structures with multiple paths from one node a to node b, counting any match in the subtree rooted at b twice. Fortunately, different paths are present in separate trees and so we can assume that they are mutually exclusive. If we constrain our search to match nodes in only one path and we match the union to a tree, we are guaranteed not to count the same subtree multiple times. Interestingly, this constraint can be merged with the obtainability constraint: we say that a match is *obtainable* if for each node v there cannot be two children a and b and a node c so that there is a path, possibly of length 0, from a to c and one from b to c. This constrain reduces to obtainability for trees when $c = b$, but it also prevents a and b from belonging two to separate paths joining at c. Hence from a node where multiple paths fork, we can extract children matches from one path only.

It is worth noting that this approach can be extended to match two union structures, as long as at most one has multiple paths to a node. To do this we iterate through each pair of weights drawn from the two sets, that is, we define the utility as:

$$\mathcal{U}(M) = \sum_{t \in T_1, t' \in T_2} \sum_{<i,j> \in M} \min(w_i^t, w_j^{t'}),$$

where $M \subset \mathcal{N}^{(T_1^\cup)} \times \mathcal{N}^{(T_2^\cup)}$ is the set of matches between the nodes of the union structures $T_1^\cup$ and $T_2^\cup$. The requirement that no more than one union has multiple paths to a node is required to avoid double counting. Solving the modified weighted clique problems we obtain the correspondence between the nodes of the trees in in the two sets.

To be able to calculate the utility we need to keep, for each node in the union structure, the weights of the matched nodes. A way to do this is to assign to each node in the union a vector of dimension equal to the number of trees in the set. The ith coordinate of this vector will be the weight of the corresponding node in the ith tree, 0 if the ith tree doesn't have a node matching to the current node. This representation also allows us to easily obtain the coordinate of each tree in the set in the embedding space induced by the union: the ith weight of the jth node is the jth coordinate of the ith tree.

In order to increase the accuracy of the approximation, we want to merge trees with smaller distance first. This is because we can be reasonably confident that, if the distance is small, the extracted correspondences are correct. We could start with the set of trees, merge the closest two and replace them with the union and reiterate until we end up with only one structure. Unfortunately, since we have no guarantees that the edit-union is a tree, we might end up trying to merge two graphs with multiple paths to a node. For this reason, if merging two trees give a union that is not a tree, we discard the union and try with the next-best match. When no trees can be merged without duplicating paths, we merge the remaining structures always merging the new nodes to the same structure. This way we are guaranteed to merge at each step at most one multi-path graph.

4 Experimental Results

We evaluate the new approach on the problem of shock tree matching. In order to asses the quality of the approach we compare the obtained embeddings with those described in [10,6]. In particular, we compare the the first two principal components of the embedding generated joining purely structural skeletal representations, with 2D multi-dimensional scaling of the pairwise distances of the shock-trees weighted with some geometrical information. The addition of matching consistency across shapes allows the embedding to better capture the structural information present in the shapes, yielding embedding comparable to those provided by localized geometrical information.

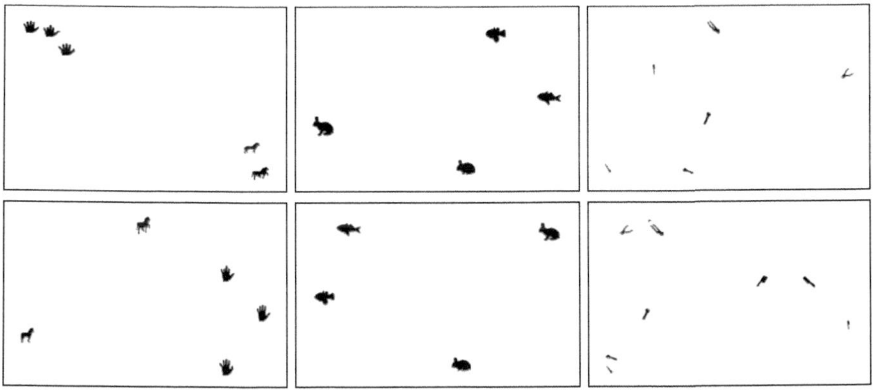

Fig. 4. Top: Embedding through union. Bottom: 2D MDS of pairwise distance

We run three experiments with 4, 5, and 9 shapes each. In each experiment the shapes belong to two or more distinct visual clusters. In order to avoid scaling effect due to difference in the number of nodes, we normalize the embedding vectors so that they have L1 norm equal to 1, and then we extract the first 2 principal components.

Figure 4 shows a comparison between embedding obtained through edit-union of shock trees and through multi-dimensional scaling of the pairwise distances. The first column shows a clear example where the pairwise edit-distances approach underestimate the distance while edit-union keep the clusters well separated. The second and third column show examples where the distance in shape space is not big enough to observe the described behavior, yet the embedding obtained through union fares well against the pairwise edit-distance, especially taking into account the fact that it uses only structural information while the edit-distance matches weight the structure with geometrical information. In particular, the third column shows a better ordering of shapes, with brushes being so tightly packed that they overlap. It is interesting to note how the union embedding puts the monkey wrench (top-center) somewhere in-between pliers and wrenches: the algorithm is able to consistently match the head to the heads of the wrenches, and the handles to the handles of the pliers.

Figure 5 plots the distances obtained through edit union of weighted shock trees (x axis) versus the corresponding pairwise edit distances (y axis). The plot clearly highlights that the pairwise distance approach tends to underestimate the distances between shapes.

4.1 Synthetic Data

To augment these real world experiments, we have performed the embedding on synthetic data. The aim of the experiments is to characterize the ability of the approach to generate a shape space. To meet this goal we have randomly generated some prototype trees and, from each tree, we generated five or ten structurally perturbed copies. The procedure for generating the random trees was as follows: we commence with an empty tree (i.e. one with no nodes) and we iteratively add the required number of nodes. At each iteration nodes are added as children of one of the existing nodes. The parents are randomly selected with uniform probability from among the existing nodes. The weight of the newly added nodes are selected at random from an exponential distribution with mean 1. This procedure will tend to generate trees in which the branch ratio is highest closest to the root. This is quite realistic of real-world situations, since shock trees tend to have the same characteristic. To perturb the trees we simply add nodes using the same approach.

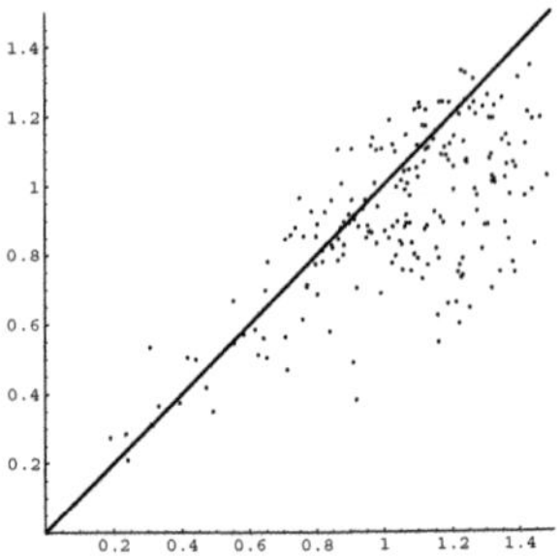

Fig. 5. Edit-union vs. pairwise edit distances

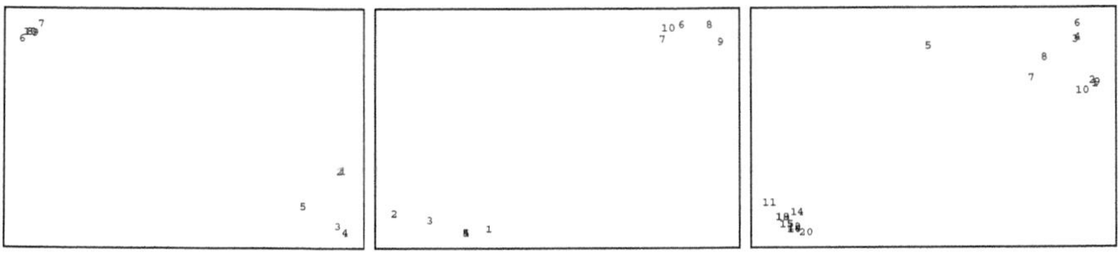

Fig. 6. Synthetic clusters

In our experiments the size of the prototype trees varied from 5 to 20 nodes. As we can see from Figure 6, the algorithm was able to clearly separate the clusters of trees generated by the same prototype. Figure 6 shows three experiments with synthetic data. The first and second images are produced embedding 5 structurally perturbed trees per prototype: trees 1 to 5 are perturbed copies of the first prototype, 6 to 10 of the second. The last image shows the result of the experiment with 10 structurally perturbed trees per prototype: 1 to 10 belong to one cluster, 11 to 20 to the other. In each image the clusters are well separated.

5 Conclusions

In this paper we investigated a technique to extend the tree edit distance framework to allow the simultaneous matching of multiple tree structures. With this approach we can impose a consistency of node correspondences between matches, avoiding the underestimation of the distance typical of pairwise edit-distances approaches. Furthermore through this methods we can get a "natural" embedding space of tree structures that can be used to analyze how tree representations vary in our problem domain.

In a set of experiments we apply this algorithm to match shock graphs, a graph representation of the morphological skeleton. The results of these experiments are very encouraging, showing that the algorithm is able to group similar shapes together in the generated embedding space.

Our future plans are to extend the framework reported in this paper by using the apparatus of variational inference to fit a mixture of trees, rather than a union tree, to the training data. Here we will perform learning by minimizing the Kullback divergence between the training data and the mixture model.

References

1. H. G. Barrow and R. M. Burstall, Subgraph isomorphism, matching relational structures and maximal cliques, *Inf. Proc. Letter*, Vol. 4, pp.83, 84, 1976. 116
2. H. Bunke and A. Kandel, Mean and maximum common subgraph of two graphs, *Pattern Recognition Letters*, Vol. 21, pp. 163-168, 2000.
3. T. F. Cootes, C. J. Taylor, and D. H. Cooper, Active shape models - their training and application, *CVIU*, Vol. 61, pp. 38-59, 1995.

4. T. Heap and D. Hogg, Wormholes in shape space: tracking through discontinuous changes in shape, *ICCV*, pp. 344-349, 1998.
5. T. Sebastian, P. Klein, and B. Kimia, Recognition of shapes by editing shock graphs, in *ICCV*, Vol. I, pp. 755-762, 2001.
6. B. Luo, et al., Clustering shock trees, in *CVPR*, Vol 1, pp. 912-919, 2001. 113, 119
7. M. Pelillo, K. Siddiqi, and S. W. Zucker, Matching hierarchical structures using association graphs, *PAMI*, Vol. 21, pp. 1105-1120, 1999.
8. S. Sclaroff and A. P. Pentland, Modal matching for correspondence and recognition, *PAMI*, Vol. 17, pp. 545-661, 1995.
9. K. Siddiqi et al., Shock graphs and shape matching, *Int. J. of Comp. Vision*, Vol. 35, pp. 13-32, 1999. 113
10. A. Torsello and E. R. Hancock, Efficiently computing weighted tree edit distance using relaxation labeling, in *EMMCVPR*, LNCS 2134, pp. 438-453, 2001 115, 116, 119
11. K. Zhang, R. Statman, and D. Shasha, On the editing distance between unorderes labeled trees, *Inf. Proc. Letters*, Vol. 42, pp. 133-139, 1992.

A Comparison of Algorithms for Maximum Common Subgraph on Randomly Connected Graphs

Horst Bunke[1], Pasquale Foggia[2], Corrado Guidobaldi[1,2], Carlo Sansone[2], and Mario Vento[3]

[1] Institut für Informatik und angwandte Mathematik,Universität Bern
Neubrückstrasse 10, CH-3012 Bern, Switzerland
bunke@iam.unibe.ch
[2] Dipartimento di Informatica e Sistemistica, Università di Napoli "Federico II"
via Claudio, 21 I-80125 Napoli, Italy
{foggiapa,cguidoba,carlosan}@unina.it
[3] Dipartimento di Ingegneria dell'Informazione ed Ingegneria Elettrica
Università di Salerno, via P.te Don Melillo, 1 I-84084 Fisciano (SA), Italy
mvento@unisa.it

Abstract. A graph g is called a *maximum common subgraph* of two graphs, g_1 and g_2, if there exists no other common subgraph of g_1 and g_2 that has more nodes than g. For the maximum common subgraph problem, exact and inexact algorithms are known from the literature. Nevertheless, until now no effort has been done for characterizing their performance. In this paper, two exact algorithms for maximum common subgraph detection are described. Moreover a database containing randomly connected pairs of graphs, having a maximum common graph of at least two nodes, is presented, and the performance of the two algorithms is evaluated on this database.

1 Introduction

Graphs are a powerful and versatile tool useful in various subfields of science and engineering. There are applications, for example, in pattern recognition, machine learning and information retrieval, where one needs to measure the similarity of objects. If graphs are used for the representation of structured objects, then measuring the similarity of objects becomes equivalent to determining the similarity of graphs.

There are some well-known concepts that are suitable graph similarity measures. Graph isomorphism is useful to find out if two graphs have identical structure[1]. More generally, subgraph isomorphism can be used to check if one graph is part of another [1,2]. In two recent papers [3,4], graph similarity measures based on maximum common subgraph and minimum common supergraph have been proposed.

Detection of the maximum common subgraph (MCS) of two given graphs is a well-known problem. In [5], such an algorithm is described and in [6] the use of this algorithm in comparing molecules has been discussed. In [7] a MCS algorithm that uses a backtrack search is introduced. A different strategy for deriving the MCS first

T. Caelli et al. (Eds.): SSPR&SPR 2002, LNCS 2396, pp. 123-132, 2002.
© Springer-Verlag Berlin Heidelberg 2002

obtains the association graph of the two given graphs and then detects the *maximum clique* (*MC*) of the latter graph [8,9].
Both, MCS and MC detection, are NP-complete problems [10]. Therefore many approximate algorithms have been developed. A survey of such algorithms, including an analysis of their complexity, and potential applications is provided in [11].

Although a significant number of MCS detection algorithms have been proposed in the literature, until now no effort has been spent for characterizing their performance. Consequently, it is not clear how the behaviour of these algorithms varies as the type and the size of the graphs to be matched changes from an application to another. The lack of a sufficiently large common database of graphs makes the task of comparing the performance of different MCS algorithms difficult, and often an algorithm is chosen just on the basis of a few data elements.

In this paper we present two exact algorithms that follows different principles. The first algorithm searches for the MCS by finding all common subgraphs of the two given graphs and choosing the largest [7]; the second algorithm builds the association graph between the two given graphs and then searches for the MC of the latter graph [12]. Moreover we present a synthetically generated database containing pairs of randomly connected pairs of graphs, in which each pair has a known MCS.
The remainder of the paper is organized as follows. In Section 2 basic terminology is introduced and the first algorithm to be compared is described. The second algorithm to be compared is described in Section 3. In Section 4 the database used is presented, while experimental results are reported in Section 5. Finally future work is discussed and conclusions are drawn in Section 6.

2 A Space State Search Algorithm for Detecting the MCS

The two following definitions will be used in the rest of the paper:

Definition 2.1: A *graph* is a 4-tuple $g = (V, E, \alpha, \beta)$, where

- V is the finite set of vertices (also called nodes)
- $E \subseteq V \times V$ is the set of edges
- $\alpha : V \rightarrow L$ is a function assigning labels to the vertices
- $\beta : E \rightarrow L$ is a function assigning labels to the edges
- Edge (u,v) originates at node u and terminates at node v.

Definition 2.2: Let $g_1 = (V_1, E_1, \alpha_1, \beta_1)$ and $g_2 = (V_2, E_2, \alpha_2, \beta_2)$ be graphs. A *common subgraph* of g_1 and g_2, $cs(g_1, g_2)$, is a graph $g = (V, E, \alpha, \beta)$ such that there exist subgraph isomorphisms from g to g_1 and from g to g_2. We call g a *maximum common subgraph* of g_1 and g_2, $mcs(g_1,g_2)$, if there exists no other common subgraph of g_1 and g_2 that has more nodes than g.

Notice that, according to Definition 2, $mcs(g_1,g_2)$, is not necessarily unique for two given graphs. We will call the set of all MCS of a pair of graphs their *MCS set.*

According to the above definition of MCS, it is also possible to have graphs with isolated nodes in the MCS set. This is in contrast with the definition given in [7], where a MCS of two given graphs is defined as the common subgraph which contains the maximum number of edges (we could call it *edge induced MCS*, in contrast with the method described in this paper, which is *node induced*). Then the case of a MCS containing unconnected nodes is not considered in [7]. Consequently, the algorithm proposed in this section, although derived from the one described in [7] by McGregor, is more general. It can be suitably described through a State Space Representation [13]. Each state `s` represent a common subgraph of the two graphs under construction.

```
Procedure MCS(s,n1,n2)
Begin
   if (NextPair(n1,n2)) then
      begin
         if (IsFeasiblePair(n1,n2)) then
           AddPair(n1,n2);
         CloneState(s,s');
         while(s' is not a leaf of the search tree)
         begin
            MCS(s',n1,n2);
            BackTrack(s');
         End
         Delete(s');
      End
End procedure
```

Fig. 1. Sketch of the space state search for maximum common subgraph detection

This common subgraph is part of the MCS to be eventually formed. In each state a pair of nodes not yet analyzed, the first belonging to the first graph and the second belonging to the second graph, is selected (whenever it exists) through the function `NextPair(n1,n2)`. The selected pair of nodes is analyzed through the function `IsFeasiblePair(n1,n2)` that checks whether it is possible to extend the common subgraph represented by the actual state by means of the this pair. If the extension is possible, then the function `AddPair(n1,n2)` actually extends the current partial solution by the pair `(n1,n2)`. After that, if the current state `s` is not a leaf of the search tree, it copies itself through the function `CloneState(s,s')`, and the analysis of this new state is immediately started. After the new state has been analyzed, a *backtrack* function is invoked, to restore the common subgraph of the previous state and to choose a different new state. Using this search strategy, whenever a branch is chosen, it will be followed as deeply as possible in the search tree until a leaf is reached. It is noteworthy that every branch of the search tree has to be followed, because - except for trivial examples - is not possible to foresee if a better solution exists in a branch that has not yet been explored. It is also noteworthy that, whenever a state is not useful anymore, it is removed from the memory through the function `Delete(s)`. The first state is the empty-state, in which two null-nodes are analyzed. A pseudo-code description of the MCS detection algorithm is shown in Fig 1.

Let N_1 and N_2 be the number of nodes of the first and the second graph, respectively, and let $N_1 \leq N_2$. In the worst case, i.e. when the two graphs are completely connected with the same label on each node and the same label on each edge, the number of states `s` examined by the algorithm is:

$$S = N_2! \cdot \left(\frac{1}{(N_2 - N_1)!} + \ldots + \frac{1}{(N_2 - 1)!} \right) \tag{1}$$

For the case $N_1 = N_2 = N$ and $N >> 1$, eq.(1) can be approximated as follows:

$$S \cong e \cdot N! \tag{2}$$

Notice that only $O(N_1)$ space is needed by the algorithm.

3 An MCS Algorithm Based on Clique Detection

The Durand-Pasari algorithm is based on the well known reduction of the search of the MCS between two graph to the problem of finding a MC in a graph [12]. The first step of the algorithm is the construction of the *association graph*, whose vertices corresponds to pair of vertices of the two starting graphs having the same label. The edges of the association graph represent the compatibility of the pair of vertices to be included; hence, MCS can be obtained by finding the MC in the association graph.
The algorithm for MC detection generates a list of vertices that represents a clique of the association graph using a depth-first search strategy on a search tree, by systematically selecting one vertex at a time from successive levels, until it is not possible to add further vertices to the list. A sketch of the algorithm is in Fig. 2.

```
Procedure MCS_DP(vert_list)
Begin
   Level = length(vert_list);
   null_count = count_null_vertices(vert_list);
   clique_length = level - null_count;
   if (null_count >= best_null_count_so_far) then
      return;
   else if (level == max_level) then
      save(vert_list);
      best_null_count_so_far = null_count;
   else
      P = set of vertices (n1,n2) having n1==level;
      Foreach(v in P ∪ { NULL_VERTEX })
      Begin
         if (is_legal(v, vert_list)) then
            MCS_DP(vert_list + v);
         end if
      end
   end if
end procedure
```

Fig. 2. Sketch of the maximum clique detection algorithm

When a vertex is being considered, the forward search part of the algorithm first checks to see if this vertex is a *legal* vertex, and if it is the algorithm next checks to see if the size of the new clique formed is as large or larger than the current largest clique, in which case it is saved. A vertex is legal if it is connected to every other vertex already in the clique. At each level l, the choice of the vertices to consider is limited to the ones which correspond to pairs (n_1, n_2) having $n_1=l$. In this way the algorithm ensures that the search space is actually a tree, i.e. it will never consider twice the same list of vertices. After considering all the vertices for level l, a special vertex, called the *null vertex*, is added to the list. This vertex is always considered legal, and can be added more than once to the list. This special vertex is used to carry the information that no mapping is associated to a particular vertex of the first graph being matched. When all possible vertices (including the null vertex) have been considered, the algorithm backtracks and tries to expand along a different branch of the search tree. The length of the longest list (excluding any null vertex entries) as well as its composition is maintained. This information is updated, as needed. If N_1 and N_2 are the number of vertices of the starting graphs, with $N_1 \leq N_2$, the algorithm execution will require a maximum of N_1 levels. Since at each level the space requirement is constant, the total space requirement of the algorithm is $O(N_1)$. To this, however, the space needed to represent the association graph must be added. In the worst case the association graph can be a complete graph of $N_1 \cdot N_2$ nodes. In the worst case the algorithm will have to explore (N_2+1) vertices at level 1, N_2 at level 2, up to $(N_2 - N_1+2)$ at level N_1. Multiplying these numbers we obtain a worst case number of states

$$S=(N_2+1)(N_2)\ldots(N_2-N_1+2)=\frac{(N_2+1)!}{(N_2-N_1+1)!} \qquad (3)$$

which, for $N_1=N_2$ reduces to $O(N \cdot N!)$.

4 The Database

During the last years the pattern recognition community recognized the importance of benchmarking activities for validating and comparing the results of proposed methods. Within the Technical Committee 15 of the International Association for Pattern Recognition (IAPR-TC15) the characterization of the performance achieved by graph matching algorithms revealed to be particularly important due to the growing need of using matching algorithms dealing with large graphs. To this concern, two artificially generated databases have been presented at the last IAPR-TC15 workshop [14, 15]. The first one [14] describes the format and the four different categories contained in a database of 72,800 pairs of graphs, developed for graph and subgraph isomorphism benchmarking purposes. The graphs composing the whole database have been distributed on a CD during the 3rd IAPR-TC15 and are also publicly available on the web at the URL: *http://amalfi.dis.unina.it/graph.* A different way for building a graph database has been proposed in [15]. Here the graphs are obtained starting from images synthetically generated by means of a set of attributed plex grammars. Different classes of graphs are therefore obtained by considering different plex grammars.

The databases cited above are not immediately usable for the purpose of benchmarking algorithms for MCS. In fact, the first database can be used with graph (or subgraph) isomorphism algorithms and provides graphs with no labels. Also the second database has not been developed for generating graphs to be used in the context of MCS algorithms. To overcome these problems we decided to generate another database of synthetic graphs with random values for the attributes, since any other choice requires making assumptions about the application dependent model of the graphs to be generated. In particular, we assumed, without any loss of generality, that attributes are represented by integer numbers with a uniform distribution over a certain interval. In fact, the purpose of the attributes in our benchmarking activity is simply to restrict the possible node or edge pairings; hence there is no need to have structured attributes. The most important parameter characterizing the difficulty of the matching problem is the number M of different attribute values: obviously the higher this number, the easier is the matching problem. Therefore, it should be important to have different values of M in a database. In order to avoid the need to have several copies of the database with different values of M, we chose to generate each attribute as a 16-bit value, using a random number generation. In this way, a benchmarking activity can be made with any M of the form 2^k, for k not greater than 16, just by using, in the attribute comparison function, only the first k bits of the attribute. As regards the kind of graphs, we chose to include in the database *randomly connected graphs,* i.e. graphs in which it is assumed that the probability of an edge connecting two nodes is independent on the nodes themselves. The same model as proposed in [1] has been adopted for generating these graphs: it fixes the value η of the probability that an edge is present between two distinct nodes n and n'. The probability distribution is assumed to be uniform. According to the meaning of η, if N is the total number of nodes of the graph, the number of its edges will be equal to $\eta N \cdot (N-1)$. However, if this number is not sufficient to obtain a connected graph, further edges are suitably added until the graph being generated becomes connected.

Table 1. The database of randomly connected graphs for benchmarking algorithms for MCS

η	# of nodes (N)	# of nodes of the MCS	# of pairs
0.05	20	2, 6, 10, 14, 18	500
	25	2, 7, 12, 17, 22	500
	30	3, 9, 15, 21, 27	500
0.1	10	3, 5, 7, 9	400
	15	4, 7, 10, 13	400
	20	2, 6, 10, 14, 18	500
	25	2, 7, 12, 17, 22	500
	30	3, 9, 15, 21, 27	500
0.2	10	3, 5, 7, 9	400
	15	5, 7, 10, 13	400
	20	2, 6, 10, 14, 18	500
	25	2, 7, 12, 17, 22	500
	30	3, 9, 15, 21, 27	500

The generated database is structured in pairs of graphs having a MCS of at least two nodes. In particular, three different values of the edge density η have been considered: 0.05, 0.1 and 0.2. For each value of η, graphs of different size N, ranging from 10 to 30 have been taken into account. Values of N equal to 10 and 15 have not been considered for η=0.05, since in these cases it was not possible to have connected graphs without adding a significant number of extra edges.

Five different percentages of the values of N have been considered for determining the size of the MCS, namely 10%, 30%, 50%, 70% and 90%. This choice allows us to verify the behavior of the algorithms as the ratio between the size of the MCS and the value of N varies. Then, for each value of N and for each chosen percentage, 100 pairs of graph have been generated, giving rise to a total of 6100 pairs. Note that for values of N equal to 10 and 15, the 10% value was not considered as it would determine a MCS size less than two nodes. Table 1 summarizes the characteristic of the graphs composing the database. The MCS size refers to the case in which $M=2^{16}$.

5 Experimental Results

In order to make an unbiased comparison of the two algorithms presented in Sections 2 and 3, we have developed an implementation of both in C++, using the VFLib class library available at *http://amalfi.dis.unina.it/graph*. The code has been compiled using the gcc 2.96 compiler, with all optimizations enabled. The machines used for the experiments are based on the Intel Celeron processor (750MHz), with 128 MB of memory; the operating system is a recent Linux distribution with the 2.4.2 kernel version. A set of Python scripts have been used to run the two algorithms on the entire database and to collect the resulting matching times.

As we have explained in the previous section, the database contains 16 bits attributes, that can be easily employed to test the algorithms with different values of the parameter M. Since both algorithms have a time complexity that grows exponentially with the number of nodes, it would be impractical to attempt the matching with a too low value of M. We have chosen to employ values of M proportional to the number of nodes in the graphs being matched, in order to keep the running times within a reasonable limit. In particular, we have tested each graph pair in the database with three M values equal to 33%, 50% and 75% of the number of nodes.

The resulting matching times are shown in Fig 3. Notice that one of the database parameters, the number of nodes of the generated MCS, does not appear in the figure. In fact, in order to reduce the number of curves to be displayed, we have averaged the times over the different MCS sizes. It should be also considered that, for different values of M, the actual size of the MCS may vary. In fact if M is large, some node pairs are excluded from the MCS because of their attribute values; if M is small, the same pairs may become feasible for inclusion. Hence, by not reporting separately the times for different MCS sizes it becomes easier to compare the results corresponding to different values of M. Examining the times reported in the figure, it can be noted that while both algorithms exhibit a very rapidly increasing time with respect to the number of nodes, they show a behavior quite different from each other with respect to the other two considered parameters that is, M and the graph density η. As regards M, it can be seen that the matching time decreases when M gets larger. But while for low values of M the Durand-Pasari algorithm performs usually better than the McGregor one, for high values of M the situation is inverted. This can be explained by the fact that the Durand-Pasari algorithm is based on the construction of an association graph, which helps reducing the computation needed by the search algorithm when the search space is large (small M) because the compatibility tests are, in a sense, "cached" in the structure of the association graph; on the other hand, the association graph construction imposes a time (and space) overhead, that is not repaid when the search space is small (large M). For the graph density, we notice that the dependency of the matching time on η is opposite for the two algorithms. In fact, while the time for Durand-Pasari decreases for larger values of η, the time for McGregor increases. An explanation of this difference is that, for Durand-Pasari, an increase in the graph density enables the algorithm to prune more node pairs on the basis of the node connections. In the McGregor algorithm, instead, this effect is compensated by the increase of the number of edge compatibility tests that must be performed at each state, to which the Durand-Pasari is immune because of the use of the association graph.

a)

b)

c)

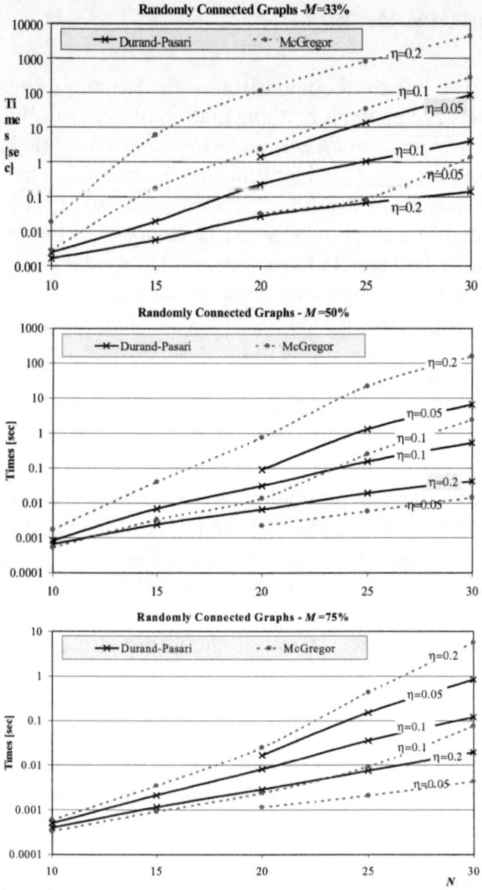

Fig. 3. Results obtained for M=33% a),50% b),75% c) of N, as a function of N and η

6 Conclusions and Perspectives

In this paper, two exact algorithms for MCS detection have been described. Moreover a database containing randomly connected pairs of graphs having a MCS of at least two nodes has been presented, and the performance of the two algorithms has been evaluated on this database. Preliminary comparative tests show that for graphs with a low density it is more convenient to search for the MCS by finding all the common subgraphs of the two given graphs and choosing the largest, while for high edge density, it is efficient to build the association graph of the two given graphs and then to search for the MC of the latter graph. At present the database presented in the paper contains 6100 pairs of randomly connected graphs. A further step will be the expansion of the database through the inclusion of pairs of graphs with more of nodes. Besides, the inclusion of other categories graphs, such as regular meshes (2-dimensional, 3-dimensional, 4-dimensional), irregular meshes, bounded valence graphs, and irregular bounded graphs will be considered. Moreover further algorithms for MCS will be implemented and their performances characterized on this database. A more precise measure of the performance could be obtained with a further parameter in the database, namely the size *s* of the MCS in each pair of graphs.

References

1. J.R. Ullmann, "An Algorithm for Subgraph Isomorphism", Journal of the Association for Computing Machinery, vol. 23, pp. 31-42, 1976.
2. L.P. Cordella, P. Foggia C. Sansone, M. Vento, "An Improved Algorithm for Matching Large Graphs", Proc. of the 3rd IAPR-TC-15 International Workshop on Graph-based Representations, Italy, pp. 149-159, 2001.
3. H. Bunke X. Jiang and A. Kandel, "On the Minimum Supergraph of Two Graphs", Computing 65, Nos. 13 - 25, pp. 13-25, 2000.
4. H. Bunke and K.Sharer, "A Graph Distance Metric Based on the Maximal Common Subgraph", Pattern Recognition Letters, Vol. 19, Nos. 3-4, pp. 255-259, 1998.
5. G. Levi, "A Note on the Derivation of Maximal Common Subgraphs of Two Directed or Undirected Graphs", Calcolo, Vol. 9, pp. 341-354, 1972.
6. M. M. Cone, Rengachari Venkataraghven, and F. W. McLafferty, "Molecular Structure Comparison Program for the Identification of Maximal Common Substructures", Journal of Am. Chem. Soc., 99(23), pp. 7668-7671 1977.
7. J.J. McGregor, "Backtrack Search Algorithms and the Maximal Common Subgraph Problem", Software Practice and Experience, Vol. 12, pp. 23-34, 1982.
8. C. Bron and J. Kerbosch, "Finding All the Cliques in an Undirected Graph", Communication of the Association for Computing Machinery 16, pp. 575-577,1973.
9. B. T. Messmer, "Efficient Graph Matching Algorithms for Preprocessed Model Graphs", Ph.D. Thesis, Inst. of Comp. Science and Appl. Mathematics, University of Bern, 1996.
10. M. R. Garey, D. S. Johnson, "Computers and Intractability: A Guide to the Theory of NP-Completeness", Freeman & Co, New York, 1979.
11. I. M. Bomze, M. Budinich, P. M. Pardalos, and M. Pelillo, "The Maximum Clique Problem", Handbook of Combinatorial Optimization, vol. 4, Kluwer Academy Pub., 1999.

12. P. J. Durand, R. Pasari, J. W. Baker, and Chun-che Tsai, "An Efficient Algorithm for Similarity Analysis of Molecules ", Internet Journal of Chemistry, vol. 2, 1999.
13. N. J. Nilsson, "Principles of Artificial Intelligence", Springer-Verlag, 1982.
14. P. Foggia, C. Sansone, M.Vento, "A Database of Graphs for Isomorphism and Sub-Graph Isomorphism Benchmarking", Proc. of the 3rd IAPR TC-15 International Workshop on Graph-based Representations, Italy, pp. 176-187, 2001.
15. H. Bunke, M. Gori, M. Hagenbuchner C. Irniger, A.C. Tsoi, "Generation of Images Databases using Attributed Plex Grammars", Proc. of the 3rd IAPR TC-15 International Workshop on Graph-based Representations, Italy, pp. 200-209, 2001.

Inexact Multisubgraph Matching Using Graph Eigenspace and Clustering Models*

Serhiy Kosinov and Terry Caelli

Department of Computing Science
Research Institute for Multimedia Systems (RIMS)
The University of Alberta, Edmonton, Alberta, CANADA T6G 2H1

Abstract. In this paper we show how inexact multisubgraph matching can be solved using methods based on the projections of vertices (and their connections) into the eigenspaces of graphs - and associated clustering methods. Our analysis points to deficiencies of recent eigenspectra methods though demonstrates just how powerful full eigenspace methods can be for providing filters for such computationally intense problems. Also presented are some applications of the proposed method to shape matching, information retrieval and natural language processing.

1 Introduction

Inexact graph matching is a fundamental task in a variety of application domains including shape matching, handwritten character recognition, natural language processing, to name a few. Naturally, there exist numerous both general and application-specific approaches for solving the problem of inexact graph matching. However, the task still presents a substantial challenge, and there still is room for improvement in some of the existing methods. Our work attempts to demonstrate the power of combining eigenspace graph decomposition models with clustering techniques to solve this problem. But before providing a detailed description of the proposed method, it is beneficial to put our work briefly into the context of previously developed solutions.

A rather generalized view point adopted by Bunke[1] poses the task of inexact graph matching as a problem of structural pattern recognition. In this work, the author has studied error-tolerant graph matching using *graph edit distance*, a concept that provides a measure of dissimilarity of two given entities and has its origins in the domain of strings. Here, a pair of graphs is compared by finding a sequence of edit operations, such as edge/vertex deletion, insertion or substitution, that transforms one graph into the other, whereas the dissimilarity, or distance, of the two graphs is said to be the minimum possible cost of such a transformation. Other important notions developed by Bunke are the weighted mean and generalized median of a pair of graphs[5], which allow a range of well-established techniques from statistical pattern recognition, such as clustering with self-organizing maps, to be applied in the domain of graphs. In a

* This project was funded by a grant from the NSERC Canada.

T. Caelli et al. (Eds.): SSPR&SPR 2002, LNCS 2396, pp. 133–142, 2002.
© Springer-Verlag Berlin Heidelberg 2002

way similar to the work of Bunke is the effort of Tirthapura et al.[14], who successfully deployed the classical Levenshtein distance in matching shock graphs that represent 2D shapes.

Another elegant and theoretically well-grounded approach to subgraph matching is that developed by Hancock et al.[6], who, instead of going further with goal-directed search, adopt a probabilistic framework and use optimization methods to solve the graph matching problem. That is, by modelling the correspondence errors encountered during graph matching with the aid of the Bernoulli probability distribution, the authors are able to devise a graph matching likelihood function that allows one to estimate the conditional likelihood of one graph given the other and recover the best possible graph node correspondence by means of Expectation-Maximization (EM) and Singular Value Decomposition (SVD).

There also exists a whole family of graph matching techniques, generally known as spectral methods, that seek to represent and distinguish structural properties of graphs using eigenvalues and eigenvectors of graph adjacency matrices. The most valuable characteristics of such methods include being invariant to edge/vertex reordering, ability to map a graph's structural information into lower-dimensional spaces and stability under minor perturbations. On top of that, the eigendecomposition technique itself is far less computationally expensive as compared to the advanced combinatorial search procedures. Among recent developments in this field are the Umeyama's[15] formulation for same-size graph matching that derives the minimum difference permutation matrix via eigendecomposition techniques, Shapiro and Brady's[10] method for comparing graphs according to the corresponding values of the rearranged eigenvectors of graph adjacency matrices, and the work of Dickinson et al.[11] on indexing hierarchical structures with topological signature vectors obtained from the sums of adjacency matrix eigenvalues.

Similarly to the above contributions, our work borrows heavily from graph eigendecompositions. The proposed model is based upon the fundamental idea that graph matching need not be posed as a combinatorial matching problem but, rather, as one of *clustering common local relational structures* between different graphs. This results in a natural grouping between vertices of quite different graphs which share similar relational properties. We show how to do this using projection principles as used in SVD where vertex vectors from different graphs can be projected into common eigenvector subspaces.

2 Graph Eigenspace Methods

2.1 Eigenspectra and Eigenvectors of Graphs

As mentioned above, the basic technique deployed in the majority of spectral methods is eigendecomposition. In general, for undirected graphs, it is expressed as follows:

$$A = VDV^T \tag{1}$$

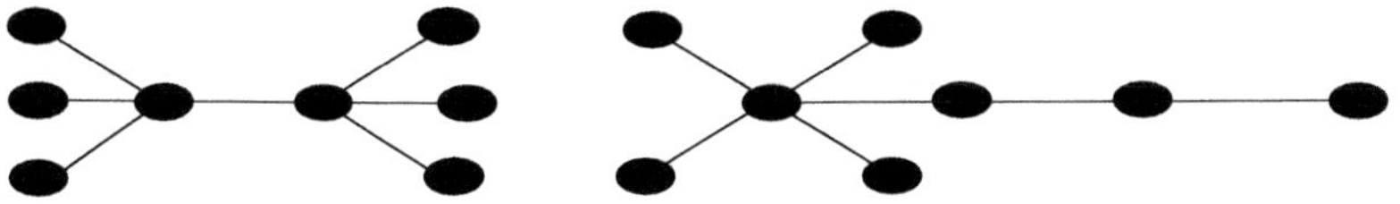

Fig. 1. Two different graphs with identical eigenspectra

where A is the square symmetric adjacency matrix of a graph, whose entry a_{ij} at the place (i,j) is equal to one if there exists an edge that connects vertex i with vertex j, and zero otherwise; V is an orthogonal matrix whose columns are normalized eigenvectors of A, and D is a diagonal matrix containing the eigenvalues λ_i of matrix A. The set of the eigenvalues found on the diagonal of matrix D is called the spectrum of A, and hence the common name for the family of methods.

One of the most well-known properties of eigendecomposition, and the one that has attracted researchers' attention for the purpose of solving inexact graph matching task in the first place, is that an eigenvalue spectrum of a matrix is invariant with respect to similarity transformations, i.e. for any non-singular matrix P, the product matrix PAP^{-1} has the same eigenvalues as A. From the view point of the graph matching problem, this means that the derived spectrum of a graph represented by its adjacency matrix is not affected by any arbitrary vertex reorderings, whose influence, or rather lack thereof, is in essence captured by the above vertex permutation matrix P.

Still, regardless of however elegant the possible graph matching problem solutions seemed at first in terms of graph eigenspectra, it was proven early on that the spectra of graphs are not unique. An obvious example that dates back to as far as 1957 was discovered by Collatz and Sinogowitz[2], and is shown in Figure 1.

The above figure depicts two non-isomorphic graphs, that are nevertheless co-spectral, i.e. the sets of eigenvalues of their adjacency matrices are identical, and therefore the two graphs cannot be distinguished by relying exclusively on their spectra. Furthermore, Schwenk[9] demonstrated that as the number of vertices gets large, the probability of occurrence of a non-isomorphic co-spectral subgraph pair in any two graphs being compared asymptotically approaches unity. This means that pure spectral methods based solely on eigenvalues are generally not rich enough to fully represent graph structure variability.

Naturally, the above arguments do not add support for spectral methods. However, it is not so difficult to see that this lack of uniqueness can be easily overcome by using graph spectra together with the set of associated eigenvectors, or even by relying on the eigenvectors alone (see Equation 1). Another drawback usually attributed to the spectral methods is that they are not extendible to matching graphs of different sizes. For example, the method developed by Umeyama[15] applies only for graphs of the same size. Nevertheless, these shortcomings can be eliminated by applying normalization and projection operations - the topic of the following section.

2.2 Normalizations and Projections

Subspace projection methods, in the principal component analysis (PCA) literature, are conventionally used to reduce the dimensionality of data, while minimizing the information loss due to the decreased number of dimensions. It is performed in the following way. The dataset covariance matrix Σ is first decomposed into the familiar eigenvalue/eigenvector matrix product (see Eq. 1):

$$\Sigma = U \Lambda U^T \tag{2}$$

where U is a matrix of eigenvectors ("principal components" of the data), and Λ is a diagonal matrix of eigenvalues. The original data is then projected onto a smaller number of the most important (i.e., associated with the largest eigenvalues) principal components as specified in the below equation (and thus, the data's dimensionality is reduced):

$$\hat{x} = U_k^T x \tag{3}$$

Here, $\hat{x}$ is the computed projection, U_k^T is the matrix of k principal components in a transposed form, and x is an item from the original data.

Taking the very same approach, we can project vertex connectivity data from a graph adjacency matrix onto a smaller set of its most important eigenvectors. The projection coordinates obtained in this way would then represent the relational properties of individual vertices relative to the others in the lower-dimensional eigenspace of a given graph. In this eigenvector subspace, structurally similar vertices or vertex groups would be located close to each other, which can be utilized for approximate comparison and matching of graphs.

However, in order to be able to use the outlined above projection method for graph matching, it is necessary to resolve the following issues: first, how many dimensions to choose for vertex eigenspace projections? Second, how to ensure the comparability of the derived projections for graphs with a different number of vertices?

The first is answered by the relative sizes of the eigenvalues associated with each dimension or eigenvector with non-zero eigenvalue signalling the redundancy of the associated subspaces. That is, for a given pair of graphs one should choose the k most important eigenvectors as the projection components, where k is the smaller value of the ranks of adjacency matrices of the two graphs being compared[1] , i.e. $k = min(rank(A_{Graph_1}), rank(A_{Graph_2}))$.

As for the second question, the empirical evidence suggests that an extra step of renormalization of the projections may suffice. Here, the idea is that for the purpose of comparing two arbitrary graphs we need not consider the values of the projections as such, but instead should look at how they are positioned and oriented *relative to each other* in their eigenvector subspace. That is, if

[1] However, in order to make the following examples more illustrative, without a loss of generality in the further discussion we will use only 2-dimensional projections, which can be easily depicted in the 2D plane.

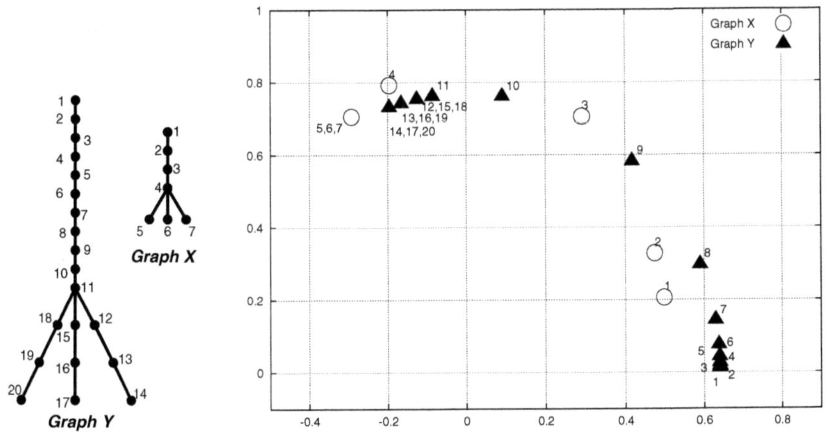

(a) Graphs X, Y (b) Projections of graphs X and Y into 2D eigenvector subspace

Fig. 2. Example 1: Graphs and their projections

projections are themselves viewed as vectors, we disregard their magnitudes, while only paying attention to their direction and orientation. And this is exactly what projection coordinate renormalization helps us to do: in the end all of the projections are unit-length vectors that can only be distinguished by their orienation, and not by their length. In addition to that, we also carry out a dominant sign correction of the projection coordinates of either of the two graphs being matched so as to align one set of graph vertex projections against the other. This corresponds to setting the direction of the axes in such a way to result in the most compatible alignment between the vertex data using the dominant sign test.

In order to provide an illustration for the described above propositions, let us consider an example with two graphs X and Y depicted in Figure 2(a). Although different in size, the two graphs are nevertheless quite similar to each other. In fact, one may see graph Y as an enlarged version of graph X. The result of projecting the two graphs into the normalized 2D eigenvector subspace shown in Figure 2(b) demonstrates the following two important features of the proposed method: firstly, the projections of vertices of both graphs follow a similar pattern, which means that it is possible to determine overall structural similarity of graphs with different number of vertices, and secondly, one may also see (by examining the juxtaposition of the projected vertices of both graphs) that graph vertices with similar relational properties tend to get projected into the areas that are close to each other. These properties are quite valuable, and, as such, have the potential to prove useful in solving the graph matching problem. The latter conjecture is confirmed by the experimental results which show that an overall graph similarity can be estimated by comparing the vertex projection distributions with the aid of multi-dimensional extension of Kolmogorov-Smirnov (K-S) statistical test. However, the K-S test becomes a rather computationally expensive procedure if applied to high-dimensional data. Also, it does not help us much to resolve another important issue of the graph matching problem, namely, the one of recovering structurally similar vertex correspondence in a pair of graphs being compared. To this end, we use clustering methods - as follows.

2.3 Clustering in Graph Eigenspaces and Inexact Solutions to Subgraph Matching

This eigenvector subspace method allows us to determine the overall similarity of a pair of graphs by the positioning of the vertex projections of both graphs relative to each other. The only remaining step for solving the graph matching problem is to find the correspondence among the vertices that have similar relational properties. The main advantage of using clustering to solve this problem is that it can equally well discover correspondence relationships of various types, i.e. it is not limited to finding the best one-to-one matches of vertices from one graph to the other, but it can also identify the whole sub-graphs and vertex groups that possess similar structural properties.[2] In order to realize this, we deploy a standard agglomerative clustering routine with only two necessary modifications: first, the algorithm gives a higher priority for clustering the candidate vertex projections that belong to different graphs, rather than the same one; second, the clustering procedure stops as soon as all of the vertex projections have been associated with a certain cluster. Once the clustering is completed, a simple customized cluster validity index that takes into account the number of obtained clusters and their quality based on the Dice[3] coefficient formula[3] is used to measure the similarity (or distance) of a pair of graphs. Figure 3 illustrates the result of vertex projection clustering (Figure 3(b)) of two sample graphs Z and T with 18 and 6 vertices respectively, that recovers a natural correspondence among the groups of vertices in these two graphs (Figure 3(a)).

3 Application

For the purpose of initial testing the proposed graph matching method, two application areas were chosen: first being the matching of shapes represented by shock trees, and second - information retrieval with sentence parse tree analysis.

In the first application area, shock tree matching (described in detail in [13,4,12]), a small set of shapes documented in [8] was used. The graphical representation of the dataset and the similarity matrix for the tested shapes, as calculated according to the aforementioned cluster validity index for measuring the similarity among the clustered eigenvector subspace projections, are shown

[2] This quality can be very important when the two graphs have substantially different number of vertices.

[3] analogous to the well-known "intersection-over-union" measure of set similarity.

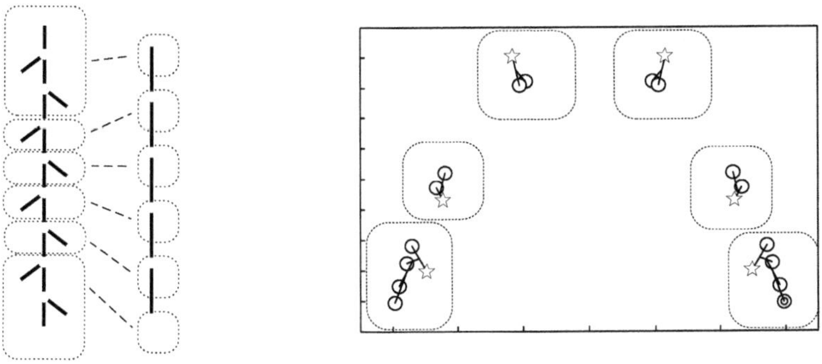

(a) Graphs Z, T

(b) Clustered projections of graphs Z and T (the formed clusters of vertex correspondence are circled)

Fig. 3. Example 2: Clustering of vertex projections of sample graphs Z and T

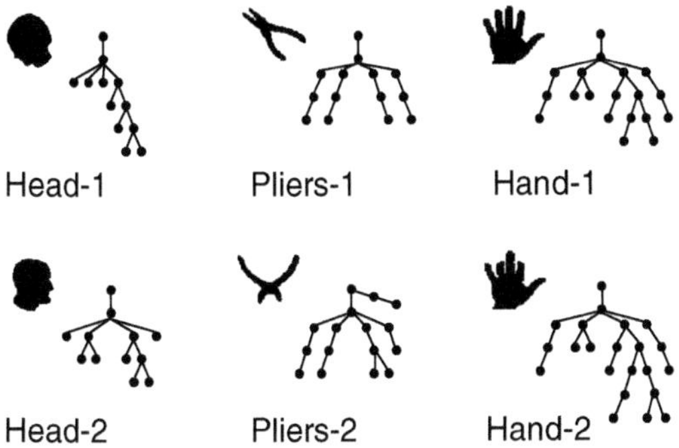

Fig. 4. The subset of shapes and their shock graph representations, from [8]

in Figure 4 and Table 1 respectively (where the best matching shape similarity values are in bold font).

In the second application area, a subset of queries and documents from ADI text collection (ftp://ftp.cs.cornell.edu/pub/smart/adi/) was parsed into a group of dependency trees. Subsequently, a standard keyword-based information retrieval system[7] was modified so as, on one hand, to restrict the keyword matching process only to words that have similar structural properties in both query and document sentence dependency trees, and, on the other hand, to allow for more flexibility in individual word comparisons by letting a direct within-cluster part of speech correspondence count as a partial match. As a result, the overall performance indicators improved, which can be illustrated by the following example.

Table 1. The similarity matrix obtained for the subset of shapes shown in Figure 4

	Head-1	Head-2	Pliers-1	Pliers-2	Hand-1	Hand-2
Head-1		**0.5536**	0.1936	0.4392	0.2000	0.1280
Head-2	**0.5536**		0.2373	0.3978	0.3133	0.1270
Pliers-1	0.1936	0.2373		**0.4857**	0.2087	0.2006
Pliers-2	0.4392	0.3978	**0.4857**		0.2126	0.1612
Hand-1	0.2000	0.3133	0.2087	0.2126		**0.3777**
Hand-2	0.1280	0.1270	0.2006	0.1612	**0.3777**	

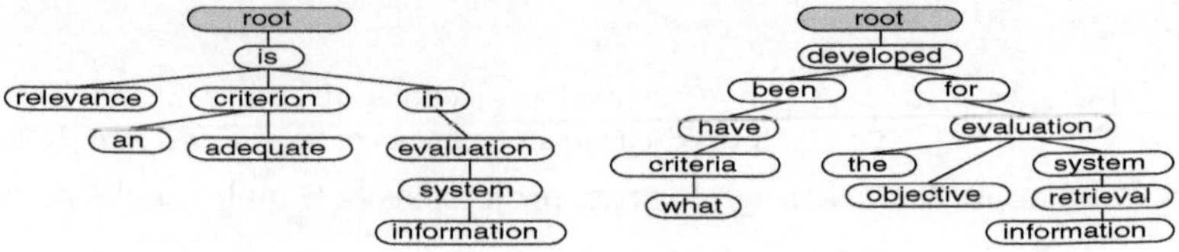

(a) Document sentence parse tree. (b) Query sentence parse tree.

Fig. 5. Parse trees of sample sentences from document 27 and query 13

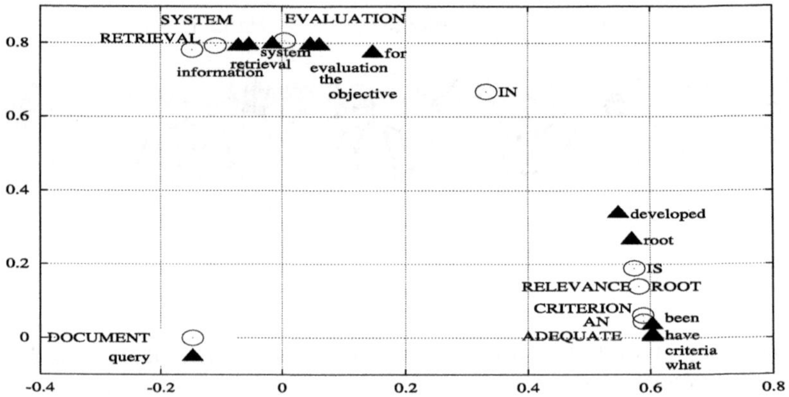

Fig. 6. Comparison of two sentence parse tree projections: an application in natural language processing

Both query 13 and document 27 in the ADI text collection have a substantial keyword overlap[4] , however, a conventional keyword-based information retrieval system does not recognize this pair as the best match. Instead such a system

[4] The sample sentences considered are: *"What criteria have been developed for the objective evaluation of information retrieval and dissemination systems?"*, and *"Is relevance an adequate criterion in retrieval system evaluation?"*.

ranks high some other "relevant" documents which share a lot of keywords with the query, even though these common keywords are quite inappropriate if one considers their context conveyed by the sentence syntactic structure. The use of the proposed eigenvector subspace projection method allowed us to take into account the parse tree structure in addition to the keyword information, which lead to improved results. The parse trees (after conjunction expansion and prepositional post-modifier normalization) of the sample sentences from the above document and query are depicted in Figure 5; their projections, that were used to estimate syntactic structural similarity of individual keywords, are shown in Figure 6.

4 Conclusion

In this paper, we have described an approach for inexact multisubgraph matching using the technique of projection of graph vertices into the eigenspaces of graphs in conjunction with standard clustering methods. The two most important properties of the proposed approach are, first, its ability to match graphs of considerably different sizes, and, second, its power to discover correspondence relationships among subgraphs and groups of vertices, in addition to the "one-to-one" type of vertex correspondence that the majority of previously developed solutions of the graph matching problem mostly focused on. In addition to that, we have also explored two potential areas for practical application for the described approach - matching of shapes represented by shock trees and natural language processing, and obtained results encouraging further research of the method.

References

1. H. Bunke. Recent advances in structural pattern recognition with application to visual form analysis. *IWVF4, LNCS*, 2059:11–23, 2001. 133
2. L. Collatz and U. Sinogowitz. Spektren endlicher grafen. *Abh. Math. Sem. Univ. Hamburg*, 21:63–77, 1957. 135
3. L. Dice. Measures of the amount of ecologic association between species. *Ecology*, 26:297–302, 1945. 138
4. P. Dimitrov, C. Phillips, and K. Siddiqi. Robust and efficient skeletal graphs. *Conference on Computer Vision and Pattern Recognition*, june 2000. 138
5. X. Jiang, A. Munger, and H. Bunke. On median graphs: properties, algorithms, and applications. *IEEE Trans. PAMI*, 23(10):1144–1151, October 2001. 133
6. B. Luo and E. Hancock. Structural graph matching using the em algorithm and singular value decomposition. *IEEE Trans. PAMI*, 23(10):1120–1136, October 2001. 134
7. N. Maloy. Successor variety stemming: variations on a theme. 2000. project report (unpublished). 139
8. M. Pelillo, K. Siddiqi, and S. Zucker. Matching hierarchical structures using association graphs. *IEEE Trans. PAMI*, 21(11), November 1999. 138, 139
9. A. Schwenk. *Almost all trees are cospectral.* Academic Press, New York - London, 1973. 135

10. L. Shapiro and J. Brady. Feature-based correspondence - an eigenvector approach. *Image and Vision Computing*, 10:268–281, 1992. 134
11. A. Shokoufandeh and S. Dickinson. A unified framework for indexing matching hierarchical shape structures. *IWVF4, LNCS*, 2059:67–84, 2001. 134
12. K. Siddiqi, S. Bouix, A. Tannebaum, and S. Zucker. Hamilton-jacobi skeletons. *To appear in International Journal of Computer Vision.* 138
13. K. Siddiqi, A. Shokoufandeh, S. Dickinson, and S. Zucker. Shock graphs and shape matching. *International Journal of Computer Vision*, 30:1–24, 1999. 138
14. S. Tirthapura, D. Sharvit, P. Klein, and B. Kimia. Indexing based on edit-distance matching of shape graphs. *Multimedia Storage and Archiving Systems III*, 3527(2):25–36, 1998. 134
15. S. Umeyama. An eigen decomposition approach to weighted graph matching problems. *IEEE Trans. PAMI*, 10:695–703, 1998. 134, 135

Optimal Lower Bound for Generalized Median Problems in Metric Space

Xiaoyi Jiang[1] and Horst Bunke[2]

[1] Department of Electrical Engineering and Computer Science
Technical University of Berlin
Franklinstrasse 28/29, D-10587 Berlin, Germany
jiang@cs.tu-berlin.de
[2] Department of Computer Science, University of Bern
Neubrückstrasse 10, CH-3012 Bern, Switzerland
bunke@iam.unibe.ch

Abstract. The computation of generalized median patterns is typically an NP-complete task. Therefore, research efforts are focused on approximate approaches. One essential aspect in this context is the assessment of the quality of the computed approximate solutions. In this paper we present a lower bound in terms of a linear program for this purpose. It is applicable to any pattern space. The only assumption we make is that the distance function used for the definition of generalized median is a metric. We will prove the optimality of the lower bound, i.e. it will be shown that no better one exists when considering all possible instances of generalized median problems. An experimental verification in the domain of strings and graphs shows the tightness, and thus the usefulness, of the proposed lower bound.

1 Introduction

The concept of average, or mean, is useful in various contexts. In sensor fusion, multisensory measurements of some quantity are averaged to produce the best estimate. Averaging the results of several classifiers is used in multiple classifier systems in order to achieve more reliable classifications. In clustering and machine learning, a typical task is to represent a set of (similar) objects by means of a single prototype. Interesting applications of the average concept have been demonstrated in dealing with shapes [6], binary feature maps [10], 3D rotation [3], geometric features (points, lines, or 3D frames) [15], brain models [4], anatomical structures [17], and facial images [13].

In structural pattern recognition symbolic structures, such as strings, trees, or graphs, are used for pattern representation. One powerful tool in dealing with these data structures is provided by the generalized median. Given a set S of input patterns, the generalized median is a pattern that has the smallest sum of distances to all patterns in S (see Section 2 for a formal definition).

T. Caelli et al. (Eds.): SSPR&SPR 2002, LNCS 2396, pp. 143–151, 2002.
© Springer-Verlag Berlin Heidelberg 2002

The computation of generalized median of symbolic structures is typically an NP-complete task. Therefore, research efforts are focused on approximate algorithms. One essential aspect in this context is the assessment of the quality of the computed approximate solutions. Since the true optimum is unknown, the quality assessment is not trivial in general. In this paper we present an optimal lower bound in terms of a linear program for this purpose. It is applicable to any pattern space. The only assumption we make is that the distance function used for the definition of generalized median is a metric.

The outline of the paper is as follows. In Section 2 we first introduce the generalized median of patterns. Then, we present the LP-based lower bound and discuss its optimality in Sections 3 and 4. The results of an experimental verification in the domains of strings and graphs are reported in Section 5 to show the usefulness of the lower bound. And finally, some discussion conclude the paper.

2 Generalized Median of Patterns

Assume that we are given a set S of patterns in an arbitrary representation space U and a distance function $d(p,q)$ to measure the dissimilarity between any two patterns $p, q \in U$. An important technique for capturing the essential information of the given set of patterns is to find a pattern $\overline{p} \in U$ that minimizes the sum of distances to all patterns from S, i.e.

$$\overline{p} = \arg\min_{p \in U} \sum_{q \in S} d(p,q).$$

Pattern $\overline{p}$ is called a *generalized median* of S. If the search is constrained to the given set S, the resultant pattern

$$\hat{p} = \arg\min_{p \in S} \sum_{q \in S} d(p,q)$$

is called a *set median* of S. Note that neither the generalized median nor the set median is necessarily unique.

Independent of the underlying representation space we can always find the set median of N patterns by means of $N(N-1)/2$ distance computations. The computational burden can be reduced if the distance function is a metric [9]. For non-metric distance functions an approximate set median search algorithm has been reported recently [12]. Note that the generalized median is the more general concept and therefore usually a better representation of the given patterns than the set median.

If U is the universe of real numbers and the distance function $d(p,q)$ is the absolute (squared) difference of p and q, then the generalized median simply corresponds to the scalar median (mean) known from statistics. Scalar median represents a powerful technique for image smoothing. Its extension to vector spaces [1,2] provides a valuable image processing tool for multispectral/color images and optical flow.

In dealing with strings, the popular Levenshtein edit distance is usually used. Under this distance function the set median string problem is solvable in polynomial time. However, the computation of generalized median strings turns out to be NP-complete [5,16]. Several approximate approaches have been reported in the literature; see [8] for a discussion. In [7] the concept of generalized median graphs is defined based on graph edit distance. Also here we are faced with an NP-complete computation problem.

An approximate computation method gives us a solution $\tilde{p}$ such that

$$\mathrm{SOD}(\tilde{p}) = \sum_{q \in S} d(\tilde{p}, q) \geq \sum_{q \in S} d(\overline{p}, q) = \mathrm{SOD}(\overline{p})$$

where SOD stands for sum of distances and $\overline{p}$ represents the (unknown) true generalized median. The quality of $\tilde{p}$ can be measured by the difference $\mathrm{SOD}(\tilde{p}) - \mathrm{SOD}(\overline{p})$. Since $\overline{p}$ and $\mathrm{SOD}(\overline{p})$ are unknown in general, we resort to a lower bound $\Gamma \leq \mathrm{SOD}(\overline{p})$ and measure the quality of $\tilde{p}$ by $\mathrm{SOD}(\tilde{p}) - \Gamma$. Note that the relationship

$$0 \leq \Gamma \leq SOD(\overline{p}) \leq SOD(\tilde{p})$$

holds. Obviously, $\Gamma = 0$ is a trivial, and also useless, lower bound. We thus require Γ to be as close to $\mathrm{SOD}(\overline{p})$ as possible. In the next two sections we present such a lower bound and prove its optimality (in a sense to be defined later). The tightness of the proposed lower bound will be experimentally verified in Section 5 in the domain of strings and graphs.

It is worth pointing out that a lower bound is not necessarily needed to compare the relative performance of different approximate methods. But it is very useful to indicate the closeness of approximate solutions to the true optimum. Such an absolute performance comparison is actually the ultimate goal of performance evaluation.

3 LP-Based Lower Bound

We assume that the distance function $d(p, q)$ be a metric. Let the set S of input patterns be $\{q_1, q_2, \ldots, q_n\}$. The generalized median $\overline{p}$ is characterized by:

$$\begin{aligned} &\text{minimize } \mathrm{SOD}(\overline{p}) = d(\overline{p}, q_1) + d(\overline{p}, q_2) + \cdots + d(\overline{p}, q_n) \text{ subject to} \\ &\forall i, j \in \{1, 2, \ldots, n\}, i \neq j, \begin{cases} d(\overline{p}, q_i) + d(\overline{p}, q_j) \geq d(q_i, q_j) \\ d(\overline{p}, q_i) + d(q_i, q_j) \geq d(\overline{p}, q_j) \\ d(\overline{p}, q_j) + d(q_i, q_j) \geq d(\overline{p}, q_i) \end{cases} \\ &\forall i \in \{1, 2, \ldots, n\}, \ d(\overline{p}, q_i) \geq 0 \end{aligned}$$

Note that the constraints except the last set of inequalities are derived from the triangular inequality of the metric $d(p, q)$. By defining n variables x_i, $i = 1, 2, \ldots, n$, we replace $d(\overline{p}, q_i)$ by x_i and obtain the linear program LP:

minimize $x_1 + x_2 + \cdots + x_n$ subject to

$$\forall i,j \in \{1,2,\ldots,n\}, i \neq j, \begin{cases} x_i + x_j \geq d(q_i, q_j) \\ x_i + d(q_i, q_j) \geq x_j \\ x_j + d(q_i, q_j) \geq x_i \end{cases}$$

$$\forall i \in \{1,2,\ldots,n\},\ x_i \geq 0$$

If we denote the solution of LP by Γ, then we have:

Theorem 1. *The true generalized median $\overline{p}$ satisfies $\Gamma \leq SOD(\overline{p})$. That is, Γ is a lower bound for $SOD(\overline{p})$.*

Proof: In the initial characterization the quantities $d(\overline{p}, q_i)$ are dependent of each other. The linear program LP results from replacing $d(\overline{p}, q_i)$ by x_i and is defined in contrast by n totally independent variables x_i. Consequently, LP poses less conditions than the initial characterization and its solution Γ thus must be smaller than or equal to SOD($\overline{p}$). QED

The linear program LP has $\frac{3n^2-n}{2}$ inequality constraints and we may apply the popular simplex algorithm [14] to find out the solution. Note that, despite its exponential worst-case computational complexity, the simplex algorithm turns out to be very efficient in practice and is used to routinely solve large-scale linear programming problems.

4 Optimality Issue

For a fixed n value, any set S of n patterns specifies $N = \frac{n(n-1)}{2}$ distances $d(p,q)$, $p, q \in S$, and can be considered as a point in the N-dimensional real space $\Re^N$. Due to the triangular inequality required by a metric, all possible sets of n patterns only occupy a subspace $\Re^N_*$ of $\Re^N$. Abstractly, any lower bound is therefore a function $f : \Re^N_* \rightarrow \Re$. The lower bound Γ derived in the last section is such a function.

Does a lower bound exist that is tighter than Γ? This optimality question is interesting from both a theoretical and a practical point of view. The answer and the implied optimality of the LP-based lower bound Γ is given by the following result.

Theorem 2. *There exists no lower bound that is tighter than Γ.*

Proof: Given a point $b \in \Re^N_*$, we denote the solution of the corresponding linear program LP by $(x_1, x_2, \ldots, x_n)$. We construct a problem instance of $n+1$ abstract patterns $q_1, q_2, \ldots, q_n, q_{n+1}$. The $\frac{n(n-1)}{2}$ distances $d(q_i, q_j)$, $1 \leq i, j \leq n$, are taken from the coordinates of b. The remaining distances are defined by $d(q_{n+1}, q_i) = x_i, 1 \leq i \leq n$. The distance function d is clearly a metric. Now we compute the generalized median $\overline{p}$ of $\{q_1, q_2, \ldots, q_n\}$. Since $\Gamma = x_1 + x_2 + \cdots + x_n$ is a lower bound, we have SOD($\overline{p}$) $\geq \Gamma$. On the other hand, the pattern q_{n+1} satisfies:

$$\begin{aligned} \text{SOD}(q_{n+1}) &= d(q_{n+1}, q_1) + d(q_{n+1}, q_2) + \cdots + d(q_{n+1}, q_n) \\ &= x_1 + x_2 + \cdots + x_n \\ &= \Gamma \end{aligned}$$

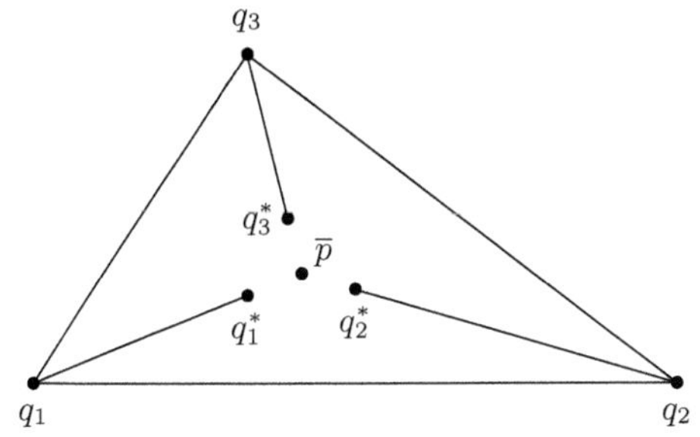

Fig. 1. The lower bound Γ cannot be reached by the generalized median $\overline{p}$

Consequently, q_{n+1} is a generalized median of $\{q_1, q_2, \ldots, q_n\}$. This means that, for each point in $\Re_*^N$, we can always construct a problem instance where the lower bound Γ is actually reached by the generalized median. Accordingly, no lower bound can exist that is more tight than the LP-based lower bound Γ. QED

At this point two remarks are in order. For most problems in practice it is likely that the lower bound Γ cannot be reached by the generalized median. The first reason is a fundamental one and is illustrated in Figure 1, where we consider points in the plane. The distance function is defined to be the Euclidean distance of two points. Let $\overline{p}$ be the true generalized median of q_1, q_2, and q_3. Then, $x_i = |q_i\overline{p}|$, $i = 1, 2, 3$, satisfy the constraints of the linear program LP. Now we select a point q_i^* on the line segment $q_i\overline{p}$ such that $|q_i^*\overline{p}| = \epsilon$ (an infinitely small number). Due to the small amount of ϵ, $x_i^* = |q_i q_i^*|$ satisfy the constraints of LP as well. But in this case we have $x_1^* + x_2^* + x_3^* < x_1 + x_2 + x_3 = \mathrm{SOD}(\overline{p})$. As a consequence, the solution of LP, i.e. the lower bound Γ, is constrained by:

$$\Gamma \;\leq\; x_1^* + x_2^* + x_3^* \;<\; \mathrm{SOD}(\overline{p})$$

and therefore not reached by the generalized median $\overline{p}$. Fundamentally, this example illustrates the decoupled nature of the quantities x_i in LP in contrast to $d(\overline{p}, q_i)$ in the original problem of generalized median computation. By doing this, however, the solution x_i of LP may not be physically realizable through a single pattern $\overline{p}$.

The special property of a concrete problem may also imply that the lower bound Γ is not reached by the generalized median. We consider again points in the plane, but now with integer coordinates only. The distance function remains the Euclidean distance. An example is shown in Figure 2 with four points q_1, q_2, q_4, and q_4. The lower bound Γ turns out to be $2\sqrt{34}$ corresponding to $x_1 = x_2 = x_3 = x_4 = \frac{\sqrt{34}}{2}$. This lower bound is satisfied by $p(\frac{5}{2}, \frac{3}{2})$, which is unfortunately not in the particular space under consideration. Any point with integer coordinates will result in a SOD value larger than Γ.

It is important to point out that Theorem 2 only implies that we cannot specify a better lower bound than the solution of LP, when considering *all* possible instances of generalized median problems. An improved lower bound may

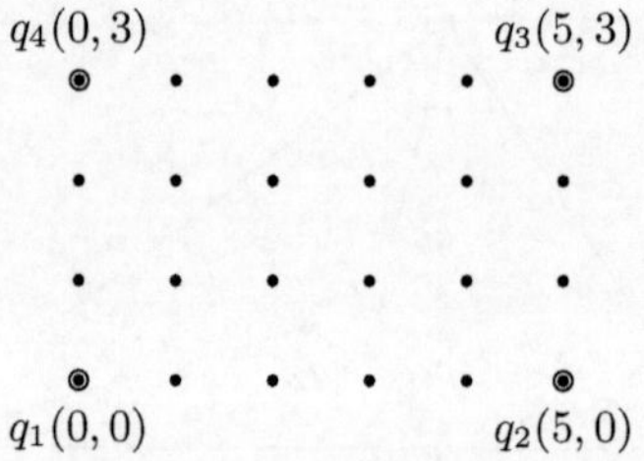

Fig. 2. The point $p(\frac{5}{2}, \frac{3}{2})$ reaching the lower bound is not in the problem space

still be computed for a *particular* problem instance. For the problem in Figure 2, for example, the constraint $x_1 + x_3 \geq d(q_1, q_3) = \sqrt{34}$ can be replaced by $x_1 + x_3 \geq \sqrt{34} + \Delta$ for some $\Delta > 0$. The reason is that no point with integer coordinates lies on the line segment q_1q_3 and the corresponding constraint can thus be made tighter. The constraint $x_2 + x_4 \geq d(q_2, q_4) = \sqrt{34}$ can be modified in a similar manner. As a final result, the modified constraints may lead to a tighter lower bound.

5 Experimental Verification

A lower bound is only useful if it is close to $\mathrm{SOD}(\overline{p})$ where $\overline{p}$ represents the (unknown) true generalized median pattern. In this section we report the results of an experimental verification in the domain of strings and graphs to show the tightness, and thus the usefulness, of the proposed lower bound. We used the MATLAB package to solve the linear program LP.

5.1 Median Strings

The median concept can be used in OCR to combine multiple classification results for achieving a more reliable final classification [11]. In doing so we may obtain multiple classification results either by applying different classifiers to a single scan of a source text or by applying a single classifier to multiple scans of the text.

To verify the usefulness of the LP-based lower bound in this context we conducted a simulation by artificially distorting the following text which consists of 448 symbols (including spaces):

> There are reports that many executives make their decisions by flipping a coin or by throwing darts, etc. It is also rumored that some college professors prepare their grades on such a basis. Sometimes it is important to make a completely 'unbiased' decision; this ability is occasionally useful in computer algorithms, for example in situations where a fixed decision made each time would cause the algorithm to run more slowly.
> Donald E. Knuth

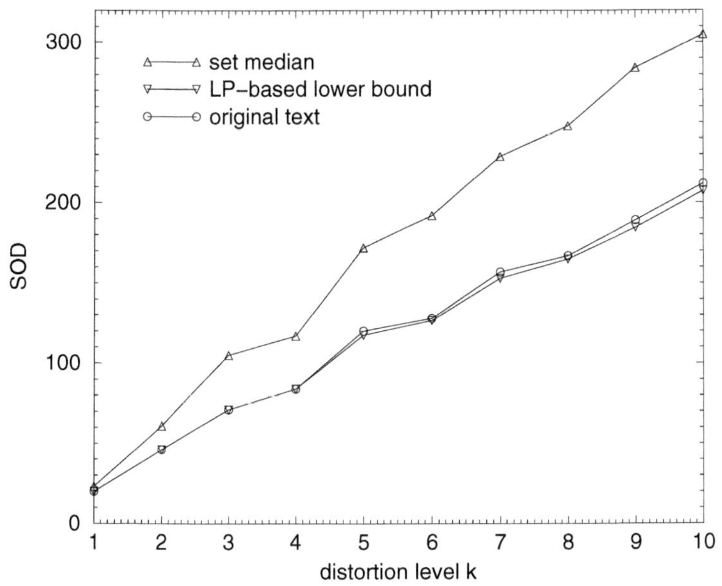

Fig. 3. Verification of lower bound for strings

Totally, ten distortion levels are used, producing $k\%$ $(k = 1, 2 \cdots, 10)$ letters in the text to be changed. For each k, five distorted samples of the text are generated. We use the Levensthein edit distance and set the insertion, deletion, and substitution cost each to be one.

Figure 3 summarizes the results of this test series. As a comparison basis, SOD of the original text is also given. The SOD of the (unknown) true generalized median string $\overline{p}$ must be between this curve and the lower bound curve. Clearly, the LP-based lower bound is a very good estimate of SOD($\overline{p}$). In addition the results confirm that the generalized median string is a more precise abstraction of a given set of strings than the set median. It has a significantly smaller SOD value, which corresponds to the representation error.

5.2 Median Graphs

The concept of generalized median graphs was introduced in [7]. We study the LP-based lower bound in this domain by means of random graphs generated by distorting a given initial graph. The initial graph g_0 contains k nodes and $2k$ edges. The node and edge labels are taken from $\{A, B, C, D, E\}$ and $\{F\}$, respectively. Both the graph structure and the labeling of g_0 are generated randomly. The distortion process first randomly changes the labels of 50% of the nodes in g_0. Then, up to two nodes are inserted or deleted in g_0. In case of an insertion the new node is randomly connected to one of the nodes in g_0. If a node in g_0 is deleted, all its incident edges are deleted as well. This way a col-

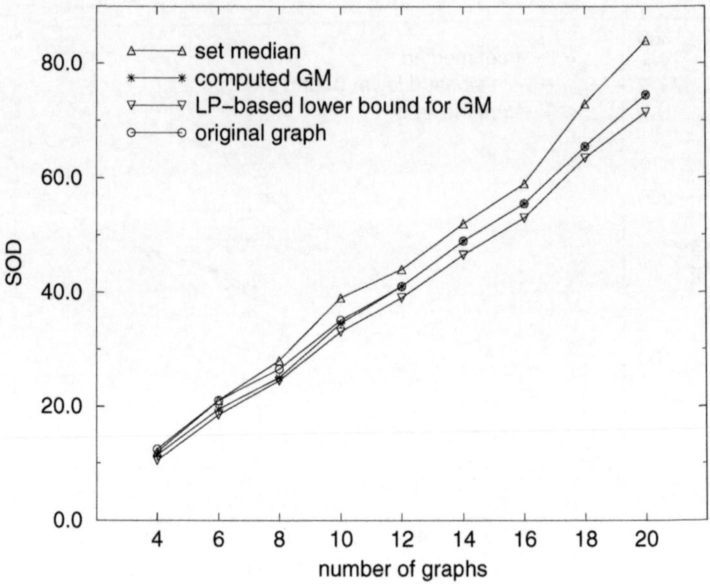

Fig. 4. Verification of lower bound for graphs

lection of 20 distorted graphs are generated for g_0 associated with a particular k value. Based on this procedure, we conducted a series of experiments by using $n \in \{4, 6, \ldots, 20\}$ out of the 20 graphs to test the lower bound. The distance function of two graphs is defined in terms of graph edit operations; see [7] for details.

The results of this test series for $k = 6$ are summarized in Figure 4. As an upper bound for $\text{SOD}(\overline{g})$ of the (unknown) true generalized median graph $\overline{g}$, we give the SOD of the original graph g_0 and an approximate solution found by the method from [7]. Clearly, $\text{SOD}(\overline{g})$ must be between the minimum of these two curves and the lower bound curve. Also here the LP-based lower bound demonstrates a high predication accuracy.

6 Conclusions

The computation of generalized median patterns is typically an NP-complete task. Therefore, research efforts are focused on approximate approaches. One essential aspect in this context is the assessment of the quality of the computed approximate solutions. In this paper we have presented an optimal lower bound in terms of a linear program for this purpose. It is applicable to any metric pattern space. An experimental verification in the domain of strings and graphs has shown the tightness, and thus the usefulness, of the proposed lower bound.

Acknowledgments

The authors would like to thank J. Crisik for valuable discussions on the topic of this paper.

References

1. J. Astola, P. Haavisto, and Y. Neuvo, Vector median filters, Proceedings of the IEEE, 78(4): 678–689, 1990. 144
2. F. Bartolini, V. Cappellini, C. Colombo, and A. Mecocci, Enhancement of local optical flow techniques, Proc. of 4th Int. Workshop on Time Varying Image Processing and Moving Object Recognition, Florence, Italy, 1993. 144
3. C. Gramkow, On averaging rotations, Int. Journal on Computer Vision, 42(1/2): 7–16, 2001. 143
4. A. Guimond, J. Meunier, and J.-P. Thirion, Average brain models: A convergence study, Computer Vision and Image Understanding, 77(2): 192–210, 2000. 143
5. C. de la Higuera and F. Casacuberta, Topology of strings: Median string is NP-complete, Theoretical Computer Science, 230(1-2): 39–48, 2000. 145
6. X. Jiang, L. Schiffmann, and H. Bunke, Computation of median shapes, Proc. of 4th. Asian Conf. on Computer Vision, 300–305, Taipei, 2000. 143
7. X. Jiang, A. Münger, and H. Bunke, On median graphs: Properties, algorithms, and applications, IEEE Trans. on PAMI, 23(10): 1144–1151, 2001. 145, 149, 150
8. X. Jiang, H. Bunke, and J. Csirik, Median strings: A review, 2002. (submitted for publication) 145
9. A. Juan and E. Vidal, Fast median search in metric spaces, in A. Amin and D. Dori (eds.), *Advances in Pattern Recognition*, Springer-Verlag, 905–912, 1998. 144
10. T. Lewis, R. Owens, and A. Baddeley, Averaging feature maps, Pattern Recognition, 32(9): 1615–1630, 1999. 143
11. D. Lopresti and J. Zhou, Using consensus sequence voting to correct OCR errors, Computer Vision and Image Understanding, 67(1): 39-47, 1997. 148
12. L. Mico and J. Oncina, An approximate median search algorithm in non-metric spaces, Pattern Recognition Letters, 22(10): 1145–1151, 2001. 144
13. A. J. O'Toole, T. Price, T. Vetter, J. C. Barlett, and V. Blanz, 3D shape and 2D surface textures of human faces: The role of "averages" in attractiveness and age, Image and Vision Computing, 18(1): 9–19, 1999. 143
14. C. H. Papadimitriou and K. Steiglitz, *Combinatorial Optimization: Algorithms and Complexity*, Prentice-Hall, Inc., 1982. 146
15. X. Pennec and N. Ayache, Uniform distribution, distance and expectation problems for geometric features processing, Journal of Mathematical Imaging and Vision, 9(1): 49–67, 1998. 143
16. J. S. Sim and K. Park, The consensus string problem for a metric is NP-complete, Journal of Discrete Algorithms, 2(1), 2001. 145
17. K. Subramanyan and D. Dean, A procedure to average 3D anatomical structures, Medical Image Analysis, 4(4): 317–334, 2000. 143

Structural Description to Recognising Arabic Characters Using Decision Tree Learning Techniques

Adnan Amin

School of Computer Science, University of New South Wales
Sydney, 2052, Australia
amin@cse.unsw.edu.au

Abstract: Character recognition systems can contribute tremendously to the advancement of the automation process and can improve the interaction between man and machine in many applications, including office automation, cheque verification and a large variety of banking, business and data entry applications. The main theme of this paper is the automatic recognition of hand-printed Arabic characters using machine learning. Conventional methods have relied on hand-constructed dictionaries which are tedious to construct and difficult to make tolerant to variation in writing styles. The advantages of machine learning are that it can generalize over the large degree of variation between writing styles and recognition rules can be constructed by example.The system was tested on a sample of handwritten characters from several individuals whose writing ranged from acceptable to poor in quality and the correct average recognition rate obtained using cross-validation was 87.23%.

Keywords: Pattern Recognition, Arabic characters, Hand-printed characters, Parallel thinning, Feature extraction, Structural classification, Machine Learning C4.5

1 Introduction

Character recognition is commonly known as Optical Character Recognition (OCR) which deals with the recognition of optical characters. The origin of character recognition can be found as early as 1870 [1] while it became a reality in the 1950's when the age of computer arrived [2]. Commercial OCR machines and packages have been available since the mid 1950's. OCR has wide applications in modern society: document reading and sorting, postal address reading, bank cheque recognition, form recognition, signature verification, digital bar code reading, map interpretation, engineering drawing recognition, and various other industrial and commercial applications.

Much more difficult, and hence more interesting to researchers, is the ability to automatically recognize handwritten characters [3]. The complexity of the problem is greatly increased by noise and by the wide variability of handwriting as a result of the mood of the writer and the nature of the writing. Analysis of cursive scripts requires

T. Caelli et al. (Eds.): SSPR&SPR 2002, LNCS 2396, pp. 152-158, 2002.
© Springer-Verlag Berlin Heidelberg 2002

the segmentation of characters within the word and the detection of individual features. This is not a problem unique to computers; even human beings, who possess the most efficient optical reading device (eyes), have difficulty in recognizing some cursive scripts and have an error rate of about 4% on reading tasks in the absence of context [4].

Different approaches covered under the general term 'character recognition' fall into either the on-line or the off-line category, each having its own hardware and recognition algorithms.

Many papers have been concerned with Latin, Chinese and Japanese characters, However, although almost a third of a billion people worldwide, in several different languages, use Arabic characters for writing, little research progress, in both on-line and off-line has been achieved towards the automatic recognition of Arabic characters. This is a result of the lack of adequate support in terms of funding, and other utilities such as Arabic text database, dictionaries, etc..[5]

This paper proposes a structural method for the extraction of line and curve primitive features. Such features will then be represented in attribute/value form, which is then input to the inductive learning system, C4.5 [6], to generate a decision tree. This decision tree can then be used to predict the class of an unseen character.

Fig. 1 depicts a block diagram of the system.

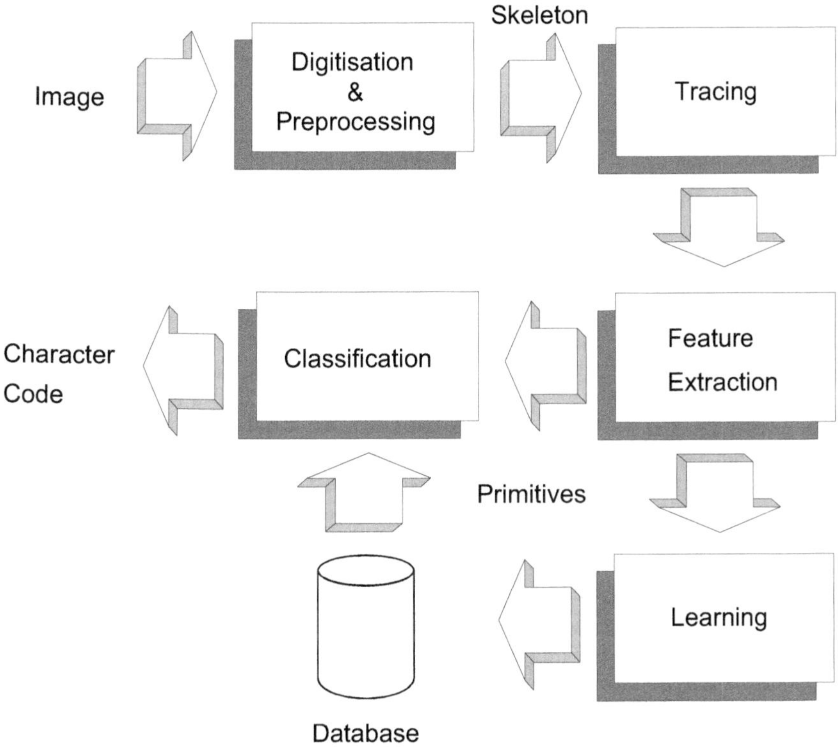

Fig. 1. Block diagram of the system

2 Feature Extraction

The characters are digitized using a 300 dpi scanner, pre-processed and thinned using one pass parallel thinning algorithm [7] The simple structural information such as lines, curves and loops that describe the characters are extracted by tracing the thinned image with some pre-defined primitives as shown in Fig. 2. A detailed description of the feature extraction process is given in [8]. The characters are then classified using a primary classifier and an exception classifier. The primary classifier uses a machine learning program (C 4.5) which uses an induction algorithm for the generation of the classification rules.

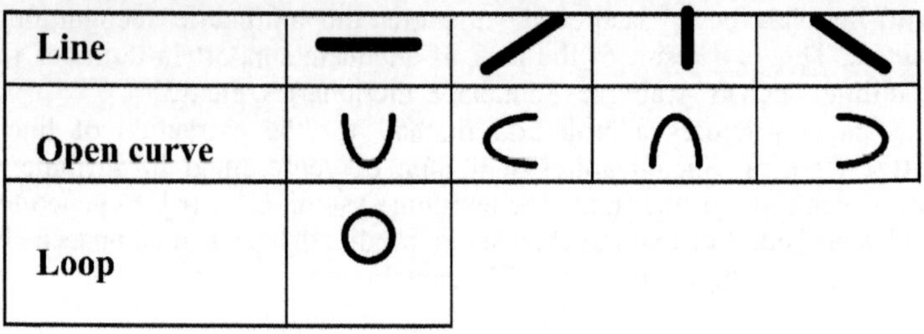

Fig. 2. Primitive features used in this system

3 Classification Using C4.5

C4.5 [6, 9], is an efficient learning algorithm that creates decision trees to represent classification rules. The data input to C4.5 form a set of examples, each labeled according to the class to which it belongs. The description of the example is a list of attribute/value pairs. A node is a decision tree represents a test on a particular attribute. Suppose its color, size and shape describe an object where color may have values red, green or blue; size may be large or small and shape may be circle or square. If the root node of the tree is labeled color then it may have three branches for each color value. Thus if we wish to test an objects color and it is red, the object descends the red branch. Leaf nodes are labeled with class names. This when an object reaches a leaf node, it is classified according to the name of the leaf node.

Building a decision tree proceeds as follows. The set of all examples forms an initial population. An attribute is chosen to split the population according to the attribute's values. Thus, if color is chosen then all red objects descend the red branch, all green objects descend the green branch, etc. Now the population has been divided into sub-populations by color. For each sub-population, another attribute value is chosen to split the sub-population. This continues as long as each population contains a mix of examples belonging to different classes. Once a uniform population has been obtained, a leaf node is created and labeled with the name of the class of the population.

The key to the success of a decision tree learning algorithm depends on the criterion used to select the attribute to use for splitting. If attribute is a strong indicator of an example's class value, it should appear as early in the tree as possible. Most decision tree learning algorithms use a heuristic for estimating the best attribute. In

C4.5, Quinlan uses a modified version of the entropy measure from information theory. For our purposes, it is sufficient to say that this measure yields a number between 0 and 1 where 0 indicates a uniform population and 1 indicates a population where there is equal likelihood of all classes being present. The splitting criterion seeks to minimize the entropy.

A further refinement is required to handle noisy data. Real data sets often contain examples that are misclassified or which have incorrect attribute values. Suppose decision tree building has constructed a node which contains 99 examples from class 1 and only one example from class 2. According to the algorithm presented above, a further split would be required to separate the 99 from the one. However, the one exception may be misclassified, causing an unnecessary split. Decision tree learning algorithms have a variety of methods for "pruning" unwanted subtrees. C4.5 grows a complete tree, including nodes created as a result of noise. Following initial tree building, the program proceeds to select suspect subtrees and prunes then, testing the new tree on a data set, which is separate from the initial training data. Pruning continues as long as the pruned trees yield more accurate classifications on the test data.

The C4.5 system requires two input files, the *names* and the *data* files. The *names* file contains the names of all the attributes used to describe the training examples and their allowed values. This file also contains the names of the possible classes. The classes are the Arabic words.

The C4.5 data files contains the attributes values for example objects in the format specified by the name file, where each example is completed by including the class to which it belongs.

Every Arabic character can be composed of at most five segments. A segment can be a complementary character, line, curve or loop. A 12-bit segment coding scheme is used to encode these segments. The scheme is depicted in Fig. 3.

The interpretation of the last seven bits of the code depends on the first five bits. However, only one of the first five should be 1. If zero or more than one bit is 1, the system automatically rejects the segment. For a segment to be misidentified, two bits would need to be incorrectly transmitted. Including the one that identifies the segment. Therefore, the encoding is relatively robust.

4 Experimental Results Using Cross-Validation

The first decision required in setting up a machine learning experiment is how the accuracy of the decision tree will be measured. The most reliable procedure is *cross-validation.* N-fold cross-validation refers to a testing strategy where the training data are randomly divided into *N* subsets. One of the *N* subsets is withheld as a *test* set and the decision tree is trained on the remaining *N-1,* subsets. After the decision tree has been built, its accuracy is measured by attempting to classify the examples in the test set. Accuracy simply refers to the percentage of examples whose class is correctly predicted by the decision tree. The error rate is the accuracy subtracted from 100%. To compensate for sampling bias, the whole process is repeated *N* times, where, in each iteration, a new test set is withheld and the overall accuracy is determined by averaging the accuracies found at each iteration.

Fu [10] describes the cross-validation process as "K-fold cross-validation (Stone[11]) repeats K times for a sample set randomly divided into disjoint subset, each time leaving one out for testing and others for training" The value of K = 10 is usually recommended [12]. Cross-validation requires that the original data set id split in K disjoint sets. At any one time, 90% of the data is used for training and the system is tested on the remaining 10%. At the end of 10 folds, all data has been tested. In every fold therefore the training and test patterns remain different. In brief, the cross-validation procedure for our purposes involves the following:

begin

1. For a total of 120 classes and 6000 patterns, interleave the patterns so that the pattern of class *j* is followed by a pattern of class *j*+1.
2. Segment data into K sets (k_1, k_2, ... k_K) of equal size, e.g. in our case for K = 10, we have 600 pattern in each set..
3. Train the C.4.5 with K-1 sets and test the system on the remaining one data set. Repeat this cycle K times, each time with a training set which is distinct from the test set.

   For i = 1 to K **do**
   Begin
   Train with data from all partitions k_n where $1 \leq n \leq K$, and $i \neq n$.
   Test with data from partition k_i.
   End;

4. Determine the recognition performance each time and take the average over a total of K performances.

end.

Table 1 shows the error rates performance using ten-fold cross-validation. It is important to note here that the system performs extremely well with recognition rates ranging between 85 % and 90% on different folds and the overall recognition is 87%. This is a very good performance taking into account the fact that we have a limited number of samples in each class.

Table 1. Error rates performance using ten fold cross validation

Fold	Error-Rate % Testing
1	12.35
2	13.89
3	14.23
4	10.14
5	11.56
6	12.90
7	11.67
8	13.76
9	14.20
10	11.45
Average	12.67

5 Conclusion

This paper presents a new technique for recognizing printed Arabic text and as indicated by the experiments performed, the algorithm resulted in a 87.23% recognition rate using the C4.5 machine learning system.

Moreover, the system used a structural approach for feature extraction (based on structure primitives such as curves, straight lines and loops in similar manner to which human begins describe characters geometrically). This approach is more efficient for feature extraction and recognition.

The study also shows that machine learning algorithms such as C4.5 are capable of learning the necessary features needed to recognize printed Arabic text achieving a best average recognition rate of 87.23% using ten fold cross-validation. The use of machine learning has removed the tedious task of manually forming rule-based dictionaries for classification of unseen characters and replaced it with an automated process which can cope with the high degree of variability which exists in printed and handwritten characters. This is very attractive feature and, therefore, further exploration of this application of machine learning is well worthwhile.

In the area of recognition, a structural approach has been previously used. This approach is sufficient to deal with ambiguity without using contextual information. This area remains undeveloped due to the immaturity of vital computational principles for Arabic character recognition.

References

1. V. Govindan and A. Shivaprasad, Character recognition- a review, Pattern Recognition, 23(7), pp. 671-683, 1990.
2. S. Mori, C. Y. Suen and K. Yamamoto, Historical review of OCR research and development, *Proceedings of the IEEE 80* (7), pp. 1029-1058, 1992.
3. E. Lecolinet E. and O. Baret, Cursive word recognition: Methods and strategies, Fundamentals in Handwriting Recognition, S. Impedovo, Ed. Springer-Verlag, 1994, pp. 235–263.
4. C. Y. Suen, R. Shingal and C. C. Kwan, Dispersion factor: A quantitative measurement of the quality of handprinted characters, Int. Conference of Cybernetics and Society, 1977, p.681–685.
5. A. Amin, Off_line Arabic characters Recognition: The State Of the Art, *Pattern Recognition 31(5),* 517-530, 1998.
6. Quilan J. R., C4.5 : programs for machine learning, San Mateo CA, Morgan Kauffman, 1993.
7. B.K. Jang and R.T. Chin, One-pass parallel thinning: analysis, properties, and quantitative evaluation, *IEEE Trans. Pattern Anal. Mach. Intell.* PAMI-14, pp. 1129-1140, 1992.
8. A. Amin, H. Al-Sadoun and S. Fischer, Hand-Printed Arabic Characters Recognition System Using an Artifial Network, *Pattern Recognition 29(4)*, pp. 663-675, 1996.
9. J. R. Quilan, Discovering rules for a large collection of examples, Edinburgh University Press, 1979.

10. L. Fu, Neural Networks in Computer Intelligence, McGraw-Hill, Singapore, pp. 331-348, 1994.
11. M. Stone, Cross-validatory choice and assessment of statistical predictions, *Journal of the Royal Statistical Society 36(1)*, pp. 111-147, 1974.
12. S. M. Weiss and G. E. Kulikowski, Computer systems that learn, Kauffman, CA, 1991.

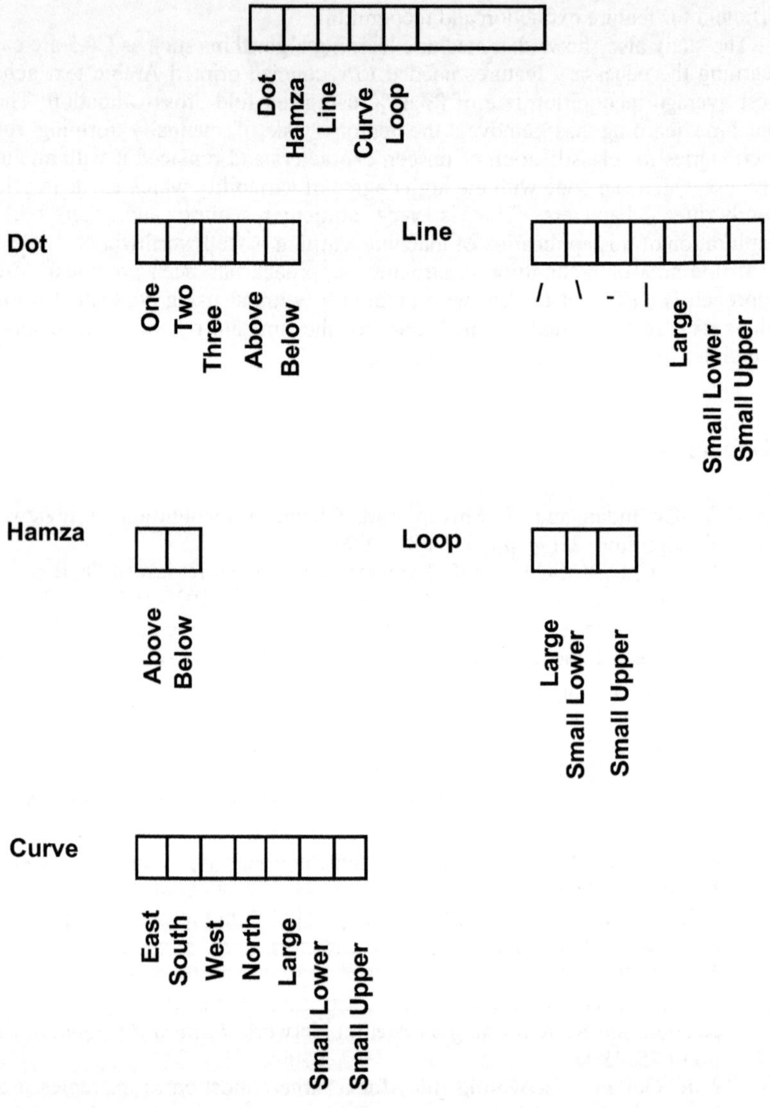

Fig. 3. Segment encoding for C4.5 input layer (each square cell holds a bit which acts as a flag)

Feature Approach for Printed Document Image Analysis

Jean Duong[1,2], Myrian Côté[1], and Hubert Emptoz[2]

[1] Laboratoire d'Imagerie Vision et Intelligence Artificielle (LIVIA)
Ecole de Technologie Supérieure, H3C 1K3 Montréal, (Québec) Canada
{duong,cote}@livia.etsmtl.ca
[2] Laboratoire de Reconnaissance de Formes et Vision (RFV)
Institut National des Sciences Appliquées (INSA) de Lyon
Bâtiment 403 (Jules Vernes)
20 Avenue Albert Einstein, 69621 Villeurbanne CEDEX
{duong,emptoz}@rfv.insa-lyon.fr

Abstract. This paper presents advances in zone classification for printed document image analysis. It firstly introduces entropic heuristic for text separation problem. Then a brief recall on existing texture and geometric discriminant parameters proposed in a previous research is done. Several of them are chosen and modified to perform statistical pattern recognition. For each of these two aspects, experiments are done. A document image database with groundtruth is used. Available results are discussed.

Introduction

In spite of the wide spread use of computers and other digital facilities, paper document keeps occupying a central place in our everyday life. Conversely to what was expected, the amount of paper produced presently is larger than ever. Important institutions like administrations, libraries, archive services, etc. are heavy paper producers and consumers. To some point of view, paper is one of the most reliable information supports. Unlike numerical records, it is not constrained by format compatibility question, or device needs.

On the other side, document storage for safety or accessibility considerations is a very tricky problem. Research is presently done in such a direction. The primary goal of document analysis and recognition is to transform a paper document into a digital file with as less information loss as possible. Many successive tasks are needed to achieve this purpose. A document image has to be produced and processed for graphic enhancement. Then physical regions of interest have to be found, labelled according to their type (text, graphic, image, etc.), ordered (hierarchically and spatially). Finally, various kinds of information may be retrieved in different ways within certain regions. For example, text can be found via optical character recognition (OCR) in text regions and stored as ASCII data while images may be compressed.

T. Caelli et al. (Eds.): SSPR&SPR 2002, LNCS 2396, pp. 159–167, 2002.
© Springer-Verlag Berlin Heidelberg 2002

Here we are concerned with printed document images. We assume that preprocessing is done and zones of interest are found. We focus on document zone classification task. This paper introduces some entropic heuristics (section 1) to achieve it. A recall about some features proposed by different authors in order to label regions physically (section 2) is done. For the most commonly used ones, a relevance study based on statistical considerations is conducted. Experiments are done on the MediaTeam Document Database and the UWI document database to validate our views.

1 Entropic Features

1.1 Text/Non-text Separation

In previous works [5,4], we introduced entropy heuristics to separate text zones from non-text ones in a black and white printed document image. As stated in [6,7], text areas will have rather regular horizontal projections while non-text elements will give projections more like random distributions (see Fig. 1). These projections are commonly stored as histograms. Thus, it is possible to compute their entropy values. Let H be the histogram representing the horizontal projection for a given region. Its entropy will be

$$E(H) = \sum_{i=1}^{n} \frac{H[i]}{n} \ln\left(\frac{H[i]}{n}\right) \quad (1)$$

assuming the index for histogram entries runs from 1 to n. If entropy is computed for every zone of interest in a given document image, this will result in low and high values for text and non-text areas respectively. Exploiting this last remark, we have been able to discriminate rather efficiently text elements from other regions of interest in various documents. Thus, entropy on horizontal projection is considered as a potentially valuable feature. To validate this assumption, we have performed some experiments which will be discussed in section 3.1.

1.2 Extensions

We have developed an adaptative binarization method to be performed on greyscale document images: within each zone of interest, we gather grey levels in two groups for low and high values via a deterministic variant of the *k-means* algorithm. Pixels with moderate or large grey scales are respectively set to black or white. Our binarization procedure is implicitly based on grey levels distribution. This histogram may carry useful information for region labelling: a text area is likely to have its grey values distribution more regular (in general bimodal) than a graphic zone. Thus, its entropy is estimated and may be considered as an interesting feature for further experiments.

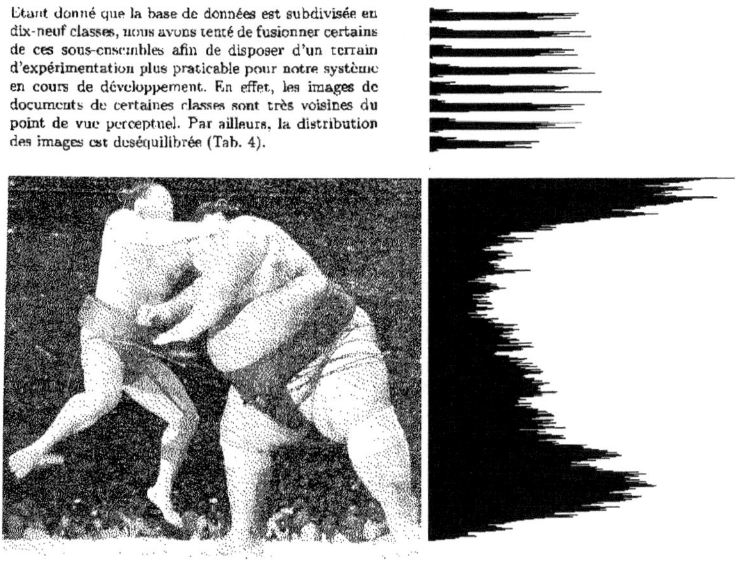
Etant donné que la base de données est subdivisée en
dix-neuf classes, nous avons tenté de fusionner certains
de ces sous-ensembles afin de disposer d'un terrain
d'expérimentation plus praticable pour notre système
en cours de développement. En effet, les images de
documents de certaines classes sont très voisines du
point de vue perceptuel. Par ailleurs, la distribution
des images est déséquilibrée (Tab. 4).

Fig. 1. Example of horizontal projection histograms for a text block and an image

More generally, entropy calculus is a convenient way to measure approximately the information conveyed in a distribution. It allows us to map a vector of a priori unknown size to a scalar. For this reason, we also use it to "compress" vertical projection, north south east and west profiles.

2 Document Zone Classification

Several "classical" features are well known in handwriting recognition. Namely, concavities, surface, profile, etc. Conversely, in printed document analysis, such a common research background does not exist. Early use of features for document zone classification can be found in [16]. A set of commonly used characteristics can be derived from recent surveys [8,10,13,11]. Most of the systems compute values for a certain set of features and perform a rule-based labelling of document zones. Many thresholds appearing in classification rules are set empirically or experimentally and tuned separately from each other. Thus, a global qualitative study remains to be done. To achieve this, we have selected the features that most frequently appear in publications.

A given document area is defined by its bounding box. Values for the entropic features introduced in precedent section are estimated. After binarization, the following measures are also computed for each region of interest.

- Eccentricity (fraction of the width to the height).
- Black pixels (ratio of black pixels population to the surface of the region).
- Horizontal relative cross-count (number of "white to black" and "black to white" transitions in horizontal direction divided by the surface of the region).
- Vertical relative cross-count.
- Mean length of horizontal black segments normalized by the region's width.
- Mean length of Vertical black segments normalized by the region's height.
- Connected components population to the region's surface.

Except for the two first ones, all these features were actually found in coarser versions (i.e. without normalization) in various works.

From now, regions of interest are represented as real vectors with fourteen components. Features can then be compared in term of relevance. Data analysis procedures are particularly well suited for this task.

3 Experiments

We ran most of our experiments with the MediaTeam Document Database[12]. This database is a collection of color or greyscale printed document images with groundtruth for physical segmentation.

3.1 Validation for Entropy Heuristic

The purpose in this first set of experiments is to test the relevance of entropy on horizontal projection histogram as a feature to separate text areas from the others. Regions of interest are retrieved via a coarse segmentation based on gradient image. Given a document image, entropy on horizontal projection is computed for each zone. Areas with low or high entropy values are labeled text or non-text respectively. To achieve this separation, a deterministic variant of the *k-means* algorithm (see Appendix) is performed over the entropy values for the image. Experimental results are presented in [5,4].

3.2 Feature Analysis

Using only one feature, we have been able to discriminate text from non-text in noisy and complicated printed document images with decent performance. Actually, the classification procedure worked locally and assumed the existence of one text zone and one non-text zone in every document image at least. This hypothesis is not fulfilled for all document images of MediaTeam. We now explore the feasability of text/non-text separation involving many characteristics. In this set of experiments, all the regions of interest (given by the database groundtruth) over all the images in MediaTeam are mapped to fourteen dimensional pattern

Table 1. Patterns distribution in MediaTeam Document Database

Pattern type	Samples
Text	4811
Graphics	735
Image	161
Composite	219

vectors. We obtained 5926 of such vectors distributed in four classes, as shown in Table 1. We try to improve the accuracy and the generality of our classification in term of text/non-text separation.

Our data are obviously insufficient and too badly distributed to train and test a neural network or a markovian process [9,3]. Thus we have decided to use classical data analysis and support vector machines for our experimental purposes. These tools are well suited to deal with the kind of data we dispose of. Due to unbalanced classes (see Table 1), classical learning machines as neural networks may lead to overfitting problems for certain classes. SVM classifiers, on the other hand, have shown robust behavior against overfitting phenomena caused by unbalanced data distribution.

Support Vector Machines Let us consider the following set of data for a two class problem: n feature vectors are called X_i with $i \in \{1, \ldots, n\}$ and $X_i \in \mathbb{R}^d, \forall i \in \{1, \ldots, n\}$. Each d-dimensional vector X_i is labelled y_i where $y_i \in \{-1, +1\}, \forall i \in \{1, \ldots, n\}$. According to its label, a vector will be said to be a negative or a positive example. A support vector machine (SVM) works seeking for optimal decision regions between the two classes. In the original formulation, the SVM searches for a linear decision surface by maximizing the margin between positive and negative examples.

Unfortunately, in most of the real-life classification problems, data are not linearly separable. They are mapped in higher dimension space via a non-linear application ϕ called *kernel* (see Table 2 for most commonly used kernels). With an appropriate kernel operating from the original feature space to a sufficiently high dimension space, data from two classes can always be separated [15,14]. The final decision function will be of type $f(X) = \sum_{i=1}^{n} \alpha_i y_i k\left(X_i, X\right)$ where k is a given kernel, $\alpha_i \in [0, C], \forall i \in \{1, \ldots, n\}$. Vector X_i is said to be a *support vector* if the corresponding α_i is non-null. C is a cost parameter. Allowing C to tend to infinity leads to optimal separation of the data, at the price of increasing processing time.

Table 2. Classical kernels for support vector machines

Kernel	Formula
Linear	$k(X,Y) = X.Y$
Sigmoïd	$k(X,Y) = tanh(\alpha X.Y + \beta)$
Polynomial	$k(X,Y) = (1 + X.Y)^d$
Radial Basis Function (RBF)	$k(X,Y) = e^{-\alpha\|X-Y\|^2}$
Exponential RFB	$k(X,Y) = e^{-\alpha\|X-Y\|}$

Table 3. Results for text/non-text separation using a SVM with RBF kernel. Vectors not used in training set are all considered test patterns

Learning set		Cost	α	Support vectors	Learning accuracy	Classification accuracy
Text	Non_text					
2000	500	1	$\frac{1}{14}$	816	90.44%	89.87%
		10		586	94.88%	93.11%
		100		450	96.76%	93.81%
		1000		383	98.12%	93.49%
		100	0.1	447	97.36%	93.46%
		100	0.01	556	93.68%	92.38%
3000	500	100	$\frac{1}{14}$	513	96.94%	91.67%
2000	750	100	$\frac{1}{14}$	559	95.60%	94.89%

Two-Class Separation Improvement Here our purpose is to improve the text/non-text separation using fourteen features. "Graphics", "Image" and "Composite" patterns are gathered in one "Non_text" class. To choose the most fitted approach, we first have to estimate the hardness of our task.

We perform a linear discriminant analysis (LDA) with all the pattern vectors for both training and validation. Observed classification accuracy is 67.09%. This leads us to conclude that our problem may be non-linearly separable (Theoretically, the problem can be considered linearly separable if the obtained accuracy is 100% with LDA.). A trial with a linear support vector machine (SVM), supposed to be the most powerful linear classifier [2], confirms this assumption: obtained classification accuracy is only 87.54%.

Since many types of SVMs exist, different sets of experiments have to be done to determine the best suited classifier. Finally, the SVM with RBF kernel shows the best performances. A preliminary collection of experimental results presented in Table 3 is available. Some subtle tradeoff between the size of the learning set, its distribution, values for kernel parameter, cost threshold remains to be found.

Table 4. "*Separability* estimation using a support vector machine with RBF kernel

Cost parameter	α	Support vectors	Accuracy
100	$\frac{1}{14}$	945	95.83%
500	$\frac{1}{14}$	861	97.11%
1000	$\frac{1}{14}$	804	97.43%
10000	$\frac{1}{14}$	705	**98.48%**

Comments When considering figures presented in Table 3, one must take into account the context of the classification task. As we did to examine the linearity of the problem, we use all the pattern vectors to train a support vector machine with RBF kernel. Different experiments (see Table 4) show that our problem is everything but trivial.

3.3 Recent Advances

Many problems arise, while using the MediaTeam document database. The corpus is not sufficiently large to deploy most of the statistical learning techniques. Moreover, proposed documents are from very different types (nineteen document classes are found in the database, some of them with less than a dozen of images). Thus, we have decided to perform another set of experiments with a more specific document database.

We have computed the above-presented characteristics over the regions of interest proposed in UWI document database. This collection consists of 1000 pages from different english journals. Since the document images are binary in this database, we dropped the grey distribution entropy characteristic. Our calculus resulted in 10573 patterns vectors. These 13-dimensional vectors are distributed as following: 9307 samples for text regions and the other 1266 ones for non-text regions.

We used SVM classifier with the KMOD kernel, newly designed by Ayat et al. [1]. This kernel's specification is given by the equation

$$kmod\,(x, y) = a \left[exp \left(\frac{\gamma}{\|x - y\|^2 + \sigma^2} \right) - 1 \right] \tag{2}$$

$\sigma = 0.01$ and $\gamma = 0.001$ are two parameters that jointly control the behavior of the kernel function. σ is a space scale parameter that define a gate surface around zero whereas γ controls the decreasing speed around zero. In other words, σ measures the spread of the kernel function, and γ describes the shape of this function within this domain. We set empirically $\sigma = 0.01$ and $\gamma = 0.001$ [1]. The normalization constant a is defined as

$$a = \frac{1}{e^{\frac{\gamma}{\sigma}} - 1} \tag{3}$$

Table 5. Results using SVM with KMOD kernel. Parameters σ and γ (in formula 2) are set to 0.01 and 0.001 respectively

Cost	0.1	1	10	100
Accuracy	91.10	97.06	97.34	97.34

UWI is both more homogeneous and more voluminous than MediaTeam. To avoid the problem of designing training and test sets, we performed a five-fold cross-validation on our data set: we divided the data into five subsets of (approximately) equal size. We trained the classifier five times, each time leaving out one of the subsets from training, and using it for test. Accuracy is defined as the mean value over the five obtained performance score for tests. Table 5 shows results for such experimental setting. Statistical examination should be conducted on data to establish an efficient preprocessing. The features we chose were interesting since they are frequently used by different authors. But some of them may be strongly correlated (redundant), due to a lack of standardization. Experiments are currently in progress to perform feature selection and extraction. Results will be presented in further publications.

Conclusion

This paper was intended to show some recent developments in printed document image analysis and several gaps in related features normalization. It also proposes a way to fill these lacks. Application to document zone classification is presented. Some basic features are selected for their simplicity and a statistical examination is performed.

The use of common statistical tools and support vector machines has been proved to be adequate for this kind of problem. Other experiments are in progress to optimize learning parameters. SVM paradigm is still under development in machine learning research community [1]. As a consequence, better results for document region classification may be obtained in a near future. The following step will be to investigate multiple class discrimination for thinner document zone classification. This should help document logical labelling process. Many other characteristics have to be jointly tested. Some will surely have to be dropped. This will be part of our further work.

Acknowledgements

We wish to thank our colleague N.E. Ayat and Professor M. Cheriet for very instructive discussions about SVM paradigm and helpful advice for experimentations.

References

1. Nedjem E. Ayat, Mohamed Cheriet, and Ching Y. Suen. Kmod-a two parameter svm kernel for pattern recognition, 2002. To appear in ICPR 2002. Quebec city, Canada, 2002. 165, 166
2. C. Cortes and V. Vapnik. Support vector networks. *Machine Learning*, 20, 1995. 164
3. Richard O. Duda, Peter E. Hart, and David G. Stork. *Pattern Classification*. Wiley Interscience, 2001. 163
4. Jean Duong, Myriam Côté, and Hubert Emptoz. Extraction des régions textuelles dans les images de documents imprimés. In *Reconnaissance de Formes et Intelligence Artificielle (RFIA)*, Angers (France), Janvier 2002. 160, 162
5. Jean Duong, Myriam Côté, Hubert Emptoz, and Ching Y. Suen. Extraction of text areas in printed document images. In *ACM Symposium on Document Engineering (DocEng)*, pages 157–165, Atlanta (Georgia, USA), November 2001. 160, 162
6. K.C. Fan, C.H. Liu, and Y.K. Wang. Segmentation and classification of mixed text/graphics/image documents. *Pattern Recognition Letters*, 15:1201–1209, 1994. 160
7. K.C. Fan and L.S. Wang. Classification of document blocks using density feature and connectivity histogram. *Pattern Recognition Letters*, 16:955–962, 1995. 160
8. Robert M. Haralick. Document image understanding: Geometric and logical layout. In *IEEE Computer Society Conference on Computer Vision and Pattern Recognition (CVPR)*, volume 4, pages 384–390, 1994. 161
9. Anil K. Jain, Robert P. W. Duin, and Jianchang Mao. Statistical pattern recognition: A review. *IEEE Transactions on Pattern Analysis and Machine Intelligence (PAMI)*, 22(1):4–37, Januray 2000. 163
10. Anil K. Jain and Bin Yu. Document representation and its application to page decomposition. *IEEE Transaction on Pattern Analysis and Machine Intelligence (PAMI)*, 20(3):294–308, March 1998. 161
11. George Nagy. Twenty years of document image ananlysis in pami. *IEEE Transactions on Pattern Analysis and Machine Intelligence (PAMI)*, 22(1):38–62, January 2000. 161
12. University of Oulu (Finland). Mediateam document database, 1998. 162
13. Oleg Okun, David Dœrmann, and Matti Pietikäinen. Page segmentation and zone classification: The state of the art, November 1999. 161
14. B. Scholkopf, C. Burges, and A. Smola. *Advances in Kernel Methods: Support Vector Learning*, chapter 1. MIT Press, 1999. 163
15. Vladimir Vapnik. *The nature of Statistical Learning Theory*. Springer Verlag, New-York (USA), 1995. 163
16. Kwan Y. Wong, Richard G. Casey, and Friedrich M. Wahl. Document analysis system. *IBM Journal of Research and Developpment*, 26(6):647–656, November 1982. 161

Example-Driven Graphics Recognition

Liu Wenyin

Dept of Computer Science, City University of Hong Kong, Hong Kong SAR, PR China
csliuwy@cityu.edu.hk

Abstract. An example-driven graphics recognition scheme is presented, which is an extension of the generic graphics recognition algorithm we presented years ago. The key idea is that, interactively, the user can specify one or more examples of one type of graphic objects in an engineering drawing image, an the system then learn the constraint rules among the components in this type of graphic objects and recognize similar objects in the same drawing or similar drawings by matching the constraint rules. Preliminary experiments have shown that this is a promising way for interactive graphics recognition.

Keywords: graphics recognition, rule-based approach, case-based reasoning, SSPR.

1 Introduction

As a pattern recognition problem, graphics recognition requires that each graphic pattern be known, analyzed, defined, and represented prior to the recognition (matching with those patterns in the image) process. This is especially true for those approaches (e.g., Neural Network based approaches [1]) that require large sets of pattern samples for pre-training. Similarly, the syntaxes and structures of the patterns should also be pre-defined before recognition in syntactic and structural approaches (e.g., [2]) and the knowledge about the patterns should also be pre-acquired before recognition in knowledge-based approaches (e.g., [2] and [8]). For example, if the task is to recognize lines from images, the attributes or features of the line patterns should be analyzed such that appropriate representations and algorithm can be designed and implemented. Through pattern analysis we know that line patterns in the image space correspond to peaks in the Hough transformed space. Therefore, these peaks are pre-defined features for detecting lines in the Hough Transform based approaches [4]. Usually, the features, syntaxes, and other knowledge about the patterns, e.g., the graphic geometry, are hard-coded in the recognition algorithms. Hence, currently, each graphics recognition algorithm only deals with a limited set of specific, known graphic patterns, e.g., dimension-sets [2], shafts [8]. Once implemented and incorporated in a graphics recognition system, these features, syntaxes, and knowledge cannot be changed. The system can only be used for these pre-defined patterns and cannot be applied to other previously unknown patterns or new patterns. In order to recognize new patterns, the same analysis-design process should be repeated. Hence, these approaches are not flexible to changeable environments.

T. Caelli et al. (Eds.): SSPR&SPR 2002, LNCS 2396, pp. 168-176, 2002.
© Springer-Verlag Berlin Heidelberg 2002

It is fine to hard-code for those very common graphic primitives, e.g., lines, arcs, characters in the recognition algorithms. However, there are many different classes of graphic symbols/patterns or higher level graphic objects in many different domains of drawings. Even within a single domain, e.g., mechanical drawings or architectural drawings, the number of symbols or commonly re-usable component patterns can be very large. Hence, it is non-realistic to hard-code all of them in the recognition algorithms. A generic method that can automatic learn new or updated patterns for run-time or just-in-time recognition is strongly desired.

In this paper, we propose a new scheme of graphics recognition, which is example-driven. That is, the user provides to the system with a selected set of representative examples for the graphic pattern to be recognized and the system learns the knowledge (attributes/constraints of the components, etc.) about the pattern from these examples and recognizes all graphic patterns similar (in terms of those attributes and constraints) to these examples. In this way, the system does not need to know any pre-defined patterns before the system is built. The knowledge of patterns can be learnt at run-time. The underlying support for this example-driven scheme is the generic graphic recognition algorithm (GGRA) [6, 7] implemented using a rule-based approach. Due to the vector-based nature of GGRA, pre-segmentation of graphic patterns is not required. In this paper, we first briefly explain GGRA and then present the rule-based framework for graphics recognition. Preliminary experiments and concluding remarks are also presented.

2 The Generic Graphics Recognition Algorithm

The Generic Graphics Recognition Algorithm (GGRA) [6, 7] was proposed and constructed based on the observation that all graphic patterns consist of multiple components satisfying (subject to) a set of constraints. For instance, a rectangle comprises a closed sequence of four connected straight lines with four right angles at the four connection points. Even a solid line may consist of several connected and collinear vectorized line fragments.

Most existing graphics recognition algorithms cluster all the potential constituent components at once, while the graphics attributes are determined later. This blind search procedure usually introduces inaccuracies in the grouping of the components, which ultimately account for inaccurate graphics recognition. Moreover, each class of graphic objects requires a particular detection algorithm. In spite of many graphics recognition algorithms reported, no research report has yet proposed to detect all classes of graphics by a generic, unifying algorithm.

The Generic Graphics Recognition Algorithm (GGRA) [6, 7] we previously proposed is a more flexible and adaptive scheme that constantly checks the graphic object's syntax rules and updates the object's parameters while grouping its components. This generic graphics recognition methodology takes vectors as input. These vectors can be produced by any vectorization algorithm, in particular our sparse pixel vectorization algorithm (SPV) [5]. As shown in **Fig. 1**, which is the C++ code illustration of the framework, GGRA (in **runWith**(...)) consists of two main phases based on the hypothesis-and-test paradigm. The first step is the hypothesis generation, in which the existence of a graphic object of the class being detected is assumed by finding its first key component from the graphics database (by calling prm = gobj->**find FirstComponentFrom**(gdb)). The second step is the hypothesis test, in which the presence of such graphic object is proved by successfully constructing it from its first key component and serially extending it to its other components. In the second step, an empty graphic object is first filled with the first key component found in the first step (by calling gobj->**fillWith**(prm)). The graphic object is further extended as far as possible in all possible directions (d<=gobj->**numOfExtensionDirections**()) in the extension process—a stepwise recovery of its other components (**extend**(d, gdb)). After the current graphic object is extended to all extension directions, a final credibility test (gobj->**isCredible**()) prevents the inclusion of false positives due to accumulative error. If the extended graphic object passes the test, it is recognized successfully and added to the graphics database (gdb), otherwise all found components are rejected as being parts of the anticipated object which should be deleted. Regardless of whether the test is successful or not, the recognition process proceeds to find the next key component, which is used to start a new hypothesis test.

```
template <class AGraphicClass>
class DetectorOf
{
        DetectorOf() {}
        void runWith(GraphicDataBase& gdb) {
          while (1) {
            AGraphicClass* gobj = new AGraphicClass();
            Primitive* prm = gobj->findFirstComponentFrom(gdb);
            if (prm == null) return;
            if (!gobj->fillWith(prm)) continue;
            for (int d=0;d<=gobj->numOfExtensionDirections(); d++)
              while (gobj->extend(d, gdb));
            if (!gobj->isCredible()) delete gobj;
            else gobj->addTo(gdb)
          }
        }
        boolean extend(int direction, GraphicDataBase& gdb) {
          Area area = extensionArea(direction);
          PrimitiveArray& candidates = gdb.search(area);
          for (int i=0; i < candidates.getSize(); i++) {
            if (!extensible(candidates[i])) continue;
            updateWith(candidates[i]);
            break;
          }
          if (i < candidates.getSize()) return true;
          return false;
        }
};
```

Fig. 1. Outline of the C++ implementation of GGRA

In the extension procedure (**extend**(…)), an extension area is first defined at the current extension direction according to the object's current state, e.g., the most recently found component (by calling area = **extensionArea**(direction)). All candidates of possible components that are found in this area and pass the candidacy test are then inserted into the candidate list, sorted by their nearest distance to the current graphic object being extended (by calling candidates = gdb.**search**(area)). The nearest candidate undergoes the extendibility test (**extensible**(candidates[i])). If it passes the test, the current graphic object is extended to include it (by calling **updateWith**(candidates[i])). Otherwise, the next nearest candidate is taken for the extendibility test, until some candidate passes the test. If no candidate passes the test, the extension process stops. If the graphic object is successfully extended to a new component, the extension process is iterated with the object's updated state.

Since in the first phase we find the first key component of the object to be recognized, making the correct hypothesis is crucial, and should be properly constrained. If it is over-constrained, only few objects will be found, while under-constraining it would lead to too many false alarms. If no key component can be found, no more objects of the type being sought can be detected and the recognition process (**runWith**(…)) stops.

The generic object recognition algorithm can be instantiated for the recognition process of a variety of objects. Especially, GGRA has been successfully applied to detection of various types of lines [9], text, arrowheads, leaders, and dimension-sets, hatched areas. However, in these applications, the rules (defining the graphic classes) are hard-coded in the overridden member functions of their classes. In this paper, GGRA is further generalized and applied to detection of user-defined types of graphic objects by implementing the abstract functions (in bold fonts in **Fig. 1**) in GGRA using the rule-based approach.

3 The Rule-Based Graphics Recognition Framework

Due to GGRA's generalized and stepwise nature, it is a good candidate to serve as the basis for the recognition framework for the graphic classes that are previously unknown but specified or defined at run-time. In this paper, we extend GGRA to such recognition framework by implementing the abstract functions (in bold fonts in **Fig. 1**) in GGRA using the rule-based approach. That is, the rule-based algorithms (and the code) in these functions are the same for all graphics classes. Each graphics class is specified using a set of rules (attributes and constraints), which are stored in the knowledge database. In the recognition process for a particular class, its rules are taken for testing and execution in the same algorithms. The knowledge base is managed separately from the main algorithms, which are fixed for all graphic classes. Hence, to make the work for a new graphics class, the only thing we need to do is to add the rules, which specify the components and their attributes/constraints, to the knowledge base. The rules are also updated when new positive/negative examples are provided for existing graphic classes.

In this section, we present how the rules for a particular graphic class are represented, learnt, and used in the recognition process.

Knowledge Representation Scheme for Graphics Classes

In order to specify a graphics class, we design the representation scheme for a graphics class as follows. Each object of such graphic class should have the following attributes or features.

1. The ID for this class, which can be specified by the user or an automatic program.
2. The components (in sequence) of the class, which can be any previously known graphic classes. Currently, we use lines, arrowheads, and textboxes as primitive types, whose attributes are known. Once new graphics classes, which can either be manually specified by the user or be automatically learnt from examples, are added, they can also be used as the types of components of future graphics classes.
3. The attributes of each individual component, which can be used to filter out those graphic objects that cannot be candidates for the component. The graphic type for this component is the most important attribute for the component. The attributes for each type can be different. For examples, the attributes for a line segment can include its shape (which can be straight, circular, free formed, etc.) and style (which can be one of the pre-defined styles: continuous, dashed, dash-dotted, dash-dot-dotted, etc.), line width, length, angle, etc. An attribute can be specified with tolerances. For example, a line width can be 5±1 pixels and an angle can be 45±5°. An attribute, e.g., the graphics type, can also be fixed. Most often, if a textbox is required, a line is usually not allowed. Sometimes, line shape and style are also not flexible attributes for a component.
4. The constraints between each individual component and the entire object or other components that are in previous position in the component sequence. For example, the relative location (or angle) of the component in the entire object is a constraint between the component and the entire object. A constraint between two components can be intersection/connection/perpendicularity/parallelism (width a tolerated distance) between two straight lines, or concentricity/tangency between two arcs, or positional (above/under, left/right, or inside/outside) between two rectangles, and so on. Tolerances are also necessary due to many reasons including drawing digitization and vectorization.

The types of attributes and constraints can also be expanded to include new ones while a few primitive types of attributes and constraints are defined initially. For examples, the connection of two lines (of any shape and any style) is defined as that the minimum distance of an endpoint of one line to an endpoint of the other line is less than a tolerance (e.g., half of the line width).

Knowledge Acquisition for a Particular Graphics Class from Examples

Knowledge acquisition is the process in which the rules (mainly, the attributes and constraints) to represent particular graphics classes are obtained. Admittedly, a user can write all the rules manually. However, to enable example-driven graphic recognition, automatic acquisition of the rules is indispensable. Hence, we implement an automatic learning process for a particular graphics class from the examples provided by the user.

In the automatic learning process, we need to determine the ID of the class, its components and sequence, especially, the first key components, which is critical in starting the recognition process (as shown in **Fig. 1**). More importantly, we need to determine the attributes of individual components and constraints among components. While the ID can be obtained quite easily (as we discussed in the last sub-section), determination of other things, however, is not-trivial.

First of all, the first key component and the sequence of the remaining components should be determined. Although there are multiple choices for the sequence, a good sequence can greatly reduce the complexity of the constraints and speedup the searching process for component candidates. We define the following heuristic rules for determination of the component sequence.

1. The components within an example are first sorted according to the priorities of their graphic types. The priority of a particular graphic type is determined as inversely proportional to the occurrence frequency of this type of objects in all graphic drawings, which can be statistically obtained. The lower the frequency, the higher the priority. The reason is that the graphic objects of those common types can be quickly filtered out during the candidacy test to save much time in later constraints checking, in which this sequence can filter out those non-promising combination of components more quickly than other possible sequences. Hence, the priority list can be sorted in the decreasing order of the occurrence frequencies of the graphic types. For example, solid lines are the most dominant graphic type in engineering drawing and hence this type is of the lowest priority.
2. If two components are of the same type, other attributes, e.g., length, size, can be used to sort their priorities.
3. When the first key component is determined, the sequence of the remaining components can be done similarly according to the type priorities. Or alternatively, the nearest principle can be used to find the next components. If multiple components are the nearest, positional sequences, e.g., from left to right, from top-down, can be used.
4. If multiple examples have been provided, the alignment (or correspondence) between the components of each example should be done. The most conformable sequence is chosen as the final sequence of components for this graphic class.
5. Optionally, the user's interaction can also be used as a method to specify the sequence. For example, we can ask experienced users to pick the key components in his examples first when the examples are provided.

After the sequence is determined, the attributes of each component is also determined. If a single example is provided, the values of the attributes can be directly calculated from the example. For example, the relative position/angle, length, angel, etc., can be calculated for line types. If permitted, a tolerance can also be added for each attributes. If multiple examples are provided, the values of the attribute for the same component all examples are used to determined the range of values that the attribute can take.

Then the constraints between the current components and each of the previous components in the sequence are determined. For each pair of component, each possible constraint in a candidate constraint list (as we discussed in the previous sub-section) is tested. If the constraint passed the test, then this constraint is valid for this pair of components. Otherwise, the final constraint list for this graphic class does not include this constraint. If only a single example is provided, the tolerance of the constraint can be set strictly. If multiple examples are provided, the tolerance should be set to include all possibilities in the examples. If one among the many examples given for this graphic pattern violates the constraint, then this constraint is not a mandatory for this pattern and should not be included in the final constraint list. Even more, if more examples, especially for those negative examples (e.g., false alarms removed by the user), are provided later, the tolerance should be updated or even the entire constraints should become invalid.

Matching for Recognition of a Particular Graphics Class

Once the rules for a particular graphic class are known, its recognition process mainly consists of searching the current graphics database for its components with the rules using GGRA (as shown in **Fig. 1**). In this sub-section, we only discuss the main functions that should be implemented. Implementations of others are intuitive. We start the process by finding its first key component, whose attributes should conform to those of the first one in the component sequence for this graphics class. Starting with the first key component found by the **findFirstComponentFrom**(…) function, in which all attribute requirements for the first component are met, we find the other components for this graphic object one by one using the **extend**(direction, …) function. The **numOfExtensionDirections**() function returns the number of components for this graphics class. The "direction" parameter in **extend**(direction, …) function specifies which component the current extension procedure is searching for. The **search**(…) function returns those candidates, which pass all attribute requirements for the current component. Each candidate undergoes further tests in **extensible**(candidates[i]) function, in which all constraints between this components and others are checked. The first candidate that passes the tests is used as the current component. If such a component can be found the graphic object is successfully extended to this component and the extension to the next component in the sequence will begin until all components are successfully found and the entire graphic object is successfully recognized. Otherwise, it means failure of recognizing the graphic object.

4 Experiments

We have implemented the rule-based graphics recognition algorithm for simple graphic patterns that consists of only various types of lines and use it to implement our strategy of example-driven graphics recognition. An example of the graphic pattern that we want to recognize is specified by clicking all of its components. For example, as shown in **Fig. 2**, the user can click the solid circles and the concentric dashed circles as the example of the pattern we want to recognize. The system automatically determines the dashed circle as the first key component and concentricity is the main constraint. The system then automatically finds other similar graphic patterns which contain the same kinds of components and are constrained similarly in the drawing. Including the example, four objects of this class have been recognized due to that we used a larger tolerance to the central angle of the arcs in the implementation. If we had selected the top-left or bottom-right pattern as the example, only three objects could have been recognized. This is due to that each such example contains a partial arc that cannot be successfully matched during recognition. Anyway, the experiment has proved that the current implementation has already been able to do example-driven graphics recognition.

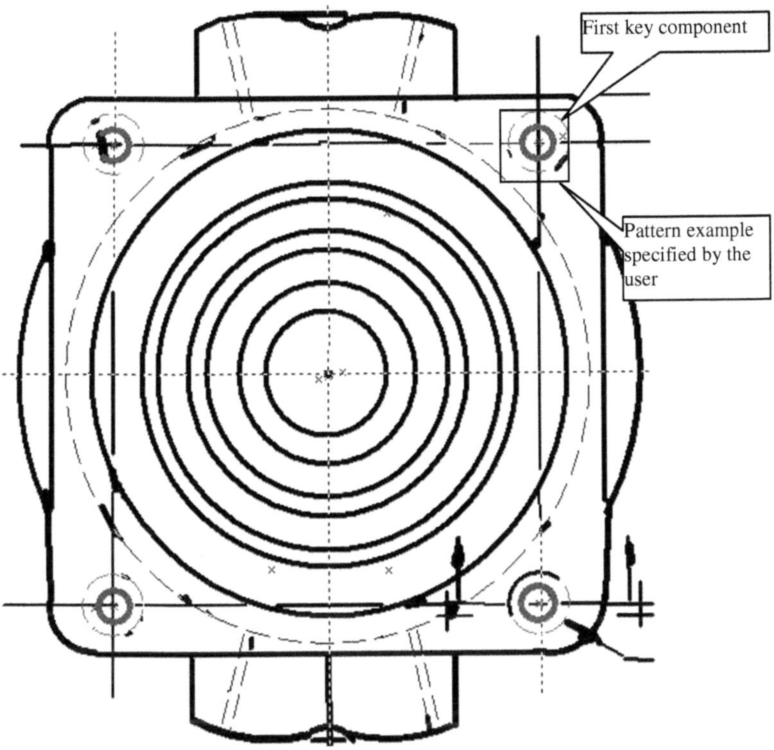

Fig. 2. Results of example-driven graphics recognition

Current, only single example can be used in our experiments. Recognition based on multiple examples, especially negative examples (from user's manual correction of false alarms), will be soon implemented. We will also test the algorithm on more complex graphic patterns, e.g., including arrowheads and textboxes in the future.

5 Summary and Future Work

We have presented a rule-based graphics recognition framework, which is based on the generic graphics recognition algorithm [6]. We have also applied it to build the example-driven graphics recognition scheme and obtained preliminary but promising results.

The scheme features a manual user interface for providing examples for particular graphic patterns, a rule-based representation of graphic patterns, and an automatic learning process for the constraint rules. This scheme provides a flexible approach, which are suitable for recognition of those graphic patterns that are unknown previously before the recognition system has been built. Such interactive graphic recognition scheme is especially useful in the current stage when automatic recognition cannot always produce reliable results. This scheme can also be used as efficient way for automatic knowledge acquisition for graphic patterns.

Although we currently only use single examples for learning graphic patterns, we believe that the scheme can also fit the cases of multiple examples from both positive and negative perspectives. Especially, the user's feedback, e.g., manual correction to those misrecognitions can be good resource to correctly learn the graphic patterns.

6 References

1. Cheng, T., Khan, J., Liu, H., Yun, D.Y.Y.: A symbol recognition system. In: Proc. ICDAR93 (1993).
2. den Hartog J.E., ten Kate T.K., and Gerbrands J.J.: Knowledge-Based Interpretation of Utility Maps. Computer Vision and Image Understanding 63(1) (1996) 105-117.
3. Dori D.: A syntactic/geometric approach to recognition of dimensions in engineering machine drawings. Computer Vision, Graphics, and Image Processing 47(3) (1989) 271-291
4. Dori D.: Orthogonal Zig-Zag: an Algorithm for Vectorizing Engineering Drawings Compared with Hough Transform. Advances in Engineering Software 28(1) (1997) 11-24
5. Dori D. and Liu W.: Sparse Pixel Vectorization: An Algorithm and Its Performance Evaluation. IEEE Trans. on Pattern Analysis and Machine Intelligence 21(3) (1999) 202-215.
6. Liu W. and Dori D.: Genericity in Graphics Recognition Algorithms. In: Graphics Recognition: Algorithms and Systems, eds. K. Tombre and A. Chhabra, Lecture Notes in Computer Science, Vol. 1389, pp. 9-21, Springer (1998).
7. Liu W. and Dori D.: A Generic Integrated Line Detection Algorithm and Its Object-Process Specification. Computer Vision and Image Understanding 70(3) (1998) 420-437
8. Vaxiviere P. and Tombre K.: Celestin: CAD Conversion of Mechanical Drawings. IEEE Computer Magazine 25(7) (1992) 46-54

Estimation of Texels for Regular Mosaics Using Model-Based Interaction Maps

Georgy Gimel'farb*

CITR, Department of Computer Science, Tamaki Campus, University of Auckland
Private Bag 92019, Auckland 1, New Zealand
g.gimelfarb@auckland.ac.nz

Abstract. Spatially homogeneous regular mosaics are image textures formed as a tiling, each tile replicating the same texel. Assuming that the tiles have no relative geometric distortions, the orientation and size of a rectangular texel can be estimated from a model-based interaction map (MBIM) derived from the Gibbs random field model of the texture. The MBIM specifies the structure of pairwise pixel interactions in a given training sample. The estimated texel allows us to quickly simulate a large-size prototype of the mosaic.

1 Introduction

Image texture is usually thought of as being formed by spatial replication of certain pixel neighbourhoods called texels [6]. Replicas of each texel may differ providing their visual similarity is not effected. We restrict our consideration to a limited number of regular textures such as translation invariant tilings, or mosaics formed with a single rectangular texel. The texel-based description holds considerable promise for fast simulation of large-size samples of these textures.

A significant advance in realistic texture simulation has been made recently by approximating particular pixel neighbourhoods of a given training sample with the neighbourhoods of the simulated image [2,3,4,5,7,9,10,11]. The chosen neighbourhoods preserve the deterministic spatial structure of signal interactions in the training sample. The approximation extrapolates the training structure to images of other size and provides random deviations from the training sample.

In most cases the pixel neighbourhoods are implicitly accounted for by using certain spatial features of a multi-resolution image representation, e.g. the top-to-bottom signal vectors along a Laplacian or steerable wavelet image pyramid [2,10], or the explicit neighbourhoods such as squares 7×7 pixels are chosen in a heuristic way to take account of conditional relative frequency distributions of the multi-resolution signals over these small close-range neighbourhoods [9]. The explicit heuristic neighbourhoods are used also in the non-parametric texture sampling [3,7] where each new pixel or small rectangular patch is chosen equiprobably from the pixels or patches having closely similar neighbourhoods

* This work was supported by the University of Auckland Research Committee grants 9343/3414113 and 9393/3600529 and the Marsden Fund grant UOA122.

T. Caelli et al. (Eds.): SSPR&SPR 2002, LNCS 2396, pp. 177–186, 2002.
© Springer-Verlag Berlin Heidelberg 2002

in the training sample or in the already simulated part of the goal image. The most characteristic explicit pixel neighbourhood describing a single-resolution texture is analytically estimated for the Gibbs random field models with multiple pairwise pixel interactions [4,5,11].

Although these techniques are efficient in simulating different natural textures, most of them involve too considerable amounts of computations per pixel to form large-size samples of the texture. The patch-based non-parametric sampling [7] is much faster than other approaches but the quality of simulation depends on the heuristic choice of the patches and their neighbourhoods.

In this paper we consider possibilities of estimating texels for certain regular mosaics using the explicit spatial structure of multiple pairwise pixel interactions in the Gibbs random field model found for the training sample. Then the large-size prototypes of the mosaic can be obtained very fast by replicating the texel.

2 Model-Based Interaction Maps and Texels

Let $g : \mathbf{R} \rightarrow \mathbf{Q}$ where $\mathbf{R}$ and $\mathbf{Q}$ denote an arithmetic lattice and a finite set of image signals, respectively, be a digital image. Let the set $\mathbf{A}$ of inter-pixel shifts specify a translation invariant neighbourhood $\mathbf{N}_{i,\mathbf{A}} = \{(i+a) : a \in \mathbf{A}\}$ of each pixel $i \in \mathbf{R}$. As shown in [4], the approximate partial Gibbs energy $E_{a,0}(\hat{g})$ of interactions over the family of the similar pixel pairs $\mathbf{C}_a = \{(i, i{+}a) : i \in \mathbf{R}\}$ in a given training sample $\hat{g}$ is easily obtained from the relative frequency distribution $\mathbf{F}_a(\hat{g}) = \{F_a(q,s|\hat{g}) : q,s \in \mathbf{Q}\}$ of the signal co-occurrences $(\hat{g}_i = q, \hat{g}_{i+a} = s)$:

$$E_{a,0}(\hat{g}) = \sum_{(q,s)\in\mathbf{Q}^2} F_a(q,s|\hat{g}) \left(F_a(q,s|\hat{g}) - \frac{1}{|\mathbf{Q}|^2} \right) \tag{1}$$

The interactions specified by a large search set $\mathbf{W}$ of the inter-pixel shifts a can be ranked by their partial energies, and the characteristic interaction structure $\mathbf{A}$ is estimated by a parallel or sequential selection of the top-rank energies [4,5,11].

Figure 1 shows training samples 128×128 of the natural image textures from [1,8] and the scaled images 81×81 of their model-based interaction maps (MBIM) $\mathbf{E}(\hat{g}) = \{E_{a,0}(\hat{g}) : a \in \mathbf{W}\}$ showing the structure of pairwise pixel interactions. Every spatial position $a \equiv (x,y)$ of the depicted MBIM indicates the partial interaction energy $E_{a,0}(\hat{g})$ for the inter-pixel shift $a \in \mathbf{W}$, the diametrically opposite shifts (x,y) and $(-x,-y)$ representing the same family $\mathbf{C}_a$. In these examples the set $\mathbf{W}$ contains all the relative inter-pixel x- and y-shifts in the range $[-40, 40]$ specifying $|\mathbf{W}| = 3280$ families of the pixel pairs. By the chosen greyscale coding, the larger the energy, the darker the dot.

As is easy to see in Fig. 1, the MBIMs reflect to a large extent the repetitive pattern of each texture. Replicate spatial clusters of the larger Gibbs energies indicate periodic parts of the texture, and relative positions of and pitches between the clusters relate to the overall rectangular or hexagonal shapes, spatial arrangement, and orientation of the parts. The choice of a single texel is not unique because the same MBIM defines many equivalent periodic partitions with different biases with respect to the lattice.

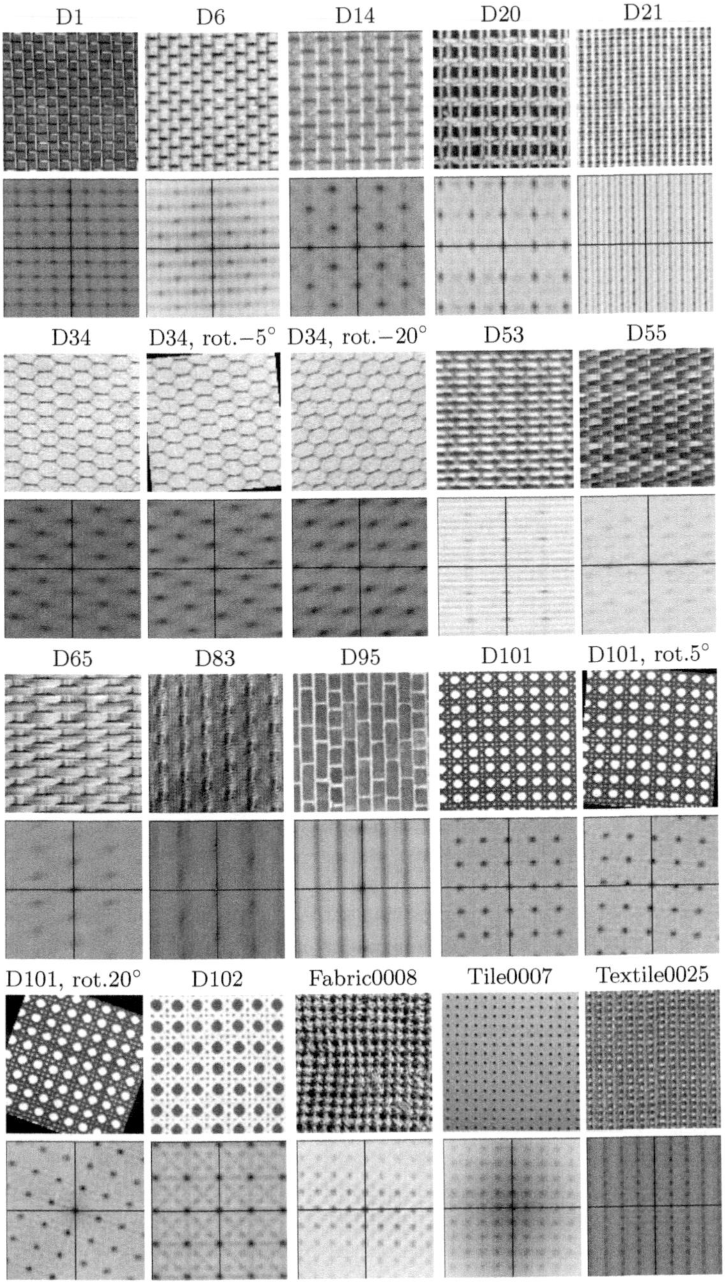

Fig. 1. Training samples and their MBIMs

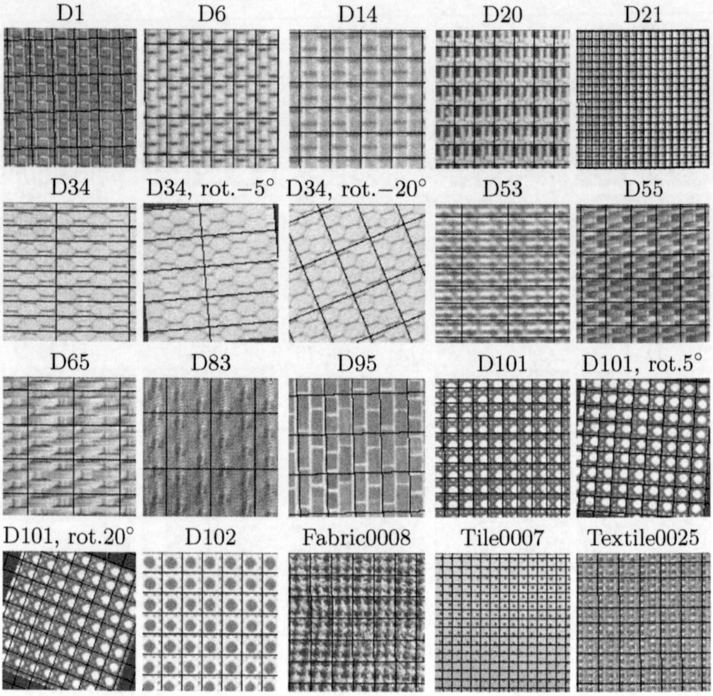

Fig. 2. Estimated partitions of the training samples

Generally it is difficult to derive a minimum-size texel from the repetitive pattern of the MBIM because some of the energy clusters may arise from the periodic fine details of the texel itself or from the secondary interactions between the distant similar parts. The shape and scale of the tiles representing the single texel as well as their photometric characteristics may also vary for different training samples and even within the same sample (e.g. D55, D95, D101, or Fabrics0008 in Fig. 1).

For simplicity, our consideration is restricted to only a rectangular texel with an arbitrary but fixed orientation and size. The central cluster of the most energetic close-range interactions in the MBIMs relates mainly to a uniform background of the image. But a repetitive pattern of the peripheral clusters (if it exists) is produced by the characteristic long-range similarities between the pixel pairs so that a single texel can be in principle estimated from the clearly defined peripheral energy clusters placed around of and closest to the central cluster.

Figure 2 demonstrates partitions of the training samples where each rectangular tile represents the texel. The partitions are estimated using a simplified heuristic approach that determines spatial clusters of the Gibbs energies by thresholding the MBIM with the threshold $E^* = \bar{E} + c \cdot \sigma_E$ where c is an empiri-

Table 1. Parameters of the rectangular texel estimated by detecting the first and second top-rank energy clusters in the MBIMs with the thresholding factor $c = 2.5$ (F08, T07, T25 stand for Fabrics0008, Tile0007, and Textile0025, respectively)

Texture:	D1	D6	D14	D20	D21	D34			D53	D55
						0°	-5°	-20°		
Texel x-size, pix.:	21.5	21.0	29.0	19.0	7.0	70.0	70.26	34.47	44.0	24.0
Texel y-size, pix.:	33.02	34.0	23.0	18.0	7.0	14.0	28.24	42.15	16.0	22.0
Orientation, °:	-1.74	0.0	0.0	0.0	0.0	0.0	-4.9	-22.31	0.0	0.0

Texture:	D65	D83	D95	D101			D102	F08	T07	T25
				0°	5°	20°				
Texel x-size, pixels:	44.0	42.0	25.96	14.0	15.1	15.2	19.0	20.0	9.0	20.0
Texel y-size, pixels:	32.0	52.0	36.76	14.0	14.04	13.92	19.0	20.0	8.0	14.0
Orientation, °:	0.0	0.0	-1.64	0.0	3.81	19.65	0.0	0.0	0.0	0.0

cally chosen factor and $\bar{E}$ and σ_E denote the mean value and standard deviation of the energies $\mathbf{E}(\hat{g})$, respectively. If the MBIM has no peripheral clusters in addition to the central cluster around the origin $a = (0, 0)$, then the texture is aperiodic and has no single texel. Otherwise each peripheral cluster is described by its maximum energy and the inter-pixel shift yielding the maximum. Two clusters with the largest energy and with the second largest energy that is not occluded by the first one from the origin of the MBIM are selected to estimate the texel. The orientation is given by the smallest angular inter-pixel shift with respect to the x-axis of the MBIM. The size is found by projecting both inter-pixel shifts for the clusters to the Cartesian axes of the texel.

Table 1 gives parameters of the texels and partitions in Fig. 2. Changes of the thresholding factor c in the range $1 \leq c \leq 3$ yield quite similar results for most of the textures used in the experiments.

This approach gives sufficiently accurate and stable estimates of the orientation angle for the rectangular and hexagonal patterns of the energy clusters in the above MBIMs. The estimates can be further refined by processing linear chains of the repetitive clusters, e.g. by finding the least scattered projections of the chains onto the coordinate axes. But the refined estimates are less stable for the hexagonal structures of the MBIMs such as for the textures D34 or D65. Generally the MBIM should be processed more in detail for finding adequate shapes, sizes, and orientations of the texels. Our approach shows only a feasibility of relating the texels to spatial periodicity of the energy clusters in the MBIMs.

Figures 3 – 5 show the simulated prototypes of the size 800×170 pixels, each prototype being obtained by replicating a single tile picked arbitrarily from the partitions in Fig. 2. Of course, the singularities of the chosen tile are replicated verbatim. But all the tiles in the partition of the training sample can be jointly

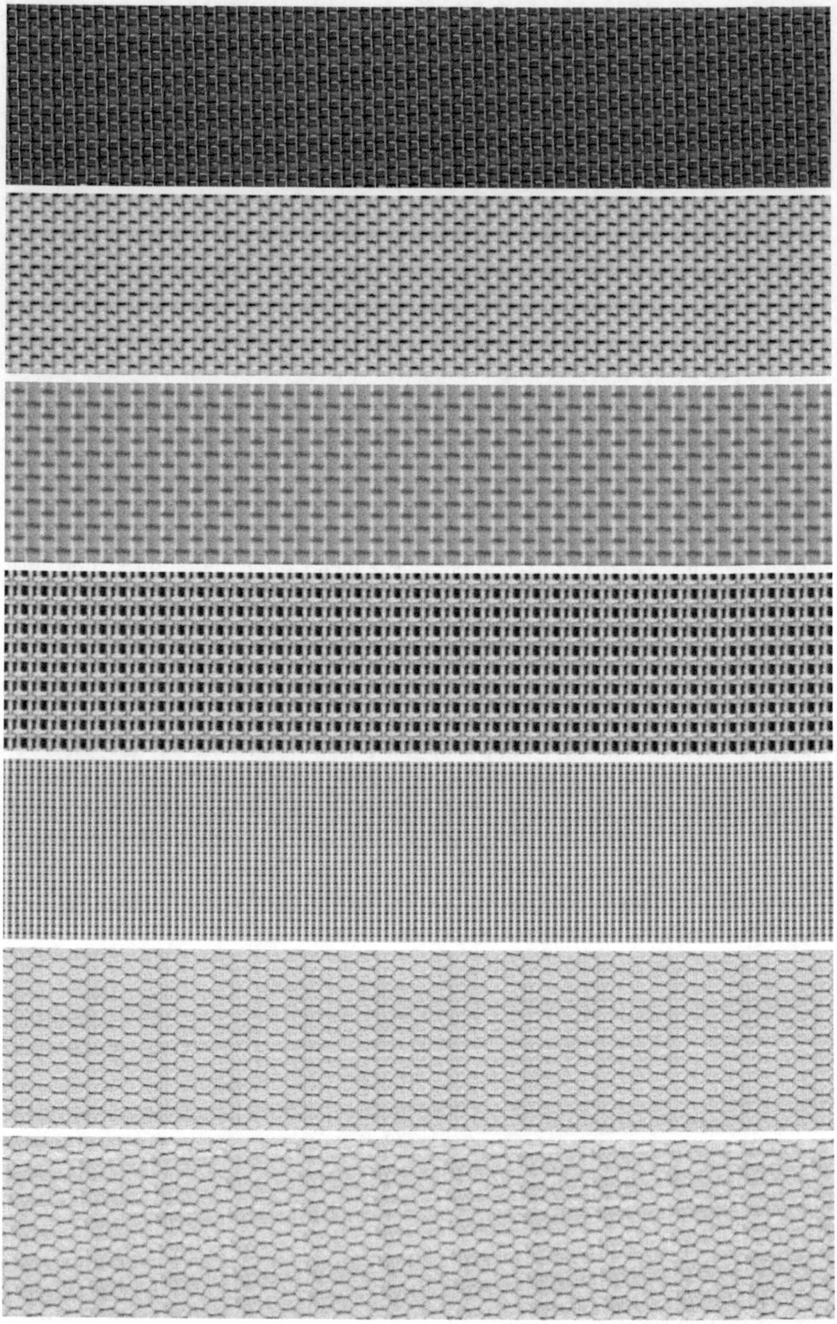

Fig. 3. Prototypes D1, D6, D14, D20, D21, and D34 (rotated 0°, −5°)

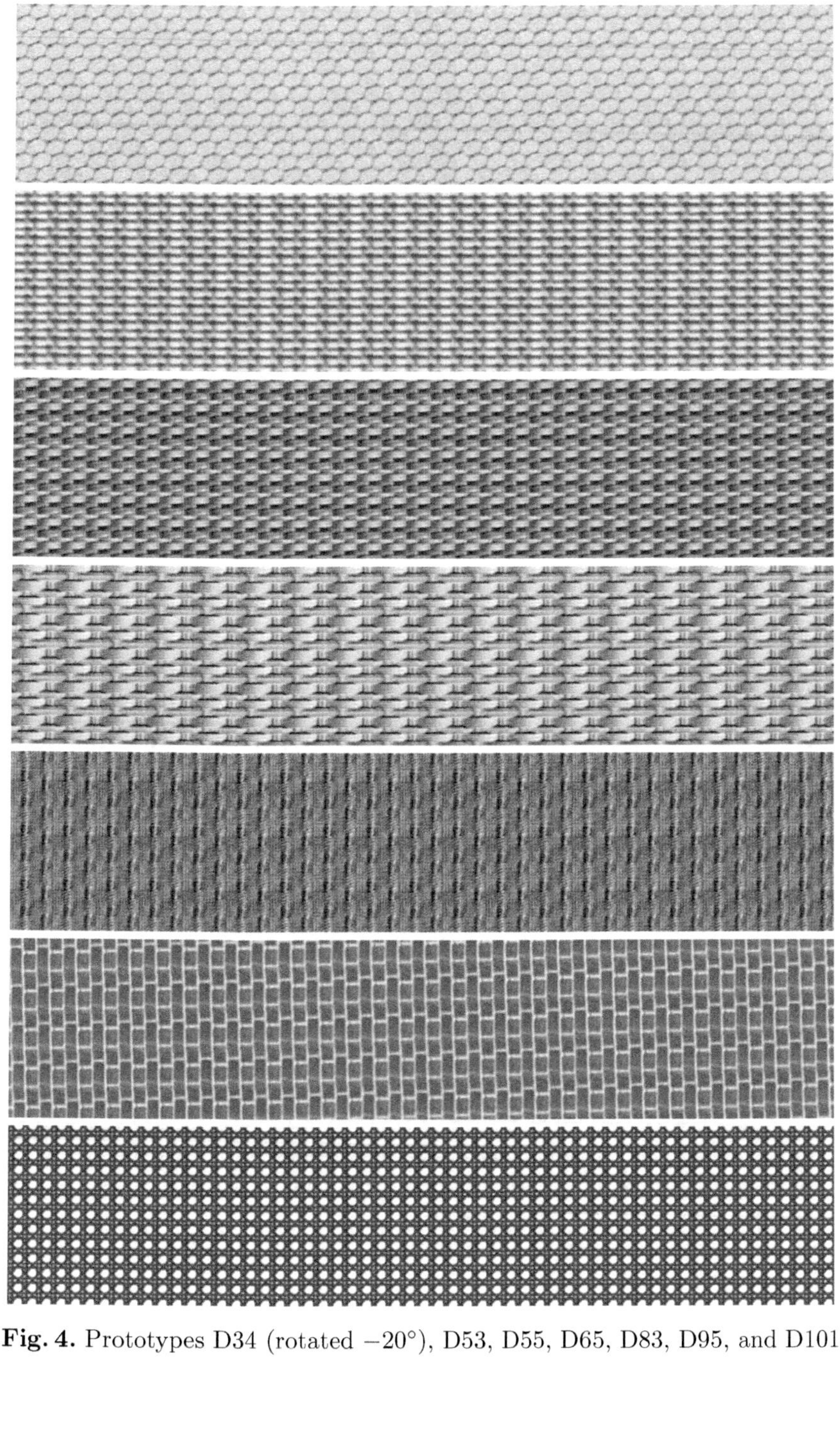

Fig. 4. Prototypes D34 (rotated $-20°$), D53, D55, D65, D83, D95, and D101

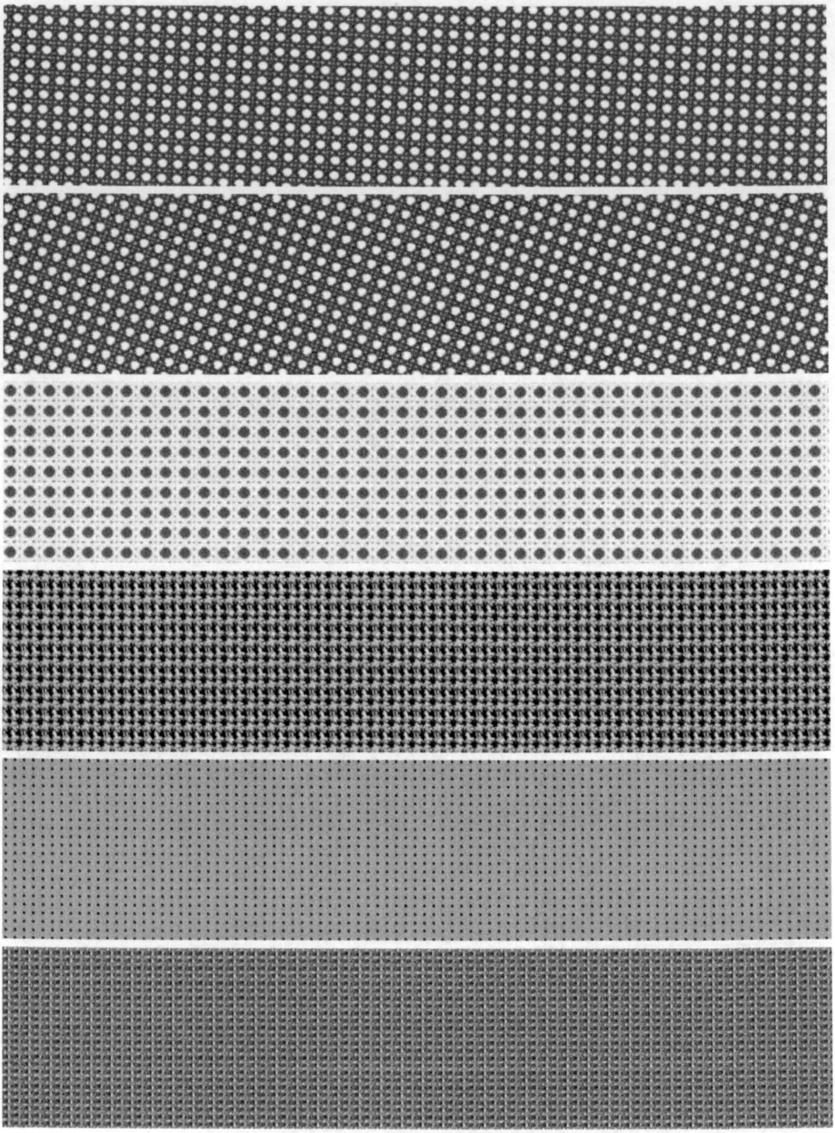

Fig. 5. Prototypes D101 (rotated 5°, 20°), D102, Fabric0008, Tile0007, and Textile0025

processed to exclude their relative distortions and produce an idealised texel. The texture prototype can easily be converted into a realistic sample by mutually agreed random photometric and geometric transformations of the adjacent tiles.

3 Concluding Remarks

Our experiments show that the orientation and size of a rectangular texel can be estimated from the structure of pairwise pixel interactions reflected in the MBIMs. Each tile obtained by partitioning the training sample can act as a provisional texel. But to obtain the ideal texel, the tiles should be jointly processed in order to exclude their geometric and photometric distortions. Replication of the texel forms a prototype of the texture.

The texel-based description is not adequate for all irregular (stochastic) textures with the MBIMs containing no repetitive peripheral energy clusters. But this description is practicable for many translation invariant regular mosaics. Because of computational simplicity of the texel estimation, the simulation of the large-size prototypes of such mosaics is considerably accelerated.

References

1. Brodatz, P.: Textures: A Photographic Album for Artists and Designers. Dover Publications: New York (1966) 178
2. De Bonet, J. S.: Multiresolution sampling procedure for analysis and synthesis of texture images. In: Proc. ACM Conf. Computer Graphics SIGGRAPH'97 (1997) 361–368 177
3. Efros, A. A., Leung, T. K.: Texture synthesis by non-parametric sampling. In: Proc. IEEE Int. Conf. Computer Vision ICCV'99, Greece, Corfu, Sept. 1999, vol.**2** (1999) 1033–1038 177
4. Gimel'farb, G. L.: Image Textures and Gibbs Random Fields. Kluwer Academic: Dordrecht (1999) 177, 178
5. Gimel'farb, G.: Characteristic interaction structures in Gibbs texture modeling. In: Blanc-Talon, J., Popescu, D. C. (Eds.): Imaging and Vision Systems: Theory, Assessment and Applications. Nova Science: Huntington, N. Y. (2001) 71–90 177, 178
6. Haralick, R. M., Shapiro, L. G.: Computer and Robot Vision, vol.**2**. Addison-Wesley: Reading (1993) 177
7. Liang, L., Liu, C., Xu, Y., Guo, B., Shum, H. Y.: Real-Time Texture Synthesis by Patch-Based Sampling. MSR-TR-2001-40. Microsoft Research (2001) 177, 178
8. Pickard, R., Graszyk, S., Mann, S., Wachman, J., Pickard, L., Campbell, L.: VisTex Database. MIT Media Lab.:Cambridge, Mass. (1995) 178
9. Paget, R., Longstaff, I. D.: Texture synthesis via a noncausal nonparametric multiscale Markov random field. IEEE Trans. on Image Processing **7** (1998) 925–931 177
10. Portilla, J., Simoncelli, E. P.: A parametric texture model based on joint statistics of complex wavelet coefficients. Int. Journal on Computer Vision **40** (2000) 49–71 177
11. Zalesny, A., Van Gool, L.: A compact model for viewpoint dependent texture synthesis. In: Pollefeys, M., Van Gool, L., Zisserman, A., Fitzgibbon, A. (Eds.): 3D Structure from Images (Lecture Notes in Computer Science **2018**). Springer: Berlin (2001) 124–143 177, 178

Using Graph Search Techniques for Contextual Colour Retrieval*

Lee Gregory and Josef Kittler

Centre for Vision Speech and Signal Processing, University Of Surrey
Guildford, Surrey. United Kingdom
L.Gregory@eim.surrey.ac.uk
http://www.ee.surrey.ac.uk/Personal/L.Gregory

Abstract. We present a system for colour image retrieval which draws on higher level contextual information as well as low level colour descriptors. The system utilises matching through graph edit operations and optimal search methods. Examples are presented which show how the system can be used to label or retrieve images containing flags. The method is shown to improve on our previous research, in which probabilistic relaxation labelling was used.

1 Introduction

The increasing popularity of digital imaging technology has highlighted some important problems for the computer vision community. As the volume of the digitally archived multimedia increases, the problems associated with organising and retrieving this data become ever more acute.

Content based retrieval systems such as ImageMiner [1], Blobworld [3], VideoQ [4], QBIC [13], Photobook [14] and others [11] were conceived to attempt to alleviate the problems associated with manual annotation of databases.

In this paper we present a system for colour image retrieval which draws on higher level contextual information as well as low level colour descriptors. To demonstrate the method we provide examples of labelling and retrieval of images containing flags. Flags provide a good illustration of why contextual information may be important for colour image retrieval. Also, flags offer a challenging test environment, because often they contain structural errors due to non rigid deformation, variations in scale and rotation. Imperfect segmentation may introduce additional structural errors.

In previous work [9], the problem was addressed using probabilistic relaxation labelling techniques. The shortcomings with the previous method in the presence of many structural errors motivated the current research which is based on optimal search and graph edit operations [12]. The method still retains the invariance to scale and rotation, since only colour and colour context are used

* This work was supported by an EPSRC grant GR/L61095 and the EPSRC PhD Studentship Program.

T. Caelli et al. (Eds.): SSPR&SPR 2002, LNCS 2396, pp. 186–194, 2002.
© Springer-Verlag Berlin Heidelberg 2002

in the matching process. In addition the examples show how this method performs in the presence of structural errors and ambiguous local regions in the images/models.

Graph representations are well suited to many computer vision problems, however matching such graphs is often very computationally expensive and may even be intractable. Non optimal graph matching methods may be much less expensive than optimal search methods, but often perform badly under conditions where structural errors prevail. Such non optimal methods include probabilistic and fuzzy relaxation labelling [5], genetic search, and eigendecomposition [10] based approaches.

Other work has focused on making optimal graph search methods more suitable for database environments. Messmer and Bunke [12], presented a decomposition approach also based on A* search, which removes the linear time dependency when matching many graphs within a database. More recently the work of Berretti et al [2] formalised metric indexing within the graph matching framework.

2 Methodology

In this section we present the details of the adopted method. First, the notation for the graph matching problem is defined and the system implementation is then described in detail. Consider an attributed relational graph (ARG) $G = \{\Omega, E, \mathbf{X}\}$, where $\Omega = \{\omega_1, \omega_2, \cdots, \omega_N\}$ denotes the set of nodes. E represents the set of edges between nodes, where $E \subseteq \Omega \times \Omega$, and $\mathbf{X} = \{x_1, x_2, \cdots, x_n\}$ defines a set of attributes associated to the nodes in Ω, where x_i denotes the attributes (features) for node ω_i.

2.1 Matching

The matching problem is often formulated by defining a model graph, representing a query, which is matched to at least one scene graph. Let $G = \{\Omega, E, \mathbf{X}\}$ and $G' = \{\Omega', E', \mathbf{X}'\}$ denote the model and scene graphs respectively.

Now consider an injective function $f : \Omega \mapsto \Omega'$ which specifies mappings from the nodes Ω in the model graph G to the nodes $X \subseteq \Omega'$ contained in some subgraph of the scene G'. Such a function represents an error correcting subgraph isomorphism, since any mapped subgraph of the scene, can be isomorphic with the model graph, subject to an appropriate set of graph edit operations. The edit operations required to achieve such an isomorphism, represent the errors of an error correcting subgraph isomorphism. These errors are quantified, and are used to guide the graph search process. Error correcting subgraph isomorphism, is well suited for computer vision tasks, where noise and clutter may distort the scene graphs.

Error correcting subgraph isomorphism matches any graph to any other given graph, since an appropriate set of graph edit operations is able to transform any graph arbitrarily. It is therefore essential to define costs for the graph edit operations. Defining such costs allows state space search methods to seek the lowest cost (best matching) mapping between any pair of graphs, given the costs for permissible edit operations. In this implementation, the following traditional graph edit operations are used. Each edit operation λ has an associated cost $C(\lambda)$.

$$\lambda : \omega_i \mapsto \omega'_j : \text{ map the model node } \omega_i \text{ to scene node } \omega'_j \tag{1}$$

$$\lambda : \omega_i \mapsto \emptyset : \text{ map the model node } \omega_i \text{ to the null attractor } \emptyset \tag{2}$$

$$\lambda : e_i \mapsto e'_j : \text{ map the model edge } e_i \text{ to the scene edge } e'_j \tag{3}$$

$$\lambda : e_i \mapsto \emptyset : \text{ map the model edge } e_i \text{ to the null attractor } \emptyset \tag{4}$$

$$\lambda : e'_j \mapsto \emptyset : \text{ map the scene edge } e_j \text{ to the null attractor } \emptyset \tag{5}$$

Note that the symbol $\emptyset$ represents a null attractor which is used to express missing edges and vertices.

Graph matching algorithms which employ state space search strategies, recursively expand partial mappings to grow error correcting subgraph isomorphisms in the state space. Our implementation uses the A* algorithm for optimal search. For any given partial mapping $f : \Omega \mapsto \Omega'$ there exists a set of graph edit operations $\Delta_f = \{\lambda_1, \lambda_2, \cdots, \lambda_N\}$ which transform the mapped scene nodes into a subgraph isomorphism with the partial model. Hence the search through the state space can be guided by the costs of the graph edit operations required for each partial mapping.

The state space search starts from the root state which is the top node in the search tree. From this node, child nodes are generated by allowing the first model node ω_1 to be mapped to each available input node in turn $\{\omega'_1, \omega'_2, \cdots, \omega'_N, \emptyset\}$. Also a child state for a missing vertex is added by mapping the model node to the null attractor.

Each leaf of the tree now represents an error correcting subgraph isomorphism $f_k : \Omega \mapsto \Omega'$ from a partial model graph to the scene graph. The cost of these graph mappings are computed as $C(\Delta_{f_k})$, and the leaf with the lowest cost is expanded. This process continues until the model is fully mapped and the isomorphism with the least cost is found. For the sake of efficiency, the graph edit distance for a given leaf node in the search tree, is computed incrementally from its parent node.

The complexity of the described state space search, is in the worst case exponential, although in practice the actual complexity is data dependent and the optimal search often becomes tractable. To further prune the search space and reduce the complexity, lookahead terms are often used when computing the costs for a given state. The lookahead term, computes an estimate of the future cost of any proceeding mappings based on the current partial interpretation. The exact computation of a minimal future mapping is itself an error correcting subgraph isomorphism problem, and therefore has a worst case exponential complexity.

Hence an estimate is used instead. To prevent false dismissals, such an estimate must provide a lower bound on future cost for any proceeding mappings. To provide such a lower bound for future mapping cost, we consider each unmapped node independently, therefore breaking the exponential complexity of the lookahead. Tests show that a lower bound which ignores edge constraints is faster than a more refined lookahead scheme which considers the edge costs. The lookahead function $L(f : \Omega \mapsto \Omega')$ is defined as

$$L(f : \Omega \mapsto \Omega') = \sum_{\omega_i \in \emptyset_M} \min_{\omega_j \in \emptyset_I} (C(\lambda : \omega_i \mapsto \omega'_j)) \tag{6}$$

where $\emptyset_M$ and $\emptyset_I$ denotes the set of model and input nodes which are not mapped in the current partial interpretation. This result is in agreement with Berretti et al[2] where a faster less accurate lookahead was shown to outperform a more complex scheme. This does not affect the optimality, since any lower bound estimate will not allow false dismissals.

2.2 Pre-processing

We now explain how the images are initialised for the graph matching. During the pre-processing stage, images are segmented so that a region adjacency graph can be built. Each pixel in the image is represented as a 5D vector, the first three dimensions are the RGB colour values for the pixel and the last two dimensions are the pixel co-ordinates. The feature space is then clustered using the mean shift algorithm [6][7]. The mean shift algorithm is an iterative procedure which seeks the modes of the feature distribution. The algorithm is non-parametric and does not assume any prior information about the underlying distributions, or the number of clusters. This is an important implication because it allows the algorithm to operate unsupervised. In practice only a window size and co-ordinate scale are needed by the algorithm. Every pixel is given a label corresponding to the cluster which it has been classified to. The region labels correspond to homogeneous colour regions within the image. A connected component analysis stage ensures that only connected pixels may be assigned the same label.

The segmented image can now be expressed as an ARG. The attributes X are defined as follows.

$$x_{i,1} = n_i \tag{7}$$

$$x_{i,2} = \bar{\mathbf{R}}_i = \frac{1}{n_i} \sum_{p \in P_i} R_p \tag{8}$$

$$x_{i,3} = \bar{\mathbf{G}}_i = \frac{1}{n_i} \sum_{p \in P_i} G_p \tag{9}$$

$$x_{i,4} = \bar{\mathbf{B}}_i = \frac{1}{n_i} \sum_{p \in P_i} B_p \tag{10}$$

where n_i is the number of pixels within region (node) ω_i and P_i denotes the set of pixels in region ω_i. R_p, G_p, B_p denote the red, green and blue pixel values respectively for pixel p. The segmentation is further improved by merging adjacent nodes which have a small number of pixels, or 'similar' feature space representation. Consider a node ω_i which has a set of neighbouring nodes N_i. The best possible candidate $\omega_{j_{best}}$, for merging with node ω_i, is given by the following equation:-

$$j_{best} = \arg \min_{j \in N_i} \left\{ \sqrt{(\bar{\mathbf{R}}_i - \bar{\mathbf{R}}_j)^2 + (\bar{\mathbf{G}}_i - \bar{\mathbf{G}}_j)^2 + (\bar{\mathbf{B}}_i - \bar{\mathbf{B}}_j)^2} \right\} \tag{11}$$

Node ω_i is only merged with node $\omega_{j_{best}}$ if the following criterion is satisfied:-

$$\sqrt{(\bar{\mathbf{R}}_i - \bar{\mathbf{R}}_{j_{best}})^2 + (\bar{\mathbf{G}}_i - \bar{\mathbf{G}}_{j_{best}})^2 + (\bar{\mathbf{B}}_i - \bar{\mathbf{B}}_{j_{best}})^2} \leq \tau_c \tag{12}$$

where τ_c is some pre-specified threshold which controls the degree of merging for similarly coloured homogeneous regions. In a second merging stage, each node ω_i is merged with node $\omega_{j_{best}}$ if the following criterion is satisfied:-

$$\tau_s \geq \frac{n_i}{\sum_{\forall j} n_j} \tag{13}$$

In practice τ_s controls how large, relative to the size of the image, the smallest region is allowed to be. It is expressed as a fraction (typically 1%) of the total number of image pixels. The resulting graph provides an efficient representation for the images within the system.

2.3 Contextual Colour Retrieval

In order to match a model image with a set of given scene images, the attributed graphs are created from the segmented images generated by the pre-processing. Edges in the attributed graph correspond to adjacent regions within the image. In contrast to the pre-processing stage, the double hexicone HLS colour space [8] is used for attribute measurements. The attributes of a vertex ω_j, are: mean hue H_j, mean lightness L_j and mean saturation S_j.

The conical bounds of the space limit the saturation according to lightness. This is intuitively better than some other polar colour spaces, which allow extremes in lightness (black and white) to be as saturated as pure hues, which is obviously not a desired trait. We define a colour distance measure $d_{i,j}$ between two vertices's ω_i and ω_j as :-

$$d_{ij} = \begin{cases} \Delta H_{i,j} & : \; s_i > \tau_{sat}, s_j > \tau_{sat} \\ \sqrt{\frac{1}{4}\Delta_x^2 + \frac{1}{4}\Delta_y^2 + \Delta L_{i,j}^2} & : \; \text{otherwise} \end{cases} \tag{14}$$

where

$$\Delta_x = S_i \cos(H_i) - S_j \cos(H_j) \tag{15}$$
$$\Delta_y = S_i \sin(H_i) - S_j \sin(H_j) \tag{16}$$
$$\Delta L_{i,j} = L_i - L_j \tag{17}$$
$$\Delta H_{i,j} = H_i - H_j \tag{18}$$

where τ_{sat} is a threshold which determines a boundary between chromatic and achromatic colours. Colour comparisons are often hindered by varying illumination and intensity. For this reason the difference in huc $\Delta H_{i,j}$ is chosen as the measurement criterion for chromatic colours. However, difference in hue is not an appropriate measurement for achromatic colours since hue is meaningless for colours with low saturation. In these cases, the more conventional euclidean distance type measurements are used.

The colour measurement defined above forms the basis of the vertex assignment graph edit operation. The assignment of a model vertex to the null attractor is defined to have a constant cost, as is the assignment of model or scene edges to the null attractor. In this implementation, edges are not attributed and therefore edge substitutions have zero cost (since all edges have the same attributes). More formally:

$$C(\lambda : \omega_i \mapsto \omega'_j) = 1 - N_\sigma(d_{i,j}) \tag{19}$$
$$C(\lambda : \omega_i \mapsto \emptyset) = \zeta_m \tag{20}$$
$$C(\lambda : e_i \mapsto e'_j) = 0 \tag{21}$$
$$C(\lambda : e_i \mapsto \emptyset) = \eta_m \tag{22}$$
$$C(\lambda : e_j \mapsto \emptyset) = \eta_i \tag{23}$$

where ζ_m is the cost for a missing node (0.5 typical), η_m is the cost for a missing edge (0.5 typical) and η_i is the cost for an inserted edge (0.1 typical). $N_\sigma()$ represents a Gaussian probability distribution

$$N_\sigma(x) = e^{-(\frac{1}{2})(\frac{x}{\sigma})^2} \tag{24}$$

where sigma has a typical value of 0.5. The shape of the assumed distribution does affect the efficiency of the search process. The distribution helps to discriminate between well and poorly matched attributes better. This allows the graph matching algorithm to expand deeper into the search tree before backtracking is necessary.

3 Experimental Results

The experimental results contained within this section were obtained using a C++ implementation running on an Athlon 1400XP with 512 Mb of RAM.

Fig. 1. Examples of synthetic models

Synthetic flags shown in figure 1 were considered as models. In contrast to the relaxation labelling approach in the previous work [9], the method is able to correctly self-label any of the given synthetic models, although in some cases (UK model) the symmetric regions were labelled arbitrarily. This is expected since the optimal search should always find the zero cost solution. Examples of such synthetic models are shown below.

Examples are presented which show how the system is able to label real images from synthetic models. The images in figure 2 show the interpretations for models of the Canadian and German flags, being matched to real images containing targets for these models. The first image in each example shows the target(scene) image. The second image shows the segmentation and hence graph structure, and the third image shows the interpretation results when matched to the corresponding synthetic model.

Note in each of these images the labelling is in complete agreement with ground truth data by manual annotation. The other example in figure 2 shows how the system is able to label quite complex model and scene images. The example of the USA flag shows how the system again is able to label with 100% accuracy the complex USA image in the presence of over segmentation errors.

In the previous example the model graph contains 15 nodes, and the scene graph contains 76 nodes. Even with this complexity, the system completed the match in 6.4 seconds of CPU time. The simple examples shown in figure 2 were typically matched in 0.8 seconds of CPU time including feature extraction and graph creation.

Retrieval performance can be evaluated by matching a given model to every scene image in the database. For each match, the sum of the graph edit operation costs is used as a similarity measure, from which the database can be ordered. A diverse database containing approximately 4000 images from mixed sources (including Internet, television and landscapes) was used as the experimental testbed. Ground truth data was created by manually identifying a set of target images $\mathbf{T}_m$ for a given synthetic model m.

In order to calculate the system performance Q_m, for a given model m, effective rank is introduced. Effective rank $\mathbf{R}(I_{j(i)})$ for a target image $I_{j(i)} \in \mathbf{T}_m$, is defined as the ranking of the target image $I_{j(i)}$ relative only to images which are not themselves target images ($I_k \notin \mathbf{T}_m$). This scheme is intuitive since the rank of a target should not be penalised by other targets with higher rank. The Effective rank is only penalised by false retrievals which have a higher database

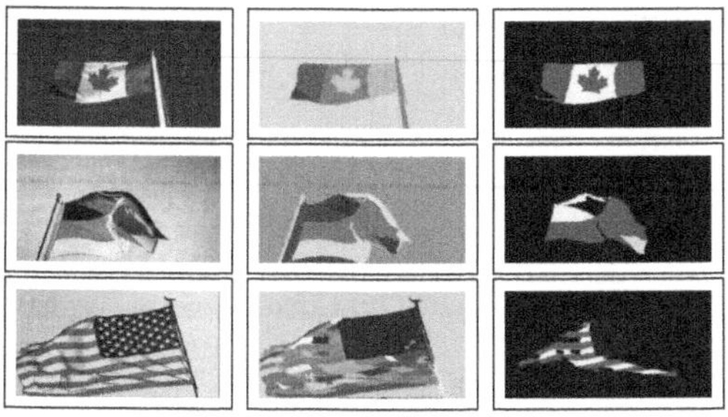

Fig. 2. Labelling Examples

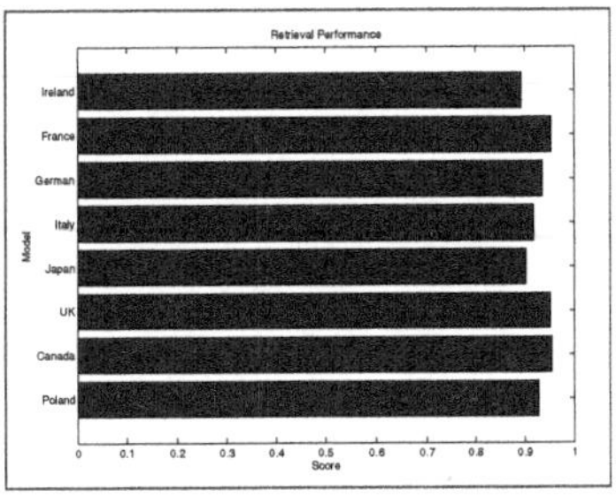

Fig. 3. Retrieval Performance

rank than the target image. Based upon the effective rank $\mathbf{R}(I_{j(i)})$, a model score $0 \leq Q_m \leq 1$ is defined as

$$Q_m = \frac{q_m}{q_{max}} \tag{25}$$

$$q_m = \sum_{I_{j(i)} \in \mathbf{T}_m} (N - \mathbf{R}(I_{j(i)}) + 1) \tag{26}$$

$$q_{max} = \sum_{i=1}^{N_{T_m}} (N - i + 1) \tag{27}$$

where N is the number of total images in the database and N_{T_m} is the total number of target images for model m. This performance evaluation criterion would yield a score of unity if all target images were ranked at the top of the database.

The system has performed well for each synthetic model. On all synthetic models, the average effective rank for the corresponding target images was always within the top 10% (approximately) of the image database.

4 Conclusion

We have presented a system for contextual colour retrieval based on graph edit operations and optimal graph search. Examples have demonstrated the performance of this system when applied to image labelling and image retrieval. Since the system uses only colour and adjacency information, it remains invariant to scale and rotation.

The results show that the adopted methods performs well in both labelling and retrieval domains. The method clearly outperforms our previous work [9]. The method is still exponential in the worst case, however the results show that for small models, the problem is quite tractable.

Future work on this system may include the incorporation of other measurements into the graph matching framework. This should improve the accuracy of labellings and the precision of retrieval. More measurement information would also push back the computational boundary since the search process would be better informed.

References

1. P. Alshuth, T. Hermes, L. Voigt, and O. Herzog. On video retrieval: content analysis by imageminer. In *SPIE-Int. Soc. Opt. Eng. Proceedings of Spie - the International Society for Optical Engineering*, volume 3312, pages 236–47, 1997. 186
2. S. Berretti, A. Del Bimbo, and E. Vicario. Efficient matching and indexing of graph models in content-based retrieval. In *IEEE Transactions on Pattern Analysis and Machine Intelligence*, volume 23, October 2001. 187, 189
3. C. Carson, S. Belongie, H. Greenspan, and J. Malik. Region-based image querying. In *Proceedings. IEEE Workshop on Content Based Access of Image and Video Libraries (Cat. No.97TB100175). IEEE Comput.Soc.*, pages 42–9, June 1997. 186
4. S-F. Chang, W. Chen, HJ. Meng, H. Sundaram, and D. Zhong. Videoq: an automated content-based video search system using visual cues. In *Proceedings ACM Multimedia 97. ACM.*, pages 313–24, USA, 1997. 186
5. W. J. Christmas, J. V. Kittler, and M. Petrou. Structural matching in computer vision using probabilistic relaxation. *IEEE Transactions on Pattern Analysis and Machine Intelligence*, 17:749–764, 8 1995. 187
6. Dorin Comaniciu and Peter Meer. Robust analysis of feature spaces: Color image segmentation. In *Proceedings of IEEE Conference on Computer Vision and Pattern Recognition*, pages 750–755, San Juan, Puerto Rico, June 1997. 189
7. Dorin Comaniciu and Peter Meer. Mean shift analysis and applications. In *IEEE Int'l Conf. Computer Vision (ICCV'99)*, pages 1197–1203, Greece, 1999. 189
8. James Foley, Andries van Dam, Steven Feiner, and John Hughes. *Computer Graphics*. Addison Wesley Longman Publishing Co, 2nd edition, 1995. 190
9. L. Gregory and J. Kittler. Using contextual information for image retrieval. In *11th International Conference on Image Analysis and Processing ICIAP01*, pages 230–235, Palermo, Italy, September 2001. 186, 192, 193
10. B. Lou and E. Hancock. A robust eigendecomposition framework for inexact graph-matching. In *11th International Conference on Image Analysis and Processing ICIAP01*, pages 465–470, Palermo, Italy, September 2001. 187
11. K Messer and J Kittler. A region-based image database system using colour and texture. *Pattern Recognition Letters*, pages 1323–1330, November 1999. 186
12. B. Messmer and H. Bunke. A new algorithm for error tolerant subgraph isomorphism detection. In *IEEE Transactions on Pattern Analysis and Machine Intelligence*, pages 493–504,, May 1998. 186, 187
13. W. Niblack, R. Barber, W. Equitz, M. Flickner, E. Glasman, D. Petkovic, P. Yanker, C. Faloutsos, and G. Taubin. The qbic project: querying images by content using color, texture, and shape. In *Proceedings of Spie - the International Society for Optical Engineering*, volume 1908, pages 173–87, Feb 1993. 186
14. A. Pentland, RW. Picard, and S. Sclaroff. Photobook: content-based manipulation of image databases. *International Journal of Computer Vision*, 18(3):233–54, June 1996. 186

Comparing Shape and Temporal PDMs

Ezra Tassone, Geoff West, and Svetha Venkatesh

School of Computing, Curtin University of Technology
GPO Box U1987, Perth 6845 Western Australia
Ph: +61 8 9266 7680 Fax: +61 8 9266 2819
{tassonee,geoff,svetha}@computing.edu.au

Abstract. The Point Distribution Model (PDM) has been successfully used in representing sets of static and moving images. A recent extension to the PDM for moving objects, the temporal PDM, has been proposed. This utilises quantities such as velocity and acceleration to more explicitly consider the characteristics of the movement and the sequencing of the changes in shape that occur. This research aims to compare the two types of model based on a series of arm movements, and to examine the characteristics of both approaches.

1 Introduction

A number of computer vision techniques have been devised and successfully used to model variations in shape in large sets of images. Such models are built from the image data and are capable of characterising the significant features of a correlated set of images. One such model is the Point Distribution Model (PDM) [1] which builds a deformable model of shape for a set of images based upon coordinate data of features of the object in the image. This is then combined with techniques such as the Active Shape Model [2] to fit the model to unseen images which are similar to those of the training set.

The PDM has been used on both static and moving images. In [3], a B-spline represents the shape of a walking person, and a Kalman filter is used in association with the model for the tracking of the person. PDMs have also been used in tracking people from moving camera platforms [4], again representing the body with a B-spline and using the Condensation algorithm to achieve the tracking. The movements of agricultural animals such as cows [5] and pigs [6] have also been described by PDMs.

Reparameterisations of the PDM have also been achieved, such as the Cartesian-Polar Hybrid PDM which adjusts its modelling for objects which may pivot around an axis [7]. Active Appearance Models extend the PDM by including the grey-level of the objects [8]. Other research has characterised the flock movement of animals by adding parameters such as flock velocity and relative positions of other moving objects in the scene to the PDM [9]. Finally purely temporal PDMs have been used to classify arm motions [10].

The aim of this research is to compare and contrast the shape PDM with the temporal PDM. The temporal PDM relies upon the sequencing of the object's

T. Caelli et al. (Eds.): SSPR&SPR 2002, LNCS 2396, pp. 195–204, 2002.
© Springer-Verlag Berlin Heidelberg 2002

motion and how this movement can be modelled on frame by frame basis. The basic shape model does not account for sequence and instead is constructed purely from spatial coordinate data. By examining the performance of both models with a classification problem, features unique to both models should become apparent. This paper will describe the derivation of both the shape and temporal models, the process used for classification and a set of experimental results.

2 The Point Distribution Model

2.1 Standard Linear PDM

The construction of the PDM is based upon the shapes of images contained within a training set of data [1]. Each shape is modelled as a set of n "land-mark" points on the object represented by xy-coordinates. The points indicate significant features of the shape and should be marked consistently across the set of shapes to ensure proper modelling. Each shape is represented as a vector of the form:

$$\mathbf{x} = (x_1, y_1, x_2, y_2, x_3, y_3, \ldots, x_n, y_n)^T \tag{1}$$

To derive proper statistics from the set of training shapes, the shapes are aligned using a weighted least squares method in which all shapes are translated, rotated and scaled to correspond with each other. This technique is based upon Generalised Procrustes Analysis [11]. The mean shape $\overline{\mathbf{x}}$ is calculated from the set of aligned shapes, where N_s is the number of shapes in the training set:

$$\overline{\mathbf{x}} = \frac{1}{N_s} \sum_{i=1}^{N_s} \mathbf{x}_i \tag{2}$$

The difference $\mathbf{dx}_i$ of each of the aligned shapes from the mean shape is taken and the covariance matrix $\mathbf{S}$ derived:

$$\mathbf{S} = \frac{1}{N_s} \sum_{i=1}^{N_s} \mathbf{dx}_i \mathbf{dx}_i^T \tag{3}$$

The modes of variation of the shape set are found from the derivation of the unit eigenvectors, $\mathbf{p}_i$, of the matrix $\mathbf{S}$:

$$\mathbf{S}\mathbf{p}_i = \lambda_i \mathbf{p}_i \tag{4}$$

The most significant modes of variation are represented by the eigenvectors aligned with the largest eigenvalues. The total variation of the training set is calculated from the sum of all eigenvalues with each eigenvalue representing a fraction of that value. Therefore the minimal set of eigenvectors that will describe a certain percentage (typically 95% or 99%) of the variation is chosen. Hence any shape, $\mathbf{x}$, in the training set can be estimated by the equation:

$$\mathbf{x} = \overline{\mathbf{x}} + \mathbf{Pb} \tag{5}$$

where $\mathbf{P} = (\mathbf{p}_1\mathbf{p}_2 \ldots \mathbf{p}_m)$ is a matrix with columns containing the m most significant eigenvectors, and $\mathbf{b} = (\mathbf{b}_1\mathbf{b}_2 \ldots \mathbf{b}_m)^T$ is the set of linearly independent weights associated with each eigenvector. For a shape x, b can thus be estimated:

$$\mathbf{b} = \mathbf{P}^T(\mathbf{x} - \overline{\mathbf{x}}) \tag{6}$$

The set of weights may also be used as parameters to produce other shapes which are possible within the range of variation described by the PDM. As the variance of each $\mathbf{b}_i$ is λ_i, the parameters would generally lie in the limits:

$$-3\sqrt{\lambda_i} \leq b_i \leq 3\sqrt{\lambda_i} \tag{7}$$

2.2 Modified PDM for Motion Components

While prior research has shown it is possible to use the standard PDM for constructing models based upon a temporal sequence of images, this paper instead proposes a reparameterisation of the PDM. The modified version of the model does not directly use image coordinates of the body but instead processes this data and derives other measures for input.

To construct the PDM, a number of frames of the object in motion are taken, and the boundary of the object extracted. A subset of n points is selected for use in developing the model. The movement of the body from frame to frame and the subsequent boundary extraction generates a new image for input and processing. The temporal sequencing of the shapes and the relative movement of the points on the shapes is what is then used to reparameterise the PDM.

To achieve this a set of three temporally adjacent frames is considered at a time with the (x, y) movement of a point from the first to the second frame being the vector $\mathbf{v}_a$ and the movement from the second frame to the third being the vector $\mathbf{v}_b$ as in Figure 1. These vectors are measured as the Euclidean norm between the (x, y) coordinates of the points. From these vectors, the relevant motion components and thus the input parameters for the PDM can be calculated:

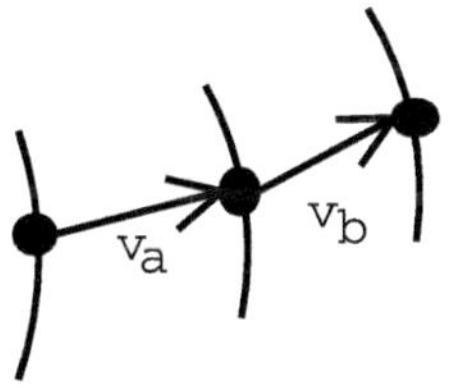

Fig. 1. Frame triple and its vectors for modified PDM

1. **Angular velocity,** $\Delta\theta$ — the change in angle between the vectors, with a counter-clockwise movement considered a positive angular velocity and a clockwise movement a negative angular velocity.
2. **Acceleration,** a — the difference in the Euclidean norm between the vectors $\| \mathbf{v}_b \| - \| \mathbf{v}_a \|$.
3. **Linear velocity,** v — this is the norm of the second vector $\| \mathbf{v}_b \|$.
4. **Velocity ratio,** r — the ratio of the second vector norm to the first vector norm, $\| \mathbf{v}_b \| / \| \mathbf{v}_a \|$. For a constantly accelerating body this measure will remain constant.

These parameters are calculated for every one of the n points of the object leading to a new vector representation for the PDM:

$$\mathbf{x} = (\Delta\theta_1, a_1, v_1, r_1, \Delta\theta_2, a_2, v_2, r_2, \ldots, \Delta\theta_n, a_n, v_n, r_n)^T$$

The user may also choose to focus on only one parameter for each point reducing the vector size and complexity of the model. This process is repeated for all triples of consecutive frames in the sequence. In this way information from all N frames in the sequence is included, however this reduces the number of temporal component shapes in the training set to be $N - 2$. After this reparameterisation of the model, the PDM can be built in the standard way. This characterisation encapsulates the temporal sequencing of the motion with the changes in parameters modelled on a frame to frame basis. This differs from the standard PDM which incorporates no temporal information in the model and encodes only variations in shape.

3 Combining Models for Classification

3.1 Video Capture and Image Processing

Image preprocessing is performed in the same way for both the shape and temporal models. The initial images are captured at a rate of 25 frames per second via one video camera parallel to the movements. As the backgrounds of the actions are not complex, simple thresholding can be applied to segment the moving object from the image yielding a binary image.

The binary images are then chaincoded to produce the boundary of the object, which generally produces a boundary of a few hundred points. Both models require a more minimal set of points for model building and hence a set of n points from these boundary points is derived. In the first frame, the subset of points is derived by choosing points from the initial boundary so the points are spaced equally. Points are then chosen in the next frames by their correspondence with points in previous frames as is typical when examining motion.

While more complex schemes are possible and could be utilised in the future, correspondence is achieved in this research by examining a specific region on the boundary of the object in the frame and choosing the point that is closest in terms of Euclidean distance to the previously found point. A further check is incorporated by using Sobel edge detection on the region to check that the found point shares the same orientation as the prior point. Experimentation demonstrates that this scheme provides reasonable correspondences as the movement between frames is not typically large and thus the likelihood of finding a suitable match is increased.

3.2 Point Distribution Models

The above scheme yields a collection of N shapes in the form of (x, y) coordinates. As described in prior research, these can be reparameterised into the motion components for the modified PDM for all points on all images. In this instance, as the motions were stable over time, the linear velocity parameter is used to build the model. After reparameterisation, this gives vectors of this form for each of the N shapes:

$$\mathbf{t} = (v_1, v_2, v_3, \ldots, v_n)^T \tag{8}$$

Standard shape PDMs also require (x, y)coordinate shapes for direct input into the modelling process, in the form of the landmark points. In order to avoid manual labelling these points or other time-consuming processes, the coordinate shapes generated from the image processing phase were used as the landmark points of the PDM. Again as the motions are relatively slow and constant, these points should provide adequate input for the shape model. This yields vectors of the form:

$$\mathbf{s} = (x_1, y_1, x_2, y_2, x_3, y_3, \ldots, x_n, y_n)^T \tag{9}$$

After having derived the data for input, both versions of the PDM can be computed in the standard way.

3.3 Movement Classification

To classify movements using both types of PDM, models are matched against test sets of data (preprocessed into (x, y) shapes as described previously). These data sets are not a part of the sequence from which the PDM was built, but are taken from the same general sequences of motion and hence provide spatial and temporal characteristics similar to those found in the models.

For the temporal PDM, the shapes are reparameterised into vectors of motion components and then the model tracked against these vectors. This is achieved through adjustment of the b values in order to determine the composition of parameters that best match the temporal "shape". The limits of these values are set to lie within three standard deviations.

The Active Shape Model [ ] is a standard iterative technique for fitting an instance of a PDM model to an example shape. However, this research uses a more general optimisation technique of the multidimensional version of Powell's method [ ]. It will attempt to minimise the error between the required vector and the vector predicted by the model, which will be measured for each vector that is tracked. Any b values that do not fall within the specified limits are adjusted to fit and so the matching will restrict the predicted motion to fall within the bounds of the PDM.

The shape PDM will also be tracked against a test set of vectors, in this case the original (x, y) shapes derived from image processing. As these form a reasonable approximation of the object's shape, these will form a "ground truth" and the model will attempt to adjust its parameters to match these new shapes. As with the temporal model, errors will be measured as to the difference between the actual and predicted shapes.

All data sets were matched against several models, one of which is part of the overall sequence of the test data. For both model types, the model which produced the lowest matching error at the end of the tracking phase would then classify the test motion as being of same type of the model. Models built from the same movement as the test set should ideally provide temporal and spatial features similar to those of the unknown sequence and hence most accurately match the motion. The characteristics of each model classification can then be compared.

4 Experimental Results

4.1 Motions and Their Models

The sequences of motion consisted of six distinct arm movements repeatedly performed by the same subject and using the same camera angle. These are illustrated with diagrams in Figure 2. A few hundred representative frames of each motion were captured, with the first 200 (or more) reserved for building the PDMs. The last 200 frames of the sequences were reserved for the test data sets. A boundary of 20 points was selected to build the temporal PDM and for input into the shape PDM. Both models were trained to describe 95% of the variation present in the training sets of data and b vector limits being $\pm 3\sigma$. An illustration of four of the modes of variation for the shape model of motion B is shown in Figure 3 from the most significant mode to the least significant mode,

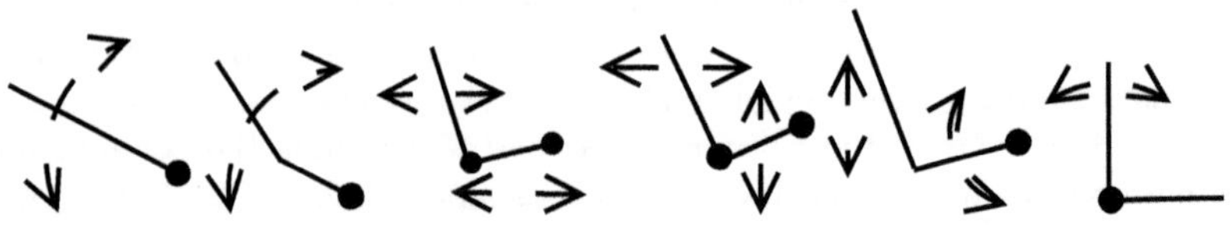

(a) A (b) B (c) C (d) D (e) E (f) H

Fig. 2. Six arm movements where each blob denotes a point of rotation. Arrows show allowable movement

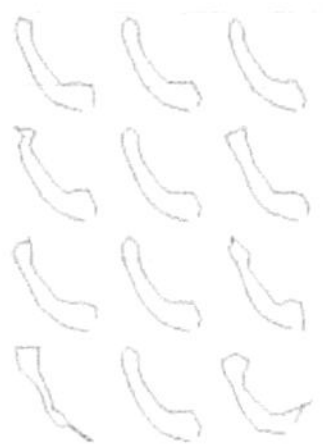

Fig. 3. Four modes of variation for a shape model of motion B

with the middle shape being the mean figure and others representing the range of variation present in each mode.

4.2 Model Classifications and Comparison

Both types of models can be examined separately for determining the lowest match error and hence the classification of the motion. Ideally all test motions should match with the prior model of their motions ie. all of the lowest errors should take place at the end of the error graph sequence or equivalently on the diagonal of the error matrix.

Figure 4 shows the progress of classification for both the temporal and spatial models for a test set of motion D. For both model types, motion D has been correctly matched to its motion model. However, it can be seen that deviations from the correct model are more pronounced for the temporal model and hence these models would seem to be more distinctive. It is also significant that the temporal model provides greater consistency in its error measurements than the shape model. As the criteria of classification is based upon choosing the model with the lowest error at the end of the sequence, this would imply ending the

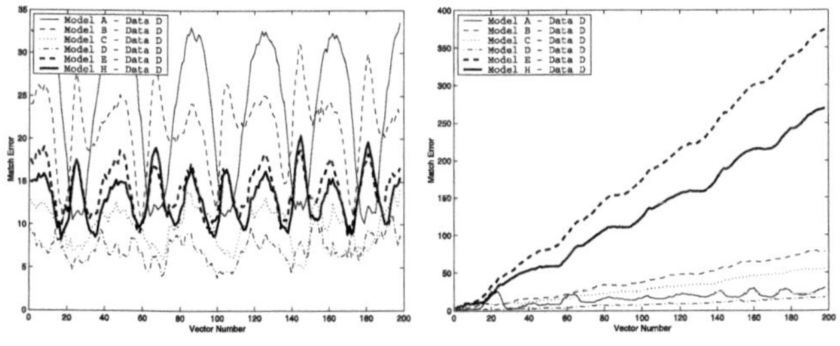

(a) Shape Model (b) Temporal Model

Fig. 4. Error plots for motion D

Table 1. Error matrix for temporal model

Data	Models					
	A	B	C	D	E	H
A	13.27	43.51	74.04	54.59	217.89	145.99
B	96.28	32.09	62.85	40.68	329.27	279.02
C	424.77	93.55	38.26	67.18	235.62	130.03
D	29.97	78.12	55.66	17.31	373.74	269.44
E	131.62	100.62	107.32	101.72	24.88	72.73
H	331.60	249.56	114.10	93.12	68.40	78.81

sequence at a different point could result in an incorrect classification using the shape model. In the error graphs of the shape model, model C is often very close and intersecting with the errors of model D particularly in the latter stages of the sequence. While model A is close to the error of D in the graphs of the temporal model, D quickly establishes that it is the correct model with the lowest error.

The error matrix for the temporal model is shown in Table 1 and that for the spatial model in Table 2. Only one misclassification occurs with the temporal model, that of motion H being matched to E. No misclassifications occur with the spatial model.

In this instance, the shape model has marginally outperformed by having no classification errors. However, inspecting the error matrix for the temporal model shows that model H provided the second lowest match error for the test set of motion H and thus a completely correct classification was very close to be attained. It may also be true that motion H (the "wave") is a less distinctive motion and hence difficult to classify temporally.

Examining the error matrices for both models, the matches provided by the shape model generally have lower levels of error. This would suggest that it can better capture the characteristics of certain types of motion. However is also likely that the shape model is more sensitive to errors in correspondence and segmentation ie. the placement of the landmark points.

Table 2. Error matrix for shape model

Data	Models					
	A	B	C	D	E	H
A	8.88	28.00	11.77	14.21	14.43	18.10
B	26.04	10.49	15.56	14.71	17.60	25.57
C	23.84	20.57	4.82	7.38	14.41	11.03
D	33.54	22.79	14.17	8.72	16.64	15.54
E	31.36	13.64	8.74	2.67	3.44	8.12
H	40.88	28.07	19.14	13.71	17.83	6.71

The temporal model also has the advantage that restrictions on possible model shapes are implicitly encoded into the model. The range of variation that it provides ensures that only those transitions which were possible in the original motions are able to be derived from the model. The shape model may also place restrictions on the movement but these are put in place after the model building and require further computation. Temporal PDMs may also be more appropriate when dealing, for example, with motions with non-uniform acceleration and velocity. A shape model will only consider the coordinate data regardless of the movement and would produce the same model, whereas the temporal PDM will be able to represent the velocities and accelerations in its model.

5 Conclusion

This paper has presented a preliminary comparison of shape and temporal PDMs. The performance of the models was similar, with only one misclassification for the temporal model and none for the shape model. The shape model provided for lower match errors than the temporal model, although the temporal models appear to be more discriminatory than the shape models. The temporal PDM also provides temporal sequencing within the model itself rather then having to be added as an additional constraint as in the case of the shape model. This provides for it to better represent the changing movements of the objects. The shape model would be unlikely to discriminate between two movements done at different velocities or accelerations, but the temporal model can cope with such data. Further work will use more of the other parameters of the temporal model and also investigate combining the models for classification.

References

1. T. F. Cootes, C. J. Taylor, D. H. Cooper, and J. Graham. Training models of shape from sets of examples. In *Proceedings of the British Machine Vision Conference, Leeds, UK*, pages 9–18. Springer-Verlag, 1992. 195, 196
2. T. F. Cootes and C. J. Taylor. Active shape models - 'smart snakes'. In *Proceedings of the British Machine Vision Conference, Leeds, UK*, pages 266–275. Springer-Verlag, 1992. 195, 199
3. A. M. Baumberg and D. C. Hogg. An efficient method of contour tracking using Active Shape Models. In *1994 IEEE Workshop on Motion of Non-rigid and Articulated Objects*, 1994. 195
4. Larry Davis, Vasanth Philomin, and Ramani Duraiwami. Tracking humans from a moving platform. In *15th. International Conference on Pattern Recognition, Barcelona, Spain*, pages 171–178, 2000. 195
5. D. R. Magee and R. D. Boyle. Spatio-temporal modeling in the farmyard domain. In *Proceeding of the IAPR International Workshop on Articulated Motion and Deformable Objects, Palma de Mallorca, Spain*, pages 83–95, 2000. 195
6. R. D. Tillett, C. M. Onyango, and J. A. Marchant. Using model-based image processing to track animal movements. *Computers and Electronics in Agriculture*, 17:249–261, 1997. 195

7. Tony Heap and David Hogg. Extending the Point Distribution Model using polar coordinates. *Image and Vision Computing*, 14:589–599, 1996. 195
8. T. F. Cootes, G. J. Edwards, and C. J. Taylor. Active appearance models. In *Proceedings of the European Conference on Computer Vision*, volume 2, pages 484–498, 1998. 195
9. N. Sumpter, R. D. Boyle, and R. D. Tillett. Modelling collective animal behaviour using extended Point Distribution Models. In *Proceedings of the British Machine Vision Conference, Colchester, UK*. BMVA Press, 1997. 195
10. Ezra Tassone, Geoff West, and Svetha Venkatesh. Classifying complex human motion using point distribution models. In *5th Asian Conference on Computer Vision, Melbourne, Australia*, 2002. 195
11. J. C. Gower. Generalized Procrustes Analysis. *Psychometrika*, 40(1):33–51, March 1975. 196
12. William H. Press, Saul A. Teukolsky, William T. Vetterling, and Brian P. Flannery. *Numerical Recipes in C: The Art of Scientific Computing*. Cambridge University Press, second edition, 1992. 199

Linear Shape Recognition with Mixtures of Point Distribution Models

Abdullah A. Al-Shaher and Edwin R. Hancock

Department of Computer Science
University of York,York YO1 5DD, UK.
{abdullah,erh}@minster.cs.york.ac.uk

Abstract. This paper demonstrates how the EM algorithm can be used for learning and matching mixtures of point distribution models. We make two contributions. First, we show how to shape-classes can be learned in an unsupervised manner. Second, we show how recognition by alignment can be realised by fitting a mixture of linear shape deformations. We evaluate the method on the problem of learning class-structure and recognising Arabic characters.

1 Introduction

Deformable models have proved to be both powerful and effective tools in the analysis of objects which present variable shape and appearance. There are many examples in the literature. These include the point distribution model of Cootes and Taylor [1], Sclaroff and Pentland's [2] finite element method, and, Duta and Jain's [3] elastic templates. There are two issues to be considered when designing a deformable model. The first of these is how to represent the modes of variation of the object under study. The second is how to train the deformable model. One of the most popular approaches is to allow the object to undergo linear deformation in the directions of the modal variations of shape. These modes of variation can be found by either performing principal components [4], or independent components analysis on the covariance matrix for a set of training examples [5], or by computing the modes of elastic vibration [6].

Recently, there have been attempts to extend the utility of such methods by allowing for non-linear deformations of shape [7]. Here there are two contrasting approaches. The first of these is to use a non-linear deformation model. The second approach is to use a combination of locally linear models. In this paper we focus on this latter approach.

In this paper, our aim is to explore how point-distribution models can be trained and fitted to data when multiple shape classes or modes of shape-variation are present. The former case arises when unsupervised learning of multiple object models is attempted. The latter problem occurs when shape variations can not be captured by a single linear model. Here we show how both learning and model fitting can be effected using the apparatus of the EM algorithm.

T. Caelli et al. (Eds.): SSPR&SPR 2002, LNCS 2396, pp. 205–215, 2002.
© Springer-Verlag Berlin Heidelberg 2002

In the learning phase, we use the EM algorithm to extract a mixture of point-distribution models from the set of training data. Here each shape-class is represented using a Gaussian distribution with its own mean-shape and covariance matrix. From the estimated parameters of the Gaussian mixtures, the point-distribution model can be constructed off-line by performing Principal Component Analysis (PCA) [8] on the class covariance matrices. In the model fitting phase, we fit a mixture of Point Distribution Models (PDM's) [9] using an architecture reminiscent of the hierarchical mixture of experts algorithm of Jordan and Jacobs [10]. Here each of the class-dependant PDM's identified in the learning step is treated as an expert. The recognition architecture is as follows. Each point in the test pattern may associated to each of the landmark points in each of the class-dependant PDM's with an *a posteriori* probability. In addition, we maintain a set of alignment parameters between the test pattern and each of the PDM's.

We experiment with the method on Arabic characters. Here we use the new methodology to learn character classes and perform recognition by alignment. This is a challenging problem since the data used exhibits a high degree of variability.

2 Point Distribution Models

The point distribution model of Cootes and Taylor commences from a set training patterns. Each training pattern is a configuration of labelled point co-ordinates or landmarks. The landmark patterns are collected as the object in question undergoes representative changes in shape. To be more formal, each landmark pattern consists of L labelled points whose co-ordinates are represented by the set of position co-ordinates $\{X_1, X_2, ..., X_l\} = \{(x_1, y_1), ...(x_L, y_L)\}$. Suppose that there are T landmark patterns. The tth training pattern is represented using the long-vector of landmark co-ordinates $X_t = (x_1, y_1, x_2, y_2, \cdots, x_L, y_L)^T$, where the subscripts of the co-ordinates are the landmark labels. For each training pattern the labelled landmarks are identically ordered. The mean landmark pattern is represented by the average long-vector of co-ordinates $Y = \frac{1}{T}\sum_{t=1}^{T} X_t$. The covariance matrix for the landmark positions is $\Sigma = \frac{1}{T}\sum_{t=1}^{T}(X_t - Y)(X_t - Y)^T$. The eigenmodes of the landmark covariance matrix are used to construct the point-distribution model. First, the unit eigenvalues λ of the landmark covariance matrix are found by solving the eigenvalue equation $|\Sigma - \lambda I| = 0$ where I is the $2L \times 2L$ identity matrix. The eigen-vector ϕ_i corresponding to the eigenvalue λ_i is found by solving the eigenvector equation $\Sigma\phi_i = \lambda_i\phi_i$. According to Cootes and Taylor [9], the landmark points are allowed to undergo displacements relative to the mean-shape in directions defined by the eigenvectors of the covariance matrix Σ. To compute the set of possible displacement directions, the K most significant eigenvectors are ordered according to the magnitudes of their corresponding eigenvalues to form the matrix of column-vectors $\Phi = (\phi_1|\phi_2|...|\phi_K)$, where $\lambda_1, \lambda_2, ..., \lambda_K$ is the order of the magnitudes of the eigenvectors. The landmark points are allowed to move in a direction which is a linear combination of the eigenvectors. The updated landmark positions are given by $\hat{X} = Y + \Phi\gamma$, where γ is a vector of modal co-efficients. This vector represents the free-parameters of the global shape-model.

3 Learning Mixtures of PDM's

In Cootes and Taylor's method [9], learning involves extracting a single covariance matrix from the sets of landmark points. Hence, the method can only reproduce variations in shape which can be represented as linear deformations of the point positions. To reproduce more complex variations in shape either a non-linear deformation or a series of local piecewise linear deformations must be employed.

In this paper we adopt an approach based on mixtures of point-distributions. Our reasons for adopting this approach are twofold. First, we would like to be able to model more complex deformations by using multiple modes of shape deformation. The need to do this may arise in a number of situations. The first of these is when the set of training patterns contains examples from different classes of shape. In other words, we are confronted with an unsupervised learning problem and need to estimate both the mean shape and the modes of variation for each class of object. The second situation is where the shape variations in the training data can not be captured by a single covariance matrix, and a mixture is required.

Our approach is based on fitting a Gaussian mixture model to the set of training examples. We commence by assuming that the individual examples in the training set are conditionally independent of one-another. We further assume that the training data can be represented by a set of shape-classes Ω. Each shape-class ω has its own mean point-pattern Y_ω and covariance matrix Σ_ω. With these ingredients, the likelihood function for the set of training patterns is

$$p(X_t, t = 1, ..., T) = \prod_{t=1}^{T} \sum_{\omega \in \Omega} p(X_t | Y_\omega, \Sigma_\omega) \tag{1}$$

where $p(X_t | Y_\omega, \Sigma_\omega)$ is the probability distribution for drawing the training pattern X_t from the shape-class ω.

According to the EM algorithm, we can maximise the likelihood function above, by adopting a two-step iterative process. The process revolves around the expected log-likelihood function

$$Q_L(C^{(n+1)} | C^{(n)}) = \sum_{t=1}^{T} \sum_{\omega \in \Omega} P(t \in \omega | X_t, Y_\omega^{(n)}, \Sigma_\omega^{(n)}) \ln p(X_t | Y_\omega^{(n+1)}, \Sigma_\omega^{(n+1)}) \tag{2}$$

where $Y_\omega^{(n)}$ and $\Sigma_\omega^{(n)}$ are the estimates of the mean pattern-vector and the covariance matrix for class ω at iteration n of the algorithm. The quantity $P(t \in \omega | X_t, Y_\omega^{(n)}, \Sigma_\omega^{(n)})$ is the *a posteriori* probability that the training pattern X_t belongs to the class ω at iteration n of the algorithm. The probability density for the pattern-vectors associated with the shape-class ω, specified by the estimates of the mean and covariance at iteration $n+1$ is $p(X_t|Y_\omega^{(n+1)}, \Sigma_\omega^{(n+1)})$. In the M, or maximisation, step of the algorithm the aim is to find revised estimates of the mean pattern-vector and covariance matrix which maximise the expected log-likelihood function. The update equations depend on the adopted model for the class-conditional probability distributions for the pattern-vectors.

In the E, or expectation, step the *a posteriori* class membership probabilities are updated. This is done by applying the Bayes formula to the class-conditional density. At iteration $n+1$, the revised estimate is

$$P(t \in \omega|X_t, Y_\omega^{(n)}, \Sigma_\omega^{(n)}) = \frac{p(X_t|Y_\omega^{(n)}, \Sigma_\omega^{(n)})\pi_\omega^{(n)}}{\sum_{\omega \in \Omega} p(X_t|Y_\omega^{(n)}, \Sigma_\omega^{(n)})\pi_\omega^{(n)}} \tag{3}$$

where

$$\pi_\omega^{(n+1)} = \frac{1}{T}\sum_{t=1}^{T} P(t \in \omega|X_t, Y_\omega^{(n)}, \Sigma_\omega^{(n)}) \tag{4}$$

3.1 Mixtures of Gaussians

We now consider the case when the class conditional density for the training patterns is Gaussian. Here we assume that the pattern vectors are distributed according to the distribution

$$p(X_t|Y_\omega^{(n)}, \Sigma_\omega^{(n)}) = \frac{1}{(2\pi)^L\sqrt{|\Sigma_\omega^{(n)}|}} \exp\left[-\frac{1}{2}(X_t - Y_\omega^{(n)})^T(\Sigma_\omega^{(n)})^{-1}(X_t - Y_\omega^{(n)})\right] \tag{5}$$

At iteration $n+1$ of the EM algorithm the revised estimate of the mean pattern vector for class ω is

$$Y_\omega^{(n+1)} = \sum_{t=1}^{T} P(t \in \omega|X_t, Y_\omega^{(n)}, \Sigma_\omega^{(n)})X_t \tag{6}$$

while the revised estimate of the covariance matrix is

$$\Sigma_\omega^{(n+1)} = \sum_{t=1}^{T} P(t \in \omega|X_t, Y_\omega^{(n)}, \Sigma_\omega^{(n)})(X_t - Y_\omega^{(n)})(X_t - Y_\omega^{(n)})^T \tag{7}$$

When the algorithm has converged, then the point-distribution models for the different classes may be constructed off-line using the procedure outlined in Section 2. For the class ω, we denote the eigenvector matrix by Φ_ω.

4 Recognition by Alignment

Once the set of shape-classes and their associated point-distribution models has been learnt, then they can be used for the purposes of alignment or classification. The simplest recognition strategy would be to align each point-distribution model in turn and compute the associated residuals. This may be done by finding the least-squares estimate of the modal co-efficient vector for each class in turn. The test pattern may then be assigned to the class of whose vector gives the smallest alignment error. However, this simple alignment and recognition strategy can be criticised on a number of grounds. First, it is difficult to apply if the training patterns and the test pattern contain different numbers of landmark points. Second, certain shapes may actually represent genuine mixtures of the patterns encountered in training.

To overcome these two problems, in this Section we detail how the mixture of PDM's can be fitted to data using a variant of the hierarchical mixture of experts (HME) algorithm of Jordan and Jacobs [10]. We view the mixture of point-distribution models learnt in the training phase as a set of experts which can preside over the interpretation of test patterns. Basic to our philosophy of exploiting the HME algorithm is the idea that every data-point can in principle associate to each of the landmark points in each of stored class shape-models with some *a posteriori* probability. This modelling ingredient is naturally incorporated into the fitting process by developing a mixture model over the space of potential matching assignments.

The approach we adopt is as follows. Each point in the test pattern is allowed to associate with each of the landmark points in the mean shapes for each class. The degree of association is measured using an *a posteriori* correspondence probability. This probability is computed by using the EM algorithm to align the test-pattern to each mean-shape in turn. This alignment process is effected using the point-distribution model to each class in turn. The resulting point alignment errors are used to compute correspondence probabilities under the assumption of Gaussian errors. Once the probabilities of individual correspondences between points in the test pattern and each landmark point in each mean shape are to hand, then the probability of match to each shape-class may be computed.

4.1 Landmark Displacements

Suppose that the test-pattern is represented by the vector $W = (\boldsymbol{w}_1, \boldsymbol{w}_2, ..., \boldsymbol{w}_D)^T$ which is constructed by concatenating D individual co-ordinate vectors $\boldsymbol{w}_1, ...\boldsymbol{w}_D$. However, here we assume that the labels associated with the co-ordinate vectors is unreliable, i.e. we can not use the order of the components of the test-pattern to establish correspondences. We hence wish to align the point distribution model for each class in turn to the unlabelled set of D point position vectors $W = \{\boldsymbol{w}_1, \boldsymbol{w}_2, ..., \boldsymbol{w}_D\}$. The size of this point set may be different to the number of landmark points L used in the training. The free parameters that must be adjusted to align the landmark points with W are the vectors modal co-efficients γ_ω for each component of the shape-mixture learnt in training.

The matrix formulation of the point-distribution model adopted by Cootes and Taylor allows the global shape-deformation to be computed. However, in order to develop our correspondence method we will be interested in individual point displacements. We will focus our attention on the displacement vector for the landmark point indexed j produced by the eigenmode indexed λ of the covariance matrix of the shape-mixture indexed ω. The two components of displacement are the elements eigenvectors indexed $2j-1$ and $2j$. For each landmark point the set of displacement vectors associated with the individual eigenmodes are concatenated to form a displacement matrix. For the jth landmark of the mixing component indexed ω the displacement matrix is

$$\Delta_j^\omega = \begin{pmatrix} \Phi_\omega(2j-1,1) & \Phi_\omega(2j-1,2) & .... & \Phi_\omega(2j-1,K) \\ \Phi_\omega(2j,1) & \Phi_\omega(2j,2) & .... & \Phi_\omega(2j,K) \end{pmatrix} \tag{8}$$

The point-distribution model allows the landmark points to be displaced by a vector amount which is equal to a linear superposition of the displacement-vectors associated with the individual eigenmodes. To this end let γ_ω represent a vector of modal superposition co-efficients for the different eigenmodes. With the modal superposition co-efficients to hand, the position of the landmark j is displaced by an amount $\Delta_j^\omega \gamma$ from the mean-position $\boldsymbol{y}_j^\omega$.

To develop a useful alignment algorithm we require a model for the measurement process. Here we assume that the observed position vectors, i.e. $\boldsymbol{w}_i$ are derived from the model points through a Gaussian error process. According to our Gaussian model of the alignment errors,

$$p(\boldsymbol{w}_i|\boldsymbol{y}_j^\omega, \gamma_\omega) = \frac{1}{2\pi\sigma} \exp\left[-\frac{1}{2\sigma^2}(\boldsymbol{w}_i - \boldsymbol{y}_j^\omega - \Delta_j^\omega\gamma_\omega)^T(\boldsymbol{w}_i - \boldsymbol{y}_j^\omega - \Delta_j^\omega\gamma_\omega)\right] \tag{9}$$

where σ^2 is the variance of the point-position errors which for simplicity are assumed to be isotropic.

4.2 Mixture Model for Alignment

We make a second application of the EM algorithm, with the aim of estimating the matrix of alignment parameters $\Gamma^{(n)} = (\gamma_1^{(n)}|\gamma_2^{(n)}|....|\gamma_{|\Omega|}^{(n)})$ is the matrix of vectors of modal alignment parameters for each of the point-distribution models residing in memory. Under the assumption that the measurements of the individual points in the test-patterns are conditionally independent of one-another, the matrix maximises the expected log-likelihood function

$$Q_A(\Gamma^{(n+1)}|\Gamma^{(n)}) = \sum_{\omega\in\Omega}\sum_{i=1}^{D}\sum_{j=1}^{L} P(\boldsymbol{y}_j^\omega|\boldsymbol{w}_i, \gamma_\omega^{(n)}) \ln p(\boldsymbol{w}_i|\boldsymbol{y}_j^\omega, \gamma_\omega^{(n+1)}) \tag{10}$$

With the displacement model developed in the previous section, maximisation of the expected log-likelihood function Q_A reduces to minimising the weighted square error measure

$$\mathcal{E}_A = \sum_{i=1}^{D}\sum_{j=1}^{L} \zeta_{ij\omega}^{(n)} (\boldsymbol{w}_i - \boldsymbol{y}_j^\omega - \Delta_j^w \gamma_\omega^{(n+1)})^T (\boldsymbol{w}_i - \boldsymbol{y}_j^\omega - \Delta_j^w \gamma_\omega^{(n+1)}) \tag{11}$$

where we have used the shorthand notation $\zeta_{ij\omega}^{(n)}$ to denote the *a posteriori* correspondence probability $P(\boldsymbol{y}_j^\omega|\boldsymbol{w}_i, \gamma_\omega^{(n)})$.

4.3 Maximisation

Our aim is to recover the vector of modal co-efficients which minimize this weighted squared error. To do this we solve the system of saddle-point equations which results by setting $\frac{\partial \mathcal{E}_A}{\partial \gamma_\omega^{(n+1)}} = 0$. After applying the rules of matrix differentiation and simplifying the resulting saddle-point equations, the solution vector is

$$\gamma_\omega^{(n+1)} = (\sum_{j=1}^{L} \Delta_j^{\omega T} \Delta_j^{\omega})^{-1} \{\sum_{i=1}^{D}\sum_{j=1}^{L} \zeta_{ij\omega}^{(n)} \boldsymbol{w}_i^T \Delta_j^{\omega} - \sum_{j=1}^{L} \boldsymbol{y}_j^{\omega T} \Delta_j^{\omega}\} \tag{12}$$

4.4 Expectation

In the expectation step of the algorithm, we use the estimated alignment parameters to update the *a posteriori* matching probabilities. The *a posteriori* probabilities $P(\boldsymbol{y}_j^\omega|\boldsymbol{w}_i, \gamma_\omega^{(n)})$ represent the probability of match between the point indexed i and the landmark indexed j from the shape-mixture indexed ω. In other words, they represent model-datum affinities. Using the Bayes rule, we re-write the *a posteriori* matching probabilities in terms of the conditional measurement densities

$$P(\boldsymbol{y}_j^\omega|\boldsymbol{w}_i, \gamma_\omega^{(n)}) = \frac{\beta_\omega^{(n)} \alpha_{j,\omega}^{(n)} p(\boldsymbol{w}_i|\boldsymbol{y}_j^\omega, \gamma_\omega^{(n)})}{\sum_{\omega' \in \Omega} \sum_{j'=1}^{L} \beta_{\omega'}^{(n)} \alpha_{j',\omega'}^{(n)} p(\boldsymbol{w}_i|\boldsymbol{y}_{j'}^{\omega'}, \gamma_{\omega'}^{(n)})} \tag{13}$$

The landmark mixing proportions for each model in turn are computed by averaging the *a posteriori* probabilities over the set of points in the pattern being matched, i.e. $\alpha_{j,\omega}^{(n+1)} = \frac{1}{D} \sum_{i=1}^{D} P(\boldsymbol{y}_j^\omega|\boldsymbol{w}_i, \gamma_\omega^{(n)})$. The *a posteriori* probabilities for the components of the shape mixture are found by summing the relevant set of point mixing proportions, i.e. $\beta_\omega^{(n+1)} = \sum_{j=1}^{L} \alpha_{j,\omega}^{(n+1)}$. In this way the *a posteriori* model probabilities sum to unity over the complete set of models. The probability assignment scheme allows for both model overlap and the assessment of ambiguous hypotheses. Above we use the shorthand notation $\alpha_{j,\omega}^{(n)}$ to represent the mixing proportion for the landmark point j from the model ω. The overall proportion of the model ω at iteration n is $\beta_\omega^{(n)}$. These quantities provide a natural mechanism for assessing the significance of the individual landmark points within each mixing component in explaining the current data-likelihood. For instance if $\alpha_{j,\omega}^{(n)}$ approaches zero, then this indicates that there is no landmark point in the data that matches the landmark point j in the model ω.

5 Experiments

We have evaluated our learning and recognition method on sets of Arabic characters. Here the landmarks used to construct the point-distribution models have

Table 1. Recognition Rate for shape-classes 1-7

		Single PDM		Mixture of PDM's	
Model No.	**No. of Samples**	**Correct**	**Wrong**	**Correct**	**Wrong**
Shape-Class 1	50	45	5	49	1
Shape-Class 2	50	48	2	50	0
Shape-Class 3	50	48	2	50	0
Shape-Class 4	50	45	5	49	1
Shape-Class 5	50	47	3	50	0
Shape-Class 6	50	48	2	50	0
Shape-Class 7	50	41	9	48	2
Recognition Rate	350	92.0%	8.0%	98.7%	1.3%

been positioned by distributing points uniformly along the length of the characters. In practice we use 20 landmarks per character in 2D space. In total there are 16 different classes of character. We use 45 samples of each character for the purposes of training. In Figure 1, we show the mean-shapes learnt in training. In the left column of the figure, we show the ground-truth mean shapes. The right column shows the learnt shapes. The two are in good agreement.

We now turn our attention to the results obtained when the shape-mixture is used for the purposes of recognition by alignment. In Figures 2 and 3 we compare the fitting of a mixture of PDM's and a single PDM to a character retained from the training-set. The different images in the sequence show the fitted PDM's as a function of iteration number. The shape shown is the one with the largest a posteriori probability. Figure 2 shows the results obtained when a single PDM is trained on the relevant set of example patterns and then fitted to the data. Figure 3 shows result obtained when training is performed using a mixture of Gaussians. The best fit is obtained when the training is performed using a mixture of Gaussians. In Figure 4 we show the alignments of the subdominant shape-components of the mixture. These are all very poor and fail to account for the data.

In Figure 5 we show the *a posteriori* probabilities β_ω for each of the mixing components on convergence. The different curves are for different shape-classes. A single dominant shape hypothesis emerges after a few iterations. The probabilities for the remaining shape-classes falls towards zero. Note that initially the different classes are equiprobable, i.e. we have not biased the initial probabilities towards a particular shape-class.

Finally, we measure the recognition rates achievable using our alignment method. Here we count the number of times the maximum a posteriori probability shape, i.e. the one for which $\omega = \arg\max \beta_\omega$, corresponds to the hand-labelled class of the character. This study is performed using 350 hand-labelled characters. Table 1 lists the recognition rates obtained in our experiments. The table

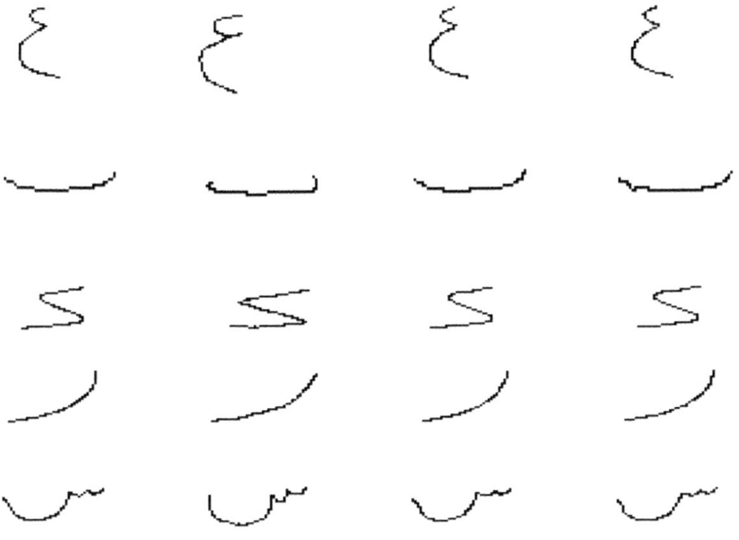

(a) (b) (c) (d)

Fig. 1. (a) Actual mean shapes, (b) EM Initialization, (c) diagonal covariance matrices, (d) non-diagonal covariance matrices

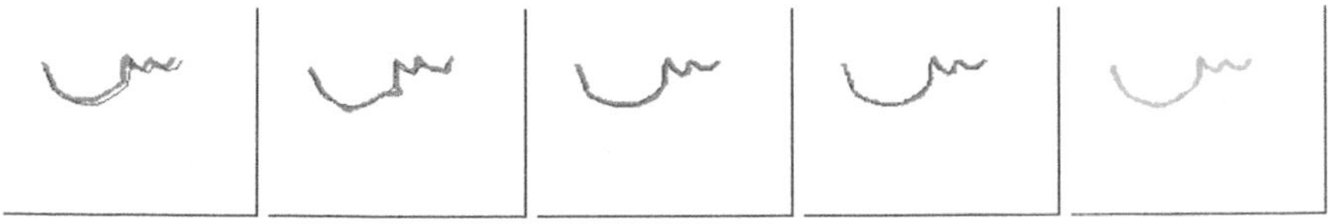

(a) (b) (c) (d) (e)

Fig. 2. Model alignment to data using Single PDM: (a) iteration 1, (b) iteration 2, (c) iteration 3, (d) iteration 5, (e) iteration 7

lists the numbers of characters recognised correctly and incorrectly for each of the shape-classes; the results a given for both single PDM's and a mixture of PDM's. The main conclusions to be drawn from the table are as follows. First, the mixture of PDM's gives a better recognition rate than using separately trained single PDM's for each class. Hence, recognition can be improved using a more complex model of the shape-space.

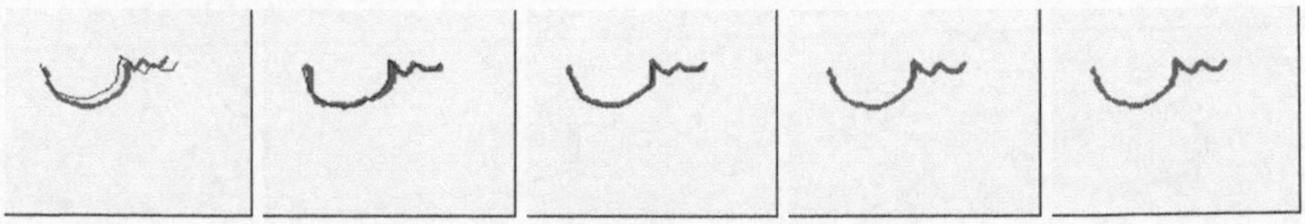

(a) (b) (c) (d) (e)

Fig. 3. Model alignment to data using mixtures of Gaussian PDM's: (a) iteration 1, (b) iteration 2, (c) iteration 3, (d) iteration 5, (e) iteration 7

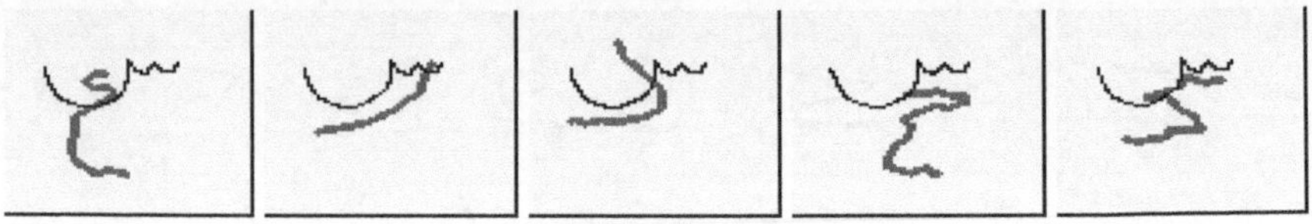

Fig. 4. Sub dominant model alignment to data using mixture of PDM's

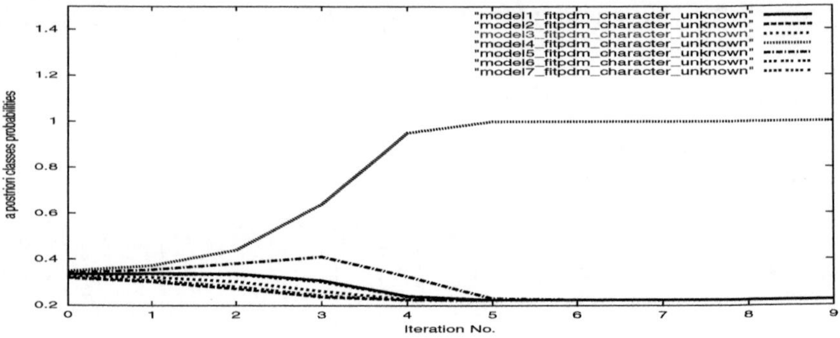

Fig. 5. Model fitting with a mixture of PDM's

6 Conclusion

In this paper, we have shown how mixtures of point-distribution models can be learned and then subsequently used for the purposes of recognition by alignment. In the training phase, we show how to use the method to learn the class-structure of complex and varied sets of shapes. In the recognition phase, we show how a variant of the hierarchical mixture of experts architecture can be used to perform detailed model alignment.

We present results on sets of Arabic characters. Here we show that the mixture of PDM's gives better performance than a single PDM. In particular we are able to capture more complex shape variations.

Our future plans revolve around developing a hierarchical approach to the shape-learning and recognition problem. Here we aim to decompose shapes into strokes and to learn both the variations in stroke shape, and the variation in stroke arrangement. The study is in hand, and results will be reported in due course.

References

1. Cootes T., Taylor C., Cooper D., Graham J. (1992). *Trainable method of parametric shape description.* Image and Vision Computing, Vol 10, no. 5, PP. 289-294. 205
2. Sclaroff S., Pentland A. (1995). *Model Matching for correspondence and recognition.* IEEE Transactions on Pattern Analysis and Machine Intelligence, Vol 17, no. 6, PP. 545-561, 1995. 205
3. Duta N., Jain A., Dubuisson P. (1999). *Learning 2D Shape Models.* International Conference on Computer Vision and pattern Recognition, Vol 2, PP. 8-14, 1999. 205
4. Cootes T., Taylor C., Cooper D., Graham J. (1995). *Active Shape Models-Their Training and Application.* Computer Vision and Image Understanding, Vol 61, no. 1, PP. 38-59. 205
5. Duda R., Hart P. (1973) *Pattern Classification and Scene Analysis.* Wiley. 205
6. Martin J., Pentland A., Sclaroff S., Kikinis R. (1998) *Characterization of neuropathological shape deformations.* IEEE transactions on Pattern Analysis and Machine Intelligence, Vol 20, no. 2, pp. 97-112, 1998. 205
7. Bowden R.; Mitchel T.; Sarhadi M. (2000). *Non-Linear statistical models for the 3D reconstruction of human pose and motion from monocular image sequences.* Image and Vision Computing, Vol 18, no. 9, PP. 729-737, 2000. 205
8. I. T. Jolliffe. (1986). *Principal Component Analysis.* Springer-Verlag, 1986. 206
9. Cootes T., Taylor C. (1999) *A mixture models for representing shape variation.* Image and Vision Computing, Vol 17, PP. 403-409, 1999. 206, 207
10. Jordan M.; Jacobs R. (1994). *Hierarchical mixtures of experts and the EM algorithm.* Neural Computation, Vol 6, PP. 181-214. 206, 209

Curvature Weighted Evidence Combination for Shape-from-Shading

Fabio Sartori and Edwin R. Hancock

Department of Computer Science
University of York, York YO10 5DD, UK

Abstract. This paper describes a new curvature consistency method for shape-from-shading. Our idea is to combine evidence for the best surface normal direction. To do this we transport surface normals across the surface using a local estimate of the Hessian matrix. The evidence combination process uses the normal curvature to compute a weighted average surface normal direction. We experiment with the resulting shape-from-shading method on a variety of real world imagery.

1 Introduction

The recovery of surface shape from shading patterns is an under constrained problem. Various constraints, including surface smoothness and occluding boundary positions can be used to render the process computationally tractable. In principal, curvature consistency constraints may also improve the quality of the recovered surface information. Despite proving alluring, there have been relatively few successful uses of curvature information in shape-from-shading, and these have overlooked much of the richness of the differential surface structure. For instance, Ferrie and Lagarde [4] have used the Darboux frame consistency method of Sander and Zucker [10]. This method uses a least-squares error criterion to measure the consistency of the directions of neighbouring surface normals and principal curvature directions. This method is applied as a postprocessing step and does not directly constrain the solution of the image irradiance equation. Worthington and Hancock have used method which enforces improved compliance with the image irradiance equation [11]. They constrain the surface normals to fall on the irradiance cone defined by Lambert's law. Subject to this constraint, the field of surface normals is smoothed using a method suggested by robust statistics. The degree of smoothing employed depends on the variance of the Koenderink and Van Doorn shape-index [8]. This is a scalar quantity which measures topography, but is not sensitive to finer differential structure, such as the directions of principal curvature.

The observation underpinning this paper is that although there is a great deal of information residing in the local Darboux frames, there has been relatively little effort devoted to exploiting this in shape-from-shading. In particular, we aim to use the field of principal curvature directions as a source of constraints that can be used to improve the shape-form-shading process. Our approach is

T. Caelli et al. (Eds.): SSPR&SPR 2002, LNCS 2396, pp. 216–224, 2002.
© Springer-Verlag Berlin Heidelberg 2002

an evidence combing one. As suggested by Worthington and Hancock [11], we commence with the surface normals positioned on their local irradiance cone and aligned in the direction of the local image gradient. From the initial surface normals, we make local estimates of the Hessian matrix. Using the Hessian, we transport neighbouring normals across the surface. This allows us to accumulate a sample of surface normals at each location. By weighting this sample by the normal curvature of the transport path and the resulting brightness error, we compute a revised surface normal direction.

2 Shape-from-Shading

Central to shape-from-shading is the idea that local regions in an image $E(x, y)$ correspond to illuminated patches of a piecewise continuous surface, $z(x, y)$. The measured brightness $E(x, y)$ will depend on the material properties of the surface, the orientation of the surface at the coordinates-ordinates (x, y), and the direction and strength of illumination.

The *reflectance map*, $R(p, q)$ characterises these properties, and provides an explicit connection between the image and the surface orientation. Surface orientation is described by the components of the surface gradient in the x and y direction, i.e. $p = \frac{\partial z}{\partial x}$ and $q = \frac{\partial z}{\partial y}$. The shape from shading problem is to recover the surface $z(x, y)$ from the intensity image $E(x, y)$. As an intermediate step, we may recover the needle-map, or set of estimated local surface normals, $\mathbf{Q}(x, y)$.

Needle-map recovery from a single intensity image is an under-determined problem [1,2,7,9] which requires a number of constraints and assumptions to be made. The common assumptions are that the surface has ideal Lambertian reflectance, constant albedo, and is illuminated by a single point source at infinity. A further assumption is that there are no inter-reflections, i.e. the light reflected by one portion of the surface does not impinge on any other part.

The local surface normal may be written as $\mathbf{Q} = (-p, -q, 1)^T$, where $p = \frac{\partial z}{\partial x}$ and $q = \frac{\partial z}{\partial y}$. For a light source at infinity, we can similarly write the light source direction as $\mathbf{s} = (-p_l, -q_l, 1)^T$. If the surface is Lambertian the reflectance map is given by $R(p, q) = \mathbf{Q} \cdot \mathbf{s}$. The image irradiance equation [6] states that the measured brightness of the image is proportional to the radiance at the corresponding point on the surface; that is, just the value of $R(p, q)$ for p, q corresponding to the orientation of the surface. Normalising both the image intensity, $E(x, y)$, and the reflectance map, the constant of proportionality becomes unity, and the image irradiance equation is simply $E(x, y) = R(p, q)$. Lambert's equation provides insufficient information to uniquely determine the surface normal direction.

3 Differential Surface Structure

In this paper we are interested in using curvature consistency information to constrain the recovery of shape-from-shading. Our characterisation of the differential structure of the surface is based on the Hessian matrix which can be computed from the currently available field of surface normals, or Gauss-map, in the following manner

$$\mathcal{H}_o = \begin{bmatrix} \frac{\partial}{\partial x}(\boldsymbol{Q}_o)_x & \frac{\partial}{\partial x}(\boldsymbol{Q}_o)_y \\ \frac{\partial}{\partial y}(\boldsymbol{Q}_o)_x & \frac{\partial}{\partial y}(\boldsymbol{Q}_o)_y \end{bmatrix} = \begin{bmatrix} h_{11} \, h_{12} \\ h_{21} \, h_{22} \end{bmatrix} \tag{1}$$

where the directional derivatives are extracted using first differences of the surface meshes in the approximation of the pixel lattice. The two eigenvalues of the Hessian matrix are the maximum and minimum curvatures:

$$K_o^{max} = -\frac{1}{2}(h_{11} + h_{22} - S) \;\; , \;\; K_o^{min} = -\frac{1}{2}(h_{11} + h_{22} + S) \tag{2}$$

where $S = \sqrt{(h_{11} - h_{22})^2 + 4(h_{21}h_{12})}$. The eigenvector associated with the maximum curvature K_o^M is the principal curvature direction. On the tangent-plane to the surface, the principal curvature direction is given by the 2-component vector

$$\boldsymbol{e}_o^{max} = \begin{bmatrix} (h_{12}, -\frac{1}{2}(h_{11} - h_{22} + S))^T & h_{11} \geq h_{22} \\ (\frac{1}{2}(h_{11} - h_{22} - S), h_{21})^T & h_{11} < h_{22} \end{bmatrix} \tag{3}$$

In this paper we are interested in using the local estimate of the Hessian matrix to provide curvature consistency constraints for shape from-shading. Our aim is to improve the estimation of surface normal direction by combining evidence from both shading information and local surface curvature. As demonstrated by both Ferrie and Lagarde [4] and Worthington and Hancock [11], the use of curvature information allows the recovery of more consistent surface normal directions. It also provides a way to control the over-smoothing of the resulting needle maps. Ferrie and Lagarde [4] have addressed the problem using local Darboux frame smoothing. Worthington and Hancock [11], on the other hand, have employed a curvature sensitive robust smoothing method. Here we adopt a different approach which uses the equations of parallel transport to guide the prediction of the local surface normal directions.

Suppose that we are positioned at the point $\boldsymbol{X}_o = (x_o, y_o)^T$ where the current estimate of the Hessian matrix is $\mathcal{H}_o$. Further suppose that $\mathbf{Q}_m$ is the surface normal at the point $\boldsymbol{X}_m = (x_m, y_m)^T$ in the neighbourhood of $\boldsymbol{X}_o$. We use the local estimate of the Hessian matrix to transport the vector $\mathbf{Q}_m$ to the location $\boldsymbol{X}_o$. The first-order approximation to the transported vector is

$$\mathbf{Q}_m^o = \mathbf{Q}_m + \mathcal{H}_o(\boldsymbol{X}_m - \boldsymbol{X}_o) \tag{4}$$

This procedure is repeated for each of the surface normals belonging to the neighbourhood R_o of the point o. In this way we generate a sample of alternative surface normal directions at the location o.

We would like to associate with the transported surface normals a measure of certainty based on the curvature of the path $\Gamma_{o,m}$ from the point m to the point o. The normal curvature at the point o in the direction of the transport path is approximately

$$\kappa_{o,m} = (\boldsymbol{T}_{o,m}.\boldsymbol{e}_o^{max})^2(K_o^{max} - K_o^{min}) + K_o^{min} \tag{5}$$

where $\boldsymbol{T}_{o.m} = \frac{\boldsymbol{X}_m - \boldsymbol{X}_o}{|\boldsymbol{X}_m - \boldsymbol{X}_o|}$ is the unit vector from o to m.

4 Statistical Framework

We would like to exploit the transported surface-normal vectors to develop an evidence combining approach to shape-from-shading. To do this we require a probabilistic characterisation of the sample of available surface normals. This is a two-component model. Firstly, we model the data-closeness of the transported surface normals with Lambert's law. We assume that the observed brightness E_o at the point $\boldsymbol{X}_o$ follows a Gaussian distribution. As a result the probability density function for the transported surface normals is

$$p(E_o|\mathbf{Q}_m^o) = \frac{1}{\sqrt{2\pi}\sigma_E}\exp\left[-\frac{(E_o - \mathbf{Q}_m^o \cdot \mathbf{s})^2}{2\sigma_E^2}\right] \tag{6}$$

where σ_E^2 is the noise-variance of the brightness errors. The second model ingredient is a curvature prior. Here we adopt a model in which we assume that the sample of transported surface normals is drawn from a Gaussian prior, which is controlled by the normal curvature of the transport path. Accordingly we write

$$p(\mathbf{Q}_m^o) = \frac{1}{\sqrt{2\pi}\sigma_k}\exp\left[-\frac{1}{2\sigma_k^2}k_{o.m}^2\right] \tag{7}$$

With these ingredients, the weighted mean for the sample of transported surface normals

$$\hat{\mathbf{Q}}_o = \sum_{m \in R_o} \mathbf{Q}_m^o p(E_o|\mathbf{Q}_m^o)p(\mathbf{Q}_m^o) \tag{8}$$

where R_o is the index set of the surface normals used for the purposes of transport. Substituting for the distributions,

$$\hat{\mathbf{Q}}_o = \frac{\sum_{m \in R_o} \mathbf{Q}_{o,m} \exp\left[-\frac{1}{2}\left(\frac{(E_o - \mathbf{Q}_m^o \cdot \mathbf{s})^2}{\sigma_E^2} + \frac{\kappa_{o,m}^2}{\sigma_k^2}\right)\right]}{\sum_m \exp\left[-\frac{1}{2}\left(\frac{(E_o - \mathbf{Q}_m^o \cdot \mathbf{s})^2}{\sigma_E^2} + \frac{\kappa_{o,m}^2}{\sigma_k^2}\right)\right]} \tag{9}$$

and the predicted brightness is $\hat{E}_o = \hat{\mathbf{Q}}_o \cdot \mathbf{s}$

This procedure is repeated at each location in the field of surface normals. We iterate the method as follows:

- 1: At each location compute a local estimate of the Hessian matrix $\mathcal{H}_o$ from the currently available surface normals $\mathbf{Q}_o$.

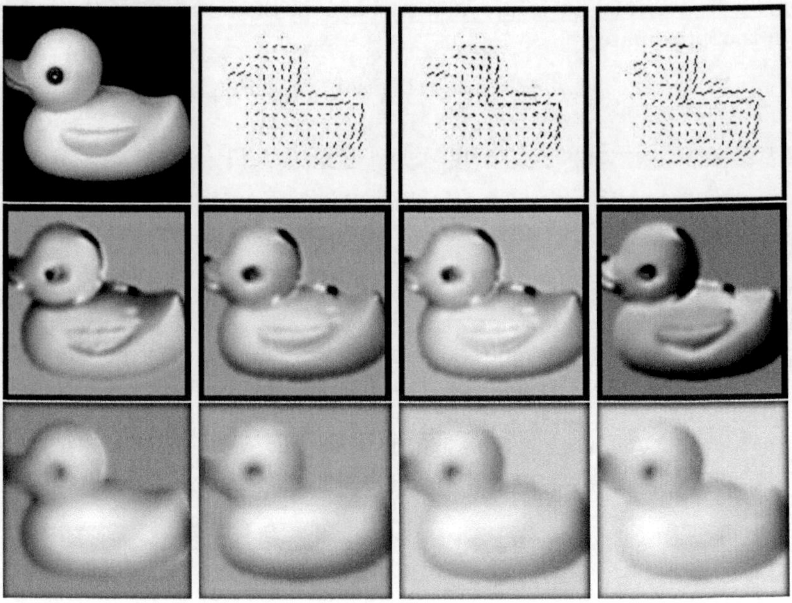

Fig. 1. Toy duck: original image, initial needle-map, final needle-maps and reconstructed images from different illumination directions

- 2: At each image location $\boldsymbol{X}_o$ obtain a sample of surface normals $N_o = \{\mathbf{Q}^o_m | m \in R_o\}$ by applying parallel transport to the set of neighbouring surface normals whose locations are indexed by the set R_o.
- 3: From the set of surface normals N_o compute the expected brightness value $\hat{E}_o$ and the updated surface normal direction $\hat{\mathbf{Q}}_o$. Note that the measured intensity E_o is kept fixed throughout the iteration process and is not updated.
- 4: With the updated surface normal direction to hand, return to step 1, and recompute the local curvature parameters.

To initialise the surface normal directions, we adopt the method suggested by Worthington and Hancock [11]. This involves placing the surface normals on the irradiance cone whose axis is the light-source direction $\mathbf{s}$ and whose apex angle is $\cos^{-1} E_o$. The position of the surface normal on the cone is such that its projection onto the image plane points in the direction of the local image gradient, computed using the Canny edge detector. When the surface normals are initialised in this way, then they satisfy the image irradiance equation.

5 Experiments

We have experimented with our new shape-form-shading method on a variety of real world imagery and have compared it with the Horn and Brooks method [7].

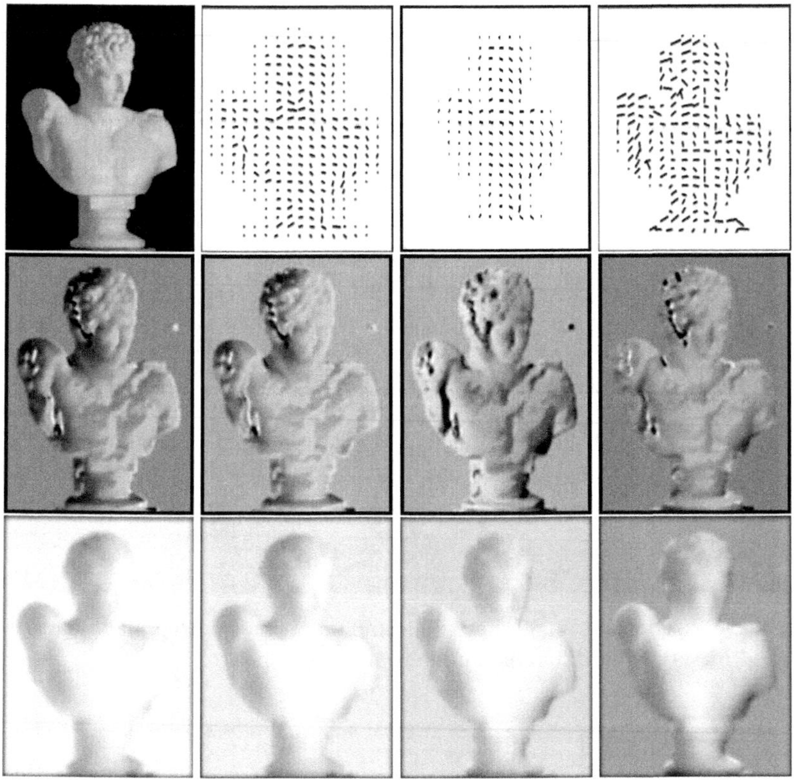

Fig. 2. Hermes: original image, initial needle-map, final needle-maps and reconstructed images from different illumination directions

The images are taken from a variety of sources including the Columbia COIL database and web epositories of images of statues.

In Figure 1 we show a sequence of results for a toy duck image from the Columbia COIL data-base. In the top row, from left to right, we show the original image, the needle-map obtained using the Horn and Brooks method [5] and the needle-map obtained using our new method. The needle-map obtained using our new method is more detailed than that obtained using the Horn and Brooks method. In the subsequent two rows of the figure, we show the results of re-illuminating the needle-map from various directions. The reilluminations reveal that the method recovers fine surface detail, especially around the wings and the head. The bottom row of the figure shows the result of reillumination using the Horn and Brooks method. This is blurred and does not convey the surface detail delivered by our new method.

Figures 2, 3 and 4 and repeat this sequence for different images. In Figure 2 we study an image of a statue of Hermes, in Figure 3 an image of a terracotta

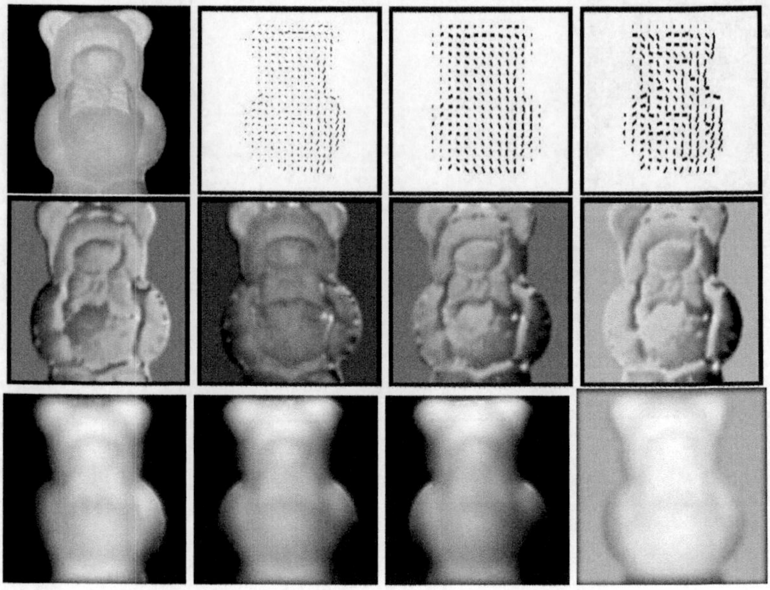

Fig. 3. Bear: Original image, initial needle-map, final needle-maps and reconstructed images from different illumination directions

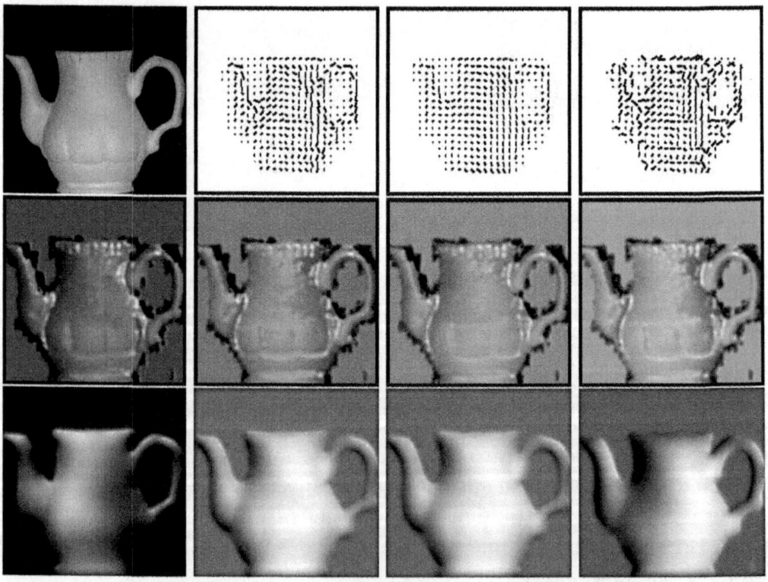

Fig. 4. Pot: Original image, initial needle-map, final needle-maps and reconstructed images from different illumination directions

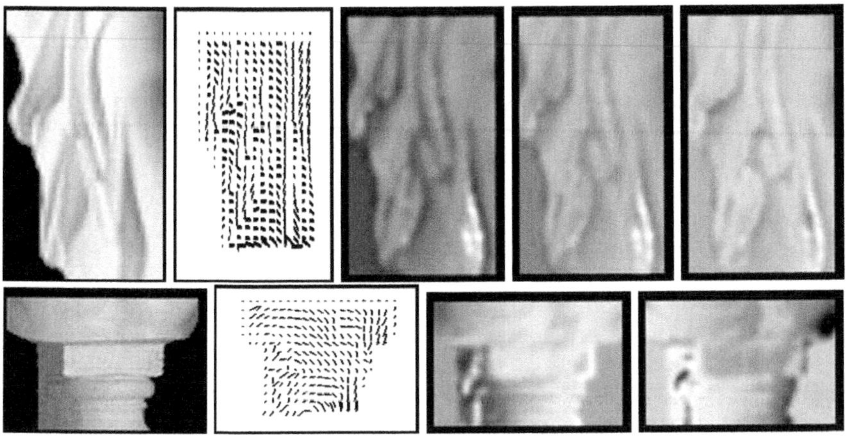

Fig. 5. Details from Vinus and Hermes: needle-map and reillumination examples

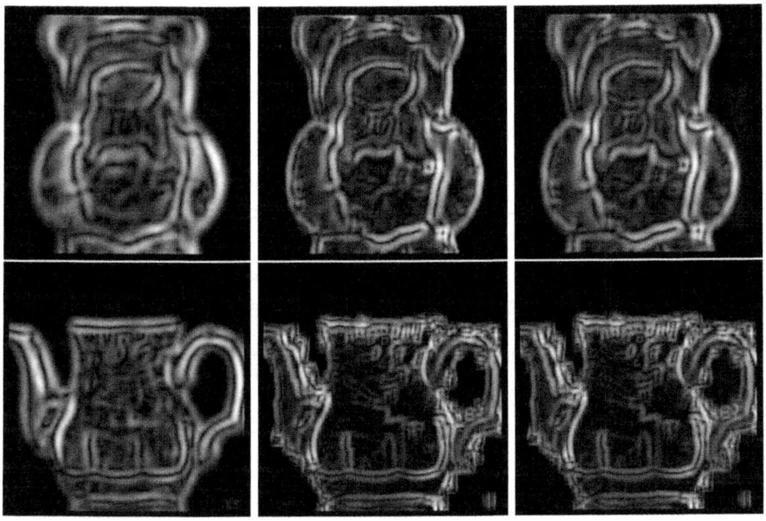

Fig. 6. Evolution of the curvedness during the minimisation process

bear, and in Figure 4 an image of a terracotta tea pot. The detail in the re-illuminations of the needle-maps obtained with our new method are much clearer than those obtained with the Horn and Brooks method.

In Figure 5 we show the results for images of highly structured surfaces. The top row shows the results for a detail in the folds of the drapery for the Venus de Milo statue. The bottom row shows the results for a detail around the plinth of the Hermes statue shown earlier. The images in each row of the figure are from left to right, the original image, the needle-map, and some example reilluminations. The results show that our method is able to recover quite fine surface detail, including high curvature structure.

Finally, we focus on the iterative qualities of the algorithm. In Figure 6, we show the curvedness $K = \sqrt{(K_o^{max})^2 + (K_o^{min})^2}$ as a function of iteration number. It is clear that the method has the effect of sharpening the curvature detail as it iterates.

6 Conclusions

In this paper we have described a curvature consistency method for shape-from-shading. The idea underpinning this work is to compute a weighted average of linearly transported surface normals. The transport is realized using a local estimate of the Hessian matrix and the weights are computed using the normal curvature of the transport path. The method proves effective on a variety of real world images.

References

1. Belhumeur, P. N. and Kriegman, D. J. (1996) What is the Set of Images of an Object Under All Possible Lighting Conditions? *Proc. IEEE Conference on Computer Vision and Pattern Recognition,* pp. 270-277. 217
2. Bruckstein, A. M. (1988) On Shape from Shading, *CVGIP,* Vol. 44, pp. 139-154. 217
3. Do Carmo M. P. (1976), Differential Geometry of Curves and Surfaces, *Prentice Hall.*
4. Ferrie, F. P. and Lagarde, J. (1990)Curvature Consistency Improves Local Shading Analysis, *Proc. IEEE International Conference on Pattern Recognition,* Vol. I, pp. 70-76. 216, 218
5. Horn, B. K. P. and Brooks, M. J. (1986) The Variational Approach to Shape from Shading, *CVGIP,* Vol. 33, No. 2, pp. 174-208. 221
6. Horn, B. K. P. and Brooks, M. J. (eds.), *Shape from Shading,* MIT Press, Cambridge, MA, 1989. 217
7. Horn, B. K. P. (1990) Height and Gradient from Shading, *IJCV,* Vol. 5, No. 1, pp. 37-75. 217, 220
8. Koenderink, J. J. (1990) *Solid Shape,* MIT Press, Cambridge MA. 216
9. Oliensis, J. and Dupuis, P. (1994) An Optimal Control Formulation and Related Numerical Methods for a Problem in Shape Reconstruction, *Ann. of App. Prob.* Vol. 4, No. 2, pp. 287-346. 217
10. Sander, P. and Zucker, S. (1990) Inferring surface trace and differential structure from 3-d images, *PAMI,* 12(9):833–854. 216
11. Worthington, P. L. and Hancock, E. R. (1999) New Constraints on Data-closeness and Needle-map consistency for SFS, *IEEE Transactions on Pattern Analysis,* Vol. 21, pp. 1250-1267. 216, 217, 218, 220

Probabilistic Decisions in Production Nets: An Example from Vehicle Recognition

Eckart Michaelsen and Uwe Stilla

FGAN-FOM Research Institute for Optronics and Pattern Recognition
Gutleuthausstr. 1, 76275 Ettlingen, Germany
{mich,stilla}@fom.fgan.de
http://www.fom.fgan.de

Abstract. A structural knowledge-based vehicle recognition method is modified yielding a new probabilistic foundation for the decisions. The method uses a pre-calculated set of hidden line projected views of articulated polyhedral models of the vehicles. Model view structures are set into correspondence with structures composed from edge lines in the image. The correspondence space is searched utilizing a 4D Hough-type accumulator. Probabilistic models of the background and the error in the measurements of the image structures lead to likelihood estimations that are used for the decision. The likelihood is propagated along the structure of the articulated model. The system is tested on a cluttered outdoor scene. To ensure any-time performance the recognition process is implemented in a data-driven production system.

1 Introduction

Vehicle recognition from oblique high resolution views has been addressed by several authors [2][7][6]. Hoogs and Mundy [7] propose to use region and contour segmentation techniques and rely on dark regions of certain size and form, that may be a vehicle shadow, and on simple features like parallel contours, that some vehicles display in a variety of perspectives. Shadows can be exploited, if the pictures are taken in bright sunlight of known direction. Omni-directional ambient lighting causes a shadowed region directly underneath the vehicle. This is visible in oblique views of vehicles but may be occluded, e.g. by low vegetation. Parallel contours are a cue to vehicles, but they are present in many environments around vehicles, too (e.g. in roads, buildings, ploughed fields).

A possibility to avoid this difficulties is to use the geometrical shape of the vehicles themselves. Viola and Wells [12] render object models and compare characteristic properties of the gray value function of the rendered graphic and the image using mutual information. Hermitson et al. [6] utilize this approach to oblique vehicle recognition. Rendering requires assumptions about the lighting and surface properties of the model. If this is not available one has to work with contours on the more abstract geometric level. Dickinson et al. [3] proposed generalized cylinder

T. Caelli et al. (Eds.): SSPR&SPR 2002, LNCS 2396, pp. 225-233, 2002.
© Springer-Verlag Berlin Heidelberg 2002

models with part-of hierarchies for contour based object recognition. Binfort and Levitt [2] applied this to vehicle recognition tasks. Generalized cylinder models capture the coarse structure of a vehicle. For details of vehicles such models are not appropriate.

Grimson [5] proposed polyhedron models and straight line segments. This has a high potential discriminative power, because many geometric properties and constraints of the targets are exploited. For the reduction of the computational effort indexing methods like generalized Hough transform as well as restricting the vehicles in position and rotation to the ground plane are proposed [10]. Some vehicles can not be covered by one rigid polyhedron alone, because they are composed of parts, that are connected by pivots or bearings (e.g. truck and trailer systems or tanks). Such objects can be captured by articulated models [11]. The appearance of polyhedrons is affected by self occlusion. This may be treated by aspect graphs [4], or by linear combination of characteristic views [11]. We use an equidistantly sampled set of views for each model [8]. In this contribution we incorporate probabilistic calculations into a structural approach.

Section 2 presents the accumulator method to solve the problem of vehicle recognition from single oblique views. The probabilistic model is described in Sect. 3. A result of an experiment on a difficult scene is given in Sect. 4. In Sect. 5 a discussion of pros and cons of the approach and an outlook on future work are given.

2 View-Based Recognition of Vehicles

View-based object recognition matches the model to the data in the 2D image space. For this purpose 2D views of the 3D model parts are constructed. It is possible to use structured models with part-of hierarchies. Then the consistency check for correct mutual positioning requires back projection.

A set of 2D lines constructed by perspective hidden-line projection from a polyhedron is called a view. In contrast to this an aspect is a line graph. Changes in the view that don't change the topology provide the same aspect [11].

2.1 The Space of Views

The space of views is originally continuous and has dimension six (three rotations and three translations). Vehicle recognition from oblique imagery constraints the distance to an interval and the spatial rotation to one off-image plane rotation (the azimuth). Depending on the focal length translations of the model may lead to geometric distortions at the margins of the image. Due to the long focal lengths used here this effect can be neglected and the same view model can be used all over the image. The model is positioned such that it appears centered in the principal point and the azimuth and distance are varied stepwise in an appropriate step width yielding finite 2D view space containing some hundred views per model. Fig. 1 shows some example views.

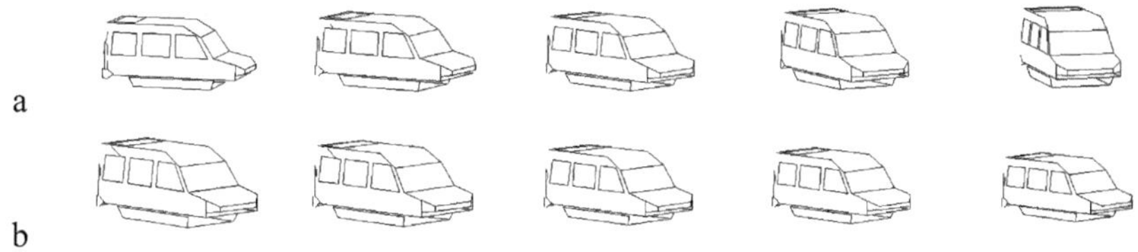

Fig. 1. Selected set of 2D models projected from a 3D polyhedron model: a) varying azimuth with $\Delta\alpha=15^o$; b) varying distance with $\Delta dis=8m$

2.2 Matching an Image to the Views

Object contours in the image are extracted using a gradient operator and morphologic thinning. The contours are approximated by short line segments. A line prolongation process improves the orientation estimation of the line objects. The set of such line objects can be matched with the lines in the views. For this task we use a generalized Hough transformation [1].

To decrease the computational complexity of the correspondence search we use L-shaped objects constructed from the lines. The L-shaped objects in all the model views are constructed off-line. As key to establish the correspondence between image and model structures the two orientations of the sides of the L-shaped objects are utilized. A structure in the image supports a part of a view if both orientations are sufficiently similar. The position of the reference point of the view is obtained subtracting the position of the part in the model view from the position in the image.

2.3 Robustness through Accumulation

Often not all modeled structures are present in images of outdoor scenes. Therefore, as much evidence as possible has to be merged from consistent cues to one specific pose. While a single cue may result from background or clutter multiple consistent cues from different structures of a specific view probably result from the presence of the modeled object in the corresponding pose. Therefore all cues are inserted in a 4D accumulator at their image position, azimuth, and distance. Resulting from different errors (modeling, imaging, feature extraction) consistent cues form a fuzzy cluster in the accumulator. For the detection of vehicles we search for dominant clusters of cues in the accumulator.

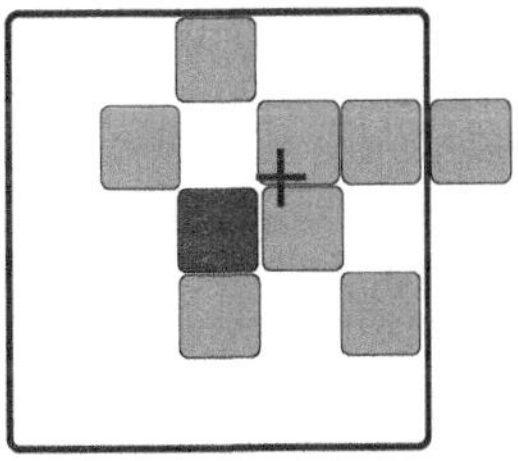

Fig.2. Searching for a proper subset in the accumulator

Each cuc locates a 4D search area. The size of this area results from the maximal expected errors. Cues within a search area are a candidate subset for a cluster object and are used to estimate the center of mass. The center of mass locates a new search area and a new subset. Such calculations are performed until convergence occurs. Fig. 2 exemplifies such procedure in 2D where the dark square indicates the position of a cue and the black square shows the corresponding search area. While the leftmost cue is missed in the first attempt it will be included in a later step, because the position of the new search area is determined by the center of mass indicated by the cross.

2.4 Part-of Hierarchies and Articulated Models

Not all vehicles are adequately described by a single shape fixed polyhedron model. Parts of a vehicle may be mutually connected and constrained by hinges or pivots (truck-trailer systems, tanks). Therefore we consider 3D models of vehicles that have a part-of hierarchy. Such a model is described by a directed graph where each basic part is a polyhedron. If the parts have mutual degrees of freedom in rotation such a model is called articulated model [11]. The resulting constraints are used by recognition process. For the consistency test the parts are projected back to the 3D scene. If a pivot or hinge is not located at the reference position of a model part, then auxiliary position attributes are used to define the search areas for partner clusters. E. g., the 2D position of the trailer hitch of a vehicle view depends on its pose. These auxiliary position attributes locate the search area for possible partners.

The information on which auxiliary attribute of which part of the model connects to which attribute of which other part, and which azimuth angle differences are permitted at this connection is given by the user in a standardized format in addition to the polyhedron models.

2.5 Production Nets and Implementation

We describe structural relations of the object models by productions. A production defines how a given configuration of objects is transformed into a single more complex object (or a configuration of more complex objects). In the condition part of a production geometrical, topological, and other relation or attributes of objects are examined. If the condition part of a production holds, an object specific generation function is executed to generate a new object. Such productions operate on sets of objects instead of graphs, strings etc. The organization of object concepts and productions can be depicted by a production net [9] which displays the part-of hierarchies of object concepts.

Our production nets are implemented in a blackboard architecture. Blackboard-systems consists of a global data base (blackboard), a set of processing modules (knowledge sources), and a control unit (selection module). The productions are implemented in the processing modules, which test the relations between objects and generate new objects. Starting with primitive objects the searched target objects are composed step by step by applying the productions. The system works in an accumulating way, this means a replaced initial configuration will not be deleted in the database. Thus all generated partial results remain available during the analysis to pursue different hypotheses. The classical backtracking in search-trees is not necessary.

3 Probabilistic Error Models

A critical issue is the choice of the optimal size of the search areas in the accumulator. With rising distance of a cue from the center of a cluster the likelihood for its membership decreases. A cue with a large distance from the cluster is probably due to background or clutter. Wells [14] used Gaussian distributions for the error of features that are in correct correspondence to the model and equal densities for background and clutter features. While he uses contour primitives attributed by their location, orientation and curvature we operate in the 4D accumulator.

3.1 Probabilistic Calculations in the Cluster Formation

Applying Wells theory we first have to estimate a reward term λ as contribution of each single cue which replaces the entry into the accumulator. From a representative training-set where the features are labeled either as correctly matched or as background or prior information λ is set to

$$\lambda = \ln\left(\frac{1}{(2\pi)^2 m} \cdot \frac{(1-B)}{B} \cdot \frac{W_1 \cdots W_4}{\sqrt{|\psi|}}\right). \tag{1}$$

The middle factor in this product is calculated from the ratio between the probability B that a feature is due to the background, and the probability $(1-B)/m$ that it corresponds to a certain model feature, where m is the number of features in the model. The rightmost factor in the product is given by the ratio between the volume of the whole feature domain $W_1 \ldots W_4$ and the volume of a standard deviation ellipsoid of the covariance matrix ψ for the correctly matched features. As feature domain we set $\beta^T=(x,y,\alpha,dis)$. Locally our accumulator domain may be treated as linear, justifying the application of this theory and its error models. The objective function L is calculated for each cluster of cues:

$$L = \sum_j \left[\lambda - \operatorname*{Min}_{\Gamma_i=j}\left[\frac{1}{2}(Y_i - \hat{\beta})^T \psi^{-1} (Y_i - \hat{\beta})\right]\right]. \tag{2}$$

Y_i is the position of the i-th cue in the accumulator domain. The pose β is estimated as mean $\hat{\beta}^T = (\hat{x}, \hat{y}, \hat{\alpha}, \hat{dis})$ of the poses of the member cues of the cluster. The correspondence Γ is coded as an attribute of the cues. For each model feature j put into correspondence in the cluster the closest cue i to the mean is taken as representative of the set of all cues i corresponding to j. This is done, because we regard multiple cues to the same model feature as not being mutual independent.

Recall that the maximization must not take those Γ into account, that include negative terms into the sum. Fig. 4 displays the 1D case: Full reward λ is only given for a precise match. With rising error the reward is diminished by a negative parabola. Finally it reaches zero level. At this point Γ is changed, setting the feature in correspondence to the background. This condition gives a new way to infer the threshold parameters for the search region in the cluster process. In 1D the covariance matrix reduces to a single variance σ and the single threshold parameter d is given by the root of λ/σ. For higher dimensional cases (e.g. 4D) the bounding box of the ellipsoid is used, that is determined by the covariance Σ and λ.

Wells rejects scenes as non recognizable, if λ turns out to be negative according to Eq. 1. This gives a profound criterion for the applicability of the approach to a task for which a test data set is provided.

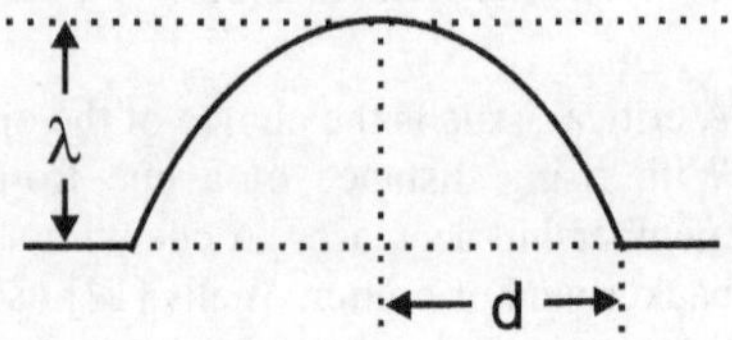

Fig. 3. Reward function after Wells [11]

3.2 Propagation of Likelihood along the Part-of Structure

The cues have an auxiliary attribute Y_1 for the position, where the partner cue should connect, e.g. the trailer hitch. This attribute is calculated by inverse projection into the scene, proper rotation of the 3D model, and projection into the image again. A search area is constructed around Y_1. For each partner cue with position Y_2 in this area the aggregation is regarded as valid and the two accumulator values are summed up yielding the accumulator value for the new aggregate object. Its position Y_n is calculated as weighted mean. This neglects the quality of the fit.

For the probabilistic setting the likelihood L is propagated along the links of the part of hierarchy. If the position Y_2 of the partner cue exactly matches the auxiliary position Y_1, we infer that there is independent evidence for the aggregate from both parts. This justifies multiplication of probabilities or adding the likelihood values. Otherwise some of the predecessors of the cue clusters may be contradicting. Lacking the precise knowledge of the distribution, the evidence for each part is assumed to be equally distributed over its search volume. Fig. 4 shows the 1D case.

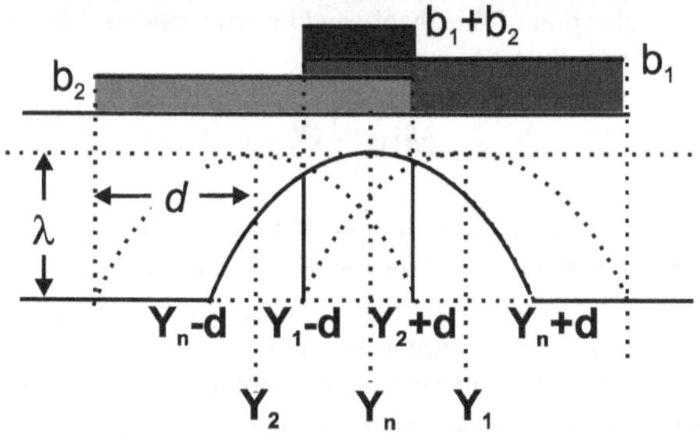

Fig. 4. Combining evidence from two different parts of a model into evidence of an aggregate: Below reward function; above density estimations

The two cue clusters with centers Y_1 and Y_2 and their error parabolas displayed in dashed lines include mutually affirming evidence, if their distance is smaller than *3d*. We indicate their evidence densities b_1 and b_2 by the differently shaded piecewise constant functions. In the overlapping region evidences b_1, b_2 are added. The reward function integrates the piecewise densities using the error parabola of the new position estimate Y_n yielding a sum of three integrals:

$$L_n = \int_{Y_n-d}^{Y_1-d} b_2(x-Y_n)^2\sigma\,dx + \int_{Y_1-d}^{Y_2+d} (b_1+b_2)(x-Y_n)^2\sigma\,dx + \int_{Y_2+d}^{Y_n+d} b_1(x-Y_n)^2\sigma\,dx \tag{3}$$

In the 2D case the upper and lower border of the integral in the middle are replaced by circular sections in the attribute domain and the parabola is replaced by a paraboloid. In the case of rigid connections the hyper ellipsoid on which the reward paraboloid is constructed is 4D, namely (x,y,α,dis). In the case of an articulated connection the azimuth α is free contributing no error. Therefore the domain is 3D containing only (x,y,dis).

4 An Experiment

Fig. 5a shows a section of a gray level image containing a vehicle with a small trailer. Input to the experiments are the lines extracted in a preprocessing stage (Fig. 5b). Both decision criteria, the maximum accumulator value as well as the maximum likelihood (ML) work. In the cluttered image region on the left (branches of a tree) and in the fairly homogenous region in the center accumulator and likelihood field are empty.

The pose estimation of the maximal elements is roughly correct. Fig. 5c shows the ML result. The interesting section of the likelihood field is enlarged in Fig 5d. The white blobs on the left correspond to correct localization. Some less significant false evidence is found on the right. Both the discrimination and the pose estimation are slightly better for the likelihood criterion.

5 Discussion

In this contribution we demonstrated the inclusion of probabilistic calculations into a structural method. Compared to previous experiments [8] the discriminative power of the accumulator on cluttered regions, e.g. in the left part of the image, has much improved due to a better parameter setting. The new settings were obtained from the probabilistic considerations. We occasionally experienced better performance of the accumulator compared to the likelihood. This occurred when the model did not fit exactly to the vehicle. The likelihood approach is more sensitive to errors in the model. Fig. 5c shows that the pose is not optimal (see nose of the vehicle). EM type optimizations including a top down search in the correspondence space can improve the result [13].

The probabilistic calculations of Wells rest on certain assumptions on the distribution of the data. Background features are assumed to be equally distributed all over the picture. Such assumption is valid only in special situations or if nothing else is known about the background [9]. If additional information is given, e.g. on certain preferences on the orientations of the lines (e.g. vertical or horizontal), this can be included in the probabilistic model. The features in correspondence to the target are modeled with a Gaussian distributed additive error. If knowledge about the error sources is available, other error models may be considered.

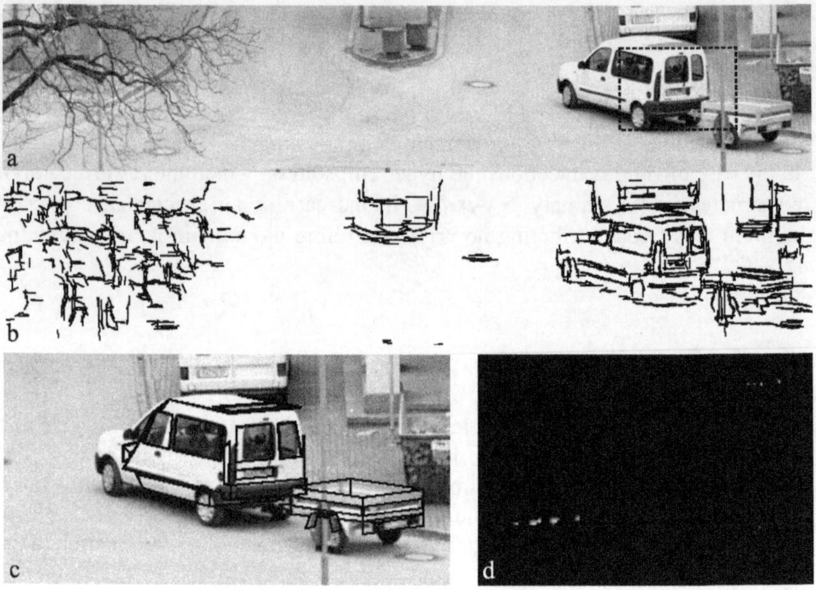

Fig. 5. Localization of an aggregate consisting of a vehicle and a small trailer. a) Image section (1000x200 pixel), b) extracted line objects; c) overlaid articulated model of ML-result, d) section of the likelihood field corresponding to the dashed box in a)

As shown in Fig 5 the evidence for the two partners of an aggregate is estimated as being equally distributed over the search volume. The evidence for the new aggregate has a stepwise constant density (lower, high and then lower again). If we include such an aggregate as a part of a higher aggregate using the same calculations, we permit a systematic estimation error. For shallow hierarchies like the one presented here this is not important. For deep hierarchies such effect has to be estimated.

In our approach all possible model views are approximated by views valid for the principal point only. This is justified for long focal lengths but will pose severe problems for views near the image margin of wide angle pictures. These are distorted by systematic errors.

Still the preliminary experiments presented in Chapter 4 yielded promising results, so that we are confident in combining statistical and structural methods.

References

1. Ballard D. H., Brown C. M.: Computer Vision. Prentice Hall, Englewood Cliffs, New Jersey, (1982).
2. Binfort T. O., Levitt T. S.: Model-based Recognition of Objects in Complex Scenes. In: ARPA (ed.). Image Understanding Workshop 1994. Morgan Kaufman, San Francisco (1994), 149-155.

3. Dickinson S. J., Pentland A. P., Rosenfeld A.: From Volumes to Views: An Approach to 3-D Object Recognition. CVGIP:IU, Vol. 55, No. 2 (1992), 130-154.
4. Eggert D. W., Bowyer K. W., Dyer C.R.: Aspect Graphs: State-of-the-Art and Applications in Digital Photogrammetry. ISPRS-XXiX, Vol. 5, Com V, (1992) 633-645.
5. Grimson W. E. L.: Object Recognition by Computer: The Role of Geometric Constraints. MIT Press, Cambridge, Mass., (1990).
6. Hermitson K. J., Booth D. M., Foulkes S. B., Reno A. L.: Pose Estimation and Recognition of Ground Vehicles in Aerial Reconnaissance Imagery. ICPR 1998, Vol. 1, IEEE, Los Alamitos, California, (1998), 578-582.
7. Hoogs A., Mundy J.:An Integrated Boundary and Region Approach to Perceptual Grouping. ICPR 2000, Vol. 1, IEEE, Los Alamitos, California, (2000), 284-290.
8. Michaelsen E., Stilla U.: Ansichtenbasierte Erkennung von Fahrzeugen. In: Sommer G., Krüger N., Perwas C. (eds.): Mustererkennung 2000, Springer, Berlin, (2000) 245-252.
9. Michaelsen E., Stilla U.: Assessing the Computational Effort for Structural 3D Vehicle Recognition. In: Ferri F.J., Inesta J.M., Amin A., Pudil P. (eds): Advances in Pattern Recognition (SSPR-SPR 2000), Springer, Berlin, (2000) 357-366.
10. Tan T., Sullivan G., Baker K.: Model-Based Localisation and Recognition of Road Vehicles. . Int. Journ. of Comp. Vision, Vol. 27, (1998) 5-25.
11. Wang P. S. P.: Parallel Matching of 3D Articulated Object Reccognition. Int. Journ. of Pattern Recognition and Artificial Intelligence, Vol. 13, (1999) 431-444.
12. Viola P., Wells W. M. III: Alignment by Maximalization of Mutual Information. Int. Journ. of Comp. Vision, Vol. 24, (1997) 137-154.
13. Wells W. M. III: Statistical Approaches to Feature-Based Object Recognition. IJCV, 21, (1997) 63-98.

Hierarchical Top Down Enhancement of Robust PCA*

Georg Langs[1], Horst Bischof[2], and Walter G. Kropatsch[1]

[1] Pattern Recognition and Image Processing Group 183/2
Institute for Computer Aided Automation, Vienna University of Technology
Favoritenstr. 9, A-1040 Vienna, Austria
{langs,krw}@prip.tuwien.ac.at
[2] Institute for Computer Graphics and Vision, TU Graz
Inffeldgasse 16 2.OG, A-8010 Graz, Austria
bischof@icg.tu-graz.ac.at

Abstract. In this paper we deal with performance improvement of robust PCA algorithms by replacing regular subsampling of images by an irregular image pyramid adapted to the expected image content. The irregular pyramid is a structure built based on knowledge gained from the training set of images. It represents different regions of the image with different level of detail, depending on their importance for reconstruction. This strategy enables us to improve reconstruction results and therefore the recognition significantly. The training algorithm works on the data necessary to perform robust PCA and therefore requires no additional input.

1 Introduction

The human visual system takes advantage of the ability to distinguish between interesting regions and less relevant regions in the field of view. By using this knowledge it is able to improve its performance considerably. [1] and [2] describe two strategies to obtain and apply information about the importance of different regions of an image when simulating the human visual system. *Bottom-up* methods retrieve their features only from the present input image [3]. *Top-down* methods are driven by knowledge which is available before getting the input. Experiments [1] have shown that human vision and particularly the scan paths of the eyes, called saccades, are not only dependent on the input image, but largely on previous knowledge i.e. *top-down* expectations. In this paper we propose a method to incorporate a top-down strategy in a robust PCA algorithm [4] for object recognition. In our approach, instead of performing sequential saccades, we change the initial representation of the images with respect to the relevance of different regions. We demonstrate that by using this top-down knowledge we are able to significantly improve the recognition results.

* This research has been supported by the Austrian Science Fund (FWF) under grant P14445-MAT and P14662-INF

T. Caelli et al. (Eds.): SSPR&SPR 2002, LNCS 2396, pp. 234–243, 2002.
© Springer-Verlag Berlin Heidelberg 2002

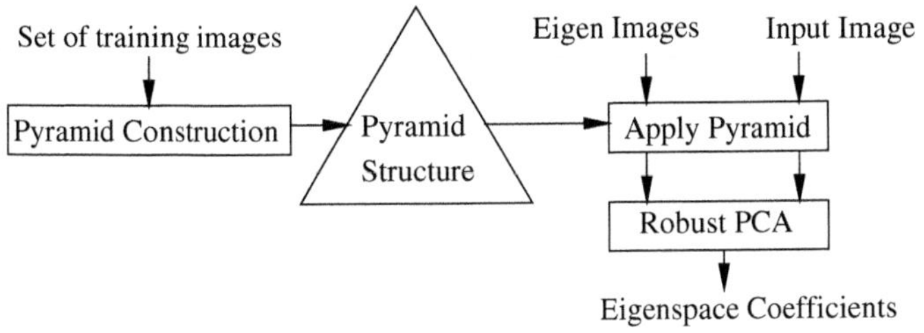

Fig. 1. The basic concept of our algorithm. It is divided into a training- (left) and a recognition phase (right)

The paper is organized as follows: In section 2 an overview of the algorithm is presented. The pyramid structure constructed by the algorithm is presented in section 3. The *training phase* is explained in section 4. Section 5 describes the application of the pyramid during the *reconstruction phase*. Finally we present experimental results in section 6 and give a conclusion in section 7.

2 Our Approach

The approach presented in this paper aims to enhance robust recognition based on eigenimages ([4]). We deal with an input, which can be an image or any other signal, that is represented as a vector of pixel values. Instead of performing sequential saccades we change the initial representation in order to stay abreast of the regions of interest. Each region is represented to an extend that corresponds to its importance for the reconstruction. The modified representation is used as input for robust PCA. Robust PCA represents training images in the eigenspace. To recognize an input image the coefficients of the eigenvectors are determined by solving an overdetermined system of linear equations in a robust manner. Further robustness is achieved by randomly selecting subsets of pixels.

The new representation has the following advantages over the unprocessed image, where all regions are represented to the same extend: Regions with small importance for the reconstruction or recognition are usually similar on different images of the training set, therefore if used for robust recognition they support almost all hypotheses. A large set of irrelevant pixels that are consistent with almost all hypotheses strongly interferes with the method in three ways: (1) It causes huge equation systems in the fitting step, which are numerically unstable (2) it wastes time because useless hypotheses are built and (3) the difference between good and bad hypotheses is likely to become smaller.

Our approach is divided into 2 main phases (Figure 1): During a *training phase* the computer is taught the importance of each pixel position as well as the dependencies between neighbouring pixels i.e. how much their values correlate in the training data. The algorithm builds an irregular *pyramid structure* that represents a given input image with different levels of detail. This pyramid structure is build by contracting the initial regular image template. Irregular image pyramids represent images as a set of graphs with decreasing number of nodes. During contraction consecutive levels are built by choosing a set of surviving vertices, and assigning them a set of sons, the receptive field. [5] gives detailed explanations of the concept of irregular image pyramids.

During the *recognition phase* different levels of the pyramid structure are applied on the input image as well as on the eigenimages of the database. The resulting vectors are used as input for robust PCA.

3 The Pyramid Structure

The result of our algorithm is a *pyramid structure* that can be applied to any input image of a given size. In 3.1 we describe the structure and give its exact definition, in 3.2 we explain an example.

3.1 Definition of the Pyramid Structure and Its Application

The pixels of the input image can be indexed by $i = 1, ..., N_1$ where N_1 is the number of pixels in the image. To convert a rectangular grid (the image) to a vector we use the transformation

$$i_{vec} = (i_{arr} - 1)n + j_{arr} \tag{1}$$

where i_{vec} indicates the index in the vector and (i_{arr}, j_{arr}) the vertical and horizontal coordinates of a pixel in an $m \times n$ image. Each pyramid level P_k consists of a vector of nodes, each of them representing a set of pixels, its *receptive field*.

$$P_k = \langle n_{1,k}, ..., n_{N_k,k} \rangle \qquad \forall i,k : n_{i,k} \in \mathfrak{P}(\{1, ..., N_1\}) \tag{2}$$

The receptive fields are not overlapping and the union of the receptive fields together with a set r covers the whole image. r is the set of pixels, that have $weight = 0$. They are irrelevant or interfering with the reconstruction and are therefore ignored.

In the first level each node represents one pixel i.e. $n_{i,1} = i$. During contraction a node in the level $k+1$ takes over the indices from its sons in level k:

$$n_{i,k+1} = \bigcup_{j:n_{j,k}\text{son of } n_{i,k+1}} n_{j,k} \tag{3}$$

The final pyramid structure with L levels consists of L vectors of nodes $P_k, k = 1, ..., L$. Each node represents a receptive field in the base level.

We define the procedure how to apply a pyramid structure P on an image [1]: The structure can be applied to an input image independently for each level, i.e. one can construct a certain level of the pyramid directly from the input image without constructing the levels in between.

[1] Note that an extension to other data representable in vector form is straightforward

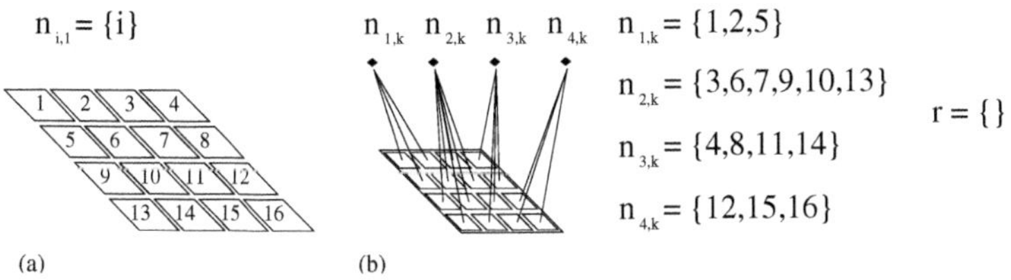

Fig. 2. An example of (a) the base-level and (b) a pyramid level P_k with 4 nodes

Definition 1. *Let B_1 be an image with pixel values $\langle b_{i,1}\rangle_{i=1,...,N_1}$ (in our experiments we used an image size of $N_1 = 128 \times 128 = 16384$ pixels). To calculate the k-th level $B_k = \langle b_{1,k}, ..., b_{N_k,k}\rangle$ of the pyramid, for each node the mean value of the pixel values in the receptive field is calculated:*

$$\forall i : b_{i,k} = \frac{\sum_{j \in n_{i,k}} b_{j,1}}{|n_{i,k}|} \tag{4}$$

Note that P is an 'empty' structure, in the sense that the nodes don't have attributes like gray values assigned to them. Only when calculating B_k of an input-image, gray values are assigned to the nodes of B_k according to P_k.

3.2 Example of a Pyramid Structure

To illustrate the pyramid structure we give an example of 2 levels P_1 and P_k. Figure 2(a)shows the base-level: The size of the images in the training set is $4 \times 4 = 16$ and for all nodes in the base-level $n_{i,1} = i$ (calculated according to equation 1) holds. (b) shows a level with 4 nodes, each of them representing a set of pixels in the base-level. The set r is empty in this example.

4 Training Phase

We assume that we are given a set of n images that represent one or more objects with different pose, and are of the same size. All these images are represented in a single eigenspace. From this set of training images we can retrieve the following information:

1. The eigenimages, eigenvalues and the coordinates of the training images in the eigenspace.
2. The variance of the value in each pixel over the training set.
3. The dependencies between pixels or receptive fields over the training set: For a given pixel or node i the pixel values in each training image $\{v_1, ..., v_n\}$ form a vector of values $(v_{i,1}, ..., v_{i,n})$, the *value profile*. In figure 3(b) value profiles of two neighbouring nodes in 3 images are depicted. Each node represents a receptive field. Two value profiles can be compared by calculating

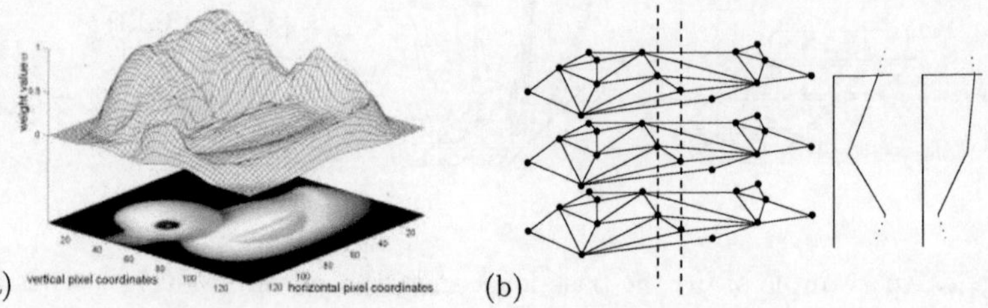

Fig. 3. (a)Weight map based on variance for a training set, consisting of 36 images of a rotating duck. (b) Value profiles of neighbouring nodes in the training set

their correlation $corr(v_i, v_j)$. By contracting two nodes with highly correlated value profiles the loss of information is expected to be smaller than the loss caused by contracting two pixels with more independent behavior.

4.1 Weight Target Contraction

During the contraction process we represent a given image with decreasing precision. In levels $2, 3, \ldots$ we deal no longer with individual pixels, but with nodes that represent a set of pixels in the original image. A node $n_{i,l}$ in level l is assigned the weight

$$\omega_{n_{i,l}} = z \cdot \sum_{x_i \in \text{ receptive field of } n_{i,l}} f(\omega_i) \tag{5}$$

We initiate z with $z_o = 1$ and define a *weight target* $1 - \tau$. P_1 is the base level of the pyramid and its neighborhood relation $\hat{N}_1 \subseteq P_1 \times P_1$ is defined according to the input data. If level P_i and $\hat{N}_i \subseteq P_i \times P_i$ have been built then we build level P_{i+1} according to the following rules:

1. Perform stochastic decimation [6] on P_i thus every node is assigned a status *survivor* or *non-survivor*. All nodes with $\omega_{i,j} > (1 - \tau)$ become *survivors.*
2. A *non-survivor* n_{i,j_0} chooses a father from all neighbouring survivors $\{n_{i,j_1}, \ldots, n_{i,j_k}\}$. $n_{i,j_{father}}$ becomes father if its *value profile* is most correlated with the *value profile* of the *non-survivor* n_{i,j_0} i.e. if the distance $(d_i)_{j_0, j_{father}}$ is minimal. d_i is the *distance map* of level i;
3. If the *weight* of the *non-survivor* $\omega_{n_{i,j_0}}$ and the *weight* of the chosen *father* $\omega_{n_{i,j_{father}}}$ sum up to a value

$$\omega_{n_{j,j_0+1}} = \omega_{n_{i,j_0}} + \omega_{n_{i,j_{father}}} > (1 + \tau) \tag{6}$$

then do not merge and change status n_{i,j_0} to *survivor*;

4. Define the neighborhood relation of the new level according to stochastic decimation [6].
5. If contraction terminates set $z_{new} = z/2$.

The algorithm proceeds with the following major steps: *Step 1* decides which receptive fields are merged with other receptive fields when constructing the successor of the level. After performing stochastic decimation algorithm [6] we modify the initial partition according to the weight map (This is an opposite strategy to [7]). *Step 2* chooses a neighbouring receptive field to merge with. All *sons* (or *grandchildren* resp.) in the base level P_1 of one *father* in an arbitrary level build its receptive field. If the resulting receptive field does not meet certain requirements defined in step (3) then steps (2) and (3) are canceled. The contraction process proceeds until no further merging is possible (Equation 6). The contraction stops, z is decreased and again contraction is performed until convergence.

While the 1st priority is to merge receptive fields with high correlation (search for father) the 2nd is to merge them only until they reach a certain weight according to (6). This strategy leads to a more balanced distribution of weights compared to a Gaussian pyramid or a pyramid built by plain stochastic decimation.

Experiments (Section 6) show that with $f(\omega_i) = \omega_i^s$ in (5) there is no exponent s that performs best on all resolutions resp. levels. A function defined in equation 8 resulted in the smallest reconstruction error.

Theorem 1. *Let* $(\omega_i)_{i=1,...,N}$ *be a* weight map *with* $0 \leq \omega_i \leq 1$ *for all* $i = 1,...,N$. *Let* P_i *denote the set of nodes and* $\hat{N}_i \in P_i \times P_i$ *be the neighborhood relation in level* i. *Let* d_i *be* distance maps *i.e. functions from* N_i *in* $[-1,1]$. *Then method 2 converges to a single node.*

A proof of Theorem 1 is given in [8].

In addition it is possible to estimate the size of a receptive field, if we are given an interval for the weights of the pixels lying in the receptive field. This is helpful during search of a monotonously ascending function $f(\omega_i) : [0,1] \rightarrow [0,1]$. Let $n_{l,j}$ be a node with a receptive field $p_i, i = 1,...,N$ and let τ denote the target tolerance, then

$$\omega_{n_{l,j}} = z \sum_{i=1}^{N} f(\omega_i) \leq (1+\tau) \quad (7)$$

holds. We assume that $\forall i = 1,...,N : \omega_i \in [\bar{\omega} - \delta, \bar{\omega} + \delta]$ and get the estimation $N \cdot f(\bar{\omega} - \delta) \leq \sum_{i=1}^{N} f(\omega_i) \leq N \cdot f(\bar{\omega} + \delta)$ and finally $N \leq \frac{\frac{1}{z}(1+\tau)}{f(\bar{\omega}-\delta)}$. Figure 4(b) gives an impression of the expected influence of $f(\omega_i)$ on the size of the receptive fields. The function *logsig* is a modified log-sigmoid transfer function. It is defined by modifying the function $ls(\omega) = \frac{1}{(1+e^{-\omega})}$ in order to get a function $logsig : [0,1] \rightarrow [0,1]$

$$logsig(\omega) = \frac{ls(l \cdot (\omega - t)) - ls(-l \cdot t)}{ls(l \cdot (1-t)) - ls(-l \cdot t)} \quad (8)$$

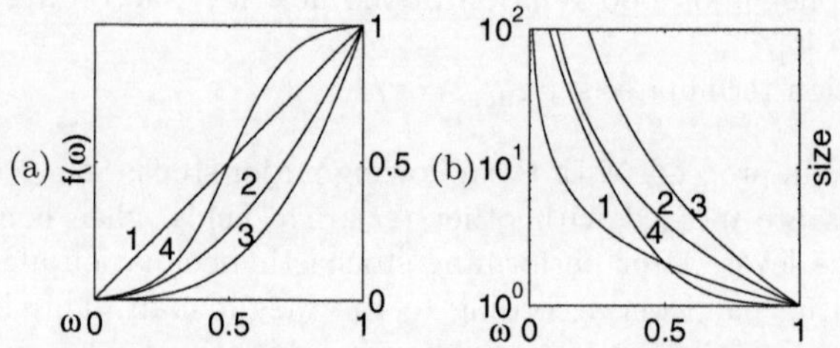

Fig. 4. Some functions (a) $f_1(\omega_i) = \omega$, $f_2(\omega_i) = \omega_i^2$, $f_3(\omega_i) = \omega_i^3$, $f_4(\omega_i) = logsig(\omega_i)$; (b): a comparison of the corresponding sizes of the receptive fields

l and t are parameters. l controls the steepness of the curve, while t shifts the steepest part of the curve along the x-direction.

5 Reconstruction

The following steps reconstruct or recognize a given image using a given eigenspace with a base consisting of N eigenimages, a pyramid P and level i. P is an empty structure as it is described in section 3. We calculate B_i according to *definition* 1:

1. For all eigenimages of the eigenspace: calculate level B_i according to the pyramid P. This results in N vectors $\{e_1, ..., e_N\}$
2. Calculate pyramid level $B_i^{inputimage}$ of the input image

The coefficients of the training images and the input image do not change [4]. The resulting vectors $\{e_1, ..., e_N, B_i^{inputimage}\}$ are input to robust PCA. Note that the 1st point is performed during the training phase. During reconstruction only one level based on the input image has to be calculated. Computational expensive steps i.e. the contraction of an image template to a pyramid structure takes place entirely during the training phase.

6 Experiments

Experiments were performed on a dataset of gray level images of different objects. The database (COIL-20 [9]) contains images of 20 objects, each object rotated around its vertical axis with images taken in 5° steps. Our training set consists of 36 images (i.e. 10° steps) of one object taken from the database. The size is $128 \times 128 = 16384$ pixels. The test set consists of the same 36 images, each 50% occluded. Target tolerance is $\tau = 0.1$. After calculation of B_i the test input images are reconstructed by unconstraint robust PCA [4]. Figure 5(c) shows a comparison of the mean squared reconstruction error. The horizontal line in figure 5(c) represents the error gained with full resolution i.e. without processing before PCA. The modified *logsig* function ($l = 9, t = 0.8$) performs best

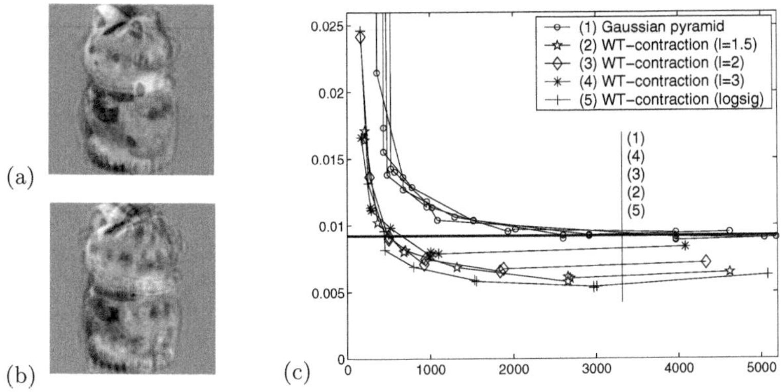

Fig. 5. (a) image of a cat, reconstructed after irregular downsampling and (b) after regular downsampling (c) Mean squared reconstruction error for pyramids constructed using different contraction algorithms

on almost all levels. Note that in figure 4 $f(\omega_i) = logsig(\omega_i)$ provides smallest receptive fields for important regions and the steepest increase of receptive field size for decreasing weight. For $f(\omega_i) = \omega_i^s$ at high resolutions lower values outperform higher values for s. This also corresponds to smaller receptive field sizes at higher weights. In figure 6 each dot represents a pixel. The x-coordinate represents its weight, the y-coordinate the size of the receptive field it lies in. (a) shows the diagram for $s = 2$ (1842 nodes) and (b) for $s = 3$ (1559 nodes). Figure 6(c) shows randomly colored receptive fields constructed by WT-contraction on training images of a rotating duck (weight map in Figure.3). Note the small receptive field size in regions where the head gives most information about the pose. For extremely low resolutions higher s-values slightly outperform lower ones. The reason is the possibility to build larger fields for pixels with low weight. This leaves more nodes for more important regions. $f(\omega_i) = logsig(\omega_i)$ attempts to combine both advantageous features.

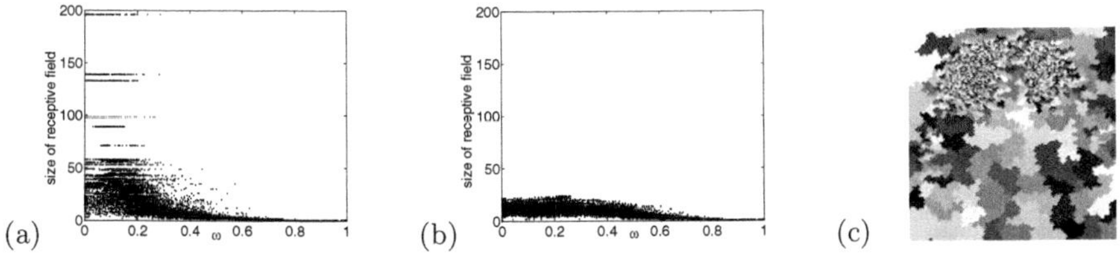

Fig. 6. (a,b) Plot of pixel weights vs. size of receptive fields; (c) randomly colored receptive fields, level of a pyramid based on the weightmap shown in Figure 3

Extensive experiments on all images of the COIL-20 database show that WT-contraction is able to significantly improve for 55% of the objects the reconstruction error (on average to 81% of the error with full resolution). Compared to a Gaussian pyramid the reconstruction error was improved to 61% of the error achieved on Gaussian pyramid levels with similar resolution. Experiments showed that contracting an input image by our algorithm to a number of ~ 3000 nodes ($\sim 18\%$ of full resolution) can decrease the mean squared reconstruction error down to $\sim 53\%$ of the error achieved with full resolution (~ 16384 pixels). For extremely low numbers of nodes few remaining small receptive fields allow stabilization: with less than $\sim 3\%$ of initial 16384 nodes the error is $\sim 2\%$ of the error achieved when the image is contracted with a regular Gaussian pyramid.

7 Conclusion

We present an approach to enhance robust PCA for object recognition and reconstruction. The algorithm simulates human vision, in particular: top-down processing and saccades by building irregular pyramid structures during a training phase. This structures are applied to an input image before robust PCA is performed. During our experiments we decreased the reconstruction error of robust PCA significantly. To represent regions of an image according to their relevance turns out to be crucial, not only to save computation time but also to improve and stabilize reconstruction and recognition results. The presented algorithm is able to meet this goal without a need for additional input. Future work will include optimization of $f(\omega_i)$ to specific tasks and a study of connection between the *distance*- and the *weight* map.

References

1. Lawrence W. Stark and Claudio M. Privitera. Top-down and bottom-up image processing. In *Int. Conf. On Neural Networks*, volume 4, pages 2294 –2299, 1997. 234
2. D.A. Chernyak and L.W. Stark. Top-down guided eye movements. *SMC-B*, 31(4):514–522, August 2001. 234
3. Claudio M. Privitera and Lawrence W. Stark. Algorithms for defining visual regions-of-interest: Comparison with eye fixations. *IEEE Trans. on PAMI*, 22(9), 2000. 234
4. Aleš Leonardis and Horst Bischof. Robust recognition using eigenimages. *CVIU*, 78:99–118, 2000. 234, 235, 240
5. Walter G. Kropatsch. Irregular pyramids. Technical Report PRIP-TR-5, Institute for Automation, Pattern Recognition and Image Processing Group, University of Technology, Vienna. 236
6. Peter Meer. Stochastic image pyramids. *CVGIP*, 45:269 – 294, 1989. 238, 239
7. Jean-Michel Jolion. Data driven decimation of graphs. In *Proc. Of GbR'01, 3rd IAPR Int. Workshop on Graph Based Representations*, pages 105–114, 2001. 239
8. Georg Langs, Horst Bischof, and Walter G. Kropatsch. Irregular image pyramids and robust appearance-based object recognition. Technical Report PRIP-TR-67, Institute for Automation, Pattern Recognition and Image Processing Group, University of Technology, Vienna. 239
9. S.A. Nene, S.K. Nayar, and H. Murase. Columbia object image library (COIL20). Technical Report CUCS-005-96, Columbia University, New York, 1996. 240

An Application of Machine Learning Techniques for the Classification of Glaucomatous Progression

Mihai Lazarescu, Andrew Turpin, and Svetha Venkatesh

Department of Computer Science, Curtin University
GPO Box U1987, Perth 6001, Australia
{lazaresc,andrew,svetha}@computing.edu.au

Abstract. This paper presents an application of machine learning to the problem of classifying patients with glaucoma into one of two classes:stable and progressive glaucoma. The novelty of the work is the use of new features for the data analysis combined with machine learning techniques to classify the medical data. The paper describes the new features and the results of using decision trees to separate stable and progressive cases. Furthermore, we show the results of using an incremental learning algorithm for tracking stable and progressive cases over time. In both cases we used a dataset of progressive and stable glaucoma patients obtained from a glaucoma clinic.

1 Introduction

Machine learning techniques have been used successfully in a number of fields such as engineering [2] and multimedia [6]. Another important field where machine learning has been applied with considerable success is in medicine for tasks such as patient diagnosis [3]. Glaucoma is a disease that affects eye sight, and is the third most common cause of blindness in the developed world, effecting 4% of people over the age of 40 [10]. The vision loss associated with glaucoma begins in peripheral vision, and as the disease progresses vision is constricted until tunnel vision and finally blindness results. Patients diagnosed with the disease usually undergo treatment which may prevent further deterioration of their vision. In some cases the treatment is successful, resulting in the patient's vision being stabilized. Unfortunately in some cases the treatment is not successful, and the visual field continues to constrict over time. One aim of the research in this field is to distinguish between stable glaucoma patients and progressive glaucoma patients as early in the life of the disease as possible. This allows ophthalmologists to determine if alternate treatments should be pursued in order to preserve as much of the patient's sight as possible. With current techniques, vision measurements must be taken at regular intervals for four to five years before progression can be determined [5].

Most research on automatic diagnosis of glaucoma has concentrated on using statistical methods such as linear discrimination functions or Gaussian classi-

T. Caelli et al. (Eds.): SSPR&SPR 2002, LNCS 2396, pp. 243–251, 2002.
© Springer-Verlag Berlin Heidelberg 2002

fiers [4, and references therein]. In this previous work, the data available combines information from several standard opthalmologic tests that assess both vision and damage to the optic nerve over time. Typical patient data has a series of observations collected at 6 monthly or yearly intervals, with each observation containing over 50 attributes. Because of the complexity of the data, many of the methods used for classification focus on the time series of a single attribute. [9] presents a comparison of several machine learning techniques such as linear support vector machines and point-wise linear-regression for data covering 8 sets of observations. [4] describes a study of 9 classification techniques but with data that covered only a single set of observations (no temporal data was available). A different approach that is based on the use of the optical disk features extracted from 3D eye scans is described in [1].

In this paper we present an approach that uses two types of learning: one-step learning that involves the use of decision trees; and incremental learning that uses a concept tracking algorithm. In both cases the processing involves a pre-processing step, where a number of new attributes are extracted from the data, and then application of one of the two learning approaches. We describe our method to extract the features from the data in detail below, and present the results obtained from applying the two learning methods to a medical dataset. The main contributions of this paper are the formulation of new features that enables the application of decision trees to separate out stable and progressive glaucoma, and the use of incremental learning to track changes in the two classes of patients.

The paper is organized as follows. In Section 2 we describe the data used in the research. Section 3 presents the features extracted from the data. In Section 4 we present our results, and conclusions are presented in Section 5.

2 The Data

The data we used in our experiments consisted of the raw visual field measurements for glaucoma patients. A visual field measurement records the intensity of a white light that can be just seen by a patient in 76 different locations of the visual field. In order to collect the data, the patient is instructed to fixate on a central spot, and lights of varying intensities are flashed throughout the 76 locations in the visual field. The patient is instructed to push a button whenever they see a light. Figure 1(b) shows the output of the machine that was used to collect the data for this data set (Humphrey Field Analyzer I, Humphrey Systems, Dublin CA). A high number (30 or above) indicates good vision in that location of the field. A score of zero at a location indicates blindness at that location. If a patient's glaucoma is progressing over time, therefore, we would expect a decrease in some of the numbers in their visual field measurement. For each patient, 6 visual field measurements are available each of which is made at an interval of 6 months. All measurements were adjusted prior to our processing to represent right eyes of 45 year old patients. The data was provided by Dr. Chris Johnson from Devers Eye Institute, Portland, Oregon.

			17	17	18	18			
		17	17	17	17	18	19		
	16	16	16	16	17	18	19	20	
15	15	15	15	15	14	16	19	20	20
14	14	14	14	13	12	13	0	21	21
8	8	8	8	9	10	10	0	1	1
7	7	7	7	7	8	7	3	2	2
	6	6	6	6	5	4	3	3	
		5	5	5	4	4	3		
			5	4	4	4			

(a)

			31	30	29	30			
		2	30	31	32	29	29		
	5	4	32	32	30	27	29	28	
29	31	31	32	28	31	32	30	29	30
28	27	29	30	29	31	31	0	27	30
28	29	30	28	31	30	31	0	29	31
27	31	30	29	29	29	27	31	28	27
	32	29	30	29	31	30	29	29	
		29	28	28	31	32	31		
			27	32	32	29			

(b)

Fig. 1. (a) A map identifying the nerve fiber bundles for each visual field location. (b) A sample visual field measurement

We also made use of a map of the nerve fiber bundles as they are arranged in the retina [11], as shown in Figure 1(a). Each location of our visual field is numbered on the map according to which nerve fiber bundle it belongs. If glaucomatous damage occurs to one nerve fiber bundle, it is expected that all locations on that bundle will have decreased visual field thresholds.

All the data had been previously labeled as either stable or progressive by experts. A fact worth noting is that the data is very noisy, as is widely reported in the opthalmologic literature [8, and references therein]. One obvious source of the noise is that human responders make mistakes: they press the button when they don't see a light, they don't press the button when they do see a light, or they do not fixate on the central spot. Another more insidious source of noise is that the patient's criteria for pushing the button (that is, determining if they have "seen" the light or not) can change during a test, and between different tests. Hence the patients with the same condition will likely produce different responses.

It is worth noting here that there are other types of data that could be used to classify patients. These include other measures of the visual field and structural measures of the retina and optic nerve. At present there is no universally accepted definition of glaucomatous progression, and various on-going clinical drug trials use their own set of criteria and definitions to determine outcomes. However, all current schemes for determining progression have one thing in common: they use visual field data such as the type described above.

3 The Features Extracted

Several approaches have been used in the past to classify visual field data. One such approach is to consider each individual visual field measurement for each observation as a separate attribute, thus creating a total of 456 attributes (76x6) [9].

A popular approach is to consider the data to have 76 attributes and then to use statistical methods such as linear regression to determine whether the visual field response decreases over time.

Our initial attempts also used the latter approach but the results were disappointing when combined with decision trees. Therefore, we decided to derive a new set of features. Unlike previous approaches, that used trends in the individual visual field measurements, we attempted to derive a set of features that can be obtained with ease, and that can be easily understood by glaucoma researchers. To obtain meaningful features we made extensive use of common knowledge about glaucoma, including:

- a consistent decrease in the response from a nerve fiber bundle indicates that the disease is progressing;
- a large decrease in the overall response indicates that the disease is progressing
- an increase in the anomalous readings of the eye over time indicates that disease is progressing;
- a low response for nerve fiber bundle 0 does not indicate glaucoma as it is the blind spot for the eye;
- nerve fiber bundles can be grouped to gain a better picture of progression;
- the nerve fiber bundles closest to the nose are more likely to show early loss due to glaucoma;

We extracted two types of features. Type 1 (seven features) uses information from only a single observation while type 2 (five features) describes the temporal aspects of the data.

The features of type 1 are described below.

- Feature 1–*Overall eye response.* Computed by obtaining the average of the 76 visual field measurements.
- Feature 2–*Existence of an anomaly*–for each location in the visual field, compare its value to the median value in a 3x3 neighborhood. If the value is greater or equal with the median, then no anomaly is indicated and feature 2 is set to 0. If the value is smaller than the median, and the difference between the pixel and at least 6 of the neighbors is larger than a threshold, then an anomaly is deemed to exist and feature 2 is set to 1. The neighborhood is set to 2x2 at the border of the visual field.
- Feature 3—*Number of anomalies per optic nerve bundle.* Each pixel in the visual field corresponds to one of the 21 nerve fiber bundles. Feature 2 is now amalgamated to get the number of anomalies per optic nerve.
- Feature 4—*Quadrant anomalies.* The 10x10 region of the eye is divided into 4 quadrants (5x5 each) and Feature 2 is amalgamated for each quadrant.

- Feature 5—*Eye response per quadrant.* Computed by averaging the visual field measurement values for each of the 4 quadrants.
- Feature 6—*Blind spot data.* The value of the amalgamated score (from feature 3) for nerve fiber bundle 0 is indicative of the blind spot data. Feature 6 is set to this value.
- Feature 7—*Number of anomalies for 3 quadrants.* The bottom right quadrant for this dataset was ignored because of the blind spot. The blind spot position varies from measurement to measurement because the data is noisy (patient's may not sit in exactly the same spot for every test). Hence any anomaly discovered in this quadrant may in fact indicate the blind spot.

Type 2 features were designed to capture the progression of the visual field responses over time. We specifically searched for consistency in the individual nerve fiber responses, the grouped nerve fiber responses, the overall change in the eye response and the number of anomalous visual responses in the eye. These include:

- Feature 8—*Average difference in eye response.* Computed by taking the difference between the overall eye response at two time instances.
- Feature 9—*Change in anomalies per optic nerve bundle.* This feature indicates whether or not a net difference has occurred for a nerve fiber bundle between two time instances. The time instances need not be consecutive.
- Feature 10—*Change in eye response for 3 quadrants.* This feature indicates whether or not a net difference has occurred for the three quadrants (feature 7) between two time instances.
- Feature 11—*Difference in anomalies per optic nerve.* This feature is computed by taking the difference between the number of anomalies for two consecutive times instances for the same optic nerve.
- Feature 12—*Difference in anomalies per quadrant.* This is computed by taking the difference between the the anomalies for each quadrant for two consecutive time instances.

4 Results

4.1 Classification Using Decision Trees

The instances containing the raw visual field measurements were processed to extract the 12 features and the data was split into 2 sets: a training and a test set. To classify and test the data we used the C4.5 (Release 8) software. The process was repeated 50 times and each time a different random training and test set was used. C4.5 generated decision trees that on average had 15 nodes

Table 1. Classification accuracy of the data using C4.5 and our 12 features

CLASS	STABLE	PROGRESSIVE	TOTAL
FALSE POSITIVE	18	108	126
FALSE NEGATIVE	93	129	222
TRUE POSITIVE	299	432	731
RECALL %	95	83	89
PRECISION %	83	72	77

using 7 features consistently (all features were used but 5 of the features were not used with any consistency). The results are shown in Table 4.1.

It can be seen that both the precision and recall for the stable class (95% and 82.5%) are higher than for the progressive class (83% and 72%).

We examined the decision trees to determine the features that best defined the stable and the progressive cases. We found that features that were consistently used in all decision trees were: average difference in eye response (feature 8), change in eye response for 3 quadrants (feature 10), number of anomalies for 3 quadrants (feature 7), quadrant anomalies (feature 4) and difference in anomalies per quadrant (feature 12).

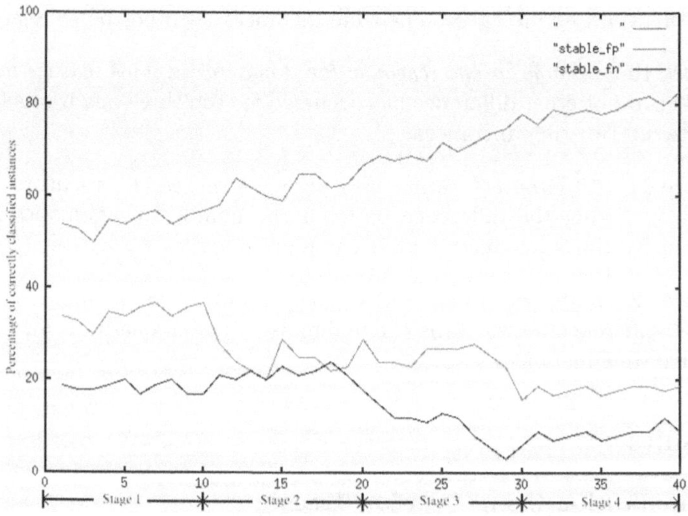

Fig. 2. Plot of the true-positive, false-positive and false-negative over the last four stages of the experiment for the stable class (each stage is equal to 10 time units)

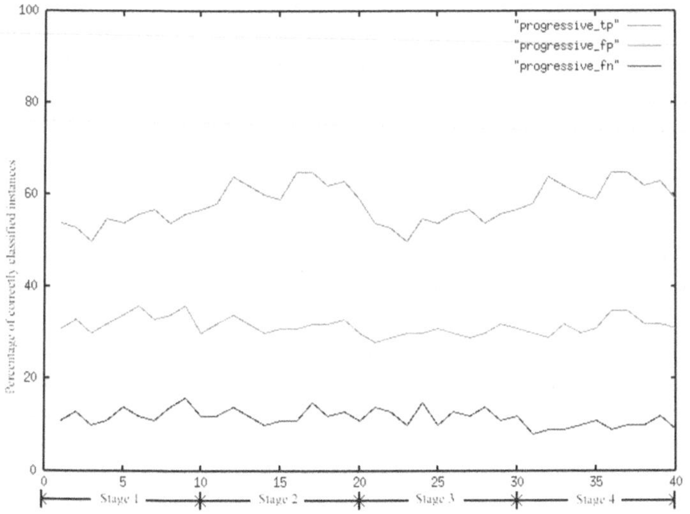

Fig. 3. Plot of the true-positive, false-positive and false-negative over the last four stages of the experiment for the progressive class (each stage is equal to 10 time units)

4.2 Using Incremental Learning to Track Patient Condition

We also investigated whether incremental learning could be used to classify a patient's condition over time. To track the patient's condition we used an incremental learning algorithm [7] that uses multiple windows to track the change in the data and adjust concept definitions. Because of the use of multiple windows, the system has the advantage that it can track changes more accurately.

In data set for consisted of pairings of time instances: (1,2), (2,3), (3,4), (4,5) and (5,6). The pairings described the patient's condition over a period of 3 years. The data is divided into 2 sets: a training set and a test set. Using the first pairing, the system builds the concepts for the progressive and stable class. These are concepts are progressively refined over four stages using subsequent samples from the training data. The system used the time pairing (2,3) for stage 1, (3,4) for stage 2, (4,5) for stage 3 and (5,6) for stage 4. The accuracy of the concepts is tested over the four stages using data from the test set. The resulting true-positive, false-positive and false-negative classification for the two classes in shown in Figures 3 and 2.

The results obtained show that the system's true-positive performance improved over time for the stable class. In the first two stages, the system classified the progressive and stable cases with 60% accuracy. In the last two stages the system averaged around 70% accuracy for the stable class. This was expected as the data covering the progressive patient is a lot more noisy than the stable cases. When comparing the performance of the incremental learning method to the one-step learning using C4.5, the latter method is more accurate. However, the difference between the methods is not large which indicates that incremental learning could be used for the classification task especially as the method we used was based on simple k-means clustering.

5 Conclusion

In this paper we present an application of machine learning techniques to the problem of classifying patients that have either stable or progressive glaucoma. The work described in this paper involves both one-step and incremental learning. We extract 12 features based on knowledge gleaned from domain experts, and applied simple machine learning techniques (decision trees and incremental learning) to solve the problem of interpreting complex medical data.

The approach described in this paper shows particular promise and that decision trees can be used to classify a patient's condition as stable or progressive using some very simple features. The features are obtained from raw visual field measurements, and do not involve any significant processing to extract. The results indicate that by using features that do not concentrate on individual visual field measurements, a good classification performance can be obtained.

References

1. D. Broadway, M. Nicolela, and S. Drance. Optic disk appearances in primary open-angle glaucoma. *Survey of Ophthalmology*, Supplement 1:223–243, 1999. 244
2. P. Clark. Machine learning: Techniques and recent developments. In A. R. Mirzai, editor, *Artificial Intelligence: Concepts and Applications in Engineering*, pages 65–93. Chapman and Hall, 1990. 243
3. R. Dybowski, P. Weller, R. Chang, and V. Gant. Prediction of outcome in critically ill patients using artificial neural networks synthesised by genetic algorithms. *Lancet*, 347:1146–1150, 1996. 243
4. M. Goldbaum, P. Sample, K. Chan et.al. Comparing machine learning classifiers for diagnosing glaucoma from standard automated perimetry. *Investigative Ophthalmology and Visual Science*, 43:162–169, 2002. 244
5. J. Katz, A. Sommer, D.E. Gaasterland, and D.R. Anderson. Analysis of visual field progression in glaucoma. *Archives of Ophthomology*, 109:1684–1689, 1991. 243
6. M. Lazarescu, S. Venkatesh, and G. West. Incremental learning with forgetting (i.l.f.). In *Proceedings of ICML-99 Workshop on Machine Learning in Computer Vision*, June 1999. 243
7. M. Lazarescu, S. Venkatesh, G. West, and H.H. Bui. Tracking concept drift robustly. In *Proceedings of AI2001*, pages 38–43, February 2001. 249
8. P.G.D. Spry, C.A. Johnson, A.M. McKendrick, and A. Turpin Determining progressiong in glaucoma using visual fields. In *Proceedings of the 5th Asia-Pacific Conference on Knowledge Discovery and Data Mining (PAKDD2001*, April 2001. 245
9. A. Turpin, E. Frank, M. Hall, I. Witten, and C.A. Johnson. Determining progressiong in glaucoma using visual fields. In *Proceedings of the 5th Asia-Pacific Conference on Knowledge Discovery and Data Mining (PAKDD2001*, April 2001. 244, 245

10. J.J. Wang, P. Mitchell, and W. Smith. Is there an association between migraine headache and open-angle glaucoma? Findings from the Blue Mountains Eye Study. *Ophthalmology*, 104(10):1714-9, 1997. 243
11. J. Weber, and H. Ulrich. A perimetric nerve fibre bundle map. *International Ophthalmology*, 15:193–200, 1991. 245

Estimating the Joint Probability Distribution of Random Vertices and Arcs by Means of Second-Order Random Graphs

Francesc Serratosa[1], René Alquézar[2], and Alberto Sanfeliu[3]

[1] Universitat Rovira i Virgili
Dept. d'Enginyeria Informàtica i Matemàtiques, Spain
Francesc.Serratosa@etse.urv.es
http://www.etse.urv.es/~fserrato
[2] Universitat Politècnica de Catalunya
Dept. de Llenguatges i Sistemes Informàtics, Spain
alquezar@lsi.upc.es
[3] Universitat Politècnica de Catalunya
Institut de Robòtica i Informàtica Industrial, Spain
sanfeliu@iri.upc.es

Abstract. We review the approaches that model a set of Attributed Graphs (AGs) by extending the definition of AGs to include probabilistic information. As a main result, we present a quite general formulation for estimating the joint probability distribution of the random elements of a set of AGs, in which some degree of probabilistic independence between random elements is assumed, by considering only 2nd-order joint probabilities and marginal ones. We show that the two previously proposed approaches based on the random-graph representation (First-Order Random Graphs (FORGs) and Function-Described Graphs (FDGs)) can be seen as two different approximations of the general formulation presented. From this new representation, it is easy to derive that whereas FORGs contain some more semantic (partial) 2nd-order information, FDGs contain more structural 2nd-order information of the whole set of AGs. Most importantly, the presented formulation opens the door to the development of new and more powerful probabilistic representations of sets of AGs based on the 2nd-order random graph concept.

1 Introduction

There are two major problems that practical applications using structural pattern recognition are confronted with. The first problem is the computational complexity of comparing two AGs. The time required by any of the optimal algorithms may in the worst case become exponential in the size of the graphs. The approximate algorithms, on the other hand, have only polynomial time complexity, but do not guarantee to find

T. Caelli et al. (Eds.): SSPR&SPR 2002, LNCS 2396, pp. 252-262, 2002.
© Springer-Verlag Berlin Heidelberg 2002

the optimal solution. For some applications, this may not be acceptable. The second problem is the fact that there is more than one model AG that must be matched with an input AG, what means that the conventional error-tolerant graph matching algorithms must be applied to each model-input pair sequentially. As a consequence, the total computational cost is linearly dependent on the size of the database of model graphs. For applications dealing with large databases, this may be prohibitive.

To alleviate these problems, some attempts have been made to try to reduce the computational time of matching the unknown input patterns to the whole set of models from the database. Assuming that the AGs that represent a cluster or class are not completely dissimilar in the database, only one structural model is defined from the AGs that represent the cluster, and thus, only one comparison is needed for each cluster [3,6,7,8].

In this paper, we review the approaches that model a set of AGs by extending the definition of graphs to include probabilistic information [3,4,6,8]. The resulting model, called *random graph* (RG) representation, is described in the most general case through a joint probability space of random variables ranging over pattern primitives (graph vertices) and relations (graph arcs). It is the union of the AGs in the cluster, according to some synthesis process, together with its associated probability distribution. In this manner, a structural pattern can be explicitly represented in the form of an AG and an ensemble of such representations can be considered as a set of outcomes of the RG.

In the following section, we introduce the formal definitions used throughout the paper. In section 3, we recall First-Order Random Graphs (FORGs) [6,8] and Function-Described Graphs (FDGs) [1,3,5,9], which are the two main approximations of the general RG concept proposed in the literature. The approach presented in the paper by Sengupta *et al.* [4] can be regarded as similar to the FORG approach. In section 4, we give a quite general formulation for estimating the joint probability of the random elements in a RG synthesised from a set of AGs. In sections 5 and 6, we show respectively that the FORG and FDG approaches can be seen as different simplifications of the general formulation given in section 4. Finally, in the last section we provide some discussion about our contribution and its future implications.

2 Formal Definitions of Random-Graph Representation

Definition 1. Let Δ_v and Δ_e denote the domains of possible values for attributed vertices and arcs, respectively. These domains are assumed to include a special value Φ that represents a *null value* of a vertex or arc. An AG *G* over (Δ_v,Δ_e) is defined to be a four-tuple $G=(\Sigma_v,\Sigma_e,\gamma_v,\gamma_e)$, where $\Sigma_v=\{v_k \mid k=1,...,n\}$ is a set of vertices (or nodes), $\Sigma_e=\{e_{ij} \mid i,j\in\{1,...,n\},\ i\neq j\}$ is a set of arcs (or edges), and the mappings $\gamma_v:\Sigma_v\rightarrow\Delta_v$ and $\gamma_e:\Sigma_e\rightarrow\Delta_e$ assign attribute values to vertices and arcs, respectively.

Definition 2. A *complete AG* is an AG with a complete graph structure (Σ_v,Σ_e), but possibly including null elements. An AG *G* of order *n* can be *extended* to form a complete AG G' of order k, $k \geq n$, by adding vertices and arcs with null attribute values Φ. We call G' the *k-extension* of G.

Definition 3. Let Ω_v and Ω_e be two sets of random variables with values in Δ_v (random vertices) and in Δ_e (random arcs), respectively. A *random-graph structure R* over (Δ_v, Δ_e) is defined to be a tuple $(\Sigma_v, \Sigma_e, \gamma_v, \gamma_e, P)$, where $\Sigma_v = \{\omega_k \mid k = 1,...,n\}$ is a set of vertices, $\Sigma_e = \{\varepsilon_{ij} \mid i, j \in \{1,...,n\}, i \neq j\}$ is a set of arcs, the mapping $\gamma_v : \Sigma_v \to \Omega_v$ associates each vertex $\omega_k \in \Sigma_v$ with a random variable $\alpha_k = \gamma_v(\omega_k)$ with values in Δ_v, and $\gamma_e : \Sigma_e \to \Omega_e$ associates each arc $\varepsilon_{ij} \in \Sigma_e$ with a random variable $\beta_k = \gamma_e(\varepsilon_{ij})$ with values in Δ_e. And, finally, P is a joint probability distribution $P(\alpha_1, \ldots, \alpha_n, \beta_1, \ldots, \beta_m)$ of all the random vertices $\{\alpha_i \mid \alpha_i = \gamma_\omega(\omega_i), 1 \leq i \leq n\}$ and random arcs $\{\beta_j \mid \beta_j = \gamma_\varepsilon(\varepsilon_{kl}), 1 \leq j \leq m\}$.

Definition 4. A *complete RG* is a RG with a complete graph structure (Σ_v, Σ_e), but possibly including null random elements (its probability of instantiation to the null value is one, i.e. $\Pr(\alpha = \Phi) = 1$ or $\Pr(\beta = \Phi) = 1$.). A RG R of order n can be *extended* to form a complete RG R' of order k, $k \geq n$, by adding null random vertices and null random arcs. We call R' the *k-extension* of R. Note that both R' and R represent the same model.

Definition 5. Any AG obtained by instantiating all random vertices and random arcs of a RG in a way that satisfies all the structural relations is called an *outcome graph* of the RG. Hence, a RG represents the set of all possible AGs that can be outcome graphs of it, according to an associated probability distribution.

Definition 6. For each outcome graph G of a RG R, the joint probability of random vertices and arcs is defined over an instantiation that produces G, and such instantiation is associated with a structural isomorphism $\mu : G' \to R$, where G' is the extension of G to the order of R. Let G be oriented with respect to R by the structurally coherent isomorphism μ; for each vertex ω_i in R, let $\mathbf{a}_i = \gamma_v(\mu^{-1}(\omega_i))$ be the corresponding attribute value in G', and similarly, for each arc ε_{kl} in R (associated with random variable β_j) let $\mathbf{b}_j = \gamma_e(\mu^{-1}(\varepsilon_{kl}))$ be the corresponding attribute value in G'. Then the *probability of G according to* (or given by) *the orientation* μ, denoted by $P_R(G|\mu)$, is defined as

$$P_R(G|\mu) = \Pr\left(\wedge_{i=1}^{n}(\alpha_i = \mathbf{a}_i) \wedge \wedge_{j=1}^{m}(\beta_j = \mathbf{b}_j)\right) = p(\mathbf{a}_1, \ldots, \mathbf{a}_n, \mathbf{b}_1, \ldots, \mathbf{b}_m) \qquad (1)$$

3 Approximating Probability Distributions in the Literature

When estimating the probability distribution of the structural patterns from an ensemble, it is impractical to consider the high order probability distribution $P(\alpha_1,\ldots,\alpha_n,\beta_1,\ldots,\beta_m)$ where all components and their relations in the structural patterns are taken jointly (eq. 1). For this reason, some other more practical approaches have been presented that propose different approximations [1,3,4,5,6]. All of them take into account in some manner the incidence relations between attributed vertices and arcs, i.e. assume some sort of dependence of an arc on its connecting vertices. Also, a common ordering (or labeling) scheme is needed that relates vertices and arcs of all the involved AGs, which is obtained through an optimal graph mapping process called synthesis of the random graph representation. In the following sections, we comment the two main such approaches, FORGs and FDGs.

3.1 First-Order Random Graphs (FORGs)

Wong and You [6] proposed the First-Order Random Graphs (FORGs), in which strong simplifications are made so that RGs can be used in practice. They introduced three suppositions about the probabilistic independence between vertices and arcs:

1) The random vertices are mutually independent;
2) The random arcs are independent given values for the random vertices;
3) The arcs are independent of the vertices except for the vertices that they connect.

***Definition** 7* A FORG *R* is a RG that satisfies the assumptions 1, 2, 3 shown above. Based on these assumptions, for a FORG *R*, the probability $P_R(G|\mu)$ becomes

$$P_R(G|\mu)=\prod_{i=1}^{n}p_i(\mathbf{a}_i)\prod_{j=1}^{m}q_j(\mathbf{b}_j \mid \mathbf{a}_{j1},\mathbf{a}_{j2}) \tag{2}$$

where $p_i(\mathbf{a})\triangleq\Pr(\alpha_i=\mathbf{a})$, $1\le i\le n$, are the marginal probability density functions for vertices and $q_j(\mathbf{b}\mid \mathbf{a}_{j1},\mathbf{a}_{j2})\triangleq\Pr(\beta_j=\mathbf{b}\mid \alpha_{j1}=\mathbf{a}_{j1},\alpha_{j2}=\mathbf{a}_{j2})$, $1\le j\le m$, are the conditional probability functions for the arcs, where α_{j1},α_{j2} refer to the random vertices for the endpoints of the random arc β_j.

The storage space of FORGs is $O(nN+mMN^2)$ where *N* and *M* are the number of elements of the domains Δ_v and Δ_e.

3.2 Function-Described Graphs (FDGs)

The FORG approach, although simplifies the representation considerably, continues to be difficult to apply in real problems where there is a large number of vertices in the AGs and their attributes have an extensive domain. The main cause of this problem is the dependence of the arc attributes with respect to the attributes of the vertices that the arc connects (assumption 3). Although this supposition is useful to constrain the generalisation of the given set of AGs, it needs a huge amount of data to estimate the probability density functions and bears a high computational cost. On the other hand, an important drawback of FORGs, which is due to the probability independence assumptions 1 and 2, is that the structural information in a sample of AGs is not well preserved in the FORG synthesised from them. This is, an FORG represents an over-generalised prototype that may cover graph structures quite different from those in the sample.

With the aim of offering a more practical approach, Serratosa *et al.* [1,3,5,9] proposed the Function-Described Graphs (FDGs), which lead to another approximation of the joint probability P of the random elements. On one hand, some independence assumptions (a) are considered, but on the other hand, some useful 2nd-order functions (b) are included to constrain the generalisation of the structure.

(a) Independence assumptions in the FDGs

1) The attributes in the vertices are independent of the other vertices and of the arcs.
2) The attributes in the arcs are independent of the other arcs and also of the vertices. However, it is mandatory that all non-null arcs be linked to a non-null vertex at each extreme in every AG covered by an FDG. In other words, any outcome AG of the FDG has to be structurally consistent.

(b) 2nd-order functions in the FDGs

In order to tackle the problem of the over-generalisation of the sample, the *antagonism*, *occurrence* and *existence* relations are introduced in FDGs, which apply to pairs of vertices or arcs. In this way, random vertices and arcs are not assumed to be mutually independent, at least with regards to the structural information, since the above relations represent a *qualitative information* of the 2nd-order joint probability functions of a pair of vertices or arcs.

To understand these 2nd-order relations it is convenient to split the domain of the joint probabilities in four regions (see figure 1.a). The first one is composed by the points that belong to the Cartesian product of the sets of actual attributes of the two elements, corresponding to the cases where both elements are defined in the initial non-extended AG and therefore their value is not null. The second and third regions are both straight lines in which only one of the elements has the null value. This covers the cases when one of the two elements does not belong to the initial AG and has been added in the extending process. Finally, the fourth region is the single point where both elements are null, which includes the cases when none of them appear in the initial AG.

The 2nd-order relations on the vertices are defined as follows (the 2nd-order relations on the arcs are defined in a similar way [3,5]):

Antagonism relations: Two vertices of the FDG are *antagonistic* if the probabilities in the first region are all zero,

$$A_\omega(\omega_i,\omega_j)=\begin{cases}1 & \text{if } \Pr(\alpha_i\neq\Phi\wedge\alpha_j\neq\Phi)=0\\ 0 & \text{if } \Pr(\alpha_i\neq\Phi\wedge\alpha_j\neq\Phi)>0\end{cases}\tag{3}$$

which means that, although these vertices are included in the prototype as different elementary parts of the covered patterns, they have never taken place together in any AG of the reference set used to synthesise the FDG. Figure 1.b shows the joint probabilities of the vertices ω_i and ω_j defined as antagonistic.

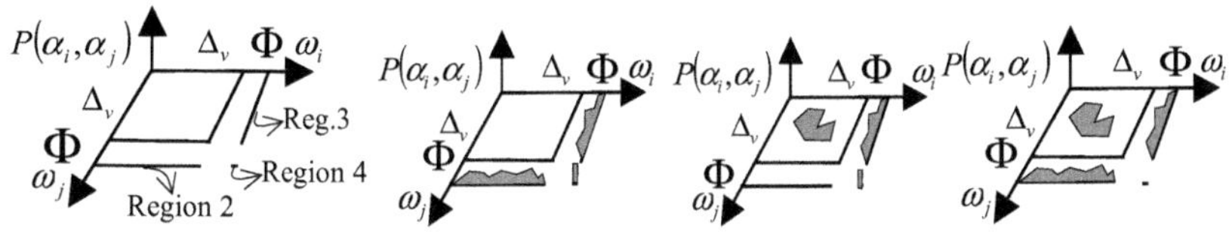

Fig. 1. (a) Split of the joint domain of two random vertices in four regions. 2nd-order density function of (b) two antagonistic, (c) occurrent and (d) existent vertices

Occurrence relations: There is an *occurrence relation* from ω_i to ω_j if the joint probability function equals zero in the second region,

$$O_\omega(\omega_i,\omega_j)=\begin{cases}1 & \text{if } \Pr(\alpha_i \neq \Phi \wedge \alpha_j = \Phi)=0\\ 0 & \text{if } \Pr(\alpha_i \neq \Phi \wedge \alpha_j = \Phi)>0\end{cases} \tag{4}$$

That is, it is possible to assure that if ω_i does appear in any AG of the reference set then ω_j must appear too. The case of the third region is analogous to the second one with the only difference of swapping the elements. Figure 1.c shows the joint probabilities of vertices ω_i and ω_j, with an occurrence from ω_i to ω_j.

Existence relations: Finally, there is an *existence* relation between two vertices if the joint probability function equals zero in the fourth region,

$$E_\omega(\omega_i,\omega_j)=\begin{cases}1 & \text{if } \Pr(\alpha_i = \Phi \wedge \alpha_j = \Phi)=0\\ 0 & \text{if } \Pr(\alpha_i = \Phi \wedge \alpha_j = \Phi)>0\end{cases} \tag{5}$$

that is, all the objects in the class described by the FDG have at least one of the two elements. Figure 1.d shows the joint probabilities of two vertices ω_i and ω_j satisfying an existence relation.

Definition 8 A *Function-Described Graph F* is a RG that satisfies the assumptions 1 and 2 shown above and contains the information of the 2nd-order relations of *antagonism*, *occurrence* and *existence* between pairs of vertices or arcs.

Based on these assumptions, for an FDG *F*, the probability $P_R(G|\mu\ )$ becomes

$$P_F(F|\mu\ )=\prod_{i=1}^{n} p_i(\mathbf{a}_i)\ \prod_{j=1}^{m} q_j(\mathbf{b}_j) \tag{6}$$

where $p_i(\mathbf{a})$ is defined as in FORGs and $q_j(\mathbf{b}) \triangleq \Pr(\beta_j = \mathbf{b} \mid \alpha_{j1} \neq \Phi, \alpha_{j2} \neq \Phi)$. Note that, due to the structural consistency requirements, there is no need to store the conditional probabilities $\Pr(\beta_j = \mathbf{b} \mid \alpha_{j1} = \Phi \vee \alpha_{j2} = \Phi)$ in the FDGs since, by definition $\Pr(\beta_j = \Phi \mid \alpha_{j1} = \Phi \vee \alpha_{j2} = \Phi) = 1$.

Moreover, the isomorphism μ not only has to be structurally coherent but also has to fulfil the 2nd-order constraints shown in (7). The basic idea of these constraints is the satisfaction by an AG to be matched of the antagonism, occurrence and existence relations inferred from the set of AGs used to synthesise the FDG. However, the 2nd-order relations caused by FDG null vertices should not be taken into account, since they are artificially introduced in the extension of the FDG (see [1] for more details).

$$
\begin{aligned}
&\overbrace{\left(A_\omega(\omega_i,\omega_j)=1 \wedge p_i(\Phi)\neq 1 \wedge p_j(\Phi)\neq 1\right)}^{\text{FDG}} \Rightarrow \overbrace{\left(\mathbf{a}_i=\Phi \vee \mathbf{a}_j=\Phi\right)}^{\text{AG}}\\
&\left(O_\omega(\omega_i,\omega_j)=1 \wedge p_i(\Phi)\neq 1 \wedge p_j(\Phi)\neq 1\right) \Rightarrow \left(\mathbf{a}_i=\Phi \vee \mathbf{a}_j\neq\Phi\right)\\
&\left(E_\omega(\omega_i,\omega_j)=1 \wedge p_i(\Phi)\neq 1 \wedge p_j(\Phi)\neq 1\right) \Rightarrow \left(\mathbf{a}_i\neq\Phi \vee \mathbf{a}_j\neq\Phi\right)
\end{aligned}
\qquad (7)
$$

The storage space of FDGs is $O\left(nN + mM + n^2 + m^2\right)$ where N and M are the number of elements of the domains Δ_v and Δ_e, respectively.

4 Second-Order Random-Graph Representation

We show next that the joint probability of an instantiation of the random elements in a RG can be approximated as follows:

$$
p(\mathbf{d}_1,,,\mathbf{d}_s) \approx \prod_{i=1}^{s} p_i(\mathbf{d}_i) \prod_{i=1}^{s-1} \prod_{j=i+1}^{s} r_{ij}(\mathbf{d}_i,\mathbf{d}_j) \qquad (8)
$$

where $p_i(\mathbf{d}_i)$ are the marginal probabilities of the random elements γ_i, (vertices or arcs) and r_{ij} are the Peleg compatibility coefficients [2] that take into account both the marginal and 2nd-order joint probabilities,

$$
r_{ij}(\mathbf{d}_i,\mathbf{d}_j) = \frac{\Pr(\gamma_i = \mathbf{d}_i \wedge \gamma_j = \mathbf{d}_j)}{p_i(\mathbf{d}_i)\, p_j(\mathbf{d}_j)} \qquad (9)
$$

The Peleg coefficient, with a non-negative range, is related to the "degree" of dependence between two random variables. If they are independent, the joint probability is defined as the product of the marginal ones, thus, $r_{ij} = 1$ (or a value close to 1 if the probability functions are estimated). If one of the marginal probabilities is null, the joint probability is also null. In this case, the indecisiveness *0/0* is solved as 1, since this do not affect the global joint probability, which is null.

Eq. (8) is obtained by assuming independence in the conditional probabilities (section 4.1) and rearranging the joint probability expression with the Bayes rule (section 4.2)

4.1 Conditional Probabilities

The conditional density probability $p(\gamma_i/(\gamma_{i+1},\ldots,\gamma_s))$ of a random element γ_i is used to compute the joint density probability $p(\gamma_1,\ldots,\gamma_s)$. Applying the Bayes rule to the conditional probability, the following expression holds,

$$p(\gamma_i/(\gamma_{i+1},\ldots,\gamma_s)) = \frac{p(\gamma_i)\cdot p((\,_{i+1},\ldots,\,_s)/\,_i)}{p(\gamma_{i+1},\ldots,\gamma_s)} \tag{10}$$

Due to the fact that this n-order probability can not be stored in practice, we have to suppose at this point that the conditioning random variables γ_{i+1} to γ_n are independent to each other. In that case, an estimate is given by

$$p(\gamma_i/(\gamma_{i+1},\ldots,\gamma_s)) = p(\gamma_i)\cdot\prod_{j=i+1}^{s}\frac{p(\gamma_j/\gamma_i)}{p(\gamma_j)} = p(\gamma_i)\cdot\prod_{j=i+1}^{s}\frac{p(\gamma_j,\gamma_i)}{p(\gamma_j)\cdot p(\gamma_i)} \tag{11}$$

Thus, if we use the Peleg compatibility coefficients then the conditional probability is,

$$\text{prob}(\gamma_i=\mathbf{d}_i/(\gamma_{i+1}=\mathbf{d}_{i+1},\ldots,\gamma_s=\mathbf{d}_s)) = p_i(\mathbf{d}_i)\cdot\prod_{j=i+1}^{s} r_{ij}(\mathbf{d}_i,\mathbf{d}_j) \tag{12}$$

4.2 Joint Probability

Using the Bayes theorem, the joint probability density function $p(\gamma_1,,,\gamma_s)$ can be split into the product of another joint probability function and a conditional one,

$$p(\gamma_1,,,\gamma_s) = p(\gamma_2,,,\gamma_s)\cdot p(\,_1/(\,_2,,,\,_s)) \tag{13}$$

and applying n-1 times the same theorem on the remaining joint probability,

$$p(\gamma_1,,,\gamma_s) = p(\gamma_s)\cdot\prod_{i=1}^{s-1} p(\,_i/(\,_{i+1},,,\,_s)) \tag{14}$$

If we use equation (12) to estimate the conditional probabilities, then the joint probability $p(\mathbf{d}_1,,,\mathbf{d}_s)$ can be estimated as $p^*(\mathbf{d}_1,,,\mathbf{d}_s)$ where,

$$p^*(\mathbf{d}_1,,,\mathbf{d}_s) = p_s(\mathbf{d}_s)\cdot\prod_{i=1}^{s-1}\left[p_i(\mathbf{d}_i)\cdot\prod_{j=i+1}^{s} r_{ij}(\mathbf{d}_i,\mathbf{d}_j)\right] \tag{15}$$

and introducing the first factor into the productory, we have

$$p^*(\mathbf{d}_1,,,\mathbf{d}_s)=\prod_{i=1}^{s}p_i(\mathbf{d}_i)\prod_{i=1}^{s-1}\prod_{j=i+1}^{s}r_{ij}(\mathbf{d}_i,\mathbf{d}_j) \tag{16}$$

In the approximations of the joint probability in the FDG and FORG approaches, random vertices and random arcs are treated separately, for this reason the above expression can be split considering vertices and arcs separately as follows

$$p^*(\mathbf{a}_1,,\mathbf{a}_n,\mathbf{b}_1,,\mathbf{b}_m)=\prod_{i=1}^{n}p_i(\mathbf{a}_i)\prod_{i=1}^{m}p_i(\mathbf{b}_i)\prod_{i=1}^{n-1}\prod_{j=i+1}^{n}r_{ij}(\mathbf{a}_i,\mathbf{a}_j)\prod_{i=1}^{n}\prod_{j=1}^{m}r_{ij}(\mathbf{a}_i,\mathbf{b}_j)\prod_{i=1}^{m-1}\prod_{j=i+1}^{m}r_{ij}(\mathbf{b}_i,\mathbf{b}_j) \tag{17}$$

5 Approximation of the Joint Probability by FORGs

In the FORG approach, the Peleg coefficients between vertices and between arcs do not influence on the computation of the joint probability. That is, by assumption 1 and 2 (section 3.1), $r_{ij}(\mathbf{a}_i,\mathbf{a}_j)=1$ and $r_{ij}(\mathbf{b}_i,\mathbf{b}_j)=1$ for all the vertices and arcs, respectively. On the contrary, assumption 3 (sec 3.1) makes that the probability on the arcs be conditioned on the values of the vertices that the arc connects, $q_j(\mathbf{b}_j \mid \mathbf{a}_{j1},\mathbf{a}_{j2})$. In a similar deduction to that of section 4.3, and considering assumption 1, we arrive at the equivalence:
$q_j(\mathbf{b}_j \mid \mathbf{a}_{j1},\mathbf{a}_{j2})=p_j(\mathbf{b}_j)r_{j_1 j}(\mathbf{a}_{j_1},\mathbf{b}_j)r_{j_2 j}(\mathbf{a}_{j_2},\mathbf{b}_j)$. Thus,

$$P_R(G|\mu)=\prod_{i=1}^{n}p_i(\mathbf{a}_i)\prod_{j=1}^{m}p_j(\mathbf{b}_j)\prod_{j=1}^{m}\prod_{i=j_1,j_2}r_{ij}(\mathbf{a}_i,\mathbf{b}_j) \tag{18}$$

6 Approximation of the Joint Probability by FDGs

In the FDG approach, the 2nd-order probabilities between vertices can be estimated from the marginal probabilities and the 2nd-order relations as follows (a similar expression is obtained for the arcs),

$$\begin{aligned}&\Pr(\alpha_i=\mathbf{a}_i\wedge\alpha_j=\mathbf{a}_j)=0\ \text{ if Condition } *_{2nd}\\&\Pr(\alpha_i=\mathbf{a}_i\wedge\alpha_j=\mathbf{a}_j)\approx p_i(\mathbf{a}_i)p_j(\mathbf{a}_j)\ \text{ otherwise}\end{aligned} \tag{19}$$

where the Condition $*_{2nd}$ is

$$*_{2nd}:\begin{cases}\left(A_\omega(\omega_i,\omega_j)\wedge\mathbf{a}_i\neq\Phi\wedge\mathbf{a}_j\neq\Phi\right)\vee\left(O_\omega(\omega_i,\omega_j)\wedge\mathbf{a}_i\neq\Phi\wedge\mathbf{a}_j=\Phi\right)\vee\\\left(O_\omega(\omega_j,\omega_i)\wedge\mathbf{a}_i=\Phi\wedge\mathbf{a}_j\neq\Phi\right)\vee\left(E_\omega(\omega_i,\omega_j)\wedge\mathbf{a}_i=\Phi\wedge\mathbf{a}_j=\Phi\right)\end{cases} \tag{20}$$

Note that, in the first case, it can be assured that the joint probability is null, but in the second case, we assume that the random elements are independent and the probability is estimated as a product of the marginal ones.

Thus, the Peleg coefficients are simplified as r_{ij}', using equation (7) and (19),

$$r_{ij}'(\mathbf{a}_i,\mathbf{a}_j)=\begin{cases}0 \text{ if } *_{2nd} \wedge p_i(\mathbf{a}_i)\neq 0 \wedge p_j(\mathbf{a}_j)\neq 0\\ 1 \text{ otherwise}\end{cases} \tag{21}$$

Moreover, due to the independence assumption 2 (sec 3.2), it is not possible to have a non-null arc and a null vertex as one of its endpoints in an outcome graph. Thus, we have $p(\alpha_{j1}=\Phi\wedge\beta_j\neq\Phi)=0$ and $p(\alpha_{j2}=\Phi\wedge\beta_j\neq\Phi)=0$. In the other cases, by assumption 1, they are assumed to be independent and so computed as the product of the marginal ones. The Peleg coefficients between vertices and arcs are simplified as

$$r_{ij}''(\mathbf{a}_i,\mathbf{b}_j)=\begin{cases}0 \text{ if } (i=j_1\vee j_2)\wedge \mathbf{a}_i=\Phi\wedge\mathbf{b}_j\neq\Phi\\ 1 \text{ otherwise}\end{cases} \tag{22}$$

The final expression of the joint probability of an outcome AG with respect to an FDG is

$$P_R(G|\mu)=\prod_{i=1}^{n}p_i(\mathbf{a}_i)\prod_{j=1}^{m}p_j(\mathbf{b}_j)\prod_{i=1}^{n}\prod_{j=1}^{n}r_{ij}'(\mathbf{a}_i,\mathbf{a}_j)\prod_{i=1}^{m}\prod_{j=1}^{m}r_{ij}'(\mathbf{b}_i,\mathbf{b}_j)\prod_{j=1}^{m}\prod_{i=j_1,j_2}r_{ij}''(\mathbf{a}_i,\mathbf{b}_j) \tag{23}$$

7 Conclusions and Future Work

We have presented a general formulation of an approximation of the joint probability of random elements in a RG, that describes a set of AGs, based on 2nd-order probabilities and marginal ones. We have seen that the FORG and FDG approaches are two specific cases (simplifications) of this general 2nd-order formulation. In both cases, the marginal probabilities of the random vertices and arcs are considered but the difference between them is in how are the 2nd-order relations between vertices or arcs estimated. FORGs keep only the 2nd-order probability between arcs and their extreme vertices, since the other joint probabilities are estimated as a product of the marginal ones. On the contrary, FDGs keep only a qualitative and structural information of the 2nd-order probabilities between all the vertices and arcs. If we compare both methods, FORGs have local (arc and endpoint vertex) 2nd-order semantic knowledge of the set of AGs but do not use any 2nd-order structural information of the set of the AGs. FDGs do not keep any 2nd-order semantic information but include the 2nd-order structural information of all the set of AGs. For this reason, the storage space of FORGs increases to the square on the size of the random-element domain but the FDGs increases to the square on the number of vertices and arcs.

However, the most important implication of the given general formulation of the 2nd-order random graph representation is that it opens the door to the development of other probabilistic graph approaches, either full 2nd-order or not. In addition, it is interesting to study empirically the relation between the amount of data to be kept in the model and the recognition ratio and run time in several applications. That is, to know in which applications is worthwhile to use explicitly the 2nd-order probabilities through the Peleg coefficients or is enough to estimate them by other ways less costly in space requirements, such as FORGs and FDGs. Moreover, a distance between the structural model and an AG can be defined. This is left for future research.

References

1. R. Alquézar, F. Serratosa, A. Sanfeliu, "Distance between Attributed Graphs and Function-Described Graphs relaxing 2nd order restrictions". *Proc. SSPR'2000 and SPR'2000,* Barcelona, Spain, Springer LNCS-1876, pp. 277-286, 2000.
2. S. Peleg and A. Rosenfeld, "Determining compatibility coefficients for curve enchancement relaxation processes", *IEEE Transactions on Systems, Man and Cybernetics*, vol. 8, pp. 548-555, 1978.
3. Sanfeliu, F. Serratosa and R. Alquezar, "Clustering of attributed graphs and unsupervised synthesis of function-described graphs", *Proceedings ICPR'2000, 15th Int. Conf. on Pattern Recog.*, Barcelona, Spain, Vol.2, pp. 1026-1029, 2000.
4. K. Sengupta and K. Boyer, "Organizing large structural modelbases", *IEEE Trans. on Pattern Analysis and Machine Intelligence*, vol. 17, pp. 321-332, 1995.
5. F. Serratosa, R. Alquezar and A. Sanfeliu, "Efficient algorithms for matching attributed graphs and function-described graphs", in *Proceedings ICPR'2000, 15th Int. Conf. on Pattern Recognition*, Barcelona, Spain, Vol.2, pp. 871-876, 2000.
6. A.K.C. Wong and M. You, "Entropy and distance of random graphs with application to structural pattern recognition", *IEEE Trans. on PAMI.*, vol. 7, pp. 599-609, 1985.
7. H. Bunke, "Error-tolerant graph matching: a formal framework and algorithms". *Proc. Workshops SSPR'98 & SPR'98,* Sydney, Australia, Springer LNCS-1451,pp.1-14, 1998.
8. D.S. Seong, H.S. Kim & K.H. Park, "Incremental Clustering of Attributed Graphs", *IEEE Transactions on Systems, Man and Cybernetics*, vol. 23, pp. 1399-1411, 1993.
9. Sanfeliu, R. Alquézar, J. Andrade, J.Climent, F. Serratosa and J.Vergés, "Graph-based Representations and Techniques for Image Processing and Image Analysis", *Pattern Recognition*, vol. 35, pp: 639-650, 2002.

Successive Projection Graph Matching

Barend Jacobus van Wyk[1,3], Michaël Antonie van Wyk[2], and Hubert Edward Hanrahan[3]

[1] Kentron, a division of Denel
Centurion, South Africa
ben.van.wyk@kentron.co.za
[2] Rand Afrikaans University
Johannesburg, South Africa
mavw@ing.rau.ac.za
[3] University of the Witwatersrand
Johannesburg, South Africa
h.hanrahan@ee.wits.ac.za

Abstract. The *Successive Projection Graph Matching* (SPGM) algorithm, capable of performing full- and sub-graph matching, is presented in this paper. *Projections Onto Convex Sets* (POCS) methods have been successfully applied to signal processing applications, image enhancement, neural networks and optics. The SPGM algorithm is unique in the way a constrained cost function is minimized using POCS methodology. Simulation results indicate that the SPGM algorithm compares favorably to other well-known graph matching algorithms.

1 Introduction

In image processing applications, it is often required to match different images of the same object or similar objects based on the structural descriptions constructed from these images. If the structural descriptions of objects are represented by attributed relational graphs, different images can be matched by performing *Attributed Graph Matching* (AGM). Because of the combinatorial nature of the AGM problem, it can be efficiently solved by an exhaustive search only when dealing with extremely small graphs.

According to [1], graph matching algorithms can be divided into two major approaches. In general, the first approach constructs a state-space which is searched using heuristics to reduce complexity. Examples of algorithms belonging to this group are those proposed by You and Wong [2], Tsai and Fu [3, 4], Depiero *et al.* [5], Eshera and Fu [6], Bunke and Shearer [7], Bunke and Messmer [8] and Allen *et al.* [9].

The second approach, which is also our approach, is based on function optimization techniques. This approach includes earlier techniques such as the symmetric *Polynomial Transform Graph Matching* (PTGM) algorithm of Almohamad [10], the *Linear Programming Graph Matching* (LPGM) algorithm of Almohamad and Duffuaa [?] and the *Eigen-decomposition Graph Matching*

T. Caelli et al. (Eds.): SSPR&SPR 2002, LNCS 2396, pp. 263–271, 2002.
© Springer-Verlag Berlin Heidelberg 2002

(EGM) method of Umeyama [11]. More recent techniques include a multitude of Bayesian, genetic, neural network and relaxation-based methods.

The *Graduated Assignment Graph Matching* (GAGM) algorithm of Gold and Rangarajan [1] proved to be very successful. It combines graduated non-convexity, two-way assignment constraints and sparsity. The literature on optimization methods for graph matching has also been complemented by the work of Hancock and his associates [13–17]. Their work builds on a relational consistency gauged by an exponential probability distribution.

This paper focuses on matching fully-connected, undirected attributed graphs using a *Projections Onto Convex Sets* (POCS) method. POCS methods have been successfully applied to signal processing applications, image enhancement, neural networks and optics [22]. In this paper, the *Successive Projection Graph Matching* (SPGM) algorithm is presented. This algorithm is unique in the way a constrained cost function is minimized using POCS methodology. Although the algorithm of Gold and Rangarajan [1] also uses a successive approximation approach, our algorithm is significantly different. We do not use graduated non-convexity, and we do not enforce constraints using repeated row and column normalization. Instead, constraints are enforced by mapping onto appropriate convex sets.

The outline of the presentation is as follows: In section 2 we introduce a constrained cost function, and in section 3 we show how successive projections can be used to obtain a constrained optimum. Numerical results, obtained during the evaluation of our algorithm, are presented in section 4.

2 Cost Function Formulation

The focus of this paper is on matching graphs where a duplicate graph, say

$$G = \left(V, E, \{\mathbf{A}_g\}_{g=1}^{r}, \{\mathbf{B}_h\}_{h=1}^{s}\right) \tag{1}$$

is matched to a reference graph, say

$$G' = \left(V', E', \{\mathbf{A}'_g\}_{g=1}^{r}, \{\mathbf{B}'_h\}_{h=1}^{s}\right) \tag{2}$$

where $\mathbf{A}_g \in \mathbb{R}^{n\times n}$, $\mathbf{B}_h \in \mathbb{R}^{n\times 1}$, $\mathbf{A}'_g \in \mathbb{R}^{n'\times n'}$ and $\mathbf{B}'_h \in \mathbb{R}^{n'\times 1}$ represent the edge attribute adjacency matrices and vertex attribute vectors respectively. The reference and duplicate graphs each have r edge attributes and s vertex attributes. The number of vertices of G' (respectively, G) is $n' := |V'|$ (respectively, $n := |V|$). Here we consider the general case of sub-graph matching. *Full-graph Matching* (FGM) refers to matching two graphs having the same number of vertices (i.e. $n' = n$) while *Sub-graph Matching* (SGM) refers to matching two graphs having a different number of vertices (i.e. $n' > n$).

We say that G is *matched* to some sub-graph of G' if there exists a matrix $\mathbf{P} \in \mathrm{Per}(n, n')$, where $\mathrm{Per}(n, n')$ is the set of all $n\times n'$ permutation sub-matrices, such that

$$\mathbf{A}_g = \mathbf{P}\mathbf{A}'_g\mathbf{P}^T, \qquad g = 1, ..., r \tag{3}$$

and

$$\mathbf{B}_h = \mathbf{P}\mathbf{B}'_h, \qquad h = 1, ..., s. \tag{4}$$

As shown in [19], the *Attributed Graph Matching* (AGM) problem can be expressed as a combinatorial optimization problem. However, due to the difficulty in solving this combinatorial optimization problem, we construct an approximate solution. Following an approach similar to [1], we can express the undirected AGM problem as finding the matrix $\overline{\mathbf{P}}$, such that the objective function,

$$J(\mathbf{p}) = -\mathbf{p}^T\mathbf{X}\mathbf{p} - \mathbf{y}^T\mathbf{p}, \tag{5}$$

is minimized, where $\mathbf{p} = \text{vec}\left(\overline{\mathbf{P}}\right)$ is subject to

$$0 \leq \overline{P}_{ij} \leq 1, \tag{6}$$

$$\sum_{j=1}^{n'} \overline{P}_{ij} = 1, \qquad i = 1, .., n \tag{7}$$

and

$$\sum_{i=1}^{n} \overline{P}_{ij} \leq 1, \qquad j = 1, ..., n'. \tag{8}$$

Here $\overline{\mathbf{P}} := \left(\overline{P}_{ij}\right)$ and vec$(\cdot)$ denote the vectorization operation from linear algebra. The elements of the matrix $\mathbf{X} \in \mathbb{R}^{nn' \times nn'}$ are given by

$$X_{kl} = \frac{\alpha}{\left(\sum_{g=1}^{r} \left| A^g_{l-\lfloor \frac{l-1}{n} \rfloor n,\, k-\lfloor \frac{k-1}{n} \rfloor n} - A'^g_{\lfloor \frac{l-1}{n} \rfloor +1,\, \lfloor \frac{k-1}{n} \rfloor +1} \right| + \alpha\right)}$$

except when $k = l$, $\lfloor \frac{k-1}{n} \rfloor + 1 = \lfloor \frac{l-1}{n} \rfloor + 1$ or $k - \lfloor \frac{k-1}{n} \rfloor n = l - \lfloor \frac{l-1}{n} \rfloor n$, in which case $X_{kl} = 0$. Here $\mathbf{A}_g := (A^g_{ij}) \in \mathbb{R}^{n \times n}$, $\mathbf{A}'_g := (A'^g_{ij}) \in \mathbb{R}^{n' \times n'}$, $|\cdot|$ denotes the absolute value, $k = 1, ..., nn'$, $l = 1, ..., nn'$ and α is a parameter controlling the steepness of the compatibility function, normally chosen equal to one. The elements of the vector $\mathbf{y} \in \mathbb{R}^{nn' \times 1}$ are given by

$$y_k = \frac{\alpha}{\left(\sum_{h=1}^{s} \left| B^m_{k-\lfloor \frac{k-1}{n} \rfloor n} - B'^m_{\lfloor \frac{k-1}{n} \rfloor +1} \right| + \alpha\right)}$$

where $\mathbf{B}_h := (B^h_i) \in \mathbb{R}^{n \times 1}$, $\mathbf{B}'_h := (B'^h_j) \in \mathbb{R}^{n' \times 1}$, $k = 1, ..., nn'$.

Relaxation methods such as [17] and [18] in general only enforce the constraint given by Eq. 6 and the row constraint given by Eq. 7. In addition, the SPGM algorithm also enforces the column constraint given by Eq. 8. Similar to our algorithm, the GAGM algorithm [1] also enforces a column constraint, but uses a significantly different approach. Central to our method is the projection of a vector onto the intersection of two convex sets formed by the row and column constraints. As noted by [17], the row constraints form a closed convex set, which can be expressed as

$$\mathbf{C}_r = \left\{ \mathbf{p}_r \in \mathbb{R}^{nn'} : \begin{array}{l} \mathbf{p}_r = [\bar{\mathbf{p}}_1, ..., \bar{\mathbf{p}}_n], \\ \bar{\mathbf{p}}_i = [\overline{P}_{i1}, ..., \overline{P}_{in'}], \\ \sum_{j=1}^{n'} \overline{P}_{ij} = 1, \\ 0 \leq \overline{P}_{ij} \leq 1 \end{array} \right\}.$$

In a similar manner, the column constraints can also be expressed as the set,

$$\mathbf{C}_c = \left\{ \mathbf{p}_c \in \mathbb{R}^{nn'} : \begin{array}{l} \mathbf{p}_c = [\bar{\mathbf{p}}_1, ..., \bar{\mathbf{p}}_{n'}], \\ \bar{\mathbf{p}}_j = [\overline{P}_{1j}, ..., \overline{P}_{nj}], \\ \sum_{i=1}^{n} \overline{P}_{ij} \leq 1, \\ 0 \leq \overline{P}_{ij} \leq 1 \end{array} \right\},$$

which is also closed and convex. Note that the intersection of the convex sets $\mathbf{C}_r$ and $\mathbf{C}_c$, denoted by $\mathbf{C}_0 = \mathbf{C}_r \cap \mathbf{C}_c$, is non-empty.

3 Projected Successive Approximations

3.1 The SPGM Algorithm

The following pseudo-code describes the SPGM algorithm with index $k > 0$:

while $(k < I \text{ and } \delta > \epsilon)$ or $(k < 3)$

$\widetilde{\mathbf{p}}^{k+1} = \mathbf{p}^k - \frac{1}{s_0^k}\left[\nabla J\left(\mathbf{p}^k\right)\right]$

$\mathbf{p}^{k+1} = T_0\left(\widetilde{\mathbf{p}}^{k+1}\right)$

$\delta = \left(\mathbf{p}^{k+1} - \mathbf{p}^k\right)^T\left(\mathbf{p}^{k+1} - \mathbf{p}^k\right)$

$k = k + 1$

end

The vector $\mathbf{p}^0$ is initialized to $\left[\frac{1}{n'}, ..., \frac{1}{n'}\right]^T$. $T_0\left(\widetilde{\mathbf{p}}^{k+1}\right)$ denotes the projection of $\widetilde{\mathbf{p}}^{k+1}$ onto $\mathbf{C}_0$, which will be discussed in section 3.2. We obtain $\widetilde{\mathbf{p}}^{k+1}$ by approximating Eq. 5, using a spherical function given by

$$\widetilde{J}\left(\widetilde{\mathbf{p}}^{k+1}\right) = J\left(\mathbf{p}^k\right) + \nabla^T J\left(\mathbf{p}^k\right)\left(\widetilde{\mathbf{p}}^{k+1} - \mathbf{p}^k\right) + \frac{1}{2}\left(\widetilde{\mathbf{p}}^{k+1} - \mathbf{p}^k\right)^T S_o^k\left(\widetilde{\mathbf{p}}^{k+1} - \mathbf{p}^k\right), \tag{9}$$

where $S_o^k = \text{diag}\left(s_0^k, ..., s_0^k\right) = s_0^k\mathbf{I}$, $\nabla J\left(\mathbf{p}^k\right)$ denotes the gradient vector of Eq. 5, given by $-2\mathbf{X}\mathbf{p}^k - \mathbf{y}$, and s_0^k is a curvature parameter. The vector $\widetilde{\mathbf{p}}^{k+1}$ is calculated as the minimum of the spherical function, occurring where

$$\nabla^T J\left(\mathbf{p}^k\right) + \left(\widetilde{\mathbf{p}}^{k+1} - \mathbf{p}^k\right)^T S_o^k = 0. \tag{10}$$

Since $S_o^k = s_0^k\mathbf{I}$, we obtain the simple update rule $\widetilde{\mathbf{p}}^{k+1} = \mathbf{p}^k - \frac{1}{s_0^k}\left[\nabla J\left(\mathbf{p}^k\right)\right]$. Although the curvature parameter, s_0^k, can be varied every iteration, the best results were obtained by keeping it constant.

The convergence parameter, ϵ, is normally chosen $< 10^{-3}$. The iteration parameter, I, limits the maximum number of iterations. Once the algorithm has terminated, devec($\mathbf{p}$), is used to obtain an estimate to $\mathbf{P} \in \mathrm{Per}(n, n')$, by setting the maximum value in each row equal to one and the rest of the values in each row equal to zero. Here devec($\cdot$) denotes the inverse of vec($\cdot$), the matrix vectorization operation from linear algebra.

3.2 Projection onto $\mathbf{C}_0$

Once we obtain $\widetilde{\mathbf{p}}^{k+1}$, our objective is to find $\mathbf{p}^{k+1} \in \mathbf{C}_0$ such that

$$\left\|\widetilde{\mathbf{p}}^{k+1} - \mathbf{p}^{k+1}\right\| = \min_{\mathbf{z} \in \mathbf{C}_0} \left\|\widetilde{\mathbf{p}}^{k+1} - \mathbf{z}\right\|. \tag{11}$$

From the POCS theory [21], Eq. 11 implies that $\mathbf{p}^{k+1} = T_0\left(\widetilde{\mathbf{p}}^{k+1}\right)$, the projection of $\widetilde{\mathbf{p}}^{k+1}$ onto the set $\mathbf{C}_0$. By applying certain fundamental POCS results, detailed in [22–23] to our problem, we can construct a sequence that converges to $T_0\left(\widetilde{\mathbf{p}}^{k+1}\right)$. The algorithm for obtaining $T_0\left(\widetilde{\mathbf{p}}^{k+1}\right)$ is described by the following pseudo-code:

Calculating $T_0\left(\widetilde{\mathbf{p}}^{k+1}\right)$:
initialization: $k_C = 1,\ \delta_{\mathbf{C}_0} > 0.05$
while $(\delta_C > 0.05$ and $k_C < n')$
  $\mathbf{p}_h = \mathbf{p}$
  $\mathbf{p} = T_r(\mathbf{p})$
  $\mathbf{p} = T_c(\mathbf{p})$
  $\delta_{\mathbf{C}_0} = (\mathbf{p}_h - \mathbf{p})^T (\mathbf{p}_h - \mathbf{p})$
  $k_C = k_C + 1$
end

$T_r(\mathbf{p})$ denotes the projection of $\mathbf{p}$ onto the set $\mathbf{C}_r$ and $T_c(\mathbf{p})$ denotes the projection of $\mathbf{p}$ onto the set $\mathbf{C}_c$, given by the following pseudo-code. The superscript $k+1$ has been omitted for simplicity. The notation $\mathbf{p}(i : j : k)$ indicates that we select every j–th element from the vector $\mathbf{p}$, starting with the i–th element and ending with the k–th element. The operation $[\mathbf{s}, \mathbf{d}] = \mathrm{sort}[\bar{\mathbf{p}}]$ indicates that we sort the elements of $\bar{\mathbf{p}}$ in ascending order, where $\mathbf{s}$ is the sorted vector and $\mathbf{d}$ is a vector containing the pre-sorted positions of the sorted elements.

Calculating $T_r(\mathbf{p})$:
for $i = 1 : n$
  $\phi = \mathrm{sum}\left[\mathbf{p}(i : n : n(n'-1)+i)\right]$
  $\sigma = n'$
  $\bar{\mathbf{p}} = \mathbf{p}(i : n : n(n'-1)+i)$
  $[\mathbf{s}, \mathbf{d}] = \mathrm{sort}\left[\bar{\mathbf{p}}\right]$
  for $j = 1 : n'$
    $\mathbf{s}(j) = \mathbf{s}(j) + \frac{1-\phi}{\sigma}$
    if $\mathbf{s}(j) < 0$
      $\mathbf{s}(j) = 0$
      $\phi = \phi - \bar{\mathbf{p}}(\mathbf{d}(j))$

$\sigma = \sigma - 1$
end
$\bar{\mathbf{p}}(\mathbf{d}(j)) = \mathbf{s}(j)$
end
$\mathbf{p}(i : n : n(n' - 1) + i) = \bar{\mathbf{p}}$
end

When $n = n'$, the approach used to calculate $T_c(\mathbf{p})$ is similar to that which we used to calculate $T_r(\mathbf{p})$. The pseudo-code below is for the case $n < n'$:

Calculating $T_c(\mathbf{p})$**:**
for $j = 1 : n'$
$\phi = \text{sum}\,[\mathbf{p}(n(j-1)+1 : nj)]$
if $\phi > 1$
$\sigma = n$
$\bar{\mathbf{p}} = \mathbf{p}(n(j-1)+1 : nj)$
$[\mathbf{s}, \mathbf{d}] = \text{sort}\,[\bar{\mathbf{p}}]$
for $i = 1 : n$
$\mathbf{s}(i) = \mathbf{s}(i) + \frac{1-\phi}{\sigma}$
if $\mathbf{s}(i) < 0$
$\mathbf{s}(i) = 0$
$\phi = \phi - \bar{\mathbf{p}}(\mathbf{d}(i))$
$\sigma = \sigma - 1$
end
$\bar{\mathbf{p}}(\mathbf{d}(i)) = \mathbf{s}(i)$
end
$\mathbf{p}(n(j-1)+1 : nj) = \bar{\mathbf{p}}$
end
end

4 Simulation Results

In order to evaluate the performance of the SPGM algorithm, the following procedure was used: Firstly, the parameters n', n, r and s were fixed. For every iteration, a reference graph G' was generated randomly with all attributes distributed between 0 and 1. An $n \times n'$ permutation sub-matrix, $\mathbf{P}$, was also generated randomly, and then used to permute the rows and columns of the edge attribute adjacency matrices and the elements of the vertex attribute vectors of G'. Next, an independently generated noise matrix (vector, respectively) was added to each edge attribute adjacency matrix (vertex attribute vector, respectively) to obtain the duplicate graph G. The element of each noise matrix/vector was obtained by multiplying a random variable —uniformly distributed on the interval $[-1/2, 1/2]$— by the noise magnitude parameter ε. Different graph matching algorithms were then used to determine a permutation sub-matrix which approximates the original permutation sub-matrix $\mathbf{P}$.

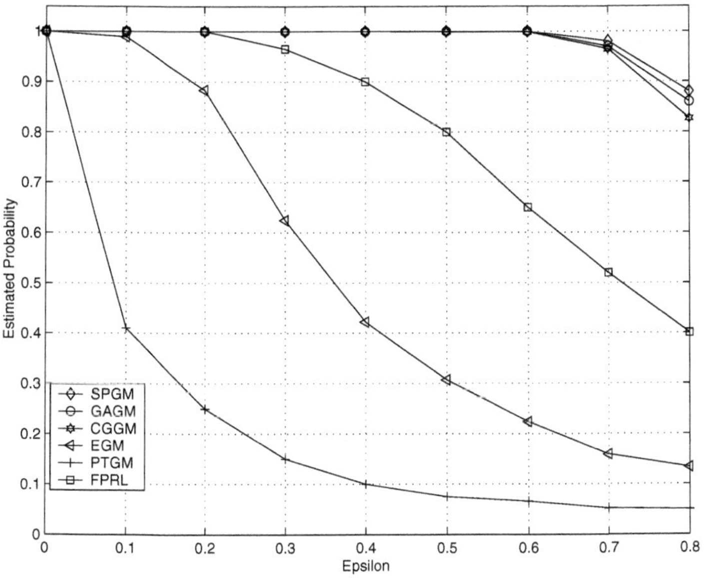

Fig. 1. Matching of (30,3,3) attributed graphs: Estimated probability of correct vertex-vertex matching versus ε

In figure 1, the performance of the SPGM algorithm is compared to the performance of the GAGM [1], PTGM [10], EIGGM [11] and CGGM [20] algorithms for $n' = 30$, $n = 30$, $r = 3$ and $s = 3$. The performance of the SPGM algorithm is also compared to the performance of the well-known *Faugeras-Price Relaxation Labelling* (FPRL) method [18]. The EIGGM algorithm has been adapted for attributed graph matching by calculating separate permutation sub-matrices for each attribute, and then selecting the permutation sub-matrix associated with the minimum cost. The FPRL algorithm was implemented using a step-size parameter of 0.1. The probability of a correct vertex-vertex assignment was estimated for a given value of ε after every 300 trials. From a probabilistic point of view, this provides us with an approximation of how well the proposed algorithm performs for a given noise magnitude.

In figure 2, the sub-graph matching performance of the SPGM is compared to the performances of the GAGM, CGGM, and FPRL algorithms for $n' = 20$, $n = 5$, $r = 3$ an $s = 3$. The EIGGM and PTGM algorithms are not suitable for performing sub-graph matching. The performance of the CGGM algorithm severely degrades when more than half the nodes are missing. The GAGM algorithm was implemented using the default parameters described in [1].

From the results it is evident that the SPGM algorithm is an extremely robust algorithm for performing full- and sub-graph matching. For $n' = 30$ and $n = 30$, the algorithm took on average 5.5 iterations to converge for $\epsilon = 10^{-3}$, $\varepsilon < 0.5$ and $s_0 = 30$. For $n' = 20$ and $n = 5$, the algorithm took on average 13.7

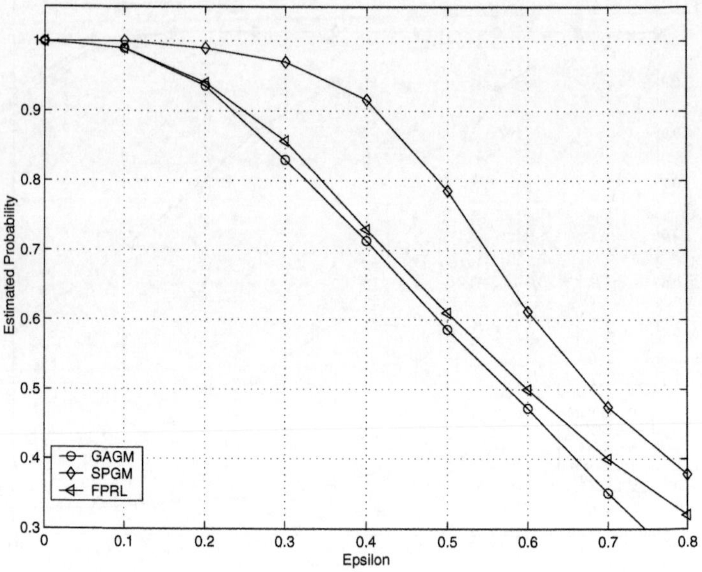

Fig. 2. Matching of (20/5,3,3) attributed graphs: Estimated probability of correct vertex-vertex matching versus ε

iterations to converge for $\epsilon = 10^{-3}$, $\varepsilon < 0.5$ and $s_0 = 30$. The complexity of the SPGM algorithm is $O(n^4)$ per iteration.

5 Conclusion

A novel algorithm for performing attributed full- and sub-graph matching was presented. The SPGM algorithm is unique in the way a constrained cost function is minimized using POCS methodology. Simulation results indicate that the SPGM algorithm is very robust against noise and performs as well or better than the algorithms it was compared against. The SPGM algorithm incorporates a general approach to a wide class of graph matching problems based on attributed graphs, allowing the structure of the graphs to be based on multiple sets of attributes.

References

1. Gold, S., Rangarajan, A.: A Graduated Assignment Algorithm for Graph Matching, IEEE Trans. Patt. Anal. Machine Intell, Vol. 18 (1996) 377–388 263, 264, 265, 269
2. You, M., Wong, K. C.: An Algorithm for Graph Optimal Isomorphism, Proc. ICPR. (1984) 316–319 263
3. Tsai, W.-H., Fu, K.-S.: Error-Correcting Isomorphisms of Attributed Relation Graphs for Pattern Recognition, IEEE Trans. Syst. Man Cybern., Vol. 9 (1997) 757–768

4. Tsai, W.-H., Fu, K.-S.: Subgraph Error-Correcting Isomorphisms for Syntactic Pattern Recognition, IEEE Trans. Systems, Man, Cybernetics, Vol. 13 (1983) 48–62
5. Depiero, F., Trived, M., Serbin, S.: Graph Matching using a Direct Classification of Node Attendance, Pattern Recognition, Vol. 29, No. 6, (1996) 1031–1048 263
6. Eshera, M. A., Fu, K.-S.: A Graph Distance measure for Image Analysis, IEEE Trans. Systems, Man, Cybernetics, Vol. 13 (1984) 398–407 263
7. Bunke, H., Shearer, K.: A Graph Distance Metric Based on the Maximal Common Subgraph, Pattern Recognition Letters, Vol. 19 (1998) 255–259 263
8. Bunke, H., Messmer, B.: Recent Advances in Graph Matching, Int. J. Pattern Recognition Artificial Intell. Vol. 11, No. 1 (1997) 169–203 263
9. Allen, R., Cinque, L., Tanimoto, S., Shapiro, L., Yasuda, D.: A Parallel Algorithm for Graph Matching and Its MarPlas Implementation, IEEE Trans. Parallel and Distb. Syst., Vol. 8, No. 5 (1997) 490–501 263
10. Almohamad, H. A. L.: Polynomial Transform for Matching Pairs of Weighted Graphs, Appl. Math. Modelling, Vol. 15, No. 4 (1991) 216–222 263, 269
11. Umeyama, S.: An Eigendecomposition Approach to Weighted Graph Matching Problems, IEEE Trans. Patt. Anal. Machine Intell., Vol. 10, No. 5 (1988) 695–703 264, 269
12. Cross, A. D. J., Wilson, C., Hancock, E. R.: Inexact Matching Using Genetic Search, Pattern Recognition, Vol. 30, No. 6 (1997) 953–970
13. Finch, A. M., Wilson, R. C., Hancock, R.: Symbolic Matching with the EM Algorithm
, Pattern Recognition, Vol. 31, No. 11 (1998) 1777–1790
14. Williams, M. L., Wilson, R. C., Hancock, E. R.: Multiple Graph Matching with Bayesian Inference, Pattern Recognition Letters, Vol. 18 (1997) 1275–1281
15. Cross, A. D. J., Hancock, E. R.: Graph Matching with a Dual Step EM Algorithm, IEEE Trans. Patt. Anal. Machine Intell., Vol. 20, No. 11 (1998) 1236–1253
16. Wilson, R. C., Hancock, E. R.: A Bayesian Compatibility Model for Graph Matching, Pattern Recognition Letters, Vol. 17 (1996) 263–276.
17. Hummel, R. A., Zucker, S. W.: On the Foundations of Relaxation Labelling Processes, IEEE Trans. Patt. Anal. Machine Intell., Vol. 5, No. 3 (1983) 267–286 265
18. Faugeras, O. D., Price, K. E.: Semantic Description of Aerial Images Using Stochastic Labeling, IEEE Trans. Patt. Anal. Machine Intell., Vol. 3, No. 6 (1981) 633–642 265, 269
19. van Wyk, M. A., Clark, J.: An Algorithm for Approximate Least-Squares Attributed Graph Matching, in Problems in Applied Mathematics and Computational Intelligence, N. Mastorakis (ed.), World Science and Engineering Society Press, (2001) 67-72 265
20. van Wyk, B. J., van Wyk, M. A., Virolleau, F.: The CGGM Algorithm and its DSP implementation, Proc. 3rd European DSP Conference on Education and Research, ESIEE-Paris, 20-21 September (2000) 269
21. Stark, H., Yang, Y.: Vector Space Projections: A Numerical Approach to Signal and Image Processing, Neural Nets and Optics., John Wiley and Sons (1998) 267
22. Youla, D. C.: Mathematical theory of image restoration by the method of convex projections, Chapter 2, in Image Recovery: Theory and Applications, H. Stark (ed.), Academic Press, Orlando, FL (1987)
23. Garey, M. R., Johnson, D. S.: Computers and Intractability: A Guide to the Theory of NP-Completeness, W. H. Freeman (1979)

Compact Graph Model of Handwritten Images: Integration into Authentification and Recognition

Denis V. Popel

Department of Computer Science, Baker University
Baldwin City, KS 66006-0065, USA
popel@ieee.org

Abstract. A novel algorithm for creating a mathematical model of curved shapes is introduced. The core of the algorithm is based on building a graph representation of the contoured image, which occupies less storage space than produced by raster compression techniques. Different advanced applications of the mathematical model are discussed: recognition of handwritten characters and verification of handwritten text and signatures for authentification purposes. Reducing the storage requirements due to the efficient mathematical model results in faster retrieval and processing times. The experimental outcomes in compression of contoured images and recognition of handwritten numerals are given.

1 Introduction

It is essential to reduce the size of an image, for example, for analysis of features in automatic character recognition or for creation of a database of authentic representations in secure applications. One can estimate that the amount of memory needed to store raster images is greater than for vector ones [5]. The problem arises if the contoured image like handwritten text or signature is scanned and only bitmap of this shape is produced. Hence, many algorithms were developed in the past years to compress raster images (LZW, JPEG,..., etc.), as well as for vector representation [3],[8]. From the other side, the extensive growth of portable devices requires developing techniques for efficient coding and transmitting audio-video information through wireless networks. Therefore, the simplification of graphical data is the first essential step towards increasing data transmission rate and enhancing services of cell phones and PDA computers [2]. The structural model discussed in this paper allows compressing images and creates a range of features to start automatic processing of contoured images, e.g. recognition.

Authentification and personal identification of financial documents are main tasks that guarantee sufficient safety in business activities. Nowadays there is a growth of crime in the area of forgery of signatures and falsifying handwritten documents. The current approach of solving this problem is using semiautomatic/manual client identification systems based on signature and handwritten

T. Caelli et al. (Eds.): SSPR&SPR 2002, LNCS 2396, pp. 272–280, 2002.
© Springer-Verlag Berlin Heidelberg 2002

text identification on personal cheques, credit card receipts, and other documents. Often only signatures and handwritten text can transform an ordinary paper into legal document. It would be ideal if the client personal characteristics like signature and handwriting could be identified by computer in full automatic mode [4],[7]. Now manual visual comparison of customer signatures and handwriting is widely used, where images are represented from the customer's document and from a database. For automatic authentification and for effective storage in a computer database, the new methods should be introduced which will lead to proper utilization of human and computing resources. This paper presents such an algorithm to build a mathematical model of handwritten characters and signatures and shows how to compress any contoured images. The reduced amount of memory is required to store this model which can be generally used for automatic analysis and identification of images, for instance, during automatic identification of a person using signature or recognition of handwritten text [1],[7].

2 Identification Process as a Motivation of Our Study

In a process of checking of signature or handwritten text authenticity from the paper document, a system should be equipped with the following components: (i) *a scanner* with image retrieving program; (ii) *a program* to compress images, which is integrated with a database management system; (iii) *a program* to reproduce compressed image.

The process of addition of an authentic signature or handwriting to the database consists of the following steps: scanning of image from a document, building mathematical model, and placing it as a separate record in the database. Afterwards the model is extracted by some database key (for example, by client account number in bank applications), and the reproduction program restores the original view of the signature or handwriting. The image to be examined should be also scanned and displayed on the screen. Thus, a bank clerk has an opportunity to access the original client signature or handwriting and to compare it with the current image. The information stored in such way is insufficient to match authentification expertise, which required from 6 to 20 original objects (depending on signature's or handwriting complexity). In this paper, we propose an algorithm to construct generalized mathematical model that stores essential and invariable features for a person and can be used as a basis to provide automatic authenticity confirmation of handwritten text or signatures.

3 Graph Model

In our approach, the transformation of raster image consists of the following stages: (1) image thinning; (2) image representation as a graph; (3) shape smoothing; (4) graph minimization; (5) shape compression.

3.1 Image Thinning

Skeletonization is an iterative procedure of bitmap processing, when all contour lines are transformed to singular pixel ones. The modified Naccache - Shinghal algorithm [6] is used in the described approach. The modification utilizes arithmetic differential operators to find the skeleton and provides good results for handwritten text and signatures. The average width of lines is measured during the skeletonization stage.

3.2 Transformation of Bitmap to Graph-Like Representation

We introduce the notation *pixel index* which is the number of neighboring image pixels in 3×3 window around the current point. Pixel with index 1 named endpoint (the beginning or the end of curve), with index 2 - line point, and pixels with indexes 3 and 4 are nodes points (junction and crossing of lines). The graph is represented in memory using two lists: the list of nodes and the list of the descriptors of contour branches. The element of the list of nodes contains X,Y coordinates of a bitmap pixel and pointers to contour branches, which started from this node. The descriptor of contour branch is a *chain code* (Freeman code [3]), where each element carries 8-bit code of the next contour direction (see Figure 3(a) for details). At this stage, the thinned bitmap is transformed to a graph description, where nodes are pixels with indexes 1,3 or 4, and arcs are contour branches. The starting node for the looped contours is selected arbitrary.

3.3 Shape Smoothing

As the scanned image has different distortions and noise, thinning shape contains some false nodes. These false nodes (treated as defects) do not have serious influence on the quality of the restored image, but their description requires a lot of additional memory. The shape is smoothed to eliminate these defects. This operation (a,b) erases false nodes, (c) smoothes of all broken contours, (d) eliminates all nodes with index 4, and (e) erases some nodes with index 3 (Figure 1). The stage of smoothing can be omitted if the lossless representation is required by applications. The suggested algorithm supports both lossy and lossless representation strategies. Eliminating nodes with index 4 does not change the shape of processing image (lossless strategy), and after the smoothing stage, graph contains only nodes with index 1 and 3.

3.4 Graph Minimization

The graph description obtained at the second stage is redundant, therefore it is possible to conduct graph minimization by eliminating some nodes and connecting corresponding contour branches. Finally, each branch has only one j-th node with index I_j assigned to the beginning of the chain of pixels.

Lemma 1. *Before minimization, each branch of the graph interconnects two nodes with indexes 1, 3 or 4.*

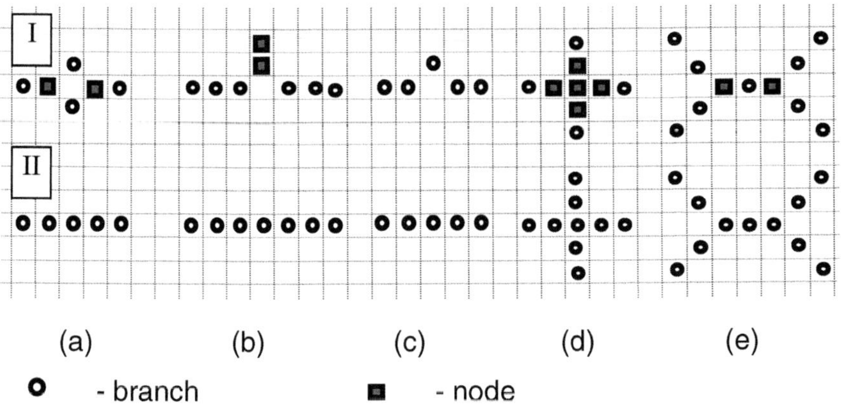

Fig. 1. Smoothing of image: (I) source shape, (II) smoothed shape

Theorem 1. *The number of branches B of the graph with no loops can be determined as follows:*

$$B = \frac{I_1 + I_2 + ... + I_N}{2}, \tag{1}$$

where N is the number of nodes of the source graph, I_j - the index of j-th node (excluding nodes with index 4).

Corollary 1. *It follows from the Theorem 1 that the number of branches for the looped contour is*

$$B = \frac{N_1 + 3 \cdot N_3}{2} + N_{loops}, \tag{2}$$

where N_1 is the number of nodes with index 1, N_3 - number of nodes with index 3, and N_{loops} is the number of arbitrary selected nodes with index 1 to cover all loops.

Graph minimization can eliminate nodes with indexes 1 and 3. It minimizes the number of branches in the graph while reducing number of nodes with index 3. The Corollary 1 can be reformulated for the minimal number of nodes and branches.

Corollary 2. *The number of branches in minimized graph equals*

$$B^{min} = \frac{N_1^{min} + 3 \cdot N_3^{min}}{2} + N_{loops}, \tag{3}$$

where N_1^{min} and N_3^{min} are the minimal number of nodes with indexes 1 and 3 correspondingly.

The graph minimization technique resolves spanning tree problem through the following steps:

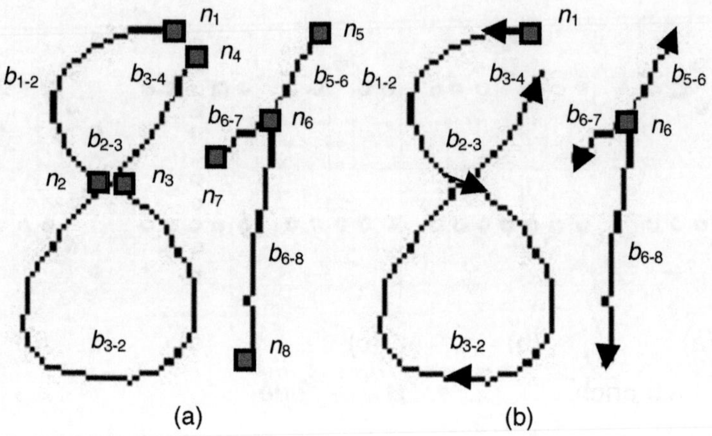

Fig. 2. Handwritten digits as a graph: (a) before smoothing and minimization; (b) after minimization (lossless smoothing)

Step 1. `Select a node with index 3. Trace three uncovered outgoing branches following links from the selected node until tracing reaches already covered node or a node with index 1. Mark the node and all branches as covered.`
Step 2. `Repeat Step 1 until all nodes with index 3 are covered.`
Step 3. `Select a node with index 1 and trace outgoing branch. Mark the node and branch as covered.`
Step 4. `Repeat Step 3 until all nodes with index 1 are covered.`

Example 1. Figure 2 shows the connected graph with eight nodes $n_1, \ldots, n_8$ and seven branches $b_{1-2}, b_{2-3}, \ldots, b_{6-8}$. Nodes n_2 and n_3 are removed by smoothing operation, and the branches b_{1-2}, b_{2-3}, b_{3-2} and b_{3-4} are transformed into one branch which covers the entire contour. Before minimization the graph description contains six nodes and four branches. Minimal graph has two nodes n_1, n_6 and four branches. So the number of nodes is reduced from eight to two.

3.5 Contour Compression

After thinning and smoothing stages, the contour does not have sharp shifts and distortions. Therefore the next contour point in the branch description has only three positions related to the current one (see Figure 3(b)). This property allows us to represent the branch dynamically using relative coordinates $(-1, 0, 1)$, which provide $L \cdot \log_2 3$ $(1.6 \cdot L)$ bits instead of $8 \cdot L$ bits for the chain of length L. Thus, the size of the branch description can be reduced. In addition, all branches are compressed by modification of widely used *Run-Length Encoding algorithm*.

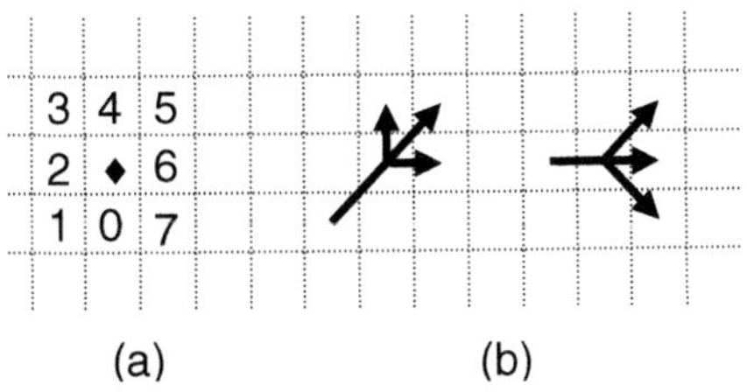

Fig. 3. Contour directions

4 The Reproduction of Image

The process of reproduction is based on mathematical vector model and should be executed in two stages: skeleton reproduction; contour thickening. The skeleton reproduction is fulfilled by traversing through graph nodes and transforming of contour branches into bitmaps. Obviously the described method does not allow us to reproduce the initial image with pixel accuracy (Figure 4). However as analysis shows this aspect does not influence on image comparison in the context of the authenticity problem as well as character recognition task discussed above. If exact image reproduction is necessary, it is possible to use a modified description, where line width is assigned to each point of the contour.

5 Experimental Results

The algorithm described above is realized and experiments were conducted on signatures, handwritten text (various languages) and contoured images.

In the first series of experiments, the described above algorithm was used directly to resolve the storage problem. Experiments show that the compression ratio for (a) signatures is 10-20, (b) handwritten text – between 5 and 12, and (c) curved shapes – between 8 and 17, (see Table 1). In some cases, the results exceed a compressing degree produced by well-known archiving programs. For example, the initial image with handwriting in BMP format has 1962 bytes, compressed by ZIP - 701 bytes, compressed using presented algorithm - 154 bytes. The contour compression ratio is 11.9, that is 4.3 times effective than using raster compression. Figure 5 compares the result of our algorithm with the outcome of CorelTrace program. Second series of experiments covers recognition of handwritten numerals. Twenty distinctive features where extracted from the mathematical model: relative distances, number of different nodes, number of loops, straight strokes,. . . , etc. These features are invariant to rotation between -45 and 45 degrees and scaling. The experiments were fulfilled on MNIST[1] database. The estimated error rate of recognition algorithm is about 5.9%.

[1] http://www.research.att.com/˜ yann/exdb/mnist/index.html

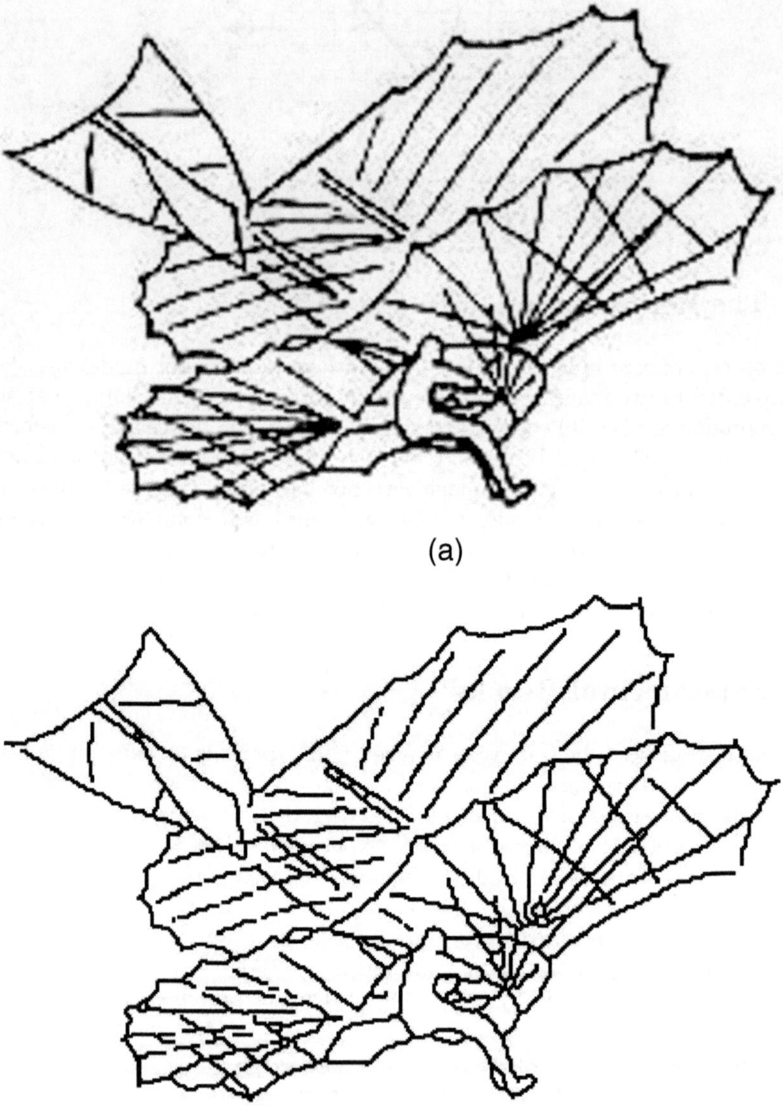

Fig. 4. Example of lossy compression and reproduction: (a) initial image – Da Vinci drawing, (b) reconstructed image

Table 1. Compression characteristics (images were scanned at 100 dpi)

Initial image and size in bytes	Restored image and compressed size in bytes	Compression rate
 3742	 193	19.3
 4158	 210	19.8
 914	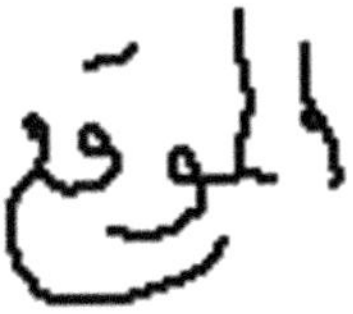 157	5.82
 10021	 683	14.67

6 Concluding Remarks and Ongoing Research

In this paper, we presented a novel algorithm to create a generalized mathematical model for contoured images. Several related issues, such as automatic authenticity confirmation and recognition of handwritten characters based on this model, were discussed. We are planning to extend the proposed mathematical model of contoured images and integrate it in authentification systems. This extension will reflect probability characteristics of image attributes and structural features of handwritten objects. Another feasible application of the graph model is the compression of large *Computer Aided Design* (CAD) and *Geographic Information System* (GIS) images.

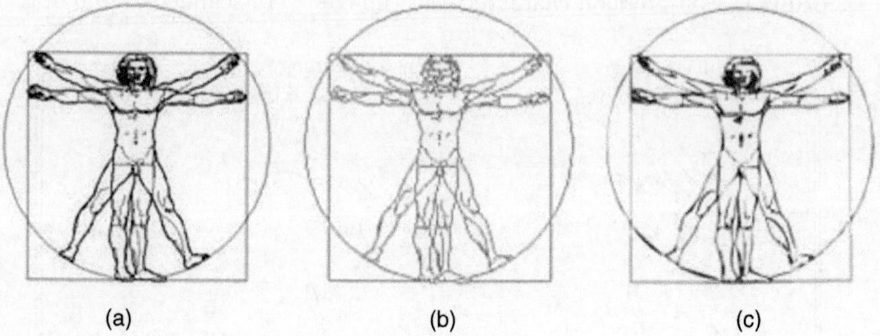

Fig. 5. Da Vinci drawing: (a) its original shape, (b) processed by suggested algorithm and (c) vector representation obtained by CorelTrace program

References

1. Al-Emami, S., Usher, M.: On-Line Recognition of Handwritten Arabic Characters, IEEE Trans. on Pattern Analysis and Machine Intelligence, vol. 12, no. 7 (1990) 704–709 273
2. Ansorge, M., Pellandini, F., Tanner, S., Bracamonte, J., Stadelmann, P., Nagel, J.-L., Seitz, P., Blanc, N., Piguet, C.: Very Low Power Image Acquisition and Processing for Mobile Communication Devices, Proc. IEEE Int. Symp. on Signals, Circuits and Systems - SCS'2001 (2001) 289–296 272
3. Freeman, H.: On the Encoding of Arbitrary Geometric Configurations. IEEE Trans. Elect. Computers, vol. ES-10 (1961) 260–268 272, 274
4. Gazzolo, G., Bruzzone, L.: Real Time Signature Recognition: A Method for Personal Identification, Proc. Int. Conf. on Document Analysis and Recognition (1993) 707–709 273
5. Gilewski, J., Phillips, P., Popel, D., Yanushkevich, S.: Educational Aspects: Handwriting Recognition - Neural Networks - Fuzzy Logic, Proc. IAPR Int. Conf. on Pattern Recognition and Information Processing - PRIP'97, vol. 1 (1997) 39–47 272
6. Naccache, N. J., Shingal, R.: SPTA: A Proposed Algorithm for Thinning Binary Patterns, IEEE Trans. Systems. Man. Cybern., SMC - 14, no. 3 (1984) 409–418 274
7. Popel, D., Ali Muhammed, T., Hakeem, N., Cheushev, V.: Compression of Handwritten Arabic Characters Using Mathematical Vector Model, Proc. Int. Workshop on Software for Arabic Language as a part of IEEE Int. Conf. on Computer Systems and Applications (2001) 30–33 273
8. Song, J., Su, F., Chen, J., Tai, C. L., Cai, S.: Line net global vectorization: an algorithm and its performance analysis, Proc. IEEE Conf. on Computer Vision and Pattern Recognition (2000) 383–388 272

A Statistical and Structural Approach for Symbol Recognition, Using XML Modelling

Mathieu Delalandre[1], Pierre Héroux[1], Sébastien Adam[1],
Eric Trupin[1], Jean-Marc Ogier[2]

[1] Laboratory PSI, University of Rouen, 76 821 Mont Saint Aignan, France
[2] Laboratory L3I, University of La Rochelle, 17042 La Rochelle, France

Abstract. This paper deals with the problem of symbol recognition in technical document interpretation. We present a system using a statistical and structural approach. This system uses two interpretation levels. In a first level, the system extracts and recognizes the loops of symbols. In the second level, it relies on proximity relations between the loops in order to rebuild loop graphs, and then to recognize the complete symbols. Our aim is to build a generic device, so we have tried to outsource models descriptions and tools parameters. Data manipulated by our system are modelling in XML. This gives the system the ability to interface tools using different communication data structures, and to create graphic representation of process results.

1 Introduction

The current improvements of intranet structures allow large companies to develop internal communications between services. The representation of the heritage of huge companies like a network managers firm is often represented through paper documents, which can be either graphic or textual. As a consequence, the sharing of these kind of information will stay very difficult as long as the storage format will not be digital. This explains the current development of studies concerning the automatic analysis of cartographic or engineering documents, which comes as a result of the growing needs of industries and local groups in the development and use of maps and charts. The aim of the interpretation of technical maps is to make the production of documents easier by proposing a set of steps to transform the paper map into interpreted numerical storage [1][2][3][4]. An important step of this conversion process consists in the recognition of symbols, which often appear on technical documents.
We present in this document a symbol recognition system. It is based on a statistical and structural approach combination. In the second section, we will briefly describe the classical approaches for symbol recognition. Then, we will present our approach. Finally, we will give conclusions and propose some perspectives for our future works.

T. Caelli et al. (Eds.): SSPR&SPR 2002, LNCS 2396, pp. 281-290, 2002.
© Springer-Verlag Berlin Heidelberg 2002

2 Classical Approaches for Symbol Recognition

Symbols constitute an important informative source on technical documents (geographical maps, utility maps, architectural drawings…). Good states of the art dealing with such a problem can be found in [5][6].

These surveys show that structural approaches are generally chosen for symbol recognition. Such approaches begin with a graphical primitives extraction step. These primitives can be either structural [7][8][9] or statistical-structural [10]. After this first step, primitives and their relations are then represented into a graph, which is used in a process of sub-graph matching in a reference graph. Nowadays, such structural symbol recognition systems are generally efficient for specific applications but cannot be generalized. Only the works of Messmer [11], Schettini [12] and Pasternak [7] can be considered as generic approaches. Indeed, these authors propose generic symbol description tools. Symbols are described by different primitives obtained through the use of low-level operators, and by the association relations between these primitives. However, some problems are not solved in these systems:

- Limitations appear when symbols integrate important variability, or when they are represented by elements, which are closed but not connected.
- Very few works propose a correction step of the primitives extracted by low-level operators. Yet, this point is important, essentially in the case of damaged documents, for which low-level operators can be disrupted by noise.
- Symbol representations are generally dedicated to existing tool libraries and to specific applications. A system allowing a more global representation of symbols does not exist.

3 Our Statistical and Structural Approach for Symbol Recognition

3.1 Introduction

Our approach may be decomposed into 3 steps:

- Extraction of loops
- Extraction of orientation invariant features and statistical recognition of loops
- Reconstruction of loop graphs, and structural symbol recognition

The system relies on proximity relations between the loops in order to recognize the symbols. Our aim is to build a generic device. So, we have tried to outsource from algorithms model descriptions and tool parameters. In this way, this system is evolutionary and can be used in practice for different applications. Until now, it has been exploited only for symbol recognition on France telecom (a French telecommunication operator) utility maps, and for meteorological symbol recognition.

The France Telecom symbols represent technical equipments permitting connections on telephonic network: Concentration Points "Point de Concentration" and room "*chambre*". They are composed of a variable number of loops belonging to 5 different classes. The Fig. 1.a shows these symbols. On the top one can see from left to right: « *chambre, PC paires sur bornes, PC paires sur appui* ». Just below one can see the 5 loops classes named from left to right: « *chambre, ellipse, triangle, cercle, portion* ».

The meteorological symbols represent cloud cover rate. They are composed of a variable number of loops belonging to 4 different classes. These symbols are shown on Fig.1.b. On the top one can see from left to right: « *aucun nuage, 2/10 à 3/10 du ciel couvert, 4/10 du ciel couvert, 5/10 du ciel couvert, 6/10 du ciel couvert, 6/10 à 7/10 du ciel couvert, ciel obscurci* ». Just below one can see the 4 loops classes named from left to right: « *cercle_25, cercle_50, cercle_75, cercle_100* ».

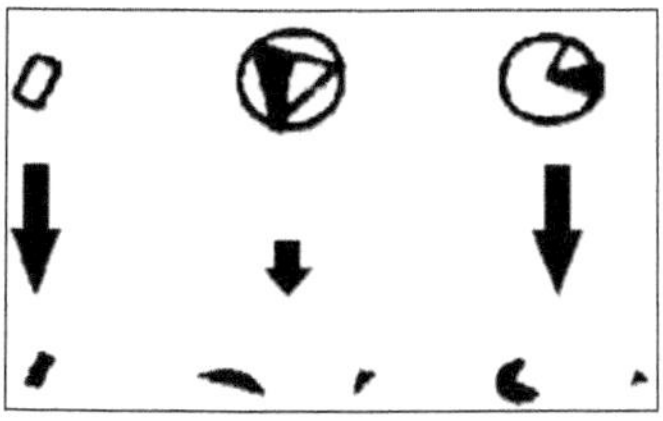

Fig. 1.a) France Telecom utility map symbols and their loops

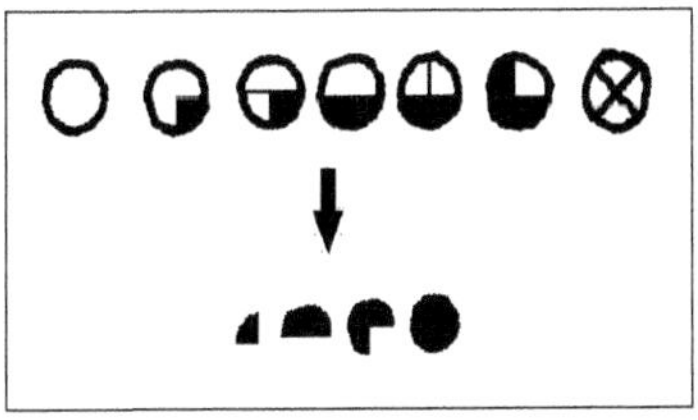

Fig. 1.b) Meteorological symbols and their loops

In the following, we will present successively each of the 3 processing steps. We will present succinctly the two first steps (loop extraction and classification) to develop more extensively the structural recognition step. Then, we will present in subsection 3.5 the strategy used for the application and the obtained results. Finally, we will present in subsection 3.6 XML use in our system.

3.2 Extraction of Loops from Symbols

An image of loops is obtained through the application of a classical connected components extraction on image. The Fig. 2.a and 2.b show a part of France Telecom utility map as well as the loop extraction result.

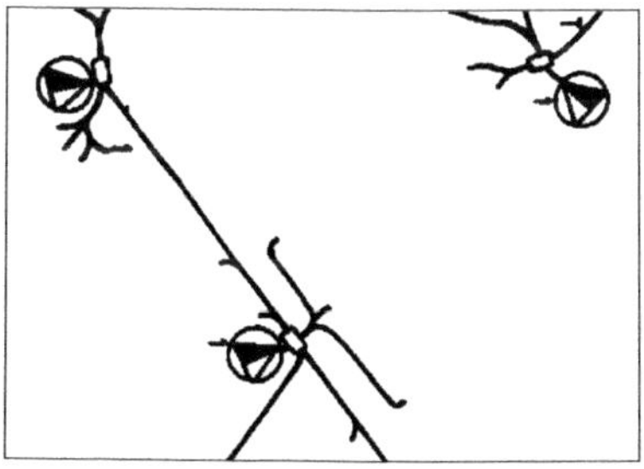

Fig. 2.a) A part of France Telecom utility map

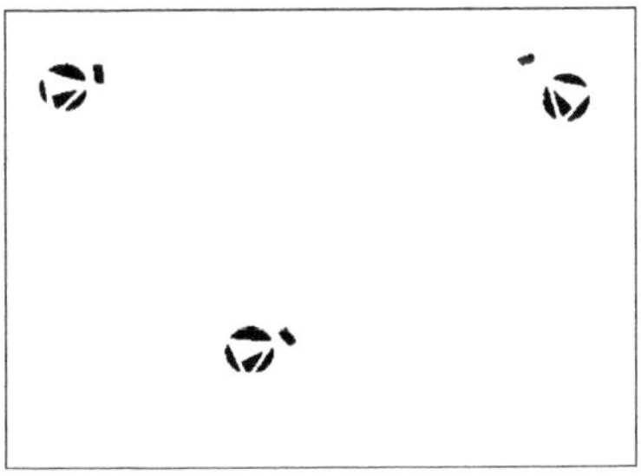

Fig. 2.b) Result after extraction of loops

3.3 Extraction of Orientation Invariant Features and Statistical Recognition of Loops

This processing step constitutes the statistical approach of our system. A feature vector is extracted for each loop on the image, using three outstanding and complementary tools: the Zernike moments, the Fourier-Mellin invariants, and the circular probes [13]. These features enable to constitute a vector set describing loops. This description is invariant to scale and orientation changes. We have constituted a test and a training set using France Télécom utility map loops, with a size of fifty loops each. Then, with the help of the k nearest neighbors (knn, with k=1) classifier, using Euclidian distance, we got recognition results that are presented in the Fig.3 for each feature extraction tool.

Characteristics extraction tools	Recognition rates
Zernike moments	97.77 %
Fourier-Mellin invariants	86.66 %
Circular probes	86.66 %

Fig. 3. Results of loop recognition

These results show that Zernike moments are the most adapted for the recognition of this loop type. Results are quite the same for meteorological symbols. Of course, these recognition rates computed on a test set of weak size are not representative of a real problem, but they indicate promising perspectives of recognition on test sets of big size.

3.4 Reconstruction of Loop Graphs, then Structural Symbol Recognition

This processing step constitutes the structural approach of our system. It may be divided into two steps.

The first step is a model reconstruction step in the sequential processing between the statistical classifier and the structural classifier showed in the Fig. 4.

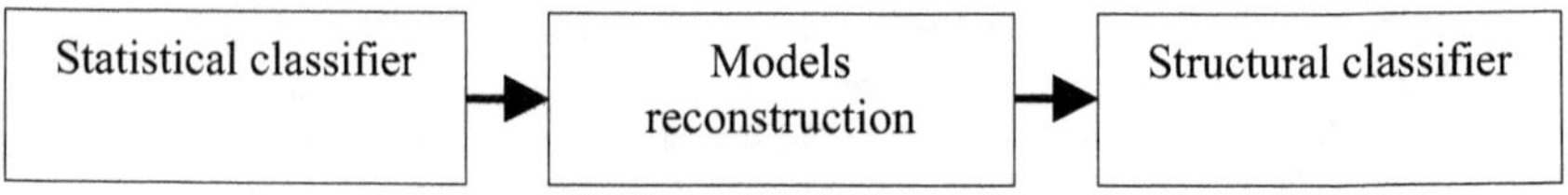

Fig. 4. Sequential processing of statistical and structural classification

The models reconstruction tool we used rebuilds some graphs under connection and/or distance constraints. This reconstruction uses the results of loop statistical recognition. The distance constraint permits to control inter-connection of the graphs corresponding to the symbols. It is thus possible to create a graph in which image loops are completely inter-connected, or to isolate each of the symbols of the image.

These connections constraints are defined according to the nature of loops. The maximal connections number is specified for each loop class. The symbol description is *a priori* considered in model reconstruction strategy. The Fig. 5.a shows an example of France Telecom utility map. Model reconstruction of this map is achieved by using constraints concerning connection and distance constraints. The distance constraint permits to detect all the 4 graphs of loops. The Fig. 5.b shows the graphic representation of the model reconstruction of the symbol located below on the right of the map. This graphic representation uses the information obtained from the statistical classification, from the model reconstruction, from a step of contours detection and polygonisation of the loop image. Here, we have a 4-connections constraint for the loop "*triangle*", and a 1-connection constraint for the loops "*ellipse*" and for the loop "*chambre*".

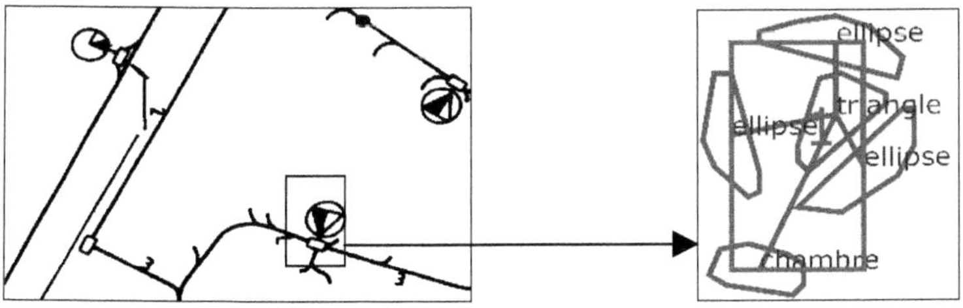

Fig. 5.a) Example of France Telecom utility map

Fig. 5.b) Graphic representation of the model reconstruction of the right low symbol

The second step is the structural recognition of symbols. It consists in submitting the graphs obtained from the model reconstruction phase to the structural classifier. Our graph matching tool [14] allows a graph edges and nodes typing possibility (integer, float, character, string, object). It permits to compute similarity criterion between graphs, based on the overlap between a candidate graph and a model graph. This overlap corresponds to their common sub-graph. This common sub-graph is searched in three times by matching candidate graph and model graph. In a first time, a filtering step aims at suppressing in the two graphs the nodes and their edges unable to be matched. This concerns the nodes whose label is not common in the two graphs nodes lists. This step has for purpose to reduce the algorithm temporal complexity. In a second time, a research of matching edges is done. The edges are matched if they are equal and if their extremities are equal. The nodes corresponding to the edges extremities are also matched during the edge matching. In a third time, the remaining nodes are matched. Two similarity criteria can finally be computed according to the number of common elements, either on the nodes (1), either on the edges (2). In these equations, n1, n2, and nc and e1, e2, and ec respectively represent the number of nodes and edges of the graph 1, the graph 2, and the common sub-graph.

$$\delta n(g1, g2) = \frac{n1 \times n2}{nc^2} - 1 \quad (1) \qquad \delta e(g1, g2) = \frac{e1 \times e2}{ec^2} - 1 \quad (2)$$

It is possible to control the combination of the results obtained out of (1) and (2) by average or variance computation. The similarity criterion computation tool is also parameterized in order to take or not into account the types and their attributes. It can be done on the nodes and/or on the edges. For example, it is possible to compute the graph matching taking into account the graph topology, or the graph types. Finally, it is possible to combine the whole similarity criteria obtained by the computation of weighted average, in order to get global similarity criterion.
The Fig. 6 is a graphic representation of a model extraction (under connection and distance constraints) and a structural recognition, the loop graphs have been submitted to the structural classifier with a similarity computation taking into account the types and their attributes based on an average of (1) and of (2).

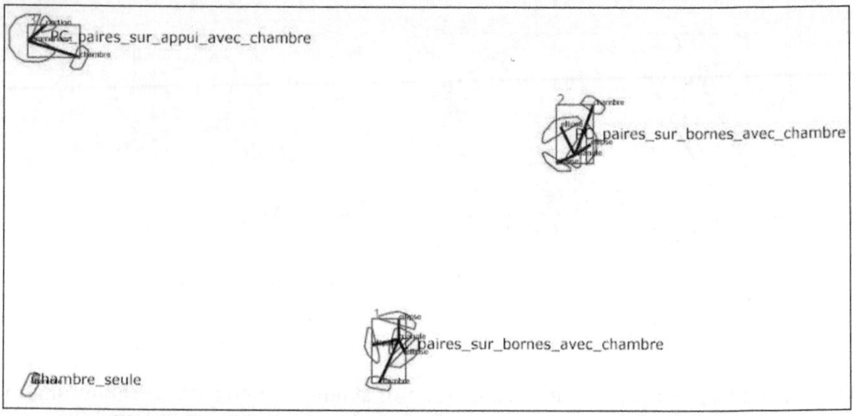

Fig. 6. Graphic representation of treatment steps of the Fig. 5.a

Fig. 7.a shows a meteorological symbol image. The Fig 7.b shows a graphic representation (of model extraction "only under distance constraint" and a structural recognition) superposition with meteorological symbol image. The loop graphs have been submitted to the structural classifier with a similarity computation taking into account the types and their attributes only based on (1).

Fig. 7. a) Example of meteorological symbol image

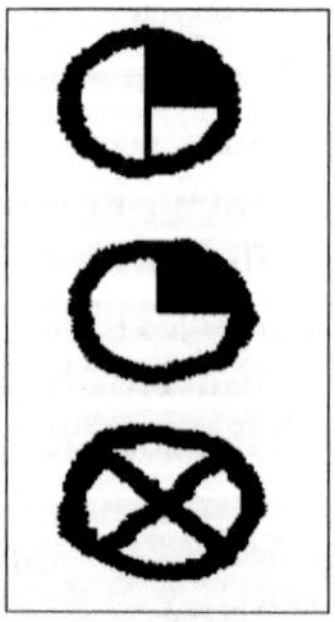

Fig. 7.b) Graphic representation of the model reconstruction and the structural recognition of Fig. 7.a

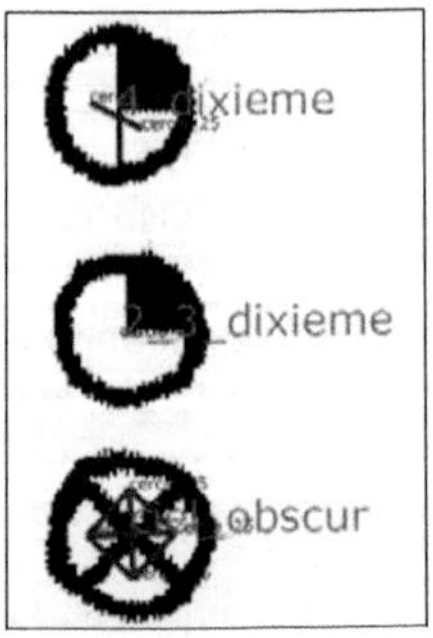

The matching graph tool used does not allow the localization and therefore the manipulation of sub-graphs in a candidate graph. Thus, it is impossible to exploit a candidate graph representing the whole image loops. For that reason, it is impossible to distinguish the symbols "*chambre*" when they are connected to the symbols "*PC paires sur appui*" and "*PC paires sur bornes*". Indeed the "*chambre*" are closely connected to this PC (Fig. 5.a and Fig. 2.a), and a too strict distance constraint could split up the PC symbol in several symbols. To overcome this problem, we have considered the symbol "*PC paires sur support + chambre*" and "*PC paires sur appui + chambre*" as complete symbols. We didn't have this problem with meteorological symbol because distance between loops is sufficiently long.

3.5 Strategy and Results

In the setting of our application, we have tested several symbol recognition strategies. The most efficient strategy uses the models only taking into account the distance constraint. Indeed, recognition errors in statistical classification inevitably generate some errors in the models reconstruction, if one takes into account connection constraints. With this strategy, we obtained completely inter-connected graph for every symbol. The matching tool is then parameterized to only take into account the global similarity criterion between nodes (Fig. 7.b). This global criterion is a weighted average of two similarity criteria, the first on topology graphs, and the second on exact graphs. The weighted average is computed with a more important coefficient for similarity criterion on graphs topology.
In term of results, we have tested this approach on 29 symbols, constituted of about hundred loops, and distributed on 9 plan extracts. We have constituted a training set of feature vectors describing the loops. This training set has been used as test set, in order to obtain a 100% statistical recognition results. We obtained 100% of recognition on symbols.
Obviously, we are interested to test the correction ability of structural recognition step on statistical recognition results. We voluntary altered statistical training set of France Telecom symbols loops in order to reduce the loops recognition rate. Tests carried out on 22 symbols composed of 74 loops give a loops statistical recognition rate of 55.4% and a symbol structural recognition rate of 86.86%. Among these symbols we have up to 75% of a wrong statistical recognition loops. These results prove the ability of structural recognition step to correct the statistical recognition step. But, taking into account similarity criterion on topology graphs increases these results. Indeed, only the node number used alone permits to distinguish France Telecom symbols (1 for "*chambre*", 2 for the "*PC paires sur appui*", 3 for the "*PC paires sur appui + chambre*", 4 for the "*PC paires sur bornes*", 5 for the "*PC paires sur bornes + chambre*"). We realized similar tests on the meteorological symbols in order to prove the importance of similarity criterion on topology graphs in symbol recognition. Tests on 56 symbols composed of 96 loops extracted from the same image rotated into 8 different directions gives 44% for loops statistical recognition and 55% symbol structural recognition. Indeed, only symbol "*ciel obscurci*" (Fig. 2) can be recognized by its loops number. Nevertheless, we have corrected 33% of symbol in which loops were badly recognized by statistical step.

3.6 XML Modelling

Data manipulated by our system are modelled in XML [15]. The use of this data description language offers several advantages. First, XML seems to become a reference concerning data description languages. This guarantees in the future a durability of our tools and especially a possibility to exploit tools provided by the scientific community. Secondly, the properties of XML can be used in a recognition system. Among these properties, we used the data transformation and the specialized sub-languages.

Data transformation tools: XML permits to use of XSLT processors (for example, the Xalan processor [16]). These processors transform a XML data flow with the help of a XSLT script program [17], which permits an easy data transformation. If data have a XML type flag format, we can have two tools using different communication data structures. In the same way, it is also possible to merge data stemming from several tools.

The specialized sub-languages: XML is described as a meta-language because it is a root language, which permits to define specialized sub-languages. For example the SVG [18] language permits a data graphic description. We used this language in order to rebuild a graphic representation of all our steps processing (Fig. 5.b, 6). Moreover, we can superpose the image with the graphic representation of our process result (Fig. 7.b). With the help of the XSLT processor, we merged and transformed information of our different tools in SVG format. We used tools provided by computer science community (SVG viewer [19], Batik [20]) for the recognition visualization.

4 Conclusions and Perspectives

In conclusion, we have presented in this document a symbol recognition system combining statistical and structural approaches. We have exploited these approaches in order to recognize some technical symbols composed of loops. These symbols have proximity relations between theirs loops. We exploited statistical approach in a first interpretation level in order to recognize loops found in symbols. In a second interpretation level, we exploited proximity relations between these loops by a structural approach, in order to recognize the complete symbols. The first results are encouraging. On one hand, a perfect statistical recognition gives a perfect symbol structural recognition. On the other hand, the statistical recognition results can be corrected by structural recognition step. The efficiency of this correction is a function of similarity between symbols (differences in topology symbols, share of loops classes between symbols). The model set and the configuration information of tools have been outsourced from algorithms. This confers to the system a generic aspect. Data manipulated by our system are modelled in XML. This gives to system the ability to interface tools using different data format, and to create graphic representation of each treatment step.

Our first perspective is to extend the statistical/structural serial combination in a parallel combination. Indeed, we hope will integrate structural model extraction tools for compare with statistical model extraction tolls actually used. These structural tools allow extracting connected component structure (contrary to statistical model extraction tolls). This property gives ability to extract part of a connected component. Among these structural model extraction tools we hope will integrate:

- squeletonization method - skeleton structuring - and mathematics approximations [21]
- follow line method - structuring method of follow line [22]
- line adjacency graph method [23]

The second perspective consists in improving our structural classifier. In a first time we plan to realize localization and manipulation of sub-graphs in a candidate graph, in order to exploit loops graph representing several symbols on image. It will permit to treat the neighbor symbols that can't be isolated by simple distance constraint (as it's the case of the France Télécom symbol "*chambre*"). In a second time we hope will compute inexact graph matching in order to allow a tolerance between node and edge values of candidate and model graphs during matching process. It would permit for example to take into account the distances between loops given by reconstruction models tool. Finally, in a third time, we wish to exploit the confusion matrix of our statistical classifier. The goal is to weight similarity computation between graphs according to confidence degrees of node labels (provided by the statistical classifier).
Finally, the third perspective is to integrate our tools in a knowledge-based approach. In a first time, the aim is to create a common knowledge set in XML, for the whole tools of our system. A simultaneous use of XSLT and XML-QL [24] (norm for the XML data set management) will permit management and adaptation of this knowledge set for the whole tools of our system. It will be necessary to define a representation formalism of a generic model for the entire recognition tools. In a second time we wish to develop a supervision program of our recognition system, permitting the combination of our different tools (classifiers and models extraction tools). The aim is to control the whole processing from the image processing to the classifier combination. We wish to control our recognition system by a processing scenario, in order to adapt easily and quickly the system to a new recognition objective.

The authors would like to thank Joël Gardes (France telecom R&D) for his contribution to this work.

References

1. L. Boatto and al, An interpretation system for land register maps, IEEE Computer Magazine, 25(7), pp 25-33, 1992.
2. S.H. Joseph, P. Pridmore, Knowledge-directed interpretation of line drawing images, IEEE Trans. on PAMI, 14(9), pp 928-940, 1992.
3. J.M. Ogier, R. Mullot, J. Labiche and Y. Lecourtier, Multilevel approach and distributed consistency for technical map interpretation: application to cadastral maps, Computer Vision and Image Understanding (CVIU), 70, pp 438-451, 1998.
4. P. Vaxivière, K. Tombre, CELESTIN : CAD conversion of Mechanical Drawings, IEEE Computer Magazine, 25, pp 46-54, 1992.

5. K. Chhabra, Graphic Symbol Recognition: An Overview, Lecture Notes in Computer Science, vol. 1389, pp 68-79, 1998.
6. J.Lladós, E. Valveny, G. Sánchez, E.Martí, Symbol recognition: current advances an perspectives, 4th IAPR International Workshop on Graphics Recognition (GREC'01), Kingston, Canada, 1:109128, 2001.
7. B. Pasternak, B. Neumann, Adaptable drawing interpretation using object oriented and constrained-based graphic specification, in proc. Second International Conference on Document Analysis and Recognition, Tsukuba, Japan, pp 359-364, 1995.
8. N.A. Langrana, Y. Chen, A.K. Das, Feature identification from vectorized Mechanical drawings, Computer Vision and Image Understanding, 68(2), pp 127-145, 1997.
9. G. Myers, P. Mulgaonkar, C. Chen, J. Decurting, E. Chen, Verification-based approach for automated text and feature extraction from raster-scanned maps, in Proc. of IAPR International Workshop on Graphics Recognition, Penn State Scanticon, USA, pp 90-99, 1995.
10. S.W. Lee, Recognized hand-drawn electrical circuit symbols with attributed graph matching, in H.S. Baird, H. Bunke, K. Yamamoto, eds., Structured Document Analysis, Springer Verlag, pp 340-358, 1992.
11. B. Messmer, H. Bunke, Automatic learning and recognition of graphical symbols in engineering drawing, in R. Katsuri and K tombre, eds., Lecture Notes In Computer Science, volume 1072, pp 123-134, 1996.
12. R. Schettini, A general purpose procedure for complex graphic Symbols Recognition, Cybernetic and System, 27, pp 353-365, 1996.
13. S. Adam, J.M. Ogier, C. Cariou, J. Gardes, Y. Lecourtier, Combination of invariant pattern recognition primitive on technical documents, Graphic Recognition – Recent Advances, A.K. Chabbra D. Dori eds., Lecture notes in Computer Science, Springer Verlag, vol 1941, pp 29-36, 2000.
14. P. Héroux, S Diana, E. Trupin, Y. Lecourtier, A structural classification for retrospective conversion of document, Lecture Notes in Computer Sciences, Springer Verlag, vol 1876, pp 154-162, 2000.
15. World Wide Web Consortium, eXtensible Markup Language (XML) 1.0, http://www.w3.org/TR/2000/REC-xml-20001006 , 2000.
16. Apache XML projects, Xalan processor 2.2 D14, http://xml.apache.org/xalan-j/index.html
17. World Wide Web Consortium, eXtensible Style-sheet Language Transformation (XSLT) 1.0, http://www.w3.org/TR/xslt , 1999.
18. World Wide Web Consortium, Scalable Vector Graphic (SVG) 1.0, http://www.w3.org/TR/SVG/ , 2001.
19. Adobe, Svg Viewer 3.0, http://www.adobe.com/svg/
20. Apache XML projects, Batik SVG toolkit 1.1, http://xml.apache.org/batik/
21. X. Hillaire, K. Tombre, improving the accuracy of skeleton-based vectorisation, IAPR International Workshop on Graphic Recognition (GREC), Kingston, Canada, 2001.
22. J.M. Ogier, C. Olivier, Y. Lecourtier, Extraction of roads from digitized maps, in processing of the sixth EUSIPCO (European Signal Processing Conference), Brussels, Belgium, pp 619-623,1992.
23. S. Di Zenzo, L. Cinque, S. Leviadi, Run-based algorithms for binary image analysis and processing, IEEE Trans. on PAMI, 18(1): 83-89, p56, 1996.
24. World Wide Web Consortium, XQuery 1.0 an XML query language, http://www.w3.org/TR/xquery/ , 2001.

A New Algorithm for Graph Matching with Application to Content-Based Image Retrieval

Adel Hlaoui and Shengrui Wang*

DMI, University de Sherbrooke
Sherbrooke (Quebec), J1K 2R1, Canada
{Hlaoui,Wang}@dmi.usherb.ca

Abstract. In this paper, we propose a new efficient algorithm for the inexact matching problem. The algorithm decomposes the matching process into K phases, each exploiting a different part of solution space. With most plausible parts being searched first, only a small number of phases is required in order to produce very good matching (most of them optimal). A Content-based image retrieval application using the new matching algorithm is described in the second part of this paper.

1 Introduction

With advances in the computer technologies and the advent of the Internet domain, the task of finding visual information is increasingly important and complex. Many attempts have been reported in the literature using low-level features such as colour, texture, shape and size. We are interested in the use of graph representation and graph matching [1] [2] for content-based image retrieval. The graph allows representation of image content by taking advantage of object/region features and their interrelationships. Graph matching [3] makes it possible to compute similarity between images. Given a database of images, retrieving images similar to a query image amounts to determining the similarity between graphs.
Many algorithms have been proposed for computing similarity between graphs by finding graph isomorphism or sub-graph isomorphism [4]. However, the algorithms for optimal matching are combinatorial in nature and difficult to use when the size of the graphs is large. The goal of this work is to develop a general and efficient algorithm that can be used easily to solve practical graph matching problems. The proposed algorithm is based on an application independent search strategy and can be run in a time-efficient way and, under some very general conditions, provides even optimal matching between graphs. We will show that the new algorithm can be effectively applied to content-based image retrieving. More importantly, this algorithm could help in alleviating the complexity problem in graph clustering, which is a very important step towards bridging the cap between structural pattern recognition and statistical pattern recognition [11].

* Dr. S. Wang is currently with School of Computer Science, University of Windsor, Windsor, Ontario, N9B 3P4, Canada

T. Caelli et al. (Eds.): SSPR&SPR 2002, LNCS 2396, pp. 291-300, 2002.
© Springer-Verlag Berlin Heidelberg 2002

2 The New Graph Matching Algorithm

In this section, we present a new algorithm for the graph-matching problem. Given two graphs, the goal is to find the best mapping between their nodes that leads to the smallest matching error. The matching error between the two graphs is a function of the dissimilarity between each pair of matched nodes and the dissimilarity between the corresponding edges. It can be viewed as the distance between the two graphs [5]. The basic idea of the new algorithm is iterative exploration of the best possible node mappings and selection of the best mapping at each iteration phase by considering both the error caused by node matching as well as that caused by corresponding edge mapping. The underlying hypothesis of this algorithm is that a good mapping between two graphs likely match similar nodes. The advantage of this algorithm is that this iterative process often allows finding the optimal mapping within a few iterations by searching only the most plausible regions of solution space. In the first phase, the algorithm selects the best possible mapping(s) that minimize the error induced by node matching only. Of these mappings, those that also give the smallest error in terms of edge matching are retained. In the second phase, the algorithm examines the mappings that contain at least one second-best mapping between nodes and then again computes those mappings that give rise to the smallest error in terms of edge matching. This process continues through a predefined number of phases.

2.1 Algorithm Description

We suppose that distance measures associated with the basic graph edit operations have been defined; i.e. costs have already been associated with substitution of nodes and edges, deletion of nodes and edges, etc. The technique proposed here is inspired by both Ullman's [1] algorithm and the error-correcting sub-graph isomorphism procedure [4],[6],[9],[10]. The new algorithm is designed for substitution operations only. It can easily be extended to deal with deletion and insertion operations by considering some special cases. For example, matching a node to a special (non-) node can perform deletion of the node. The algorithm is designed to find a graph isomorphism when both graphs have the same number of nodes and a sub-graph isomorphism when one has fewer nodes than the other.

Given two graphs $G_1 = (V_1, E_1, \mu_1, \nu_1)$ and $G_2 = (V_2, E_2, \mu_2, \nu_2)$, a $n \times m$ matrix $P = (p_{ij})$ is introduced, where n and m are the numbers of nodes in the first and the second graph, respectively. Each element p_{ij} in P denotes the dissimilarity between node i in G_1 and node j in G_2. We also use a second $n \times m$ matrix $B = (b_{ij})$.

The first step is to initialize matrix P by setting $p_{ij} = d(\mu_1(v_i), \mu_2(v_j))$. The second step consists of initializing B by setting $b_{ij} = 0$. The third (main) step contains K phases. In the first phase $(Current\_Phase = 1)$, the elements of B corresponding to the minimum elements in each row of matrix P are set to 1, $(b_{ij} = 1)$. Then, for each possible mapping extracted from B, the algorithm computes the error induced by nodes and the error induced by edges. The mapping that gives the smallest matching error will be recorded. In the second phase ($Current\_Phase = 2$), the algorithm will set the value to 1 those elements of *B* corresponding to the second-smallest elements in each row of matrix *P*. The algorithm will extract the mappings from matrix *B* that contain at least one node-to-node mapping added to matrix *B* at this phase. Of these mappings and the mappings obtained in the first phase, those with the smallest cost are retained. The algorithm then proceeds to the next phase, and so on.

A direct implementation of the above ideas would result in redundant extraction and testing of mappings, since any mapping extracted from matrix *B* at a given time will also be extracted from any subsequent matrix *B*. To solve this problem, a smart procedure has been designed. First, a matrix *B'* is introduced to contain all the possible node-to-node mappings considered by the algorithm so far. *B* is used as a 'temporary' matrix. At each phase (except the first), each of the *n* rows of *B* is examined successively. For each row *i* of *B*, all of the previous rows of *B* will contain all of the possible node-to-node mappings examined so far. The row *i* contains only the possible node-to-node mapping in the present phase. Finally, all of the following rows of *B* will contain only the possible node-to-node mappings examined in the previous phases. Such a matrix *B* guarantees that the mappings extracted as the algorithm progresses will never be the same and that all of the mappings that need to be extracted at each phase will indeed be extracted.

To illustrate the algorithm, we present a detailed example. Fig. 1 shows the weights attributed to nodes and edges in the input and the model graphs respectively. The first step in the proposed algorithm computes a *P* matrix. Each row in *P* represents a node in the model graph and the columns represent nodes in the input graph. The *P* matrix is given in Table 1. The second step of the algorithm computes the *B* matrix. Each element b_{ij} in this matrix is set to 1 if the corresponding p_{ij} has the smallest value in the ith row of *P*, to 0 otherwise. At this stage, there is no possible matching. This step can be interpreted as level one or $Current\_Phase = 1$. Next the algorithm enters its second phase, exploring mappings containing at least one node-to-node matching which corresponds to the second-smallest value in a row of the matrix P. Table 4 illustrates the possible mappings extracted from the current *B*.

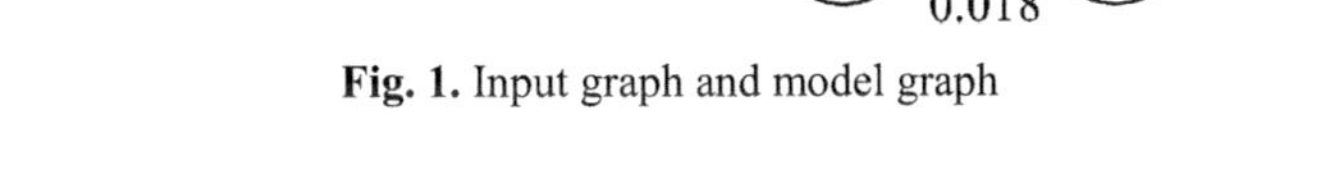

Fig. 1. Input graph and model graph

Table 1. Matrix *P*

0.225	0.068	0.645	0.19
0.232	0.075	0.638	0.183
0.377	0.22	0.493	0.038

Table 2. Matrix B (first phase)

0	1	0	0
0	1	0	0
0	0	0	1

Table 3. Matrix B (second phase)

1	1	0	0
0	1	0	0
0	0	0	1

Table 4. Best mataching with $Current\_Phase = 2$

Mappings	Matching error
(1,1) (2,2) (3,4)	0.711

2.2 Algorithm and Complexity

Input: two attributed graphs G_1 and G_2 .
Output: matching between nodes in G_1 and G_2, from the smaller graph (e.g., G_1) to the larger (e.g., G_2)

1. **Initialize** P as follows: For each p_{ij}, set $p_{ij} = d(\mu_1(v_i), \mu_2(v_j))$.
2. **Initialize** B as follows: For each b_{ij}, $i = 1,...,n$ and $j = 1,...,m$, set $b_{ij} = 0$.
3. **While** $Current\_Phase < K$

   If $Current\_Phase = 1$, **Then**

   For $i = 1,...,n$

   Set the value 1 to elements of B corresponding to the smallest value in i^{th} row of P;

   Call Matching_Nodes(B).

   Else For all $i = 1,...,n$

   Set $B' = B$

   For all $j = 1,...,m$ set $b_{ij} = 0$

   Select the element with the smallest value in P that is not marked 1 in B' and set it to 1 in B and B';

   Call Matching_Nodes(B);

   Set $B = B'$.

   If all the elements in B are marked 1, **Then**

   Set $Current\_Phase = K$

   Else add 1 to *Current_Phase*.

Matching_Nodes(B)

For each valid mapping in *B*

1. Compute the matching error induced by nodes.
2. Add the error induced by the corresponding edges to the matching error.
3. Save the actual matching if the matching error is minimal.

The major parameter *K* defines the number of phases to be performed in order to find the best matching. Suppose, without loss of generality, that the size of the two graphs satisfies the following condition $n = |V_1| \leq |V_2| = m$, then the worst case complexity of the new algorithm is $O(n^2 K^n)$. This is to compare with $O(n^2 m^n)$, the complexity for Ullman's algorithm [1] and the A*-based error-correcting sub-graph isomorphism algorithm [4],[6]. In general, the new algorithm reduces the number of steps in the error-correcting algorithm by the factor of about $(m/K)^n$. This can be very significant when matching large graphs. Table 5 shows a comparison with the A*-based error-correcting algorithm over 1000 pairs of graphs generated randomly. The size of each graph is between 2 and 10 nodes. The experiment was run on a Sun Ultra 60 workstation (450 MHz CPUs). From the table, one can notice that the new algorithm performs extremely well in computing the optimal matching while maintaining very low average CPU times. For instance, when using $K = 4$, the algorithm finds the optimal matching in 971 cases while using only 11 seconds in average. The A*-based algorithm needs 186 seconds in average although it guarantees to find the optimal matching. It is to be remarked that due to its complexity, the A*-based algorithm is generally not usable when the graphs to be matched have more than 10 nodes. The new algorithm does not suffer this limit. For example, matching two graphs of 11 and 30 nodes with $K = 5$ takes about 100 seconds. Details about the deduction of the complexity and about the performance of the algorithm can be found in our technical report [8]. The new algorithm does not require the use of heuristics. It can be used to find good matchings (usually optimal) in a short time. In this sense, it can be categorised in the class of approximate algorithms.

Table 5. Comparison with the error-correcting sub-graph isomorphism algorithm

Number of phases *K*	1	2	3	4	5	Error-Correcting (A*)
Optimal matchings reached by the proposed algorithm	609	827	940	971	1000	1000
Average time in seconds	2.14	3.69	6.14	11.04	16.28	186.57

3 Image Retrieval Based on the New Graph Matching Algorithm

The aim of this section is to show how graph matching contributes to image retrieval. In particular, we would like to show how the new matching algorithm could be used. For this purpose, we have generated an artificial image database so that extraction of

objects and representation of the content by a graph are simplified. Our work is divided into two parts. First, we build an image database and define a graph model to represent images. Second, we make use of the new matching algorithm to derive a retrieval algorithm for retrieving similar images.
The advantage of using a generated database is that it allows us to evaluate a retrieval algorithm in a more systematic way. We suppose that each image in the database contains regular shapes such as rectangles, squares, triangles, etc. An algorithm has been developed to build such a database. Only the number of images needs to be given by the user. The algorithm randomly generates all the other parameters. These random parameters define the number of object, the shape, color, size and position of each object in the image.
For easy manipulation of the database, only the description of the image is stored in a text file and a subroutine is created to save an image from and restore it to this text file. The description includes following variables: the numerical index of each image, the number of objects in the image, the shape of an object represented by a value between 1 and 5 (a square is represented by 1, a rectangle by 2, etc.), the size of the object; its color; its position; and its dimension.
The second step in the process is to use graphs to represent the contents of images. Each node represents an object in an image and an edge represents the relation between two objects. In our work, three features describe a node: the shape, size and color of the object. Two features describe an edge: the distance between two objects and their relative position. These features are represented, respectively, using S, Z, C, D, and RP. The values of the first three features figure in the database. The Hausdorff distance [7] is computed for D. The relative position RP is a discrete value describing the location of objects with respect to each other [8].

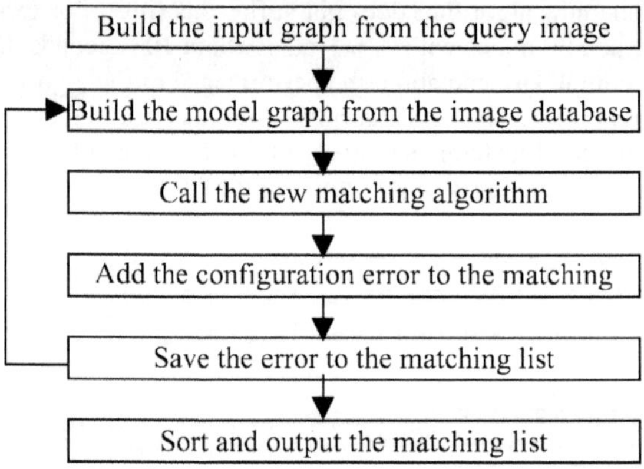

Fig. 2. The flow diagram of the retrieval algorithm

3.1 The Retrieval Algorithm

In this section, we adapt the matching algorithm described in Section 2 for retrieving images by content using graphs. Given a query image, the algorithm computes a matching error for each image in the data base, finds the best matching between the query image and any of the images in the database and extracts the similar images from the database. Fig.2 gives the schema of the retrieval algorithm. Obviously, if the database is very large, such a retrieval algorithm may not be appropriate. Organization of the database indices would be required so that the matching process will be done only on those images that are most likely similar to the query image. Graph clustering is one of the issues that we plan to investigate in the near future.

The retrieval algorithm has six steps. The construction of the input and model graphs from the query and database images is done in the first and the second steps respectively. The new matching algorithm is then called in the third step to compute the matching error. To perform this task, the algorithm should compute f_n, the error induced by the node-to-node matching, and f_e, the error induced by the edge-to-edge matching. Since a node includes multiple features, f_n must combine them using a weighting scheme. It is formulated as follows:

$$f_n = \alpha\ e_s(S_I, S_B) + \beta\ e_z(Z_I, Z_B) + \gamma\ e_c(C_I, C_B) \tag{1}$$

Where I and B represent the input and the database graph respectively, and α, β, γ are the weighting coefficients for the shape, color and size. Similarly, f_e is defined as:

$$f_e = \delta\ \ e_p(PR_I, PR_B) + \varepsilon\ \ e_d(D_I, D_B) \tag{2}$$

The error related to the shape e_s is set to zero if the two objects have the same shape; otherwise it is set to 1. Similarly, the error related to the relative position e_p is set to zero if the pair of objects have the same value according to this feature; otherwise the error is set to 1. The respective errors related to the size, the color and the distance between two objects, e_z, e_c and e_d, are defined by the following formulas:

$$e_z(Z_I, Z_B) = \frac{|Z_I - Z_B|}{(Z_I + Z_B)} \tag{3}$$

$$e_c(C_I, C_B) = \sqrt{(CL_I - CL_B)^2 + (CU_I - CU_B)^2 + (CV_I - CV_B)^2} \tag{4}$$

$$e_d(D_I, D_B) = \frac{|D_I - D_B|}{(D_I + D_B)} \tag{5}$$

In the fourth step, the retrieval algorithm computes a configuration error f_c associated to the image that does not have the same number of objects or of edges as the query image. This error is effectively added to the matching error if the coefficient c is greater than zero.

$$f_c = c\ \left(\left||V_I| - |V_B|\right| + \left||E_I| - |E_B|\right|\right) \tag{6}$$

$$matching\_error = f_n + f_e + f_c \tag{7}$$

Here $|V_I|, |E_I|, |V_B|$ and $|E_B|$ are the number of objects and edges in the query and the database images respectively.

In the next step, the algorithm saves the matching error and the corresponding mappings into a matching list. This process will be repeated for each image in the database. Finally, the algorithm sorts the matching list and outputs the most similar images. The different parameters α, β, γ, δ, ε, and *c* provide a variety of possibilities for the users to control the query.

3.2 The Experimental Results

In this section, we present some image retrieval experiments performed using the new retrieval algorithm. The aim of these experiments is to show that the algorithm can indeed retrieve expected images similar to the query image and that such retrieval can be performed according to various needs of the user. We have conducted the retrieval with the generated database containing 1000 images. The number of objects in each image varies between 2 and 9. For each experiment, the specification of the query will be detailed and the first three similar images will be showed. For these experiments, the query image itself is not a member of the database.

3.2.1 Image Retrieval by Shape

In this experiment, the user is searching for images that contain three objects. Only the shape (two triangles and a square) is important to the user. For this purpose, the parameters in the two dissimilarity functions should be set as follows: $\alpha = 1$, $c = 1$ and all other parameters are set to zero.

Query image

Image: 528 Error: 0

Image: 7 Error: 1

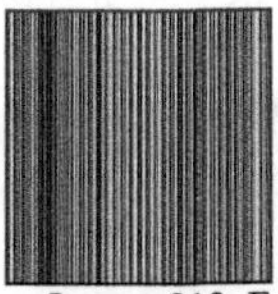
Image: 213 Error: 5

The image 528 has exactly the same objects as the query image according to the shape. In the second image only two objects can be matched and thus the error is not null. The third image has four objects and only two objects can be matched.

3.2.2 Image Retrieval by Shape and Relative Position

In this experiment, the same query image is used. The user is searching for images that contain objects having the same shape and relative position as in the query image. For this purpose, the parameters in the two dissimilarity functions should be set as follows: $\alpha = 0.5$, $\delta = 0.5$, $c = 1$ and all other parameters are set to zero.

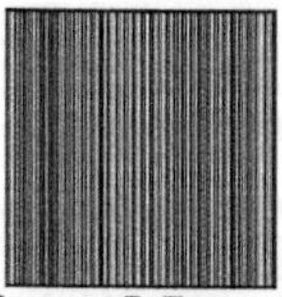
Image : 7 Error : 1

Image : 184 Error : 1

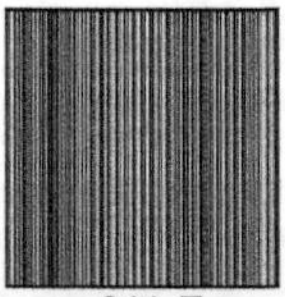
Image : 244 Error : 1.5

The algorithm is able to find the similar images considering both criteria. The image 7 is one of the two closest ones to the query image. The result is appealing visually. The (minimum) error of 1 is caused by two factors. One is the presence of a square object in the image 7 instead a triangle in the query image. The other one is the difference between the relative position square-triangle(big) in the query image and relative position Square-Square in the image 7.

4 Conclusion and Perspectives

The new graph-matching algorithm presented in this paper performs the search process in K phases. The promising mappings are examined in early phases. This allows computation of good matching with a small number of phases and increased computational efficiency. The new algorithm compares extremely well to the A*-based error correcting algorithm on randomly generated graphs. The new matching algorithm will be part of our content-based image retrieval system. A preliminary retrieval algorithm based on the new graph-matching algorithm has been reported here. Investigation is underway to discover cluster structures in the graphs so that the retrieval process can be focused on a reduced set of model graphs.

Acknowledgement

This work has been supported by a Strategic Research Grant from Natural Sciences and Engineering Research Council of Canada (NSERC) to the team composed of Dr. F. Dubeau, Dr. J. Vaillancourt, Dr. S. Wang and Dr. D. Ziou. Dr. S. Wang is also supported by NSERC via an individual research grant.

References

1. J. R. Ulmann. An algorithm for subgraph isomorphism, Journal of the association for Computing Machinery, vol. 23, no 1, January 1976, pp. 31-42.
2. D. G. Corneil and C. G. Gotlieb. An Efficient Algorithm for Graph Isomorphism, Journal of the Association for Computing Machinery, vol. 17, no. 1, January 1970, pp. 51-64.
3. J. Lladós. Combining Graph Matching and Hough Transform for Hand-Drawn Graphical Document Analysis. http://www.cvc.uab.es/~josep/articles/tesi.html.
4. B. T. Messmer and H. Bunke. A New Algorithm for error-Tolerant Subgraph Isomorphism Detection, IEEE Trans on PAMI, vol. 20, no. 5, May 1998.
5. Sanfeliu and K.S. Fu, A Distance Measure Between Attributed Relational Graphs for Pattern Recognition. IEEE Trans. on SMC, vol. 13, no. 3. May/June 1983.
6. W.H. Tsai and K.S. Fu. Error-Correcting Isomorphisms of Attributed Relational Graphs for Pattern Analysis. IEEE Trans. on SMC, vol. 9, no. 12. December 1979.
7. Hausdorff. Hausdorff distance http://cgm.cs.mcgill.ca/~godfried/teaching/cg-projects/98/normand/main.html

8. Hlaoui and S. Wang. Graph Matching for Content-based Image Retrieval Systems. Rapport de Recherche, No. 275, Département de mathématiques et d'informatique, Université de Sherbrooke, 2001.
9. Y. Wang, K. Fan and J. Horng. Genetic-Based Search for Error-Correcting Graph Isomorphism. IEEE Trans. on SMC, Part B, vol. 27, no. 4. August 1997.
10. Huet, A. D. J. Cross And E. R. Hancock. Shape Retrieval by Inexact Graph Matching. ICMCS, vol. 1, 1999, pp. 772-776. http://citeseer.nj.nec.com/325326.html
11. X. Jiang, A. Munger, and H. Bunke. On Median Graphs: Properties, Algorithms, and Applications. IEEE Trans on PAMI, vol. 23, no. 10, October 2001.

Efficient Computation of 3-D Moments in Terms of an Object's Partition

Juan Humberto Sossa Azuela[1], Francisco Cuevas de la Rosa[2,*] and Héctor Benitez[1]

[1] Centro de Investigación en Computación del IPN
Av. Juan de Dios Bátiz esquina con M. Othón de Mendizábal
Colonia Nueva Industrial Vallejo, México, D. F. 07738, México
[2] Centro de Investigaciones en Óptica+
Apdo. Postal 1-948, León, Gto. México
{hsossa,fjcuevas}@cic.ipn.mx
hbenitez75@hotmail.com

Abstract. The method here proposed is based on the idea that the object of interest is first decomposed in a set of cubes under d_∞. This decomposition is known to form a partition. The required moments are computed as a sum of the moments of the partition. The moments of each cube can be computed in terms of a set of very simple expressions using the center of the cube and its radio. The method provides integral accuracy by applying the exact definition of moments over each cube of the partition. One interesting feature of our proposal is that once the partition is obtained, moment computation is faster than with earlier methods.

1 Introduction

The two-dimensional moment (for short 2D moment) of a 2D object *R* is defined as [1]:

$$M_{pq} = \iint_R x^p y^q f(x,y)\,dxdy \tag{1}$$

where $f(x,y)$ is the characteristic function describing the intensity of *R*, and $p+q$ is the order of the moment. In the discrete case, the double integral is often replaced by a double sum giving as a result:

$$m_{pq} = \sum\sum_R x^p y^q f(x,y) \tag{2}$$

with $f(x,y)$, *p and q* defined in equation 1, where $(x,y) \in \mathbf{Z}^2$.

* Francisco Cuevas is in a post-doctoral stay at the Centro de Investigación en Computación of the Instituto Politécnico Nacional.

T. Caelli et al. (Eds.): SSPR&SPR 2002, LNCS 2396, pp. 301-309, 2002.
© Springer-Verlag Berlin Heidelberg 2002

The tri-dimensional geometric moment (for short 3D moment) of order $p+q+r$ of a 3D object is defined as [2]:

$$M_{pqr} = \iiint_R x^p y^q z^r f(x,y,z)\,dxdydz \tag{3}$$

where R is a 3D region. In the discrete case, the triple integral is often replaced by the triple sum giving as a result:

$$m_{pqr} = \sum\sum_R\sum x^p y^q z^r f(x,y,z) \tag{4}$$

with $f(x,y,z)$, p,q y r defined in equation (3), where $(x,y,z) \in \mathbf{Z}^3$.

In the binary case, the characteristic function takes only values 1 or 0, assuming that for the volume of interest $f(x,y,z)=1$. When we replace this value in equation (4) we get the equation to compute the moments of order $(p+q+r)$ of a 3-D image R as

$$m_{pqr} = \sum\sum_R\sum x^p y^q z^r \tag{5}$$

with $(x,y,z) \in \mathbf{Z}^3$ y $p, q, r = 0,1,2,3,...$

The world around us is generally three-dimensional, and 3D shape information for an object can be obtained by computer tomographic reconstruction, passive 3D sensors, and active range finders. As 2D moments, 3D moments have been used in 3D image analysis tasks including movement estimation [3], shape estimation [4], and object recognition [2].

Several methods have been proposed to compute the 3D moments. In [6], Li uses a polyhedral representation of the object for the computing of its 3D moments. The number of required operations is a function of the number of edges of the surfaces of the polyhedral. The methods of Cyganski *et al.* [5], Li and Shen [7] and Li and Ma [8] use a voxel representation of the object. The difference among these methods is the way to compute the moments. Cyganski *et al.* uses the filter proposed in [9]. Li and Shen use a transformation based on Pascal triangle for the computation of the monomials and only additions are used for the computation of the moments. In the other hand, Li and Ma relate 3D moments with the so-called LT moments that are easier to evaluate. Although these methods allow to reduce the number of operations to compute the moments, they require a computation of $O(N^3)$. Recently, Yang *et al.* [10] propose to use the so called discrete divergence theorem to compute the 3D moments of an object. It allows a reduction in the number of operations to $O(N^2)$.

In this note we present an efficient method to compute the 3-D moments of a binary object in $\mathbf{Z}^3$. The method is an extension of the method recently introduced in [11] to compute the 2D moments of an object. It provides integral accuracy (see [12] for the details) on the values obtained by applying the original definition of moments (equation (3)) instead of that one using triple sums (equation (4) or (5)). This could not happen if equation (5) was used.

The object is first partitioned into convex cubes which moments evaluation can be reduced to the computation of very simple formulae instead of using triple integrals. The desired 3D moments are obtained as the sum of the moments of each cube of the partition, given that the intersection among cubes is empty.

2 Moments of a Cube

In the last section we mentioned that to compute the desired moments of a 3D object, it should be first decomposed into a set of cubes. Then we also said that a set of simple expressions should be applied to get the desired values. In this section, this set of expressions is provided.

Depending on the definition of moments used, the set of expressions obtained might differ resulting in some differences. This situation was first studied in [12] and recently re-discussed in [13], both in the 2-D case. As stated in [13], if M_{pq} are the 2D moments obtained by means of equation (1) and m_{pq} those obtained in terms of equation (2) an error $|M_{pq} - m_{pq}|$ is introduced due to the approximations and numeric integration of $x^p y^q$ over each pixel. As we will next see, this also happens with 3D moments.

To derive the set of expressions needed to accurately compute the desired 3D moments, let us consider a cube centered in (X_c, Y_c, Z_c), with radius t and coordinates of its vertices in $(X_c - t, Y_c - t, Z_c - t)$, $(X_c + t, Y_c - t, Z_c - t)$, $(X_c - t, Y_c + t, Z_c - t)$, $(X_c - t, Y_c - t, Z_c + t)$, $(X_c + t, Y_c + t, Z_c - t)$, $(X_c + t, Y_c - t, Z_c + t)$, $(X_c - t, Y_c + t, Z_c + t)$ and $(X_c + t, Y_c + t, Z_c + t)$. The characteristic function of this block is

$$f(x,y,z) = \begin{cases} 1 & \text{if} \quad (x,y,z) \in (a,b) \times (c,d) \times (e,f) \\ 0 & \text{otherwise} \end{cases}$$

with

$$\begin{aligned} a &= X_c - t - 0.5 \\ b &= X_c + t + 0.5 \\ c &= Y_c - t - 0.5 \\ d &= Y_c + t + 0.5 \\ e &= Z_c - t - 0.5 \\ f &= Z_c + t + 0.5 \end{aligned}$$

According to equation (3), the exact moments of a cube are given as

$$\begin{aligned} M_{pqr} &= \int_{-\infty}^{\infty}\int_{-\infty}^{\infty}\int_{-\infty}^{\infty} x^p y^q z^r f(x,y,z)\,dx\,dy\,dz \\ &= \frac{1}{(p+1)(q+1)(r+1)} \left[b^{p+1} - a^{p+1}\right] \cdot \left[d^{q+1} - c^{q+1}\right] \cdot \left[f^{r+1} - e^{r+1}\right] \end{aligned} \tag{6}$$

The reader can easily verify that the first 20 expressions for the moments are:

$$M_{000} = (2t+1)^3 \qquad M_{100} = M_{000}X_c$$
$$M_{010} = M_{000}Y_c \qquad M_{001} = M_{000}Z_c$$
$$M_{200} = \frac{M_{000}}{3}\left(3X_c^2 + t(t+1) + 0.25\right) \qquad M_{020} = \frac{M_{000}}{3}\left(3Y_c^2 + t(t+1) + 0.25\right)$$
$$M_{002} = \frac{M_{000}}{3}\left(3Z_c^2 + t(t+1) + 0.25\right) \qquad M_{110} = M_{100}Y_c = M_{000}X_cY_c$$
$$M_{101} = M_{100}Z_c = M_{000}X_cZ_c \qquad M_{011} = M_{001}Y_c = M_{000}Y_cZ_c$$
$$M_{300} = M_{000}X_c\left(X_c^2 + t(t+1) + 0.25\right) \tag{7}$$
$$M_{030} = M_{000}Y_c\left(Y_c^2 + t(t+1) + 0.25\right)$$
$$M_{003} = M_{000}Z_c\left(Z_c^2 + t(t+1) + 0.25\right)$$
$$M_{120} = \frac{M_{000}}{3}X_c\left(3Y_c^2 + t(t+1) + 0.25\right)$$
$$M_{210} = \frac{M_{000}}{3}Y_c\left(3X_c^2 + t(t+1) + 0.25\right)$$
$$M_{102} = \frac{M_{000}}{3}X_c\left(3Z_c^2 + t(t+1) + 0.25\right)$$
$$M_{201} = \frac{M_{000}}{3}Z_c\left(3X_c^2 + t(t+1) + 0.25\right)$$
$$M_{012} = \frac{M_{000}}{3}Y_c\left(3Z_c^2 + t(t+1) + 0.25\right)$$
$$M_{021} = \frac{M_{000}}{3}Z_c\left(3Y_c^2 + t(t+1) + 0.25\right)$$
$$M_{111} = M_{000}X_cY_cZ_c$$

The reader can be easily verify that the same set of 20 expressions obtained through equation (5) is the following:

$$m_{000} = (2t+1)^3 \qquad m_{100} = m_{000}X_c$$
$$m_{010} = m_{000}Y_c \qquad m_{001} = m_{000}Z_c$$
$$m_{200} = \frac{m_{000}}{3}\left(3X_c^2 + t(t+1)\right) \qquad m_{020} = \frac{m_{000}}{3}\left(3Y_c^2 + t(t+1)\right)$$
$$m_{002} = \frac{m_{000}}{3}\left(3Z_c^2 + t(t+1)\right) \qquad m_{110} = m_{100}Y_c = m_{000}X_cY_c$$
$$m_{101} = m_{100}Z_c = m_{000}X_cZ_c \qquad m_{011} = m_{001}Y_c = m_{000}Y_cZ_c$$
$$m_{300} = m_{000}X_c\left(X_c^2 + t(t+1)\right) \tag{8}$$
$$m_{030} = m_{000}Y_c\left(Y_c^2 + t(t+1)\right)$$
$$m_{003} = m_{000}Z_c\left(Z_c^2 + t(t+1)\right)$$

$$m_{120} = \frac{m_{000}}{3} X_c \left(3Y_c^2 + t(t+1)\right)$$

$$m_{210} = \frac{m_{000}}{3} Y_c \left(3X_c^2 + t(t+1)\right)$$

$$m_{102} = \frac{m_{000}}{3} X_c \left(3Z_c^2 + t(t+1)\right)$$

$$m_{201} = \frac{m_{000}}{3} Z_c \left(3X_c^2 + t(t+1)\right)$$

$$m_{012} = \frac{m_{000}}{3} Y_c \left(3Z_c^2 + t(t+1)\right)$$

$$m_{021} = \frac{m_{000}}{3} Z_c \left(3Y_c^2 + t(t+1)\right)$$

$$m_{111} = m_{000} X_c Y_c Z_c$$

One might would like know which accuracy provides the proposed approach. It is know that for pixel or voxel represented objects, the computed moment values have mainly two types of accuracy. One of them is obtained by exactly performing the double or triple sum, given by equation (2) or (4). The another is obtained by assuming that a pixel is a square and a voxel is a cube, and computing the moments as an integral over the area covered by the small pixel squares, or the volume covered by the small cubes. None of the above approaches gives the true values of the moments. It is not possible to obtain the true moment values if digitalization is done. Our proposal provides integral accuracy. This was very well studied in [12] for the 2-D case.

One might think that because each cube has its own center located at (X_c, Y_c, Z_c), the summing of moments computed from different cubes of different centers is not possible. The summing is possible even if each cube has its own center. This is due to M_{pqr} are expressed in terms of t and one or more of the coordinates of the center. These last terms introduce the needed values to compensate the fact that each cube has its own center.

3 Discussion and Comparison

While equation (3) yields exact results, equation (5) provides some moments with small errors due to the zero-order approximation for numerical integration when using sums. We will always find $M_{pqr} \geq m_{pqr}$. The error $M_{pqr} - m_{pqr}$ depends directly on p, q and r. Your can easily verify that:

$$M_{200} - m_{200} = \frac{m_{000}}{12} \qquad M_{020} - m_{020} = \frac{m_{000}}{12} \qquad M_{002} - m_{002} = \frac{m_{000}}{12}$$

$$M_{300} - m_{300} = \frac{m_{100}}{4}. \qquad M_{030} - m_{030} = \frac{m_{100}}{4}. \qquad M_{003} - m_{003} = \frac{m_{100}}{4}.$$

On the other hand, for some moments both methods produce exact results:

$$M_{000} - m_{000} = 0 \qquad M_{100} - m_{100} = 0$$
$$M_{010} - m_{010} = 0 \qquad M_{001} - m_{001} = 0$$
$$M_{110} - m_{110} = 0 \qquad M_{101} - m_{101} = 0$$
$$M_{011} - m_{011} = 0 \qquad M_{111} - m_{111} = 0$$

On the main features of our method is that once the partition is obtained, moment computation is much faster than in the case of earlier methods. For this, let us take the next simple example. Let us suppose that the an object is composed of $N \times N \times N$ pixels, with t as its radius.

The number of operations required by one of the fastest methods (for example the method of Yang, Albregtsen and Taxt, [10]) to compute all the moments of order $(p+q+r)$ up to some K, let say $K=3$ from a discrete image of $N \times N \times N$ pixels is: $2KN^2$ multiplications and $\left(\frac{1}{2}K^2 + \frac{7}{2}K + 3\right)N^2$ additions (for the details, refer to [10]).

The number of operations required by our proposal **once the partition has been obtained** will depend basically on the radius t of the object: $26t$ multiplications and $10t$ additions.

4 A Method to Compute the Desired Object Moments

To compute the desired moments we could use the same idea already used in [11], this is :

1. Decompose the object into the union of disjoint cubes.
2. Compute the geometric moments for each of these cubes, and
3. Obtain the final moments as a sum of the moments computed for each cube.

The key problem to apply this idea is how to obtain the desired partition, i.e. the union of disjoint cubes. For this we can use the same morphological approach used in [11] (extended to the 3D case). According to [11] there two main variants to compute the desired moments:

4.1 Method Based on Iterated Erosions

The following method to compute the geometric moments of a 3D object $R \subset Z^3$, using morphological operations is an extension of the one described in [11]. It s composed of the following steps:

1. Initialize 20 accumulators $C_i=0$, for i=1,2,...,20, one for each geometric moment.
2. Make $A=R$ and $B=\{(\pm a, \pm b, \pm c) / a,b,c \in \{-1,0,1\}\}$, B is a 3x3x3 pixel neighborhood in $\mathbf{Z}^3$.
3. Assign $A \leftarrow A \,\theta\, B$ iteratively until the next erosion results in $\varnothing$ (the null set). The number of iterations of the erosion operation before set $\varnothing$ appears, is the

radius r of the maximal cube completely contained in the original region R. The center of this cube is found in set A just before set $\varnothing$ appears.
4. Select one of the points of A and given that the radius r of the maximal cube is known, we use the formulae derived in the last section to compute the moments of this maximal cube, the resulting values are added to the respective 20 accumulators, C_i, for 1,2,3,...,20.
5. Eliminate this ball from region R, and assign this new set to R.
6. Repeat steps 2 to 5 with the new R until it becomes $\varnothing$.

The method just described gives us as a result the true values of the geometric moments of order $(p+q+r) \leq 3$, using only erosions and the formulae developed in Section 2.

4.2 Method Based on Iterated Erosions and Parallel Processing

This method is a brute force method. A considerable enhancement can obtained if steps 4 and 5 are replaced by:

1. Select those points in A at a distance among them greater than $2t$ and use the formulae given by equation (7), to compute the geometric moments of these maximal cubes, and add these values to the respective accumulators.
2. Eliminate the maximal cubes from region R, and assign this new set to R.

The enhancement consists in processing all maximal cubes of the same radius in just a step, coming back to the iterated erosions until the value of the radio t is to be changed. At this step it is very important to verify that the eliminated cubes do not intersect with those just eliminated, for one of the important conditions is that the set of maximal cubes forms a partition of the image. Thus one has to guarantee that these maximal cubes be disjoint sets.

5 Results

Suppose we use the proposed two variants described in the last section to compute the desired object's moments. Because both variants are not designed to work in a conventional computer, the processing times are only significant to compare the method eliminating a cube at the time against the method eliminating, at the same step, all the non-intersecting maximal cubes at the same time.

Both variants were tested on several hundreds of images. All of them are binary and $101\times101\times101$ voxel sized. These images were obtained by generating at random P touching and overlapping cubes of different sizes inside the $101\times101\times101$ cubical image. At the beginning all the locations of the $101\times101\times101$ cube are zero.

The original method takes on average 150 seconds to compute all moments of order $(p+q+r) \leq 3$; while the enhanced method requires only about 25 seconds onto 233 Mhz PC based system to compute the same moments.

6 Conclusions and Present Research

In this note an extended version of the recently proposed method in [11] to compute accurately the 3D geometric moments for a object has been presented. Initially, the object is partitioned in a set of convex cubes whose moment evaluation can be reduced to the computation of very simple formulae.

These expressions were derived from the original definition of moments given by equation (3). This gives more accurate values for the moments. This would not happen if equation (5) would be used. An error is introduced due to zero-order approximation and numeric integration of $x^p y^q z^r$ over each voxel. The resulting shape moments are finally obtained by addition of the moments of each cube forming the partition, giving that the intersections are empty.

As implemented until now the proposed approach is very slow, for and image of 100×100×100 voxels, 150 seconds in the case of the first variant and 25 in the case of the second variant.

To make our proposal really competitive with classical sequential algorithms we need a better way the obtain the desired partition. Apparently, the fast distance transform (see [14]) could be an excellent option. In this case, the idea here would be to first decompose the image into a set of disjoint cubes by means of the fast tri-dimensional distance transform, which would provide the necessary information of all the maximal cubes covering the image. We would then apply the simple formulae given by equation (7) to obtain the exact moments for each cube. We would finally get the final desired moments of the image by summing the partial results from all the cubes.

One of the huge advantages of the fast distance transform is that it can be efficiently programmed in a sequential machine. At this moment, we are working on the development of a suitable algorithm.

Acknowledgments

The authors would like to thank the CIC-IPN and the CONACYT under project 34880-A for their economical support to develop this work.

References

1. M. K. Hu, Visual pattern recognition by moment invariants, IRE Transactions on Information Theory, 179-187, 1962.
2. C. H. Lo and H. S. Don, 3-D moment forms: Their construction and application to object identification and positioning, IEEE Transactions on Pattern Analysis and Machine Intelligence, 11:1053-1064, 1989.
3. S. C. Pei and L. G. Liou, Using moments to acquire the motion parameters of a deformable object without correspondences, Image Vision and Computing, 12:475-485, 1994.

4. J. Shen and B. C. Li, Fast determination of center and radius of spherical surface by use of moments, in Proceedings of the 8th Scandinavian Conference on Image Analysis, Tromso, Norway, pp. 565-572, 1993.
5. D. Cyganski, S. J. Kreda and J. A. Orr, Solving for the general linear transformation relating 3-D objects from the minimum moments, in SPIE Intelligent Robots and Computer Vision VII, Proceedings of the SPIE, Vol. 1002, pp. 204-211, Bellingham, WA, 1988.
6. B. C. Li, The moment calculation of polyhedra, Pattern Recognition, 26:1229-1233, 1993.
7. B. C. Li and J. Shen, Pascal triangle transform approach to the calculation of 3D moments, CVGIP: Graphical Models and Image Processing, 54:301-307, 1992.
8. B. C. Li and S. D. Ma, Efficient computation of 3D moments, in Proceedings of 12 the International Conference on Pattern Recognition, Vol 1, pp. 22-26, 1994.
9. Z. L. Budrikis and M. Hatamian, Moment calculations by digital filters, AT&T Bell Lab. Tech. J. 63:217-229, 1984.
10. L. Yang and F. Albregtsen and T. Taxt, Fast computation of three-dimensional geometric moments using a discrete divergence theorem and a generalization to higuer dimensions, CGVIP: Graphical models and image processing, 59(2):97-108, 1997.
11. H. Sossa, C. Yañez and J. L Díaz, Computing geometric moments using morphological erosions, Pattern Recogntition, 34(2), 2001.
12. M. Dai, P. Baylou and M. Najim, An efficient algorithm for computation of shape moments from run-length codes or chain codes, Pattern Recognition, 25(10):1119-1128, 1992.
13. J. Flusser, Refined moment calculation using image block representation, IEEE Transactions on Image Processing, 9(11):1977-1978, 2000.
14. J. D. Díaz de León and J. H. Sossa, Mathematical Morphology based on linear combined metric spaces on Z^2 (Part I): Fast distance transforms, Journal of Mathematical Imaging and Vision, 12:137-154, 2000.

A Visual Attention Operator Based on Morphological Models of Images and Maximum Likelihood Decision

Roman M. Palenichka

Université du Québec, Dept. of Computer Science
Hull, Québec, Canada
palenich@uqah.uquebec.ca

Abstract. The goal of the image analysis approach presented in this paper was two-fold. Firstly, it is the development of a computational model for visual attention in humans and animals, which is consistent with the known psychophysical experiments and neurology findings in early vision mechanisms. Secondly, it is a model-based design of an attention operator in computer vision, which is capable to detect, locate, and trace objects of interest in images in a fast way. The proposed attention operator, named image relevance function, is an image local operator that has local maximums at the centers of locations of supposed objects of interest or their relevant parts. This approach has several advantageous features in detecting objects in images due to the model-based design of the relevance function and the utilization of the maximum likelihood decision.

1 Introduction

Time-effective detection and recognition of objects of interest in images is still a matter of intensive research in computer vision community because the artificial vision systems usually fail to outperform the detection results by a human being. The detection problem is complicated when objects of interest have low contrast and various sizes or orientations and can be located on noisy and inhomogeneous background with occlusions. In many practical applications, the real-time implementation of object detection algorithms in such natural conditions is a matter of great concern. The results of numerous neurophysiological and psychophysical investigation of human visual system (HVS) indicate that the human vision can successfully cope with these complex situations because of using a visual attention mechanism associated with a model-based image analysis [1,2]. The goal of presented here investigation was not the simulation of human visual perception but the incorporation of its advantageous features into computer vision algorithms. Besides many remarkable properties of HVS like the mentioned model-based visual attention, the HVS has also some disadvantages such as visual illusions while detecting and identifying objects [3].

T. Caelli et al. (Eds.): SSPR&SPR 2002, LNCS 2396, pp. 310-319, 2002.
© Springer-Verlag Berlin Heidelberg 2002

Several models of attention mechanism in HVS in the context of reliable and time-effective object detection in static scenes have been proposed in the literature. They are mostly based on the generalization of edge and line detection operators and on the utilization of a multi-resolution image analysis including the wavelet theory [4-8]. Very good results of attention modeling have been reported by the application of symmetry operators to images [8]. There are known attention operators, which combine both the multi-resolution approach and the symmetry operators. Attention operators based on the wavelet image analysis also showed great potential, especially when integrating such novel types of wavelets as curvelets and ridgelets [7]. In contrast to the standard isotropic image analysis, they incorporate a multi-scale analysis of anisotropy of objects of interest in images.

The feature extraction approach *per se* is also a method for selecting regions of interest although it is a generic approach and requires explicit defining of relevant features. It can be considered as an intermediate stage between pre-attentive vision and post-attentive vision. Recently, a method for directed attention during visual search has been developed based on the maximum likelihood strategy [9]. It is suitable to detection of objects of interest of a particular class by pairing certain image features with the objects of interest but is restricted only to the detection task.

However, a few work has been done toward designing a model-based attention mechanism which is quite general and based on an image model of low and intermediate levels (description of object regions and their shape), and can yield an optimal detection and segmentation performance with respect to the underlying model. The low-level image modeling is requested because the attention mechanism in its narrow sense is a *bottom-up* image analysis process based on quite general intensity and shape properties in order to respond to various unknown *stimuli* as well as to provide a reasonable response when no object of interest is present.

In this paper, a new model of visual attention based on the concept of a multi-scale relevance function is proposed as a mathematical representation of some generally recognized results regarding the explanation of HVS mechanisms. The introduced relevance function is an image local operator that has local maxima at centers of location of supposed objects of interest or their parts if the objects have complex or elongated shapes. The visual attention mechanism based on the relevance function provides several advantageous features in detecting objects of interest due to the model-based approach used. While detecting objects, it provides a quick location of the objects of interest with various sizes and orientations. Besides some other advantages, the operating with the property map as an intermediate image representation enhances the possibility to treat images with inhomogeneous backgrounds and textured appearance of objects.

2 Representation of Planar Shapes of Objects

Analysis of images and detection of local objects of interest in images can be efficiently performed by using object-relevant image properties. There are considered properties of object planar shape as well as intensity properties within a region of interest containing an object on the background. Such properties have to be computed in each image point in order to be able to perform the image segmentation, which provides object and background regions for the object recognition. This results in the computation of a *property map* as input data for further image analysis including the detection of objects of interest. It is assumed that in the general case the image intensity is represented by a vector of n primary features $\boldsymbol{x}=[x_1,\ldots,x_n]$. For example, pixels of a color image are three-component vectors. The primary features can be extracted from a gray-scale image on the basis of one feature vector per pixel. This includes the case of texture features when a set of local features is computed in each image point. Some examples of used primary features are given in the section of experimental results. In fact, the vector $\boldsymbol{x}=[x_1,\ldots,x_n]$ describes the parameters (properties) of an image intensity model.

On the second step, one final property z is computed by a linear clustering transformation:

$$z=a_1\cdot x_1+a_2\cdot x_2+,\ldots,+a_n\cdot x_n\ , \tag{2.1}$$

where the coefficient vector $\boldsymbol{a}=[a_1,\ldots,a_n]$ have to be computed in such a way that a separability measure between object and background pixels will be maximized in the new data space of z. One such reasonable choice is the well-known (in mathematical statistics) Fisher's linear discriminant based on sample mean value vectors within class and scatter vectors computed over object pixels and background pixels during the learning (estimation) of the property map transformation by Eq. (2.1).

In order to consider both the intensity and shape description of objects of interest in images simultaneously, a structural image model of the property map is used which is an intermediate image representation. This is a piecewise constant representation of the property map by the function $f(i,j)$, which is a certain linear function of components of the primary feature vector $\boldsymbol{x}=[x_1,\ldots,x_n]$ in point (i,j), i.e. $f(i,j)=z(i,j)$. It is also supposed that a zero-mean perturbation term $\lambda\cdot\nu(i,j)$ with a unit variance is present in the property map $f(i,j)$:

$$f(i,j)=h(i,j)*[\lambda\cdot\nu(i,j)+\sum_{l=0}^{1}\tau_l\cdot\varphi_l(i,j)]\,, \tag{2.2}$$

where $\{\tau_l,\ l=0,1\}$ are two constant intensity values of image plane segments corresponding to the background and objects of interest, respectively, $\varphi_1(i,j)$ is the binary map for objects, $h(i,j)$ is the smoothing kernel of a linear smoothing filter denoted by the convolution sign $*$, λ is the noise standard deviation. The function $\varphi_1(i,j)$ is equal to zero in the whole image plane Π except for the points belonging to objects of interest, whereas the function $\varphi_0(i,j)=1-\varphi_1(i,j)$ for $\forall\ (i,j)\in\Pi$.

The planar shape modeling is aimed at a concise shape representation of possible objects of interest whose property map satisfies the model by Eq. (2.1-2.2). It consists of a description of shape constraints for the representation of object binary map $\varphi_1(i,j)$ in Eq. (2.2). An efficient approach to describe the planar shape is the multi-scale morphological image modeling which defines objects of interest by using structuring elements and piecewise-linear skeletons [10]. In the underlying morphological model, one initial structuring element S_0 of minimal size as a set of points on the image grid is selected that determines the size and resolution of the objects. The structuring element at the scale m in a uniform scale system is formed as a consecutive binary dilation (denoted by ⊕) by S_0, $S_m = S_{m-1} \oplus S_0$, m=1,2,...,K-1, where K is the total number of scales. The generation of the planar shape of a simple object can be modeled in the continuous case by a growth process along *generating lines* [10].

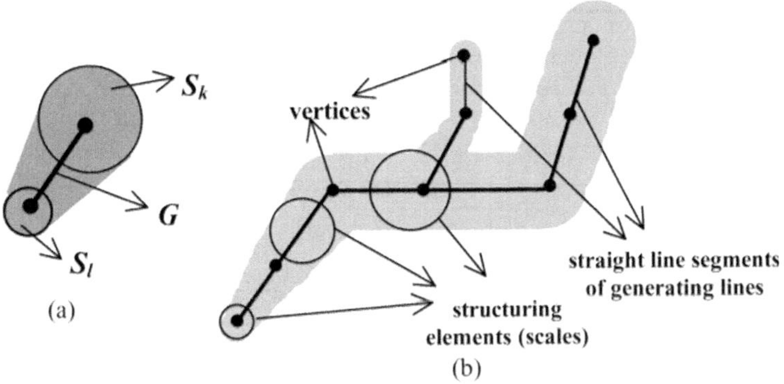

Fig. 1. Multi-scale formation of a simple object of interest (b) by the concatenation of blob-like objects (a)

A *local scale* value is assigned to each vertex point and generating lines are represented as concatenations of straight-line segments. A *blob-like* object defined by its two vertices is formed by two structuring elements S_k and S_l corresponding to the end vertices of a given straight-line segment G (see Fig. 1a) The domain region U of a blob-like object is formed by using the operation of dilation of a generating straight line segment (set) G with a variable in size structuring element (scale), $S(G)$:

$$U = G \oplus S(G) = \bigcup_{(i,j)\in G} S_{m(i,j)}(i,j)\text{, and } m(i,j) = \alpha_k(i,j)\cdot k + \alpha_l(i,j)\cdot l\text{,} \quad (2.3)$$

where $S_m(i,j)$ is the structuring element with a variable size m, k and l are the sizes of the structuring elements S_k and S_l, $\alpha_k(i,j)$ and $\alpha_l(i,j)$ are the two ratios of distances of the current point (i,j) to the end points of the segment G. A simple model is adopted for multi-scale object formation using the blob-like objects at different scales: an object of interest is formed from blob-like objects by a concatenation of their vertices, start and end points (see Fig. 1b). Finally, this morphological planar shape model is coupled with the model of image property map by Eq.(2.2) in such a way that the function $\varphi_1(i,j)$ in Eq. (2.2) satisfies the described morphological model.

3 Multi-scale Relevance Function of Images

3.1 Definition of the Relevance Function

Here, an improved model-based relevance function is presented as a modification of the relevance function approach that was initially described in [11]. First of all, it is considered as applied not to the initial image $g(i,j)$ but to the property map $f(i,j)$ represented by Eq. (2.1-2.2). The point on the image plane located on an object generating line, which corresponds to the maximal value of the likelihood function, allows optimal localization of the object of interest. Two basic local characteristics (constraints) of the image property $f(i,j)$ are involved in the definition of the relevance function: local object-to-background contrast, x, and homogeneity of object, y. Considering a single scale S_k , let the object sub-region $O(i,j)$ be a symmetric structuring element centered at point (i,j) and the sub-region $B(i,j)$ be a ring around it generated by the background structuring element, i.e. $O=S_k$ and $B=S_{k+1}\setminus S_k$ (see Fig. 2). The local contrast can be defined as the difference between mean value of object with a disk structuring element and background intensity within a ring around it:

$$x=\frac{1}{|O|}\sum_{(m,n)\in O(i,j)} f(m,n)-\frac{1}{|B|}\sum_{(m,n)\in B(i,j)} f(m,n).$$

The homogeneity of object y is measured by the difference between an object *intensity of reference, a,* and local (current) estimated intensity. The two constraints x and y take into account all object's potential scales in the definition of the multi-scale relevance function:.

$$\frac{1}{|O|}\sum_{(m,n)\in O(i,j)} f(m,n)\leftrightarrow\frac{1}{K}\sum_{k=0}^{K-1}\left(\frac{1}{|S_k|}\sum_{(m,n)\in S_k} f(m,n)\right),$$

where the object mean intensity is averaged over all K scales $\{S_k\subseteq O(i,j)\}$, $|S_k|$ denotes the number of points in S_k (see Fig. 2a). Similarly, the multi-scale estimation of the background intensity is made by averaging over K single-scale background regions (see Fig. 2b).

The object position, a *focus of attention* (i_f, j_f), is determined as the point in which the joint probability $P(x,y/\text{object})$ will be maximal provided the object point is being considered:

$$(i_f,j_f)=\arg\max_{(m,n)\in A}\{P(x(m,n)/object)P\{(y(m,n)/object)\}\tag{3.1}$$

where A is a region of interest, which might be the whole image plane. In the proposed definition of the relevance function, it was supposed that $P(y/\text{object})$ follows a Gaussian distribution $\mathbf{N}(0;\sigma_y^2)$ and $P(y/\text{object})$ is also approximated by a normal distribution law $\mathbf{N}(h;\sigma_x^2)$, where h is the mean value of the object local contrast. It can be easily proved that the maximization of joint probability by Eq. (3.1) in the conditions of the assumed model is reduced to the maximization of the image relevance function:

$$R(i,j)=\left(\frac{1}{|O|}\sum_{(m,n)\in O(i,j)} f(m,n)-\frac{1}{|B|}\sum_{(m,n)\in B(i,j)} f(m,n)\right)^2-\alpha\left(a-\frac{1}{|O|}\sum_{(m,n)\in O(i,j)} f(m,n)\right)^2\tag{3.2}$$

For the case of the assumed model without noise the relevance function takes the maximum at the start or end point of a blob-like object. Insignificant shift in the location might be introduced by the present noise depending on the noise variance.

The relevance function $R\{f(i,j)\}$ have to be computed within a region of interest A and takes its maximal value in the focus of attention (i_f, j_f).

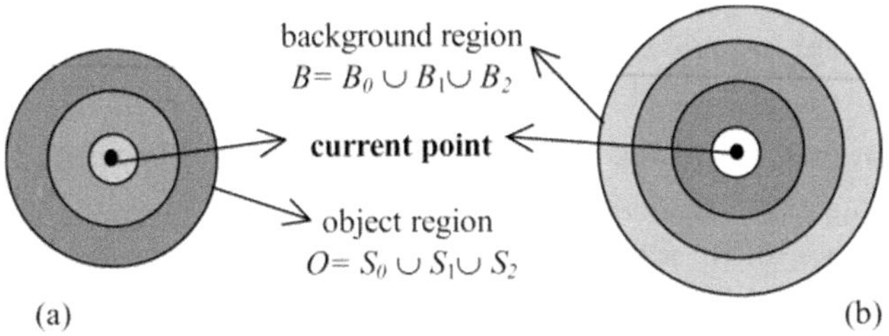

Fig. 2. An illustration to the definition of a three-scale relevance function. Kernel functions for the estimation of object intensity (a) and background intensity (b) are shown as gray levels

3.2 Robust Anisotropic Estimation of Object Intensity

The approach of relevance function is more suitable for large in size objects and low level of noise in the model of property map by Eq. (2.2). Often, thin and elongated low-contrast objects to be detected appear in real images. The simple estimation of average object intensity yields poor results since the object-to-background contrast x in Eq. (3.1) will be low. The remedy to such a situation is the anisotropic estimation of object intensity at certain expenses of the computational complexity. It is based on the morphological image model and the notion of so-called *object structuring regions*. The lth object structuring region V_l^k, l=1,..,L, at scale k is a sub-region of a dilation of a straight-line segment of generating lines with slope θ_l by the scale S_k. Object structuring region V_0^k at scale k coincides with the kth structuring element, i.e. it is a disk of radius r_k. Some examples of object structuring regions are given in Fig. 3. The concept of structuring regions and their derivation from the object morphological model was first introduced in the context of adaptive intensity estimation and image filtering [10]. The object intensity is estimated adaptively depending on the object orientation for the case of elongated object parts and edges.

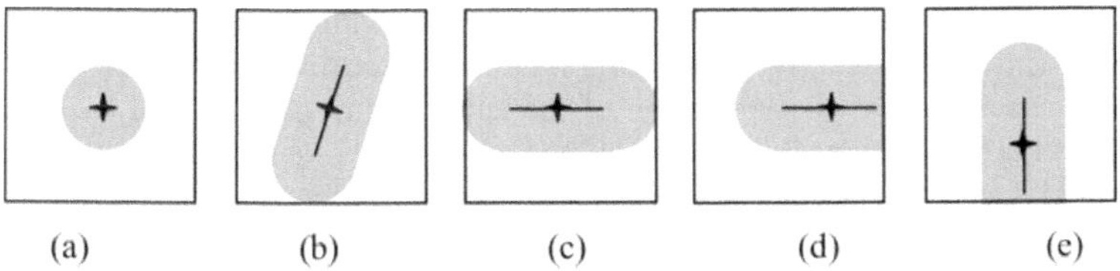

Fig. 3. Examples of object structuring regions (shaded areas) used in the robust estimation of object intensity

This approach can be successfully applied for the robust parameter estimation when computing the relevance function. First, average object intensities $\{q_l\}$ and local variances $\{s_l\}$ are computed inside all the object structuring regions. The average intensity value in region V_μ^k is selected as the result of intensity estimation, where μ

is the structuring region with minimal variances among all L regions. It is clear that such a decision coincides with the maximum likelihood estimation of intensity when assuming Gaussian distributions for point-wise deviations of intensities from the mean value inside respective structuring regions.

3.3 Estimation of Local Scales and Extraction of Planar Shapes

The location of an object of interest as the focus of attention point is followed by the determination of its potential scale and orientation in order to ensure a size-invariant recognition. On the other hand, such a preliminary estimation of scale simplifies the further image analysis provided the estimation is computationally simple. For example, the potential object scale is determined by the maximal value of absolute difference of intensities within a disk S_k and a ring around it, $R_k = S_{k+1} \setminus S_k$, for all $k=0,1,\ldots,K-1$.

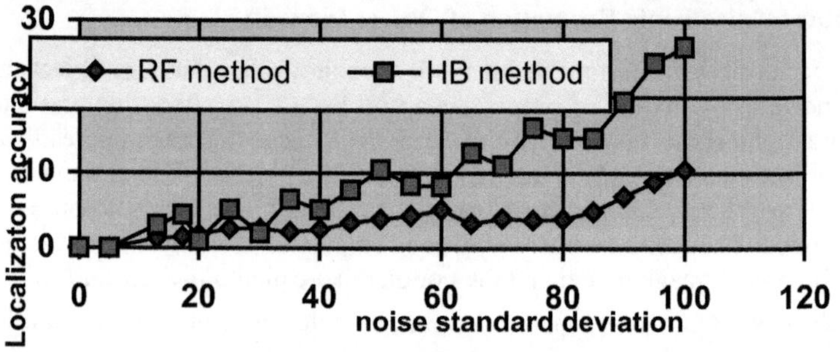

Fig. 4. Localization accuracy (in pixels) vs. noise deviation for two methods: relevance function (RF) and histogram-based binarization (HB)

The proposed model of visual attention mechanism can be successfully applied to time-efficient detection of objects of interest and its shape description by binarization and piecewise-linear skeletonization. In this framework, the object detection consists of several (many) consecutive stages of a multi-scale local image analysis while each of them is aimed at the determination of the next salient maximum of the relevance function [11]. A statistical hypothesis, the so-called *saliency hypothesis*, is first formulated and tested concerning whether an object is present or not with respect to the current local maximum of the relevance function. Statistically, the estimated value of actual object contrast x in Eq. (3.1) is tested on its significance. For this purpose, the result of scale estimation is used in order to estimate the contrast value in a better way. If the hypothesis testing result is positive then the current point is selected as a vertex of object skeleton. The image fragment in the neighborhood (region of attention) of the current attention point is binarized in order to have local binary shape of detected object of interest [11]. If using the property map as an input image, the threshold value is the mean value of object intensity and the background intensity.

(a) (b)

Fig. 5. Experiments with the illusion of Kanisza triangle: (a) - initial noisy image of an imaginary triangle; (b) - result of attention mechanism (maximum points of the relevance function) starting at large scales

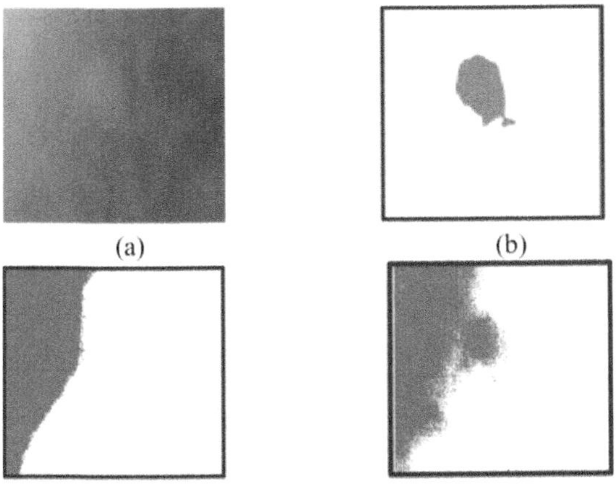

(a) (b) (c) (d)

Fig. 6. Results of lesion detection and segmentation in an X-ray image fragment of lungs with a significant slope of intensity: (b) - using the multi-scale relevance function; (c) and (d) using the histogram-based binarization [12]

4 Experimental Results

The relevance function approach to the detection of objects has been tested on synthetic and real images from industrial and medical diagnostic imaging. The main purpose of testing on synthetic images was the performance evaluation during the localization of low-contrast and noisy images. For example, the graph in Fig. 4 shows the experimental dependence of the location bias on the noise level for the noisy image of a bar-like object of known position. For comparison, the object center has been determined in the result of a wavelet transform [7] followed by a histogram-based binarization [12] with subsequent computation of the image centroid.

Several shape and intensity illusions can be modeled (i.e. explained) by the above described visual attention mechanism. Such known examples of illusions connected to the planar shape of objects are the Kanizsa figures (see Fig. 5) [3]. The application of the relevance function at larger scales yields the focus of attention at the centers of the illusionary triangle in Fig. 5a. The next three local maximums of the relevance function are located at the corners of the Kanisza figures (Fig. 5b). After the local binarization, the local fragments in the respective regions of attention are then identified as corners and the whole object as an illusionary triangle.

The proposed object detection method using relevance function has been tested on real images from diagnostic imaging where the visual attention model has its suitable application areas. The objects of interest are defect indications (quality inspection) or abnormalities of a human body (medical diagnostics), which are usually small in size, low-contrast and located on inhomogeneous backgrounds. One such example is related to lesion detection in radiographic images of lungs (see Fig. 6). Here, the property map has been obtained by a linear clustering transformation of such primary features as three polynomial coefficients of linear polynomial intensity. Such polynomial model is an adequate representation of image intensity when a significant slope value is presenting the background intensity. The result of lesion detection and binarisation is shown in Fig. 6b. The application of the method of histogram-based binarisation [12] gives poor results of shape extraction because of the significant slope in the background intensity even after making a correction to the threshold position on the histogram (see Fig. 6c and 6d for comparison).

5 Conclusions

A model for visual attention mechanisms has been proposed in the context of object detection and recognition problems. A multi-scale relevance function has been introduced for time-effective and geometry-invariant determination of object position. As compared to known visual attention operators based on the standard multi-resolution analysis and wavelet transform, this method has several distinctive features. Firstly, it is a model-based approach, which incorporates some structural features of sought objects in the design of the relevance function. Secondly, it provides a tracking capability for the case of large and elongated objects with complex shape due to the constraint of object homogeneity. The third advantage of this approach is the possibility to treat images with inhomogeneous backgrounds and textured appearance of objects because of working with the property map as an intermediate image representation. It exhibits a high localization accuracy at the same computation time as compared to the multi-resolution approach.

References

1. V Cantoni, S. Levialdi and V. Roberto, Eds., *Artificial Vision: Image Description, Recognition and Communication,* Academic Press, (1997).
2. L. Yarbus, *Eye movement and vision*, Plenum Press, N. Y., (1967).
3. M. D. Levine, *Vision in Man and Machine*, McGraw-Hill, (1985).

4. T. Lindeberg, "Detecting salient blob-like image structures and their scale with a scale-space primal sketch: a method for focus of attention", *Int. Journal of Computer Vision*, Vol. 11, (1993) 283-318.
5. L. Itti, C. Koch, and E. Niebur, "A model of saliency-based visual attention for rapid scene analysis", *IEEE Trans., Vol. PAMI-20,* No. 11, (1998) 1254-1259.
6. J. K. Tsosos *et al.*, "Modeling visual attention via selective tuning", *Artificial Intelligence*, Vol. 78, No. 1-2, (1995) 507-545.
7. J. L. Starck, F. Murtagh, and A. Bijaoui, *Image Processing and Data Analytsis: the Multiscale Approach*, Cambridge University Press, Cambridge, (1998).
8. D. Reisfeld *et. al.,* "Context-free attentional operators: the generalized symmetry transform", *Int. Journal of Computer Vision*, Vol. 14, (1995) 119-130.
9. H. D. Tagare, K. Toyama, and J.G. Wang, "A maximum-likelihood strategy for directing attention during visual search", *IEEE Trans., Vol. PAMI-2,* No. 5, (2001) 490-500.
10. R. M. Palenichka and P. Zinterhof, "A fast structure-adaptive evaluation of local features in images", *Pattern Recognition*, Vol. 29, No. 9, (1996) 1495-1505.
11. R. M. Palenichka and M. A. Volgin, "Extraction of local structural features in images by using multi-scale relevance function", *Proc. Int. Workshop MDML'99*, LNAI 1715, Springer, (1999) 87-102.
12. P. K. Sahoo *et al.*, "A survey of thresholding techniques", *Computer Vision, Graphics and Image Process.,* Vol. 41, (1988) 233 -260.

Disparity Using Feature Points in Multi Scale*

Ilkay Ulusoy[1], Edwin R. Hancock[2], and Ugur Halici[1]

[1] Computer Vision and Artificial Neural Networks Lab.
Middle East Technical University, Ankara, Turkey
{ilkay}@metu.edu.tr
http://vision1.eee.metu.edu.tr/~halici/

[2] Department of Computer Science, University of York
York, Y01 5DD, UK

Abstract. In this paper we describe a statistical framework for binocular disparity estimation. We use a bank of Gabor filters to compute multiscale phase signatures at detected feature points. Using a von Mises distribution, we calculate correspondence probabilities for the feature points in different images using the phase differences at different scales. The disparity map is computed using the set of maximum likelihood correspondences.

1 Introduction and Motivation

For many species with frontally located eyes including humans, binocular disparity provides a powerful and highly quantitative cue to depth. For primates, it has been shown that different neurons in a number of visual cortical areas signal distinct ranges of binocular disparities [1,2,3,4]. This observation has lead to the use of Gabor filters to model the phase differences for the receptive fields and to act as disparity decoders. However, although promissing this Gabor model of complex cell responses has a number of shortcomings. First, a phase selective complex cell model can not uniquely signal a given retinal disparity. Second, they can not signal disparities beyond the quarter cycle limit of the input. Qian [12,13,14] has improved the complex cell model so that it can uniquely signal definite disparities. Furthermore, the experimental data of Anzai et. al. suggest that there may be a possibility of positional differences in disparity encoding [1]. Complex Gabor filters have also been used for finding disparity from the region-based phase differences between the left and right images [15]. Potential problems with the use of phase as a disparity encoder have been identified by Jenkin and Jepson [6,7,8]. If the stereo images are subjected to affine image deformations such as scaling or shifting with respect to one-another, at certain locations phase may not be stable through scale. Since there is extensive physiological and psychophysical evidence which indicates the frequency selectivity of cortical receptive fields, many algorithms incorporate spatial filters of multiple scale or size to model the shift in peak spatial frequency. For instance, Pollard

* This study is partially supported by TUBITAK BDP and METU AFP 2000.07.04.01

T. Caelli et al. (Eds.): SSPR&SPR 2002, LNCS 2396, pp. 320–328, 2002.
© Springer-Verlag Berlin Heidelberg 2002

et. al. refine stereo correspondences by checking their behaviour through scale [11]. Sanger combines disparities at different scales using a weighting method [15]. Fleet simply sums the energy responses at different scales [5], Qian has a simple method which averages over different scales [12]. Marr et. al argue for a coarse to fine search procedure [10].

The observation underpinning this paper is that there is considerable scope for combining multiscale phase information to improve the estimation of disparity. Our approach is as follows: We commence from feature points detected using the method of Ludtke, Wilson and Hancock [9]. Next, a phase vector is calculated for each feature point. Correspondences are estimated using the similarity of orientation and phase at multiple scales. In this way we avoid the singular points encountered in the method of Jenkin and Jepson [6]. After calculating disparity from the positional difference between corresponding points, fine-tuning is performed using the phase difference information. This is done using a probabilistic model based on a von Mises distribution for the phase difference. The outline of the paper is follows. Extraction of features and their usage is explained in Section 2. In Section 3 we discuss the use of multiple scales for correspondence. The probabilistic phase difference model is explained in Section 4. In section 5 the results are discussed.

2 Extraction of Features Used in the Correspondence Algorithm

Gabor filters are well known models of simple cells:

$$g_{cos}(x,y,\omega_0,\theta) = \exp[-(\frac{x^2}{2\sigma_x^2}+\frac{y^2}{2\sigma_y^2})]\cos[2\pi\omega_0(xcos(\theta)+ysin(\theta))] \tag{1}$$

$$g_{sin}(x,y,\omega_0,\theta) = \exp[-(\frac{x^2}{2\sigma_x^2}+\frac{y^2}{2\sigma_y^2})]\sin[2\pi\omega_0(xcos(\theta)+ysin(\theta))] \tag{2}$$

where σ_x, σ_y express width of 2D Gaussian envelope along x and y direction, ω_0 is the spatial frequency and θ gives the orientation in space. Experiments show that adjacent simple cells have the same orientation and spatial frequency, but are in quadrature pairs (i.e. they differ in spatial phase by 90°) [4]. Thus a simple cell pair can be expressed by a complex Gabor filter:

$$g(x,y,\omega_0,\theta) = \exp\left\{-[\frac{x^2}{2\sigma_x^2}+\frac{y^2}{2\sigma_y^2}] + i2\pi\omega_0[x\cos\theta + y\sin\theta]\right\} \tag{3}$$

In this paper we use a bank of 8 complex Gabor filters of different orientation. From the output of the filter-bank, we compute a population vector [9]:

$$\mathbf{p}(x,y) = \begin{bmatrix} p_x(x,y) \\ p_y(x,y) \end{bmatrix} = \sum_{i=1}^{n} G(x,y,\omega_0,\theta_i)\mathbf{e}_i \tag{4}$$

where (x,y) is the position of the pixel in the image, n is the number of different orientation states, $G(x, y, \omega_0, \theta_i)$ is the response (energy) of a quadrature pair of Gabor filters with orientation θ_i and $\mathbf{e}_i = (\cos\theta_i, \sin\theta_i)^T$ is the unit vector in the direction θ_i. Here, the population vector is the vector sum of the n=8 filter response vectors and the resultant orientation is given by $\theta_{pop}(x, y) = \arctan[p_y(x, y)/p_x(x, y)]$. When compared to the tuning width of a single Gabor filter, the orientation estimate returned by the population vector is very accurate even though a relatively limited number of filters is used.

In our study, the feature points used for correspondence analysis are the locations where the length of population vector is locally maximum (see [9] for details). These points are located on object boundaries. In Figure 1a,b we show stereo images with numbered feature points on right image. Figure 1c,d shows the feature points from the images with the estimated orientation encoded as a grey-level.

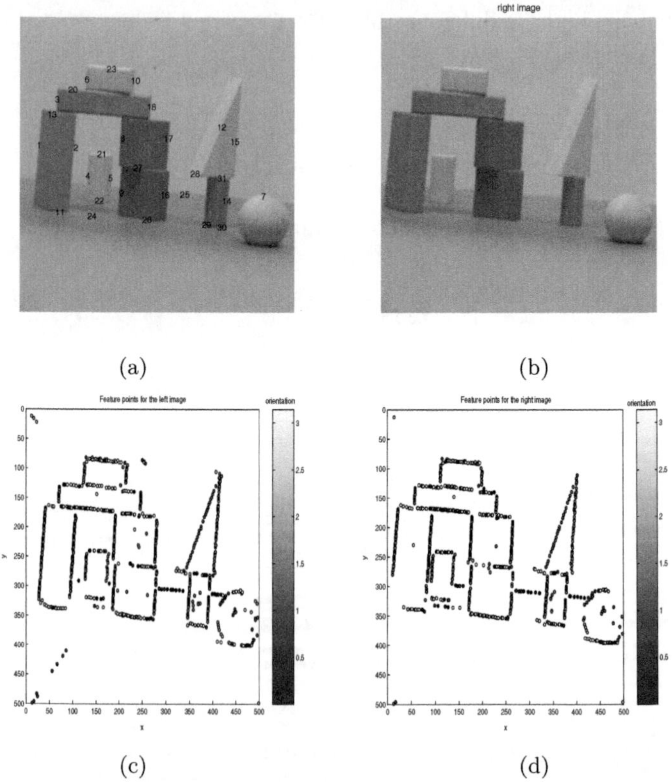

Fig. 1. (a) Right image of the stereo pair. (b)Left image of the stereo pair. (c) Feature points for right image. (d) Feature points for left image

3 Finding Corresponding Pairs and Disparity Using Multi-phase

The attributes used for the correspondence matching of feature points are orientation and phase. It is well known that phase based methods for disparity estimation are successful except in the neighbourhood of singularities [6]. In particular phase is stable with respect to geometric deformations and contrast variations between the left and right stereo views. In this paper, disparity is estimated from the region-based phase differences between the left and right images. Our estimate is obtained by first filtering the raw image data with a complex Gabor filter and computing the quantity

$$\phi_w(x, y) = \arctan\left[\frac{G_{sin}(x, y, w_0, \theta)}{G_{cos}(x, y, w_0, \theta)}\right] \tag{5}$$

where $G_{cos}(x, y, w_0, \theta)$ and $G_{sin}(x, y, w_0, \theta)$ are the cosine-phase and sine-phase filter responses of the image. We use the phase measurements for Gabor filters of different width, i.e. different scales, to locate correspondences. We use three filters each separated by one octave. The width of the narrowest filter is 6 pixels. For each feature point at the right image, we search over a window for feature points of similar orientation and phase in the left image. Let $\Phi_i = (\phi_1, \phi_2, \phi_3)^T$ be a vector of phase estimates obtained using the three filters. We measure the similarity of the phase-vectors by weighting the different components using the method described by Sanger [15]. Let C be the weighting matrix. The candidate j which has the closest weighted phase to the feature point i is the one that satisfies the condition $j = arg\left\{\min\{\Phi_i C^{-1} \Phi_j^T\}\right\}$. The disparity is the distance between corresponding feature points. In performing this, position shift between the receptive fields of binocular disparity selective cells are mimicked [4].

The matching algorithm explained above is cross checked for left-right correspondences and righ-left correspondences. In this way we may discard occluded feature points. For the stereo shown in Figure 1a,b we find correspondences for 537 of the 980 feature points in the right-hand image (Figure 1b). The final disparity values are displayed as gray scale values in Figure 2a and height plot in Figure 2b. Also in Figure 3 three main depth layers are shown separately. Out of the 537 matched feature points only 62 are in error, hence the succes rate is 90%. Most of the errors are for feature points having a population vector orientation in the disparity direction. In order to obtain subpixel accuracy, a phase shift model of binocular cell receptive fields can be used [4]. Here, the subpixel disparity is calculated from the interocular phase differences for between corresponding points using the quantity $\Delta d = \frac{\phi_{ij}\lambda}{2\pi}$, where Δd is the fine tuning in disparity, $\phi_{ij} = \phi_i - \phi_j$ is the measured phase difference, i and j are the left and the right feature point indenties respectively. In this way, the rough disparity estimate found by using only the position shift model is tuned by the phase shift model. As an example, the rough disparities on the edge-segment numbered 12 in Figure 1a shows a stair shaped structure (see Figure 2c top plot (*)). After fine tuning, the disparity varies more smoothly (see Figure 2c top plot (line)).

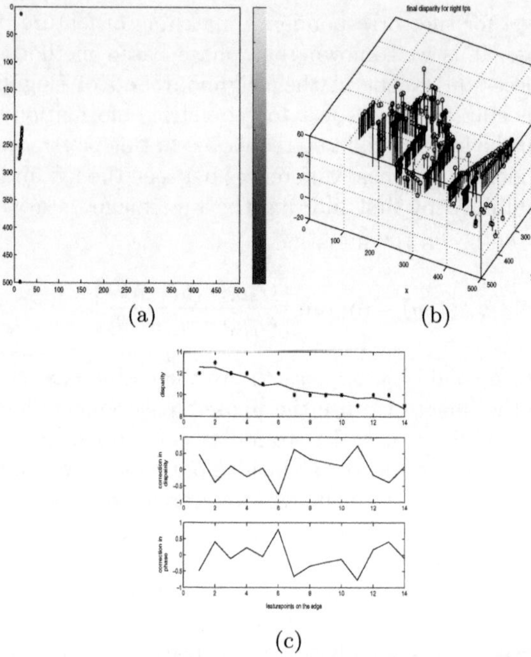

Fig. 2. (a),(b) Disparity. (c)Fine tuning result. Top: Coarse disparity (*), and fine disparity (line). Middle: Subpixel disparity. Bottom: Phase difference

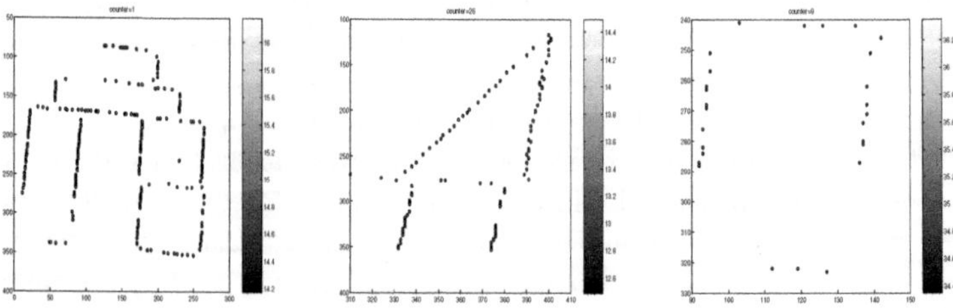

Fig. 3. Different depth layers

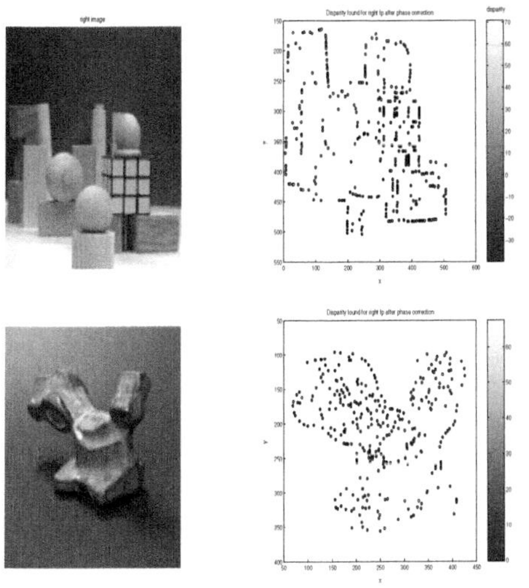

Fig. 4. Left: right images; Right: disparities

Disparity results for other image pairs are shown in Figure 4. Although the shapes in the images have very different characteristics, the results are still satisfactory.

4 Probabilistic Model of the Disparity Algorithm

After finding correspondences and computing the associated disparities, we refine the correspondences using a probabilistic model for the distribution of phase differences. This model is based on the assumption that the measured phase follows the von Mises distribution:

$$p(\phi_{ij}|\kappa, \mu) = \frac{1}{2\pi I_0(\kappa)} \exp\left[\kappa \cos(\phi_{ij} - \mu)\right] \tag{6}$$

where the distribution width or standard deviation is κ, the mean is μ and I_0 is the zero order Bessel function. For each scale, we fit a mixture of von Mises distributions to the measured phase differences. We use the EM algorithm to estimate the parameters of the mixture components κ_w and μ_w. At iteration $n+1$ of the algorithm the expected log likelihood functions for the estimation process is

$$Q = \sum_{i=1}^{N}\sum_{j=1}^{N}\sum_{w=1}^{W} P(w|\phi_{i,j}, \kappa_w^{(n)}, \mu_w^{(n)}) \ln p(\phi_{ij}|\kappa_w^{(n+1)}, \mu_w^{(n+1)}) P(w) \tag{7}$$

where N is the total number of phase difference measurements, and W is the total number of von Mises distributions in the mixture model. In the E or expectation

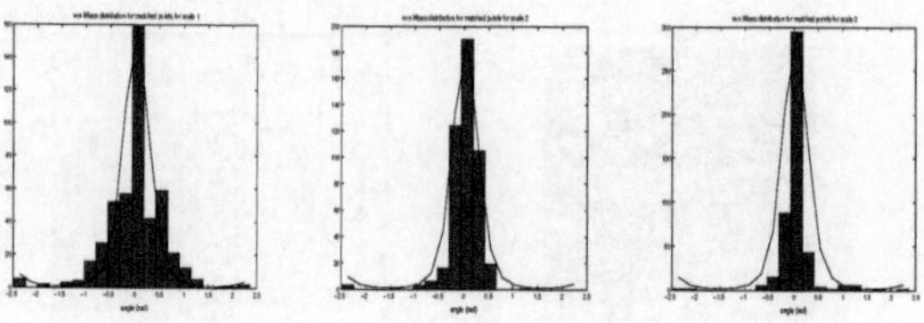

Fig. 5. Von Mises distributions fitted at three of the scales

step we compute the updated a posteriori probabilities

$$P(w|\phi_{i,j},\kappa_w^{(n)},\mu_w^{(n)}) = P_{i,j}^{(n)}(w) = 1/N\sum_{ij}^{N} p(\phi_{i,j}|\kappa_w^{(n)},\mu_w^{(n)}) \tag{8}$$

In the M-step, the distribution means are given by

$$\mu_w^{(n+1)} = 1/2\arctan\left[\frac{\sum_{i,j}^{N} P_{i,j}^{(n)}(w)\sin(2\phi_{ij})}{\sum_{ij}^{N} P_{i,j}^{(n)}(w)\cos(2\phi_{ij})}\right] \tag{9}$$

The distribution widths are more difficult to obtain, and involve computing the quantity

$$R = \frac{I_1(\kappa_w^{(n+1)})}{I_0(\kappa_w^{(n+1)})} = \frac{\sum_{ij}^{N} p(\kappa_w^{(n)},\mu_w^{(n)}|\phi_{ij})\cos(2(\phi_{ij}-\mu_w^{(n)}))}{p(\kappa_w^{(n)},\mu_w^{(n)}|\phi_{ij})} \tag{10}$$

For small values of R $\kappa_w^{(n+1)} \simeq (1/6)R(12+6R^2+5R^4)$ while when R is large $\kappa_w^{(n+1)} \simeq 1/(2(1-R)-(1-R^2)-(1-R^3))$. The result of fitting the von Mises mixture at different scales is shown in Figure 5.

With the parameters of the mixture model to hand, we can estimate correspondence probabilities from the phase differences. The correspondence probabilities are taken to be a posteriori probability of the mixture with the smallest mean μ_{min} at convergence of the EM algorithm. Suppose that $S_{i,j}^s$ is the a posteriori correspondence probability for scale s. The overall correspondence probability is the product of correspondence probabilities computed at the different scales, i.e. $q_{i,j} = \prod_{s=1}^{3} S_{i,j}^s$. The correspondences are taken so as to maximise $q_{i,j}$.

Applying the correspondences located in this way the computed disparities were very similar to those found using the method described in the previous section. The main differences are at horizontal edges as can be seen in the Figure 6.

5 Conclusion

We have presented a stereo correspondence method which is motivated by physiological and biological information. To do this we have modelled visual cortex

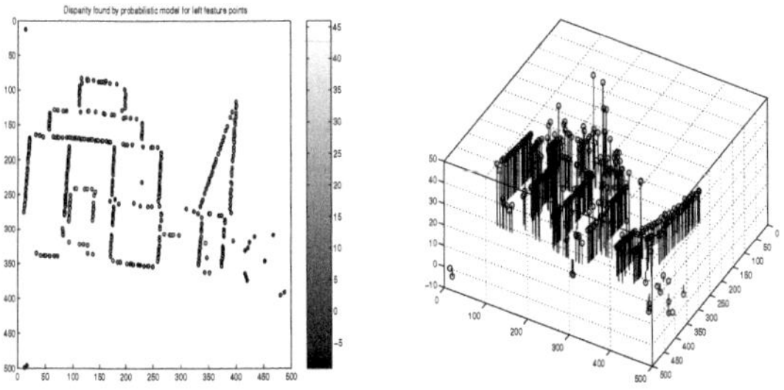

Fig. 6. Disparity found by probabilistic model

cell receptive fields using Gabor functions. Hypercolumns are encoded using population vectors. Thus, instead of calculating disparities using oriented Gabor filters and pooling the results over different orientations, a single orientation for each feature is obtained prior to disparity computation. The population vector estimate of stimulus orientation found using this method is very accurate given the small number of filters used. Although the feature points are sparse, since they are the points of high contrast edges that define the bounding contours of objects, they still prove to be informative. Correspondences between similarly oriented feature points are located using the phase information. This idea is also biologically grounded. The reason for this is that simple binocular cells occur in pairs that are in quadrature phase. Also, phase is sensitive to spatial differences, and hence it provides fine image detail which is helpful in discriminating neighbouring image regions. Phase is also robust to small scale differences. Unfortunately, there are image locations where phase is singular and can not be reliably used. In this study, by performing phase comparisons at multiple scales and by using confidence information we overcome these difficulties. We use the confidence weighting to augment phase information with information concerning the magnitude of the population vector to improve the correspondence method. Our use of multiple scales is also biologically plausable. The reason for this is that disparity encoding binocular cells are sensitive to different spatial wavelengths. We explore two routes to locating feature-point correspondences. Using the position shift model, rough disparity values are obtained and a large range of disparities can be calculated, but to a limited accuracy. Using the phase shift model, fine tuning is performed without encountering the quarter cycle limit. This tuning scheme also allows a continium of disparity estimates to be obtained. The algorithm proves to be effective for textureless images, especially at depth boundaries. The next step is to use the computed disparity values for surface reconstruction.

References

1. Anzai, A., Ohzawa, I., Freeman, R. D.: Neural mechanisms for encoding binocular disparity: Receptive field position vs. phase. Journal of Neurophysiology, vol. 82, no. 2, pp. 874-890, 1999. 320
2. Anzai, A., Ohzawa, I., Freeman, R. D.: Neural mechanisms for processing binocular information I. Simple cells. Journal of Neurophysiology, vol. 82, no. 2, pp. 891-908, 1999. 320
3. Anzai, A., Ohzawa, I., Freeman, R. D.: Neural mechanisms for processing binocular information II. Complex cells. Journal of Neurophysiology, vol. 82, no. 2, pp. 909-924, 1999. 320
4. DeAngelis, G.: Seeing in three dimension: the neurophysiology of stereopsis. Trends in Cognitive Science, vol. 4, no. 3, pp. 80-89, 2000. 320, 321, 323
5. Fleet, D. J., Wagner, H.,Heeger, D. J.: Neural encoding of binocular disparity: Energy models, position shifts and phase shifts. Vision Research, vol. 36, no. 12, pp. 1839-1857, 1996. 321
6. Jenkin, M. R. M., Jepson, A. D.: Recovering local surface structure through local phase difference measurements. CVGIP: Image Understanding, vol. 59, no. 1, pp. 72-93, 1994. 320, 321, 323
7. Jepson, A. D., Fleet, D. J.: Scale space singularities. Lecture Notes in Computer Science, vol. 427, pp. 50-55, 1990. 320
8. Jepson, A. D., Fleet, D. J.: Phase singularities in scale space. Image and Vision Computing, vol. 9, no. 5, pp. 338-343, 1991. 320
9. Ludtke, N., Wilson, R. C., Hancock, E. R.: Tangent fileds from population coding. Lecture Notes in Computer Science, vol. 1811, pp. 584-593, 2000. 321, 322
10. Marr, D., Poggio, T.: A computational theory of human stereo vision. Proceedings of the Royal Society of London, B207, pp. 187-217, 1979. 321
11. Pollard, S. B., Mayhew, J. E. W., Frisby, J. P.: PMF: A stereo correspondence algorithm using a disparity gradient limit. Perception, vol. 14, pp. 449-470, 1985. 321
12. Qian, N.: Computing stereo disparity and motion with known binocular cell properties. Neural Computation, vol. 6, no. 3, pp. 390-404, 1994. 320, 321
13. Qian, N., Zhu, Y.: Physiological computation of binocular disparity. Vision Research, vol. 37, no. 13, pp. 1811-1827, 1997. 320
14. Qian, N.: Relationship between phase and Energy methods for disparity computation. Neural Computation, 12, pp. 279-292, 2000. 320
15. Sanger, T. D.: Stereo disparity computation using Gabor filters. Biol. Cybern., 59, pp. 405-418, 1988. 320, 321, 323

Detecting Perceptually Important Regions in an Image Based on Human Visual Attention Characteristic

Kyungjoo Cheoi and Yillbyung Lee

Dept. of Computer Science, Yonsei University
134 Sinchon-Dong, Seodaemun-Gu, Seoul, 120-749, Korea
{kjcheoi,yblee}@csai.yonsei.ac.kr

Abstract. In this paper a new method of automatically detecting perceptually important regions in an image is described. The method uses bottom-up components of human visual attention, and includes the following three components : i) several feature maps known to influence human visual attention, which are computed in parallel directly from the original input image, ii) importance maps, each of which has the measure of "perceptual importance" of local regions of pixels in each corresponding feature map, and are computed based on lateral inhibition scheme, iii) single saliency map, integrated across multiple importance maps based on a simple iterative non-linear mechanism which uses statistical information and local competence of pixels in importance maps. The performance of the system was evaluated over some synthetic and complex real images. Experimental results indicate that our method correlates well with human perception of visually important regions.

1 Introduction

We can say that the main problem in computer vision lies in its limited ability which is caused by enlarging the size of a given image, and the computational complexity followed by it. Actually, computer vision system receives vast amount of visual information, and real-time image capturing at any useful image resolution yields prodigious quantities of visual information. Therefore, analyzing all of inputted visual information for high-level process, such as object recognition, is actually impossible, and is also unnecessary in aspects of using limited computational resources efficiently. Therefore, the mechanism of selecting and analyzing only the information "most" relevant to the current visual task is needed to computer vision system.

It is known that human visual system does not handle all visual information received by the eye but selects and processes only the information essential to the task at hand while ignoring a vast flow of irrelevant details [7]. Many existent experimental evidences about primate report that there are a lot of mechanisms related to the function of "visual selection", and visual attention belongs to this mechanism.

T. Caelli et al. (Eds.): SSPR&SPR 2002, LNCS 2396, pp. 329-338, 2002.
© Springer-Verlag Berlin Heidelberg 2002

Visual attention is one of the primate's most important intellectual ability that maximizes visual information processing capability by rapidly finding the portion of an image with which the information is most relevant to the current goals(See Colby's work [4] for a neurophysiological review). From these, one usable method of reducing prodigious quantities of visual information of input image is deploying the function of human visual attention within the system. That is, extract the regions of interest from the image which usually constitute a considerably lesser proportion of the whole image, and discard the rest, non-interest regions [12].

This paper describes a new method of automatically detecting salient regions in an image based on the bottom-up human visual attention characteristic. The proposed method can be explained by following three stages (See Fig. 1). First, the input image is represented in several independent feature maps, two chromatic feature maps and one achromatic feature map. Second, all feature maps are converted into corresponding number of importance map by lateral inhibition scheme. The importance map has the measure of "perceptual importance" of local regions of pixels in feature map. Third, all importance maps are combined into a single representation, saliency map. Iterative non-linear mechanism using statistical information and local competence of pixels in importance map is processed on all of importance maps and the output is just simply summed. The saliency map represents the saliency of the pixels at every location in an image by a scalar quantity in relation to its perceptual importance, and guides the selection of attended regions [7].

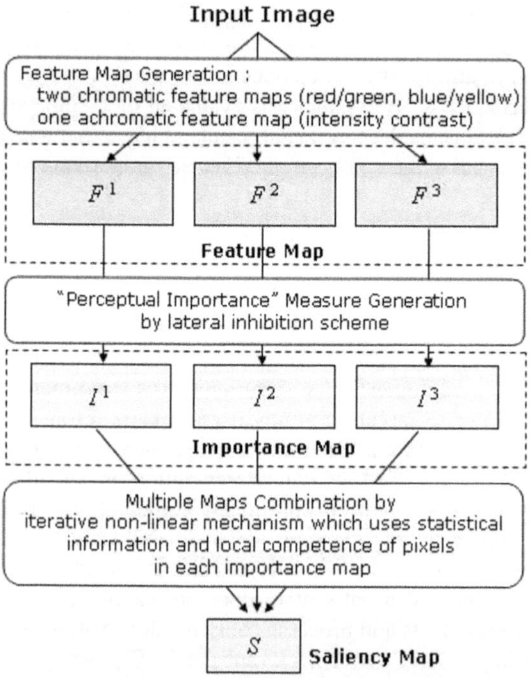

Fig. 1. Overall Architecture of the System Proposed

The organization of the paper is as follows. Related works are given in Section 2. In Section 3, the proposed method for detecting salient regions is explained. Experimental results are shown in Section 4, and concluding remarks are made in Section 5.

2 Related Works

Researches related with visual attention have been studied in two primary approaches according to the ways of directing attention, that is, the *bottom-up*(or *data-driven*) approach and the *top-down*(or *model-driven*) approach.

In bottom-up approach, the system selects regions of interest by bottom-up cues obtained by extracting various elementary features of visual stimuli [3,6,7,13]. And in top-down approach, the system uses top-down cues obtained from a-priori knowledge about current visual task [8]. A hybrid approach combining both bottom-up and top-down cues has also been reported [2,5,9,10].

As bottom-up models do not use any kind of priori knowledge about the given task, they can be employed on a variety of applications without major changes in the architecture. Also the most of what is known about human visual attention is related to the bottom-up cues. Meanwhile, almost all previous top-down systems neglect bottom-up cue, so they are very useful to match specific patterns whose high-level information was presented to the system. In such cases, the system needs training process and also needs partial interaction with the recognition system. Therefore, it is very difficult to extend the top-down system to other applications. Because of these reasons, relatively few studies have been made to provide quantitative top-down systems although the importance of top-down cues of attention has long been emerged.

Treisman's "Feature Integration Theory" [13] which proposed to explain strategies of human visual search has been very influential theory of attention. The first biological plausible computational model for controlling visual attention was proposed by Koch and Ullman [7]. Many successful computational models for bottom-up control of visual attention have the common stages of computing several feature maps and single saliency map. The differences between those models are the differences of the strategies used to create feature maps and the saliency map. Among existing computational models, our system is built at the basis of [6] and [9]. Itti *et al.* proposed the purely bottom-up attention model that consists of saliency map and winner-take-all network [6], and Milanese's model [9] extracts regions of interest by integrating the bottom-up and the top-down cues by a non-linear relaxation technique using energy minimization-like procedure.

At least two main remarks can be made about most of the systems reviewed in this section. The first remark is that, the most of existing systems are in the progress of establishing the concept of visual attention, and they put too much emphasis on the theoretical aspects of human visual attention, not the real aspects. The second remark is that, in many cases, the performance of most of the systems have been evaluated over just synthetic or simple simulated images, so they yield a rare example of a system that can be applied to natural color images.

From these two remarks, we can conclude that the existent systems are not general-purpose enough or widely applied to real actual problem of visual world yet. Our method proposed here is designed to extend the capabilities of previous systems. In doing so it proved that our system was suitable for applications to real color images including noisy images.

3 The System

Our system detects regions of interest by properties of the input image without any a-priori knowledge. As shown in Fig. 1, our system has three main components, the feature map, the importance map, and the saliency map. In this section, these three components are described in detail.

3.1 The Feature Maps

Two kinds of topographic feature maps known to influence human visual attention are generated from the input image : two chromatic feature maps for color contrast and one achromatic feature map for intensity contrast.

The chromatic information is one of the biggest properties of human vision that discriminates an object from others, and psychophysical results also show that it is available for pre-attentive selection. In human vision, the spectral information is represented by the collective responses of the three main types of cones(R, G, B) in retina. These responses are projected to the ganglion cells, and then to the LGN, and to the visual cortex. In this way, we can get both chromatic and achromatic information about the input objects. In V1, there exist three types of cells with center-surround receptive fields, homogeneous receptive fields, and more complex receptive fields which combine above mentioned two types. Among them, the cells with homogeneous receptive fields respond the highest when both the center and the surround receives the same stimuli of a specific wavelength, and this means that they are not spatially selective but responds very strongly to color contrast. From these, two chromatic feature maps which simulate the effect of two types of color opponency exhibited by the cells with homogeneous receptive fields are generated. The process of generating two kinds of chromatic feature maps is as follows.

First, red, green and blue components of the original input image are extracted as R, G, and B, and four broadly tuned color channels are created by

$$r=R-(G+B)/2,\quad g=G-(R+B)/2,\quad b=B-(R+G)/2,\quad y=R+G-2(|R-G|+2) \tag{1}$$

where r, g, b, and y denote red, green, blue, and yellow channels respectively. Each channel yields maximal response for pure, fully saturated hue to which it is tuned, and yields zero response both for black and for white inputs.

Second, based on above color channels, two chromatic feature maps are created by

$$F^1 = r - g,\quad F^2 = b - y \tag{2}$$

F^1 is generated to account for red/green color opponency, and F^2 for blue/yellow color opponency.

If no chromatic information is available, the *gray-level*(or *intensity*) image can be used as an achromatic feature map. Gray-level information can be obtained from the chromatic information of the original color input image as $I = (R+G+B) / 3$, and is used as an achromatic feature map F^3.

$$F^3 = I \tag{3}$$

These generated multiple independent feature maps are then normalized in the range of 0~1 in order to eliminate across-modality differences due to dissimilar feature extraction mechanisms, and to simplify further processing of the feature maps.

3.2 The Importance Maps

Since each of the computed feature maps has the special meaning at every locations of input image, we have to assign measure of importance to each of the feature maps in order to detect salient regions based on this. We used center-surround operator, based on the *DOOrG*(Difference-Of-Oriented-Gaussians) model [9] to generate corresponding number of importance maps. This operator is also based on lateral inhibition scheme which compares local values of the feature maps to their surround and enhances those values strongly different from their surroundings' while inhibiting the others. Aguilar and Ross have suggested that the regions of interest are those regions which differ the most [1]. With this operator, the system also can have the effect of reducing noises. The processing of generating importance maps for each available feature map is as follows.

First, construct filter bank h at 8 orientations (Fig. 2) by

$$h_{x',y'}(\theta) = \left| DOOrG_{x',y'}(\sigma, r_{x'/y'}, r_{on/off}) \right| \tag{4}$$

where $DOOrG_{x',y'}(\cdot,\cdot,\cdot)$ denotes 2-D *DOOrG* function. The *DOOrG* model is defined by the difference of two Gaussians of different sizes with the width of positive Gaussian being smaller than the width of the negative one. The two Gaussians may have an elliptic shape characterized by different width of the two Gaussians while the *DoG*(Difference-of-Gaussian) model has isotropic shape of Gaussians. See [9] for more details. If we change a coordinate, it is possible to extend the canonical *DOOrG* model to vary the orientation of the filter. In our system, θ is fixed as $\theta \in \{0, \pi/8, 2\pi/8, \cdots, 7\pi/8\}$ and the values of other parameters are as follows: $\sigma=5.5$, $r_{x'/y'}=1/9$, $r_{on/off}=4.76$, $K_1=1/6$, $K_2=17/60$.

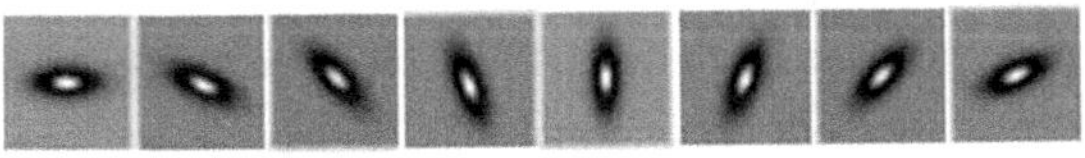

Fig. 2. Generated filter bank $h(\theta)$

Second, for each importance map, convolution is processed over the map with eight $h(\theta)$ filters, and then the results are squared to enhance the contrast. Finally, all computed maps are summed to factor out θ.

Since the importance map computed in this section used filter bank based on the *DOOrG* model at 8 orientations, the system can have the ability of detecting orientation.

3.3 The Saliency Map

In general, the system extracts perceptually important regions based on importance measures provided by importance maps. The difficulty of using these measures resides in the fact that importance maps are derived from different types of feature maps. Since importance map provides different measures of importance for same image, each map may guide different regions as a salient region. To settle down this problem, different measures must be integrated in order to obtain a single global measure of importance for each location of image. In this case, computed global measure could guide the detection of the final salient regions. But, combining information across multiple importance maps is not an easy work. In principle, this could be done by taking high activity pixels over all information [2,11] or by weighted sum of all information [3,14]. However, in the former case, there is no reason why a high intensity region should be more interesting than a low intensity one. And in the latter case, the results highly depend on the appropriate choice of the weights. Also, both do not consider the fact that each importance map represents different measures of importance about the "same" input image.

Here, we propose a simple iterative non-linear combination mechanism which uses statistical information and local competence of pixels in importance maps. Our method promotes those maps in which small number of meaningful high activity areas present while suppressing others. The saliency map has been generated through following three steps.

At the first step, important maps I^k (k=1,2,3), computed in section 3.2, are inputted to the system. Each importance map is convolved with the large size of the *LoG* filter and the result is added with the original input one. Iterate this procedure several times, and corresponding number of IT^k maps are generated as a result. This procedure causes the effect of short-range cooperation and long-range competition among neighboring values of the map. And we can also have the advantage of reducing noises of an image. The *LoG* function we used is given by

$$LoG(x,y) = \frac{1}{\pi\sigma^4}\left[1 - \frac{x^2+y^2}{2\sigma^2}\right] \cdot e^{-\left(\frac{x^2+y^2}{2\sigma^2}\right)} \tag{6}$$

where σ denotes the scale of the Gaussian. We set σ as 3.6.

At the second step, each IT^k map is evaluated iteratively by statistical information of the map to enhance the values associated with strong peak activities in the map while suppressing uniform peak activities. For each IT^k map, update the map by

$$\mathrm{IT}^k = \mathrm{IT}^k \times (\mathrm{GMax}^k - \mathrm{Ave}^k)^2 \tag{7}$$

where $GMax^k$ denotes the global maximum value of the map and Ave^k denotes the average value of the map. After this, normalization is processed on each of computed IT^k maps by

$$IT^k = \frac{IT^k - IT_{\min}}{IT_{\max} - IT_{\min}} \tag{8}$$

where $IT_{\min}$ and $IT_{\max}$ denote the global minimum and the maximum value out of all IT^k maps. Through this, relative importance of an IT^k map with respect to other ones would be remained, and irrelevant information extracted from ineffective IT^k maps would be suppressed. Iterate the procedure of this step several times. Here, we iterated 4 times.

At the third step, computed IT^k maps are summed and normalized to a fixed range of 0~1 to generate saliency map S.

4 Experimental Results and Analysis

To evaluate the performance of our system, we used three kinds of experimental images. In this section, we will describe what kinds of images are we used, and the experimental results, in detail. As explained already, our system was developed in order to solve several problems caused by enlarging the size of input image in computer vision system, through selecting regions of interest which humans think to be perceptually important. Therefore, it is useless to use images which contain only the target. So, we used images of not only the target, but images including complex background or other objects photographed at a great distance. By the way, many previous researchers have been concentrated more on the evaluation of their system's performance on simple synthetic images, not real images. However, this is not proper from the fact that computer vision system actually operates in the real world. Therefore, we cannot neglect the system's performance on complex real images. So, we used various images from simple synthetic images to complex real images. Besides, many images of real visual world may have lots of noises caused by the properties of images themselves, or added through the image acquiring processes. For these reasons, we included our testing with noisy images. With the images selected by above mentioned three criteria, we tested our system. And through experimental results, we've found that our system detects the interest part of an image that a human is likely to attend to, and it has following three properties. First, the system was able to reproduce human performance for a number of pop-out tasks [13], using images of the type shown in Fig. 3(a). A target defined by a simple and unique feature such as color, orientation, size, contrast, etc. distinguishing it without any ambiguity, or isolated, is easily detected at almost constant time independent from the number of the other stimuli. To evaluate the system's performance of this paradigm, we used various images that differed in orientations by 30°, 45°, 90°, in colors by red, blue, green, white, black, yellow, in sizes, and in intensity contrasts. Also, the system was tested with the images of which the background has lighter contrast than those of the target, and vice versa. The system detected the target properly, and some results of these tasks were shown in Fig. 3(a), Fig 4(a). Second, the system could be successfully applied to complex real images. The system was tested with complex real color and gray-level images such as signal lamp image of the type shown in Fig. 3(b) and various images of traffic sign, food, animal, and natural scenes. See Fig 4(b) for example. One major difficulty of deciding whether the result is good or not is that each person may choose a different region as the most salient region. However, if we follow the assumption that the most salient region to which attention goes is an object of interest, the results for complex real images are successful. Third, the system was very strong to noises. See Fig. 4(a) for example.

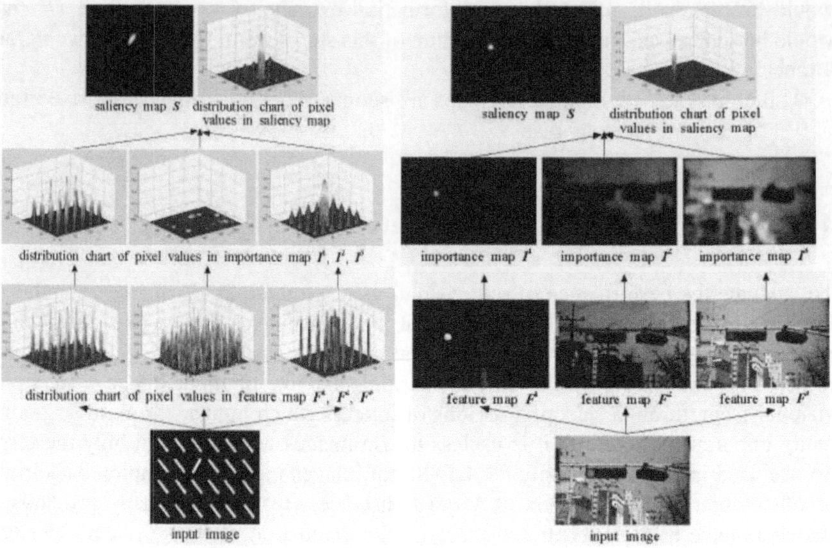

(a) (b)

Fig. 3. Examples of experimental results for synthetic and real images : (a) orientation pop-out task. Orientation is detected in importance map, and this feature wins among other features through the procedure of saliency map generation (b) detects red signal lamp

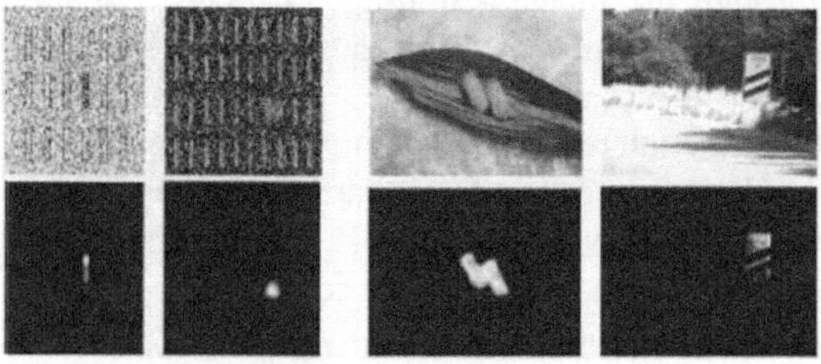

(a) (b)

Fig. 4. Some more examples. (a) Noisy Image : (left) color pop-out task(middle:blue, the very left upside and right downside:light green, remainder:yellow):detects blue bar, (right) size pop-out task:detects circle shaped object (b) Non-Noisy Image : (left) detects red sliced raw fish, (right) detects yellow traffic sign

5 Concluding Remarks

In this paper, we proposed a new method of detecting salient regions in an image in order to solve several problems caused by enlarging the size of input image in computer vision system. The proposed method uses only bottom-up components of human visual attention. As shown in experimental results, the performance of the system is very strong to not only synthetic images but also complex real images, although the system employed very simple mechanisms in feature extraction and combination. Also our system can be extended to other vision applications such as arbitrary target detection tasks through just simply modifying feature maps. However, our method needs more experiments and analysis with more complex real and noisy images in order to confirm whether our system can be applicable to other various actual problems. And we are currently doing this kind of job with more complex real and noisy images. In addition, as human visual attention actually depends on both bottom-up and top-down controls, researches to integrate the proposed method with top-down cue still has to be carried out.

References

1. Aguilar, M., Ross, W.:Incremental art:A neural network system for recognition by incremental feature extraction. Proc. of WCNN-93 (1993)
2. Cave, K., Wolfe, J.: Modeling the Role of Parallel Processing in Visual Search. Cognitive Psychology 22 (1990) 225-271
3. Chapman, D.: Vision, Instruction, and Action. Ph.D. Thesis, AI Laboratory, Massachusetts Institute of Technology (1990)
4. Colby:The neuroanatomy and neurophysiology of attention. Journal of Child Neurology 6 (1991) 90-118
5. Exel, S., Pessoa, L.:Attentive visual recognition. Proc. of Intl. Conf. on Pattern Recognition 1 (1998) 690-692
6. Itti, L., Koch, C., Niebur, E.: Model of saliency-based visual attention for rapid scene analysis. IEEE Transactions on Pattern Analysis and Machine Intelligence (1998) 1254-1259
7. Koch, C., Ullman, S.: Shifts in Selective Visual Attention : Towards the Underlying Neural Circuitry. Human Neurobiology 4 (1985) 219-227
8. Laar, P., Heskes, T., Gielen, S.:Task-Dependent Learning of Attention. Neural Networks 10, 6 (1997) 981-992
9. Milanese, R., Wechsler, H., Gil, S., Bost, J.,Pun, T.: Integration of Bottom-up and Top-down Cues for Visual Attention Using Non-Linear Relaxation. Proc. of IEEE Conf. on Computer Vision and Pattern Recognition (1994) 781-785
10. Olivier, S., Yasuo, K., Gordon, C.:Development of a Biologically Inspired Real-Time Visual Attention System. In:Lee, S.-W.,Buelthoff, H.-H., Poggio, T.(eds.):BMCV 2000.Lecture Notes in Computer Science, Vol. 1811. Springer-Verlag, Berlin Heidelberg New York (2000) 150–159
11. Olshausen, B., Essen, D., Anderson, C.: A neurobiological model of visual attention and Invariant pattern recognition based on dynamic routing of information. NeuroScience 13 (1993) 4700-4719
12. Stewart, B., Reading, I., Thomson, M., Wan, C., Binnie, T.: Directing attention for traffic scene analysis. Proc. of Intl. Conf. on Image Processing and Its Applications (1995) 801-805

13. Treisman, A.-M., Gelade, G.-A.: A Feature-integration Theory of Attention. Cognitive Psychology 12 (1980) 97-136
14. Yagi, T., Asano, N., Makita, S., Uchikawa, Y.:Active vision inspired by mammalian fixation mechanism. Intelligent Robots and Systems (1995) 39-47

Development of Spoken Language User Interfaces: A Tool Kit Approach

Hassan Alam, Ahmad Fuad Rezaur Rahman, Timotius Tjahjadi, Hua Cheng, Paul Llido, Aman Kumar, Rachmat Hartono, Yulia Tarnikova, and Che Wilcox

Human Computer Interaction Group, BCL Technologies Inc.
990 Linden Drive, Suite #203, Santa Clara, CA 95050,USA
fuad@bcltechnologies.com

Abstract. This paper introduces a toolkit that allows programmers with no linguistic knowledge to rapidly develop a Spoken Language User Interface (SLUI) for various applications. The applications may vary from web-based e-commerce to the control of domestic appliances. Using the SLUI Toolkit, a programmer is able to create a system that incorporates Natural Language Processing (NLP), complex syntactic parsing, and semantic understanding. The system has been tested using ten human evaluators in a specific domain of a web based e-commerce application. The evaluators have overwhelmingly endorsed the ease of use and applicability of the tool kit in rapid development of speech and natural language processing interfaces for this domain.

1 Introduction

Automatic Speech Recognition (ASR) technology is making significant advancements, and voice recognition software is becoming more and more sophisticated. However, because computers are unable to understand the meaning of the words they identify, the practical use of this technology is severely restricted. This paper describes a Spoken Language User Interface (SLUI) Toolkit that allows programmers to rapidly develop spoken language input for computer applications. The Toolkit will allow programmers to generate SLUI front-ends for new and existing applications by a program-through-example method. The programmer will specify a set of sample input sentences for each task, and the SLUI Toolkit will generate a large set of semantically equivalent sentences. The programmer will then select the sentences needed for the task and the SLUI Toolkit will generate code that will take a users spoken request and execute a command on an application. The work reported here is based on a contract awarded by the Advanced Technology Program (ATP) of National Institute of Standards and Technology (NIST) in the USA [1].

T. Caelli et al. (Eds.): SSPR&SPR 2002, LNCS 2396, pp. 339-347, 2002.
© Springer-Verlag Berlin Heidelberg 2002

2 Overview

As hand-held computers merge with cellular telephones with limited keyboard and pointing capabilities, a user interface that allows spoken language will become the input of choice [2,3]. Current spoken language interface systems mimic menu driven Graphical User Interfaces (GUI). This does not exploit the full power of *naturally spoken* language that allows users to express commands at a higher level of abstraction than is possible with current GUI or command-line interfaces. However, to develop a spoken language interface such as this, the programmer needs to learn a number of different technologies. These include: *Automatic Speech Recognition (ASR),* for transcribing spoken language, *Syntactic Parser,* for transforming the transcribed text into a Lexical Conceptual Form, *Semantic Analyzer,* to understand the meaning of the sentence and *Dialog Manager,* to manage the interaction between the user and the computer. Most computer programmers do not have the knowledge to develop these components. While Commercial Off The Shelf (COTS) ASR systems are available, a programmer needs to understand linguistics and human discourse theory to write an effective SLUI system that incorporates complex syntactic parsing, semantic understanding and dialog. This situation is similar to the pre X-Windows GUI where programmers had to develop custom GUI after learning graphics programming. This clearly hampered adoption of the technology. If anything, linguistic theory is more complex than learning graphical routines. The proposed SLUI Tool Kit will solve this problem.

The underlying technology for Natural Language Processing is being developed for over 30 years. One of the earliest command-and-control like integrated systems developed by Woods [4] used a QLF like formalism. Some of the more recent systems using robust semantic interpretation include the Core Language Engine developed at SRI Cambridge aimed as an interactive advisor system, the Rosetta automatic translation system, the SQUIRREL portable natural language front-end to databases, the Tacitus system developed at SRI International by Jerry Hobbs, et al, the TRAINS system at the University of Rochester for planning a railroad freight system and the Verbmobil system for translation in face-to-face dialogs. At NIST, a spoken language interface to libraries was developed using COTS ASR, and parsers to retrieve from a database.

The computational linguistics field of parsing is mature in terms of available implementations. The main challenges being faced now are robust parsing and increasing coverage of the different parsers. A lot of work is also being done in automatic learning of parser grammar rules based on corpora. Recently shallow parsing methods have received more attention because of their success in text-extraction. But application of shallow-parsing methods to command-and-control-like sentences have not been a focus. There is a body of work on automatically acquiring semantic concepts knowledge through statistical techniques. Knowledge Representation is an old field in Artificial Intelligence. A lot of the current research is focused on building reusable ontology. Also lexical semantics deals with better representations of the lexical knowledge itself.

Most of the current research on generation focuses on how to make efficient, large-scaled feature based generation systems for dialogs. FUF/SURGE and PENMAN/KPML are two of the most widely used systems.

The Agents field has received widespread interest from Computational Linguists in recent years. Natural Language systems are being built as Agent interface technologies. A good overview of the field can be found from the proceedings of the annual conference on autonomous agents. Although agents architecture is receiving attention, most of the current commercially available production systems are implemented as client-server solutions.

A Spoken Language User Interface (SLUI) is the newest method for controlling a computer program. Using simple Natural Language (NL) input, a user can interact with a program through a SLUI. The SLUI Toolkit is a suite of applications and tools that allow programmers to develop SLUI enabled programs that can process NL input. Using the Toolkit, an NL interface can be incorporated into a new or existing program.

3 Methodology

This Section briefly describes the methodology of the development of SLUI.

- Query Inputter. A GUI with relevant functionality is implemented to easily specify test queries, and to map related application parameters. This is based on [5].
- Syntax Recognizer. A lot of the queries can come with specific types of embedded phrases. Examples of these may include embedded dates, emails, documents, currency and many more. Within specific domains, this list can be a lot longer and richer. We harvest these patterns using regular expression (RE) matching and replace them with standard, pre-selected parameters.
- Spell Checker. At this stage, a Commercial-Off-the-Shelf (COTS) spell checker was used [6].
- Sentence Tokenizer. A sentence tokenizer is implemented to tag various parts of speech (POS).
- Parser. We adopted the MiniPar [7] parser at this stage. We also developed a shallow parser to work as a back-up parser. We implemented a framework where these two parsers can be optimally combined to provide maximum robustness.
- Anaphora Resolver. We implement an anaphora resolver at this stage [8].
- Translator. We implemented a translator module to translate the parsed tree to a semantic tree using an internal data structure. This translator transforms the parse tree into a semantic representation that captures its head, its arguments and all its modifiers and complements.
- Frame Generator. We implemented a frame generator to generate a semantic internal representation of the translated tree.
- Frame Handler. We implemented a module to validate the generated frame.

4 Functionality of SLUI Tool Kit

The SLUI Toolkit assists programmers in creating SLUI enabled programs that recognize NL input. Essentially, The SLUI Toolkit creates a channel of communication between the user and the program through a SLUI. The Toolkit handles syntax and semantic processing with minimal direction, and removes the need for an in depth understanding of linguistics and human discourse theory (although basic understanding of English grammar is helpful) [9]. Using the SLUI Toolkit, a programmer is able to create a system that incorporates Natural Language Processing (NLP), complex syntactic parsing, and semantic understanding.

The SLUI Tool Kit works in the following steps:

- Toolkit begins to create a SLUI by using NLP to create semantic representations of sample input sentences provided by the programmer.
- These representations are expanded using synonym sets and other linguistic devices, and stored in a Semantic Frame Table (SFT). The SFT becomes a comprehensive database of all the possible commands a user could request a system to do.
- The Toolkit then creates methods for attaching the SLUI to the program.
- When the SLUI enabled program is released, a user may enter a NL sentence. The sentence is translated into a semantic frame, and the SFT is searched for an equivalent frame. If a match is found, the program executes the action linked to this frame.

Fig. 1. shows a schematic of the SLUI Tool Kit. The SLUI, Program, User and Programmer are the basic components of the system, and each plays a necessary role in the system as a whole. In this system, the user provides SLUI with relevant inputs, and provides responses to dialog questions. The programmer provides SLUI with setup information. SLUI then uses this information and executes code relevant to the desired action.

5 SLUI Toolkit System Process

All of the functions of the SLUI Toolkit can be categorized into 4 processes. Each process consists of several operations that must be completed by the programmer or the system in order to create a SLUI enabled program. The first 3 processes (Input, Create SFT, and Debug SFT) are executed in the Toolkit's main UI. Once these 3 processes have been completed, the programmer must tie the system together with several external functions in order to deploy the SLUI enabled program.

Fig. 2. describes the basic purpose of the system processes.

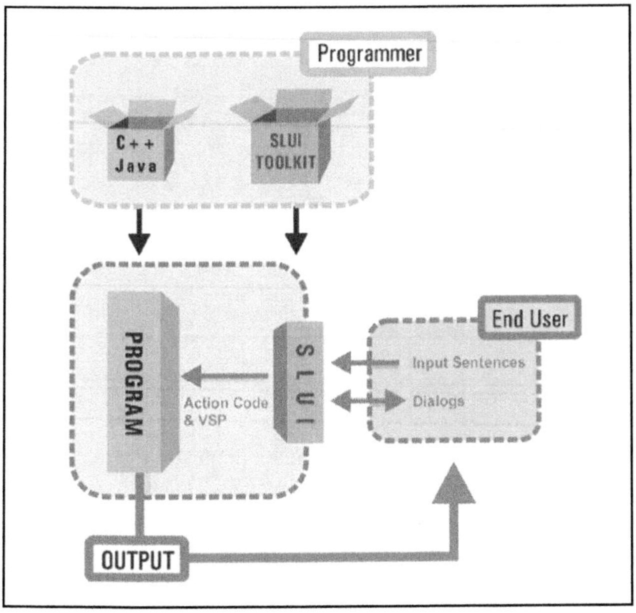

Fig. 1. Component Interaction

5.1 Input Setup Information

The programmer provides the Toolkit with sets of sample input sentences that map to specific actions. The programmer also provides a domain specific lexicon. The Toolkit produces semantic representations of the sample sentences. The SLUI Toolkit needs this information so that it can build a database of all the possibilities the user may input. These are the 3 basic processes within Input Setup Information:

- Provide sample input sentences
  - "When was our last shipment delivered?"
  - "I would like an update on my status."
- Provide lexicon
  - A sample lexicon is provided.
  - The SLUI Toolkit contains tools that can modify this lexicon to fit many different needs.
- Process sample sentences
  - The Toolkit uses the lexicon to extract critical words from the sample input sentences and create semantic representation of them.

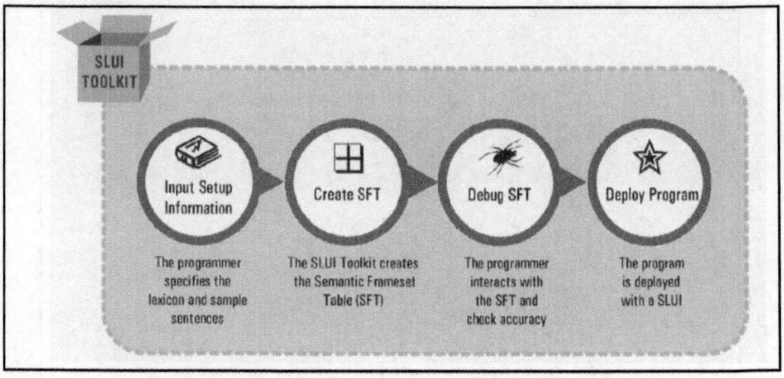

Fig. 2. System Processes

5.2 Create SFT

The SLUI Toolkit uses synonym sets and other linguistic devices to expand the semantic representations of the sample input sentences. These representations are individual frames, and they are stored in a Semantic Frame Table (SFT). The SFT becomes a comprehensive database of all the possible commands, with all possible variations, a user could request a system to do. The programmer then assigns action codes to each frame of the table. These are the 4 basic processes within Create SFT:

- Provide variable sentence parameters (VSP)
  - If the sample input sentence contains a variable, then each variable possibility must be specified:
  - For example: "I would like a <<COLOR>> shirt.", where <<COLOR>> = red, orange, yellow, green, blue, purple etc.
- Expand sample sentences
  - Expand the sentences using synonym sets, Variable Sentence Parameters and the lexicon.
- Assign Action Codes
  - For example, 345: Search for shipment date, 456: Display user status, etc.
- Semantic Frame Table
  - Create the SFT. Fig. 3. SFT structure
  - It shows the structure of a Semantic Frame Table.

Action Code	Sentence Type	Predicate	Argument 1	Argument 2	Argument 3	Head 1	Modifier 1

Fig. 3. SFT structure

5.3 Debug SFT

The programmer may debug the output of the Build Process. It is critical that the sample input sentences have correct semantic representations in the SFT. The programmer does not need a linguistic background to check the values in the SFT. If an SFT element looks as though it would map to an inappropriate User Input sentence, then it should be deleted.

- Analyze SFT
  - The programmer has the opportunity to confirm the results in the Semantic Frame Table.
  - The SLUI Toolkit contains tools that allow the programmer to verify that NL test input will match the proper frames.
- Attach SLUI to Program
  - In order to create a SLUI enabled program, the programmer needs to direct User Input to the SLUI and connect the SLUI to the program.
- COTS ASR Engines
  - Commercial Off The Shelf (COTS) Automatic Speech Recognition (ASR) engines are used to direct User Input to the SLUI.
- Application Agents and Wrappers
  - These utilities assist in binding the SLUI to the C++ code that will be executed in the program.

5.4 Deploy SLUI

When the Toolkit deploys the SLUI enabled program, the user may interact with it. The user enters an NL sentence, and if it has been well anticipated by the programmer, the input sentence maps to the appropriate frame in the SFT. The SLUI sends an action code and the VSP value to the program, and the program executes the appropriate task.

Table 1. Evaluation of SLUI

Subject	Response Scale 1-10
Overall Intuitivity in Using the UI	7.6 / 10
Locating Functionalities	7.5 / 10
Flexibility	6.8 / 10
Modification of Code	7.0 / 10
Integration with an Application	7.6 / 10
VSP Usability	7.5 / 10
Data Organization	8.6 / 10

6 Performance

The system has been tested using ten evaluators. The evaluators have overwhelmingly endorsed the ease of use and applicability of the tool kit in rapid development of speech and natural language processing interfaces for a web based e-commerce application. Table 1 shows the summary of the evaluation. The basic evaluation was on the usability of the tool kit, how easy it was to integrate with existing applications, how easy it is to write a new interface from scratch and how east it is to modify the interface once it has been designed. The numbers in the second column shows the score in a scale of 1-10, 1 being "poor" and 10 being "excellent".

7 Discussion

The SLUITK is designed to recognize the main concepts in sentences and match User Input sentences with sample sentences provided by the developer. This Natural Language recognition technique works especially well with mono-clausal sentences and common requests. However, currently, the SLUI can only recognize one sense of a word at a time [8]. For example, "Run" can either mean: "Run around the block" or "Run the application." Using a domain specific lexicon will determine which sense of the word is most common. It is critical that the correct sense of a word is chosen to be expanded during the creation of the semantic frame tables. Therefore, we recommend that the programmer verify the results of the Build Process before launching the program.

The biggest advantage of this system is its robustness against variations in the queries made using natural language. A keyword-based approach is easy to develop for a small domain, but as the domain is expanded, the maintenance and updating of such a keyword driven system becomes impossible. SLUI, on the other hand, is very easy to maintain, as the system can be continually updated automatically based on new unseen inputs.

This paper discusses the current status of an ongoing research. We are now working on improving the parsing techniques and automatically grouping similar queries together. Work is also underway to improve automatic rule generation for the parsers and word sense disambiguation for better understanding of the NL inputs.

References

1. Spoken Language user Interface (SLUI) Toolkit. NIST Award #70NANB9H3025.
2. Rahman, A. F. R., Alam, H., Hartono, R. and Ariyoshi, K. Automatic Summarization of Web Content to Smaller Display Devices, 6th Int. Conf. On Document Analysis and Recognition, ICDAR01, USA (2001) pages 1064-1068,
3. Alam, H., Rahman, A. F. R., Lawrence, P., Hartono, R., Ariyoshi, K. Automatic Extraction, Display and Processing of Forms from Web Content to Various Display Devices. U.S. Patent Application pending.
4. Woods, W. Semantics and quantification in natural language question answering. In Advances in Computers. 17(187), (1977).
5. Alam, H. Spoken language generic user interface (SLGUI). Technical Report, AFRL-IF-RS-TR-2000-58, Air Force Research Laboratory, Rome, NY, (2000).

6. Sentry Spelling Checher Engine. Wintertree Software, Nepean, Ontario, Canada K2J 3N4.
7. Scholkopf, B. Dumais, S. T., Osuna, E. and Platt, J. Support Vector Machine. In IEEE Intelligent Systems Magazine, Trends and Controversies, Marti Hearst, ed., 13(4), pages 18-28, (1998).
8. DeKang Lin. University of Manitoba. http://www.cs.ualberta.ca/~lindek/minipar.htm.
9. Mitkov, R. "The latest in anaphora resolution: going robust, knowledge-poor and multilingual". Procesamiento del Lenguaje Natural, No. 23, 1-7, (1998).
10. Burton, A. and Steward, A. P. Effects of linguistic sophistication on the usability of a natural language interface. Interacting with Computers, 5(1), pages 31-59, (1993).

Document Image De-warping for Text/Graphics Recognition

Changhua Wu and Gady Agam

Department of Computer Science
Illinois Institute of Technology
Chicago, IL 60616
{agam,wuchang}@iit.edu

Abstract. Document analysis and graphics recognition algorithms are normally applied to the processing of images of 2D documents scanned when flattened against a planar surface. Technological advancements in recent years have led to a situation in which digital cameras with high resolution are widely available. Consequently, traditional graphics recognition tasks may be updated to accommodate document images captured through a hand-held camera in an uncontrolled environment. In this paper the problem of perspective and geometric deformations correction in document images is discussed. The proposed approach uses the texture of a document image so as to infer the document structure distortion. A two-pass image warping algorithm is then used to correct the images. In addition to being language independent, the proposed approach may handle document images that include multiple fonts, math notations, and graphics. The de-warped images contain less distortions and so are better suited for existing text/graphics recognition techniques.

Keywords: perspective correction, document de-warping, document pre-processing, graphics recognition, document analysis, image processing.

1 Introduction

Document analysis and graphics recognition algorithms are normally applied to the processing of images of 2D documents scanned when flattened against a planar surface. Distortions to the document image in such cases may include planar rotations and additional degradations characteristic of the imaging system [3]. Skewing is by far the most common geometric distortion in such cases, and have been treated extensively [2].

Technological advancements in recent years have led to a situation in which digital cameras with high resolution are widely available. Consequently, traditional graphics recognition tasks may be updated to accommodate document images captured through a hand-held camera in an uncontrolled environment. Examples of such tasks include analysis of documents captured by a digital camera, OCR in images of books on bookshelves [12], analysis of images of outdoor

T. Caelli et al. (Eds.): SSPR&SPR 2002, LNCS 2396, pp. 348–357, 2002.
© Springer-Verlag Berlin Heidelberg 2002

signs [7], license plate recognition [13], and text identification and recognition in image sequences for video indexing [10]. Consequently, distortions characteristic of such situations should be addressed.

Capturing a document image by a camera involves perspective distortions due to the camera optical system, and may include geometric distortions due to the fact that the document is not necessarily flat. Rectifying the document in order to enable its processing by existing generic graphics recognition algorithms require the cancellation of perspective and geometric structural distortions.

A treatment of projective distortions in a scanned document image have been proposed [9] for the specific case of scanner optics and a thick bound book modeled by a two parameter geometric model. Extending this approach to more general cases requires the generation of parametric models for specific cases, and the development of specific parameter estimation techniques. A more general approach that is capable of handling different document deformations by using structured light projection in order to capture the geometric structure of the document is described independently in [4] and [5]. The disadvantage of these approaches lie in the need for additional equipment and calibration needs. In [1] a method is described for correcting geometric and perspective distortions based on structure inference from two uncalibrated views of a document. In this approach the structure recovery may be in some cases inaccurate and lead to distortions in the recovered image. Finally in [15] small deformations of a document image of a thick bound book obtained by a scanner are treated by rotating and aligning segmented words. Entities other than words such as math notations or graphics cannot be handled by this approach.

Contrary to the above described approaches, the proposed approach is not coupled to specific structural models and does not depend on external means for structure recovery. Instead, the document structure distortion is inferred from a single document image obtained by a hand held camera. In addition to being language independent, the proposed approach may handle document images that include multiple fonts, math notations, and graphics.

In the proposed approach, the restoration of the document image so as to reduce structural and perspective distortions in the acquired image, depends on a reconstruction of a target mesh which is then used to de-warp the original image. This target mesh is generated from a single image of the document based on the assumption that text lines in the original document are straight. The detection of the starting position and orientation of curved text lines is described in Section 2. The tracing of curved text lines in outlined in Section 3. The mesh reconstruction and document de-warping presented in Section 4. Section 5 concludes the paper.

2 Detecting the Starting Position and Orientation of Curved Text Lines

Let $F \equiv \{f_p\}_{p \in \mathbb{Z}_m \times \mathbb{Z}_n}$ be an $m \times n$ image in which the value of each pixel is represented by an s-dimensional color vector $f_p \in \mathbb{Z}^s$ where in this expression $\mathbb{Z}_m$ represents the set of non-negative integers $\{0, \ldots, m-1\}$. Without loss of generality, for efficiency reasons, we assume that the input image is binarized [11] so as to produce $G \equiv \{g_p\}_{p \in \mathbb{Z}_m \times \mathbb{Z}_n}$ where $g_p \in \mathbb{Z}_2$ and black pixels have intensity of 0.

In the proposed approach the user interactively specifies the four corner points p_{lt}, p_{lb}, p_{rt}, p_{rb} of a quadrilateral containing the portion of the image that has to be rectified. It is assumed that the user specification of the corner points is such that text lines in the document are in a general direction which is approximately perpendicular to the left edge $\overline{p_{lb}p_{lt}}$ of the quadrilateral. While the identification of these corners may be done automatically under certain assumptions, interactive point specification is simple and so we did not find it necessary to address this problem. Without loss of generality we assume in the rest of this paper that the orientation of the left edge $\overline{p_{lb}p_{lt}}$ is approximately vertical.

Based on the above assumptions the starting position of non-indented text lines should be approximately along the line segment $\overline{p_{lb}p_{lt}}$ and their orientation should be approximately perpendicular to that line segment. Detecting the starting point of text lines is a common problem in document analysis that is normally treated after skew correction by detecting the extremum points in the graph of a cumulative horizontal projection [8]. It should be noted, however, that the performance of this approach strongly depends on the preliminary skew distortion correction. As the distortion in our case is non-linear, skew correction will not suffice. Furthermore, the required correction is the overall target of the proposed approach and can not be used at this stage. In order to solve this problem, the cumulative projection that is used in the proposed approach is constructed from a local neighborhood of the left edge $\overline{p_{lb}p_{lt}}$ which due to is locality is assumed to be less distorted. As the text lines are not necessarily perpendicular the the left edge $\overline{p_{lb}p_{lt}}$ directions adjacent to the horizontal direction should be checked as well, and the maximal projection should be retained.

Consequent to the above description, the graph of a cumulative horizontal projection is replaced in the proposed approach by a graph of local adaptive cumulative projection. This graph is constructed by computing the local adaptive cumulative projection $\Phi(p)$ at each possible starting point $p \equiv (x, y)$ starting with p_{lb} and progressing along $\overline{p_{lb}p_{lt}}$ towards p_{lt}. The value of $\Phi(p)$ is defined by:

$$\Phi(p) \equiv \min\{\Phi_\beta(p) \quad | \quad \theta - \alpha < \beta < \theta + \alpha\} \tag{1}$$

where $\Phi_\beta(p)$ is the local cumulative projection in the direction of β at p, the angle θ if the angle that produced the minimal projection of the previous starting point $(x, y-1)$ and α is a preset angular limit (see Figure 1). The use of θ is designed to promote a smoothing constraint. Its initial value at p_{lb} is taken as 0. The angle β that produced the minimal value of $\Phi_\beta(p)$ is the estimated starting orientation of the text line emanating from p. It is stored for later use. The local cumulative projection $\Phi_\beta(p)$ is computed by the sum:

$$\Phi_\beta(p) \equiv \sum_{p \in R(p,\beta)} g_p \tag{2}$$

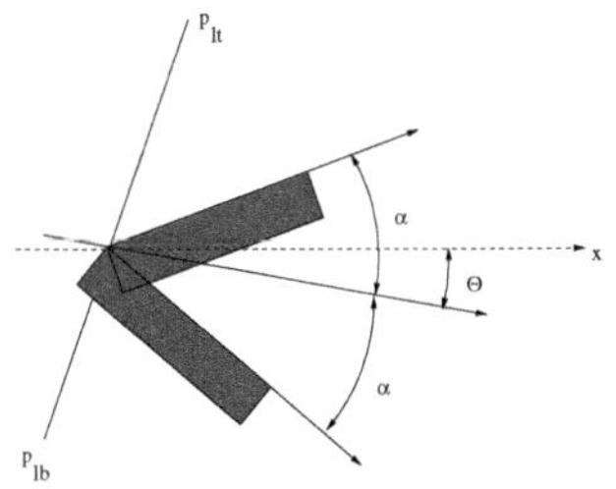

Fig. 1. Constructing the local adaptive cumulative projection

where $R(p,\beta)$ is the set of pixels contained within a rectangle emanating from p in the direction of β. Based on simple geometric considerations, the corner points of this rectangle may be computed by:

$$p_1 \equiv (x_{lt} + (x_{lb} - x_{lt})(y - y_{lt})/(y_{lb} - y_{lt}), y) \tag{3}$$
$$p_2 \equiv (x + h\cos(\beta), y + h\sin(\beta)) \tag{4}$$
$$p_3 \equiv (x + w\cos(\beta - \pi/2) + h\cos(\beta), y + w\sin(\beta - \pi/2) + h\sin(\beta)) \tag{5}$$
$$p_4 \equiv (x + w\cos(\beta - \pi/2), y + w\sin(\beta - \pi/2)) \tag{6}$$

where w and h are preset parameters corresponding to the width and height of the rectangle respectively. By using the corner points p_1, p_2, p_3, p_4, the pixels belonging to the rectangle are obtained by using a standard scan-line filling algorithm [6]

Once obtaining the local adaptive cumulative projection graph, extremum points in it may be used to separate text lines. Minimum points in particular are used to detect the beginning of text lines. In order to reduce errors due noise this graph is smoothed by using a low-pass filter prior to the extremum points detection. In addition, in cases of several detected minimum points in close proximity to each other, only the smallest one is kept. Figure 2 presents the smoothed local adaptive cumulative projection graph of the Chinese document image in Figure 3. Minimum points in this graph correspond to starting points of text lines. The identified starting points of text lines together with the estimated starting orientation are overlaid in gray on the binarized document image in Figure 3. As can be observed the fact that the text lines are curved does not mislead the outlined detection algorithm. It should be noted that in the experiments we conducted, the described algorithm was equally successful in detecting the starting position and orientation of text lines in images of English and Chinese documents in accordance with the observation in [8].

3 Tracing Curved Text Lines

After obtaining the starting position and orientation of each text line, the complete text lines may be traced in a similar way. That is, given a point and

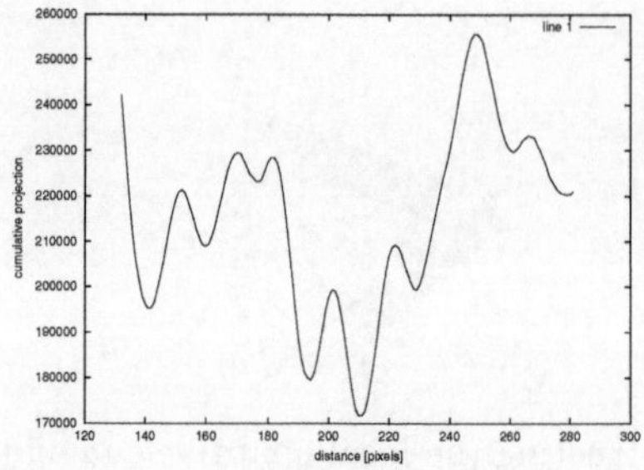

Fig. 2. The smoothed local adaptive cumulative projection graph of the Chinese document in Figure 3. Minimum points in this graph correspond to starting points of text lines. The values on the x-axis of this graph were multiplied by 255

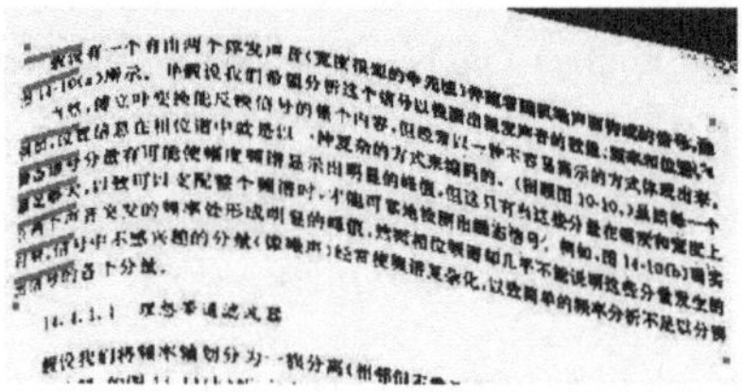

Fig. 3. Identified starting position and orientation of text lines (by using the proposed approach) overlaid in gray on the binarized image of a Chinese document. As can be observed the fact that the text lines are curved does not mislead the detection algorithm

orientation on a text line the next point on that text line is selected by evaluating the cumulative projection in a local neighborhood in a range of directions around the given direction. The next point is selected as the one for which the cumulative projection is minimal. More formally, given the j-th point on the i-th traced text line p_{ij} and the text line orientation θ_{ij} at that point (see Figure 4), the next point on that line $p_{i,j+1}$ is obtained as:

$$p_{i,j+1} \equiv p_{ij} + h \cdot s_i \cdot (\cos(\theta_{i,j+1}), \sin(\theta_{i,j+1})) \tag{7}$$

where h is the length of the rectangular neighborhood as defined for Equation 4, the angle $\theta_{i,j+1}$ is the angle that minimizes $\Phi(p)$ in Equation 1, and s_i is a scale factor.

The scale factor s_i is introduced in order to produce a similar number of points on each text line when the specified quadrilateral $(p_{lt}, p_{lb}, p_{rt}, p_{rb})$ is not rectangular. Let $L_t \equiv ||p_{lt} - p_{rt}||$ and $L_b \equiv ||p_{lb} - p_{rb}||$ be the length of the top and bottom edges of the quadrilateral respectively. Assuming that the step length on the top edge is 1, the step length on the bottom edge is taken to be $\frac{L_b}{L_t}$.

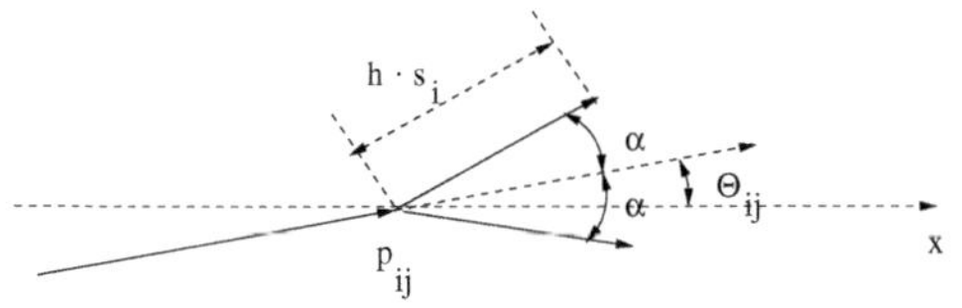

Fig. 4. Tracing a text line. Given a point p_{ij} and orientation θ_{ij} the next point on that text line is searched in an angular range of $\pm\alpha$ around θ_{ij}. The length of the step is adjusted by a scale factor s_i

The step length in an intermediate line i can be then interpolated by:

$$s_i \equiv (1 - \eta_i) + \eta_i \frac{L_b}{L_t} \tag{8}$$

where $\eta_i \equiv (y_{lt} - y_{i0})/(y_{lt} - y_{lb})$.

The tracing of a curved text line is stopped at the point p_{ij} if black pixels are not found in any of the projection rectangles of that point or if any of the projection rectangles of that point intersects the right edge $\overline{p_{rb}p_{rt}}$ of the quadrilateral. Due to the non-uniformity of characters in the document the traced lines may contain small variations. These variations are eliminated by low pass filtering the traced curves.

The angular range searched in the process of tracking a curved text line is normally smaller than the one used for the detection of the starting point of text lines. The angular range should be small enough in order to prevent possible crossings to neighboring text lines, and large enough in order to facilitate the tracing of curved text lines. In order to reduce crossings between text lines while maintaining a larger angular range search area, the local cumulative projection $\Phi_\beta(p)$ is modulated by a weight factor $W(\beta)$ which is inversely proportional to the angular deviation $(\beta - \theta)$:

$$W(\beta) \equiv 1 + |\tan(\beta - \theta)|/\mu \tag{9}$$

where μ is a constant and it is assumed that $|\beta - \theta| < \pi/4$.

The above description of modulation of the local cumulative projection assists in reducing the number of crossings between text lines, but do not eliminate them. Consequently, a consistency constraint is introduced in order to remove such crossings. For that purpose the average orientation in each column is computed by: $\overline{\theta}_j \equiv \sum_i \theta ij$, and lines containing any points with orientation deviating by more than a preset threshold τ from the average θ_j are removed. Text lines not intersecting the right edge $\overline{p_{rb}p_{rt}}$ do not contribute to the generation of a regular grid, and so they are removed as well. It should be noted that as the proposed approach does not rely on a dense grid of lines, incomplete/inaccurate traced lines may be removed instead of attempting to correct them. Figure 5 presents the result of the line removal stage where Figures 5-a and 5-b display the traced lines before and after correcting them respectively.

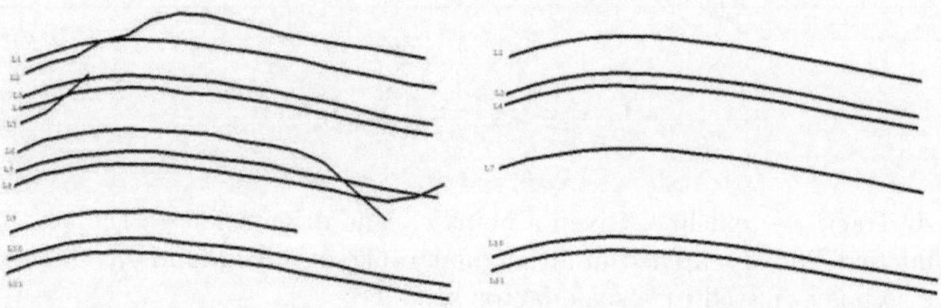

(a) (b)

Fig. 5. Demonstration of the line removal stage. **(a)** – **(b)** The traced lines before and after correction respectively. As can be observed, lines L_2, L_5, L_6, L_8 are removed due to the orientation consistency constraint whereas lines L_5, L_9 are removed due to the fact that they do not intersect the right edge

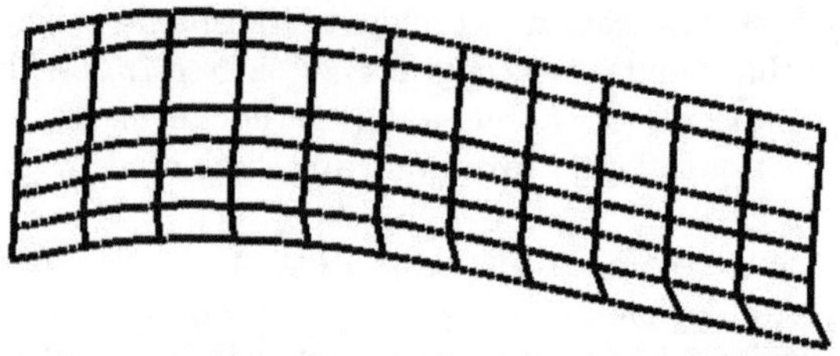

Fig. 6. The reconstructed source mesh for the document image in Figure 3. As can be observed the reconstructed mesh corresponds to the structural deformation in that document

4 De-warping the Document Image

For the purpose of de-warping the document image a source and target rectangular meshes should be produced. The source mesh contains curved lines corresponding to the structural distortion in the document image, whereas the target mesh should be rectilinear so as to represent the document structure without distortion. The source mesh is produced based on the traced lines obtained as described in the previous section. The horizontal lines of that mesh are the traced lines whereas the vertical lines are generated by subdividing each traced line into a fixed number of uniform length segments. Figure 6 presents the reconstructed source mesh for the document image in Figure 3. As can be observed the reconstructed mesh corresponds to the structural deformation in that document.

The target mesh is generated based on the source mesh and the assumption that the text lines in the document were straight before going through structural deformation. The rectilinear target mesh is generated with the same number of

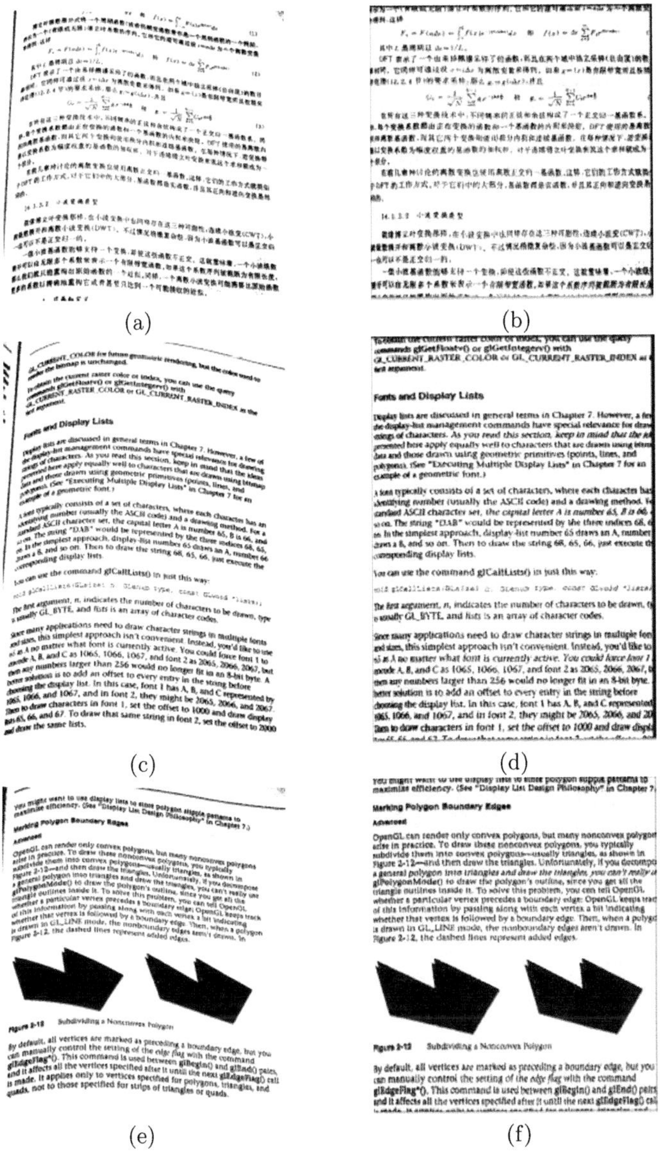

Fig. 7. Document image de-warping obtained by the proposed approach. The left column presents the input document images whereas the right column presents the rectification results obtained automatically by the proposed approach. As can be observed the proposed approach is capable of handling documents in different languages which include graphics, math notations, and different fonts

rows and columns as the source mesh. The distance between neighboring rows in the target mesh is set to the average distance between the corresponding rows in the source mesh multiplied by a uniform scale factor which is used to scale the size of the rectified image. The distance between neighboring columns in the target mesh is set to be uniform. This distance is selected as the uniform segment length on the longest row in the source mesh. It should be noted that, in general, the distance between neighboring columns in the target mesh should not be uniform due to perspective foreshortening. More specifically, the distance between columns of the document corresponding to an area of the document closer to the camera should be smaller. In future work we intend to estimate this non-uniform length based on character density estimation in each column.

Given the reconstructed source and target meshes the de-warping of the document image is done by a 2-pass image warping algorithm as described in [14]. This image warping algorithm is particularly suitable for our case as it is based on a rectangular mesh structure which is inherent to document images and as it prevents foldovers.

5 Results

The results of document image de-warping obtained by the proposed approach are presented in Figure 7. In this figure, the left column presents the input document images whereas the right column presents the rectification results obtained automatically by the proposed approach. As can be observed the proposed approach is capable of handling documents in different languages which include graphics, math notations, and different fonts. This is due to the fact that only a sparse mesh grid has to be reconstructed for the rectification.

As mentioned earlier in this work we do not yet take care of generating non-uniform columns in the target mesh. Consequent to that it is possible to observe that characters in parts of the document image which were originally closer to the camera appear to have a slightly larger width.

References

1. G. Agam. Perspective and geometric correction for graphics recognition. In *Proc. GREC'01*, pages 395–407, Kingston, Ontario, 2001. 349
2. A. Amin, S. Fischer, A.F. Parkinson, and R. Shiu. Comparative study of skew algorithms. *Journal of Electronic Imaging*, 5(4):443–451, 1996. 348
3. H. Baird. Document image defect models. In *Proc. SSPR'90*, pages 38–46, 1990. 348
4. M.S. Brown and W.B. Seales. Document restoration using 3d shape: a general deskewing algorithm for arbitrarily warped documents. In *Proc. ICCV'01*, pages 367–374, Vancouver, BC, Jul. 2001. IEEE. 349
5. A. Doncescu, A. Bouju, and V. Quillet. Former books digital processing: image warping. In *Proc. Workshop on Document Image Analysis*, pages 5–9, San Juan, Puerto Rico, Jun. 1997. 349

6. David F.Rogers. *Procedural elements for computer graphics.* McGraw-Hill, second edition, 1998. 351
7. H. Fujisawa, H. Sako, Y. Okada, and S. Lee. Information capturing camera and developmental issues. In *Proc. ICDAR'99*, pages 205–208, 1999. 349
8. D.J. Ittner and H.S. Baird. Language-free layout analysis. In *Proc. ICDAR'93*, pages 336–340, Tsukuba, Japan, 1993. 350, 351
9. T. Kanungo, R. Haralick, and I. Philips. Global and local document degradation models. In *Proc. ICDAR'93*, pages 730–734, 1993. 349
10. H. Li, D. Doermann, and O. Kia. Automatic text detection and tracking in digital video. *IEEE Trans. Image Processing*, 9(1):147–156, 2000. 349
11. J. Sauvola, T. Seppanen, S. Haapakoski, and M. Pietikainen. Adaptive document binarization. In *Proc. ICDAR'97*, pages 147–152, Ulm, Germany, Aug. 1997. 350
12. M. Sawaki, H. Murase, and N. Hagita. Character recognition in bookshelf images by automatic template selection. In *Proc. ICPR'98*, pages 1117–1120, Aug. 1998. 348
13. M. Shridhar, J.W.V. Miller, G. Houle, and L. Bijnagte. Recognition of license plate images: issues and perspectives. In *Proc. ICDAR'99*, pages 17–20, 1999. 349
14. G. Worlberg. *Digital Image Warping.* IEEE Computer Society Press, Los Alamitos, California, 1990. 356
15. Z. Zhang and C.L. Tan. Recovery of distorted document images from bound volumes. In *Proc. ICDAR'01*, pages 429–433, Seattle, WA, 2001. 349

A Complete OCR System for Gurmukhi Script

G. S. Lehal[1] and Chandan Singh[2]

[1] Department of Computer Science and Engineering
Thapar Institute of Engineering & Technology, Patiala, India
[2] Department of Computer Science and Engineering
Punjabi University, Patiala, India

Abstract. Recognition of Indian language scripts is a challenging problem. Work for the development of complete OCR systems for Indian language scripts is still in infancy. Complete OCR systems have recently been developed for Devanagri and Bangla scripts. Research in the field of recognition of Gurmukhi script faces major problems mainly related to the unique characteristics of the script like connectivity of characters on the headline, characters in a word present in both horizontal and vertical directions, two or more characters in a word having intersecting minimum bounding rectangles along horizontal direction, existence of a large set of visually similar character pairs, multi-component characters, touching characters which are present even in clean documents and horizontally overlapping text segments. This paper addresses the problems in the various stages of the development of a complete OCR for Gurmukhi script and discusses potential solutions.

1 Introduction

Research on Devanagri, Tamil and Telugu optical text recognition started around mid 70s[1-4] and recently complete OCR systems for Indian scripts such as Devanagri and Bangla[5-6] have been developed. The research work for Gurmukhi OCR has started only in mid 90s and is still in its infancy stage. To the best of our knowledge this is the first time that a complete OCR solution for Gurmukhi script has been developed and presented.

The word 'Gurmukhi' literally means from the mouth of the Guru. Gurmukhi script is used primarily for the Punjabi language, which is the world's 14th most widely spoken language. Gurmukhi script like most of other Indian language scripts is written in a nonlinear fashion. The width of the characters is also not constant. The vowels getting attached to the consonant are not in one (or horizontal) directions, they can be placed either on the top or the bottom of consonant. Some of the properties of the Gurmukhi script are:

- Gurmukhi script is cursive and the Gurmukhi script alphabet consists of 41 consonants and 12 vowels and 3 half characters, which lie at the feet of consonants (Fig 1).

T. Caelli et al. (Eds.): SSPR&SPR 2002, LNCS 2396, pp. 358-367, 2002.
© Springer-Verlag Berlin Heidelberg 2002

- Most of the characters have a horizontal line at the upper part. The characters of words are connected mostly by this line called head line and so there is no vertical inter-character gap in the letters of a word and formation of merged characters is a norm rather than an aberration in Gurmukhi script
- A word in Gurmukhi script can be partitioned into three horizontal zones (Fig 2). The upper zone denotes the region above the head line, where vowels reside, while the middle zone represents the area below the head line where the consonants and some sub-parts of vowels are present. The middle zone is the busiest zone. The lower zone represents the area below middle zone where some vowels and certain half characters lie in the foot of consonants.
- The bounding boxes of 2 or more characters in a word may intersect or overlap vertically.
- The characters in the lower zone frequently touch the characters in the middle zone.

ੳ	ਅ	ੲ	ਸ	ਹ	ਕ	ਖ	ਗ	ਘ	ਙ	
ਚ	ਛ	ਜ	ਝ	ਞ	ਟ	ਠ	ਡ	ਢ	ਣ	
ਤ	ਥ	ਦ	ਧ	ਨ	ਪ	ਫ	ਬ	ਭ	ਮ	
ਯ	ਰ	ਲ	ਵ	ੜ	ਸ਼	ਜ਼	ਖ਼	ਫ਼	ਗ਼	ਲ਼
ੋ	ੇ	ੈ	ੌ	ੱ	ੰ	ਂ	ਿ	ੀ	ਾ	
ੁ	ੂ	੍ਰ	੍	੍ਵ						

Fig. 1. Character set of Gurmukhi script

Fig. 2. Three zones of a word in Gurmukhi script

In our current work, after digitization of the text, the text image is subjected to pre-processing routines such as noise removal, thinning and skew correction. The thinned and cleaned text image is then sent to the text segmenter, which segments the text into connected components. Next these connected components are recognized and combined to form back characters. Finally post processing is carried out to refine the results.

2 Text Segmentation

Gurmukhi script is a two dimensional composition of consonants, vowels and half characters which require segmentation in vertical as well in horizontal directions. Thus the segmentation of Gurmukhi text calls for a 2D analysis instead of the commonly used one-dimensional analysis for Roman script. Besides the common segmentation problems faced in Indian language scripts, Gurmukhi script has other typical problems such as horizontally overlapping text segments and touching characters in various zonal positions in a word.

To simplify character segmentation, since it is difficult to separate a cursive word directly into characters, a smaller unit than a character is preferred. In our current work, we have taken an 8-connected component as the basic image representation throughout the recognition process and thus instead of character segmentation we have performed *connected component segmentation.* The segmentation stage breaks up a word and characters which lie above and below the headline into connected components and the classifier has been trained to recognize these connected components or sub-symbols (Table 1). It is to be noted that the headline is not considered the part of the connected component.

Table 1. Sub-symbols of Gurmukhi script used for segmentation and recognition

Symbol	**Sub-symbols**	**Symbol**	**Sub-symbols**	**Symbol**	**Sub-symbols**
ੳ	ਚ and ⌒	ਖ਼	ਖ and .	ਿ	। and ⌒
ਗ	ਰ and ।	ਫ਼	ਫ and .	ੀ	। and ⌒
ਸ਼	ਸ and .	ਗ਼	ਰ , । and .	=	- and -
ਜ਼	ਜ and .	ਲ਼	ਲ and .	ਓ	ਚ and ⌐
Gurmukhi Characters in upper zone	Same shapes retained	Gurmukhi Characters in lower zone	Same shapes retained	Rest of Gurmukhi characters in middle zone	Gurmukhi characters with their headlines stripped off

A combination of statistical analysis of text height, horizontal projection and vertical projection and connected component analysis is performed to segment the text image into connected components. We have employed a 5 phased segmentation scheme. These phases, which are described in detail in [7] are:

1) Dissect the text image into text strips using valleys in the horizontal projection profiles. Each of these strip could represent either one text row or only the upper or lower zones of a text row or more than one text row (Fig. 3).

2) Perform statistical analysis to automatically label the text strips as multi strip, core strip, upper strip or lower strip, depending on if the text strip contains more than one text row, one text row, upper zone or lower zone of a text row respectively. As for example, in Fig. 3 strip nos. 2 and 3 are lower strips, strip no. 1 is core strip, strip no. 12 is upper strip and strip no. 15 is multi strip.
3) Decompose the text strips into smaller components such as words and connected components using vertical projection profile analysis. In case of multi strip, the strip is first split into individual text rows using the statistics based on the average height of a core strip and next each text row is split into words. In case of upper and lower strips we just have sub parts of upper and lower zone vowels respectively. A connected component analysis is carried out to obtain the connected components in these strips.
4) Split words into connected components in case of core strip and multi strip. For obtaining the connected components the headline is rubbed off and after segmentation it is restored back.
5) Detect and segment touching characters in connected components. This phase is explained briefly in the following subsection.

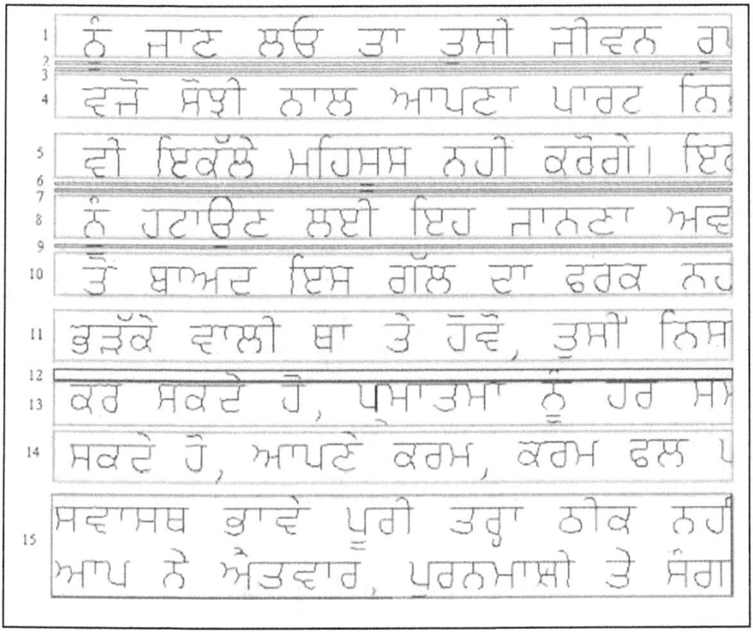

Fig. 3. A sample image split into text strips by horizontal projection

2.1 Touching Characters

It was observed that touching characters were frequently present even in clean machine printed texts. As already mentioned, segmentation process for Gurmukhi script proceeds in both x and y directions since two or more characters of a word may be sharing the same x coordinate. Therefore, for the segmentation of touching characters in Gurmukhi script, the merging points of the touching characters have to be determined both along the x and y axes. These touching characters can be categorized as follows:

(a) Touching characters in upper zone
(b) Touching characters in middle zone
(c) Lower zone characters touching with middle zone characters
(d) Lower zone characters touching with each other

Fig. 4 shows examples of touching characters for these categories. The existing techniques for detecting and segmenting touching characters were used and certain heuristics were developed to solve the segmentation problem for Gurmukhi characters. The details are discussed elsewhere[7]. Table 2 displays the percentage frequency of occurrence of the touching characters in the three zones and the accuracy rate of detection and correction.

Fig. 4. Examples of touching characters a) touching characters in upper zone, b)touching characters in middle zone, c) Lower zone characters touching with middle zone characters, d) Lower zone characters touching with each other

Table 2. Accuracy rate of detecting and segmenting touching characters

Type of touching characters	% of occurrence	% of correct detection and segmentation
Touching/merging upper zone vowels	6.90%	92.5%
Touching middle zone consonants	0.12%	72.3%
Touching middle zone and lower zone characters	19.11%	89.3%
Touching lower zone characters	0.03%	95.2%

3 Recognition Stage

The main phases of the recognition stage of OCR of Gurmukhi characters in our present work are:

1. Feature extraction.
2. Classification of connected components using extracted features and zonal information.
3. Combining and converting the connected components to form Gurmukhi symbols.

3.1 Feature Extraction

After a careful analysis of shape of Gurmukhi characters for different fonts and sizes, two sets of features were developed. The first feature set called primary feature set is made up of robust and font and size invariant features. The purpose of primary feature set is to precisely divide the set of characters lying in middle zone into smaller subsets which can be easily managed. The cardinality of these subsets varies from 1 to 8. The Boolean valued features used in the primary feature set are:

1. Is the number of junctions with the headline one
2. Is a sidebar present
3. Is there a loop
4. Is a loop formed with headline

The second feature set, called secondary feature set, is a combination of local and global features, which are aimed to capture the geometrical and topological features of the characters and efficiently distinguish and identify the character from a small subset of characters. The secondary feature set consists of following features:

1. Number of endpoints and their location (S_1)
2. Number of junctions and their location (S_2)

3. Horizontal Projection Count (S_3)
4. Right Profile depth (S_4)
5. Left Profile Upper Depth (S_5)
6. Left Profile Lower Depth (S_6)
7. Left and Right Profile Direction Code (S_7, S_8)
8. Aspect Ratio (S_9) :
9. Distribution of black pixels about the horizontal mid line (S_{10})

3.2 Classification

In our present work, we have used a multi-stage classification in which the binary tree and nearest neighbour classifiers have been used in a hierarchical fashion. The classification scheme for the Gurmukhi characters proceeds in the following 3 stages:

(a) Using zonal information, we classify the symbol into one of the three sets, lying either in upper zone, middle zone or in lower zone. The upper zone and lower symbols are assigned to set nos. 11 and 12 of Table 3 respectively.

(b) If the symbol is in the middle zone, then we assign it to one of the first ten sets of Table 3 using primary features and binary classifier tree. At the end of this stage the symbol has been classified into one of 12 sets including the sets for characters in upper and lower zones.

(c) Lastly, the symbol classified to one of the 12 sets of Table 3 is recognized using nearest neighbour classifier and the feature set of secondary features assigned for that particular set.

Table 3. Secondary feature set for classification of character sets

Set #	Character Set	Primary Feature Vector	Classification features
1	ਚ ਰ	[1, 1, 1, X]	S_1 S_2 S_3
2	ਹ ਜ ।	[1, 1, 0, X]	S_1 S_2 S_3
3	ਕ ਙ ਛ ਠ ਡ ਢ ਫ ਭ	[1, 0, 1, X]	S_1 S_2 S_3 S_4 S_5 S_6 S_7 S_8
4	ਟ ਣ ਤ ਦ ਨ ਵ ਝ	[1, 0, 0, X]	S_1 S_2 S_3 S_4 S_5 S_6 S_7 S_8 S_{10}
5	ਖ	[0, 1, 1, 1]	-
6	ਥ ਬ	[0, 1, 1, 0]	S_5 S_8
7	ਅ ਘ ਪ ਮ	[0, 1, 0, 1]	S_1 S_2 S_3 S_5
8	ਸ ਧ ਯ	[0, 1, 0, 0]	S_1 S_2 S_3 S_5
9	ੳ	[0, 0, 1, X]	-
10	ੲ ਞ ੜ ਲ	[0, 0, 0, X]	S_1 S_2 S_3 S_4 S_7 S_8
11	ਿ ੀ ੇ ੈ ੱ ੰ ਂ	[X, X, X, X]	S_1 S_7 S_8 S_{10}
12	ੁ ੂ ੍	[X, X, X, X]	S_8 S_9

The complete feature set used for classification is tabulated in Table 3. The primary feature vector is obtained from binary classifier tree and the i_{th} component of the vector is 1 or 0 depending on if the P_i primary feature is true or false for that character set. X denotes that the feature is not needed for classification. Thus for example, the primary feature vector for set number 1 is [1, 1, 1, X], which means that all the characters in this set have one junction with the headline, have a side bar and have a loop.

3.3 Merging Sub-symbols

In this last stage of recognition of characters, the information about coordinates of bounding box of sub-symbols and context is used to merge and convert the sub-symbols to Gurmukhi characters. It is to be noted that most of the sub-symbols can as such be converted to equivalent character (Table 1). It is only in some typical cases where a character may be broken into more than one sub-symbol that some rules have to be devised to merge these sub-symbols. For example, if the sub-symbol in middle zone is ਰ and the next sub-symbols in middle and upper zones are | and ⌒respectively, then if the upper sub-symbol is vertically overlapping with one or more middle zone sub-symbols, then these sub-symbols might represent one of the character combinations ਰੀ, ਗ or ਰਾਂ.The information regarding the overlapping of the upper and middle zone connected components (CCs) is used to identify the characters represented by the CCs. Thus, if ⌒ is overlapping with both ਰ and | then the CCs combine to form ਰੀ . If ⌒ is overlapping with only | then the CCs combine to form ਗ and if ⌒ is overlapping only with only ਰ then the CCs combine to form ਰਾਂ.

4 Post Processing

For the post processing we have used a Punjabi corpus, which serves the dual purpose of providing data for statistical analysis of Punjabi language and also checking the spelling of a word. Punjabi grammar rules are also incorporated to check for illegal character combinations such as presence of two consecutive vowels or a word starting with a forbidden consonant or vowel. The main steps in the post processing phase are:

1. Create word frequency list from the Punjabi corpus. The list stores the frequency of occurrence of all words present in the corpus.
2. Partition the word frequency list into smaller sub lists based on the word size. We have created 7 sub-lists corresponding to word sizes of two, three, four, five, six, seven and greater than seven characters.
3. Generate from each of the sub-list an array of structures which is based on visually similar characters. We say that two characters are visually similar if they belong to the same set of Table 3. Similarly two words are visually similar if each character is corresponding position is visually similar. Thus the words m^C and p>r are visually similar since the first character in both the words belongs to set 7, the second character belongs to set 11 and third character belongs to set 1. This array of visually similar words records the percentage frequency of occurrence of character in all the positions of these visually similar words. This list is combined with the confidence rate of recognition of the recognizer to correct the mistakes of the recognizer.

4. Store the twenty most commonly occurring words. Any word which is visually similar to any of these words and which is not recognized with high confidence is automatically converted to the nearest visually similar word.
5. Use Punjabi grammar rules to eliminate illegal character combinations.

These steps are explained in detail in[8].

5 Experimental Results and Conclusion

We tested our OCR on about 25 Gurmukhi text documents consisting of about 30000 characters. The documents were pages from good quality books and laser print outs in multiple sizes and fonts. We tested on font sizes in the range 10-24 point size and 8 fonts were used.

It was found that seven characters (ਹ ਖ ਘ ਙ ਞ ਢ ਧ) with a combined frequency of occurrences of 5.67% were recognized with almost 100% accuracy. Out of these the character ਹ has a high frequency of occurrence (4.2%) but in the subset 2 (Table 3), there are only two other characters for resolving the confusion and their shapes are quite different so ਹ is not confused with them. Twenty two characters with cumulative frequency of occurrences of 44.69% are recognized with more than 98% accuracy. On the lower end, eleven characters (ਲ਼, ਜ਼ ੱ ਂ ਸ਼ ਬ ਚ ੂ ਖ਼ ਫ਼ ਗ਼) with a cumulative frequency of occurrences of 10.08% have a low recognition rate of 80% or less. It is these characters which are the main bottlenecks in the performance of the OCR. It can be seen that majority of these characters are the characters with dot at their feet. The reason for this inaccuracy is that during the thinning either the dot is deleted or it gets merged with the character. Even among the characters with dot at their feet the characters ਗ਼ ਖ਼ and ਫ਼ have a far more poor recognition accuracy as compared to characters ਲ਼ ਸ਼ and ਜ਼. The reason for this is that the dot is positioned in centre for characters ਲ਼ ਸ਼ and ਜ਼ while for characters ਗ਼ ਖ਼ and ਫ਼ the dot is positioned very close to the character and so it gets easily merged on thinning. The characters ਰ and ਬ have low recognition accuracy as they are very closely resembling with charactersਹ and ਥ respectively and are often confused with them. The characters ੱ and ੂ , have their strokes often joined together or touching with other characters which makes it difficult to recognize them. The character, ਂ (*bindi*), which is similar to a dot and is present in the upper zone is also difficult to recognize. There were two type of errors produced: a) Deletion - The character *bindi* would be removed during the scanning and binarization process or by the thinning algorithm. In many cases the *bindi* character would be merged with other symbols in the upper zone and vanish. b)Insertion - The noise present in the upper zone would be confused with *bindi*. Sometimes an upper zone vowel would be broken into smaller components, which would generate extra *bindi* characters. The above statistics are obtained without the application of the post processor. The recognition accuracy of the OCR without post processing was 94.35%, which was increased to 97.34% on applying the post processor to the recognized text.

This is the first time that a complete multi-font and multi-size OCR system for Gurmukhi script has been developed. It has been tested on good quality images from books and laser print outs and has recognition accuracy of more than 97%. We are now working for testing and improving the performance of the OCR on newspapers and low quality text.

References

1. Govindan, V. K., Shivaprasad, A. P.: Character recognition-A review. Pattern Recognition. Vol. 23. (1990) 671-683.
2. S. N. S. Rajasekaran, S. N. S., Deekshatulu, B. L.: Recognition of printed Telugu characters. Computer Graphics and Image Processing. Vol. 6. (1977) 335-360.
3. G. Siromoney, G., Chandrasekaran, R., Chandrasekaran, M.: Machine recognition of printed Tamil characters. Pattern Recognition. Vol. 10. (1978) 243-247.
4. Sinha, R. M. K., Mahabala, H. N.: Machine recognition of Devanagari script. IEEE Trans on Systems, Man and Cybernetics. Vol. 9. (1979) 435-449.
5. Chaudhuri, B. B., Pal, U.: A complete printed Bangla OCR system. Pattern Recognition. Vol. 31. (1998) 531-549.
6. Bansal, V.: Integrating knowledge sources in Devanagri text recognition. Ph.D. thesis. IIT Kanpur (1999).
7. Lehal, G. S., Singh, C.: Text segmentation of machine printed Gurmukhi script. Document Recognition and Retrieval VIII. Paul B. Kantor, Daniel P. Lopresti, Jiangying Zhou (eds.), Proceedings SPIE, USA. Vol. 4307. (2001) 223-231.
8. Lehal, G. S., Singh, C.: A shape based post processor for Gurmukhi OCR. Proceedings 6th International Conference on Document Analysis and Recognition, Seattle, USA. (2001) 1105-1109.

Texprint: A New Algorithm to Discriminate Textures Structurally*

Antoni Grau[1], Joan Climent[1], Francesc Serratosa[1], and Alberto Sanfeliu[2]

[1]Dept Automatic Control, Technical University of Catalonia UPC
c/ Pau Gargallo, 5 E-08028 Barcelona, Spain
{agrau,climent}@esaii.upc.es
Universitiy Rovira i Virgili Tarragona
[2] Institute for Robotics, UPC/CSIC
c/ Lloren s i Artigues, 4-6 E-08028 Barcelona, Spain
asanfeliu@iri.upc.es

Abstract. In this work a new algorithm for texture analysis is presented. Over a region with size NxN in the image, a texture print is found by means of counting the number of changes in the sign of the derivative in the gray level intensity function by rows and by columns. These two histograms (**Hx** and **Hy**) are represented as a unique string R of symbols. In order to discriminate different texture regions a distance measure on strings based on minimum-cost sequences of edit operations is computed.

1 Introduction

Texture provides an important feature to characterize and discriminate regions. In general, textural features extraction methods can be classified in statistical and structural methods [5]. In statistical approach, texture is characterized by a set of statistics that are extracted from large ensemble of local properties representing interpixel relationships. On the other hand, structural methods are based on the model that texture is made of a set of elements arranged with some regular placement rule. In [10] sets of connected pixels with similar gray level as elements are extracted and characterized by size, directionality and shape. In [11] texture elements by a region growing procedure are analyzed and the spatial arrangement among them by regularity vectors is described. For texture discrimination, in [7] a syntactic model for the generation of structured elements is proposed. In this work we present a new algorithm to generate the texture print (*texprint*) over regions in an image. This texture print will be the basis for a comparison between texture images and a further discrimination step among texture regions. Because texture is not a perfect pattern repeated along images with similar texture, it is not possible to use an exact matching algorithm to compare *texprints*. To perform such comparison the Levenshtein distance

* This work has been funded by Spanish Department of Science & Technology, DPI2001-2223.

T. Caelli et al. (Eds.): SSPR&SPR 2002, LNCS 2396, pp. 368-377, 2002.
© Springer-Verlag Berlin Heidelberg 2002

between *texprints*, which are represented by strings, will be found. If the distance is short enough, these *texprints* correspond to similar textures. We propose the use of string-to-string correction problem as placement rules of elements (primitives) obtained statistically. This technique can be applied to pattern recognition [6], [9] and [11]. This paper is organized as follows: in Section 2 we present the algorithm to extract the *texprint*. In Section 3 we propose the use of Levenshtein distance as a measure to discriminate textures with different *texprint*. Experimental results are shown in Section 4.

2 Algorithm for Extracting the *Texprint*

In this Section the algorithm used to compute and generate the *texprint* from a textured image is described. First, the original image is normalized in order to make further steps invariant to illumination changes. Then, the input image is divided into regions, or windows, with *N*x*N*-pixels size. For each region *W*, the algorithm finds two histograms. First histogram, **Hx**, is calculated from pixels in the *W* region in X axis. Second histogram, **Hy**, is calculated in Y axis. Each position in both histograms is incremented after the evaluation of a condition over a row (for **Hx**) or over a column (for **Hy**). The condition is defined as: there will be an increment in a position of the histogram, **Hx**(i) or **Hy**(i), if there is a change in the sign of the first derivative of the gray level intensity in its row or in its column, respectively. This is a measure of the texture appearing in the image. We have seen that images with no texture (smooth texture) present some non-null histograms, for this reason a second condition is defined: the difference between pixels with different derivative has to be greater than a certain threshold *T*. The use of a threshold is due to the fluctuation in the image during the acquisition process.

The algorithm, for **Hx**, can be formalized as follows.

```
For each row, i
  For all the pixels in this row, k
    Evaluate condition 1
    Evaluate condition 2
    If (condition 1 AND condition 2) then Hx(i)++
  endfor
endfor
```

Here, *Condition* 1 is "(S = Sign $(k+1)$ * Sign $(k+2)$) < 0"
where Sign $(k+1)$ = I(i, k) - I$(i,k+1)$; and
Sign $(k+2)$ = I$(i, k+1)$ - I$(i, k+2)$.

The image function is represented by I, indexed by rows and colums (r,c). Following, we can define *Condition* 2: "ABS(S)>Threshold *T*", where ABS is the absolute value.

If *condition* 1 is false there is no change in the increasing or decreasing of the gray level values. For the **Hy** histogram, conditions are similar with the unique difference that the pixels to be evaluated are taken by columns (Y axis). These histograms represent the *texprint* of the evaluated window *W*.

3 Discrimination Step

Since we propose an structural approach, the prints represented by **Hx** and **Hy** can be considered as two strings of symbols (characters) (R_{Hx} and R_{Hy}). If these strings are concatenated ($\oplus$) a new string R with double length is obtained ($R = R_{Hx} \oplus R_{Hy}$). Therefore, the characteristic that defines the texture print over a region is the string R (the new texture print). The problem in the discrimination step is now reduced and it is defined as follows: two texture regions p and q are similar iif their texture prints R_p and R_q approximate match.

The problem of string-matching can generally be classified into exact matching and approximate matching. For exact matching, a single string is matched against a set of strings and this is not the purpose of our work. For approximate string matching, given a string v of some set V of possible strings, we want to know if a string u approximately matches this string, where u belongs to a subset U of V. In our case, V is the global set of texture prints and u and v are texture prints obtained from different texture images. Approximate string matching is based on the string distances that are computed by using the editing operations: substitution, insertion and deletion [12].

Let Σ be a set of symbols and let Σ^* be the set of all finite strings over Σ. Let Λ denote the *null string*. For a string $A = a_1a_2...a_n \in \Sigma^*$, and for all $i, j \in \{1, 2,..., n\}$, let $A<i, j>$ denote the string $a_ia_{i+1}...a_j$, where, by convention $A<i, j> = \Lambda$ if $i > j$.
An *edit operation* s is an ordered pair $(a, b) \neq (\Lambda,\Lambda)$ of strings, each of length less than or equal to 1, denoted by $a \rightarrow b$. An edit operation $a \rightarrow b$ will be called an *insert* operation if $a = \Lambda$, a *delete* operation if $b = \Lambda$, and a substitution operation otherwise.
We say that a string B results from a string A by the edit operation $s = (a \rightarrow b)$, denoted by $A \rightarrow B$ via s, if there are strings C and D such that $A = CaD$ and $B= CbD$. An *edit sequence* $S:= s_1s_2...s_k$ is a sequence of edit operations. We say that S *takes* A *to* B if there are strings $A_0, A_1, ..., A_k$ such that $A_0 = A$, $A_k = B$ and $A_{i-1} \rightarrow A_i$ via s_i for all $i \in \{1, 2, ..., k\}$.
Now let γ be a cost function that assigns a nonnegative real number $\gamma(s)$ to each edit operation s. For an edit sequence S as above, we define the cost $\gamma(S)$ by $\gamma(S):= \Sigma_{i=1,...,k}\gamma(s_i)$. The *edit distance* $\delta(A, B)$ from string A to string B is now defined by $\delta(A, B):= \min\{\gamma(S) \mid S$ is an edit sequence taking A to $B\}$. We will assume that $\gamma(a \rightarrow b)= \delta(A, B)$ for all edit operations $a \rightarrow b$. The key operation for string matching is the computation of edit distance. Let A and B be strings, and $D(i,j)= \delta(A(1, i), B(1, j))$, $0 \leq i \leq m$, $0 \leq j \leq n$, where m and n are the lengths of A and B respectively, then:

$$D(i,j)= \min\{ D(i\text{-}1,j\text{-}1) + \gamma(A(i) \rightarrow B(j)), D(i\text{-}1,j) + \gamma(A(i) \rightarrow \Lambda), D(i,j\text{-}1) + \gamma(\Lambda \rightarrow B(j)) \} \quad (1)$$

for all $1 \leq i \leq m$, $1 \leq j \leq n$. Determining $\delta(A, B)$ in this way can in fact be seen as determining a minimum weighted path in a weighted directed graph. Note that the arcs of the graph correspond to insertions, deletions and substitutions. The Levenshtein distance (metric) is the minimum-cost edit sequence taking A to B from vertices $v(0,0)$ to $v(n,m)$. In our case both strings have the same length (N) and the algorithm used is $O(N^2)$, [2].

4 Results

Free parameters *N* and *T* (size of the region and threshold, respectively) can not be found in a formal way and it is only from experiments and empirical proofs that a set of optimal values we can obtained to best discriminate textures. First, we show the texture images used in this experiment. These images have been obtained from [1] representing an universal accepted set of textures in many works. We used 20 Brodatz images, figure 1, highly representative form natural textures.

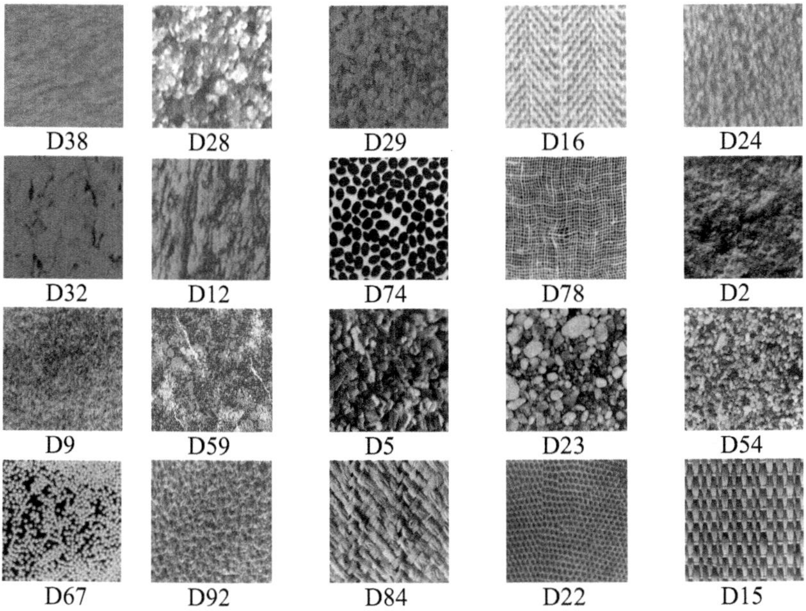

Fig. 1. Brodatz's texture images used in this experiment and their indexes, from [1]

Due to the uncertain parameters *N* and *T*, firstly we will find visually the *texprints* searching for some information 'at a glance'. In figure 2, the shape of the *texprint* can be observed for texture D67, pellets. We choose three values for the threshold *T* (0, 8 and 16, every column in figure 2) and five values for the regions of exploration (8, 16, 32, 48 and 64, every row in figure 2). In each plot, histograms **Hx** and **Hy** have been already concatenated. Their shape was predictable: the bigger the region of exploration, the higher the histogram values, that is, there are more changes of sign in the derivative. Respect the consequences of the threshold, it can be seen its attenuative effect, reducing the number of changes. In such a situation, it is not necessary to normalize the histogram values because, once a region size is chosn, this will be the unique size for all the experiments.

Visually, it is not possible to discriminate any texture from the shape of the *texprint*, but intuitively it contains some outstanding information about the texture of the image. Therefore, a numerical method is needed in order to evaluate the differences between *texprints*. The Levenshtein distance will be the measure of how different the prints are.

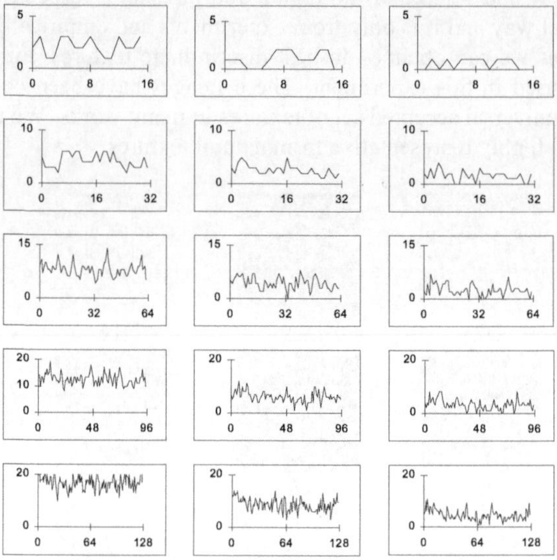

Fig. 2. Some texprints for texture "pellets", D67

As the visual observation of plots do not supply rellevant information, the next step is: for each texture image we compute the distances among *texprints* in the same image. For this proof, different region sizes have been evaluated (*N* ranging from 5 to 64), taken randomly over the image. In a similar way, different values for the threshold *T* have been evaluated (*T* ranging from 1 to 20). The computed distances are an average of 50 distances for a given *N* and *T*.

The distances between *texprints* obtained from the same texture are low and they follow a pattern: when the region of exploration grows and the threshold is low, the distance between *texprints* is higher. This result was predictable: the accumulated number of variations depends directly on the size of the region, while the attenuative effect of the threshold *T* disappears when its value is low. A sample of this proof can be seen in figure 3. For different values N and T a surface map indicates the distance between texture regions in D38 texture (left) and D15 texture, right of figure 3. The costs for insertion and deletion are constants, while the cost of substitution is the difference between the symbols of the string to be substituted. The maximum distances can be found in the upper right corner of each map indicating, with with values for the threshold *T* and low size of the regions, *N*. In the rest of the *T* and *N* values the distance is lower than 2.

The next step in texture discrimination is to observe the distances between different textures and to evaluate whether they are significative enough. We compute the distances among the whole set of textures by pairs of textures. Once more, we use different values for the region of exploration (N=5 to 64) and for the threshold (T=0 to 20). Each distance is averaged for 50 distances with N and T fixed.

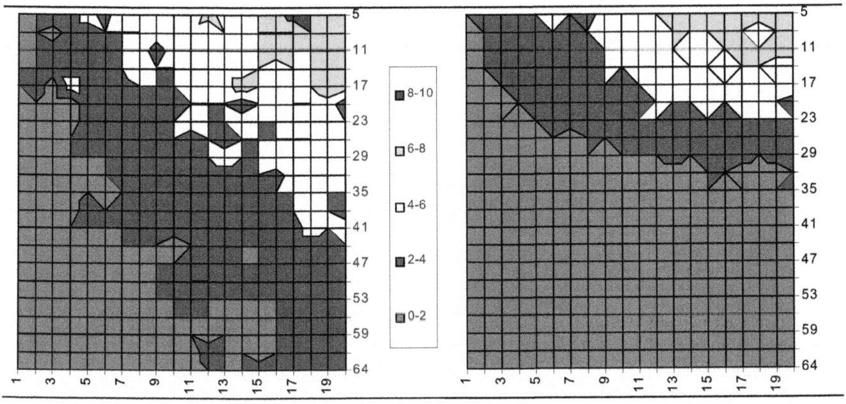

Fig. 3. Left. Distance between texture D38 and D38. Right. Distance between D15 and D15

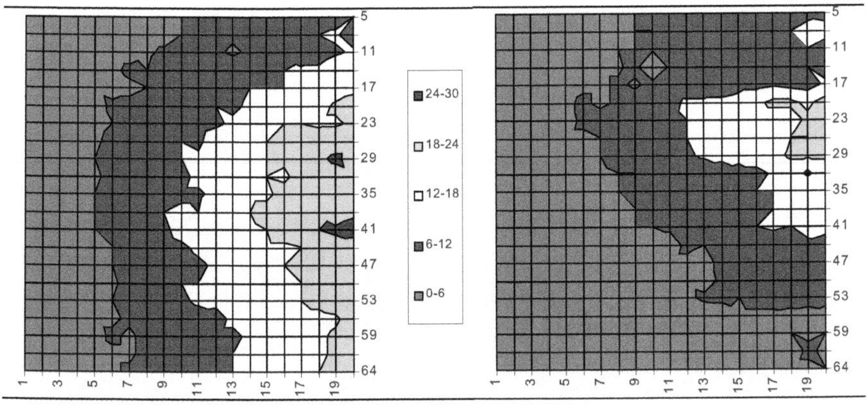

Fig. 4. Left. Distance between texture D38 and D78. Right. Distance between D74 and D92

In figure 4, the distance map of different textures can be seen. The maximum distances has moved to the right center of the map, with high values of the threshold and a medium size for the exploration region. The distances range from 10 to 30 in a big area of the surface map.

Comparing the results between figure 3 and figure 4, the distances between different textures are higher than the distances between a same texture and this effect indicates that the discrimination can be possible. But, a further step is needed in order to find an better value for the size of the regions and the value of the threshold. These values are not always the same for any texture and it is necessary to find the values that best discriminate. For this reason, the best (suboptimal) values can be found by equation (2).

$$\text{Best}(N, T) = \max \frac{\text{Dist}(W_{TexA}, W_{TexB})}{\text{Dist}(W_{TexA}, W_{TexA}) + 1} \tag{2}$$

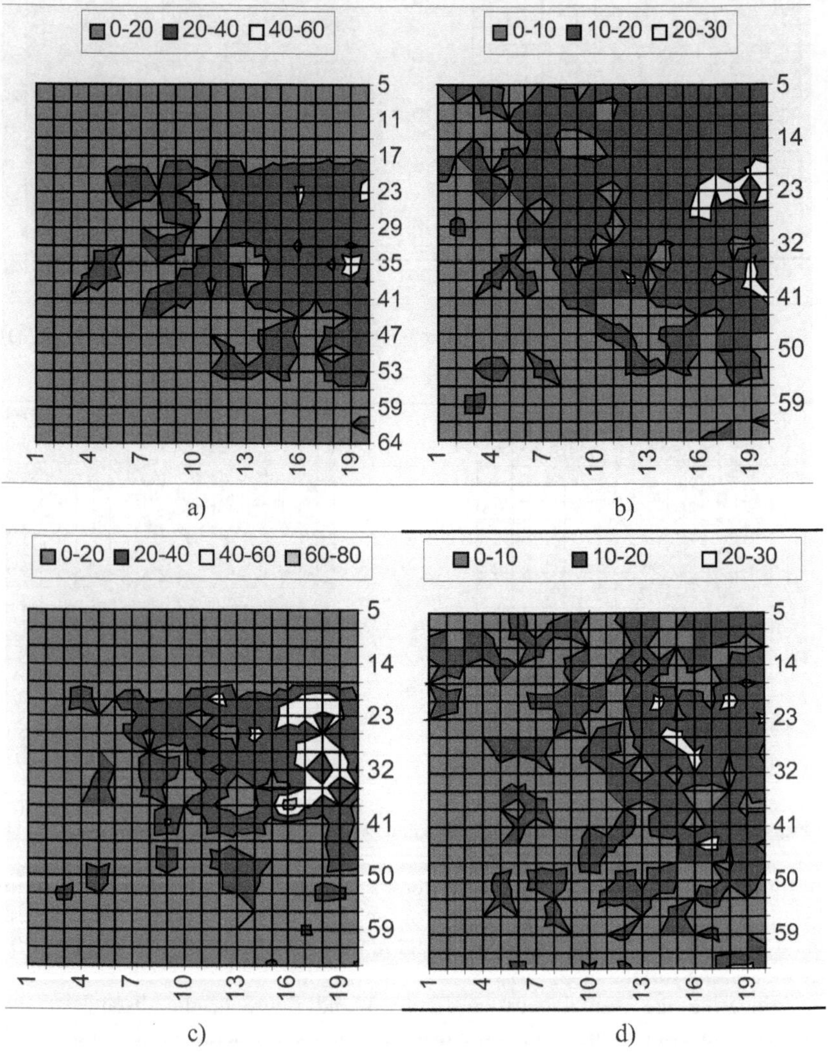

Fig. 5. Plots for the Best N and T parameters, using textures a) D74 and D15; b) D74 and D24; c) D9 and D15; d) D67 and D15

The values that most contribute to the discrimination between texture are those that maximize the relation in equation (2). This quotient is the distance between different texture divided by the distance between similar textures.

In figure 5, a few combinations for best finding the N and T parametres are shown. The values that maximize the equation (2) are located at the right center of the plot, that means, values ranging from 24 to 40 for the N (size of the exploration region) and values from 14 to 19 for T, the threshold value. Therefore, we choose N=32 and T=16, both values inside the intervals and also powers of 2, thinking about present and future hardware implementations for reducing the algorithm cost-time, [3] and [4].

Thus, there is only a last important experiment. Once the values for N and T are already fixed, we compute the distances among the whole set of available textures. In table 1, these distances are shown and they are normalized respect the size of the region, N, because the distance depends on the amount of nodes in the graph.

Table 1. Distances among textures

D38	2																			
D28	16	7																		
D29	33	29	14																	
D16	19	14	23	5																
D24	14	11	27	12	3															
D32	8	15	32	15	12	4														
D12	18	12	30	12	10	13	5													
D74	5	41	30	16	11	12	13	1												
D78	23	46	21	14	18	22	18	21	10											
D2	11	41	27	11	10	10	12	96	17	4										
D9	18	40	27	15	12	16	10	13	17	12	6									
D59	9	42	30	13	11	9	15	7	20	8	14	5								
D5	10	42	29	15	10	7	12	7	20	10	12	8	5							
D23	12	44	28	13	11	10	13	8	19	9	13	8	9	5						
D54	15	41	27	13	11	12	11	11	17	12	11	10	11	10	7					
D67	14	39	29	11	11	16	11	11	17	10	13	14	10	12	10	6				
D92	19	40	23	10	12	14	12	16	15	14	12	16	15	15	12	11	8			
D84	16	41	26	19	11	16	12	15	20	11	11	12	12	11	10	10	11	7		
D22	8	37	29	12	10	7	15	7	20	9	11	6	7	8	10	11	10	11	3	
D15	16	42	29	14	11	12	11	12	18	12	10	11	13	13	12	12	12	11	12	5
Textures	D38	D28	D29	D16	D24	D32	D12	D74	D78	D2	D9	D59	D5	D23	D54	D67	D92	D84	D22	D15

The indexes of table 1 correspond to all the textures used in this experiment. The distances between similar textures (the diagonal of table 1) are, in the whole cases, lower than the distance between different textures. For this reason, we can affirm that a discrimination between textures is achieved using, as a characteristics, histograms **Hx** and **Hy** treated structurally. The quadratic order of the algorithm is not problematic because the length of the strings are short enough to achieve good discrimation results in less than 1 second in a 500-Mhz Pentium PC, for 512x512-pixel input images.

5 Conclusions

We have seen that it is possible to demonstrate empirically certain questions when their formalization is difficult to carry out. Texture has something to do with variations of gray levels in the image and, it is under this assumption that we propose an algorithm for generating a texture print of a region into a image. This print is related to the changes in the sign of the derivative-gray-level intensity function by rows and by columns. These accountings are represented by a string *R* that will be approximate matched with strings obtained from other texture images. The result of this matching will be measured as a distance based on minimum-cost sequences of edit operations. This approximate matching is translation invariant. Through the results presented above, we verify that texture is an implicit characteristic in the images represented by its gray levels and, moreover, it is possible to discriminate regions with different textures. As a future work, we can consider the cyclic string-to-string correction problem as approximate matching reaching an important improve: the comparison between texture prints will be invariant to rotations but, in the other hand, the algorithm is $O(N^2 \log N)$. Another challenge is to implement this algorithm with a specific architecture to be used in, i.e., robot navigation.

References

1. P. Brodatz, *Textures: A Photographic Album for Artist and Designers*, Dover Publishing Co., New York, 1966.
2. H. Bunke and A. Sanfeliu, *Syntactic and Structural Pattern Recognition Theory and Applications*, Series in Computer Science, Vol. 7, World Scientific Publ., 1990.
3. J. Climent, A. Grau, J. Aranda and A. Sanfeliu, "Low Cost Architecture for Structure Measure Distance Computation", *ICPR'98*, Australia, pp. 1592-1594, August 1998.
4. J. Climent, A. Grau, J. Aranda and A. Sanfeliu ,"Clique-to-Clique Distance Computation Using a Specific Architecture", *SSPR'98*, Sydney, Australia, pp. 405-412, August 1998.
5. R.M. Haralick, "Statistical and Structural Approaches to Texture", *Proc. of the IEEE* 67, No. 5, pp. 786-804, 1979.
6. H.-C. Liu and M.D. Srinath, "Classification of partial shapes using string-to-string matching", *Intell. Robots and Comput. Vision*, SPIE Proc. Vol. 1002, pp. 92-98, 1989.
7. S.Y. Lu and K.S. Fu, "A Syntactic Approach to Texture Analysis", *Computer Graphics & Im. Proc.*, Vol. **7**, No. 3, 1978.
8. T. Matsuyama, K. Saburi and M. Nagao, "A Structural Analyzer for Regularly Arranged Textures", *Computer Graphics and Image Processing*, Vol. **18**, pp. 259-278, 1982.
9. D. Sankoff and J.B. Kruskal, eds, Time Warps, String Edit and Macromolecules: The Theory and Practice of Sequence Comparison, Addison-Wesley, Reading, MA, 1983.

10. F. Tomita and S. Tsuji, *Computer Analysis of Visual Textures*, Kluwer Academic Publishers, 1990.
11. W.H. Tsai and S.S. Yu, "Attributed string matching with merging for shape recognition", *IEEE Trans. Patt. Anal. Mach. Intell.* **7**, No. 4, pp. 453-462, 1985.
12. R.A. Wagner et al., "The string-to-string correction problem", *J. Ass. Comput. Mach.* 21, No 1, pp. 168-173, 1974.

Optical Music Interpretation

Michael Droettboom, Ichiro Fujinaga, and Karl MacMillan

Digital Knowledge Center, Milton S. Eisenhower Library
Johns Hopkins University, Baltimore, MD 21218
{mdboom,ich,karlmac}@peabody.jhu.edu

Abstract. A system to convert digitized sheet music into a symbolic music representation is presented. A pragmatic approach is used that conceptualizes this primarily two-dimensional structural recognition problem as a one-dimensional one. The transparency of the implementation owes a great deal to its implementation in a dynamic, object-oriented language. This system is a part of a locally developed end-to-end solution for the conversion of digitized sheet music into symbolic form.

1 Introduction

For online databases of music notation, captured images of scores are insufficient to perform musically meaningful searches (Droettboom et al. 2001) and analyses on the musical content itself (Huron 1999). Such operations require a logical representation of the musical content of the score. To date, creating those logical representations has been very expensive. Methods of input include manually entering data in a machine-readable format (Huron and Selfridge-Field 1994) or hiring musicians to play scores on MIDI keyboards (Selfridge-Field 1993). Optical music recognition (OMR) technology promises to accelerate this conversion by automatically producing the musical content directly from a digitized image of the printed score.

2 The Lester S. Levy Collection of Sheet Music

The present system is being developed as part of a larger project to digitize the Lester S. Levy Collection of Sheet Music[1] (Milton S. Eisenhower Library, Johns Hopkins University) (Choudhury et al. 2001). The Levy Collection consists of over 29,000 pieces of popular American music. Phase One of the digitization project involved optically scanning the music in the collection and cataloging them with metadata such as author, title, and date. Currently, Phase Two of the project involves using OMR to derive the musical information from the score images. The OMR system being developed for this purpose must be flexible and extensible enough to deal with the diversity of the collection.

[1] http://levysheetmusic.mse.jhu.edu

T. Caelli et al. (Eds.): SSPR&SPR 2002, LNCS 2396, pp. 378–387, 2002.
© Springer-Verlag Berlin Heidelberg 2002

3 Overview

For the purposes of this discussion, the problem of optical music recognition is divided into two subproblems: a) the classification of the symbols on the page, and b) the interpretation of the musical semantics of those symbols. The first subproblem has been thoroughly explored and implemented by Fujinaga (1996) in the Adaptive Optical Music Recognition (AOMR) system. The second subproblem, Optical Music Interpretation (OMI), builds on this work and is the subject of this paper, discussed in greater detail in Droettboom (2002).[2]

The AOMR system proceeds through a number of steps. First, the staff lines are removed from the input image file to separate the individual symbols that overlap them. Lyrics are also removed using various heuristics. Commonly occurring symbols, such as stems and noteheads, are then identified and removed using simple filtering techniques. The remaining musical symbols are segmented using connected-component analysis. A set of features, such as width, height, area, number of holes, and low-order central moments, is stored for each segmented graphic object and used as the basis for the adaptive recognition system. The recognition itself is exemplar-based and built around the k-nearest-neighbor (k-NN) algorithm (Cover and Hart 1967). The accuracy of the k-NN database can be improved offline by adjusting the weights of different feature dimensions using a genetic algorithm (GA) (Holland 1975).

Recently, the AOMR part of the system has been extended into a more general and powerful system currently under active development: *Gamera* (MacMillan et al. 2002).

4 Background

In general, OMI involves identifying the relationships between symbols by examining their identities and relative positions, and is therefore a structural pattern recognition problem. From this information, the semantics of the score (e.g. the pitches and durations of notes) can be derived.

A number of approaches to OMI use two-dimensional graph grammars as the central problem-solving mechanism (Fahmy and Blostein 1993; Couasnon and Camillerapp 1994; Baumann 1995). Fahmy and Blostein use a novel approach, called graph-rewriting, whereby complex syntactic patterns are replaced with simpler ones until the desired level of detail is distilled. Graph grammar systems may not be the best fit for the present problem, however, since notated music, though two-dimensional on the page, is essentially a one-dimensional stream. It is never the case that musical objects in the future will affect objects in the past. This property can be exploited by sorting all the objects into a one-dimensional list before performing any interpretation. Once sorted, all necessary operations for interpretation can be performed on the objects quite conveniently. Any errors in the ordering of symbols, often cited as a major difficulty in OMI, in fact tend

[2] All of the software discussed here is open source and licensed under the GNU General Public License, and runs on Microsoft Windows, Apple MacOS X, and Linux.

to be quite localized and simple to resolve. Therefore, graph grammars are not used as part of the present implementation.

Another approach to OMI is present in the underlying data structure of a music notation research application, *Nutator* (Diener 1989). Its T-TREES (temporal trees) are object-oriented data structures used to group objects in physical space and time. Each symbol in a score is composed of a type name, an (x, y) coordinate and a z ordering. Collectively, this object is referred to as a *glyph*. Glyphs exist in a "two-and-a-half dimensional space" and thus can be stacked on top of each other. Glyphs in the foreground communicate with glyphs in the background in order to determine their semantics. For instance, a note would determine its pitch by communicating with the staff underneath it and the clef on top of that staff. This paradigm of communication between glyphs is used heavily throughout the present system. The advantage of this approach is that glyphs can be edited throughout the process at run-time and the results of those changes can be determined very easily.

5 Procedure

In general, the OMI system proceeds in a linear, procedural fashion, applying heuristic rules, and is therefore not a learning system. However, some amount of feedback-based improvement is provided by consistency-checking. In general, due to the diversity and age of our collection, *ad hoc* rules for music notation are used, which are not necessarily those laid out in music notation texts (e.g. Gerou and Lusk 1996).

The OMI system moves through the following steps: input and clean-up, sorting, reference assignment, metric correction, and output. Optionally, the system itself can be tested using an interactive self-debugging system. Each phase of execution is discussed below.

5.1 Input and Clean-Up

The output from AOMR used by OMI is an eXtensible Markup Language (XML) description of the glyphs identified on the page. Each `<glyph>` entry contains a classification, a bounding box, and a list of features. Object instances are created from the input based on a class name, therefore new classes of glyphs can be easily added to the system.

Glyphs that were separated by poor printing or improper binary thresholding are then joined together using heuristic rules.

5.2 Sorting

Since the glyphs are output from AOMR in an arbitrary order, the sorting phase must put them into a useful order for interpretation, i.e. that in which they would be read by a musician. This ordering makes many of the algorithms both easier to write and maintain as well as more efficient.

Contextual information, such as clefs and time signatures, must carry over from one page to the next. The easiest way to deal with this problem is to treat multi-page scores as one very long page. Each page is input in sequence and the bounding boxes are adjusted so that each page is placed physically below the previous one. In this way, multi-page scores are not a special case: they can be interpreted exactly as if they were printed on a single page.

In common music notation, events are read from left to right on each staff. Therefore, before the glyphs can be put into this order, they must first *belong* to a staff. Each glyph will have a reference to exactly one staff. Staff assignment is determined by starting with glyphs that physically overlap a staff and then moving outward to include other related glyphs. Once glyphs have been assigned to staves, those staves need to be grouped into systems (a set of staves performed simultaneously), and then each staff in each system is assigned to a part (a set of notes played by a single performer, or set of performers).

Lastly, in the sorting phase, glyphs are put into musical order. Glyphs are sorted first by part, then voice (see Section 5.3), and then staff. Next, the glyphs are sorted in temporal order from left to right. Finally, glyphs that occur at the same vertical position are sorted top to bottom. This sorted order has a number of useful properties. Most inter-related glyphs, such as noteheads and stems, appear very close together in the list. Finding relationships between these objects requires only a very localized search. `staff` glyphs serve to mark system breaks and `part` glyphs mark the end of the entire piece for each part. Lastly, this ordering is identical to that used in many musical description languages, including GUIDO (Hoos and Hamel 1997), Mudela (Nienhuys and Nieuwenhuizen 1998) and MIDI (MIDI 1986), and therefore output files can be created with a simple linear traversal of the list.

5.3 Reference Assignment

The purpose of this phase is to build the contextual relationships between glyphs to fully obtain their musical meaning. For instance, to fully specify the musical meaning of a notehead, it must be related to a staff, stem, beam, clef, key signature, and accidentals (Figure 1). This is the core of OMI.

Reference assignment proceeds through a number of categories of symbols: pitch, duration, voice, chord, articulation, and text. Most of these processes are performed using a single linear scan through the sorted glyphs, much like a Turing machine.

Fig. 1. References to other glyphs (shaded in grey) are required to fully determine the meaning of a notehead (marked by ×)

Class Hierarchy All glyph classes are members of an object-oriented class hierarchy based on functionality. In this style, most abstract subclasses can be named by adjectives describing their capabilities. For instance, all symbols that can have their duration augmented by dots are subclasses of `DOTTABLE`. This allows new classes of glyphs to be added to the system simply by combining the functionalities of existing classes. It also means that reference-assignment algorithms using these classes can be as abstract as possible. This general design would be much more difficult to implement in more static languages, such as C++, where type modification at run-time is difficult and cumbersome. All of the reference assignment operations described below make extensive use of this class hierarchy.

Pitch OMI has a three-tiered hierarchy of pitch: staff line (which requires a reference to a staff), white pitch (which requires a reference to a clef) and absolute pitch (which requires references to key signatures and accidentals). Each level adds more detail and requires more information (i.e. references to more classes of glyphs) in order to be fully specified. These three different levels are used so that the functionality can be shared between glyphs that use all three, such as notes, and those that only use a subset, such as accidentals.

Determining the correct staff line location of notes on the staff is relatively easy, since most scores have relatively parallel staff lines, pitch can be determined by a simple distance calculation from the center of the staff. However, one of the most difficult problems in determining pitch is dealing with notes outside the staff. Such notes, which require the use of short "ledger" lines, are often placed very inaccurately in hand-engraved scores (Figure 2). The most reliable method to determine the pitches of these notes is to count the number of ledger lines between the notehead and the staff, as well as determining whether a ledger line runs through the middle of the notehead.

Duration Durations are manipulated throughout the system as rational (fractional) numbers. Operations upon `Rational` objects preserve the full precision (e.g. triplet eighth notes are represented as exactly $\frac{1}{3}$).

Assigning stems to noteheads, the single most important step in determining the duration of a note, is a difficult problem since stems are visually identical to

Fig. 2. An example of poorly aligned ledger lines. The grey lines are perfectly horizontal and were added for emphasis

barlines, although they serve a very different purpose. Height alone is not enough information to distinguish between the two, since many stems may be taller than the staff height, particularly if they are part of a chord. Instead, vertical lines are dealt with by a process of elimination.

1. Any vertical lines that touch noteheads are assumed to be stems.
2. Any remaining vertical lines taller than the height of a staff are assumed to be barlines.
3. The remaining vertical lines are likely to be vertical parts of other symbols that have become broken, such as sharps or naturals.

If the guesses made about stem/barline identity turn out to be wrong, they can often be corrected later in the metric correction stage (Section 5.4).

The direction of the stem is determined based on the horizontal location of the stem. If the stem is on the right-hand side, the stem direction is assumed to be up. If the stem is on the left-hand side, the stem direction is down. Stem direction can not be determined based on the vertical position of the stem because the notehead may be part of a chord, in which case the notehead intersects the stem somewhere in the middle. This method must be superseded by a more complex approach for chords containing second (stepwise) intervals, since some of the noteheads are forced to the other side of the stem.

Voices Multi-voicing, where multiple parts are written on the same staff, often occurs in choral music or compressed orchestral scores to conserve space. Just as in multi-page scores, the approach here is to massage the data into a form where it no longer is a special case. Therefore, each voice is split into a separate logical part (Figure 3). Note that some glyphs exist in all logical parts (such as clefs and time signatures) whereas others are split (notes). Determining whether to split a measure into multiple parts is determined automatically.

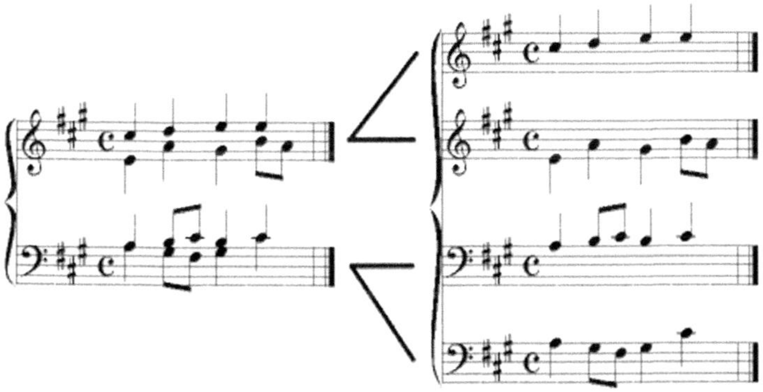

Fig. 3. Splitting multi-voiced scores

5.4 Metric Correction

Physical deterioration of the input score can cause errors at the recognition (AOMR) stage. Missing or erroneous glyphs cause voices to have the wrong number of beats per measure. These errors are quite serious, since they accumulate over time, and parts become increasingly out of synchronization. Fortunately, many of these errors can be corrected by exploiting a common feature of typeset music: notes that occur at the same time are aligned vertically within each system (set of staves) of music. Unfortunately, some poorly typeset scores do not exhibit this feature. In that case, metric correction fails consistently, and is automatically bypassed.

The score is examined, one measure at a time, across all parts simultaneously. A number of approaches are then applied to that measure to correct the durations of notes and rests and barline placement. The primary goal is to ensure that the length of the measure across all parts is the same before moving to the next measure, and to make any corrections in the most intelligent way possible.

At present, there are seven approaches to metric correction that are attempted. For each, a particular change is made, and then the consistency check is performed again. If the change does not improve the measure, the change is undone and the next approach is tried. a) Measures containing only a single rest are adjusted to the length of the entire measure. b) Whole rests and half rests, which are visually identical, are traded and checked for consistency. c) Specks of ink or dust on the page can be confused for augmentation dots. Therefore, augmentation dots are ignored. d) Stems that are too far from a notehead may be interpreted as a barline. These barlines are reexamined as if they were stems. e) Barlines can be missed entirely, and new ones are inserted based on the locations of barlines in other parts. f) Flags and beams can be misread. In this case, the duration of notes is estimated by examining their horizontal position in relation to notes in other parts (Figure 4). g) As a worst case scenario, empty durational space is added to the end of the measure so that all parts have the same duration. This does not usually produce an elegant solution, but it still prevents the errors of one measure to accumulate across an entire piece.

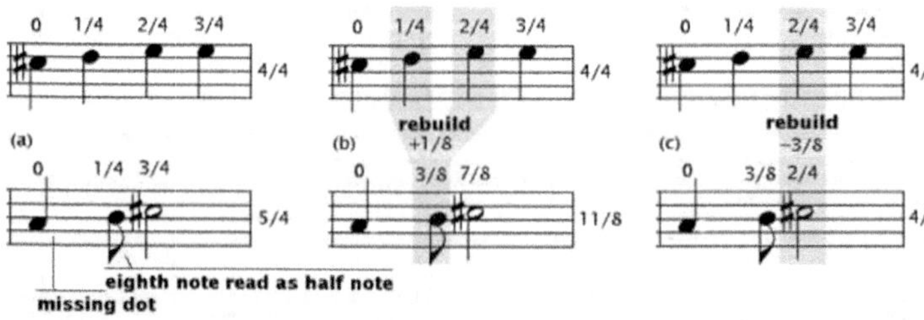

Fig. 4. Adjusting the durations of notes based on the durations in other parts

Metric correction works best in scores with many parts, because there is a large amount of information on which to base the corrections. It is also in multi-part scores where metric correction is most crucial. However, many of the algorithms can improve the accuracy of single-part scores as well.

5.5 Output

Unfortunately, there is no single accepted standard for symbolic musical representation (Selfridge-Field 1997). It is therefore necessary for the present system to support different output formats for different needs. Relying on external converters, as many word processors do, is not ideal, since many musical representation formats have radically different ordinal structures and scope. For example, GUIDO files are organized part by part, whereas Type 0 MIDI files interleave the parts together by absolute time (a temporal stream). To handle this, OMI uses pluggable back-ends that map from OMI's internal data structure, a list of glyphs, to a given output file format. Presently, output to GUIDO and MIDI is implemented, but other musical representation languages such as Lilypond Mudela are planned.

5.6 Interactive Self-Debugger

The ability to interact with the data of a running program, using a scripting language such as Python, greatly reduces the length of the develop-test cycle. However, manipulating graphical data, such as that in OMI, is quite cumbersome using console-based tools. For example, selecting two-dimensional coordinates with a mouse is much easier than entering them numerically. For this reason, a graphical, interactive debugger was implemented that allows the programmer to examine the data structures of a running OMI session and execute arbitrary Python code upon it. The interactive self-debugger proved to be an invaluable tool when developing the OMI application. While extra development effort was expended to create it, those hours were easily made up by the ease with which it allows the programmer to examine the state of the data structures.

6 Conclusion

The system presented here represents a number of pragmatic solutions to the problem, providing a useful tool that is effective on a broad range of scores. In the near future, it will allow for the creation large online databases of symbolic musical data: a valuable resource for both musicologists and music-lovers alike.

Acknowledgements

The second phase of the Levy Project is funded through the NSF's DLI-2 initiative (Award #9817430), an IMLS National Leadership Grant, and support from the Levy Family.

References

[1995] Baumann, S.: A simplified attributed graph grammar for high-level music recognition. International Conference on Document Analysis and Recognition. (1995) 1080–1083 379

[2001] Choudhury, G. S., DiLauro, T., Droettboom, M., Fujinaga, I., MacMillan, K.: Strike up the score: Deriving searchable and playable digital formats from sheet music. D-Lib Magazine. **7(2)** (2001) 378

[1994] Couasnon, B., and Camillerapp, J.: Using grammars to segment and recognize music scores. International Association for Pattern Recognition Workshop on Document Analysis Systems. (1994) 15–27 379

[1967] Cover, T., Hart, P.: Nearest neighbor pattern classification. IEEE Transactions on Information Theory. **13(1)** (1967) 21–27 379

[1989] Diener, G.: TTREES: A tool for the compositional environment. Computer Music Journal. **13(2)** (1989) 77–85 380

[2001] Droettboom, M., Patton, M., Warner, J. W., Fujinaga, I., MacMillan, K., DiLauro, T., Choudhury, G. S.: Expressive and efficient retrieval of symbolic musical data. International Symposium on Music Information Retrieval. (2001) 163–172 378

[2002] Droettboom, M.: Selected Research in Computer Music. Master's thesis. (2002) The Peabody Institute of the Johns Hopkins University. 379

[1993] Fahmy, H., D. Blostein.: A graph grammar programming style for recognition of music notation. Machine Vision and Applications. **6(2)** (1993) 83–99 379

[1996] Fujinaga, I.: Adaptive Optical Music Recognition. (1996) Ph. D. thesis, McGill University. 379

[1996] Gerou, T., L. Lusk.: Essential Dictionary of Music Notation. (1996) Alfred, Los Angeles. 380

[1975] Holland, J. H.: Adaptation in Natural and Artificial Systems. (1975) University of Michigan Press, Ann Arbor. 379

[1997] Hoos, H. H., Hamel, K.: GUIDO Music Notation Version 1.0: Specification Part I, Basic GUIDO. (1997) Technical Report TI 20/97, Technische Universität Darmstadt. 381

[1999] Huron, D.: Music Research Using Humdrum: A User's Guide. (1999) Center for Computer Assisted Research in the Humanities, Menlo Park, CA. 378

[1994] Huron, D., Selfridge-Field, E.: Research notes (the J. S. Bach Brandenburg Concertos). (1994) Computer software. 378

[2002] MacMillan, K., Droettboom, M., Fujinaga, I.: Gamera: A Python-based toolkit for structured document recognition. Tenth International Python Conference. (2002) (In press) 379

[1986] MIDI Manufacturers Association Inc.: The Complete MIDI 1.0 specification. (1986) 381

[2000] Musitek.: MIDISCAN. Computer Program (Microsoft Windows).

[2000] Neuratron. Photoscore. Computer Program (Microsoft Windows, Apple MacOS).

[1998] Nienhuys, H., Nieuwenhuizen. J.: LilyPond User Documentation (Containing Mudela Language Description). (1998) 381

[2000] Van Rossum, G., Drake, F. L.: Python Tutorial. (2000) iUniverse, Campbell, CA.

[1993] Selfridge-Field, E.: The MuseData universe: A system of musical information. Computing in Musicology **9** (1993) 11–30 378

[1997] Selfridge-Field, E. Beyond MIDI: The Handbook of Musical Codes. (1997) MIT Press, Cambridge, MA. 385

On the Segmentation of Color Cartographic Images

Juan Humberto Sossa Azuela, Aurelio Velázquez, and Serguei Levachkine

Centro de Investigación en Computación – IPN
Av. Juan de Dios Bátiz s/n, Esq. Miguel Othón de Mendizábal
UPALM-IPN Zacatenco, México. D. F. C.P. 07738
avela@cic.ipn.mx, hsossa@cic.ipn.mx, palych@cic.ipn.mx

Abstract. One main problem in image analysis is the segmentation of a cartographic image into its different layers. The text layer is one of the most important and richest ones. It comprises the names of cities, towns, rivers, monuments, streets, and so on. Dozens of segmentation methods have been developed to segment images. Most of them are useful in the binary and the gray level cases. Not to many efforts have been however done for the color case. In this paper we describe a novel segmentation technique specially applicable to raster-scanned color cartographic color images. It has been tested with several dozen of images showing very promising results.

1 Introduction

Color cartographic images are very important commercially speaking but complex at the same time. They contain many information usually divided in layers: the text layer, the river layer, the symbol layer, and so on. Segmentation of an image like these into its layers is a complex task because in general the information from the different layers is mixed. Letters might share the same color as text, river traces or street paths.

One of the most important layers in any cartographic map is the text layer. It can help us to identify cities, towns, rives, lakes, monuments, streets, etcetera. One might apply a simple thersholding technique trying to separate the text layer from a color cartographic image. Thresholding has been applied with success for many years to isolate objects of interest from their background. Dozens of techniques have been described in the literature. Refer for example to [1]-[6] for isolated examples and to [7] and [8] for good surveys.

Image thresholding is mostly applicable to the case of high contrasted images (images where objects contrast strongly with respect to the background). In general, however, this is not the case because objects in most gray-level images normally share the same gray-levels. Almost any simple thresholding method will not adequately (as requested by humans beings) isolate the desired objects. Intuitively, trying to do the same with color images will be even more complicated. Several segmentation techniques have developed during these last years for the color images. For a very good survey see [9].

T. Caelli et al. (Eds.): SSPR&SPR 2002, LNCS 2396, pp. 387-395, 2002.
© Springer-Verlag Berlin Heidelberg 2002

In this paper we introduce a novel technique specially designed to work with color images. The proposed technique has been tested with several raster-scanned color cartographic images with excellent results.

2 Preliminaries

Most of the existing color-based image segmentation techniques use the R, G and B information contained in the image to accomplish the task. Our proposal also uses this information plus the average of the combination of two of them, the average of the three, and the transformation of the color image to an intensity image, transforming it into an image YIQ (Y - luminance, I - In-phase and Q - Quadratic) by selecting only the luminance component. The eight images considered by our segmentation algorithm are thus the following:

M1="R",

M2="G",

M3="B",

M4=int(("R"+"G")/2),

M5=int(("R"+"B")/2),

M6=int(("G"+"B")/2),

M7=int(("R"+"G"+"B")/3),

M8=color_to_gray("RGB").

Let us denote these images as the 8 M-images of a standard color image.

3 The Novel Technique

The proposed technique to segment raster-scanned color cartographic images is divided in three main stages: Image pre-processing, image processing and image post-processing. Each one of these three stages is next described.

3.1 Image Preprocessing

During this stage, the goal it to accentuate the pixels of the objects of interest in the image: the text pixels in our case, those coming from alpha-numeric characters. This is done by first decomposing the test mage into its 8 M-images. Fig. 1 shows an image and its eight M-sub-images. Each M-sub-image is next adjusted in contrast fit the range of 0 to 255 gray levels. The same original 8 M-sub-image are histogram-equalized. This results in 24 sub-images (8 original M-sub-images, 8 normalized sub-images and 8 equalized sub-images). Figures 2 and 3 show, respectively the 8 normalized sub-images and the 8 equalized sub-images of the 8 M-sub-images of Fig. 1(b).

3.2 Image Processing

The goal here is to further emphasize the pixels of interest. This stage is divided in two steps: Image pre-threholding and image post-thresholding. These steps are next explained.

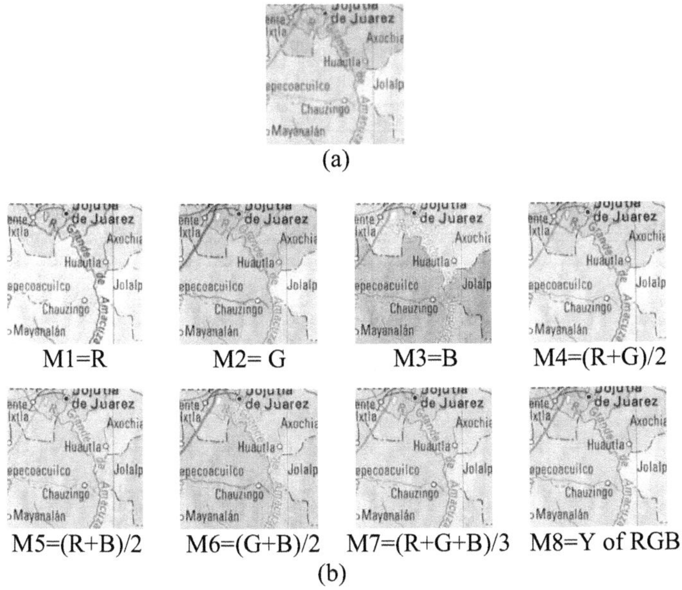

Fig. 1. (a) A color cartographic image and (b) its 8 M-images.

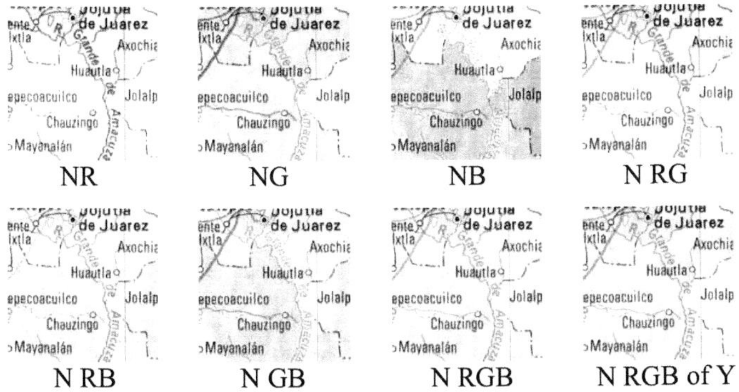

Fig. 2. Normalized sub-images of the eight M sub-images of Fig. 1(b).

3.2.1 Image Pre-segmentation

The 24 images shown in Figs. 1(b), 2 and 3 are next thresholded to get their 24 binary versions. The goal as mentioned before is to strongly emphasize the presence of text in the image. Next we describe how to automatically select the desired threshold.

Before describing how to choose the desired threshold let us say some words. It is well known that threshold selection is not only a critical but also a very difficult task in the general case. A bad threshold would result in very bad results; a good threshold would provide in the contrary very nice results.

One very well known way to obtain the threshold is by using the histogram of the image (see for example [10]). We have used this approach at the beginning of our research with very bad results. We have experimented with several algorithms. After many tries we have retained the following one:

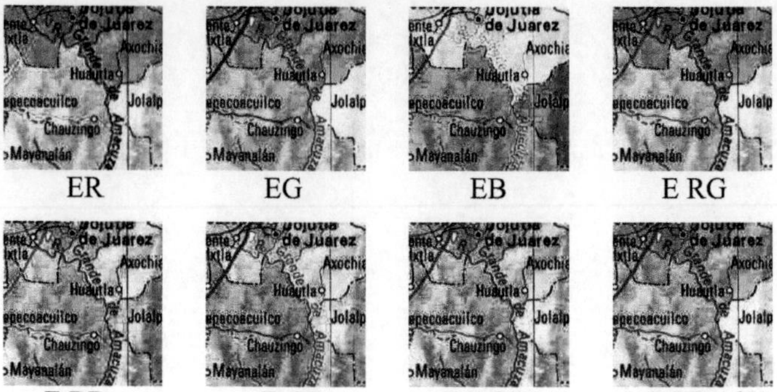

ER EG EB E RG

E RB E GB E RGB E RGB of Y

Fig. 3. Histogram equalized sub-images of the eight sub-images of Fig. 1(b)

For each of the 24 for sub-images:

Step 1. Convolve sub-image with the following modified Prewit's edge masks in both the x and the y directions:

1/6

-1	0	1
-1	2	1
-1	0	1

1/6

-1	-1	-1
0	2	0
1	1	1

Step 2. Instead of computing the magnitude of the gradient as usual, add pixel-by-pixel the two resulting images to get once again one image, say g.

Step 3. Add all the gray-level values of this image to get just one number.

Step 4. Divide this number by the total size of the image (the number of pixels of the image) to get the desired threshold u. If g has M rows and N columns then,

$$u = \frac{1}{MN} \sum_{i=1}^{M} \sum_{j=1}^{N} g(i, j) \tag{1}$$

Step 5. Threshold the sub-image with this u.

Repeat steps 1 to 5 for the 24 sub-images to get the desired binary version.

We have found that a threshold computed like this provides very good results. We could also use another edge detector such as the Roberts or Sobel operators with very similar results.

In the case of histogram-equalized images the applied threshold is the original one u (obtained by means of Eq. (1)) but divided by seven. If the original threshold was applied directly to an equalized image a completely black image would result. This factor was found empirically. Fig. 4 shows the 24 binary versions of the 24 sub-images shown in Figs. 1(b), 2 and 3.

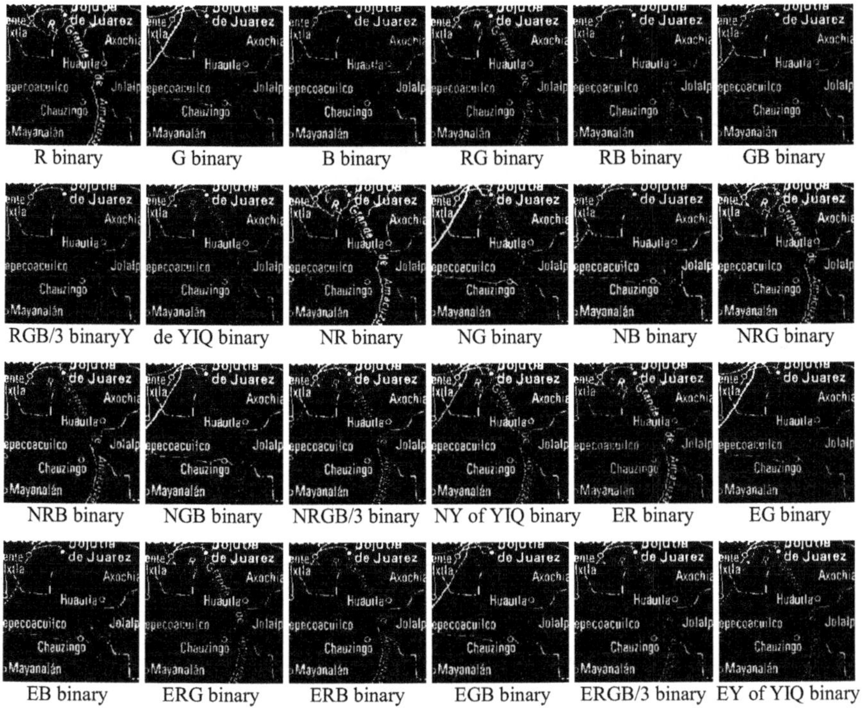

Fig. 4. The 24 binary versions of the 24 sub-images shown in Figs. 1(b), 2 and 3

A comment. We have used an edge detector to get the desired threshold due to letters and numerals are associated in their frontiers with abrupt changes with respect to the image's background. The use of an edge detector would thus accentuate letters and numerals at least in theirs frontiers. The application of a method like this would also emphasize the presence of rivers, country borders and so on. These, however, could be later eliminated if desired, for example, to only separate the alphanumeric layer of the image.

3.2.2 Image Post-segmentation and Region Labeling

The white pixels in each of the 24 binary sub-images are representative of strokes of a letters or numerals, the border of a river, the limit between cities, and so on. They appear however fragmented and not clustered to form connected regions of letters or numerals, in some images appear incomplete, in some others they appear complete.

To solve this problem we have used the following heuristic method. We first obtain one image from all the 24 binary sub-images by simple pixel-by-pixel addition of all the 24 images. The value of a pixel in this new image will oscillate between 0 and 24. To obtain again a binary image we verify if the sum at a given (x,y) position is greater that 21 (a manually selected value). If it is true we put a 1 into the buffer image and zero otherwise. We have thus another binary image with 1's in the most probably positions representing letters, numerals and other symbols and 0's in the backgrounds areas. Fig. 5(a) shows the resultant binary sub-image.

Fig. 5. (a) Binary sub-image after thresholding the image resulting when adding pixel-to-pixel the 24 sub-images shown in Fig 4. (b) Resulting image after filling the gaps on (a). (c) Resulting image after eliminating the pixels on Fig. 1 as explained in the text.

We next apply a standard labeling algorithm [10] to this image to get adjacent regions representing the desired elements (letters, numerals, and other symbols). Let's call this labeled image the P-image.

3.3 Image Post-processing

If you take a look to Fig. 5(a), the adjacent connected regions in this image obtained with the procedure just described appear not well emphasized as desired and yet fragmented.

The gaps between isolated regions must be filled to get the complete desired regions. These gaps are filled by means of the following procedure:

Procedure FILLING GAPS

The input is a P-image.

1. Select a connected region of pixels in the P-image. Dilate this region with a 3x3 structural element, from 3 to 15 times, depending on the number of pixels of the region. Each 10 pixels add one morphological-dilation to the process. This process gives as a result a square window whose size is given by the number of dilations. Let us call this resulting square window the D-mask.
2. AND-Image-mask (see [11], pp. 50 for the details) the 8 M-sub-images with the D-mask obtained in step 1 in order to compute the average gray-level and the standard deviation of this region. Only the pixel values under the 1's in the D mask are taken into account.
3. Turn off all pixels in the D-mask if the corresponding gray-level value if any of the M-sub-image is greater than the gray-level average plus the standard deviation value obtained in step 2. This allows, on the one hand, to eliminate the undesired background pixels added to the alphanumeric character during the

dilation process. On other hand, this step permits to aggregate missing pixels of the character.

4. Apply steps 1-3 to all labeled regions inside the P-image.

The output is another image (the F-image) with the isolated characters with gaps filled.

Fig. 5(b) shows the resulting image after applying the filling gap procedure just described to the image shown in Fig. 5(a). Note how the gaps between isolated regions have disappeared, and the letters and numerals now appear more complete.

Fig 5(c) shows the resulting image after the segmented pixels (those appearing in Fig. 5(b)) were eliminated from the original image (Fig. 1(a)). Fig. 5(c) was obtained by using Fig. 5(b) as a mask. If a pixel in Fig. 5(b) is 1 (white), its corresponding pixel in Fig. 1(a) is substituted by an interpolated value obtained by averaging the surrounding pixels of the pixel.

From Figs. 5(b) and 5(c), you can see that some letters, rivers and other small objects did not were separated from the original image (Fig. 1(a)). They do not appear in white in Fig. 5(b), they do appear however in Fig 5(c). We would like of course these objects to be also segmented. To accomplish this we have applied our technique to other RGB combinations: RG, RB and GB. As we will next see the remaining objects of interest will be accentuated and thus also segmented.

The M-images for an image are not 8 but 5. For the RG combination the 5 M-images are: M1, M2, M4, M7 and M8. For the RB combination were M1, M3, M5, M7 and M8. For the GB combination were M2, M3, M6, M7 and M8. For a given combination now five normalized and five equalized images were obtained. Fifteen thresholded images as explained in Section 3.2.1 were obtained from each of the three combinations and their 10 images. Each set of ten images was processed as explained in Sections 3.3.2 and 3.3 to get the final images with gaps filled. Fig. 6 shows the three resulting images. Note how the missing letters in Fig. 5(a) (appearing in Fig. 5(c)) are now well emphasized in Fig. 6(a). The path of the river under "R. Grande Amacuza", also missing in Fig 5(a) appears also accentuated in Fig. 6(a). Note also how the three rivers missing in Fig. 5(a) (appearing in red in Fig. 5(c)) are well emphasized in Fig. 6(c).

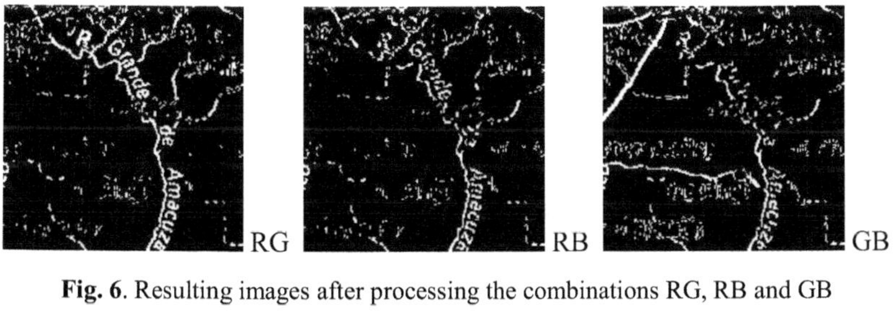

Fig. 6. Resulting images after processing the combinations RG, RB and GB

One important feature of our proposal is that each combination emphasizes different elements of the original image. The RGB combination strongly emphasizes black objects. The RG combination emphasizes mostly blue objects such as river names and their paths and the GB combination the red elements such as roads. The RB combination in this case appears to be useless.

As we are interested in text characters, we can now take each character from each resulting image along with its position and stock them into a file for further processing.

4 Results

In this section additional testing results are shown. Figure 7 (top) shows six raster-scanned color cartographic images. All of these were processed as described in the previous sections to get their final segmented versions. Only the RGB combination was used. Figure 7 (down) shows these final versions. You can appreciate that the final results are good enough. The methodology has been tested until now with more than 100 images with very promising results.

Fig. 7. Other images (top) and their segmented results only on their RGB combinations

5 Conclusions

In this short paper a novel threshoding approach applicable for color images has been proposed. It incorporates three main stages: Image preprocessing, image processing and image post-processing.

This approach has been tested with many raster-scanned color cartographic images giving promising results.

The resulting images are being now analyzed by another module that will allow to separate alphanumeric characters from other objects. The goal is to isolate as much as possible each alphanumeric character and determine its identity by a trained classifier.

Acknowledgments

The authors would like to express their acknowledgment to CONACYT under grants 34880-A and 32019-A and to the Centro de Investigación en Computación of the IPN from Mexico for their support for the development of this research.

References

1. N. Otsu, A threshold selection method from gray-level histograms, IEEE Transactions on Systems, Man and Cybernetics, 9(1): 62-66, 1979.
2. J. N. Kapur, P. S. Sahoo and A. K. C. Wong, A new method for gray-level picture thresholding using entropy of the histogram, Computer Graphics and Image Processing, 29:273-285, 1985
3. J. Kittler and J. Illingworth, Minimun error thresholding, Pattern Recognition, 19:41-47, 1986.
4. P. Sahoo, C. Wilkings and J. Yeager, Threshold selection using Renyi's entropy, Pattern Recognition, 30(1):71-84, 1997.
5. L. Li, J. Gong and W. Chen, Gray-level image thresholding based on Fisher linear projection of two-dimensional histogram, Pattern Recognition, 30(5):743-749, 1997.
6. X. J. Wu, Y.J. Zhang and L. Z. Xia, A fast recurring two-dimensional entropic thresholding algorithm, Pattern Recognition, 32:2055-2061, 1999.
7. J. S. Weska, A survey of threshold selection techniques, Computer Graphics and Image Processing, 7:259-265, 1978.
8. P. S. Sahoo, . Soltani, A. K. C. Wong and Y. Chen, A survey of thresholding techniques, Computer Graphics and Image Processing, 41:233-260, 1988.
9. H. D. Cheng, X. H. Jiang, Y. Sun and J. Wang, Color image segmentation: advances and prospects, Pattern recognition, 34(12):2259-2281, 2001.
10. R. C. Gonzalez and R. E. Woods, Digital image processing, Addison Wesley Pub. Co. 1993.
11. S. E. Umbaugh, Computer Vision and Image Processing: A practical Approach using CVIPtools, Prentice Hall PTR, NJ, 1998.

Projection Pursuit Fitting Gaussian Mixture Models*

Mayer Aladjem

Department of Electrical and Computer Engineering
Ben-Gurion University of the Negev
P.O.B. 653, 84105 Beer-Sheva, Israel
http://www.ee.bgu.ac.il/~aladjem/

Abstract. *Gaussian mixture models* (GMMs) are widely used to model complex distributions. Usually the parameters of the GMMs are determined in a *maximum likelihood* (ML) framework. A practical deficiency of ML fitting of the GMMs is the poor performance when dealing with high-dimensional data since a large sample size is needed to match the numerical accuracy that is possible in low dimensions. In this paper we propose a method for fitting the GMMs based on the *projection pursuit* (PP) strategy. By means of simulations we show that the proposed method outperforms ML fitting of the GMMs for small sizes of training sets.

1 Introduction

We consider the problem of modeling an n-variate probability density function $p(\mathbf{x})$ ($\mathbf{x} \in R^n$) on the basis of a training set

$$X = \{\mathbf{x}_1, \mathbf{x}_2, \ldots, \mathbf{x}_N\}. \tag{1}$$

Here $\mathbf{x}_i \in R^n$; $i = 1, 2, \ldots, N$ are data points drawn from that density. We require a normalization of the data, called *sphering* [7] (or *whitening* [4]). For the sphered X the sample covariance matrix becomes the identity matrix and the sample mean vector is a zero vector. In the remainder of the paper, all operations are performed on the sphered data.

In this paper we seek a *Gaussian mixture model* (GMM) of $p(\mathbf{x})$, which is a linear combination of M Gaussian densities

$$\hat{p}(\mathbf{x}) = \sum_{j=1}^{M} \omega_j \phi_{\mathbf{\Sigma}_j}(\mathbf{x} - \mathbf{m}_j). \tag{2}$$

Here, ω_j are the mixing coefficients which are non-negative and sum to one, and $\phi_{\mathbf{\Sigma}_j}(\mathbf{x} - \mathbf{m}_j)$ denotes $N(\mathbf{m}_j, \mathbf{\Sigma}_j)$ density in the vector $\mathbf{x}$.

* This work was supported in part by the Paul Ivanier Center for Robotics and Production Management, Ben-Gurion University of the Negev, Israel.

T. Caelli et al. (Eds.): SSPR&SPR 2002, LNCS 2396, pp. 396–404, 2002.
© Springer-Verlag Berlin Heidelberg 2002

The mixture model is widely applied due to its ease of interpretation by viewing each fitted Gaussian component as a distinct cluster in the data. The clusters are centered at the means $\mathbf{m}_j$ and have geometric features (shape, volume, orientation) determined by the covariances $\mathbf{\Sigma}_j$.

The problem of determining the number M of the clusters (Gaussian components) and the parameterization of $\mathbf{\Sigma}_j$ is known as model selection. Usually several models are considered and an appropriate one is chosen using some criterion, such as the *Bayesian information criterion* (BIC) [6].

In this paper we study GMMs with *full* (unrestricted) covariance matrices $\mathbf{\Sigma}_j$, *spherical* $\mathbf{\Sigma}_j$ with a single parameter for the whole covariance structure and *diagonal* $\mathbf{\Sigma}_j$. The use of GMMs with full covariance matrices leads to a large number of parameters for high-dimensional input vectors and presents the risk of over-fitting. Therefore $\mathbf{\Sigma}_j$ are often constrained to be spherical and diagonal. The latter parameterizations do not capture correlation of the variables and cannot match the numerical accuracy that is possible using unrestricted $\mathbf{\Sigma}_j$. Additionally, the diagonal GMMs are strongly dependent on the rotation of the data. An attractive compromise between these parameterizations is the recently introduced mixture of latent variable models. In this paper we study a latent variable model, called a mixture of *probabilistic principal component analyses* (*PPCA*) [5]

$$\hat{p}(\mathbf{x}) = \sum_{j=1}^{M} \omega_j \phi_{(\sigma_j^2 \mathbf{I} + \mathbf{W_j W_j^T})}(\mathbf{x} - \mathbf{m}_j), \tag{3}$$

where W_j is a $(n \times q)$ matrix. The dimension q is called the latent factor. For $q < n$ an unrestricted $\mathbf{\Sigma}_j$ (not spherical or diagonal) can be captured using only $(1 + nq)$ parameters instead of the $(n(n+1)/2)$ parameters required for the full covariance matrix. Usually the parameters of the conventional and latent GMMs are determined in a *maximum likelihood* (ML) framework [4], [5].

In this paper we propose a method for fitting GMMs based on the *projection pursuit* (PP) density estimation [7], [8]. By means of simulations we show that our method outperforms the ML methods for small sizes of the training samples.

2 Projection Pursuit Fitting GMMs

We propose to set the parameters of GMM (2) using the *projection pursuit* (PP) density estimation [7], [8] proposed by Friedman. In Section 2.1 we summarize the original method of Friedman and in Section 2.2 we present our method for fitting GMMs.

2.1 Friedman's PP Density Estimation

Friedman [7], [8] proposed to approximate the density $p(\mathbf{x})$ by multiplication of K univariate functions $f_k(.)$

$$\hat{p}(\mathbf{x}) = \phi(\mathbf{x}) \prod_{k=1}^{K} f_k(\mathbf{a}_k^T \mathbf{x}), \tag{4}$$

where $\phi(\mathbf{x})$ is $N(\mathbf{0},\mathbf{I})$ density in the vector $\mathbf{x}$ (the standard normal n-variate probability density function) , $\mathbf{a}_k$ is an unit vector specifying a direction in R^n and f_k is

$$f_k(y) = \frac{\hat{p}_k(y)}{\phi(y)}. \tag{5}$$

Here $\phi(y)$ denotes $N(0,1)$ density in the variable y and $\hat{p}_k(y)$ is a density function along $\mathbf{a}_k$. Friedman approximate/estimate $\hat{p}_k(y)$ using the Legendre polynomial expansion of the density along $\mathbf{a}_k$. The directional vectors $\mathbf{a}_k$ are set by the projection pursuit strategy explained in Appendix A.

2.2 GMM Expansion of the PP Density Estimation

In order to expand (4) to the multivariate GMM we model $\hat{p}_k(y)$ in (5) by a mixture of the univariate normals

$$\hat{p}_k(y) = \sum_{j=1}^{M_k} \omega_{kj}\phi_{\sigma_{kj}}(y-\mu_{kj}). \tag{6}$$

Here $\phi_{\sigma_{kj}}(y-\mu_{kj})$ denotes $N(\mu_{kj},\sigma_{kj})$ density in the variable y and ω_{kj} are the mixing coefficients for $j=1,2,\ldots M_k$. After manipulations of (5) using (6) $f_k(y)$ becomes

$$f_k(y) = \sum_{j=1}^{M_k} \tilde{\omega}_{kj}\phi_{\tilde{\sigma}_{kj}}(y-\tilde{\mu}_{kj}), \tag{7}$$

with

$$\tilde{\omega}_{kj} = \omega_{kj}\sqrt{\frac{2\pi}{1-{\sigma_{kj}}^2}}\, exp\left(\frac{{\mu_{kj}}^2}{2(1-{\sigma_{kj}}^2)}\right), \tag{8}$$

$$\tilde{\mu}_{kj} = \frac{\mu_{kj}}{1-{\sigma_{kj}}^2}, \tag{9}$$

$$\tilde{\sigma}_{kj} = \frac{\sigma_{kj}}{\sqrt{1-{\sigma_{kj}}^2}}. \tag{10}$$

Substituting (7) into (4), we have

$$\hat{p}(\mathbf{x}) = \phi(\mathbf{x})\prod_{k=1}^{K}\left[\sum_{j=1}^{M_k}\tilde{\omega}_{kj}\phi_{\tilde{\sigma}_{kj}}(\mathbf{a}_k^T\mathbf{x}-\tilde{\mu}_{kj})\right]. \tag{11}$$

Finally, we employ the identity

$$\phi_{\mathbf{\Sigma}}(\mathbf{x}-\mathbf{m})\phi_\sigma(\mathbf{a}^T\mathbf{x}-\mu) = \alpha\phi_{\tilde{\mathbf{\Sigma}}}(\mathbf{x}-\tilde{\mathbf{m}}), \tag{12}$$

with $\mathbf{x}$, $\mathbf{m}$, $\mathbf{a}\in R^n$; $\mathbf{a}^T\mathbf{a}=1$ and

$$\tilde{\mathbf{\Sigma}} = \mathbf{\Sigma} - \frac{\frac{1}{\sigma^2}\mathbf{\Sigma}\mathbf{a}\mathbf{a}^T\mathbf{\Sigma}}{1+\frac{1}{\sigma^2}\mathbf{a}^T\mathbf{\Sigma}\mathbf{a}}, \tag{13}$$

$$\tilde{\mathbf{m}} = \tilde{\mathbf{\Sigma}}\mathbf{\Sigma}^{-1}\mathbf{m} + \frac{\mu}{\sigma^2}\tilde{\mathbf{\Sigma}}\mathbf{a}, \tag{14}$$

$$\alpha = \frac{|\tilde{\mathbf{\Sigma}}|^{\frac{1}{2}}}{\sqrt{2\pi}\sigma|\mathbf{\Sigma}|^{\frac{1}{2}}} exp\{\tfrac{\mu^2}{2\sigma^2}(\tfrac{1}{\sigma^2}\mathbf{a}^T\tilde{\mathbf{\Sigma}}\mathbf{a} - 1) + \tfrac{1}{2}\mathbf{m}^T(\mathbf{\Sigma}^{-1}\tilde{\mathbf{\Sigma}}\mathbf{\Sigma}^{-1} - \mathbf{\Sigma}^{-1})\mathbf{m} + \tfrac{\mu}{\sigma^2}\mathbf{a}^T\tilde{\mathbf{\Sigma}}\mathbf{\Sigma}^{-1}\mathbf{m}\}. \tag{15}$$

The proof of formulae (12) - (15) will be included in an extended version of this paper.

The identity (12) shows that the multiplication of any n-variate normal density $\phi_{\mathbf{\Sigma}}(\mathbf{x}-\mathbf{m})$ by any univariate normal density $\phi_\sigma(\mathbf{a}^T\mathbf{x}-\mu)$ along a directional vector $\mathbf{a} \in R^n$ implies an n-variate normal density $\phi_{\tilde{\mathbf{\Sigma}}}(\mathbf{x} - \tilde{\mathbf{m}})$ scaled by a constant α. After an iterative application of the identity (12) into (11), Friedman's approximation (4) becomes the form of an GMM

$$\hat{p}(\mathbf{x}) = \sum_{j=1}^{\tilde{M}} \tilde{\omega}_j \phi_{\tilde{\mathbf{\Sigma}}_j}(\mathbf{x} - \tilde{\mathbf{m}}_j) \tag{16}$$

having $\tilde{M} = \prod_{i=1}^{K} M_i$ Gaussian components. We name (16) the GMM expansion of the PP density estimation (4). Here $\tilde{\omega}_j$, $\tilde{\mathbf{\Sigma}}_j$ and $\tilde{\mathbf{m}}_j$ denote the parameter values implied by the iterative application of (12) - (15) into (11). The GMM expansion (16) can be simplified, i.e. the number $\tilde{M}$ of the Gaussian components can be reduced by suitable replacement of the similar components with a single normal. The latter is out of the scope of this paper and is subject of our current research.

2.3 Fitting Strategy

In the previous Section 2.2 we showed that Friedman's approximation (4) implies a GMM model (16) for the specific choice (6) of $\hat{p}_k(y)$. For this scenario the purpose of the *PP* fitting is to choose K and $\mathbf{a}_k$ of the model (4), and to set the parameters M_k, ω_{kj}, μ_{kj} and σ_{kj} of the univariate mixture density $\hat{p}_k(y)$ (6).

We compute K and $\mathbf{a}_k$ by a method of Friedman, called *projection pursuit* (PP). We summarize the PP method in the Appendix A. The PP method computes each $\mathbf{a}_k$ for a specific data set $X^{(k)} = \{\mathbf{x}_1^{(k)}, \mathbf{x}_2^{(k)}, \dots, \mathbf{x}_N^{(k)}\}$ (18). In the next explanation we refer to $X^{(k)}$.

Our strategy for setting the parameters M_k, ω_{kj}, μ_{kj} and σ_{kj} of $\hat{p}_k(y)$ (6) is based on the *maximum likelihood* (ML) technique [4, pages 65-72] and the *Bayesian information criterion* (BIC) [6]. In summary, it is as follows. First we project the data points $\mathbf{x}_i^{(k)} \in X^{(k)}$, $i = 1, 2, \dots, N$ onto $\mathbf{a}_k$. We denote the projections $y_i = {\mathbf{a}_k}^T\mathbf{x}_i^{(k)}$. Then for $M_k = 1, 2, \dots, M_{max}$ we fit $\hat{p}_k(y)$ to the data points $y_i, i = 1, 2, \dots, N$ by the ML technique. The maximal number M_{max} of the components of $\hat{p}_k(y_i)$ is set by the user (in our experiments described in Section 3 we set $M_{max} = 10$). For each M_k we compute the value of the *log-likelihood function* L_{M_k} ($L_{M_k} = \sum_{i=1}^{N} ln\hat{p}_k(y_i)$) at the maximized values of the parameter ω_{kj}, μ_{kj} and σ_{kj}. Then we compute the values $BIC_{M_k} = 2L_{M_k} - (3M_k - 1)ln(N)$ [6] and plot them for $M_k = 1, 2, \ldots, M_{max}$. Finally, following [6], we select the model having the number M_k giving rise to a decisive first local maximum of the BIC values. In the case of monotonically decreasing BIC values we drop $\mathbf{a}_k$ from Friedman's approximation (4).

3 Comparative Studies

In this section, we compare the performance of the *maximum likelihood* (ML) [4], [5] and the *projection pursuit* (PP) (Section 2) fittings the GMMs. We study a wide spectrum of situations in terms of the size N of the training samples (1) drawn from 15-dimensional trimodal densities $p_{\mathbf{IK}}(\mathbf{x})$, $p_{\mathbf{JK}}(\mathbf{x})$, $p_{\mathbf{IJ}}(\mathbf{x})$ and $p_{\mathbf{IJK}}(\mathbf{x})$ set in Appendix B. We ran experiments for $N = 50, 100, 150, \ldots, 700$.

An experiment for a given combination of particular setting, density function and size of the training sample consisted of 10 replications of the following procedure. We generated training data of size N from an appropriate distribution. Then we normalized (sphered [7]) the data and rotated the coordinate system randomly in order not to favor the rotating dependent GMMs. Using this data

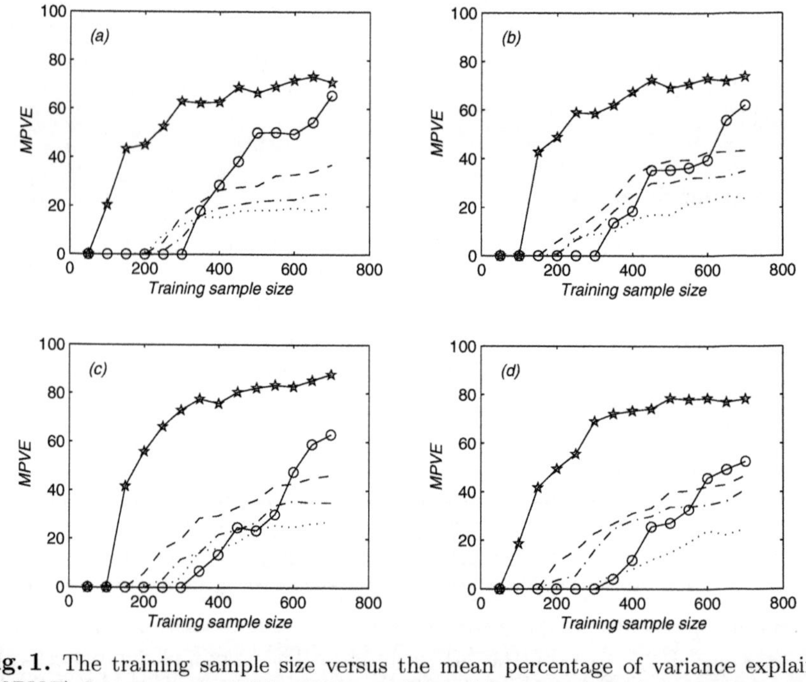

Fig. 1. The training sample size versus the mean percentage of variance explained ($MPVE$) for our method ($\star$), GMMs with full ($\circ$), diagonal (-.-) and spherical (...) covariance matrices, and the mixture of PPCAs [5] (- -). Comparison on the 15-dimensional data sets drawn from densities: a) $p_{\mathbf{IJK}}(\mathbf{x})$, b) $p_{\mathbf{IK}}(\mathbf{x})$, c) $p_{\mathbf{JK}}(\mathbf{x})$, d) $p_{\mathbf{IJ}}(\mathbf{x})$.

we fitted the GMMs by our method (Section 2) and PPCA [5]. The number q of the latent factors for PPCA and the number M of the components of the GMMs were varied $q = 1, 2, 3, 4$ and $M = 3, 4, 5, 6, 7, 8$. For the same data we fitted the GMMs with full, diagonal and spherical covariance matrices by the ML technique. The EM algorithm [4, pages 65-72] was used as a local optimizer of the likelihood of the GMMs for the training data. A k-mean clustering technique [4, page 187] was used to set the starting GMM parameter values for the EM algorithm. In order not to favor our method the starting point for the optimization (17) was set by the k-mean clustering, as well. The number M of the components of the GMM (2) was set $M = 3$ for the GMMs with full covariance matrices, and M was varied for GMMs with diagonal ($M = 3, 4, 5, 6, 7, 8$) and spherical ($M = 3, 4, \ldots, 14$) covariance matrices. For each GMM a performance criterion named the *percentage of variance explained* (PVE) (Appendix C) was computed. Finally we calculated the mean of the PVE values over the 10 replications and denoted it by *mean percentage of variance explained* ($MPVE$). In Fig. 1 we show the largest $MPVE$ values among the variation of q and M.

The results in Fig. 1 show that our method ($\star$) outperforms all the methods for $N = 150 - 700$. We succeeded to explain 50-80% of the variance, while the other methods explain 0-40% only. The GMMs with full covariance matrices ($\circ$) were highly sensitive to over-fitting ($MPVE \approx 0\%$) for $N = 50 - 300$. The mixture of PPCAs (- -) was better than sphered and diagonal GMMs for all variations of N, and better than full GMMs for $N < 400$. The latter results are consistent with the observations in [10].

4 Summary and Conclusion

We proposed a method for fitting GMMs based on the *projection pursuit* (PP) strategy proposed by Friedman [7]. The results obtained by means of simulations (Section 3) show that the PP strategy outperforms the *maximal likelihood* (ML) fitting of the GMMs for small sizes of the training sets.

In Section 2.2 we showed that the PP density estimation implies a GMM model for a specific setting of the Friedman's approximation. The formulae (12)-(15) derived allow us to set the parameters of the GMM implied by the PP estimation. This allows simple exact computation of the performance (PVE, Appendix C) in the simulations with normal mixture densities. The exact calculation of the PVE of the GMMs is carried out by direct matrix computations instead of a complicated Monte-Carlo evaluation of the n-fold integrals of PVE provided in [8] and [9]. The exact computation of the PVE is possible for a high-dimensional input space $n >> 10$, while the Monte-Carlo evaluation of the PVE is restricted to $n < 10$.

In our previous works we employed the PP strategy successfully in the discriminant analysis [1], [2] and for training neural networks for classification [3]. In this paper we showed that the PP strategy is an attractive choice for fitting GMMs using small sizes of the training sets.

Appendices

A Projection Pursuit

A.1 Computation the Directions $\mathbf{a}_1$, $\mathbf{a}_2$, ..., $\mathbf{a}_K$

Following Friedman [7] we choose $\mathbf{a}_1$, $\mathbf{a}_2$, ..., $\mathbf{a}_K$ by solving a sequence of *nonlinear programming* (NP) problems

$$\mathbf{a}_k = \underset{\mathbf{a}}{arg\ max} \left\{ I(\mathbf{a}|X^{(k)}) \right\} \text{ for } k = 1, 2, \ldots, K \qquad \text{subject to } \mathbf{a}^T\mathbf{a} = 1. \tag{17}$$

Here $I(\mathbf{a}|X^{(k)})$ is an objective function, named *projection pursuit* (*PP*) index (see Section A.2). It depends implicitly on a specific data set, denoted by

$$X^{(k)} = \left\{ \mathbf{x}_1^{(k)}, \mathbf{x}_2^{(k)}, \ldots, \mathbf{x}_N^{(k)} \right\}. \tag{18}$$

Here, $\mathbf{x}_1^{(k)}$, $\mathbf{x}_2^{(k)}$, ..., $\mathbf{x}_N^{(k)}$ are n-dimensional vectors. The data sets $X^{(k)}$, $k = 1, 2, \ldots, K$ are constructed in a sequential way, explained in Section A.3.

For solving the *nonlinear programming* (NP) problems (17) we employ a hybrid optimization strategy proposed in [11].

A.2 PP Index

The *PP* index $I(\mathbf{a}|X^{(k)})$ is defined in the following way. We project the data points $\mathbf{x}_i^{(k)} \in X^{(k)}$ onto $\mathbf{a}$ (an arbitrary n-dimensional vector having unit length) and obtain the projections $y_i^{(k)} = \mathbf{a}^T\mathbf{x}_i^{(k)}$. Obviously the shape of the density of these projections depends on the direction of $\mathbf{a}$. Friedman [7] defined the PP index as a measure of the departure of that density from N(0,1). He constructed the PP index based on an J-term Legendre polynomial expansion of the L_2 distance between the densities [7, pages 250-252]

$$I(\mathbf{a}|X^{(k)}) = \sum_{j=1}^{J} \frac{2j+1}{2} \left[\frac{1}{N} \sum_{i=1}^{N} P_j(r_i^{(k)}) \right]^2, \tag{19}$$

with

$$r_i^{(k)} = 2\Phi(y_i^{(k)}) - 1. \tag{20}$$

Here Φ denotes the standard normal (cumulative) distribution function and the Legendre polynomials P_j are defined as follows:

$$\begin{aligned} &P_0(r) = 1,\ P_1(r) = r,\ P_2(r) = \tfrac{1}{2}(3r^2 - 1), \\ &P_j(r) = \tfrac{1}{j}\left\{(2j-1)rP_{j-1}(r) - (j-1)P_{j-2}(r)\right\}, j = 3, 4, \ldots \ . \end{aligned} \tag{21}$$

If the projected density onto $\mathbf{a}$ is N(0,1) then PP index (19) achieves its minimum value (≈ 0). The solution of the NP problem (17) defines direction $\mathbf{a}_k$ which manifests non-normal projected density as much as possible.

Following Friedman [7] we set $J = 6$ in (19). We used the value of $I(\mathbf{a}_k|X^{(k)})$ to set the number K in the approximation (4). If $I(\mathbf{a}_k|X^{(k)}) < \epsilon$ then we dropped $\mathbf{a}_k$ from (4). In our experiments in Section 3 we set $\epsilon = 0.0001$.

A.3 Computation the Data Sets $X^{(1)}$, $X^{(2)}$, ..., $X^{(K)}$

Following Friedman [7] we compute the data sets $X^{(1)}$, $X^{(2)}$, ..., $X^{(K)}$ by the following successive transformation of the original training data set X (1).

For $k = 1, 2, \ldots, K$

We assign $X^{(k)} = X$ (X is the original data set (1) for $k = 1$).

We compute $\mathbf{a}_k$ solving (17).

We transform $X^{(k)}$ into $\tilde{X}^{(k)}$. We require $\tilde{X}^{(k)}$ to have $N(0,1)$ onto $\mathbf{a}_k$, and the same data structure as $X^{(k)}$ into an $(n-1)$-dimensional subspace orthogonal to $\mathbf{a}_k$. By this means we eliminate the maximum value of the PP index for $\tilde{X}^{(k)}$ at the point $\mathbf{a}_k$ ($I(\mathbf{a}_k|\tilde{X}^{(k)}) = 0$). The transformed data $\tilde{X}^{(k)}$ is computed by a method [7, pages 253-254], called *structure removal.*

We assign X to be the transformed data $\tilde{X}^{(k)}$ ($X = \tilde{X}^{(k)}$) and continue.

End

B Density Used to Generate the Training Data Sets

We generated training data sets from 15-dimensional density functions

$$p_{\mathbf{IK}}(\mathbf{x}) = [\textstyle\sum_{j=1}^{3} \alpha_j g_{\mathbf{I}_j}(x_1, x_2) g_{\mathbf{K}_j}(x_3, x_4)] \prod_{k=5}^{15} \phi(x_k),$$
$$p_{\mathbf{JK}}(\mathbf{x}) = [\textstyle\sum_{j=1}^{3} \alpha_j g_{\mathbf{J}_j}(x_1, x_2) g_{\mathbf{K}_j}(x_3, x_4)] \prod_{k=5}^{15} \phi(x_k),$$
$$p_{\mathbf{IJ}}(\mathbf{x}) = [\textstyle\sum_{j=1}^{3} \alpha_j g_{\mathbf{I}_j}(x_1, x_2) g_{\mathbf{J}_j}(x_3, x_4)] \prod_{k=5}^{15} \phi(x_k),$$
$$p_{\mathbf{IJK}}(\mathbf{x}) = [\textstyle\sum_{j=1}^{3} \alpha_j g_{\mathbf{I}_j}(x_1, x_2) g_{\mathbf{J}_j}(x_3, x_4) g_{\mathbf{K}_j}(x_5, x_6)] \prod_{k=7}^{15} \phi(x_k).$$

Here $\mathbf{x} = (x_1, x_2, \ldots x_{15})^T$, $\phi(x_k)$ is $N(0,1)$ density in the variable x_k and $g_{\mathbf{I}_j}(x_1, x_2)$, $g_{\mathbf{J}_j}(x_3, x_4)$, $g_{\mathbf{K}_j}(x_5, x_6)$ for $j = 1, 2, 3$ are bivariate normal densities from [12, Table 1]. We set the mixing coefficients $\alpha_1 = \alpha_2 = \frac{9}{20}$ and $\alpha_3 = \frac{1}{10}$. The structure of $p_{\mathbf{IK}}(\mathbf{x})$, $p_{\mathbf{JK}}(\mathbf{x})$ and $p_{\mathbf{IJ}}(\mathbf{x})$ lies in the first four variables, and the structure of $p_{\mathbf{IJK}}(\mathbf{x})$ lies in the first six variables. The remaining variables only add noise (variables having $N(0,1)$ densities). Note that the data sets drawn from these densities were normalized (sphered [7]), and randomly rotated in the runs discussed in Section 3.

C Percentage of Variance Explained (PVE)

In Section 3 we evaluated the performance of the GMMs by *percentage of variance explained* $PVE = 100(1 - \frac{ISE}{var})$ [8], where $ISE = \int_{R^n}(\hat{p}(\mathbf{x}) - p(\mathbf{x}))^2 d\mathbf{x}$ is the *integrated squared error* of the GMM $\hat{p}(\mathbf{x})$ and $var = \int_{R^n}(p(\mathbf{x}) - \frac{1}{vol(E)})^2 d\mathbf{x}$ is a normalization. Here $p(\mathbf{x})$ is the true underlying density and $vol(E)$ denotes the volume of a region E in space containing most of the mass of $p(\mathbf{x})$. We set $E = \{(-5 < x_i < 5), i = 1, 2, \ldots 15\}$. We employed a closed-form solution of the n-fold integrals ISE and var, which is available within the class of the normal mixture densities [13]. The latter allows us to compute the PVE for the densities $p_{\mathbf{IJK}}(\mathbf{x})$, $p_{\mathbf{IK}}(\mathbf{x})$, $p_{\mathbf{JK}}(\mathbf{x})$ and $p_{\mathbf{IJ}}(\mathbf{x})$ (Appendix B) exactly by direct matrix calculations. The formulae for the latter calculations will be included in an extended version of this paper.

References

1. Aladjem, M. E.: Linear discriminant analysis for two-classes via removal of classification structure. IEEE Trans. Pattern Anal. Mach. Intell. **19** (1997) 187–192 401
2. Aladjem, M. E.: Non-parametric discriminant analysis via recursive optimization of Patrick-Fisher distance. IEEE Trans. on Syst., Man, Cybern. **28B** (1998) 292–299 401
3. Aladjem, M. E.: Recursive training of neural networks for classification. IEEE Trans. on Neural Networks. **11** (2000) 488–503 401
4. Bishop, C. M.: Neural Networks for Pattern Recognition. Oxford University Press Inc., New York (1995) 396, 397, 399, 400, 401
5. Bishop, C. M.: Latent variable models. In: Jordan, M. I. (ed.): Learning in Graphical Models. The MIT Press, London (1999) 371–403 397, 400, 401
6. Fraley, C., Raftery, A. E.: How many clusters? Which clustering method? Answers via model-based cluster analysis. The Computer Journal. **41** (1998) 578–588 397, 399, 400
7. Friedman, J. H.: Exploratory projection pursuit. Journal of the American Statistical Association. **82** (1987) 249–266 396, 397, 400, 401, 402, 403
8. Friedman, J. H., Stuetzle, W., Schroeder, A.: Projection pursuit density estimation. Journal of the American Statistical Association. **79** (1984) 599–608 397, 401, 404
9. Hwang, J. N., Lay, S. R., Lippman, A.: Nonparametric multivariate density estimation: A comparative study. IEEE Trans. on Signal Processing. **42** (1994) 2795–2810 401
10. Moerland, P.: A comparison of mixture models for density estimation. In: Proceedings of the International Conference on Artificial Neural Networks (1999) 401
11. Sun, J.: Some practical aspects of exploratory projected pursuit. SIAM J. Sci. Comput. **14** (1993) 68–80 402
12. Wand, M. P., Jones, M. C.: Comparison of smoothing parameterizations in bivariate kernel density estimation. Journal of the American Statistical Association. **88** (1993) 520–528 403
13. Wand, M. P., Jones, M. C.: Kernel Smoothing. Charman & Hall/CRC (1995) 404

Asymmetric Gaussian and Its Application to Pattern Recognition

Tsuyoshi Kato, Shinichiro Omachi, and Hirotomo Aso

Graduate School of Engineering, Tohoku University, Sendai-shi, 980-8579 Japan
{kato,machi,aso}@aso.ecei.tohoku.ac.jp

Abstract. In this paper, we propose a new probability model, 'asymmetric Gaussian(AG),' which can capture spatially asymmetric distributions. It is also extended to mixture of AGs. The values of its parameters can be determined by Expectation-Conditional Maximization algorithm. We apply the AGs to a pattern classification problem and show that the AGs outperform Gaussian models.

1 Introduction

Estimation of a probability density function(pdf) of the patterns in given data set is a very important task for pattern recognition [1], data mining and so on. Single Gaussian and mixtures of Gaussians are most popular probability models, and they are used for many applications [2]. However, they do not always fit any distribution of patterns, so it is meaningful to provide another probability model which can be chosen instead of single/mixture Gaussian model.

In this paper, we propose a new probability model, 'asymmetric Gaussian(AG),' which is an extension of Gaussian. The AG can capture spatially asymmetric distributions. In the past, 'Asymmetric Mahalanobis Distance(AMD), ' was introduced [3] and it was applied to handwritten Chinese and Japanese character recognition. The AMD can measure a spatially asymmetrical distance between an unknown pattern and the mean vector of a class and shows excellent classification performance. However, the AMD is suitable only for an unimodal distribution, so the range of its application is necessarily somewhat limited. Meanwhile, since our model is formulated by a density function, it is easy to be extended to mixture model, which can capture multi-modal distributions. Moreover, due to its probabilistic formulation, we can develop a wide variety of extensions in a theoretically well-appointed setting.

The remainder of the paper is organized as follows. In the next section, we introduce the concept of latent variable model of single Gaussian model. In section 3 we then propose the AG model by extending the framework to the asymmetric version. Next we extend the AG to mixture models in Section 4. Section 5 presents its maximum likelihood estimation algorithm. In section 6 we show empirically that the mixture of AGs captures clusters of patterns, each of which are distributed asymmetrically. In section 7 we apply AG models to pattern recognition and present results using a real-world data sets. The final section presents our conclusions.

T. Caelli et al. (Eds.): SSPR&SPR 2002, LNCS 2396, pp. 405-413, 2002.
© Springer-Verlag Berlin Heidelberg 2002

2 A View of Single Gaussian

In this section we introduce a view of single Gaussian by a latent variable model. The goal of the latent variable model is to extend the representation for asymmetric distribution. We consider that single Gaussian has a d-dimensional latent variable $\boldsymbol{z}$ related to an observed data $\boldsymbol{x}$ in d-dimensional space. The i-th element of the latent variable, z_i, is distributed according to the following normal distribution $\mathcal{N}(z_i; \mu_i^z, \sigma_i^2)$ with mean μ_i^z and variance σ_i^2:

$$\mathcal{N}(z_i; \mu_i^z, \sigma_i^2) = \frac{1}{\sqrt{2\pi\sigma_i^2}} \exp\left(-\frac{(z_i - \mu_i^z)^2}{2\sigma_i^2} \right). \tag{1}$$

The Gaussian-distributed observed data vector $\boldsymbol{x}$ is generated by rotating $\boldsymbol{z}$ via an orthonormal matrix $\Phi = [\boldsymbol{\phi}_1, \cdots, \boldsymbol{\phi}_d] \in \mathcal{R}^{d \times d}$ as follows:

$$\boldsymbol{x} = \Phi \boldsymbol{z}. \tag{2}$$

The pdf of the observed variable $\boldsymbol{x}$ is consequently given by:

$$p(\boldsymbol{x}) = \int p(\boldsymbol{x}|\boldsymbol{z}) \prod_{i=1}^{d} \mathcal{N}(z_i; \mu_i^z, \sigma_i^2) dz_i \tag{3}$$

$$= \prod_{i=1}^{d} \mathcal{N}(\boldsymbol{\phi}_i^T \boldsymbol{x}; \mu_i^z, \sigma_i^2). \tag{4}$$

The last equality follows because the conditional density of $\boldsymbol{x}$ given $\boldsymbol{z}$ is $p(\boldsymbol{x}|\boldsymbol{z}) = \prod_{i=1}^{d} \delta(\boldsymbol{\phi}_i^T \boldsymbol{x} - z_i)$ where $\delta(\cdot)$ is the Dirac's delta function.

Next, we show an arbitrary Gaussian can be represented by the latent variable model. The observation variable $\boldsymbol{x}$ is assumed to be distributed according to a Gaussian $\mathcal{N}(\boldsymbol{\mu}^x, \Sigma^x)$ (the mean is $\boldsymbol{\mu}^x$ and the covariance matrix is Σ^x). The pdf of the Gaussian can be rewritten as

$$\mathcal{N}(\boldsymbol{x}; \boldsymbol{\mu}^x, \Sigma^x) = \prod_{i=1}^{d} \mathcal{N}(\boldsymbol{\psi}_i^T \boldsymbol{x}; \boldsymbol{\psi}_i^T \boldsymbol{\mu}^x, \lambda_i) \tag{5}$$

where λ_i and $\boldsymbol{\psi}_i$ denote i-th eigenvalue of the covariance matrix Σ^x and the corresponding eigenvector, respectively. By comparison between the formulae (4) and (5), it is shown that the above-mentioned latent variable model represents any Gaussian distribution by letting $\boldsymbol{\phi}_i = \boldsymbol{\psi}_i$, $\mu_i^z = \boldsymbol{\psi}_i^T \boldsymbol{\mu}^x$, $\sigma_i^2 = \lambda_i$.

3 Asymmetric Gaussian

We now introduce an asymmetric Gaussian(AG) model by extending the latent variable model.

In the same manner as Gaussian, the d-dimensional AG has a latent variable $\boldsymbol{z} \in \mathcal{R}^d$ and the observation variable $\boldsymbol{x}$ is modeled using $\boldsymbol{z}$ and an orthonormal

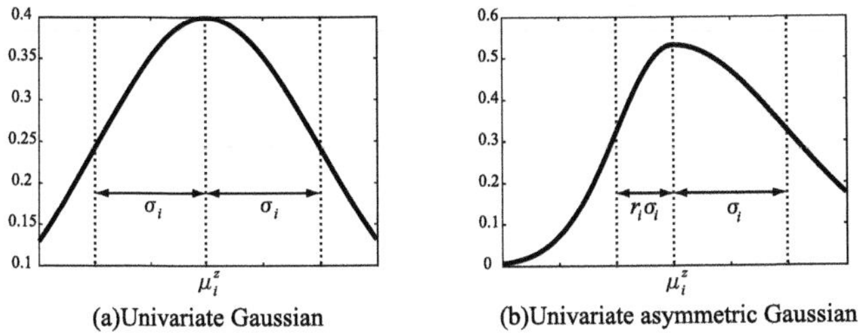

Fig. 1. Univariate Gaussian and univariate asymmetric Gaussian.

matrix $\Phi \in \mathcal{R}^{d \times d}$: $\boldsymbol{x} = \Phi \boldsymbol{z}$. The different point between the AG and the Gaussian is the distribution of the latent variable $\boldsymbol{z}$. We choose the following distribution of each element of $\boldsymbol{z}$:

$$\mathcal{A}(z_i; \mu_i^z, \sigma_i^2, r_i) \equiv \frac{2}{\sqrt{2\pi}} \frac{1}{\sqrt{\sigma_i^2}(r_i+1)} \begin{cases} \exp\left(-\frac{(z_i-\mu_i^z)^2}{2\sigma_i^2}\right) & \text{if } z_i > \mu_i^z, \\ \exp\left(-\frac{(z_i-\mu_i^z)^2}{2r_i^2\sigma_i^2}\right) & \text{otherwise,} \end{cases} \tag{6}$$

where μ_i^z,σ_i^2 and r_i are parameters of $\mathcal{A}(z_i; \mu_i^z, \sigma_i^2, r_i)$. We term the density model (6) 'univariate asymmetric Gaussian(UAG).' It is shown that UAG have an asymmetric distribution by the Figure 1(b) where the density function is plotted. In addition, UAG is an extension of Gaussian since UAG with $r_i = 1$ is equivalent to Gaussian.

The pdf of AG is given by:

$$p(\boldsymbol{x}) = \mathcal{A}(\boldsymbol{x}; \Theta) \equiv \int p(\boldsymbol{x}|\boldsymbol{z}) \prod_{i=1}^{d} \mathcal{A}(z_i; \mu_i^z, \sigma_i^2, r_i) dz_i \tag{7}$$

$$= \prod_{i=1}^{d} \mathcal{A}(\boldsymbol{\phi}_i^T \boldsymbol{x}; \mu_i^z, \sigma_i^2, r_i), \tag{8}$$

where $\Theta = \{\boldsymbol{\phi}_i, \mu_i^z, \sigma_i^2, r_i\}_{i=1}^d$ is the set of the adaptive parameters.

4 Mixture of Asymmetric Gaussians

Due to the definition of the density model, it is straightforward to consider a mixture of AG, which is able to model complex data structures with a linear combination of local AGs. The overall density of the K-component mixture model is written by

$$p(\boldsymbol{x}) = \sum_{k=1}^{K} \pi_k \mathcal{A}(\boldsymbol{x}; \Theta^{(k)}) \tag{9}$$

where $\mathcal{A}(\boldsymbol{x};\Theta^{(k)})$ is the kth local AG, with its own set of independent parameters, $\Theta^{(k)} = \{\boldsymbol{\phi}_{i,k}, \mu^z_{i,k}, \sigma^2_{i,k}, r_{i,k}\}_{i=1}^d$, and $\{\pi_k\}_{k=1}^K$ are mixing proportions satisfying $0 \le \pi_k \le 1$ and $\sum_{k=1}^K \pi_k = 1$.

5 The EM Algorithm for Maximum Likelihood Estimation

Optimal values of the parameters of each local AG, $\{\Theta^{(k)}\}$, and mixing proportions $\{\pi_k\}$ are unable to be obtained in the closed form, and here we describe the formulae using Expectation-Maximization(EM) algorithm [4], [5] which provides a numerical method for estimating these maximum likelihood parameters. Given a data set $\{\boldsymbol{x}^n\}_{n=1}^N$, the log-likelihood function is given by

$$\mathcal{L} = \sum_{n=1}^{N} \log \sum_{k=1}^{K} \left(\pi_k \mathcal{A}(\boldsymbol{x}^n;\Theta^{(k)}) \right). \tag{10}$$

The maximization of the log-likelihood can be regarded as a missing-data problem in which the identity k of the component that has generated each pattern $\boldsymbol{x}^n$ is unknown.

In the E-step, we compute the posterior probability h_{nk}, called responsibility, of each local AG component k for generating pattern $\boldsymbol{x}^n$ using the current values of $\Theta^{(k)}$ and π_k:

$$h_{nk} = \hat{P}(k|\boldsymbol{x}^n) = \frac{\pi_k \mathcal{A}(\boldsymbol{x}^n;\Theta^{(k)})}{\sum_{k'} \pi_{k'} \mathcal{A}(\boldsymbol{x}^n;\Theta^{(k')})}. \tag{11}$$

In the M-step, the quantity of the expected complete-data log-likelihood which is given by

$$\langle \mathcal{L}_{\text{comp}} \rangle = \sum_{n=1}^{N} \sum_{k=1}^{K} h_{nk} \left(\log \mathcal{A}(\boldsymbol{x}^n;\Theta^{(k)}) + \log \pi_k \right) \tag{12}$$

is maximized with respect to $\{\Theta^{(k)}, \pi_k\}_{k=1}^K$. The following updates of $\{\pi_k\}$ maximize the quantity of the term containing $\{\pi_k\}$ in (12) with subject to the constraint $\sum_{k=1}^K \pi_k = 1$:

$$\pi_k = \frac{1}{N} \sum_{n=1}^{N} h_{nk}. \tag{13}$$

Although the parameter set of each local AG, $\Theta^{(k)} = \{\boldsymbol{\phi}_{i,k}, \mu^z_{i,k}, \sigma^2_{i,k}, r_{i,k}\}_{i=1}^d$, must also be found so that it maximizes the expected complete-data log-likelihood in the standard EM algorithm, it is not tractable to compute both Φ^k and the other parameters simultaneously. We therefore use a two-stage procedure. In the first stage of the M-step, $\bigcup_{i,k}\{\mu^z_{i,k}, \sigma^2_{i,k}, r_{i,k}\}$ is held constant, and the orthonormal matrix $\Phi^k = \{\phi_{i,k}\}$ is updated so as to increase $\langle \mathcal{L}_{\text{comp}} \rangle$ in (12). In the second stage, we find the optimal parameters of each UAG in each local AG, $\mu^z_{i,k}$,$\sigma^2_{i,k}$ and $r_{i,k}$, keeping the orthonormal matrix Φ^k constant. This procedure performs only partial maximization, however, the partial maximization of $\langle\mathcal{L}_{\text{comp}}\rangle$ also guarantees the log-likelihood not to decrease during each iteration. Such a strategy is called generalized Expectation-Maximization(GEM) algorithm [4], [6]. The proposed maximum likelihood(ML) estimation scheme is an example of Expectation-Conditional Maximization(ECM) algorithm [7], which is a subclass of GEM algorithms. Further details concerning the two-stage procedure can be seen in Appendix.

The ML estimation algorithm is summarized as follows:

begin
  repeat
  { E-step }
  Evaluate responsibilities (11);
  { M-step }
  Update mixing proportions using (13);
  foreach $\forall k$ **begin**
    Update the orthonormal matrix Φ^k with $\{\mu^z_{i,k}, \sigma^2_{i,k}, r_{i,k}\}^d_{i=1}$ fixed;
    Find the optimal values of $\{\mu^z_{i,k}, \sigma^2_{i,k}, r_{i,k}\}^d_{i=1}$ with Φ^k fixed
  end;
  until the convergence of $\mathcal{L}$
end.

6 Simulations

We applied the ML estimation algorithm mentioned in the previous section for AG model to a problem involving 229 hand-crafted data points in the 2-dimensional space shown in Figure 2.

Figure 2(b) shows the results using three components. We also fitted the mixture of (standard) Gaussians for comparison (Figure 2(a)). The ellipse in (a) denotes the set of points that have the same Mahalanobis distance from the mean of each component, and the cross point in each ellipse lies on the mean. Similarly the loop in (b) denotes the set of points satisfying the values of the exponent of each local AG equal to one, and the cross point in each ellipse lies on the point $(\mu^z_{1,k}, \mu^z_{0,k})$. The AG captures the asymmetric distribution, which cannot be done by the Gaussian intrinsically. Although it might seem that AG tends to over-fit to the data set, we expect that this problem could be overcome by evidence framework [8].

7 Application to Pattern Recognition

In this section, we first present how to apply mixture of AGs to pattern recognition, and then show the experimental results on character recognition problem.

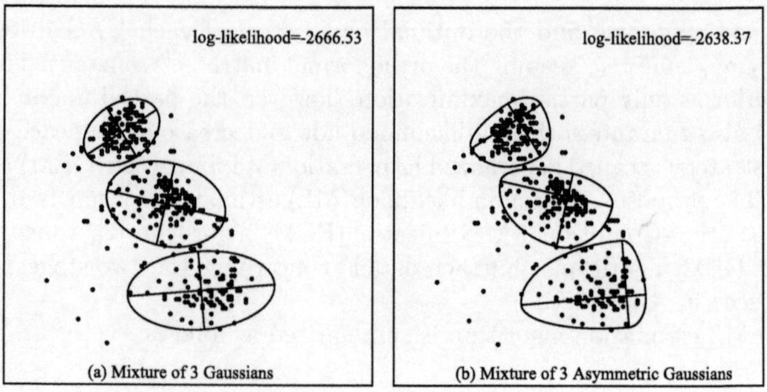

Fig. 2. Comparison between Mixture of Gaussians and Mixture of Asymmetric Gaussians.

In the training stage, we estimate the density function of each class w, $p(\boldsymbol{x}|w)$, using the ML estimation algorithm. In the classification stage, we find the class which has the largest posterior class probability:

$$P(w|\boldsymbol{x}) = \frac{p(\boldsymbol{x}|w)P(w)}{\sum_{w'} p(\boldsymbol{x}|w')P(w')} \tag{14}$$

where the prior class probability $P(w)$ is assumed to be non-informative.

We have tested the method in the public database 'Letter' [9] obtained from the UCI Machine Learning repository. The data contain 20,000 instances extracted from character images. Each of them has 16 features. The number of classes is 26. The database is partitioned into five almost equal subsets. In rotation, four subsets are used to train the AG parameters of each class and the trained AGs are tested on the remaining subset. In this experiment, we choose $K = 1$ for each class, that is, non-mixture AG models are used. For comparison, we also test Gaussians.

The accuracy on each subset is plotted in Figure 3. The 'average' in the figure is obtained by the ratio of the sum of the numbers classified correctly on each subset to the number of all instances. AGs improve in classification performance on every subset and AGs obtain 88.14% 'average' accuracy while Gaussians obtain 87.71%. It can be considered that AGs capture the distribution of patterns more precisely than Gaussians.

8 Conclusion

In this paper, we proposed a new probability density model, asymmetric Gaussian, which can fit the spatially asymmetric distribution, and extended it to mixture model. We also developed an algorithm of the maximum likelihood estimation for mixture of AGs using the Expectation-Conditional Maximization technique and it was applied to a two-dimensional problem. We also applied the

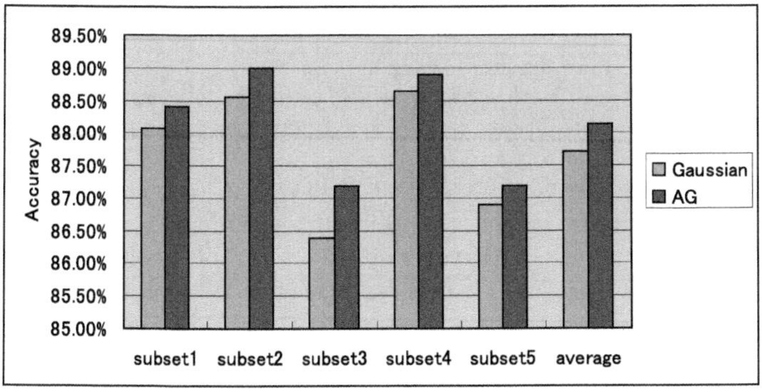

Fig. 3. Experimental results on the database 'Letter'.

AGs to character classification problem and showed that the AGs outperform Gaussian models.

Appendix: M-Step in the EM Algorithm

We now describe the details about how to update the parameters of mixture of AGs, $\Theta^{(k)}$, in the M-step. We use a two-stage procedure to update $\Theta^{(k)}$ which increases the expected complete-data log-likelihood function. The two-stage procedure runs as follows: (1) Update the orthonormal matrix Φ^k with remaining parameters $\{\mu^z_{i,k}, \sigma^2_{i,k}, r_{i,k}\}^d_{i=1}$ fixed. (2) Update $\{\mu^z_{i,k}, \sigma^2_{i,k}, r_{i,k}\}^d_{i=1}$ with Φ^k fixed.

(1) Update Φ^k with $\{\mu^z_{i,k}, \sigma^2_{i,k}, r_{i,k}\}^d_{i=1}$ fixed

We compute Φ^k_{new} as follow:

$$\Phi^k_{\text{new}} = \Phi^k_{\text{old}} + \eta \left. \frac{\partial \langle \mathcal{L}_{\text{comp}} \rangle}{\partial \Phi^k} \right|_{\Phi^k = \Phi^k_{\text{old}}} \tag{15}$$

where η is the learning constant and Φ^k_{old} denotes the old value of Φ^k. Note that there is no constraint to ensure that Φ^k_{new} in (15) will result in an orthonormal matrix. Therefore, after updating, we modify Φ^k_{new} by using Gram-Schmidt orthonormalization procedure. Then the log-likelihood $\mathcal{L}$ using Φ^k_{new} is evaluated. If $\mathcal{L}$ improves, Φ^k_{new} is chosen as the new value of Φ^k. If not, Φ^k is not updated.

(2) Update $\{\mu^z_{i,k}, \sigma^2_{i,k}, r_{i,k}\}$ with Φ^k fixed

The expected complete-data log-likelihood function can be factorized by $Q_{i,k}$'s:

$$\langle \mathcal{L}_{\text{comp}} \rangle = \left(\sum_{k=1}^{K} \sum_{i=1}^{d} Q_{i,k} \right) + \left(\sum_{n=1}^{N} \sum_{k=1}^{K} h_{nk} \log \pi_k \right), \tag{16}$$

where

$$Q_{i,k} = \sum_{n=1}^{N} h_{nk} \log \mathcal{A}\left((\boldsymbol{\phi}_i^k)^T \boldsymbol{x}^n;\ \mu^z_{i,k}, \sigma^2_{i,k}, r_{i,k} \right). \tag{17}$$

Note that $Q_{i,k}$ depends only on three parameters, $\mu^z_{i,k}$, $\sigma^2_{i,k}$ and $r_{i,k}$. The above factorization permits us to find the optimal values of $\{\mu^z_{i,k}, \sigma^2_{i,k}, r_{i,k}\}$ separately so that $Q_{i,k}$ is maximized. However, it is intractable to maximize $Q_{i,k}$ with respect to the triple $\{\mu^z_{i,k}, \sigma^2_{i,k}, r_{i,k}\}$ simultaneously. So each of $Q_{i,k}$ is maximized sequentially with respect to each of parameters by the following iterative scheme:

begin
  repeat
    Find the optimal value of $\mu^z_{i,k}$ with $\sigma^2_{i,k}$ and $r_{i,k}$ fixed;
    Find the optimal value of $r_{i,k}$ with $\sigma^2_{i,k}$ and $\mu^z_{i,k}$ fixed;
    Find the optimal value of $\sigma^2_{i,k}$ with $\mu^z_{i,k}$ and $r_{i,k}$ fixed;
  until the convergence of $Q_{i,k}$
end.

Each maximization step is performed by finding the value of $\mu^z_{i,k}, \sigma^2_{i,k}, r_{i,k}$ so that $\frac{\partial Q_{i,k}}{\partial \mu^z_{i,k}} = 0$ $\frac{\partial Q_{i,k}}{\partial r_{i,k}} = 0$ and $\frac{\partial Q_{i,k}}{\partial \sigma^2_{i,k}} = 0$ are satisfied, respectively. It is straightforward to maximize $Q_{i,k}$ with respect to $\mu^z_{i,k}, \sigma^2_{i,k}$ because the equations are linear. $r_{i,k}$ is optimized by Newton-Raphson method [10] since the equation $\frac{\partial Q_{i,k}}{\partial r_{i,k}} = 0$ is non-linear.

References

1. T. Kato, S. Omachi and H. Aso: "Precise hand-printed character recognition using elastic models via nonlinear transformation", Proc. 15th ICPR, Vol. 2, pp. 364–367 (2000).
2. Z. R. Yang and M. Zwolinski: "Mutual information theory for adaptive mixture models", IEEE Trans. PAMI, **23**, 4, pp. 396–403 (2001).
3. N. Kato, M. Suzuki, S. Omachi, H. Aso and Y. Nemoto: "A handwritten character recognition system using directional element feature and asymmetric Mahalanobis distance", IEEE Trans. PAMI, **21**, 3, pp. 258–262 (1999).
4. A. P. Dempster, N. M. Laird and D. B. Rubin: "Maximum likelihood from incomplete data via the EM algorithm", J.R. Statistical Society, Series B, **39**, pp. 1–38 (1977).
5. C. M. Bishop: "Neural network for pattern recognition", Oxford, England: Oxford University Press (1995).
6. R. M. Neal and G. E. Hinton: "A view of the EM algorithm that justifies incremental, sparse, and other variants", Learning in Graphical Models (Ed. by M. I. Jordan), Kluwer Academic Publishers, pp. 355–368 (1998).
7. X. L. Meng and D. B. Rubin: "Recent extensions of the EM algorithms", Bayesian Statistics (Eds. by J. M. Bernardo, J. O. Berger, A. P. Dawid and A. F. M. Smith), Vol. 4, Oxford (1992).
8. D. J. C. MacKay: "Bayesian interpolation", Neural Computation, **4**, 3, pp. 415–447 (1992).

9. P. W. Frey and D. J. Slate: "Letter recognition using holland-style adaptive classifiers", Machine Learning, **6**, 2 (1991).
10. W. H. Press, S. A. Teukolski, W. T. Vetterling and B. P. Flannery: "Numerical Recipes in C", Cambridge University Press (1988).

Modified Predictive Validation Test for Gaussian Mixture Modelling

Mohammad Sadeghi and Josef Kittler

Centre for Vision, Speech and Signal Processing
School of Electronics, Computing and Mathematics, University of Surrey
Guildford GU2 7XH, UK
{M.Sadeghi,J.Kittler}@surrey.ac.uk
http://www.ee.surrey.ac.uk/CVSSP/

Abstract. This paper is concerned with the problem of probability density function estimation using mixture modelling. In [7] and [3], we proposed the `Predictive Validation`, PV, technique as a reliable tool for the Gaussian mixture model architecture selection. We propose a modified form of the PV method to eliminate underlying problems of the validation test for a large number of test points or very complex models.

1 Introduction

Consider a finite set of data points $\mathbf{X} = \mathbf{x}_1, \mathbf{x}_2, \ldots \mathbf{x}_N$, where $\mathbf{x}_i \in \Re^d$ and $1 \leq i \leq N$, that are identically distributed samples of the random variable $\mathbf{x}$. We wish to find the function that describes the data, i.e. its *pdf*, $p(\mathbf{x})$. Building such a model has many potential applications in pattern classification, clustering and image segmentation.

There are basically two major approaches to density estimation: parametric and non-parametric. The parametric approach involves assuming a specific functional form for the data distribution and estimating its parameters from the data set with a likelihood procedure. If the selected form is correct, it leads to an accurate model. In contrast, non-parametric methods attempt to perform an estimation without constraints on the global structure of the density function. The problem with this approach is that the number of parameters in the model quickly grows with the size of the data set. This leads to a huge computational burden, even with today's most capable processors. Semi-parametric techniques offer a successful compromise between parametric and non parametric methods. A finite mixture of functions is assumed as the functional form but the number of free parameters are allowed to vary which motivates a more complex and adaptable model. The number of free parameters does not depend upon the size of data set. The most widely used class of density functions for mixture modelling are Gaussian functions, which are attractive because of their isotropic and unimodal nature, along with their capability to represent distribution by a mean vector and covariance matrices.

An important problem of Gaussian mixture modelling approaches is selection of the model structure, i.e. the number of components. In [7] and [3], we

T. Caelli et al. (Eds.): SSPR&SPR 2002, LNCS 2396, pp. 414–423, 2002.
© Springer-Verlag Berlin Heidelberg 2002

proposed the `Predictive Validation`, PV, technique as a reliable solution to this problem. The PV method provides an absolute measure of goodness of the model which is based on the `calibration` concept: a density function is calibrated if, for the set of events they try to predict, the predicted frequencies match the empirical frequencies derived from the data set. The agreement between the predicted and empirical frequencies is checked using the `chi-squared` statistic. The main problem with this goodness of fit test is that it usually rejects almost everything for a large number of test points. As more accurate models can be built only with a large number of samples, we face a fundamental contradiction which could be resolved only by accepting compromise solutions. Furthermore, in some applications, the data distribution is not exactly a mixture of Gaussian functions and it is not possible to model the data accurately with a finite mixture of such functions. However, a model with a reasonable goodness of fit works practically well.

In this article, we revisit the PV test and eliminate the underlying problem of model validation for a large number of test samples or a very complex model. We show that with a modified test, we can obtain a well behaving measure of goodness of fit which identifies the best structure of the mixture. If the data set can truly be modelled by a finite mixture of Gaussian functions, the method succeeds in finding it. Otherwise, it tries to find the `best` estimation. By best, we mean the simplest model which describes the data distribution well. Also, an important problem in *pdf* modelling approaches is model initialisation. We demonstrate that the PV technique is also quite useful for dealing with this problem.

The rest of this paper is organised as follows. In the next section we define Gaussian mixture models and review the PV technique used to obtain the mixture structure. The problem of the goodness of fit test and our solution to the problem is detailed in Section 3. In Section 4, the use of the validation test to aid the model initialisation is shown. The experimental results are given in Section 5. Finally, some conclusions are drawn and possible directions for future research are suggested in Section 6.

2 Gaussian Mixture Modelling

A mixture model is defined by equation (1). The mixture components, $p(\mathbf{x}|j)$, satisfy the axiomatic property of probability density functions, $\int p(\mathbf{x}|j)d\mathbf{x} = 1$ and the coefficients P_j, the mixing parameters, are chosen such that $\sum_{j=1}^{M} P_j = 1$ and $0 \leq P_j \leq 1$.

$$p(\mathbf{x}) = \sum_{j=1}^{M} p(\mathbf{x}|j)P_j \tag{1}$$

A well known group of mixture models is Gaussian mixture in which

$$p(\mathbf{x}|j) = \frac{1}{\sqrt{(2\pi)^d|\mathbf{\Sigma}_j|}} \exp\left\{-\frac{1}{2}(\mathbf{x}-\boldsymbol{\mu}_j)^T\mathbf{\Sigma}_j^{-1}(\mathbf{x}-\boldsymbol{\mu}_j)\right\} \tag{2}$$

where $\boldsymbol{\mu}_j$ is the d-dimensional mean vector of component j and $\boldsymbol{\Sigma}_j$ the covariance matrix. For a given number of components, P_j, $\boldsymbol{\mu}_j$, and $\boldsymbol{\Sigma}_j$ are estimated via a standard maximum likelihood procedure using the *EM* algorithm [2,3]. An initial guess of the Gaussian mixture parameters is made first. The parameter values are then updated so that they locally maximise the log-likelihood of the samples. Unfortunately, the EM algorithm does not guarantee to find a global maximum. It can easily get confined to a local maximum or saddle point. For this reason, different initialisations of the algorithm have to be considered which may give rise to different models being obtained. Also, the most important problem which is examined under model selection is that prior knowledge of the number of components is rarely available.

2.1 Model Selection

There are several methods for selecting the architecture, i.e. the number of components, M. The simplest approach is to select the model which optimises the likelihood of the data given model [8]. However, this method requires a very large data set which is rarely available. Moreover, this model building process is biased to selecting more complex models than actually required, with the risk of over-fitting the data. Information criteria attempt to remove this bias using an auxiliary term which penalises the log-likelihood by the number of parameters required to define the model (AIC) [1] or by a factor related to the sample size (BIC) [8]. The main advantage of information criteria is their simplicity. The downside is that the chosen penalty term depends on the problem analysed. If the function is complex and the penalty is too strong the model will be under-fitted. We advocated the use of the `predictive validation` method. The goal is to find the least complex model that gives a satisfactory fit to the data. The model selection algorithm using the *PV* technique is a bottom up procedure which starts from the simplest model, a one component model and keeps adding components until the model is validated [3].

The basis of the validation test is that a good model can predict the data. Suppose that a model M_j with j components has been computed for data set $\mathbf{X}$. The validation test is performed by placing hyper-cubic random size windows in random places of the observation space and comparing the empirical and predicted probability. The former is defined as $p_{emp}(\mathbf{x}) = \frac{N_W}{N}$, where N_W is the number of training points falling within window W, and the latter $p_{pred}(\mathbf{x}) = \int_W p(\mathbf{x})d\mathbf{x}$. Although the window size is selected randomly, more stable results can be achieved by controlling it so that p_{emp} falls within a limited range[3]. The agreement between the empirical and predicted frequency is checked by a weighted linear least square fit of p_{emp} against p_{pred} .

3 Weighted Least Squares Fit

If the estimated *pdf* model is good the empirical and predicted frequencies should be approximately equal. Making repeated observations of p_{emp} and p_{pred} permits a weighted linear least square fit between p_{emp} and p_{pred} to be formed

$$p_{emp} = a + b \cdot p_{pred} \tag{3}$$

where a is the intercept and b is the gradient. If the model is good then it should be possible to fit a linear model to the data points. Furthermore, the fitted line should lie close to the line $y = x$. To fit the straight line to the set of points and to check whether the fitted line is close to the desired line the chi-square statistic is used. In these statistical procedures, measurement error plays a crucial role. The chi-square statistics is defined as

$$\chi^2 = \sum_i \left[\frac{1}{\sigma^{(i)}} (y^{(i)} - a - bx^{(i)}) \right]^2 \tag{4}$$

where $\sigma^{(i)}$ is the standard deviation of the measurement error in the y coordinate of the ith point. If the measurement errors are normally distributed then this function will give the maximum likelihood parameter estimation of a and b. To determine a and b, equation (4) is minimised [3].

To check whether a linear model can be applied to the data correctly, a goodness-of-fit measure, $Q(\chi^2|\nu)$, is computed. This is done via the incomplete gamma function, Γ [5]. If the goodness-of-fit test fails, the validation test also fails and it proceeds no further. Since, the line parameters are estimated by minimising equation (4), from this equation, we can see that the relative sizes of $\sigma^{(i)}$ do not affect the placement of the fitted line. They do affect the value of the χ^2 statistic which we use to test the linear model's validity. This is why it is imperative to calculate $\sigma^{(i)}$ correctly.

After the best fit line has been found we need to check whether this line is statistically close to the $y = x$ line. This can be done again by making use of the chi-squared statistic. For the data set we have found a minimum value of χ^2_{min} for our estimated parameters, a and b. If these values are perturbed then the value of χ^2 increases. The change in the chi-squared value, $\Delta\chi^2 = \chi^2 - \chi^2_{min}$, defines an elliptical confidence region around the point $[a, b]^T$.

$$\Delta\chi^2 = \begin{bmatrix} \delta a \\ \delta b \end{bmatrix}^T \cdot \begin{bmatrix} \sigma_a^2 & \sigma_{ab}^2 \\ \sigma_{ab}^2 & \sigma_b^2 \end{bmatrix}^{-1} \cdot \begin{bmatrix} \delta a \\ \delta b \end{bmatrix} \tag{5}$$

where δa and δb are the changes in the line parameters, σ_a^2 and σ_b^2 are the variances in the estimates of a and b respectively and σ_{ab}^2 is the covariance of a and b [5,3]. In the original *PV* test, to accept a model we computed the 99.0% confidence region around $[a, b]^T$ and checked whether our true parameter value vector $[0, 1]^T$ is encompassed within this elliptical region [3], i.e. the model is accepted if for $[\delta a \;\; \delta b] = [0 - a \;\; 1 - b]$

$$\Delta\chi^2 \leq \Delta\chi^2_\nu(p) \tag{6}$$

where ν is the degree of freedom in $\Delta\chi^2$, i.e. the number of parameters and p is the desired confidence interval. For a confidence level of 99.0% with two degrees of freedom, the value of $\Delta\chi^2_\nu(p)$ is 9.21.

Our further investigations showed that this test is very hard to pass when it is performed using too many test points. In [3] we checked experimentally the assumption of the un-correlatedness between measurement errors which affects the value of the degree of freedom and we found that this assumption is justifiable. The choice of the standard deviation of the measurement errors, $\sigma^{(i)}$, is an important issue in the test. An overestimated value helps the test to pass, but it may lead to an under-fitted model and very small error make the χ^2 test difficult to pass. So, to deal with the problem of modelling using a large data set, the choice of the measurement error is studied more accurately.

3.1 Measurement Uncertainty

In the *PV* method [7,3], standard deviation of the measurement error, $\sigma^{(i)}$, is estimated using a binomial distribution. Considering the random sized window, W, in the feature space, the probability of finding a point within the window is p. The probability of finding it outside W is $q = 1 - p$. In other words the number of points falling inside W, N_W, is a stochastic variable which is binomially distributed. The standard deviation of a binomial distribution is given by

$$\sigma_{binom}^{(i)} = \sqrt{\frac{p^{(i)}(1 - p^{(i)})}{N}} \tag{7}$$

$p^{(i)}$ is estimated by the empirical probability value within the window, $p_{emp}^{(i)}$.

By considering p_{pred} as the measurement without uncertainty (x coordinate), equation (7) makes a good approximation of the measurement error on the empirical probability value, p_{emp}. However, the effect of some other error sources like the effects of sampling and the integration error on p_{pred} need to be studied. If such errors are important a bias term has to be added to the measurement error.

To investigate the effect of the bias experimentally, we built a single component Gaussian model. This model was then used to generate 500 samples. The empirical and predicted probabilities were then calculated within randomly placed windows using the data and the true model. Finally, the mean and variance of $p_{diff} = p_{emp} - p_{pred}$ were calculated. Obviously, in the ideal conditions these values should be zero. This experiment was repeated for the different number of Gaussian components, data samples and window placements. The experimental results showed that the variance is almost independent of the number of components and the number of window placements and highly dependent on the data set size. Moreover, the experiments showed that σ_{diff} changes in a very similar manner to σ_{binom} when the number of test points changes. Therefore, σ_{binom} describe the sampling error well and integration error is negligible and no additional term as the bias error needs to be taken into account.

3.2 F Test

Consider a specified number of Gaussian components, M. As the number of data samples increases, the variance of the binomial distribution, equation (7), and therefore, the value of the elements of the covariance matrix in equation (5) reduce. At the same time, if the Gaussian model has not been improved significantly, the change in the chi-squared value, $\Delta\chi^2$, increases which makes the test more difficult to pass. In fact, in the modelling process, if the data distribution is a perfect mixture of normal functions, a model with the same number of Gaussian components would become more accurate as the number of samples increases. Eventually, its parameters would become identical with the true distribution parameters. So, in the validation test, although σ_{binom} and the variances of the line parameters decrease, the difference between the estimated and the true line parameters also decrease, so $\Delta\chi^2$ will not increase noticeably.

However, in a number of practical applications, the distribution is not an exact Gaussian mixture model. In such a conditions, when the number of data samples increases, although a more accurate model is achieved, the resulting effect on the improvements of the line parameters is not as significant as the effect caused by the reduction of σ_{binom}. Now, even when the number of components is increased, the improvement in $\Delta\chi^2$ is not significant enough to meet the condition 6.

Figure 1(a) shows the value of $\Delta\chi^2$ versus the number of components, M, when the method is used to model 1000 and 10000 samples generated by a mixture of 5 Gaussian components while figure 1(b) shows the results of the same experiments for a face image data set. Figure 2 also shows the logarithm of $\Delta\chi^2$ versus the number of data points for different number of Gaussian components. As we expect, when the size of the `class5` data set increases, although $\Delta\chi^2$ for the incorrect structures ($M < 5$) becomes larger, for the correct one ($M = 5$), it is even smaller than the value for the smaller size data set which emphasises that using more data points, more accurate model is achieved. For the face data set, the problem is not the same. Using about 1 percent of the image samples, 1000 samples, an eight components model is validated. When 10000 samples are used to train and validate the model, as the number of components is increased, a better model is built and $\Delta\chi^2$ is reduced accordingly. For the models with more than 13 Gaussian components, although $\Delta\chi^2$ is very close to the acceptance threshold, it is not reduced noticeably. In the PV method, we are seeking the simplest model which predicts the data well. So, it seems that, the model selection process has to be controlled using a more intelligent test.

The simplest solution to this problem is to avoid using large data sets and instead use a few samples, especially in the model validation stage. Such a solution may lead to an inaccurate model. The other solution is to define an acceptable structural error and add a term as the residual error to equation 7. However, selection of such an error is an important and difficult problem. In different applications and for different data sets, this term has to be selected carefully. As plot 1(b) suggests, the best solution is to check whether adding more components to the model improves the prediction ability of the model or not. Since, very high structural error is also not desirable, the absolute value of $\Delta\chi^2$ also has to be taken into account. As we mentioned earlier in the validation test the 99.0% confidence region around the estimated line parameters is considered as the re-

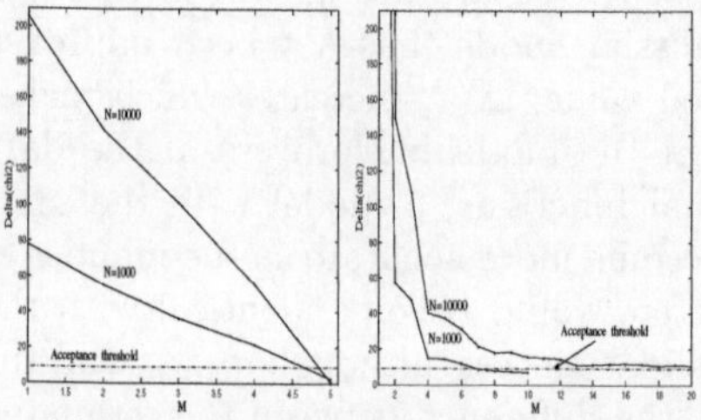

(a) class5 data set (b) A face data set

Fig. 1. $\Delta\chi^2$ versus the number of components using 1000 and 10000 samples

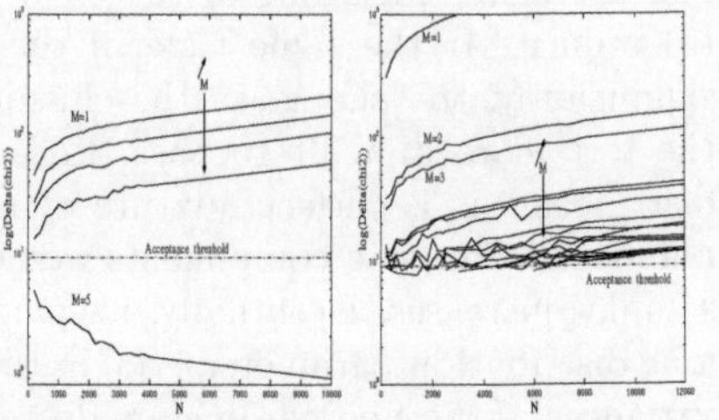

(a) class5 data set (b) A face data set

Fig. 2. $log(\Delta\chi^2)$ versus the number of samples for various M

quired confidence limit. Our experiments demonstrate that if the condition 6 is satisfied with such a confidence limit, the Gaussian model is absolutely reliable. Now, we propose that if the true parameters are within the 99.9% confidence area and more complex models do not improve $\Delta\chi^2$ significantly, the model is acceptable. In order to check the $\Delta\chi^2$ value variations, we apply the `F-test`.

The `F-test` is usually applied to check whether two distributions have significantly different variances. It is done by trying to reject the null hypothesis that the variances are consistent. The statistic `F` which is the ratio of the variances indicates very significant differences if either $F >> 1$ or $F << 1$ [5]. If we consider $\Delta\chi^2_M$ and $\Delta\chi^2_{M-1}$ as the change in the chi-squared value of the model M and $M-1$, since they have χ^2 distribution, their ratio obeys the Fisher's F-distribution law with (ν_1, ν_2) degrees of freedom where $\nu_1 = \nu_2 = 2$, i.e. the number of the line parameters. The distribution of `F` in the null case is calculated using equation 8.

$$Q(F|\nu_1, \nu_2) = I_{\frac{\nu_1}{\nu_2+\nu_1 F}}\left(\frac{\nu_2}{2}, \frac{\nu_1}{2}\right) \tag{8}$$

where $I_x(\alpha, \beta)$ is incomplete beta function [5]. In the validation process, if the true line parameters are between the 99.0% and 99.9% confidence area of the estimated parameters, to check whether $\Delta\chi^2_M$ and $\Delta\chi^2_{M-1}$ are consistent, F is considered as the ratio of the larger value to the smaller one. Then $p = 2 \cdot Q(F|\nu_1, \nu_2)$ is calculated. If the value of p is very close to one ($p > 0.99$), the null hypothesis is accepted [5] and the model is validated.

4 Model Initialisation

As we mentioned earlier, model initialisation is an important problem in the mixture modelling and different initialisations may lead to different models. We adopted our *PV* technique to select the best initialised model. In the model selection algorithm, for a given number of components, M, different models are built using the *EM* algorithm with different initialisation. During the validation step, the change in the chi-squared value, $\Delta\chi^2$, is calculated and the model with the minimum $\Delta\chi^2$ value is selected as the best M components model. If this minimum value satisfies the PV tests conditions also, the model is accepted.

5 Experiments

Two groups of experiments are reported here. In the first experiments the performance of the modified PV technique is compared with the original one. Then, the improvement achieved in a specific application, lip tracker initialisation, is shown.

5.1 Comparison of the Model Selection Methods

These experiments were performed on the `class5` data set, the face data set and the lip area of the face data set. The first row of figure 3 contains the experiments results using the information criteria methods, AIC and BIC, while the next row shows the results of the same experiments using the PV methods. In these plots the results using the original validation method (say *M1*), the results when the model initialisation is checked by the PV technique (*M2*) and the results when the modified test is also applied (*M3*) have been shown. Figures 3(a) and (d) contain plots of the number of components accepted versus the sample size considering the `class5` data set. As one can see, the AIC and BIC methods usually select over-fitted models. A five component model is always built using the *M2* and *M3* methods. Apparently, in such a cases, no structural error needs to be taken into account. Figures 3(b, e) and (c, f) show the results when performing the same experiments considering samples generated from the face and lip data sets. Although more stable results are obtained when the model is initialised intelligently, the effect of the `F` test on the model validation is noticeable. The test offers a compromise solution between the model accuracy and the model complexity.

5.2 Lip Tracker Initialisation

In [6], Gaussian Mixture Modelling using the PV technique along with a Gaussian components grouping algorithm was used to aid an un-supervised classification of lip data. The lip pixel classification experiments were performed on 145 colour images taken from the xm2vts database [4]. The first column of figure 4 shows two examples of the rectangular colour mouth region image blocks. The second and the third columns show the associated segmentation results using the original and modified algorithm. The segmentation error was calculated using ground truth images. The average error decreases from 7.12% using the original method to 6.87% after modifying the test.

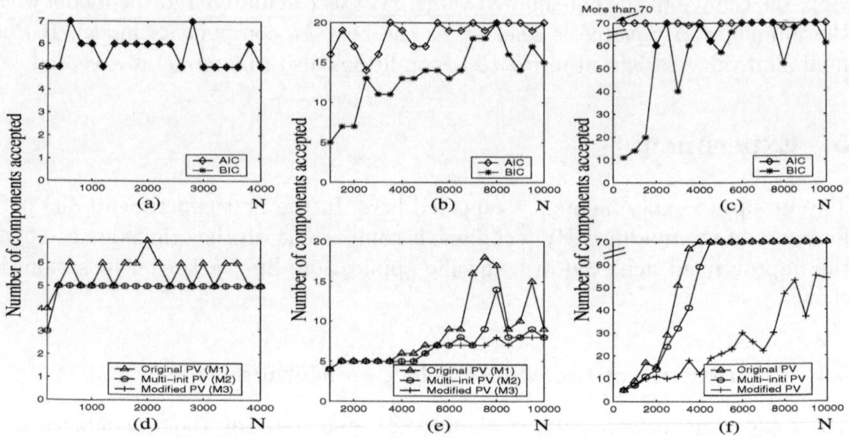

Fig. 3. The number of components accepted versus the number of samples using (top) AIC and BIC, (below) the original, multi-initialised and modified *PV* methods.(left: `class5` data set, middle: A face data set, right: Lip area data set)

6 Conclusions

In this paper we modified our proposed `Predictive Validation` algorithm in order to eliminate underlying problems of the model validation test for a large number of test points or very complex Gaussian mixture model. We demonstrated that `F` test avoids uncontrolled growth of the model complexity when more complex models do not improve the model calibration. It was also demonstrated that the *PV* technique is quite useful for dealing with the problem of model initialisation.

Even using the modified test, when we are dealing with a huge data set to avoid computational complexity of the *PV* test, it is desirable to place the vali-

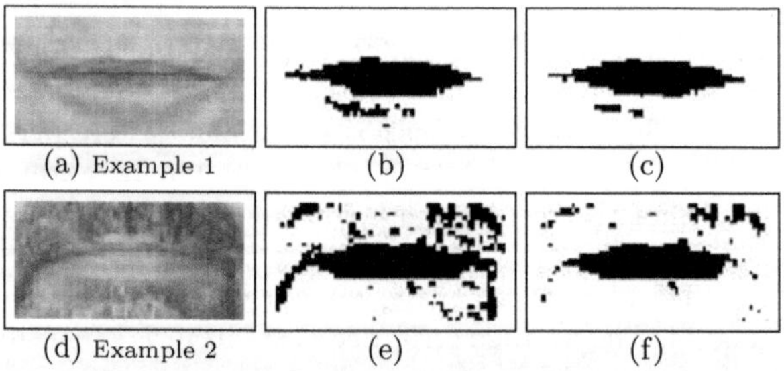

Fig. 4. *(Left:)*Two examples of the rectangular blocks taken from the xm2vts database images. *(Middle:)* The segmentation results using the original method.*(Right:)* The segmentation results using the modified method

dation windows over a sub-samples of the data set. The effective selection of the number of windows is a matter of interest in the future works.

Acknowledgements

The financial support from the EU project Banca and from the Ministry of Science, Research and Technology of Iran is gratefully acknowledged.

References

1. H. Akaike. A new look at the statistical model identification. *IEEE trans. on Automatic Control*, AC-19(6):716–723, 1974. 416
2. A. Dempster, N. Laird, and D. Rubin. Maximum likelihood from incomplete data via the em algorithm. *Journal of the Royal Statistical Society*, 39(1):1–38, 1977. 416
3. J. Kittler, K. Messer, and M. Sadeghi. Model validation for model selection. In S. Singh, N. Murshed, and W. Kropatsch, editors, *Proceedings of International Conference on Advances in Pattern Recognition ICAPR 2001*, pages 240–249, 11-14 March 2001. 414, 416, 417, 418
4. K. Messer, J. Matas, J. Kittler, J. Luettin, and G. Maitre. Xm2vtsdb: The extended m2vts database. In *Second International Conference on Audio and Video-based Biometric Person Authentication*, March 1999. 421
5. W. Press, B. Flanney, S. Teukolsky, and W. Vetterling. *Numerical Recipes in C: The Art of Scientific Computing.* Cambridge University Press, Cambridge, 2nd edition, 1992. 417, 420
6. M. Sadeghi, J. Kittler, and K. Messer. Segmentation of lip pixels for lip tracker initialisation. In *Proceedings IEEE International Conference on Image Processing, ICIP2001*, volume I, pages 50–53, 7-10 October 2001. 421
7. L. Sardo and J. Kittler. Model complexity validation for pdf estimation using gaussian mixtures. In S. V. A.K. Jain and B. Lovell, editors, *International Conference on Pattern Recognition*, pages 195–197, 1998. 414, 418
8. G. Schwarz. Estimating the dimension of a model. *The Annals of Statistics*, 6(2):461–464, 1978. 416

Performance Analysis and Comparison of Linear Combiners for Classifier Fusion

Giorgio Fumera and Fabio Roli

Dept. of Electrical and Electronic Eng., University of Cagliari
Piazza d'Armi, 09123 Cagliari, Italy
{fumera,roli}@diee.unica.it

Abstract. In this paper, we report a theoretical and experimental comparison between two widely used combination rules for classifier fusion: simple average and weighted average of classifiers outputs. We analyse the conditions which affect the difference between the performance of simple and weighted averaging and discuss the relation between these conditions and the concept of classifiers' "imbalance". Experiments aimed at assessing some of the theoretical results for cases where the theoretical assumptions could not be hold are reported.

1 Introduction

In the past decade, several rules for fusion of classifiers outputs have been proposed [10]. Some theoretical works also investigated the conditions which affect the performance of specific combining rules [1,2,3]. For the purposes of our discussion, the combining rules proposed in the literature can be classified on the basis of their "complexity". Simple rules are based on fixed combining methods, like the majority voting [1] and the simple averaging [2,3]. Complex rules use adaptive or trainable techniques, like the weighted voting [4] and the Behaviour Knowledge Space rule [5]. Researchers agree that simple combining rules work well for ensembles of classifiers exhibiting similar performance ("balanced" classifiers). On the other hand, experimental results showed that complex combining rules can outperform simple ones for ensembles of classifiers exhibiting different performance ("imbalanced" classifiers), supposed that a large and independent validation set is available for training such rules [10]. From the application viewpoint, it would be very useful to evaluate the maximum performance improvement achievable by trained rules over fixed ones for a classifier ensemble exhibiting a certain degree of imbalance. If such improvement is not significant for the application at hand, the use of a trained rule could be not worth, since the quality and the size of the training set can strongly reduce the theoretical improvement. However, no theoretical framework has been developed so far, which allows a clear quantitative comparison between different combining rules.

In this paper, we focus on two widely used combining rules, namely, simple and weighted averaging of classifiers outputs. Weighted averaging is often claimed to

T. Caelli et al. (Eds.): SSPR&SPR 2002, LNCS 2396, pp. 424-432, 2002.
© Springer-Verlag Berlin Heidelberg 2002

perform better than simple averaging for unbalanced classifier ensembles. However, to the best of our knowledge, no work clearly analysed the conditions which affect the difference between the performance of simple and weighted averaging. The performance improvement achievable by weighted averaging was not clearly quantified so far [2,3,11]. Moreover, experimental results, for instance, the ones reported in [6], showed a small improvement.

In the following, we report a theoretical and experimental comparison between weighted averaging and simple averaging. For our theoretical comparison, we used an analytical framework developed by Tumer and Ghosh [2,3] for the simple averaging rule, and extended it to the weighted averaging rule (Section 2). In Section 3, we quantify the theoretical performance improvement achievable by weighted averaging over simple averaging. We also discuss the conditions under which such improvement can be achieved, and the connection with the concept of classifier "imbalance". In Section 4, experiments aimed at assessing some of the theoretical results for cases where the theoretical assumptions could not be hold are reported.

2 An Analytical Framework for Linear Combiners

Following the work of Tumer and Ghosh [2,3], the outputs of an individual classifier approximating the a posteriori probabilities can be denoted as:

$$\hat{p}_i(x) = p_i(x) + \varepsilon_i(x) \ , \tag{1}$$

where $p_i(x)$ is the "true" posterior probability of the i-th class, and $\varepsilon_i(x)$ is the estimation error. We consider here a one-dimensional feature vector x. The multi-dimensional case is discussed in [7]. The main hypothesis made in [2,3] is that the decision boundaries obtained from the approximated a posteriori probabilities are close to the Bayesian decision boundaries. This allows focusing the analysis of classifier performance around the decision boundaries. Tumer and Ghosh showed that the expected value of the added error (i.e., the error added to the Bayes one due to estimation erros), denoted as E_{add}, can be expressed as:

$$E_{add} = \frac{1}{2s} E\left\{\left(\varepsilon_i(x_b) - \varepsilon_j(x_b)\right)^2\right\}, \tag{2}$$

where $E\{\}$ denotes the "expected" value, and s is a constant term depending on the values of the probability density functions in the optimal decision boundary. Let us assume that the estimation errors $\varepsilon_i(x)$ on different classes are i.i.d. variables [2,3], with zero mean (note that we are not assuming that the estimated a posteriori probabilities sum up to 1). Denoting their variance with σ_ε^2, we obtain from Eq. 2:

$$E_{add} = \frac{\sigma_\varepsilon^2}{s} \ . \tag{3}$$

Let us now evaluate the expected value E_{add}^{ave} of the added error for the weighted averaging of the outputs of an ensemble of N classifiers. We consider the case of normalised weights w_k:

$$\sum_{k=1}^{N} w_k = 1, \quad w_k \geq 0 \quad k = 1,\ldots,N \ . \tag{4}$$

The outputs of the combiner can be expressed as:

$$\hat{p}_i^{ave}(x) = \sum_{k=1}^{N} w_k \hat{p}_i^k(x) = \sum_{k=1}^{N} w_k \left(p_i(x) + \varepsilon_i^k(x)\right) = p_i(x) + \bar{\varepsilon}_i(x) \ , \tag{5}$$

where

$$\bar{\varepsilon}_i(x) = \sum_{k=1}^{N} w_k \varepsilon_i^k(x) \tag{6}$$

is the estimation error of the combiner. By proceeding as shown above for an individual classifier, one obtains the following expression for E_{add}^{ave}:

$$E_{add}^{ave} = \frac{1}{2s} E\left\{\left(\bar{\varepsilon}_i(x_b^{ave}) - \bar{\varepsilon}_j(x_b^{ave})\right)^2\right\}, \tag{7}$$

where x_b^{ave} denotes the decision boundary estimated by the combiner. We assume again that, for any individual classifier, the estimation errors $\varepsilon_i^k(x)$ on different classes are i.i.d. variables with zero mean, and denote their variances with $\sigma_{\varepsilon^k}^2$. We also assume that the errors $\varepsilon_i^m(x)$ and $\varepsilon_i^n(x)$ of different classifiers on the same class are correlated [2,3], with correlation coefficient ρ_i^{mn}, while they are uncorrelated on different classes. Under these assumptions, we obtain from Eq. 7:

$$E_{add}^{ave} = \frac{1}{s}\sum_{k=1}^{N} \sigma_{\varepsilon^k}^2 w_k^2 + \frac{1}{s}\sum_{m=1}^{N}\sum_{n \neq m} \left(\rho_i^{mn} + \rho_j^{mn}\right)\sigma_{\varepsilon^m}\sigma_{\varepsilon^n} w_m w_n \ . \tag{8}$$

This expression generalises the result obtained in [2,3] for simple averaging to the case of weighted averaging. For the purposes of our discussion, let us assume that the correlation coefficients of the different classes are equal: $\rho_i^{mn} = \rho_j^{mn} = \rho^{mn}$. From Eq. 3 it follows that E_{add}^{ave} can be rewritten as follows:

$$E_{add}^{ave} = \sum_{k=1}^{N} E_{add}^k w_k^2 + \sum_{m=1}^{N}\sum_{n \neq m} 2\rho^{mn} \sqrt{E_{add}^m E_{add}^n}\, w_m w_n \ . \tag{9}$$

Let us now analyse Eq. 9. We first consider the case of uncorrelated estimation errors (i.e., $\rho^{mn}=0$ for any $m \neq n$). In this case Eq. 9 reduces to:

$$E_{add}^{ave} = \sum_{k=1}^{N} E_{add}^k w_k^2 \ . \tag{10}$$

Taking into account the constraints of Eq. 4, it is easy to see that the optimal weights of the linear combination, that is, the ones which minimise the above E_{add}^{ave}, are:

$$w_k = \left(\sum_{m=1}^{N} \frac{1}{E_{add}^m}\right)^{-1} \frac{1}{E_{add}^k} \ . \tag{11}$$

Eq. 11 shows that the optimal weights are inversely proportional to the expected added error of the corresponding classifiers. Accordingly, for equal values of the expected added error, the optimal weights are $w_k=1/N$. This means that simple averaging is the optimal combining rule in the case of classifiers with equal performance ("balanced" classifiers).

Consider now the case of correlated estimation errors (Eq. 9). In this case it is not easy to derive an analytical expression for the optimal weights. However, from Eq. 9 it turns out that the optimal weights are $w_k=1/N$ if all classifiers exhibit both equal average performance and equal correlation coefficients. Otherwise, different weights are needed to minimise the expected added error E_{add}^{ave} of the combiner. This means that even for equal average accuracies, simple averaging is not the optimal rule, if the estimation errors of different classifiers exhibit different correlations.

3 Performance Analysis and Comparison

In this section, we quantitatively evaluate the theoretical improvement achievable by weighted averaging over simple averaging. To this end, we use the theoretical model described in Sect. 2. In the following we denote with ΔE_{add}^{ave} the difference between the expected added error achieved by simple averaging and the one achievable by weighted averaging using the optimal weights given in Eq. 11. Without loss of generality, we consider the N classifiers ordered for decreasing values of their expected added error E_{add}^{k}, so that $E_{add}^{1} \geq E_{add}^{2} \geq \ldots \geq E_{add}^{N}$.

3.1 Combining Uncorrelated Classifiers

Let us first consider the case of uncorrelated estimation errors (i.e., $\rho^{mn}=0$ for any $m \neq n$). According to Eq. 10, ΔE_{add}^{ave} can be written as:

$$\Delta E_{add}^{ave} = \frac{1}{N^2} \sum_{k=1}^{N} E_{add}^{k} - \left(\sum_{k=1}^{N} \frac{1}{E_{add}^{k}} \right)^{-1} . \tag{12}$$

By a mathematical analysis of Eq. 12 we proved that, for any given value of the difference $E_{add}^{1} - E_{add}^{N}$, the maximum of ΔE_{add}^{ave} is achieved when the N-2 classifiers 2,…,N-1 exhibit the same performance of the worst individual classifier, that is, $E_{add}^{1} = E_{add}^{2} = \ldots = E_{add}^{N-1} > E_{add}^{N}$. For the sake of brevity, we omit this proof. According to our model, this is therefore the condition under which, for a given value of the difference $E_{add}^{1} - E_{add}^{N}$, the advantage of weighted averaging over simple averaging is maximum. Hereafter we will denote this condition as performance "imbalance".

Under the above condition, in Fig. 1 we reported the values of ΔE_{add}^{ave} for values of $E_{add}^{1} - E_{add}^{N}$ ranging from 0 to 25%. Three different values of E_{add}^{N} for the best classifier were considered (1%, 5%, and 10%), and two values of the ensemble size, N=3,5. From Fig. 1, two conclusions can be drawn. First, weighted averaging significantly outperforms simple averaging (say, more than 1%) only if the performance of the individual classifiers are highly imbalanced (that is, for high values of $E_{add}^{1} - E_{add}^{N}$), and if the performance of the best individual classifier is very high (that is, for low values of E_{add}^{N}). Moreover, the advantage of weighted averaging decreases for increasing values of N (note that in practice it is unlikely to have a high number of uncorrelated classifiers [8]).

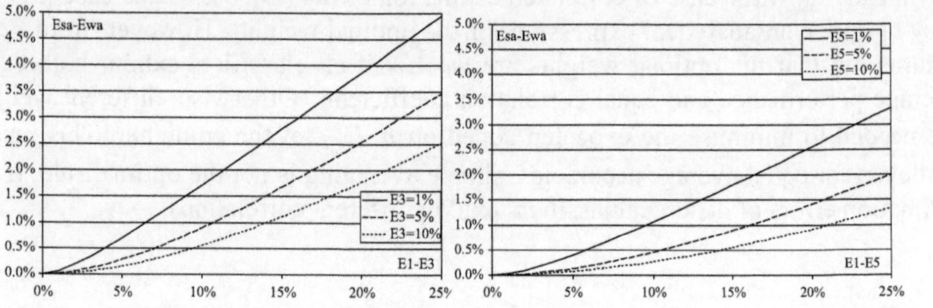

Fig. 1. Values of ΔE_{add}^{ave} (denoted as E^{sa}-E^{wa}) for uncorrelated classifiers, for N=3 (left) and N=5 (right). The values of E_{add}^{i} are denoted as E_i

Consider now the optimal weights given in Eq. 11. It is easy to see that the highest weight is assigned to the best individual classifier. Moreover, the weights of classifiers 1,...,N-1 are equal, as these classifiers have equal values of the expected added error. Their weight is reported in Fig. 2, plotted against $E_{add}^{1} - E_{add}^{N}$, for the same values of E_{add}^{N} and N as in Fig. 1.

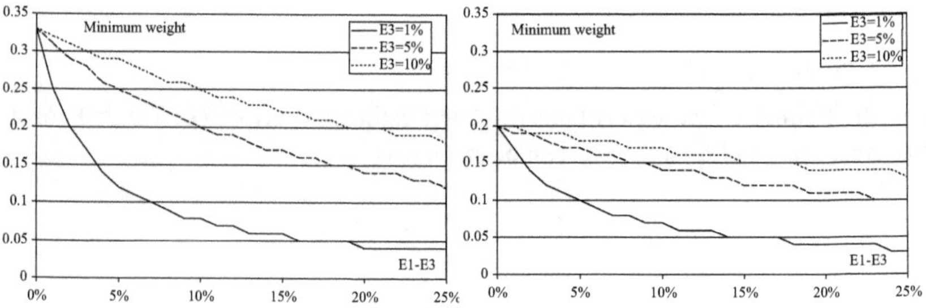

Fig. 2. Values of the minimum of the optimal weights, for N=3 (left) and N=5 (right)

The comparison of Figs. 1 and 2 shows that higher values of ΔE_{add}^{ave} correspond to lower weights for classifiers 1,...,N—1. In particular, if the best individual classifier performs very well (i.e., E_{add}^{N} is close to 0), a value of ΔE_{add}^{ave} greater than 1% can be achieved only by assigning to the other classifiers a weight lower than 0.1. This means that the performance of weighted averaging gets close to the one of the best individual classifier, as the other classifier are discarded.

To sum up, the theoretical model predicts that weighted averaging can significantly outperform simple averaging only if a classifier with very high performance is combined with few other classifiers exhibiting much worse performance. However, in this case, using only the best classifier could be a better choice than combining.

3.2 Combining Correlated Classifiers

Let us now consider the case of correlated estimation errors (Eq. 9). We evaluated ΔE_{add}^{ave} by first computing numerically the optimal weights from Eq. 9. As in the case of uncorrelated errors, it turned out that, for any given value of $E_{add}^{1} - E_{add}^{N}$, the maximum of ΔE_{add}^{ave} is achieved for $E_{add}^{1} = E_{add}^{2} = \ldots = E_{add}^{N-1}$. Under this condition, in Fig. 3 we report the values of ΔE_{add}^{ave} for N=3, and for values of ρ^{mn} in the range [—0.4, 0.8]. Fig. 3 shows that the advantage of weighted averaging over simple averaging is greater than in the case of uncorrelated errors. However, note that achieving a significant advantage still requires that the performance of the individual classifiers are highly imbalanced. Moreover, it turns out that the weight of one of the worst classifiers is always zero.

Let us now consider the values of the correlation coefficients. For given values of E_{add}^{3} and $E_{add}^{1} - E_{add}^{3}$, it turned out that the maximum of ΔE_{add}^{ave} is achieved when the best individual classifier is little correlated with one of the others (in our case, ρ^{13} = —0.4), while the other correlation coefficients must be as high as possible ($\rho^{12} = \rho^{23}$ = 0.8). It seems therefore that the correlations must be imbalanced in an analogous way as the performance.

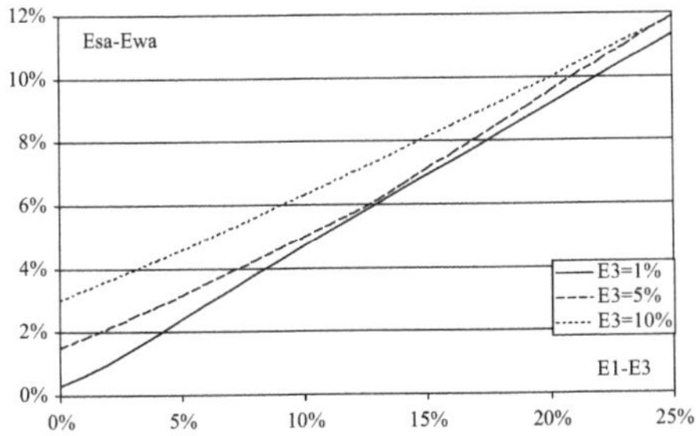

Fig. 3. Values of ΔE_{add}^{ave} for correlated classifiers, for N=3

To better analyse the effects of correlation, we evaluated ΔE_{add}^{ave} for varying values of the correlations coefficients. We considered imbalanced values in the sense defined above, that is, $\rho^{13}<\rho^{12}=\rho^{23}$. In Fig. 4 the values of ΔE_{add}^{ave} are plotted against the value of ρ^{12}—ρ^{13}, for three different values of ρ^{13}. Two different cases are considered for the expected added errors: $E_{add}^{1} = E_{add}^{2} = E_{add}^{3} = 5\%$ and $E_{add}^{1} = E_{add}^{2} = 10\%,\ E_{add}^{3} = 5\%$.

Fig. 4 shows that imbalanced correlations significantly affect the performance of simple averaging only when the individual classifiers have imbalanced performance. Moreover, it turned out that the weight assigned to one of the worst individual classifiers drops to zero as the imbalancing in performance or in correlation increases. This means that discarding such classifier would not affect the performance of weighted averaging, while simple averaging would perform significantly better. We found that the classifier whose weight is minimum is the one highly correlated with the best individual classifier. In Fig. 5 the value of the lowest of the optimal weights is reported, with reference to the cases shown in Fig. 4.

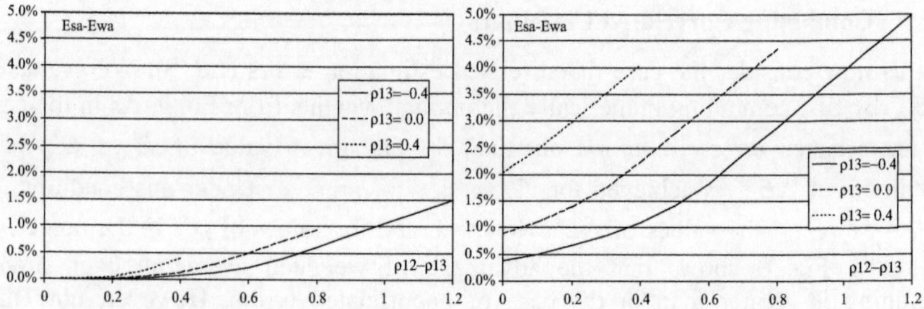

Fig. 4. Values of ΔE_{add}^{ave} for $E_{add}^{1} = E_{add}^{2} = E_{add}^{3} = 5\%$ (balanced performance, left), and $E_{add}^{1} = E_{add}^{2} = 10\%, E_{add}^{3} = 5\%$ (imbalanced performance, right)

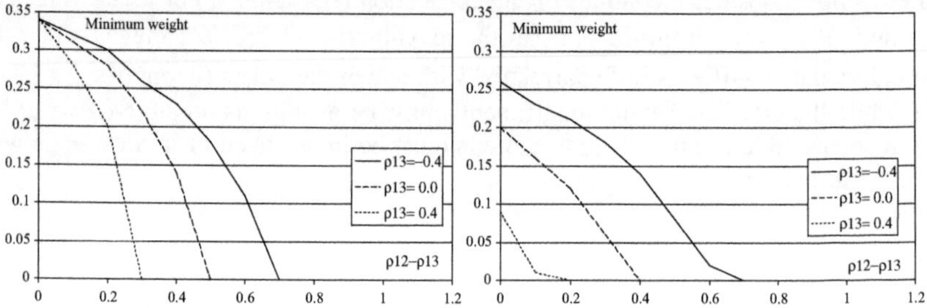

Fig. 5. Values of the minimum of the optimal weights for balanced (left), and imbalanced (right) performance, as in Fig. 5

To sum up, according to our model, weighted averaging significantly improves the performance of simple averaging only for ensembles of classifiers with highly imbalanced performance and correlations. However, such improvement is often achieved by discarding one of the worst classifiers, that is, assigning to it a weight close to zero. When the optimal weights are significantly greater than zero, the advantage of weighted averaging over simple averaging is quite small. It is worth noting that the advantage of weighted averaging over simple averaging is smaller than one can think of. This conclusion is in agreement with experimental results recently reported [6]. Moreover, it should be noted that, in practical applications, it can be very difficult to obtaining good estimates of the optimal weights.

4 Experimental Results

In this section, we report experiments aimed at comparing the performance of simple averaging and weighted averaging for ensembles with different degrees of imbalance.

We used a data set of remote-sensing images related to an agricultural area near the village of Feltwell (UK) [9]. This data set consists of 10,944 pixels belonging to five agricultural classes. It was randomly subdivided into a training set of 5,820 pixels, and a test set of 5,124 pixels. Each pixel is characterised by fifteen features, corresponding to the brightness values in the six optical bands, and over the nine radar channels considered.

We considered ensembles made up of a k-nearest neighbours classifier (k-NN), with a value of k equal to 15, and two multi-layer perceptron (MLP1 and MLP2) neural networks. Two ensembles were selected so that the performance of individual classifiers were imbalanced as defined in Sect. 3.1. We used MLPs with 15 hidden units for ensemble 1, and two hidden units for ensemble 2. The test set error percentages of the individual classifiers are shown in Table 1. The difference between the error percentages of the worst and the best classifier is shown in the last column as E_1—E_3. These values are averaged over ten runs corresponding to ten training set / validation set pairs, obtained by sampling without replacement from the original training set. The validation set contained the 20% of patterns of the original training set, and was used for the stopping criterion of the training phase of the MLPs.

Table 1. Error percentages of the individual classifiers on the test set. E_1-E_3 indicates the difference between the error percentages of the worst and the best classifier

	k-NN	MLP1	MLP2	E_1-E_3
ensemble 1	10.01	18.20	18.00	7.99
Ensemble 2	10.01	25.97	26.23	16.22

Table 2. Error percentages of weighted averaging (E^{wa}) and simple averaging (E^{sa}) on the test set

	ensemble performance			optimal weights		
	E^{sa}	E^{wa}	E^{sa}-E^{wa}	k-NN	MLP1	MLP2
Ensemble 1	12.09	9.69	2.40	0.689	0.080	0.231
Ensemble 2	16.81	9.79	7.02	0.838	0.006	0.156

In both ensembles, the k-NN was the best classifier. The two MLPs exhibited a similar error probability, which was higher than the one of the k-NN of about 8% (ensemble 1) and 16% (ensemble 2). As in these experiments we were not interested in the problem of weight estimation, the optimal weights of the linear combination were computed on the test set by "exhaustive" search. The average performance of simple and weighted averaging are reported in Table 2, respectively as E^{sa} and E^{wa}, together with the values of the optimal weights.

For ensemble 1, where the difference E_1—E_3 between the performance of the best and the worst classifier is about 8%, weighted averaging outperforms simple averaging of 2.4%. This value increases to about 7% for ensemble 2, where E_1-E_3 is about 16%. For both cases the weight of MLP1 is close to 0, and the performance of weighted averaging is very close to the one of the best individual classifier. These preliminary results are in agreement with the theoretical predictions. They show that weighted averaging significantly outperforms simple averaging only for highly imbalanced classifiers, and only by discarding one of the worst classifiers.

References

1. Lam, L., Suen, C. Y.: Application of Majority Voting to Pattern Recognition: An Analysis of Its Behavior and Performance. IEEE Trans. on Systems, Man and Cybernetics - Part A 27 (1997) 553-568

2. Tumer, K., Ghosh, J.: Analysis of Decision Boundaries in Linearly Combined Neural Classifiers. Pattern Recognition 29 (1996) 341-348
3. Tumer, K., Ghosh, J.: Linear and Order Statistics Combiners for Pattern Classification. In: Sharkey, A. J. C. (ed.): Combining Artificial Neural Nets. Springer (1999) 127-161
4. Xu, L., Krzyzak, A., Suen, C. Y.: Methods of Combining Multiple Classifiers and Their Applications to Handwriting Recognition. IEEE Trans. on Systems, Man, and Cybernetics 22 (1992) 418-435
5. Huang, Y. S., Suen, C. Y.: A Method of Combining Multiple Experts for the Recognition of Unconstrained Handwritten Numerals. IEEE Transactions on Pattern Analysis and Machine Intelligence 17 (1995) 90-94
6. Ueda, N.: Optimal Linear Combination of Neural Networks for Improving Classification Performance. IEEE Trans. on Pattern Analisys and Machine Int. 22 (2000) 207-215
7. Tumer, K.: Linear and Order Statistics Combiners for Reliable Pattern Classification. PhD thesis. The University of Texas, Austin, TX (1996)
8. Perrone, M., Cooper, L. N.: When Networks Disagree: Ensemble Methods for Hybrid Neural Networks. In: Mammone, R.J. (ed.): Neural Networks for Speech and Image Processing. Chapman-Hall, New York (1993)
9. Roli, F.: Multisensor Image Recognition by Neural Networks with Understandable Behaviour. Int. J. of Pattern Recognition and Artificial Intelligence 10 (1996) 887-917
10. Kittler, J., Roli, F. (eds.): Proc. of the 1st and 2nd Int. Workshop on Multiple Classifier Systems. Springer-Verlag, LNCS, Vol. 1857 (2000), and Vol. 2096 (2001)
11. Tumer, K., Ghosh, J.: Robust Combining of Disparate Classifiers through Order Statistics. To appear in: Pattern Analysis and Applications, special issue on "Fusion of Multiple Classifiers"

Comparison of Two Classification Methodologies on a Real-World Biomedical Problem

Ray Somorjai[1], Arunas Janeliunas[2], Richard Baumgartner[1], and Sarunas Raudys[2]

[1] Institute for Biodiagnostics, NRCC
435 Ellice Avenue, Winnipeg, MB, Canada, R3B 1Y6
{ray.somorjai,richard.baumgartner}@nrc.ca
[2] Department of Mathematics and Informatics, Vilnius University
Naugarduko 24, LT, 2006 Vilnius, Lithuania
raudys@ktl.mii.lt
arunas.janeliunas@verslas.com

Abstract. We compare two diverse classification strategies on real-life biomedical data. One is based on a genetic algorithm-driven feature extraction method, combined with data fusion and the use of a simple, single classifier, such as linear discriminant analysis. The other exploits a single layer perceptron-based, data-driven evolution of the optimal classifier, and data fusion. We discuss the intricate interplay between dataset size, the number of features, and classifier complexity, and suggest different techniques to handle such problems.

1 Introduction

Many modern pattern classification and data mining problems are characterized by hundreds or thousands of attributes and huge amounts of data records. However, for most spectroscopy-based biomedical classification problems, although the number of attributes is large, data scarcity is the rule rather than the exception. Hence, relations between classifier complexity, feature space dimensionality and sample set size continue to be among the major research topics in pattern classification and data analysis.

Traditionally, complexity/ feature space dimensionality/sample size interrelations are tackled by first reducing the number of features, using some feature extraction/selection method [12]. An alternative approach is to adjust the classification rule to the number of training samples and feature space dimensionality.

A third way employs multiple classification systems (MCSs). In using MCS, the designer divides the attributes into non-intersecting or intersecting subsets and uses each subset of attributes to design a corresponding simple classifier ("expert"). Then the individual decisions of the experts are combined to arrive at the final decision. In each separate procedure, considerably fewer features may be required. In a modification of this approach, the designer divides the training records into separate non-intersecting or intersecting subsets and uses each subset to design a simple expert

T. Caelli et al. (Eds.): SSPR&SPR 2002, LNCS 2396, pp. 433-441, 2002.
© Springer-Verlag Berlin Heidelberg 2002

classification rule instead of a complex one. This latter approach does need large sample sizes. In the last decade, the MCS approach has received considerable attention in the pattern recognition literature [1].

The existence of many similar approaches to solve the complexity/sample size/ dimensionality problem necessitates reviewing "the fundamental problems arising from this challenge" [2]. Notwithstanding theoretical attempts and advances, truly decisive tests of practical relevance can only be arrived at by experimental comparisons of the successes of different approaches on real-world problems. In the present paper, we conduct such comparisons of two strategies: MCS/classifier complexity regularization with a 3-stage feature-extraction-based statistical classification strategy (SCS) [9,10]. SCS was devised specifically for the classification of biomedical spectra. (The SCS's third stage is closely related conceptually to MCS.) The particular real-world example we use is a two-class biomedical data classification problem of typical difficulty. The observation vectors to be classified are magnetic resonance (MR) spectra of biofluids obtained from normal subjects and cancer patients. The dataset consists of 140 samples with 300 spectral features (the intensities at 300 frequencies). There are 71 spectra (31 healthy, 40 cancerous) in the training set and 69 (30 healthy, 39 cancerous) in the validation set. MR spectra of the 140 samples were acquired on a Bruker 360 MHz spectrometer. The MR magnitude spectra were preprocessed by normalizing each spectrum to unit spectral area. These are the data analyzed by both strategies.

2 A Feature-Selection-Based 3-Stage Statistical Classification Strategy (SCS)

The first stage of the SCS is feature selection. For MR magnitude spectra, the original N features are the intensity values at the different spectral frequencies. The feature selector algorithm we have used is an optimal region selector (ORS) [8]. ORS searches for spectral regions (frequency intervals) that are maximally discriminatory. ORS is guided by a genetic algorithm (GA), explicitly optimized for preprocessing spectra. GA is particularly appropriate for spectra, since the latter are naturally representable as "chromosomes", vectors of length N, with 1s indicating the presence, 0s the absence of features. The GA's input is M, the maximum number of features i.e., distinct spectral subregions required, the type of feature space-reducing operation/transformation (typically averaging) to be carried out, the population size, the number of generations and a random seed. The operations comprise the standard GA options: mutation and crossover. To ensure robust classification, the number of features M is typically kept much smaller than the sample size,. ORS begins searching the entire feature space, i.e. the complete spectrum. The output is the set of (averaged) spectral regions that optimally separate the classes.

For a limited number N of original features, exhaustive search (ES) for the best subset(s) is feasible. For larger N, we developed a dynamic programming (DP) based algorithm [8] that often produces near-optimal solutions, in feasible computer times.

Once M << N good features have been found, the second stage, a *crossvalidated classifier development* follows, with appropriately selected *training*, *test* and *validation* sets. We have developed an approach (RBS, Robust "BootStrap") that was inspired by the conventional nonparametric bootstrap [15]. RBS proceeds by randomly selecting approximately half the spectra from each class and using these to *train* a classifier, usually linear discriminant analysis (LDA). The resulting classifier is then used to *validate* the remaining half of the spectra. This process is repeated B times (*with replacement*), and every time the optimized LDA coefficients are saved. (B is typically 500-1000.) The *weighted average* of these B sets of coefficients produces the final, single *W-weighted classifier*. The weight for the *m*th set is $W_m = \kappa_m C_m^{1/2}$, m = 1,...,B, where $0 \le C_m \le 1$ is the *crispness* (for two classes, the fraction of samples with class probability ≥ 0.75), and $\sim 0 \le \kappa_m \le 1$ is Cohen's [16] chance-corrected *measure of agreement*, $\kappa_m = 1$ signifying perfect classification. The B weight values W_m are those obtained for the less optimistic bootstrap *test sets*. Classifier outcome is reported as a *class probability*.

For difficult classification problems the third stage is activated. At this stage, the outcomes of several classifiers are combined into an overall classifier via classifier fusion methods [13,14]. We use Wolpert's Stacked Generalizer [17] (WSG) for classifier combination. For the ultimate classifier to be developed, the *input features* for WSG are the *output* class *probabilities* obtained by the individual classifiers. For 2-class problems (since the *two* class probabilities are not independent, $p_1 + p_2 = 1$), the number of such features is *one* probability per sample (spectrum). The overall classification quality of the fused classifier is generally higher than that of the individual classifiers. In particular, the *crispness* of this final classifier is invariably greater. This is important in a clinical setting, because greater class assignment certainty means that fewer patients will have to be re-examined.

3 Controlling Classifier Complexity by Training a Single Layer Perceptron

There are many pattern classification strategies. They may differ conceptually, in the assumptions used to establish the design procedure, in the way the parameters of the classifier are estimated, and in the complexity of the decision boundary. When the design set is small, one promising approach is to use a linear classifier obtained while training a single layer perceptron (SLP). An important characteristic of an SLP-based classifier is that while training the perceptron, a number of standard pattern classification algorithms of differing complexity can be obtained by simply changing certain conditions [3,4]. Moreover, prior to training the perceptron, one can use sample estimates of statistical parameters of the training data to perform data transformations (rotation and scaling) such that various statistical models (i.e., prior information about the problem) may be incorporated into the perceptron design [4].

We followed the procedure described in [4] and prior to training the perceptron:

- The data centre $\hat{\boldsymbol{M}} = ½\,(\hat{\boldsymbol{M}}_1 + \hat{\boldsymbol{M}}_2)$ was zeroed, and all single features were scaled to unit variance by their sample standard deviations s_i;
- An estimate $\mathbf{S}_e$ of the pooled covariance matrix (CM) of the training data was constructed, followed by a singular-value-decomposition of $\mathbf{S}_e$, to linearly

transform the data to $\boldsymbol{Y} = \mathbf{F}(\boldsymbol{X} - \hat{\boldsymbol{M}})$, where $\mathbf{F} = \boldsymbol{\Lambda}^{-1/2}\,\boldsymbol{\Phi}^T$, and Λ, $\boldsymbol{\Phi}^T$ are eigenvalues and eigenvectors of the $\mathbf{S}_e$;
- Training started from zero initial weights;
- The standard sum of squares cost function and total gradient training (batch mode) was used.

To be able to work in the original 300-dimensional spectral feature space, given the very small training sets, it was necessary to use some prior information about the data. We assumed that all mutual correlations among all features are equal, and depend only on the noise variance. Hence, the correlation matrix $\mathbf{S}_e$ will be characterized by a single parameter, the correlation coefficient ρ.

This model, describing the dependence among the data features, is called the *additive noise model* Accordingly, after subtracting the mean vector $\hat{\boldsymbol{M}}$ and scaling to unity the variances of all features, we calculated the correlation matrix and used the average value of the correlation coefficient to obtain $\mathbf{S}_e$. Then we used this matrix to perform the singular value decomposition and to rotate the input feature vectors. In the new, transformed feature space, the main dependency between the components of the feature vector is already taken into account and the training process is faster. Optimal stopping of the iterative training process, control of target values, and the addition of an antiregularization term to the cost function can help balancing the complexity of the classification rule and the training set size.

4 Multiple Classifier Systems

In order to design an MCS, we partitioned the MR magnitude spectral features into 12 non-overlapping subsets. Then on each of these subsets, we constructed a simple classification rule by training a single-layer perceptron with the exponential threshold function $f(x) = 1/(1 + e^{-x})$. The outputs (i.e., the values of f(x)) of these single expert classifiers actually served as new input features for the "governor" SLP training.

For a small design set, it is very difficult to determine the optimal number of SLP training iterations. In such cases, one must use the same training set both to validate classifier performance and to define a stopping point for SLP training. This method of classification performance estimate is called the resubstitution error estimate.

Another approach is to create the independent validation data from random noise vectors, by augmenting the training set with them. One usually injects "white" noise vectors from a Gaussian distribution $N(0, \lambda \boldsymbol{I})$, where $\boldsymbol{I}$ is the identity matrix and λ is some scalar. However, further improvement is obtainable for high-dimensional problems, by adding instead *k-NN*-directed "colored" noise [6]. We used this technique to produce the validation dataset for training the expert classifiers. The outputs of these expert classifiers for the validation data provided the validation dataset for governor-SLP training.

We performed several experiments with different expert-SLP training techniques in order to compare the resulting classification performances. In our first attempt to design expert-SLP classifiers, we simply followed the recommendations in [4]. We scaled and rotated the original training data vectors and then trained the SLP classifier, starting from zero weights. No other assumption about the data was used. We will refer to this set of expert classifiers as the "simple" SLP experts.

During training, the single-layer perceptron tends to adapt to the specific training dataset (often called overtraining). When we use the resubstitution error to estimate classifier performance, each expert classifier "boasts" to the fusion rule, i.e., overestimates its own accuracy. Thus, if the same data are used to train both the classifiers and the combiner, the outcomes of the classifiers to the training data vectors create an optimistically biased training data for the fusion rule. We call this "boasting bias". Therefore, our next attempt was to improve the classification performance of our MCS by correcting the boasting bias of the simple SLP experts.

Assuming that the simple SLP experts are similar to Fisher discriminant functions (FDF), we have applied FDF-type theoretical corrections to the outputs of the SLP experts. Using the mean and variance values of the outputs of FDF classifiers given in [7], we define the following boasting-bias-correcting (BBC) transformation of the outputs of expert classifiers:

$$\tilde{O}_i = N/(N-p_i)\, O_i + (-1)^j N p_i(\delta_i^2+4)/2(N-p_i)^2 \qquad (1)$$

where O_i is the original output of the i-th classifier, N is the total number of training vectors, p_i is the dimensionality of the training data for the i-th classifier, δ_i^2 is the squared Mahalonobis distance between the two classes for the i-th classifier, and j is the class number. These corrections change the experts' means and variances. Thus, we produced a new, corrected training dataset for the governor-SLP, anticipating that it will help us obtain a better classifier fusion rule.

However, the simple SLP experts are obviously not FDF classifiers and such FDF-oriented BBC may not be appropriate. A more suitable BBC rule is the one that most affects the common distribution parameters of the expert classifier outputs. Hence, we tried a simpler BBC technique that affects only the means of the expert outputs, as shown in [7]:

$$\tilde{O}_i = O_i + (-1)^j\, 2p_i/(N-p_i), \qquad (2)$$

This was the second corrected training dataset for the governor-SLP training.

The second group of SLP experts was designed using the additive noise model. We used the previous splitting of data features into 12 subsets, and then we trained the individual SLP classifiers using the complexity control techniques described in Section 3. We assumed equal mutual correlations among data features, hence introduced into the expert SLP training additional information about the data structure.

We also trained the governor-SLP using the additive noise model. The outputs (or corrected outputs) of the SLP experts for the training data were used as input features for governor-SLP training, and the outputs for the noise data were used to stop the governor-SLP training. However, the role of colored noise vectors in the training of expert classifiers can also be changed. If we assume that we can create as many noise vectors as we need, and that the *k-NN*-directed colored noise retains information about the data configuration, we can use it as the training data for the governor-SLP. Furthermore, we can use the real training data for a more reliable determination of the number of governor-SLP training iterations than is possible by the random noise vectors.

5 Comparison Experiments

For the experiments with both the SCS and the CCR & MCS, we partitioned the data into two subsets. There are 71 spectra (31 healthy, 40 cancerous) in Set 1, and 69 (30 healthy, 39 cancerous) in Set 2. We then performed two runs of experiments:

Run 1: Train on Set 1, validate on Set 2
Run 2: Train on Set 2, validate on Set 1.

5.1 Experiments with the SCS: Dynamic Programming (DP) & Exhaustive Search (ES) on the Original Attributes

Because the number of attributes is relatively small (300), we did not need to use the more sophisticated, genetic algorithm-driven optimal region selector preprocessor; instead, we applied the DP-based feature selection algorithm.

Run 1: We first applied DP to Set 1. Because this is a 2-class problem, we could use a classifier that is a robust equivalent of LDA (we employed least-trimmed-squares regression, 10% trimming), with leave-one-out (LOO) crossvalidation. For the attribute selection, we used an objective function F that simultaneously minimizes the squared classification error and maximizes the crispness. In the range 2-13 of requested number of attributes, the minimum F for Set 2, *the validation set*, was obtained by the 8 attributes 8, 15, 18, 29, 115, 124, 151, 265. Because DP is a suboptimal feature selector, and to avoid overfitting, we used ES to select the best 2-7 subsets of these 8 attributes. The best of these, comprising only 3 attributes (8, 124, 151) gave a misclassified percentage of 13.0% for Set 2. The crisp result was 5.3% (38 of 69, 55.1% of total). A total of (12+6) = 18 models were tested.

Run 2: From the original 300 attributes DP selected 30 (using 6 tries, i.e., models). From these 30, the best 5 chosen by ES were 26, 151, 164, 165 and 248 (8 models). The misclassification percentage for the switched validation set (Set 1) was 14.1%, the crisp result 14.3% (70 of 71, 98.6%). (6+8) = 14 models were tested.

Averaging the two runs yielded 13.6%; based only on the crisp assignments, the average was 9.8% (108 of 140, 77.1% of total).

5.2 Experiments with CCR & MCS

For each run, we designed two sets of SLP-experts: 12 simple SLP classifiers and 12 SLP classifiers with the additive noise model. We used *3-NN*-directed colored noise vectors as the independent validation data set. For each training vector we added 300 colored noise vectors with variance $\lambda = 0.5$.

The governor-SLP was trained both on the training data and on the noise data. The optimal number of training iterations was determined by using the real validation data vectors. When the noise data (or the real training data, when we used noise data for training) was used to stop the governor-SLP training, we have the non-optimally stopped governor-SLP.

In the table below, we present the percentages of misclassified validation data vectors. To produce the results, 6 different feature distributions were tested for the

SLP experts (3 using the information measure, 3 the correlation matrix). For each of these, 3 different values of the correlation coefficient were tried in the additive noise model. From this total, (3+3)*3 = 18 different versions, the best results are listed in the table.

Expert classifiers	Average % performance of expert classifiers			Real training data (Noise training data)					
				Optimally stopped governor-SLP			Non-optimally stopped governor-SLP		
	1st run	2nd run	Avg	1st run	2nd run	Avg	1st run	2nd run	Avg
Simple SLP	39.6	36.6	38.1	18.8 (27.5)	26.8 (22.5)	22.8 (25.0)	26.1 (29.0)	28.2 (26.8)	27.1 (27.9)
Correction Eq. (1)	-	-	-	20.3 (26.1)	28.2 (25.4)	24.2 (25.7)	27.5 (29.0)	28.2 (26.8)	27.9 (27.9)
Correction Eq. (2)	-	-	-	17.4 (21.7)	26.8 (22.5)	22.1 (22.1)	29.0 (31.9)	26.8 (26.8)	27.9 (29.3)
Additive noise model	34.3	0.4	37.3	14.5 (13.0)	22.5 (21.1)	18.5 (17.1)	18.8 (21.7)	28.2 (22.5)	23.5 (22.1)

6 Discussion and Concluding Remarks

For the first time, two comprehensive classification strategies, developed independently by the Vilnius and Winnipeg groups, were compared on a real-world dataset not commonly accessible to the machine intelligence community.

The differences between the results obtained by the two strategies are not significant statistically. Nevertheless, the two strategies arrived at these results by quite different routes, using somewhat different philosophies. Comparing individually the misclassified test samples produced by the two best classifiers (13.0% both for the SCS-based approach and for the MCS method, the latter for the additive noise model with noise training data and optimally stopped governor-SLP) shows that the two approaches misclassified the same number of, but not the same individual samples.

The focus of the SCS is selecting maximally discriminatory features, by various preprocessing approaches. Extensive experience with biomedical spectra (see references in [10]) indicates that when the number of appropriate features extracted from the spectra is ~1/5th – 1/10th of the number of spectra per class, even a simple linear classifier, such as LDA will be reliable, once properly crossvalidated. The third stage, classifier fusion, is invoked only if the outcome probabilities are low. This was not done for this study.

In contrast, the MCS approach starts with a combination of implicit feature selection and classifier fusion (governor-SLP). (Unlike in the SCS, where classifier fusion is via the output class probabilities that serve as input features for the ultimate classifier, MCS uses the outputs O_i of the SLP experts as inputs to the governor-SLP.) However, the overall strategy's most distinguishing aspect is the use of a data-driven selection of the optimal classifier (CCR), which may range from the simplest (Euclidean distance classifier) to the most complex (support vector machines).

For the MSC approach, we tested several versions of the strategy (different subsets of experts, optimal stopping, etc). For the SCS, we selected the best attributes, based on the validation set's classification accuracy. Therefore, in both cases we adapted to the validation data, leading to optimistically biased results. Ideally, one would need a sufficiently large (hence statistically significant), completely independent validation set that was never used in the actual classifier development. This, at least for biomedical or genomics (microarray) applications, is unrealistic, and reliable methods for augmenting an originally sparse dataset (e.g., by noise injection) become particularly relevant.

Clearly, there is no best universal strategy or classifier! Hence, one cannot decide in advance what classification strategy and/or classifier to use. For each particular situation and dataset, comparative experimentation will be necessary. For real-life data, the designer will have to test several different models to decide which is best. This is an important consideration for the final assessment of the classifier(s). An essential requirement of success is that the classification strategy be comprehensive and sufficiently flexible to adapt to the peculiarities of the data. This is highlighted in this study: although the emphasis may be different, both strategies rely, explicitly or implicitly, both on feature selection and on classifier fusion. (Note that the explicit feature selection stage of the SCS, designed to create features that retain spectral identity, is driven by the biomedical imperative to understand the biochemical origin of the diseases.) However, the SCS and MCS place different emphasis on the components of the strategies, and carry them out in a different order. The major differences are SCS's reliance on feature selection vs. MCS's data-tuned classifier development.

We recommend having a toolbox of different classification strategies, and that the user experiment to select the most appropriate for the task. In the present context, a fusion of the best components of the two strategies used above seems promising. We shall report on these experiments in a future communication.

In general, to understand better the intricate interplay between dataset size, the number of features, and classifier complexity, it is highly desirable that several different classification experiments be performed on many different *types* of real-life datasets (beyond those archived in a few machine intelligence-oriented databases). Furthermore, both details of positive and negative results should be reported.

References

1. Kittler J., Roli F. (eds): Multiple Classifier Systems. Springer Lecture Notes in Computer Science, Springer Vol. 1857 (2000), Vol. 2096 (2001)
2. Ho T.K.: Data complexity analysis for classifier combination. In: Multiple Classifier Systems. J. Kittler and F. Roli (eds). Springer Lecture Notes in Computer Science, Springer Vol. 2096, (2001), 53-67
3. Raudys S.: Evolution and generalization of a single neuron. I. SLP as seven statistical classifiers. Neural Networks 11, 1998, 283–96

4. Raudys S.: Statistical and Neural Classifiers: An integrated approach to design. Springer, London, (2001) 312
5. Pivoriunas V.: The linear discriminant function for the identification of spectra. In: S Raudys (editor), Statistical Problems of Control 27, (1978), 71–90. Institute of Mathematics and Informatics, Vilnius (in Russian)
6. Skurichina M., Raudys S., Duin R.P.W.: K-nearest neighbours directed noise injection in multilayer perceptron training. IEEE Trans. On Neural Networks. 11(2) (2000), 504-511
7. Janeliūnas A.: Bias correction of linear classifiers in the classifier combination scheme. In: Proceedings of the 2nd International Conference on Neural Networks and Artificial Intelligence, BSUIR, Minsk, (2001) 91-98
8. Nikulin A., Dolenko B., Bezabeh T., Somorjai R.: Near-optimal region selection for feature space reduction: novel preprocessing methods for classifying MR spectra. NMR in Biomedicine, 11 (1998) 209-216
9. Mountford C., Somorjai R., Gluch L., Malycha P., Lean C., Russell P., Bilous M., Barraclough B., Gillett D., Himmelreich U., Dolenko B., Nikulin A., Smith I.: MRS on breast fine needle aspirate biopsy determines pathology, vascularization and nodal involvement. Br. J. Surg. 88 (2001) 1234-1240
10. Somorjai R.L., Dolenko B., Nikulin A., Nickerson P., Rush D., Shaw A., de Glogowski M., Rendell J., Deslauriers R. (2002) Distinguishing normal allografts from biopsy - proven rejections: application of a three - stage classification strategy to urine MR and IR spectra. Vibrational Spectroscopy 28: (1) 97-102
11. Zhilkin P., Somorjai R.: Application of several methods of classification fusion to magnetic resonance spectra. Connection Science 8(3) (1996) 427-442
12. Jain A.K., Duin R.P.W., Mao J.: Statistical pattern recognition: A review. IEEE Transactions on Pattern Analysis and Machine Intelligence 22 (2000) 4-37
13. Somorjai R.L., Nikulin A.E., Pizzi N., Jackson D., Scarth G., Dolenko B., Gordon H., Russell P., Lean C.L., Delbridge L., Mountford C.E., Smith I.C.P.: Computerized consensus diagnosis: a classification strategy for the robust analysis of MR spectra. I. Application to 1H spectra of thyroid neoplasms. Magn. Reson. Med. 33 (1995) 257-263
14. Somorjai R.L., Dolenko B., Nikulin A.E., Pizzi N., Scarth G., Zhilkin P., Halliday W., Fewer J., Hill N., Ross I., West M., Smith I., Donnelly M., Kuesel A., Brière K.: Classification of 1H MR spectra of human brain biopsies: The influence of preprocessing and computerized consensus diagnosis on classification accuracy. J Magn Reson Imaging 6 (1996) 437-444
15. Efron B., Tibshirani R.: An Introduction to the Bootstrap. Monographs on Statistics and Applied Probability, Cox D., Hinkley D., Reid N., Rubin D. and Silverman B.W. (General Eds.) Vol. 57 Chapman & Hall, London (1993)
16. Cohen J.: Weighted kappa: Nominal scale agreement with provision for scaled disagreement or partial credit. Psychological Bulletin 70 (1968) 213-220
17. Wolpert D.H.: Stacked generalization. Neural Networks 5 (1992) 241-259

Evidence Accumulation Clustering Based on the K-Means Algorithm

Ana Fred[1] and Anil K. Jain[2]

[1] Instituto de Telecomunicações
Instituto Superior Técnico, Lisbon, Portugal
afred@lx.it.pt

[2] Department of Computer Science and Engineering
Michigan State University, USA
jain@cse.msu.edu

Abstract. The idea of evidence accumulation for the combination of multiple clusterings was recently proposed [7]. Taking the K-means as the basic algorithm for the decomposition of data into a large number, k, of compact clusters, evidence on pattern association is accumulated, by a voting mechanism, over multiple clusterings obtained by random initializations of the K-means algorithm. This produces a mapping of the clusterings into a new similarity measure between patterns. The final data partition is obtained by applying the single-link method over this similarity matrix. In this paper we further explore and extend this idea, by proposing: (a) the combination of multiple K-means clusterings using variable k; (b) using cluster lifetime as the criterion for extracting the final clusters; and (c) the adaptation of this approach to string patterns. This leads to a more robust clustering technique, with fewer design parameters than the previous approach and potential applications in a wider range of problems.

1 Introduction

Clustering algorithms can be categorized into hierarchical methods and partitional methods [3,12]. A partitional structure organizes patterns into a small number of clusters. The K-means is one of the simplest clustering algorithms in this class: it is computationally efficient and does not require the specification of many parameters. Hierarchical methods propose a nesting of clusterings, providing additional information about data structure, represented graphically as a dendrogram. A particular partition is obtained by cutting the dendrogram at some level. The single link algorithm is one of the most popular methods in this class [12].

A large number of clustering algorithms exist [12,13]. Examples of different classes of algorithms are model-based techniques [8,18,23], non-parametric density estimation based methods [21], central clustering [2], square-error clustering [19], and graph theoretical based [4,26] methods. Each handles differently the issues related to cluster validity [1,10,20,8], number of clusters [15,25], and

T. Caelli et al. (Eds.): SSPR&SPR 2002, LNCS 2396, pp. 442–451, 2002.
© Springer-Verlag Berlin Heidelberg 2002

structure imposed on the data [6,24,16]; yet, no single algorithm can adequately handle all sorts of cluster shapes and structures.

Inspired by the work in sensor fusion and classifier combination techniques in pattern recognition [14], Fred [7] proposed a combination of clusterings in order to devise a consistent data partition. It follows a split and merge strategy. First, the data is split into a large number of small clusters, using the K-means algorithm; with fixed k, different clusterings are produced by an arbitrary initialization of cluster centers. The clustering results are combined using a voting mechanism, leading to a new similarity matrix between patterns. The final clusters are obtained by applying the single-link (SL) method on this matrix, thus merging small clusters produced in the first stage of the method.

In this paper we further analyze the above method and propose three main refinements/extensions: the use of cluster lifetime as a criterion for the identification of the final data partition from the dendrogram produced by the SL method, instead of fixed level thresholding; the combination of clusterings with different values of k in a reasonably large range; adaptation of this approach to process string patterns. These modifications improve the previous strategy in terms of robustness and simplicity of the method, with fewer parameters to be defined.

Section 2 discusses the method in [7]. Refinements and extensions of the method are proposed in section 3. The performance of the new method is illustrated through a set of experimental results given in section 4, followed by the conclusions.

2 Evidence Accumulation Clustering

The idea of evidence accumulation clustering is to combine the results of multiple clusterings into a single data partition, by viewing each clustering result as an independent evidence of data organization. Fred [7] used the K-means algorithm as the basic algorithm for decomposing the data into a large number, k, of compact clusters; evidence on pattern association is accumulated, by a voting mechanism, over N clusterings obtained by random initializations of the K-means algorithm. This produces a mapping of the clusterings into a new similarity measure between patterns, summarized in the matrix co_assoc, where $co\_assoc(i,j)$ indicates the fraction of times the pattern pair (i,j) is assigned to the same cluster among N clusterings. The final data partition is obtained by applying the single-link method over this similarity matrix, using a fixed threshold, t.

The method has two design parameters: k, the number of clusters for the K-means algorithm; and t, the threshold on the dendrogram produced by the SL method. We discuss these parameters using the half-rings data set example, depicted in figure 1(a). This data set is composed of 400 two-dimensional patterns (upper cluster - 100 patterns; lower cluster - 300 patterns). Due to the particular cluster shapes, the K-means algorithm by itself is unable to identify the two natural clusters (see figure 1(b)). The uneven data sparseness of the two

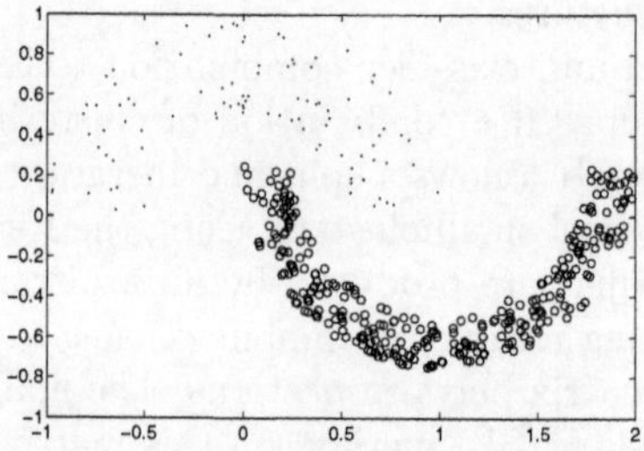

(a) Half-rings shaped clusters

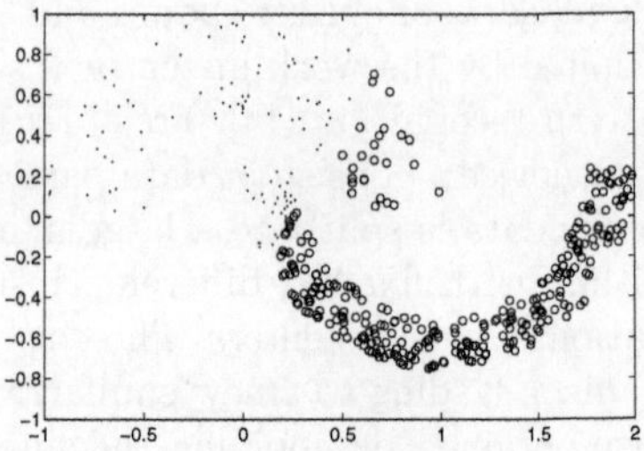

(b) K-means clustering, $K = 2$

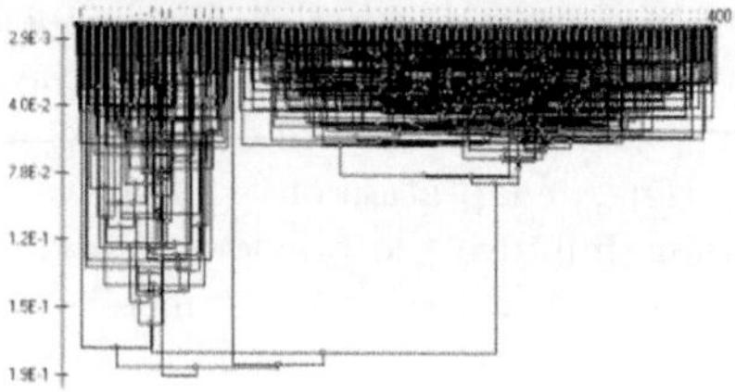

(c) Single-link method. Thresholding this graph splits the upper ring cluster into several small clusters

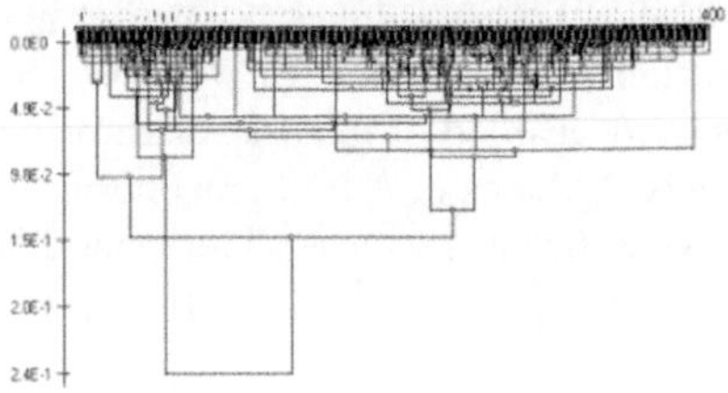

(d) Evidence accumulation clustering, $k = 5$

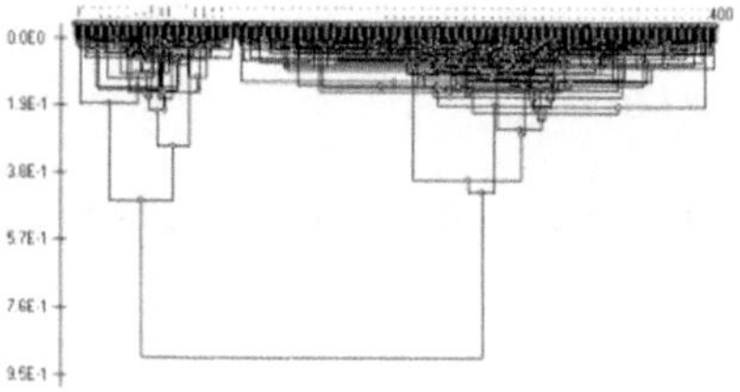

(e) Evidence accumulation clustering, $k = 15$

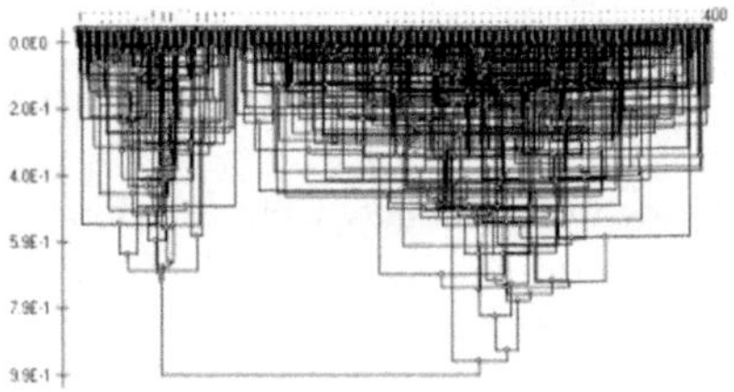

(f) Evidence accumulation clustering, $k = 80$

Fig. 1. Half-rings data set. Vertical axis on dendrograms (d) to (f) corresponds to distances, $d(i,j)$, with $d(i,j) = 1 - co\_assoc(i,j)$

clusters also prevents the SL method to produce the correct data partition, as shown by the associated dendrogram (figure 1(c)). Figures 1(d)- 1(f) plot the dendrograms produced by the evidence accumulation algorithm after 200 runs ($N = 200$) of the K-means algorithm, for different values of k. The K-means algorithm can be seen as performing a decomposition of the data into a mixture of Gaussians. k is the critical parameter in this decomposition: low values of k are not enough to capture the complexity of the data, while large values may produce an over-fragmentation of the data (in the limit, each pattern forming a cluster). By using the method in [5] the data set is decomposed into 10 gaussian components. This should be a lower bound on the value of k to be used with the K-means, as this algorithm imposes spherical shaped clusters, and therefore a higher number of components may be needed for evidence accumulation. This is in agreement with the dendrograms in figures 1(d)- 1(f). As shown in figure 1(d), although the two-cluster structure starts to emerge in the dendrogram for $k = 5$, the two natural clusters cannot yet be identified. A clear cluster separation is present in the dendrogram for $k = 15$ (fig. 1(e)). As k increases, similarity values between pattern pairs decrease, and links in the dendrograms progressively form at higher levels, causing the two natural clusters to be less clearly defined (see fig. 1(f) for $k = 80$). The same conclusions can be drawn by analyzing table 1, showing the number of clusters identified for different values of k and of t. The lifetime of a cluster in the dendrogram for a given k (distance gap between two successive merges) can be evaluated on the corresponding line in this table. As shown, using a fixed threshold, the range of k values for which the true number of clusters is identified is limited and depends on t. Using the longest lifetime (clusters persisting for the largest range of t) as the criterion for identifying the final number of clusters, leads to the values on the rightmost column of table 1, with the identification of the true clusters for a larger k range.

Table 1. Number of clusters identified as a function of k and t for the half-rings data set ($N = 200$). The 2^* notation indicates that, although the correct number of clusters is identified, this does not correspond to the correct data partition. The rightmost column indicates the final number of clusters according to the largest lifetime criterion

$k \backslash t$	.05	.10	.15	.20	.25	.30	.35	.40	.45	.50	.55	.60	.65	.70	.75	.80	.85	.90	.95	NC
2	1	1	1	1	1	1	1	1	1	1	1	1	1	1	1	1	2^*	5	7	1
5	1	1	1	1	1	1	1	1	1	1	1	1	1	1	1	2^*	2^*	5	20	2^*
10	1	1	1	1	1	2	2	2	2	2	2	2	2	3	5	7	8	18	71	2
15	1	2	2	2	2	2	2	2	2	2	3	5	5	6	8	13	33	64	134	2
20	1	2	2	2	2	2	2	2	3	5	6	6	7	9	18	34	61	94	171	2
25	1	2	2	2	2	2	2	2	4	6	7	10	14	19	31	54	84	121	215	2
30	1	1	2	2	2	2	3	5	6	8	9	14	20	25	40	71	99	145	252	2
40	1	2	2	2	2	2	3	7	9	11	18	25	34	46	67	95	137	197	279	2
80	3	4	4	5	7	14	19	26	31	47	65	85	97	130	157	188	227	276	334	4

3 Evidence Accumulation Clustering with Varying k and Dynamic Threshold

As noted in the previous section, cluster lifetime is a better criterion for identifying the natural clusters than a fixed threshold, as the dendrogram scales up with increasing values of k. On the other hand, in order to determine an adequate value or range for k, one should use some *a priori* information (for instance, by applying a mixture decomposition method for determining the number of components in the mixture). Otherwise, several values of k should be tested, the final number of clusters being the most stable solution found.

The evidence of a clear cluster separability on the dendrograms associated with a large range for k (see figures 1(e), 1(f)) suggests a combination of K-means clusterings with variable k. Our hypothesis is that the combined evidence will reinforce the intrinsic data structure, diluting the effect produced by low values of k (while combined with other values, low k values contribute to a scaling up of similarity measures - lower values on the dendrograms); high values of k produce random, high granularity data partitions, so they should also not be disruptive of the structure imposed by more adequate k values, scaling down the similarity values. We therefore propose a combination of multiple K-means clusterings with varying k, the final data partition being obtained as the cluster configuration with the highest lifetime in the dendrogram produced by the SL method over the similarity matrix, *co_assoc*. The proposed evidence accumulation clustering method is summarized below:

Data clustering using Evidence Accumulation.

Input:

n $d-$dimensional patterns;

k_min - minimum initial number of clusters;

k_max - maximum initial number of clusters;

N - number of clusterings.

Output: Data partitioning.

Initialization: Set *co_assoc* to a null $n \times n$ matrix.

1. Do N times:

1.1. Randomly select k in the interval $[k\_min; k\_max]$.

1.2. Randomly select k cluster centers.

1.3. Run the K-means algorithm with the above k and initialization, and produce a partition P.

1.4. Update the co-association matrix: for each pattern pair, (i, j), in the same cluster in P, set $co\_assoc(i, j) = co\_assoc(i, j) + \frac{1}{N}$.

2. Detect consistent clusters in the co-association matrix using the SL technique: compute the SL dendrogram and identify the final clusters as the ones with the highest lifetime.

4 Experimental Results

4.1 Vector Representations: Artificial Data Sets

The proposed evidence accumulation clustering method was applied to the half-rings data set, used as the illustrative example in section 2. Several ranges for k were tested in order to evaluate the robustness of the method. Dendrograms for some of these tests are plotted in figure 2. The number of clusterings used were $N = 600$; experiments with $N = 200$ and lower values led to similar results, since the method converges for values of N around 50 (see figure 2(d)).

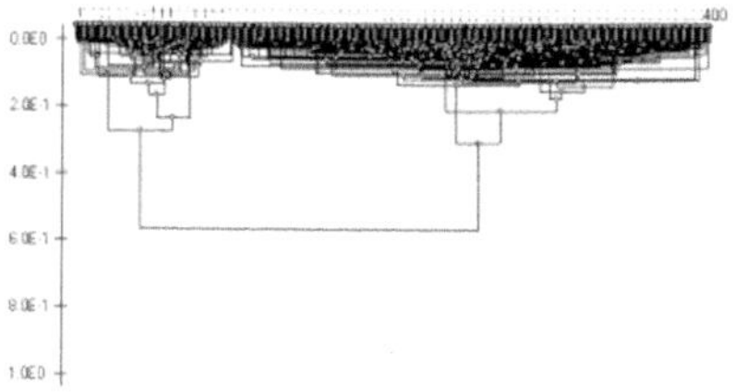

(a) $k \in [2; 20]$.

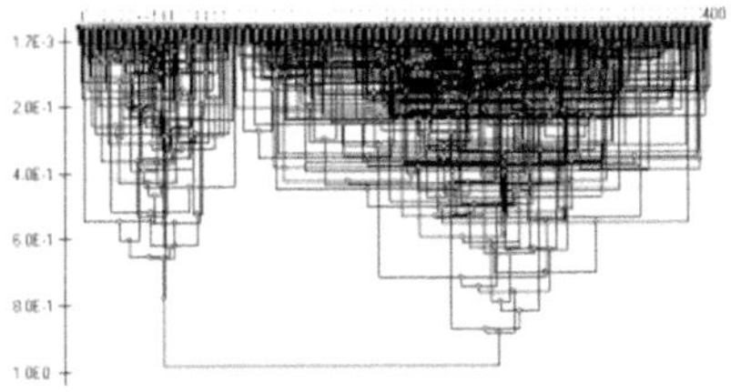

(b) $k \in [60; 90]$.

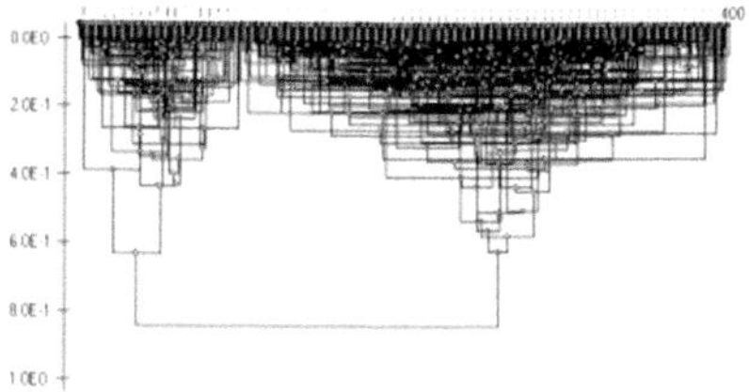

(c) $k \in [2; 80]$.

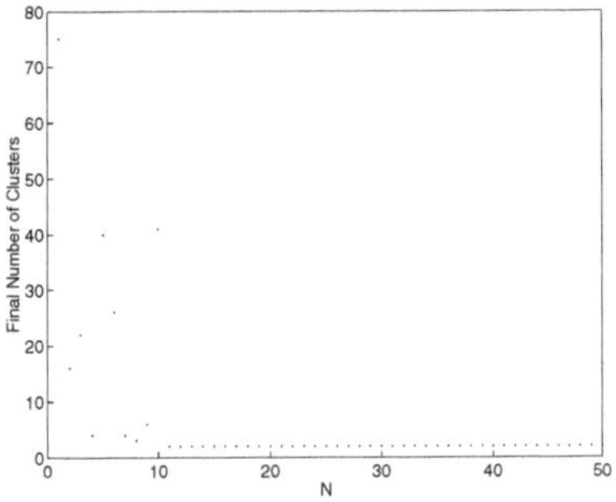

(d) Convergence curve ($k \in [2; 80]$).

Fig. 2. Combining 600 K-means clusterings, with varying k for data in fig 1(a). Dendrograms (a) to (c) illustrate the wide range of k values with a clear cluster separation, showing the robustness of the combination technique. (d)- Convergence curve of the final number of clusterings as a function of N, the number of clusterings, for $k \in [2; 80]$

Table 4 summarizes the experiments and the number of clusters (NC) obtained. As shown, all ranges for k, except the ones completely below the minimum number of mixture components, 10, (first two columns), lead to the correct identification of the natural clusters, demonstrating the robustness of the method.

The spiral data set (fig. 3(a)) is another example of complex shaped clusters. Using the method of [7], the two natural clusters are identified for values of k in the interval $[25; 70]$ for $t = 0.5$ or $t = 0.6$. In all the tests with the proposed method, the true clusters were identified for all the intervals considered (values of $k > 90$ were not tested as the number of training patterns is only 200), except for ranges totally in the interval $[2; 20]$, as this is lower than the minimum number of components required to decompose the data (the method in [5] identifies 24 gaussian components).

We also performed tests on uni-modal random data (gaussian and uniform distributions) in order to assess if the proposed clustering technique imposes

Table 4. Evidence accumulation clustering with varying k for the half-rings data set

	k-range							
	[2; 5]	[2; 10]	[2; 20]	[5; 20]	[10; 30]	[30; 60]	[60; 90]	[2; 80]
NC	1	1	2	2	2	2	2	2

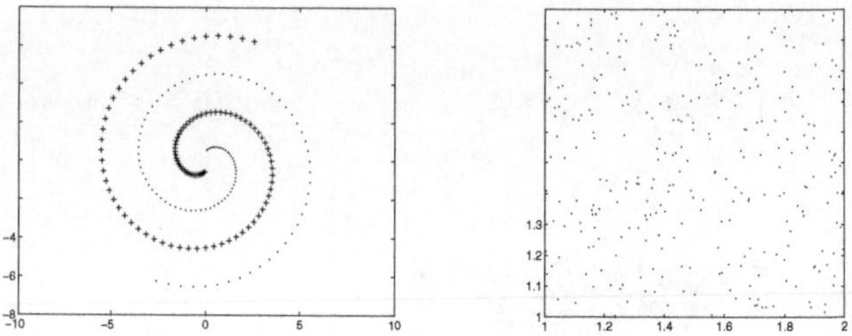

(a) (b)

Fig. 3. Artificial data sets. (a)- Spiral data set (100 samples per class). (b)- 2-D projection of 300 patterns uniformly distributed in a 5-dimensional hypercube

some structure on data. In all the tests performed (an example of uniform data set is illustrated in figure 3(b)), a single cluster was identified, no matter what interval for k was considered.

4.2 String Patterns: Clustering of Contour Images

We have applied the proposed technique to the classification of string descriptions of contour images of 2D shapes. The data set consists of 126 images from three types of tools (42 patterns per class); sample images are shown in figures 4(a) to 4(c). Each image was segmented to separate the object from the background and the object boundary was sampled at 50 equally spaced points; object shapes were encoded using an 8-directional differential chain code [9,11]. In order to apply the cluster combination technique, similarity between all pattern pairs was calculated using the Levensthein distance normalized by the length of the editing path [17,22]. The K-means algorithm was adapted in order to handle string patterns: cluster centroids are selected as the training pattern with the minimum average distance to the remaining patterns within a cluster; therefore, the algorithm simply needs a similarity/dissimilarity matrix between pattern pairs as input.

As shown in figure 4(d), a direct application of the SL method to the string patterns using the normalized string edit distance does not produce a correct partitioning of the data. With the proposed method, a good separation of the three clusters is obtained, for instance with $k \in [2; 30]$ and N=200.

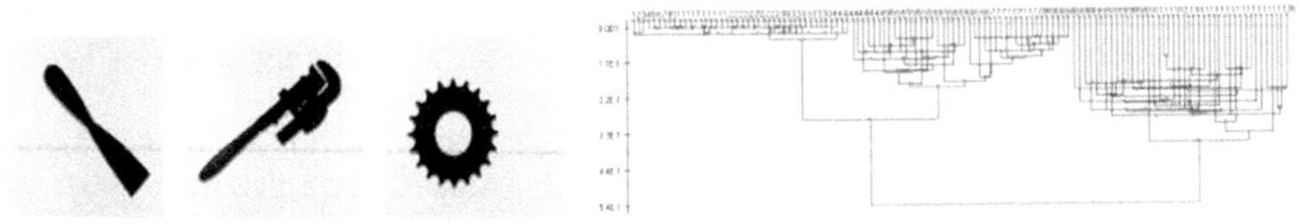

(a) T1. (b) T2. (c) T3. (d) SL dendrogram.

Fig. 4. Hardware tools data set

5 Conclusions

We have proposed a novel algorithm for evidence accumulation clustering. The method introduced in [7] was extended/modified by: (1) using cluster lifetime as a criterion for determining the final number of clusters; (2) proposing the formation of clustering ensembles by using the K-means algorithm with random initialization and arbitrary k values within a large interval. Furthermore, the adaptation of the K-means algorithm by using cluster median patterns, and thus simply requiring as input a similarity or dissimilarity matrix, extended the potential use of this technique to a wider range of applications, namely those based on string descriptions. The new method enhances the previous approach in terms of robustness and simplicity of evaluations, with fewer parameters being defined. The ability of the clustering method to correctly identify well separated clusters with complex shapes has been demonstrated on a set of artificial and real data, using both vector and string descriptions of patterns. Moreover, tests on unimodal/uniform data showed that the method does not impose any structure on data, a single cluster being identified for this data. Further tests are needed for touching clusters.

Acknowledgments

This work was partially supported by the Portuguese Foundation for Science and Technology (FCT), Portuguese Ministry of Science and Technology, and FEDER, under grant POSI/33143/SRI/2000, and ONR grant no. N00014-01-1-0266.

References

1. T. A. Bailey and R. Dubes. Cluster validity profiles. *Pattern Recognition*, 15(2):61–83, 1982. 442
2. J. Buhmann and M. Held. Unsupervised learning without overfitting: Empirical risk approximation as an induction principle for reliable clustering. In Sameer Singh, editor, *International Conference on Advances in Pattern Recognition*, pages 167–176. Springer Verlag, 1999. 442

3. R. O. Duda, P. E. Hart, and D. G. Stork. *Pattern Classification.* Wiley, second edition, 2001. 442
4. Y. El-Sonbaty and M. A. Ismail. On-line hierarchical clustering. *Pattern Recognition Letters*, pages 1285–1291, 1998. 442
5. M. Figueiredo and A. K. Jain. Unsupervised learning of finite mixture models. *IEEE Trans. Pattern Analysis and Machine Intelligence*, 24(3):381–396, 2002. 445, 447
6. B. Fischer, T. Zoller, and J. Buhmann. Path based pairwise data clustering with application to texture segmentation. In M. Figueiredo, J. Zerubia, and A. K. Jain, editors, *Energy Minimization Methods in Computer Vision and Pattern Recognition*, volume 2134 of *LNCS*, pages 235–266. Springer Verlag, 2001. 443
7. A. L. Fred. Finding consistent clusters in data partitions. In Josef Kittler and Fabio Roli, editors, *Multiple Classifier Systems*, volume LNCS 2096, pages 309–318. Springer, 2001. 442, 443, 447, 449
8. A. L. Fred and J. Leitão. Clustering under a hypothesis of smooth dissimilarity increments. In *Proc. of the 15th Int'l Conference on Pattern Recognition*, volume 2, pages 190–194, Barcelona, 2000. 442
9. A. L. Fred, J. S. Marques, and P. M. Jorge. Hidden markov models vs syntactic modeling in object recognition. In *ICIP'97*, 1997. 448
10. M. Har-Even and V. L. Brailovsky. Probabilistic validation approach for clustering. *Pattern Recognition*, 16:1189–1196, 1995. 442
11. A. Jain. *Fundamentals of Digital Image Processing.* Prentice-Hall, 1989. 448
12. A. K. Jain and R. C. Dubes. *Algorithms for Clustering Data.* Prentice Hall, 1988. 442
13. A.K. Jain, M. N. Murty, and P.J. Flynn. Data clustering: A review. *ACM Computing Surveys*, 31(3):264–323, September 1999. 442
14. J. Kittler, M. Hatef, R. P Duin, and J. Matas. On combining classifiers. *IEEE Trans. Pattern Analysis and Machine Intelligence*, 20(3):226–239, 1998. 443
15. R. Kothari and D. Pitts. On finding the number of clusters. *Pattern Recognition Letters*, 20:405–416, 1999. 442
16. Y. Man and I. Gath. Detection and separation of ring-shaped clusters using fuzzy clusters. *IEEE Trans. Pattern Analysis and Machine Intelligence*, 16(8):855–861, August 1994. 443
17. A. Marzal and E. Vidal. Computation of normalized edit distance and applications. *IEEE Trans. Pattern Analysis and Machine Intelligence*, 2(15):926–932, 1993. 448
18. G. McLachlan and K. Basford. *Mixture Models: Inference and Application to Clustering.* Marcel Dekker, New York, 1988. 442
19. B. Mirkin. Concept learning and feature selection based on square-error clustering. *Machine Learning*, 35:25–39, 1999. 442
20. N. R. Pal and J. C. Bezdek. On cluster validity for the fuzzy c-means model. *IEEE Trans. Fuzzy Systems*, 3:370–379, 1995. 442
21. E. J. Pauwels and G. Frederix. Fiding regions of interest for content-extraction. In *Proc. of IS&T/SPIE Conference on Storage and Retrieval for Image and Video Databases VII*, volume SPIE Vol. 3656, pages 501–510, San Jose, January 1999. 442
22. E. S. Ristad and P. N. Yianilos. Learning string-edit distance. *IEEE Trans. Pattern Analysis and Machine Intelligence*, 20(5):522–531, May 1998. 448
23. S. Roberts, D. Husmeier, I. Rezek, and W. Penny. Bayesian approaches to gaussian mixture modelling. *IEEE Trans. Pattern Analysis and Machine Intelligence*, 20(11), November 1998. 442

24. D. Stanford and A. E. Raftery. Principal curve clustering with noise. Technical report, University of Washington, http://www.stat.washington.edu/raftery, 1997. 443
25. H. Tenmoto, M. Kudo, and M. Shimbo. MDL-based selection of the number of components in mixture models for pattern recognition. In Adnan Amin, Dov Dori, Pavel Pudil, and Herbert Freeman, editors, *Advances in Pattern Recognition*, volume 1451 of *Lecture Notes in Computer Science*, pages 831–836. Springer Verlag, 1998. 442
26. C. Zahn. Graph-theoretical methods for detecting and describing gestalt structures. *IEEE Trans. Computers*, C-20(1):68–86, 1971. 442

A Kernel Approach to Metric Multidimensional Scaling

Andrew Webb

QinetiQ, St. Andrews Road, Malvern, WR14 3PS
webb@signal.qinetiq.com

Abstract. The solution for the parameters of a nonlinear mapping in a metric multidimensional scaling by transformation, in which a stress criterion is optimised, satisfies a nonlinear eigenvector equation, which may be solved iteratively. This can be cast in a kernel-based framework in which the configuration of training samples in the transformation space may be found iteratively by successive linear projections, without the need for gradient calculations. A new data sample can be projected using knowledge of the kernel and the final configuration of data points.
Keywords. multidimensional scaling; kernel representation; nonlinear feature extraction;

1 Introduction

Multidimensional scaling by transformation (MST) describes a class of procedures that implements a nonlinear, dimension-reducing mapping that minimises a criterion, stress, in the output or *representation* space with the aim of retaining the structure and important relationships within the original dataset defined in the data or *observation space*. Often, these mappings are characterised by feed-forward neural network models (for example, multilayer perceptrons [6,7,9], or radial basis functions [8,16]) whose parameters are adjusted to optimise the stress. The stress criterion is strongly related to the conventional metric multidimensional scaling objective function [3], sometimes termed Sammon mappings in the pattern recognition literature after (Sammon, 1969 [14]), with generalisations to include subjective information [8] and class information [5,16].

Many *linear* methods of feature extraction are based on matrices of first and second order statistics. These include transformations based on principal components analysis and linear discriminant analysis and variants [18], and lead to eigenvector / generalised eigenvector equations for the weights of the transformation.

Many methods of *nonlinear* feature extraction, and also methods for discrimination and regression, are linear models. The nonlinear transformation from the data space to the output space is expressed as a linear combination of basis functions. This linear transformation can be determined by first explicitly mapping the input data to the feature space defined by the outputs of the basis functions and then choosing the linear transformation that optimises a criterion defined in

T. Caelli et al. (Eds.): SSPR&SPR 2002, LNCS 2396, pp. 452–460, 2002.
© Springer-Verlag Berlin Heidelberg 2002

the output space. This criterion may be, for example, a least squares error, stress (in multidimensional scaling) or variance and, in the case of nonlinear principal components analysis, leads to eigenvector equations for the weights.

An alternative to the approach of *explicitly* mapping data to an intermediate feature space and then determining the nonlinear transformation is to design nonlinear data processing algorithms using linear techniques in an *implicit* feature space induced by kernel functions defined on the data space. These algorithms include support vector machines, kernel principal components analysis [15] and kernel multidimensional scaling [20]. In this paper we show that MST in which a stress function is minimised (and so differs from [20] which is the application of classical scaling in the feature space) can also be expressed in such a framework.

2 Statement of the Problem

Let $\boldsymbol{X} = \{\boldsymbol{x}_1, \ldots, \boldsymbol{x}_N \in \mathbb{R}^n\}$ denote the dataset of N n-dimensional measurement vectors, $\boldsymbol{x}_i, i = 1, \ldots, N$. Each measurement vector $\boldsymbol{x}_i$ may also have an associated class label, $z_i \in \{1, \ldots, C\}$, the class to which $\boldsymbol{x}_i$ belongs, where C is the number of classes. MST seeks a transformation $\boldsymbol{f} : \mathbb{R}^n \to \mathbb{R}^m$ from n-dimensional space to m-dimensional space ($m < n$) such that a loss function of the form,

$$\sigma^2 = \sum_{i=1}^{N}\sum_{j=1}^{N} \alpha_{ij}(q_{ij} - d_{ij}(\boldsymbol{X}))^2 \tag{1}$$

is minimised, where $\alpha_{ij}, i, j = 1, \ldots, N$ are positive weights that may depend on the classes of $\boldsymbol{x}_i$ and $\boldsymbol{x}_j$; the term $d_{ij}(\boldsymbol{X})$ given by

$$d_{ij}(\boldsymbol{X}) = |\boldsymbol{x}_i - \boldsymbol{x}_j|$$

is a distance in the observation space and

$$q_{ij} = |\boldsymbol{f}(\boldsymbol{x}_i) - \boldsymbol{f}(\boldsymbol{x}_j)|$$

is a distance in the representation space. The distances are usually taken to be Euclidean, but other forms may be considered [17]. Minimisation of the loss, σ^2, is performed with respect to the functional form of the transformation, $\boldsymbol{f}$. A convenient choice for $\boldsymbol{f}$ is a basis function expansion of the form

$$\boldsymbol{f}(\boldsymbol{x}) = \sum_{j=1}^{l} \boldsymbol{w}_j \phi_j(\boldsymbol{x}) \tag{2}$$

for a set of basis functions, $\{\phi_j, j = 1, \ldots, l\}$, where $\{\boldsymbol{w}_j \in \mathbb{R}^m, j = 1, \ldots, l\}$ is a set of weights optimised by the procedure. Equation (2) may be written

$$\boldsymbol{f}(\boldsymbol{x}) = \boldsymbol{W}^T \boldsymbol{\phi}(\boldsymbol{x}) \tag{3}$$

where $\boldsymbol{W}$ is the $l \times m$ matrix with (i, j) elements w_{ij} and $\boldsymbol{\phi} = [\phi_1(\boldsymbol{x})| \ldots |\phi_l(\boldsymbol{x})]^T$ is the l-dimensional vector of nonlinear responses.

3 Iterative Solution

The minimisation of σ^2 (Equation (1)) with respect to $\boldsymbol{W}$, the parameters of the nonlinear transformation $\boldsymbol{f}$, may be performed using standard nonlinear optimisation schemes [12] that require evaluation of the gradient of σ^2 with respect to $\boldsymbol{W}$. It may also be performed without gradient calculations [16,19] using an *iterative majorisation* approach. Given an initial starting point for the parameters, a *majorising function* is specified that touches the loss function to be minimised at this point, but everywhere else lies above it. The majorising function is simple to minimise and the position of the minimum is used as the starting point for the next iteration (see Figure 1).

It is shown in [16] that the weights $\boldsymbol{W}$ may be determined iteratively using the equation

$$\boldsymbol{S}\boldsymbol{W}^{(t+1)} = 2\boldsymbol{R}(\boldsymbol{W}^{(t)})\boldsymbol{W}^{(t)} \tag{4}$$

for symmetric $l \times l$ matrices $\boldsymbol{S}$ and $\boldsymbol{R}$ given by

$$\begin{aligned} \boldsymbol{S} &= 2\sum_i \sum_j \alpha_{ij}(\boldsymbol{\phi}(\boldsymbol{x}_i) - \boldsymbol{\phi}(\boldsymbol{x}_j))(\boldsymbol{\phi}(\boldsymbol{x}_i) - \boldsymbol{\phi}(\boldsymbol{x}_j))^T \\ &= 4N\boldsymbol{\Phi}^T\left(I - \frac{1}{N}\mathbf{1}\mathbf{1}^T\right)\boldsymbol{\Phi} \end{aligned} \tag{5}$$

where the α_{ij} are taken to be unity in the last equality and $\boldsymbol{\Phi} = [\boldsymbol{\phi}(\boldsymbol{x}_1)| \dots |\boldsymbol{\phi}(\boldsymbol{x}_N)]^T$ is the $N \times l$ matrix of nonlinear features; and

$$\boldsymbol{R}(\boldsymbol{W}^{(t)}) = 2\boldsymbol{\Phi}^T(\tilde{\boldsymbol{C}} - \boldsymbol{C})\boldsymbol{\Phi} \tag{6}$$

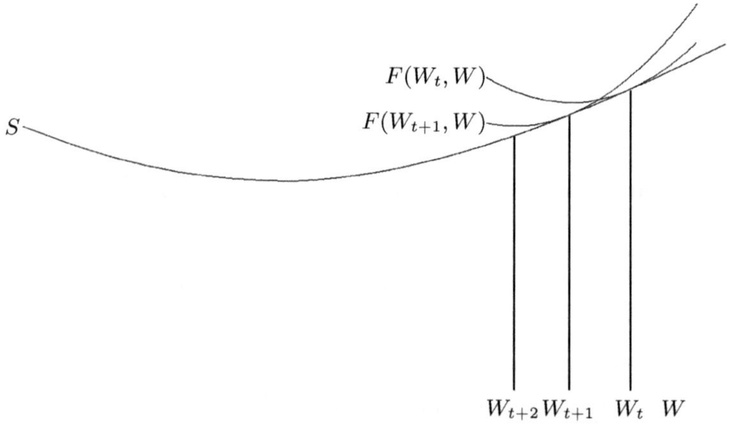

Fig. 1. Illustration of iterative majorisation principle: minimisation of $S(W)$ is achieved through successive minimisations of the majorisation functions, F

where $\boldsymbol{C} = \boldsymbol{C}(\boldsymbol{W}^t)$ is the $N \times N$ matrix that depends on the configuration at stage t through the q_{ij} with (i, j) element

$$c_{ij} = \begin{cases} d_{ij}(\boldsymbol{X})/q_{ij}(\boldsymbol{W}^{(t)}) & q_{ij}(\boldsymbol{W}^{(t)}) > 0 \\ 0 & q_{ij}(\boldsymbol{W}^{(t)}) = 0 \end{cases}$$

and $\tilde{\boldsymbol{C}} = \text{Diag}\{\boldsymbol{C}\boldsymbol{1}\}$, the diagonal matrix with (i, i) element $(\boldsymbol{C}\boldsymbol{1})_i$.

An alternative derivation for the iterative equation for $\boldsymbol{W}$ may be obtained by showing that $\boldsymbol{W}$ satisfies a *nonlinear eigenvector equation* with an algorithm for its solution based on the inverse iteration method for the ordinary eigenvector equation [19].

4 Kernel Representation

4.1 Iterative Solution Using Kernels

We now re-cast the iterative solution for the weights as an iterative solution for the final configuration in the transformed space that requires the specification of a kernel defined on the data space.

Defining $\boldsymbol{H}$ as the $N \times N$ idempotent centring matrix

$$\boldsymbol{H} = \left(I - \frac{1}{N}\boldsymbol{1}\boldsymbol{1}^T\right)$$

so that $\boldsymbol{H}\boldsymbol{H} = \boldsymbol{H}$ and $\boldsymbol{H} = \boldsymbol{H}^T$, and using (5) and (6) Equation (4) may be written

$$N(\boldsymbol{H\Phi})^T(\boldsymbol{H\Phi})\boldsymbol{W}^{(t+1)} = (\boldsymbol{H\Phi})^T(\tilde{\boldsymbol{C}} - \boldsymbol{C})(\boldsymbol{H\Phi})\boldsymbol{W}^{(t)} \quad (7)$$

Writing $\boldsymbol{P}^{(t)} = (\boldsymbol{H\Phi})\boldsymbol{W}^{(t)}$, the $N \times m$ matrix of centred data coordinates in the projected space at stage t, then Equation (7) may be written

$$N(\boldsymbol{H\Phi})^T\boldsymbol{P}^{(t+1)} = (\boldsymbol{H\Phi})^T(\tilde{\boldsymbol{C}} - \boldsymbol{C})\boldsymbol{P}^{(t)} \quad (8)$$

Taking the pseudo-inverse of $(\boldsymbol{H\Phi})^T$, we can express $\boldsymbol{P}^{(t+1)}$ as

$$N\boldsymbol{P}^{(t+1)} = [(\boldsymbol{H\Phi})(\boldsymbol{H\Phi})^T]^{\dagger}(\boldsymbol{H\Phi})(\boldsymbol{H\Phi})^T(\tilde{\boldsymbol{C}} - \boldsymbol{C})\boldsymbol{P}^{(t)} \quad (9)$$

Equation (9) above provides an iterative equation for the coordinates of the transformed data samples. This is the procedure followed in standard approaches to multidimensional scaling [4]. The difference here is that constraints on the form of the nonlinear transformation describing the multidimensional scaling projection are incorporated into the procedure through the $N \times N$ matrix $(\boldsymbol{H\Phi})(\boldsymbol{H\Phi})^T$.

The matrix $(\boldsymbol{H\Phi})(\boldsymbol{H\Phi})^T$ depends on dissimilarities in feature space and may be written

$$(\boldsymbol{H\Phi})(\boldsymbol{H\Phi})^T = \boldsymbol{HFH} \quad (10)$$

where the $N \times N$ matrix, $\boldsymbol{F}$ has (i,j) element $f_{ij} = -\frac{1}{2}\hat{\delta}_{ij}^2$, where

$$\hat{\delta}_{ij}^2 = (\boldsymbol{\phi}(\boldsymbol{x}_i) - \boldsymbol{\phi}(\boldsymbol{x}_j))^T(\boldsymbol{\phi}(\boldsymbol{x}_i) - \boldsymbol{\phi}(\boldsymbol{x}_j))$$

is the squared Euclidean distance in feature space. Denoting the inner product $\boldsymbol{\phi}^T(\boldsymbol{x}_j)\boldsymbol{\phi}(\boldsymbol{x}_i)$ by the kernel function $K(\boldsymbol{x}_i, \boldsymbol{x}_j)$, then $\boldsymbol{F}$ may be written

$$\boldsymbol{F} = \boldsymbol{K} - \frac{1}{2}\boldsymbol{k}\boldsymbol{1}^T - \frac{1}{2}\boldsymbol{1}\boldsymbol{k}^T \tag{11}$$

where $\boldsymbol{K}$ is the matrix with (i,j) element $K(\boldsymbol{x}_i, \boldsymbol{x}_j)$ and the ith element of the vector $\boldsymbol{k}$ is $K(\boldsymbol{x}_i, \boldsymbol{x}_i)$. Substituting for $\boldsymbol{F}$ from Equation (11) into (10) gives

$$(\boldsymbol{H\Phi})(\boldsymbol{H\Phi})^T = \boldsymbol{HKH} \tag{12}$$

Thus, the iterative procedure for the projected data samples depends on the kernel function K, which must satisfy the usual conditions [2] to ensure that it is an inner product on some feature space. Example kernels are polynomials and gaussians.

However, note that if the pseudo-inverse is used to calculate $\boldsymbol{P}^{(t+1)}$ from $\boldsymbol{P}^{(t)}$ in (9), then the only influence of the kernel is through the space spanned by the (non-zero) eigenvectors of $\boldsymbol{HKH}$. That is, if we write $\boldsymbol{HKH}$ as its singular value decomposition, $\boldsymbol{U}_r\boldsymbol{\Sigma}_r\boldsymbol{U}_r^T$, for $N \times r$ matrix of eigenvectors $\boldsymbol{U}_r$ and $\boldsymbol{\Sigma}_r = \text{Diag}\{\sigma_1, \ldots, \sigma_r\}$ for non-zero singular values $\sigma_i, 1 \leq i \leq r$, then

$$(\boldsymbol{HKH})^\dagger(\boldsymbol{HKH}) = \boldsymbol{U}_r\boldsymbol{U}_r^T$$

and

$$N\boldsymbol{P}^{(t+1)} = \boldsymbol{U}_r\boldsymbol{U}_r^T(\tilde{\boldsymbol{C}} - \boldsymbol{C})\boldsymbol{P}^{(t)} \tag{13}$$

Thus, the new coordinates comprise a transformation of the coordinates at step t followed by a projection onto the subspace defined by the columns of $\boldsymbol{U}_r$. The final solution for the coordinates, which we denote by $\tilde{\boldsymbol{P}}$, must lie in the subspace defined by $\boldsymbol{U}_r$.

The matrix $\boldsymbol{C} = \boldsymbol{C}(\boldsymbol{P}^{(t)})$ depends on the configuration of points in the transformed space and is given by

$$c_{ij} = \begin{cases} \alpha_{ij} d_{ij}(\boldsymbol{X})/q_{ij}(\boldsymbol{P}^t) & q_{ij}(\boldsymbol{P}^{(t)}) > 0 \\ 0 & q_{ij}(\boldsymbol{P}^{(t)}) = 0 \end{cases}$$

where $q_{ij}(\boldsymbol{P}^{(t)})$ is the distance between transformed points i and j at stage t.

4.2 Projection of New Data Samples

The iterative procedure described above finds a configuration of the data samples in the transformed space. We would also like to determine where in the transformed space a new sample maps to without having to calculate a weight vector explicitly.

Let $\tilde{\boldsymbol{P}}$ denote the final $N \times m$ matrix of coordinates of the N data samples in the projected m-dimensional space. For a data sample $\boldsymbol{x}$, the new projection, $\boldsymbol{z}$, is given by

$$\boldsymbol{z} = \boldsymbol{W}^T \phi(\boldsymbol{x})$$

and using the solution for the weights ($\boldsymbol{W} = (\boldsymbol{H\Phi})^\dagger \tilde{\boldsymbol{P}}$) we have

$$\begin{aligned} \boldsymbol{z} &= \tilde{\boldsymbol{P}}^T [(\boldsymbol{H\Phi})^\dagger]^T \phi(\boldsymbol{x}) \\ &= \tilde{\boldsymbol{P}}^T [\boldsymbol{HKH}]^\dagger \boldsymbol{H\Phi}\phi(\boldsymbol{x}) \\ &= \tilde{\boldsymbol{P}}^T [\boldsymbol{HKH}]^\dagger \boldsymbol{Hl} \end{aligned} \tag{14}$$

where $\boldsymbol{l} = [l_1, \ldots, l_N]^T$ and $l_i = k(\boldsymbol{x}, \boldsymbol{x}_i)$. Thus, a new projection can be expressed using the kernels only, and not the feature space representation and is a weighted sum of the final training data projections, $\tilde{\boldsymbol{P}}$.

5 Choice of Kernel

We adopt a Gaussian kernel of the form

$$K(\boldsymbol{x}_i, \boldsymbol{x}_j) = \exp\left(-\theta(\boldsymbol{x}_i - \boldsymbol{x}_j)^T(\boldsymbol{x}_i - \boldsymbol{x}_j)\right)$$

with inverse scale parameter θ.

As $\theta \to \infty$, the matrix $\boldsymbol{K} \to \boldsymbol{I}$ and $\boldsymbol{HKH} \to \boldsymbol{H} = \boldsymbol{I} - \boldsymbol{1}\boldsymbol{1}^T/N$, which is independent of the training data.

As $\theta \to 0$, the matrix $\boldsymbol{K} \to \boldsymbol{1}\boldsymbol{1}^T - \theta\boldsymbol{D}$, where $\boldsymbol{D}$ is the matrix of squared distances in the data space, $\boldsymbol{D}_{ij} = |\boldsymbol{x}_i - \boldsymbol{x}_j|^2$. The matrix $\boldsymbol{HKH} \to -\theta\boldsymbol{HDH}$, showing that the kernel is equivalent to the quadratic kernel[1] $K(\boldsymbol{x}_i, \boldsymbol{x}_j) = |\boldsymbol{x}_i - \boldsymbol{x}_j|^2$, which does not depend on θ.

6 Illustration

The technique is illustrated using a simulated dataset comprising 500 points uniformly distributed over a part-sphere with added noise.

$$\begin{aligned} x_1 &= A\cos(\psi)\sin(\phi) + n_1 \\ x_2 &= A\cos(\psi)\cos(\phi) + n_2 \\ x_3 &= A\sin(\psi) + n_3 \end{aligned}$$

where $\phi = 2\pi u$, $\psi = \sin^{-1}(v(1 + \sin(\psi_{\max})) - 1)$ and u and v are uniformly-distributed on $[0, 1]$; $A = 1$ and n_1, n_2 and n_3 are normally-distributed with variance 0.1. A value of $\pi/4$ was taken for $\psi_{\max}$, so that the surface covers the

[1] Any scaling of a kernel does not affect the final configuration of training samples or the point to which a test pattern is projected.

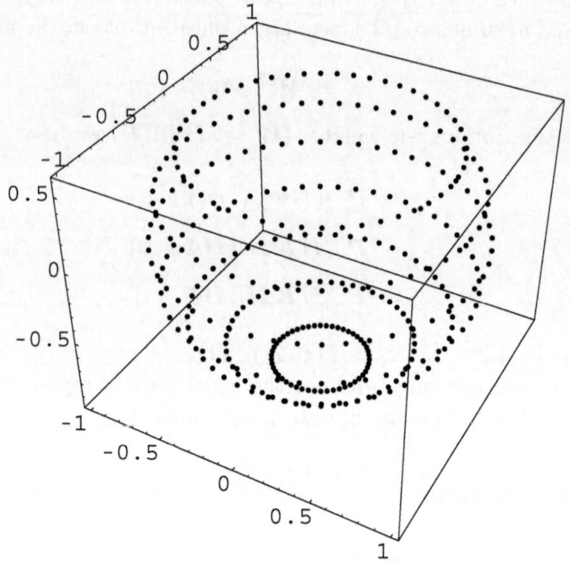

Fig. 2. Lines of latitude on underlying surface

lower hemisphere ($-\pi/2 \leq \psi \leq 0$), together with the upper hemisphere up to a latitude of 45 degrees. Figure 2 shows lines of latitude on the underlying sphere.

The algorithm is trained to convergence on the noisy sphere data and a projection to two dimensions is sought. Figure 3 plots the normalised stress (after the algorithm has converged),

$$\sigma^2 = \frac{\sum_{i=1}^{N}\sum_{j=1}^{N}(q_{ij} - d_{ij}(\boldsymbol{X}))^2}{\sum_{i=1}^{N}\sum_{j=1}^{N} d_{ij}(\boldsymbol{X})^2} \tag{15}$$

as a function of θ for a test dataset generated using the same distribution as the training data.

For small values of θ, there is very little variation in the stress, showing that a quadratic kernel is close to optimal. Figure 4 give a two-dimensional plot of the training data and the points on the underlying surface (lines of latitude on the sphere) for a value of $\theta = 1.0$. We see that the transformation has 'opened out' the sphere to produce a two dimensional projection.

7 Summary

The main results of this paper can be summarised as follows.

1. The solution for the weights of a generalised linear model that minimise a stress criterion can be obtained using an iterative algorithm (Equations (4)).

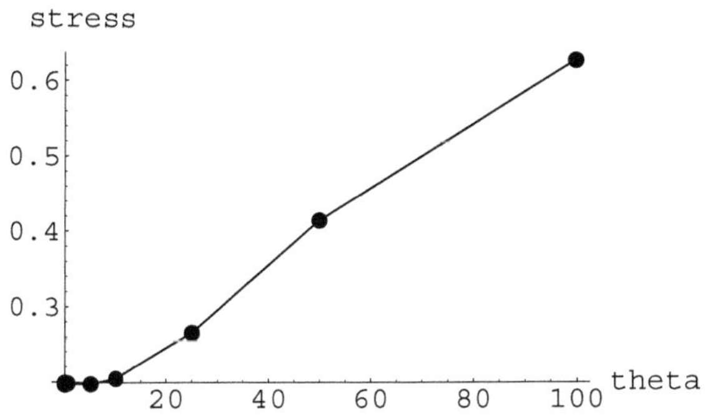

Fig. 3. Normalised stress on test set as a function of θ

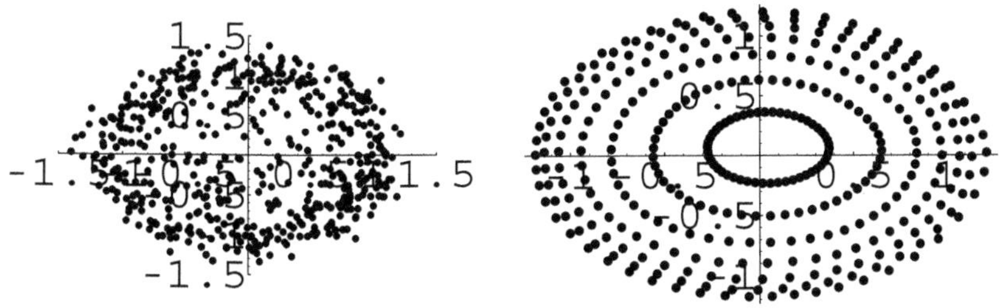

Fig. 4. Projection of training data (left) and points on the underlying surface (right) for the noisy sphere dataset

2. The iterative algorithm for the weights may be re-expressed as an iterative algorithm for the projected data samples (Equation (9)), which depends on a kernel function defined in the data space.
3. For a Gaussian kernel, there is one model selection parameter, θ, that can be determined using a validation set.
4. The projection of new data points may be achieved using the solution for the projected training samples (Equation (14)). The projection is a weighted sum of the projected training samples.

Acknowledgments

This research was sponsored by the UK MOD Corporate Research Programme.

References

1. J. M. Borwein and A. S. Lewis. *Convex Analysis and Nonlinear Optimisation. Theory and Examples.* Springer-Verlag, New York, 2000.
2. N. Cristianini and J. Shawe-Taylor. *An introduction to support vector machines.* Cambridge University Press, Cambridge, 2000. 456
3. W. R. Dillon and M. Goldstein. *Multivariate Analysis Methods and Applications.* John Wiley and Sons, New York, 1984. 452
4. W. J. Heiser. Convergent computation by iterative majorization: theory and applications in multidimensional data analysis. In W. J. Krzanowski, editor, *Recent Advances in Descriptive Multivariate Analysis*, pages 157–189. Clarendon Press, Oxford, 1994. 455
5. W. L. G. Koontz and K. Fukunaga. A nonlinear feature extraction algorithm using distance information. *IEEE Transactions on Computers*, 21(1):56–63, 1972. 452
6. B. Lerner, H. Guterman, M. Aladjem, and I. Dinstein. A comparative study of neural network based feature extraction paradigms. *Pattern Rccognition Letters*, 120:7–14, 1999. 452
7. B. Lerner, H. Guterman, M. Aladjem, I. Dinstein, and Y. Romem. On pattern classification with Sammon's nonlinear mapping – an experimental study. *Pattern Recognition*, 31(4):371–381, 1998. 452
8. D. Lowe and M. Tipping. Feed-forward neural networks and topographic mappings for exploratory data analysis. *Neural Computing and Applications*, 4:83–95, 1996. 452
9. J. Mao and A. K. Jain. Artificial neural networks for feature extraction and multivariate data projection. *IEEE Transactions on Neural Networks*, 6(2):296–317, 1995. 452
10. R. Mathar and R. Meyer. Algorithms in convex analysis to fit l_p-distance matrices. *Journal of Multivariate Analysis*, 51:102–120, 1994.
11. R. Meyer. Nonlinear eigenvector algorithms for local optimisation in multivariate data analysis. *Linear Algebra and its Applications*, 264:225–246, 1997.
12. W. H. Press, B. P. Flannery, S. A. Teukolsky, and W. T. Vetterling. *Numerical Recipes. The Art of Scientific Computing.* Cambridge University Press, Cambridge, second edition, 1992. 454
13. R. T. Rockafellar. *Convex Analysis.* Princeton University Press, Princeton, New Jersey, 1970.
14. J. W. Sammon. A nonlinear mapping for data structure analysis. *IEEE Transactions on Computers*, 18(5):401–409, 1969. 452
15. B. Schölkopf, A. Smola, and K.-R. Müller. Nonlinear component analysis as a kernel eigenvalue problem. *Neural Computation*, 10:1299–1319, 1998. 453
16. A. R. Webb. Multidimensional scaling by iterative majorisation using radial basis functions. *Pattern Recognition*, 28(5):753–759, 1995. 452, 454
17. A. R. Webb. Radial basis functions for exploratory data analysis: an iterative majorisation approach for Minkowski distances based on multidimensional scaling. *Journal of Classification*, 14(2):249–267, 1997. 453
18. A. R. Webb. *Statistical Pattern Recognition.* Arnold, London, 1999. 452
19. A. R. Webb. A kernel approach to metric multidimensional scaling. In preparation, 2002. 454, 455
20. C. K. I. Williams. On a connection between kernel pca and metric multidimensional scaling. *Machine Learning*, 46(1/3):11–19, 2001. 453

On Feature Selection with Measurement Cost and Grouped Features

Pavel Paclík[1], Robert P.W. Duin[1],
Geert M.P. van Kempen[2], and Reinhard Kohlus[2]

[1] Pattern Recognition Group, Delft University of Technology
The Netherlands
{pavel,duin}@ph.tn.tudelft.nl
[2] Unilever R&D Vlaardingen, The Netherlands
Geert-van.Kempen@unilever.com
Reinhard.Kohlus@unilever.com

Abstract. Feature selection is an important tool reducing necessary feature acquisition time in some applications. Standard methods, proposed in the literature, do not cope with the measurement cost issue. Including the measurement cost into the feature selection process is difficult when features are grouped together due to the implementation. If one feature from a group is requested, all others are available for zero additional measurement cost. In the paper, we investigate two approaches how to use the measurement cost and feature grouping in the selection process. We show, that employing grouping improves the performance significantly for low measurement costs. We discuss an application where limiting the computation time is a very important topic: the segmentation of backscatter images in product analysis.

1 Introduction

Feature selection is usually used to choose a feature subset with the best possible performance [,]. The acquisition cost of selected features is, however, also an important issue in some applications. We can mention, for example, medical diagnosis or texture segmentation. In this paper, we are interested in cases where well-performing cheap features are preferred over the expensive ones delivering just slightly better results.

Due to the implementation efficiency, features are often produced in groups. The computation of time-consuming intermediate results is performed just once and then used for the generation of a number of different features. Examples may be Fourier descriptors or various texture features. If the traditional feature selection technique is used in such a case, the resulting feature subset will often require unnecessarily long acquisition time when classifying of new objects. In this paper, we discuss a strategy how to include the information about the feature grouping into the feature selection process and thereby save the time.

An application that enabled our interest in the feature selection with measurement cost is the segmentation of backscatter images (BSE) in the analysis

T. Caelli et al. (Eds.): SSPR&SPR 2002, LNCS 2396, pp. 461–469, 2002.
© Springer-Verlag Berlin Heidelberg 2002

of laundry detergents [4]. Let us use it as an example of a problem where the speeding-up of the feature computation is an important design issue. The development of laundry detergents is based on structural analysis of BSE images. For each powder type, the batch of BSE images is acquired, segmented and analyzed. Image segmentation is performed by supervised statistical pattern recognition algorithm using a number of mainly texture features [5]. Feature selection is run on a single training image. The batch of BSE images from the same powder formulation is then segmented by a trained texture classifier. An important point is that the feature selection is performed for each new batch of images due to variable magnification and type of detergent structures to be labeled. Feature acquisition is computationally intensive problem as the image pixels of high-resolution images are treated as individual data samples. Taking into account also the number of processed images within a batch, the feature selection method should optimize both the performance and the feature computation time. From the implementation point of view, features form several groups. Our intention is to use feature grouping in the feature selection process to find a time-effective feature set.

In the next section, we explain two strategies how the feature grouping may be employed in the feature selection. In the section 3, we discuss experiments on two different problems: handwritten digit recognition and backscatter image segmentation. Finally, in the last section, we give conclusions.

2 Feature Selection with Grouped Features

Feature selection algorithm searches for the best subset of d features from a complete set of D measurements, $d < D$. Several searching strategies have been proposed in the literature [2,3]. In the following, we use the sequential forward feature selection algorithm (SFS). Selection of features is based on a criterion function. In the paper, we use performance of a classifier on an evaluation set as a criterion. It predicts the performance degradation caused by the use of weak features in high dimensionalities (curse of dimensionality).

Standard feature selection algorithms do not take into account measurement cost. Therefore, expensive features may be selected while a number of weaker features is available at low cost. Measurement cost may be combined into the selection criterion in several different ways. In this paper, we consider a criterion C of the following form:

$$C = \frac{\Delta P}{\Delta T}. \tag{1}$$

Here, ΔP stands for the increase of performance and ΔT denotes the increase of measurement cost between two algorithm steps. This criterion favors cheap features offering a small performance gain before better but expensive ones. If linear weighting of performance and measurement cost is of interest, different criterion might be a better choice.

In reality, implementation often defines grouping of features. Group G of N features is computed at once. If one feature from the group is used, all others are available for zero additional measurement cost. If time optimization is of interest, adding descriptive features with zero measurement cost should be preferred. Unfortunately, zero increase of the measurement cost poses a problem for the selection algorithm using criterion (1).

We propose to change the selection strategy and choose the features on the *per-group* basis. It means, that the feature selection algorithm runs at two levels. At the higher level, it operates over feature groups. For each group, a convenient feature subset is found. The performance of the selected subset in the group is used to choose the best group. We have been investigating two variants of this approach.

2.1 Group-Wise Forward Selection (GFS)

In this method, forward feature selection is run for each group. A group is judged based on the performance of its *all* features. For the group with the best score, all the features are included to the list of selected features. The method, which is fast and easy to implement, is appropriate in cases where including all the features from the group does not dramatically decrease the system performance. In the following algorithm, function `getcost(subset)` returns relative measurement cost of the feature subset and `getperf(data,subset)` returns subset performance.

2.2 Group-Wise Nested Forward Selection (GNFS)

The main idea of this method is to use the best feature subset in the group instead of all the group's features. In order to identify such a subset, nested feature selection search is launched within each group. The group is judged on the basis of its best feature subset. Features that were not selected in one step may be used later for a zero additional measurement cost.
GNFS algorithm keeps track of group-specific information (structure *group*, lines 6-11). Newly computed features are judged by the criterion (1) while features from already computed groups are judged solely by their performance. If a subset of already computed and therefore cheap features may be found, which improves the performance, it is used preferably to features from a new group offering a bigger performance gain. This decision is made on the line 18 of the GNFS algorithm and its implementation was omitted for the sake of simplicity.
If just single feature groups are present, both proposed algorithms perform sequential forward selection with criterion (1).

3 Experiments

3.1 Handwritten Digit Recognition

In the first experiment, we use the proposed methods on the handwritten digit `mfeat` dataset from [1]. The dataset contains 10 digit classes with 200 samples per class and six different feature sets (649 features). In order to lower the computational requirements in this illustrative example, we have reduced the number of features in all of the six feature sets. The dataset used in this experiment contains 78 features. The set with 100 samples per class was used for training; the other 100 samples per class were used as the evaluation set for the feature selection criterion (error of the linear discriminant classifier assuming normal densities).

Experimental results are presented in Figure 1. Performance of selected feature subset on the evaluation set is estimated as a function of the measurement cost. The measurement cost is expressed on the relative scale $\langle 0, 1.0 \rangle$, where 1.0 corresponds to the computational time of all the features. Because the computational time of individual feature groups is not known for this dataset, we assume equal measurement cost for all the features. Measurement cost of particular group is then a sum of measurement costs for the group's features.

The solid line in the graph represents results of the forward feature selection algorithm not using the feature grouping. The dashed line with cross markers corresponds to GFS, and dash-dotted line with circles to GNFS algorithm.

Points on the curves denote steps of the corresponding feature selection algorithms. While one step represents one added feature for SFS method, for GFS is that adding of all and for GNFS adding of the subset of the group's fea-

Algorithm 1 Group-wise Forward Selection (GFS)

```
input: data, features F = {1, ..., D}, feature groups G = {1, ..., N}
C_max = 0; F_best = {};               // best criterion and subset found
G_sel = {}; F_sel = {};               // selected groups, selected features
currperf = 0;                         // performance of the current subset F_sel
currcost = 0;                         // meas.cost of the current subset F_sel
while length(G_sel)<N
    C = {}; perf = {};                // criteria values, performance values
    for i=1 to length(G)
        A ← all features from group G(i);
        mincr =getcost({F_sel∪A})−currcost;        // measurement cost increase
        perf(i) =getperf(data,{F_sel ∪ A});
        C(i) = (perf(i) − currperf)/mincr;          // criterion
    end
    m=max(C);                         // maximum criterion value
    imax=argmax(C);                   // index of the best group
    F_sel ← features from group G(imax);
    currperf = perf(imax);
    currcost =getcost(F_sel);
    G_sel ← G(imax);                  // add group to list
    G(imax) = ∅;                      // remove group from list G
    if m > C_max
        C_max = m; F_best = F_sel;    // adjust the best achieved criterion and subset
    end
end
output: the best subset found F_best
```

tures. Note also the vertical curve segment on the solid line which is caused by adding features with a zero measurement cost (they come from already computed groups).

It can be seen, that both methods using the feature grouping reach a very good result (0.028) already for one third of the computation time. The standard method achieves the similar performance for 48% of the measurement cost.

The lower graph in the Figure 1 presents the number of used features as a function of the relative measurement cost for all three methods. It can be seen, that methods using feature grouping perform better than the standard selection due to larger number of employed features at the same measurement cost.

3.2 Backscatter Image Segmentation

In the second experiment, we apply proposed methods in order to speed-up the feature acquisition process in the backscatter image segmentation. For the sake of feature selection, a dataset with 3000 samples, three classes, and 95 features

Algorithm 2 Group-wise Nested Forward Selection (GNFS)

```
input: data, features F = {1,...,D}, feature groups G = {1,...,N}
C_max = 0; F_best = {};              // best criterion and subset found
F_sel = {};                          // selected features
currperf = 0;                        // performance of the current subset F_sel
currcost = 0;                        // measurement cost of the current subset F_sel
for i=1 to N
    group(i).perf = 0;               // set-up auxiliary group structure
    group(i).computed = 0;           // is this group already computed?
    group(i).F_sel = {};             // best found subset in the group
    group(i).F ← features from group G(i);
end
while length(F_sel)<D
    for i=1 to N                     // perform nested search for each group
        group(i).perf = 0;
        A = group(i).F;              // assign not-yet-used features
        group(i) ← find best subset and its performance on A;
    end
    m, imax ← choose the best subset;
    F_sel ← group(imax).F_sel                  // added features
    group(imax).F = group(imax).F \ group(imax).F_sel;     // remove selected
    group(imax).computed = 1;                  // toggle group as computed
    currperf = group(imax).perf;
    currcost =getcost(F_sel);
    if m > C_max
        C_max = m; F_best = F_sel;   // adjust the best achieved criterion and subset
    end
end
output: the best subset found F_best
```

was computed from a BSE image. Six different types of features were used: intensity features, cooccurrence matrices (CM), gray-level differences (SGLD), local binary patterns (LBP), features based on Discrete Cosine Transform (DCT), and Gabor filters. More details regarding feature types and segmentation algorithm can be found in [5].

The Table 2 summarizes the actual feature grouping defined by the used implementation. It follows from the table, that each of eight DCTs and 24 Gabor filters is computed apart forming a separate group. The last column in the table indicates a relative cost to compute the group (1.0 is the total cost using all groups).

The experimental results for four different backscatter images are presented in Figure 3. A complete dataset with 95 features was computed for each image. Then, three feature selection methods were run (standard forward selection not using feature grouping and two presented methods employing grouping information). Once again, the error of the linear discriminant classifier assuming normal densities on the independent evaluation set was used as the criterion. Evaluation

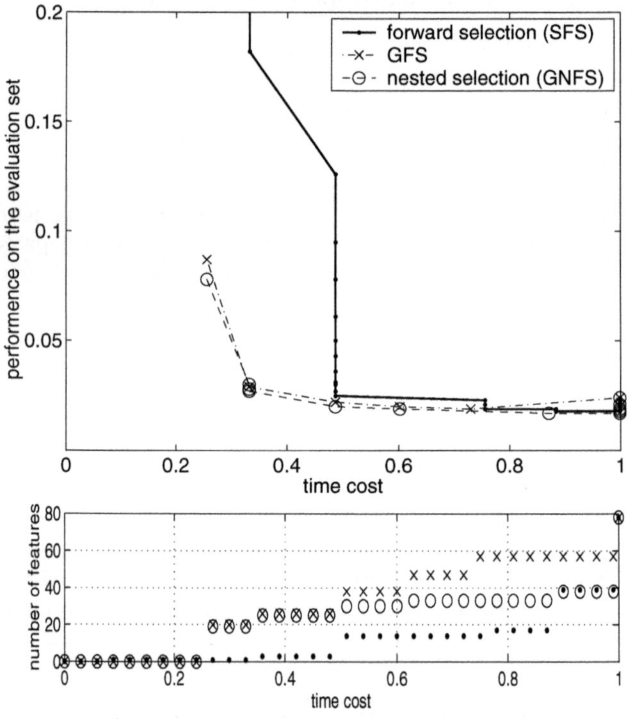

Fig. 1. Performance as a function of the measurement cost for handwritten digit dataset (upper plot). Number of features as a function of the measurement cost (lower plot)

group number	*feature type*	*features per group*	*group cost*
1	intensity	4	0.0091
2	CM	4	0.0422
3	SGLD	4	0.0118
4	LBP1	17	0.0694
5	LBP2	25	0.1009
6	LBP3	9	0.0434
7-14	DCT filters	1	0.0349
15-38	Gabor filters	1	0.0185

Fig. 2. Feature groups in the backscatter segmentation experiments

set consists of different 3000 samples from the same BSE image. All the three lines end up in the same point (performance of the complete feature set).

Maximum performance for all methods is summarized in the Table 4. It appears, that the best achieved performance is similar in all cases. Proposed methods utilizing feature grouping lower the measurement cost of feature computation for all but the last image. Further examination of performance-cost curves in Figure 3 suggests possible better choice of operating points with lower measurement costs.

It is interesting to note areas where the standard feature selection algorithm finds better solutions than both proposed methods (images 2 and 3 in Figure 3). We think, that the reason is in the fine-grained approach of the standard algorithm. Both proposed algorithms outperform standard forward feature selection for low measurement costs which is our area of interest. In general, nested feature selection (GNFS) works better than adding all the group's features (GFS) but is computationally more intensive.

4 Conclusions

We investigate ways how the information about feature grouping may be used in the feature selection process for finding well-performing feature subset with low measurement cost. The problem arises in many real applications where the feature acquisition cost is of importance and feature grouping is defined by the implementation.

We show, that it is beneficial to perform feature selection on the per-group basis. Different strategies may be then chosen to select appropriate feature subset within the groups. We have investigated two such approaches – adding all features from the group (GFS) and Group-wise nested feature selection (GNFS).

It follows from our experiments on handwritten digits and backscatter images, that proposed methods outperform standard feature selection algorithm in the low measurement cost area. We also conclude, that using nested feature selection is better strategy than adding all group's features but it is computationally more intensive.

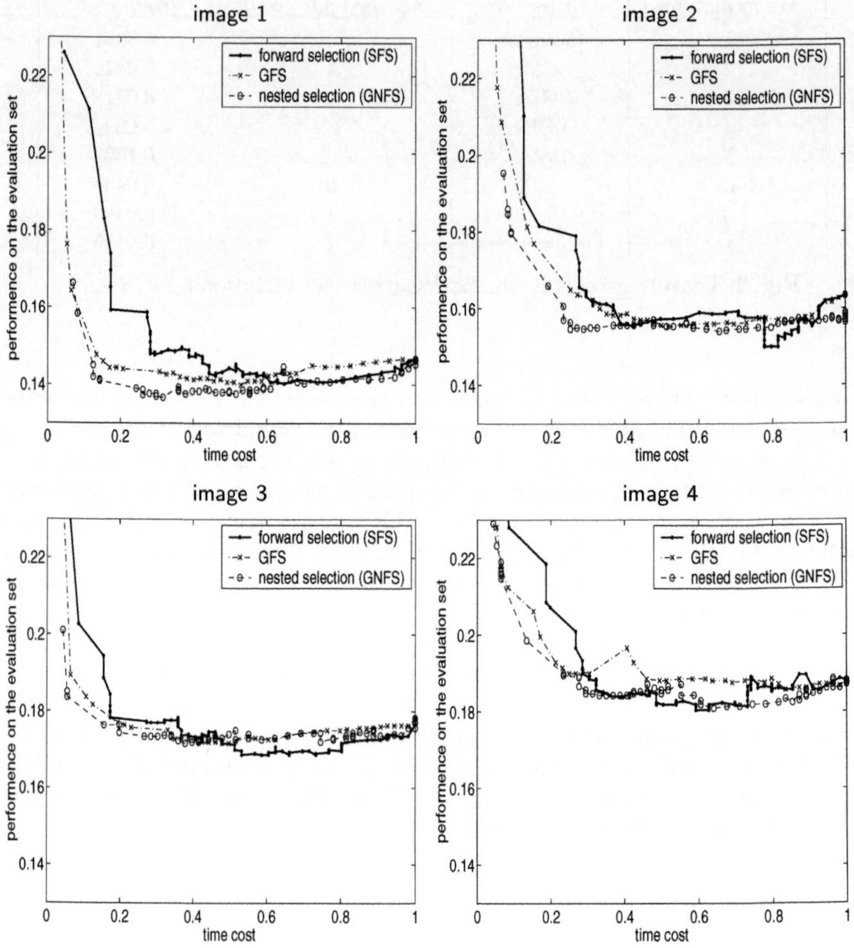

Fig. 3. Performance as a function of the measurement cost for backscatter segmentation experiment

image	forward selection (SFS)	GFS	GNFS
1	0.140 at 0.65 (77)	0.142 at 0.30 (59)	0.138 at 0.32 (50)
2	0.150 at 0.78 (57)	0.158 at 0.56 (77)	0.154 at 0.65 (69)
3	0.169 at 0.53 (60)	0.174 at 0.28 (49)	0.172 at 0.38 (48)
4	0.181 at 0.59 (48)	0.181 at 0.83 (63)	0.181 at 0.64 (71)

Fig. 4. Best performances of feature subsets in backscatter segmentation experiment. The first number is the best performance, the second is corresponding measurement cost and the number if parentheses is the feature count

Presented methods are based on simple forward feature selection algorithm. It is possible to replace forward selection by more powerful methods like floating search [6]. Computation time of the feature selection may, however, increase considerably what may be not acceptable in some applications.

References

1. C. L. Blake and C. J. Merz. UCI repository of machine learning databases, 1998. `http://www.ics.uci.edu/ |mlearn/MLRepository.html`. 463
2. F. Ferri, P. Pudil, M. Hatef, and J. Kittler. Comparative study of techniques for large-scale feature selection, 1994. 461, 462
3. Anil Jain and Douglas Zongker. Feature selection: Evaluation, application, and small sample performance. *IEEE Trans. Pattern Analysis and Machine Inteligence*, 19(2):153–158, February 1997. 461, 462
4. Pavel Paclík, Robert P.W. Duin, and Geert M.P. van Kempen. Multi-spectral Image Segmentation Algorithm Combining Spatial and Spectral Information. In *Proceedings of SCIA 2001 conference*, pages 230–235, 2001. 462
5. Pavel Paclík, Robert P.W. Duin, Geert M.P. van Kempen, and Reinhard Kohlus. Supervised segmentation of backscatter images for product analysis. accepted for International Conference on Pattern Recognition, ICPR2002, Quebec City, Canada, August 11-15, 2002. 462, 466
6. P. Pudil, J. Novovičová, and Kittler J. Floating search methods in feature selection. *Pattern Recognition Letters*, 15:1119–1125, 1994. 469

Classifier-Independent Feature Selection Based on Non-parametric Discriminant Analysis

Naoto Abe[1], Mineichi Kudo[1], and Masaru Shimbo[2]

[1] Division of Systems and Information Engineering
Graduate School of Engineering
Hokkaido University, Sapporo 060-8628, Japan
{chokujin,mine}@main.eng.hokudai.ac.jp

[2] Faculty of Information Media, Hokkaido Information University
Ebetsu 069-8585, Japan
shimbo@do-johodai.ac.jp

Abstract. A novel algorithm for classifier-independent feature selection is proposed. There are two possible ways to select features that are effective for any kind of classifier. One way is to correctly estimate the class-conditional probability densities and the other way is to accurately estimate the discrimination boundary. The purpose of this study is to find the discrimination boundary and to determine the effectiveness of features in terms of normal vectors along the boundary. The fundamental effectiveness of this approach was confirmed by the results of several experiments.

1 Introduction

Feature selection is a procedure to find a feature subset that has the most discriminative information from an original feature set. In large-scale problems with over 50 features, there may exist *garbage features* that have an adverse effect on construction of classifiers. In such a case, it is expected that the performance of classifiers can be improved by removing such garbage features.

Many algorithms for feature selection have been proposed. Many references to report presenting such algorithms are given in [1]. These algorithms can be divided into two groups. One group is called *classifier-specific feature selection* algorithms [1], where the goodness of a feature subset is measured in terms of the estimated performance of a certain classifier. These algorithms are useful when it is known in advance what classifier is used. However, it is more desirable to select a feature subset that is effective for any kind of classifier. Therefore, another group of algorithms, called *classifier-independent feature selection* algorithms [2,3,4], has been studied, and their criterion functions are connected with estimates of the recognition rate of the Bayes classifier. Algorithms belonging to the latter group can further be divided into two groups: one group of algorithms designed to estimate class-conditional densities or a distributional structure [2,3] and another group of algorithms designed to estimate discrimination boundaries [4,5]. When we have only a small training sample set, estimating

T. Caelli et al. (Eds.): SSPR&SPR 2002, LNCS 2396, pp. 470–479, 2002.
© Springer-Verlag Berlin Heidelberg 2002

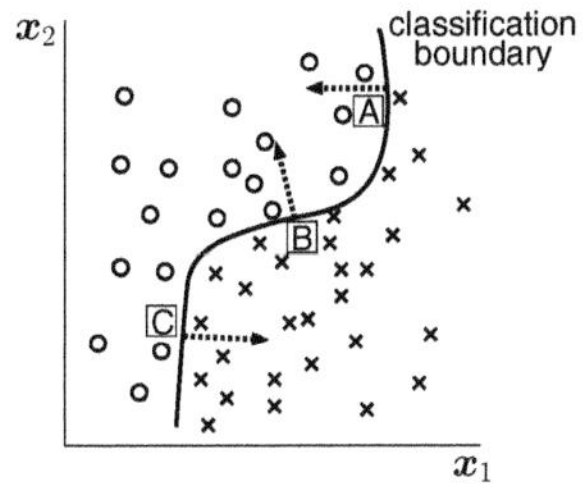

Fig. 1. Local importance of features

the discrimination boundary seems to be better than estimating densities. In this paper, we examine the effectiveness of such a trial.

2 Discriminative Information Based on Non-parametric Discriminant Analysis

2.1 Feature Importance Based on Classification Boundary

Our approach is twofold: (1) to estimate the discrimination boundary as precisely as possible and (2) to evaluate the effectiveness of features in terms of the boundary. First, we notice that the normal vectors along the discrimination boundary show which features are necessary for discriminating two classes (Fig. 1). A vector at point A or C indicates that feature x_1 is important, and feature x_2 is important at point B. Thus, it can be seen that these normal vectors reflect the importance of features locally. Since what we want is a feature subset that is effective globally, we find all necessary features by combining such local evidence.

2.2 Non-parametric Discriminant Analysis

Here, let us consider two-class (ω_1 and ω_2) problems first. The method described below can easily be extended to multi-class problems. The normal vectors along the discrimination boundary can be estimated using non-parametric discriminant analysis proposed by Fukunaga and Mantock [6]. In this analysis, k nearest unlike neighbors (NUNs) taken from the opposite class play a main role. For a sample $\boldsymbol{x} \in \omega$, k nearest unlike neighbors $\boldsymbol{y}_1, \ldots, \boldsymbol{y}_k \in \omega' \neq \omega$, where $\|\boldsymbol{y}_1 - \boldsymbol{x}\| \leq \|\boldsymbol{y}_2 - \boldsymbol{x}\| \leq \cdots \leq \|\boldsymbol{y}_k - \boldsymbol{x}\|$, are found. Then, a normal vector $\boldsymbol{v}_{\boldsymbol{x}}$ at $\boldsymbol{x}$ is formulated as

$$\boldsymbol{v}_{\boldsymbol{x}} = \frac{1}{k}\left(\sum_{i=1}^{k} \boldsymbol{y}_i\right) - \boldsymbol{x}.$$

Also, vector $\boldsymbol{v}_{\boldsymbol{x}}$ is weighted by

$$w_v = \frac{\min\{\|\boldsymbol{y}_k - \boldsymbol{x}\|, \|\boldsymbol{x}_k - \boldsymbol{x}\|\}}{\|\boldsymbol{y}_k - \boldsymbol{x}\| + \|\boldsymbol{x}_k - \boldsymbol{x}\|}. \tag{1}$$

Here, $\boldsymbol{x}_k$ is the kth NN of $\boldsymbol{x}$ taken from the same class ω. This weight takes a value close to 0.5 when $\boldsymbol{x}$ is close to the discrimination boundary and declines to zero as $\boldsymbol{x}$ moves away from the discrimination boundary. Fukunaga and Mantock used these vectors with weights to calculate a non-parametric form of a between-class covariance matrix.

3 Proposed Method

3.1 Modified Normal Vectors

There is a problem in calculation (1) of weights for the normal vectors. Such an example is shown in Fig. 2. As shown in Fig. 2(a), if we use weighting formula (1), when the distances $\|\boldsymbol{y}_k - \boldsymbol{x}\|$ and $\|\boldsymbol{x}_k - \boldsymbol{x}\|$ are almost the same, such a point $\boldsymbol{x}$ is treated as being located near the discrimination boundary, which results in $w_v \doteq \frac{1}{2}$. As a result, a normal vector for such a point can show a wrong direction with a high weight close to the maximum weight $\frac{1}{2}$. To cope with this problem, we calculate the normal vectors as

$$\boldsymbol{v}_{\boldsymbol{x}} = \sum_{i=1}^{k} e^{-\sigma\|\boldsymbol{y}_i - \boldsymbol{x}\|} \frac{\boldsymbol{y}_i - \boldsymbol{x}}{\|\boldsymbol{y}_i - \boldsymbol{x}\|}. \tag{2}$$

Here, $\sigma (\sigma > 0)$ is a control parameter. In this way, as shown in Fig. 2(b), vectors $(\boldsymbol{y}_i - \boldsymbol{x})$'s cancel their bad influence by themselves. Two examples of normal vectors calculated in this way are shown in Fig. 3.

3.2 Evaluation of Features

To combine the local importance of features, we take a simple way. That is, for calculated normal vectors $\boldsymbol{v}_{\boldsymbol{x}} = (v_{1x}, v_{2x}, \ldots, v_{Dx})$ of class ω_c, we sum up the absolute values of ith component v_{ix} to have the importance measure $f_i^c = 1/n_c \sum_{\boldsymbol{x}\in\omega_c} |v_{ix}|$, and take the average over classes as $f_i = \sum_{c=1}^{C} P_c f_i^c$. Here, D is the dimensionality, n_c is the number of samples from ω_c, and P_c is a priori probability of class ω_c which is estimated by $P_c = n_c / \sum_{c=1}^{C} n_c$. Last, we normalize f_i as $f_i \leftarrow f_i / \sum_{i=1}^{D} f_i$.

3.3 Determination of Size

In classifier-independent feature selection, it is important to remove only garbage features. If the number of training samples is sufficiently large, the feature importance measure of a garbage feature is expected to take a very small value.

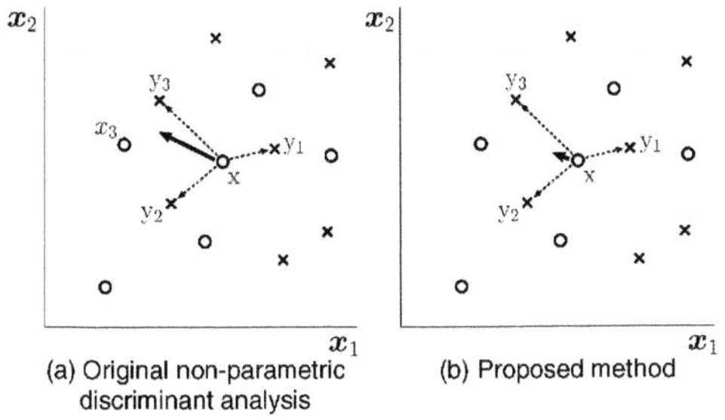

Fig. 2. Calculation of weights for normal vectors when $k = 3$

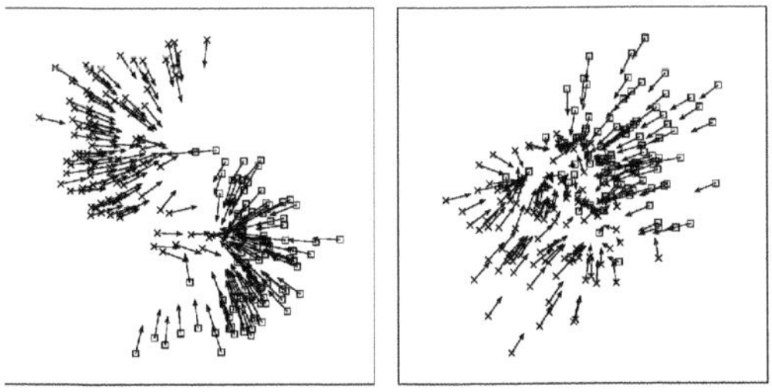

Fig. 3. Two examples of normal vectors calculated by the proposed method (k=4)

Therefore, we use a function $J(d) = 1 - \min_{1 \le i \le d} f_i$ for evaluating the goodness of a feature subset of size d ($d = 1, 2, \ldots, D$). In an ideal case in which d^* ($d^* < D$) features are garbage and the remaining $D - d^*$ features contribute equally to the performance, we have a criterion curve as shown in Fig. 4(a). In this case, we can remove all garbage features with $\theta < 1/d^*$. However, in a practical case in which the number of training samples is limited, even garbage features may show a little contribution. In this case, we use a threshold θ in order to determine the size d_θ. That is, we find the θ-degradation point of the maximum value $J(D)$ (Fig. 4(b)). The proposed algorithm is summarized as follows:

Step 0: Set the value of threshold $\theta > 0$. Let d=D.
Step 1: Calculate the normal vectors $\boldsymbol{v_x}$ with respect to every sample $\boldsymbol{x}$ by Eq. (2).
Step 2: Calculate feature importance f_i ($i = 1, 2, \ldots, d$). If $d = D$, set $J(D)$ for $J_{\max}$.
Step 3: Remove the worst feature $f_i = \arg\min_i f_i$.
Step 4: If $J_d < (1 - \theta)J_{\max}$, terminate the algorithm and output the subset. Otherwise, with $d = d - 1$, go to step 1.

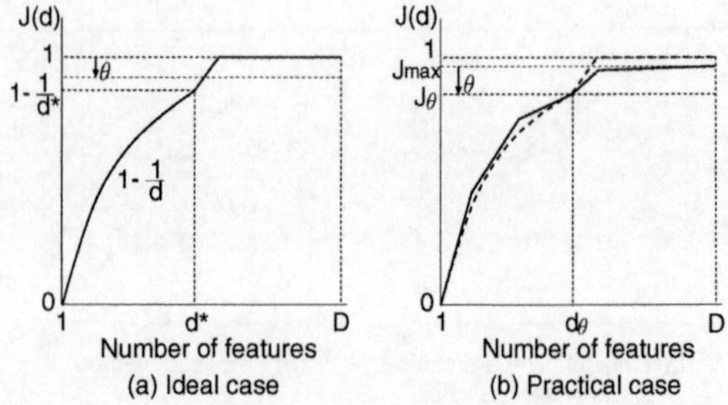

Fig. 4. Determination of size

4 Experiments

Two artificial datasets and one real dataset were used in the experiments [9]. In the experiments, θ was taken as 1% or 5%. Here, $\theta = 1\%$ was chosen from the viewpoint of removing only garbage features. The value of 5% was taken from the practical viewpoint of choosing a smaller feature subset at the expense of a slight degradation of discriminative information. A simple experiment was carried out to determine the appropriate values of k and σ. A value two-times greater than the number of features was chosen for k and 0.1 was chosen for σ. Six classifiers were used to evaluate the goodness of a selected feature subset: the Bayes linear classifier (LNR), the Bayes quadratic classifier (QDR), the C4.5 decision tree classifier [7] (C45), the one-nearest neighbor (1NN) classifier, the subclass classifier [8] (SUB), and the neural network classifier (NRL). The recognition rates were calculated by the 10-fold cross validation technique.

1. *Friedman*: A Friedman database [4].

   In this database, there are two classes, ω_1 and ω_2. The samples of ω_1 were generated according to a Gaussian with a unit covariance matrix and zero mean. The samples of ω_2 surround those of ω_1 in the first four features that are distributed uniformly within a four-dimensional spherical slab centered at the origin with an inner radius 3.5 and outer radius 4.0 in the samples of ω_2. The last six features of ω_2 are distributed as a Gaussian with a unit covariance matrix and zero mean. Each class has 100 samples.

2. *Waveform*: A waveform database [9].

   In this database, there are three classes and 40 features with continuous values between 0 and 6. Each class is a random combination of two triangular waveforms with noise added. Each class is generated from a combination using two of three base waves. The instances are generated such that features $1, 2, \ldots, 21$ are necessary for class separation, whereas features $22, 23, \ldots, 40$ have random variables. The numbers of samples were 67 for class 1, 54 for class 2, and 79 for class 3, respectively.

3. *Sonar*: A sonar database [9].

The task was to discriminate between sonar signals bounced off a metal cylinder and those bounced off a roughly cylindrical rock using 60 features, each of which describes the energy within a particular frequency band, integrated over a certain period of time. The database consists of 111 patterns obtained by bouncing sonar signals off a metal cylinder at various angles and under various conditions and 97 patterns obtained from rocks under similar conditions.

The results were evaluated mainly from the following two viewpoints.

- Compared with the case in which all original features are used, if all garbage features are successfully removed, the recognition rates of all classifiers would be improved or maintained. We examined whether this is the case or not.
- Since the classifier that shows the best performance among all classifiers can be considered as being closest to the Bayes classifier, we examined whether the number of selected features is larger than that corresponding to the peak of the best classifier or not.

Unlike classifier-specific feature selection, what we hope for is a minimal feature subset that includes all discriminative features in classifier-independent feature selection. Thus, a slightly larger feature subset that includes them is acceptable. For comparison with the density estimation approaches, we carried out the same experiments using the divergence method [10], which is based on an axis-parallel Gaussian mixture model. In the divergence method, a feature subset of a given size that maximizes Kullback-Leibler divergence is selected on the basis of estimated densities. Changing the size from D down to 1, we found a sequence of feature subsets and selected a feature subset using a threshold ($\alpha = 1\%$ or 5%) as the proposed method does. For details, see [10].

The results of the experiments are shown in Figs. 5-7. Compared with the case that all the features were used, the recognition rates of the six classifiers were all either improved or maintained by using the proposed method, as well as the divergence method. Therefore, the proposed method is effective in the first viewpoint of evaluation. In addition, the results with $\theta = 1\%$ were satisfactory even from the second viewpoint of evaluation.

Next, we examined the difference between the recognition rates using the proposed method and the divergence method. The difference between the recognition rates is shown in Fig. 8. In this figure, only the best three classifiers are shown. For the *Friedman* dataset, there was no notable difference between recognition rates. However, for the *Waveform* and *Sonar* datasets, a slight improvement in the recognition rate was obtained by using the proposed method.

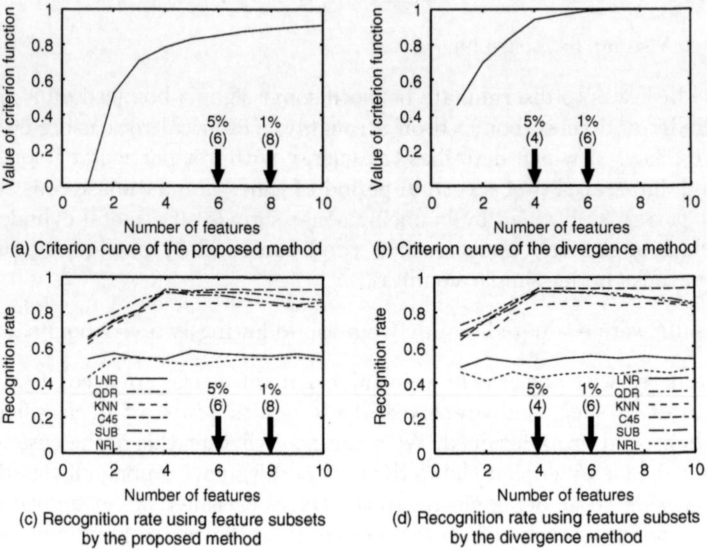

Fig. 5. Results of *Friedman* data. The values in parentheses are the numbers of selected features

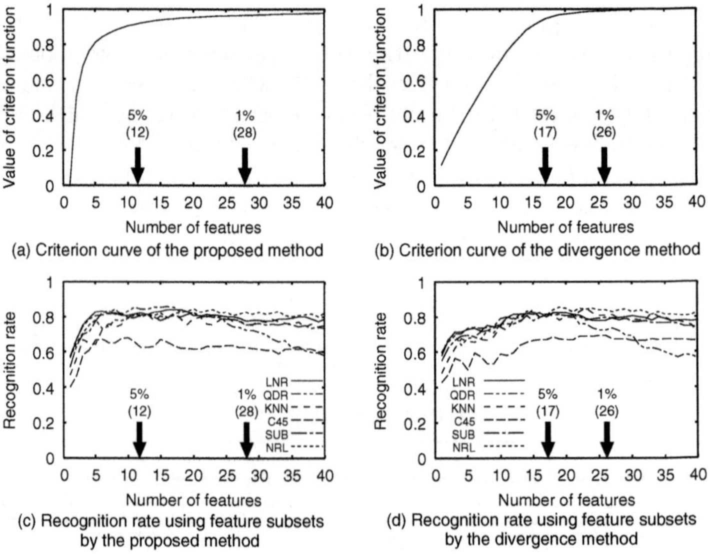

Fig. 6. Results of *Waveform* data. The values in parentheses are the numbers of selected features

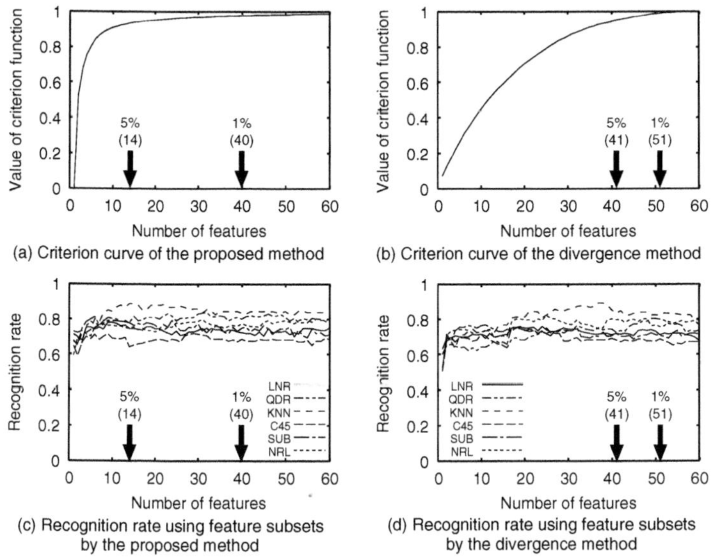

Fig. 7. Results of *Sonar* data. The values in parentheses are the numbers of selected features

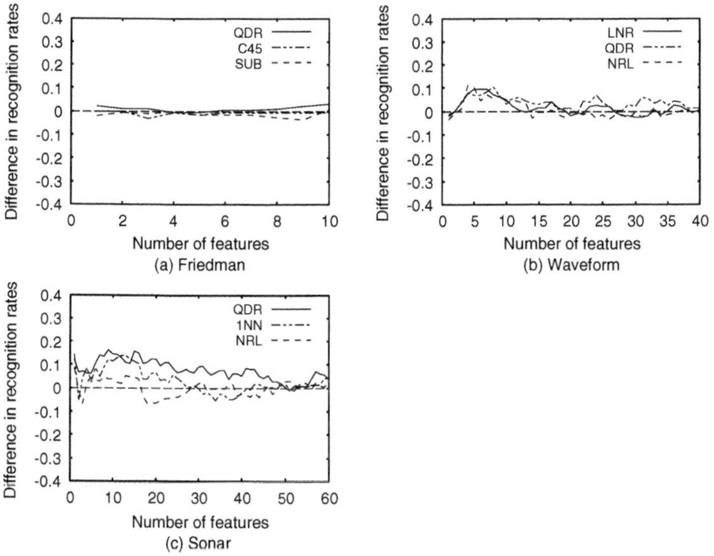

Fig. 8. Difference in recognition rates (A)-(B), where (A) is the recognition rate of the proposed method and (B) is that of the divergence method. Only the best three classifiers are shown

5 Discussion

The curve of the criterion function $J(d)$ $(d = 1, 2, \ldots, D)$ obtained by the proposed method is almost proportional to that of the recognition rates of many classifiers. This means that the proposed method is better than the divergence method for evaluation of feature subsets. For example, see Fig. 6. In the divergence method, many parameters must be estimated appropriately in the probability density. On the other hand, the proposed method requires only two parameters, a control parameter of the normal vectors and a threshold for determining the size of a feature subset to be selected. In addition, the discrimination boundary is expected to be learned faster than the densities of classes. Unfortunately, we have not carried out any experiments to confirm this point, it is worth examining if this is true or not by dealing different sizes of training samples.

6 Conclusion

We have proposed an algorithm for classifier-independent feature selection using non-parametric discriminant analysis. The fundamental effectiveness of the proposed method was confirmed by results of experiments using two artificial datasets and one real dataset. Overall, the effectiveness of the proposed method is comparable with that of the divergence method and, the proposed method is superior in simplicity of the parameter setting.

References

1. Kudo, M., and Sklansky J.: Comparison of Algorithms that Select Features for Pattern Classifiers. Pattern Recognition **33**-1 (2000) 25–41 470
2. Holz, H. J., and Loew, M. H.: Relative Feature Importance: A Classifier-Independent Approach to Feature Selection. In: Gelsema E. S. and Kanal L. N. (eds.) Pattern Recognition in Practice IV, Amsterdam: Elsevier (1994) 473–487 470
3. Novovičová, J., Pudil, P., and Kittler, J.: Divergence Based Feature Selection for Multimodal Class Densities. IEEE Transactions on Pattern Analysis and Machine Intelligence **18** (1996) 218–223 470
4. Kudo, M., and Shimbo, M.: Feature Selection Based on the Structural Indices of Categories. Pattern Recognition **26** (1993) 891–901 470, 474
5. Egmont-Petersen, M., Dassen, W. R. M., and Reiber, J. H. C.: Sequential Selection of Discrete Features for Neural Networks - A Bayesian Approach to Building a Cascade. Pattern Recognition Letters **20** (1999) 1439–1448 470
6. Fukunaga, K., and Mantock, J. M.: Nonparametric Discriminant Analysis. IEEE Transactions on Pattern Analysis and Machine Intelligence **5** (1983) 671–678 471
7. Quinlan, J. R.: C4.5: Programs for Machine Learning. Morgan Kaufmann San Mateo CA (1993) 474
8. Kudo, M., Yanagi, S., and Shimbo, M.: Construction of Class Region by a Randomized Algorithm: A Randomized Subclass Method. Pattern Recognition **29** (1996) 581–588 474

9. Murphy, P. M., and Aha, D. W.: UCI Repository of machine learning databases [Machine-readable data repository]. University of California Irvine, Department of Information and Computation Science (1996) 474, 475
10. Abe, N., Kudo, M., Toyama, J., and Shimbo, M.: A Divergence Criterion for Classifier-Independent Feature Selection. In: Ferri, F. J., Inesta, J. M., Amin, A., and Pudil, P. (eds.) Advances in Pattern Recognition, Lecture Notes in Computer Science, Alicante, Spain, (2000) 668–676 475

Effects of Many Feature Candidates in Feature Selection and Classification

Helene Schulerud[1,2] and Fritz Albregtsen[1]

[1] University of Oslo
PB 1080 Blindern, 0316 Oslo, Norway
[2] SINTEF
PB 124 Blindern, 0314 Oslo, Norway
hsc@sintef.no

Abstract. We address the problems of analyzing many feature candidates when performing feature selection and error estimation on a limited data set. A Monte Carlo study of multivariate normal distributed data has been performed to illustrate the problems. Two feature selection methods are tested: *Plus-1-Minus-1* and *Sequential Forward Floating Selection.* The simulations demonstrate that in order to find the correct features, the number of features initially analyzed is an important factor, besides the number of samples. Moreover, the sufficient ratio of number of training samples to feature candidates is not a constant. It depends on the number of feature candidates, training samples and the Mahalanobis distance between the classes. The two feature selection methods analyzed gave the same result. Furthermore, the simulations demonstrate how the leave-one-out error estimate can be a highly biased error estimate when feature selection is performed on the same data as the error estimation. It may even indicate complete separation of the classes, while no real difference between the classes exists.

1 Introduction

In many applications of pattern recognition, the designer finds that the number of possible features which could be included in the analysis is surprisingly high and that the number of samples available is limited. High-dimensional functions have the potential to be much more complicated then low-dimentional ones, and those complications are harder to discern. Evaluating many features on a small set of data is a challenging problem which has not yet been solved. In this paper some pitfalls in feature selection and error estimation in discriminant analysis on limited data sets will be discussed. It is well known that the number of training samples affects the feature selection and the error estimation, but the effect of the number of feature candidates initially analyzed is not much discussed in the pattern recognition literature.

The goal of the feature selection is to find the subset of features which best characterizes the differences between groups and which is similar within the groups. In pattern recognition literature there is a large amount of papers addressing the problem of feature selection [4,5]. In this study two commonly used

T. Caelli et al. (Eds.): SSPR&SPR 2002, LNCS 2396, pp. 480–487, 2002.
© Springer-Verlag Berlin Heidelberg 2002

suboptimal feature selection methods are analyzed, *Stepwise Forward Backward selection (SFB)* [11], also called *Plus-1-Minus-1*, and *Sequential Forward Floating Selection (SFFS)* [7]. The SFB method was chosen since it is commonly used for exploratory analyses and is available in statistical packages, such as SAS and BMDP. The SFFS method has been reported as the best sub-optimal feature selection method [5] and was therefore included.

An important part of designing a pattern recognition system is to evaluate how the classifier will perform on future samples. There are several methods of error estimation like leave-one-out and holdout. In the leave-one-out method [6], one sample is omitted from the dataset of n samples, and the $n-1$ samples are used to design a classifier, which again is used to classify the omitted sample. This procedure is repeated until all the samples have been classified once. For the holdout method, the samples are divided into two mutually exclusive groups (training data and test data). A classification rule is designed using the training data, and the samples in the test data are used to estimate the error rate of the classifier.

The leave-one-out error estimate can be applied in two different ways. The first approach is to first perform feature selection using all data and afterwards perform leave-one-out to estimate the error, using the same data. The second approach is to perform feature selection and leave-one-out error estimation in one step. Then one sample is omitted from the data set, feature selection is performed and a classifier is designed and the omitted sample is classified. This procedure is repeated until all samples are classified.

The goal of this study is to demonstrate how the number of correctly selected features and the performance estimate depends on the number of feature candidates initially analyzed.

2 Study Design

A Monte Carlo study was performed on data generated from two 200 dimensional normal distributions regarded as class one and two. The class means were $\mu_1 = (0, ..., 0)$ and $\mu_2 = (\mu_1', \mu_2', \mu_3', \mu_4', \mu_5', 0, ..., 0)$, $\mu_j' = (\delta/\sqrt{r})$, $r = 5$ being the number of features separating the classes and δ^2 being the Mahalanobis distance between the classes. The data sets consisted of an equal number of observations from each class. We used the Stepwise Forward-Backward (SFB) feature selection method, also called Plus-1-Minus-1, with Wilk's λ as quality criterion ($\alpha - to - enter = \alpha - to - stay = 0.2$) [1], from the SAS statistical package. Sequential Forward Floating Selection (SFFS) was also analyzed, using the MATLAB based toolbox PRTools from Delft [3]. Sum of Mahalanobis distance was used as quality criterion.

Bayesian minimum error classifier [2] was applied, assuming Gaussian distributed probability density functions with common covariance matrix and equal apriori class probabilities. The covariance matrix is equal to the identity matrix. The Bayesian classification rule then becomes a linear discriminant function. The values of parameters tested are given in Table 1. For each set of parameters,

Table 1. Values of the different parameters tested

Symbol	Design variable	Values
n^{Tr}	No. of training samples	20, 50, 100, 200, 500, 1000
n^{Te}	No. of test samples	20, 100, 200, 1000
D	No. of feature candidates	10, 50, 200
δ^2	Mahalanobis distance	0, 1, 4

100 data sets were generated and the expected error rate, $\hat{P}_e^i$, and variance were estimated for i equal to the leave-one-out (L) and the holdout (H) method, using 50 % of the data as test samples. The expected number of correctly selected features was estimated by the mean number of correctly selected features of the k simulations, and is denoted $\hat{F}$.

3 Experimental Results

3.1 Feature Selection

The simulations show that the number of correctly selected features increases when the Mahalanobis distance between the classes increases, the number of samples increases and the number of feature candidates decreases, as shown in Figure 1 and 2. Normally we do not know the Mahalanobis distance between the classes, so we need to analyze the number of training samples (n^{Tr}) and feature candidates (D) and their relation.

Figure 1 shows the results of applying stepwise forward-backward (SFB) selection. Figure 1 left shows the average number of correctly selected features as a function of the number of training samples for three different values of the number of feature candidates. In Figure 1 right, the average number of correctly selected features for four different values of the ratio n^{Tr}/D is shown. Some additional simulations using 500 feature candidates were performed in order to complete this Figure.

Figure 2 shows the results of stepwise forward-backward (SFB) and sequential forward floating selection (SFFS) when the Mahalanobis distance equals 1 (left) and 4 (right).

We observe that:

- If the number of samples is low (less than 200), the number of feature candidates is of great importance, in order to select the correct features.
- When the number of training samples increases, the number of correctly selected features increases.
- The optimal ratio, n^{Tr}/D, depends on the Mahalanobis distance, the number of training samples and feature candidates. Hence, recommending an optimal ratio is not advisable.
- The performance of the two feature selection methods analyzed is almost the same.

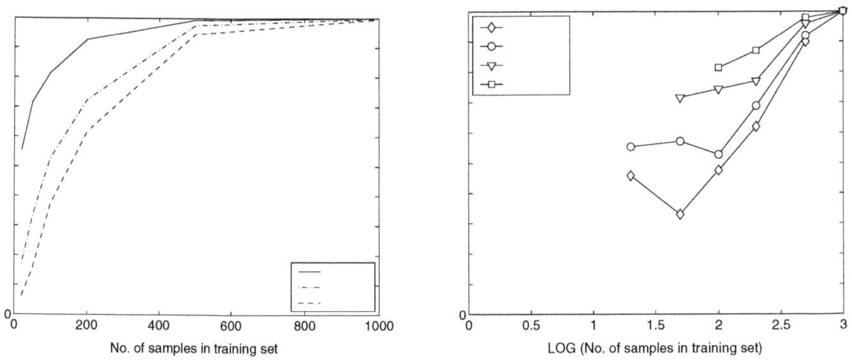

Fig. 1. The average number of correctly selected features, $\hat{F}$, when selecting 5 features and the Mahalanobis distance is 1.**Left:** $\hat{F}$ as a function of training samples for three different numbers of feature candidates. **Right:** $\hat{F}$ as a function of constant ratio

3.2 Performance Estimation

The bias of the resubstitution error estimate introduced by estimating the parameters of the classifier and the error rate on the same data set, is avoided in the leave-one-out, since the sample to be tested is not included in the training process. However, if all data are first used in the feature selection process and then the same data are used in error estimation using e.g. the leave-one-out method ($\hat{P}_e^L$), a bias is introduced. To avoid this bias, feature selection and leave-one-out error estimation can be performed in one process ($\hat{P}_e^{L2}$). We have analyzed the bias and variance of these two variants of the leave-one-out error estimate and of the holdout error estimate.

Figure 3 left shows the bias and variance of the two leave-one-out error estimates when there is no difference between the classes and we select 5 out of 200 feature candidates using SFB. The simulations show that when the number of samples is low (less than 200), the $\hat{P}_e^L$ estimate tends to give a highly optimistic error estimate. Moreover, when analyzing many features on a small data set, the $\hat{P}_e^L$ estimate can indicate complete separation of the classes, while no real difference between the classes exists. As the number of samples increases, the $\hat{P}_e^L$ approaches the true error. The number of samples necessary to get a good estimate of the true error depends on the Mahalanobis distance between the classes and the number of feature candidates. However, the simulation results show that if the number of training samples is greater than 200, the bias of the leave-one-out estimate is greatly reduced.

Performing feature selection and leave-one-out error estimation in one process results in an almost unbiased estimate of the true error, but the $\hat{P}_e^{L2}$ estimate has a high variance, see Figure 3 left. When the number of samples is less than 200, the $\hat{P}_e^{L2}$ gives a clearly better estimate of the true error than $\hat{P}_e^L$. The bias and variance of the holdout error estimate ($\hat{P}_e^H$) were analyzed under the same

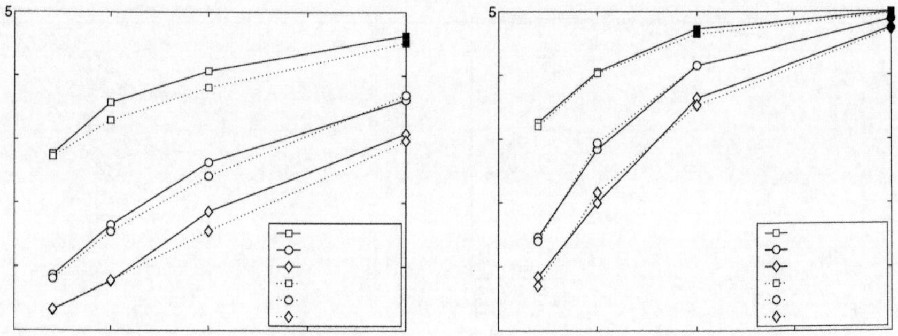

Fig. 2. The average number of correctly selected features as a function of training samples and feature candidates, when the Mahalanobis distance is 1 (left) and 4 (right) for Sequential Forward Backward (SFB) and Sequential Forward Floating Selection (SFFS). D = number of feature candidates

conditions as the leave-one-out estimates, see Figure 3 right. The holdout error estimate is also an unbiased estimate of the true error, but with some variance.

The bias of the three error estimates as a function of the number of feature candidates are shown in Figure 4 left. The Figure shows how the bias of the $\hat{P}_e^L$ error estimate increases with increasing number of feature candidates, while the two other estimates are not affected. Figure 4 right shows the bias of the $\hat{P}_e^L$ estimate as a function of Mahalanobis distance and number of training samples. The Figure shows how the bias of the $\hat{P}_e^L$ estimate increases when the Mahalanobis distance decreases. We note that for a small number of training samples (less than 200), this leave-one-out error estimate has a significant bias, even for high class distances.

4 Discussion

Our experiments are intendant to show the potential pitfalls of analyzing a large number of feature candidates on limited data sets. We have analyzed how the number of feature candidates and training samples influence the number of correctly selected features and how they influence different error estimates. Monte Carlo simulations have been performed in order to illustrate the problems.

The simulations show that when the number of training samples is less than 200, the number of feature candidates analyzed is an important factor and affects the number of correctly selected features. Moreover, few of the correct features are found when the number of samples is low (less than 100). To find most of the correct features the ratio n^{Tr}/D (number of training samples/number of feature candidates) differs between 1 and 10, depending on the Mahalanobis distance, the number of feature candidates and the number of training samples. Hence, to give a recommended general ratio n^{Tr}/D is not possible. However,

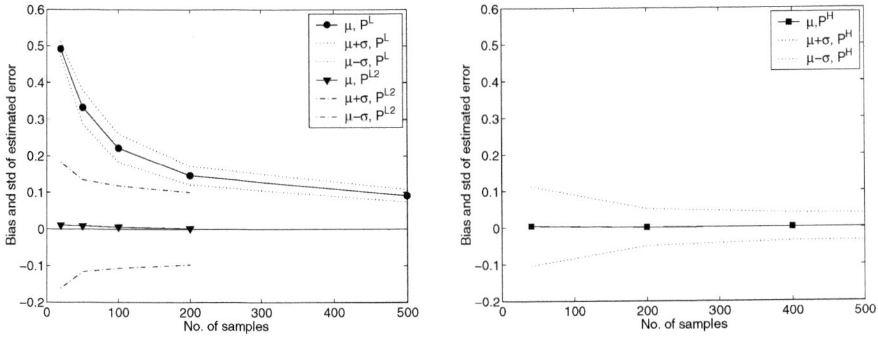

Fig. 3. Bias and variance of error estimates when the Mahalanobis distance between the classes is zero. **Left:** Leave-one-out error estimates. **Right:** Holdout error estimate with 50 % left out

Figure 2 could be used to indicate if the given number of samples and feature candidates used in a stepwise feature selection is likely to find the features which separate the classes. This result corresponds only partially to previous work by Rencher and Larson [8]. They state that when the number of feature candidates exceeds the degrees of freedom for error $[D > (n^{Tr} - 1)]$ in stepwise discriminant analysis, spurious subsets and inclusion of too many features can occur. Rutter et al. [9] found that when the ratio of sample size to number of feature candidates was less than 2.5, few correct features were selected, while if the ratio was 5 or more, most of the discriminative features were found. The two feature selection methods analyzed, *Stepwise Forward Backward selection (SFB)* and *Sequential Forward Floating Selection (SFFS)*, gave the same result.

Furthermore, the simulation results demonstrate the effect of performing feature selection before leave-one-out error estimation on the same data. If the classes are overlapping, the number of training samples is small (less than 200) and the number of feature candidates is high, the common approach of performing feature selection before leave-one-out error estimation on the same data $(\hat{P}_e^L)$ results in a highly biased error estimate of the true error. Performing feature selection and leave-one-out error estimation in one process $(\hat{P}_e^{L2})$ gives an unbiased error estimate, but with high variance. The holdout error estimate is also an unbiased estimate, but with less variance than $\hat{P}_e^{L2}$.

The following conclusions can be drawn based on the simulation results:

- The number of feature candidates analyzed statistically is critical when the number of training samples is small.
- Perform feature selection and error estimation on separate data, $(\hat{P}_e^{L2}, \hat{P}_e^H)$, for small sample sizes.
- In order to find the correct features the n^{Tr}/D ratio differs depending on the number of training samples, feature candidates and the Mahalanobis distance.

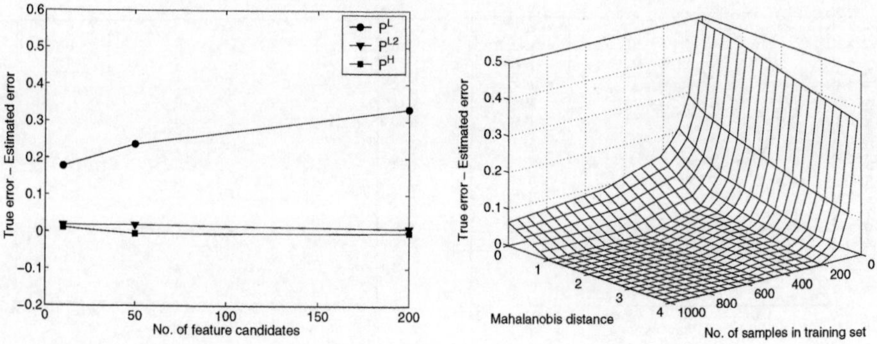

Fig. 4. Left: Bias of error estimates as a function of the number of feature candidates analyzed. **Right**: Bias of the $\hat{P}_e^L$ error estimate as a function of the Mahalanobis distance between the classes and the number of samples, when selecting 5 out of 200 feature candidates

- The traditional *Stepwise Forward Backwardselection (SFB)* gave the same results as the more advanced *Sequential Forward Floating Selection (SFFS).*

A method often used to eliminate feature candidates is to discard one of a pair of highly correlated features. However, this is a multiple comparison test, comparable to the tests performed in the feature selection process. So, the number of feature candidates analyzed will actually not be reduced. If the n^{Tr}/D ratio is low for a given sample size, one should either increase the sample size or reduce the number of feature candidates using non-statistical methods.

In a previous work [10] the bias and variance of different error estimates have been analyzed in more detail and some of the main results from this study are included here.

Some of the results presented here may be well known in statistical circles, but it is still quite common to see application papers where a small number of training samples and/or a large number of feature candidates render the conclusion of the investigation doubtful at best. Statements about the unbiased nature of the leave-one-out error estimate are quite frequent, although it is seldom clarified whether the feature selection and the error estimation are performed on the same data ($\hat{P}_e^L$) or not ($\hat{P}_e^{L2}$). Finally, comparison between competing classifiers, feature selection methods and so on are often done without regarding the heightened variance that accompanies the proper unbiased error estimate, particularly for small sample sizes. The key results of this study are the importance of the number of feature candidates and that the proper n^{Tr}/D ratio in order to select the correct features is not a constant, but depends on the number of training samples, feature candidates and the Mahalanobis distance.

Acknowledgment

This work was supported by the Norwegian Research Council (NFR).

References

1. M. C. Constanza and A. A. Afifi. Comparison of stopping rules in forward stepwise discriminant analysis. *Journal of the American Statistical Association*, 74:777–785, 1979. 481
2. R. O Duda and P. E Hart. *Pattern classification and scene analysis*. A Wiley-interscience publication, first edition, 1973. 481
3. R. P. W. Duin. A matlab toolbox for pattern recognition. Technical Report Version 3.0, Delft University of Technology, 2000. 481
4. K. S. Fu, P. J. Min, and T. J. Li. Feature selection in pattern recognition. *IEEE Trans on Syst Science and Cybern - Part C*, 6(1):33–39, 1970. 480
5. A. Jain and D. Zongker. Feature selection: Evaluation, application, and small sample performance. *IEEE Trans Pattern Anal Mach Intell*, 19(2):153–158, 1997. 480, 481
6. P. A. Lachenbruch and M. R. Mickey. Estimation of error rates in discriminant analysis. *Techometrics*, 10(1):1–11, 1968. 481
7. P. Pudil, J. Novovicova, and J. Kittler. Floating search methods in feature selection. *Pat Rec Let*, 15:1119–1125, 1994. 481
8. A. C. Rencher and S. F. Larson. Bias in Wilks' lambda in stepwise discriminant analysis. *Technometrics*, 22(3):349–356, 1980. 485
9. C. Rutter, V. Flack, and P. Lachenbruch. Bias in error rate estimates in discriminant analysis when setpwise variable selection is employed. *Commun. Stat., Simulation Comput*, 20(1):1–22, 1991. 485
10. H. Schulerud. The influence of feature selection on error estimates in linear discriminant analysis. *Submittet to Pattern Recognition.* 486
11. S. D. Stearns. On selecting features or pattern classifiers. *Proc. Third Intern. Conf. Pattern Recognition*, pages 71–75, 1976. 481

Spatial Representation of Dissimilarity Data via Lower-Complexity Linear and Nonlinear Mappings

Elżbieta Pekalska and Robert P. W. Duin

Pattern Recognition Group, Department of Applied Physics
Faculty of Applied Sciences, Delft University of Technology
Lorentzweg 1, 2628 CJ Delft, The Netherlands
{ela,duin}@ph.tn.tudelft.nl

Abstract. Dissimilarity representations are of interest when it is hard to define well-discriminating features for the raw measurements. For an exploration of such data, the techniques of multidimensional scaling (MDS) can be used. Given a symmetric dissimilarity matrix, they find a lower-dimensional configuration such that the distances are preserved. Here, Sammon nonlinear mapping is considered. In general, this iterative method must be recomputed when new examples are introduced, but its complexity is quadratic in the number of objects in each iteration step. A simple modification to the nonlinear MDS, allowing for a significant reduction in complexity, is therefore considered, as well as a linear projection of the dissimilarity data. Now, generalization to new data can be achieved, which makes it suitable for solving classification problems. The linear and nonlinear mappings are then used in the setting of data visualization and classification. Our experiments show that the nonlinear mapping can be preferable for data inspection, while for discrimination purposes, a linear mapping can be recommended. Moreover, for the spatial lower-dimensional representation, a more global, linear classifier can be built, which outperforms the local nearest neighbor rule, traditionally applied to dissimilarities.

1 Introduction

An alternative to the feature-based description is a representation based on dissimilarity relations between objects. Such representations are useful when features are difficult to obtain or when they have little discriminative power. Such situations are encountered in practice, especially when shapes, blobs, or some particular image characteristics have to be recognized [6,8]. The use of dissimilarities is, therefore, dictated by the application or data specification.

For an understanding of dissimilarity data, techniques of multidimensional scaling (MDS) [1,10] can be used. MDS refers to a group of methods mainly used for visualizing the structure in high-dimensional data by mapping it onto a 2- or 3-dimensional space. The output of MDS is a spatial representation of the data, i.e. a configuration of points, representing the objects, in a space. Such a

T. Caelli et al. (Eds.): SSPR&SPR 2002, LNCS 2396, pp. 488–497, 2002.
© Springer-Verlag Berlin Heidelberg 2002

display is believed to allow for a better understanding of the data, since similar objects are represented by close points.

In the basic approach, MDS is realized by Sammon mapping [1,10]. This nonlinear, iterative projection minimizes an error function between original dissimilarities and Euclidean distances in a lower-dimensional space. For n objects, it requires computation of $\mathcal{O}(n^2)$ distances in each iteration step and the same memory storage. However, for a lower, m-dimensional representation, only mn variables should be determined, which suggests that a number of $\mathcal{O}(n^2)$ constraints on distances is redundant and, therefore, could be neglected. The leads to the idea that only distances to the, so-called, representation set R (a subset of all objects), could be preserved, for which a modified version of the Sammon mapping should be considered. A similar reduction of complexity can be applied to a linear projection of dissimilarity data, being an extension of Classical Scaling, i.e. the linear MDS technique [1].

In this paper, we compare the linear and nonlinear projection methods, reduced in complexity, for data visualization and classification. Our experiments show that for dissimilarity data of smaller intrinsic dimensionality, its lower-dimensional spatial representation allows for building a classifier that significantly outperforms the nearest neighbor (NN) rule, traditionally used to discriminate between objects represented by dissimilarities. The NN rule, based on local neighborhoods, suffers from sensitivity to noisy objects. The spatial representation of dissimilarities, reflecting the data structure, is defined in a more global way, and therefore, better results can be achieved.

The paper is organized as follows. Sections 2 and 3 give insight into linear and nonlinear projections of the dissimilarity data. Section 4 explains how the reduction of complexity is achieved. Section 5 describes the classification experiments conducted, presents some $2D$ projection maps and discusses the results. Conclusions are summarized in section 6.

2 Linear Projection of the Dissimilarity Data

Non-metric distances may arise when shapes or objects in images are compared e.g. by template matching [8,6]. For projection purposes, the symmetry condition is necessary, but for any symmetric distance matrix, an Euclidean space is not 'large enough' for a distance-preserving linear mapping onto the specified dimensionality. It is, however, always possible [4] for a pseudo-Euclidean space.

The Pseudo-Euclidean Space A pseudo-Euclidean space $\mathcal{R}^{(p,q)}$ of the signature (p,q) [5,4] is a real linear vector space of dimension $p+q$, composed of two Euclidean subspaces, $\mathcal{R}^p$ and $\mathcal{R}^q$, such that $\mathcal{R}^{(p,q)}=\mathcal{R}^p \oplus \mathcal{R}^q$ and the inner product $\langle \cdot, \cdot \rangle$ is positive definite on $\mathcal{R}^p$ and negative definite on $\mathcal{R}^q$. The inner product w.r.t. the orthonormal basis is defined as $\langle \boldsymbol{x}, \boldsymbol{y} \rangle = \sum_{i=1}^{p} x_i y_i - \sum_{j=p+1}^{p+q} x_j y_j = \boldsymbol{x}^T M \boldsymbol{y}$, $M = \begin{bmatrix} I_{p\times p} & 0 \\ 0 & -I_{q\times q} \end{bmatrix}$, where I is the identity matrix. Using the notion of inner product, $d^2(\boldsymbol{x},\boldsymbol{y})=||\boldsymbol{x}-\boldsymbol{y}||^2=\langle \boldsymbol{x}-\boldsymbol{y},\,\boldsymbol{x}-\boldsymbol{y}\rangle=(\boldsymbol{x}-\boldsymbol{y})^T M(\boldsymbol{x}-\boldsymbol{y})$, can be positive, negative or zero. Note that an Euclidean space $\mathcal{R}^p$, is a pseudo-Euclidean space $\mathcal{R}^{(p,0)}$.

Linear Projection and Generalization to New Objects Let T consists of n objects. Given a symmetric distance matrix $D(T,T)\in\mathcal{R}^{n\times n}$, a configuration $X_{red}\in\mathcal{R}^{n\times m}$ $(m<n)$ in a pseudo-Euclidean space can be found, up to rotation and translation, such that the distances are preserved as well as possible. Without loss of generality, a linear mapping is constructed such that the origin coincides with the mean. X is then determined, based on the relation between distances and inner products. The matrix of inner products B can be expressed only by using the square distances $D^{(2)}$ [4,9]:

$$B=-\frac{1}{2}JD^{(2)}J, \quad J=I-\frac{1}{n}\mathbf{1}\mathbf{1}^T\in\mathcal{R}^{n\times n}, \tag{1}$$

where J takes care that the final configuration has a zero mean. By the eigendecomposition of $B=XMX^T$, one obtains: $B=Q\Lambda Q^T=Q|\Lambda|^{\frac{1}{2}}\begin{bmatrix}M & \\ & 0\end{bmatrix}|\Lambda|^{\frac{1}{2}}Q^T$, where $|\Lambda|$ is a diagonal matrix of first, decreasing p positive eigenvalues, then decreasing absolute values of q negative eigenvalues, and finally zeros. Q is the matrix of corresponding eigenvectors and $M\in\mathcal{R}^{k\times k}$, $k=p+q$, is defined as before (or it is equal to $I_{k\times k}$ if $\mathcal{R}^k$ is Euclidean). X is then represented in the space $\mathcal{R}^k$ as $X=Q_k\,|\Lambda_k|^{\frac{1}{2}}$ [4]. Note that X is an uncorrelated representation, i.e. given w.r.t. the principal axes. The reduced representation $X_{red}\in\mathcal{R}^{n\times m}$, $m<k$, is, therefore, determined by largest p' positive and smallest q' negative eigenvalues, i.e. $m=p'+q'$, and it is found as [4,9]:

$$X_{red}=Q_m\,|\Lambda_m|^{\frac{1}{2}}, \tag{2}$$

New objects can be orthogonally projected onto the space $\mathcal{R}^m$. Given the matrix of square distances $D_n^{(2)}\in\mathcal{R}^{s\times n}$, relating s new objects to the set T, a configuration X_{red}^n is then sought. Based on the matrix of inner products $B^n\in\mathcal{R}^{s\times n}$:

$$B^n=-\frac{1}{2}(D_n^{(2)}J-UD^{(2)}J), \quad U=\frac{1}{s}\mathbf{1}\mathbf{1}^T\in\mathcal{R}^{s\times n}, \tag{3}$$

$$X_{red}^n=B^nX_{red}|\Lambda_m|^{-1}M_m \quad \text{or} \quad X_{red}^n=B^nB^{-1}X_{red}. \tag{4}$$

Classifiers For a pseudo-Euclidean configuration, a linear classifier $f(\boldsymbol{x})=\langle \boldsymbol{v},\boldsymbol{x}\rangle+v_0=\boldsymbol{v}^TM\boldsymbol{x}+v_0$ can be constructed by addressing it as in the Euclidean case, i.e. $f(\boldsymbol{x})=\langle \boldsymbol{w},\boldsymbol{x}\rangle_{Eucl}+v_0=\boldsymbol{w}^T\boldsymbol{x}+v_0$, where $\boldsymbol{w}=M\boldsymbol{v}$; see [4,9].

3 Nonlinear Projection of the Dissimilarity Data

Sammon mapping [10,1] is the basic MDS technique used. It is a nonlinear projection onto an Euclidean space, such that the distances are preserved. For this purpose, an error function, called *stress*, is defined, which measures the difference between the original dissimilarities and Euclidean distances of the configuration X (consisting of n objects) in an m-dimensional space. Let D be the given dissimilarity matrix and $\tilde{D}$ be the distance matrix for the projected configuration X. A variant of the Sammon stress is here considered [3,10]:

$$S = \frac{1}{\sum_{i=1}^{n-1}\sum_{j=i+1}^{n} d_{ij}^2} \sum_{i=1}^{n-1}\sum_{j=i+1}^{n} (d_{ij} - \tilde{d}_{ij})^2 \tag{5}$$

and it is chosen since it emphasizes neither large nor small distances. To find a Sammon representation, one starts from an initial configuration of points for which all the pairwise distances are computed and the stress value is calculated. Next, the points are adjusted such that the stress will decrease. This is done in an iterative manner, until a configuration corresponding to a (local) minimum of S is found. Here, the scaled conjugate gradients algorithm is used to search for the minimum of S. It is important to emphasize that the minimum found depends on the initialization. In this paper, the principal component projection of the dissimilarity data is used to initialize the optimization procedure.

4 Reduction of Complexity

For our (non-)linear projection, although X has the dimensionality m, it is still determined by n objects. In general, such a space can be defined by $m+1$ linearly independent objects. If they were lying one in the origin and the others on the axes, they would determine our space exactly. Since this is unlikely to happen, the space retrieved will be an approximation of the original one. When more objects are used, the space becomes more filled and, therefore, better defined. The question now arises how to select the representation set $R \subseteq T$ of the size $r > m$, on which the (non-)linear mapping could be based. Following [2], we choose objects, lying in the areas of higher density, i.e. with relatively many close neighbors. For a dissimilarity representation $D(T, T)$, a natural way to proceed is the K-centers algorithm. It looks for K center objects, i.e. examples that minimize the maximum of the distances over all objects to their nearest neighbors, i.e. it minimizes the error $E_{K-cent} = \max_i (\min_k d_{ik})$. It uses a forward search strategy, starting from a random initialization. (Note that the K-means [3] cannot be used since no potential feature representation is assumed.)

For a chosen R, the linear mapping onto m-dimensional space is defined by formulas (1)–(2) based on $D(R, R)$. The remaining objects $D(T \backslash R, R)$ can then be added by the use of (3) and (4). In this way, the complexity is reduced from $\mathcal{O}(mn^2)$ (computing m eigenvectors and eigenvalues) to $\mathcal{O}(mr^2) + \mathcal{O}(nr)$.

In case of the Sammon mapping, a modified version should be defined, which generalizes to new objects. Following [2], first the Sammon mapping of $D(R, R)$ onto the space $\mathcal{R}^m$ is performed, yielding the configuration X_R.The remaining objects can be mapped to this space, while preserving the dissimilarities to the set R, i.e. $D' = D(T\backslash R, R)$. This can be done via an iterative minimization procedure of the modified stress S_M, using the found representation X_R:

$$S_M = \frac{1}{\sum_{i=1}^{n}\sum_{j=1}^{r}(d'_{ij})^2}\sum_{i=1}^{n}\sum_{j=1}^{r}(d'_{ij}-\tilde{d}'_{ij})^2 \tag{6}$$

This procedure allows for adding objects to an existing map, which can now be used for classification purposes. Its complexity reduces from $\mathcal{O}(mn^2)$, computing $\mathcal{O}(n^2)$ distances in the $\mathcal{R}^m$ space, to $\mathcal{O}(nmr+nr^2)$ in each iteration step.

5 Experiments

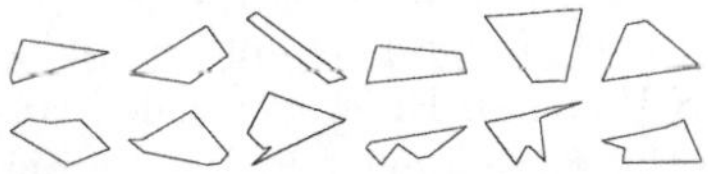

Fig. 1. Examples of the polygons

Two datasets are used in our study. The first data consists of randomly generated polygons (see Figure 1): 4-edge convex polygons and 7-edge convex and non-convex polygons. The polygons are first scaled and then the modified Hausdorff distance [8] is computed. The second data describes the NIST digits [11], represented by 128×128 binary images. Here, the symmetric dissimilarity, based on deformable template matching, as defined by Zongker and Jain [7], is used.

The experiments are performed 50/10 times for the polygon/digit data and the results are averaged. In each run, both datasets are randomly split into equally sized the training and testing sets. Each class is represented by 50/100 objects (i.e. n=100/1000) for the polygon/digit data. In each experiment, first the dimensionality m of the projection is established. In case of the linear mapping, one may predict the intrinsic dimensionality based on the number of significant eigenvalues [4,9] (similarly to the principal component analysis [3]). However, this might be different for Sammon mapping. Therefore, a few distinct dimensionalities are used. For the dimensionality m, representation sets of the size r, varying from $m+1$ to n are considered. Each set R is selected by the K-centers algorithm, except for R equal to the training set T (i.e. $r = n$). Next, an approximated space, defined by objects from R is determined (i.e. the (non-)linear mapping is based on $D(R, R)$). The remaining $T\backslash R$ objects are then mapped to this space, as described in section 4 and the Fisher linear classifier (FLC) is trained on *all* n objects (a quadratic classifier has also been used, but the linear one performs better). The test data is then projected to the space and the classification error is found. For a new object, only r distances have to be computed and the complexity of the testing stage becomes $\mathcal{O}(mr)$ for the linear projection and $\mathcal{O}(\max(mr, r^2))$ in each iteration step for Sammon mapping.

The results of our experiments on the polygon/digit data are presented in Figure 2. For the polygon data, the best performance of the FLC is achieved when the dimensionality of the projected space is 15 for Sammon mapping or 20 for the linear mapping. For the set R consisting of only 20 training objects, the FLC built in both linear and nonlinear projected spaces (i.e. using distances to

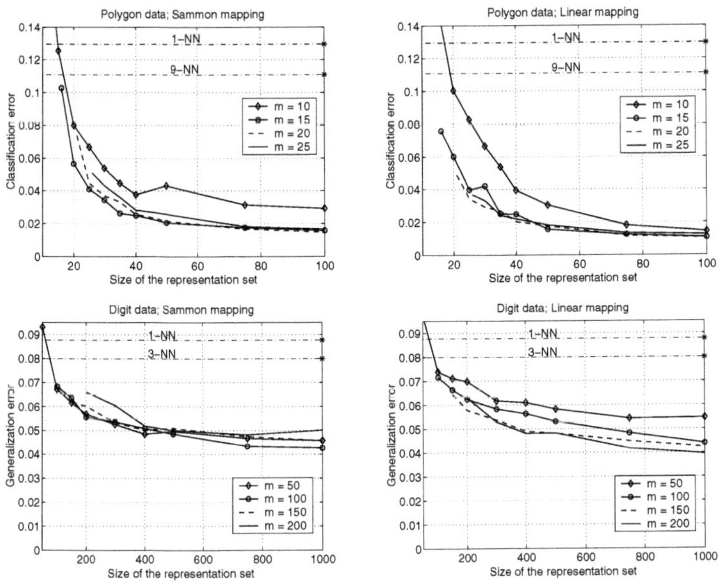

Fig. 2. The NN rule on dissimilarities (marked by '*') and the FLC on the spatial representations for the polygon data (top) and the digit data (bottom)

the set R only), outperforms the 1-NN rule and the best 9-NN rule, both based on 100 objects. This shows that by making use of the structure information present in the data, a less noise-sensitive decision rule than, the NN method, can be constructed. When R contains $30 - 40\%$ of the data, the error of nearly 0.02 is reached, which is close to the error of $0.015 - 0.018$ gained when $R{=}T$.

For the digit data, the best accuracy is found when $m = 100$ or $m = 200$ for the Sammon mapping or the linear projection, respectively. For the set R, consisting of 10% of the training objects only, the FLC built in both nonlinear and linear 50-dimensional spaces, outperforms the 1-NN rule and the best 3-NN rule, both based on all 1000 objects. When $r = 400$ objects are chosen to the set R, an error of 0.05 can be reached; when $R{=}T$, an error of 0.04 is achieved.

In Figure 3, one can also observe that for both data, the stress S changes only slightly when R is larger than half of the training set. For the linear mapping, the stress values are not shown for $r{=}m{+}1$, since some of the pseudo-Euclidean distances are negative and S becomes complex. For larger r, the imaginary part of S becomes nearly zero and can, therefore, be neglected. The stress is, of course, relatively large for the linear mapping, but this does not disturb a good classification performance. Apparently, the variance present in the data, revealed by the linear projection, is good enough for discrimination purposes, since major differences in classes are captured.

In summary, in terms of the stress, a nonlinear configuration preserves the data structure better than the linear one. The nonlinear mapping requires less

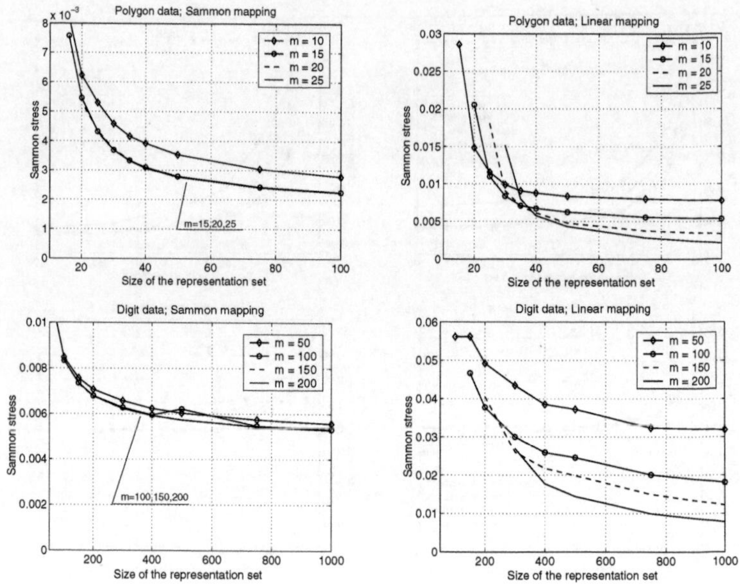

Fig. 3. Sammon stress for the spatial representations of the polygon data (top) and the digit data (bottom)

dimensions for about the same performance of the FLC than the linear mapping, although for the latter, a bit higher accuracy can overall be reached.

Visualization From all the linear mappings of the fixed dimensionality, our linear projection preserves the distances in the best way [1,4]. Since it is constructed to explain the maximum of the (generalized) variance in the data, some details in the structure might remain unrevealed. When data lies in a nonlinear subspace, Sammon mapping is preferred since it provides an additional information.

The difference between the original (non-)linear $2D$ maps and the maps based on smaller representation sets can be observed in Figure 4, where the results for four datasets are shown. The first two examples are illustrative: banana dataset is an artificial $2D$ dataset for which the theoretical, nearly-Euclidean distance is found; for the $4D$ Iris dataset, the Euclidean distance is considered. The last two datasets refer to data from our classification experiments.

Each subfigure presents plots for the linear and nonlinear projections. Those plots show the difference between the original (non-)linear maps and the maps, constructed while preserving the dissimilarities to the set R only. From Figure 4, one can observe that the (non-)linear maps, based on a smaller R resemble well the original maps, based on all objects. The Sammon stress computed for those configurations reveals the loss up to 20%. This is reasonable, given that the

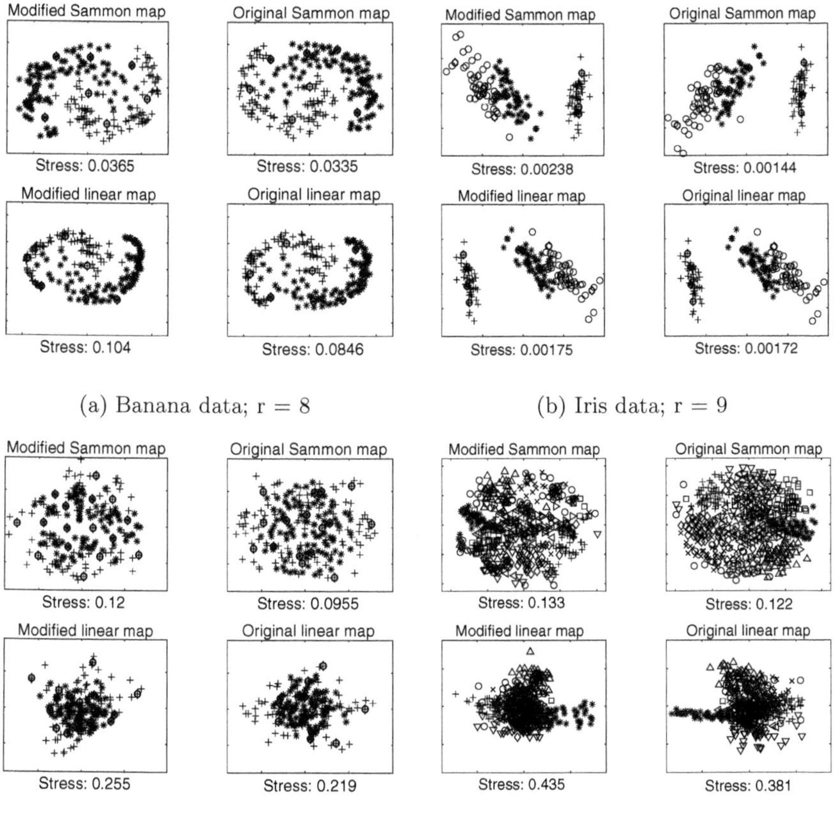

(a) Banana data; r = 8 (b) Iris data; r = 9

(c) Polygon data; r = 16 (d) Digit data; r = 50

Fig. 4. Linear and nonlinear 2D maps; set R marked in black 'o' when feasible

chosen R consists of less than 10% of all objects, which means that around 90% of distances are not taken into account during the mapping process.

6 Discussion and Conclusions

The presented mappings of finding a faithful spatial configuration do not make use of class labels. So, the class separability could potentially be enhanced by using such information. This remains an open issue for further research.

To reduce noise in the data, in a mapping process, the distances are preserved approximately. By this, the class separability may be somewhat improved, although, in general, it is reflected in a similar way as given by all the dissimilarity relations. The advantage of building e.g. a linear classifier in such a projected space over the k-NN is that the data information is used in a more complex and comprehensive way, based on relations between a number of objects both in the mapping process and in the classifier construction. Since the k-NN rule is locally noise sensitive, for dissimilarity data, noisy in local neighbourhoods, our approach can be beneficial. It is important to emphasize, however, that the generality of our approach holds for data of a lower intrinsic dimensionality.

A number of conclusions can be drawn from our study. First of all, the modified Sammon algorithm allows for adding new data to the existing map. Secondly, the (non-)linear mapping onto m dimensions, based on the set R of the size r, reduces its complexity both in the training and testing stage. For an evaluation of a novel object, only r dissimilarities have to be computed, and for the linear mapping $\mathcal{O}(mr)$ operations are needed, while for the Sammon mapping, $\mathcal{O}(\max(mr, r^2))$ operations are necessary in each iteration step.

Thirdly, the projections considered, allow for obtaining a spatial configuration of the dissimilarity data, which can be beneficial for the classification task. Our experiments with dissimilarity representations of the polygon and digit data show that such spaces offer a possibility to build decision rules that significantly outperform the NN method. Based on the set R consisting of 45% of the training objects, the FLC, constructed in a projected space defined by the dissimilarities to R only, reaches an error of 0.02/0.05, while the best NN rule makes an error of 0.11/0.088 and makes use of all objects.

Next, the 2D spatial representations of dissimilarity data, obtained by the linear and modified Sammon projections, resemble the original maps. A similar structure is revealed in the data when R consists of 10% of objects, chosen by the K-centers algorithm, as well as of all of them. These approaches are especially useful when dealing with large datasets. In general, Sammon maps provide an extra insight into the data and can be preferred for visualization. Our experience shows also that the use of the K-centers is not crucial; what is important is the choice of significantly *different* objects to represent the variability in the data.

Finally, the FLC built on the linear configuration yields about the same (somewhat better) classification results as the FLC on the modified-Sammon representation, but in a space of a larger dimensionality than for the nonlinear case. However, since no iterations are involved for an evaluation of novel examples, the linear projection can be recommended for the classification task.

Acknowledgments

This work is supported by the Dutch Organization for Scientific Research (NWO). The authors thank prof. Anil Jain for the NIST dissimilarity data.

References

1. I. Borg and P. Groenen. *Modern Multidimensional Scaling.* Springer-Verlag, New York, 1997. 488, 489, 490, 494
2. D. Cho and D. J. Miller. A Low-complexity Multidimensional Scaling Method Based on Clustering. *concept paper*, 2002. 491

3. R. O. Duda, P. E. Hart, and D. G. Stork. *Pattern Classification.* John Wiley & Sons, Inc., 2nd edition, 2001. 491, 492
4. L. Goldfarb. A new approach to pattern recognition. In L.N. Kanal and A. Rosenfeld, editors, *Progress in Pattern Recognition*, volume 2, pages 241–402. Elsevier Science Publishers B.V., 1985. 489, 490, 492, 494
5. W. Greub. *Linear Algebra.* Springer-Verlag, 1975. 489
6. D. W. Jacobs, D. Weinshall, and Y. Gdalyahu. Classification with Non-Metric Distances: Image Retrievaland Class Representation. *IEEE Trans. on PAMI*, 22(6):583–600, 2000. 488, 489
7. A. K. Jain and D. Zongker. Representation and recognition of handwritten digits using deformable templates. *IEEE Trans. on PAMI*, 19(12):1386–1391, 1997. 492
8. Dubuisson M. P. and Jain A. K. Modified Hausdorff distance for object matching. In *12th Int. Conf. on Pattern Recognition*, volume 1, pages 566–568, 1994. 488, 489, 492
9. E. Pekalska, P. Paclík, and R. P. W. Duin. A Generalized Kernel Approach to Dissimilarity Based Classification. *J. of Mach. Learn. Research*, 2:175–211, 2001. 490, 492
10. J. W. Sammon. A nonlinear mapping for data structure analysis. *IEEE Transaction on Computers*, C-18:401–409, 1969. 488, 489, 490, 491
11. C. L. Wilson and M. D. Garris. Handprinted character database 3. Technical report, National Institute of Standards and Technology, February 1992. 492

A Method to Estimate the True Mahalanobis Distance from Eigenvectors of Sample Covariance Matrix

Masakazu Iwamura, Shinichiro Omachi, and Hirotomo Aso

Graduate School of Engineering, Tohoku University
Aoba 05, Aramaki, Aoba-ku, Sendai-shi, 980-8579 Japan
{masa,machi,aso}@aso.ecei.tohoku.ac.jp

Abstract. In statistical pattern recognition, the parameters of distributions are usually estimated from training sample vectors. However, estimated parameters contain estimation errors, and the errors cause bad influence on recognition performance when the sample size is not sufficient. Some methods can obtain better estimates of the eigenvalues of the true covariance matrix and can avoid bad influences caused by estimation errors of eigenvalues. However, estimation errors of eigenvectors of covariance matrix have not been considered enough. In this paper, we consider estimation errors of eigenvectors and show the errors can be regarded as estimation errors of eigenvalues. Then, we present a method to estimate the true Mahalanobis distance from eigenvectors of the sample covariance matrix. Recognition experiments show that by applying the proposed method, the true Mahalanobis distance can be estimated even if the sample size is small, and better recognition accuracy is achieved. The proposed method is useful for the practical applications of pattern recognition since the proposed method is effective without any hyper-parameters.

1 Introduction

In statistical pattern recognition, the Bayesian decision theory gives a decision to minimize the misclassification probability as long as the true distributions are given. However, the true distributions are unknown in most practical situations. The forms of the distributions are often assumed to be normal and the parameters of the distributions are estimated from the training sample vectors. It is well known that the estimated parameters contain estimation errors and the errors cause bad influence on recognition performance when there are not enough training sample vectors.

To avoid bad influence caused by estimation errors of eigenvalues, there are some methods to obtain better estimates of the true eigenvalues. Sakai et al. [1,2] proposed a method to rectify the sample eigenvalues (the eigenvalues of the sample covariance matrix), which is called RQDF. James and Stein indicated that the conventional sample covariance matrix is not admissible (which means there are some better estimators). They proposed an improved estimator of the

T. Caelli et al. (Eds.): SSPR&SPR 2002, LNCS 2396, pp. 498–507, 2002.
© Springer-Verlag Berlin Heidelberg 2002

sample covariance matrix (James-Stein estimator) [3] by modifying the sample eigenvalues.

However, estimation errors of eigenvectors of covariance matrix have not been considered enough and still an important problem. In this paper, we aim to achieve high-performance pattern recognition without many training samples and any hyper-parameters. We present a method to estimate the true Mahalanobis distance from the sample eigenvectors. First of all, we show the error of the Mahalanobis distance caused by estimation errors of eigenvectors can be regarded as the errors of eigenvalues. Then, we introduce a procedure for estimating the true Mahalanobis distance by deriving the probability density function of estimation errors of eigenvectors. The proposed method consists of two-stage modification of the sample eigenvalues. At the first stage, estimation errors of eigenvalues are corrected using an existing method. At the second stage, the corrected eigenvalues are modified to compensate estimation errors of eigenvectors. The effectiveness of the proposed method is confirmed by recognition experiments. This paper is based on the intuitive sketch [4] and formulated with statistical and computational approaches.

2 A Method to Estimate the True Mahalanobis Distance

2.1 The Eigenvalues to Compensate Estimation Errors of Eigenvectors

If all the true parameters of the distribution are known, the true Mahalanobis distance is obtained. Let $\boldsymbol{x}$ be an unknown input vector, $\boldsymbol{\mu}$ be the true mean vector, $\boldsymbol{\Lambda} = \mathrm{diag}(\lambda_1, \lambda_2, \ldots, \lambda_d)$ and $\boldsymbol{\Phi} = (\boldsymbol{\phi}_1\ \boldsymbol{\phi}_2 \cdots \boldsymbol{\phi}_d)$, where λ_i and $\boldsymbol{\phi}_i$ are the ith eigenvalue and eigenvector of the true covariance matrix. All the eigenvalues are assumed to be ordered in the descending order in this paper. The true Mahalanobis distance is given as

$$d(\boldsymbol{x}) = (\boldsymbol{x} - \boldsymbol{\mu})^{\mathrm{T}} \boldsymbol{\Phi} \boldsymbol{\Lambda}^{-1} \boldsymbol{\Phi}^{\mathrm{T}} (\boldsymbol{x} - \boldsymbol{\mu}). \tag{1}$$

In general, the true eigenvectors are unknown and only the sample eigenvectors $\left\{\hat{\boldsymbol{\phi}}_i\right\}$ are obtained. Let $\hat{\boldsymbol{\Phi}} = (\hat{\boldsymbol{\phi}}_1\ \hat{\boldsymbol{\phi}}_2 \cdots \hat{\boldsymbol{\phi}}_d)$. The Mahalanobis distance using $\hat{\boldsymbol{\Phi}}$ is

$$\hat{d}(\boldsymbol{x}) = (\boldsymbol{x} - \boldsymbol{\mu})^{\mathrm{T}} \hat{\boldsymbol{\Phi}} \boldsymbol{\Lambda}^{-1} \hat{\boldsymbol{\Phi}}^{\mathrm{T}} (\boldsymbol{x} - \boldsymbol{\mu}). \tag{2}$$

Of course, $d(\boldsymbol{x})$ and $\hat{d}(\boldsymbol{x})$ differ. Now, let $\hat{\boldsymbol{\Psi}} \equiv \boldsymbol{\Phi}^{\mathrm{T}} \hat{\boldsymbol{\Phi}}$ be *estimation error matrix* of eigenvectors. Since both $\boldsymbol{\Phi}$ and $\hat{\boldsymbol{\Phi}}$ are orthonormal matrices, $\hat{\boldsymbol{\Psi}}$ is also an orthonormal matrix. Substituting $\hat{\boldsymbol{\Phi}} = \boldsymbol{\Phi} \hat{\boldsymbol{\Psi}}$ into Eq. (2), we obtain

$$\hat{d}(\boldsymbol{x}) = (\boldsymbol{x} - \boldsymbol{\mu})^{\mathrm{T}} \boldsymbol{\Phi} \left(\hat{\boldsymbol{\Psi}} \boldsymbol{\Lambda} \hat{\boldsymbol{\Psi}}^{\mathrm{T}}\right)^{-1} \boldsymbol{\Phi}^{\mathrm{T}} (\boldsymbol{x} - \boldsymbol{\mu}). \tag{3}$$

Comparing Eq. (3) and Eq. (1), $\left(\hat{\boldsymbol{\Psi}} \boldsymbol{\Lambda} \hat{\boldsymbol{\Psi}}^{\mathrm{T}}\right)^{-1}$ in Eq. (3) corresponds to $\boldsymbol{\Lambda}^{-1}$ (the true eigenvalues) in Eq. (1). If we can ignore the non-orthogonal elements of

$\left(\hat{\boldsymbol{\Psi}}\boldsymbol{\Lambda}\hat{\boldsymbol{\Psi}}^{\mathrm{T}}\right)^{-1}$, the error of the Mahalanobis distance caused by the estimation errors of eigenvectors will be regarded as the errors of eigenvalues. This means that even if eigenvectors have estimation errors, we can estimate the true Mahalanobis distance using certain eigenvalues. Now, let $\tilde{\boldsymbol{\Lambda}}$ be a diagonal matrix which satisfies $\boldsymbol{\Lambda}^{-1} \approx \left(\hat{\boldsymbol{\Psi}}\tilde{\boldsymbol{\Lambda}}\hat{\boldsymbol{\Psi}}^{\mathrm{T}}\right)^{-1}$. Namely, $\tilde{\boldsymbol{\Lambda}}$ is defined as

$$\tilde{\boldsymbol{\Lambda}} = \mathrm{D}\left(\hat{\boldsymbol{\Psi}}^{\mathrm{T}}\boldsymbol{\Lambda}\hat{\boldsymbol{\Psi}}\right), \tag{4}$$

where D is a function which returns diagonal elements of the matrix. $\tilde{\boldsymbol{\Lambda}}$ is the eigenvalues which compensate estimation errors of eigenvectors. The justification of ignoring the non-diagonal elements of $\hat{\boldsymbol{\Psi}}^{\mathrm{T}}\boldsymbol{\Lambda}\hat{\boldsymbol{\Psi}}$ is confirmed by the experiment in Sect. 3.1.

$\tilde{\boldsymbol{\Lambda}}$ is defined by the true eigenvalues ($\boldsymbol{\Lambda}$) and estimation errors of eigenvectors ($\hat{\boldsymbol{\Psi}}$). $\hat{\boldsymbol{\Psi}}$ is defined by using the true eigenvectors ($\boldsymbol{\Phi}$). Since we assume that $\boldsymbol{\Phi}$ are unknown, we can not observe $\hat{\boldsymbol{\Psi}}$. However, we can observe the probability density function of $\hat{\boldsymbol{\Psi}}$ because the probability density function of $\hat{\boldsymbol{\Psi}}$ depends only on the dimensionality of feature vectors, sample size, the true eigenvalues and the sample eigenvalues, and does not depend on the true eigenvectors (See Appendix). Therefore, the expectation of $\hat{\boldsymbol{\Psi}}$ is observable even if the true eigenvectors are unknown. Let $\acute{\boldsymbol{\Psi}}$ be random estimation error matrix of eigenvectors. Eq. (4) is rewritten as

$$\acute{\tilde{\boldsymbol{\Lambda}}} = \mathrm{D}\left(\acute{\boldsymbol{\Psi}}^{\mathrm{T}}\boldsymbol{\Lambda}\acute{\boldsymbol{\Psi}}\right), \tag{5}$$

where $\acute{\tilde{\boldsymbol{\Lambda}}}$ is a diagonal matrix of the random variables representing the eigenvalues for the compensation.

The conditional expectation of Eq. (5) given $\hat{\boldsymbol{\Lambda}}$ is calculated as

$$\begin{aligned}\tilde{\tilde{\boldsymbol{\Lambda}}} &= \mathrm{E}\left[\acute{\tilde{\boldsymbol{\Lambda}}}\middle|\hat{\boldsymbol{\Lambda}}\right] = \mathrm{E}\left[\mathrm{D}\left(\acute{\boldsymbol{\Psi}}^{\mathrm{T}}\boldsymbol{\Lambda}\acute{\boldsymbol{\Psi}}\right)\middle|\hat{\boldsymbol{\Lambda}}\right] \\ &= \mathrm{D}\left(\mathrm{E}\left[\acute{\boldsymbol{\Psi}}^{\mathrm{T}}\boldsymbol{\Lambda}\acute{\boldsymbol{\Psi}}\middle|\hat{\boldsymbol{\Lambda}}\right]\right),\end{aligned} \tag{6}$$

where $\tilde{\tilde{\boldsymbol{\Lambda}}} = \mathrm{diag}\left(\tilde{\tilde{\lambda}}_1, \tilde{\tilde{\lambda}}_2, \ldots, \tilde{\tilde{\lambda}}_d\right)$. The ith diagonal element of Eq. (6) is

$$\tilde{\tilde{\lambda}}_i = \mathrm{E}\left[\sum_{j=1}^{d}\left\{\acute{\psi}_{ji}\right\}^2\lambda_j\middle|\hat{\boldsymbol{\Lambda}}\right] \tag{7}$$

$$= \sum_{j=1}^{d}\mathrm{E}\left[\left\{\acute{\psi}_{ji}\right\}^2\middle|\hat{\boldsymbol{\Lambda}}\right]\lambda_j. \tag{8}$$

Letting

$$\left\{\tilde{\psi}_{ji}\right\}^2 = \mathrm{E}\left[\left\{\acute{\psi}_{ji}\right\}^2\middle|\hat{\boldsymbol{\Lambda}}\right], \tag{9}$$

we obtain

$$\tilde{\tilde{\lambda}}_i = \sum_{j=1}^{d} \left\{ \tilde{\psi}_{ji} \right\}^2 \lambda_j. \tag{10}$$

2.2 Calculation of Eq. (10)

We show a way to calculate Eq. (10). We will begin by generalizing the conditional expectation of Eq. (9). Let $f(\acute{\boldsymbol{\Psi}})$ be an arbitrary function of $\acute{\boldsymbol{\Psi}}$. The integral representation of the conditional expectation of $f(\acute{\boldsymbol{\Psi}})$ is given as

$$\mathrm{E}\left[f(\acute{\boldsymbol{\Psi}}) \middle| \hat{\boldsymbol{\Lambda}} \right] = \int_{\acute{\boldsymbol{\Psi}}} f(\acute{\boldsymbol{\Psi}}) \mathrm{P}(\acute{\boldsymbol{\Psi}} | \hat{\boldsymbol{\Lambda}}) d\acute{\boldsymbol{\Psi}}, \tag{11}$$

where $\mathrm{P}(\acute{\boldsymbol{\Psi}}|\hat{\boldsymbol{\Lambda}})$ is the probability density function of estimation errors of eigenvectors. Obtaining exact value of $\mathrm{P}(\acute{\boldsymbol{\Psi}}|\hat{\boldsymbol{\Lambda}})$ is difficult especially for large d. In this paper, Eq. (11) is estimated by Conditional Monte Carlo Method [5]. By assuming $f(\acute{\boldsymbol{\Psi}}) = \left\{ \acute{\psi}_{ji} \right\}^2$ in Eq. (11), $\left\{ \tilde{\psi}_{ji} \right\}^2$ of Eq. (9) is obtained. Therefore, we can calculate $\tilde{\tilde{\lambda}}_i$ in Eq. (10).

To carry out Conditional Monte Carlo Method, we deform the right side of Eq. (11). For the preparation of the deformation, let $\acute{\boldsymbol{\Sigma}}$ be a random symmetric matrix and $\acute{\boldsymbol{\Lambda}}$ be a random diagonal matrix that satisfies $\acute{\boldsymbol{\Sigma}} = \acute{\boldsymbol{\Psi}} \acute{\boldsymbol{\Lambda}} \acute{\boldsymbol{\Psi}}^{\mathrm{T}}$. Since the probability density function of estimation errors of eigenvectors ($\acute{\boldsymbol{\Psi}}$) is independent of the true eigenvectors ($\boldsymbol{\Phi}$), $\boldsymbol{\Phi} = \boldsymbol{I}$ is assumed without loss of generality, and $\acute{\boldsymbol{\Phi}} = \acute{\boldsymbol{\Psi}}$ immediately. Therefore, $\acute{\boldsymbol{\Sigma}} = \acute{\boldsymbol{\Phi}} \acute{\boldsymbol{\Lambda}} \acute{\boldsymbol{\Phi}}^{\mathrm{T}}$. Hence the probability density of $\acute{\boldsymbol{\Sigma}}$ is given as the Wishart distribution (See Appendix). We have $\mathrm{P}(\acute{\boldsymbol{\Sigma}}) = \mathrm{P}(\acute{\boldsymbol{\Psi}} \acute{\boldsymbol{\Lambda}} \acute{\boldsymbol{\Psi}}^{\mathrm{T}}) = \mathrm{P}(\acute{\boldsymbol{\Psi}}, \acute{\boldsymbol{\Lambda}}) J(\acute{\boldsymbol{\Psi}}, \acute{\boldsymbol{\Lambda}})$, where Jacobian $J(\acute{\boldsymbol{\Psi}}, \acute{\boldsymbol{\Lambda}}) = \frac{d\acute{\boldsymbol{\Psi}} d\acute{\boldsymbol{\Lambda}}}{d\acute{\boldsymbol{\Sigma}}}$. We also have $\mathrm{P}(\acute{\boldsymbol{\Psi}} \hat{\boldsymbol{\Lambda}} \acute{\boldsymbol{\Psi}}^{\mathrm{T}}) = \mathrm{P}(\acute{\boldsymbol{\Psi}}, \hat{\boldsymbol{\Lambda}}) J(\acute{\boldsymbol{\Psi}}, \hat{\boldsymbol{\Lambda}})$ since $\hat{\boldsymbol{\Lambda}}$ is a realization of random variable $\acute{\boldsymbol{\Lambda}}$. Let $g(\acute{\boldsymbol{\Lambda}})$ be an arbitrary function and $G = \int_{\acute{\boldsymbol{\Lambda}}} g(\acute{\boldsymbol{\Lambda}}) d\acute{\boldsymbol{\Lambda}}$.

Based on the preparation above, the right side of Eq. (11) can be deformed as

$$\begin{aligned} &\int_{\acute{\boldsymbol{\Psi}}} f(\acute{\boldsymbol{\Psi}}) \mathrm{P}(\acute{\boldsymbol{\Psi}} | \hat{\boldsymbol{\Lambda}}) d\acute{\boldsymbol{\Psi}} \\ &= \int_{\acute{\boldsymbol{\Psi}}} f(\acute{\boldsymbol{\Psi}}) \frac{\mathrm{P}(\acute{\boldsymbol{\Psi}} | \hat{\boldsymbol{\Lambda}})}{G} \left[\int_{\acute{\boldsymbol{\Lambda}}} g(\acute{\boldsymbol{\Lambda}}) d\acute{\boldsymbol{\Lambda}} \right] d\acute{\boldsymbol{\Psi}} \end{aligned}$$

$$= \int_{\acute{\boldsymbol{\Psi}}\times\acute{\boldsymbol{\Lambda}}} f(\acute{\boldsymbol{\Psi}}) \frac{\mathrm{P}(\acute{\boldsymbol{\Psi}},\hat{\boldsymbol{\Lambda}})}{\mathrm{P}(\hat{\boldsymbol{\Lambda}})} \frac{g(\acute{\boldsymbol{\Lambda}})}{G} d\acute{\boldsymbol{\Psi}} d\acute{\boldsymbol{\Lambda}}$$

$$= \frac{1}{\mathrm{P}(\hat{\boldsymbol{\Lambda}})} \int_{\acute{\boldsymbol{\Sigma}}} f(\acute{\boldsymbol{\Psi}}) \frac{\mathrm{P}(\acute{\boldsymbol{\Psi}}\hat{\boldsymbol{\Lambda}}\acute{\boldsymbol{\Psi}}^{\mathrm{T}})}{J(\acute{\boldsymbol{\Psi}},\hat{\boldsymbol{\Lambda}})} \frac{g(\acute{\boldsymbol{\Lambda}})}{G} J(\acute{\boldsymbol{\Psi}},\acute{\boldsymbol{\Lambda}}) d\acute{\boldsymbol{\Sigma}}$$

$$= \frac{1}{\mathrm{P}(\hat{\boldsymbol{\Lambda}})} \int_{\acute{\boldsymbol{\Sigma}}} f(\acute{\boldsymbol{\Psi}}) w_0(\acute{\boldsymbol{\Sigma}};\hat{\boldsymbol{\Lambda}}) \mathrm{P}(\acute{\boldsymbol{\Sigma}}) d\acute{\boldsymbol{\Sigma}}, \tag{12}$$

where

$$w_0(\acute{\boldsymbol{\Sigma}};\hat{\boldsymbol{\Lambda}}) = \frac{\mathrm{P}(\acute{\boldsymbol{\Psi}}\hat{\boldsymbol{\Lambda}}\acute{\boldsymbol{\Psi}}^{\mathrm{T}})}{J(\acute{\boldsymbol{\Psi}},\hat{\boldsymbol{\Lambda}})} \frac{J(\acute{\boldsymbol{\Psi}},\acute{\boldsymbol{\Lambda}})}{\mathrm{P}(\acute{\boldsymbol{\Psi}}\acute{\boldsymbol{\Lambda}}\acute{\boldsymbol{\Psi}}^{\mathrm{T}})} \frac{g(\acute{\boldsymbol{\Lambda}})}{G}. \tag{13}$$

Eq. (12) means that the expectation of $f(\acute{\boldsymbol{\Psi}})$ with probability density $\mathrm{P}(\acute{\boldsymbol{\Psi}}|\hat{\boldsymbol{\Lambda}})$ is the same as the expectation of $f(\acute{\boldsymbol{\Psi}})w_0(\acute{\boldsymbol{\Sigma}};\hat{\boldsymbol{\Lambda}})\frac{1}{\mathrm{P}(\hat{\boldsymbol{\Lambda}})}$ with probability density $\mathrm{P}(\acute{\boldsymbol{\Sigma}})$. Therefore, Eq. (11) can be calculated using the random vectors following normal distribution. By substituting Eq. (19) and Eq. (20) into Eq. (13), we have

$$w_0(\acute{\boldsymbol{\Sigma}};\hat{\boldsymbol{\Lambda}}) = \frac{|\hat{\boldsymbol{\Lambda}}|^{\frac{1}{2}(n-p-2)} \prod_{i<j}^{d}(\hat{\lambda}_i - \hat{\lambda}_j) \exp\left(-\frac{n-1}{2} \operatorname{tr} \boldsymbol{\Lambda}^{-1}\acute{\boldsymbol{\Psi}}\hat{\boldsymbol{\Lambda}}\acute{\boldsymbol{\Psi}}^{\mathrm{T}}\right)}{|\acute{\boldsymbol{\Lambda}}|^{\frac{1}{2}(n-p-2)} \prod_{i<j}^{d}(\acute{\lambda}_i - \acute{\lambda}_j) \exp\left(-\frac{n-1}{2} \operatorname{tr} \boldsymbol{\Lambda}^{-1}\acute{\boldsymbol{\Psi}}\acute{\boldsymbol{\Lambda}}\acute{\boldsymbol{\Psi}}^{\mathrm{T}}\right)} \frac{g(\acute{\boldsymbol{\Lambda}})}{G}. \tag{14}$$

Since calculating $\mathrm{P}(\hat{\boldsymbol{\Lambda}})$ is hard, though the formula of $\mathrm{P}(\hat{\boldsymbol{\Lambda}})$ is known, we show an another approach to obtain $\mathrm{P}(\hat{\boldsymbol{\Lambda}})$. When $f(\acute{\boldsymbol{\Psi}}) = 1$ is assumed in Eq. (12), $\frac{1}{\mathrm{P}(\hat{\boldsymbol{\Lambda}})} \int_{\acute{\boldsymbol{\Sigma}}} w_0(\acute{\boldsymbol{\Sigma}};\hat{\boldsymbol{\Lambda}})\mathrm{P}(\acute{\boldsymbol{\Sigma}})d\acute{\boldsymbol{\Sigma}} = 1$. Then, we obtain a calculatable solution

$$\mathrm{P}(\hat{\boldsymbol{\Lambda}}) = \int_{\acute{\boldsymbol{\Sigma}}} w_0(\acute{\boldsymbol{\Sigma}};\hat{\boldsymbol{\Lambda}})\mathrm{P}(\acute{\boldsymbol{\Sigma}})d\acute{\boldsymbol{\Sigma}}. \tag{15}$$

Let us assume $f(\acute{\boldsymbol{\Psi}}) = \left\{\acute{\psi}_{ji}\right\}^2$. Since Eq. (12) is the right side of Eq. (9), by substituting Eq. (15) into Eq. (12), Eq. (9) is obtained as $\left\{\tilde{\psi}_{ji}\right\}^2 = \frac{\int_{\acute{\boldsymbol{\Sigma}}}\left\{\acute{\psi}_{ji}\right\}^2 w_0(\acute{\boldsymbol{\Sigma}};\hat{\boldsymbol{\Lambda}})\mathrm{P}(\acute{\boldsymbol{\Sigma}})d\acute{\boldsymbol{\Sigma}}}{\int_{\acute{\boldsymbol{\Sigma}}} w_0(\acute{\boldsymbol{\Sigma}};\hat{\boldsymbol{\Lambda}})\mathrm{P}(\acute{\boldsymbol{\Sigma}})d\acute{\boldsymbol{\Sigma}}}$. By replacing the integrals of both numerator and denominator with the averages of $\acute{\boldsymbol{\Sigma}}_k, k = 1,\ldots,t$, and erasing $\frac{1}{t}$ of both numerator and denominator, we obtain

$$\left\{\tilde{\psi}_{ji}\right\}^2 = \frac{\sum_{k=1}^{t}\left\{\acute{\psi}_{k,ji}\right\}^2 w_0(\acute{\boldsymbol{\Psi}}_k\acute{\boldsymbol{\Lambda}}_k\acute{\boldsymbol{\Psi}}_k^{\mathrm{T}};\hat{\boldsymbol{\Lambda}})}{\sum_{k=1}^{t} w_0(\acute{\boldsymbol{\Psi}}_k\acute{\boldsymbol{\Lambda}}_k\acute{\boldsymbol{\Psi}}_k^{\mathrm{T}};\hat{\boldsymbol{\Lambda}})}, \tag{16}$$

where $\acute{\boldsymbol{\Psi}}_k$ is the eigenvectors of $\acute{\boldsymbol{\Sigma}}_k$ and $\acute{\psi}_{k,ji}$ is the ji element of $\acute{\boldsymbol{\Psi}}_k$.

Algorithm 1 Estimation of $\tilde{\tilde{\lambda}}_i$

1: Create nt sample vectors $\boldsymbol{X}_1, \ldots, \boldsymbol{X}_{nt}$ that follows normal distribution $\mathrm{N}(\boldsymbol{0}, \breve{\boldsymbol{\Lambda}})$ by using random numbers.
2: **for** $k = 1$ to t **do**
3: Estimate the sample covariance matrix $\acute{\boldsymbol{\Sigma}}_k$ from $\boldsymbol{X}_{n(k-1)+1}, \ldots, \boldsymbol{X}_{nk}$.
4: Obtain the eigenvalues $\acute{\boldsymbol{\Lambda}}_k$ and the eigenvectors $\acute{\boldsymbol{\Psi}}_k$ of $\acute{\boldsymbol{\Sigma}}_k$.
5: **end for**
6: Calculate $\left\{\tilde{\Psi}_{ji}\right\}^2$ in Eq. (16).
7: Calculate $\tilde{\tilde{\lambda}}_i$ in Eq. (10)

From the discussion above, we have an algorithm to estimate $\tilde{\tilde{\lambda}}_i$ in Eq. (10). The algorithm is shown in Algorithm 1. n is the available sample size for training and $\breve{\boldsymbol{\Lambda}}$ is the matrix which represents the true eigenvalues or the corrected eigenvalues by an existing method for correcting estimation errors of eigenvalues. $g(\acute{\boldsymbol{\Lambda}}_k) = \frac{1}{t}$ in this paper.

2.3 Procedure for Estimating the True Mahalanobis Distance from the Sample Eigenvectors

When n sample vectors are available for training, we can estimate the true Mahalanobis distance by the following procedure.

1. The sample eigenvalues ($\hat{\boldsymbol{\Lambda}}$) and the sample eigenvectors ($\hat{\boldsymbol{\Phi}}$) are calculated from available n sample vectors.
2. The estimation errors of the sample eigenvalues are corrected by an existing method, e.g. Sakai's method [1,2] or James-Stein estimator [3].
3. $\tilde{\tilde{\lambda}}_i$ is calculated by Algorithm 1.
4. Use $\tilde{\tilde{\lambda}}_i$ as eigenvalue with the sample eigenvectors for recognition.

3 Performance Evaluation of the Proposed Method

3.1 Estimated Mahalanobis Distance

The first experiment was performed to confirm that the proposed method has the ability to estimate the true Mahalanobis distance correctly from the sample eigenvectors. To show the ability, t_{ij}, e_{ij} and p_{ij} were calculated. t_{ij} is the true Mahalanobis distance between the jth input vector of class i and the true mean vector of the class to which the input vector belongs. e_{ij} was the calculated Mahalanobis distance from the true mean vectors, the true eigenvalues and the sample eigenvectors. p_{ij} was the one from the true mean vectors, the eigenvalues modified by the proposed method and the sample eigenvectors. Then, the average ratios of the Mahalanobis distances to the true ones $r_{\mathrm{e}} = \frac{1}{cs}\sum_{i=1}^{c}\sum_{j=1}^{s}\frac{e_{ij}}{t_{ij}}$

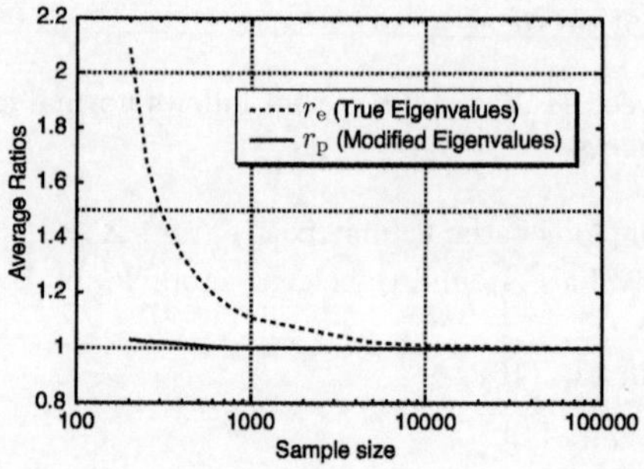

Fig. 1. The average ratios of the Mahalanobis distances

and $r_\mathrm{p} = \frac{1}{cs}\sum_{i=1}^{c}\sum_{j=1}^{s}\frac{p_{ij}}{t_{ij}}$ were compared, where c was the number of classes and s was the number of samples for testing. Here, $c = 10$ and $s = 1000$.

The experiments were performed on artificial feature vectors. The vectors of class i followed normal distribution $\mathrm{N}(\boldsymbol{\mu}_i, \boldsymbol{\Sigma}_i)$. $\boldsymbol{\mu}_i$ and $\boldsymbol{\Sigma}_i$ were calculated from feature vectors of actual character samples. The feature vectors of actual character samples were created as follows: The character images of digit samples in NIST Special Database 19 [6] were normalized nonlinearly [7] to fit in a 64×64 square, and 196-dimensional Directional Element Features [8] were extracted. The digit samples were sorted by class, and shuffled within the class in advance. $\boldsymbol{\mu}_i$ and $\boldsymbol{\Sigma}_i$ of class i were calculated from 36,000 feature vectors of the actual character samples of class i.

The parameter t of Algorithm 1 was 10,000. The average ratios of the Mahalanobis distances are shown in Fig. 1. r_e are far larger than one for small sample sizes. However, r_p are almost one for all sample sizes. This means the true Mahalanobis distance is precisely estimated by the proposed method. Moreover, it shows the justification of the approximation of ignoring the non-diagonal elements of $\hat{\boldsymbol{\Psi}}^\mathrm{T}\boldsymbol{\Lambda}\hat{\boldsymbol{\Psi}}$ described in Sect. 2.

3.2 Recognition Accuracy

The second experiment was carried out to confirm the effectiveness of the Mahalanobis distance estimated by the proposed method as a classifier. Character recognition experiments were performed by using three kinds of dictionaries "Control," "True eigenvalue" and "Proposed method." The dictionaries had common sample mean vectors, common sample eigenvectors and different eigenvalues. "Control" had the sample eigenvalues, "True eigenvalue" had the true eigenvalues, and "Proposed method" had the eigenvalues modified by the proposed method. The recognition rates of the dictionaries were compared.

The experiments were performed on the feature vectors of actual character samples and the artificial feature vectors described in Sect. 3.1. Since the true eigenvalues are not available for feature vectors of actual character samples,

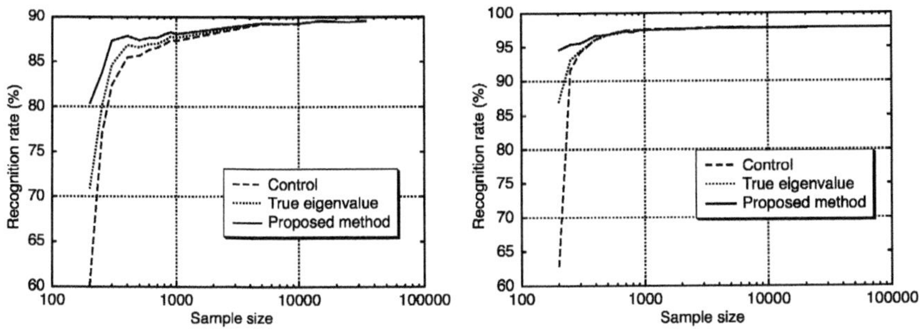

(a) (b)

Fig. 2. The recognition rates. (a)On the use of the feature vectors of actual character samples. (b)On the use of the artificial feature vectors

the eigenvalues corrected by Sakai's method [1,2] were used in place of the true eigenvalues. The parameter t of Algorithm 1 was 10,000. 1,000 samples were used for testing. The recognition rates of the experiment on the use of feature vectors of actual character samples and artificial feature vectors are shown in Fig.2(a) and Fig.2(b) respectively. Both of the figures show that the recognition rate of "Proposed method" is higher than that of "True eigenvalue." Therefore, the effectiveness of the proposed method is confirmed. Although the feature vectors of actual character samples does not follow normal distribution, the proposed method is effective. When the number of training samples is small, the difference between the recognition rates of "True eigenvalue" and "Proposed method" is large. The difference decreases as the sample size increases. This seems to depend on the amount of estimation errors of eigenvectors. The figures also show that the recognition rate of "True eigenvalue" is higher than that of "Control."

4 Conclusion

In this paper, we aimed to achieve high-performance pattern recognition without many training samples. We presented a method to estimate the true Mahalanobis distance from the sample eigenvectors. Recognition experiments show that the proposed method has the ability to estimate the true Mahalanobis distance even if the sample size is small. The eigenvalues modified by the proposed method achieve better recognition rate than the true eigenvalues. The proposed method is useful for the practical applications of pattern recognition since the proposed method is effective without any hyper-parameters, especially when the sample size is small.

References

1. Sakai, M., Yoneda, M., Hase, H.: A new robust quadratic discriminant function. In: Proc. ICPR. (1998) 99–102 498, 503, 505
2. Sakai, M., Yoneda, M., Hase, H., Maruyama, H., Naoe, M.: A quadratic discriminant function based on bias rectification of eigenvalues. Trans. IEICE **J82-D-II** (1999) 631–640 498, 503, 505
3. James, W., Stein, C.: Estimation with quadratic loss. In: Proc. 4th Berkeley Symp. on Math. Statist. and Prob. (1961) 361–379 499, 503
4. Iwamura, M., Omachi, S., Aso, H.: A modification of eigenvalues to compensate estimation errors of eigenvectors. In: Proc. ICPR. Volume 2., Barcelona, Spain (2000) 378–381 499
5. Hammersley, J. M., Handscomb, D. C.: 6. In: Monte Carlo Methods. Methuen, London (1964) 501
6. Grother, P. J.: NIST special database 19 handprinted forms and characters database. Technical report, National Institute of Standards and Technology (1995) 504
7. Yamada, H., Yamamoto, Saito, T.: A nonlinear normalization method for handprinted kanji character recognition — line density equalization —. Pattern Recognition **23** (1990) 1023–1029 504
8. Sun, N., Uchiyama, Y., Ichimura, H., Aso, H., Kimura, M.: Intelligent recognition of characters using associative matching technique. In: Proc. Pacific Rim Int'l Conf. Artificial Intelligence (PRICAI'90). (1990) 546–551 504
9. Muirhead, R. J.: Aspects of Multivariate Statistical Theory. John Wiley & Sons, Inc., New York (1982) 506, 507

A Probability Density Function of Estimation Errors of Eigenvectors

Let the d-dimensional column vectors $\boldsymbol{X}_1, \boldsymbol{X}_2, \ldots, \boldsymbol{X}_n$ be random sample vectors from normal distribution $\mathrm{N}(\mathbf{0}, \boldsymbol{\Sigma})$. The distribution of the random matrix $\boldsymbol{W} = \sum_{t=1}^{n} \boldsymbol{X}_t \boldsymbol{X}_t^{\mathrm{T}}$ is the Wishart distribution $\boldsymbol{W}_d(n, \boldsymbol{\Sigma})$. The probability density function of $\boldsymbol{W}$ is

$$\mathrm{P}(\boldsymbol{W}|\boldsymbol{\Sigma}) = \frac{v(d,n)|\boldsymbol{W}|^{\frac{1}{2}(n-d-1)}}{|\boldsymbol{\Sigma}|^{\frac{1}{2}n}} \exp\left(-\frac{1}{2}\operatorname{tr}\boldsymbol{\Sigma}^{-1}\boldsymbol{W}\right), \tag{17}$$

where $v(d,n) = \frac{1}{2^{\frac{1}{2}nd}\pi^{\frac{1}{4}d(d-1)}\prod_{i=1}^{d}\boldsymbol{\Gamma}\left[\frac{1}{2}(n+1-i)\right]}$.

Let $\acute{\boldsymbol{\Sigma}} = \frac{1}{n-1}\sum_{t=1}^{n}(\boldsymbol{X}_t - \acute{\boldsymbol{\mu}})(\boldsymbol{X}_t - \acute{\boldsymbol{\mu}})^{\mathrm{T}}$ be the sample covariance matrix and $\acute{\boldsymbol{\mu}} = \frac{1}{n}\sum_{t=1}^{n}\boldsymbol{X}_t$ be the sample mean vector. The distribution of $\acute{\boldsymbol{\Sigma}}$ is given as $\boldsymbol{W}_d(n-1, \frac{1}{n-1}\boldsymbol{\Sigma})$ and the probability density function is as follows [9]:

$$\mathrm{P}(\acute{\boldsymbol{\Sigma}}|\boldsymbol{\Sigma}) = (n-1)^{\frac{1}{2}(n-1)d} v(d, n-1) \frac{|\acute{\boldsymbol{\Sigma}}|^{\frac{1}{2}(n-d-2)} \exp\left(-\frac{n-1}{2}\operatorname{tr}\boldsymbol{\Sigma}^{-1}\acute{\boldsymbol{\Sigma}}\right)}{|\boldsymbol{\Sigma}|^{\frac{1}{2}(n-1)}}. \tag{18}$$

$\boldsymbol{\Sigma}$ and $\acute{\boldsymbol{\Sigma}}$ are decomposed into $\boldsymbol{\Phi}\boldsymbol{\Lambda}\boldsymbol{\Phi}^{\mathrm{T}}$ and $\acute{\boldsymbol{\Phi}}\acute{\boldsymbol{\Lambda}}\acute{\boldsymbol{\Phi}}^{\mathrm{T}} = \boldsymbol{\Phi}\acute{\boldsymbol{\Psi}}\acute{\boldsymbol{\Lambda}}\acute{\boldsymbol{\Psi}}^{\mathrm{T}}\boldsymbol{\Phi}^{\mathrm{T}}$. Eq. (18) is deformed as

$$\mathrm{P}(\boldsymbol{\Phi}\acute{\boldsymbol{\Psi}}\acute{\boldsymbol{\Lambda}}\acute{\boldsymbol{\Psi}}^{\mathrm{T}}\boldsymbol{\Phi}^{\mathrm{T}}|\boldsymbol{\Phi}\boldsymbol{\Lambda}\boldsymbol{\Phi}^{\mathrm{T}}) = (n-1)^{\frac{1}{2}(n-1)d}v(d,n-1) \cdot \frac{|\acute{\boldsymbol{\Lambda}}|^{\frac{1}{2}(n-d-2)} \exp\left(-\frac{n-1}{2}\operatorname{tr}\boldsymbol{\Lambda}^{-1}\acute{\boldsymbol{\Psi}}\acute{\boldsymbol{\Lambda}}\acute{\boldsymbol{\Psi}}^{\mathrm{T}}\right)}{|\boldsymbol{\Lambda}|^{\frac{1}{2}(n-1)}} . \quad (19)$$

Since the right side of Eq. (19) is independent of $\boldsymbol{\Phi}$, the left side of Eq. (19) is simplified as $\mathrm{P}(\acute{\boldsymbol{\Psi}}\acute{\boldsymbol{\Lambda}}\acute{\boldsymbol{\Psi}}^{\mathrm{T}}|\boldsymbol{\Lambda})$, and denoted as $\mathrm{P}(\acute{\boldsymbol{\Psi}}\acute{\boldsymbol{\Lambda}}\acute{\boldsymbol{\Psi}}^{\mathrm{T}})$ by omitting the condition.

Finally, the probability density function of estimation errors of eigenvectors are given as $\mathrm{P}(\acute{\boldsymbol{\Psi}}|\acute{\boldsymbol{\Lambda}}) = \frac{\mathrm{P}(\acute{\boldsymbol{\Psi}},\acute{\boldsymbol{\Lambda}})}{\mathrm{P}(\acute{\boldsymbol{\Lambda}})} = \frac{\mathrm{P}(\acute{\boldsymbol{\Psi}}\acute{\boldsymbol{\Lambda}}\acute{\boldsymbol{\Psi}}^{\mathrm{T}})}{\mathrm{P}(\acute{\boldsymbol{\Lambda}})J(\acute{\boldsymbol{\Psi}},\acute{\boldsymbol{\Lambda}})}$, where Jacobian $J(\acute{\boldsymbol{\Psi}},\acute{\boldsymbol{\Lambda}})$ is as follows [9]:

$$J(\acute{\boldsymbol{\Psi}},\acute{\boldsymbol{\Lambda}}) = \frac{\prod_{i=1}^{d}\boldsymbol{\Gamma}\left[\frac{1}{2}(d+1-i)\right]}{2^{d}\pi^{\frac{1}{4}d(d+1)}} \frac{1}{\prod_{i<j}^{d}(\acute{\lambda}_i - \acute{\lambda}_j)} . \quad (20)$$

Non-iterative Heteroscedastic Linear Dimension Reduction for Two-Class Data

From Fisher to Chernoff

Marco Loog[1] and Robert P. W. Duin[2]

[1] Image Sciences Institute, University Medical Center Utrecht
Utrecht, The Netherlands
marco@isi.uu.nl

[2] Pattern Recognition Group, Department of Applied Physics
Delft University of Technology, Delft, The Netherlands

Abstract. Linear discriminant analysis (LDA) is a traditional solution to the linear dimension reduction (LDR) problem, which is based on the maximization of the between-class scatter over the within-class scatter. This solution is incapable of dealing with heteroscedastic data in a proper way, because of the implicit assumption that the covariance matrices for all the classes are equal. Hence, discriminatory information in the difference between the covariance matrices is not used and, as a consequence, we can only reduce the data to a single dimension in the two-class case. We propose a fast non-iterative eigenvector-based LDR technique for heteroscedastic two-class data, which generalizes, and improves upon LDA by dealing with the aforementioned problem. For this purpose, we use the concept of directed distance matrices, which generalizes the between-class covariance matrix such that it captures the differences in (co)variances.

1 Introduction

Probably the most well-known approach to supervised linear dimension reduction (LDR), or feature extraction, is *linear discriminant analysis* (LDA). This traditional and simple technique was developed by Fisher [6] for the two-class case, and extended by Rao [16] to handle the multi-class case. In LDA, a $d \times n$ transformation matrix that maximizes the *Fisher criterion* is determined. This criterion gives, for a certain linear transformation, a measure of the between-class scatter over the within-class scatter (cf. [7,9]). An attractive feature of LDA is the fast and easy way to determine this optimal linear transformation, merely requiring simple matrix arithmetics like addition, multiplication, and eigenvalue decomposition. A limitation of LDA is its incapability of dealing with heteroscedastic data, i.e., data in which classes do not have equal covariance matrices.

This paper focusses on the generalization of the Fisher criterion to the heteroscedastic case in order to come to heteroscedastic linear dimension reduction

T. Caelli et al. (Eds.): SSPR&SPR 2002, LNCS 2396, pp. 508–517, 2002.

© Springer-Verlag Berlin Heidelberg 2002

(HLDR). We restrict our attention to two-class data, e.g. where pattern classes can typically be divided into good or bad, 0 or 1, benign or malignant, on or off, foreground or background, yin or yang, etc. With this kind of data the limitation of LDA is very obvious: A reduction to only a single dimension is possible (cf. [7]).

Our generalization takes into account the discriminatory information that is present in the difference of the covariance matrices. This is done by means of directed distance matrices (DDMs) [12], which are generalizations of the between-class covariance matrix. This between-class covariance matrix, as used in LDA, merely takes into account the discriminatory information that is present in the differences between class means and can be associated with the Euclidean distance.

The specific heteroscedastic generalization of the Fisher criterion, we study more closely in Section 2, is based on the *Chernoff distance* [2,3]. This measure of affinity of two densities considers mean differences as well as covariance differences—as opposed to the Euclidean distance—and can be used to extend LDA, while retaining the attractive feature of fast and easily determining a dimension reducing transformation. Furthermore, we are able to reduce the data to any dimension d smaller than n and not only to a single dimension. We call our HLDR criterion the *Chernoff criterion.*

Several alternative approaches to HLDR are known, of which we mention the following ones. See also [14].

In the two-class case, under the assumptions that both classes are normally distributed and that one wants a reduction to one dimension, Kazakos [10] reduces the LDR problem to a one-dimensional search problem. Finding the optimal solution for this search problem, is equivalent to finding the optimal linear feature. The work of Kazakos is closely related to [1].

Three other HLDR approaches for two-class problems, that generalize upon Fisher, were proposed in [13], [4], and [5], of which the latter is also applicable in the multi-class case. [13] uses scatter measures different to the one used in LDA. In [4] and [5] the criterions to be optimized utilize the Bhattacharyya distance (cf. [7]) and the Kullback divergence, respectively. The drawback of these criteria is that the maximization of them needs complex or iterative optimization procedures.

Another iterative multi-class HLDR procedure, which is based on a maximum likelihood formulation of LDA, is studied in [11]. Here LDA is generalized by dropping the assumption that all classes have equal within-class covariance matrices, and maximizing the likelihood for this model.

A fast HLDR method based on a singular value decomposition (svd) was developed in [19] by Tubbs et al. We discuss this method in more detail in Section 3, where we also compare our non-iterative method to theirs. The comparison is done on three artificial and seven real-world data sets.

Section 4 presents the discussion and the conclusions.

2 From Fisher to Chernoff

2.1 The Fisher Criterion

LDR is concerned with the search for a linear transformation that reduces the dimension of a given n-dimensional statistical model to d ($d < n$) dimensions, while maximally preserving the discriminatory information for the several classes within the model. Due to the complexity of utilizing the Bayes error as the criterion to optimize, one resorts to suboptimal criteria. LDA is such a suboptimal approach. It determines a linear mapping $\mathbf{L}$, a $d \times n$-matrix, that maximizes the so-called *Fisher criterion* J_{F} [7,9,12,13,16]:

$$J_{\mathrm{F}}(\mathbf{A}) = \mathrm{tr}((\mathbf{A}\mathbf{S}_W\mathbf{A}^t)^{-1}(\mathbf{A}\mathbf{S}_B\mathbf{A}^t)) \,. \tag{1}$$

Here $\mathbf{S}_B := \sum_{i=1}^{K} p_i(\mathbf{m}_i - \bar{\mathbf{m}})(\mathbf{m}_i - \bar{\mathbf{m}})^T$ and $\mathbf{S}_W := \sum_{i=1}^{K} p_i\mathbf{S}_i$ are the between-class and the average within-class scatter matrix, respectively; K is the number of classes, $\mathbf{m}_i$ is the mean vector of class i, p_i is its a priori probability, and the overall mean $\bar{\mathbf{m}}$ equals $\sum_{i=1}^{K} p_i\mathbf{m}_i$. Furthermore, $\mathbf{S}_i$ is the within-class covariance matrix of class i.

From Equation (1) we see that LDA maximizes the ratio of between-class scatter to average within-class scatter in the lower-dimensional space.

Our focus is on the two-class case, in which case we have $\mathbf{S}_B = (\mathbf{m}_1 - \mathbf{m}_2)(\mathbf{m}_1 - \mathbf{m}_2)^t$ [7,12], $\mathbf{S}_W = p_1\mathbf{S}_1 + p_2\mathbf{S}_2$, and $p_1 = 1 - p_2$. Optimizing (1) comes down to determining an eigenvalue decomposition of $\mathbf{S}_W^{-1}\mathbf{S}_B$, and taking the rows of $\mathbf{L}$ to equal the d eigenvectors corresponding to the d largest eigenvalues [7].

Note that the rank of $\mathbf{S}_B$ is 1 in the two-class case, assuming unequal class means, and so we can only reduce the dimension to 1: According to the Fisher criterion there is no discriminatory information in the features, apart from this single dimension.

2.2 Directed Distance Matrices

We now turn to the concept of *directed distance matrices* (DDMs) [12], by which means, we are able to generalize LDA in a proper way.

Assume that the data is linearly transformed such that the within-class covariance matrix $\mathbf{S}_W$ equals the identity matrix, then $J_{\mathrm{F}}(\mathbf{A})$ equals $\mathrm{tr}((\mathbf{A}\mathbf{A}^t)^{-1}$ $(\mathbf{A}(\mathbf{m}_1 - \mathbf{m}_2)(\mathbf{m}_1 - \mathbf{m}_2)^t\mathbf{A}^t))$, which is maximized by taking the eigenvector $\mathbf{v}$ associated with the largest eigenvalue λ of the matrix $(\mathbf{m}_1 - \mathbf{m}_2)(\mathbf{m}_1 - \mathbf{m}_2)^t$. As pointed out earlier, this matrix has only one nonzero eigenvalue and we can show that $\mathbf{v} = \mathbf{m}_1 - \mathbf{m}_2$ and $\lambda = \mathrm{tr}((\mathbf{m}_1 - \mathbf{m}_2)(\mathbf{m}_1 - \mathbf{m}_2)^t) = (\mathbf{m}_1 - \mathbf{m}_2)^t(\mathbf{m}_1 - \mathbf{m}_2)$. The latter equals the squared Euclidean distance between the two *class means*, which we denote by ∂_E.

The matrix $(\mathbf{m}_1 - \mathbf{m}_2)(\mathbf{m}_1 - \mathbf{m}_2)^t$, which we call $\mathbf{S}_E$ from now on, not only gives us the distance between two distributions, but it also provides the direction, by means of the eigenvectors, in which this specific distance can be found. As a matter of a fact, if both classes are normally distributed and have equal covariance matrix, there is only distance between them in the direction $\mathbf{v}$ and this distance equals λ. All other eigenvectors have eigenvalue 0, indicating that there is no distance between the two classes in these directions. Indeed, reducing the dimension using one of these latter eigenvectors results in a complete overlap of the classes: There is no discriminatory information in these directions, the distance equals 0.

The idea behind DDMs is to give a generalization of $\mathbf{S}_E$. If there is discriminatory information present because of the heteroscedasticity of the data, then this should become apparent in the DDM. This extra distance because of the heteroscedasticity, is, in general, in different directions then the vector $\mathbf{v}$, which separates the means, and so DDMs have more than one nonzero eigenvalue.

The specific DDM we propose is based on the Chernoff distance ∂_C. For two normally distributed densities, it is defined as[1] [2,3]

$$\begin{aligned}\partial_C =&(\mathbf{m}_1-\mathbf{m}_2)^t(\alpha\mathbf{S}_1+(1-\alpha)\mathbf{S}_2)^{-1}(\mathbf{m}_1-\mathbf{m}_2)\\&+\frac{1}{\alpha(1-\alpha)}\log\frac{|(\alpha\mathbf{S}_1+(1-\alpha)\mathbf{S}_2)|}{|\mathbf{S}_1|^\alpha|\mathbf{S}_2|^{1-\alpha}}.\end{aligned} \tag{2}$$

Like ∂_E, we can obtain ∂_C as the trace of a positive semi-definite matrix $\mathbf{S}_C$. Simple matrix manipulation [18] shows that this matrix equals[2] (cf. [12])

$$\begin{aligned}\mathbf{S}_C :=&\mathbf{S}^{-\frac{1}{2}}(\mathbf{m}_1-\mathbf{m}_2)(\mathbf{m}_1-\mathbf{m}_2)^t\mathbf{S}^{-\frac{1}{2}}\\&+\frac{1}{\alpha(1-\alpha)}(\log\mathbf{S}-\alpha\log\mathbf{S}_1-(1-\alpha)\log\mathbf{S}_2),\end{aligned} \tag{3}$$

where $\mathbf{S} := \alpha\mathbf{S}_1 + (1-\alpha)\mathbf{S}_2$. Now, before we get to our HLDR criterion, we make the following remarks. (Still assume that $\mathbf{S}_W$ equals the identity matrix.) We want our criterion to be a generalization of Fisher's, so if the data is homoscedastic, i.e., $\mathbf{S}_1 = \mathbf{S}_2$, we want $\mathbf{S}_C$ to equal $\mathbf{S}_E$. This suggests to set α equal to p_1, from which it directly follows that $1-\alpha$ equals p_2. In this case $\mathbf{S}_C = \mathbf{S}_E$, and we obtain the same dimension reducing linear transform via an eigenvalue decomposition on either.

Now assume that $\mathbf{S}_1$ and $\mathbf{S}_2$ are diagonal—$\mathrm{diag}(a_1,\ldots,a_n)$ and $\mathrm{diag}(b_1,\ldots,b_n)$, respectively—but not necessarily equal. Furthermore, let $\mathbf{m}_1 = \mathbf{m}_2$. Now because $\alpha = p_1$, and hence $\alpha\mathbf{S}_1 + (1-\alpha)\mathbf{S}_2 = \mathbf{I}$, we have

$$\mathbf{S}_C = \frac{1}{p_1p_2}\mathrm{diag}(\log\frac{1}{a_1^{p_1}b_1^{p_2}},\ldots,\log\frac{1}{a_n^{p_1}b_n^{p_2}}). \tag{4}$$

---

[1] Often, the Chernoff distance is defined as $\frac{\alpha(1-\alpha)}{2}\partial_C$, this constant factor, however, is of no essential influence on the rest of our discussion.

[2] We define the function f, e.g. some power or the logarithm, of a symmetric positive definite matrix $\mathbf{A}$, by means of its eigenvalue decomposition $\mathbf{RVR}^{-1}$, with eigenvalue matrix $\mathbf{V} = \mathrm{diag}(v_1,\ldots,v_n)$. We let $f(\mathbf{A})$ equal $\mathbf{R}\,\mathrm{diag}(f(v_1),\ldots,f(v_n))\,\mathbf{R}^{-1} = \mathbf{R}(f(\mathbf{V}))\mathbf{R}^{-1}$. Although generally $\mathbf{A}$ is nonsingular, determining $f(\mathbf{A})$ might cause problems, because the matrix is close to singular. Most of the times, alleviation of this problem is possible by using the svd instead of an eigenvalue decomposition, or by properly regularizing $\mathbf{A}$.

On the diagonal of $\mathbf{S}_C$ are the Chernoff distances of the two densities if the the dimension is reduced to one in the associated direction, e.g., linearly transforming the data by the n-vector $(0,\ldots,0,1,0,\ldots,0)$, where only the dth entry is 1 and all the other equal 0, would give us a Chernoff distance of $\frac{1}{p_1 p_2}\log\frac{1}{a_d^{p_1} b_d^{p_2}}$ in the one-dimensional space. Hence, determining a LDR transformation by an eigenvalue decomposition of the DDM $\mathbf{S}_C$, means that we determine a transform which preserves as much of the Chernoff distance in the lower dimensional space as possible.

In the two cases above, where in fact, we considered $\mathbf{S}_1 = \mathbf{S}_2$ and $\mathbf{m}_1 = \mathbf{m}_2$ respectively, we argued that our criterion gives eligible results. We also expect reasonable results if we do not necessarily have equality of means or covariance matrices, because in this case we obtain a solution that is approximately optimal with respect to the Chernoff distance. In conclusion: The DDM $\mathbf{S}_C$, captures differences in covariance matrices in a certain way and indeed generalizes the homoscedastic DDM $\mathbf{S}_E$.

2.3 Heteroscedasticization of Fisher: The *Chernoff Criterion*

If $\mathbf{S}_W = \mathbf{I}$, $J_\mathrm{F}(\mathbf{A})$ equals $\mathrm{tr}((\mathbf{A}\mathbf{A}^t)^{-1}(\mathbf{A}\mathbf{S}_E\mathbf{A}^t))$. Hence in this case, regarding the discussion in the foregoing subsection, we simply substitute $\mathbf{S}_C$ for $\mathbf{S}_E$, to obtain a heteroscedastic generalization of LDA, because optimizing this criterion is similar to optimizing J_F: Determine an eigenvalue decomposition of $\mathbf{S}_C$, and take the rows of the transform $\mathbf{L}$ to equal the d eigenvectors corresponding to the d largest eigenvalues.

In case $\mathbf{S}_W \neq \mathbf{I}$, note that we can first transform our data by $\mathbf{S}_W^{-\frac{1}{2}}$, so we *do* have $\mathbf{S}_W = \mathbf{I}$. In this space we then determine our criterion, which for LDA equals $\mathrm{tr}((\mathbf{A}\mathbf{A}^t)^{-1}(\mathbf{A}\mathbf{S}_W^{-\frac{1}{2}}\mathbf{S}_B\mathbf{S}_W^{-\frac{1}{2}}\mathbf{A}^t))$, and then transform it back to our original space using $\mathbf{S}_W^{\frac{1}{2}}$, giving the criterion $\mathrm{tr}((\mathbf{A}\mathbf{S}_W^{\frac{1}{2}}\mathbf{S}_W^{\frac{1}{2}}\mathbf{A}^t)^{-1}(\mathbf{A}\mathbf{S}_B\mathbf{A}^t))$. Hence for LDA, this procedure gives us just Criterion (1), as if it was determined directly in the original space. For our heteroscedastic *Chernoff criterion* J_C the procedure above gives the following:

$$\begin{aligned} J_\mathrm{C}(\mathbf{A}) :=& \mathrm{tr}((\mathbf{A}\mathbf{S}_W\mathbf{A}^t)^{-1}(\mathbf{A}(\mathbf{m}_1-\mathbf{m}_2)(\mathbf{m}_1-\mathbf{m}_2)^t\mathbf{A}^t \\ &- \mathbf{A}\mathbf{S}_W^{\frac{1}{2}}\frac{p_1\log(\mathbf{S}_W^{-\frac{1}{2}}\mathbf{S}_1\mathbf{S}_W^{-\frac{1}{2}}) + p_2\log(\mathbf{S}_W^{-\frac{1}{2}}\mathbf{S}_2\mathbf{S}_W^{-\frac{1}{2}})}{p_1p_2}\mathbf{S}_W^{\frac{1}{2}}\mathbf{A}^t)) . \end{aligned} \tag{5}$$

This is maximized by determining an eigenvalue decomposition of

$$\mathbf{S}_W^{-1}(\mathbf{S}_B - \mathbf{S}_W^{\frac{1}{2}}\frac{p_1\log(\mathbf{S}_W^{-\frac{1}{2}}\mathbf{S}_1\mathbf{S}_W^{-\frac{1}{2}}) + p_2\log(\mathbf{S}_W^{-\frac{1}{2}}\mathbf{S}_2\mathbf{S}_W^{-\frac{1}{2}})}{p_1p_2}\mathbf{S}_W^{\frac{1}{2}}) , \tag{6}$$

and taking the rows of the transform $\mathbf{L}$ to equal the d eigenvectors corresponding to the d largest eigenvalues.

3 Experimental Results: Comparing Chernoff to Fisher and Svd

This section compares the performance of the HLDR transformations obtained by means of the Chernoff criterion with transformations obtained by the traditional Fisher criterion, and by the svd method as discussed in, e.g., [19]. The latter method determines a dimension reducing transform—in the two-class case—by constructing an $n \times (n+1)$-matrix $\mathbf{T}$ that equals $(\mathbf{m}_2 - \mathbf{m}_1, \mathbf{S}_2 - \mathbf{S}_1)$, then performing an svd on $\mathbf{T}\mathbf{T}^t = \mathbf{U}\mathbf{S}\mathbf{V}^t$, and finally choosing the row vectors from $\mathbf{U}$ associated with the largest d singular values as the HLDR transformation.

Tests were performed on three artificial [7] (cf. [12])—labelled (a) to (c)—and seven real-world data sets [8,15]—labelled (d) to (j). To be able to see what discriminatory information is retained in using a HLDR, classification is done with a quadratic classifier assuming the underlying distributions to be normal. Results obtained with the svd-based approach, and the Chernoff criterion are presented in the Figures 1(a) to 1(j), and indicated by gray and black lines, respectively. Figures 1(a) to (j) are associated to data sets (a) to (j). The dimension of the subspace is plotted horizontally and the classification error vertically. Results of reduction to a single dimension, mainly for comparison with LDA, are in Table 1.

In presenting the results on the real-world data sets, we restricted ourselves to discussing the main results, and to the most interesting observations. The p-values stated in this part are obtained by comparing the data via a signed rank test [17].

3.1 Fukunaga's Heteroscedastic Two-Class Data and Two Variations

Fukunaga [7] describes a heteroscedastic model consisting of two classes in eight dimensions. The classes are normally distributed with $\mathbf{m}_1 = (0, \ldots, 0)^t$, $\mathbf{S}_1 = \mathbf{I}$, and

$$\mathbf{m}_2 = (3.86, 3.10, 0.84, 0.84, 1.64, 1.08, 0.26, 0.01)^t \,, \tag{7}$$

$$\mathbf{S}_2 = \mathrm{diag}(8.41, 12.06, 0.12, 0.22, 1.49, 1.77, 0.35, 2.73) \,. \tag{8}$$

Furthermore, $p_1 = p_2 = \frac{1}{2}$.

The first test (a) on artificial data uses these parameters. Two variants are also considers. In the first variant (b), the two means are taken closer to each other to elucidate the performance of the Chernoff criterion, when most of the discriminatory information is in the difference in covariances. For this variant we take the mean of the second class to equal $\frac{1}{10}\mathbf{m}_2$ (cf. [12]). The second variant (c) is a variation on the first, where we additionally set $p_1 = \frac{1}{4}$ and $p_2 = \frac{3}{4}$. This is to elucidate the effect of a difference in class priors, something the svd approach does not account for.

Tests are carried out using Monte Carlo simulation in which we take a total of 1,000,000 instances from the two classes, proportional to the values p_1 and p_2, and determine the error by classifying these instances.

The results from Figures 1(a)–(c) are clear: For a reduction to one dimension, the LDR by means of the Chernoff criterion is as good as, or unmistakably better than LDR by the Fisher criterion or the svd approach. Furthermore, when reducing the data to several dimensions, our approach is also preferable to the svd approach, which merely outperforms our approach in reducing the dimension to 2 in experiment (b).

Table 1. Information w.r.t. the 10 data sets including (mean) classification errors for comparison of the three considered LDR techniques for reduction to one dimension (last three columns). Best results are in boldface

data set		n	size	N	Fisher	Chernoff	svd
Fukunaga's two-class data	(a)	8	-	-	**0.054**	**0.054**	0.140
Variation one	(b)	8	-	-	0.415	**0.231**	**0.231**
Variation two	(c)	8	-	-	0.245	**0.159**	0.240
Wisconsin breast cancer	(d)	9	682	350	0.028	**0.027**	0.031
Wisconsin diagnostic breast cancer	(e)	30	569	500	0.035	**0.029**	0.086
Bupa liver disorders	(f)	6	345	200	**0.364**	0.407	0.466
Cleveland heart-disease	(g)	13	297	200	0.174	**0.172**	0.463
Pima indian diabetes	(h)	8	768	576	0.230	**0.229**	0.342
Musk "Clean2" database	(i)	166	6598	6268	**0.056**	0.061	0.152
Lung/non-lung classification	(j)	11	72000	36000	0.223	0.225	**0.217**

3.2 Real-World Data

Six tests in this subsection are on data sets from the UCI Repository of machine learning databases [15], a seventh test is on the chest radiograph database used in [8]. For a description of the first six data sets refer to [15]. The seventh database consists of 115 chest radiograph images and the classification of their pixels in lung or non-lung. For our purpose, we took 20 images and sub-sampled them from 256×256 to 64×64 pixels. To come to a lung/non-lung classification of a pixel, we used its gray value, its eight neighboring gray values, and its x- and y-coordinate as features, which finally gives us 72000 instances[3] in an 11-dimensional feature space.

The seven tests are carried out by randomly drawing N instances from the data for training and using the remaining instances for testing. (If provided, the value N is taken from the UCI Repository.) This procedure is repeated 100 times and the mean classification error is given. Most of the time, the 100 repetitions give us enough measurements to reliably decide whether one approach consistently outperform the other. This decision is based on the signed rank test [17] for which the p-values are provided.

[3] There are only 72000 instances, because in building the feature vector, we excluded pixels that were too close to the border of the image resulting in $60 \times 60 \times 20$ instances.

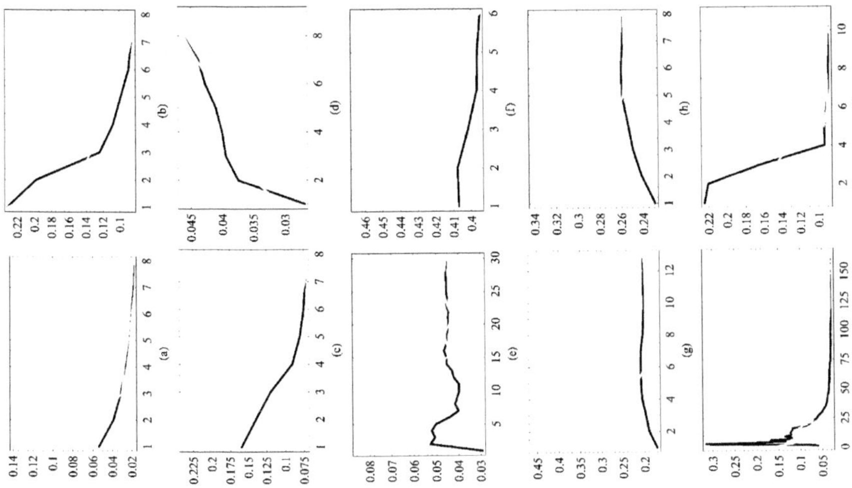

Fig. 1. Plots of feature dimension (vertically) versus classification error (horizontally) for comparison of HLDR via the Chernoff criterion and via the svd-based approach. The grey lines give the results obtained by svd, the black lines provide results obtained by using the Chernoff criterion

Considering Tables 1 and 2, and Figures 1(d) to 1(j), we can generally conclude that the Chernoff criterion improves upon the Fisher criterion. Even though the Chernoff criterion clearly needs more than one dimension, about 25, to outperform LDA in case of data set (i), it dramatically improves upon LDA in case of a dimension greater than, say, 50, with its best result at $d = 106$. Fisher is clearly better for data set (f), however for all other data sets Chernoff is, in general, the one to be preferred although its improvement w.r.t. Fisher for data sets (d) and (h) are considered insignificant.

Concerning the comparison of Chernoff with the svd approach, we can be brief: In case of reducing data set (j) to one or two dimensions, the svd approach is clearly preferable, however, in all other cases the use of the Chernoff criterion is preferable to the svd approach.

See also the captions of Tables 1 and 2.

4 Discussion and Conclusions

We proposed a new heteroscedastic linear dimension reduction (HLDR) criterion for two-class data, which generalizes the well-known Fisher criterion used in LDA. After noting that the Fisher criterion can be related to the Euclidean distance between class means, we used the concept of directed distance matrices (DDMs) to replace the matrix that incorporates the Euclidean distance by one

Table 2. Results w.r.t. the 7 real-world data sets—(d) to (j). Included are the best results over all dimensions ($< n$) for the three approaches (for LDA this, of course, equals $d = 1$). The dimension for which this result is obtained is denoted by d. For comparison of our approach (Chernoff) to both other approaches, the p-values are provided, which are indicated between the compared approaches. Best overall results are in boldface

data set	Fisher	p-value	Chernoff	d	p-value	svd	d
(d)	0.028	*0.070*	**0.027**	1	*0.000*	0.031	1
(e)	0.035	*0.006*	**0.029**	1	*0.000*	0.043	16
(f)	**0.364**	*0.000*	0.396	5	*0.000*	0.416	5
(g)	0.174	*0.033*	**0.172**	1	*0.000*	0.195	7
(h)	0.230	*0.721*	**0.229**	1	*0.000*	0.256	6
(i)	0.056	*0.000*	**0.030**	106	*0.035*	0.031	165
(j)	0.223	*0.000*	**0.089**	9	*0.000*	0.090	10

incorporating the *Chernoff distance.* This distance takes into account the difference in the covariance matrices of both groups, which, by means of a DDM, can be used to find an LDR transformation that takes such differences into account. In addition, it enables us to reduce the dimension of the two-class data to more than a single one.

An other important property of our *Chernoff criterion* is that it is computed in a simple and efficient way, merely using standard matrix arithmetics and not using complex or iterative procedures. Hence its computation is almost as easy as determining an LDR transform using LDA. Furthermore, it should be noted that, although it was used in the derivation of our criterion, it is not necessary that both classes are normally distributed. The Chernoff criterion only uses the first and second order central moments of the class distribution in a way that is plausible for dimension reduction, whether the data is normally distributed or not.

In ten experiments, we compared the performance of the Chernoff criterion to that of LDA and an other simple and efficient approach based on the svd. Three of these experiments were on artificial data and seven on real-world data. The experiments clearly showed the improvement that is possible when utilizing the Chernoff criterion instead of the Fisher criterion or the svd-based approach and we can generally conclude that our method is clearly preferable to the others.

Finally, it is of course interesting to look at the possibility of extending our criterion to the multi-class case. [7] and [12] offer ideas for doing so. They suggest a certain averaged criterion that takes into account all multi-class discriminatory information at once. Future investigations will be on these kinds of extensions of the Chernoff criterion.

References

1. T. W. Anderson and R. R. Bahadur. Classification into two multivariate normal distributions with different covariance matrices. *Annals of Mathematical Statistics*, 33:420–431, 1962. 509
2. C. H. Chen. On information and distance measures, error bounds, and feature selection. *The information scientist*, 10:159–173, 1979. 509, 511
3. J. K. Chung, P. L. Kannappan, C. T. Ng, and P. K. Sahoo. Measures of distance between probability distributions. *Journal of mathematical analysis and applications*, 138:280–292, 1989. 509, 511
4. H. P. Decell and S. K. Marani. Feature combinations and the Bhattacharyya criterion. *Communications in Statistics. Part A. Theory and Methods*, 5:1143–1152, 1976. 509
5. H. P. Decell and S. M. Mayekar. Feature combinations and the divergence criterion. *Computers and Mathematics with Applications*, 3:71–76, 1977. 509
6. R. A. Fisher. The use of multiple measurements in taxonomic problems. *Annals of Eugenics*, 7:179–188, 1936. 508
7. K. Fukunaga. *Introduction to Statistical Pattern Recognition*. Academic Press, New York, 1990. 508, 509, 510, 513, 516
8. B. van Ginneken and B. M. ter Haar Romeny. Automatic segmentation of lung fields in chest radiographs. *Medical Physics*, 27(10):2445–2455, 2000. 513, 514
9. A. K. Jain, R. P. W. Duin, and J. Mao. Statistical pattern recognition: A review. *IEEE Transactions on Pattern Analysis and Machine Intelligence*, 22(1):4–37, 2000. 508, 510
10. D. Kazakos. On the optimal linear feature. *IEEE Transactions on Information Theory*, 24:651–652, 1978. 509
11. N. Kumar and A. G. Andreou. Generalization of linear discriminant analysis in a maximum likelihood framework. In *Proceedings of the Joint Meeting of the American Statistical Association*, 1996. 509
12. M. Loog. *Approximate Pairwise Accuracy Criteria for Multiclass Linear Dimension Reduction: Generalisations of the Fisher Criterion*. Number 44 in WBBM Report Series. Delft University Press, Delft, 1999. 509, 510, 511, 513, 516
13. W. Malina. On an extended Fisher criterion for feature selection. *IEEE Transactions on Pattern Analysis and Machine Intelligence*, 3:611–614, 1981. 509, 510
14. G. J. McLachlan. *Discriminant Analysis and Statistical Pattern Recognition*. John Wiley & Sons, New York, 1992. 509
15. P. M. Murphy and D. W. Aha. UCI Repository of machine learning databases. `[http://www.ics.uci.edu/~mlearn/mlrepository.html]`. 513, 514
16. C. R. Rao. The utilization of multiple measurements in problems of biological classification. *Journal of the Royal Statistical Society. Series B*, 10:159–203, 1948. 508, 510
17. J. A. Rice. *Mathematical Statistics and Data Analysis*. Duxbury Press, Belmont, second edition, 1995. 513, 514
18. G. Strang. *Linear algebra and its applications*. Harcourt Brace Jovanovich, third edition, 1988. 511
19. J. D. Tubbs, W. A. Coberly, and D. M. Young. Linear dimension reduction and Bayes classification. *Pattern Recognition*, 15:167–172, 1982. 509, 513

Some Experiments in Supervised Pattern Recognition with Incomplete Training Samples

Ricardo Barandela[1], Francesc J. Ferri[2], and Tania Nájera[1]

[1] Lab for Pattern Recognition, Instituto Tecnológico de Toluca, México
C.P. 52140, Metepec, Estado de México
rbarandela@hotmail.com
[2] Dept. d'Informàtica, Universitat de València
46100 Burjassot, València, Spain

Abstract. This paper presents some ideas about automatic procedures to implement a system with the capability of detecting patterns arising from classes not represented in the training sample. The procedure aims at incorporating automatically to the training sample the necessary information about the new class for correctly recognizing patterns from this class in future classification tasks. The Nearest Neighbor rule is employed as the central classifier and several techniques are added to cope with the peril of incorporating noisy data to the training sample. Experimental results with real data confirm the benefits of the proposed procedure.

1 Introduction

Supervised pattern recognition methods are based on the information supplied by a training sample (TS). It is usually assumed that this TS satisfies two important conditions:

1. Each member is a pattern that has been identified as representative of one of the classes and is actually a member of that class.
2. The *c* classes represented in the training sample span the entire set of the classes that need to be considered in the problem at hand.

Experience in practical applications has shown that these assumptions do not always hold and that violation of one or both of them may seriously degrade classification accuracy of the recognition system. Particularly sensitive to these deficiencies are nonparametric classifiers whose learning is not based upon any assumption about probability density functions.

In several real domains, class identification of the training patterns is a difficult and very costly task. Also, the fact that one (or some) of the classes is not known a priori leads to situations lying between supervised and unsupervised methods that have already been coined as partially exposed environments [2].

T. Caelli et al. (Eds.): SSPR&SPR 2002, LNCS 2396, pp. 518-527, 2002.
© Springer-Verlag Berlin Heidelberg 2002

In this paper, the situation in which one of the classes is missing from the original TS is taken into account. This situation is common in some fields as Remote Sensing in which data from one zone is used to classify other zones (or the same zone but at a different time) in which new classes may appear. Another similar situation arises in Banknote classification when trying to detect forgery. Although patterns from missing classes can always be treated as outliers (with regard to the available information) it is quite clear that this is not the best option in real situations where a classification system is continuously running and receiving new data that has potentially the ability of giving us information about possible new classes.

The problem statement is challenging in itself and has a straightforward real application. In this paper, we try to contribute with a preliminary solution based on prototype selection techniques for the nearest neighbor rule. To this end, we must first detect the situation in which a running system misclassifies patterns that may belong in fact to a class about which no information is available. Second, we must put the information collected in this first classification stage in such a way that the system can update the TS to include particular samples belonging to the new class. Both of this are challenging and difficult problems to be solved in general. On the other hand, there are several different ways for dealing with these situations. In this paper, some of these will be used and put together to deal with a simplified version of the problem.

The paper is organized as follows. In section 2, basic ideas about NN rules are introduced. Several proposals to the problem of extracting information about missing classes are introduced in sections 3 and 4. Section 5 explains the experiments carried out and the paper ends with the conclusions in section 6.

2 The NN Rule and Some Related Techniques

The Nearest Neighbor (NN) rule is one of the oldest and better-known algorithms for nonparametric classification. The entire TS is stored in the computer memory. To classify a new instance, its distance is computed to each one of the stored training cases. The new instance is then assigned to the class represented by its nearest neighboring training pattern. The NN rule is very popular because of: a) conceptual simplicity, b) easy implementation, c) known error rate bounds and d) potentiality to compete favorably with other classification methods in real data applications.

The performance of the NN rule can be improved in practice by using prototype selection techniques. Wilson [3] proposed a procedure (Editing) that consists of applying the k-NN classifier to estimate the class label of all prototypes in the TS and discard those instances whose class label does not agree with the class associated with the largest number of the k neighbors. Most prototype selection techniques share the general idea of discarding prototypes that significantly deviate from the general tendency in their class and also prototypes lying in the overlapping zones between classes. In this way, the processed TS is able to classify new patterns in a close-to-optimal way. Apart from its optimal behavior, editing prototypes should be almost compulsory in situations in which an optimal (or at least fair) labeling of the training is impossible because it relies in a very difficult or costly (usually manual) task. These situations have been previously referred to as imperfectly supervised [1]. Many theoretical and empirical results have corroborated the convenience of selecting training patterns in this way [4-7].

The idea of discarding can be extended to consider changing the class labels apart from removing patterns (Generalized Edition, [9]). In this way, the obtained TS may still perform well (or at least better) in the small sample size situation, which is one of the drawbacks of editing in practice. In short, Generalized Edition looks for modifications of the training sample structure through changes of the labels (re-identification) of some of the patterns and elimination (edition) of some others. The decisions about whether re-label or discard depends upon a qualified majority voting among the corresponding k-neighborhood (more than k' out of k) [9].

On the other hand, pattern re-labeling may introduce new problems because not all discarded prototypes have the same probability of belonging to one of the other classes that may lead to instabilities and unbalanced situations between neighboring classes. To better cope with imperfectly supervised environments, Barandela and Gasca [8] have proposed a combined methodology (Decontamination) in which both class re-labeling and prototype deletion are used in a more intensive way. Decontamination consists of several applications of the Generalized Edition scheme followed by Wilson's editing [3] also reiterated. Decontamination has proved profitable by correcting the training data and cleaning the errors both in the input features and in the class labels [8].

3 Proposals to Detect Patterns Arising from Missing Classes

The usual way of dealing with situations in which patterns may exhibit tendencies significantly different from the ones a priori represented in the TS is through outlier detection. In this work, it is intended to go one step further by identifying new classes (or at least new homogeneous groups of patterns). The process will then be divided into two stages. First, in an off-line classification phase, a given TS will be used to classify new patterns but with the ability of adding some of them to the TS. The updated TS (containing one or more newly identified classes) will then be used to classify patterns in the definitive online classification phase. This situation is summarized in Fig. 1.

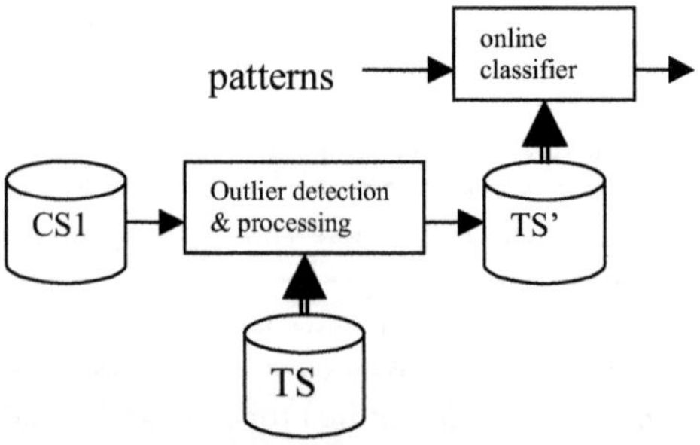

Fig. 1. Schematic representation of the proposed technique. CS1 contains a control set of (unlabelled) patterns possibly containing classes not present in TS

There are a number of different techniques that are able to detect outliers. Virtually any classifier with a rejection option can be used for this purpose. The main problem is that we have to look for a convenient rejection rate in order to be able to identify new classes into the rejected patterns. Apart from the trivial option of using the *k*-NN rule with a variable rejection rate that is not reported here, a slight modification of the ALIEN technique [2] has been adopted in this study. This method consists of assigning a threshold or weight to each training pattern that relates to the maximum distance at which neighbors can be located with regard to any pattern. That is, given *x*, any neighbor of *x* farther than its threshold is not considered and will not cast any vote to classify *x*. The procedure consists of two phases, the first one to compute thresholds (learning) and the second one to classify new patterns with reject. The details of both phases are as follows:

Learning phase:

For each x_i in TS do:

a) Find the *k*-NNs of x_i in TS - $\{x_i\}$
b) **If** there is a majority of NNs from the same class as x_i
  then set d^i = max $\{d(x_i, x_m)\}$ for all NNs x_m from the same class of x_i **else** set $d^i = 0$.

Classification phase:

For every new pattern *X* to be classified do:

c) Let $X_u = \{x_i$ in TS / $d(X, x_i) \le d^i\}$
d) **If** $|X_u| = 0$, **then** reject *X*.
  else, assign *X* to the most voted class in X_u.

The above technique can be seen as an implicit edition of prototypes. A modification of this is proposed in this paper by effectively removing the prototypes on one hand and substituting individual thresholds of patterns by an averaged threshold for the whole class. The resulting technique, which will be referred to as Averaged Distance in this paper, can schematically be explained as follows:

Learning phase:

For each class *j* do:

For each x_i in TS from class *j* do:
e) Find the *k*-NN of x_i in TS - $\{x_i\}$
f) **If** there is a majority of NNs from class j
  then set d^i = mean value $\{d(x_i, x_t)\}$, for all NNs x_t from class *j* **else** mark x_i for removal.

Let T_j = mean value $\{d^i)\}$, for all x_i in TS from class *j*.

Classification phase (all patterns marked are removed from TS):

For every new pattern *X* to be classified do:
a) Find the *k*-NN of *X* in TS
b) Let *q* be the most voted class into these NNs.

c) Set d^X = mean value $\{d(X, x_m)\}$, for all NNs x_m from class q
d) **If** $d^X \leq T_q$, **then** X is assigned to the class q **else**, reject X.

According to our experimentation, this modification of the ALIEN procedure exhibits a more stable behavior and, moreover, can be more efficiently implemented.

4 Looking for New Classes among Rejected Patterns

Once a subset of patterns that cannot possibly belong to any of the classes represented in the TS has been identified, we are interested in processing these to see whether some of them are grouped together and can be of interest for future operation of the classifier. In its full generality the problem is very complex but we will make appropriate simplifications in order to carry out this empirical study. First, we will suppose in our experiments that there is always a natural class that has been intentionally omitted in the TS. We will suppose that all classes (including the missing one) have a (normal-like) unimodal distribution.

With these assumptions in mind, the proposed procedure begins the second phase with the application of a clustering algorithm (*k*-means; [14]). A fixed and reduced number of patterns from each of the classes represented in the TS are joined to the set of the rejected patterns and this augmented set is partitioned into $c+1$ groups (the TS includes c classes). Once the clusters have been identified, we need to match each one of them to the c known classes. In particular, as clusters include some labeled (and representative) prototypes, the majority of these labels are used to assign a particular cluster to a class.

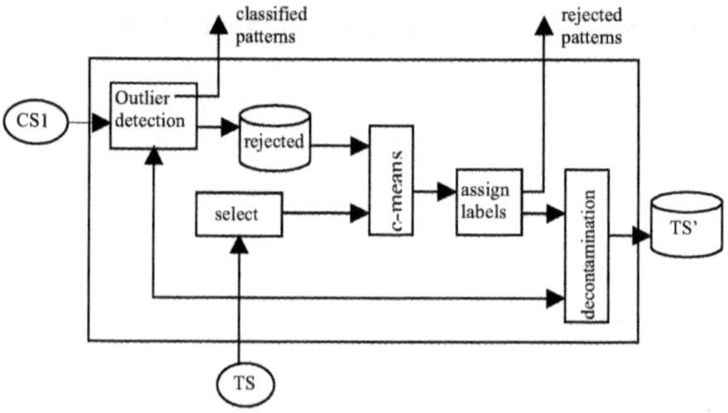

Fig. 2. Schematic description of the proposed method. Given a training set (TS) and patterns to be classified (CS1), the method classifies some of them, rejects other and produces prototypes from a new class and adds them to TS giving rise to TS'

The newly obtained labeled set may contain some patterns close to the boundaries that can increase the classification risk. A further application of the decontamination technique is used then in order to discard some of these prototypes and changing labels of others in order to obtain a "clean" and definitive training set, TS', that can be hopefully used to classify new patters from any of the $c+1$ classes. The whole proposed procedure including the first step of detecting potential members of unknown classes is summarized in the block diagram of Figure 2.

It is worth noting that the method splits the input set CS1 into (definitively) rejected patterns, classified patterns and patterns to be added to TS. It is possible as well (as shown in the experiments) to re-classify the rejected patterns using the updated training set TS'. Although this technique has been presented as a procedure taking two input sets and giving one updated TS, the obvious extension is to have a system of this kind operating continuously in an online learning basis.

5 Experimental Results

For this empirical study, 24 real datasets taken from the UCI Repository [12] have been considered. In all experiments, one of the classes of each dataset was removed, in turn, to simulate a partially exposed situation.

First, the different techniques to be applied in the first step (outlier detection) were compared. To this end, five replications (different random partitions) were realized for each dataset and for every situation in which one of the classes was eliminated from the training sample. In each of these replications, 80% of the dataset (but without patterns from one of the classes) was used as training set and the 20% of the data as the control set. Since the data contained numerical as well as non-numerical features, the Heterogeneous Euclidean Overlap Metric [13] was employed. The parameter value k=2 was used for Alien and Averaged Distance. The different databases used are shown in Table 1.

Table 1. Description of the datasets employed in the experiments of phase one. Figures in colums 4 and 9 (TS size) are averaged. Real numbers depend on the class being removed

Dataset	class	feature	TS size	CS size	Dataset	Class	feature	TS size	CS size
Iris	3	4	80	30	Wine	3	13	96	34
Glass	6	9	150	40	Bridge	7	11	78	19
Zoo	7	16	64	15	Image	7	19	314	82
Vehicle	4	18	504	168	Pima	2	8	215	153
Cancer	2	9	200	139	Liver	2	6	160	69
Australia	2	14	243	137	CRX	2	15	246	137
Cleveland	2	13	110	60	Cardiology	2	9	30	15
Hungarian	2	13	110	58	Beach	2	13	75	40
Heart	2	13	120	54	Swiss	2	13	49	24
Voting	2	16	135	86	Hepatitis	2	19	62	30
Horse	2	23	121	59	Promoter	2	57	43	20
Ionosphere	2	34	141	70	Sonar	2	60	78	41

The most important characteristic to assess in this part of the work is the capacity of each technique to detect (reject) patterns that do not belong to any of the classes considered in the training sample. These are the results (in percentage) presented in Table 2: mean values (and standard deviation) of each technique for each dataset. Results are analyzed in two aspects: rejected patterns from new classes (those classes not included in the current training sample) and rejected patterns from the known classes (those classes represented in the current training sample). Of course, the optimum values would be 100% of rejected patterns from the new classes and, at the same moment, 0% of rejected patterns from the known classes.

Unfortunately, these optimum goals are far from being simultaneously reached as results in Table 2 show. The dilemma is that, as the number of patterns from new classes being rejected grows (good sign) it is also increased the number of patterns from known classes remaining unidentified (undesirable effect). In this context, Averaged Distance presents a more convenient tradeoff between these two effects. It is true that this technique also produces a considerable amount of patterns from known classes that remains without classification but this effect is especially good if we aim at extracting information from these rejected patterns. Experiments with other values for the parameter *k* are not reported here since they did not offer better results. Preprocessing of the training sample through Edition [3] produced some improvements but not in a significant amount.

Table 2. Experimental results –mean values (and standard deviations)- of the percentage of patterns rejected (from the new and the known classes) in each dataset and for each technique assessed

	New Classes		Known Classes	
Dataset	Alien	Av. Dist.	Alien	Av. Dist.
Iris	82.7 (3.7)	98.7 (1.8)	15.3 (6.4)	34.7 (13.0)
Wine	76.5 (5.5)	100.0 (0)	22.7 (4.2)	52.7 (9.0)
Glass	59.0 (9.5)	88.5 (3.8)	48.1 (6.8)	72.4 (5.8)
Bridge	84.2 (9.9)	95.8 (4.4)	57.0 (12.5)	80.7 (9.9)
Zoo	94.7 (3.0)	100.0 (0)	9.6 (4.1)	50.4 (6.2)
Image	90.0 (2.5)	99.5 (1.1)	22.6 (7.6)	44.1 (11.7)
Vehicle	56.7 (1.8)	96.7 (0.3)	36.3 (3.5)	82.7 (2.1)
Pima	22.1 (5.7)	48.8 (2.9)	14.9 (2.8)	37.3 (2.8)
Cancer	52.7 (17.4)	85.9 (5.0)	11.8 (3.3)	35.7 (8.7)
Liver	21.7 (9.0)	43.8 (17.5)	18.0 (5.8)	39.1 (17.4)
Australian	32.9 (7.2)	89.1 (2.1)	17.1 (7.1)	49.9 (8.1)
CRX	28.9 (8.0)	89.6 (3.0)	20.0 (3.9)	51.4 (6.5)
Cleveland	32.0 (9.1)	80.3 (6.2)	19.0 (9.3)	49.3 (6.3)
Cardiology	77.3 (13)	96.0 (4)	24.0 (14)	38.7 (20)
Hungarian	27.6 (9.8)	83.5 (4.3)	31.4 (12.6)	39.3 (8.7)
Beach	20.5 (8.6)	57.5 (15)	16.5 (8.6)	46.5 (16.6)
Heart	29.3 (3.3)	82.2 (8.6)	18.5 (5.4)	45.9 (5.8)
Swiss	25.8 (21.9)	62.5 (29)	27.5 (23.7)	61.7 (19.9)
Voting	58.8 (7.9)	99.5 (0.6)	14.0 (7)	48.8 (11.3)
Hepatitis	20.7 (9.6)	66.0 (15.9)	18.0 (6.5)	53.3 (11.8)
Horse	26.8 (5.4)	69.8 (3)	19.7 (1.9)	54.2 (7.3)
Promoter	46.0 (6.5)	89.0 (7.4)	8.0 (5.7)	56.0 (9.6)
Ionosphere	44.6 (5.5)	44.6 (4.7)	37.8 (10.3)	49.7 (16.5)
Sonar	59.5 (5.9)	81.0 (8)	19.5 (5.7)	46.3 (12.2)
Average	48.8 (2.1)	81.2 (2.6)	22.8 (1.8)	50.9 (1.1)

Experiments about the whole proposed technique (Figure 2) were carried out using the Averaged Distance method only. Four datasets were selected since it was necessary to employ only those with more than two classes (to correctly simulate a partially exposed environment) and with a sufficient number of patterns in each class. Each dataset was divided into three groups: 50% for the TS, 25% for CS1 (the control set to be used in the first step of the procedure) and 25% for CS2 (control set for second step). Five different random partitions of this kind were performed and their corresponding results averaged in all the experiments. The details of these datasets are shown in Table 3.

Table 3. Description of the real datasets employed in the experiments of Phase two. Figures of TS size are averaged. Real numbers depend on the class being removed

Datasets	Classes	Features	TS size	CS1 size	CS2 size
Iris	3	4	50	38	37
Wine	3	13	59	45	45
Glass	6	9	88	54	53
Vehicle	4	18	316	212	211

For each partition of the data set into three subsets as explained above, four different methods have been taken into account.

1. The proposed method as shown in Figure 2 and re-classifying the rejected patters from CS1 by using TS'.
2. The same as above but preprocessing the initial TS by using decontamination.
3. External supervision I: which corresponds to using the same *k*-NN rule to classify an independent set corresponding to 50% of the patterns from known classes and 25% from the missing one, using a training sample with the same composition.
4. External supervision II: which corresponds to an ideal behavior of the proposed technique. The data sets are split in the same way but the patterns from the missing class in CS1 are identified by hand.

The two last methods are taken into account for comparison purposes. Both methods represents in some way the ideal situation the proposed method is trying to mimic.

Results in Table 4 indicate that the two variants of the proposed procedure excel, in classification accuracy, the behavior of the External Supervision II method. They can also compete in some cases with the External Supervision I, which optimally uses the same amount of information about the missing class available for the proposed method. In fact, in Glass dataset, both variants are more accurate than Ext. Supervision I. In Wine and Vehicle, the variant with initial Decontamination obtains more classification errors than variant 1. These two datasets present a rather low size/dimensionality ratio that has been reduced even more by the Decontamination methodology. It is important to remark that these variants of the proposed procedure work in a complete automatic way, not requiring the help of a human expert.

Table 4. Experimental results of the two variants of the proposed procedure. Mean values and standard deviations of the percentage of misclassification

Iris	Misclassification			Vehicle	Misclassification		
Procedure	New	Known	Total	Procedure	New	Known	Total
Ext. Sup I	1.4	4.5	5.9	Ext. Sup. I	6.5	13.3	19.9
	(0.3)	(1.3)	(1.6)		(0.4)	(0.4)	(0.6)
Ext Sup. II	17.8	3.2	21.0	Ext. Sup. II	18.1	19.6	37.6
	(0.3)	(1.2)	(1.4)		(0.4)	(1.2)	(0.9)
Variant 1	5.8	1.9	7.7	Variant 1	12.2	13.0	25.1
	(4.6)	(0.6)	(4.7)		(1.0)	(0.5)	(0.6)
Variant 2	4.4	2.1	6.6	Variant 2	12.4	13.7	26.1
	(4.2)	(1.1)	(3.4)		(1.7)	(1.2)	(2.0)
Glass	Misclassification			Wine	Misclassification		
Procedure	New	Known	Total	Procedure	New	Known	Total
Ext. Sup I	3.7	31.6	35.3	Ext. Sup. I	1.2	3.1	4.3
	(0.5)	(3.0)	(3.0)		(0.7)	(0.9)	(1.4)
Ext Sup. II	11.8	27.8	39.5	Ext. Sup. II	17.6	2.3	19.9
	(0.5)	(3.0)	(2.9)		(0.6)	(0.3)	(0.8)
Variant 1	9.5	17.4	26.9	Variant 1	7.1	2.1	9.2
	(0.6)	(1.6)	(1.6)		(6.3)	(0.5)	(6.2)
Variant 2	9.4	17.3	26.7	Variant 2	12.0	1.9	13.9
	(0.8)	(1.9)	(1.4)		(7.4)	(0.8)	(7.2)

6 Some Final Comments

The initial experimental results above reported exhibit the usefulness of the proposed procedure to handle classification situations with incomplete training samples. Working in an automated way, the recognition accuracy of the procedure can compete with the results obtained when help of a human supervisor is available.

The work presented will be extended and improved by retaking some of the decisions and simplifications that have been done here. In particular, the way representative patterns from TS are taken and the identification between classes and clusters when *k*-means is applied have to be revised. Many well studied methods to fully automate our technique and making it more generally applicable exists. Following with this, research is to be done also to adapt our procedure to those cases when two or more classes are not represented in the initial training sample. We intend now to study how to incorporate this idea to an ongoing learning system, for continuously increasing the knowledge of the supervised system, not only with the perception of new classes but also with a better understanding of the already defined classes (see [15]). In these cases, the more difficult problem of distinguishing between pure outliers or noise and rejected patterns that can belong to new classes will arise.

Acknowledgements

This work has been partially supported by grants 32016-A from Conacyt and 744.99-P from Cosnet, both from Mexico and by the Spanish project TIC2000-1703-C03-03.

References

1. Dasarathy, B.V.: All you need to know about the neighbors. *Proceedings International Conference on Cybernetics and Society*, Denver (1979)
2. Dasarathy, B.V.: "There goes the neighborhood": An Alien identification approach to recognition in partially exposed environments. *Proceedings of the 5th International Conference on Pattern Recognition* (1980)
3. Wilson, D.L.: Asymptotic properties of Nearest Neighbor rules using edited data sets. *IEEE Trans. Systems, Man and Cybernetics,* SMC-2 (1972) 408-421
4. Barandela, R.: La regla NN con muestras de entrenamiento no balanceadas. *Investigación Operacional*, X (1) (1990) 45-56
5. Sanchez, J.S., F. Pla and F. Ferri: Prototype selection for the Nearest Neighbor rule through Proximity Graphs. *Pattern Recognition Letters*, 18, 6 (1997) 507-513
6. Barandela, R.: Una metodología para el reconocimiento de patrones en tareas geólogo-geofísicas. *Geofísica Internacional,* 34 (4) (1995) 399-405
7. Gasca, E. and R. Barandela: Influencia del preprocesamiento de la muestra de entrenamiento en el poder de generalización del Perceptron Multicapa. *6th Brazilian Symposium on Neural Networks*. Río de Janeiro (2000)
8. Barandela, R. and E. Gasca: Decontamination of training samples for supervised pattern recognition methods. In: *Advances in Pattern Recognition, Lecture Notes in Computer Science* 1876, Springer Verlag (2000) 621-630
9. Koplowitz, J. And T.A. Brown: On the relation of performance to editing in nearest neighbor rules. *Proc. 4th Int. Joint Conf. on Pattern Recognition,* Japan (1978)
10. Dasarathy, B.V.: Adaptive decision systems with extended learning for deployment in partially exposed environments. *Optical Engineering*, 34, 5, (1995) 1269-1280
11. Tax, D.M.J. and R.P.W. Duin: Outlier Detection using Classifier Instability. In: A. Amin et al. (eds.), *Advances in Pattern Recognition, Lecture Notes in Computer Science*, vol 1451, Springer, Berlin (1998)
12. Merz, C.J. and P.M. Murphy: *UCI Repository of machine learning databases.* University of California at Irvine. http://www.csi.uci.edu/~mlearn (1998)
13. Wilson, D.R. and T.R. Martinez: Reduction Techniques for Instance-Based Learning Algorithms. *Machine Learning*, 38, 3 (2000) 257-286
14. Duda, R.O. and P.E. Hart: *Pattern Classification and Scene Analysis.* Wiley, New York (1973)
15. Barandela, R. and M. Juarez: Ongoing learning for Supervised Pattern Recognition. *Proceedings of the XIV SIBGRAPI*, Brazil, IEEE Press (2001) 51-58

Recursive Prototype Reduction Schemes Applicable for Large Data Sets

Sang-Woon Kim[1] and B. J. Oommen[2]

[1] Member IEEE.
Div. of Computer Science and Engineering, Myongji University
Yongin, 449-728 Korea
kimsw@mju.ac.kr
[2] Senior Member IEEE.
School of Computer Science, Carleton University
Ottawa, ON, K1S 5B6, Canada
oommen@scs.carleton.ca

Abstract. Most of the Prototype Reduction Schemes (PRS), which have been reported in the literature, process the data in its entirety to yield a subset of prototpyes that are useful in nearest-neighbour-like classification. Foremost among these are the Prototypes for Nearest Neighbour (PNN) classifiers, the Vector Quantization (VQ) technique, and the Support Vector Machines (SVM). These methods suffer from a major disadvantage, namely, that of the excessive computational burden encountered by processing *all* the data. In this paper, we suggest a recursive and computationally superior mechanism. Rather than process all the data using a PRS, we propose that the data be recursively subdivided into smaller subsets. This recursive subdivision can be arbitrary, and need not utilize any underlying clustering philosophy. The advantage of this is that the PRS processes *subsets* of data points that effectively sample the entire space to yield smaller subsets of prototypes. These prototypes are then, in turn, gathered and processed by the PRS to yield more refined prototypes. Our experimental results demonstrate that the proposed recursive mechansim yields classification comparable to the *best* reported prototype condensation schemes to-date, for both artificial data sets and for samples involving real-life data sets. The results especially demonstrate the computational advantage of using such a recursive strategy for *large* data sets, such as those involved in data mining and text categorization applications.

1 Introduction

In non-parametric pattern classification like the nearest neighbour (NN) or the k-nearest neighbour ($k-$NN) rule, each class is described using a set of sample prototypes, and the class of an unknown vector is decided based on the identity of the closest neighbour(s) which are found among all the prototypes [1]. For applications, involving large data sets, such as those involved in data mining, financial forecasting, retrieval of multimedia databases and biometrics, it

T. Caelli et al. (Eds.): SSPR&SPR 2002, LNCS 2396, pp. 528–537, 2002.
© Springer-Verlag Berlin Heidelberg 2002

is advantageous to reduce the number of training vectors while simultaneously insisting that the classifiers that are built on the reduced design set perform as well, or nearly as well, as the classifiers built on the original data set. Various prototype reduction schemes, which are useful in nearest-neighbour-like classification, have been reported in the literature [2].

One of the firsts of its kind is the Condensed Nearest Neighbour (CNN) rule [3]. The reduced set produced by the CNN, however, customarily includes "interior" samples, which can be completely eliminated, without altering the performance of the resultant classifier. Accordingly, other methods have been proposed successively, such as the Reduced Nearest Neighbour (RNN) rule [4], the Prototypes for Nearest Neighbour (PNN) classifiers [5], the Selective Nearest Neighbour (SNN) rule [6], two modifications of the CNN [7], the Edited Nearest Neighbour (ENN) rule [8], and the non-parametric data reduction method [9]. Besides these, in [10], the Vector Quantization (VQ) and the Bootstrap [11] techniques have also been reported as being extremely effective approaches to data reduction. Recently, Support Vector Machines (SVM) [12] have proven to possess the capability of extracting vectors that support the boundary between the two classes. Thus, they have been used satisfactorily to represent the global distribution structure. Recently, we have proposed a new hybrid scheme, the Kim_Oommen Hybridized Technique, [13], which is based on the philosophy of invoking *creating* and *adjusting* phases to the prototype vectors. First, a reduced set of initial prototype vectors is chosen by any of the previously mentioned methods, and then their optimal positions are learned with an LVQ3-type algorithm, thus, minimizing the average classification error. The details of this are omitted here.

All the PRS methods reported in the literature (including the one proposed in [13]) are practical as long as the size of the data set is not "too large". The applicability of these schemes for large-sized data set is limited because they all suffer from a major disadvantage – they incur an excessive computational burden encountered by processing *all the data points*. To overcome this disadvantage for large-sized data sets, in this paper, we suggest a recursive mechanism. Rather than process all the data using a PRS, we propose that the data be recursively subdivided into smaller subsets. *We emphasize that the smaller subsets need* **not** *represent sub-clusters of the original data set.* After this recursive subdivision, the smaller subsets are reduced with a PRS. The resultant sets of prototypes obtained are, in turn, gathered and processed at the higher level of the recursion to yield more refined prototypes. This sequence of divide-reduce-coalesce is invoked recursively to ultimately yield the desired reduced prototypes. We refer to the algorithm presented here as the *Kim_Oommen_Recursive PRS*.

The main contribution of this paper is the demonstration that the speed of data condensation schemes can be increased by recursive computations - which is crucial in large-sized data sets. This has been done by introducing the *Kim_Oommen_Recursive PRS*, and demonstrating its power in both speed and accuracy. We are not aware of any similar reported recursively-motivated PRS.

The first outstanding advantage of this mechanism is that the PRS can select more refined prototypes in significantly less time. This is achieved without sacrificing the classification accuracy and the prototype reduction rate. This is primarily because the new scheme does **not** process all the data points at *any* level of the recursion. A second advantage of this scheme is that the recursive subdivision can be arbitrary, and *need not utilize any clustering philosophy.* Furthermore, *all* subsequent recursive partitionings, also, need not involve clustering. Finally, the higher level PRS invocations do not involve any points interior to the Voronoi space because they are eliminated at the leaf levels.

The reader should observe that this philosophy is *quite distinct* from the partitioning using clustering methods which have recently been proposed in the literature to solve the Travelling Salesman Problem (TSP) [15]. The differences between these two philosophies can be found in [16].

The experimental results on synthetic and real-life data prove the power of these enhancements. The real-life experiments include three "medium-size" data sets, and two *large* data sets with a fairly high dimensionality. The results are conclusive.

2 Prototype Reduction Schemes

As mentioned previously, various Prototype Reduction Schemes (PRS) have been proposed in the literature - a survey of which is found in [2]. The most pertinent ones are reviewed here by two groups: the conventional methods and a newly proposed hybrid method. Among the conventional methods, the CNN and the SVM are chosen as representative schemes of *selecting* methods. The former is one of first methods proposed, and the latter is more recent. As opposed to these, the PNN and VQ (or SOM) are considered to fall within the family of prototype-*creating* algorithms. The reviews of these methods is not attempted here. However, it should be emphasized that our new recursive method can utilize *any one* of the reported PRS as an atomic building block – thus extending them. In that light, many of the PRS are *briefly* surveyed in the unabridged version of the paper [16]. This present study (and so the review in [16]) includes the CNN rule [3], the PNNs [5], the VQ and SOM methods [14], and the SVM [12].

The unabridged paper [16] also describes a newly-reported hybrid scheme, the Kim_Oommen Hybridized Technique, [13], which is based on the philosophy of invoking *creating* and *adjusting* phases. First, a reduced set of initial prototypes or code-book vectors is chosen by any of the previously mentioned methods, and then their optimal positions are learned with an LVQ3-type algorithm, thus, minimizing the average classification error.

3 Recursive Invocations of PRS

3.1 The Rationale of the Recursive Algorithm

Since prototypes near the boundary play more important roles than the interior ones for designing NN classifiers, the points near the boundary are more important in selecting the prototypes. In all the currently reported PRS, however, points in the interior of the Voronoi space are processed for, apparently, no reason. Consequently, all reported PRS suffer from an excessive computational burden encountered by processing all the data, which becomes very prominent in "large" data sets.

To overcome this disadvantage, in this section, we propose a recursive mechanism, where the data set is sub-divided recursively into smaller subsets to filter out the "useless" internal points. Subsequently, a conventional PRS processes the smaller subsets of data points that effectively sample the entire space to yield *subsets* of prototypes – one set of prototypes for each subset. The prototypes, which result from each subset, are then coalesced, and processed again by the PRS to yield more refined prototypes. In this manner, prototypes which are in the interior of the Voronoi boundaries, and are thus ineffective in the classification, are eliminated at the subsequent invocations of the PRS.

A direct consequence of eliminating the "redundant" samples in the PRS computations, is that the processing time of the PRS is *significantly* reduced. This will be clarified by the example below.

To illustrate the functioning of the recursive process, we present an example for the two-dimensional data set referred to as "Random". Two data sets, namely the training and test sets, are generated randomly with a uniform distribution, but with irregular decision boundaries. The training set of 200 sample vectors is used for computing the prototypes, and the test set of 200 sample vectors is used for evaluating the quality of the extracted prototypes.

To demonstrate the power of the mechanism, we first select prototypes from the whole training set using the CNN method. This is also repeated after randomly dividing the training set into two subsets of equal size – each with 100 vectors. Fig. 1 shows the whole set of the "Random" training data set, the divided subsets and the prototypes selected with the CNN and the recursive PRS methods, respectively. Observe that in this example, to render the presentation brief, we have invoked the recursive procedure *only* twice.

In Fig. 1, the set of prototypes of (e), which is extracted from the whole set of (a), consists of 36 points and has a classification accuracy of 96.25 % . The prototypes of (f) and (g), selected from the subsets of (b) and (c), *both* consist of the 21 vectors, and have accuracies of 96.00 % and 94.50 % respectively.

On the other hand, the set of prototypes of (h), which is extracted from a data set obtained by combining two prototype sets of (f) and (g), consists of 27 points, and has the accuracy of 97.00 %. Moreover, it should be pointed out that the time involved in the prototype selection of (h), is *significantly* less than that of (e), because the number of sample vectors of the combined sets of (f) and (g) together, is smaller than that of the whole set of (a).

The formal algorithm is omitted here in the interest of brevity, but can be found in [16]. However, a *brief* explanation of the algorithm is not out of place. If the size of the original data set is smaller than K, a traditional PRS is invoked to get the reduced prototypes. Otherwise, the original data set is recursively subdivided into J subsets, and the process continues down towards

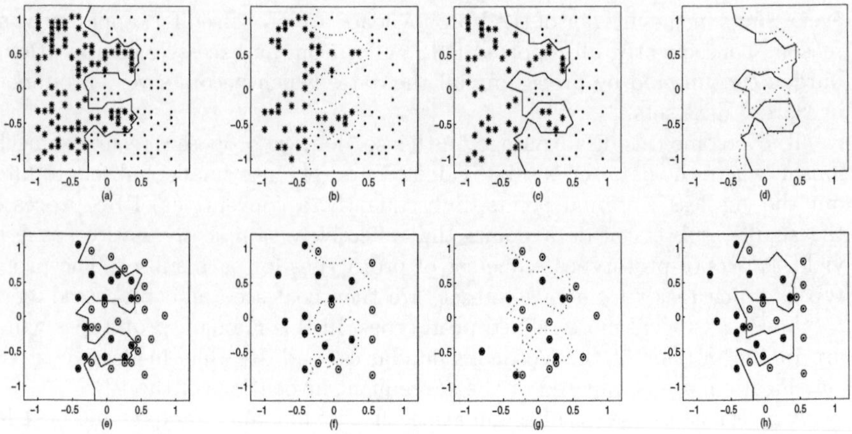

Fig. 1. The entire set, the divided subsets and the prototypes of the "Random" data set, where the vectors of each class are represented by '$*$' and '$\cdot$', and the selected prototype vectors are indicated by the circled '$*$' and '$\cdot$' respectively. The details of the figure are found in the text

the leaf of the recursive tree. Observe that a traditional PRS is invoked only when the corresponding input set is "small enough". Finally, at the tail end of the recursion, the resultant output sets are merged, and if *this* merged set is greater than K the procedure is again recursively invoked.

It should be also noted that the traditional PRS, which can be otherwise time consuming for large data sets, is never invoked for *any* sets of cardinality larger than K. It is invoked *only* at the leaf levels when the sizes of the sets are "small", rendering the entire computation very efficient.

4 Experimental Results

4.1 Experimental Results: Medium-Sized Data Sets

The Kim_Oommen_Recursive PRS has been rigorously tested and compared with many conventional PRS. This was first done by performing experiments on a number of "medium-sized" data sets, both real and artificial. The data set named "Non_normal (Medium-size)", which has been also employed in [9], [10] and [11] as a benchmark experimental data set, was generated from a mixture of four 8-dimensional Gaussian distributions as follows:

1. $p_1(x) = \frac{1}{2}N(\mu_{11}, I_8) + \frac{1}{2}N(\mu_{12}, I_8)$ and
2. $p_2(x) = \frac{1}{2}N(\mu_{21}, I_8) + \frac{1}{2}N(\mu_{22}, I_8)$,

where $\mu_{11} = [0, 0, \cdots, 0]$, $\mu_{12} = [6.58, 0, \cdots, 0]$, $\mu_{21} = [3.29, 0, \cdots, 0]$ and $\mu_{22} = [9.87, 0, \cdots, 0]$. Here, I_8 is the *8*-dimensional *Identity* matrix.

The "Sonar" data set contains 208 vectors. Each sample vector, of two classes, has 60 attributes which are all continuous numerical values. The "Arrhythmia" data set contains 452 patterns with 279 attributes, 206 of which are real-valued, and the rest are nominal. The details of these sets are found in [16] and omitted here in the interest of brevity. However, we mention that both the data sets "Sonar" and "Arrhythmia" are real benchmark data sets, cited from the UCI Machine Learning Repository [17].

In the above data sets, all of the vectors were normalized within the range $[-1, 1]$ using their standard deviations. Also, for every class j, the data set for the class was randomly split into two subsets, $T_{j,t}$ and $T_{j,V}$, of equal size. One of them was used for choosing initial code-book vectors and training the classifiers as explained above, and the other subset was used in the validation (or testing) of the classifiers. The roles of these sets were later interchanged.

In this case, because the size of the sets was not excessively large, the recursive versions of CNN, PNN, VQ and SVM, were all invoked only for a depth of *two*. The experimental results of the CNN, PNN, VQ and SVM methods implemented with the recursive mechanism, for the "Non_normal (Medium-size)" data sets is shown in Table 1. The results for the data sets "Sonar" and "Arrhythmia" are not included here in the interest of brevity. They can be found in [16].

Table 1. The experimental results of the recursive CNN, PNN, VQ and SVM methods for the "Non_normal (Medium-size)" data set. Here, *DS*, *CT*, *NP* and *Acc* are the data set size (the number of sample vectors), the processing CPU-time (in seconds), the number of prototypes, and the classification accuracy rate (%), respectively

Methods	*DS1*	*CT1*	*NP1*	*Acc1*	*DS2*	*CT2*	*NP2*	*Acc2*
	500	0.61	64	92.60	500	0.63	66	91.20
CNN	250	0.13	34		250	0.15	37	
	250	0.15	34		250	0.12	29	
	68	0.11	54	91.60	66	0.11	54	90.00
	500	81.74	56	92.40	500	208.55	380	91.60
PNN	250	7.58	31		250	7.55	34	
	250	7.54	29		250	7.54	26	
	60	0.22	46	89.20	60	0.26	50	88.80
	500	0.23	4	95.60	500	0.41	4	94.80
VQ	250	0.11	4		250	0.15	4	
	250	0.16	4		250	0.13	4	
	8	0.06	4	95.60	8	0.07	4	94.40
	500	0.86	62	95.40	500	0.97	57	94.40
SVM	250	0.24	32		250	0.25	35	
	250	0.23	30		250	0.24	24	
	62	0.07	60	95.60	59	0.06	57	94.40

The Kim_Oommen_Recursive PRS can be compared with the non-recursive versions using three criteria, namely, the processing CPU-time (*CT*), the classification accuracy rate (*Acc*), and the prototype reduction rate (*Re*).

We report below a summary of the results obtained for the case when one subset was used for training and the second for testing. The results when the roles of the sets are interchanged are almost identical. From Table 1, we can see that the *CT* index (the processing CPU-time) of the pure CNN, PNN, VQ and SVM methods can be reduced significantly by merely employing the recursive philosophy.

Consider the PNN method for the "Non_normal (Medium-size)" data set. If the 500 samples were processed non-recursively, the time taken is 81.74 seconds, the size of the reduced set is 56, and the resulting classification accuracy is 92.4%. However, if the 500 samples are subdivided into two sets of 250 samples each, processing each subset involves only 7.58 and 7.54 seconds leading to 31 and 29 reduced prototypes respectively. When *these* 60 samples are, in turn, subjected to a pure PNN method, the number of prototype samples reduced to 46 in just 0.22 seconds and yielded an accuracy of 89.2 %. If we reckon that the recursive computations can be done in parallel, the time required is only *about one-tenth* of the time which the original PNN would take. Even if the computations were done serially, the advantage is marked. Such results are typical, as can be seen from [16].

4.2 Experimental Results: Large-Sized Data Sets

In order to further investigate the advantage gained by utilizing the proposed recursive PRS for more computationally intensive sets, we conducted experiments on "large-sized" data sets, which we refer to as the "Non_normal (Large-size)" and "Adult" sets, which consisted of 20,000 patterns and 8 dimensions, and 33,330 samples and 14 dimensions respectively. In this case, because the size of the sets was reasonably large, the recursive version of the SVM was invoked to a depth of *four*.

As in the case of the "Non_normal (Medium-size)" data set, the data set "Non_normal (Large-size)" was generated randomly with the normal distributions. The "Adult" data set, which had been extracted from a census bureau database[1], has also been obtained from the UCI Machine Learning Repository [17]. The aim of the pattern recognition task here was to distinguish the income into two groups, in the first group the salary is more than 50K dollars, and in the second group the salary is less than or equal to 50K dollars. Each sample vector has fourteen attributes. Some of the attributes, such as the age, hours-per-week, etc., are continuous numerical values. The others, such as education, race, etc., are nominal symbols. In the experiments, the nominal attributes were replaced with numeric zeros.

[1] http://www.census.gov/ftp/pub/DES/www/welcome.html

Table 2. The experimental results of the recursive SVM for the "Adult" data set. Here, *Depth*(i) means the depth at which the data set is sub-divided into 2^{i-1} subsets. *DS*, *CT*, *SV* and *Acc* are the data set size (the number of sample vectors), the processing CPU-time (in seconds), the number of support vectors, and the classification accuracy rate (%), respectively

Depth(i)	*DS1*	*CT1*	*SV1*	*Acc1*	*DS2*	*CT2*	*SV2*	*Acc2*
1	16665	3825.46	6448	82.84	16665	2365.16	6218	82.63
	2084	19.19	843		2084	17.07	801	
	2083	17.82	819		2083	15.04	814	
	2083	15.94	825		2083	14.01	771	
	2083	20.17	821		2083	19.32	825	
4	2083	15.94	836		2083	29.01	810	
	2083	21.75	807		2083	12.10	802	
	2083	15.26	853		2083	17.01	808	
	2083	18.39	814		2083	16.20	775	
	6621	256.80	6270	81.28	6406	219.46	6059	79.49

Although the experimental results for the "Non_normal (Large-size)" and the "Adult" data sets are given in [16], we briefly cite the results for the latter in Table 2.

Consider Table 2. At the *depth* 1, the data set of 16,665 samples was processed requiring a computation time of 3,825.46 seconds, and gave an accuracy of 82.84 % with a reduction of 61.31 %. However, if the 16,665 samples are subdivided into eight subsets of 2,083 each at the *depth* 4, processing each of these involves only the times given in the third column whose average is 18.06 seconds, leading to 843, 819, 825, 821, 836, 807, 853 and 814 reduced prototypes, respectively. When *these* 6,621 samples are, in turn, subjected to a pure SVM method, the number of reduced samples reduced to 6,270 in 256.80 seconds and yielded an accuracy of 81.28 %. As the recursive computations were done serially, the time required was 401.26 seconds, which is only 10.5 % of the time which the original SVM would take. The power of the newly-introduced recursive philosophy is obvious!

5 Conclusions

Conventional PRS (Prototype Reduction Schemes) suffer from a major disadvantage, namely that of the excessive computational burden encountered by processing all the data, even though the sample data in the interior of the Voronoi space is typically processed for no reason. In this paper, we proposed a recursive mechanism, where the data sets are recursively sub-divided into smaller subsets, and the prototype points which are ineffective in the classification are eliminated for the subsequent invocations of the PRS. These prototypes are, in turn, gathered and processed by the PRS to yield more refined prototypes.

The proposed method was tested on both artificial and real-life benchmark data sets (of both medium and large sizes), and compared with the reported conventional methods, and the superiority has been clearly demonstrated both with regard to the required CPU time and the classification accuracy. The results obtained are conclusive and prove that it is futile to invoke a PRS on any large data set. Rather, it is expedient to recursively split the data, and invoke the PRS on the smaller subsets. Undoubtedly, the advantage of the recursive invocations increases with the size of the data set.

Acknowledgements

The work of the first author was done while visiting with the School of Computer Science, Carleton University, Ottawa, Canada. The work of the second author was partially supported by the NSERC of Canada. The authors are very grateful to Luis Rueda who helped us with the manuscript.

References

1. A. K. Jain, R. P. W. Duin and J. Mao.: Statistical pattern recognition: A review. *IEEE Trans. Pattern Anal. and Machine Intell.*, **PAMI-22(1)**:4–37, 2000. 528
2. D. V. Dasarachy.: *Nearest Neighbor (NN) Norms: NN Pattern Classification Techniques.* IEEE Computer Society Press, Los Alamitos, 1991. 529, 530
3. P. E. Hart.: The condensed nearest neighbor rule. *IEEE Trans. Inform. Theory,* **IT-14**: 515–516, May 1968. 529, 530
4. G. W. Gates.: The reduced nearest neighbor rule. *IEEE Trans. Inform. Theory,* **IT-18**: 431–433, May 1972. 529
5. C. L. Chang.: Finding prototypes for nearest neighbor classifiers. *IEEE Trans. Computers,* **C-23(11)**: 1179–1184, Nov. 1974. 529, 530
6. G. L. Ritter, H. B. Woodruff, S. R. Lowry and T. L. Isenhour.: An algorithm for a selective nearest neighbor rule. *IEEE Trans. Inform. Theory,* **IT-21**: 665–669, Nov. 1975. 529
7. I. Tomek.: Two modifcations of CNN. *IEEE Trans. Syst., Man and Cybern.,* **SMC-6(6)**: 769–772, Nov. 1976. 529
8. P. A. Devijver and J. Kittler.: On the edited nearest neighbor rule. *Proc. 5th Int. Conf. on Pattern Recognition,* 72–80, Dec. 1980. 529
9. K. Fukunaga.: *Introdction to Statistical Pattern Recognition, Second Edition.* Academic Press, San Diego, 1990. 529, 532
10. Q. Xie, C. A. Laszlo and R. K. Ward.: Vector quantization techniques for nonparametric classifier design. *IEEE Trans. Pattern Anal. and Machine Intell.,* **PAMI-15(12)**: 1326–1330, Dec. 1993. 529, 532
11. Y. Hamamoto, S. Uchimura and S. Tomita.: A bootstrap technique for nearest neighbor classifier design. *IEEE Trans. Pattern Anal. and Machine Intell.,* **PAMI-19(1)**:73–79, Jan. 1997. 529, 532
12. C. J. C. Burges.: A tutorial on support vector machines for pattern recognition. *Data Mining and Knowledge Discovery,* **2(2)**:121–167, 1998. 529, 530
13. S.-W. Kim and B. J. Oommen.: Enhancing prototype reduction schemes with LVQ3-type algorithms. To appear in *Pattern Recognition.* 529, 530

14. T. Kohonen.: *Self-Oganizing Maps.* Berlin, Springer - Verlag, 1995. 530
15. N. Aras, B. J. Oommen and I. K. Altinel.: The Kohonen network incorporating explicit statistics and its application to the travelling salesman problem. *Neural Networks,* 1273–1284, Dec. 1999. 530
16. S.-W. Kim and B. J. Oommen.: Recursive prototype reduction schemes applicable for large data sets. *Unabridged version of this paper.* 530, 531, 533, 534, 535
17. http://www.ics.uci.edu/mlearn/MLRepository.html. 533, 534

Combination of Tangent Vectors and Local Representations for Handwritten Digit Recognition

Daniel Keysers[1], Roberto Paredes[2], Hermann Ney[1], and Enrique Vidal[2]

[1] Lehrstuhl für Informatik VI - Computer Science Department
RWTH Aachen - University of Technology
52056 Aachen, Germany
{keysers,ney}@informatik.rwth-aachen.de
[2] Instituto Tecnológico de Informática
Universidad Politécnica de Valencia
Camino de Vera s/n, 46022 Valencia, Spain
{rparedes,evidal}@iti.upv.es

Abstract. Statistical classification using tangent vectors and classification based on local features are two successful methods for various image recognition problems. These two approaches tolerate global and local transformations of the images, respectively. Tangent vectors can be used to obtain global invariance with respect to small affine transformations and line thickness, for example. On the other hand, a classifier based on local representations admits the distortion of parts of the image. From these properties, a combination of the two approaches seems very likely to improve on the results of the individual approaches. In this paper, we show the benefits of this combination by applying it to the well known USPS handwritten digits recognition task. An error rate of 2.0% is obtained, which is the best result published so far for this dataset.

1 Introduction

Transformation tolerance is a very important aspect in the classification of handwritten digits because of individual writing styles, pen properties and clutter. Among the relevant transformations we can distinguish the two cases of

- global transformations of the image, e.g. scale, rotation, slant, and
- local transformations of the image, e.g. clutter or missing parts.

These types of transformations do not change the class of the object present in the image and therefore we are interested in classifiers that can tolerate these changes, in order to improve classification accuracy. There exists a variety of ways to achieve invariance or transformation tolerance of a classifier, including e.g. normalization, extraction of invariant features and invariant distance measures.

In this work, we present two classification methods that are particularly suitable for the two types of transformations: a statistical classifier using tangent vectors for global invariance and a classifier based on the nearest neighbor

T. Caelli et al. (Eds.): SSPR&SPR 2002, LNCS 2396, pp. 538–547, 2002.
© Springer-Verlag Berlin Heidelberg 2002

technique and local representations of the image, which tolerates local changes. Because these two methods deal with different types of transformations it seems especially useful to combine the results of the classifiers.

The combination of the classifiers is evaluated on the well known US Postal Service database (USPS), which contains segmented handwritten digits from US zip codes. There are many results for different classifiers available on this database and the combined approach presented here achieves an error rate of 2.0% on the test set, which is the best result reported so far.

2 The Statistical Classifier Using Tangent Distance

First, we will describe the statistical classifier used. To classify an observation $x \in \mathrm{IR}^D$, we use the Bayesian decision rule

$$x \longmapsto r(x) = \underset{k}{\operatorname{argmax}} \{p(k) \cdot p(x|k)\} .$$

Here, $p(k)$ is the *a priori* probability of class k, $p(x|k)$ is the *class conditional* probability for the observation x given class k and $r(x)$ is the decision of the classifier. This decision rule is known to be optimal with respect to the expected number of classification errors if the required distributions are known [1]. However, as neither $p(k)$ nor $p(x|k)$ are known in practical situations, it is necessary to choose models for the respective distributions and estimate their parameters using the training data. The class conditional probabilities are modeled using *kernel densities* in the experiments, which can be regarded as an extreme case of a mixture density model, since each training sample is interpreted as the center of a Gaussian distribution:

$$p(x|k) = \frac{1}{N_k} \sum_{n=1}^{N_k} \mathcal{N}(x|x_{kn}, \Sigma),$$

where N_k is the number of training samples of class k, x_{kn} denotes the n-th reference pattern of class k and here we assume $\Sigma = \alpha\sigma^2 I$, i.e. we use variance pooling over classes and dimensions and apply a factor α to determine the kernel width.

2.1 Overview of Tangent Distance

In this section, we first give an overview of an invariant distance measure, called *tangent distance* (TD), which was introduced in [2]. In the following section, we will then show how it can be effectively integrated into the statistical classifier presented above. An invariant distance measure ideally takes into account transformations of the patterns, yielding small values for patterns which mostly differ by a transformation that does not change class-membership.

Let $x \in \mathrm{IR}^D$ be a pattern and $t(x, \alpha)$ denote a transformation of x that depends on a parameter L-tuple $\alpha \in \mathrm{IR}^L$, where we assume that t does not

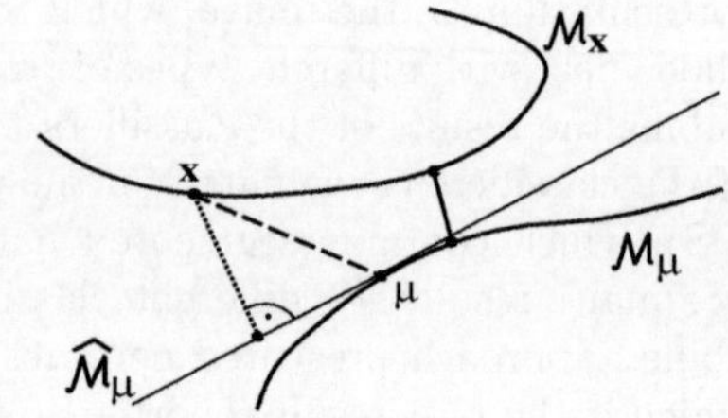

Fig. 1. Illustration of the Euclidean distance between an observation x and a reference μ (dashed line) in comparison to the distance between the corresponding manifolds (plain line). The tangent approximation of the manifold of the reference and the corresponding (one-sided) tangent distance is depicted by the thin line and the dotted line, respectively

affect class membership (for small α). The set of all transformed patterns now is a manifold $\mathcal{M}_x = \{t(x,\alpha) \ : \ \alpha \in \mathrm{IR}^L\} \subset \mathrm{IR}^D$ in pattern space. The distance between two patterns can then be defined as the minimum distance between the manifold $\mathcal{M}_x$ of the pattern x and the manifold $\mathcal{M}_\mu$ of a class specific prototype pattern μ. This manifold distance is truly invariant with respect to the regarded transformations (cf. Fig. 1). However, the distance calculation between manifolds is a hard non-linear optimization problem in general. These manifolds can be approximated by a *tangent subspace* $\widehat{\mathcal{M}}$. The *tangent vectors* x_l that span the subspace are the partial *derivatives* of the transformation t with respect to the parameters α_l $(l = 1,\ldots,L)$, i.e. $x_l = \partial t(x,\alpha)/\partial \alpha_l$. Thus, the transformation $t(x,\alpha)$ can be approximated using a Taylor expansion at $\alpha = 0$:

$$t(x,\alpha) = x + \sum_{l=1}^{L} \alpha_l x_l + \sum_{l=1}^{L} \mathcal{O}(\alpha_l^2)$$

The set of points consisting of the linear combinations of the tangent vectors x_l added to x forms the tangent subspace $\widehat{\mathcal{M}}_x$, a first-order approximation of $\mathcal{M}_x$:

$$\widehat{\mathcal{M}}_x = \{x + \sum_{l=1}^{L} \alpha_l x_l \ : \ \alpha \in \mathrm{IR}^L\} \subset \mathrm{IR}^D$$

Using the linear approximation $\widehat{\mathcal{M}}_x$ has the advantage that distance calculations are equivalent to the solution of linear least square problems or equivalently

Fig. 2. Example of first-order approximation of affine transformations and line thickness. (Left to right: original image, ± horizontal translation, ± vertical translation, ± rotation, ± scale, ±axis deformation, ± diagonal deformation, ± line thickness)

projections into subspaces, which are computationally inexpensive operations. The approximation is valid for small values of α, which nevertheless is sufficient in many applications, as Fig. 2 shows for examples of USPS data. These examples illustrate the advantage of TD over other distance measures, as the depicted patterns all lie in the same subspace and can therefore be represented by one prototype and the corresponding tangent vectors. The TD between the original image and any of the transformations is therefore zero, while the Euclidean distance is significantly greater than zero. Using the squared Euclidean norm, the TD is defined as:

$$d_{2S}(x,\mu) = \min_{\alpha,\beta\in\mathbb{R}^L} \{||(x + \sum_{l=1}^{L} \alpha_l x_l) - (\mu + \sum_{l=1}^{L} \beta_l \mu_l)||^2\}$$

This distance measure is also known as two-sided tangent distance (2S) [ ]. To reduce the effort for determining $d_{2S}(x,\mu)$ it may be convenient to restrict the tangent subspaces to the derivatives of the reference (or the observation). The resulting distance measure is called one-sided tangent distance.

2.2 Integration Into the Statistical Approach

The considerations presented above are based on the Euclidean distance, but equally apply when using the Mahalanobis distance in a statistical framework. The result of the integration of one-sided tangent distance into the densities is a modification of the covariance matrix of each kernel in the kernel densities [ ]:

$$p(x|k) = \frac{1}{N_k} \sum_{n=1}^{N_k} \mathcal{N}(x|x_{kn}, \Sigma'_{kn}), \qquad \Sigma'_{kn} = \Sigma + \gamma^2 \sum_{l=1}^{L} \mu_{knl} \mu_{knl}^T$$

Here, the parameter γ denotes the variance of the coefficients α in the tangent subspace. The resulting distances (i.e. the values of the exponent in the Gaussian distribution) approach the conventional Mahalanobis distance for $\gamma \rightarrow 0$ and the TD for $\gamma \rightarrow \infty$. Thus, the incorporation of tangent vectors adds a corrective term to the Mahalanobis distance that only affects the covariance matrix which can be interpreted as *structuring* Σ_{kn}.

2.3 Virtual Data

In order to obtain a better approximation of $p(x|k)$, the domain knowledge about invariance can be used to enrich the training set with shifted copies of the given training data. In the experiments displacements of one pixel in eight directions were used. Although the tangent distance should already compensate for shifts of that amount, this approach still leads to improvements, as the shift is a transformation following the true manifold, whereas the tangents are a linear approximation. As it is possible to use the knowledge about invariance for the training data by applying both tangent distance and explicit shift, this is true for the test data as well. The resulting method is called *virtual test sample method* [4]. When classifying a given image, shifted versions of the image are generated and independently classified. The overall result is then obtained by combining the individual results using the sum rule.

3 The Nearest Neighbor Classifier Using Local Features

As the second method, the nearest neighbor (NN) paradigm is used to classify handwritten characters. To use the NN algorithm, a distance measure between two character images is needed. Usually, the solution is to represent each image as a feature vector obtained from the entire image, using the appearance-based approach as above (each pixel corresponds to one feature) or some type of feature extraction. Finally, using vector space dissimilarity measures, the distance between two character images is computed.

In the handwritten characters classification problem, there usually appear clear differences between handwritten versions of the same character. This is an important handicap to the NN classification algorithm if the feature vector is obtained from the entire image. But it is possible to find *local* parts of the characters that seem to be unchanged, that is, the distance between them is low in two handwritten version of the same character. This leads to the idea of using a local feature approach, where each character is represented by *several* feature vectors obtained from parts of the image.

In a classical classifier [1], each object for training and test is represented by a feature vector, and a discrimination rule is applied to classify a test vector. In the handwritten characters scenario, the estimation of posterior class probabilities from the whole object seems to be a difficult task, but taking local representations we obtain simpler features to learn the posterior probabilities. Moreover, we obtain a model that is invariant with respect to horizontal and vertical translations.

3.1 Extraction of Local Features

Many local representations have been proposed, mainly in the image database retrieval literature [5,6]. In the present work, each image is represented by several (possibly overlapping) square windows of size $w \times w$, which correspond to a set of "local appearances" (cf. Fig. 3).

To obtain the local feature vectors from an image, a selection of windows with highly relevant and discriminative content is needed. Although a number of methods exist to detect such windows [7], most of them are not appropriate for handwritten images or they are computationally too expensive.

In this work, the grey value of the pixels is used as selection criterion. Dark pixels (with low grey value) are selected in order to determine points on the trace of the handwritten character. The surrounding window of each selected pixel is used as one of the local features for the representation of the whole character.

The possibly high dimensionality of w^2 vector components is reduced using a principal component analysis on the set of all local features extracted from the training set.

3.2 Classification Using Local Features

Given a training set, for each image of the training set a set of feature vectors is obtained. The size of these sets may be different, depending on the number of local features chosen. Each local feature vector has the same class label associated as the image it was obtained from. All these feature vectors are then joined to form a new training set.

Given a test image x, we obtain M_x feature vectors, denoted by $\{x_1, \ldots, x_{M_x}\}$. Then, to solve the problem of classification of a test object represented by local features, the sum rule is used to obtain the posterior probability of the object from the posterior probabilities of its local representations [8]:

$$r(x) = \underset{k}{\operatorname{argmax}}\, P(k|x) \approx \underset{k}{\operatorname{argmax}} \sum_{m=1}^{M_x} P(k|x_m)$$

And to model the posterior probability of each local feature, a κ-NN is used:

$$P(k|x_m) \approx \frac{v_k(x_m)}{\kappa},$$

where $v_k(x_m)$ denotes the number of votes from class k found for the feature x_m among the κ nearest neighbors of the new training set. We adopt the sum rule as an approximation for the object posterior probabilities and the k-NN estimate is used to approximate each local feature posterior probability, yielding:

$$r(x) = \underset{k}{\operatorname{argmax}} \sum_{m=1}^{M_x} \frac{v_k(x_m)}{\kappa} = \underset{k}{\operatorname{argmax}} \sum_{m=1}^{M_x} v_k(x_m) \qquad (1)$$

In words, the classification procedure is summarized as follows: for each local feature of the test image, the k-nearest neighbor algorithm gives a fraction of votes to each class, which is an approximation of the posterior probability that each local feature belongs to each class. As each of the vectors obtained from the test image can be classified into a different class, a joint decision scheme is required to finally decide on a single class for the entire test image. The probabilities obtained from each local feature are combined using the sum rule to obtain the overall posterior probability for the entire image for each class. The

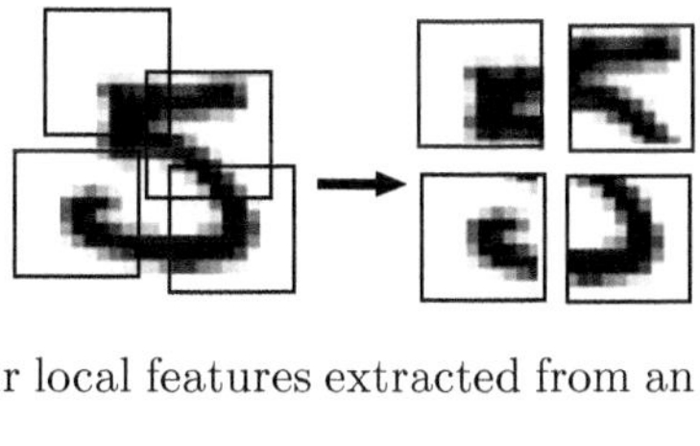

Fig. 3. Example of four local features extracted from an image of a handwritten digit

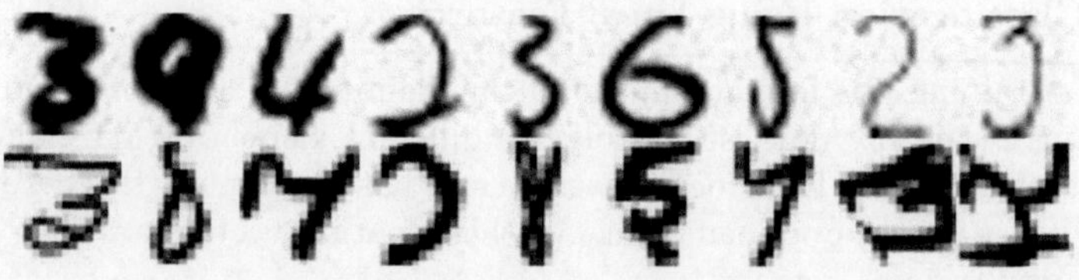

Fig. 4. Examples of digits misclassified by the local feature approach, but correctly classified by the tangent distance classifier (first row, note the variation in line thickness and affine changes) and vice versa (second row, note the missing parts and clutter)

test image is assigned to the class with highest posterior probability. According to Eq. (1), this decision corresponds to the most voted class counting all votes from all local features of the test image [8].

3.3 Computational Considerations

Representing objects by several local features involves a computational problem if the number of local features to represent one object is very large. The k-NN algorithm needs to compare every local feature of a test object with every local feature of every training object. This high computational cost is considerably reduced by using a fast approximate k-nearest neighbor search technique [9].

4 Experimental Results on the USPS Database

All the results presented here were obtained using the well known US Postal Service handwritten digits recognition corpus (USPS). It contains normalized grey scale images of size 16×16, divided into a training set of 7291 images and a test set of 2007 images. A human error rate estimated to be 2.5% shows that it is a hard recognition task. Some (difficult) examples of the test are shown in Fig. 4. Several other methods have been tried on this database and some results are included in Table 1.

Observing that the two classifiers described here led to different errors on the USPS data, this situation seemed to be especially suited for the use of classifier combination in order to improve the results [10]. For example, tangent distance is able to cope with different line thicknesses very well, while the local feature approach can tolerate missing parts (like segmentation errors) or clutter. Fig. 4 shows some of the errors, which were different between the two classifiers.

Therefore, the experimental setup was comparably simple. The best result obtained so far (2.2% error rate) was already based on classifier combination on the basis of class posterior probabilities. Hence, it was only necessary to include the results of the local feature approach (which yielded an error rate of 3.0%) in the combiner. We used the decision based on the local features with two votes, one statistical classifier with one-sided tangent distance and two statistical

Table 1. Summary of results for the USPS corpus (error rates, [%])

*: training set extended with 2,400 machine-printed digits

method		ER[%]
human performance	[SIMARD et al. 1993] [2]	2.5
relevance vector machine	[TIPPING et al. 2000] [11]	5.1
neural net (LeNet1)	[LECUN et al. 1990] [12]	4.2
invariant support vectors	[SCHÖLKOPF et al. 1998] [13]	3.0
neural net + boosting	[DRUCKER et al. 1993] [12]	*2.6
tangent distance	[SIMARD et al. 1993] [2]	*2.5
nearest neighbor classifier	[14]	5.6
mixture densities	[15] baseline	7.2
	+ LDA + virtual data	3.4
(1) kernel densities	[14] tangent distance, two-sided	3.0
	+ virtual data	2.4
	+ classifier combination	**2.2**
(2) k-nearest neighbor, local representations		**3.0**
classifier combination using methods (1) and (2)		**2.0**

classifiers with two-sided tangent distance. Using majority vote as combination rule, ties were arbitrarily broken by choosing the class with the smallest class number k. With this approach, we were able to improve the result from 2.2% to 2.0%. Table 1 shows the error rates in comparison to those of other methods, which are mainly single classifier results.

Note that the improvement from 2.2% to 2.0% is not statistically significant, as there are only 2007 test samples in the test set (the 95% confidence interval for the error rate on this experiment is $[1.4\%, 2.8\%]$). Furthermore, it must be admitted that these improvements seem to result from "training on the testing data". Against this impression we may state several arguments: On the one hand, only few experiments using classifier combination were performed here. Secondly, there exists no development test set for the USPS dataset. Therefore, all the results presented on this dataset (cf. e.g. Table 1) must be considered as training on the testing data to some degree and therefore a too optimistic estimation of the real error rate. This adds some fairness to the comparison. Despite these drawbacks, the presented results are interesting and important in our opinion, because the combination of two classifiers, which are able to deal with different transformations of the input (cf. Fig. 4), was able to improve on a result which was already very optimized.

5 Conclusion

In this work, the combination of two different approaches to handwritten character classification was presented. These two methods are complementary in the transformations of the images that are tolerated and thus in the sets of misclassified images. Therefore, the application of a combined classifier based on these two techniques is a suitable approach. In the experiments carried out, it was observed that the combination improves the results of the previously best classifier on the USPS corpus from 2.2% to 2.0%. Although this is not a statistically significant improvement, qualitatively, the advantages of the combination become clear when regarding Fig. 4. This shows the benefits of the applied combination, which will possibly be helpful for image classification tasks in the future.

References

1. R. Duda, P. Hart, and D. Stork. *Pattern Classification.* John Wiley & Sons, New York, 2nd edition, 2001. 539, 541, 542
2. P. Simard, Y. Le Cun, and J. Denker. Efficient Pattern Recognition Using a New Transformation Distance. In S. Hanson, J. Cowan, and C. Giles, editors, *Advances in Neural Information Processing Systems 5.* Morgan Kaufmann, pages 50–58, 1993. 539, 545
3. D. Keysers, W. Macherey, J. Dahmen, and H. Ney. Learning of Variability for Invariant Statistical Pattern Recognition. In *ECML 2001, 12th European Conference on Machine Learning*, volume 2167 of *Lecture Notes in Computer Science*, Springer, Freiburg, Germany, pages 263–275, September 2001. 541
4. J. Dahmen, D. Keysers, and H. Ney. Combined Classification of Handwritten Digits using the 'Virtual Test Sample Method'. In *MCS 2001, 2nd International Workshop on Multiple Classifier Systems*, volume 2096 of *Lecture Notes in Computer Science*, Springer, Cambridge, UK, pages 109–118, May 2001. 542
5. C. Schmid and R. Mohr. Local grayvalue invariants for image retrieval. *IEEE Transactions on Pattern Analysis and Machine Intelligence*, 19(5):530–535, 1997. 542
6. C. Shyu, C. Brodley, A. Kak, A. Kosaka, A. Aisen, and L. Broderick. Local versus Global Features for Content-Based Image Retrieval. In *Proc. of the IEEE Workshop on Content-Based Access of Image and Video Libraries*, Santa Barbara, CA, pages 30–34, June 1998. 542
7. R. Deriche and G. Giraudon. A Computational Approach to Corner and Vertex Detection. *Int. Journal of Computer Vision*, 10:101–124, 1993. 542
8. R. Paredes, J. Perez-Cortes, A. Juan, and E. Vidal. Local Representations and a Direct Voting Scheme for Face Recognition. In *Workshop on Pattern Recognition in Information Systems*, Setúbal, Portugal, pages 71–79, July 2001. 543, 544
9. S. Arya, D. Mount, N. Netanyahu, R. Silverman, and A. Wu. An optimal algorithm for approximate nearest neighbor searching fixed dimensions. *Journal of the ACM*, 45:891–923, 1998. 544
10. J. Kittler, M. Hatef, and R. Duin. Combining Classifiers. In *Proceedings 13th International Conference on Pattern Recognition*, Vienna, Austria, pages 897–901, August 1996. 544
11. M. Tipping. The Relevance Vector Machine. In S. Solla, T. Leen, and K. Müller, editors, *Advances in Neural Information Processing Systems 12.* MIT Press, pages 332–388, 2000. 545
12. P. Simard, Y. Le Cun, J. Denker, and B. Victorri. Transformation Invariance in Pattern Recognition — Tangent Distance and Tangent Propagation. In *Neural*

Networks: Tricks of the Trade, volume 1524 of *Lecture Notes in Computer Science*, Springer, Heidelberg, pages 239–274, 1998. 545
13. B. Schölkopf, P. Simard, A. Smola, and V. Vapnik. Prior Knowledge in Support Vector Kernels. In M. Jordan, M. Kearns, and S. Solla, editors, *Advances in Neural Inf. Proc. Systems*, volume 10. MIT Press, pages 640–646, 1998. 545
14. D. Keysers, J. Dahmen, T. Theiner, and H. Ney. Experiments with an Extended Tangent Distance. In *Proceedings 15th International Conference on Pattern Recognition*, volume 2, Barcelona, Spain, pages 38–42, September 2000. 545
15. J. Dahmen, D. Keysers, H. Ney, and M. O. Güld. Statistical Image Object Recognition using Mixture Densities. *Journal of Mathematical Imaging and Vision*, 14(3):285–296, May 2001. 545

Training Set Expansion in Handwritten Character Recognition

Javier Cano, Juan-Carlos Perez-Cortes, Joaquim Arlandis, and Rafael Llobet

Instituto Tecnologico de Informatica, Universidad Politecnica de Valencia *
Camino de Vera, s/n 46071 Valencia (SPAIN)
{jcano,jcperez,jarlandi,rllobet}@iti.upv.es

Abstract. In this paper, a process of expansion of the training set by synthetic generation of handwritten uppercase letters via deformations of natural images is tested in combination with an approximate $k-$Nearest Neighbor ($k-$NN) classifier. It has been previously shown [11] [10] that approximate nearest neighbors search in large databases can be successfully used in an OCR task, and that significant performance improvements can be consistently obtained by simply increasing the size of the training set. In this work, extensive experiments adding distorted characters to the training set are performed, and the results are compared to directly adding new natural samples to the set of prototypes.

1 Introduction

The $k-$NN classifier has received renewed attention in the last years and is being successfully used in many pattern recognition tasks, in particular in Handwritten Character Recognition [11] [10]. The k Nearest Neighbors Rule is a classical statistical method which offers consistently good results, ease of use and certain theoretical properties related to the expected error. Being a memory-based technique that does not build a reduced model of the data but stores every prototype to compare it to the test observations, it can benefit from very large sets of training data. In [11], for example, consistent performance improvements of a k-NN classifier in a character recognition task are reported when the original training-set size is increased. The practical use of large databases is being increasingly allowed by advances in hardware speed and memory capacity.

In fact, the spatial cost of storing all the training vectors is often not a problem for current computers even in the case of huge data-sets. The temporal cost, however, can be a limiting factor in many cases. This problem can be approached from two points of view: trying to reduce the number of prototypes without degrading the classification power, or using a fast nearest neighbors search algorithm.

The first approach has been widely studied in the literature. Editing [12], [3], condensing [6], and their multiple variations are well known representatives. These methods have shown good results equaling or sometimes improving the

* Work partially supported by the Spanish CICYT under grant TIC2000-1703-CO3-01

T. Caelli et al. (Eds.): SSPR&SPR 2002, LNCS 2396, pp. 548–556, 2002.
© Springer-Verlag Berlin Heidelberg 2002

classification rates of the $k-$NN rule. Their power resides in the smoother discrimination surfaces they yield by eliminating redundant and noisy prototypes. Cleaner discrimination surfaces reduce the risk of over-training. In a pure $k-$NN classifier with a large database, an adequate choice of k provides the same effect. The best value of k often grows with the size of the database.

The second approach has also been extensively analyzed and many techniques have been proposed to reduce the cost of the exhaustive search of the prototypes set to find the k nearest neighbors to a test point. Most of them make use of $kd-$trees [4], [2], [8] or similar data structures. A $kd-$tree is a binary tree where each node represents a region in a $k-$dimensional space. Each internal node also contains a hyper-plane (a linear subspace of dimension $k-1$) dividing the region into two disjoint sub-regions, each inherited by one of its sons. Most of the trees used in the context of our problem divide the regions according to the points that lay in them. This way, the hierarchical partition of the space can either be carried out to the last consequences to obtain, in the leaves, regions with a single point in them, or can be halted in a previous level so as each leaf node holds b points in its region.

In a $kd-$tree, the search of the nearest neighbor of a test point is performed starting from the root, which represents the whole space, and choosing at each node the sub-tree that represents the region of the space containing the test point. When a leaf is reached, an exhaustive search of the b prototypes residing in the associated region is performed. But the process is not complete at this point, since it is possible that among the regions defined by the initial partition, the one containing the test point does not contain the nearest prototype. If this can happen, the algorithm backtracks as many times as necessary until it is sure to have checked all the regions that can hold a prototype nearer to the test point than the nearest one in the original region. The resulting procedure can be seen as a Branch-and-Bound algorithm.

In most pattern recognition applications, there is no need to guarantee that the exact nearest neighbors of a test observation are found. Fast approximate nearest neighbors search algorithms allow for efficient classification without losing significant classification performance.

In a $kd-$tree, if a guaranteed exact solution is not needed, the backtracking process can be aborted as soon as a certain criterion is met by the current best solution. In [1], the concept of $(1+\epsilon)-$approximate nearest neighbor query is introduced. A point p is a $(1+\epsilon)-$approximate nearest neighbor of q if the distance from p to q is less than $1+\epsilon$ times the distance from p to its nearest neighbor. An algorithm based on these concepts have been used on conventional $kd-$trees in the experiments.

The rest of the paper is organized as follows: Section 2 describes the parameterization, classification and the database expansion procedures proposed. In Section 3, the data and the deformation patterns used are presented. In Section 4, extensive empirical results are reported, and the conclusions are presented in Section 5.

2 Parameterization and Character Classification

The preprocessing and feature extraction methods used are simple and can be performed easily and quickly. The character images were first sub-sampled from their original 128x128 binary pixels into 14x14 gray value representations by first computing the minimum inclusion box of each character, keeping the original aspect ratio, and then accumulating into each of the 14x14 divisions the relative area taken by black pixels, to obtain a continuous value between 0 and 1. Other grid sizes from 8x10 to 16x16 have been tested in this task with similar results. Principal Component Analysis (or Karhunen-Loeve Transform) was then performed on the image representations to reduce their dimensionality to 45, the value which showed the best results in preliminary experiments.

The $k-$NN classifier with the Euclidean Distance in the 45-dimensional space described above has been used in the experiments, employing a fast $(1+\epsilon)-$approximate search with $\epsilon = 2$ in every case. This was the largest value for ϵ empirically found to keep, in our task, almost the same error rates as an exact search. The temporal cost of the classification grows approximately with the logarithm of the size of the training set. Given this slow increase in the search times, an interesting approach to improve the accuracy, keeping at the same time high recognition speeds, is to insert new prototypes into the training set. Of course, making larger the original database is an evident way to do it, but the manual or semi-automated segmentation and labeling procedures needed to build a good, large database are very time-consuming. Therefore, a possible approach to exploit the information of a given database as much as possible is to perform controlled deformations on the characters to insert them into a new larger training set.

Improving a handwritten character recognizer using appropiately distorted samples has been proposed in [5] with good results. In [10], expanding the training set by including the distorted characters is proposed as a faster and cleaner option to include the distorted characters in the training set instead of distorting the test character in several ways and carrying out the classification of each deformed pattern. In this work, we perform extensive experimentation to validate that approach.

3 Databases and Transformations

Two different sets of databases have been employed to perform the experiments. One is public and has been widely used in the literature, and the other one was a locally acquired image database that was used only for one experiment where a larger training set was required.

The first image databases are the well-known NIST Special Database 3, for training, and Special Database 7 for test. Only the upper-case images have been used, with a total of 44951 images for training set and 11941 for test set.

The other database is a locally acquired set of images with slightly over 1 million upper-case letters acquired from forms written by different writers.

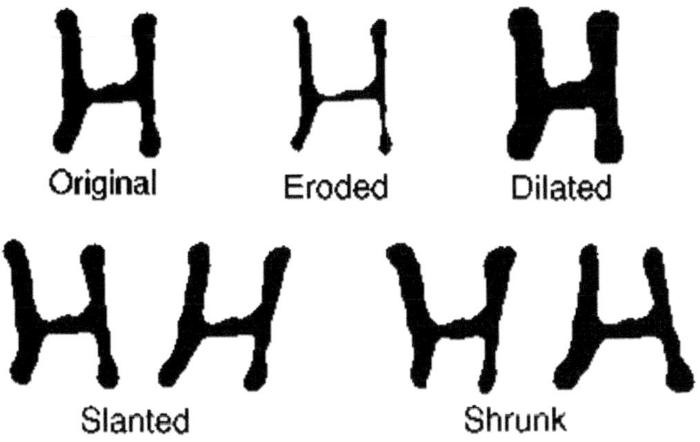

Fig. 1. Families of transformations tested

A database of this size was required to compare the performance of two systems trained with increasingly large sets composed by the same initial core and expanded with real and distorted images respectively.

Four simple kinds of image transformations have been tested (see Figure 1): slant and shrink, to cope with geometric distortions of the writing, and erosion and dilation to account for different writing implements, acquisition conditions, etc. [9] [7]. The transformations were first tested separately and then the one offering the best results (slant) was applied first, expanding the training set so that the rest of transformations were incrementally applied to the original plus the slanted characters. The details are presented in Section 4

All the transformations were applied to the original binary images before scaling them to 14x14 pixels. The slant transformation consisted of right- or left-shifting each row an integer number of positions. The 4 different slants performed were equivalent to rotations of a vertical segment by -26°, -9°, 9°, and 26°. The central row was never shifted, and the amount of shift increased linearly from there to the top and bottom rows, in opposite directions. The new pixels entering the area due to the shifts were set to white.

The shrink transformation consisted of two proportional scalings of the image, keeping the top or bottom line at its original size and then gradually scaling each line until the opposite line is reached and scaled to a fixed smaller size (50%). In the case of erosion and dilation, the classical binary morphological operations were applied by one pixel.

4 Experiments

The first experiment was targeted to determine the value of each of the different distortions, used alone, to improve the accuracy of a $k-$NN classifier. The results,

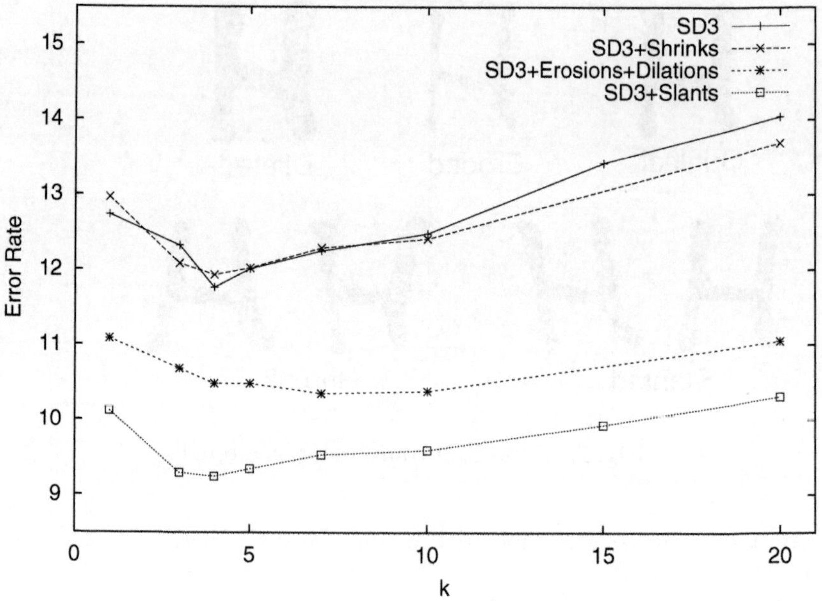

Fig. 2. Error rates of a k−NN classifier for a range of values of k and different training sets composed of real and synthetic images of the NIST database

using a holdout estimation, for the 44951 upper case letters from SD3 as the training set and the 11941 images of SD7 as the test set are shown in Figure 2. The insertion of artificially slanted, eroded and dilated images to the training set produced significant improvements. However, the inclusion of shrunk images did not improve the results. Therefore, this last transformation has not been included in the subsequent experiments. The best value of k was 4 in all cases but one.

In the second experiment, with the same training and test data, the goal was to expand the training set in the best possible way to improve the classification accuracy. To do that, the best transformation, slant, was first used to expand the core of original images, obtaining 224755 images $(44951 + 4 \times 44951)$. Then, the next best transformations, erosion and dilation, have been sequentially applied to the previous set obtaining training sets of 449160 (224755×2) and 673740 $(449160 + 224755)$ respectively. The results are shown in Figure 3.

Another effect that can be noticed in the results of these experiments is that the best value of k seems to gradually increase as the database grows, which is in accordance with what could be expected.

To find out if the accuracy improvement achieved by artificially expanding a core database of images is comparable to using extra real data, another experiment was carried out. In this case, a large locally acquired real database was used, allowing increasingly large training sets up to 674265 images, equivalent to

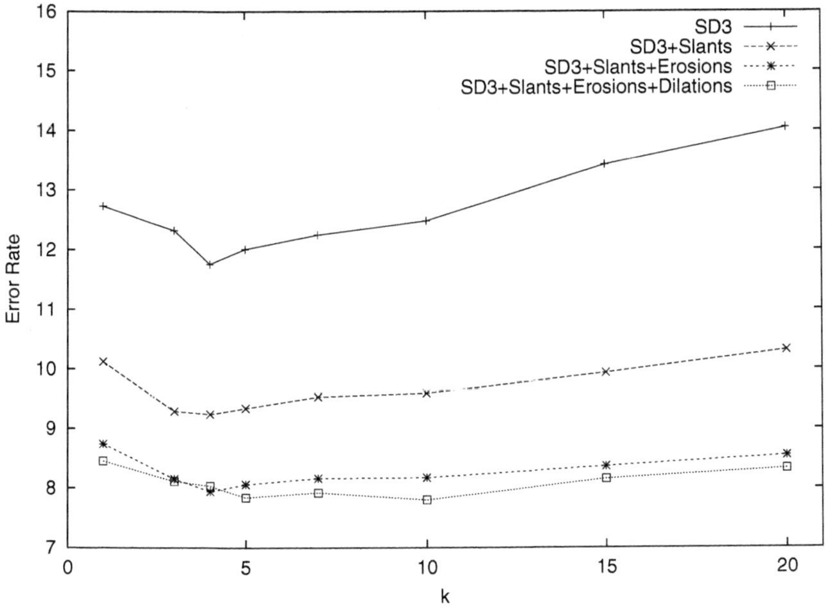

Fig. 3. Error rates of a k−NN classifier for a range of values of k. The original data set was incrementally expanded using deformed images of the NIST database

the size of SD3 expanded with all the synthetic images generated in the previous experiments.

A core of 44951 images (the same size of SD3) from this local database was randomly selected. On the one hand, this core set was made larger by adding new real images from the rest of the local database, and on the other hand, the proposed expansion of the core using deformations was applied. In Figure 4, the results of this experiment on the local database using a k−NN classifier are shown. The value of k used was 4.

From this figure we can see that a reduction from 11.77% to 7.77% on the error rate has been achieved in the experiments with NIST SD3/SD7 by adding synthetic images to the training set. Similarly, an error-rate reduction from 4.08% to 2.85% has been achieved in the experiments with the local database. But an even higher reduction of the error rate, from 4.08% to 1.88%, has been found in the local database when the expansion of the training set is performed using additional real data, instead of deformed images.

These results suggest that, with the synthetic image generation methods employed, it is preferable to make use of extra real data. Of course, this is only possible if a very large database is available.

In Figure 5, the results of the largest real and synthetic local databases, both of similar size, around 675.000 images, are plotted for a range of values of k.

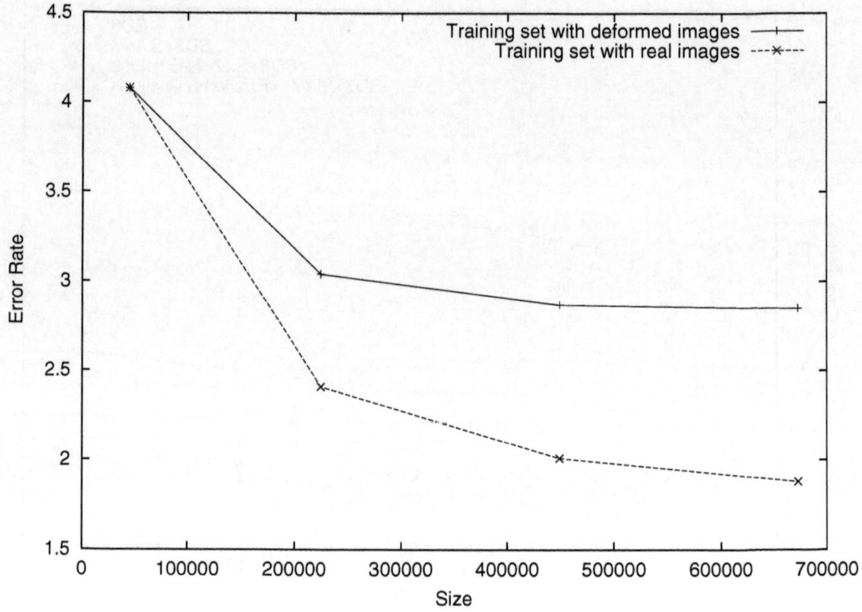

Fig. 4. Comparison of error rates of a k−NN classifier for increasingly large training sets composed of real-only and real+synthetic images from the local database

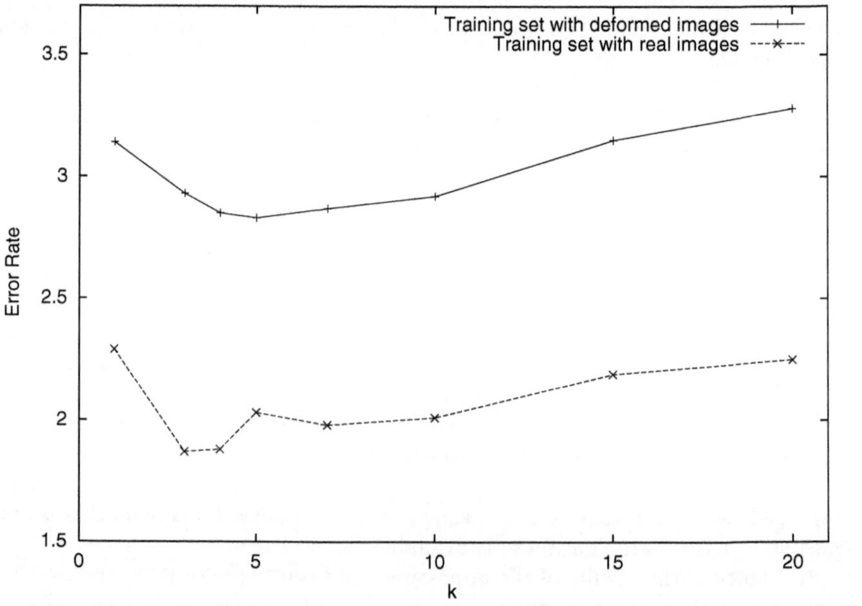

Fig. 5. Error rates of a k−NN classifier for several values of k. A training set comprising the full local database (real images) is compared to one composed by a core subset of real images plus those images deformed in various ways

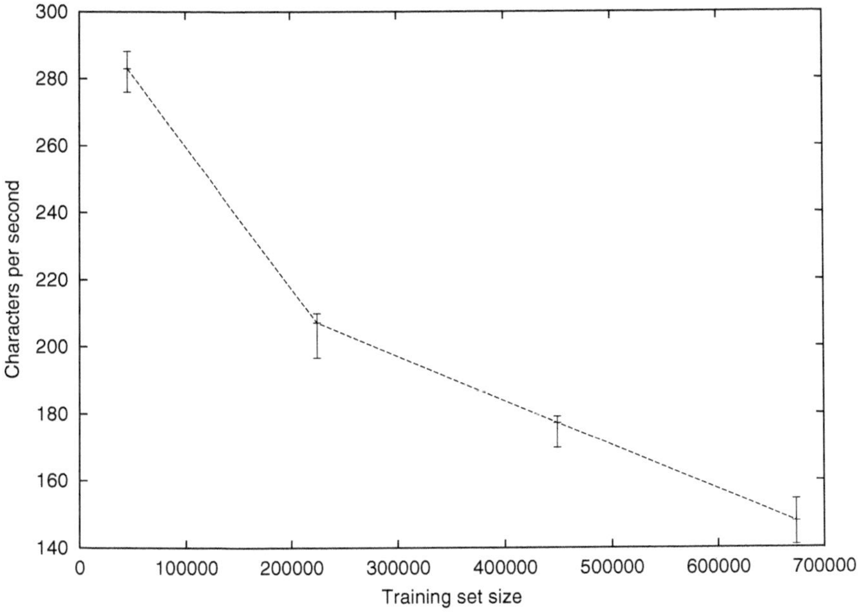

Fig. 6. Recognition speed against traininig set size with real data from the local database in an AMD Athlon PC at 1.2 Ghz. The average, maximum and minimum of 10 experiments are shown for each point

As can be seen from Figure 6, the improvements obtained when the database is expanded only incur a slight decrease on the average recognition speed.

5 Conclusions

In this work, large image databases stored in $kd-$tree structures and approximate nearest neigbour search have been used to obtain competitive error rates and recognition speed.

A recognition rate improvement of 4.0% has been achieved in the upper-case letters classification problem with the NIST-SD3/SD7 (Figure 3), and a 2.20% improvement in the local database (Figure 4) has been achieved by increasing the database size fifteen times (from 44951 original images to a final set of 674265 images).

However, this training set size increase does not seriously affect the processing time requirements of the recognition method. For instance, with the original training set, a search is performed in 3.6 ms (280 cps) with an increase of about 3 ms (to 150 cps) for the whole training set with 674265 images in a computer with an AMD Athlon processor and 1.2 Ghz clock frequency.

A comparison between artificial database expansion and real enlargement of the training set has also been performed. The results suggest that, although both approaches provide significant improvements, the latter is clearly preferable.

References

1. S. Arya, D. M. Mount, N. S. Netanyahu, R. Silverman, and A. Wu. An optimal algorithm for approximate nearest neighbor searching. *Journal of the ACM*, 45:891–923, 1998. 549
2. J. L. Bentley, B. W. Weide, and A. C. Yao. Optimal expected time algorithms for closest point problems. *ACM Trans. on Math. Software*, 6:563–580, 1980. 549
3. P. A. Devijver and J. Kittler. On the edited nearest neighbour rule. In *Proceedings of the 5th International Conference on Pattern Recognition*, pages 72–80. IEEE Computer Society Press, Los Alamitos, CA, 1980. 548
4. J. H. Friedman, J. L. Bentley, and R. A. Finkel. An algorithm finding best matches in logarithmic expected time. *ACM Trans. Math. Software*, 3:209–226, 1977. 549
5. T. M. Ha and H. Bunke. Off-line, handwritten numeral recognition by perturbation method. *IEEE Trans. on PAMI*, 19(5):535–539, May 1997. 550
6. P. E. Hart. The condensed nearest neighbor rule. *IEEE Trans. on Information Theory*, 125:515–516, 1968. 548
7. A. Jain. Object matching using deformable templates. *IEEE Transactions on Pattern Analysis and Machine Intelligence*, 18:268–273, 1996. 551
8. B. S. Kim and S. B. Park. A fast k nearest neighbor finding algorithm based on the ordered partition. *IEEE Trans. on PAMI*, 8:761–766, 1986. 549
9. Jianchang Mao. Improving ocr performance using character degradation models and boosting algorithm. *Pattern Recognition Letters*, 18:1415–1419, 1997. 551
10. J. C. Perez-Cortes, J. Arlandis Arlandis, and R. Llobet. Fast and accurate handwritten character recognition using approximate nearest neighbours search on large databases. In *Workshop on Statistical Pattern Recognition SPR-2000*, Alicante (Spain), 2000. 548, 550
11. S. J. Smith. Handwritten character classification using nearest neighbor in large databases. *IEEE Trans. on PAMI*, 16(9):915–919, September 1994. 548
12. D. L. Wilson. Asymptotic properties of nearest neighbor rules using edited data. *IEEE Trans. on Systems, Man and Cybernetics*, 2:408–420, 1972. 548

Document Classification Using Phrases*

Jan Bakus and Mohamed Kamel

Pattern Analysis and Machine Intelligence Lab
Department of Systems Design Engineering, University of Waterloo
Waterloo, Ont., Canada
{jbakus,mkamel}@pami.uwaterloo.ca

Abstract. This paper presents a Bayes document classifier using phrases as features. The phrases are extracted using a grammar that iteratively applies the rules to the sequence of words in the document. This grammar is generated from a training set using statistical word association. We report an improvement in the classification over the "bag of words" representation.

1 Introduction

Growing number of text information available on the internet is raising interest in automatic analysis of text data. A number of machine learning approaches to sort and organize this data have been proposed, such as document classification, document clustering, and information retrieval [7]. Many of these approaches are based on the "bag of words" approach, where the entire document is represented as a list of all the unique words found in the document ignoring the order of the words. This paper presents a document classifier that takes the word order into account by extracting phrases from the documents and using them as features in the classifier.

One of the first approaches to extract these collocations was by Church and Hanks [2]. They use the word associations to build more comprehensive dictionaries that better reflect actual language use. An improvement on this technique by Smadja presents the Xtract lexicographical tool in [10]. This tool extracts phrases by first identifying bigrams using a technique similar to the one presented by Church and Hanks and then analyzes all the sentences containing the bigrams by examining the distribution of words and parts of speech of the words surrounding the bigram pair in each sentence where the pair occurs.

A phrase extraction technique presented by Ahonen-Myka et. al. in [1] finds a set of all maximal frequent sentences, i.e. such sequences of words that are frequent in the document collection and are not contained in any other longer frequent sequence. This technique works in two phases. The first phase extracts all the frequent phrases, where a sentence is considered frequent if it occurs in at least σ documents. In the second phase, co-occurrences of the maximal frequent set of sentences are found by discovering which sequences tend to co-occur together.

* This work was partially supported by NSERC strategic grant and TL-NCE.

T. Caelli et al. (Eds.): SSPR&SPR 2002, LNCS 2396, pp. 557–565, 2002.
© Springer-Verlag Berlin Heidelberg 2002

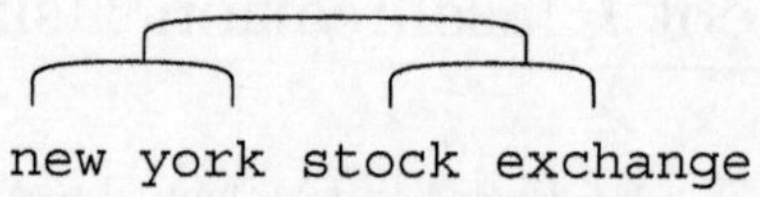

Fig. 1. Structure of the phrase "new york stock exchange"

A number of approaches have been used to extract phrases for the purpose of machine learning tasks. Fagan [5] presents two methods to extract phrases for an information retrieval application: a statistical method based on simple text characteristics and a syntactic method that generates phrases from syntactic parse trees. He reports that the syntax based method is better than the statistical method, however the performance of both methods was found to be variable.

Mladenić and Gorbelnik [8] extract phrases for text categorization. They enumerate all possible phrases up to a length of five words and use feature extraction to prune out the irrelevant phrases. The remaining phrases are used in a document classification task using a Naive Bayes classifier.

This paper presents an algorithm for extraction of phrases using the statistical collocation by building a hierarchical grammar structure of sub-phrases. The hierarchical structure allows extraction of phrases of any length by simply adding more rules, as well as extraction of all the constituent sub phrases. For example, the the phrase "new york stock exchange" contains two sub phrases "new york" and "stock exchange". The phrase forms a hierarchical structure, as shown in Figure 1, which can be captured by the presented hierarchical grammar.

To extract the phrases, a grammar is first constructed from a training corpus. This grammar is iteratively built by merging adjacent words tokens together to form a rule, which can be further merged together in subsequent iterations. The merging is performed according to mutual information association measure. To extract the phrases, the grammar rules are applied in the order of the association weights to a test document until no more rules can be applied. The extracted phrases are tested on a document classification task and is shown to improve the performance of a Naive Bayes document classifier.

This paper is organized as follows. Section 2 describes document classification using Naive Bayes classifier, Section 3 outlines the grammar based phrase extraction technique. Section 4 gives the results of the classification experiments, and Summary and Future Research is presented in Section 5.

2 Document Classification

The objective of document classification is to assign a document into one or more classes. This is typically achieved using one of the machine learning techniques, which learns the classification decision from a training set of examples. We use the Reuters document data set, where the categories are not mutually exclusive and may overlap. As a result, the classification task is treated as a number of binary classification tasks. A different classifier is constructed for each category, and each of these classifiers determines if the document belongs in the corresponding category.

Using the machine learning techniques, the objective is to learn from a training set of examples to assign a document into one or more categories. since the categories may overlap, each class is treated as a binary classification problem.

The first step in the document classification is to transform each document into a feature vector that can be used in a machine learning approach. Each document is represented as a feature vector of distinct words (also called terms) t_i that appear in the document. To reduce the dimensionality of the feature vector, the words in the document are processed by stemming each word to its base form, and removing a list of very frequent words, called "stop words". Most document classification approaches use the individual words as the features and ignore the order of the words in the document.

This paper presents a technique that extracts phrases as well as words from each document, and uses both as the features to classify the documents. The Naive Bayes classifier is used to classify both the base case using only words as well as the test case using words and phrases.

2.1 Naive Bayes Classifier

To classify the documents we use a Naive Bayes classifier as presented in [6]. The assumption behind Naive Bayes classifier is that the features are independent of each other, and this is certainly not the case with words and phrases as features. However, according to [4] the lack of independence does not necessarily mean that the classifier will perform poorly. Empirical evidence suggests that a Naive Bayes classifier performs surprisingly well despite the lack of feature independence [6,8].

The classifier is constructed form the training data to estimate the probability of each class given the document feature vector. Using the Bayes theorem, we can estimate the probability that a document with a feature vector $\mathbf{d}$ belongs to a class c_j as:

$$P(c_j|\mathbf{d}) = \frac{P(c_j)P(\mathbf{d}|c_j)}{P(\mathbf{d})} \tag{1}$$

The denominator of the above equation does not differ between categories and can therefore be left out. Using the assumption that the features of the vector $\mathbf{d} = (t_1, \ldots, t_M)$ are conditionally independent and (1) becomes.

$$P(c_j|\mathbf{d}) = P(c_j)\prod_{i=1}^{M} P(t_i|c_j) \tag{2}$$

An estimate $\hat{P}(c_j)$ for the class probability $P(c_j)$ is calculated as a fraction of the training documents assigned to the class c_j as:

$$\hat{P}(c_j) = \frac{N_j}{N} \tag{3}$$

where N_j is the number of documents assigned to class c_j, and N is the total number of documents. To estimate the probability $P(t_i|c_j)$, we use the Laplace estimator:

$$\hat{P}(t_i|c_j) = \frac{1 + N_{ij}}{M + \sum_{k=1}^{M} N_{kj}} \tag{4}$$

where M is the feature vector size, and N_{ij} is the number of times feature i occurred within documents from class c_j. Despite the fact that the assumptions of conditional independence is generally not true for word appearances in documents, the Naive Bayes classifier is surprisingly effective.

2.2 Feature Selection

Using all the available features results in a very large feature vector. In order to reduce the size of the feature vector we use the χ^2 feature extraction [11]. For each feature t_i and class c_j we calculate the χ^2 statistic of the features relevance to the category.

$$\chi^2(t_i, c_j) = \frac{N(AC - CB)^2}{(A + C)(B + D)(A + B)(C + D)} \tag{5}$$

A is the number of times t_i and c_j co-occur, B is the number of times t_i occurs without c_j, C is the number of times c_j occurs without t_i, D is the number of times neither c_j nor t occurs, and N is the total number of documents. The features that occurred at least 3 times in the training corpus and have χ^2 statistic greater than 3 are selected.

3 Phrase Grammar

The phrase extraction is performed by iteratively merging bigrams together using a similar approach as Ries et. al. in [9]. The first step is to create a frequency table of all the bigrams in a training corpus, and calculate the mutual information for all bigrams. The bigram with the highest positive mutual information is replaced in this corpus with a new symbol, and a rule is entered in the grammar. This new symbol may be merged with other adjacent words or symbols in subsequent iterations. The frequency counts and association weights for all the bigrams are recalculated to reflect the merge operation and the procedure is repeated until there are no more bigrams with positive mutual information. To extract the phrases, a similar iterative algorithm is used. First, a list of bigrams from the input word sequence is collected and similar iterative bigram merging algorithm is applied. However, instead of calculating the mutual information from the occurrences, the weights extracted from the training corpus are used.

3.1 Grammar Generation Algorithm

The grammar generation algorithm is shown in Figure 2. The algorithm inputs a training word sequence **L** and collects occurrence counts for all the words and

```
input training sequence L
fill the word counts C and pair counts M from seq. L
for for every non-zero entry of M[i,j] do
  X[i,j] = association measure between i and j
end for
while do
  i,j = indices of maximum entry in X
  if X[i,j] ≤ 0 then
    break
  end if
  g = next available grammar rule index
  add the rule {X[i,j], g ← i, j} to the grammar
  for ever occurrence of adjacent pair i,j in seq. L do
    replace the pair i,j with g
    update counts M, C and recalculate X from counts
  end for
end while
return grammar G
```

Fig. 2. Grammar extraction algorithm

adjacent word pairs. The occurrence count of each word is stored in a vector **C** and the occurrence count for each adjacent pair of words is stored in a count matrix **M**.

From these counts, we can calculate the probability $P(t_i)$ of a token t_i and the probability $P(t_i, t_j)$ that a token t_i is followed by token t_j. With the probabilities, the association measure for all possible token pairs is calculated and stored in an association matrix **X**. The algorithm then selects two adjacent tokens t_i and t_j with the highest association measure, $w = \mathbf{X[i,j]}$, and forms a new rule:

$$w : t_k \leftarrow t_i, t_j$$

The rule is added to the grammar and all occurrences of the adjacent tokens t_i and t_j in the sequence **L** are replaced with the new token t_k. The occurrence count vector **C** and matrix **M** are updated to reflect the fact that there is now a new token t_k and fewer of the child tokens t_i and t_j, and then the association matrix **X** is recalculated. The algorithm iterates until there is no pair of adjacent tokens with positive association weights.

An example of the generated grammar for the phrase "`new york stock exchange`" extracted from the Reuters corpus using mutual information as the association measure is shown in Figure 3. This grammar contains four terminal rules and three non terminal rules. The terminal rules have no association weight and simply map each of the input word strings to of the tokens: t_1, t_2, t_3, and t_4. Each of the three non terminal rules combines together a pair of adjacent tokens with a positive association measure.

The association measure is calculated by first counting the number of times two tokens t_i and t_j occur independently and the number of times they occur together, and from these the corresponding probabilities $P(t_i)$, $P(t_j)$, and $P(t_i, t_j)$

Weight	Rule	Expanded Phrase
–	$t_1 \leftarrow$ `new`	`new`
–	$t_2 \leftarrow$ `york`	`york`
–	$t_3 \leftarrow$ `stock`	`stock`
–	$t_4 \leftarrow$ `exchange`	`exchange`
7.681	$t_5 \leftarrow t_1\ t_2$	`new york`
5.260	$t_6 \leftarrow t_3\ t_4$	`stock exchange`
8.513	$t_7 \leftarrow t_5\ t_6$	`new york stock exchange`

Fig. 3. Resulting grammar from the phrase `new york stock exchange` using the mutual information association measure

are calculated. Each of the measures then compares the actual probability of the two tokens occurring together with the estimated probability of co-occurrence assuming the tokens are independent.

Mutual information is used as the association measure, and is defined as the measure of the amount of information that one event contains about another event [3]. Positive mutual information indicates that if the first event is observed, the probability of observing the second event is increased. Given two tokens t_i and t_j, then the mutual information between them is given as:

$$I(t_i;t_j) = \log_2 \frac{P(t_i,t_j)}{P(t_i)P(t_j)} \tag{6}$$

where $P(t_i)$ and $P(t_j)$ are the probabilities of the occurrence of each of the tokens and $P(t_i,t_j)$ is the probability that the token t_i is followed by token t_j.

3.2 Phrase Extraction Algorithm

Given a grammar of rules and an input sequence of word tokens, the phrases contained in the text sequence are to be extracted. This algorithm, as shown in Figure 4, inputs a sequence **L** and a grammar **G** and outputs a vector **output** containing counts of all the applicable rules from the grammar in the test sequence. These rules are expanded to form the extracted phrases.

The initialization is similar to the grammar generation, where occurrence count of each adjacent word pair is stored in a matrix **M**. Rather than using association measure the weights in the grammar are used. For each non-zero entry **M[l,r]**, that also has the corresponding rule in the grammar, the weight of the rule is used in the corresponding entry **X[i,j]** in the association matrix.

Similar iterative loop is used to find the greatest weight in **X**, but instead of creating new rule, a corresponding index of the vector **output** is incremented. At the end of the algorithm the vector **output** contains the counts of all the applicable rules, which can be expanded to the extracted strings. The update of the sequence **L**, count matrix **M** and the association matrix **X** is done the same way as in the grammar generation algorithm. The algorithm iterates until there are no more applicable rules, i.e. there are no more positive entries in the **X** matrix.

```
input test sequence L and grammar G
fill the pair counts M from sequence L
for every non-zero entry M[i,j] do
  if there exists rule {weight, g ← i, j} then
    X[i,j] = weight
  end if
end for
while do
  i,j = indices of maximum entry in X
  if X[i,j] ≤ 0 then
    break
  end if
  find the rule {weight, g ← i, j} in the grammar
  increment output[g]
  for ever occurrence of adjacent pair i,j in seq. L do
    replace the pair i,j with g
    update counts M and recalculate X from grammar
  end for
end while
return output
```

Fig. 4. Phrase extraction algorithm

4 Results

The classifier was tested using the Reuters corpus. From all the available documents, 6 000 were randomly selected as the test set. From the remaining documents 10 000 were randomly selected as the phrase extraction training set, and from this set 6 000 documents were selected to train the classifier. Total of 6 experiments were carried out, each with a different random split and the results were averaged.

The effectiveness of the classification is summarized in four contingency table values:

a = the number of documents *correctly assigned* to the class
b = the number of documents *incorrectly assigned* to the class
c = the number of documents *correctly rejected* from the class
d = the number of documents *incorrectly rejected* from the class

To evaluate the performance of classifiers, precision and recall are calculated as:

$$\begin{aligned} precision &= \frac{a}{a+c} \\ recall &= \frac{a}{a+b} \end{aligned} \tag{7}$$

Examining precision or recall alone is misleading, as there is a trade-off between the two values. Therefore each classifier is tuned, such that the precision and

Table 1. Break-even values for the Reuters test corpus

category	Naive Bayes words	Naive Bayes phrases	Difference
australia	46.98 %	50.30 %	3.31 %
belgium	46.46 %	48.10 %	1.63 %
brazil	72.83 %	74.28 %	1.45 %
canada	56.24 %	59.92 %	3.67 %
china	66.47 %	67.15 %	0.67 %
france	58.82 %	62.18 %	3.36 %
japan	67.81 %	72.71 %	4.90 %
switzerland	51.87 %	61.26 %	9.39 %
uk	69.24 %	72.65 %	3.41 %
usa	91.34 %	92.38 %	1.03 %
ussr	62.36 %	65.25 %	2.89 %
west-germany	66.11 %	71.20 %	5.08 %
earn	88.23 %	88.77 %	0.54 %
acq	68.14 %	70.68 %	2.54 %
money-fx	73.07 %	76.54 %	3.47 %
grain	75.62 %	75.44 %	-0.17 %
crude	74.42 %	76.66 %	2.24 %
trade	62.08 %	63.61 %	1.53 %
interest	63.04 %	68.21 %	5.17 %
ship	70.54 %	71.22 %	0.67 %
wheat	66.97 %	68.93 %	1.96 %
corn	55.77 %	57.44 %	1.66 %

recall are equal, resulting in a break-even point. Table 1 shows the break-even points for both the classifier using words only and classifier using phrases (and all sub phrases) for the tested categories. We apply all the extracted phrases and sub-phrases to the classifier, and let the feature selection "decide" which phrases are best suited for the classification task. The phrase classifier consistently improves the break-even point by several percent over the word classifier.

5 Summary and Future Research

We presented a method of phrase extraction based on the mutual information association measures of adjacent words and reported on the use of the phrases in a document classification task. We reported consistent improvement of the break-even point classification performance, however the improvement is on the order of a few percent. Examining the extracted phrases manually showed that many of the extracted phrases are very good, however we attribute this poor performance to two factors.

The first factor is the total number of extracted phrases is significantly less than the total number of extracted words. Typically, only about 10% – 15% of the extracted features are phrases and the rest are words. In the future we hope to increase the number of phrases extracted by increasing the training set to generate the grammar. Using a small phrase extraction training set, we are able to only extract phrases that occur within the small training set.

The second factor in the poor performance is the fact that there are many phrases that correspond to the same feature, yet have a different sequence of words. Consider the phrases: `president bush`, `president george bush`, and `president george w. bush`. All these phrases would be considered as unique features, however they all refer to the same person and should be treated as the same feature. To improve this, in the future we propose that the extracted words and phrases should be clustered together based on the the context they occur in.

References

1. Ahonen-Myka, H., Heinonen, O., Klemettinen, M., and Verkamo, A. I. Finding co-occurring text phrases by combining sequence and frequent set discovery. In *Proceedings of 16th International Joint Conference on Artificial Intelligence* (1999). 557
2. Church, K. W., and Hanks, P. Word association norms, mutual information and lexicography. *Computational Linguistics 16*, 1 (Mar. 1990), 22–29. 557
3. Cover, T. M., and Thomas, J. A. *Elements of Information Theory.* Wiley and Sons, Inc., 1991. 562
4. Domingos, P., and Pazzani, M. On the optimality of the simple bayesian classifier under zero-one loss. *Machine Learning 29* (1997), 103 – 130. 559
5. Fagan, J. L. *Experiments in Automatic Phrase Indexing for Document Retrieval: A Comparison of Syntactic and Non-Syntactic Methods.* PhD thesis, Department of Computer Science, Cornell University, Ithaca, USA, 1997. 558
6. Joachims, T. A probabilistic analysis of the rocchio algorithm with tfidf for text categorization. In *International Conference on Machine Learning (ICML)* (1997). 559
7. Kosala, R., and Blockeel, H. Web mining research: A survey. *ACM SIGKDD Explorations Newsletter 2*, 1 (June 2000). 557
8. Mladenić, D., and Grobelnik, M. Word sequences as features in text-learning. In *Proceedings of the 17th Electrotechnical and Computer Sciences Conference (ERK-98)* (Ljubljana, Slovenia, 1998). 558, 559
9. Ries, K., Buø, F., and Waibel, A. Class phrase models for language modelling. In *Proc. of the 4th International Conference on Spoken Language Processing (ICSLP'96)* (1996). 560
10. Smadja, F. Retrieving collocations form text: Xtract. *Computational Linguistics 19*, 1 (1993), 143–177. 557
11. Yang, Y., and Pedersen, J. O. A comparative study on feature selection in text categorization. In *Text Categorization Proceedings of the Fourteenth International Conference on Machine Learning (ICML-97)* (1997). 560

Face Detection by Learned Affine Correspondences*

Miroslav Hamouz, Josef Kittler, Jiri Matas, and Petr Bílek**

Centre for Vision, Speech and Signal Processing
University of Surrey, United Kingdom
{m.hamouz,j.kittler,g.matas}@eim.surrey.ac.uk

Abstract. We propose a novel framework for detecting human faces based on correspondences between triplets of detected local features and their counterparts in an affine invariant face appearance model. The method is robust to partial occlusion, feature detector failure and copes well with cluttered background. Both the appearance and configuration probabilities are learned from examples. The method was tested on the XM2VTS database and a limited number of images with cluttered background with promising results – 2% false negative rate – was obtained.

1 Introduction

Human face detection and precise localisation play a major role in face recognition systems, significantly influencing their overall performance. In spite of the considerable past research effort it still remains an open problem. The challenge stems from the fact that face detection is an object-class recognition (categorisation) problem, where an object to be recognised is not just a previously seen entity under different viewing conditions, but rather an instance from a class of objects sharing common properties such as symmetry and shape structure. Existing face detection algorithms may be classified according to different criteria, but for our purpose the two following categories are appropriate.

Holistic Face Models In this approach, a representation of the image function defined over manually selected image region containing a face is learned from examples. During detection, probability of an image patch belonging to the face class is evaluated (or the patch is passed into a face - non-face classifier). This probability or classification must be computed for all possible positions, scales and rotations of the face. Typical examples of this method are the work of Moghaddam and Pentland [MP96, MP95] and Sung and Poggio [SP98]. Holistic methods suffer from at least three problems. During detection, a computationally exhaustive search through several scales and rotations has to be carried out in order not to miss any instances of a face in the image. Secondly, the human face is as a whole a highly variable object (class) from the appearance point of

* This work was supported by the EU Project BANCA

** Currently at CMP, Czech Technical University, Prague, bilek@cmp.felk.cvut.cz

T. Caelli et al. (Eds.): SSPR&SPR 2002, LNCS 2396, pp. 566–575, 2002.
© Springer-Verlag Berlin Heidelberg 2002

view thus making modelling of photometric effects difficult. Thirdly, effects of occlusion, e.g. by glasses, hair, beards or other objects are difficult to overcome.

Local Feature Face Models In this framework, local feature detectors are used. A face is represented by a shape (configuration) model together with models of local appearance. The most popular of such methods has been the Dynamic Link Architecture, where the preferred shape is defined by an energy function and local appearance is captured in the so-called jets, i.e. responses of Gabor filters [LVB+93, KP97]. Typically, the positions of local features are chosen manually and the appearance models are learned from examples. The work of Burl and Perona [BLP95, BP96, BWP98] is a rare example where an attempt is made to learn the local feature detectors automatically. Other approaches in this group include [VS00, SK98].

Some local methods formulate face detection as a search (minimisation) in the (in principle) continuous pose space. However, the problem can be formulated as a combinatorial search for correspondence between the (hopefully few) responses of the local detectors and the face model. Especially if the local feature detectors produce a small number of false positives, this could be significantly less computationally expensive.

In this paper we present a novel method which addresses these issues. We introduce a detection framework which exploits the advantages of both approaches and at the same time avoids their drawbacks. We use small local detectors to promote fast processing. Face location hypotheses are generated by the correspondence between the detected features and the model. In hypothesis verification all the available photometric information is exploited. The crucial concept of our method is a face space in which all the faces are geometrically normalised and consequently photometrically correlated. In such space, both the natural biological shape variability of faces and distortions introduced by scene capture are removed. As a result of face normalisation, the face features become tightly distributed and the whole face class very compact. This greatly simplifies the face detection process.

In contrast to holistic methods, our search for a face instance in the image is navigated by the evidence coming from the local detectors. By using correspondences between the features in the face space and features found by local detectors in the image a full affine invariance is achieved. This is substantially different from the holistic methods, where face patch is not affinely aligned and the detector has to be trained to cope with all the variability. Additionally, the search through subsets of features which invoke face hypotheses is reduced using geometric feature configuration constraints learned from the training set. The method is robust to occlusion since any triple of features is sufficient to instantiate a face hypothesis. In contrast to existing local methods, as the feature distributions in the face space are very compact, the geometric constraints become very selective. Moreover, all the photometric information is used for hypotheses verification. In other words any geometrically consistent configuration of features gets a favourable score, but serves only as a preliminary evidence of

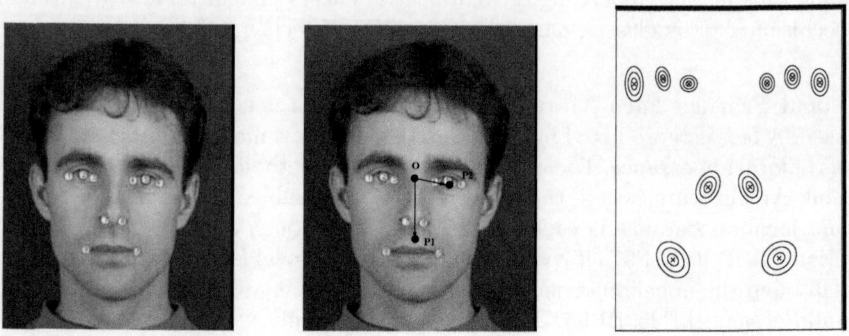

(a) Manually selected features

(b) Co-ordinate system

(c) Variability of features in the face space

Fig. 1. Construction of face space co-ordinate system

a face being present. The final decision whether a face is present or not is made by comparing the underlying photometric information (or its function) against a fully affinely aligned appearance model employed by the detector.

The aim of this paper is to present the framework of the proposed approach. First, the concept of the face space is introduced in Section 2. Next we show how geometric constraints derived from the probabilistic model learned on training data can be used to prune the list of face hypotheses significantly. The probabilistic feature configuration model and "feature location confidence regions" used for pruning are described in Section 3. The speed up is demonstrated experimentally on the commonly used XM2VTS database [MMK+99] in Section 4. The paper is concluded in Section 5.

2 Proposed Methodology

Face Space In order to reduce the inherent face variability, each face is registered in a common coordinate system. We assume that the geometric frontal human face variability can be, to a large extent, modelled by affine transformations. To determine an affine transformation from one face to another, we need to define correspondences of three reference points. As a good choice we propose the midpoint between the eye coordinates, one eye coordinate and the point on the face vertical axis of symmetry half way between the tip of the nose and the mouth as illustrated in Figure 1(b). We refer to the coordinate system defined by these facial points as "face space". We established experimentally that the total variance of the set of XM2VTS training images registered in the face space was minimised by this particular selection of reference points. Note that two of these reference points are not directly detectable. However, their choice is quite effective at normalizing the width and the length of each face. Consequently

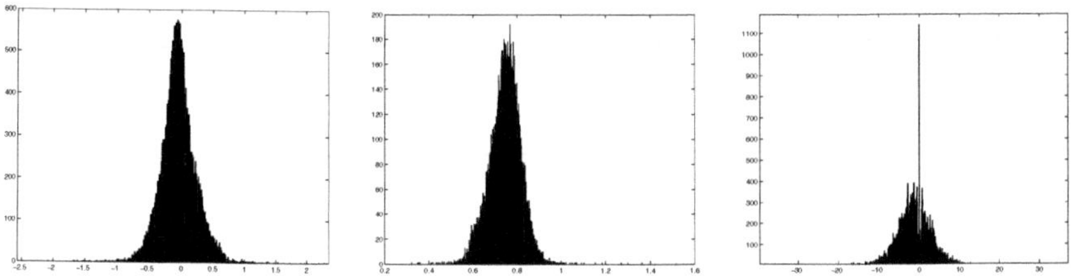

(a) Histogram of shear (b) Histogram of squeeze (c) Histogram of rotation (in degrees)

Fig. 2. Histograms of transformation parameters on XM2VTS

in the coordinate system defined by these points, the ten detectable face features, shown in Figure 1(a) (left and right eye corners, eye centres, nostrils and mouth corners) become tightly clustered. The compactness of the class of faces in the face space was verified experimentally, see Figure 1(c) where the position variance of the detectable face features is depicted.

Face Detection Method Our method uses local feature detectors, which are optimised to detect the 10 key face features shown in Figure 1(a). Any triplet of face features detected in the image allows us to determine the affine transformation that would map the features and therefore the whole face into the face space. As shown in the previous section, the distribution of the key features is very compact and therefore it can be adequately represented by their mean vectors. Mapping the image features onto these mean positions will enable us to register the hypothesised face in the face space and perform appearance-based hypothesis verification. In contrast to Schmidt [VS00], Burl [BLP95, BP96], and others, in our approach the face is not detected simply as an admissible (i.e. conforming to a model) configuration of several features found in an image, rather a face position is hypothesized by a triplet of local features and all the photometric information is used to verify whether face is present or not. It can be argued that this framework is quite close to the way humans are believed to localise faces [BY98].

In the simplest case, the matching of a hypothesised face with the prototype face could be accomplished by normalised correlation. We adopted a PCA representation of the face class in the face space. However, any appearance model, e.g. exploiting a neural network or SVMs, could be used. The novelty of our approach lies in the hypothesis generation part, not in the verification. Our hypothesis verification method is in essence similar to that of Moghaddam and Pentland, originally proposed for face recognition [MP95, MP96].

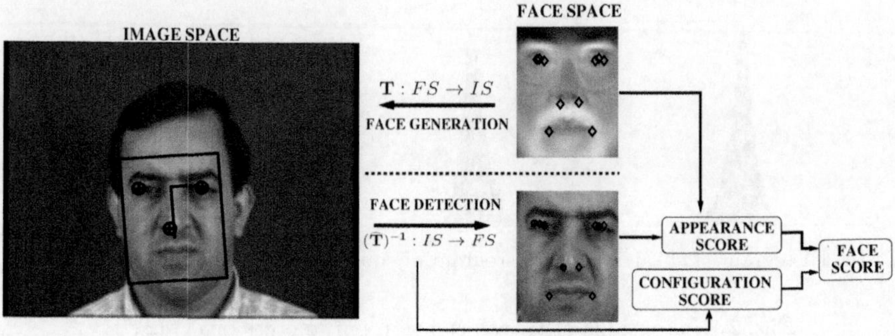

Fig. 3. Schematic diagram of the proposed method

3 Probabilistic Feature Configuration Model

An essential prerequisite of the proposed approach is that the face features in the image are detected and identified with manageable false positive and false negative rates. Clearly this is not an easy task, in spite of the redundancy. To hypothesise a face we need to detect at least three face features of different type. As there are 10 features on each face that our detectors attempt to extract, there are potentially many triplets that may generate a successful hypothesis. However, some of the configurations would be insufficient to define the mapping to the face space, such as a triplet composed of eye features. The two types of error of the feature detectors have different effects. If less than three features are detected, the face will not be detected. On the other hand, false positives will increase the number of triplets, and thus face hypotheses, that have to be verified.

Rather than setting a restrictive threshold, the number of false hypotheses is controlled by geometric pruning. This is achieved as follows. Firstly, we eliminate all the combinations of features which would lead to a degenerate solution for the affine transformation or the transform would be sensitive to errors. The condition number (ratio of the biggest and the smallest singular value) of the matrix made up by putting the face space coordinates of features as columns (homogeneous) was used to determine the well-posedness of triplets. Second, we set out to identify and filter out the triplets that would yield an unfeasible (unrealistic) affine transformation - such that are outside an approximation of the convex hull of transformation encountered in the training set. To construct the probabilistic transformation model, the 6-parameter affine transformations $\widehat{T}$ was decomposed in the following way:

$$\begin{pmatrix} a & b & c \\ d & e & f \\ 0 & 0 & 1 \end{pmatrix} = \begin{pmatrix} 1 & 0 & t_x \\ 0 & 1 & t_y \\ 0 & 0 & 1 \end{pmatrix} \begin{pmatrix} r & 0 & 0 \\ 0 & r & 0 \\ 0 & 0 & 1 \end{pmatrix} \begin{pmatrix} \cos\phi & \sin\phi & t_x \\ -\sin\phi & \cos\phi & t_y \\ 0 & 0 & 1 \end{pmatrix} \begin{pmatrix} t & 0 & 0 \\ 0 & \frac{1}{t} & 0 \\ 0 & 0 & 1 \end{pmatrix} \begin{pmatrix} 1 & n & 0 \\ 0 & 1 & 0 \\ 0 & 0 & 1 \end{pmatrix} \begin{pmatrix} 0 & 1 & 0 \\ 1 & 0 & 0 \\ 0 & 0 & 1 \end{pmatrix}^R \quad (1)$$

where R is reflection (either 0 or 1), n shear, t squeeze, ϕ rotation, r scale

Data : Local detector positions and labels, face-to-image transformation model $p(\widehat{T})$, face appearance model, input image, translation-free feature position confidence regions
Result : Frames of detected face instances in the image
for *All the three-tuples of feature types X, Y, Z defining a well-posed triplet* **do**
  for *All responses of feature detector X* **do**
    Compute feature position confidence regions for remaining features.;
    for *All responses of feature detector Y, Z in confidence regions* **do**
      estimate face to image space affine transformation $\widehat{T}$;
      compute probability $p(\widehat{T})$ of transformation ;
      if $p(\widehat{T}) > threshold1$ **then**
        map image patch into the face space using $\widehat{T}^{-1}$;
        compute probability of appearance $P(A)$;
        if $p(face) = p(\widehat{T}) \cdot p(A) > threshold2$ **then**
          Face detected, record face position defined by $\widehat{T}$;
        end
      end
    end
  end
end

Algorithm 1: Summary of the Face Detection Algorithm

and t_x, t_y translation. This decomposition provides an intuitive feel for the given mapping. We assumed that these parameters are independent and this has been confirmed experimentally. The probability of a given transformation $\widehat{T}$

$$p(\hat{T}) \approx p(n) \cdot p(t) \cdot p(\phi) \cdot p(r) \cdot p(t_x, t_y) \tag{2}$$

is used, together with the appearance model, to reject or accept a face hypothesis. The components $p(n)$, $p(t)$, and $p(\phi)$ of $p(\hat{T})$ are assumed Gaussian, which is in good agreement with experimental data, see Figure 2; $p(t_x, t_y)$ and $p(r)$ are assumed uniform. This assumption reflects the understanding that the useful information content is carried only by shear, squeeze and rotation. Scale which corresponds to the size of a face and translation that represents the position in an image are chosen to lie within a predefined interval (i.e. certain range of face sizes and positions are allowed). Such explicit modelling of probability of face to image transformations may be useful in applications. It is easy to make the system ignore e.g. small faces in the background. The final likelihood of a face hypothesis whose location is defined by a feature triplet is

$$p(face|triplet) = p(A|triplet) \cdot p(\widehat{T}|triplet) \tag{3}$$

where p(A) is the likelihood of appearance. In fact, instead of computing probabilities, we work in the log space, where the product becomes a sum and instead of probability we get a score based on Mahalanobis distance.

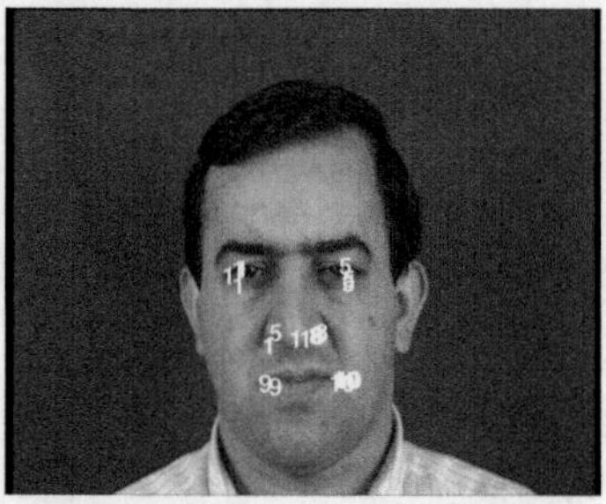

Fig. 4. Five types of detected features, roughly corresponding to a left eye corner, right eye centre, left nostril, left mouth corner and right mouth corner

Efficiency In case that all well-posed triplets would be checked (i.e. at least $\widehat{T}$ computed) the complexity of the search algorithm would be $O(n^3)$, where n is the total number of features detected in the image. In our approach, the set of face-to-image transformations from the training set is used to restrict the number of triplets tested. After picking a first feature of a triplet, a region where the other two features of the triplet may lie is established as an envelope of all positions encountered in the training set. Such regions are called *feature position confidence regions.* Since the probabilistic transformation model and these regions were derived from the same training data, triplets formed by features taken from these regions will have nonzero transformation probability. All other triplets are false alarms, since such configurations did not appear in the training data. The detection algorithm is summarised in Alg. 1. The structure of the detection process is graphically depicted in Figure 3. We assume that an instance of the face in the image is obtained by a randomly chosen appearance and transformation T with probability as in the training set (face generation). In the detection stage, first a linear estimate of T, denoted ($\widehat{T}$) is obtained from a triplet of different types of face features. Next, an image patch is mapped back from the image to face space by an inverse transformation $\widehat{T}^{-1}$ and the probability of the image function being an instance of face appearance is evaluated. The 'appearance score' and the 'configuration score' (the probability of $\widehat{T}$) is combined to make a decision on the presence of face.

4 Experiments

In the experiments reported here, local feature detectors were implemented via the PCA-based classification of neighbourhoods of local maxima of the Harris corner detector. At the detected interest points a scale-invariant face feature detectors based on Moghaddam's and Pentland's probabilistic matching [MP95, MP96] were trained. The face-space coordinate of each feature was defined as the position with the highest frequency of occurrence produced by the respective feature detectors in the training set. Details of the detection process are described

Fig. 5. Confidence regions for the second and third feature after left eye corner detection

in [MBHK02]. An example of the result of detection performed on a test image is shown in Figure. 4

To make the testing for admissible transformation fast and simple, the feature position confidence regions were approximated by bounding boxes. An illustration of confidence regions on XM2VTS data for a detected left eye corner is depicted in Figure 5.

An experiment on 400 test images from the XM2VTS database and several images with cluttered background was carried out in order to find out the detection rates and the efficiency of pruning. **The overall detection rate on XM2VTS database was 98%**. Since there can be more than 3 features detected on a face, more triplets can lead to a successful detection (at most 120 if all features are detected). In our experiment only the best face hypothesis with distance below a global threshold is taken as valid. Figure 6 shows a typical result.

The speed up achieved by search-pruning was measured. For the XM2VTS data, the pruning reduced the search by 55 per cent. **For images with cluttered background, the reduction was 92 %**, making the detection process more than 10 times faster. Clearly, the search reduction achieved on the XM2VTS database gives a very conservative estimate of potential gains as the background is homogeneous. In the presence of cluttered background, the reduction in the number of hypotheses is much more impressive, as many triplets involving false positives are present.

5 Conclusions

We proposed a novel framework for detecting human faces based on correspondences between triplets of detected local features and their counterparts in an affine invariant face appearance model. The method is robust to partial occlusion or feature detector failure since a face may be detected if only three out of ten currently used detectors succeed. Robustness with respect to cluttered

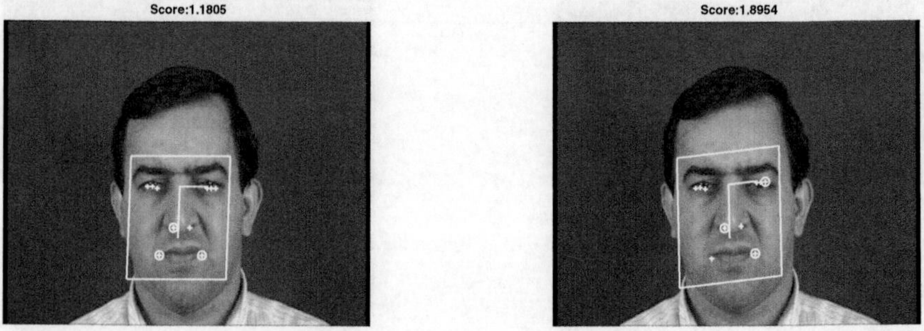

Fig. 6. Typical detection result. The best face hypothesis (left) and the worst-score face hypothesis below a distance threshold (right). Features that formed the triplets are marked. The worst face hypothesis has a quite high shear, and consequently high score because high shears were unlikely transformation in the training set

background is achieved via pruning, since background hypotheses lead to extremely unlikely face-to-image space transformations. Since both the appearance and transformation probability is learned from examples, the method can easily be tuned to incorporate application-dependent constraints. The application of the probabilistic model of feature configuration reduced the search complexity by more than 55% in uncluttered images and by 92% in images with complex background. The method was tested on the XM2VTS database and a limited number of images with cluttered background with very promising results – 2% false negative rate and 0 false positives - were obtained.

References

[BLP95] M. C. Burl, T. K. Leung, and P. Perona. Face localization via shape statistics. In *Proc. of International Workshop on Automatic Face and Gesture Recognition*, pages 154–159, 1995. 567, 569

[BP96] M. C. Burl and P. Perona. Recognition of planar object classes. In *Proc. of Computer Vision and Pattern Recognition*, pages 223–230, 1996. 567, 569

[BWP98] M. C. Burl, M. Weber, and P. Perona. A Probabilistic approach to object recognition using local photometry abd global Geometry. In *Proc. of European Conference on Computer Vision*, pages 628–641, 1998. 567

[BY98] V. Bruce and A. Young. *In the Eye of Beholder, The Science of face perception.* Oxford University Press, 1998. 569

[KP97] C. Kotropoulos and I. Pitas. Rule-based face detection in frontal views. In *Proc. of International Conference on Acoustics, Speech, and Signal Processing*, pages 21–24, 1997. 567

[LVB+93] M. Lades, J. C. Vorbrüggen, J. Buhmann, J. Lange, Ch. von der Malsburg, R. P. Würtz, and W. Konen. Distrotion invariant object recognition in the dynamic link architecture. *IEEE Trans. on Pattern Analysis and Machine Intelligence*, 42(3):300–310, 1993. 567

[MBHK02] J. Matas, P. Bílek, M. Hamouz, and J. Kittler. Discriminative regions for human face detection. In *Proceedings of Asian Conference on Computer Vision*, January 2002. 573

[MMK+99] K. Messer, J. Matas, J. Kittler, J. Luettin, and G. Maitre. XM2VTSDB: The extended M2VTS database. In R. Chellapa, editor, *Second International Conference on Audio and Video-based Biometric Person Authentication*, pages 72–77, Washington, USA, March 1999. University of Maryland. 568

[MP95] B. Moghaddam and A. Pentland. Probabilistic visual learning for object detection. In *Proc. of International Conference on Computer Vision*, pages 786–793, 1995. 566, 569, 572

[MP96] B. Moghaddam and A. Pentland. Probabilistic visual learning for object representation. In *Early Visual Learning*, pages 99–130. Oxford University Press, 1996. 566, 569, 572

[SK98] H. Schneiderman and T. Kanade. Probabilistic modeling of local appearance and spatial relationships for object recognition. In *Proc. Computer Vision and Pattern Recognition, pages 45–51. IEEE*, 1998. 567

[SP98] K. Sung and T. Poggio. Example-based learning for view-based human face detection. *IEEE Transactions on Pattern Analysis and Machine Intelligence*, 20(1):39–50, January 1998. 566

[VS00] V. Vogelhuber and C. Schmid. Face detection based on generic local descriptors and spatial constraints. In *Proc. of International Conference on Computer Vision*, pages I:1084–1087, 2000. 567, 569

Shape-from-Shading for Highlighted Surfaces

Hossein Ragheb * and Edwin R. Hancock

Department of Computer Science, University of York
York YO10 5DD, UK
{hossein,erh}@minster.cs.york.ac.uk

Abstract. One of the problems that hinders the application of conventional methods for shape-from-shading to the analysis of shiny objects is the presence of local highlights. The first of these are specularities which appear at locations on the viewed object where the local surface normal is the bisector of the light source and viewing directions. Highlights also occur at the occluding limb of the object where roughness results in backscattering from microfacets which protrude above the surface. In this paper, we consider how to subtract both types of highlight from shiny surfaces in order to improve the quality of surface normal information recoverable using shape-from-shading.

1 Introduction

Shape-from-shading (SFS) is concerned with recovering surface orientation from local variations in measured brightness [5]. The observation underpinning this paper is that although considerable effort has gone into the recovery of accurate surface geometry, existing SFS methods are confined to situations in which the reflectance is predominantly Lambertian. When the surface under study is shiny, then the estimated geometry may be subject to error. The main problem that can occur is that surface intensity highlights may lead to misestimation of surface curvature. The most familiar example here is that of surface specularities. These occur at locations on the surface where the local surface normal direction is the bisector of the light source and viewing directions For this reason, if specular highlights can be accurately located, then they can provide important cues that can be used to constrain the recovery of surface shape. However, there is a less well known effect that results in limb brightening. This is due to surface roughness and results from oblique scattering from microfacets that protrude above the limb perpendicular to the line of sight.

The problem of non-Labmertian and specular reflectance has been widely studied [6,13]. For instance, Healey and Binford [4] have shown how to simplify the Beckmann distribution [1] using a Gaussian approximation to the distribution of specular angle. This simplification can be used in conjunction with the Torrance and Sparrow model [14] to model intensity variations in the analysis of surface curvature. Several authors have looked critically at the physics underlying specular reflectance. For instance, Nayar, Ikeuchi and Kanade [9] have

* Sponsored by the university of Bu-Ali Sina, Hamedan, Iran.

T. Caelli et al. (Eds.): SSPR&SPR 2002, LNCS 2396, pp. 576–586, 2002.
© Springer-Verlag Berlin Heidelberg 2002

shown that the Torrance-Sparrow model [14] is applicable to the modelling of the specular lobe rather than the specular spike. Wolff [15] also has a model which combines diffuse and specular reflectance components, in which the parameters are chosen on the basis of the known physical properties of particular surfaces. In a series of recent papers, Lin and Lee have shown how specular reflections due to multiple light-sources can be located in multi-band imagery [7]. Finally, Nayar, Fang and Boult [10] have used polarisation filters to detect specular reflection.

There has also been a considerable body of work devoted to reflectance from rough surfaces. As noted above, this process is responsible for limb brightening. Oren and Nayar [11] have developed a model which can be used to account for reflectance from surfaces with a rough microfacet structure. Dana, Nayar, Van Ginneken and Koenderink [3] have catalogued the BRDF's for 3D surface textures. Recently, Magda, Kriegman, Zickler and Belhumeur [8] have commented on how shape can be recovered from surfaces with arbitrary BRDF's. Finally, Wolff [15] has shown how the Fresnel term can be used to model reflectance from a variety of surfaces.

In this paper our aim is to incorporate both specular and rough limb reflectance into the SFS process. This is a two-step process. First, we make estimates of the local surface normals using geometric constraints on the directions of Lambertian and specular reflectance to recover surface normal directions. The approach is a probabilistic one, which uses a mixture model to estimate the posterior mean direction of Lambertian and specular reflectance. Once the posterior mean surface normals are to hand, then we can perform photometric correction on the original image. This is again a two-step process. First, we subtract specularities using the Torrance-Sparrow (T-S) model. Second, we correct the residual non-specular component using the Oren-Nayar model. The result is a corrected Lambertian image from which both local specularities and limb-brightening effects are removed. By applying a Lamertian SFS algorithm to the corrected image, we obtain an improved estimate of the surface normal directions.

2 Reflectance Geometry

In this section we outline the geometry of the reflectance processes which underpin our SFS model. We adopt a two-component model in which the predominantly Lambertian surface reflectance exhibits local specular highlights.

In the case of Lambertian reflectance from a matte surface of constant albedo illuminated with a single collimated light-source, the observed intensity is independent of the viewing direction. Suppose that $\boldsymbol{L}$ is the unit-vector in the direction of the light source and that $\boldsymbol{N}_L(i,j)$ is the unit-vector in the surface normal at the pixel (i,j). According to Lambert's law, the observed image intensity at the pixel with coordinates (i,j) is $E(i,j) = \boldsymbol{N}_L(i,j) \cdot \boldsymbol{L}$

The second component of our reflectance process is concerned with modelling local specular highlights on the observed surface. For specular reflection the surface normal, the light source and the viewing directions are coplanar. The incidence angle is equal to the angle of specular reflectance (Figure 1a). Hence, the direction of the surface normal $\boldsymbol{N}_S^{(n)}$ is the bisector of the light source ($\boldsymbol{L}$) and the viewing ($\boldsymbol{V}$) directions and the unit-vector is $\boldsymbol{N}_S^{(n)} = \frac{(\boldsymbol{L}+\boldsymbol{V})}{\|\boldsymbol{L}+\boldsymbol{V}\|}$.

3 Bayesian Framework

The aim in this paper is to develop a Bayes-decision scheme for separating the two reflectance modes [12]. In other words, we wish to compute the *a posteriori* probabilities of specular or Lambertian reflectance. The aim is to label pixels according to the reflectance mode from which they originated. The class identity for the pixel (i,j) at iteration n is denoted by $\omega_{i,j}^{(n)}$. The class-identity may be drawn from the set $\Omega = \{S, L\}$ where S is the specular reflectance label and L is the Lambertian reflectance label. For each image location, we maintain a specular surface normal and a Lambertian surface normal which satisfy the geometric constraints outlined in Section 2. At iteration n of the algorithm the currently available estimates of the two surface normals are respectively $\boldsymbol{N}_L^{(n)}(i,j)$ and $\boldsymbol{N}_S^{(n)}(i,j)$. In the case of the specular component, the normal direction is in the direction of local specular reflection, and does not change with iteration number. In the case of Lambertian reflectance, the normal direction varies with iteration number, but is always projected to be positioned on the irradiance cone.

To develop our decision process, we require two probabilistic modelling ingredients. The first of these are separate probability density functions which can be used to represent the distributions of surface normals for the two reflectance components. We evaluate these densities at the posterior mean (PM) surface normal $\boldsymbol{M}^{(n)}(i,j)$ computed at iteration n. The reason for doing this is that the current values of the two normals are guaranteed to satisfy the geometric constraints outlined in Section 2. As a result, they will be associated with vanishing angular error. Accordingly, we let $q_{i,j}^{(n)}(L) = p(\boldsymbol{M}^{(n)}(i,j)|\omega_{i,j}^{(n)} = L)$ be the probability distribution for the PM surface normal under the Lambertian reflectance model. Similarly, we let $q_{i,j}^{(n)}(S) = p(\boldsymbol{M}^{(n)}(i,j)|\omega_{i,j}^{(n)} = S)$ denote the distribution function for the PM surface normal for the specular reflectance component.

The second probabilistic ingredient is a smoothness prior for the selected surface normal. This component of the model incorporates contextual information. Indexing the surface normals according to their pixel locations, suppose that $\Gamma_{i,j}^{(n)} = \{\boldsymbol{M}^{(n)}(k,l)|(k,l) \in G_{i,j}\}$ is the set of PM surface normals in the neighbourhood $G_{i,j}$ of the pixel (i,j). We let $P_{i,j}^{(n)}(L) = P(\boldsymbol{N}_L^{(n)}(i,j)|\Gamma_{i,j}^{(n)})$ be the conditional probability (or smoothness prior) of the Lambertian surface normal at the location (i,j) given the field of surrounding PM surface normals. With these ingredients, then according to the iterated conditional modes, the probability that the pixel (i,j) belongs to the Lambertian class at iteration n is

$$P(\omega_{i,j}^{(n)} = L|\boldsymbol{M}^{(n)}(i,j)) = \frac{q_{i,j}^{(n)}(L)P_{i,j}^{(n)}(L)}{\sum_{\Lambda\in\Omega} q_{i,j}^{(n)}(\Lambda)P_{i,j}^{(n)}(\Lambda)} \quad (1)$$

The probability that the surface normal belongs to the specular class is the complement, These probabilities can be used to separate the two reflectance modes. With these probabilities to hand, we can update the estimate of the PM surface normal in the following manner

$$\begin{aligned} \boldsymbol{M}^{(n+1)}(i,j) = \boldsymbol{N}_S^{(n)}(i,j)P(\omega_{i,j}^{(n)} &= S|\boldsymbol{M}^{(n)}(i,j)) \\ + \boldsymbol{N}_L^{(n)}(i,j)P(\omega_{i,j}^{(n)} &= L|\boldsymbol{M}^{(n)}(i,j)) \end{aligned}$$

4 Surface Normal Distribution

To apply the MAP estimation scheme outlined in the previous section, we require probability distributions for the two surface normals together with the smoothness prior. For the specular surface normals, we use the Beckmann distribution to model the angle $\alpha = \cos^{-1}(\boldsymbol{M}^{(n)}(i,j) \cdot \boldsymbol{N}_S^{(n)}(i,j))$, between the PM surface normal $\boldsymbol{M}^{(n)}(i,j)$ and the predicted direction of the specular spike $\boldsymbol{N}_S^{(n)}$. The normalized distribution is

$$q_{i,j}^{(n)}(S) = D(\alpha) = \frac{1}{\sigma_S(2+\sigma_S^2)\cos^4\alpha} \exp\left[-\left(\frac{\tan(\alpha)}{\sigma_S}\right)^2\right] \tag{2}$$

where σ_S is a parameter which controls the angular spread of the specular spike. This distribution can be used to model the shape of both the specular spike and the specular lobe. Our model of the Lambertian reflectance process assumes that the observed intensity values follow a Gaussian distribution with variance σ_L^2. The mean intensity is $\boldsymbol{M}^{(n)} \cdot \boldsymbol{L}$. Under these assumptions we can write

$$q_{i,j}^{(n)}(L) = \frac{1}{\sqrt{2\pi}\sigma_L} \exp\left[-\frac{1}{2}\left(\frac{E(i,j) - \boldsymbol{M}^{(n)}(i,j)\cdot \boldsymbol{L})}{\sigma_L}\right)^2\right] \tag{3}$$

Our model for the surface normal smoothness prior is based on the average value of the inner product of the surface normal at the location (i,j) with the surrounding field of PM surface normals. We write

$$P_{i,j}^{(n)}(\Lambda) = \frac{1}{2|G_{i,j}|}\left[|G_{i,j}| + \sum_{(k,l)\in G_{i,j}} \boldsymbol{N}_\Lambda^{(n)}(i,j)\cdot \boldsymbol{M}^{(n)}(k,l)\right] \tag{4}$$

When the PM surface normals from the neighbourhood $G_{i,j}$ are aligned in the direction of $\boldsymbol{N}_\Lambda^{(n)}(i,j)$, then $P_{i,j}^{(n)}(\Lambda) = 1$, the larger the misalignment then the smaller the value of smoothness prior.

5 Specular SFS Algorithm

Having described the Bayes framework and the associated two-mode reflectance model, we are now in a position to develop a practical SFS algorithm.

We commence by initialising the algorithm. The initial Lambertian surface normal $\boldsymbol{N}_L^{(0)}$ is constrained to lay on the irradiance cone in the direction of the image gradient. The subsequent iterative steps of the algorithm are as follows:

- 1: The field of PM surface normals $\boldsymbol{M}^{(n)}$ (initially equal to $\boldsymbol{N}_L^{(0)}$) is subjected to local smoothing. Here we use the curvature sensitive smoothing method [16]. The smoothed PM surface normal is denoted by $\boldsymbol{M}_R^{(n)}(i,j)$.
- 2: We update the current estimate of the Lambertian surface normal by projecting the smoothed PM surface normal onto the nearest location on the irradiance cone. This gives us the revised surface normal of $\boldsymbol{N}_L^{(n)}$.
- 3: With $\boldsymbol{M}_R^{(n)}(i,j)$ to hand we compute the conditional measurement densities $q_{i,j}^{(N)}(L)$ and $q_{i,j}^{(N)}(S)$ for the two reflectance modes. Taking $\boldsymbol{M}_R^{(n)}(i,j)$, $\boldsymbol{N}_L^{(n)}$ and $\boldsymbol{N}_S^{(n)}$, we compute the smoothness priors $P_{i,j}^{(n)}(L)$ and $P_{i,j}^{(n)}(S)$, and also the updated *a posteriori* probabilities for both reflectance modes.
- 4: Using $\boldsymbol{N}_L^{(n)}$ and $\boldsymbol{N}_S^{(n)}$ and the updated *a posteriori* probabilities, we compute the new PM surface normal $\boldsymbol{M}^{(n+1)}(i,j)$ and we return to step 1.

The steps of the algorithm are summarised in Figure 1b. The PM surface normals delivered by our SFS algorithm can be used for the purposes of reconstructing the specular intensity component. The reason for doing this is that the specular intensity may be removed from the original image intensity to give a corrected Lambertian image. We use the T-S model [14] to reconstruct the specular intensity component I_S. With the reconstructed specular intensity to hand, we can compute the matte reflectance component $I_L(i,j) = E(i,j) - I_S(i,j)$. By re-applying the SFS algorithm to this corrected intensity image, we aim to recover improved surface normal estimates, free of the high curvature artifacts of specular highlights.

6 Correcting for Limb-Brightening

As mentioned earlier, there may also surface brightness anomalies due to rough reflectance from the limbs of objects. Our aim in this section is to show how the Oren and Nayar model [11] can be used to further correct the images obtained by specular subtraction for limb-brightening. According to this model, for a point on a rough surface with roughness parameter of σ and illuminant incidence direction (θ_i, ϕ_i) and reflectance direction (θ_r, ϕ_r), the reflectance functions is

$$L_r(\theta_i, \theta_r, \phi_r - \phi_i; \sigma) = \frac{\rho}{\pi} E_0 \cos(\theta_i)(A + B \max\left[0, \cos(\phi_r - \phi_i)\right] \sin(\alpha) \tan(\beta))$$

where $A = 1.0 - 0.5\frac{\sigma^2}{\sigma^2+0.33}$, $B = 0.45\frac{\sigma^2}{\sigma^2+0.09}$ and $\alpha = \max[\theta_i, \theta_r]$, $\beta = \min[\theta_i, \theta_r]$. It is important to note that the model reduces to the Lambertian case when $\sigma = 0$. Here, we aim to utilize this model to deduce a corrected Lambertian reflectance image from the matte component delivered by our specular subtraction method. To do this, we assume that σ is almost constant and the reflectance measurements are obtained in the plane of incidence ($\phi_r = \phi_i = 0$). We also confine our attention to the case where the angle between the light source and the viewing directions is small, i.e. $\theta_r = \theta_i = \theta$. With these two restrictions, we can write $\cos(\phi_r - \phi_i) = 1$ and $\alpha = \beta = \theta$. Hence, the non-specular (or matte) intensity predicted by the simplified Oren-Nayar (O-N) model is

$$I_M(i,j) = A\cos\theta + B\sin^2\theta \tag{5}$$

Hence, the matte intensity consists of two components. The first of these is a Lambertian component $A\cos\theta$. The second is the non-Lambertian component $B\sin^2\theta$ which takes on its maximum value where $\theta = \frac{\pi}{2}$, i.e. close to the occluding boundary. To perform Lambertian correction, we proceed as follows. At every pixel location, we use Equation (5) to estimate the angle θ using the subtracted matte intensity and solving the resulting quadratic equation in $\cos\theta$. The solution is

$$\cos\theta = \frac{A \mp \sqrt{A^2 - 4B(I_M(i,j) - B)}}{2B} \tag{6}$$

We take the sign above which results in a value of $A\cos\theta$ which is closest to the matte intensity I_M (in the majority of cases this involves taking the solution associated with the minus sign). This hence allows us to reconstruct the corrected Lambertian reflectance image $I_L = A\cos\theta$. It also gives us an estimate of the opening angle of the Lambertian reflectance cone. This can then be used in the Worthington and Hancock SFS scheme which assumes the Lambertian reflectance model to recover improved surface normal estimates.

In Figure 2a we show the Lambertian reflectance $\cos\theta$ (equation 6) as a function of the roughness parameter σ and the matte intensity I_M. When the roughness is zero, then the Lambertian and matte intensities are equal to one another. When the roughness increases, then the departures from Lambertian reflectance become more marked. In Figure 2b we plot the ratio $\frac{I_L}{I_M}$ as a function of the incidence angle θ. The different curves are for different values of the roughness parameter σ. For zero roughness, the ratio is flat, i.e. the reflectance is purely Lambertian. As the roughness increases, then so the value of the ratio decreases with increasing incidence angle. For normal incidence, the ratio is always unity, i.e. the reflectance is indistinguishable from the Lambertian case, whatever the value of the roughness parameter is.

7 Experiments

The images used in our experiments have been captured using an Olympus 10E camera. The objects studied are made of white porcelain and are hence shiny. Each object has been imaged under controlled lighting conditions in a darkroom. The objects have been illuminated using a single collimated tungsten light source. The light source direction is recorded at the time the images are captured. To ground-truth the surface highlight removal process, we have used a pair of polaroid filters. We have placed the first filter between the light source and the illuminated object. The second filter was placed between the illuminated object and the camera. For each object we have collected a pair of images. The first of these is captured when the second filter (i.e. the one between the camera and the object) is rotated until there is maximal extinction of the observed specularities. The second image is obtained when the polaroid is rotated through 90°, i.e. there is minimal extinction of the specularities. We refer to the polarisation conditions of the former image as "uncrossed" and of the latter as "crossed".

In Figure 3 we show the results obtained for three of the objects used in our study. The objects are a porcelain bear, a porcelain vase and a porcelain urn. The top row of the Figure shows the images obtained with uncrossed polaroids while the second row shows the images obtained with crossed polaroids. The third row shows the difference between the crossed and uncrossed polaroid images. The strongest differences occur at two different locations. Firstly, there are the positions of specularities. From the uncrossed polaroid images it is clear that there are several quite small specular reflections across the surface of the bear. The vase has larger specularities on the neck and the centre of the bulb. The urn has a complex pattern of specularities around the handles. From the crossed polaroid images it is clear that most of the specular structure is removed. The second feature in the difference images are the locations of occluding object limbs, where oblique scattering occurs.

In the fifth row of Figure 3, we show the reconstructed specular intensity obtained using the T-S model, i.e. I_S, The fourth row shows the matte images I_M obtained after specularity subtraction. Turning our attention to the matte images and the specular images, it is clear that for each of the objects the specular structure is cleanly removed and the matte appearance is close to that obtained with the crossed polaroids. Also, the pattern of specularities obtained in each case corresponds to that obtained by subtracting the crossed polaroid images from the uncrossed polaroid images.

In Figure 4 we investigate the shape information recoverable. The top row shows the Lambertian images after correction for rough limb reflectance using the simplified O-N model, i.e. I_L. In the second row of the Figure we show the needle maps obtained when the SFS is applied to the corrected Lambertian images (I_L) appearing in the top row of this Figure. The third row of Figure 4 shows the difference in needle-map directions for the matte (I_M) and Lambertian images (I_L). Here the main differences occur at the limbs of the objects. The fourth row of Figure 4 show the curvedness estimated using the surface normals delivered by the corrected Lambertian images. In the case of the urn the ribbed structure emerges well. The complex surface structure of the bear, in particular the boundaries of the arms and legs, is clearly visible. For the urn the symmetric structure of the neck and the bulb is nicely preserved.

In Figure 5 we provide some analysis of the different reflectance models used in our experiments. In the left hand panel of the Figure, the solid curve is the intensity cross-section along a horizontal line crossing the uncrossed image of the neck of the vase shown in Figure 3. The dashed curve shows the matte image I_M while the dotted curve is the specular component I_S . The specularity on the neck is clearly visible as a peak in the solid curve. This peak is cleanly subtracted in the matte (dashed) curve. In the right-hand panel we focus on the corrected Lambertian image. Here the solid curve is the matte reflectance I_M . The dashed curve is the corrected Lambertian reflectance I_L. The differences between the two curves are small except at the limbs of the object. To examine the effect of the model in more detail, the dotted curve shows the ratio of corrected Lambertian and matte reflectance $\rho = \frac{I_L}{I_M}$. The ratio drops rapidly towards zero as the limbs are approached. Also shown on the plot as a dash-dot curve is the predicted value of the ratio based on the assumption that the object has a circular cross-section. If x is the distance from the centre and r is the radius of the circle, then value of the ratio at a distance x from the centre is $\rho(x) = \frac{A\sqrt{1-(\frac{x}{r})^2}}{A\sqrt{1-(\frac{x}{r})^2}+B(\frac{x}{r})^2}$. This simple model is in reasonable agreement with the empirical data.

8 Conclusions

In this paper we have shown how to use shape-from-shading to perform photometric correction of images of shiny objects. Our approach is to use estimated surface normal directions together with reflectance models for specular and rough reflectance to perform specularity removal and rough limb-correction. Specularities are modelled using the T-S model while the rough limb brightening is modelled using the O-N model. We commence by using an iterated conditional modes algorithm to extract surface normals using a mixture of specular and matte reflectance directions. The resulting surface normals are used to perform specularity subtraction. Finally, we correct the residual matte reflectance component for rough limb scattering using the O-N model. The resulting corrected Lambertian images can be used as input to a conventional shape-from-shading algorithm and result in improved recovery of object-geometry.

References

1. P. Beckmann and A. Spizzochino, *The Scattering of Electromagnetic Waves from Rough Surfaces,* Pergamon, New York, 1963. 576
2. A. Blake and H. Bulthoff, "Shape from Specularities: computation and psychophysics," *Phil Trans R. Soc. Lond. B*, Vol. 331, pp. 237-252, 1991.
3. K. Dana, S. Nayar, B. Van Ginneken, J. Koenderink, "Reflectance and Texture of Real-World Surfaces," *CVPR*, pp. 151-157, 1997. 577
4. G. Healey and T. Binford "Local shape from specularity" *ICCV*, pp. 151-160, 1987. 576
5. B. K. P. Horn and M. J. Brooks, "The Variational Approach to Shape from Shading," *CVGIP*, Vol. 33, No. 2, pp. 174-208, 1986. 576
6. C. C. J. Kuo and K. M. Lee, "Shape from Shading With a Generalized Reflectance Map Model," *CVIU*, Vol. 67, No. 2, pp. 143-160, 1997. 576
7. S. Lin and S. W. Lee, "Estimation of Diffuse and Specular Appearance," *ICCV*, pp. 855-860, 1999. 577
8. S. Magda, D. Kriegman, T. Zickler and P. Belhumeur, "Beyond Lambert: Reconstructing Surfaces with Arbitrary BRDFs," *ICCV*, Vol. 2, pp. 391-399, 2001. 577
9. S. K. Nayar, K. Ikeuchi and T. Kanade, "Surface Reflection: Physical and Geometrical Perspectives," *PAMI*, Vol. 13, No. 7, pp. 611-634, 1991. 576
10. S. K. Nayar, X. Fang and T. Boult, "Removal of specularities using color and polarization," *CVPR*, pp. 583-590, 1993. 577
11. M. Oren and S. K. Nayar, "Generalization of the Lambertian Model and Implications for Machine Vision," *IJCV*, vol. 14, No. 3, pp. 227-251, 1995. 577, 580

12. H. Ragheb, and E. R. Hancock, "Separating Lambertian and Specular Reflectance Components using Iterated Conditional Modes," *BMVC*, pp. 541-552, 2001. 578
13. H. D. Tagare and R. J. P. deFigueiredo, "A Theory of Photometric Stereo for a Class of Diffuse Non-Lambertian Surfaces," *PAMI*, Vol. 13, No. 2, pp. 133-151, 1991. 576
14. K. Torrance and E. Sparrow, "Theory for Off-Specular Reflection from Roughened Surfaces," *JOSA*, Vol. 57, pp. 1105-1114, 1967. 576, 577, 580
15. L. B. Wolff, "On The Relative Brightness of Specular and Diffuse Reflection," *CVPR*, pp. 369-376, 1994. 577
16. P. L. Worthington and E. R. Hancock, "New Constraints on Data-closeness and Needle-map consistency for SFS," *PAMI*, Vol. 21, No. 11, pp. 1250-1267, 1999. 580

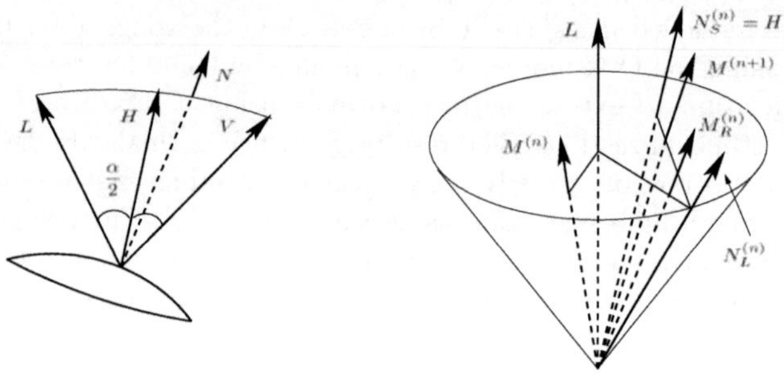

Fig. 1. Geometry of the specular reflectance and the surface normal update process

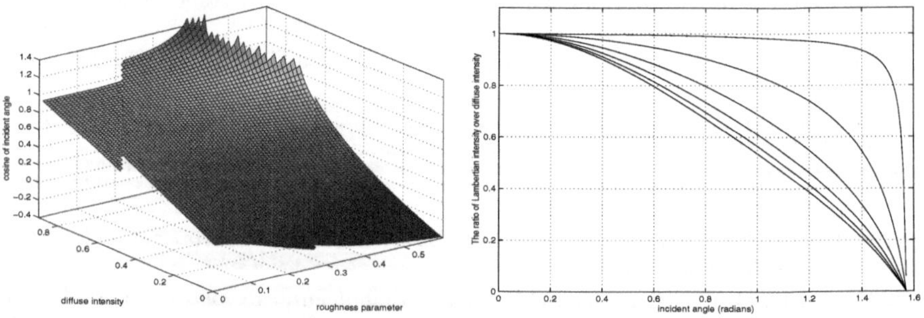

Fig. 2. Plots showing the behaviour of the Oren and Nayar model for rough surfaces

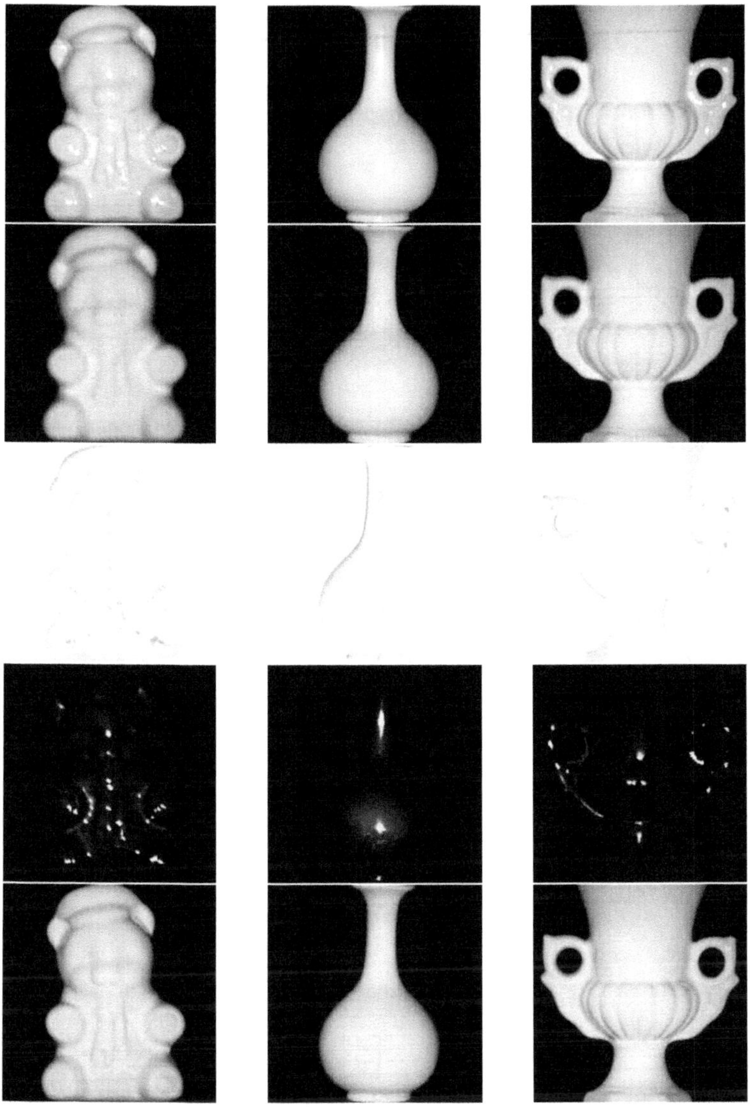

Fig. 3. Applying our specular SFS to separate the specular and matte components

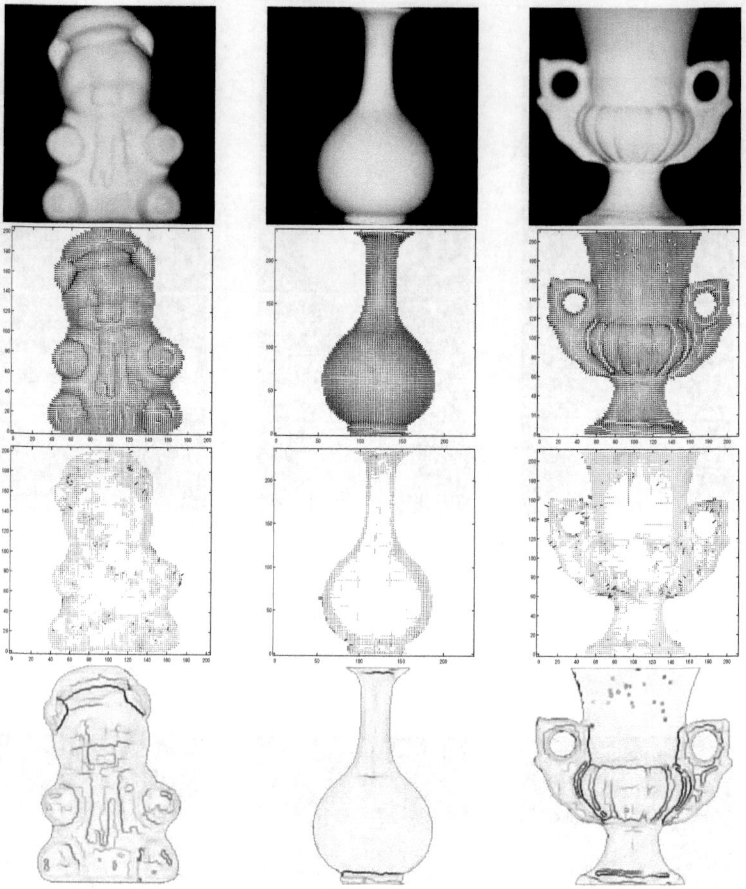

Fig. 4. Surface normals obtained by running SFS over matte and Lambertian images

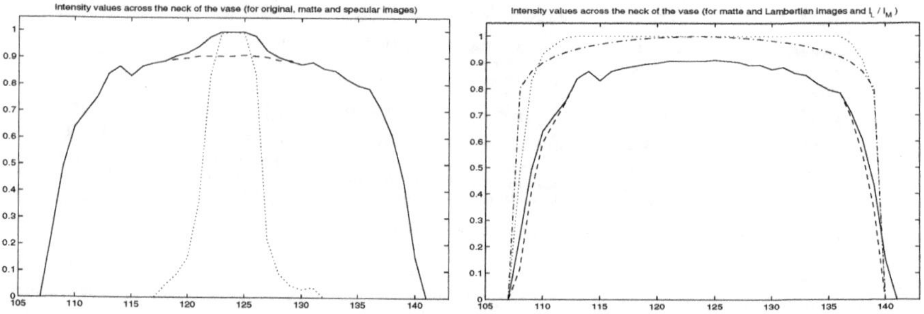

Fig. 5. Intensity plots for different reflectance components across the neck of the vase

Texture Description by Independent Components

Dick de Ridder[1], Robert P. W. Duin[1], and Josef Kittler[2]

[1] Pattern Recognition Group
Dept. of Applied Physics, Faculty of Applied Sciences, Delft University of Technology
Lorentzweg 1, 2628 CJ Delft, The Netherlands
Phone +31 15 278 1845, Fax +31 15 278 6740
http://www.ph.tn.tudelft.nl/~dick
[2] Centre for Vision, Speech and Signal Processing
School of Electronics, Computing and Mathematics, University of Surrey
Guildford GU2 7XH Surrey, United Kingdom

Abstract. A model for probabilistic independent component subspace analysis is developed and applied to texture description. Experiments show it to perform comparably to a Gaussian model, and to be useful mainly for problems in which the detection of little occurring, high-frequency image elements is important.

1 Introduction

In many applications of statistical pattern recognition techniques to image data, images (or image patches) are described as high-dimensional vectors in which each element corresponds to a pixel position in the rectangular image grid. A problem associated with this approach is that estimating parameters in such a high-dimensional space is cumbersome. Furthermore, images that make sense to human observers only form a small subset of all possible positions in this space. The vector description of images therefore contains more parameters than necessary. Irrespective of these problems, one would often like to model image information locally invariant to offset, contrast, translation, rotation and scale. However, if an image patch is just slightly brightened, contrast enhanced, translated, rotated or scaled, the distribution of the vectors will change drastically, whereas to a human observer the image content looks very similar. That is, the representation is not naturally invariant.

It can be shown that transformed versions of an image patch all lie on an m-dimensional manifold in the d-dimensional space spanned by all pixel values, where m is the number of degrees of freedom present in the transformations [10]. Although this manifold may be intrinsically low-dimensional, it is likely to be nonlinear and lie folded up in the d-dimensional space. A good representation would therefore be one which describes this manifold using a small number of parameters, thereby avoiding the estimation problems in high-dimensional spaces and enforcing invariance to elementary transformations.

T. Caelli et al. (Eds.): SSPR&SPR 2002, LNCS 2396, pp. 587–596, 2002.
© Springer-Verlag Berlin Heidelberg 2002

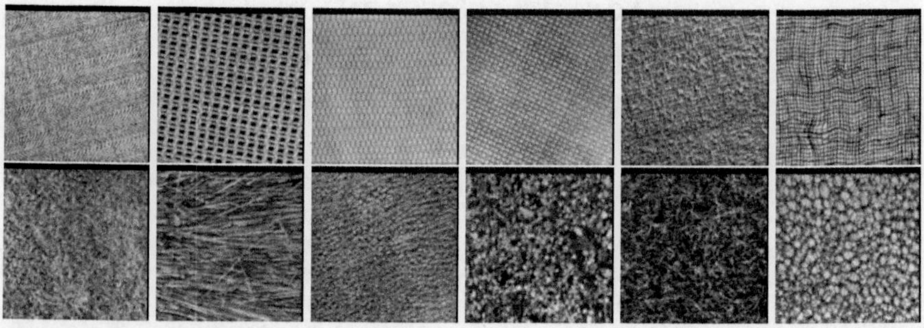

Fig. 1. The 12 Brodatz textures used in the experiments in this paper. Top: structured textures S1–S6; bottom: natural textures N1–N6

Earlier work [2,3,4] discussed the use of mixtures-of-PCA for locally modelling such manifolds, in applications to texture segmentation and image description. The main focus in this paper is on the use of independent component analysis (ICA) for describing sets of image patches. ICA is a recently proposed technique for finding subspaces in which the projected data distributions are independent. There have been reports of modelling sets of natural image patches using ICA (e.g. [6]), leading to sets of wavelet-like basisvectors. However, there have been no true applications of the subspaces thus found. In this paper, ICA will be applied as a texture description method and compared to a full Gaussian model.

The remainder of the paper is laid out thus: Sect. 2 discusses the texture image data used in the experiments. Next, in Sect. 3 the Gaussian and ICA models will be discussed. Sect. 4 describes experiments in texture description. Questions raised by these experiments with regard to ICA will be answered in Sect. 5, followed by some conclusions in Sect. 6.

2 Texture Data

For the experiments in this paper, two sets of textures were taken from the Brodatz album [1], shown in Fig. 1. It is to be expected that subspace methods will work better on the structured textures, as these exhibit stronger correlation between pixels. The original images were rescaled to make sure texel sizes (i.e. the sizes of the 'structuring elements") in the structured textures were smaller than 16×16 pixels (the image patch size used in the experiments). For each image, the grey value range was rescaled to $[0, 1]$.

As the models will be trained on image patches rather than entire images, training sets will have to be sampled from the images. To this end, 16×16 pixel windows were used. Furthermore, the notion of *episodes* introduced by Kohonen [8] was used. This artificial enlargement of the data set allows for incorporation of prior knowledge, i.e. that (subspace) models should be invariant to small amounts of translation, rotation and scaling. Next to a sample extracted at position (x_0, y_0), four translated samples are extracted at positions $(x_0 + x, y_0 + y)$, where $x \sim U(-5,5)$ and $y \sim U(-5,5)$; five samples are taken at the same position in images rotated over -45°, -22.5°, 0°, 22.5°and 45°; and five samples are taken at the same position in images scaled to 1.1×, 1.2×, 1.3×, 1.4× and 1.5× the original size. This gives a total of 15 samples per episode. Data sets of 1,500 samples *per texture* were created in this way, containing 100 episodes.

The models used in this paper cannot be fitted when certain dimensions do not contain data, i.e. when there is zero variance. Therefore, directions in which there is little or no variance (assumed to contain noise) can be removed using PCA. This is called *pre-mapping*; the data is projected onto the eigenvectors corresponding to the set of largest eigenvalues explaining more than 90% of the variance. The data is not whitened. For 16×16 windows taken out of structured textures, this leaves 50 dimensions on average; out of natural textures, on average 70 dimensions remain.

3 Models

This section discusses two basic models that can be applied to texture description: the Gaussian model and the independent component analysis (ICA) model. The basis for the discussion will be the log-likelihood, which in segmentation algorithms would be used to assign image windows to models. A data set of d-dimensional vectors $\boldsymbol{x}$ will be denoted by $\mathcal{L} = \{\boldsymbol{x}^n\}, n = 1, \ldots, N$. For the ICA model, m will denote the number of independent components.

3.1 Gaussian

The log-likelihood of a sample $\boldsymbol{z}$ belonging to a Gaussian distribution is:

$$\mathfrak{L}_{\text{Gauss}}(\boldsymbol{z}|\boldsymbol{\mu}, \boldsymbol{C}) = -\frac{d}{2}\ln(2\pi) - \frac{1}{2}\ln|\det(\boldsymbol{C})| - \frac{1}{2}(\boldsymbol{z}-\boldsymbol{\mu})^T \boldsymbol{C}^{-1}(\boldsymbol{z}-\boldsymbol{\mu}) \quad (1)$$

where $\boldsymbol{\mu}$ is the mean and $\boldsymbol{C}$ the covariance matrix, estimated on all $\boldsymbol{x} \in \mathcal{L}$. For a number of texture data sets, the inverse of $\boldsymbol{C}$ cannot be calculated in a 256D space; hence PCA pre-mapping (see the previous section) was applied.

3.2 Independent Component Analysis

Lately, independent component analysis or ICA [7] has enjoyed considerable attention. In its most basic form, this method tries to find directions in which the data is independent (rather than just decorrelated as in principal component analysis, PCA). The linear ICA model is $\boldsymbol{u} = \boldsymbol{W}(\boldsymbol{x} - \boldsymbol{\mu})$, in which $\boldsymbol{u}$ are the m-dimensional projected vectors. As ICA was originally applied to blind separation of various signals (or *sources*) $\boldsymbol{s}$ under an additive model, the matrix $\boldsymbol{W}$ is often called the *unmixing matrix* and the backprojection matrix $\boldsymbol{A}$ the *mixing matrix*. This backprojection is defined as $\boldsymbol{x} = \boldsymbol{A}\boldsymbol{s} + \boldsymbol{\mu}$. Note that in the notation here, $\boldsymbol{u}$ is used to denote an estimate of the *sources* $\boldsymbol{s}$.

Several (related) iterative algorithms have been proposed to perform ICA. For this work, an algorithm was developed as a variation on the extended infomax ICA algorithm of Lee et al. [9]. Lee's algorithm is a maximum likelihood (ML) fit of a set of distributions, in which for each component u_i it is learned whether to fit a sub-Gaussian or a super-Gaussian distribution. The number of independent components m is assumed to be equal to the number of dimensions in the data d. The log-likelihood can be expressed as [9]

$$\mathfrak{L}_{\text{ICA}}(\boldsymbol{z}|\boldsymbol{A}) = -\ln|\det(\boldsymbol{A})| + \ln p(\boldsymbol{u}) \quad (2)$$

where $\boldsymbol{u} = \boldsymbol{W}\boldsymbol{z}$ and $\boldsymbol{W} = \boldsymbol{A}^{-1}$, which is possible since $m = d$ and therefore $\boldsymbol{A}$ is a square matrix. For notational simplicity, $\boldsymbol{\mu}$ is assumed to be $\mathbf{0}$. An ICA base can be found using a generalised EM algorithm to maximise Eqn. 2, in which the estimate for $\boldsymbol{W}$ is updated by a gradient descent learning rule. This batch learning rule, for the entire dataset (where $\boldsymbol{U}$ is the matrix containing the source approximations $\boldsymbol{u}$ as its columns and $\boldsymbol{X}$ likewise contains the data vectors), is

$$\Delta \boldsymbol{W} = \eta \left[N\boldsymbol{A}^T + \phi(\boldsymbol{U})\boldsymbol{X}^T \right] \quad (3)$$

where η is a learning rate and $\phi_i(u_i) = \frac{\partial \ln p_i(u_i)}{u_i}$, with $p_i(u_i)$ a sub-Gaussian or super-Gaussian distribution:

$$p_i(u_i) = \begin{cases} f_{sub}\,(u_i) = \frac{1}{2}\left(f_{Gauss}(u_i; 1, 1) + f_{Gauss}(u_i; -1, 1)\right) \\ f_{sup}\,(u_i) = c\; f_{Gauss}\,(u_i; 0, 1)\,\text{sech}\,(u_i) \end{cases} \quad (4)$$

with $f_{Gauss}(u_i; \mu, \sigma^2)$ the Gaussian distribution function and c a normalisation constant. Rule 3 was (approximately) found by various authors (e.g. [?]); note that the assumptions made to find it were (a) that there is no noise and (b) that matrix $\boldsymbol{A}$ is full rank, i.e. there are as many sources m as there are mixtures d.

The main limitation of this ML algorithm is that it works only for $m = d$. In the case of texture description, it is not to be expected that there are that many ICs. Therefore, a new learning rule was derived for the *undercomplete* case where $m < d$, by modelling the remaining dimensions as Gaussian noise [2]. This violates the assumptions necessary for the learning rule derived above.

The revised model contains an added noise term, $\boldsymbol{x} = \boldsymbol{A}\boldsymbol{s} + \boldsymbol{\mu} + \boldsymbol{\epsilon}$, with $\boldsymbol{s}$ now an m-dimensional vector, $\boldsymbol{A}$ a $d \times m$ matrix and $\boldsymbol{\epsilon}$ Gaussian i.i.d. distributed noise with $\boldsymbol{C} = \sigma^2 \boldsymbol{I} = \frac{1}{\beta}\boldsymbol{I}$. The log-likelihood now becomes

$$\mathfrak{L}_{\text{ICA}}(\boldsymbol{z}|\boldsymbol{A}, \beta) \approx \frac{d}{2} \ln \frac{\beta}{2\pi} - \frac{\beta}{2} (\boldsymbol{x} - \boldsymbol{A}\boldsymbol{u})^T (\boldsymbol{x} - \boldsymbol{A}\boldsymbol{u}) - \frac{1}{2} \ln|\det(\boldsymbol{A}^T\boldsymbol{A})| + \ln p(\boldsymbol{u}) \quad (5)$$

and the batch learning rule changes to

$$\Delta \boldsymbol{W} = \eta \left[\beta N \boldsymbol{A}^T \boldsymbol{C} \left(\boldsymbol{I} - \boldsymbol{A}\boldsymbol{W} \right) + N\boldsymbol{A}^T + \phi(\boldsymbol{U})\boldsymbol{X}^T \right] \quad (6)$$

The first term is an orthogonalisation term which works in the space rotated and scaled by $\boldsymbol{C}$. Note that in this model $\boldsymbol{W} = \boldsymbol{A}^{-1}$ no longer holds since $\boldsymbol{A}$ is not a square matrix; instead, $\boldsymbol{W} = (\boldsymbol{A}^T\boldsymbol{A})^{-1}\boldsymbol{A}^T$.

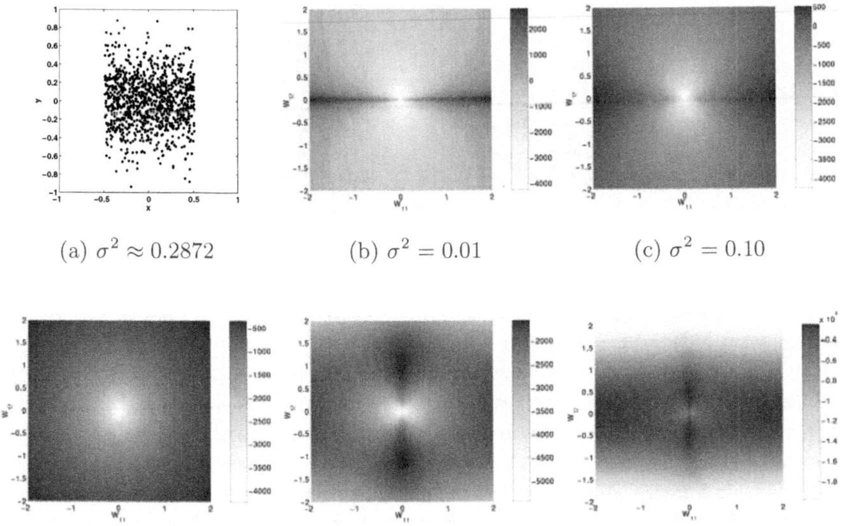

(a) $\sigma^2 \approx 0.2872$ (b) $\sigma^2 = 0.01$ (c) $\sigma^2 = 0.10$

(d) $\sigma^2 = 0.2872$ (e) $\sigma^2 = 1.0$ (f) $\sigma^2 = 2.5$

Fig. 2. (a) A simple 2D data set. (b)-(f) $\mathfrak{L}_{\text{ICA}}$ as a function of $\boldsymbol{W}$. For small σ^2, the local maxima (dark spots) correspond to the independent component (horizontal); for large σ^2, they correspond to the Gaussian component (vertical)

Eqn. 6 gives the basic learning rule for $\boldsymbol{W}$. However, the noise parameter β has to be estimated as well. This can easily be done using the GEM algorithm. In each step, estimate the current parameters in the E-step by:

$$\boldsymbol{u}^n = \boldsymbol{W}\boldsymbol{x}^n,\ n = 1, \ldots, N \quad (7) \qquad \boldsymbol{v}^n = \boldsymbol{W}^\top \boldsymbol{x}^n,\ n = 1, \ldots, N \quad (8)$$

$$\beta^{-1} = (d-m)^{-1} \sum_{i=1}^{d-m} \text{var}(v_i) \quad (9) \qquad \boldsymbol{A} = \boldsymbol{W}^T (\boldsymbol{W}\boldsymbol{W}^T)^{-1} \quad (10)$$

where $\boldsymbol{W}^\top$ is the nullspace of $\boldsymbol{W}$, i.e. β corresponds to the inverse of the average noise outside the subspace (cf. probabilistic PCA [11]). In the M-step, the log-likelihood is then maximised by applying the learning rule (Eqn. 6) to $\boldsymbol{W}$.

The estimation of β has a large influence on convergence. It controls the trade-off between finding a subspace (the 2nd term in Eqn. 5) and independent components (the 4th term). This is illustrated in Fig. 2. The data set consists of 2D samples of which the x-coordinate is drawn from a $U(-0.5, 0.5)$ distribution with $\sigma^2 \approx 0.2872$ and the y-coordinate is drawn from a Gaussian distribution with $\mu = 0$ and variable σ^2. If β is fixed at σ^{-2}, the algorithm converges for any setting of σ. However, if β is learned, this is not always the case. For various settings of σ^2, Fig. 2 indicates $\mathfrak{L}_{\text{ICA}}$ as a function of $\boldsymbol{W}$. Clearly, as σ increases, finding the independent component becomes less likely than finding the Gaussian component, depending on initialisation of $\boldsymbol{W}$.

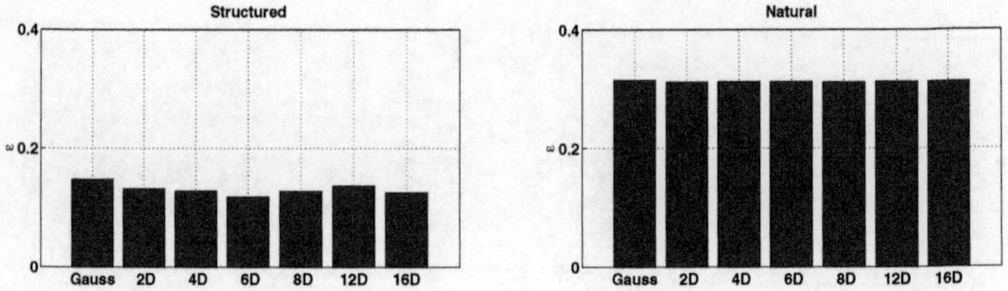

Fig. 3. Predicted error ε on texture sets, for the Gaussian model and ICA models of varying dimensionalities

This exposes a problem of most ICA algorithms: to find only independent components, the role variance plays will have to be eliminated. Although alternatives exist, the easiest method is by whitening the data before applying the ICA subspace algorithm, i.e. using $\boldsymbol{x}' = \boldsymbol{C}^{-\frac{1}{2}}\boldsymbol{x}$. This will make the data variance 1 in all directions. Pre-whitening also simplifies the algorithm in a number of ways: β can be fixed at 1; the rows of $\boldsymbol{W}$ and columns of $\boldsymbol{A}$ can be constrained to have unit length; and in the batch learning rule (Eqn. 6), the first term drops out, as $\boldsymbol{C} = \boldsymbol{I}$, so the learning rule is identical to the one originally proposed by Lee at al. (Eqn. 3). The ICA subspace model essentially changes to a Gaussian model in which m directions should contain non-Gaussian distributions.

4 Experiments

In a simple experiment, the Gaussian model and ICA were compared on their ability to describe texture[1]. Individual models M_i $(i = 1, \ldots, 6)$ were trained on the 6 individual Brodatz texture images, for both the structured and the natural texture set. After training, the negative log-likelihood of each pixel in each texture image I_j of belonging to each model M_i was calculated. This gives a set of distances F_{ij} between model M_i and texture image I_j. As these distances are approximately normally distributed[2], a simple performance measure is [5]:

$$\varepsilon = \operatorname*{med}_{i=1,\ldots,6;\ j\neq i} \frac{1}{2} - \frac{1}{2}\mathrm{erf}\left(\frac{1}{2}\sqrt{\mathcal{F}(F_{ii}, F_{ij})}\right) \tag{11}$$

where $\mathcal{F}(F_1, F_2)$ is the Fisher distance between two distributions F_1 and F_2 and $\varepsilon \in [0, 0.5]$ is the median *predicted Bayes error*, assuming normally distributed features with equal variance and equal prior probabilities[3]. The median is taken over all models and all textures to prevent outliers from having too much influence.

[1] For details on experimental settings, see [2].

[2] The distribution of a likelihood calculated in d dimensions can be approximated by a χ^2_d distribution, which for large d can be approximated by a normal distribution.

[3] Note that the per-model pre-mapping makes this measure pessimistic; in a true segmentation application, PCA pre-mapping would be performed on the entire data set rather than single textures. However, relative performance is not influenced.

The results (Fig. 3) show, firstly, that the predicted error indeed is pessimistic; in [2], the Gaussian model is applied in a mixture model setting and gives very good segmentation results. However, of interest for this paper is the fact that the Gaussian and ICA models give nearly equal performance. Furthermore, for an increasing number of independent components performance hardly changes. This raises the question as to what exactly the advantage of modelling dimensions by non-Gaussian distributions is.

5 Independent Component Analysis

Investigation of the kurtoses of the distributions of data projected onto the extracted independent components shows that non-Gaussian directions have indeed been found: the kurtoses of the projected data distributions deviated significantly from 3 for any model. The algorithm works.

If the ICA model really fits better than the Gaussian, its likelihood $\mathfrak{L}_{\mathrm{ICA}}$ (Eqn. 5) should be significantly larger than that of the Gaussian, $\mathfrak{L}_{\mathrm{Gauss}}$. However, for nearly all textures, the average increase in likelihood (over all pixels) was negligible. To get a better idea of why these increases are so low, the "straw" natural texture N2 (Fig. 1) is considered. Fig. 4 (a) shows the ICA basis vectors found after training a 16D ICA model. Clearly, they correspond to directed edge detectors, which one would expect to be of use in segmenting the straw image. To see their effect, for each pixel in the original image the likelihood of the window of which it is the center can be plotted, again as an image. Figs. 4 (b) and (c) show these likelihood images for both the 64D ICA model and the Gaussian model. At first glance, there is no difference between the two. However, there is a difference, albeit small; Fig. 4 (d)-(i) shows this difference for an increasing number of independent components in the ICA model. For presentation purposes, negative differences (which fell in the range $[-2, 0]$) have not been shown.

It now becomes obvious why the ICA model shows no improvement over the Gaussian model. In general, the independent components correspond to characteristic but more or less unique high-frequency events in images. For this image, these are the few straws that have a different orientation than the majority. The first few independent components (2D-4D models) correspond to the straws below the center of the image; as more dimensions are added, the straws at the top of the image get modelled better as well. Finally, as more and more independent components are found, all straws with non-standard orientations are singled out (relative to the Gaussian model), whereas the main structure of the texture becomes slightly more likely, but in a noise-like fashion. In fact, this happens for all textures.

6 Conclusions

A probabilistic subspace ICA model was developed and compared to a Gaussian model on the task of texture description. It was shown that data pre-whitening is necessary to be able to train the ICA model, which effectively makes the ICA

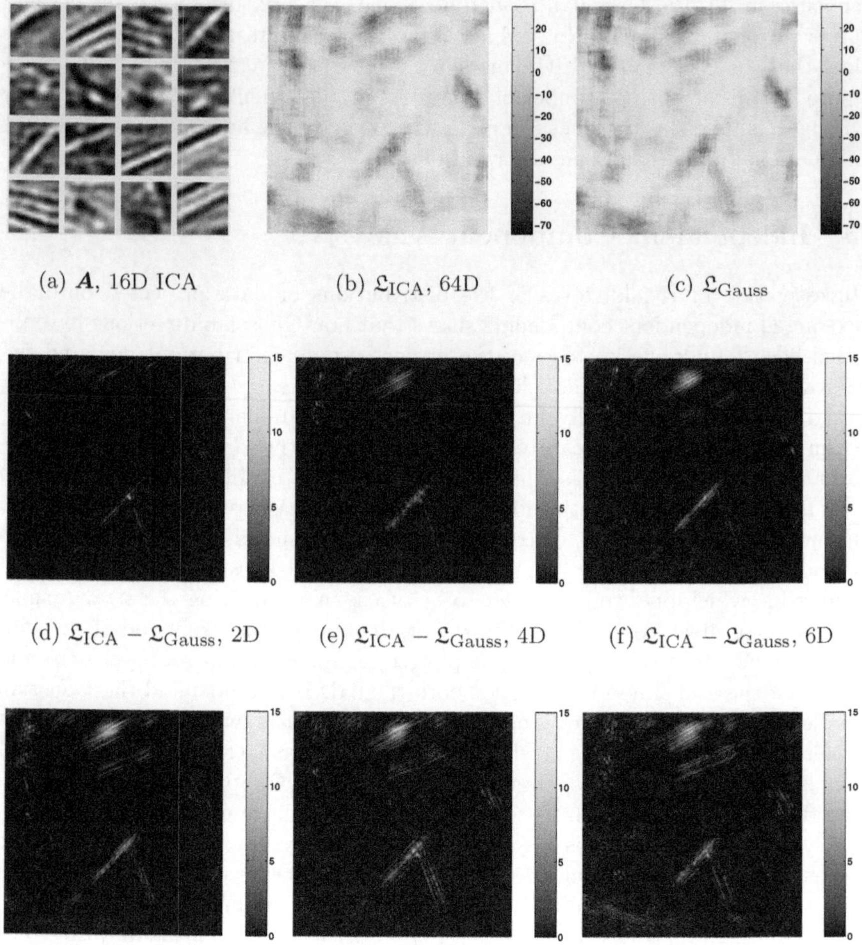

(a) $\boldsymbol{A}$, 16D ICA (b) $\mathfrak{L}_{\text{ICA}}$, 64D (c) $\mathfrak{L}_{\text{Gauss}}$

(d) $\mathfrak{L}_{\text{ICA}} - \mathfrak{L}_{\text{Gauss}}$, 2D (e) $\mathfrak{L}_{\text{ICA}} - \mathfrak{L}_{\text{Gauss}}$, 4D (f) $\mathfrak{L}_{\text{ICA}} - \mathfrak{L}_{\text{Gauss}}$, 6D

(g) $\mathfrak{L}_{\text{ICA}} - \mathfrak{L}_{\text{Gauss}}$, 8D (h) $\mathfrak{L}_{\text{ICA}} - \mathfrak{L}_{\text{Gauss}}$, 12D (i) $\mathfrak{L}_{\text{ICA}} - \mathfrak{L}_{\text{Gauss}}$, 16D

Fig. 4. (a) ICA basis vectors found on texture N2; (b) log-likelihood of each pixel belonging to a 64D ICA model; (c) same for a Gaussian model and (d)-(i) the difference between $\mathfrak{L}_{\text{ICA}}$ of various dimensionalities and $\mathfrak{L}_{\text{Gauss}}$

model a Gaussian one in which a subset of dimensions have non-Gaussian distributions. Subsequently, both models were shown to perform equally well on texture description. Experimental observations led to the conclusion that: (a) the increase in likelihood is generally so low, that the Gaussian model may be considered as good a model as ICA for a data set consisting of texture image patches; (b) ICA – given its high computational complexity – is not useful as a

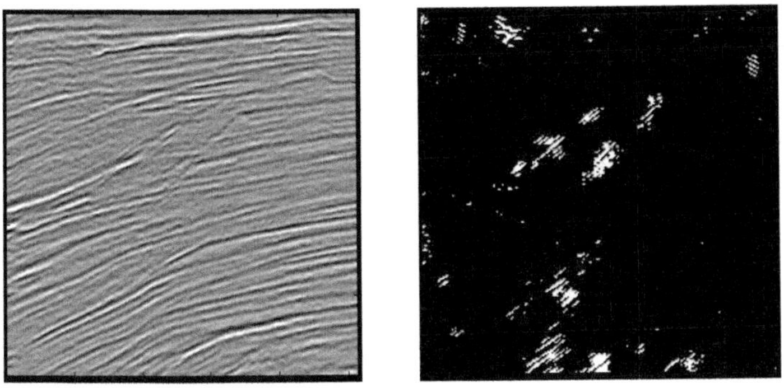

Fig. 5. A useful application of ICA: left, a seismic image; right, thresholded $(\mathfrak{L}_{\mathrm{ICA}} - \mathfrak{L}_{\mathrm{Gauss}})$, for $m = 8$

texture description tool. For segmentation, the goal is to find models which describe textures in a shift-invariant way. To this end, non-standard regions should be *ignored* rather than modelled. The Gaussian model does this by modelling the data by (co)variance only, ignoring outliers. Under the ICA model, as it focuses on non-Gaussianity alone, outliers are more probable (for super-Gaussian sources); in fact, the sources are indirectly optimised to make outliers more likely.

The main conclusion is that, where many authors have suggested ICA might be useful for image processing, its application area is limited to problems in which the detection of unique, characteristic events is of importance. As an example, consider the seismic image shown in Fig. 5. The Gaussian model will pick up only on the two dominant directions present. However, the ICA model is better at description of some faint – yet to a human observer clearly visible – discontinuities and bifurcations in the image.

Acknowledgements

This work was partly supported by the Foundation for Computer Science Research in the Netherlands (SION), the Dutch Organisation for Scientific Research (NWO), the Foundation for Applied Sciences (STW) and the Engineering and Physical Sciences Research Council (EPSRC) in the UK (grant numbers GR/M90665 and GR/L61095).

References

1. P. Brodatz. *Textures, a photographic album for artists and designers.* Dover Publications, New York, NY, 1966. 588
2. D. de Ridder. *Adaptive methods of image processing.* PhD thesis, Delft University of Technology, Delft, 2001. 588, 590, 592, 593
3. D. de Ridder, J. Kittler, O. Lemmers, and R.P.W. Duin. The adaptive subspace map for texture segmentation. In *Proc. ICPR 2000*, pages 216–220, Los Alamitos, CA, 2000. IAPR, IEEE Computer Society Press. 588

4. D. de Ridder, O. Lemmers, R.P.W. Duin, and J. Kittler. The adaptive subspace map for image description and image database retrieval. In *Proc. S+SSPR 2000*, pages 94–103, Berlin, 2000. IAPR, Springer-Verlag. 588
5. K. Fukunaga. *Introduction to statistical pattern recognition.* Electrical Science Series. Academic Press, NY, NY, 1972. 592
6. J. Hurri. Independent component analysis of image data. Master's thesis, Dept. of Computer Science and Engineering, Helsinki University of Technology, Espoo, Finland, March 1997. 588
7. A. Hyvärinen. Survey on independent component analysis. *Neural Computing Surveys*, 1(2):94–128, 1999. 589
8. T. Kohonen, S. Kaski, and H. Lappalainen. Self-organized formation of various invariant-feature filters in the Adaptive-Subspace SOM. *Neural Computation*, 9(6):1321–1344, 1997. 588
9. T.-W. Lee, M. Girolami, and T.J. Sejnowski. Independent component analysis using an extended infomax algorithm for mixed sub-Gaussian and super-Gaussian sources. *Neural Computation*, 11(2):417–441, 1999. 590
10. H. Lu, Y. Fainman, and R. Hecht-Nielsen. Image manifolds. In *Applications of Artificial Neural Networks in Image Processing III, Proceedings of SPIE*, volume 3307, pages 52–63, Bellingham, WA, 1998. SPIE, SPIE. 587
11. M.E. Tipping and C.M. Bishop. Mixtures of probabilistic principal component analyzers. *Neural Computation*, 11(2):443–482, 1999. 591

Fusion of Multiple Cue Detectors for Automatic Sports Video Annotation

Josef Kittler, Marco Ballette, W. J. Christmas,
Edward Jaser, and Kieron Messer

Centre for Vision
Speech and Signal Processing, University of Surrey
Guildford, Surrey GU2 7XH, UK
J.Kittler@eim.surrey.ac.uk

Abstract. This paper describes an aspect of a developing system named ASSAVID which will provide an automatic and semantic annotation of sports video. This annotation process segments the sports video into semantic categories (e.g. type of sport) and permits the user to formulate queries to retrieve events that are significant to that particular sport (e.g. goal, foul). The system relies upon the concept of "cues" which attach semantic meaning to low-level features computed on the video. In this paper we adopt the multiple classifier system approach to fusing the outputs of multiple cue detectors using *Behaviour Knowledge Space* fusion. Using this technique, unknown sports video can be classified into the type of sport being played. Experimental results on sports video provided by the BBC demonstrate that this method is working well.

1 Introduction

There is a vast amount of sports footage being recorded every day. For example, each year the British Broadcasting Corporation (BBC) provides coverage of the Wimbledon tennis championships. During this event up to fifteen different live feeds are being recorded simultaneously. At the end of the Wimbledon fortnight over one thousand tapes of sports related footage are brought back to the BBC headquarters. All this video data is generated from just one event. The BBC records hundreds of different sporting events each year.

Ideally, all this sports video should be annotated and the meta-data generated on it should be stored in a database along with the video data. Such a system would allow an operator to retrieve any shot or important event within a shot at a later date. Also, the level of annotation provided by the system should be adequate to facilitate simple text-based queries. For example a typical query could be: *"Retrieve the winning shot played in the last game of each set in the ladies final in the Wimbledon tennis championship 2001"*. Such a system has many uses, such as in the production of television sport programmes and documentaries.

Due to the large amount of material being generated, manual annotation is both impractical and very expensive. However, automatic annotation is a very

T. Caelli et al. (Eds.): SSPR&SPR 2002, LNCS 2396, pp. 597–605, 2002.
© Springer-Verlag Berlin Heidelberg 2002

demanding and an extremely challenging computer vision task as it involves high-level scene interpretation. It is unlikely that an efficient, fully automated video annotation system will be realised in the near future.

Perhaps the most well known automatic video annotation system reported in the literature is Virage [1]. Virage has an open framework which allows for the integration of many real time audio and video analysis tools and places the data into an industry-standard database such as Informix or Oracle. However, the number of analysis tools available is limited, although always expanding. The main problem with Virage is that no effort has been made to bridge the gap between the information provided by the low-level analysis tools and the high-level interpretation of the video, which is required for our application.

Other work, specific to some form of sports annotation includes [9] in which camera motion is used to help in the automatic annotation of basketball. Mo et al. utilize state transition models, which include both top-down and bottom-up processes, to recognise different objects in sports scenes [8]. In [10] work has been undertaken to distinguish between sports and non-sports MPEG compressed video. Finally, MIT have been working on the analysis of American football video [4].

This paper describes one aspect of the development of a novel system which will provide a semantic annotation of sports video. This annotation process segments the sports video into semantic categories (e.g. type of sport) and permits the user to formulate queries to retrieve events that are significant to that particular sport (e.g. goal, foul). The system will aid an operator in the generation of the high-level annotation for incoming sports video. The basic building blocks of the system are low-level audio and video analysis tools, which we term cue detectors. Examples of cue-detectors include: grass; swimming pool lanes; ocean; sprint frequency; referee whistle and crowd cheer. A contextual reasoning engine will then be used to analyse the output of these cue detectors and attach semantic information to the video being analysed. The generated meta-data and video data will then be stored in a database which is based on a mixture of IBM's Media360 and Informix. The system will also provide a Java graphical user interface to the database which will allow the user to browse the video, view sequences and generate story boards, formulate queries and analyse and modify the generated indices.

In general, different methods can be adopted to design cue detectors. For instance, the same cue can be detected on the basis of different image properties, such as texture, colour or shape. Having, for a given cue, more than one cue detector raises the problem of how the respective response of these multiple detectors should be combined to improve the confidence in the cue being present or absent.

In this paper we adopt the multiple classifier system approach to fusing the outputs of multiple cue detectors. We use the Behaviour Knowledge Space fusion method [3] and show that significant improvements can be gained in this way.

The paper is organised as follows. In Section 2 we briefly describe the cue detectors used in this study. The fusion technique is overviewed in Section 3.

The results of the experiments designed to demonstrate the effectiveness of the method are presented in Section 4 and 5. The paper is concluded in Section 6.

2 Cue Detectors

The objective in the automatic annotation of video material is to provide indexing material that describes as usefully as possible the material itself. In much of the previous work in this area (for example [2]), the annotation consisted of the output of various feature detectors. By itself, this information bears no semantic connection to the actual scene content — it is simply the output of some image processing algorithms. In this project we are taking the process one stage further. By means of a set of training processes, we aim to generate an association between the feature detector outputs and the occurrence of actual scene features. Thus for example we might train the system to associate the output of a texture feature detector with crowds of people in the scene. We can then use this mechanism to generate confidence values for the presence of a crowd in a scene, based on the scene texture. We denote the output of this process as a "cue". These cues can then be combined in the contextual reasoning engine to generate higher-level information, e.g. the type of sport being played.

We have developed many different cue detection methods. In this section we briefly discuss three visual cue generation methods. Each method can be used to form a number of different cue-detectors provided that suitable training data is available. These methods are neural network, multimodal neighbourhood signature and texture codes. These are briefly described in the following subsections.

2.1 Neural Network

Each cue-detector is a neural network trained on colour and texture descriptors computed at a pixel level on examples image regions of the cue of interest (see [7]) and on image regions which are known not to contain the cue. The resulting trained network is then able to distinguish between the features of cue and non-cue pixels. A high output represents the case when the feature vector of the pixel belongs to same distribution as the cue and vice-versa.

To check for the presence of a cue in a test image the same colour and texture features are computed for each test image pixel and the feature vector is passed to the neural network. If many high outputs are observed then this gives an indication of how likely it is that the given cue is present in the image. Cues suitable for this method include sky, grass, tennis court and athletics track.

2.2 Multimodal Neighbourhood Signature

In the Multimodal Neighbourhood Signature (MNS) approach (described in detail in [6]), object colour structure is represented by a set of invariant features computed from image neighbourhoods with a multimodal colour density function. The method is image-based – the representation is computed from a set of examples of the object of interest.

In the implemented method, an MNS is a set of invariant colour pairs corresponding to pairs of coordinates of the located density function modes from each neighbourhood. The MNS's of all the example images are then merged into a composite one by superposing the features (colour pairs). Considering each colour pair as an independent object descriptor (a *detector*), its discriminative ability is measured on a pre-selected *training set* consisting of positive and negative example image regions. A simple measure, the absolute difference of true and false positive percentages is computed. Finally, the n most discriminative detectors are selected to represent the object of interest. For the reported experiments n was set to 3.

For matching, we view each detector as a point in the detector space. A hyper-sphere with radius h is defined around each point. Given another image of the object, measurements are likely to lie inside the detector hyper-spheres. A binary n-tuple is computed for each test image, each binary digit assigned 1 if at least one test measurement was within the corresponding detector sphere, 0 otherwise. One of 2^n possible n-tuples are the measurements output from the matching stage. The relative frequency of each possible n-tuple over the positive and negative cue examples of the training set define an estimate of the probability of each measurement given the cue and not given the cue respectively. These 2 numbers are output to the decision making module.

2.3 Texture Codes

The texture-based cue detector consists of two components: a training phase, in which a model for the cue is created using ground truth, and the cue extractor (see [5]). In the training stage, template regions from the keyframes are selected for each cue. Several templates are needed for each cue to account for appearance variations. Textural descriptors are extracted form the templates using a texture analysis module based on Gabor filters. These descriptors, with the number of occurrences, form the model for the cue.

In the cue extractor, the whole image is presented to the texture analysis module. Then, by comparing the result with the model, a coarse detection component selects the three templates which are most likely to be visually similar to an area of the image being annotated. The similarity is evaluated using the histogram intersection. We increase the computational efficiency by hashing the meta-data; this also enables us to compute the similarity measure only for templates which share descriptors with the input image. A localisation component finally identifies the areas of the image which the selected templates match most closely, and the image location which yields the best match confidence is retained. The highest confidence, with its location, are the output for the cue.

3 Fusion Method

The fusion method we experimented with is the *Behaviour Knowledge Space* [3] which proved to be very effective in other applications. In order to describe it, let us first introduce the necessary notation. We consider the task of cue detection as a two class pattern recognition problem where pattern Z (video key frame, video sequence, audio segment) is to be assigned to one of the *2* possible classes ω_i $i = 1, 2$. Let us assume that we have L classifiers, each representing the given pattern by some measurement vector. Denote the measurement vector used by the jth classifier $\mathbf{x}_j$. The measurement vectors may be distinct, as, for example, in the case of the multimodal neighbourhood signature expert, or some components may be identical (shared) as exemplified by the texture codes and neural network experts. Each classifier computes the respective aposteriori probabilities for the two hypotheses that a cue is either present or absent. The aposteriori probability for the cue being present, computed by expert j will be denoted $P(\omega_1|\mathbf{x}_j)$. The probability that the cue is absent $P(\omega_2|\mathbf{x}_j)$ is given by $1 - P(\omega_1|\mathbf{x}_j)$.

The *Behaviour-Knowledge Space* (BKS) method proposed by Huang et al [3] also considers the support from all the experts to all the classes jointly. However, the degree of support is quantified differently than in the Decision Templates approach. Here the decisions Δ_{ji} of experts $i = 1, ..., L$ regarding the class membership ω_j, $j = 1, ..., c$ of pattern Z are mapped into an L dimensional discrete space and the BKS fusion rule is defined in this space. In order to be more specific, let us designate the decision of the jth expert about pattern Z by $\delta_j(Z)$ which can be expressed as

$$\delta_j(Z) = argmax_{i=1}^{c} P(\omega_i|\mathbf{x}_j) \tag{1}$$

Thus $\delta_j(Z)$ assumes integer values from the interval $[1, c]$. The combination of the decision outputs $\delta_j(Z)$, $j = 1, ..., L$ defines a point in the L-dimensional discrete space, referred to as the Behaviour Knowledge Space (BKS). We can consider each point in this space to index a bin (cell). The BKS fusion rule then associates a separate consensus decision with each of the bins in the BKS space.

Let us denote by $h_i(d_1, .., d_L)$ the histogram of the patterns from a set Ξ which belong to class ω_i and fall into the bin $(d_1, .., d_L)$ by virtue of the indexation process defined in (1). The BKS fusion rule then assigns a new pattern Z as follows

$$\delta(Z) = \begin{cases} argmax_{i=1}^{c} h_i(\delta_1(Z), .., \delta_L(Z)) & if \ \sum_{k=1}^{c} h_k(\delta_1(Z), .., \delta_L(Z)) > 0 \\ & and \ \frac{h_{\delta(Z)}(\delta_1(Z),..,\delta_L(Z))}{\sum_{k=1}^{c} h_k(\delta_1(Z),..,\delta_L(Z))} \geq \lambda \\ \delta_0(Z) & otherwise \end{cases} \tag{2}$$

where $\delta_0(Z)$ denotes a rejection or a decision by an alternative fusion process such as the majority *Vote*, or the fusion strategies suggested in [11] and [3]. Thus a special line of action is taken when the indexed bin of the BKS histogram is zero, or if the proportional vote held by the most representative class in the bin is below threshold λ. In our two class experiments the above conditions were always satisfied. However, a number of studies on how to find the value of the threshold have been reported (see for instance Huang and Suen [3].

In summary, the class with the greatest number of votes in each bin is chosen by the BKS method. In our experiments we considered different weights for the two different classes based on the class a priori probabilities. Thus for each combination of classifiers we divided the number of occurrences of each class by the respective numbers of samples in set Ξ.

4 Experiments

The aim of the experiments described in this section was to investigate the effect of multiple expert fusion in the context of sports video cue detection. We considered the following sports: boxing, shooting, swimming and tennis. In each frame we looked for a boxing ring, a shooting target, a swimming pool and a tennis court. These cues are considered indicative of the respective sport disciplines. The images in figure 1 show such examples for each sport and cue. The study was limited to these four cues because for each of them we had the responses from multiple cue detectors that we wished to combine. In particular, for each cue we had the outputs of three experts. These experts and the associated identifiers (codes) are : Texture Code Expert (code 0), Multimodal Neighbourhood Signature Expert (code 1), Neural Net Expert (code 2).

The experiments were conducted on a database of key frames which were manually annotated to provide ground truth information about cue presence. The database contained 517 key frames of boxing, 172 frames of shooting, 1087 frames of swimming and 469 frames of tennis. Each key frame then had the cue-outputs for each cue-detector and each cue-method (expert) computed. Thus for each expert we had 517*4 outputs for boxing, 172*4 outputs for shooting and so on. Each detector generated two scores : $p(\mathbf{x}_j|\omega_1)$ and $p(\mathbf{x}_j|\omega_2)$. These scores are the density function values for the key frame measurements $\mathbf{x}_j$ computed by the jth cue-detector when the cue is present and absent from the scene respectively. These scores are converted into aposteriori class probabilities under the assumption that, a priori, the presence and absence of a cue are equally likely, i.e.

$$P(\omega_1|\mathbf{x}_j) = \frac{p(\mathbf{x}_j|\omega_1)}{p(\mathbf{x}_j|\omega_1)+p(\mathbf{x}_j|\omega_2)} \tag{3}$$

A global thresholding was then applied in order to obtain a crisp label.

We split our data set into an evaluation set and a test set. We considered two different configurations that we called CONFIGURATION 1 (evaluation set = 20% of the total set; test set = 80% of the total set) and CONFIGURATION 2

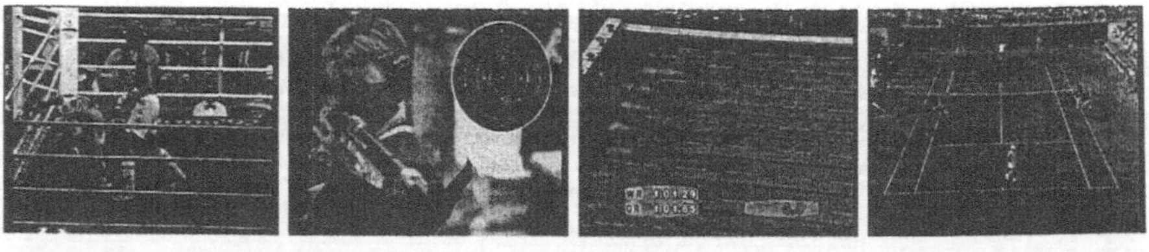

Fig. 1. Examples images selected to represent each of the four chosen sports

(evaluation set =30%; test set =70%). The evaluation set was used to find the optimal global threshold which produced the lowest total error rate. The performance of the fusion method was then evaluated on a completely independent test set.

The results of the experiments are presented in Figures 2(a) and 2(b). The false rejection and false acceptance error rates of the system were estimated separately and then averaged. This is a standard practice in detection problems as it is impossible to specify the prior probabilities of the populations of key frames which do and do not contain a particular cue accurately. The resulting error rates are then shown in the figures for the four sport disciplines.

In general, the error rates are lower for CONFIGURATION 2 than for CONFIGURATION 1. This is understandable, as CONFIGURATION 2 uses more data for training the fusion system than CONFIGURATION 1.

Boxing In the case of boxing cue detection it is interesting to note that the best pair of experts, as selected on the evaluation set, did not include the individually best cue detector based on MNS. This is reflected in the performance on the test set which is worse than that of the individually best detector. Once the size of the evaluation set is increased, better generalisation behaviour is observed. This is apparent from the monotonicity of the fusion results, i.e. as the number of detectors is increasing, the performance monotonically improves.
Shooting In the case of the shooting cue, the Texture Code and Neural Network detectors produced very high false acceptance rates. Interestingly, the MNS detector has a zero rejection rate as the shooting cue - shooting target - is a very distinctive object. There was a dramatic swing of false acceptance and false rejection rates between the two configurations. Again, CONFIGURATION 2 results exhibited better generalisation and most importantly, the benefit of multiple detector fusion was the most pronounced.
Swimming For CONFIGURATION 1, both texture codes and MNS detectors produced very high false rejection rates and all detectors showed zero false acceptance rate. The imbalance in performance was corrected, though with some overshoot, with the enhanced training under CONFIGURATION 2. Again, much better generalisation was achieved for CONFIGURATION 2.
Tennis The performance trends noted in the case of tennis were quite consistent with the previous cues. The main points are the improved generalisation when moving from CONFIGURATION 1 to CONFIGURATION 2. Most importantly, the multiple cue detector fusion consistently provides at least as good or better results than the individually best detector.

5 Results on a Relabelled Database

A detailed analysis of the frequently unbalanced error rates and the difficulty in selecting a sensible threshold reported in the previous section revealed that the problem derived primarily from the way the data set was labelled. Any key frame that was part of a video segment reporting on a particular sport was routinely

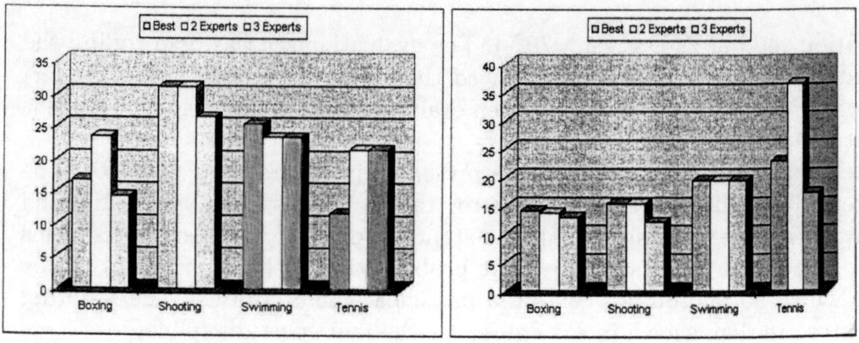

(a) Configuration 1 (b) Configuration 2

Fig. 2. Error rates on sport images using best single and multiple experts

assigned the label corresponding to that sport. Yet the individual cue detectors were trained to detect specific cues that are characteristic of the respective disciplines. For instance, the visual content of the swimming cue was an image segment of the swimming pool containing ropes delineating the swimming lanes. However, some segments of the swimming video sequence contained interviews with the competitors and the swimming pool was not visible. Often the number of such frames was significantly high and this resulted in a complete overlap of the cue present and cue absent distributions. In view of this, all the key frames were carefully re-examined and whenever appropriate re-labelled.

The experiments of the previous section were then repeated. The results are reported in Figures 3(a) and 3(b) using the same format of presentation. In general, we observe that the error rates are more balanced, although not always better than in the previous section. However, most importantly we observe dramatic improvements in the results of the multiple cue detector system for all disciplines. The results are particularly promising for CONFIGURATION 2 for which average error rates are not worse than 3.11%. In the case of shooting they drop to zero both for false rejection and false acceptance.

6 Conclusion

In this paper we have described a process for the automatic sports classification within the developing ASSAVID system. We have demonstrated that the method, based on the concept of cue detection, is working well on a set of ground-truthed static images. It has also been demonstrated that by adopting the multiple classifier system approach of *Behaviour Knowledge Space* to fusing the outputs of the multiple cue detectors can significantly improve the performance of the automatic sports classification.

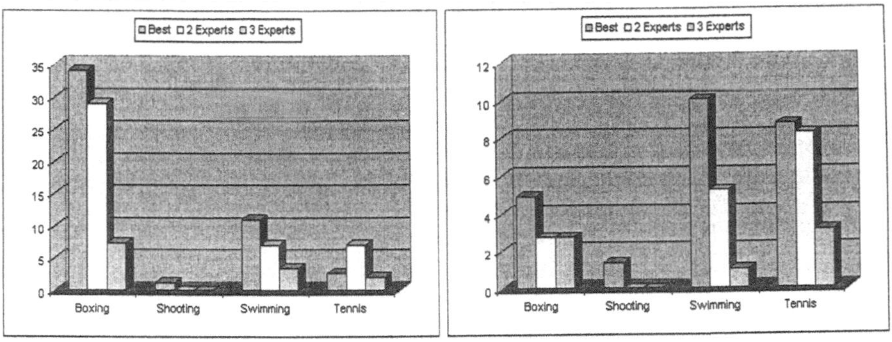

(a) Configuration 1 (b) Configuration 2

Fig. 3. Error rates on relabelled sport images using best single and multiple experts

It was also demonstrated that the sports were only recognised when a suitably trained cue for that sport was identified in the image. If that cue for that sport was not in the image the frame was incorrectly labelled. The use of more cues for each specific sport should increase the recognition accuracy and robustness. At present we are working on more cue-methods and training more cue-detectors. These include cues based on other audio and motion features.

In the final system it is intended to make a decision about the sport being played over an entire shot and not just a single frame. This will allow us to incorporate temporal information into the decision making decision and this should robustify the results further.

Acknowledgements

This work has been performed within the framework of the ASSAVID project granted by the European IST Programme.

References

1. http://www.virage.com.
2. B. V. Levienaise-Obadia, W. Christmas, J. Kittler, K. Messer, and Y. Yusoff. Ovid: towards object-based video retrieval. In *Proceedings of Storage and Retrieval for Video and Image Databases VIII (part of the SPIE/ITT Symposium: Electronic Imaging'2000)*, Jan 2000.
3. Y. Huang and C. Suen. A method of combining multiple experts for the recognition of unconstrained handwritten numerals. *IEEE Transaction on Pattern Analysis and Machine Intelligence*, 17(1), 1 1995.

4. S. S. Intille and A. F. Bobick. A framework for representing multi-agent action from visual evidence. In *Proceedings of the National Conference on Artificial Intelligence (AAAI)*, July 1999.
5. B. Levienaise-Obadia, J. Kittler, and W. Christmas. Defining quantisation strategies and a perceptual similarity measure for texture-based annotation and retrieval. In IEEE, editor, *ICPR'2000*, volume III, Sep 2000.
6. J. Matas, D. Koubaroulis, and J. Kittler. Colour Image Retrieval and Object Recognition Using the Multimodal Neighbourhood Signature. In D. Vernon, editor, *Proceedings of the European Conference on Computer Vision*, LNCS vol. 1842, pages 48–64, Berlin, Germany, June 2000. Springer.
7. K. Messer and J. Kittler. A region-based image database system using colour and texture. *Pattern Recognition Letters*, pages 1323–1330, November 1999.
8. H. Mo, S. Satoh, and M. Sakauchi. A study of image recognition using similarity retrieval. In *First International Conference on Visual Information Systems (Visual'96)*, pages 136–141, 1996.
9. D. D. Saur, Y.-P. Tan, S. R. Kulkarni, and P. J. Ramadge. Automated analysis and annotation of basketball video. In *SPIE Storage and Retrieval for Still Image and Video Databases V, Vol.3022*, pages 176–187, 1997.
10. V. Kobla, D. DeMenthon, and D. Doermann. Identifying sporst video using replay, text and camera motion features. In *SPIE Storage and retrieval for Media Database 2000*, pages 332–342, 2000.
11. K.-D. Wernecke. A coupling procedure for the discrimination of mixed data. *Biometrics*, 48:497–506, 6 1992.
12. L. Xu, A. Krzyzak, and C. Y. Suen. Methods of combining multiple classifiers and their applications to handwriting recognition. *IEEE Transaction. SMC*, 22(3):418–435, 1992.

Query Shifting Based on Bayesian Decision Theory for Content-Based Image Retrieval

Giorgio Giacinto and Fabio Roli

Dept. of Electrical and Electronic Engineering - University of Cagliari
Piazza D'Armi 09123 Cagliari, Italy
Tel: +39 070 675 5752 Fax: +39 070 675 5782
{giacinto,roli}@diee.unica.it

Abstract. Despite the efforts to reduce the so-called semantic gap between the user's perception of image similarity and feature-based representation of images, the interaction with the user remains fundamental to improve performances of content-based image retrieval systems. To this end, relevance feedback mechanisms are adopted to refine image-based queries by asking users to mark the set of images retrieved in a neighbourhood of the query as being relevant or not. In this paper, Bayesian decision theory is used to compute a new query whose neighbourhood is more likely to fall in a region of the feature space containing relevant images. The proposed query shifting method outperforms two relevance feedback mechanisms described in the literature. Reported experiments also show that retrieval performances are less sensitive to the choice of a particular similarity metric when relevance feedback is used.

1 Introduction

The availability of large image and video archives for many applications (art galleries, picture and photograph archives, medical and geographical databases, etc.) demands advanced query mechanisms that address perceptual aspects of visual information. To this end, a number of image retrieval techniques based on image content, where the visual content of images is captured by extracting low-level features based on color, texture, shape, etc., have been developed [4],[19]. Content-based queries are often expressed by visual examples in order to retrieve from the database all images that are "similar" to the examples. The retrieval process is usually performed by a *k-nn* search in the feature space using the Euclidean metric [4].

It is easy to see that the effectiveness of a content-based image retrieval system (CBIR) strongly depends on the choice of the set of visual features and on the choice of the "metric" used to model the user's perception of image similarity. The gap between user's perception of image similarity and feature-based image representation is usually small for databases related to tasks where the semantic description of the images is reasonably well defined. For example, data bases of lithographs, frontal

T. Caelli et al. (Eds.): SSPR&SPR 2002, LNCS 2396, pp. 607-616, 2002.
© Springer-Verlag Berlin Heidelberg 2002

views of faces, outdoor pictures, etc. [19]. For this kind of databases, a pair of images that the user judges as being similar to each other is often represented by two near points in the feature space.

However, no matter how suitable for the task at hand the features and the similarity metric have been designed, the set of retrieved images often fits the user's needs only partly. Typically, different users may categorise images according to different semantic criteria [1]. Thus, if we allow different users to mark the images retrieved with a given query as "relevant" or "non-relevant", different subsets of images will be marked as "relevant", and the intersection of such subsets is usually non-empty. Accordingly, the need for mechanisms to "adapt" the CBIR system response based on some "feedback" from the user is widely recognised.

A number of techniques aimed at exploiting such relevance feedback have been proposed in the literature [2],[3],[6],[7],[9],[10],[11],[12],[13],[14],[16],[18]. As discussed in Section 2, they are based on the fact that the user does not know the actual distribution of images in the feature space, nor the feature space itself, nor the similarity metric employed.

In this paper, Bayesian decision theory is used to compute a new query point based on relevance feedback from the user. The basic idea behind our proposal is the local estimation of the decision boundary between the "relevant" and "non relevant" regions of the neighbourhood of the original query. The new query is then placed at a suitable distance from such boundary, on the side of the sub-region containing relevant images. A similar query shifting mechanism was proposed by the authors in [7], where the query shifting computation was derived by heuristics. In this paper, the computation of the new query is placed in the framework of Bayesian decision theory.

In section 2, a brief overview of relevance feedback techniques for CBIR is given. The proposed relevance feedback method is described in Section 3. Experiments with two image datasets are reported in Section 4. The reported results show that the proposed method outperforms two relevance feedback mechanisms recently described in the literature. Section 4 also points out that, when relevance feedback is performed, retrieval performances are less sensitive to the choice of a particular similarity metric.

2 Relevance Feedback for CBIR

It is well known that information retrieval system performances can be improved by user interaction mechanisms. This issue has been studied thoroughly in the text retrieval field, where the relevance feedback concept has been introduced [15]. Techniques developed for text retrieval systems should be suitably adapted to content based image retrieval, on account of differences in both feature number and meaning, and in similarity measures [10],[13].

Basically, relevance feedback strategies are motivated by the observation that the user is unaware of the distribution of images in the feature space, nor of the feature space itself, nor of the similarity metric. Therefore, relevance feedback techniques proposed in the literature involve the optimisation of one or more CBIR components, e.g., the formulation of a new query, the modification of the similarity metric, or the transformation of the feature space. Query reformulation is motivated by the observation that the image used to query the database may be placed in a region of the feature space that is "far" from the one containing images that are relevant to the user. A query shifting technique for CBIR based on the well known Rocchio formula developed in the text retrieval field [15] has been proposed in [13]. The estimation of probability densities of individual features for relevant and non relevant images is used in [11] to compute a new query. The new query is determined by randomly drawing individual feature components according to the estimated distributions.

In order to optimise the similarity metric to user interests, many CBIR systems rely on parametric similarity metrics, whose parameters are optimised by relevance feedback. Theoretical frameworks involving both the computation of a new query and the optimisation of the parameters of similarity metric have been proposed in the literature [9],[14]. A linear combination of different similarity metrics, each suited for a particular feature set, has been proposed in [18]. Relevance feedback information is then used to modify the weights of the combination to reflect different feature relevance. Santini and Jain also proposed a parametrized similarity measure updated according to feedback from the user [16]. Rather than modifying the similarity metric, Frederix et al. proposed a transformation of the feature space by a logistic regression model so that relevant images represented in the new feature space exhibit higher similarity values [6]. A probabilistic feature relevance scheme has been proposed in [12], where a weighted Euclidean distance is used.

A different perspective has been followed in [3] where relevance feedback technique based on the Bayesian decision theory was first proposed. The probability of all images in the database of being relevant is estimated, and images are presented to the user according to the estimated probability.

3 Query Shifting by Bayesian Decision Theory

3.1 Problem Formulation

Let us assume first that the database at hand is made up of images whose semantic description is reasonably well defined. In these cases it is possible to extract a set of low level features, such that a pair of images judged by the user as being similar to each other is represented by two near points in the feature space. Let us also assume that the user wishes to retrieve images belonging to a specific *class*, that is, she/he is interested in performing a so-called “category” search [19]. As different users have different perceptions of similarity depending on the goal they are pursuing, for a given query, different users may identify different subsets of relevant images. According to the first hypothesis, each subset of relevant images identifies a region in the feature space. Relevance feedback is thus needed to locate the region containing relevant images for a given user.

The user marks the images retrieved by the k-nn search as being relevant or not, so that the neighbourhood of the query in the feature space is subdivided into a relevant and a non-relevant region. Our approach is based on the local estimation of the boundary between relevant and non-relevant images belonging to the neighbourhood. Then a new query is computed so that its neighbourhood is more likely to be contained in the relevant region.

In order to illustrate our approach, let us refer to the example shown in Figure 1. The boundary of the region that contains the relevant images that the user wishes to retrieve is depicted in the figure. It is worth noting that this boundary is not known a-priori because its knowledge would require the user to mark all images contained in the database. $\mathbf{Q}_0$ is the initial query provided by the user to perform the k-nn search. The neighbourhood $N(\mathbf{Q}_0)$ of $\mathbf{Q}_0$ does not fall entirely inside the region of relevant images because it contains both relevant and non-relevant images.

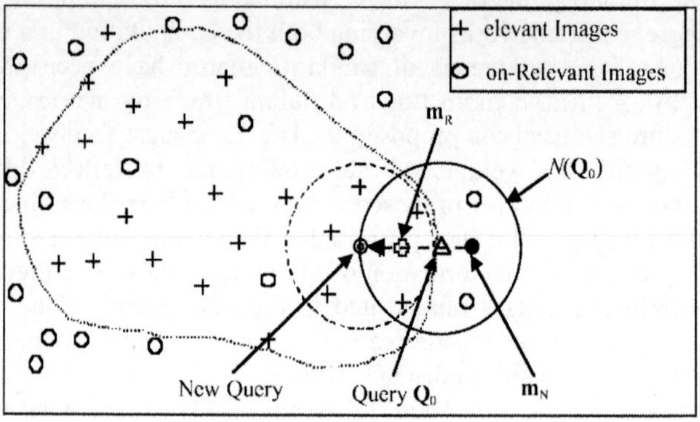

Fig. 1. The proposed query shifting method: an example. The boundary of the region containing the images that are relevant to user query $\mathbf{Q}_0$ is depicted by the dotted line. The initial query $\mathbf{Q}_0$ and the neighbourhood $N(\mathbf{Q}_0)$ related to the *k-nn* search (k = 5) are depicted. A new query computed in the $\mathbf{m}_R$ - $\mathbf{m}_N$ direction is shown such that its neighbourhood (dashed line) is contained in the relevant region. $\mathbf{m}_R$ and $\mathbf{m}_N$ are the mean vectors of relevant and non-relevant image subsets, respectively, retrieved with the initial query $\mathbf{Q}_0$

The decision boundary between relevant and non relevant images belonging to $N(Q_0)$ can be estimated as explained in the following.

Let $\mathbf{I}$ be a feature vector representing an image in a d–dimensional feature space. Let $I_R(\mathbf{Q}_0)$ and $I_N(\mathbf{Q}_0)$ be the sets of relevant and non-relevant images, respectively, contained in $N(\mathbf{Q}_0)$. The mean vectors of relevant and non-relevant image, $\mathbf{m}_R$ and $\mathbf{m}_N$, can be computed as follows

$$\mathbf{m}_R = \frac{1}{k_R} \sum_{\mathbf{I} \in I_R(\mathbf{Q}_0)} \mathbf{I}, \quad \mathbf{m}_N = \frac{1}{k_N} \sum_{\mathbf{I} \in I_R(\mathbf{Q}_0)} \mathbf{I} \tag{1}$$

where k_R and k_N are the sizes of relevant and non-relevant image sets, respectively (k_R+k_N=k). The average variance of relevant and non relevant images can be computed as follows:

$$\sigma^2 = \frac{1}{k} \left(\sum_{\mathbf{I} \in I_R(\mathbf{Q}_0)} (\mathbf{I} - \mathbf{m}_R)^t (\mathbf{I} - \mathbf{m}_R) + \sum_{\mathbf{I} \in I_N(\mathbf{Q}_0)} (\mathbf{I} - \mathbf{m}_N)^t (\mathbf{I} - \mathbf{m}_N) \right) \mathrm{y} \tag{2}$$

Let us assume that relevant and non-relevant images in $N(\mathbf{Q}_0)$ are normally distributed with means $\mathbf{m}_R$ and $\mathbf{m}_N$ and equal variance σ^2. Then, according to the Bayesian decision theory, the decision surface between these two "classes" of images is orthogonal to the line linking the means and passes through a point $\mathbf{x}_0$ defined by the following equation [5]:

$$\mathbf{x}_0 = \frac{1}{2}(\mathbf{m}_R + \mathbf{m}_N) - \frac{\sigma^2}{\left|\mathbf{m}_R - \mathbf{m}_N\right|^2} \ln\frac{P(\omega_R)}{P(\omega_N)}(\mathbf{m}_R - \mathbf{m}_N) \tag{3}$$

where the priors $P(\omega_R)$ and $P(\omega_N)$ are related to the images belonging to $N(\mathbf{Q}_0)$ and can be estimated as the fraction of relevant and non relevant images in $N(\mathbf{Q}_0)$, respectively.

When the prior probabilities are equal, i.e., half of the images of $N(\mathbf{Q}_0)$ are relevant, $\mathbf{x}_0$ is halfway between the means, while it moves away from the more likely mean in the case of different priors. In $\mathbf{x}_0$ the posterior probabilities for the two classes are equal, while points with higher values of posterior probability for class ω_R are found by moving away from $\mathbf{x}_0$ in the $(\mathbf{m}_R\text{-}\mathbf{m}_N)$ direction (if we move in the opposite direction, higher posteriors for ω_N are obtained). Therefore, as clearly shown in Figure 1, candidate query points that could improve retrieval performances, are those located on the line connecting $\mathbf{m}_R$ and $\mathbf{m}_N$. In particular the new query point should be selected in the $\mathbf{m}_R$ - $\mathbf{m}_N$ direction so that its neighbourhood is contained in the relevant region.

3.2 Query Shifting Computation

The rationale behind the query computation proposed hereafter can be briefly explained as follows. The desired result is to have the neighbourhood of the new query totally contained in the relevant region of the feature space. Therefore we first hypothesise an optimal location for the desired neighbourhood and then we will compute the query that can be associated with such a neighbourhood.

An optimal neighbourhood can be obtained by shifting the neighbourhood of $\mathbf{Q}_0$ in the $\mathbf{m}_R$ - $\mathbf{m}_N$ direction until it contains only relevant images (see Figure 1). Let $\mathbf{m}_R^{(1)}$ be the mean vector of relevant images captured by this shifted neighbourhood. We propose to use this point as the new query.

Exploiting the hypotheses of section 3.1, and following the above rationale, let us derive formally the computation of the new query. Let us define the shifted neighbourhood as the neighbourhood whose images satisfy the following properties: i) the mean vectors of relevant and non-relevant images are collinear with the mean vectors of $N(\mathbf{Q}_0)$ and their distance is constant, i.e.

$$\mathbf{m}_R^{(1)} - \mathbf{m}_N^{(1)} = \mathbf{m}_R^{(0)} - \mathbf{m}_N^{(0)} \tag{4}$$

where superscripts (0) and (1) refer to original neighbourhood position and the shifted one, respectively; ii) the average variance of relevant and non-relevant images is always equal to σ^2; iii) the boundary between relevant and non relevant images estimated according to Equation (3) for any shifted neighbourhood, coincides with the boundary computed using the original neighbourhood, i.e., $\mathbf{x}_0$ represents a point of the actual boundary between relevant and non relevant images. Accordingly, the location of the point $\mathbf{x}_0$ (Equation 3) can be computed using either neighbourhood (0) or (1). By making equal the two computations of $\mathbf{x}_0$, the following relation holds:

$$\frac{1}{2}\left(\mathbf{m}_R^{(1)}+\mathbf{m}_N^{(1)}-\mathbf{m}_R^{(0)}-\mathbf{m}_N^{(0)}\right)=$$

$$\frac{\sigma^2}{\left\|\mathbf{m}_R^{(0)}-\mathbf{m}_N^{(0)}\right\|^2}\left(\ln\frac{P^{(1)}(\omega_R)}{P^{(1)}(\omega_N)}-\ln\frac{P^{(0)}(\omega_R)}{P^{(0)}(\omega_N)}\right)\left(\mathbf{m}_R^{(0)}-\mathbf{m}_N^{(0)}\right) \tag{5}$$

It is worth noting that the two neighbourhoods capture different fractions of relevant and non relevant images, i.e., in the above formula the priors are different.

To simplify the above expression and avoid infinite results, let us substitute the logs with the following first-order approximation:

$$\ln\frac{P(\omega_R)}{P(\omega_N)}\cong\frac{k_R-k_N}{\max(k_R,k_N)} \tag{6}$$

where each prior is estimated as the fraction of relevant and non relevant images contained in the neighbourhood ($P(\omega_R)+P(\omega_N)=1$).

If we let the superscript (0) indicate the data computed from the relevance information provided by the user, Equation 5 let us compute the position of the mean $\mathbf{m}_R^{(1)}$ for which the related neighbourhood does not contain any non-relevant image, i.e. $P^{(1)}(\omega_N)$ is equal to 0. We shall select this point as the new query. By substituting Equation 6 in Equation 5 and expressing $\mathbf{m}_N^{(1)}$ as a function of $\mathbf{m}_R^{(1)}$, $\mathbf{m}_R^{(0)}$ and $\mathbf{m}_N^{(0)}$ according to Equation 4, and letting $P^{(1)}(\omega_N) = 0$, the point $\mathbf{m}_R^{(1)}$ where the new query $\mathbf{Q}_1$ should be placed can be computed as follows

$$\mathbf{m}_R^{(1)}=\mathbf{m}_R^{(0)}+\frac{\sigma^2}{\left\|\mathbf{m}_R^{(0)}-\mathbf{m}_N^{(0)}\right\|^2}\left(1-\frac{k_R^{(0)}-k_N^{(0)}}{\max\left(k_R^{(0)},k_N^{(0)}\right)}\right)\left(\mathbf{m}_R^{(0)}-\mathbf{m}_N^{(0)}\right) \tag{7}$$

Summing up, the query computed by Equation (7) coincides with $\mathbf{m}_R^{(0)}$ only when all images in $N(\mathbf{Q}_0)$ are relevant. Otherwise the larger the fraction of non-relevant images in the neighbourhood of the original query, the further the new query from the original neighbourhood.

It is worth to point out the main difference between the proposed query computation and other methods proposed in the literature. Usually the new query is computed as the solution of the minimisation of the average distance of the query with all the retrieved images, where relevant images have larger weights than non-relevant images. The new query is thus the weighted average of all retrieved images, the weights being related to the degree of relevance [14]. Therefore the new query is the "optimum" query with respect to the retrieved images. On the other hand the proposed mechanism is based on a local model of the distribution of relevant and non relevant images. This model is used to "optimise" the location of the neighbourhood of the new query with respect to the local boundary between relevant and non relevant images. To this end the new query is computed at a distance from the boundary proportional to the neighbourhood "size" expressed in terms of the ratio between the variance and the distance between means.

Further discussions on the validity of the proposed approach in comparison with other approaches proposed in the literature, are out of the scope of the present paper.

4 Experimental Results

In order to test the proposed method and compare it with other methods described in the literature, two image databases have been used: the MIT database and a database contained in the UCI repository.

The MIT database was collected by the MIT Media Lab (ftp://whitechapel.media.mit.edu/pub/VisTex). This database contains 40 texture images that have been processed as described in [13]. Images have been manually classified into fifteen classes. Each of these images has been subdivided into sixteen non-overlapping images, obtaining a data set with 640 images. Sixteen Gabor filters were used to characterise these images, so that each image is represented by a 16-dimensional feature vector .

The database extracted from the UCI repository (http://www.cs.uci.edu/mlearn/MLRepository.html) consists of 2,310 outdoor images. The images are subdivided into seven data classes (brickface, sky, foliage, cement, window, path, and grass). Nineteen colour and spatial features characterise each image. (Details are reported on the UCI web site).

For each dataset, a normalisation procedure has been performed, so that each feature takes values in the range between 0 and 1. This normalisation procedure is necessary when the Euclidean distance metric is used.

For both databases, each image is used as a query and the top twenty nearest neighbours are returned. Relevance feedback is performed by marking images belonging to the same class of the query as relevant, and all other images in the top twenty as non-relevant. This experimental set up affords an objective comparison among different methods and is currently used by many researchers [11], [12],[13].

Tables 1 and 2 report the results of the proposed method on the two selected datasets in terms of average percentage retrieval precision and Average Performance Improvement (API). Precision is measured as the ratio between the number of relevant retrievals and the number of total retrievals averaged over all the queries. API is computed averaging the following ratio over all the queries:

$$\frac{relevant\ retrievals(n+1) - relevant\ retrievals(n)}{relevant\ retrievals(n)}$$

where $n = 0, 1, \ldots$ is the number of feedbacks performed. In the reported experiments, n equals 1, because the relative performances of the compared methods does not change significantly by increasing the number of feedbacks performed.

For the sake of comparison, retrieval performances obtained with other methods recently described in the literature are also reported, namely the RFM (Relevance Feedback Method) [13] and the PFRL (Probabilistic Feature Relevance Learning) [12]. PFRL is a probabilistic feature relevance feedback method aimed at weighting each feature according to the information extracted from the relevant images. This method uses the Euclidean metric to measure the similarity between images. RFM is an implementation of the Rocchio formula for CBIR, that is, it implements the query shifting strategy. It is worth noting that RFM uses the cosine metric to compute similarity between images. Therefore, a different normalisation procedure is performed on the data sets in order to adapt features to the cosine metric.

The first columns of Tables 1 and 2 report the average percentage retrieval precision without feedback step. It is worth noting that the reported differences in performances depend on the different similarity metrics used. These results show that the cosine metric is more suited than the Euclidean metric to the MIT data set, while the reverse is true for the UCI data set. This points out that, if no relevance feedback mechanism is used, retrieval performances are highly sensitive to the selected similarity metric.

Table 1. Retrieval Performances for the MIT data set. Average percentage retrieval precision and Average Performance Improvement (API) are reported

RF mechanism	1st retrieval	2nd retrieval with RF	API
Rocchio	83.74%	90.23%	13.53
PFRL	79.24%	85.48%	12.70
Bayesian query shifting	79.24%	91.11%	28.79

Table 2. Retrieval Performances for the UCI data set. Average percentage retrieval precision and Average Performance Improvement (API) are reported

RF mechanism	1st retrieval	2nd retrieval with RF	API
Rocchio	86.39%	91.95%	15.33
PFRL	90.21%	94.56%	7.66
Bayesian query shifting	90.21%	96.24%	15.64

The second columns of Tables 1 and 2 report the average percentage retrieval precision after relevance feedback. The proposed relevance feedback method always outperformed the PFRL and the Rocchio formula. It is worth noting that while the Rocchio formula and the PFRL relies on some parameters that must be chosen by heuristics, the proposed method is based only on statistical estimates in the neighbourhood of the original query. However, the limited experimentation carried out does not allow to draw definitive conclusions.

A comparison between the PFRL and the proposed query shifting method shows that query shifting is more suited to relevance feedback than feature weighting alone. This is also confirmed by the results reported in [8], where PFRL performances are improved by combining PFRL with a query shifting mechanism.

With regard to the results on the MIT data set, it should be noted that although the method based on the Rocchio formula obtained a larger number of relevant images in the first retrieval the proposed query shifting method outperformed it when relevance feedback was used. Therefore, one can argue that retrieval performances provided by the proposed relevance feedback method are less sensitive to the choice of the similarity metric.

The above conclusions are also confirmed when comparing the average performance improvements (API). Our method provided the largest performance improvements on both data sets. In particular, the advantages of the proposed method are more evident on the MIT data set.

References

1. Bhanu, B., Dong, D.: Concepts Learning with Fuzzy Clustering and Relevance Feedback. In: Petra, P. (Ed.): Machine Learning and Data Mining in Pattern Recognition. LNAI 2123, Springer-Verlag, Berlin (2001) 102-116
2. Ciocca G, Schettini R.: Content-based similarity retrieval of trademarks using relevance feedback. Pattern Recognition, 34(8) (2001) 1639-1655
3. Cox, I.J., Miller, M.L., Minka T.P., Papathomas T.V., Yianilos, P.N.: The Bayesian image retrieval system, PicHunter: theory, implementation, and psychophysical experiments. IEEE Trans. on Image Processing 9(1) (2000) 20-37
4. Del Bimbo A.: Visual Information Retrieval. Morgan Kaufmann Pub. Inc., San Francisco, CA (1999)
5. Duda R.O., Hart P.E., Stork D.G.: Pattern Classification. J. Wiley & Sons (2000)
6. Frederix G., Caenen G., Pauwels E.J.: PARISS: Panoramic, Adaptive and Reconfigurable Interface for Similairty Search. Proc. of ICIP 2000 Intern. Conf. on Image Processing. WA 07.04, vol. III (2000) 222-225
7. Giacinto, G., Roli, F., Fumera, G.: Content-Based Image Retrieval with Adaptive Query Shifting. In: Petra, P. (Ed.): Machine Learning and Data Mining in Pattern Recognition. LNAI 2123, Springer-Verlag, Berlin, (2001) 337-346
8. Hesterkamp DR, Peng J, Dai HK.: Feature relevance learning with query shifting for content-based image retrieval. In Proc. of the 15th IEEE International Conference on Pattern Recognition (ICPR 2000), vol 4. IEEE Computer Society (2000) 250-253
9. Ishikawa Y., Subramanys R., Faloutsos C.: MindReader: Querying databases through multiple examples. In Proceedings. of the 24th VLDB Conference (1998) 433-438
10. McG Squire D, Müller W., Müller H., Pun T.: Content-based query of image databases: inspirations from text retrieval. Pattern Recognition Letters 21(13-14) (2000) 1193-1198
11. Nastar C., Mitschke M., Meilhac C.: Efficient query refinement for Image Retrieval. Proc. of IEEE Conf. Computer Vision and Pattern Recognition, CA (1998) 547-552
12. Peng J., Bhanu B., Qing S.: Probabilistic feature relevance learning for content-based image retrieval. Computer Vision and Image Understanding 75(1-2) (1999) 150-164
13. Rui Y., Huang T.S., Mehrotra S.: Content-based image retrieval with relevance feedback: in MARS. In Proceedings of the IEEE International Conference on Image Processing, IEEE Press (1997) 815-818
14. Rui Y., Huang T.S.: Relevance Feedback Techniques in Image retrieval. In Lew M.S. (ed.): Principles of Visual Information Retrieval. Springer-Verlag, London, (2001) 219-258
15. Salton G,, McGill M.J.: Introduction to modern information retrieval. McGraw-Hill, New York (1988)
16. Santini S., Jain R.: Integrated browsing and querying for image databases. IEEE Multimedia 7(3) (2000) 26-39

17. Santini S., Jain R.: Similarity Measures. IEEE Trans. on Pattern Analysis and Machine Intelligence 21(9) (1999) 871-883
18. Sclaroff S., La Cascia M., Sethi S., Taycher L.: Mix and Match Features in the ImageRover search engine. In Lew M.S. (ed.): Principles of Visual Information Retrieval. Springer-Verlag, London (2001) 219-258
19. Smeulders A.W.M., Worring M., Santini S., Gupta A., Jain R.: Content-based image retrieval at the end of the early years. IEEE Trans. on Pattern Analysis and Machine Intelligence 22(12) (2000 1349-1380

Recursive Model-Based Colour Image Restoration

Michal Haindl

Institute of Information Theory and Automation, Academy of Sciences
Prague, CZ182 08, Czech Republic
haindl@utia.cas.cz

Abstract. This paper presents a derivation of a fast recursive filter for colour image restoration if degradation obeys a linear degradation model with the unknown possibly non-homogeneous point-spread function. Pixels in the vicinity of steep discontinuities are left unrestored to minimize restoration blurring effect. The degraded image is assumed to follow a causal simultaneous multidimensional regressive model and the point-spread function is estimated using the local least-square estimate.

1 Introduction

Physical imaging systems, the recording medium, the atmosphere are imperfect and thus a recorded image represents a degraded version of the original scene. Similarly an image is usually further corrupted during its processing, transmission or storage. Possible examples are lens defocusing or aberration, noisy transmission channels, motion between camera and scene, etc. The image restoration task is to recover an unobservable image given the observed corrupted image with respect to some statistical criterion. Image restoration is the busy research area for already several decades and many restoration algorithms have been proposed. The simplest restoration method is to smooth the data with an isotropic linear or non-linear shift-invariant low-pass filter. Usual filtering techniques (e.g. median filter, Gaussian low pass filter, band pass filters, etc.) tend to blur the location of boundaries. Several methods [17] try to avoid this problem by using a large number of low-pass filters and combining their outputs. Similarly anisotropic diffusion [18],[5] addresses this problem but it is computationally extremely demanding. Image intensity in this method is allowed to diffuse over time, with the amount of diffusion at a point being inversely proportional to the magnitude of local intensity gradient. A nonlinear filtering method developed by Nitzberg and Shiota [16] uses an offset term to displace kernel centers away from presumed edges and thus to preserve them, however it is not easy to propose all filter parameters to perform satisfactory on variety of different images and the algorithm is very slow. In the exceptional case when the degradation point-spread function is known the Wiener filter [1] or deconvolution methods [12] can be used. Model-based methods use most often Markov random field type of models either in the form of wide sense Markov (regressive models) or strong

T. Caelli et al. (Eds.): SSPR&SPR 2002, LNCS 2396, pp. 617–626, 2002.
© Springer-Verlag Berlin Heidelberg 2002

Markov models. The noncausal regressive model used in [3],[4] has the main problem in time consuming iterative solution based on the conjugate gradient method. Similarly Markov random field based restoration methods [7], [6], [13] require time consuming application of Markov chain Monte Carlo methods. Besides this both approaches have solve the problem when to stop these iterative processes. A similar combination of causal and non-causal regressive models as in this paper was used in [14]. However they assume the homogeneous point-spread function and they identify all parameters simultaneously using extremely time consuming iterations of the EM algorithm which is not guaranteed to reach the global optimum. This work generalizes our monospectral restoration method [8] for the multispectral (e.g., colour) images. It is seldom possible to obtain a degradation model analytically from the physics of the problem. More often a limited prior knowledge supports only some elementary assumptions about this process. Usual assumption, accepted also in this work, is that the corruption process can be modeled using a linear degradation model.

2 Image Model

Suppose Y represents a true but unobservable colour image defined on finite rectangular $N \times M$ underlying lattice I. The observable data are X, a version of Y distorted by noise independent of the signal. We assume knowledge of all pixels elements from the reconstructed scene. For the treatment of the more difficult problem when some date are missing see [10], [11]. The image degradation is supposed to be approximated by the linear discrete spatial domain degradation model

$$X_r = \sum_{s \in I_r} H_s Y_{r-s} + \epsilon_r \tag{1}$$

where H is a discrete representation of the unknown point-spread function and X_r, Y_{r-s} corresponding $d \times 1$ multispectral pixels. The point-spread function is assumed to be either homogeneous or it can be non-homogeneous but in this case we assume its slow changes relative to the size of an image. I_r is some contextual support set, and a noise vector ϵ is uncorrelated with the true image, i.e.,

$$E\{Y \epsilon\} = 0 \ . \tag{2}$$

The point-spread function is unknown but such that we can assume the unobservable image Y to be reasonably well approximated by the expectation of the corrupted image

$$\hat{Y} = E\{X\} \tag{3}$$

in regions with gradual pixel value changes. The above method (3) changes all pixels in the restored image and thus blurs discontinuities present in the scene although to much less extent than the classical restoration methods due to adaptive restoration model (10). This excessive blurring can be avoided if pixels with steep step discontinuities are left unrestored, i.e.,

$$\hat{Y}_r = \begin{cases} E\{X_r\} & \text{if (5) holds} \\ X_r & \text{otherwise} \end{cases} , \tag{4}$$

where the adaptive condition (5) is

$$|\,E\{X_r\} - X_r\,| < \frac{1}{n_s} \sum_s |\,E\{X_{r-s}\} - X_{r-s}\,| \ . \tag{5}$$

The expectation (3) can be expressed as follows:

$$E\{X\} = \int x\, p(x)\, dx = \int \begin{pmatrix} x_1 & x_2 & \dots & x_M \\ x_{M+1} & x_{M+2} & \dots & x_{2M} \\ \vdots & \vdots & \ddots & \vdots \\ x_{NM-M+1} & x_{NM-M+2} & \dots & x_{NM} \end{pmatrix} \prod_{r=1}^{NM} p(x_r \,|\, X^{(r-1)})\, dx_1 \dots dx_{NM} \tag{6}$$

where $X^{(r-1)} = \{X_{r-1}, \dots, X_1\}$ is a set of noisy pixels in some chosen but fixed ordering. For single matrix elements in (6) it holds

$$\begin{aligned} E\{X_j\} &= \int x_j \prod_{r=1}^{NM} p(x_r \,|\, x^{(r-1)})\, dx_1 \dots dx_{NM} \\ &= \int X_j \prod_{r=1}^{j} p(X_r \,|\, X^{(r-1)})\, dX_1 \dots dX_j \\ &= \int E\{X_j \,|\, X^{(j-1)}\} \prod_{r=1}^{j-1} p(X_r \,|\, X^{(r-1)})\, dX_1 \dots dX_{j-1} \\ &= E_{X^{(j-1)}}\{\, E_{X_j}\{X_j \,|\, X^{(j-1)}\}\,\} \end{aligned} \tag{7}$$

Let us approximate after having observed $x^{(j-1)}$ the $\hat{Y}_j = E\{X_j\}$ by the $E\{X_j \,|\, X^{(j-1)} = x^{(j-1)})$ where $x^{(j-1)}$ are known past realization for j. Thus we suppose that all other possible realization $x^{(j-1)}$ than the true past pixel values have negligible probabilities. This assumption implies conditional expectations approximately equal to unconditional ones, i.e., then the expectation (7) is

$$E\{X_j\} \approx E\{X_j \,|\, X^{(j-1)}\} \ , \tag{8}$$

and

$$\hat{Y} = E\{X\} \approx \begin{pmatrix} E\{X_1 \,|\, x^{(0)}\} & \dots & E\{X_M \,|\, x^{(M-1)}\} \\ E\{X_{M+1} \,|\, x^{(M)}\} & \dots & E\{X_{2M} \,|\, x^{(2M-1)}\} \\ \vdots & \ddots & \vdots \\ E\{X_{NM-M+1} \,|\, x^{(NM-M)}\} & \dots & E\{X_{NM} \,|\, x^{(NM-1)}\} \end{pmatrix} \tag{9}$$

Suppose further that the noisy image can be represented by an adaptive causal simultaneous autoregressive model

$$X_r = \sum_{s \in I_r^c} A_s X_{r-s} + \epsilon_r \ , \tag{10}$$

where ϵ_r is a white Gaussian noise vector with zero mean, and a constant but unknown covariance matrix Σ. The noise vector is uncorrelated with data from a causal neighbourhood I_r^c, but noise vector components can be mutually correlated. The model adaptivity is introduced using the standard exponential forgetting factor technique in parameter learning part of the algorithm. The model can be written in the matrix form

$$X_r = \gamma Z_r + \epsilon_r \ , \tag{11}$$

where

$$\gamma = [A_1, \dots, A_\eta] \ , \tag{12}$$

$$\eta = card(I_r^c) \tag{13}$$

is a $d \times d\eta$ parameter matrix and Z_r is a corresponding vector of X_{r-s}. To evaluate conditional mean values in (9) the one-step-ahead prediction posterior density $p(X_r \,|\, X^{(r-1)})$ is needed. If we assume the normal-Wishart parameter prior for parameters in (10) (alternatively we can assume the Jeffreys parameter prior) this posterior density has the form of d-dimensional Student's probability density

$$p(X_r|X^{(r-1)}) = \frac{\Gamma(\frac{\beta(r)-d\eta+d+2}{2})}{\Gamma(\frac{\beta(r)-d\eta+2}{2}) \ \pi^{\frac{d}{2}} \ (1 + Z_r^T V_{z(r-1)}^{-1} Z_r)^{\frac{d}{2}} \ \lambda_{(r-1)}^{\frac{1}{2}}}$$

$$\left(1 + \frac{(X_r - \hat{\gamma}_{r-1} Z_r)^T \lambda_{(r-1)}^{-1} (X_r - \hat{\gamma}_{r-1} Z_r)}{1 + Z_r^T V_{z(r-1)}^{-1} Z_r}\right)^{-\frac{\beta(r)-d\eta+d+2}{2}} \ , \tag{14}$$

with $\beta(r) - d\eta + 2$ degrees of freedom, where the following notation is used:

$$\beta(r) = \beta(0) + r - 1 = \beta(r-1) + 1 \ , \tag{15}$$

$$\beta(0) > 1 \ ,$$

$$\hat{\gamma}_{r-1}^T = V_{z(r-1)}^{-1} V_{zx(r-1)} \tag{16}$$

$$V_{r-1} = \tilde{V}_{r-1} + I \ ,$$

$$\tilde{V}_{r-1} = \begin{pmatrix} \tilde{V}_{x(r-1)} & \tilde{V}_{zx(r-1)}^T \\ \tilde{V}_{zx(r-1)} & \tilde{V}_{z(r-1)} \end{pmatrix} \ , \tag{17}$$

$$\tilde{V}_{x(r-1)} = \sum_{k=1}^{r-1} X_k X_k^T \ , \tag{18}$$

$$\tilde{V}_{zx(r-1)} = \sum_{k=1}^{r-1} Z_k \, X_k^T \ , \tag{19}$$

$$\tilde{V}_{z(r-1)} = \sum_{k=1}^{r-1} Z_k \, Z_k^T \ , \tag{20}$$

$$\lambda_{(r)} = V_{x(r)} - V_{zx(r)}^T \, V_{z(r)}^{-1} \, V_{zx(r)} \ . \tag{21}$$

If $\beta(r-1) > \eta$ then the conditional mean value is

$$E\{X_r | X^{(r-1)}\} = \hat{\gamma}_{r-1} Z_r \tag{22}$$

and it can be efficiently computed using the following recursion

$$\hat{\gamma}_r^T = \hat{\gamma}_{r-1}^T + (1 + Z_r^T V_{z(r-1)}^{-1} Z_r)^{-1} \, V_{z(r-1)}^{-1} \, Z_r (X_r - \hat{\gamma}_{r-1} Z_r)^T \ .$$

3 Optimal Contextual Support

The selection of an appropriate model support (I_r^c) is important to obtain good restoration results. If the contextual neighbourhood is too small it can not capture all details of the random field. Inclusion of the unnecessary neighbours on the other hand add to the computational burden and can potentially degrade the performance of the model as an additional source of noise. The optimal Bayesian decision rule for minimizing the average probability of decision error chooses the maximum posterior probability model, i.e., a model M_i corresponding to $\max_j \{p(M_j | X^{(r-1)})\}$. If we assume uniform prior for all tested support sets (models) the solution can be found analytically. The most probable model given past data is the model M_i $(I_{r,i}^c)$ for which $i = \arg\max_j \{D_j\}$.

$$D_j = -\frac{d}{2} \ln |V_{z(r-1)}| - \frac{\beta(r) - d\eta + d + 1}{2} \ln |\lambda_{(r-1)}| + \frac{d^2\eta}{2} \ln \pi$$
$$\sum_{i=1}^{d} \left[\ln \Gamma(\frac{\beta(r) - d\eta + d + 2 - i}{2}) - \ln \Gamma(\frac{\beta(0) - d\eta + d + 2 - i}{2}) \right] . \tag{23}$$

4 Global Estimation of the Point-Spread Function

Similarly with (11) the degradation model (1) can be expressed in the matrix form

$$X_r = \psi W_r + \epsilon_r \ , \tag{24}$$

where $\nu = \ card(I_r)$,

$$\psi = [H_1, \ldots, H_\nu] \ , \tag{25}$$

and W_r is a corresponding vector of Y_{r-s}. The unobservable $\nu \times 1$ image data vector W_r is approximated using (3), (8),(22), i.e.,

$$\hat{W}_r = [\hat{\gamma}_{r-s-1} Z_r]^T_{s \in I_r} \ . \tag{26}$$

In contrast to the model (10) the degradation model (1) is non-causal and hence it has no simple analytical Bayesian parameter estimate. Instead we use the least square estimate

$$\hat{\psi} = \min_{\psi} \left\{ \sum_{\forall r \in I} (X_r - \psi_r \hat{W}_r)^T (X_r - \psi_r \hat{W}_r) \right\} \ . \tag{27}$$

The optimal estimate is $\hat{\psi}^T = V^{-1}_{\hat{W}} V_{\hat{W}X}$ where the data gathering matrices $V_{\hat{W}}$, $V_{\hat{W}X}$ are corresponding analogies with the matrices (18),(19).

5 Local Estimation of the Point-Spread Function

If we assume a non-homogeneous slowly changing point-spread function, we can estimate its local value using the local least square estimate

$$\hat{\psi}_r = \min_{\psi_r} \left\{ \sum_{\forall r \in J_r} (X_r - \psi_r \hat{W}_r)^T (X_r - \psi_r \hat{W}_r) \right\} \ . \tag{28}$$

The locally optimal estimate is $\hat{\psi}^T_r = \tilde{V}^{-1}_{\hat{W}} \tilde{V}_{\hat{W}X}$. The matrices $\tilde{V}_{\hat{W}}$, $\tilde{V}_{\hat{W}X}$ are computed from subwindows $J_r \subset I$. This estimator can be efficiently evaluated using the fast recursive square-root filter introduced in [9].

Table 1. Comparison of the presented method and median filter restoration results for different noise levels

	Cymbidium image - Gaussian noise							
SNR $[dB]$	66.5	27	24.5	17.5	15.8	13.3	9.1	7.8
σ^2	0.001	9	16	81	121	225	625	900
MAD - AR	1.7	3.2	3.4	5.2	5.6	5.9	6.7	9.5
SNR_{imp}	3.3	-1.4	-0.6	1.7	2.8	6.1	8.4	7.5
MAD - median	3.1	3.9	4.1	5.3	5.9	6.0	8.8	10.3
SNR_{imp}	-7.8	-3.3	-0.7	1.6	2.6	6	6.6	7.3

Fig. 1. Original and corrupted ($\sigma^2 = 900$ white noise) Cymbidium image

6 Results

The test image of the Cymbidium orchid (Fig.1-left), quantized at 256 levels per spectral band, was corrupted by the white Gaussian noise with $\sigma^2 \in \langle 0.001; 900 \rangle$ Fig.1-right ($\sigma^2 = 900$). The signal-to-noise ratio for these corrupted images is

$$SNR = 10 \log \left(\frac{var(X)}{\sigma^2} \right) \qquad dB \ . \tag{29}$$

The resulting reconstructed image using our method is on the Fig.2-left ($\sigma^2 = 900$) while the image Fig.2-right shows reconstruction using identical model but without differentiating discontinuity pixels. Visual comparison of both reconstructed images demonstrates deblurring effect of the presented algorithm.
The performance of the both methods is compared on artificially degraded images (so that the unobservable data are known) using the criterion of mean absolute difference between undegraded and restored pixel values

$$MAD = \frac{1}{MNd} \sum_{r_1=1}^{M} \sum_{r_2=1}^{N} \sum_{r_3=1}^{d} |Y_{r_1,r_2,r_3} - \hat{Y}_{r_1,r_2,r_3}| \tag{30}$$

and the criterion SNR_{imp} which denotes the improvement in signal-to-noise ratio

$$SNR_{imp} = 10 \log \left(\frac{\mu(X)}{\mu(\hat{Y})} \right) \qquad dB \tag{31}$$

where $\mu(X)$ is the mean-square error of X.
Both proposed methods are superior over the classical methods using both criteria (30),(31). The edge preserving version of the restoration method demonstrates visible deblurring effect Fig.2-left without significantly affecting numerical complexity of the method.

Fig. 2. The reconstructed Cymbidium image using (4),(5) and (3) (right), respectively

The Tab. 1 demonstrates the influence of noise increasing on the performance of our and median filter methods. The proposed method is clearly superior for noisy images.

Fig. 3. The reconstructed Cymbidium image using the median filter

7 Conclusions

The proposed recursive blur minimizing reconstruction method is very fast (approximately five times faster than the median filter) robust and its reconstruction results surpasses some standard reconstruction methods. Causal models such as

(10) have obvious advantage to have the analytical solution for parameter estimation, prediction, or model identification tasks. However, this type of models may introduce some artifacts in restored images. These undesirable effects are diminished by introducing adaptivity into the model. This novel formulation allow us to obtain extremely fast adaptive restoration and / or local or global point-spread function estimation which can be easily parallelized. The method can be also easily and naturally generalized for multispectral (e.g. colour, multispectral satellite images) or registered images which is seldom the case for alternative methods. Finally, this method enables to estimate homogeneous or slowly changing non-homogeneous degradation point-spread function.

Acknowledgments

This research was supported by the GAČR grant no. 102/00/0030 and partially supported by the GAČR grant no. 106/00/1715.

References

1. Andrews, H. C., Hunt, B.: Digital Image Restoration. Prentice-Hall, Englewood Cliffs (1977) 617
2. Chalmond, B.: Image restoration using an estimated markov model. Signal Processing **15** (1988) 115–129
3. Chellappa, R., Kashyap, R.: Digital image restoration using spatial interaction models. IEEE Trans. Acoustics, Speech and Sig. Proc. **30** (1982) 284–295 618
4. Deguchi, K., Morishita, I.: Two-dimensional auto-regressive model for the representation of random image fields. In: Proc.ICPR Conf., IEEE, Munich (1982) 90–93 618
5. Fischl, B., Schwartz, E.: Learning an integral equation approximation to nonlinear anisotropic diffusion in image processing. IEEE Trans. Pattern Anal. Mach. Int. **19** (1997) 342–352 617
6. Geman, D.: Random fields and inverse problems in imaging. Springer, Berlin, 1990 618
7. Geman, S., Geman, D.: Stochastic relaxation , gibbs distributions and bayesian restoration of images. IEEE Trans. Pattern Anal. Mach. Int. **6** (1984) 721–741 618
8. Haindl, M.: Recursive model-based image restoration. In: Proc. of the 13th ICPR Conf. vol. III, IEEE Press, Barcelona (2000) 346–349 618
9. Haindl, M.: Recursive square-root filters. In: Proc. of the 13th ICPR Conf. vol. II, IEEE Press, Barcelona (2000) 1018–1021 622
10. Haindl, M., Šimberová, S.: A high - resolution radiospectrograph image reconstruction method. Astronomy and Astrophysics, Suppl.Ser. **115** (1996) 189–193 618
11. Haindl, M., Šimberová, S.: A scratch removal method. Kybernetika **34** (1998) 423–428 618
12. Hunt, B.: The application of constraint least square estimation to image restoration by digital computer. IEEE Trans. Computers **22** (1973) 805–812 617

13. Jeffs, B., Pun, W.: Simple shape parameter estimation from blurred observations for a generalized Gaussian mrf image prior used in map restoration. In: Proc. IEEE CVPR Conf., IEEE, San Francisco (1996) 465–468 618
14. Lagendijk, R., Biemond, J., Boekee, D.: Identification and restoration of noisy blurred images using the expectation-maximization algorithm. IEEE Trans. on Acoust., Speech, Signal Processing **38** (1990) 1180–1191 618
15. Marroquin, J., Poggio, T.: Probabilistic solution of ill-posed problems in computational vision. J. Am. Stat. Assoc. **82** (1987) 76–89
16. Nitzberg, M., T. Shiota, T.: Nonlinear image filtering with edge and corner enhancement. IEEE Trans. Pattern Anal. Mach. Int. **16** (1992) 826–833 617
17. Perona, P.: Deformable kernels for early vision. IEEE Trans. Pattern Anal. Mach. Int. **17** (1995) 488–489 617
18. Perona, P., Malik, J.: Scale-space and edge detection using anisotropic diffusion. IEEE Trans. Pattern Anal. Mach. Int. **12** (1990) 629–639 617
19. Reeves, S., Mersereau, R.: Identification by the method of generalized cross-validation. IEEE Trans. Im. Processing **1** (1992) 301–311

Human Face Recognition with Different Statistical Features

Javad Haddadnia[1], Majid Ahmadi[1], and Karim Faez[2]

[1] Electrical and Computer Engineering Department, University of Windsor
Windsor, Ontario, Canada, N9B 3P4
{javad,ahmadi}@uwindsor.ca
[2] Electrical Engineering Department, Amirkabir University of Technology
Tehran, Iran, 15914
kfaez@cic.aku.ac.ir

Abstract. This paper examines application of various feature domains for recognition of human face images to introduce an efficient feature extraction method. The proposed feature extraction method comprised of two steps. In the first step, a human face localization technique with defining a new parameter to eliminate the effect of irrelevant data is applied to the facial images. In the next step three different feature domains are applied to localized faces to generate the feature vector. These include Pseudo Zernike Moments (PZM), Principle Component Analysis (PCA) and Discrete Cosine Transform (DCT). We have compared the effectiveness of each of the above feature domains through the proposed feature extraction for human face recognition. The Radial Basis Function (RBF) neural network has been utilized as classifier. Simulation results on the ORL database indicate the effectiveness of the proposed feature extraction with the PZM for human face recognition.

1 Introduction

In recent years there has been a growing interest in machine recognition of faces due to potential commercial application such as film processing, law enforcement, person identification, access control systems, etc. A recent survey of the face recognition systems can be found in reference [1]. The ultimate goal of designing human face recognition systems is to develop different feature extractions and classification schemes that achieve the best possible recognition performance.

A complete conventional human face recognition system should include three stages. The first stage involves detecting the location of face in arbitrary images. Although many researchers tried to solve this problem [2-3], however, detecting the location of a face is still difficult and complicated due to unknown position, orientation and scaling of face in an arbitrary image. The second stage requires extraction of pertinent features from the face image. Two main approaches to feature extraction have been extensively used by other researchers [4]. The first one is based on extract-

T. Caelli et al. (Eds.): SSPR&SPR 2002, LNCS 2396, pp. 627-635, 2002.
© Springer-Verlag Berlin Heidelberg 2002

ing structural and geometrical facial features that are local structure of face images, for example, the shapes of the eyes, nose and mouth. The structural-based approaches deal with local data instead of global data. It has been shown that structural-based approaches by explicit modeling of facial features have been troubled by the unpredictability of face appearance and environmental conditions. The second method is statistical-based approaches that extract features from the whole image and therefore use global data instead of local data. Since the global data of an image are used to determine the feature elements, data that are irrelevant to facial portion such as hair, shoulders and background may contribute to creation of erroneous feature vectors that can affect the recognition results [5]. Finally the third stage involves classification of facial images based on the derived feature vector. Neural networks have been employed and compared to conventional classifiers for a number of classification problems. The results have shown that the accuracy of the neural network approaches are equivalent to, or slightly better than, other methods. Also, due to the simplicity, generality and good learning ability of the neural networks, these types of classifiers are found to be more efficient [6].

In this paper a new feature extraction technique is developed. This algorithm is based on the face localization using the shape information [3] and a definition of a new parameter for eliminating the irrelevant data from arbitrary face images. This parameter is named Correct Information Ratio (CIR). We have shown how CIR can improve the recognition rate. Once the face localization process was completed, a subimage is created and then PZM, PCA and DCT are computed on the subimage to generate the feature vector associated with each image. These feature vectors are sent to classifier, which is RBF neural network. The recognition performance of each feature domain is subsequently analyzed and compared. The organization of this paper is as follows: Section 2 presents face localization and feature extractionmethods. In section 3, feature domains are presented. Classifier techniques are described in section 4 and finally, section 5 and 6 presents the experimental results and conclusions.

2 Face Localization and Feature Extraction

The ultimate goal of the face localization is finding an object in an image as a face candidate that its shape resembles the shape of a face. Many researchers have concluded that an ellipse generally can approximate the face of a human [2-4]. Considering the elliptical shape of a face in general, it is convenient to search for connected components using a region-growing algorithm and fit an ellipse to every connected component of nearly elliptical shape. A technique is presented in [3], which finds the best-fit ellipse to enclose the facial region of the human face in a frontal view of facial image.

The aim of the feature extractor is to produce a feature vector containing all pertinent information about the face to be recognized. The feature vector generation is very important in any high accuracy pattern recognition system. In this paper global data from a facial image are used to derive the feature vector. It is important that in this phase all irrelevant data pertaining to the face images such as hair, shoulders and background be eliminate and to keep only the important data about the face images.

Our feature extraction method therefore has two different steps. In the first step a subimage is created to enclose only the important information about the face in an ellipse. In the second step, feature vector elements are determined by computing PZM, PCA and DCT on the derived subimage.

The subimage encloses all pertinent information around the face candidate in an ellipse while pixel values outside the ellipse are set to zero. Unfortunately through creation of the subimage with the best-fit ellipse many unwanted regions of the face image may still appear in this subimage, as shown in Fig. (1). These include hair portion, neck and part of the background. To overcome this problem, instead of using the best-fit ellipse for creating subimage we have defined another ellipse. The proposed ellipse has the same orientation and center as the best-fit ellipse but the length of the major (A) and the minor (B) axis are calculated as follows:

$$A = \rho.\alpha \quad , \quad B = \rho.\beta \tag{1}$$

Where α and β are the length of the major and minor axis of the best-fit ellipse [3] and the coefficient ρ that we have named Correct Information Ratio (CIR) varies from 0 to 1. Fig. (2) shows the effect of changing CIR while Fig. (3) shows the corresponding subimages. Our exprimental results with 400 images show that the best value for CIR is around 0.87. By using a subimage with CIR parameter, data that are irrelevant to facial portion are disregarded. Also the speed of computing feature domains is increased due to smaller nonzero pixels content of the subimages.

Fig. 1. Face localization method

$\rho = 1.0$ $\rho = 0.7$ $\rho = 0.4$

Fig. 2. Different ellipses with related CIR value

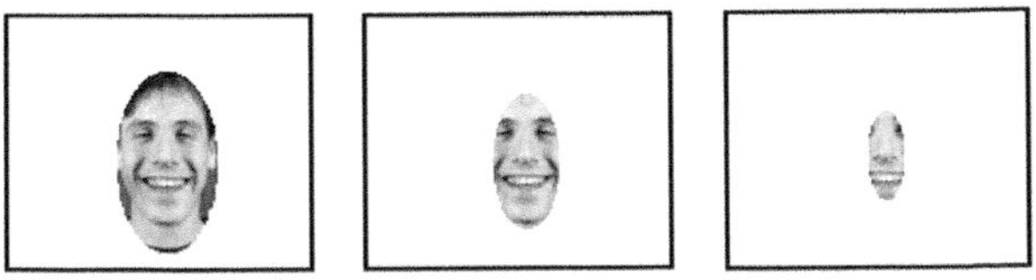

Fig. 3. Creating subimage based on CIR value

3 Feature Domains

In order to design a good face recognition system, the choice of feature domains is very crucial. To design a system with low to moderate complexity the feature vectors should contain the most pertinent information about the face to be recognized. In this paper different feature domains are extracted from the derived subimages. These include PZM, PCA and DCT.

3.1 Pseudo Zernike Moment (PZM)

The advantages of considering orthogonal moments are that they are shift, rotation and scale invariant and very robust in the presence of noise. The PZM of order n and repetition m can be computed using the scale invariant central moments and the radial geometric moments that defined in reference [8] as follows:

$$PZM_{nm} = \frac{n+1}{\pi} \sum_{(n-m-s)even, s=0}^{n-|m|} D_{n,|m|,s} \sum_{a=0}^{k} \sum_{b=0}^{m} \binom{k}{a} \binom{m}{b} (-j)^b CM_{2k+m-2a-b, 2a+b} + \frac{n+1}{\pi} \sum_{(n-m-s)odd, s=0}^{n-|m|} D_{n,|m|,s} \sum_{a=0}^{d} \sum_{b=0}^{m} \binom{d}{a} \binom{m}{b} (-j)^b RM_{2d+m-2a-b, 2a+b} \quad (2)$$

where $k = (n-s-m)/2$, $d = (n-s-m+1)/2$, $CM_{i,j}$ is the Central moments, $RM_{i,j}$ is the Radial moments [8] and $D_{n,|m|,s}$ is defined as:

$$D_{n,|m|,s} = (-1)^s \frac{(2n+1-s)!}{s!(n-|m|-s)!(n-|m|-s+1)!} \quad (3)$$

3.2 Principle Component Analysis (PCA)

PCA is a well-known statistical technique for feature extraction [9]. Each $M \times N$ image in the training set was row concatenated to form $MN \times 1$ vectors x_i. Given a set of N_T training images $\{x_i\}_{i=0,1,\ldots N_T}$ the mean vector of the training set was obtained as:

$$\overline{x} = \frac{1}{N_T} \sum_{i=1}^{N_T} x_i \quad (4)$$

A $N_T \times MN$ training set matrix $X = [x_i - \overline{x}]$ can be built. The basis vectors are obtained by solving the eigenvalue problem:

$$\Lambda = V^T \Sigma_X V \quad (5)$$

where $\Sigma_X = XX^T$ is the covariance matrix, V is the eigenvector matrix of Σ_X and Λ is the corresponding diagonal matrix of eigenvalues.

As PCA has the property of packing the greatest energy into the least number of principal components, in PCA the eigenvectors corresponding to the k largest eigenvalues are selected to form a lower dimensional subspace. It is proven that the residual reconstruction error generated by dismissing the $N_T - k$ components are low even for small k [9].

3.3 Discrete Cosine Transform (DCT)

The DCT transforms spatial information to decoupled frequency information in the form of DCT coefficients. Also it exhibits excellent energy compaction. The definition of DCT for an $N \times N$ image is [10]:

$$DCT_{uv} = \frac{1}{N^2} \sum_{x=0}^{N-1} \sum_{y=0}^{N-1} f(x,y) \cos\left[\frac{(2x+1)u\pi}{2N}\right] \cos\left[\frac{(2y+1)v\pi}{2N}\right] \quad (6)$$

where $f(x,y)$ is $N \times N$ image pixels.

4 Classifier Design

Radial Basis Function (RBF) neural networks have found to be very attractive for many engineering problem because: (1) they are universal approximators, (2) they have a very compact topology and (3) their learning speed is very fast because of their locally tuned neurons. Therefore RBF neural networks serve as an excellent candidate for pattern applications and attempt to make the learning process in this type of classification faster than normally required for the multi-layer feed forward neural networks [11-12]. The construction of the RBF neural network involves three different layers with feed forward architecture. The input layer of this network is fully connected to the hidden layer. Connections between the input and hidden layers have unit weights and, as a result, do not have to be trained. The hidden units are also fully connected to the output layer. The goal of the hidden units is to cluster the data and reduces its dimensionality with a nonlinear transformation and maps the input data to a new space. Therefore the transformation from the input space to the hidden space is nonlinear, whereas the transformation from the hidden space to the output space is linear.

The RBF neural network is a class of neural networks, where the activation function of the hidden units is determined by the distance between the input vector and a prototype vector. The activation function of the hidden units is expressed as [11-12]:

$$R_i(x) = R_i\left(\frac{\| x - c_i \|}{\sigma_i}\right) \quad , \quad i=1,2,\ldots,r \quad (7)$$

Where x is an n-dimensional input feature vector, c_i is a n-dimensional vector called the center of the hidden unit, σ_i is the width of hidden unit and r is the number of the hidden units. Typically the activation function of the hidden units is chosen as a Gaussian function with mean vector c_i and variance vector σ_i as follows:

$$R_i(x) = \exp(-\frac{\| x - c_i \|^2}{\sigma_i^2}) \quad (8)$$

Note that σ_i^2 represents the diagonal entries of covariance matrix of Gaussian function. The output units are linear and therefore the response of the j-th output unit for input x is given as:

$$y_j(x) = b(j) + \sum_{i=1}^{r} R_i(x) w_2(i, j) \quad (9)$$

Where $w_2(i, j)$ is the connection weight of the i-th hidden unit to the j-th output node and $b(j)$ is the bias of the j-th output. The bias is omitted in this network in order to reduce network complexity. Therefore:

$$y_j(x) = \sum_{i=1}^{r} R_i(x) \times w_2(i, j) \quad (10)$$

Training RBF neural network can be made faster than the methods used to train multi-layer neural networks. In this paper a Hybrid Learning Algorithm (HLA) based on [12] was used to estimate the width and center of hidden units and synaptic weights.

5 Experimental Results

To check the suitability of each feature domain experimental studies are carried out on the ORL database images of Cambridge University. 400 face images from 40 individuals in different states from the ORL database have been used to evaluate the performance of the each feature domain. A total of 200 images have been used to train and another 200 for test. Each training set consists of 5 randomly chosen images from the same individual in the training stage. In the first step after subimage creation, the classifier is trained separately based on each feature domains using HLA technique [12]. In the second step, recognition performance is evaluated. Each test image is projected onto the feature extraction and sent to the classifier. This procedure was repeated for each feature domain. Also in each feature domain the number of feature elements used to represent feature vectors was varied.

The results of the experiments are summarized in Fig. (4) to (7). Fig. (4) shows the average error rate for each feature domains computed for the 40 runs as a function of the number of feature elements. The average error rate curves show that for the PCA a minimum error rate of 1.5% with 60 elements feature vector is obtainable. Also Fig. (4) shows that the DCT with 40 elements feature vector can yield 3.9% error rate. It is also clear from this figure the PZM with the proposed feature extraction outperforms the PCA and DCT by providing a minimum error rate of 0.3% with 21 elements feature vector which is derived from the moments of order 9 and 10. It is interesting to note that although the feature elements of PZM are far less than the other two feature domains, the recognition rate is far superior.

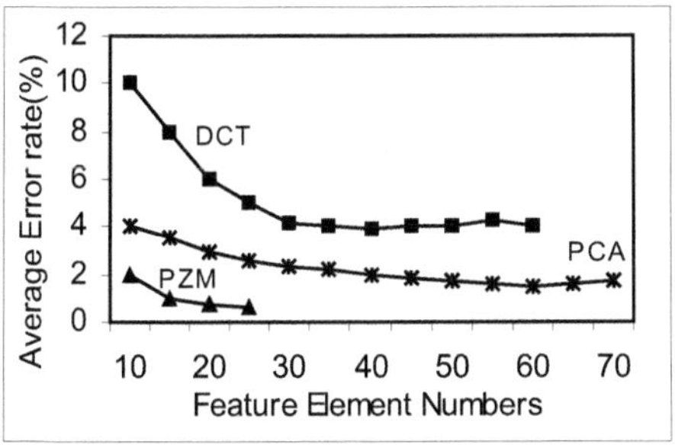

Fig. 4. Average error rate based on feature elements

Another important result revealed by the experiments is shown in Fig. (5). These curves present the standard deviation of the error rate computed for 40 runs for each feature domains. These graphs indicate sensitivity of results to the choice of the training and testing sets. The higher order of the PZM through the proposed feature extraction presents again the lowest standard deviation.

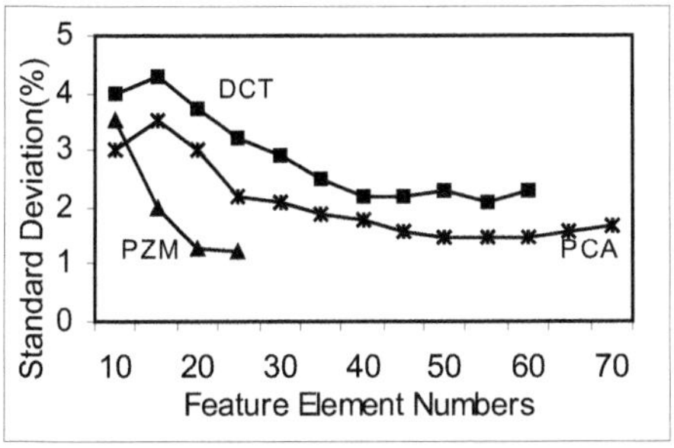

Fig. 5. Standard deviation based on feature elements

As a complement to Fig. (5), the minimum performance among 40 runs as a function of the number of feature elements is plotted in Fig. (6). The PZM presents again the best performance in the experiments. These results show that higher orders of PZM contain more and useful information for face recognition process.

It should be noted that when the PCA was applied to the entire facial image as reported in [13] an error rate of 3% with the same database was obtained. For the purpose of evaluating how the non-face portion of a face image such as hair, neck, shoulder and background will influence the recognition results we have chosen the PZM with 21 elements, the PCA with 60 elements and the DCT with 40 elements for feature extraction. We have also used RBF neural network with HLA learning algorithm as classifier. We varied the CIR value and evaluated the recognition rate. Fig. (7) shows the effect of CIR on the error rate. As Fig. (7) shows by variation of the CIR value error rate also changes. By defining and using CIR parameter we have obtained a better recognition rate which is 99.3% for PZM, 98.5% for PCA and 96.1% for DCT. The above results were obtained for CIR=0.87 that it is the optimum value for CIR.

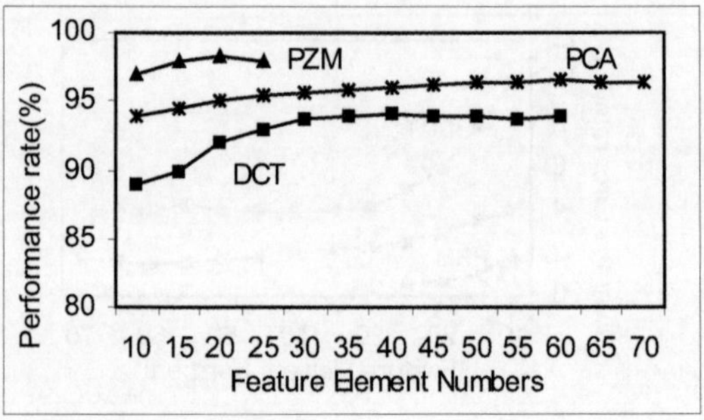

Fig. 6. Minimum performance rate for each feature domain

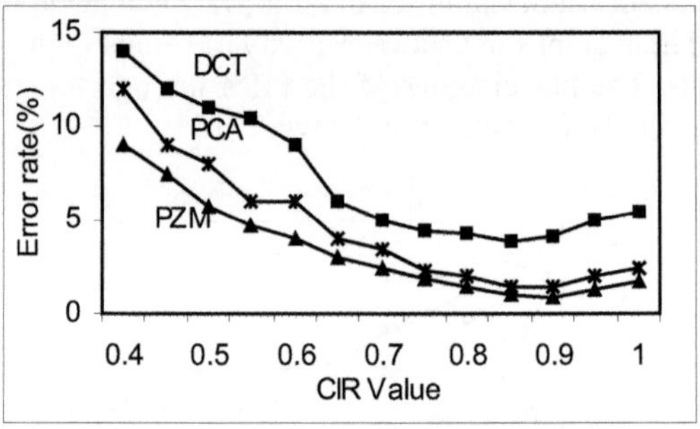

Fig. 7. Error rate based on CIR value

6 Conclusion

This paper presented a feature extraction method for the recognition of human faces in 2-Dimensional digital images. The proposed technique utilizes a modified feature extraction technique, which is based on a flexible face localization algorithm followed by various feature domains. This paper has compared several feature domains for human face recognition. These include PZM, PCA and DCT. Also we have introduced CIR parameter for efficient and robust feature extraction technique. It was shown through experimentation the effect of varying this parameter in recognition rate. We have also indicated the optimum value for the CIR for the best recognition results through exhaustive experimentation. We have shown that high order PZM contains very useful information about the facial images. The highest recognition rate of 99.3% with ORL database was obtained using this proposed algorithm.

References

1. Grudin, M. A.: On Internal Representation in face Recognition Systems. Pattern Recognition. Vol. 33, No. 7 (2000) 1161-1177
2. Yung M. H., Kreigman D. J. and Ahuja N., Detecting Face in Images: A Survay, IEEE Trans. on Patt. Anal. and Mach. Intel., Vol. 34, No. 1 (2002) 34-58
3. Haddadnia J., Faez K.: Human Face Recognition Based on Shape Information and Pseudo Zernike Moment. 5th Int. Fall Workshop Vision, Modeling and Visualization, Saarbrucken, Germany, Nov. 22-24 (2000) 113-118
4. Daugman, J.: Face Detection: A Survey. Computer Vision and Image Understanding, Vol. 83, No. 3, Sept. (2001) 236-274
5. Chen L. F., Liao H. M., Lin J. and Han C.: Why Recognition in a statistic-based Face Recognition System should be based on the pure Face Portion: A Probabilistic decision-based Proof. Pattern Recognition, Vol. 34, No. 7 (2001) 1393-1403
6. Zhou W.: Verification of the nonparametric characteristics of backporpagation neural networks for image classification. IEEE Transaction On Geoscience and Remote Sensing, Vol. 37, No. 2, March (1999) 771-779
7. Haddadnia J., Faez K., Moallem P.: Neural Network Based Face Recognition with Moments Invariant. IEEE International Conference On Image Processing, Vol. I, Thessaloniki, Greece, 7-10 October (2001) 1018-1021
8. The C. H. and Chin R. T.: On Image Analysis by the Methods of Moments. IEEE Transaction On Pattern Analysis And Machine Intelligence, Vol. 10, No. 4, (1988) 496-513
9. Truk M. and Pentland A.: Eigenfaces for Recognition. Journal Cognitive Neuroscience, Vol. 3, No. 1 (1991) 71-86
10. Embree P. M. and Kimble B.: C language Algorithm for Digital Signal Processing. Printice Hall, New Jercy , 1991
11. Haddadnia J. and Faez K.: Human Face Recognition Using Radial Basis Function Neural Network. 3rd Int. Conf. On Human and Computer, Aizu, Japan, Sep. 6-9 (2000) 137-142
12. Haddadnia J., Ahmadi M. and Faez K., A Hybrid Learning RBF Neural Network for Human Face Recognition with Pseudo Zernike Moment Invariant, IEEE International Joint conference on Neural Network, Honolulu, HI, May 12-17 (2002), Accepted for presentation
13. Thomaz C. E., Feitosa R. Q. and Veiga A.: Design of Radial Basis Function Network as Classifier in Face Recognition Using Eigenfaces. IEEE Proceedings of Vth Brazilian Symposium on Neural Network, (1998) 118-123

A Transformation-Based Mechanism for Face Recognition

Yea-Shuan Huang and Yao-Hong Tsai

Advanced Technology Center
Computer & Communications Research Laboratories
Industrial Technology Research Institute, Chutung, Hsinchu, Taiwan
yeashuan@itri.org.tw

Abstract. This paper proposes a novel mechanism to seamlessly integrate face detection and face recognition. After extracting a human face x from an input image, not only x but also its various kinds of transformations are performed recognition. The final decision is then derived from aggregating the accumulated recognition results of each transformed pattern. From experiments, the proposed method has shown a significantly improved recognition performance compared with the traditional method on recognizing human faces.

1 Introduction

Due to the rapid advance of computer hardware and the continuous progress of computer software, we are looking forward to developing more powerful and friendly computer use models so that computers can serve people in a more active and intelligent way. The concept of "*computer will be more human*" is not required just in the scientific fiction but also is in our daily life. To this end, the computer basically needs to have a surveillance ability, which enables it to detect, track, and recognize its surrounding people so that the computer can offer various kinds of user oriented services automatically. This results in the situation that researches on face processing (including detection [1-2], tracking [3], and recognition [4-7]) are very prosperous in the last two decades.

Many promising algorithms have been proposed to deal with the basic face processing problems, such as (1) how to detect the human faces from an input image? and (2) how to recognize the people identity based on a detected face? High accuracy has already been individually reported for face detection (FD) and face recognition (FR). However, when integrating FD and FR together (as shown in Figure 1), it often results in a considerably degraded performance. For example, both of the chosen algorithms of FD and FR may have over a 90% individual correction rate, but the integrated system actually just has only 60% in correctly recognizing a face from an input image. This phenomenon mainly comes from three factors: (1) the criterion of deciding successful face detection is too rough for face recognition so that many detected face

T. Caelli et al. (Eds.): SSPR&SPR 2002, LNCS 2396, pp. 636-643, 2002.
© Springer-Verlag Berlin Heidelberg 2002

images considered to be correct in the FD stage are not good enough in the FR stage due to either partially lost important face information or contain extra non-face image; (2) the chosen FR has little generalization ability so that it is easy to obtain a wrong recognition when a face image is not perfectly detected; and (3) the training samples of FR are extracted manually so that the automatically detected face images from a FD having different properties from the manually extracted ones are prone to be misrecognized. In this paper, a novel mechanism has been devised which can seamlessly integrate FD and FR and improve the accuracy of the whole integrated system. The basic concept of this approach is that not only the detected face image x but also its various transformations are performed the face recognition operation, and the final decision is derived from the accumulated recognition results such as the decision is the class with the highest similarity among all individual recognition or the class with the highest average similarity.

This paper consists of four sections. Section 2 describes the newly proposed face recognition integration mechanism which contains four main processing steps: face detection, image transformation, image matching, and result accumulation. Section 3 then performs experiments with and without multiple transformation and investigates their performance difference on the ITRI (Industrial Technology and Research Institute) face databases. Finally, Section 4 draws our conclusions and point out the future research directions.

2 The Proposed Integration Mechanism

Figure 2 shows the newly proposed transformation-based mechanism to integrate a face recognition system. The main concept of this mechanism is that not only an extracted face image x but also its various transformations are performed recognition, and the final decision is derived from the accumulated recognition results. In this section, face detection and face matching is discussed first, then a recognition by accumulating multiple transformation is formally introduced.

2.1 Face Detection

In a face recognition system, it is essential that a face image is extracted from a processed image by containing only the portion of face image which is useful for recognition and excluding the image portions (such as background, clothes, and hair) invalid to recognition. Chen et al. [8] have shown that it is incorrect to recognize a person by using the image including face, hair, shoulder, and background as used in [9] because the trained statistical classifiers learns too much irrelevant information in identifying a person. Instead, they proposed to extract the face-only image for training and testing a face recognition system. A face-only image as shown in Figure 3 ideally is the minimal rectangle containing eyebrows. eyes, nose, and mouth of a face. Here, we adopted Han's method [10] to extract the face-only image which consists of three main steps. In the first step, a morphology-based technique is devised to perform eye-analogue segmentation. Morphological operations are applied to locate eye-analogue pixels in the original image. Then a labeling process is executed to generate the eye-analogue segments. Each eye-analogue segment is considered a candidate of one of the eyes. In the second step, the previously located eye-analogue segments are used to find meaningful eye pairs by using four geometrically matching rules to guide the merging of the eye-analogue elements into pairs. Each meaningful eye pair is further directed to specify a corresponding face region. The last step is to verify each specified face region to be a face or an non-face by a neural network. This method performs rather fast and accurate when dealing with uniformly well lit face images.

2.2 Face Matching

A three-layer feed-forward network with a Generalized Probabilistic Descent (GPD) learning rule is served as the face classifier. GPD is originally proposed by Juang [11] to train a speech classifier, which is reported to have a much better recognition performance than the well-known Back-Propagation (BP) training. However, to our best knowledge, GPD is rarely or even never used in the computer-vision community. Because GPD is based on minimizing a classification related error function, it theoretically can produce a better classification performance than the classifiers (such as BP) based on minimizing a least-mean-square error. Because the space limit, this GPD face learning approach is not introduced in this article, interested readers can find detail information in [12].

2.3 Recognition by Accumulating Multiple Transformations

Assume K subjects of people in the concerned pattern domain, $C_1,\cdots,C_K$, x denotes an inputted face-only image, and $S_k(x)$ is the possibility that *x* belongs to subject *k*. Let $F_n(x)$ be the feature of the *n*-th transformed image of *x*, where $0 \le n \le N$ and N is the total number of transformations, and $F_0(x)$ be the feature of the original *x*. Traditionally, x is recognized to subject *k* if $S_j(x)$ has the largest value among all $S_k(x)$ for $1 \le k \le K$. That is

$$D(x) = j, \quad \text{if } S_j(x) = \arg\max_{1\le k\le K} S_k(x).$$

The proposed decision by accumulating multiple transformation is

$$D(x) = j, \quad \text{if } G_j(S_j(F_0(x)),\cdots,S_j(F_N(x))) = \arg\max_{1\le k\le K} G_k(S_k(F_0(x)),\cdots,S_k(F_N(x)))$$

where $G_k(\cdots)$ is an aggregation function which specifies the appropriate way to derive the accumulated score that *x* belongs to subject *k*. Two possible selections of G_k have been proposed, they are

$$G_k(S_k(F_0(x)),\cdots,S_k(F_N(x))) = \max_{0\le n\le N} S_k(F_n(x)) \tag{1}$$

and,

$$G_k(S_k(F_0(x)),\cdots,S_k(F_N(x))) = \frac{1}{N}\sum_{n=0}^{N} S_k(F_n(x)) \tag{2}$$

Of course, there are many other possible choices of G_k. As a matter of fact, this concept can be applied to deduce a lot of variational recognition processes, such as

(1) not only the best detected object but also many other possible detected objects are performed recognition and the final decision is derived from the proposed G_k, and
(2) not only many possibly detected object but also their various transformations are applied recognition, and the final decision is derived from these tentative recognitions.

In reality, there exists various kinds of image transformations such as image rotation, affine transform, boundary shifting, lighting compensation and so on. In this paper, two kinds of commonly-used transformations (image rotation and boundary shifting) are described as follows: Let *I* be an image with *M* horizontal and *N* vertical pixels, *I(m,n)* be one image pixel locating at the *m*-th horizontal and the *n*-th vertical pixels ($1 \le m \le M$ and $1 \le n \le N$). Let *T* denote a segmented target image, *T(x,y)* be one pixel of *T*, and T' be a transform of *T*. The rotation transform is

$$\begin{pmatrix} x' \\ y' \end{pmatrix} = \begin{bmatrix} \cos\theta & \sin\theta \\ -\sin\theta & \cos\theta \end{bmatrix} \begin{pmatrix} x \\ y \end{pmatrix}$$

where θ is the rotation angle. One more generalized transformation form is the affine transform and its matrix notation is

$$\begin{pmatrix} x' \\ y' \end{pmatrix} = \begin{bmatrix} a_{11} & a_{12} \\ a_{21} & a_{22} \end{bmatrix} \begin{pmatrix} x \\ y \end{pmatrix} + \begin{pmatrix} b_1 \\ b_2 \end{pmatrix}$$

which covers translation, scaling, rotation and slant.

Another possible transformation is boundary shifting which shifts any of *T*'s four boundaries (i.e. top, bottom, left and right) in and out. Suppose *T(x,y)* be one pixel of *T* where $x_{min} \le x \le x_{max}$ and $y_{min} \le y \le y_{max}$, if T' is achieved by shifting the *T*'s left boundary *p* pixels out (i.e. the left boundary is shifted *p* pixels more left but keep the top, the bottom and the right boundaries the same), then $T'(x,y) = T(x,y)$ with $x_{min} - p \le x \le x_{max}$ and $y_{min} \le y \le y_{max}$. T' could also be achieved by shifting both *T*'s left boundary *p* pixels out and *T*'s bottom boundary *q* pixels in, then $T'(x,y) = T(x,y)$ with $x_{min} - p \le x \le x_{max}$ and $y_{min} \le y \le y_{max} - q$.

3 Experiment Results

To investigate the performance of this method, two sets of face databases were constructed and face recognition experiments were performed by training and testing a face classifier (GPD with 625-100-26 nodes) with and without adopting this proposed method respectively. The first face dataset was taken by asking 26 persons to slightly rotate their faces or change their face expressions when they stood about 1.5 meters away from the camera. The second face dataset was taken by allowing the same 26 persons to approach from 6 meters to 1 meter away from the camera. Figure 4 shows some examples of the two face datasets which reveals that (1) the face images are about the same size in the first dataset, but they have quite different sizes (more than 10 times) in the second dataset, and (2) the second dataset consists of much image difference in brightness gain and intensity uniformity.

In this experiment, face-only images are manually selected from the first face database, which are then used to train the GPD face classifier; but face-only images are automatically extracted from the second face database, which are further used to test the face recognition performance. A threshold T is defined to be $0.125 \times D$, where D is the distance between the centers of two eyes. If any location distance of the eye on the automatically extracted face-only image to its corresponding manually selected eye is larger than T, then it is counted as an invalid face detection; otherwise, it is counted as a valid one. Since a face-only image is extracted directly based on its right and left eyes, shifting the eye's locations can correspondingly generate transformed images having the rotation and boundary shifting effects to the original one. In this experiment, the locations of each right and left eyes are individually shift S pixels (here S is $0.06 \times D$) left and right. This results in totally 81 $(3 \times 3 \times 3 \times 3)$ transformed images from each detected face-only image.

After the training procedure, the test images are inputted to the trained recognizer, and the recognition decision is made respectively by using the tradition decision rule and the proposed decision rule with the first aggregating rule

$$G_k(S_k(F_0(x)),\cdots,S_k(F_N(x))) = \max_{0 \le n \le N} S_k(F_n(x)).$$

In order to analyze the effectiveness of the proposed method, the face recognition rate is computed by two different situations: recognition of validly detected faces, and recognition of invalidly detected faces. Table 1 shows respectively the recognition performance of the tradition decision rule and the proposed transformation-based decision accumulation rule. Obviously, the proposed method performs much better than the traditional one by improving the recognition accuracy from 65% to 85% in the first (valid) situation, and from 40% to 60% in the second (invalid) situation. Since the two recognition indexes are based on the same face detection result, it is clear that the better performance of the proposed method comes from the better generalization ability of the proposed method. This robustness enables the integration system to have the ability in producing the correct recognition decision even when the detected face image is not complete or good enough to the adopted face classifier.

4 Conclusions

This paper proposes a novel mechanism to seamlessly integrate face detection and face recognition. After extracting a human face x from an input image, not only x but also its various kinds of transformations are performed recognition. The final decision is then derived from aggregating the accumulated recognition results of each trans-

formed pattern. From experiments, the proposed method has shown a much better performance compared with the traditional face recognition system.
One drawback of this method is that it takes a much longer period of FD+FR (0.85 seconds/image) than the traditional one (0.2 second/image). Many speed-up methods can be easily applied to reduce the processing time, such as (1) to reduce the number of transformations, and (2) to reduce the scale in each transformation. However, it is more interesting if an appropriate way can be designed to guide the suitable transformation types and transformation scales automatically based on the extracted face information. We are currently working on this direction.

References

1. K.K. sung and T. Poggio, "*Example-Based Learning for view-Based Human Face Detection*," IEEE Trans. Patt. Anal. Machine Intell., Vol. 20, pp. 39-51, 1998.
2. H. A. Rowley, S. Baluja, and T. Kanade, "Neural network-based face detection," *IEEE transactions on PAMI.*, vol. 20, no. 1, pp. 22-38, Jan. 1998.
3. D. M. Gavrila , "The visual analysis of human movement: a survey," Computer Vision and Image Understanding, vol. 73, pp. 82-98, 1999.
4. M. Turk and A. Pentland, "*Eihenfaces for Recognition*", Journal of Cognitive Neuroscience, March, 1991.
5. R. Brunelli and T. Poggio, "*Face Recognition: Features Versus Templates*", IEEE Trans. Patt. Anal. Machine Intell. Vol. 15, No. 10, October, pp 1042-1052, 1993.
6. R. Chellappa, C. Wilson and S. Sirohey, "*Human and Machine Recognition of Faces: A Survey*", Proc. Of IEEE, Vol. 83, No. 5, May, pp 705-740, 1995.
7. A.K. Jain, R. Bolle and S. Pankanti, *Biometrics: Personal Identification in Networked Society*, Kluwer Academic Publishers, 1999.
8. L.-F. Chen, C.-C. Han, and J.-C. Lin, "Why Recognition in a Statistics-Based Face Recognition System Should be Based on the Pure Face Portion: a Probabilistic Decision-Based Proof," to appear in *Pattern Recognition*, 2001.
9. F. Goudail, E. Lange, T. Iwamoto, K. Kyuma, and N. Otsu, "Face Recognition System Using Local Autocorrelations and Multiscale Integrations," *IEEE Trans. Pattern Analysis and Machine Intelligence*, Vol. 18, pp. 1024-1028, Oct., 1996.
10. C.C. Han, H.Y. Mark, K.C. Yu, and L.H. Chen, "Fast Face Detection via Morphology-Based Pre-Processing," Pattern Recognition 33, pp. 1701-1712, 2000.
11. B.H. Juang and S. Katagiri, "*Discriminative Learning for Minimum Error Classification*", IEEE Trans. On Signal Processing, Vol. 40, No. 12, December, pp 3043-3054, 1992.
12. Yea-Shuan Huang, Yao-Hong Tsai, Jun-Wei Shieh, "Robust Face Recognition with Light Compensation," to appear in The Second IEEE Pacific-Rim Conference on Multimedia, 2001.

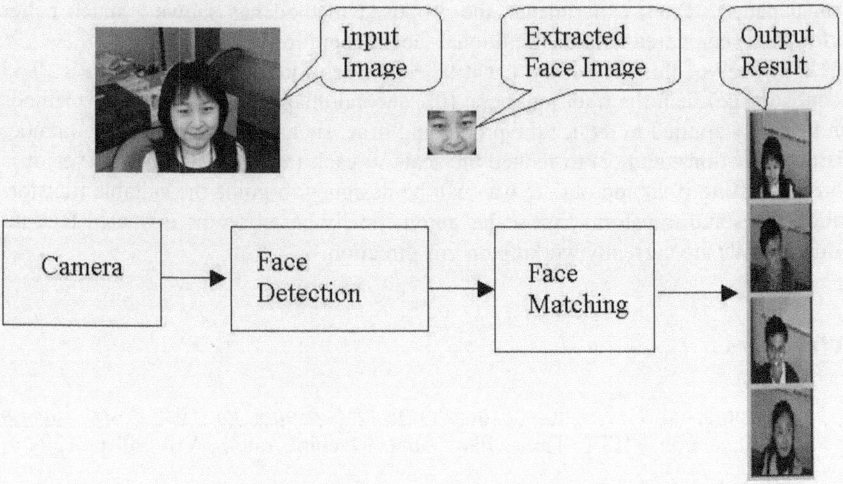

Fig. 1. A traditional face recognition system

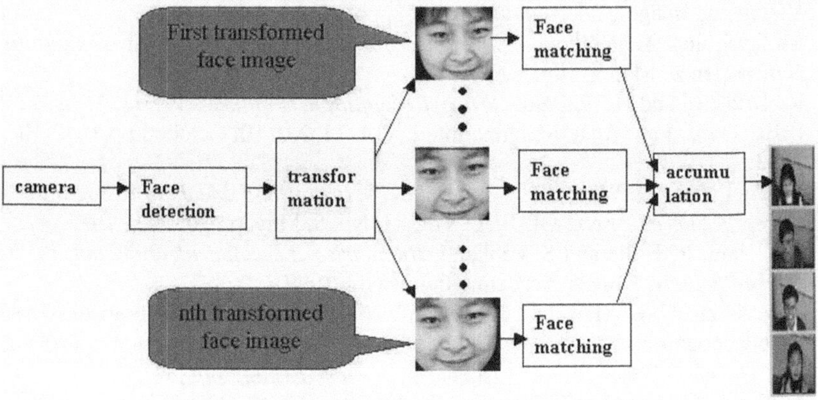

Fig. 2. The proposed transformation-based face recognition system

Fig. 3. One example of the approximate face-only image delimitated by a black-line rectangle which is extracted based on the two eyes marked in white circles

Fig. 4. Examples of two face datasets. The first row shows images of the first dataset images which were taken at a constant distance between camera and the subject, and the second row displays the images of the second dataset which were taken during a subject approached close to the camera

Table 1. This table shows (1) the performance of face detection, and (2) the recognition performance by the tradition decision rule and the corresponding recognition rate by the proposed decision rule

	Valid Faces	Invalid Faces
Sample Number	2456	345
Traditional FR	65%	40%
Proposed FR	85%	60%

Face Detection Using Integral Projection Models*

Ginés García-Mateos[1], Alberto Ruiz[1], and Pedro E. Lopez-de-Teruel[2]

[1] Dept. Informática y Sistemas
[2] Dept. de Ingeniería y Tecnología de Computadores
University of Murcia, 30.170 Espinardo, Murcia (Spain)
{ginesgm,aruiz}@um.es
pedroe@ditec.um.es

Abstract. Integral projections can be used to model the visual appearance of human faces. In this way, model based detection is done by fitting the model into an unknown pattern. Thus, the key problem is the alignment of projection patterns with respect to a given model of generic face. We provide an algorithm to align a 1-D pattern to a model consisting of the mean pattern and its variance. Projection models can also be used in facial feature location, pose estimation, expression and person recognition. Some preliminary experimental results are presented.

1 Introduction

Human face detection is an essential problem in the context of perceptual interfaces and human image processing, since a fixed location assumption is not possible in practice. It deals with determining the number of faces that appear in an image and, for each of them, its location and spatial extent [1]. Finding a fast and robust method to detect faces under non-trivial conditions is still a challenging problem.

Integral projections have already been used in problems like face detection [2, 3] and facial feature location [4]. However, most existing techniques are based on max-min analysis [2, 3], fuzzy logic [4] and similar heuristic approaches. To the best of our knowledge, no rigorous study on the use of projections has been done yet. Besides, the use of projections constitutes a minor part in the vision systems. Our proposal is to use projections as a means to create 1-dimensional face models.

The structure of this paper is the following. In Section 2, we show how projections can be used by themselves to model 3-D objects like faces. The face detection process is presented in Section 3. Section 4 focuses in the key problem of projection alignment. Some preliminary experimental results on the proposed model are described in Section 5. Finally, we present some relevant conclusions.

---

* This work has been supported by the Spanish MCYT grant DPI-2001-0469-C03-01.

T. Caelli et al. (Eds.): SSPR&SPR 2002, LNCS 2396, pp. 644-653, 2002.
© Springer-Verlag Berlin Heidelberg 2002

2 Modeling Objects with Integral Projections

Integral projections can be used to represent the visual appearance of a certain kind of object under a relatively wide range of conditions, i.e., to model object classes. In this way, object analysis can be done by fitting a test sample to the projection model. We will start this section with some basic definitions on integral projections.

2.1 One-Dimensional Projections

Let $i(x, y)$ be a grayscale image and $R(i)$ a region in this image, i.e., a set of contiguous pixels in the domain of i. The horizontal and vertical integral projections of $R(i)$, denoted by $P_{HR(i)}$ and $P_{VR(i)}$ respectively, are discrete and finite 1-D signals given by

$$P_{HR(i)} : \{x_{min}, \ldots, x_{max}\} \rightarrow \mathbf{R} \ ; \ P_{HR(i)}(x) := |Rx(i)|^{-1} \sum_{y \in Rx(i)} i(x, y) \ ; \tag{1}$$

$$P_{VR(i)} : \{y_{min}, \ldots, y_{max}\} \rightarrow \mathbf{R} \ ; \ P_{VR(i)}(y) := |Ry(i)|^{-1} \sum_{x \in Ry(i)} i(x, y) \ ; \tag{2}$$

where

$$x_{min} = \min_{(x,y)\in R(i)} x; \ x_{max} = \max_{(x,y)\in R(i)} x; \ y_{min} = \min_{(x,y)\in R(i)} y; \ y_{max} = \max_{(x,y)\in R(i)} y; \tag{3}$$

$$Rx(i) = \{y \,/\, \forall\, y, (x, y) \in R(i)\} \ ; \ Ry(i) = \{x \,/\, \forall\, x, (x, y) \in R(i)\} \ . \tag{4}$$

The sets $\{x_{min}, \ldots, x_{max}\}$ and $\{y_{min}, \ldots, y_{max}\}$ are called the domains of the horizontal and vertical integral projection, denoted by $Domain(P_{HR(i)})$ and $Domain(P_{VR(i)})$ respectively. Similarly, we can define the projection along any direction with angle α, $P_{\alpha R(i)}$, as the vertical projection of region $R(i)$ rotated by angle α. Applied on faces, vertical and horizontal projections produce typical patterns, as those in Fig. 1.

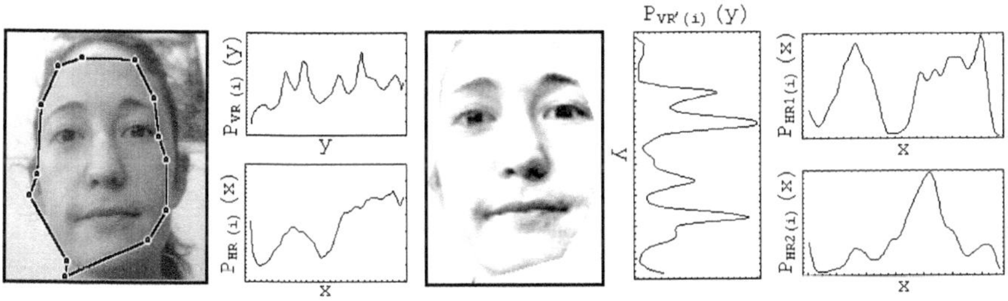

a) b) c) d) e)

Fig. 1. Vertical and horizontal integral projections. a) A face region R found using skin color analysis. b) Vertical (up) and horizontal (down) projections of R. c) Segmentation and intensity equalization of R, to produce R'. d) Vertical projection of R'. e) Horizontal projection of the upper ($R1$) and lower ($R2$) halves of R'

Integral projections give marginal distributions of gray values along one direction, so they usually involve a loss of information. An approximate reconstruction of $R(i)$ can be easily computed from $P_{HR(i)}$ and $P_{VR(i)}$. Let us suppose $i(x,y)$ is normalized to values in [0, 1], then the reconstruction is: $\hat{\imath}(x,y) = P_{HR(i)}(x) \cdot P_{VR(i)}(y)$, $\forall (x,y) \in R(i)$.

Modeling Faces

The following questions have to be dealt with when using a model of projections:

- How many projections are used to model the object class.
- For each of them, which angle and which part of the region is projected.
- What projected pixels represent, e.g., intensity, edge-level, color hue.

As we have mentioned, an approximate reconstruction can be computed from the projections and, somewhat, the similarity with respect to the original image indicates the accuracy of the representation. For instance, three reprojections of a face image using different numbers of projections are shown in Fig. 2.

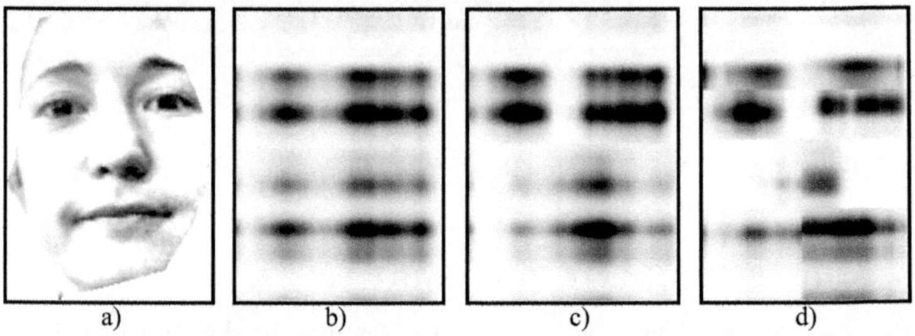

Fig. 2. Face image reconstruction by reprojection. a) Original face image, segmented and intensity equalized. b)-d) Reconstruction using vertical (*vip*) and horizontal (*hip*) integral projections: b) 1 *vip*, 1 *hip*; c) 1 *vip*, 2 *hip*; d) 2 *vip*, 4 *hip*

Obviously, the accuracy of the reprojection increases with the number of projections used. However, using a high number of projections involves an important loss of robustness and efficiency. Thus, we have chosen the representation of 1 vertical and 2 horizontal projections, shown in Fig. 2, which gives admissibly results.

To model the variability of the face class, we propose a gaussian-style representation of the one-dimensional signals. That is, for each point j in the domain of the signal, the mean value $M(j)$ and the variance $V(j)$ are computed. Summing up, the face model consists of the following 1-D signals:

- $M_{V,FACE}$, $V_{V,FACE}$: $\{1, ..., f_{max}\} \rightarrow \mathbf{R}$. Mean and variance of the vertical projection of the whole face region, respectively.
- $M_{H,EYES}$, $V_{H,EYES}$: $\{1, ..., e_{max}\} \rightarrow \mathbf{R}$. Mean and variance of the horizontal projection of the upper part of the face, from forefront to nose (not included).
- $M_{H,MOUTH}$, $V_{H,MOUTH}$: $\{1, ..., m_{max}\} \rightarrow \mathbf{R}$. Mean and variance of the horizontal projection of the lower part of the face, from nose (included) to chin.

3 Face Detection Using Projections

In a general sense, object detection using models consists of fitting a known model into an unknown pattern. If a good fitting is found, the object is said to be detected. However, as the location, scale and orientation of the object in the image is unknown, either selective attention [3, 5, 6] or exhaustive multi-scale searching [7, 8], are needed. We use the selective attention mechanism described in [3], based on connected components of skin-like color. This process was used in the experiments to extract the face and non-face candidate regions, which are the input to the face detection algorithm using the projection models. The algorithm is shown in Fig. 3.

Algorithm: Face Detection Using a Projection Model

Input

i: Input image
$M = (M_{V,FACE}, V_{V,FACE}, M_{H,EYES}, V_{H,EYES}, M_{H,MOUTH}, V_{H,MOUTH})$: Face model

Output

n: Number of detected faces
$\{R_1, ..., R_n\}$: Region of the image occupied by each face

1. Segment image *i* using connected components of skin-like color regions.

2. For each candidate region *R(i)* found in step 1, do.
2.1. Compute $P_{VR(i)}$, the vertical integral projection of *R(i)*, taking the principal direction of *R(i)* as the vertical axis.
2.2. Align $P_{VR(i)}$ to $(M_{V,FACE}, V_{V,FACE})$ obtaining $P'_{VR(i)}$.
2.3. If a good alignment was obtained in step 2.2, compute $P_{HR1(i)}$ and $P_{HR2(i)}$, the horizontal integral projections of the upper and lower parts of *R(i)* respectively, according to the results of the alignment $P'_{VR(i)}$.
2.4. Align $P_{HR1(i)}$ to $(M_{H,EYES}, V_{H,EYES})$ obtaining $P'_{HR1(i)}$, and align $P_{HR2(i)}$ to $(M_{H,MOUTH}, V_{H,MOUTH})$ obtaining $P'_{HR2(i)}$.
2.5. If good alignments were obtained in step 2.4, then *R(i)* corresponds to a face. Increment *n*, and make $R_n = R(i)$. The location of the facial features can be computed by undoing the alignment transformations in steps 2.2 and 2.4.

Fig. 3. Global structure of the algorithm for face detection using a projection model

First, the $(M_{V,FACE}, V_{V,FACE})$ part of the model is fitted into the vertical projection of the whole candidate region, which might contain parts of hair or neck. If a good alignment is obtained, the vertical locations of the facial components are known, so we can compute the horizontal projections of the eye and mouth regions, removing hair and neck. If both are also correctly aligned, a face has been found.

4 Projection Alignment

The key problem in object detection using projections is alignment. The purpose of projection alignment is to produce new derived projections where the location of the facial features is the same in all of them. Figs. 4a,b) show two typical skin-color regions containing faces, which produce unaligned patterns of vertical projections. After alignment, eyes, nose and mouth appear at the same position in all the patterns.

In this section, we describe one solution to the problem of aligning 1-D patterns, or signals, to a model consisting of the mean signal and variance at each point, as introduced in Section 2.2. Note that in the algorithm for face detection, in Section 3, the goodness of alignment is directly used to classify the pattern as face or non-face. In general, any classifier could be used on aligned patterns, thus making clear the difference between preprocessing (alignment) and pattern recognition (binary classification face/non-face). In the following, we will suppose any classifier can be used.

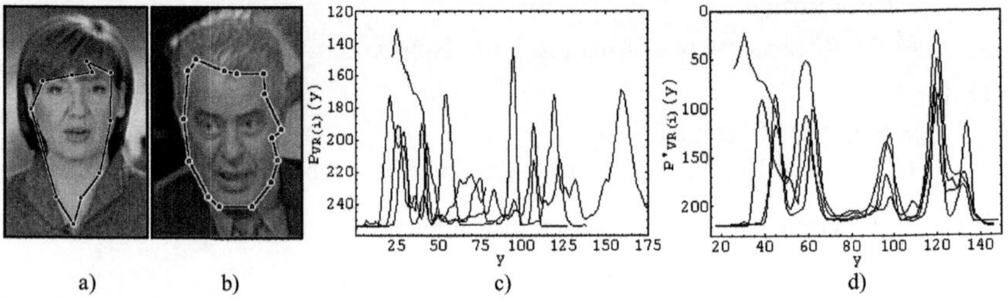

Fig. 4. Alignment of projections. a)-b) Two typical face regions, producing unaligned projections. c) Unaligned vertical projections of 4 faces. d) The same projections, after alignment

4.1 Alignment Criterion

Theoretically speaking, a good alignment method for detection should produce a representation of face patterns invariant to lighting conditions, pose, person and face expression. Let us suppose we have a set of projections $P=P_{FACE} \cup P_{NON\text{-}FACE}$, and a set of alignment transformations $A=\{a_1, ..., a_m\}$. The best alignment is the one that produces the best detection ratios, that is, a high number of detected faces and a low number of false-positives. Instead, we will work with a more practical criterion.

In order to achieve good detection results, independently from the classifier used, the alignment should minimize the variance of aligned face patterns, denoted by P^a_{FACE}, and maximize the interclass variance of $\{P^a_{FACE}, P^a_{NON-FACE}\}$. However, estimating this interclass variance is a hard problem, since no finite set $P_{NON\text{-}FACE}$ can be representative enough of everything which is not a face.

Supposing the average projection of the infinite class of non-faces is a uniform signal, the variance between face and non-face classes can be estimated with the inner variance, or energy, of the mean signal $\overline{P^a_{FACE}}$. In this way, the goodness of an alignment transformation a can be estimated with the ratio

$$RATIO\left(a,P_{FACE}\right):=\frac{VARIANCE(\overline{\overline{P^{a}_{FACE}}})}{VARIANCE(P^{a}_{FACE})} \ . \tag{5}$$

A lower value of (5) means a better alignment. In practice, we are interested in aligning patterns according to a projection model in the form (M: mean; V: variance) learnt by training. This involves that the average face projection (and, consequently, its energy) is computed in the training process. As a result, for a given pattern p, the alignment should minimize its contribution to (5), which can be expressed as

$$Distance\left(a,p,M,V\right):=\sum_{i\in Domain(M)}\frac{(p^{a}(i)-M(i))^{2}}{V(i)} \ . \tag{6}$$

4.2 Transformation Functions

A transformation is a function that takes a signal as an input and produces a derived signal. It is called an alignment, or normalization, if the transformed signals verify a given property. We are interested in parameterized functions, where the parameters of the transformation are calculated for each signal and model. It is convenient to limit the number of free parameters, as a high number could produce a problem of *over-alignment*: both face and non-face patterns could be transformed to face-similar patterns, causing many false-positives.

We will use the following family of parameterized transformation functions

$$t_{a,b,c,d,e}:(\{s_{min},\dots,s_{max}\}\rightarrow \mathbf{R})\longrightarrow(\{\left|\frac{s_{min}-e}{d}\right|,\dots,\left|\frac{s_{max}-e}{d}\right|\}\rightarrow \mathbf{R}) \ , \tag{7}$$

defined by

$$t_{a,b,c,d,e}(S)(i):=a+b{\cdot}i+c{\cdot}S(|\,d{\cdot}i+e\,|) \ . \tag{8}$$

As expressed in (8), the function $t_{a,b,c,d,e}$ makes a linear transformation both in value and in domain of the input signal S. It has five free parameters: (a, b, c) the value transformation parameters, and (d, e) the domain transformation parameters. Geometrically interpreted, (a, e) are translation parameters in value and domain, respectively; (c, d) are scale parameters, and b is a skew parameter that, in our case, accounts for a non-uniform illumination of the object.

4.3 Alignment Algorithm

For the alignment, we will use the family of transformation functions with the form $t_{a,b,c,d,e}$, defined in (7) and (8). We can obtain the objective function of alignment replacing p^a in (6) with (8). Thus, the optimum alignment of a signal S to a model (M, V) is given by the set of values (a, b, c, d, e) minimizing

$$Distance\left(a,b,c,d,e\right):=\sum_{i\in Domain(M)}\frac{(a+b{\cdot}i+c{\cdot}S(|\,d{\cdot}i+e\,|)-M(i))^{2}}{V(i)} \ . \tag{9}$$

Due to the form of the transformation, both in value and domain, standard optimization techniques can not be applied to minimize (9). Instead of it, we use an iterative two-step algorithm that, alternatively, solves for the domain parameters (*d*, *e*) and the value parameters (*a*, *b*, *c*). The algorithm is presented in Fig. 5.

Algorithm: Linear Alignment of a 1-D Pattern to a Mean/Variance Model

Input

S: Signal pattern to be aligned

M, *V*: Signal model, mean and variance respectively

Output

S': Optimum alignment of signal *S* to model (*M*, *V*)

1. Transformation initialization. Set up initial values for (*a*, *b*, *c*, *d*, *e*), e.g., locating two clearly distinguishable points in *S*. Obtain *S'* applying equation (8).
2. Repeat until convergence is reached or after MAX_ITER iterations:

2.1. Pattern domain alignment.

2.1.1. Assign each point *i* of *S'*, in *Domain(S')*$\cap$*Domain(M)*, to a point *h(i)* of *M*, in *Domain(M)*, with reliability degree *w(i)*.

2.1.2. Estimate parameters (*d*, *e*), as the linear regression parameters of the set (*i*, *h(i)*) taking into account the weights *w(i)*, for *i* in *Domain(S')*$\cap$*Domain(M)*.

2.1.3. Transform *S'* in domain, to obtain S'_d. That is, setting (*a*, *b*, *c*) = (0,0,1), make $S'_d(i) := S'(d \cdot i + e)$

2.2. Pattern value alignment.

2.2.1. Estimate value transformation parameters (*a*, *b*, *c*) as the values minimizing $\Sigma(a + b \cdot i + c \cdot S'_d(i) - M(i))^2 / V(i)$

2.2.2. Transform pattern S'_d to obtain the new *S'*, using: $S'(i) := a + b \cdot i + c \cdot S'_d(i);\ \forall\ i \in$ Domain (S') = Domain (S'_d)

Fig. 5. Structure of the algorithm to compute the optimum linear alignment of a 1-D pattern, or signal, to a mean/variance model of the signal

The algorithm is based on an assignment *h(i)* of points in *Domain(S)* with the corresponding points in *Domain(M)*. This assignment is computed as the most similar point to *S(i)* around a local proximity in *M(i)*. The similarity is a combination of position and slope likeness. The reliability degree *w(i)* is proportional to the maximum similarity and inversely proportional to the similarity of non-maximum.

5 Experimental Results

The purpose of the experiments described herein has been to assess the invariance and robustness of the aligned projection representation and its potential to discriminate between faces and non-faces. In this way, the results indicate both the strength of the alignment algorithm and the feasibility of modeling faces with projections.

The test set consists of 325 face (R_{FACE}) and 292 non-face ($R_{NON\text{-}FACE}$) regions segmented from 310 color images, using color segmentation (see step 1, in Fig. 3). These images were captured from 12 different TV channels, with samples taken from news, series, contests, documentaries, etc. The existing faces present a wide range of different conditions in pose, expression, facial features, lighting and resolution. Some of them are shown in Fig. 7. The face model was computed using a reduced set of 45 faces, not used in the test set, and is shown in Fig. 6.

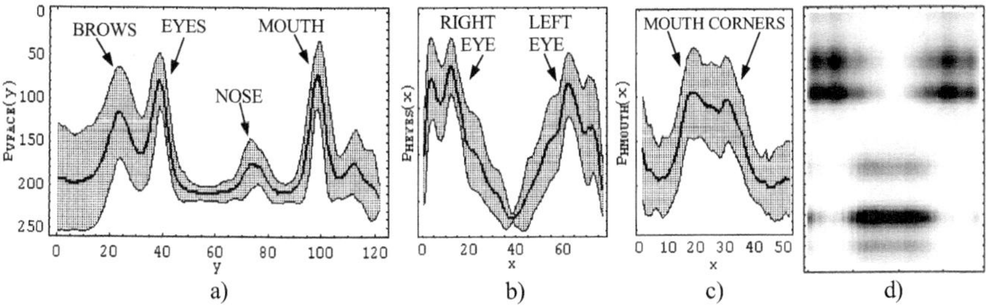

Fig. 6. Integral projection model of the face. a) $M_{V,FACE}$ and $V_{V,FACE}$. b) $M_{H,EYES}$ and $V_{H,EYES}$. c) $M_{H,MOUTH}$ and $V_{H,MOUTH}$. d) Reprojection of the face model

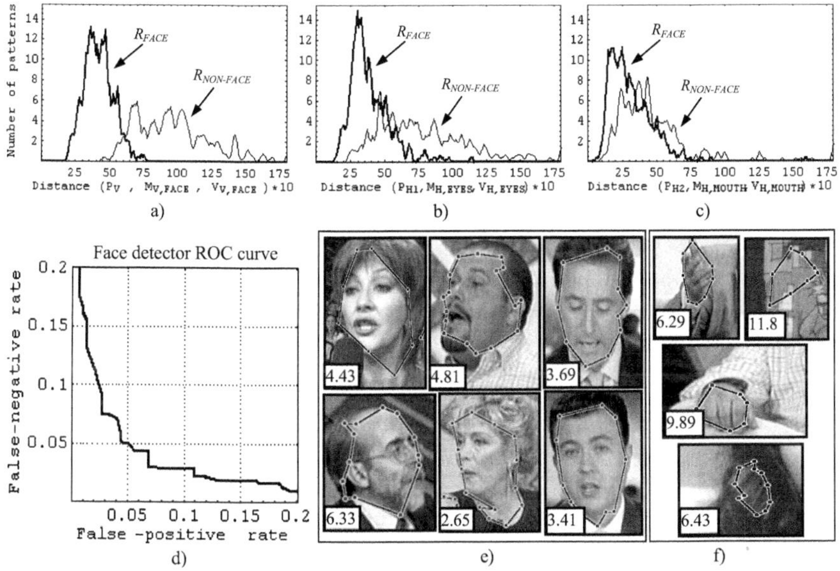

Fig. 7. Detection results. a-c) Distances of face and non-face aligned projections to the model: a) *V,FACE*; b) *H,EYES*; c) *H,MOUTH*. d) ROC curve of the face detector. e), f) Some face and non-face regions, respectively. The distances of P_V to ($M_{V,FACE}$, $V_{V,FACE}$) are shown below

As described in Section 3, detection is done according to the distance from the model to the aligned projections (P_{VR}, P_{HR1}, P_{HR2}) of each region R, using equation (9). The distances obtained for the test set are shown in Fig. 7. As expected, R_{FACE} projections yield lower distance values, although a certain overlapping exists. This overlapping is 11%, 39% and 67% for Figs. 7a), 7b) and 7c) respectively, so, by itself, the vertical projection of the whole face is the most discriminant component.

The results of the face detector, using different distance thresholds, are shown in the ROC curve in Fig. 7d). At the point with equal number of false-positives and false-negatives the detection ratio is 95.1%. A direct comparison with other methods is not meaningful, since the color-based attention process should also be taken into account. In previous experiments [3], this process showed an average detection ratio of 90.6% with 20.1% false-negatives (similar results are reported by other authors in [5]). Combined with the projection method, the faces detected are 86.2% with 0.96% of false-negatives. These results are comparable with some state-of-the-art appearance based methods (see [7] and references therein), detecting between 76.8% and 92.9% of the faces, but with higher numbers of false detections.

6 Conclusion and Future Work

This work constitutes, to the best of our knowledge, the first proposal concerning the definition and use of one-dimensional face patterns. This means that projections are used not only to extract information from max-min analysis or similar heuristic methods, but to model object classes and perform object detection and analysis. The preliminary experiments have clearly shown the feasibility of our proposal.

Somewhat, our approach could be considered equivalent to a 2-D appearance based face detection, where the implicit model is like the image shown in Fig. 6d). However, our method has several major advantages. First, working with 1-D signals involves an important improvement in computational efficiency. Second, the separation in vertical projection and then horizontal projections, makes the process very robust to non-trivial conditions or bad segmentation, without requiring exhaustive multi-scale searching. Third, the kind of projection model we have used, has proven an excellent generalization capability and invariance to pose, facial expression, facial elements and acquisition conditions.

References

1. Yang, M. H., Ahuja, N., Kriegman, D.: A Survey on Face Detection Methods. IEEE Trans. on Pattern Analysis and Machine Intelligence (to appear 2002)
2. Sobottka, K., Pitas, I.: Looking for Faces and Facial Features in Color Images. PRIA: Advances in Mathematical Theory and Applications, Vol. 7, No. 1 (1997)
3. García-Mateos, G., Vicente-Chicote, C.: Face Detection on Still Images Using HIT Maps. Third International Conference on AVBPA'2000, Halmstad, Sweden, June 6-8, (2001)

4. Yang, J., Stiefelhagen, R., Meier, U., Waibel, A.: Real-time Face and Facial Feature Tracking and Applications. In Proc. of AVSP'98, pages 79-84, Terrigal, Australia (1998)
5. Terrillon, J. S., Akamatsu, S.: Comparative Performance of Different Chrominance Spaces for Color Segmentation and Detection of Human Faces in Complex Scene Images. Vision Interface '99, Trois-Rivieres, Canada, pp.1821, (1999)
6. Gong, S., McKenna, S. J., Psarrou, A.: Dynamic Vision, From Images to Face Recognition. Ed. Imperial College Press (2000)
7. Rowley, H. A., Baluja, S., Kanade, T.: Neural Network-Based Face Detection. IEEE Trans. on Pattern Analysis and Machine Intelligence, Vol. 20, No. 1, pp. 23-38 (January 1998)
8. Moghaddam, B., Pentland, A.: Probabilistic Visual Learning for Object Detection. International Conference on Computer Vision, Cambridge, MA (1995)

Illumination Normalized Face Image for Face Recognition

Jaepil Ko, Eunju Kim, and Heyran Byun

Dept. of Computer Science, Yonsei Univ.
134, Shinchon-dong Sudaemoon-ku, Seoul, 120-749, Korea
{nonezero,outframe,hrbyun}@csai.yonsei.c.kr

Abstract. A small change in illumination produces large changes in appearance of face even when viewed in fixed pose. It makes face recognition more difficult to handle. To deal with this problem, we introduce a simple and practical method based on the multiple regression model, we call it ICR (Illumination Compensation based on the Multiple Regression Model). We can get the illumination-normalized image of an input image by ICR. To show the improvement of recognition performance with ICR, we applied ICR as a preprocessing step. We achieved better result with the method in preprocessing point of view when we used a popular technique, PCA, on a public database and our database.

1 Introduction

The visual recognition system suffers from the different appearances of objects according to the illumination conditions [1]. Especially face images are highly sensitive to the variations in illumination conditions so small change in illumination produces large changes in appearance of face [2]. That makes face recognition/verification problem more difficult to handle. The FERRET test report shows that the performance significantly drops in the case of the illumination changes [3,4]. Until now many face recognition methods have been proposed and there are several methods for dealing with illumination problem. For details, reader should consult recent survey paper [5].

The first approach to handle the effects results from illumination changes is constructing illumination model from several images acquired under different illumination condition [6]. The representative method, the illumination cone model that can deal shadow and multiple lighting sources, is introduced by [7,8]. This approach is not practical in smart card application, which can just memorize one or two presentations (prototypes) of a person and to construct the cone model for a person it needs well-controlled image capturing circumstances. The standard answer for the problem with variable lightings, the second approach, is to extract illumination invariant features, such as edges, corner, and contour, which is often considered as the basic image representation but these are insufficient to contain useful information for recognition. Furthermore, edges are susceptible to the illumination conditions for complex object

T. Caelli et al. (Eds.): SSPR&SPR 2002, LNCS 2396, pp. 654-661, 2002.
© Springer-Verlag Berlin Heidelberg 2002

and when the image has cluttered background. Instead of edge-based description, image-based description is preferred in face recognition system. The method use low-dimensional representation of image by subspace technique such as Eighenfaces and Fisherfaces [9]. In the above case, with the assumption that the first few principal components are strongly involved in illumination effects, discarding the first three principal components improves recognition performance under illumination changes [10]. However, the performance is not improved on images captured under normal illumination. Because in the case of normal lighted images discarding the first three components could also eliminate important information for recognition. Another eigenspace method was developed by [11]. The major idea was to incorporate a set of gradient based filter banks into the eigenspace recognition framework. It might strongly depend on gradient operator being charge of illumination variations. Without loosing important information in an image itself, the SSFS (symmetric shape-from-shading) algorithm as a tool to obtain illumination-normalized prototype image, which is based on shape-from-shading assuming just one image is available was proposed by [12].

In this paper, we describe more simple and practical algorithm for getting illumination-compensated face image by applying the illumination compensation method based on the multiple regression model for finding the best-fit intensity plane.

This paper is organized as follows. In the next section, we overview multiple regression model in brief. Section 3, we describe simple illumination compensation algorithm, we call it ICR (Illumination Compensation based on the multiple Regression model), for face image. The experimental results are shown in section 4. Finally, in section 5 conclusions are drawn.

2 Multiple Regression Model

In this section, we will give a brief overview of MRM (multiple regression model) well known technique in statistics. For details of MRM see the book [13]. MRM is the linear regression model for multivariate case.

The multiple regression model can be written as

$$\mathbf{Y} = \mathbf{X}\boldsymbol{\beta} + \mathbf{e} \tag{1}$$

where $\mathbf{Y}$ is an $n \times 1$ response vector, $\mathbf{X}$ is an $n \times (k+1)$ matrix for an k input variables and n samples, and $\mathbf{e}$ is an $n \times 1$ random error vector that we shall assume is normally distributed with mean 0 and variance σ^2.

The parameters $\boldsymbol{\beta}$ and σ^2 must be estimated from the samples. If we let $\mathbf{B}^T = [B_0 B_1 \cdots B_k]$ of the least square estimator of the $\boldsymbol{\beta}$. Then the least square estimator $\mathbf{B}$ are given by

$$\mathbf{B} = (\mathbf{X}^T\mathbf{X})^{-1}\mathbf{X}^T\mathbf{Y} \tag{2}$$

In finding the best-fit intensity plane, input variable is the coordinate of each pixel and response is the intensity value of the location and the number of samples are number of pixels of the image. After estimating parameters, we can get a new intensity value of the location. That is the best-fit intensity plane of the image. The next section will give you for details.

3 Illumination Compensation Based on the Multiple Regression Model (ICR)

We try to get illumination- compensated face image by itself, which is similar to the stored prototype image when the input face image is captured under different illumination conditions without any illumination model. We assume that small set of face images is available which is acquired under a single ambient light. That is practical assumption. Because in office circumstance we can easily imagine that the window sided face image is brighter than that of the other side.

To get illumination- compensated face image, we first find the best-fit intensity plane of an input image. The best-fit intensity plane can be found by the multiple regression model described in section 2. We start with a face image whose dimension is *q* from *n* x *m* face image pixels.

$$\mathbf{x} = [x_0, x_1, \cdots, x_{q-1}]^T \tag{3}$$

then, we generate *q* samples for the regression model.

$$\mathbf{z}_k = [i, j, x_k]^T, \ k = i \times m + j$$
$$i = 0,1, \cdots, n-1, j = 0,1, \cdots, m-1 \tag{4}$$

where the *i* and *j* are input values and x_k is response value for the regression model. After applying samples $\mathbf{z}_k$ to the regression, we can get the best-fit intensity plane:

$$\mathbf{y} = [y_0, y_1, \cdots, y_{q-1}]^T \tag{5}$$

The center value in the best-fit intensity plane $y_c = [\max(y_i) - \min(y_j)] / 2$, $i<j<q$, is computed. The next step is that we reverse the intensity plane against its center value:

$$\mathbf{y}' = [y_c - y_0, y_c - y_1, \cdots, y_c - y_{q-1}]^T \tag{6}$$

Finally, we can get illumination- compensated face image by adding the original input image $\mathbf{x}$ and the adjusted intensity image $\mathbf{y}'$.

The Fig. 1 shows an original input image, the best-fit intensity plane, adjusted intensity plane, and illumination- compensated image by applying our method. The final image appears a little bit uniform intensity preserving its relative intensity.
To compare the existing preprocessing techniques such as gamma correction and histogram equalization are shown in the Fig. 2.

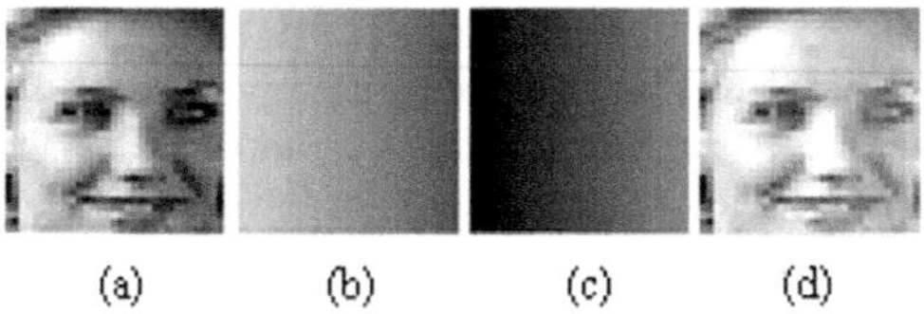

Fig. 1. (a) input image, (b) the best-fit intensity plane, (c) adjusted intensity plane, (d) final result from adding (a) and (c)

Segmenting an object in a gradient image is very difficult problem in threshold techniques. The box object in the image can be easily segmented by a single threshold in the illumination-compensated image. In the case of gamma correction, the gradient of the image was not changed when the global intensity became bright. The image with histogram equalization, it made worse segment the box object by striking the gradient of the input image.

Fig. 2. (a) original input image, (b) the best-fit intensity plane of (a), (c) adjusted the best-fit intensity plane, (d) illumination-compensated image of (a), (e) image with gamma correction, (f) histogram equalized image of (a)

4 Experimental Results

In this section, we show that the performance of ICR in preprocessing point of view by comparing the histogram equalization widely used for normalizing illumination effect in face recognition system. We also compared the standard technique that drops the first three components in PCA for mitigating illumination effect [10].

All the face images were preprocessed to normalize geometry and to remove background. The preprocessing procedure is that we first manually located the centers of the eyes for translating, rotating and scaling the faces, and then applied the histogram equalization or ICR. The next step is to scale facial pixels to have zero mean and unit variance that is required for PCA inputs to improve the performance. In the last step, we masked the face image to remove the background and hair.

We have applied ICR to both the Cambridge ORL face database and our database to demonstrate the improvement of face recognition performance. Our database contains 200 frontal face images for 200 individuals captured with single light source. PCA was performed on all the images on each database and made a probe set by flipping the face image by turning right-side left. In our experiments, we applied the nearest neighbor classifier.

Four types of preprocessed face image whose size is 32 x 32 are shown in Fig. 3. The right part of the face image is darker than that of left side due to light source on the left hand. The difference between the both sides gets tight in (b) by histogram equalization. However, the gap between them is loosed by ICR in (c). We expected that this effect would improve recognition performance.

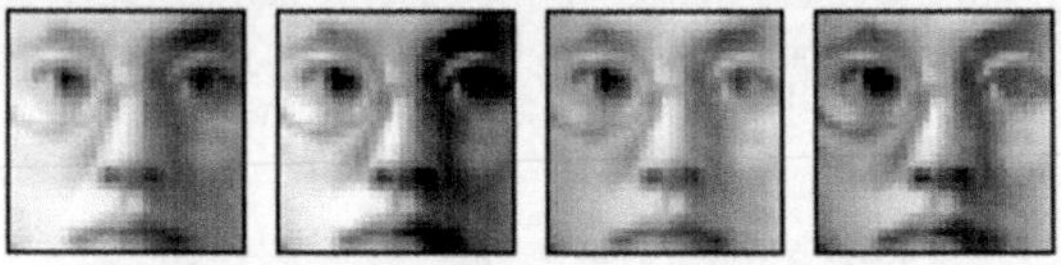

(a) (b) (c) (d)

Fig. 3. (a) Original image, (b) histogram equalized image, (c) ICR processed image, (d) histogram equalization followed by ICR

First, we experimented the effect of discarding the first few components in PCA on the ORL database. In the ORL database, in the case of applying ICR the recognition performance was better when it included the first component. However, Original and histogram equalized image showed better results without the first component.According to the result, we could carefully say that ICR efficiently removes illumination-effect contained in the first few principle components. Table 1 shows the results.

Table 1. The recognition performance on the ORL database with rank 1, original (Org), histogram equalization (HE), ICR, and histogram equalization followed by ICR (HE&ICR). The row was tested without the first n principal components

	Org	HE	ICR	HE & ICR
w/o 0	0.63	0.49	**0.80**	**0.74**
w/o 1	**0.66**	**0.77**	0.76	0.63
w/o 2	0.65	0.75	0.69	0.56
w/o 3	0.53	0.69	0.65	0.46
w/o 4	0.47	0.65	0.60	0.36

Second, we collected the best results by changing the number of principal components and listed in Table 2, 3 on the ORL database and on our database respectively.

We had the same result on both the ORL database and our database. The proposed technique, ICR, showed the best result among them but we could not achieve the same result on the subset including face images just related to the light conditions in the Yale database. On the Yale database, the HE&ICR showed better result. It means that histogram equalization is still useful when the illumination situation is worse than normal case. In addition, on the Yale database the performance of the techniques with

ICR is remarkable comparing to the others without ICR. Fig. 4 shows the results. We tested rank 1 success rate as removing the first eigenvectors. ICR, ICR followed by HE and HE followed by ICR techniques applying our proposed method show better result than ORG and HE techniques. In the case of ORG and HE, they achieved best result when it removed the first two or three eigenvectors respectively. The techniques applying both ICR and HE show the best results when they used the first eigenvector. It means that our proposed method is alleviating illumination effects like eigen-technique discarding the first n eigenvectors. Moreover, it does not remove any important information for recognition.

Table 2. The recognition performance on the ORL database from rank 1 to rank 10. For each rank, the first row is success rate and the second row is the success number of faces among 400 faces

	Org	HE	ICR	HE&ICR
rank1	0.66	0.77	**0.80**	0.74
	265	307	318	294
rank2	0.73	0.83	**0.85**	0.80
	292	332	340	320
rank3	0.78	0.85	**0.87**	0.85
	310	341	348	338
rank4	0.80	0.87	**0.89**	0.87
	321	349	355	348
rank5	0.82	0.88	**0.90**	0.88
	327	352	360	353
rank10	0.87	0.92	**0.95**	0.93
	347	368	381	372

Table 3. The recognition performance on our database from rank 1 to rank 10. For each rank, the first row is success rate and the second row is the success number of faces among 200 faces

	Org	HE	ICR	HE & ICR
rank1	0.47	0.60	**0.76**	0.62
	93	119	151	123
rank2	0.55	0.65	**0.85**	0.72
	109	130	170	144
rank3	0.57	0.71	**0.90**	0.77
	113	141	179	153
rank4	0.60	0.74	**0.91**	0.81
	119	147	182	162
rank5	0.64	0.76	**0.92**	0.84
	128	152	184	167
rank10	0.72	0.82	**0.94**	0.87
	143	163	188	174

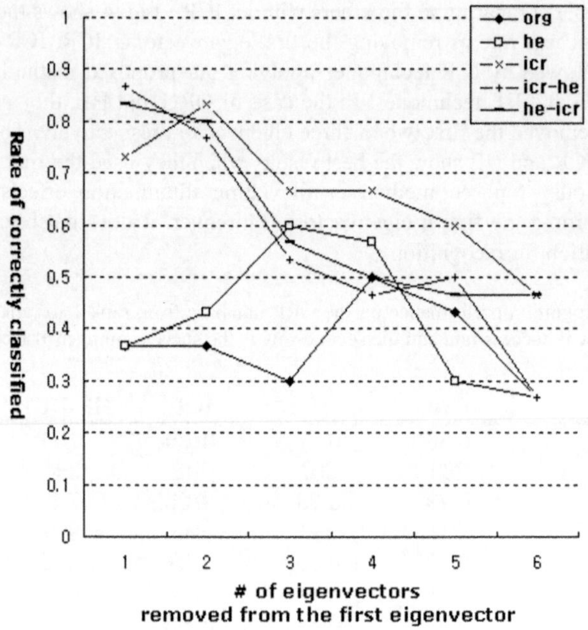

Fig. 4. Comparison of success rate according to the preprocessing technique. The techniques applying our proposed method outperform when they use the first eigenvector

5 Conclusions

Face images are highly sensitive to the variations in illumination conditions. We described a simple and practical method, ICR, to get an illumination- compensated image. The method does not need to construct illumination model from several images, also it does not remove any important information for recognition by discarding the first *n* principal components.

We showed experimentally that the possibility of the normalized face image preprocessed by ICR could be used to alleviate illumination effect in face recognition system.

We think that the ICR can be one of the preprocessing techniques in face recognition system for its simplicity and effectiveness of dealing with illumination effect. However, we assumed that the light source is single and ambient. In future work, even thought it is a practical assumption we will extend experimentation by applying ICR locally to handle the situation that several light sources exist.

Acknowledgement

This research was supported as a Brain Neuroinformatics Research Program sponsored by the Korean Ministry of Science and Technology (M1-0107-00-0008).

References

1. Michael J. Tarr, Daniel Kersten, Heinrich H. Bulthoff, : Why the visual recognition system might encode the effects of illumination, Pattern Recognition (1998)
2. Yael Adini, Yael Moses, and Shimon Ullman, : Face Reconition: The problem of Compensating for Changes in Illumination Direction, IEEE Trans. on PAMI Vol. 19, No. 7 (1997) 721-732
3. P. J. Phillips, H. Moon, P. Rauss, and S. A. Rizvi. : The FERET Evaluation Methodology for Face-Recognition Algorithms. IEEE Conference on CVPR, Puerto Rico (1997) 137-143
4. S. Rizvi, P. Phillips, and H. Moon. : The FERET verication testing protocol for face recognition algorithms. IEEE Conference on Automatic Face- and Gesture-Recognition (1998) 48-53
5. R. Chellappa and W. Zhao, : Face Recognition: A Literature Survey. ACM Journal of Computing Surveys (2000)
6. A. Yuille, D. Snow, R. Epstein, P. Belhumeur, : Determining Generative Models of Objects Under Varying Illumination: Shape and Albedo from Multiple Images Using SVD and Integrability, International Journal of Computer Vision, 35 (3)(1999) 203-222
7. P. N. Belhumeur and D. J. Kriegman. : What is the set of images of an object under all possible lighting conditions?, IEEE Conference on CVPR (1996)
8. Athinodoros S. Georghiades, David J. Kriegman, Peter N. Belhumeur, : Illumination Cones for Recognition Under Variable Lighting: Faces, IEEE Conference on CVPR (1998) 52-58
9. M.Turk and A. Pentland, : Eigenfaces for recognition. Journal of Cognitive Neuroscience,Vol 3 (1991)
10. V. Belhumeur, J. Hespanha, and D. Kriegman. : Eigenfaces vs. fisherfaces: Recognition using class specific linear projection. IEEE Trans. on PAMI (1997) 711-720
11. Bischof, H.; Wildenauer, H.; Leonardis, A.: Illumination insensitive eigenspaces, IEEE Conference on Computer Vision, Vol. 1 (2001) 233-238
12. Wen Yi Zhao; Chellappa, R. : Illumination-Insensitive Face Recognition Using Symmetric Shape-from-Shading, IEEE Conference on CVPR, Vol. 1, (2000) 286-293
13. S. M. Ross, : Introduction to Probability and Statistics for Engineers and Scientists, Wiley, New York (1987)

Towards a Generalized Eigenspace-Based Face Recognition Framework

Javier Ruiz del Solar and Pablo Navarrete

Department of Electrical Engineering
Universidad de Chile
{jruizd,pnavarre}@cec.uchile.cl

Abstract. Eigenspace-based approaches (differential and standard) have shown to be efficient in order to deal with the problem of face recognition. Although differential approaches have a better performance, their computational complexity represents a serious drawback. To overcome that, a *post-differential approach*, which uses differences between reduced face vectors, is here proposed. The mentioned approaches are compared using the Yale and FERET databases. Finally, a generalized framework is also proposed.

1. Introduction

Face Recognition is a highly dimensional pattern recognition problem. Even low-resolution face images generate huge dimensional spaces (20,000 dimensions in the case of a 100x200 pixels face image). In addition to the problems of large computational complexity and memory storage, this high dimensionality makes very difficult to obtain statistical models of the input space using well-defined parametric models. However, the intrinsic dimensionality of the face space is much lower than the dimensionality of the image space, since faces are all similar in appearance and posses significant statistical regularities. This fact is the starting point of the use of eigenspace methods to reduce the dimensionality of the input face space, which is the subject to be studied in this paper.

The main task of face recognition is the identification of a given face image among all the faces stored in a database. The result of identification corresponds to the subject that shows the most similar features to the ones of the requested face image. In this way, it is necessary to define a similarity function $S(\mathbf{x}, \mathbf{y})$ to quantify the likeness between two face feature vectors $\mathbf{x}$ and $\mathbf{y}$. Furthermore, in most face image databases exist only a few number of images per subject (usually one) and then there is not a reliable statistical knowledge about each subject or class. Thus, the immediate problem consists in finding the nearest-neighbor of the requested face image among all the face images in the database, working in a metric space defined by the similarity function. Two approaches can be utilized to implement that: (i) feature matching, in which local features (eyes, mouth, etc.) are used to compute the feature vector, and (ii) template matching, in which the whole face image is considered as the feature

T. Caelli et al. (Eds.): SSPR&SPR 2002, LNCS 2396, pp. 662-671, 2002.
© Springer-Verlag Berlin Heidelberg 2002

vector, considering each pixel value as a vector component. This last approach works but is not efficient. The huge and redundant number of components demands the application of dimensional reduction methods, leading to the eigenspace-based approaches in which we are focused.

Standard eigenspace-based approaches project input faces onto a dimensional reduced space where the recognition is carried out. In 1987 Sirovich and Kirby used Principal Component Analysis (PCA) in order to obtain a reduced representation of face images [9]. Then, in 1991 Turk and Pentland used the PCA projections as the feature vectors to solve the problem of face recognition, using the euclidean distance as the similarity function [10]. This system was the first eigenspace-based face recognition approach and, from then on, many eigenspace-based systems have been proposed using different projection methods and similarity functions. A differential eigenspace-based approach was proposed in 1997 by Pentland and Moghaddam [4], and it allows the application of statistical analysis in the recognition process. The main idea is to work with differences between face images, rather than with face images. In this way the recognition problem becomes a two-class problem, because the so-called "differential image" contains information of whether the two subtracted images are of the same class or different classes. In this case the number of training images per class increases so that statistical information becomes available. The system proposed in [4] used Dual-PCA projections and a Bayesian classifier. Following the same approach, a system using single PCA projections and a Support Vector Machine (SVM) classifier is here outlined.

In several comparisons that we have done between these two different approaches, we have realized that the "differential" approaches work better than the standard ones. However, in the differential case all the face images need to be stored in the database, which slow down the recognition process. This is a serious drawback in practical implementations. To overcome this drawback a so-called post-differential approach is here proposed. Under this new approach, differences between reduced face vectors are used instead of differences between face images. This allows a decreasing of the computations and storage required (only reduced face vectors are stored in the database), without losing the recognition performance of the differential approaches.

This paper is structured as follows. In section 2 the mentioned eigenspace-based recognition approaches are described. In this section is also proposed a generic framework in which the standard and the differential approaches can be included. In section 3 are presented some simulation results of recognition using the Yale and FERET databases, which allows to compare the different approaches. Finally, some conclusions of this work are given in section 4.

2. Eigenspace-Based Face Recognition

2.1 Standard Eigenspace Approaches

Fig. 1 shows the block diagram of a generic, standard eigenspace-based face recognition system. Standard eigenspace-based approaches approximate the face vectors (face images) by lower dimensional feature vectors. The main supposition behind this procedure is that the face space has a lower dimension than the image space, and that the recognition of the faces can be performed in this reduced space. These approaches consider an off-line phase or training, where the *projection matrix*, the one that achieve the dimensional reduction, is obtained using all the database face images. In the off-line phase are also calculated the *mean face* and the reduced representation of each database image. These representations are the ones to be used in the recognition process. Among the projection methods employed for the reduction of dimensionality, we can mention: PCA [10], Linear Discriminant Analysis (LDA) [2], and Evolutionary Pursuit (EP) [3]. Among the similarity matching criteria employed for the recognition process, they have been used: Euclidean-, Cosine- and Mahalanobis-distance, Self-Organizing Map (SOM) clustering, and Fuzzy Feature Contrast (FFC) similarity (see definitions in [6]). All this methods have been analyzed and compared in [6]. Under this standard eigenspace approach a *Rejection System* for unknown faces is implemented by placing a threshold over the similarity measure (see Fig. 1).

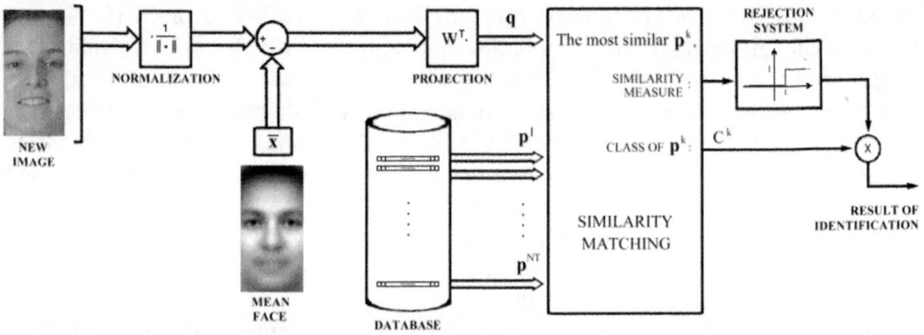

Fig. 1. Block diagram of a generic, standard eigenspace-based face recognition system

2.2 Differential Eigenspace Approaches

Fig. 2 shows the block diagram of a generic, differential eigenspace-based face recognition system. In this approach the whole face images are stored in the database. Previously the face images are centered and scaled so that they are aligned. An input face image is normalized and subtracted from each database image. The result of each subtraction is called "differential image" $\mathbf{\Delta}$ in R^N and it is the key for identification. That because it contains information of whether the two subtracted images are of the same class or different classes. In this way the original problem of *NC* classes becomes a two-class problem. The so-called differential images are projected into a reduced space using a given projection method. Thus, each image is transformed into a reduced differential vector $\boldsymbol{\delta}$ in R^m. Thereafter the classification of the reduced differential vectors is performed. The result of each classification (S_i) is negative if the subtracted images (each $\boldsymbol{\delta}$) are of different classes and positive in other case. In order to determine the class of the input face image, the reduced vector with maximum classification value is chosen, and the class of its initial database image is given as the result of identification. The rejection system acts just when the maximum classification value is negative, i.e. it corresponds to the subtraction of different classes. Dual-PCA and Single-PCA projections have been used as projection methods. The Dual-PCA projections employ two projection matrices: $\mathbf{W}_{\mathrm{I}} \in R^{N \times m_{\mathrm{I}}}$ for intra-classes $\mathbf{\Omega}_{\mathrm{I}}$ (subtractions within equal classes), and $\mathbf{W}_{\mathrm{E}} \in R^{N \times m_{\mathrm{E}}}$ for extra-classes $\mathbf{\Omega}_{\mathrm{E}}$ (subtractions between different classes). Dual-PCA projections are employed together with a Bayesian classifier in order to perform the classification of the differential images [4]. Single-PCA projection employs a single projection matrix $\mathbf{W} \in R^{N \times m}$ (standard PCA) that reduces the dimension of the differential face images, and it is used together with a SVM classifier in order to perform the classification.

2.3 Post-Differential Eigenspace Approaches

Fig. 3 shows the block diagram of the generic, post-differential eigenspace-based face recognition system here proposed. In this approach only the reduced face images are stored in the database. Input face images are normalized and then projected into a reduced space using a given projection method (we consider only Single-PCA projections). Thereafter, the new reduced face image is subtracted from each database reduced face image. The result of each subtraction is called "post-differential image" $\boldsymbol{\delta}$ in R^m. This vector contains information of whether the two subtracted vectors are of the same class or different classes, and then it works in the same way as the "differential images" once projected on the reduced space. The classification module performs the classification of the post-differential vectors. The class of the reduced database vector that has the maximum classification value gives the class of the initial input face image. If the projection module does not significantly change the topology of the differential-image space, then the differential and post-differential approaches should have very similar recognition rates. The rejection system acts just when the maximum classification value is negative, i.e. it corresponds to the subtraction of different classes. We have implemented two different systems that follow this approach, and they are described in the following subsections.

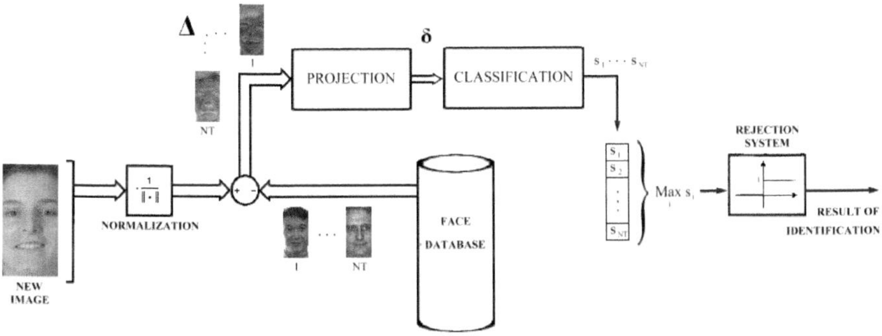

Fig. 2. Block diagram of a generic, differential eigenspace face recognition system

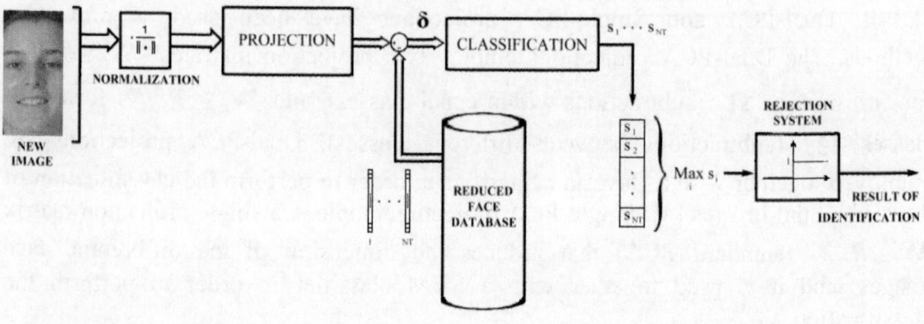

Fig. 3. Block diagram of a generic, post-differential eigenspace face recognition system

SVM Classification

SVM in its simplest form, linear and separable case, is defined as the hyperplane that separates the vector sets belonging to different classes with the maximum distance to its closest samples, called support vectors. The problem is solved using a particular Lagrange formulation in which the problem is reduced to the computation of Lagrange multipliers. SVM in its general form, non-linear and non-separable, is very similar to its simplest form. Non-separable cases are considered by adding an upper bound to the Lagrange multipliers [7], and non-linear cases are considered by replacing all the dot products like $\mathbf{x}\cdot\mathbf{y}$, by a so-called kernel function $\mathrm{K}(\mathbf{x},\mathbf{y})$. As a classification system we are using SVM over the differential reduced training vectors $\boldsymbol{\delta}\in R^m$. Thus, the system to be solved corresponds to the following [1]:

$$\underset{\alpha_i}{\mathrm{Max}}\ \mathrm{L_D}(\alpha_i) = -\tfrac{1}{2}\sum_{i=1}^{NT}\sum_{j=1}^{NT} y_i y_j \alpha_i \alpha_j \mathrm{K}(\boldsymbol{\delta}_i,\boldsymbol{\delta}_j) + \sum_{i=1}^{NT}\alpha_i \tag{1}$$

$$\text{subject to :} \qquad 0 \le \alpha_i \le C \quad i = 1\ldots\mathrm{NT}\,;\ \sum_{i=1}^{NT} y_i \alpha_i = 0 \,.$$

Then, classification rule will be $S(\boldsymbol{\delta}) = \sum \alpha_i\, y_i\, \mathrm{K}(\boldsymbol{\delta}_i,\boldsymbol{\delta}) + b$, in which the parameter b is given by the following expression:

$$b = y_k - \sum \alpha_i\, y_i\, \mathrm{K}(\boldsymbol{\delta}_i,\boldsymbol{\delta}_k)$$

$$\text{for some } k \text{ so that } \alpha_k > 0 \quad (\boldsymbol{\delta}_k : \text{support vector}). \tag{2}$$

Bayes Classification

If we suppose a normal distributed pattern for both $\boldsymbol{\delta}_\mathrm{I} \in R^m$ ($\boldsymbol{\delta}\in \boldsymbol{\Omega}_\mathrm{I}$, intra-class) and $\boldsymbol{\delta}_\mathrm{E} \in R^m$ ($\boldsymbol{\delta}\in \boldsymbol{\Omega}_\mathrm{E}$, extra-class), then the likelihood of a given $\boldsymbol{\delta}$ will be [5]:

$$P(\boldsymbol{\delta}/\boldsymbol{\Omega}) = \frac{\exp\left[-\frac{1}{2}(\boldsymbol{\delta}-\bar{\boldsymbol{\delta}})^{\mathrm{T}}\,\mathbf{R}^{-1}\,(\boldsymbol{\delta}-\bar{\boldsymbol{\delta}})\right]}{(2\pi)^{m/2}\,\left|\mathbf{R}\right|^{1/2}}\,, \tag{3}$$

with $\bar{\boldsymbol{\delta}}$ the mean differential image and $\mathbf{R}$ the correlation matrix, for a given set $\boldsymbol{\Omega}$ ($\boldsymbol{\Omega}_{\mathrm{I}}$ or $\boldsymbol{\Omega}_{\mathrm{E}}$). Thus we can compute the likelihood $P(\boldsymbol{\delta}/\boldsymbol{\Omega}_{\mathrm{I}})$ and $P(\boldsymbol{\delta}/\boldsymbol{\Omega}_{\mathrm{E}})$ in order to obtain the *a posteriori* probability using the Bayes rule:

$$P(\boldsymbol{\delta} \in \boldsymbol{\Omega}) = P(\boldsymbol{\Omega}/\boldsymbol{\delta}) = \frac{P(\boldsymbol{\delta}/\boldsymbol{\Omega})\,P(\boldsymbol{\Omega})}{P(\boldsymbol{\delta}/\boldsymbol{\Omega}_{\mathrm{I}})\,P(\boldsymbol{\Omega}_{\mathrm{I}}) + P(\boldsymbol{\delta}/\boldsymbol{\Omega}_{\mathrm{E}})\,P(\boldsymbol{\Omega}_{\mathrm{E}})} \tag{4}$$

Therefore a given $\boldsymbol{\delta}$ would be classified as an intra-class vector if $P(\boldsymbol{\Omega}_{\mathrm{I}}/\boldsymbol{\delta}) - P(\boldsymbol{\Omega}_{\mathrm{E}}/\boldsymbol{\delta}) > 0$. Using expression (4) the decision rule yields:

$$S(\boldsymbol{\delta}) = P(\boldsymbol{\delta}/\boldsymbol{\Omega}_{\mathrm{I}})\,P(\boldsymbol{\Omega}_{\mathrm{I}}) - P(\boldsymbol{\delta}/\boldsymbol{\Omega}_{\mathrm{E}})\,P(\boldsymbol{\Omega}_{\mathrm{E}})\,, \tag{5}$$

and for numerical stability the logarithm of this decision rule is computed.

2.4 Generalized Eigenspace Framework

The approaches previously presented can be thought as independent eigenspace-based systems. Nevertheless, the foundations of each approach are based on similar principles. Fig. 4 shows a generalized eigenspace-based face recognition framework from which all the previous approaches presented could be derived. The standard eigenspace approach is formed when the switches in Fig. 4 are set in the positions 1 and 3; the differential eigenspace approach is formed when the switches are set in the positions 2, 4 and 5; and the post-differential eigenspace approach is formed when the switches are set in the positions 1, 4 and 6. The main idea is that all the eigenspace approaches use a projection module that could work with original or differential images and, when differential approaches are being used, the differences could be computed before or after the dimensional reduction.

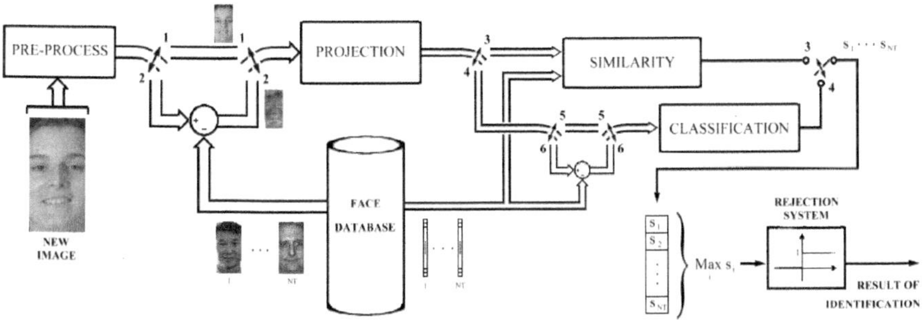

Fig. 4. Block diagram of a generalized eigenspace-based face recognition system

3. Comparison among the Approaches

3.1 Simulations Using the Yale Face Image Database

Tables 1, 2 and 3. Mean recognition rates using the Yale database and different numbers of training images per class, and taking the average of 20 different training sets. The small numbers are standard deviations. All results consider the top 1 match

Table 1. Standard Eigenspace

projection method	images per class	axes	euclidean	cos(·)	SOM	FFC	whitening euclidean	whitening cos(·)	whitening SOM	whitening FFC
PCA		56	87.9 6.2	86.0 6.8	84.6 7.0	77.1 10.1	64.7 9.4	79.3 11.6	64.7 10.5	77.1 10.1
FISHER	6	14	91.5 6.6	91.6 6.5	90.3 6.7	83.9 9.3	91.9 5.8	92.6 5.6	92.1 6.2	85.6 8.3
E.P.		15	81.2 9.0	85.3 8.7	83.7 9.8	77.2 8.0	-	-	-	-
PCA		34	88.7 3.8	87.1 4.1	86.0 5.1	78.5 8.1	69.5 8.9	83.2 9.0	66.1 10.5	78.5 8.1
FISHER	5	14	92.2 5.7	91.7 6.2	90.3 6.4	85.1 9.1	92.3 4.7	92.4 5.7	92.1 5.3	85.4 8.5
E.P.		13	84.1 5.7	87.7 6.6	86.7 7.6	78.7 6.8	-	-	-	-
PCA		46	87.3 3.9	86.7 3.9	84.8 3.6	77.6 5.2	72.9 5.5	84.4 5.6	66.7 6.5	77.6 5.2
FISHER	4	14	90.3 4.5	91.1 5.0	90.3 4.4	84.4 5.9	90.4 4.2	91.0 4.4	90.1 4.7	82.9 5.7
E.P.		18	83.6 4.6	86.9 4.7	85.0 5.0	74.7 6.0	-	-	-	-
PCA		35	86.6 4.0	85.4 3.9	82.0 5.6	77.9 4.6	75.0 5.6	84.8 5.4	67.4 6.9	77.9 4.6
FISHER	3	14	89.0 3.6	90.4 4.0	87.4 4.0	80.7 6.3	88.9 3.1	89.9 3.9	88.7 3.9	81.5 3.4
E.P.		14	81.1 4.3	86.9 3.7	82.5 3.7	75.9 4.4	-	-	-	-
PCA		26	82.7 5.9	80.8 5.9	76.2 7.9	71.1 5.9	75.6 4.9	82.1 4.6	60.8 7.3	71.1 5.9
FISHER	2	14	81.5 5.6	82.2 5.8	79.4 5.8	69.3 8.6	80.7 4.7	82.8 4.9	78.8 5.8	73.6 6.2
E.P.		14	77.8 5.6	81.2 5.3	76.0 7.3	70.0 7.4	-	-	-	

In order to compare the described approaches we have first made several simulations using the Yale University - Face Image Database [11]. We used 150 images of 15 different classes. First, we preprocessed the images manually by masking them in windows of 100 x 200 pixels and centering the eyes in the same relative places. In table 1 we show the results of several simulations for standard approaches using different kind of representations and similarity matching methods. For each simulation we used a fixed number of training images, using the same type of images per class, according with the Yale database specification. In order to obtain representative results we take the average of 20 different sets of images for each fixed number of training images. All the images not used for training are used for testing. In tables 2 and 3 we show the results of several simulations using differential approaches. We used equal *a priori* probabilities for the Bayes-based methods, $P(\Omega_I) = P(\Omega_E)$, and a penalty for non-separable cases $C = 0.01$ in the SVM classification method. The number of axes obtained with single-PCA was slightly smaller than the one obtained with standard PCA (shown in table 1). On the other hand, the number of axes obtained with dual-PCA was about the same for intra-class and extra-class images, and smaller than the number obtained with standard PCA. As it can be seen in these simulations, the differential and post-differential approaches show a slightly better performance, which increases when a low number of training images per class (2) is used. That shows that both approaches have a better generalization ability than the standard one.

Table 2. Differential Eigenspace.

images per class	Dual PCA Bayes	SVM
6	93.5 6.1	94.1 4.1
5	93.3 5.7	92.5 5.3
4	90.9 3.5	91.3 4.5
3	90.0 3.7	89.6 5.7
2	84.7 5.1	86.9 5.9

Table 3. Post-differential Eigenspace.

images per class	Bayes	SVM
6	91.6 6.2	93.5 4.8
5	90.9 6.5	92.1 4.5
4	89.8 5.1	90.2 4.1
3	88.3 4.8	89.5 6.5
2	87.5 5.5	87.0 5.8

3.2 Simulations Using FERET

In order to test the described approach using a large database, we made simulations using the FERET database [8]. We use a target set with 762 images of 254 different classes (3 images per class), and a query set of 254 images (1 image per class). Eyes' location is included in FERET database for all the images being used. Then the preprocessing consists in centering and scaling images so that eyes' position keeps in the same relative place. In table 4 we show the results of simulations for standard approaches using different kind of representations and similarity matching methods. In this table the SOM-based clustering was not included because in these tests the number of classes (254) is much larger than the number of images per class (3), and the training process is very difficult. In tables 4 and 5 we show the results of simulations using differential and post-differential approaches. The recognition rates of both approaches are better than almost all results using standard approaches, with the exception of the FLD-cosine and EP-cosine when 3 images per class were used for training. It must be noted that, when 2 images per class were used for training, the differential and post-differential approaches work better than all the standard ones. This fact shows again that differential approaches have a better generalization ability.

Tables 4, 5 and 6. Mean recognition rates for standard approaches using FERET. All results consider the top 1 match for recognition

Table 4. Standard Eigenspace

projection method	images per class	axes	euclidean	cos(·)	FFC	whitening euclidean	whitening cos(·)	whitening FFC
PCA	3	212	87.0	88.6	85.0	74.4	89.4	85.0
FISHER	3	108	91.3	94.1	92.1	85.8	92.1	92.1
E.P.	3	115	91.0	93.1	91.0	-	-	-
PCA	2	180	81.9	83.7	80.7	62.3	86.0	80.7
FISHER	2	73	79.5	88.2	85.2	79.5	88.2	85.2
E.P.	2	96	80.3	85.8		-	-	-

Table 5. Differential Eigenspace

images per class	Axes Bayes	Axes SVM	Bayes	SVM
3	148 (i) / 156 (e)	186	92.6	92.7
2	106 (i) / 128 (e)	124	88.3	90.6

Table 6. Post-differential Eigenspace

images per class	Axes Bayes	Axes SVM	Bayes	SVM
3	158	218	91.3	92.8
2	115	173	88.1	90.6

4. Conclusions

Eigenspace-based approaches have shown to be efficient in order to deal with the problem of face recognition. Although differential approaches have a better performance than the standard ones, their computational complexity represents a serious drawback in practical applications. To overcome that, a *post-differential approach*, which uses differences between reduced face vectors was here proposed.

The three mentioned approaches were compared using the Yale and FERET databases. The simulations results obtained have shown that the two differential approaches have a better generalization ability than the standard one. This is probably because of the classification techniques used in the differential and post-differential approaches take advantage of additional statistical information. This property was decisive when a low number of training images per class (2) was used. In the simulations, the Bayes and SVM implementations do not show a significant difference in their performance, so both could be used for this kind of application. The simulation results show that the here proposed post-differential approach corresponds to a very practical solution in order to obtain a good recognition performance as well as a fast processing speed.

Eigenspace decompositions can be divided in generic (e.g. generic PCA) and specific. In a specific decomposition, the faces which need to be identified are those whose images are used when computing the projection matrix. This is not the case in a generic decomposition. In the here presented simulations only specific decompositions where used. As a future work we want to perform simulations using generic decompositions to go deeper in the comparison of the described approaches. We believe that in this case differential approaches will show an even better generalization ability.

Acknowledgements

Portions of the research in this paper use the *FERET* database of facial images collected under the *FERET* program.
This research was supported by the DID (U. de Chile) under Project ENL-2001/11, and by the join "Program of Scientific Cooperation" of CONICYT (Chile) and BMBF (Germany) under the project "Face Recognition and Signature Verification using Soft-Computing".

References

1. Burges C. J. C., "A tutorial on support vector machines for pattern recognition", *Data Mining and Knowledge Discovery*, 2(2), pp. 121–167, 1998.
2. Fisher R. A., "The Use of Multiple Measures in Taxonomic Problems", *Ann. Eugenics*, vol. 7, pp. 179-188, 1936.
3. Liu C., and Wechsler H., "Evolutionary Pursuit and Its Application to Face Recognition", *IEEE Trans. Pattern Analysis and Machine Intelligence*, vol. 22, no. 6, pp. 570-582, June 2000.
4. Pentland A., and Moghaddam B., "Probabilistic Visual Learning for Object Representation", *IEEE Trans. Pattern Analysis and Machine Intelligence*, vol. 19, no. 7, pp. 696-710, July 1997.
5. Duda R. O., Hart P. E., and Stork D. G., "Pattern Classification", Second Edition, 2001.
6. Navarrete P., and Ruiz-del-Solar J., "Comparative study between different Eigenspace-based approaches for Face Recognition", *Lecture Notes in Artificial Intelligence 2275*, AFSS 2002, Springer, 178 - 184.
7. Cortes C., and Vapnik V., "Support Vector Networks", *Machine Learning*, 20, pp. 273-297, 1995.
8. Phillips P. J., Wechsler H., Huang J., and Rauss P., "The FERET database and evaluation procedure for face recognition algorithms", *Image and Vision Computing J.*, Vol. 16, no. 5, 295-306, 1998.
9. Sirovich L., and Kirby M., "A low-dimensional procedure for the characterization of human faces", *J. Opt. Soc. Amer. A*, vol. 4, no. 3, pp. 519-524, 1987.
10. Turk M., and Pentland A., "Eigenfaces for Recognition", *J. Cognitive Neuroscience*, vol. 3, no. 1, pp. 71-86, 1991.
11. Yale University Face Image Database, publicly available for non-commercial use, http://cvc.yale.edu/projects/yalefaces/yalefaces.html .

Automatic Segmentation of Speech at the Phonetic Level*

Jon Ander Gómez and María José Castro

Departamento de Sistemas Informáticos y Computación
Universidad Politécnica de Valencia, Valencia, Spain
{jon,mcastro}@dsic.upv.es

Abstract. A complete automatic speech segmentation technique has been studied in order to eliminate the need for manually segmented sentences. The goal is to fix the phoneme boundaries using only the speech waveform and the phonetic sequence of the sentences.
The phonetic boundaries are established using a Dynamic Time Warping algorithm that uses the *a posteriori* probabilities of each phonetic unit given the acoustic frame. These *a posteriori* probabilities are calculated by combining the probabilities of acoustic classes which are obtained from a clustering procedure on the feature space and the conditional probabilities of each acoustic class with respect to each phonetic unit.
The usefulness of the approach presented here is that manually segmented data is not needed in order to train acoustic models. The results of the obtained segmentation are similar to those obtained using the HTK toolkit with the "flat-start" option activated. Finally, results using Artificial Neural Networks and manually segmented data are also reported for comparison purposes.

1 Introduction

The automatic segmentation of continuous speech using only the phoneme sequence is an important task, specially if manually pre-segmented sentences are not available for training. The availability of segmented speech databases is useful for many purposes, mainly for the training of phoneme-based speech recognizers [1]. Such an automatic segmentation can be used as the primary input data to train other more powerful systems like those based on Hidden Markov Models (HMMs) or Artificial Neural Networks (ANNs).

In this work, two different Spanish speech databases composed of phonetically balanced sentences were automatically segmented. The phonetic boundaries are established using a Dynamic Time Warping algorithm that uses the *a posteriori* probabilities of each phonetic unit given the acoustic frame. These *a posteriori* probabilities are calculated by combining the probabilities of acoustic classes which are obtained from a clustering procedure on the feature space and the conditional probabilities of each acoustic class with respect to each phonetic unit.

* Work partially supported by the Spanish CICYT under contract TIC2000-1153.

T. Caelli et al. (Eds.): SSPR&SPR 2002, LNCS 2396, pp. 672–680, 2002.
© Springer-Verlag Berlin Heidelberg 2002

2 Description of the System

The core of the approach presented here is the estimation of $P(ph_u|x_t)$, that is, the *a posteriori* probability that the phonetic unit ph_u has been uttered given the feature vector x_t, obtained at every instant of analysis t. When this probability is broken down using the Bayes rule, we obtain:

$$P(ph_u|x_t) = \frac{P(ph_u) \cdot p(x_t|ph_u)}{\sum_{i=1}^{U} P(ph_i) \cdot p(x_t|ph_i)} \tag{1}$$

where U is the number of phonetic units used in the system, and $P(ph_u)$ is the *a priori* probability of ph_u. In this approach, we assume $P(ph_u) = 1/U$ for all units, so it can be removed from expression (1).

Now, we need to calculate $p(x_t|ph_u)$, which is the conditional probability density that x_t appears when ph_u is uttered. To do so, a clustering procedure to find "natural" classes or groups in the subspace of $\mathbb{R}^d$ formed by the feature vectors is done. From now on, we will refer to this subspace as "feature space". Once the clustering stage has been completed, we are able to calculate $P(w_c|x_t)$, that is, the *a posteriori* probability that a class w_c appears given an input feature vector x_t, applying the Bayes rule as follows:

$$P(w_c|x_t) = \frac{P(w_c) \cdot p(x_t|w_c)}{\sum_{i=1}^{C} P(w_i) \cdot p(x_t|w_i)} \tag{2}$$

where C is the number of "natural" classes estimated using the clustering procedure, $P(w_c)$ is the *a priori* probability of the class w_c, and $p(x_t|w_c)$ is the conditional probability density calculated as Gaussian distributions. In this work, we assume $P(w_c) = 1/C$ for all classes.

At this point, the conditional probability densities $p(x_t|ph_u)$ from equation (1) can be estimated from the models learned using the clustering procedure. The "natural" classes make a partition of the feature space which is more precise than the phoneme partition. Since we already have $p(x_t|w_c)$ from the clustering procedure, $p(x_t|ph_u)$ can be approximated as

$$p(x_t|ph_u) \approx \sum_{c=1}^{C} p(x_t|w_c) \cdot P(w_c|ph_u) \tag{3}$$

where $P(w_c|ph_u)$ is the conditional probability that the class w_c is observed when the phonetic unit ph_u has been uttered (see how to obtain these conditional probabilities in section 2.2).

Given that the *a priori* probabilities of the phonetic units $P(ph_u)$ are considered to be equal, we can rewrite equation (1) using (3) as

$$P(ph_u|x_t) = \frac{\sum_{c=1}^{C} p(x_t|w_c) \cdot P(w_c|ph_u)}{\sum_{i=1}^{U} \sum_{c=1}^{C} p(x_t|w_c) \cdot P(w_c|ph_i)}, \tag{4}$$

which is the *a posteriori* probability we were looking for.

2.1 Clustering Procedure

One of the underlying ideas of this work is that we do not know how many different acoustical manifestations can occur for each phoneme from a particular parametrization. The obtained acoustical feature vectors form a subspace of $\mathbb{R}^d$. We assume that this subspace can be modeled with a Gaussian Mixture Model (GMM), where each class or group is identified by its mean and its diagonal covariance matrix. In our case, the *a priori* probabilities of each class or group, $P(w_c)$, are considered to be equal to $1/C$. The unsupervised learning of the means and the diagonal covariances for each class have been done by maximum likelihood estimation as described in [3, chapter 10].

The number of classes C has been fixed after observing the evolution of some measures which compare the manual segmentation with the automatic one (see section 3 and Figure 1). Once the number of classes is fixed and the parameters which define the GMM are learned, we can calculate the conditional probability densities $p(x_t|w_c)$. Then, the probabilities $P(w_c|x_t)$ are obtained as shown in equation (2).

2.2 Coarse Segmentation and Primary Estimation of the Conditional Probabilities

We need a segmentation of each sentence for the initial estimation of the conditional probabilities $P(w_c|ph_u)$. This first coarse segmentation has been achieved by applying a set of acoustic-phonetic rules knowing only the phonetic sequence of the utterance. The phonetic sequence of each sentence is automatically obtained from the orthographic transcription using a grapheme-to-phoneme converter [4].

The coarse segmentation used at this stage is done by:

1. Searching for relative maxima and minima over the speech signal based on the energy.
2. Associating maxima with vowel or fricative units and minima with possible silences.
3. Estimating the boundaries between each unit by simple spectral distances.

Searching for relative maxima and minima over the speech signal based on the energy. The location of relative maxima is restricted to those instants t where the energy is greater or equal to the energy at the interval of ± 30 ms. around t. Each maximum is considered to be more or less important depending on whether its energy is greater or smaller than a threshold for maxima calculated specifically for each sentence. The importance of a maximum is used to properly weight its deletion. The location of relative minima is done by searching for intervals where the energy is under a threshold for minima, which is also calculated for each sentence. After this step, we have a list of maxima and minima $(m_1, m_2, \ldots, m_{|m|})$ for the sentence.

Association of maxima with vowel or fricative units and minima with possible silences. The association of phonetic units (vowels or fricative consonants) is performed by a Dynamic Time Warping (DTW) algorithm that aligns the list of maxima and minima $(m_1, m_2, \ldots, m_{|m|})$ with the phonetic sequence $p_1 p_2 \ldots p_{|p|}$. The DTW algorithm uses the following set of productions:

- $\{(i-1, j-1), (i, j)\}$: Location of the phonetic unit p_j around the instant which m_i occurs. If m_i is a maximum, p_j is a vowel or a fricative consonant; if it is a minimum, p_j is a silence.
- $\{(i, j-1), (i, j)\}$: Insertion of phonetic unit p_j, which is not associated with any maximum or minimum.
- $\{(i-1, j), (i, j)\}$: Deletion of maximum or minimum m_i.
- $\{(i-1, j-\delta), (i, j)\}$, with $\delta \in [2..5]$: To align consecutive vowels (such as diphthongs).

Each production is properly weighted; for instance, the weight of the insertion of possible silences between words is much lower than the weight of the insertion of a vowel. In the case of several continuous vowels, the association of this subsequence with a maximum is also allowed.

The association of vowels and fricative consonants with a maximum is weighted using a measure which is related to the first MFCC (CC1). Fricative consonants, when associated with a maximum, have a cost which is calculated by differentiation of the CC1 with a threshold for fricatives, which is estimated for each sentence. This differentiation is also used for the vowel "i", and inverted in the case of the vowels "a", "o" and "u".

Estimation of the boundaries between each phonetic unit by simple spectral distances. After the association is done, we have some phonetic units (vowels and fricative consonants) located around the instant where its associated event (maximum or minimum) was detected. For instance, we could have the following situation:

$$\begin{array}{cccccc} \ldots & m_i & & & m_{i+1} & \ldots \\ \ldots & \mathrm{a} & \mathrm{r} & \mathrm{d} & \mathrm{o} & \ldots \end{array}$$

We then take subsequences of units to locate the boundaries. These subsequences are formed by two units which are associated with an event (in the example, "a" and "o") and the units between them ("r" and "d" in the example). The boundaries are located by searching for relative maxima of spectral distances. The interval used to locate the boundaries begins at the position where the event m_i is located, and it ends where m_{i+1} is located. The Euclidean metric is used as the spectral distance, which is calculated using the feature vectors before and after instant t, as $||x_{t-1} - x_{t+1}||$.

At this point, with a segmented and labeled sentence, the estimation of each joint event (w_c, ph_u) can be carried out as its absolute frequency. The conditional probabilities $P(w_c|ph_u)$ are calculated by normalizing with respect to each ph_u. Now, we can calculate $p(x_t|ph_u)$ as in equation (3).

2.3 Conditional Probability Tuning

At this point, we can apply both a DTW algorithm, which uses the *a posteriori* probabilities $P(ph_u|x_t)$ obtained as in equation (4), and the phonetic sequence to segment a sentence. This algorithm assigns a phonetic unit ph_u to an interval of the signal in order to minimize the measure

$$\sum_{f=1}^{F} \sum_{t=t^0_{ph_f}}^{t^1_{ph_f}} -\log P(ph_f|x_t) \qquad (5)$$

where F is the number of phonetic units of the sentence, and $t^0_{ph_f}$ is the initial frame of ph_f and $t^1_{ph_f}$ the final frame.

When the DTW algorithm is used to segment all the sentences of the training corpus, we obtain a new segmentation, which is used to make a new estimation of the absolute frequency of each joint event (w_c, ph_u). Then, the conditional probabilities $P(w_c|ph_u)$ are recalculated by normalizing with respect to each ph_u. This process is repeated until the difference between all the conditional probabilities $P(w_c|ph_u)$ of two continuous iterations is smaller than an ϵ (we use $\epsilon = 0.01$).

To perform this iterative tuning process do the following:

1. Initialize the absolute frequencies to 0.
2. For each sentence of the training corpus:
   (a) Estimate $P(ph_u|x_t)$ using equation (4).
   (b) Segment minimizing equation (5).
   (c) Increment the absolute frequencies.
3. Calculate the new conditional probabilities $P(w_c|ph_u)$ from the new absolute frequencies.
4. If the difference between the conditional probabilities is smaller than ϵ, then finish, otherwise go to step 1.

3 Evaluation

The measures used to evaluate the performance of the segmentation were extracted from [5]. The percentage of correctly located boundaries (*PB*) compares the location of automatically obtained phoneme boundaries with the location of manually obtained reference boundaries. The *PB* is the percentage of boundaries located within a given distance from the manually set boundaries. Tolerance intervals of 10 ms. to 30 ms. are considered.

The second measure used in this work is the percentage of frames (*PF*) which matches both segmentations, the automatic one and the manual one. Other measures are calculated using the ratios C_{man}, and C_{aut} for each phonetic unit:

$$C_{man} = \frac{correct}{tot\text{-}man} \times 100 \qquad C_{aut} = \frac{correct}{tot\text{-}aut} \times 100$$

where *correct* is the number of frames matching both segmentations, *tot-man* is the total number of frames in the manual segmentation for each phonetic unit, and *tot-aut* is the total number of frames in the automatic segmentation.

These ratios allow us to determine the type of segmentation error for each phonetic unit. A low value of C_{man} indicates a tendency of the system to assign shorter segments than needed to the unit under consideration. A low value of C_{aut} indicates a tendency to assign longer segments than needed.

4 Experiments and Results

The experiments performed in this work were performed using two Spanish speech databases composed of phonetically balanced sentences. The first one (*Frases*) was composed of 170 different sentences uttered by 10 speakers (5 male and 5 female) with a total of 1,700 sentences (around one hour of speech). The second one was the *Albayzin* speech database [2], from which we only used 6,800 sentences (around six hours of speech) which were obtained by making subgroups from a set of 700 distinct sentences uttered by 40 different speakers.

Each acoustic frame was formed by a d-dimensional feature vector: energy, 10 mel frequency cepstral coefficients (MFCCs), and their first and second time derivatives using a ± 20 ms. window, which were obtained every 10 ms. using a 25 ms. Hamming window. A preemphasis filter with the transfer function $H(z) = 1 - 0.95z^{-1}$ was applied to the signal.

For the *Frases* database, 1,200 sentences were used for training. First of all, the feature vectors of these sentences were clustered to find "natural" classes. The phonetic sequence of each sentence was obtained using a grapheme-to-phoneme converter [4]. The coarse segmentation of each sentence was obtained using the sequence of phonetic units uttered and the acoustic-phonetic rules explained in section 2.2. The initial values of the conditional probabilities $P(w_c|ph_u)$ were estimated using this coarse segmentation and the clusters. Next, the tuning process to re-estimate the conditional probabilities is iterated by segmenting the sentences with a DTW algorithm. Finally, the segmentation of the test sentences was carried out for evaluation purposes.

A subset of 77 manually segmented sentences was used for testing. Table 1 shows the results obtained with 300 "natural" classes and the results reported in [5] using HMMs for the same corpus and the same test sentences. In the case of HMMs, the same 77 manually segmented sentences were also used for training. From Table 1, it can be observed that our automatic system performs slightly better than the HMM approach. The same segmentation task was carried out using the HTK toolkit [6] with the "flat-start" option activated. Our automatic procedure and the HTK toolkit led to similar results.

In addition, we trained a Multilayer Perceptron (MLP) with the 77 manually segmented sentences to estimate the *a posteriori* probabilities of the phonetic units given the acoustic input. In this case, no derivatives were used: the input to the MLP was composed of a context window of nine acoustic frames, each of which was formed by energy plus 10 MFCCs. An MLP of two hidden lay-

Table 1. Percentage of frames (*PF*) matching both segmentations, and percentage of correct boundaries (*PB*) within tolerance intervals of different lengths (in ms.) for the *Frases* database.

	Frases ($C = 300$)			
	PF	*PB*		
		10	*20*	*30*
Automatic	82.1	67.9	85.1	93.0
HMMs	81.7	67.7	82.4	91.1
HTK	82.2	69.8	85.1	90.1
MLP+DTW	94.2	93.1	97.2	98.3

ers of 100 units each was trained achieving a classification error of around 6% (at frame level). In order to have a biased result to compare our system to, we resegmented the same 77 sentences using the trained MLP and the DTW segmentation algorithm. The result of this experiment is also shown in Table 1. As might be expected, the results of the closed-experiment (the same manually segmented training data and testing data) using the MLP were much better than our automatic approach, which did not use manual segmentation at all.

The same procedure was applied to the phonetic corpus of the *Albayzin* database. In this case, the number of sentences used was 6,800. They were divided into two subsets, one of 5,600 sentences used for training, and the other of 1,200 sentences used for testing. These 1,200 test sentences were manually segmented.

A subset of 400 sentences was selected out of the training sentences to do the clustering and to obtain the initial conditional probabilities. All the 5,600 training sentences were used in the iterative tuning process to adjust the conditional probabilities. In order to select the number of classes C, the whole experiment was carried out for increasing values of C, from 80, 100, 120, ..., 500 (see Figure 1). From this graph, it can be seen that performance is similar for values of C above 120.

The results obtained for the *Albayzin* database are shown in Table 2 and Figure 2 (performance of our automatic procedure is given for 400 "natural" classes). As before, the same task was performed by using the HTK toolkit with the "flat-start" option activated. An experiment with a MLP was also performed using the same 1,200 manually segmented sentences for training and testing. The results were quite similar to those obtained by the other speech database. Thus, the system for automatic segmentation can be scaled to any speech database.

5 Conclusions

In this work, we have presented a completely automatic procedure to segment speech databases without the need for a manually segmented subset. This task is important in order to obtain segmented databases for training phoneme-based speech recognizers.

Table 2. Percentage of frames (*PF*) matching both segmentations, and percentage of correct boundaries (*PB*) within tolerance intervals of different lengths (in ms.) for the *Albayzin* database

	Albayzin ($C = 400$)			
	PF	*PB*		
		10	*20*	*30*
Automatic	81.3	70.5	87.1	93.4
HTK	82.8	72.9	84.3	87.7
MLP+DTW	83.8	80.6	89.0	92.5

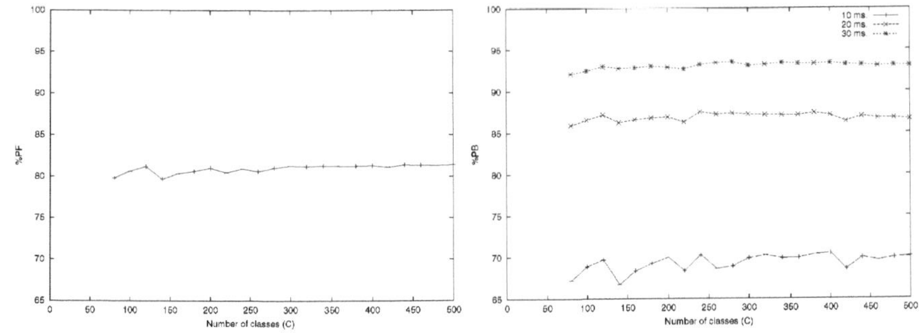

Fig. 1. *Left:* Percentage of frames (*PF*) matching manual segmentation and automatic segmentation versus the number of classes C for the *Albayzin* database. *Right:* Percentage of correct boundaries (*PB*) within tolerance intervals of different lengths (in ms.) versus the number of classes C for the *Albayzin* database

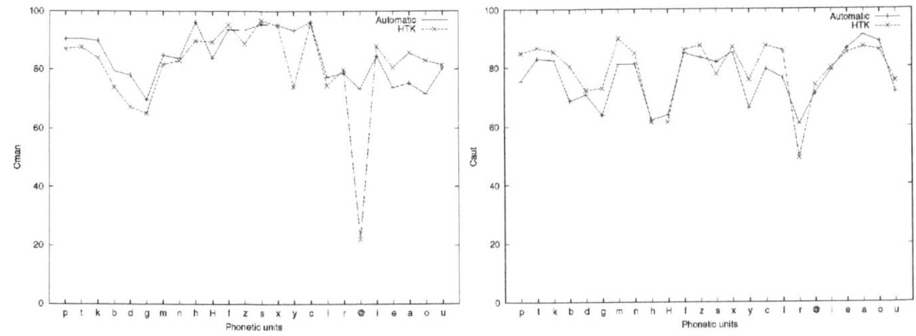

Fig. 2. C_{man} and C_{aut} for the phonetic units (SAMPA allophones) for the automatic segmentation (with 400 classes) and for the segmentation obtained using the HTK toolkit for the *Albayzin* database

As future extensions, we plan to increment the ratio of analysis from 10 ms. to 5 ms. to obtain better representations of the acoustical transitions, specially the burst of the plosive consonants. We also plan to extend the feature vectors with a contextual window of acoustic frames. We hope that the incorporation of these extensions will significantly increase the accuracy of the obtained segmentation.

References

1. B. Angelini, F. Brugnara, D. Falavigna, D. Giuliani, R. Gretter, and M. Omologo. Automatic Segmentation and Labeling of English and Italian Speech Databases. In *Eurospeech93*, volume 3, pages 653–656, Berlin (Germany), September 1993. 672
2. A. Moreno, D. Poch, A. Bonafonte, E. Lleida, J. Llisterri, J. B. Mariño, and C. Nadeu. Albayzin Speech Database: Design of the Phonetic Corpus. In *Eurospeech93*, volume 1, pages 653–656, Berlin (Germany), September 1993. 677
3. R. O. Duda, P. E. Hart, and D. G. Stork. *Pattern Classification.* John Wiley and Sons, 2001. 674
4. María José Castro, Salvador España, Andrés Marzal, and Ismael Salvador. Grapheme-to-phoneme conversion for the Spanish language. In *Proceedings of the IX National Symposium on Pattern Recognition and Image Analysis*, pages 397–402, Benicàssim (Spain), May 2001. 674, 677
5. I. Torres, A. Varona, and F. Casacuberta. Automatic segmentation and phone model initialization in continuous speech recognition. *Proc. in Artificial Intelligence*, I:286–289, 1994. 676, 677
6. Steve Young, Julian Odell, Dave Ollason, Valtcho Valtchev, and Phil Woodlan. *The HTK Book.* Cambridge University, 1997. 677

Class-Discriminative Weighted Distortion Measure for VQ-based Speaker Identification

Tomi Kinnunen and Ismo Kärkkäinen

University of Joensuu, Department of Computer Science
P.O. Box 111, 80101 JOENSUU, FINLAND
{tkinnu,iak}@cs.joensuu.fi

Abstract. We consider the distortion measure in vector quantization based speaker identification system. The model of a speaker is a codebook generated from the set of feature vectors from the speakers voice sample. The matching is performed by evaluating the distortions between the unknown speech sample and the models in the speaker database. In this paper, we introduce a weighted distortion measure that takes into account the correlations between the known models in the database. Larger weights are assigned to vectors that have high discriminating power between the speakers and vice versa.

1 Introduction

It is well known that different phonemes have unequal discrimination power between speakers [14, 15]. That is, the inter-speaker variation of certain phonemes are clearly different from other phonemes. This knowledge should be exploited in the design of speaker recognition [6] systems. Acoustic units that have higher discrimination power should contribute more to the similarity or distance scores in the matching.

The description of acoustic units in speech and speaker recognition is often done via *short-term spectral features*. Speech signal is analyzed in short segments (frames) and a representative feature vector for each frame is computed. In speaker recognition, *cepstral coefficients* [5] along with their 1st and 2nd time derivatives (Δ- and $\Delta\Delta$-coefficients) are commonly used. Physically these represent the shapes of the vocal tract and their dynamic changes [1, 2, 5], and therefore carry information about the *formant structure* (vocal tract resonant frequencies) and dynamic formant changes.

In *vector quantization* (*VQ*) based speaker recognition [3, 8, 9, 10, 16], each speaker (or *class*) is presented by a codebook which approximates his/her data density by a small number of representative *code vectors*. Different regions (*clusters*) in the feature space represent acoustically different units.

The question how to benefit from the different discrimination power of phonemes in VQ-based speaker recognition returns into question how to assign *discriminative weights* for different code vectors and how to adopt these weights into the distance or similarity calculations in the matching phase. As a motivating example, Fig. 1 shows

T. Caelli et al. (Eds.): SSPR&SPR 2002, LNCS 2396, pp. 681-688, 2002.
© Springer-Verlag Berlin Heidelberg 2002

two scatter plots of four different speakers cepstral code vectors derived from the TIMIT speech corpus. In both plots, two randomly chosen components of the 36-dimensional cepstral vectors are shown. Each speakers data density is presented as a codebook of 32 vectors. As can be seen, different classes have strong overlap. However, some speakers do have code vectors that are far away from all other classes. For instance, speakers marked by "•" and "Δ" in the rightmost plot have both such code vectors that are especially good for discriminating them from other speakers.

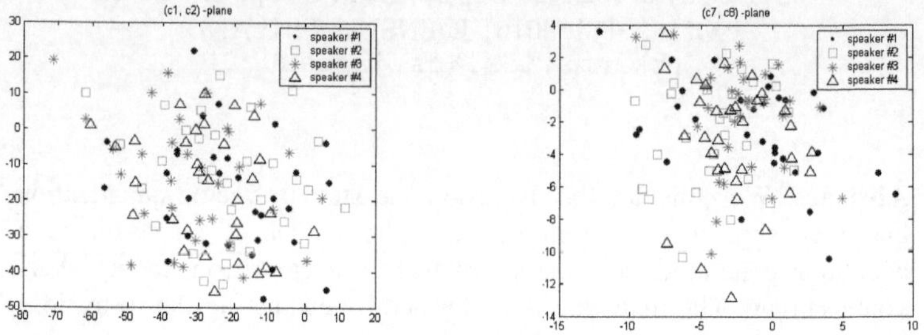

Fig. 1. Scatter plots of two randomly chosen dimensions of four speakers cepstral data from TIMIT database

There are two well-known ways for improving class separability in pattern recognition. The first one is to improve separability in the *training phase* by *discriminative training algorithms*. Examples in the VQ context are *LVQ* [12] and *GVQ* [8] algorithms. The second discrimination paradigm, *score normalization*, is used in the *decision phase*. For instance, matching scores of the client speaker in speaker verification can be normalized against matching scores obtained from a *cohort set* [3].

In this paper, we introduce a third alternative for improving class separability and apply it to speaker identification problem. For a given set of codebooks, we assign discriminative weights for each of the code vectors. In the matching phase, these weights are retrieved from a look-up table and used in the distance calculations directly. Thus, the time complexity of the matching remains the same as in the unweighted case.

The outline of this paper is as follows. In Section 2, we shortly review the baseline VQ-based speaker identification. In Section 3, we give details of the weighted distortion measure. Experimental results are reported in Section 4. Finally, conclusions are drawn in Section 5.

2 VQ-based Speaker Identification

Speaker identification is a process of finding the best matching speaker from a speaker database, when given an unknown speakers voive sample [6]. In VQ-based speaker identification [8, 9, 11, 16], vector quantization [7] plays two roles. It is used both in the training and matching phases. In the training phase, the speaker models are constructed by clustering the feature vectors in K separate clusters. Each cluster is represented by a *code vector* $\mathbf{c}_i$, which is the centroid (average vector) of the cluster. The resulting set of code vectors is called a *codebook*, and notated here by $C^{(j)} = \{\mathbf{c}_1^{(j)}, \mathbf{c}_2^{(j)}, ..., \mathbf{c}_K^{(j)}\}$. The superscript (j) denotes speaker number.

In the codebook, each vector represents a single acoustic unit typical for the particular speaker. Thus, the distribution of the feature vectors is represented by a smaller set of sample vectors with similar distribution than the full set of feature vectors of the speaker model. The codebook size should be set reasonably high since the previous results indicate that the matching performance improves with the size of the codebook [8, 11, 16]. For the clustering we use the *randomized local search* (RLS) algorithm [4] due its superiority in codebook quality over the widely used LBG method [13].

In the matching phase, VQ is used in computing a *distortion* $D(X, C^{(i)})$ between an unknown speakers feature vectors $X = \{\mathbf{x}_1, ..., \mathbf{x}_T\}$ and all codebooks $\{C^{(1)}, C^{(2)}, ..., C^{(N)}\}$ in the speaker database [16]. A simple decision rule is to select the speaker i^* that minimizes the distortion, i.e.

$$i^* = \arg\min_{1 \le i \le N} D(X, C^{(i)}) . \tag{1}$$

A natural choice for the distortion measure is the average distortion [8, 16] defined as

$$D(X,C) = \frac{1}{T} \sum_{\mathbf{x} \in X}^{T} d(\mathbf{x}, \mathbf{c}_{NN[\mathbf{x}]}), \tag{2}$$

where $NN[\mathbf{x}]$ is the index of the nearest code vector to $\mathbf{x}$ in the codebook and $d(.,.)$ is a *distance measure* defined for the feature vectors. In words, each vector from the unknown feature set is quantized to its nearest neighbor in the codebook and the sum of the distances is normalized by the length of the test sequence. A popular choice for the distance measure d is the Euclidean distance or its square. In [15] it is justified that Euclidean distance of two cepstral vectors is a good measure for the dissimilarity of the corresponding short-term speech spectra. In this work, we use squared Euclidean distance as the distance measure.

In the previous work [10] we suggested an alternative approach to the matching. Instead minimizing distortion, maximization of a *similarity measure* was proposed. However, later experiments have pointed out that it is difficult to define a natural and intuitive similarity measure in the same way as distortion (2) is defined. For that reason, we limit our discussion to distortion measures.

3 Speaker Discriminative Matching

As an introduction, consider the two speakers codebooks illustrated in Fig. 2. Vectors marked by "•" represent an unknown speakers' data. Which one is this speaker? We can see that the uppermost code vector $\mathbf{c}_2^{(1)}$ is actually the only vector which clearly turns the decision to the speaker #1. Suppose that there wasn't that code vector. Then the average distortion would be approximately same for both speakers. There are clearly three regions in the feature space which cannot distinguish these two speakers. Only the code vectors $c_2^{(1)}$ and $c_3^{(2)}$ can make the difference, and they should be given a large discrimination weight.

3.1 Weighted Distortion Measure

We define our distortion measure by modifying (2) as follows:

$$D(X,C) = \frac{1}{T}\sum_{x \in X}^{T} f(w_{NN[x]})d(x, c_{NN[x]}). \tag{3}$$

Here $w_{NN[x]}$ is the weight associated with the nearest code vector, and f is a non-increasing function of its argument. In other words, code vectors that have good discrimination (large weight) tend to decrease the distances d; vice versa, non-discriminative code vectors (small weight) tend to increase the distances. Product $f(w)d(\boldsymbol{x},\boldsymbol{c})$ can be viewed as an operator which "attracts" (decreases overall distortion) vectors $\boldsymbol{x}$ that are close to $\boldsymbol{c}$ or the corresponding weight w is large. Likewise, it "repels" (increases overall distortion) such vectors $\boldsymbol{x}$ that are far away or are quantized with small w.

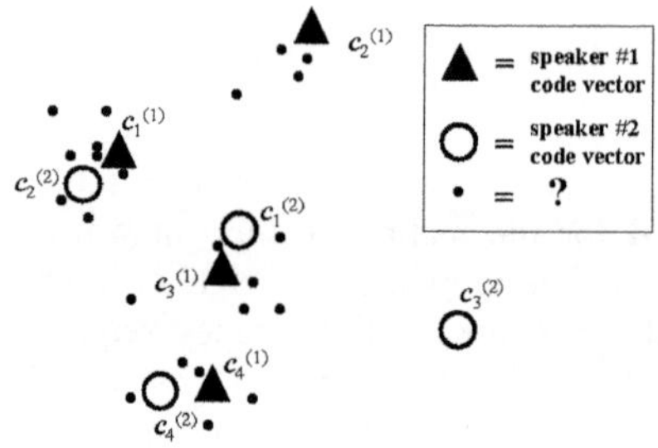

Fig. 2. Illustration of code vectors with unequal discrimination powers

An example of a quantization of a single vector is illustrated in Fig. 3. Three speakers' code vectors and corresponding weights are shown. For instance, the code vector at location (8, 4) has a large weight, because there are no other classes' presentatives in its neighborhood. The three code vectors in the down left corner, in turn, have all small weights because they all have another classes' representative near. When quantizing the vector marked by ×, the unweighted measure (2) would give the same distortion value $D \cong 7.5$ for all classes (squared Euclidean distance). However, when using the weighted distortion (3.1), we get distortion values $D_1 \cong 6.8$, $D_2 \cong 6.8$ and $D_3 \cong 1.9$ for the three classes, respectively. Thus, × is favored by the class #3 due to the large weight of the code vector. We have not yet specified two important issues in the design of the weighted distortion, namely:

- How to assign the code vector weights,
- Selection of the function f.

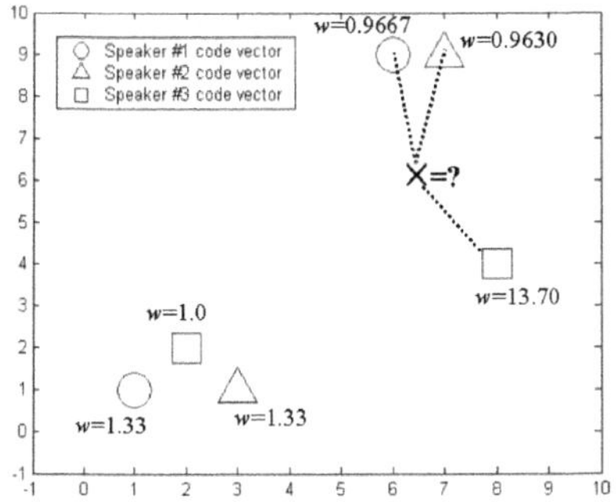

Fig. 3. Weighted quantization of a single vector

In this work, we fix the function *f* as a decaying exponential of the form

$$f(w)=e^{-\alpha w}, \tag{4}$$

where $\alpha > 0$ is a parameter that controls the rate of decay. In the above example, $\alpha=0.1$.

1.2 Assigning the Weights

The weight of a code vector should depend on the minimum distances to other classes code vectors. Let $\mathbf{c} \in C^{(j)}$ be a code vector of the *j*th speaker. Let us denote the index of its nearest neighbor in the *k*th codebook simply by $NN^{(k)}$. The weight of $\mathbf{c}$ is then assigned as follows:

$$w(\mathbf{c}) = \frac{1}{\sum_{k \neq j} 1/d(\mathbf{c}, \mathbf{c}_{NN^{(k)}})}. \tag{5}$$

In other words, nearest code vector from all other classes are found, and the inverse of the sum of inverse distances is taken. If some of the distances equals 0, we set $w(\mathbf{c}) = 0$ for mathematical convenience. The algorithm is looped over all code vectors and all codebooks.

As an example, consider the code vector located at (1,1) in Fig. 3. The distances (squared Euclidean) to the nearest code vectors in other classes are 2.0 and 4.0. Thus, the weight for this code vector is $w = 1/(1/2.0 + 1/4.0) = 1.33$. In the practical implementation, we further normalize the weights within each codebook such that their sum equals 1. Then all weights satisfy $0 \leq w \leq 1$, which makes them easier to handle and interpret.

2 Experimental Results

For testing purposes, we used a 100 speaker subset from the American English TIMIT corpus. We resampled the wave files down to 8.0 kHz with 16-bit resolution. The average duration of the training speech per speaker was approximately 15 seconds. For testing purposes we derived three test sequences from other files with durations 0.16, 0.8 and 3.2 seconds. The feature extraction was performed using the following steps:

- Pre-emphasis filtering with $H(z) = 1 - 0.97z^{-1}$.
- 12th order mel-cepstral analysis with 30 ms Hamming window, shifted by 15 ms.

The feature vectors were composed of the 12 lowest mel-cepstral coefficients (excluded the 0th coefficient). The Δ- and $\Delta\Delta$-cepstral were added to the feature vectors, thereby implying 3×12=36-dimensional feature space.

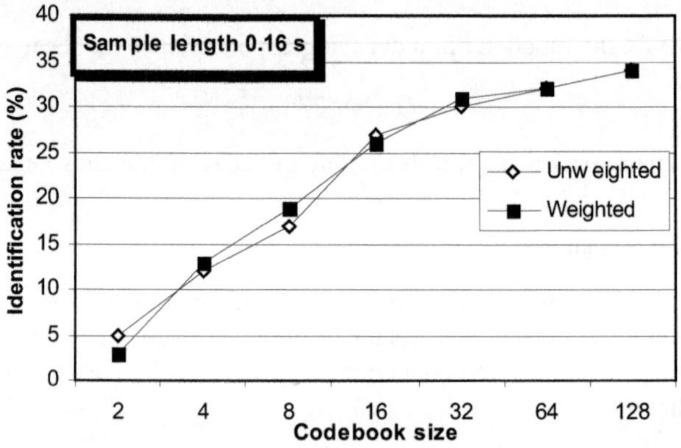

Fig. 4. Performance evaluation using ~0.16 s. speech sample (~10 vectors)

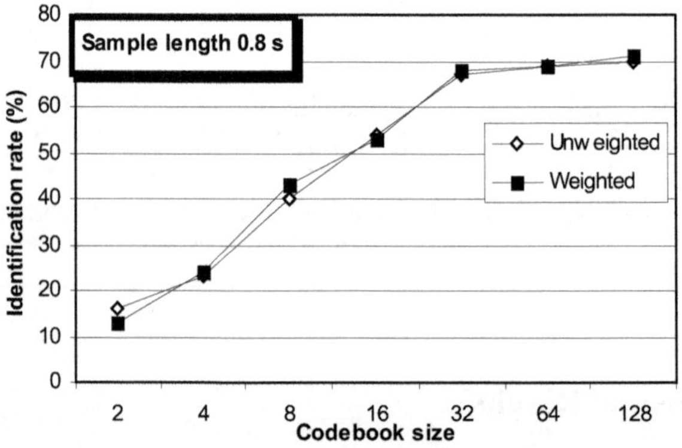

Fig. 5. Performance evaluation using ~0.8 s. speech sample (~50 vectors)

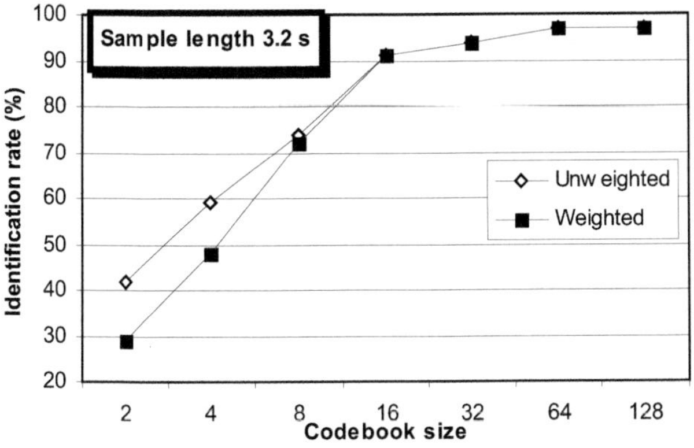

Fig. 6. Performance evaluation using ~3.2 s. speech sample (~200 vectors)

The identification rates by using the reference method (2) and the proposed method (3) are summarized through Figs. 4 - 6 for the three different subsequences by varying the codebook sizes from $K = 2$ to 128. The parameter α of (4) is fixed in all three experiments to $\alpha = 1$.

The following observations can be made from the figures. The proposed method does not perform consistently better than the reference method. In some cases the reference method (unweighted) outperforms the proposed (weighted) method, especially for low codebook sizes. For large codebooks the ordering tends to be opposite. This phenomenon is probably due to the fact that small codebook sizes give a poorer representation of the training data, and thus the weight estimates cannot be good either.

Both methods give generally better results with increasing codebook size and test sequence length. Both methods saturate to the maximum accuracy (97 %) with the longest test sequence (3.2 seconds of speech) and codebook size K=64. In this case, using codebook K=128 does not improve accuracy any more.

3 Conclusions

We have proposed a framework for improving class separability in pattern recognition and evaluated the approach in the speaker identification problem. In general, results show that with proper design VQ-based speaker identification system can achieve high recognition rates with very short test samples while model having low complexity (codebook size $K = 64$). Proposed method adapts to a given set of classes represented by codebooks by computing discrimination weights for all code vectors and uses these weights in the matching phase. The results obtained in this work show no clear improvement over the reference method. However, together with the results obtained in [10] we conclude that weighting indeed can be used to improve class separability. The critical question is: how to take full advantage of the weights in the distortion or similarity measure? In future work, we will focus on the optimization of the weight decay function f.

References

1. Deller, J. R. Jr., Hansen, J. H. L., Proakis, J. G.: *Discrete-time Processing of Speech Signals*. Macmillan Publishing Company, New York, 2000.
2. Fant, G.: *Acoustic Theory of Speech Production*. The Hague, Mouton, 1960.
3. Finan R. A., Sapeluk A. T., Damper R. I.: "Impostor cohort selection for score normalization in speaker verification," *Pattern Recognition Letters*, **18**: 881-888, 1997.
4. Fränti, P., Kivijärvi, J.: „Randomized local search algorithm for the clustering problem," *Pattern Analysis and Applications*, **3**(4): 358-369, 2000.
5. Furui, S.: "Cepstral analysis technique for automatic speaker verification," *IEEE Transactions on Acoustics, Speech and Signal Processing*, **29**(2): 254-272, 1981.
6. Furui, S.: "Recent advances in speaker recognition," *Pattern Recognition Letters*, **18**: 859-872, 1997.
7. Gersho, A., Gray, R. M., Gallager, R.: *Vector Quantization and Signal Compression*. Kluwer Academic Publishers, 1991.
8. He, J., Liu, L., Palm, G.: "A discriminative training algorithm for VQ-based speaker identification," *IEEE Transactions on Speech and Audio Processing*, **7**(3): 353-356, 1999.
9. Jin, Q., Waibel, A.: „A naive de-lambing method for speaker identification," *Proc. ICSLP 2002*, Beijing, China, 2000.
10. Kinnunen, T., Fränti, P.: "Speaker discriminative weighting method for VQ-based speaker identification," *Proc. 3rd International Conference on Audio- and Video-Based Biometric Person Authentication (AVBPA))*: 150-156, Halmstad, Sweden, 2001.
11. Kinnunen, T., Kilpeläinen, T., Fränti P.: "Comparison of clustering algorithms in speaker identification," *Proc. IASTED Int. Conf. Signal Processing and Communications (SPC)*: 222-227, Marbella, Spain, 2000.
12. Kohonen T.: *Self-Organizing Maps*. Springer-Verlag, Heidelberg, 1995.
13. Linde, Y., Buzo, A., Gray, R. M.: "An algorithm for vector quantizer design," *IEEE Transactions on Communications*, **28**(1): 84-95, 1980
14. Nolan, F.: *The Phonetic Bases of Speaker Recognition*. Cambridge CUP, Cambridge, 1983.
15. Rabiner, L., Juang B.: *Fundamentals of Speech Recognition*. Prentice Hall, 1993.
16. Soong, F. K., Rosenberg, A. E., Juang, B.-H., Rabiner, L. R.: "A vector quantization approach to speaker recognition," *AT&T Technical Journal*, **66**: 14-26, 1987.

Alive Fishes Species Characterization from Video Sequences

Dahbia Semani, Christophe Saint-Jean, Carl Frélicot, Thierry Bouwmans, and Pierre Courtellemont

L3I - UPRES EA 2118
Avenue de Marillac, 17042 La Rochelle Cedex 1, France
{dsemani,csaintje,cfrelico}@univ-lr.fr

Abstract. This article presents a method suitable for the characterization of fishes evolving in a basin. It is based on the analysis of video sequences obtained from a fixed camera. One of the main difficulties of analyzing natural scenes acquired from an aquatic environment is the variability of illumination. This disturbs every phase of the whole process. We propose to make each task more robust. In particular, we propose to use a clustering method allowing to provide species parameters estimates that are less sensitive to outliers.

1 Introduction

Segmentation of natural scenes from an aquatic environment is a very difficult issue due to the variability of illumination [17]. Ambient lighting is often insufficient as ocean water absorbs light. In addition, the appearance of non-rigid and deformable objects detected and tracked in a sequence is highly variable and therefore makes identification of these objects very complex [13]. Furthermore, recognition of these objects represent a very challenging problem in computer vision. We aim at developing a method suitable to the characterization of classes of deformable objects in an aquatic environment in order to make their online and real-time recognition easier to a vision-based system. In our application, the objects are fishes of different species evolving in a basin of the *Aquarium of La Rochelle* (France). The method we propose is composed of the following tasks:

1. *scenes acquisition*: a basin of the aquarium is filmed by a fixed CDD camera to obtain a sequence in low resolution (images of size 384 x 288);
2. *region segmentation*: color images are segmented to provide the main regions of the each scene;
3. *feature extraction and selection*: different features (e.g. color, moments, texture) are computed on each region, then selected to form pattern vectors;
4. *species characterization*: pattern vectors are clustered using a robust mixture decomposition algorithm.

T. Caelli et al. (Eds.): SSPR&SPR 2002, LNCS 2396, pp. 689–698, 2002.
© Springer-Verlag Berlin Heidelberg 2002

2 Segmentation

Image segmentation is a key step in an object recognition or scene understanding system. The main goal of this phase is to extract regions of interest corresponding to objects in the scene [9]. Obviously, this task is more difficult for *moving objects* as fishes or parts of fishes.

Under the assumption of almost constant illumination and fixed camera, the motion detection is directly connected to temporal changes in the intensity function of each pixel (x, y). Then, background substraction is usually applied to segment the moving objects from the remaining part of the scene [10][3]. By assuming that the scene background does not change over successive images, the temporal changes can be easily captured by subtracting the background frame $I_{back}(x, y)$ to the current image $I(x, y, t)$ at time t. The obtained image is denoted $I_{sub}(x, y, t)$. However, such detection of temporal changes are not robust to illumination changes and electronic noise of the camera. A solution consists in updating dynamically the background image by $I_{back}(x, y, t) = \sum_{s=1}^{t} I(x, y, s)/t$. Since obtaining a suitable background requires numerous images, $I_{back}(x, y, t = 1)$ is initialized off-line (from another available sequence). Then, thresholding the difference image provides the so-called binary difference picture:

$$I_{bin}(x, y, t) = \begin{cases} 1 \text{ if } |I_{sub}(x, y, t)| > \tau \\ 0 \text{ otherwise} \end{cases} \quad (1)$$

When color images are available, e.g. in the three dimensional color space RGB (Red, Green and Blue), one can proceed for each color plane. Three corresponding binary difference pictures $I_{bin}^{R}(x, y, t)$, $I_{bin}^{G}(x, y, t)$ and $I_{bin}^{B}(x, y, t)$ are combined to compute the segmented image:

$$I_{seg}(x, y, t) = \begin{cases} 1 \text{ if } (I_{bin}^{R}(x, y, t) = 1 \text{ or } I_{bin}^{G}(x, y, t) = 1) \text{ or } I_{bin}^{B}(x, y, t) = 1) \\ 0 \text{ otherwise} \end{cases} \quad (2)$$

Thresholds are fixed empirically according to the sequence properties. Figure 1 shows: (a) an individual frame in the sequence, (b) the reconstructed background and (c) the resulting segmented image with $\tau_R = 40, \tau_G = 30$ and $\tau_B = 35$. Note that changes in illumination due to the movement of water induce false alarms as one can see at the top right part of (c).

3 Feature Extraction and Selection

Regions issued from the segmentation process can be used as objects for the identification task. 38 features of different types, e.g. in [18], are extracted from each object:

- *Geometric* features directly relate to the objects' shape, e.g. area, perimeter, roundness ratio, elongation, orientation. Note that the wide variety of possible orientations of fishes to the camera focal axis makes geometric features inappropriate. A fish which is parallel to the image plane will exhibit its main shape while another one being orthogonal will not.

(a) An original frame $I(x, y, t)$

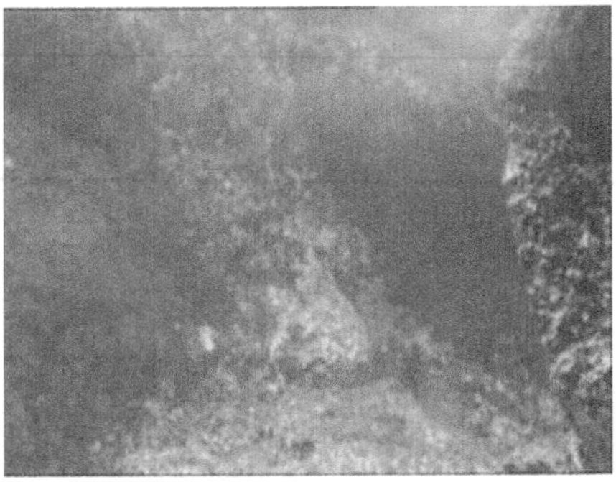

(b) Reconstructed $I_{back}(x, y, t)$

(c) Segmentation result $I_{seg}(x, y, t)$

Fig. 1. From an input image to segmented regions

- *Photometric* parameters are descriptors of the gray level distribution or the different color ones, e.g. maximum, minimum, mean and variance.
- *Texture* features are computed from the co-occurrence matrix, e.g. contrast, entropy, correlation.
- *Moments of Hu* which are known to be invariant under translation, scaling and rotation. Only the first four ones showed significant values.
- *Motion features* are computed from two consecutive frames within a sequence. Correspondence between regions from frame t and $t+1$ are established with respect to geometric and photometric features. A classical *hypothesize-and-verify* scheme [2] is used to solve this correspondence problem which is similar to the correspondence problem in stereo [11] except that a geometric constraint (a disparity-window centred around the each region's centroid) replaces the epipolar one. The extracted features are the centroid displacement and the angle of this displacement. Note that some regions do not match because of occlusion, disappearance and appearance of objects.

Feature reduction is motivated by making the characterization process easier and speeding the recognition step up to achieve a real-time processing. In order to eliminate features which are either not useful or redundant, we have selected the most pertinent features in a two-stage process:

1. **Group-based clustering:** To make sure that every features group is represented in the reduced feature space, a hierarchical clustering algorithm is applied to each group with respect to the minimization of an aggregation measure, e.g. the increase of intra-cluster dispersion for Ward's method [1]. Cutting the hierarchy to a significant value leads to a partition of the features in clusters. Among the features within a cluster, the most discriminatory powerful one is selected and the others are discarded. We recall that the discriminatory power of a feature is its usefulness in determining to which class an object belongs.

2. **Global clustering:** In order to check whether some features from different groups are similar or not, the same clustering method is globally applied to the remaining features.

4 Species Characterization

From a statistical point of view, each extracted region being described by p features can be considered as a realization x of a p-dimensional random vector X [8]. We have then to estimate the Probability Density Function (pdf) $f(x)$ from a set of realizations $\chi = \{x_1, \ldots, x_N\}$, i.e. featured regions. In mixture model approach, $f(x)$ is decomposed as a mixture of C components:

$$f(x) = \sum_{k=1}^{C} \pi_k f(x; \theta_k) \tag{3}$$

where $f(x; \theta_k)$ denotes the conditional pdf of the kth component and pairs (π_k, θ_k) are the unknown parameters associated with the parametric model of the pdf [12]. A priori probabilities π_k sum up to one. If a normal model is assumed, $\theta_k = (\mu_k, \Sigma_k)^T$ reduces to the mean μ_k and the covariance matrix Σ_k. Under the assumption of independent features of X, estimates of the model parameters $\Theta = (\pi_1, \ldots, \pi_C, \theta_1^T, \ldots, \theta_C^T)^T$ can be chosen such as the likelihood $\mathcal{L}(\Theta)$

$$\mathcal{L}(\Theta) = P(\chi|\Theta) = \prod_{i=1}^{N} \sum_{k=1}^{C} \pi_k f(x_i; \theta_k) \tag{4}$$

is maximized.

To solve this estimation problem, the EM (Expectation-Maximization) algorithm [5] has been widely used in the field of statistical pattern recognition because of its convergence. However, it is sensitive to outliers as pointed out in [15]. This is a major drawback in the context of our application because incorrectly segmented regions can disturb the estimation process. Several strategies to robust clustering are available, including:

1. contamination models of data, e.g. fitting Student distributions [14],
2. influence functions of robust statistics, e.g. using an M-estimator [6],
3. adding a class dedicated to noise, e.g. Fuzzy Noise Clustering (FNC) [4].

We propose to use a robust clustering method (based on EM algorithm) that is a combination of the first two types [15]. Each component is modelled as a mixture of two sub-components:

$$f(x;\theta_k) = \underbrace{(1-\gamma_k)\mathcal{N}(x;\mu_k,\Sigma_k)}_{(A)} + \underbrace{\gamma_k\mathcal{N}(x;\mu_k,\alpha_k\Sigma_k)}_{(B)} \tag{5}$$

where $\mathcal{N}$ stands for the gaussian multivariate pdf.
First term (A) intends to track cluster kernel points while second term (B) allows to take into account surrounding outliers via multiplicative coefficients α_k. These γ_k and α_k control respectively the combination of the two sub-components and the spread of the second one by modifying its variance. Parameters of both sub-components are estimated through different estimators so that the conditional pdf is estimated by:

$$\hat{f}(x;\theta_k) = (1-\gamma_k)\mathcal{N}(x;\tilde{\mu}_k,\tilde{\Sigma}_k) + \gamma_k\mathcal{N}(x;\hat{\mu}_k,\alpha_k\hat{\Sigma}_k) \tag{6}$$

where $\tilde{\mu}_k, \tilde{\Sigma}_k$ are robust estimates whereas $\hat{\mu}_k, \hat{\Sigma}_k$ are standard ones. Among the possible M-estimators to be used, e.g. Cauchy, Tuckey, Huber, we have chosen the Huber M-estimator [7] because it performs well in many situations [19]. It is parametrized by a constant value h that controls the size of the filtering area. Such an estimator is an influence function $\psi(y,h)$, e.g. the Huber one:

$$\psi_{Huber}(y,h) = \begin{cases} y \text{ if } |y| \leq h \\ h\ sgn(y) \text{ otherwise} \end{cases} \tag{7}$$

This function allows to associate a weight $w(y,h) = \frac{\psi(y,h)}{y}$ as a decreasing function of y, e.g. the Huber one:

$$w_{huber}(y,h) = \begin{cases} 1 \text{ if } |y| \leq h \\ \frac{h}{|y|} \text{ otherwise} \end{cases} \tag{8}$$

We apply it to the distances between each point x_i and the cluster prototypes in order to compute a weight w_i associated with each x_i. According to the equation (8), all w_i belong to $[0,1]$ and outlying points are given a zero weight (see Figure 2).
Algorithm 1 replaces the parameters updating in the M-step of the EM algorithm. The more iterations, the less points are taken into account in the estimation process, so that one needs to use a stop criterion in order to ensure

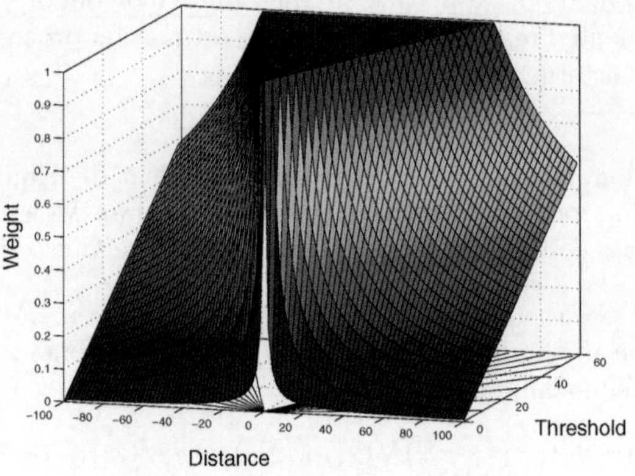

Fig. 2. Huber M-estimator weight as a function of distance y and threshold h

H 1: Iterative robust estimation of means and covariance matrices

Input: $\chi = \{x_1, \ldots, x_N\}$, $\hat{z}_{ik}$ current estimates of $P(C_k|x_i)$ from the E-Step, h the M-estimator threshold

$$\tilde{\mu}_k = \hat{\mu}_k = \frac{\sum_{i=1}^{N} \hat{z}_{ik}\, x_i}{\sum_{i=1}^{N} \hat{z}_{ik}}$$

$$\tilde{\Sigma}_k = \hat{\Sigma}_k = \frac{\sum_{i=1}^{N} \hat{z}_{ik}\, (x_i - \tilde{\mu}_k)(x_i - \tilde{\mu}_k)^T}{\sum_{i=1}^{N} \hat{z}_{ik}}$$

repeat

for $i = 1$ *to* N **do**

$d_i = (x_i - \tilde{\mu}_k)^T \tilde{\Sigma}_k (x_i - \tilde{\mu}_k)$ (Mahalanobis distance)

$w_i = w_{huber}(d_i, h)$ (Huber M-estimator weight function - see Fig. 2)

$$\tilde{\mu}_k = \frac{\sum_{i=1}^{N} w_i\, \hat{z}_{ik}\, x_i}{\sum_{i=1}^{N} w_i\, \hat{z}_{ik}}$$

$$\tilde{\Sigma}_k = \frac{\sum_{i=1}^{N} w_i\, \hat{z}_{ik}\, (x_i - \tilde{\mu}_k)(x_i - \tilde{\mu}_k)^T}{\sum_{i=1}^{N} w_i\, \hat{z}_{ik}}$$

until *Stop Criterion*;

sufficient statistics. We use a combination of maximum number of iterations and maximum elimination rate (proportion of sample having a quite zero weight). It can be shown that the property of monotonous increase of log-likelihood of the EM algorithm no more holds because the iterative estimation process yields an approximated realization of the maximum log-likelihood estimator. However, relaxing the maximum likelihood estimation principle allows to obtain more accurate estimates.

5 Experiments and Discussion

A sequence of 550 images was acquired in the *Aquarium of La Rochelle*, the filmed basin comprising 12 species. After segmentation and false alarms discarding, 5009 regions were obtained and labelled according to the different species. The first feature selection step (group-based clustering) allowed to reduce the 38 original attributes to 22 ones while the second step (global clustering) allowed to keep only 18 of them (see Table 1 for details), representing a compression rate greater than 52%.

Table 1. Summary of features selection

Number of features	Before selection	Group-based clustering	Global clustering
Geometric	10	4	4
Photometric	14	7	5
Texture	7	5	4
Moments of Hu	4	3	2
Motion	3	3	3
Total	38	22	18
Compression rate (%)		42.11%	52.63%

At least two features of each group are present in the final set of 18 selected features:

- *Geometric feautures:* width, elongation, roundness ratio and orientation.
- *Photometric feautures:* gray-level mean, minimum and variance ; blue average and minimum of the color.
- *Texture features:* entropy, contrast, homogeneity, and uniformity.
- *Moments of Hu:* second and third moments of Hu.
- *Motion features:* vector and angle of displacement.

During the labelling, we have noticed that different species were indeed subspecies members of which look very similar, e.g. subspecies *Acanthurs bahianus* and *Acanthurs chirurgus* shown in Figure 3. We decided to merge such subspecies decreasing the number of classes to 8. This choice was validated by the BIC (*Bayesian Information Criterion*) using unconstrained normal classes [16]. As labels were available and under the assumption of gaussian classes, class parameters $\theta^D = [\mu_1^D, \Sigma_1^D, \ldots, \mu_c^D, \Sigma_c^D]$ were computed directly from training 5009 samples.

Our goal was to provide as accurate as possible class parameters estimates with an unsupervised technique in order to characterize the fish species. We applied our clustering algorithm several times under random initializations. Parameters γ_k, α_k and h were fixed empirically and identical for each class ($\gamma_k = \gamma$ and $\alpha_k = \alpha$). According to the semantics of theoretical model of classes, only robust estimates $\tilde{\theta}_k = [\tilde{\mu}_k, \tilde{\Sigma}_k]$ were considered ($k = 1, c$). In order to assess the species characterization, a distance between θ^D and the final estimated parameters provided by the algorithm was calculated. Because of possible labels switching in class numbering, optimal permutation σ^* was obtained by computing the minimum over all possible permutations σ:

$$A(\theta^D, \tilde{\theta}, \sigma^*) = \min_{\sigma} \left(\sum_{k=1}^{C} dist_{\mathcal{M}}(\theta_k^D, \tilde{\theta}_{\sigma(k)}) \right) \quad (9)$$

where $dist_{\mathcal{M}}$ is the Mahalanobis distance between two normal distributions:

$$dist_{\mathcal{M}}(\theta_k, \theta_l) = dist_{\mathcal{M}}(\mu_k, \Sigma_k, \mu_l, \Sigma_l) = (\mu_k - \mu_l)^T (\Sigma_k + \Sigma_l)^{-1} (\mu_k - \mu_l) \quad (10)$$

A value of 15.07 was obtained for $A(\theta^D, \tilde{\theta}, \sigma^*)$. Using the EM algorithm, standard estimates $\hat{\theta}$ are involved, so (9) becomes $A(\theta^D, \hat{\theta}, \sigma^*)$. In this case, we obtained a value of 20.30. This clearly shows the advantage of including robust estimators as well as a contamination model.

(a) Acanthurs bahianus (b) Acanthurs chirurgus

Fig. 3. Specimens from different subspecies to be merged

6 Conclusion

In this paper, we address the problem of characterizing moving deformable objects in an aquatic environment using a robust mixture decomposition based clustering algorithm. Despite several difficulties in our application, particularly changes in illumination conditions induced by water, preliminary experiments showed that our approach provides better estimates than the EM algorithm. Further investigations will concern the automatic selection of the different coefficients involved in the model and the test of non normal models.

Acknowledgements

This work is partially supported by the region of Poitou-Charentes.

References

1. M. R. Anderberg. *Cluster Analysis for Applications.* Academic Press, 1973. 692
2. N. Ayache. *Artificial Vision for Mobile Robots : Stereo Vision and Multisensory perception.* MIT Press, Cambridge, MA, 1991. 691
3. A. Cavallaro and T. Ebrahimi. *Video object extraction based on adaptative background and statistical change detection*, pages 465–475. In Proceedings of SPIE Electronic Imaging, San Jose, California, USA, January 2001. 690
4. R. Davé and R. Krishnapuram. Robust clustering methods: A unified view. *IEEE Transactions on Fuzzy Systems*, 5(2):270–293, 1997. 693
5. A. P. Dempster, N. M. Laird, and D. B. Rubin. Maximum likelihood from incomplete data via the EM algorithm. *Journal of the Royal Statistical Society series B*, 39:1–38, 1977. 692
6. H. Frigui and R. Krishnapuram. A robust competitive clustering algorithm with applications in computer vision. *IEEE Trans. on Pattern Analysis and Machine Intelligence*, 21(5):450–465, May 1999. 693
7. P. J. Huber. *Robust Statistics.* John Wiley, New York, 1981. 693
8. A. Jain, R. Duin, and J. Mao. Statistical pattern recognition: A review. *IEEE Transactions on Pattern Analysis and Machine Intelligence*, 22(1):4–38, 2000. 692
9. A. Jain and P. Flynn. Image segmentation using clustering. *in Advances in Image Understanding, K. Bowyer and N. Ahuja (Eds), IEEE Computer Society Press*, pages 65–83, 1996. 690
10. R. Jain and H. Nagel. On the analysis of accumulative difference pictures from image sequences of real world scenes. *IEEE Transactions on Pattern Analysis and Machine Intelligence*, 1(2):206–214, 1979. 690
11. R. Jain, R. Kasturi and B. G. Schunck. *Machine Vision.* McGRAW-HILL Inc., 1995. 691
12. G. McLachlan and D. Peel. *Finite Mixture Models.* Wiley and Sons, 2000. ISBN 0-471-00626-2. 692
13. E. Meier and F. Ade. Object detection and tracking in range image sequences by separation of image features, stuttgart, germany. *In IEEE International Conference on Intelligent Vehicles*, pages 176–181, 1998. 689
14. D. Peel and G. McLachlan. Robust mixture modelling using the t distribution. *Statistics and Computing*, 10(4):339–348, October 2000. 693
15. C. Saint-Jean, C. Frélicot, and B. Vachon. *Clustering with EM: complex models vs. robust estimation*, pages 872–881. In proceedings of SPR 2000: F. J. Ferri, J. M. Inesta, A. Amin, and P. Pudil (Eds.). Lectures Notes in Computer Science 1876, Springer-Verlag, 2000. 693

16. G. Schwarz. Estimating the dimension of a model. *Annals of Statistics*, 6:461–464, 1978. 695
17. Z. Tauber, Z. Li, and M. S. Drew. *Local-based Visual Object Retrieval under Illumination Change*, volume 4, pages 43–46. In Proceedings of the 15th International Conference on Pattern Recognition, Barcelona,Spain, 2000. 689
18. S. Theodoridis and K. Koutroumbas. *Pattern Recognition.* Academic Press Inc. - ISBN 0-12-686140-4, 1999. 690
19. Z. Zhang. Parameter estimation techniques: A tutorial with application to conic fitting. Technical Report RR-2676, Inria, 1995. 693

Automatic Cut Detection in MPEG Movies: A Multi-expert Approach

Massimo De Santo[1], Gennaro Percannella[1], Carlo Sansone[2], Roberto Santoro[1], and Mario Vento[1]

[1] Dipartimento di Ingegneria dell'Informazione e di Ingegneria Elettrica
Università di Salerno - Via P.te Don Melillo,1 I-84084, Fisciano (SA), Italy
{desanto,pergen,rsantoro,mvento}@unisa.it
[2] Dipartimento di Informatica e Sistemistica
Università di Napoli "Federico II"- Via Claudio, 21 I-80125 Napoli, Italy
carlosan@unina.it

Abstract. In this paper we propose a method to detect abrupt shot changes in MPEG coded videos that operates directly on the compressed domain by using a Multi-Expert approach. Generally, costly analysis for addressing the weakness of a single expert for abrupt shot change detection and the consequent modifications would produce only slight performance improvements. Hence, after a careful analysis of the scientific literature, we selected three techniques for cut detection, which extract complementary features and operate directly in the compressed domain. Then, we combined them into different kinds of Multi-Expert Systems (MES) employing three combination rules: Majority Voting, Weighted Voting and Bayesian rule. In order to assess the performance of the proposed MES, we built up a huge database, much wider than those used in the field. Experimental results demonstrate that the proposed system performs better than each of the three single algorithms.

1 Introduction

Information filtering, browsing, searching and retrieval are essential issues to be addressed in order to allow a faster and more appealing use of video databases. Even if it does not yet exist a unique and definitive solution to the former problem, the field experts have agreed upon a position: the first step toward an effective organization of the information in video databases consists in the segmentation of the video footage in shots, defined as the set of frames obtained through a continuous camera recording. There are two different types of transitions between shots: abrupt and gradual. The difference between them relies on the number of frames involved, which are two in the case of abrupt shot changes and more than two in the case of gradual shot changes. In the latter case, different types of shot transitions may be outlined as dissolves, fades, wipes, iris, etc., according to the mathematical model used to transform the visual content from a shot to the successive one. The automatic detection of gradual transitions is much more complicated than that of abrupt shot

T. Caelli et al. (Eds.): SSPR&SPR 2002, LNCS 2396, pp. 699-708, 2002.
© Springer-Verlag Berlin Heidelberg 2002

changes and requires more complex mathematical models. However, gradual transitions are also less frequent than abrupt shot changes; therefore, in this paper we focused our investigation only on abrupt shot changes detection. It is worth to consider that automatic abrupt shot changes detection (SCD) is not a trivial task and is often complicated by some video effects, like camera or objects movements, impulsive variations of luminance signals, that may be easily confused with abrupt shot changes.

It has to be noted that video sources are often provided in compressed form according to standards like MPEG. In the recent past, many researchers have tried to face the problem of cut detection by processing videos in compressed form. In fact, the direct analysis in the coded domain offers at least two advantages: firstly, the computational efficiency of the whole process is improved; secondly, video compression is generally performed using signal processing techniques capable of deriving features for video segmentation, e.g. motion vectors in MPEG coding. Thus, such features become readily available for any parsing operation, and would have to be re-derived if a decoding step were applied. For these reasons, we perform the whole analysis of the video stream directly in the MPEG coded domain.

In the scientific literature many techniques for SCD have been proposed. However, the efforts for increasing performance of a single classifier appear, in general, unjustified, especially when the classifier has been repeatedly improved over the time, by adjusting its features, the learning procedures, the classification strategies and so on. Generally, costly analysis for addressing the weakness of a single classifier and the consequent modifications would produce only slight performance improvements. In these cases, the ensemble of rather simple experts, complementary as regards their errors, makes it possible to improve the overall performance, and often relatively little efforts are rewarded by high performance increases.

Therefore, our intention was to employ a Multi-Expert approach that can give good performance improvements with relatively few efforts. The use of a Multi-Expert System (MES) for complex classification tasks has been widely explored in the last ten years [1, 2]. The underlying idea of using a MES is to combine a set of experts in a system taking the final classification decision on the basis of the classification results provided by any of the experts involved. The rationale of this approach lies on the assumption that the performance obtained by suitably combining the results of a set of experts is better than that of any single expert. The successful implementation of a MES requires experts which use complementary features, and the definition of a combining rule for determining the most likely class a sample should be assigned to, given the class it is attributed to by each single expert.

Hence, our idea was to select three methods proposed in the scientific literature and to combine them into a MES according to the most commonly used combining rules: Majority Voting, Weighted Voting and Bayesian rules [1, 2]. We considered two principal aspects when choosing the algorithms to integrate in our MES: the complementarity of the used features and the performance declared by the authors.

For the training and testing phases, we used a database consisting of more than 130 thousands frames with a percentage of about 1% of abrupt cut frames. This is a significant amount of both frames and cuts, especially if compared to the size of the data sets usually employed in this scientific realm. The experimental results showed that the proposed MES performs better than each of the considered classifier.

The organization of the paper is the following: in section 2 we report about previous works in the field of automatic abrupt cut detection in the MPEG compressed domain; in section 3 the proposed system architecture is presented, together with some details about the cut detection algorithm implemented into the three experts; in section 4 we analyze the experimental campaign carried out in order to assess the performance of the proposed system; finally, in section 5 we draw conclusions and discuss on the future work.

2 Previous Work

In this section, we briefly report about proposed methods for automatic abrupt cut detection, which, according to our opinion, are the most representative. As mentioned in the introduction, we focus our attention on MPEG coded videos. A possible taxonomy for classifying algorithms for automatic detection of video shot transitions can be based on the required level of decoding. From this point of view, we devised four different groups of techniques sorted according to the increasing number of decoding steps required to derive the basic information needed for shot boundaries detection: bit rate, macroblock prediction type, motion vectors, DCT coefficients.

Bit rate techniques [3, 4] rely on the idea that large variation in the visual content between two consecutive frames generally results in a large variation in the amount of bits used for coding the DCT coefficients of the blocks of the respective frames. Anyway, the variations in the amount of bits may occur both when a cut or other effects like zooming, panning or dissolves are present. Being the used information trivial, there are no ways to distinguish among these cases. The idea behind macroblock prediction type techniques for SCD [5, 6] is that the visual change generated by a cut usually gives rise to specific patterns of macroblocks into the frames across the shot boundary. Therefore, recognizing these patterns means recognizing cuts. The use of features based on motion vectors [7] relies on a very simple principle: temporally adjacent frames belonging to the same shot are usually characterized by the same motion. Hence, the motion vectors in the inter-coded frames (i.e. *P* or *B*) might be used to this aim. In particular, the difference between the motion vectors of a block of two successive inter-coded frames should be small or large if the two frames are respectively in the same shot or not.Another source of information that has been often used for shot segmentation is represented by DCT coefficients [8, 9, 10, 11]. The idea is that a variation in the content of a block of 8x8 pixels results in a variation in the content of the block in the transformed domain, and so in its DCT coefficients.

Generally speaking, all the above mentioned techniques for SCD operate in a similar fashion. Each one parses a MPEG coded video frame by frame, computes the distance between each couple of successive frames and, if this difference is greater or lower than a fixed threshold, they declare or not a cut between the two frames the difference is referred to. The distinction among these techniques relies on the way they compute the difference between two frames. Therefore, each technique for SCD can be viewed as a single classifier that for each couple of frames declares the presence of a cut or not.

3 The Proposed System Architecture

According to the rationale inspiring MES, we selected three techniques for SCD (three experts) whose features were complementary and combined them according to a parallel scheme, as shown in Fig. 1. Each of the three single classifiers receives in input the MPEG coded bitstream and for each couple of frames provides its own classification (e.g. cut or not cut). Then the combination module of the MES, for each couple of frames, provides the final classification on the basis of the outputs of the three experts and of the implemented combining rule.

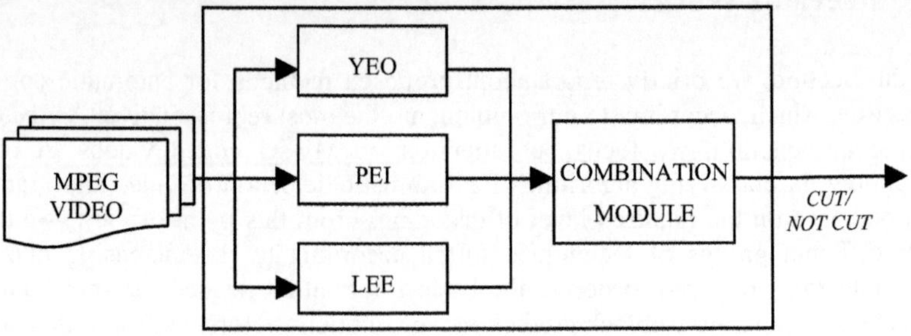

Fig. 1. The system architecture of the proposed MES

The experts used in our system implement the SCD algorithms proposed by Yeo et al. in [8], Pei et al. in [5] and Lee et al. in [9]. These techniques offers the advantage to extract complementary features (e.g. DC-image, edges and macroblock prediction type) and the performances reported by the authors are interesting, as shown in Table 1. Hereinafter, for the sake of simplicity we will refer to these three classifiers with the terms YEO, PEI and LEE, according to the name of the first author. The performance index (*PI*) used for comparing the various techniques is defined by the sum of *Precision* and *Recall*. In (1), (2) and (3), we reported the formulas of *Precision*, *Recall* and *PI*, respectively:

$$Precision = \frac{cd}{cd + f} \tag{1}$$

$$Recall = \frac{cd}{cd + m} \tag{2}$$

$$PI = Precision + Recall \tag{3}$$

where *cd* is the number of correctly detected cut, *f* is the number of false positive and *m* is the number of misses.

The SCD method proposed by Yeo et al. employs the average value of the luminance computed on each frame. For each video frame a *DC-image* is constructed; such image is obtained considering for each 8x8 pixels luminance block only the value of the DCT-DC coefficient. Therefore, a frame of 352x288 pixels is represented through 44x36 DC coefficients. For each couple of successive frames, YEO computes the distance as the sum of the absolute differences among the corresponding pixels of the two *DC-images*. Then, it considers sliding windows of m frames, computes X and Y, respectively as the first and the second maximum distances between each couple of successive frames into the window: if X is n times greater than Y, then a cut is declared between the two frames whose distance is X.

Table 1. Experimental results reported by Yeo et al., Pei et al. and Lee et al.

	Precision	Recall	PI
YEO	0.93	0.93	1.86
PEI	0.99	1	1.99
LEE	0.97	0.99	1.96

DCT coefficients have been used also in the SCD method proposed by Lee et al. In this case, for each 8x8 pixels luminance block of every frame, seven DCT coefficients are needed. Anyway in [9], the key idea is to perform cut detection on the basis of the variations of the edges. In fact, the authors developed a mathematical model to approximately characterize an eventual edge in the block by using only on the first seven DCT coefficients in the zig-zag order. The characteristics of an edge are represented in terms of its intensity (*strength*) and orientation (ϑ). The technique works as follows: for each frame, they compute the histogram of the edge strengths, *H(strength)*, and the histogram of the edge orientations, $H(\vartheta)$. Then, for each couple of successive frames, they compute the differences *D(strength)* and $D(\vartheta)$ between their $H(s)$ and $H(\vartheta)$, respectively. Finally, the interframe distance is obtained as a linear combination of $D(s)$ and $D(\vartheta)$; if such distance is higher than a fixed threshold, a cut is declared between the two frames.

The third technique for SCD we used is that proposed by Pei et al. In this case, the employed feature is the macroblock prediction type. Each macroblock consists of four 8x8 pixels blocks and can be coded as *I*, *P* or *B*. The idea is that a particular pattern of coded macroblocks should reveal the presence of a cut. As an example, typically most macroblocks of a *B* frame are coded referring both to a precedent anchor frame (*I* or *P*) and to a successive anchor frame (i.e. *forward and backward prediction*). Differently, when a cut occurs between a *B* and a successive anchor frame, most macroblocks of a *B* frame are coded only referring to the previous anchor frame (i.e. *forward prediction*). This SCD technique is very fast as the macroblock coding type is a readily available information, requiring only few MPEG decoding steps.

As regards the combiner, we implemented the most common combination rules: Majority Voting, Weighted Voting and Bayesian rule. In particular, for the Weighted Voting and the Bayesian rules, we considered the combinations of all the three algorithms and the three possible combinations of two out of three algorithms. Therefore, on the whole, we considered nine combinations.

As regards the weighted voting, it is worth to specify that the votes provided by each expert have been weighted proportionally to the percentage of correct recognition evaluated on the training set for each class. As an example, if a test set sample has been classified as a cut by the i-th expert, this vote will weight 0.8 if the percentage of correctly detected cut on the training set for the i-th classifier was 80%. As to the Bayesian rule, the set up was done according to [12].

4 Experimental Results

In this section we report the description of the database used for the experimental campaign, together with experimental results provided by the single experts and the proposed MES.

4.1 The Video Database

In the field of SCD a topic still in progress is the definition of a common database. Nowadays, each researcher involved in this field uses a different database, so making comparisons among SCD algorithms a very complicated task. Moreover, quite often, these databases consist of too few frames and cuts.

In order to carry out a significant analysis of the proposed system, we set up a database consisting of sixteen excerpts of MPEG coded movies of various genres for a total amount of 134314 frames and 1071 cuts. This is a significant amount of frames above all if compared to other databases used in the SCD field. As example, in Table 2, we reported a comparison among our database and databases used in [5, 8, 9].

Table 2. Size comparison of the databases used in [5, 8, 9] and in this paper

Databases	Number of frames	Number of cuts
Yeo et al. [8]	9900	41
Pei et al. [5]	36000	269
Lee et al. [9]	80887	611
This paper	134314	1071

With the aim to obtain reliable experimental results, we selected excerpts of videos MPEG of various genres. Furthermore, the database contains several common video effects as zooming, panning, dissolves and fades: this is important as it allows stressing the algorithms in several difficult conditions. Moreover, in the database is also present the ground truth, therefore it is available, for each frame, the information concerning the presence, or the absence, of a cut. This information is structured as follows: if the frame i is labeled as a cut, it is intended that there is a cut between the frame i and the frame i-1. This information is essential to assess the performance of every employed classifier.

The whole database has been divided into two sets (training and test, respectively) of approximately the same size: both sets contain eight video fragments. In Table 3 there are reported some details about the composition of the two sets.

Table 3. Composition of the training and test sets

	Number of frames	Number of cuts
Training Set	64343	543
Test Set	69971	528

The training set was used to fix the optimal threshold for each classifier. To this aim, the optimal threshold is intended as the threshold that allows the classifiers to maximize PI (the index defined in Section 2) on the training set.

4.2 Experimental Results

In this section we report the performance of the three single classifiers and of the proposed MES, evaluated on the test set.

Firstly, we compared the performance obtained by the three single classifiers. From Table 4, it is possible to note that performances obtained by YEO and PEI are much better than those obtained by LEE. This is due to the very high percentage of false cuts detected by LEE, which degrades the value of *Precision* and consequently the performance index. As shown in Table 4, the performance index for PEI is 1.83 versus 1.70 for YEO and 0.66 for LEE.

Table 4. A comparison among performances of the three single classifiers on the test set

Expert	%cd	%f	Precision	Recall	PI
YEO	82.00	0.08	0.88	0.82	1.70
PEI	91.60	0.07	0.91	0.92	1.83
LEE	64.40	23.50	0.02	0.64	0.66

Table 5 reports the coverage Table evaluated on the test set for the combination of the YEO, PEI and LEE experts, in terms of correctly detected cut and false positive. Table 5 is extremely useful as it allows sketching the complementarity of the employed experts. As an example, the value of *%cd* for "None" represents the percentage of correctly detected cuts which neither YEO nor PEI nor LEE are able to detect. Therefore there is no MES that by combining the three single classifiers can detect these cuts, independently from the employed combination rule. Obviously, the complement of this percentage constitutes the upper bound for the percentage of correctly detected cuts that any MES can obtain by combining YEO, PEI and LEE.

Table 5. Coverage table evaluated on the test set for the combination of the YEO, PEI and LEE experts, in terms of correctly detected cut and false positive

Number of classifiers	Classifiers	%cd		%f	
		Percentage	Sum	Percentage	Sum
3	YEO, PEI and LEE	52.65	52.65	0	0
2	YEO and PEI	27.84	35.80	0	0.06
	YEO and LEE	1.33		0.03	
	PEI and LEE	6.63		0.03	
1	Only YEO	0.19	8.52	0.05	22.11
	Only PEI	4.54		0.04	
	Only LEE	3.79		22.02	
0	None	3.03	3.03	77.83	77.83

The theoretic MES that could provide this percentage is called *oracle*. In our MES the *oracle* is characterized by a *%cd* of 96.97. Moreover, from Table 5, it is possible to deduce the performance that majority voting MES is able to gain. The sum of percentages of correctly detected cuts provided by at least two out of three single classifiers is 88.45, which is the percentage of correctly detected cuts by using the majority voting rule. Table 5 shows also the coverage percentages of false positive obtained by YEO, PEI and LEE on the test set. Here, the value obtained in case of "None" means that on 77.83% of the test set frames, neither YEO nor PEI nor LEE give a false positive. Interestingly enough, the lower bound for the percentage of false positive is 0%; while the percentage of false positive given by using the majority voting rule is 0.06.

In Table 6, a comparison among all considered classification systems is reported, sorted according to the global performance index *PI*. For simplicity we indicated the various MES as follows: MV, W and BAY stand for majority, weighted voting and Bayesian rule respectively; Y, P and L are abbreviations for YEO, PEI and LEE. As an example, BAY-YP is the MES obtained by applying the Bayesian rule to YEO and PEI classifiers.

From table 6, it can be observed that some MES performs exactly the same. This is due partly to the small number of experts (three) and mostly to the small number of classes (two). Table 6 reports in bold the MES which obtained the highest percentage of correctly detected cuts, lower percentage of false positive and higher value of performance index *PI*. In the first row of the same table, we reported also the performance of the *oracle*. Therefore, we can conclude that there are three MES (i.e. BAY, BAY-YP and W-YP) which performs better than the single expert (i.e. YEO, PEI and LEE), considered individually. It is also very interesting to note that the maximum improvement to *PI* that the multi-expert approach would give is 0.17 and the best MES are able to recover about 21% of this maximum improvement; such point constitutes a very good result.

Table 6. Parameters *%cd*, *%f*, *Precision*, *Recall* and *PI*, evaluated on test set, for all the considered classification systems

Algorithms	%cd	%f	Precision	Recall	PI
Oracle	96.97	0	1	0.97	1.97
BAY-YPL	93.20	0.05	0.93	0.93	1.86
BAY-YP	93.20	0.05	0.93	0.93	1.86
W-YP	93.20	0.05	0.93	0.93	1.86
BAY-PL	91.60	0.07	0.91	0.92	1.83
W-PL	91.60	0.07	0.91	0.92	1.83
PEI	91.60	0.07	0.91	0.92	1.83
MV	88.45	0.06	0.92	0.88	1.80
W-YPL	88.45	0.06	0.92	0.88	1.80
BAY-YL	82.00	0.08	0.88	0.82	1.70
W-YL	82.00	0.08	0.88	0.82	1.70
YEO	82.00	0.08	0.88	0.82	1.70
LEE	64.40	23.50	0.02	0.64	0.66

5 Conclusions

Automatic abrupt cut detection in the MPEG coded domain is still an open problem. In spite of efforts of many researchers involved in this field, there are not yet techniques which provide fully satisfactory performances.

In this context, our idea was to employ a Multi-Expert approach to combine some of the best techniques available in the scientific literature with the aim to improve the recognition rates. Therefore, we implemented three algorithms for cut detection, operating directly on the compressed format and combined them in a parallel MES using common combination rules. The experimental results have demonstrated that the bayesian combination of the three algorithms (also only the best two) performs better than each classifier considered individually, with respect to each of the considered performance parameters: percentage of correctly detected cuts, percentage of false positive and value of *Precision + Recall.*

At this moment, as future direction of our research we foresee from a side the integration in the MES of other cut detection techniques in order to further improve the overall performances of the system and on the other side the application of the MES approach to the detection of gradual transitions.

References

1. T.K. Ho, J.J. Hull, S.N. Srihari, Decision Combination in Multiple Classifier Systems, IEEE Transactions on Pattern Analysis and Machine Intelligence, 16(1), (1994), 66-75.
2. J. Kittler, M. Hatef, R.P. W. Duin, J. Matas, On Combining Classifiers, IEEE Transactions on Pattern Analysis and Machine Intelligence, 20(3), (1998), 226-239.
3. J. Feng, K.T. Lo and H. Mehrpour, "Scene change detection algorithm for MPEG video sequence", Proc. of the IEEE International Conference on Image Processing, vol. 2, pp. 821–824, Sept. 1996.
4. G. Boccignone, M. De Santo, and G. Percannella, "An algorithm for video cut detection in MPEG sequences," Proc. of the IS&T/SPIE International Conference on Storage and Retrieval of Media Databases 2000, pp. 523-530, Jan. 2000, San Jose, CA.
5. S.C. Pei, Y.Z. Chou, Efficient MPEG compressed video analysis using macroblock type information, in IEEE Transactions on Multimedia, 1(4), (1999), 321–333.
6. J. Nang, S. Hong, Y. Ihm, "An efficient video segmentation scheme for MPEG video stream using Macroblock information", Proc. of the ACM International Conference on Multimedia, pp. 23-26, 1999.
7. S.M. Bhandarkar, A.A. Khombhadia, "Motion-based parsing of compressed video", Proc. of the IEEE International Workshop on Multimedia Database Management Systems, pp. 80-87, Aug. 1998.
8. B.L. Yeo, B. Liu, Rapid Scene Analysis on Compressed Video, IEEE Transactions on Circuits and Systems for Video Technology, 5(6), (1995), 533-544.

9. S.W. Lee, Y.M. Kim, S.W. Choi, Fast Scene Change Detection using Direct Features Extraction from MPEG Compressed Videos, IEEE Transactions on Multimedia, 2(4), (2000), 240-254.
10. N.V. Patel, I.K. Sethi, Compressed video processing for cut detection, IEE Proceedings on Vision, Image and Signal Processing, 143(5), (1996), 315–323.
11. S.S. Yu, J.R. Liou, W.C. Chen, Computational similarity based on chromatic barycenter algorithm, IEEE Transactions on Consumer Electronics, 42(2), (1996), 216-220.
12. L. Xu, A. Krzyzak, C.Y. Suen, Methods of Combining Multiple Classifiers and Their Application to Handwritten Numeral Recognition, IEEE Transactions on Systems, Man and Cybernetics 1992; 22(3), (1992), 418-435.

Bayesian Networks for Incorporation of Contextual Information in Target Recognition Systems

Keith Copsey and Andrew Webb

QinetiQ
St Andrews Road, Malvern, Worcs, WR14 3PS, UK
K.Copsey@signal.QinetiQ.com
webb@signal.QinetiQ.com

Abstract. In this paper we examine probabilistically the incorporation of contextual information into an automatic target recognition system. In particular, we attempt to recognise multiple military targets, given measurements on the targets, knowledge of the likely groups of targets and measurements on the terrain in which the targets lie. This allows us to take into account such factors as clustering of targets, preference to hiding next to cover at the extremities of fields and ability to traverse different types of terrain. Bayesian networks are used to formulate the uncertain causal relationships underlying such a scheme. Results for a simulated example, when compared to the use of independent Bayesian classifiers, show improved performance in recognising both groups of targets and individual targets.

1 Introduction

In this paper we examine probabilistically the incorporation of contextual information and domain specific knowledge into an automatic target recognition (ATR) system for military vehicles.

In a realistic ATR scenario, after an initial detection phase, there will be a set of multiple locations which have been identified as possibly occupied by targets/vehicles. Appropriate measurements (*e.g.* radar, sonar, infra-red, *etc*) will then be taken at each of these locations, so that classifications can be made. Some of the measurements might be from real targets, while others will be false alarms from clutter objects, or just background noise. Most standard ATR techniques[7] will consider each potential target independently.

In this work, we look at how, in a military scenario, the posterior probabilities of class membership at each location, can be combined with additional domain specific knowledge. This reflects the fact that a human assigning classes to measurements would take into account contextual information as well as the data measurements themselves. The use of this sort of additional contextual information by an operator might be stronger than just having a closer look at the data measurements in certain locations; it may tip the balance towards (or

T. Caelli et al. (Eds.): SSPR&SPR 2002, LNCS 2396, pp. 709–717, 2002.
© Springer-Verlag Berlin Heidelberg 2002

away from) certain classes. Thus, two nearly identical measurements may actually be assigned to different classes, depending on their respective contextual information.

The type of contextual information that can be incorporated could include the proximity of other vehicles, recognising that military targets will often travel in groups. A human operator might also pay more attention to the extremities of fields close to hedges and woodland edges, reflecting the fact that military commanders would consider their vehicles exposed in the centre of a field and might choose to get them as close to cover as possible. Further domain specific knowledge that could be brought to the problem includes the type of terrain that surrounds the target and knowledge about the likely deployment of military vehicles, such as formations.

Simply altering our set of possible classes, based on the contextual information, is not appropriate, since there are almost always going to be uncertainties. For example, our estimates of the surrounding terrain might be in error. The most appropriate formalism for handling the possibly conflicting pieces of information in a consistent manner is probabilistic. Thus, conventional expert systems[5] are not appropriate. However, a Bayesian network[3, 4] based on the causal relationships leading to a deployment of targets within a region, can be used in a probabilistic way to integrate domain specific knowledge with the actual data measurements.

1.1 Bayesian Networks

Bayesian networks (also referred to as belief networks, Bayesian graphical models and probabilistic expert systems) can be used to model situations in which causality plays a role, but where there is a need to describe things probabilistically, since our understanding of the processes occurring is incomplete or, indeed, the processes themselves have a random element.

A Bayesian network can be illustrated graphically as a series of nodes, which may assume particular states, connected by directional arrows. Figure 1 shows such a network. The states of a node can be discrete or continuous. Every node has an associated conditional probability table or density, specifying the probability or probability density of the node being in a certain state given the states of the nodes with arrows pointing towards it. Nodes with no arrows pointing towards them are assigned prior distributions over their states. Given observations on some of the nodes, beliefs are propagated up through the network to obtain posterior probabilities for the unobserved nodes.

1.2 Problem Definition

We focus on the situation where (after an initial detection phase) there is a set of objects at estimated locations, with each object being a potential target/vehicle. Each of these objects needs to be assigned to a class, *i.e.* either a specific class of target or just clutter. A single multi-dimensional measurement of each object (*i.e.* at each location) is available, as well as estimates of the terrain of the region surrounding the objects. This terrain estimate consists of the division of the overall region into sub-regions, each with an associated local terrain (*e.g.* field, marsh, *etc*). These sub-regions are separated by boundaries, which are of unspecified type. The boundaries are allowed to split sub-regions of the same local terrain, so fields split by hedges or walls are treated as separate sub-regions.

In our work we consider a subset of the contextual or domain specific information that can be incorporated into such an ATR problem. In particular, we focus on the proximity of other vehicles, the distances to boundaries, the immediate type of terrain and known groupings of targets.

1.3 Related Work

A related approach to the work documented here is given by Blacknell[1], who looked at the incorporation of contextual information in SAR target detection, by altering the prior probabilities of targets, depending on terrain, the distances to boundaries and the proximity of other potential targets. The use of Bayesian networks to exploit contextual information for vehicle detection in infrared line-scan imagery has been reported by Ducksbury *et al*[2]. Musman and Lehner[6] use Bayesian networks to schedule a shipboard self-defence mechanism.

2 Incorporation of Domain Specific Knowledge

Our proposed Bayesian network is illustrated in Fig. 1. The nodes denote states of affairs and the arcs can be interpreted as causal connections.

The "groups" node represents the collections/groups that targets are likely to be deployed from, while the "terrain" node represents the terrain over the region

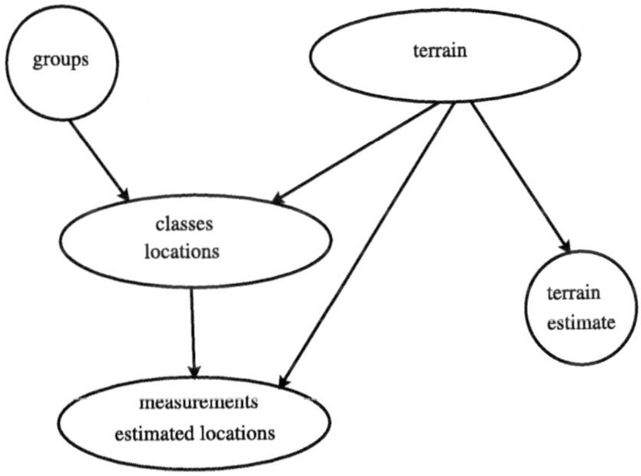

Fig. 1. Bayesian network formulation for the incorporation of contextual information in an ATR system

of interest. The "terrain estimate" node is made up of our observations of the terrain in the region of interest, based, for example, on Synthetic Aperture Radar (SAR) image analysis. The node labelled "classes" and "locations" represents the classes and locations of the objects, whereas the node labelled "measurements" and "estimated locations" contains our measurements of the objects and our estimates of the locations.

In our scenario, after an initial detection phase, we have a set of n_l potential target locations, $\hat{l} = \{\hat{l}_i; 1 \leq i \leq n_l\}$, with corresponding data measurements $x = \{x_i; 1 \leq i \leq n_l\}$.

There are $J \geq 1$ target types and these are supplemented with a noise/clutter class, giving, in total, $J+1$ classes. The measurements at the potential target locations can, therefore, be referred to as coming from a collection of n_l objects, each of which belongs to one of the $J+1$ classes.

2.1 Groups of Targets

Deployed targets are taken to come from specific groups of targets/vehicles, which we denote by the discrete random variable, G. The cover of G is the set of possible groups, which is assigned using expert knowledge. If the number of targets within a group is less than the number of potential target locations, n_l, the group is augmented with clutter measurements. The prior probabilities for the states of G would, ideally, be assigned using intelligence information.

2.2 Terrain

The random variable representing the terrain, denoted T, is a broad variable covering many aspects and is made up of both continuous and discrete elements. In our case, this includes the positions of the boundaries of the sub-regions of the area of interest and the local terrain types within each sub-region. We suppose that the local terrain types must belong to a discrete set of n_τ terrain types, such as "field", "urban", "marsh" *etc.*

Attempting to assign decent prior distributions for the constituents of T is a hard task, since the number and locations of the sub-regions are random variables. Assigning flat priors is possible, but this makes later inference awkward, since the states have to be integrated over. However, progress can be made if our conditional distribution for the observation of the terrain given the actual terrain is restrictive. This is covered in Section 2.5.

2.3 Classes and Locations

The random variable C denotes the classes at each of the n_l potential target locations. A state of C consists of a n_l-dimensional vector, $c = (c_1, \ldots, c_{n_l})$, with elements the classes for each of the n_l objects. The class allocations variable is coupled with the locations variable, L, which contains the actual locations of the objects. A state of L consists of a n_l-dimensional vector, $l = (l_1, \ldots, l_{n_l})$, with elements the locations for each of the n_l objects. A pair $(C, L) = (c, l)$ is referred to as an allocation of targets.

The conditional distribution for this node of the Bayesian network is, for ease of notation, denoted by $p(c, l|g, t)$. Its specification (Section 2.6) allows incorporation of contextual information.

2.4 Measurements and Locations

The measurements, x and estimated locations, $\hat{l}$, depend on the actual classes and locations, $(C, L) = (c, l)$ and the terrain $T = t$, via the conditional distributions, $p(x, \hat{l}|c, l, t)$.

We take the data measurements to be conditionally independent and assume:

$$p(x, \hat{l}|c, l, t) = \prod_{i=1}^{n_l} p(x_i|c_i, l_i, t)p(\hat{l}_i|l_i, t). \tag{1}$$

The distribution $p(x_i|c_i, l_i, t)$ is the measurement distribution for the class, c_i, in the terrain, t, at location, l_i. In practice we frequently take this distribution to be independent of the terrain and given by $p(x_i|c_i)$, although this is not necessary in our Bayesian network approach. Specification of these distributions is, of course, very difficult and the subject of much research interest[7].

The distribution $p(\hat{l}_i|l_i, t)$ is generally taken to be a δ-function, $p(\hat{l}_i|l_i, t) = \delta(\hat{l}_i - l_i)$, so that the measured locations are the same as the actual locations. This simplification is for computational reasons in specifying the conditional probability distributions, $p(c, l|g, T)$ and in propagating evidence in the network.

A standard Bayesian classifier comprises only the distributions $p(x_i|c_i)$, along with some very simple prior probabilities for the classes, $p(c_i)$. Classifications are made to the *maximum a posteriori* (MAP) class, as determined by the posterior class probabilities from Bayes' rule, $p(c_i|x_i) \propto p(x_i|c_i)p(c_i)$.

2.5 Terrain Estimate

The random variable $\hat{T}$ representing our estimates of the terrain, depends on the actual terrain via the conditional distribution $p(\hat{T}|T)$. Similarly to Section 2.2, the terrain estimate consists of the positions and boundaries of sub-regions, along with their respective local terrain types.

A full treatment of the possible states (and associated conditional distribution) is not feasible, because of the requirement to specify the distribution $p(c, l|g, t)$ for each allowable t. Thus, we are forced to assume that the number of sub-regions is correctly observed, as are the boundaries of these sub-regions. However, the observations/estimates of the local terrain types within these boundaries can be erroneous. Thus, the conditional distribution, $p(\hat{T}|T)$, is defined using a $n_T \times n_T$ matrix of the conditional probabilities of the terrain type estimates given the actual terrain types. This matrix would be determined by consultation with experts, who could take into account the techniques used to estimate local types of terrain.

2.6 The Conditional Distributions of Target Allocations

If suitable training data were available, the conditional distribution, $p(c, l|g, t)$, could be learnt from the data[3, 4]. However, the availability of such data is often quite limited, so we rely on expert opinion and representative values to determine the distribution.

Our conditional distribution is expressed as a product of weights:

$$p(c, l|g, t) \propto w_{\text{bndry}}(c, l, t) \times w_{\text{clust}}(c, l, t) \times w_{\text{terr}}(c, l, t) \times w_{\text{grp}}(c, l, g, t), \quad (2)$$

where $w_{\text{bndry}}(c, l, t)$ is a factor related to the distances of vehicles from boundaries, $w_{\text{clust}}(c, l, t)$ is a factor related to the clustering of targets, $w_{\text{terr}}(c, l, t)$ is a factor related to the local types of terrain at the object locations and $w_{\text{grp}}(c, l, g, t)$ is a factor relating the allocation defined by (c, l) to the group of targets g. Due to a lack of space we do not go into the details of these weighting factors.

3 Using the Bayesian Network

Having specified our Bayesian network we need to be able to calculate the required posterior distributions of the nodes of the network, based on our measurements, $(x, \hat{l}, \hat{t})$. This is referred to as updating the beliefs in a Bayesian network. There are a number of ways of making inference on Bayesian networks and these are well documented in [3, 4]. Whatever the method, the posterior node distributions of interest are $p(g|x, \hat{l}, \hat{t})$, $p(c, l|x, \hat{l}, \hat{t})$ and $p(c, l|g, x, \hat{l}, \hat{t})$. These correspond to the marginal posterior probabilities of the groups; the marginal posterior probabilities for the allocations; and the marginal posterior probabilities for the allocations conditional on the group.

In this paper we use a "long-hand" approach for belief updating, which explicitly carries out the summations required for each marginal or conditional posterior distribution of interest. This has the advantage of being quick to program and is exact. However, as the number of states in the network increases, the direct summation approach to updating beliefs becomes computationally complex (the computational cost increases exponentially with the number of objects to be examined).

3.1 Using the Probabilities

The posterior probabilities contain our updated beliefs about the objects and should be incorporated into a model that reflects the whole decision making process. By considering the expected posterior loss of decisions, account can be made of the different costs involved in making erroneous decisions. However, for the purposes of this paper, single classifications are proposed, based on the MAP probabilities of interest.

The most likely group is taken to be the group, $\hat{g}$, that maximises the probability $p(g|x, \hat{l}, \hat{t})$. The most likely allocation of classes is the allocation maximising the probability $p(c, l|x, \hat{l}, \hat{t})$. Also of interest is the allocation maximising $p(c, l|\hat{g}, x, \hat{l}, \hat{t})$.

3.2 Standard Bayesian Classifier

For the standard Bayesian classifier (described in Section 2.4), determination of the most likely group is awkward. To assess the performance of the Bayesian network in determining groups we have therefore invented a simple group assignment scheme for the standard Bayesian classifier. In particular, we assign a set of objects to belong to a specific group, if all the targets in the group are present within the set of classes from the MAP class probabilities for the objects.

4 Simulated Example

4.1 Scenario

The performance of the Bayesian network approach (compared to a standard Bayesian classifier) is illustrated with a simulated example.

We consider a problem with three target classes, namely main battle tanks (MBTs), armoured personnel carriers (APCs) and air defence units (ADUs), which on addition of the clutter class gives a four class problem. Target deployments are taken to come from three possible groupings. The 1st group, g_1, consists of 3 MBTs and an APC. The 2nd group, g_2, consists of 4 APCs and the 3rd group, g_3, consists of 2 ADUs and an APC. We consider the situation where intelligence information indicates that group g_1, is more likely than group g_2, which is, in turn, more likely than group g_3.

For demonstration purposes, the measurement distributions are taken to be given by 2-dimensional Gaussian distributions, which have been chosen to overlap. In our experiments, these measurement distributions are assumed known. This affects both the Bayesian network and the standard Bayesian classifier. It is, of course, unrealistic in an actual ATR scenario, but allows us to focus on the effects of the contextual information.

A terrain has been simulated, dividing the area of interest into five subregions (each of which has a local terrain type). Scenarios of target deployments following our expected target behaviour (*i.e.* travelling together and close to cover) for each of the three groups have been simulated, along with some additional clutter objects. These scenarios are illustrated in Fig. 2. In each case, 5 objects have been picked up in the (unspecified) initial detection phase. For scenario 1 (a deployment for g_1), from top to bottom the objects and their respective classes are O-5 (Clutter), O-4 (APC), O-1, O-3 and 0-2 (all MBTs), respectively. For scenario 2 (a deployment for g_2), from top to bottom the objects are O-5 (Clutter), O-1, O-4, O-2 and 0-3 (all APCs). Finally, for scenario 3 (a deployment for g_3), from top to bottom the objects are 0-2 (ADU), O-3 (APC), O-4 (Clutter), O-1 (ADU) and O-5 (Clutter).

For each of the three scenarios, 500 sets of measurements for the objects have been simulated. Performance within each scenario is estimated from the results for that scenario's sets of measurements.

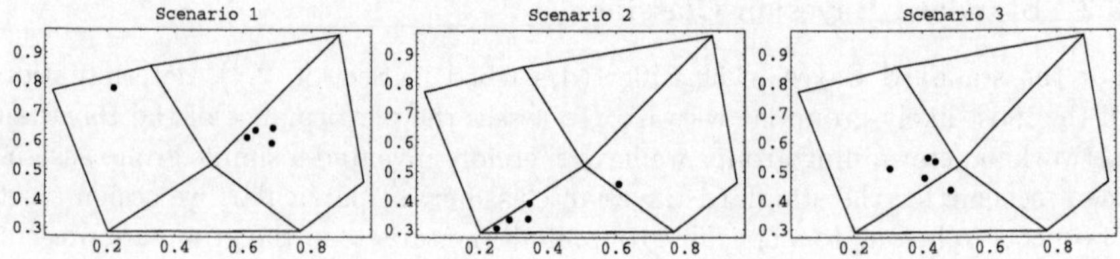

Fig. 2. Scenario for each of the three groups, with objects marked by dots

4.2 Experimental Results

The percentages of correctly identified groups for the Bayesian network are 85.0%, 90.2% and 63.0% for the 1st, 2nd and 3rd scenarios respectively. This outperforms significantly the corresponding standard Bayesian classifier results of 16.0%, 48.2% and 56.6%.

In Fig. 3 we show the classification rates for each of the five objects, in each of the three scenarios (with the underlying classes of the objects detailed in Section 4.1). For the first two scenarios, the Bayesian network approaches can be seen to out-perform the standard approach significantly. For the third scenario, the standard approach marginally out-performs the Bayesian networks for the first two objects (ADUs) and the clutter objects, but the Bayesian network approaches have better performance on the APC. There is little to choose between the two Bayesian network approaches.

The poorer performance of the network on the 3rd scenario, in terms of improvement over the the standard classifier, is an artefact of the lower prior probability assigned to the 3rd group (which comes from the simulated intelligence information that the 3rd group is less likely than the other two groups).

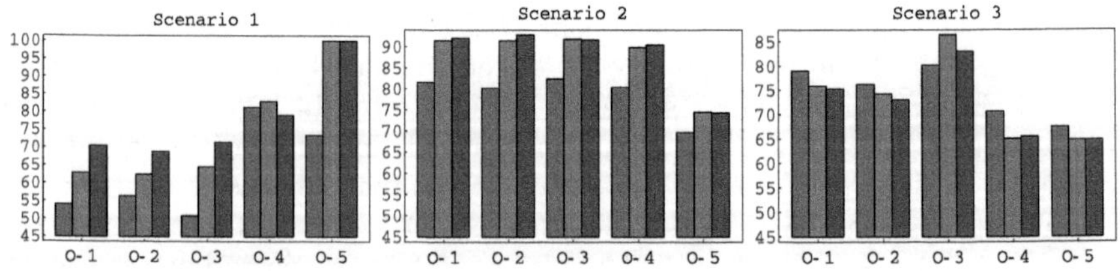

Fig. 3. The classification rates for the two techniques based on the Bayesian network, compared to the standard Bayesian classifier. For each object and scenario, from left to right, the bars correspond to the standard classifier, the Bayesian network with MAP $p(c, l|x, \hat{l}, \hat{t})$, and the Bayesian network with MAP $p(c, l|\hat{g}, x, \hat{l}, \hat{t})$

5 Summary and Discussion

In this work we have shown how Bayesian networks can be used to incorporate domain specific knowledge and contextual information into an ATR system, used for multiple target recognition of military vehicles. Given measurements on the terrain in which the targets lie, we have taken into account such factors as clustering of targets, preference to hiding next to cover at the extremities of fields and the varying abilities of vehicles to traverse different types of terrain. These have been combined in a consistent probabilistic manner, with the information contained in measurements of the targets and with prior knowledge on the groupings of targets. A potential area for further research is the incorporation of other contextual factors into the system, the major difficulty lying in the translation of the factors into appropriate conditional probability distributions.

In a simulated scenario, the Bayesian network has been shown to outperform classification using a standard Bayesian classifier (which uses only the target measurements), both in terms of recognising groups of targets and the performance at specific locations. Currently, the technique has been tested only on simulated data. Future research will need to assess the approach on real data.

Acknowledgments

This research was sponsored by the UK MOD Corporate Research Programme.
©Copyright QinetiQ 2002.

References

[1] D. Blacknell. Contextual information in SAR target detection. *IEE Proceedings - Radar, Sonar and Navigation*, 148(1):41–47, February 2001. 711

[2] P. G. Ducksbury, D. M. Booth, and C. J. Radford. Vehicle detection in infrared linescan imagery using belief networks. *Proceedings of 5th International Conference on Image processing and its applications, Edinburgh, UK*, July 1995. 711

[3] F. V. Jensen. *Introduction to Bayesian Networks.* Springer-Verlag, 1997. 710, 714

[4] M. I. Jordan. *Learning in Graphical Models.* The MIT Press, February 1999. 710, 714

[5] G. Luger and W. Stubblefield. *Artificial Intelligence: Structures and Strategies for Complex Problem Solving.* Addison Wesley, 2nd edition, 1993. 710

[6] S. Musman and P. Lehner. Real-time scheduling under uncertainty for ship self defence. *Submitted to IEEE Expert Special Issue on Real-time Intelligent Systems*, 1998. 711

[7] A. R. Webb. *Statistical Pattern Recognition.* John Wiley & Sons, Chichester, 2nd edition, August 2002. 709, 713

Extending LAESA Fast Nearest Neighbour Algorithm to Find the *k* Nearest Neighbours

Francisco Moreno-Seco, Luisa Micó, and Jose Oncina*

Dept. Lenguajes y Sistemas Informáticos
Universdad de Alicante, E-03071 Alicante, Spain
{paco,mico,oncina}@dlsi.ua.es

Abstract. Many pattern recognition tasks make use of the k nearest neighbour (k–NN) technique. In this paper we are interested on fast k–NN search algorithms that can work in any metric space i.e. they are not restricted to Euclidean–like distance functions. Only symmetric and triangle inequality properties are required for the distance.
A large set of such fast k–NN search algorithms have been developed during last years for the special case where $k = 1$. Some of them have been extended for the general case. This paper proposes an extension of LAESA (Linear Approximation Elimination Search Algorithm) to find the k-NN.

1 Introduction

The k nearest neighbour problem consists in finding the k nearest points (prototypes) from a database to a given point sample using a dissimilarity function $d(\cdot, \cdot)$. This problem appears often in computing problems and, of course, in pattern recognition tasks [2].

Usually, a brute force approach is used but, when the database is large and/or the dissimilarity function is computationally expensive, this approach results in a real bottleneck.

In this paper we are interested in fast k-NN algorithms that can work in any metric space *i.e.* the algorithm is not restricted to work with Euclidean–like dissimilarity functions, and no assumption is made about the point's data structure. It is only required that the dissimilarity function fulfils the following conditions:

- $d(x, y) = 0 \Leftrightarrow x = y$.
- $d(x, y) = d(y, x)$ (symmetry).
- $d(x, z) \leq d(x, y) + d(y, z)$ (triangle inequality).

That is, the dissimilarity function defines a *metric space*, and thus can be properly called a *distance* function.

* The authors wish to thank the Spanish CICyT for partial support of this work through project TIC2000–1703-CO3-02.

T. Caelli et al. (Eds.): SSPR&SPR 2002, LNCS 2396, pp. 718–724, 2002.
© Springer-Verlag Berlin Heidelberg 2002

Such algorithms can efficiently find the k–NN when the points are represented by structures like strings, trees or graphs and the distance functions can be some variants of the edit distance ([9], [10]).

Many general metric space k–NN fast search algorithms have been developed trough these years for the special case where $k = 1$ (Fukunaga and Narendra's [3], Kalantary and McDonald's [5], AESA [8], LAESA [7], TLAESA [6], ...). One of these algorithms has been extended for the general case (k-AESA [1]).

This paper proposes an extension of LAESA (Linear Approximation Elimination Search Algorithm) fast 1–NN search algorithm to cope with the k–NN problem.

2 The LAESA

As an evolution of AESA algorithm, LAESA is a branch and bound algorithm based on the triangle inequality.

In a preprocessing step a set of n_{bp} *base prototypes* are selected and their distances to the rest of prototypes are stored in a table.

When searching the nearest neighbour of a sample s, first a lower bound ($g[\cdot]$) of the distance from each prototype (p) to the sample is computed. This lower bound distance is based on the triangle inequality and can be computed as follows:

$$g[p] = MAX_{i=1}^{n_{bp}}(|d(b_i, p) - d(s, b_i)|) \tag{1}$$

where $d(b_i, p)$ is the precomputed distance between p and the base prototype b_i, and $d(s, b_i)$ is the actual distance between the sample s and b_i.

After that, the prototype set is traversed in ascending order of $g[\cdot]$ until a prototype with a distance to the sample lower than the lower bound distance is found. Then the traversal stops and the nearest prototype to the sample found so far is outputed as the nearest neighbour.

The performance of the algorithm depends on the number of base prototype and the way they are selected. In [7] it was shown that selecting the prototypes so that they are maximally separated is a good choice. A pseudo-algorithmic description of LAESA is shown in figure 1.

LAESA was devised to work with very time consuming distances, then, in practice, the time cost of the algorithm is dominated by the number of distance computations (n_d). In [7] it was shown that in most usual cases n_{bp} is practically independent of the database size (n) and can be set to a number much smaller that n. Nevertheless, worst-case time complexity can be expressed as $O(n + n_d \log n)$, but since n_d in practice does not grow with n, the expended time grows linearly with n. Please note that n_d includes the distances to base prototypes, so n_d is always bigger than n_{bp}. The space complexity of LAESA can be expressed as $O(n_{bp}n)$.

Preprocessing (given the number n_{bp} of base prototypes)

1. Select the n_{bp} base prototypes B maximally separated
2. Compute and store the distances $d(b, p)\ \forall b \in B\ \forall p \in P$

Classification (input: the sample s; output: the nearest neighbour of s, $p_{\min}$)

1. compute and store the distances $d(b, s)\ \forall b \in B$
2. $p_{\min} = \text{argmin}_{b \in B} d(b, s)$ and compute the lower bound $g[p]\ \forall p \in P$
3. for all p in P in ascending order of $g[p]$
   (a) if $g[p] > d(p_{\min}, s)$ stop the algorithm
   (b) compute $d(p, s)$; if $d(p, s) < d(p_{\min}, s)$ then $p_{\min} = p$

Fig. 1. The LAESA

3 Extending LAESA to *k*–LAESA

Instead of stopping the algorithm when the lower bound distance of the current prototype is bigger than its distance to the sample (line 3a), the algorithm is stopped when the lower bound distance is bigger that the kth nearest neighbour found so far.

K–LAESA must store the k nearest neighbours found up to the moment. As k is lower than n the space and time complexity does not change.

As can be expected, our experiments show that the number of distance computations increases as the value of k increases. Despite of this, the total number of distance computations remains much lower than exhaustive k–NN search, thus k–LAESA can be very useful when distance computations are very expensive.

4 Experiments

K–LAESA is intended for tasks where a very time consuming distance is required. The actual bottleneck in these tasks is the number of distance computations (n_d). Thus, the experiments reported here are focused exclusively on the number of distance computations for several tasks.

In a first set of experiments, 4 and 10 dimension spaces along with the Euclidean distance were used. Of course, there are some specially designed fast k–NN algorithms for such spaces that can beat LAESA and k–LAESA (remind that LAESA was devised for very time consuming distances). Those experiments are included just to show the behaviour of LAESA in a well known metric space.

Second, some experiments with misspelled words and the edit distance were performed to show k–LAESA's behaviour in its application field.

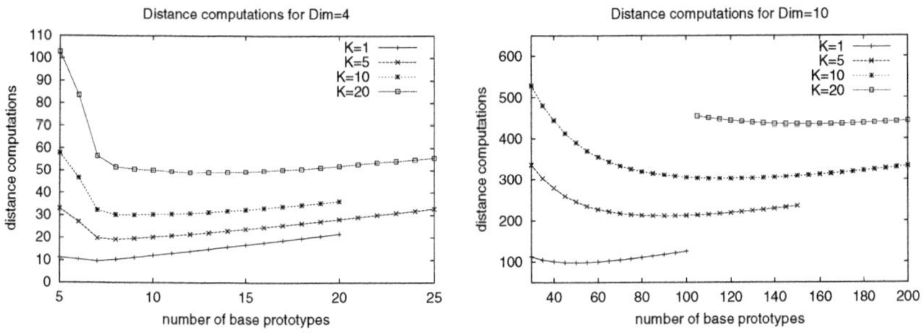

Fig. 2. Searching for optimal *number of base prototypes* in uniform distributions

Table 1. Optimal values of n_{bp} for uniform distribution data

VALUE	Dimensionality	
OF k	4	10
1	7	50
5	8	90
10	9	120
20	12	155

4.1 Experiments with the Euclidean Distance

For these experiments the well-known uniform distribution on the 4 and 10 dimension unit hipercube has been chosen as a reference.

First the optimal number of base prototypes has to be found. Then the evolution of the number of distance computations (n_d), for different values of k (1, 5, 10, 20), is studied when the number of prototypes grows.

As shown in figure 2, the optimal values of n_{bp} depends on the dimension and on the value of k. Those values (table 1) are used on the following experiments.

Next, the number of distance computations was studied as the database grows (1024, 2048, 4096, 6144 and 8192 prototypes). The test set was a collection of 1024 samples. For each database size, the experiment was repeated for 16 different pairs train/test set in order to obtain sounder results. In figure 3 it can be observed that n_d grows very slightly with respect to the database size, but, as figure 4 shows, n_d grows as the value of k increases (only the results for 2048 prototype database is plotted).

4.2 Experiments with the Edit Distance

In these experiments a dictionary of more than 60000 words was used. To obtain test words, one edition mistake (insertion, deletion or substitution) was equiprobably introduced in each word. Only results for $k = 1$ and $k = 5$ will be reported here.

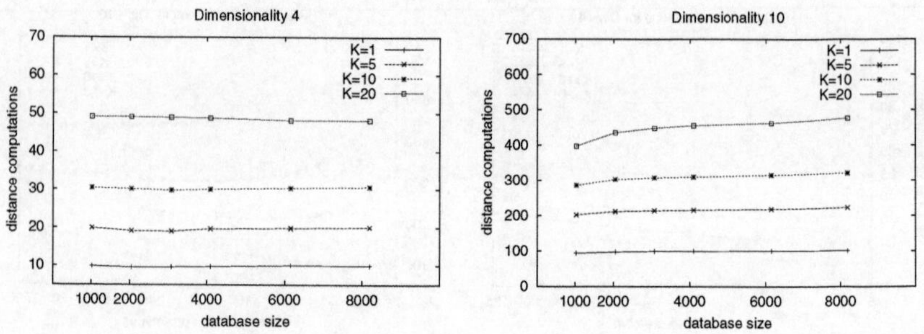

Fig. 3. Distance computations for uniform distributions

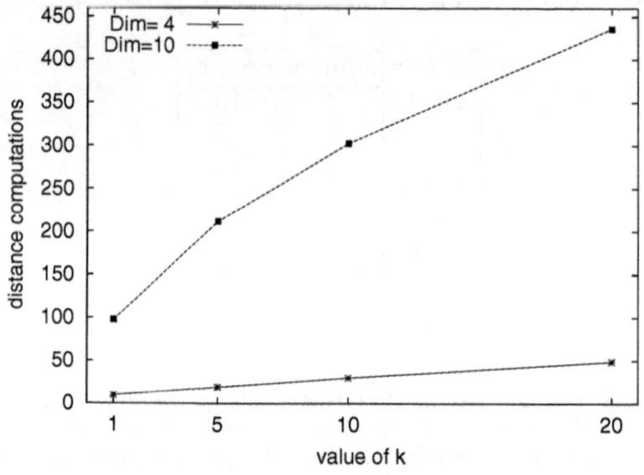

Fig. 4. Distance computations as k increases (uniform distributions)

As in previous experiments, exhaustive experiments were developed to obtain the optimal value of n_{bp} for each value of k; the results are plotted in figure 5. The optimal values obtained were 102 base prototypes for $k = 1$ and 510 base prototypes for $k = 5$. Then, experiments with databases of increasing sizes and 1000 samples were performed. The number of distance computations (n_d) obtained in these experiments is plotted in figure 6, which confirms that the increasing in n_d depends more on the value of k than on the size of the database.

5 Conclusions

We have developed an extension of LAESA to find the k nearest neighbours. This new algorithm (k–LAESA) is intended for tasks where the distance computation is very time consuming. No special data structure is required for points, only the distance is required to fulfill the symetric and the triangle inequality properties.

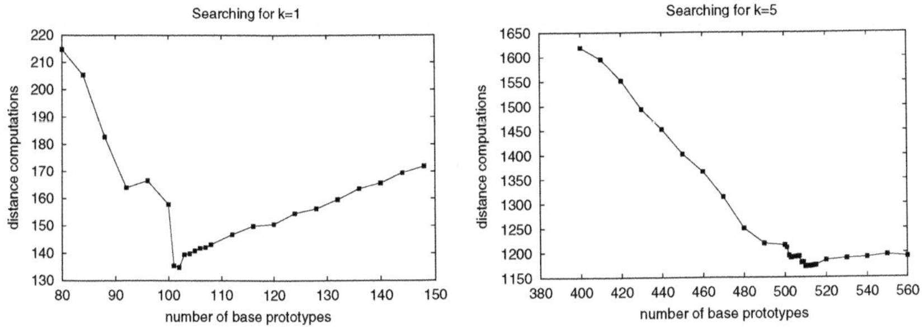

Fig. 5. Searching optimal n_{bp} for distorted words

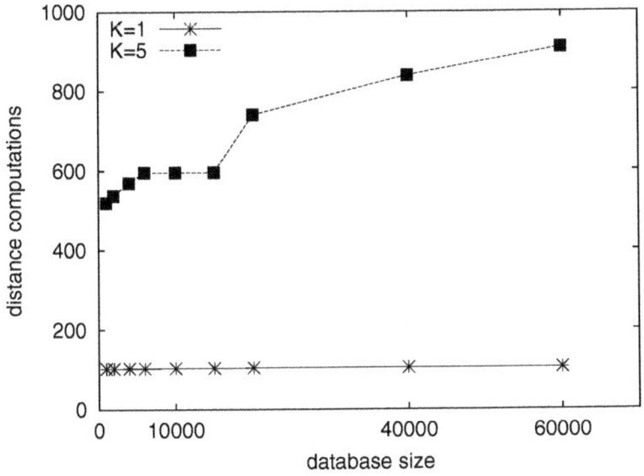

Fig. 6. Distance computations for distorted words

The experiments reported in this work show that the number of distance computations grows with the value of k, but remains always much lower than k–NN exhaustive search. This number of distance computations seems to grow very slowly with the database size. Also, the space required by this algorithm is almost linear with the database size. K–LAESA is good alternative when the distance computation is very time consuming and the database is large.

References

1. Aibar, P., Juan, A., Vidal, E.: Extensions to the approximating and eliminating search algorithm (AESA) for finding k-nearest-neighbours. New Advances and Trends in Speech Recognition and Coding (1993) 23–28 719
2. Duda, R., Hart, P.: Pattern Classification and Scene Analysis. Wiley (1973) 718
3. Fukunaga, K., Narendra, M.: A branch and bound algorithm for computing k–nearest neighbors. IEEE Trans. Computing (1975) **24** 750–753 719
4. Jain, A. K., Dubes, R. C.: Algorithms for clustering data. Prentice-Hall (1988)

5. Kalantari, I., McDonald, G.: A data structure and an algorithm for the nearest point problem. IEEE Trans. Software Engineering (1983) **9** 631–634 719
6. Micó, L., Oncina, J., Carrasco, R. C.: A fast branch and bound nearest neighbour classifier in metric spaces. Pattern Recognition Letters (1996) **17** 731–739 719
7. Micó, L., Oncina, J., Vidal, E.: A new version of the nearest neighbour approximating and eliminating search algorithm (AESA) with linear preprocessing-time and memory requirements. Pattern Recognition Letters (1994) **15** 9–17 719
8. Vidal, E.: New formulation and improvements of the Nearest-Neighbour Approximating and Eliminating Search Algorithm (AESA). Pattern Recognition Letters (1994) **15** 1–7 719
9. Wagner, R. A., Fischer, M. J.: The String-to-String Correction Problem. Journal of the Association for Computing Machinery (1974) **21**(1) 168–173 719
10. Zhang, K., Shasha, D.: Simple fast algorithms for the editing distance between trees and related problems. SIAM Journal of Computing (1989) **18** 1245–1262 719

A Fast Approximated k–Median Algorithm

Eva Gómez–Ballester, Luisa Micó, and Jose Oncina*

Universidad de Alicante, Departamento de Lenguajes y Sistemas Informáticos
{eva,mico,oncina}@dlsi.ua.es

Abstract. The *k-means* algorithm is a well–known clustering method. Although this technique was initially defined for a vector representation of the data, the set median (the point belonging to a set P that minimizes the sum of distances to the rest of points in P) can be used instead of the mean when this vectorial representation is not possible. The computational cost of the set median is $O(|P|^2)$. Recently, a new method to obtain an approximated median in $O(|P|)$ was proposed. In this paper we use this approximated median in the *k–median* algorithm to speed it up.

1 Introduction

Given a set of points P, the k-clustering of P is defined as the partition of P into k distinct sets (clusters)[2]. The partition must have the property that the points belonging to each cluster are *most similar*.

Clustering algorithms may be divided in different categories [9], although in this work we are interested in clustering algorithms based on cost function optimization. Usually, in this type of clustering, the number of clusters k is kept fixed.

A well–known algorithm based on cost function optimization is the *k–means* algorithm described in figure 1.

In particular, the *k–means* algorithm finds locally optimal solutions using as cost function the sum of the distances between each point and its nearest cluster center (the mean). This cost function can be formuled as

$$J(C) = \sum_{i \in k} \sum_{p \in C_i} ||p - m_i||^2, \quad (1)$$

where $C = \{C_1, C_2 \ldots, C_k\}$ and $m_1, m_2 \ldots m_k$ are, respectively, the clusters and the means of each cluster.

Although the *k–means* algorithm was developed for a vector representation of data (the defined cost function uses the euclidean distance), a more general definition can be used for data where a no vector representation is possible, or is not a good alternative. For example, in character recognition, speech recognition or any application of syntactic recognition, data can be represented using strings.

* The authors thank the Spanish CICyT for partial support of this work through project TIC2000–1703–CO3–02

T. Caelli et al. (Eds.): SSPR&SPR 2002, LNCS 2396, pp. 725–733, 2002.
© Springer-Verlag Berlin Heidelberg 2002

algorithm k–Means Clustering
input: P : set of points; k : number of classes;
output: $m_1, m_2 \ldots m_k$
select the initial cluster representatives $m_1, m_2 \ldots m_k$
do
  classify the $|P|$ points according to nearest m_i
  recompute $m_1, m_2 \ldots m_k$
until no change in m_i $\forall i$
end algorithm

Fig. 1. The *k–means algorithm*

For this type of problems, the median string can be used to be the representative of each class.

The median can be obtained from a constrained point set P (*set median*) or can be extended to the whole space U where the points are extracted (*generalized median*).

Given a set of points P, the *generalized median string* m is defined as the point (in the whole space U) that minimizes the sum of distances to P:

$$m = argmin_{p \in U} \sum_{p' \in P} d(p, p') \tag{2}$$

When the points are strings and the edit distance is used, this is a NP-Hard problem [4]. A simple greedy algorithm to compute an approximation to the median string was proposed in [6].

When the median is selected among the points belonging to the set, the *set median* is obtained. In this case the median is defined as the point (in the set P) that minimizes de sum of distances to P. In this case, the search of the median is constrained to a finite set of points:

$$m = argmin_{p \in P} \sum_{p' \in P} d(p, p') \tag{3}$$

Unlike the *generalized median string*, the cost of computing the *set median* (using points or strings) is for the best known algorithm $O(|P|^2)$.

When the set median is used in *k–means* algorithm instead of the mean, the new cost function is:

$$J(C) = \sum_{i \in k} \sum_{p \in C_i} d(p, m_i), \tag{4}$$

where $m_1, m_2 \ldots m_k$ are the k medians obtained with the equation (3) leading to the *k–median* algorithm.

Recently, a fast algorithm to obtain an approximated median was proposed [7]. The main characteristics of this algorithm are: 1) no assumption about the structure of the points or the distance function are made and 2) it has a linear time complexity. The experiments showed that very accurate medians can be obtained using appropriate parameters.

In this paper this approximated median algorithm has been used in the *k–median* algorithm instead of the median. So this new approximation can be used in problems, as exploratory data analysis, where data are represented by strings.

Depending on the initialization of the *k–median* algorithm the behaviour is different. Some different initializations of the algorithm have been proposed in the literature ([1,8]). In this work we have used the most simple initialization that consists in selecting randomly k cluster representatives.

In the next section this approximated median algorithm is described. Some experiments using synthetic and real data to compare the approximated and the exact set median are shown in section 3 and finally conclusions are drawn in section 4.

2 The Approximated Median Search Algorithm

Given a set of points P, the algorithm selects as median the point that minimizes the sum of distances of a subset of the complete set ([7]).

The algorithm has two main steps: 1) a subset of n_r points (*reference points*) from the set P is randomly selected. It is very important to select randomly the reference points because it is expected that they have a similar behaviour of the whole set (some experiments were made to support this conclusion in [7]). The sum of distances from each point in P to the reference points are calculated and stored. 2) The n_t points of P whose sum of distances is lowest are selected (*test points*). For each test point, the sum of distances to every point belonging to P is calculated and stored. The test point that minimizes this sum is selected as the median of the set. The algorithm is described in figure 2.

The algorithm needs two parameters: n_r and n_t. Experiments reported in [7] show that the choice $n_r = n_t$ is reasonable. Moreover, the use of a small number of n_r and n_t points is enough to obtain accurate medians.

3 Experiments

A set of experiments was made to know the behaviour of the *approximated k–median* in relation to the *k–median* algorithm.

As the objective in the *k–median* algorithm is the minimization of the cost function, the cost function and the expended time have been studied with both algorithms.

Two main groups of experiments with synthetic and real data were made. All experiments (real and synthetic) were repeated 30 times using different random initial medians. Deviations were always below 4% and are not plotted in the figures.

algorithm approximated median
input:
P : set of points; $d(\cdot,\cdot)$: distance function;
n_r : number of reference points; n_t : number of test points
output:
$m \in P$: median
var:
U : used points (reference and test); T : test points;
PS : array of $|P|$ partial sums; FS : array of $|P|$ full sums
// **Initialization**
$U = \emptyset$
$\forall p \in P$
$PS[p] = 0$
// **Selecting the reference points**
repeat n_r times
$u =$ random point in $P - U$
$U = U \cup \{u\}$
$FS[u] = PS[u]$
$\forall p \in P$
$d = d(p, u)$
$PS[p] = PS[p] + d$
$FS[u] = FS[u] + d$
// **Selecting the test points**
$T = n_t$ points in $P - U$ that minimize $PS[\cdot]$
// **Calculating the full sums**
$\forall t \in T$
$FS[t] = PS[t]$
$U = U \cup \{t\}$
$\forall p \in P - U$
$d = d(t, p)$
$FS[t] = FS[t] + d$
$PS[p] = PS[p] + d$
// **Selecting the median**
$m =$ the point in U that minimizes $FS[\cdot]$
end algorithm

Fig. 2. The approximate median search algorithm

Experiments with Synthetic Data

To generate synthetic clustered data, the algorithm proposed in [5] has been used. This algorithm generates random synthetic data from different classes (clusters) with a given maximum overlap between them. Each class follows a Gaussian distribution with the same variance and different randomly chosen means. For the experiments presented in this work, syntethic data from 4 and 8 classes was generated with dimensions 4 and 8. The overlap was set to 0.05 and the variance to 0.03.

Table 1. Expended time (in seconds) by the *k–median* and the *approximated k–median* algorithm for a set of 2048 prototypes using the 1% threshold stop criterion

DIMENSIONALITY 4				
NUMBER OF CLASSES	*k–m*	*ak–m (40)*	*ak–m (80)*	*ak–m (320)*
4	2.86	0.29	0.54	1.82
8	3.11	0.54	1.01	1.92
DIMENSIONALITY 8				
NUMBER OF CLASSES	*k–m*	*ak–m (40)*	*ak–m (80)*	*ak–m (320)*
4	2.09	0.25	0.44	1.46
8	2.16	0.46	0.86	1.74

The first set of experiments was designed to study the evolution of the cost function in each iteration for *k-median* and *approximated k–median* algorithms. 2048 points were generated using 4 and 8 clusters from spaces of dimension 4 and 8 (512 and 256 points respectively for each class) (see figure 3). In the *approximated k–median* three different sizes of the *used points* ($n_r + n_t$) were used (40, 80 and 320). As shown in this figure, the behaviour of the cost function is similar in all the cases. It is important to note that the *approximated k–median* algorithm computes much less distances than the *k–median.* For example, in the experiment with 4 classes, the number of distance computations in one iteration is around 500,000 for the *k–median* and 80,000 for the *approximated k–median* algorithm with 40 *used points.*

Note that the cost function decreases quickly and, afterwards, stays stable until the algorithm finishes. Then, the stop criterion can be relaxed to stop the iteration when the cost function changes less than a threshold. Using a 1% threshold in the experiments of figure 3, all the experiments would be stopped around the tenth iteration.

In table 1 the expended time by both algorithms are represented for a set of 2048 points. In the *approximated k–median* three different numbers of *used points* were used (40, 80 and 320) with the 1% threshold stop criterion. The time was measured on a Pentium III running at 800 MHz under a Linux system.

Table 1 show that the time expended by the *approximated k–median* using any of the three different sizes of *used points*, is much lower than the *k–median* algorithm.

In the last experiment with synthetic data, the expended time was measured with different set sizes. Figure 4 illustrates that the use of the *approximated k–median* reduces drastically the expended time when the set size increases.

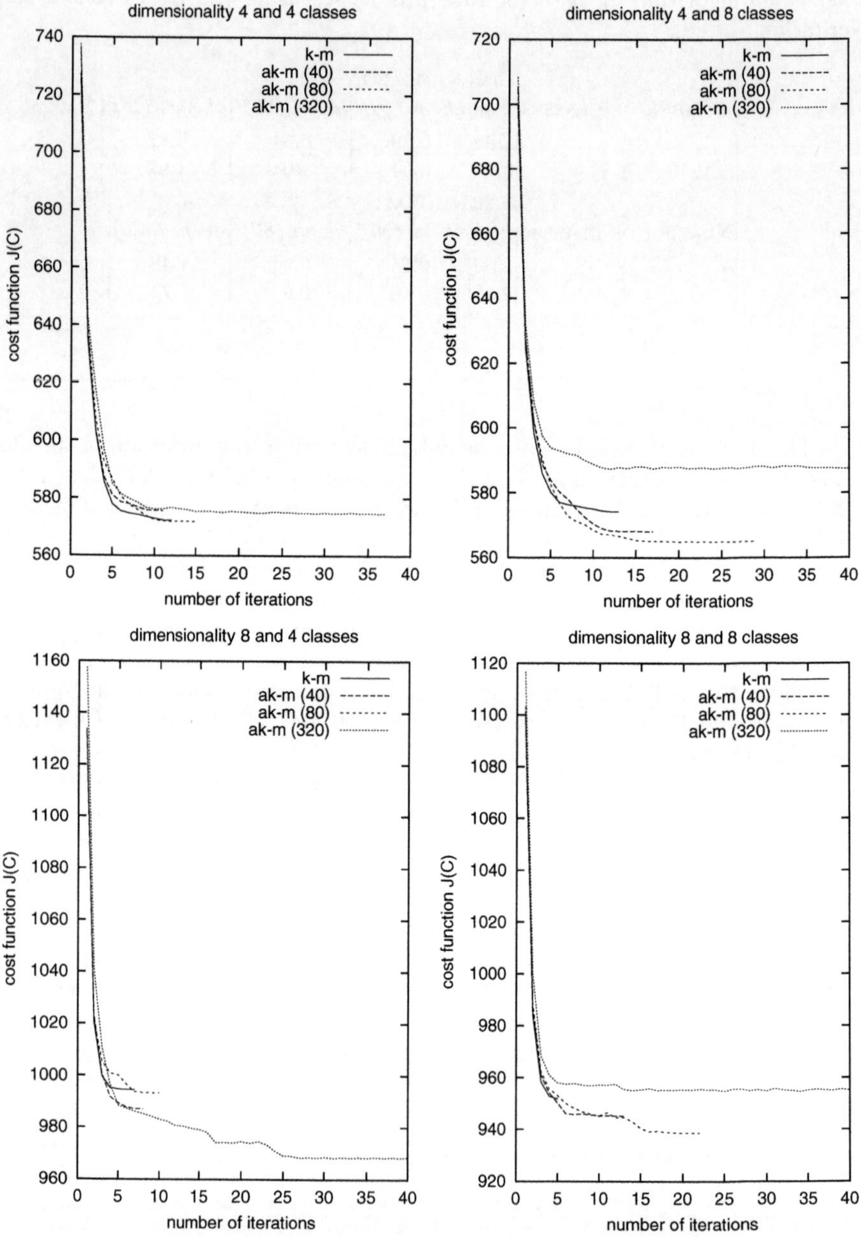

Fig. 3. Comparison of the cost function using 4 and 8 classes for a set of 2048 points with dimensionality 4 and 8

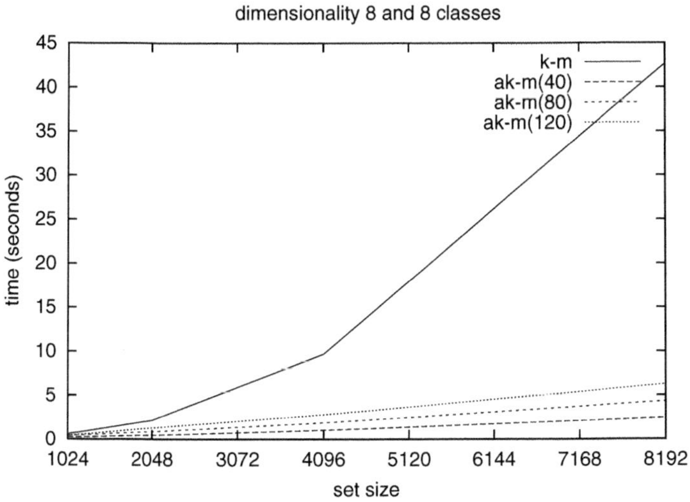

Fig. 4. Expended time (in seconds) by the *k–median* and the *approximated k–median* algorithm when the size of the set increases using the 1% threshold stop criterion

3.1 Experiments with Real Data

For real data, a chain code description of the handwritten digits (10 writers) of the NIST Special Database 3 (National Institute of Standards and Technology) was used. Each digit has been coded as an octal string that represents the contour of the image.

The edit distance [3] with deletion and insertion cost set to 1 was used. The substitution costs are proportional to the relative angle between the directions (in particular 1, 2, 3 and 4 were used).

As can be seen in figure 5, the results are similar to the synthetic data experiments. Moreover, as it was said previously for synthetic data, the process for the *approximated k–median* can also be stopped when the value of the cost function between two consecutive iterations is lower than a threshold. These results are showed in figure 6.

As figure 6 shows, a relatively low number of used points (20 and 40) is enough to obtain a similar behaviour of the cost function with different set sizes. Moreover, the expended time grows linearly with the set size in the *approximated k–median*, while this increase is quadratic for the *k–median* algorithm.

4 Conclusions

In this work an effective fast *approximated k–median* algorithm is presented. This algorithm has been obtained using an approximated set median instead of

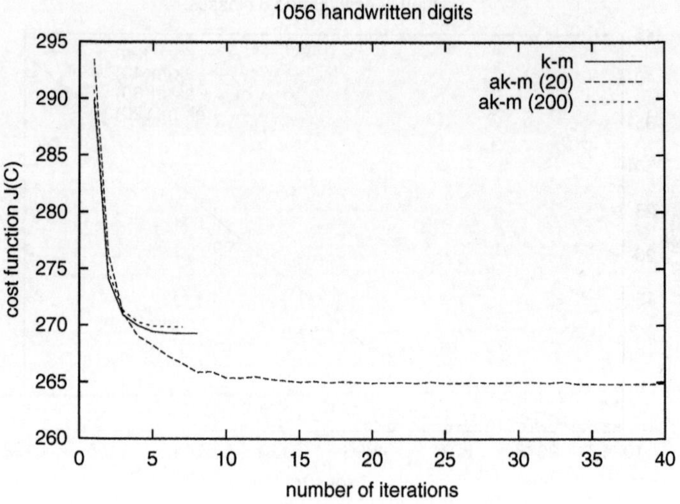

Fig. 5. Comparison of the cost function using 1056 digits

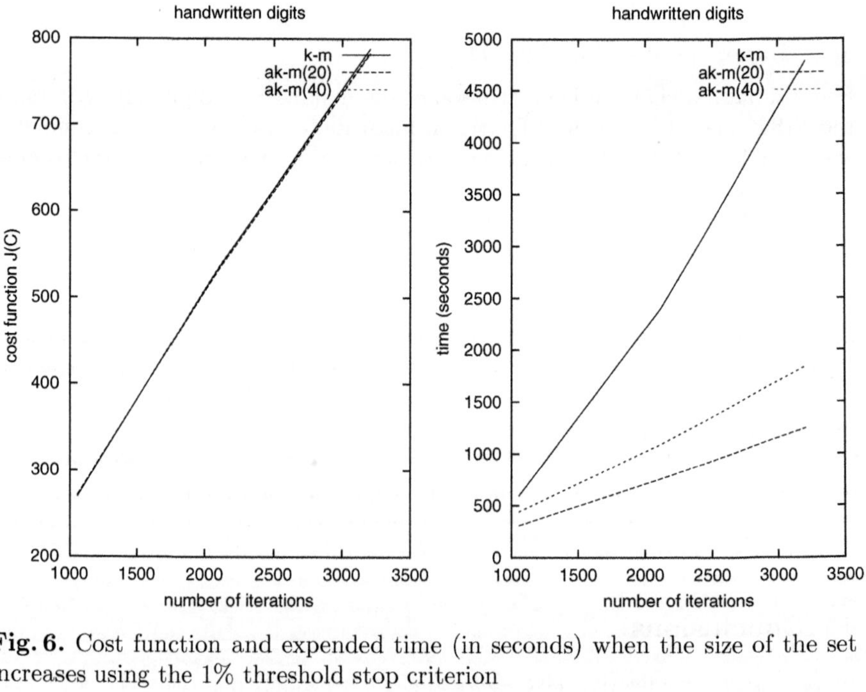

Fig. 6. Cost function and expended time (in seconds) when the size of the set increases using the 1% threshold stop criterion

the set median in the *k–median* algorithm. As the computation of approximated medians is very fast, its use speeds the *k–median* algorithm. The experiments show that a low number of *used points* in relation to the complete set is enough to obtain similar value for the cost function in the *approximated k–median.*

In future works we will develop some ideas related to the stop criterion and the initialization of the algorithm.

Acknowledgement

The authors wish to thank to Francisco Moreno–Seco for permit us the use of his synthetic data generator.

References

1. Bradley, P. S., Fayyad, U. M.: Refining Initial Points for K–Means Clustering. Proc. 15th International Conf. on Machine Learning (1998). 727
2. Duda, R., Hart, P., Stork, D.: Pattern Classification. Wiley (2001). 725
3. Fu, K. S.: Syntactic Pattern Recognition and Applications. Prentice–Hall, Englewood Cliffs, NJ (1982). 731
4. de la Higuera, C., Casacuberta, F.: The topology of strings: two NP–complete problems. Theoretical Computer Science **230** 39–48 (2000). 726
5. Jain, A. K., Dubes, R. C.: Algorithms for clustering data. Prentice-Hall (1988). 728
6. Martínez, C., Juan, A., Casacuberta, F.: Improving classification using median string and nn rules. In: Proceedings of IX Simposium Nacional de Reconocimiento de Formas y Análisis de Imágenes, 391–394 (2001). 726
7. Micó, L., Oncina, J.: An approximate median search algorithm in non–metric spaces. Pattern Recognition Letters **22** 1145–1151 (2001). 726, 727
8. Peña, J. M., Lozano, J. A., Larrañaga, P.: An empirical comparison of four initialization methods for the K–means algorithm. Pattern Recognition Letters **20** 1027–1040 (1999). 727
9. Theodoridis, S., Koutroumbas, K.: Pattern Recognition. Academic Press (1999). 725

A Hidden Markov Model-Based Approach to Sequential Data Clustering

Antonello Panuccio, Manuele Bicego, and Vittorio Murino

Dipartimento di Informatica, University of Verona
Ca' Vignal 2, Strada Le Grazie 15, 37134 Verona, Italy
{panuccio,bicego,murino}@sci.univr.it

Abstract. Clustering of sequential or temporal data is more challenging than traditional clustering as dynamic observations should be processed rather than static measures. This paper proposes a Hidden Markov Model (HMM)-based technique suitable for clustering of data sequences. The main aspect of the work is the use of a probabilistic model-based approach using HMM to derive new proximity distances, in the likelihood sense, between sequences. Moreover, a novel partitional clustering algorithm is designed which alleviates computational burden characterizing traditional hierarchical agglomerative approaches. Experimental results show that this approach provides an accurate clustering partition and the devised distance measures achieve good performance rates. The method is demonstrated on real world data sequences, i.e. the EEG signals due to their temporal complexity and the growing interest in the emerging field of Brain Computer Interfaces.

1 Introduction

The analysis of sequential data is without doubts an interesting application area since many real processes show a dynamic behavior. Several examples can be reported, one for all is the analysis of DNA strings for classification of genes, protein family modeling, and sequence alignment.

In this paper, the problem of unsupervised classification of temporal data is tackled by using a technique based on Hidden Markov Models (HMMs). HMMs can be viewed as stochastic generalizations of finite-state automata, when both transitions between states and generation of output symbols are governed by probability distributions [1]. The basic theory of HMMs was developed in the late 1960s, but only in the last decade it has been extensively applied in a large number of problems, as speech recognition [1], handwritten character recognition [2], DNA and protein modeling [3], gesture recognition [4], behavior analysis and synthesis [5], and, more in general, to computer vision problems.

Related to sequence clustering, HMMs has not been extensively used, and a few papers are present in the literature. Early works were proposed in [6,7], all related to speech recognition. The first interesting approach not directly linked to speech issues was presented by Smyth [8], in which clustering was faced by devising a "distance" measure between sequences using HMMs. Assuming each

T. Caelli et al. (Eds.): SSPR&SPR 2002, LNCS 2396, pp. 734–743, 2002.
© Springer-Verlag Berlin Heidelberg 2002

model structure known, the algorithm trains an HMM for each sequence so that the log-likelihood (LL) of each model, given each sequence, can be computed. This information is used to build a LL distance matrix to be used to cluster the sequences in K groups, using a hierarchical algorithm.

Subsequent work, by Li and Biswas [9,10], address the clustering problem focusing on the model selection issue, i.e. the search of the HMM topology best representing data, and the clustering structure issue, i.e. finding the most likely number of clusters. In [9], the former issue is addressed using standard approach, like Bayesian Information Criterion [11], and extending to the continuous case the Bayesian Model Merging approach [12]. Regarding the latter issue, the sequence-to-HMM likelihood measure is used to enforce the within-group similarity criterion. The optimal number of clusters is then determined maximizing the Partition Mutual Information (PMI), which is a measure of the inter-cluster distances. In the second paper [10], the same problems are addressed in terms of Bayesian model selection, using the Bayesian Information Criterion (BIC) [11], and the Cheesman-Stutz (CS) approximation [13]. Although not well justified, much heuristics is introduced to alleviate the computational burden, making the problem tractable, despite remaining of elevate complexity. Finally, a model-based clustering method is also proposed in [14], where HMMs are used as cluster prototypes, and Rival Penalized Competitive Learning (RPCL), with state merging is then adopted to find the most likely HMMs modeling data. These approaches are interesting from the theoretical point of view, but they are not tested on real data. Moreover, some of them are very computationally expensive.

In this paper, the idea of Smyth [8] has been extended by defining a new metric to measure the distance, in the likelihood sense, between sequences. Two clustering algorithms are proposed, one based on the hierarchical agglomerative approach, and the second based on a partitional method, variation of the K-means strategy. Particular care has been posed on the HMM training initialization by utilizing a Kalman filtering and a clustering method using mixture of Gaussians. Finally, and most important, the proposed algorithm has been tested using real data sequences, the electroencephalographic (EEG) signals. Analysis of this kind of signals became very important in the last years, due to the growing interest in the field of *Brain Computer Interface (BCI)* [15]. Among all we choose these signals for their temporal complexity, suitable for HMM modeling.

The rest of the paper is organized as follows. In Sect. 2, HMM will be introduced. Section 3 describes how the EEG signal has been modeled and the specific initialization phase of the proposed approach. The core of the algorithm is presented in Sect. 4, in which the definition of distances and the clustering algorithms will be detailed. Subsequently, experimental results are presented in Sect. 5, and, finally, conclusions are drawn in Sect. 6.

2 Hidden Markov Models

A discrete HMM is formally defined by the following elements [1]:

- *A set* $S = \{S_1, S_2, \cdots, S_N\}$ *of (hidden) states.*
- *A state transition probability distribution*, also called transition matrix $A = \{a_{ij}\}$, representing the probability to go from state S_i to state S_j.

$$a_{ij} = P[q_{t+1} = S_j | q_t = S_i] \qquad 1 \leq i, j \leq N, \quad a_{ij} \geq 0, \quad \sum_{j=1}^{N} a_{ij} = 1 \quad (1)$$

- *A set* $V = \{v_1, v_2, \cdots, v_M\}$ *of observation symbols.*
- *An observation symbol probability distribution*, also called emission matrix $B = \{b_j(k)\}$, indicating the probability of emission of symbol v_k when system state is S_j.

$$b_j(k) = P[v_k \text{ at time t } | q_t = S_j] \qquad 1 \leq j \leq N, 1 \leq k \leq M \qquad (2)$$

  with $b_i(k) \geq 0$ and $\sum_{j=1}^{M} b_j(k) = 1$.
- *An initial state probability distribution* $\pi = \{\pi_i\}$, representing probabilities of initial states.

$$\pi_i = P[q_1 = S_i] \qquad 1 \leq i \leq N, \quad \pi_i \geq 0, \quad \sum_{i=1}^{N} \pi_i = 1 \qquad (3)$$

For convenience, we denote an HMM as a triplet $\lambda = (A, B, \pi)$.

All of our discussion has considered only the case where the observation was characterized as a sequence of discrete symbols chosen from a finite alphabet. In most application, observations are continuous signals. Although it is possible to quantize such continuous signals via codebooks, it would be advantageous to be able to use HMMs with continuous observation densities. In this case the emission probability distribution B becomes

$$P(O|j) = b_j(\mathbf{O}) = \sum_{m=1}^{M} c_{jm} \mathcal{M}[\mathbf{O}, \mu_{jm}, \mathbf{\Sigma}_{jm}] \qquad (4)$$

where $\mathbf{O}$ is observation vector being modeled, c_{jm} is the mixture coefficient for the mth mixture in state j and $\mathcal{M}$ is any log-concave or elliptically symmetric density (e.g. Gaussian density). The adaption of reestimation formulas of Baum-Welch procedure for the continuous case is straightforward [16].

Although the general formulation of continuous density HMMs is applicable to a wide range of problems, there is one other very interesting class of HMMs that seems to be particularly suitable for EEG signals: the autoregressive HMMs [17]. In this case, the observation vectors are drawn from an autoregression process. In the next section it is explained how these models are applied to EEG modeling.

3 EEG Signal Modeling

Electroencephalographic (EEG) signals represent the brain activity of a subject and give an objective mode of recording brain stimulation. EEGs are an useful tool used for understanding several aspects of the brain, from diseases detection to sleep analysis and evocated potential analysis. The system used to model EEG signal is largely based on Penny and Roberts paper [18]: the key idea above this approach is to train an autoregressive HMM directly on the EEG signal, rather than use an intermediate AR representation. Each HMM state can be associated with a different dynamic regime of the signal, determined using a Kalman Filter approach [19]. Kalman filter is used to preliminary segment the signal in different dynamic regimes: these estimates are then fine-tuned with HMM model. The approach is briefly resumed in the rest of this section.

3.1 Hidden Markov AR Models

This type of models differs from those defined in Sect. 2 by the definition of observation symbol probability distribution. In this case B is defined as

$$P(y_t|q_t = S_i) = N(y_t - \mathbf{F}_t\hat{\mathbf{a}}_i, \sigma_i^2) \tag{5}$$

where $\mathbf{F}_t = -[y_{t-1}, y_{t-2}, \cdots, y_{t-p}]$, $\hat{\mathbf{a}}_i$ is the (column) vector of AR coefficients for the ith state and σ_i^2 is the estimated observation noise for the i-th state, estimated using Jazwinski method [20]. The prediction for the ith state is $\hat{y}_t^i = \mathbf{F}_t\hat{\mathbf{a}}_i$. The order of AR model is p.

The HMM training procedure is fundamentally a gradient descent approach, sensitive to initial parameters estimate. To overcome this problem, a Kalman filter AR model is passed over the data, obtaining a sequence of AR coefficients. Coefficients corresponding to low evidence are discarded. Others are then clusterized with Gaussian Mixture Models [21]. The center of each Gaussian cluster is then used to initialize the AR coefficients in each state of the HMM-AR model.

The number of clusters (i.e. the number of HMM states) and the order of autoregressive model were decided by performing a preliminary analysis of classification accuracy. Varying number of states from 4 to 10, and varying order of autoregressive model from 4 to 8, we have found that best configuration was $K = 4$ and $p = 6$. The classification accuracy obtained was about 2% superior than one obtained using Neural Network [22] on same data, showing that Hidden Markov Models are more effective in modeling EEG signals.

To initialize the transition matrix we used prior knowledge from the problem domain about average state duration densities. We use the equation $a_{ii} = 1 - \frac{1}{d}$ to let HMM remain in state i for d samples. This number is computed knowing that EEG data is stationary for a period of the order of half a second [23].

4 The Proposed Method

Our approach, inspired by [8], can be depicted by the following algorithm:

1. We train an $m-$states HMM for each sequence S_i, $(1 \leq i \leq N)$ of the dataset D. These N HMM are identified by λ_i, $(1 \leq i \leq N)$ and have been initialized with a Kalman filter AR model as described in Sect. 3.
2. For each model λ_i we evaluate its probability to generate the sequence S_j, $1 \leq j \leq N$, obtaining a measure matrix L where

$$L_{ij} = P(S_j|\lambda_i), \qquad 1 \leq i, j \leq N \tag{6}$$

3. We apply a suitable clustering algorithm to the matrix L obtaining K clusters on the data set D.

This method aims to exploits the measure defined by (6) which naturally expresses the similarity between two observation sequences. Through the use of Hidden Markov Models, that are able to describe a sequence with a simple scalar number, we could transform the difficult task of clustering sequences in the easier one of clustering points.

About step 3 we can apply several clustering algorithms but first of all we need to "symmetrize" the matrix L because the result of step 2 is not really a distance matrix. Thus we define

$$L_S^{ij} = \frac{1}{2}\left[L_{ij} + L_{ji}\right] \tag{7}$$

Another kind of HMM based measure that we applied, which remind the Kullback-Leibler information number, defines the distance L_{KL} between two HMM λ_i and λ_j, and its symmetrized version L_{KLS}, as

$$L_{KL}^{ij} = L_{ii}\left[\ln\frac{L_{ii}}{L_{ji}}\right] + L_{ij}\left[\ln\frac{L_{ij}}{L_{jj}}\right], \quad L_{KLS}^{ij} = \frac{1}{2}\left[L_{KL}^{ij} + L_{KL}^{ji}\right] \tag{8}$$

Finally, we introduced another measure, called BP metric, defined as

$$L_{BP}^{ij} = \frac{1}{2}\left\{\frac{L_{ij} - L_{ii}}{L_{ii}} + \frac{L_{ji} - L_{jj}}{L_{jj}}\right\} \tag{9}$$

motivated by the following considerations: the measure (6), defines a similarity measure between two sequences S_i and S_j as the likelihood of the sequence S_i with respect to the model λ_j, trained on S_j, without really taking into account the sequence S_j. In other words this kind of measure assumes that all sequences are modeled with the same quality without considering how well sequence S_j is modeled by the HMM λ_j: this could not always be true. Our proposed distance also considers the modeling goodness by evaluating the relative normalized difference between the sequence and the training likelihoods. About step 3 we investigated two clustering algorithms [21], namely

- Complete Link Agglomerative Hierarchical Clustering: this class of algorithms produces a sequence of clustering of decreasing number of clusters at each step. The clustering produced at each step results from the previous one by merging two clusters into one.
- Partitional Clustering: this methods obtains a single partition of the data instead of a clustering structure, such as a dendogram produced by hierarchical technique. Partitional method have advantages in application involving large data sets for which the construction of a dendogram is computationally prohibitive. In this context we developed an ad hoc partitional method described in the next section and henceforth called "DPAM".

4.1 DPAM Partitional Clustering Algorithm

The proposed algorithm shares the ideas of the well known k-means techniques. This method finds the optimal partition by evaluating at each iteration the distance between each item and each cluster descriptor, and assigning it to the nearest class. At each step, the descriptor of each cluster will be reevaluated by averaging its cluster items. A simple variation of the method, partition around medoid (PAM) [24], determines each cluster representative by choosing the point nearest to the centroid. In our context we cannot evaluate centroid of each cluster because we only have item distances and not values.

To address this problem a novel algorithm is proposed. This method is able to determine cluster descriptors in a PAM paradigm, using item distances instead of their values. Moreover, the choice of the initial descriptors could affect algorithm performances. To overcome this problem we have adopted a multiple initialization procedure, where the best resulting partition is determined by a sort of Davies-Bouldin criterion [21].

Fixed η as the number of tested initializations, N the number of sequences, k the number of clusters and L the proximity matrix characterized by previously defined distances (7), (8), (9), the resulting algorithm is the following:

- for t=1 to η
  - Initial cluster representatives θ_j are randomly chosen ($j = 1, \ldots, k$, $\theta_j \in \{1, \ldots, N\}$).
  - Repeat:
    - *Partition evaluation step:* Compute the cluster which each sequence S_i, $i = 1, \ldots, N$ belongs to; S_i lies in the j cluster for which the distance $L(S_i, \theta_j)$, $i = 1, \ldots, N$, $j = 1, \ldots k$ is minimum.
    - *Parameters upgrade:*
      - Compute the sum of the distance of each element of cluster C_j from each other element of the jth cluster
      - Determine the index of the element in C_j for which this sum is minimal
      - Use that index as new descriptor for cluster C_j

- Until the representatives θ_j values between two successive iterations don't change.
- $\mathcal{R}_t = \{C_1, C_2, \ldots, C_k\}$
- Compute the Davies–Bouldin–like index defined as:

$$\mathcal{DBL}^{(t)} = \frac{1}{k}\sum_{r=1}^{k} \max_{s \neq r} \left\{ \frac{S_c^L(C_r, \theta_r) + S_c^L(C_s, \theta_s)}{L(\theta_r, \theta_s)} \right\}$$

where S_c is an intra–cluster measure defined by:

$$S_c^L(C_r, \theta_r) = \frac{\sum_{i \in C_r} L(i, \theta_r)}{|C_r|}$$

– endfor t
– *Final solution:* The best clustering $\mathcal{R}_p$ has the minimum Davies–Bouldin–like index, viz.: $p = \arg\min_{t=1,\ldots,\eta}\{\mathcal{DBL}^{(t)}\}$

5 Experiments

In order to validate the exposed modeling technique we worked primarily on EEG data recorded by Zak Keirn at Purdue University [25]. The dataset contains EEGs signal recorded from different subjects which were asked to perform five mental tasks: a *baseline* task, for which the subjects were asked to relax as much as possible; the *math task*, for which the subjects were given nontrivial multiplications problems, such as 27*36, and were asked to solve them without vocalizing or making any other physical movements; the *letter task*, for which the subjects were instructed to mentally compose a letter to a friend without vocalizing; the *geometric figure rotation*, for which the subjects were asked to visualize a particular 3D block figure being rotated about an axis; and a *visual counting task*, for which the subjects were asked to image a blackboard and to visualize numbers being written on the board sequentially. We applied the method on a segment-by-segment basis, 1s signals sampled at 250Hz and drawn from a dataset of cardinality varying from 190 (two mental states) to 473 sequences (five mental states) where we removed segments biased by signal spikes arising human artifact (e.g. ocular blinks).

The proposed HMM clustering algorithm has been first applied to two mental states: *baseline* and *math task*, then we extend trials to all available data. Accuracies are computed by comparing the clustering results with real segment labels, percentage is merely the ratio of correct assigned label with respect to the total number of segments. First we applied the hierarchical complete link technique, varying the proximity measure: results are shown in Table 1(a), with number of mental states growing from two to five.

We note that accuracies are quite satisfactory. None of the method experimented can be considered the best one, nevertheless, measures (7) and (8) seem to be more effective. Therefore we applied the partitional algorithm to the same

Table 1. Results for (a) Hierarchical Complete Link and (b) Partitional DPAM Clustering varying the distances defined in (9) BP, (8) KL and (7) SM

(a)

	BP	KL	SM
2 natural clusters	97.37%	97.89%	97.37%
3 natural clusters	71.23%	79.30%	81.40%
4 natural clusters	62.63%	57.36%	65.81%
5 natural clusters	46.74%	54.10%	49.69%

(b)

BP	KL	SM
95.79%	96.32%	95.79%
75.44%	72.98%	65.61%
64.21%	62.04%	50.52%
57.04%	46.74%	44.80%

datasets setting the number of initializations $\eta = 5$ during all the experiments. Results are presented in Table 1(b): in this last case the BP distance is overall slightly better than the others experimented measures. A final comparison of partitional and agglomerative hierarchical algorithms underlines that there are no remarkable differences between the proposed approaches. Clearly, partitional approaches alleviates computational burden, thus they should be preferred when dealing with complex signals clustering (e.g. EEG). The comparison of clustering and classification results (obtained in earlier works) shown that the latter are just slightly better. This strengthen the quality of the proposed method, considering that unsupervised classification is inherently a more difficult task.

6 Conclusions

In this paper we addressed the problem of unsupervised classification of sequences using an HMM approach. These models, very suitable in modeling sequential data, are used to characterize the similarity between sequences in different ways. We extend the ideas exposed in [8] by defining a new metric in likelihood sense between data sequences and by applying to these distance matrices two clustering algorithms: the traditional hierarchical agglomerative method and a novel partitional technique. Partitional algorithms are generally less computational demanding than hierarchical, but could not be applied in this context without some proper adaptations, proposed in this paper. Finally we tested our approach on real data, using complex temporal signals, the EEG, that are increasing in importance due to recent interest in Brain Computer Interface. Results shown that the proposed method is able to infer the natural partitions with patterns characterizing a complex and noisy signal like the EEG ones.

References

1. Rabiner, L. R.: A tutorial on Hidden Markov Models and Selected Applications in Speech Recognition. Proc. of IEEE **77(2)** (1989) 257–286. 734, 736
2. Hu, J., Brown, M. K., Turin, W.: HMM based on-line handwriting recognition. IEEE Trans. Pattern Analysis and Machine Intelligence, **18(10)** (1996) 1039–1045. 734

3. Hughey, R., Krogh, A.: Hidden Markov Model for sequence analysis: extension and analysis of the basic method. Comp. Appl. in the Biosciences **12** (1996) 95–107. 734
4. Eickeler, S., Kosmala, A., Rigoll, G.: Hidden Markov Model based online gesture recognition. Proc. Int. Conf. on Pattern Recognition (ICPR) (1998) 1755–1757. 734
5. Jebara, T., Pentland, A.: Action Reaction Learning: Automatic Visual Analysis and Synthesis of interactive behavior. In 1st Intl. Conf. on Computer Vision Systems (ICVS'99) (1999). 734
6. Rabiner, L. R., Lee, C. H., Juang, B. H., Wilpon, J. G.: HMM Clustering for Connected Word Recognition. Proceedings of IEEE ICASSP (1989) 405–408. 734
7. Lee, K. F.: Context-Dependent Phonetic Hidden Markov Models for Speaker-Independent Continuous Speech Recognition. IEEE Transactions on Acoustics, Speech and Signal Processing **38(4)** (1990) 599–609. 734
8. Smyth, P.: Clustering sequences with HMM, in Advances in Neural Information Processing (M. Mozer, M. Jordan, and T. Petsche, eds.) MIT Press **9** (1997). 734, 735, 738, 741
9. Li, C., Biswas, G.: Clustering Sequence Data using Hidden Markov Model Representation, SPIE'99 Conference on Data Mining and Knowledge Discovery: Theory, Tools, and Technology, (1999) 14–21. 735
10. Li, C., Biswas, G.: A Bayesian Approach to Temporal Data Clustering using Hidden Markov Models. Intl. Conference on Machine Learning (2000) 543–550. 735
11. Schwarz, G.: Estimating the dimension of a model. The Annals of Statistics, **6(2)** (1978) 461–464. 735
12. Stolcke, A., Omohundro, S.: Hidden Markov Model Induction by Bayesian Model Merging. Hanson, S. J., Cowan, J. D., Giles, C. L. eds. Advances in Neural Information Processing Systems **5** (1993) 11–18. 735
13. Cheeseman, P., Stutz, J.: Bayesian Classification (autoclass): Theory and Results. Advances in Knowledge discovery and data mining, (1996) 153–180. 735
14. Law, M. H., Kwok, J. T.: Rival penalized competitive learning for model-based sequence Proceedings Intl Conf. on Pattern Recognition (ICPR) **2** (2000) 195–198. 735
15. Penny, W. D., Roberts, S. J., Curran, E., Stokes, M.: EEG-based communication: a PR approach. IEEE Trans. Rehabilitation Engineering **8(2)** (2000) 214–215. 735
16. Juang, B. H., Levinson, S. E., Sondhi, M. M.: Maximum likelihood estimation for multivariate mixture observations of Markov Chain. IEEE Trans. Informat. Theory **32(2)** (1986) 307–309. 736
17. Juang, B. H., Rabiner, L. R.: Mixture autoregressive hidden Markov models for speech signals. IEEE Trans. Acoust. Speech Signal Proc. **33(6)** (1985) 1404–1413. 736
18. Penny, W. D., Roberts, S. J.: Dynamic models for nonstationary signal segmentation. Computers and Biomedical Research **32(6)** (1998) 483–502. 737
19. Kalman, R. E.: A New Approach to Linear Filtering and Prediction Problems. Transaction of the ASME - Journal of Basic Engineering (1960) 35–45. 737
20. Jazwinski, A.: Adaptive Filtering. Automatica **5** (1969) 475–485. 737
21. Theodoridis, S., Koutroumbas, K.: Pattern Recognition. Academic Press (1999). 737, 738, 739
22. Anderson, C. W., Stolz, E. A., Shamsunder, S.: Multivariate autoregressive models for classification of spontaneous electroencephalogram during mental tasks. IEEE Transactions on Biomedical Engineering, **45(3)** (1998) 277–286. 737
23. Nunez, P. L.: Neocortical Dynamics and Human EEG Rhythms. Oxford University Press, (1995). 737
24. Kaufman, L., Rousseuw, P.: Findings groups in Data: An Introduction to Cluster Analysis. John Wiley & Sons – New York (1990). 739
25. Keirn, Z.: Alternative modes of communication between man and machine. Master's thesis. Purdue University (1988). 740

Genetic Algorithms for Exploratory Data Analysis*

Alberto Perez-Jimenez and Juan-Carlos Perez-Cortes

Departamento de Informatica de Sistemas y Computadores
Universidad Politecnica de Valencia
Camino de Vera, s/n 46071 Valencia, Spain
{aperez,jcperez}@disca.upv.es

Abstract. Data projection is a commonly used technique applied to analyse high dimensional data. In the present work, we propose a new data projection method that uses genetic algorithms to find linear projections, providing meaningful representations of the original data. The proposed technique is compared with well known methods as Principal Components Analysis (PCA) and neural networks for non-linear discriminant analysis (NDA). A comparative study of these methods with several data sets is presented.

1 Introduction

Data projection is a commonly used technique applied to exploratory data analysis [3]. By projecting high dimensional data into a 2- or 3-dimensional space, a better understanding of the structure of the data can be acquired. Characteristics such as clustering tendency, intrinsic dimensionality, similarity among families or classes, etc. can be studied on a planar or tridimensional projection, which also can help to build a classifier or another statistical tool [12][8].

Data projection methods can be divided into linear and non-linear, depending on the nature of the mapping function [7]. They can also be classified as supervised or unsupervised, depending on whether the class information is taken into account or not. The best known linear methods are Principal Component Analysis, or PCA (unsupervised), Linear Discriminant Analysis or LDA (supervised) [3], and projections pursuit [2]. Schematically, PCA preserves as much variance of the data as possible, LDA tries to group patterns of the same class, separating them from the other classes, and, finally, projection pursuit tries to search projections in which points do not distribute normally. On the other hand, well known non-linear methods are: Sammon's Mapping (unsupervised) [10] , non-linear discriminant analysis, or NDA (supervised) [8] and Kohonen's self-organising map (unsupervised) [6]. Sammon's mapping tries to keep the distances among the observations using hill-climbing or neural networks methods [8][10], NDA obtains new features from the coefficients of the hidden layers of a multi-layer perceptron (MLP) and Kohonen Maps project data trying to preserve the topology.

* Work partially supported by the Spanish CICYT under grant TIC2000-1703-CO3-01

T. Caelli et al. (Eds.): SSPR&SPR 2002, LNCS 2396, pp. 743–751, 2002.
© Springer-Verlag Berlin Heidelberg 2002

In the present paper, a new linear supervised data projection method referred to as GLP (genetic linear projection) is proposed. The goal of this method is to find a set of linear projections maximising a certain criterion function. In this work, the accuracy of a Nearest Neighbour classifier has been used as the criterion to maximise. The optimisation is performed by means of a genetic algorithm (GA) [5] [4].

In Section 2 we describe the GLP algorithm, in Section 3 a comparison between a linear method (PCA), a non-linear method (NDA) and the proposed GLP algorithm over several data sets is presented. Finally, some conclusions and further works are presented in section 4.

2 Genetic Linear Projection (GLP)

A linear projection (LP) is defined as follow,

$$LP(x) = c_1 x_1 + c_2 x_1 + \dots c_d x_d,$$

where x is a d-dimensional vector with components x_i, and c_i are the projections coefficients representing the projection axis.

The GLP searches for m (being m the projected space dimensionality) LP's at the same time, optimising the accuracy rate of a Nearest Neighbour classifier. The goal of using this criterion is to preserve the class structure of the data in the projected space. Since the projections obtained are always linear, the representation does not produce an excessive distortion of the original space and therefore the observed data is directly related to the original data.

This criterion does not impose the orthogonality of the projections, as opposed to methods such as PCA or LDA, neither forces the recomputation of the data distribution after choosing each new axis, as in Projection Pursuit.

The number of parameters to estimate by GLP is $m \times d$, since a linear projection is defined by d coefficients, being d the dimensionality of the original data, and m the dimension of the projected space. If we want to project high-dimensional data, the number of parameters to estimate will be large. For that reason, we propose a Genetic Algorithm to carry out the optimisation.

Genetic Algorithms have proved to be specially useful on large search spaces [4]. We have used a GA with the following properties:

- An individual is composed of m chromosomes representing the m LP's to search. Each chromosome contains d genes, holding each a binary string of b bits that encodes a coefficient of the LP in fixed point format.
- For the fitness function, the computed accuracy of a Nearest Neighbour classifier trained with the projected data obtained from the linear projection coded in the individual is used.
- As a genetic selection scheme, a rank-based strategy [9] has been used. In this strategy, the probability of being selected is computed from the rank position of the individuals. This method gave in our case a faster convergence than a fitness-proportionate method.

- Finally, the following setting are used for the rest of parameters: crossover probability is 0.6, mutation probability is 0.001, population size is 100, and the maximum number of generations is 300.

Finally, because to estimate the accuracy of a Nearest Neighbour classifier is a time consuming task. A micro-grain parallel GA [11] has been implemented to reduce computational time. In these algorithms several computers are used to compute individual fitness functions, obtaining a linear speedup.

3 Comparative Study

3.1 Methodology

In this section our GLP method will be compared with the well known PCA (linear, unsupervised) and NDA (non-linear, supervised) methods. The three methods will be applied to four data sets in order to obtain 2-dimensional projections. The data sets used are described below.

- *Digits.* This is a high-dimensional data set containing 3000 patterns, representing 128×128 images of hand-written digits. Each pattern is obtained by resizing images to 14×14 and using gray values as features. The dimension of the data is 196.
- *IRIS.* This data set, obtained from the UCI repository [1], consists of 150 4-dimensional pattern from 3 classes. It contains four measurements on 50 flowers from each of three species of the Iris flower.
- *Cookies.* This synthetic corpus consists of two 10-dimensional normal distributions with
$$\Sigma_1 = \Sigma_2 = \begin{bmatrix} 0.0001 & 0 \cdots 0 \\ 0 & 1 \cdots 0 \\ \vdots & \vdots \ddots \vdots \\ 0 & 0 \cdots 1 \end{bmatrix},$$
$\mu_1 = (+0.1, 0, 0, \ldots)$ and $\mu_2 = (-0.1, 0, 0, \ldots)$, having each class 1000 patterns. These distributions represent two hyperspehers flattened (like cookies) in the dimension they are separated. This data set represents a well known case in which PCA does not work well because the maximal scattered axes are not the most significant.
- *Page Blocks.* This corpus, also obtained from the UCI repository, consists of 5473 10-dimensional patterns representing block documents. Each pattern is represented by 10 features representing geometrical and image properties of the segmented blocks. Blocks are classified into 5 classes.

The performance of these methods will be first compared by means of visual judgement over the 2-dimensional projections obtained from the data sets. And then by means of the error rate of a Nearest Neighbour classifier (E_{NN}) computed for each data set in the original and projected spaces. This quantitative criterion shows how well the class structure is preserved by the projections.

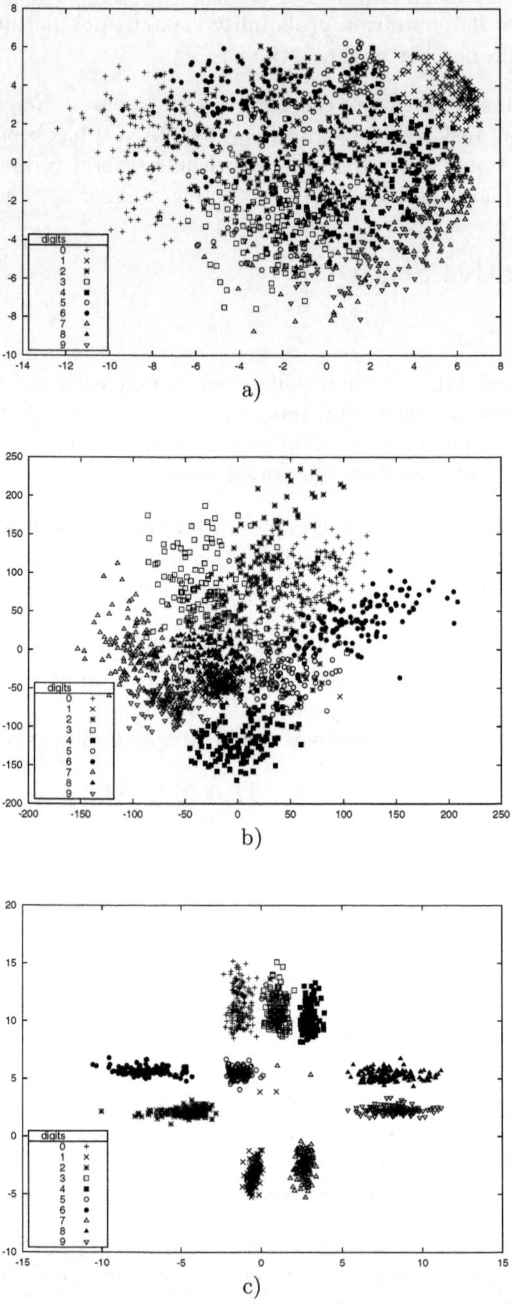

Fig. 1. *Digits* data set 2D projections using: a) PCA, b) GLP and c) NDA

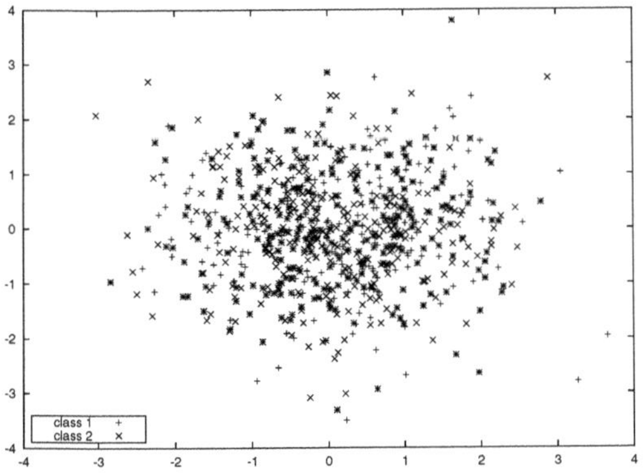

a)

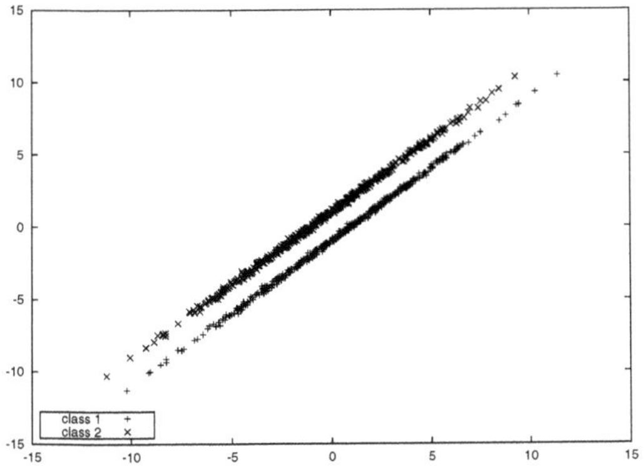

b)

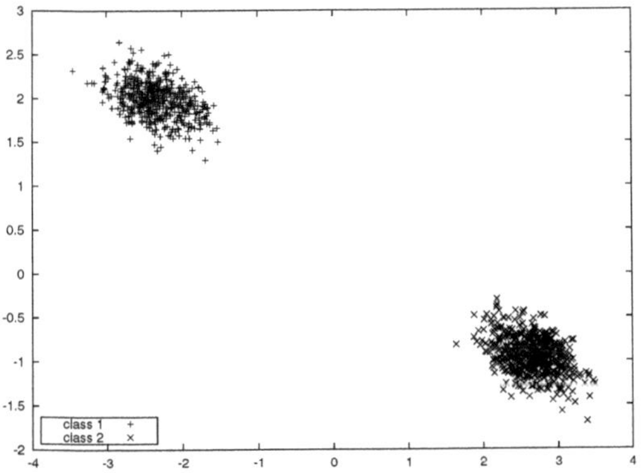

c)

Fig. 2. *Cookies* data set 2D projections using: a) PCA, b) GLP and c) NDA

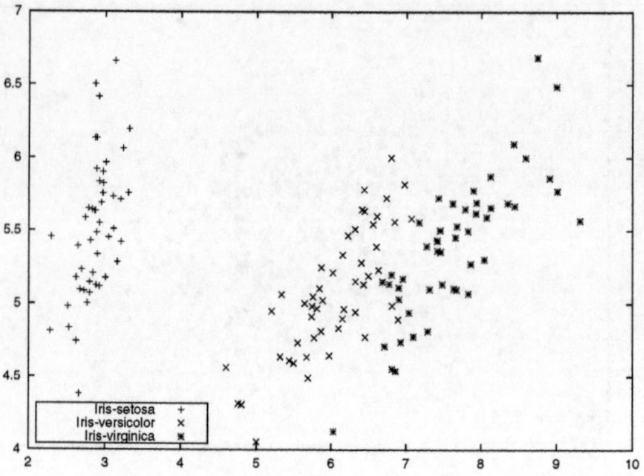

a)

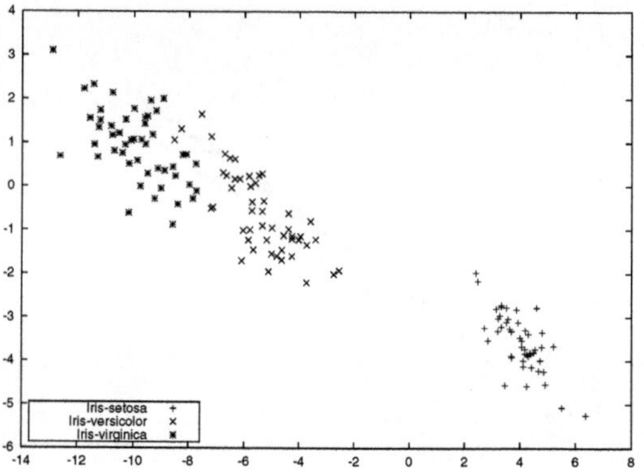

b)

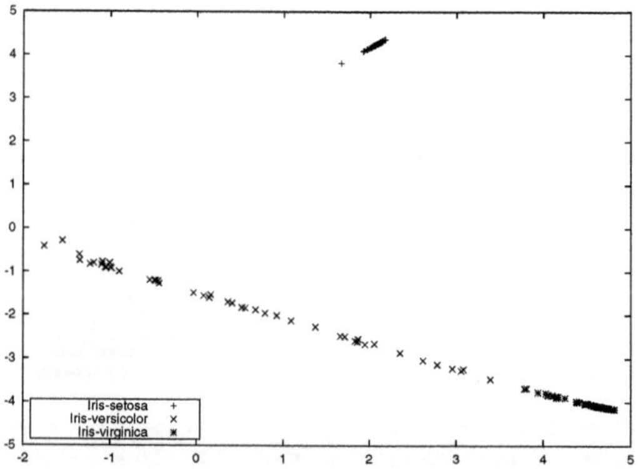

c)

Fig. 3. *Iris* data set 2D projections using: a) PCA, b) GLP and c) NDA

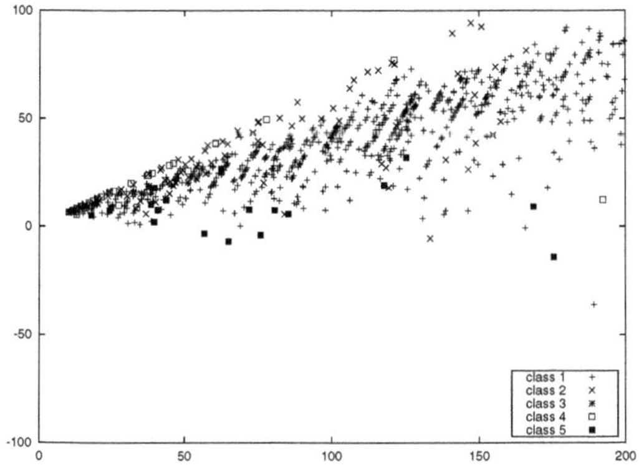

a)

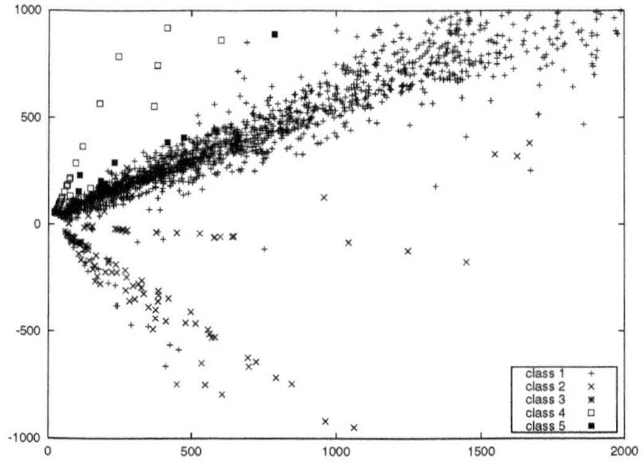

b)

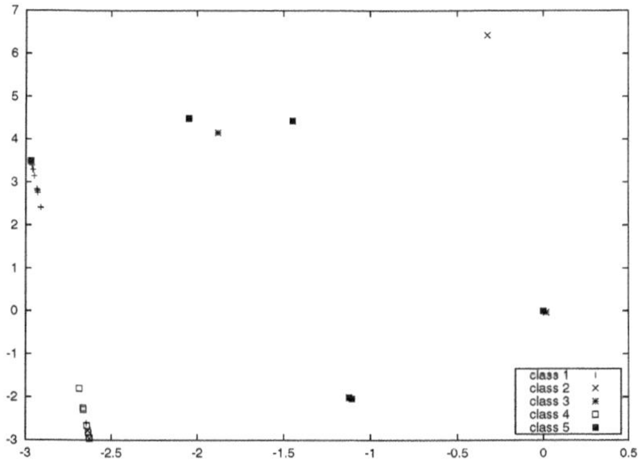

c)

Fig. 4. *Page Blocks* data set 2D projections using: a) PCA, b) GLP and c) NDA

Table 1. Average error rates (%) of the Nearest Neighbour classifier (E_{NN}) computed over the four data sets

	Digits	Iris	Cookies	Page Blocks
ORIGINAL	3.3	4.0	0.4	9.2
PCA	56.3	4.0	42.7	11.4
GLP	24.0 ± 4.4	0.6 ± 0.6	0.3 ± 0.4	3.9 ± 0.8
NDA	0.2 ± 0.4	3.7 ± 1.3	0.0 ± 0.0	8.5 ± 1.5

3.2 Results

These data sets have been projected into a 2-dimensional space. In the case of GLP and NDA methods, 10 runs have been averaged for each data set with different initialisations values. The number of generations necessary to obtain GLP convergence for the *Digits*, *Cookies*, *Iris* and *Page Blocs* data sets was 300, 50, 25 and 50 respectively.

As can be seen from Figure 1a and 2a, PCA projections are not particularly meaningful for the *Digits* and *Cookies* data sets. In them, the directions of maximal data scatter are not interesting. Nevertheless, the projections obtained for the *Iris* and *Page blocks* data sets (Figures 3a and 4a) give an interesting view of the data structure. On the other hand, while GLP projection obtains a view of the *Iris* data set (Figure 3b) similar to the PCA projection, a more interesting view of the rest of data sets is obtained because the class information is considered. In Figure 2b, the cluster structure of the *Cookies* data set appears now clearly. In the same way, a much more meaningful view of the cluster structure from the *Digits* data set (Figure 1b) can be seen. Finally, the NDA projection shows the power of a supervised non-linear method extracting the cluster structure of the data sets. In the case of the *digits* data set, an remarkable view of its strong cluster structure can be seen (Figure 1c).

On the other hand, the study of E_{NN} values (Table 1) leads to similar conclusions. PCA obtains poor results for the *Digits* data set, this is not surprising considering that the original space is 196-dimensional. Results for the *Cookies* data set are particularly bad because the projection found by PCA, completely mixes the classes. GLP outperforms clearly PCA specially for this data set because the optimal projection is found. The NDA method shows that non-linear transformations are necessary to extract the class structure of the data when the intrinsic dimensionality is higher than the projected space dimensionality, this can be shown by the results obtained for data set *Digits*. For the remaining data sets, similar E_{NN} values as in the GLP method have been obtained. In some cases, the GLP method outperforms NDA, although the GLP algorithm is oriented to optimise this criterion, and therefore small differences of E_{NN} values are not important.

4 Conclusions

From the results obtained, it can be concluded that NDA projections outperform our GLP method for high dimensional data. In these cases, the NDA projection is able to extract the class structure even in a 2-dimensional projection. Nevertheless, we consider that NDA shows two important drawbacks. In the first place, because non-linear transformations are used, an important distortion of the original space is obtained, specially when projecting into a 2-dimensional space, trying to preserve the class structure. In these situations, a synthetic view of the configuration of real clusters is obtained. Moreover, the process of training an NDA neural network is not straightforward in many cases. The GLP method uses linear transformations, producing less distorted and more meaningful views of the original space (distortion can appear because the new axes are not necessarily orthogonal). Additionally, this method does not present the convergence problems of NDA networks. The PCA method is linear and does not present convergence problems, but it is an unsupervised method and therefore, the projections computed do not always show a good view of the class structure if the discriminant axes are not the ones with the higher variance.

References

1. C. L. Blake and C. J. Merz. UCI repository of machine learning databases, 1998. http://www.ics.uci.edu/~mlearn/ MLRepository.html, University of California, Irvine. 745
2. J. H. Friedman. Exploratory projection pursuit. *Journal of the American Statistical Association*, 82(397), 1987. 743
3. K. Fukunaga. *Statistical Pattern Recognition.* Academic Press, second edition edition, 1990. 743
4. D. E. Goldberg. *Genetic Algorithms in Search, Optimization, and Machine Learning.* Addison-Wesley, 1989. 744
5. J. H. Holland. *Adaptation in Natural and Artificial Systems.* Ann Tabor: The University of Michigan Press, 1975. 744
6. T. Kohonen. The self organizing map. *Proceedings IEEE*, pages 1464–1480, 1990. 743
7. B. Lerner, H. Guterman, M. Aladjem, I. Dinstein, and Y. Romen. On pattern classification with sammon's nonlinear mapping (an experimental study). *Pattern Recognition*, 31(4):371–381, 1998. 743
8. J. Mao and A. K. Jain. Artificial neural networks for feature extraction and multivariate data projection. *IEEE Transactions on Neural Networks*, 6(2), 1995. 743
9. M. Mitchell. *An Introduction to Genetic Algorithms.* MIT Press, Cambridge, MA, 1996. 744
10. J. W. Sammon. A non-linear mapping for data structure analysis. *IEEE Transactions on Computers*, C-18(5), 1969. 743
11. L. Shyh-Chang, W. F. Punch III, and E. D. Goodman. Coarse-grain parallel genetic algorithms: Categorization and new approach. *Parallel and Distributed Processing*, 1994. 745
12. W. Siedlecki, K. Siedlecka, and J. Sklansky. An overview of mapping techniques for exploratory pattern analysis. *Pattern Recognition*, 21(5):411–429, 1988. 743

Piecewise Multi-linear PDF Modelling, Using an ML Approach

Edgard Nyssen, Naren Naik, and Bart Truyen

Vrije Universiteit Brussel, Vakgroep Elektronica en Informatieverwerking (ETRO)
Pleinlaan 2, B-1050 Brussel, Belgium
ehnyssen@etro.vub.ac.be

Abstract. This paper addresses the problem of estimating the model parameters of a piecewise multi-linear (PML) approximation to a probability density function (PDF). In an earlier paper, we already introduced the PML model and discussed its use for the purpose of designing Bayesian pattern classifiers. The estimation of the unknown model parameters was based on a least squares minimisation of the difference between the estimated PDF and the estimating PML function. Here, we show how a Maximum Likelihood (ML) approach can be used to estimate the unknown parameters and discuss the advantages of this approach. Subsequently, we briefly introduce its application in a new approach to histogram matching in digital subtraction radiography.

1 Introduction

In an earlier paper [1], we already addressed the problem of estimating the class-conditioned probability density function (PDF) $\mathrm{f}(\bar{x}|\omega \in \Omega_t)$, appearing in the expression of a Bayesian discriminant function $\mathrm{d}_t(\bar{x}) = \mathrm{P}(\omega \in \Omega_t)\mathrm{f}(\bar{x}|\omega \in \Omega_t)$. We cited different approaches [2,3,4,5,6] to the solution of this problem and proposed an alternative representation of approximated PDFs $\mathrm{f}_t(\bar{x})$ defined in a bounded domain I. In this approach, the domain is divided into cells on the basis of a multidimensional rectangular point lattice. The probability densities inside the cells are obtained by a multi-linear interpolation of function values at the lattice points (i.e. inside a cell, and along any line segment parallel to one of the main axes of the coordinate system, values are obtained by linear interpolation). In [1], we showed that in a low-dimensional feature space, this interpolation model allows a fast approximation of a PDF value in any point of I, and unlike other models, the speed of the calculations is independent of the model complexity. The piecewise multi-linear (PML) function $\mathrm{f}_t(\bar{x})$, which maps the points of the domain I to the interpolated values, and which serves as an approximation of $\mathrm{f}(\bar{x}|\omega \in \Omega_t)$, is reformulated as a weighted sum of PML basis functions. This allows the application of a procedure to optimise the approximation. In [1], we considered the minimisation of the least squares (LS) fitting criterion

$$C = \int_{\bar{x}\in I} \left(\mathrm{f}(\bar{x}|\omega \in \Omega_t) - \mathrm{f}_t(\bar{x})\right)^2 \mathrm{d}\bar{x} \ . \tag{1}$$

T. Caelli et al. (Eds.): SSPR&SPR 2002, LNCS 2396, pp. 752–760, 2002.
© Springer-Verlag Berlin Heidelberg 2002

In the present paper, we will prove that the approximating function $f_t(\bar{x})$, obtained in this way, satisfies

$$\int_{\bar{x}\in I} f_t(\bar{x})d\bar{x} = 1 \ . \tag{2}$$

We will also prove that this property holds both for the "theoretical" approximating function $f_t(\bar{x})$ — which paradoxically requires the knowledge of the exact PDF $f(\bar{x}|\omega \in \Omega_t)$ — and its estimation $\hat{f}_t(\bar{x})$, derived from the data in a learning set of patterns vectors. Although (2) suggests that the approximation has the same properties as a probability density function, another fundamental property of PDF's, namely positivity, unfortunately is not always satisfied, as is illustrated by some of the examples given in [1]. If this positivity is crucial, one may consider to search for a solution by finding a minimum of C in (1) under the constraint of positivity. Unfortunately, this solution not only is suboptimal but also, may no longer satisfy (2).

An alternative is to use another meaningful optimisation criterion. In the present paper, we suggest a Maximum Likelihood approach (see e.g. [2,7]). We show how this criterion can be applied, by reformulating the basic problem in an appropriate way, and demonstrate that the corresponding solution satisfies all properties of a PDF. We then present very briefly results obtained by a new approach to the problem of histogram matching in digital subtraction radiography, based upon the piecewise linear (PL) approximation of the PDFs underlying the histograms.

2 Fundamental Considerations

Preliminary remark: since the present paper does not relate data with pattern classes, we will omit the class index t, in our notations. The approximated PDF will be denoted by $f(\bar{x})$, its approximation will be referred to as $f'(\bar{x})$, and the estimator of this approximation will be indicated by $\hat{f}'(\bar{x})$.

2.1 Fundamental Properties of the Approximating Functions Obtained by Applying the Least Squares (LS) Fitting Criterion

The properties of the approximating functions, proven here, are based on the existence of a decomposition of any given constant function into a given set of basis functions ψ_j, $j \in \{1, \ldots, m\}$. Obviously, this is the case with any PML model.

We will prove two fundamental properties regarding the approximation of a probability function $f(\overline{x})$, defined in a finite domain I, using weighted sums of basis functions. This approximation is given by

$$f'(\overline{x}) = \sum_{j=1}^{m} \alpha_j \psi_j(\overline{x}) \ , \tag{3}$$

where the weight coefficients α_j minimise the criterion

$$C = \int_{\overline{x} \in I} (\mathrm{f}(\overline{x}) - \mathrm{f}'(\overline{x}))^2 \mathrm{d}\overline{x} \ . \tag{4}$$

Theorem 1. *If there exists a set of coefficients b_i, $i \in \{1, \ldots, m\}$, satisfying $\sum_{i=1}^{m} b_i \psi_i(\overline{x}) = 1$, $\forall \overline{x} \in I$, then the approximation $\mathrm{f}'(\overline{x})$ satisfies $\int_{\overline{x} \in I} \mathrm{f}'(\overline{x}) d\overline{x} = 1$.*

Proof. Substituting $\mathrm{f}'(\overline{x})$ in (4), by the right hand side of (3), and equating to zero the derivative of C, with respect to α_j, one obtains

$$\sum_{j=1}^{m} \alpha_j \int_{\overline{x} \in I} \psi_j(\overline{x}) \psi_i(\overline{x}) \mathrm{d}\overline{x} = \int_{\overline{x} \in I} \psi_i(\overline{x}) \mathrm{f}(\overline{x}) \mathrm{d}\overline{x} \ , \ \ \forall i \in \{1, \ldots, m\} \ . \tag{5}$$

Now, multiplying both sides of this equation by b_i, calculating the sum over $i \in \{1, \ldots, m\}$ for both expressions, and rearranging the order of summation and integration operations yields

$$\int_{\overline{x} \in I} \sum_{j=1}^{m} \alpha_j \psi_j(\overline{x}) \sum_{i=1}^{m} b_i \psi_i(\overline{x}) \mathrm{d}\overline{x} = \int_{\overline{x} \in I} \left(\sum_{i=1}^{m} b_i \psi_i(\overline{x}) \right) \mathrm{f}(\overline{x}) \mathrm{d}\overline{x} \ , \ \ \forall i \in \{1, \ldots, m\} \ .$$

Substituting the first sum by the left hand side of (3), this equation simplifies to

$$\int_{\overline{x} \in I} \mathrm{f}'(\overline{x}) \mathrm{d}\overline{x} = \int_{\overline{x} \in I} \mathrm{f}(\overline{x}) \mathrm{d}\overline{x} = 1 \ ,$$

since b_i, $i \in \{1, \ldots, m\}$, satisfy $\sum_{i=1}^{m} b_i \psi_i(\overline{x}) = 1$, $\forall \overline{x} \in I$. □

When a representative learning sample of pattern vectors $\overline{x}_l$ is available, the coefficients α_j can be estimated by replacing the expression at the right hand side of (5) — which represents the expectation of the value of the basis function $\psi_i(\overline{x})$ — by the sample mean value of this function, which gives

$$\sum_{j=1}^{m} a_j \int_{\overline{x} \in I} \psi_j(\overline{x}) \psi_i(\overline{x}) \mathrm{d}\overline{x} = \frac{1}{p} \sum_{l=1}^{p} \psi_i(\overline{x}_l) \ , \ \ \forall i \in \{1, \ldots, m\} \ . \tag{6}$$

Here, p is the sample size and a_j are the estimations of the original coefficients α_j in (3). When the coefficients α_j in the decomposition of the PDF are substituted by the coefficients a_j, the resulting function becomes an estimator for $\mathrm{f}'(\overline{x})$:

$$\widehat{\mathrm{f}'}(\overline{x}) = \sum_{j=1}^{m} a_j \psi_j(\overline{x}) \ . \tag{7}$$

A second theorem shows that the property, proven previously for $\mathrm{f}'(\overline{x})$, also holds for $\widehat{\mathrm{f}'}(\overline{x})$:

Theorem 2. *If there exists a set of coefficients b_i, $i \in \{1, \ldots, m\}$, satisfying $\sum_{i=1}^{m} b_i \psi_i(\overline{x}) = 1$, $\forall \overline{x} \in I$, then the estimated approximation $\widehat{\mathrm{f}}'(\overline{x})$ satisfies $\int_{\overline{x} \in I} \widehat{\mathrm{f}}'(\overline{x}) d\overline{x} = 1$.*

Proof. Summation over $i \in \{1, \ldots, m\}$ of both sides of (6), after multiplication with b_i, and rearranging the order of summation and integration operations, yields

$$\int_{\overline{x} \in I} \sum_{j=1}^{m} a_j \psi_j(\overline{x}) \sum_{i=1}^{m} b_i \psi_i(\overline{x}) \mathrm{d}\overline{x} = \frac{1}{p} \sum_{l=1}^{p} \left(\sum_{i=1}^{m} b_i \psi_i(\overline{x}_l) \right) , \quad \forall i \in \{1, \ldots, m\} .$$

After the substitution of the first sum by the left hand side of (7), this equation yields:

$$\int_{\overline{x} \in I} \widehat{\mathrm{f}}'(\overline{x}) \mathrm{d}\overline{x} = \frac{1}{p} \sum_{l=1}^{p} 1 = 1 ,$$

since b_i satisfy $\sum_{i=1}^{m} b_i \psi_i(\overline{x}) = 1$, $\forall \overline{x} \in I$, including the learning pattern vectors $\overline{x}_l$. □

2.2 Derivation of a Maximum Likelihood Model for Estimating the Coefficients of the PML Approximation

Let us assume that the basis functions ψ_j satisfy

$$\forall \overline{x} \in I : \psi_j(\overline{x}) >= 0 \quad \text{and} \quad \int_{\overline{x} \in I} \psi_j(\overline{x}) \mathrm{d}\overline{x} = 1 . \tag{8}$$

In other words, the ψ_j behave like probability density functions. If the second condition does not hold for basis functions ψ'_j in the decomposition

$$\mathrm{f}(\overline{x}) = \sum_{j=1}^{m} \alpha'_j \psi'_j(\overline{x}) ,$$

it is sufficient to multiply these with an appropriate scale factor s, i.e. $\psi_j(\overline{x}) = s\psi'_j(\overline{x})$, so that (8) is satisfied, and to replace the coefficients α'_j by α_j which will satisfy $\alpha_j = \alpha'_j / s$ for the solution.

Therefore, consider a probability density function, of the following form

$$\mathrm{f}(\overline{x}) = \sum_{j=1}^{m} \alpha_j \psi_j(\overline{x}) ,$$

where the $\psi_j(\overline{x})$, $j = 1, \ldots m$, satisfy (8) and α_j are the weighting coefficients of the mixture of these density functions. It is obvious that for a random vector $\overline{x}$ that satisfies the distribution $\mathrm{f}(\overline{x})$, α_j can be considered as the prior probability by which the vector will be attributed to component j of the mixture – satisfying

$$\sum_{j=1}^{m} \alpha_j = 1 . \tag{9}$$

The function ψ_j can be considered as the probability density of $\overline{x}$, conditioned by the knowledge that the vector is attributed to component j of the mixture, i.e.: $\psi_j(\overline{x}) = \mathrm{f}\left(\overline{x}|j\right)$.

When a sample $\{\overline{x}_1, \ldots, \overline{x}_p\}$ of independent observations of the random vector $\overline{x}$ is given, together with a predetermined set of probability density functions $\psi_j(\overline{x})$, $j = 1, \ldots m$, of the mixture, the weighting coefficients α_j can be estimated using a maximum likelihood approach. It is obvious that the likelihood to be maximised is

$$L = \prod_{l=1}^{p} \sum_{j=1}^{m} \hat{\alpha}_j \psi_j(\overline{x}_l) \ ,$$

or equivalently, the log-likelihood to be maximised is

$$\log L = \sum_{l=1}^{p} \log \sum_{j=1}^{m} \hat{\alpha}_j \psi_j(\overline{x}_l) \ .$$

The maximisation for the values of $\hat{\alpha}_j$ must be subject to the constraint (9), and, therefore, involves the use of a Lagrange multiplier λ. We thus search for the solution of

$$\frac{\partial}{\partial \hat{\alpha}_i} \left(\sum_{l=1}^{p} \log \sum_{j=1}^{m} \hat{\alpha}_j \psi_j(\overline{x}_l) + \lambda \sum_{j=1}^{m} \hat{\alpha}_j - \lambda \right) = 0 \ , \quad \forall i \in \{1, \ldots, m\} \ ,$$

which yields

$$-\lambda = \sum_{l=1}^{p} \frac{\psi_i(\overline{x}_l)}{\sum_{j=1}^{m} \hat{\alpha}_j \psi_j(\overline{x}_l)} \ , \quad \forall i \in \{1, \ldots, m\} \ .$$

Multiplying both sides of this equation with $\hat{\alpha}_i$, we obtain a set of equations which allow us to calculate the value of λ:

$$-\lambda \hat{\alpha}_i = \sum_{l=1}^{p} \frac{\hat{\alpha}_i \psi_i(\overline{x}_l)}{\sum_{j=1}^{m} \hat{\alpha}_j \psi_j(\overline{x}_l)} \ , \quad \forall i \in \{1, \ldots, m\} \ . \tag{10}$$

Indeed, summing both sides of these equations over index $i \in \{1, \ldots, m\}$, interchanging the order of the summations $\sum_i$ and $\sum_l$ in the right hand side, and using (9), gives: $-\lambda = \sum_{l=1}^{p} 1 = p$. Substituting this result in (10) finally yields a set of equations from which the values of $\hat{\alpha}_j$ can be solved. We have

$$\hat{\alpha}_i = \frac{1}{p} \sum_{l=1}^{p} \frac{\hat{\alpha}_i \psi_i(\overline{x}_l)}{\sum_{j=1}^{m} \hat{\alpha}_j \psi_j(\overline{x}_l)} \ , \quad \forall i \in \{1, \ldots, m\} \ . \tag{11}$$

This set of equations is immediately formulated in a form, appropriate for the application of a recursive solution procedure. In such approach, one starts with a tentative set of coefficients $\hat{\alpha}_i$, $i \in \{1, \ldots, m\}$, and plugs it in the right hand side of (11), yielding a new set of estimates for the coefficients $\hat{\alpha}_i$. This is repeated till a convergence criterion is satisfied. When the starting values of the coefficients are all positive, it is obvious from (11), that they remain positive during the whole procedure, since the functions ψ_i are also positive. It is also evident from (11) that the sum of the coefficients $\hat{\alpha}_i$ is one. For this reason and because of (8), the proposed ML solution will satisfy all properties of a PDF.

An interesting set of equations, similar to the equations derived by Duda and Hart ([2], pp. 192, 193), can be derived from (11), by replacing the functions $\psi_j(\overline{x})$ with their expression as conditional probabilities, namely f $(\overline{x}|j)$, and using the Bayes theorem:

$$\hat{\alpha}_i = \frac{1}{p} \sum_{l=1}^{p} \mathrm{P}(i|\overline{x}_l) \ .$$

We indeed see that the coefficient $\hat{\alpha}_i$ can be considered as the mean posterior probability to attribute the observed random vectors $\overline{x}_l$ to component i of the mixture probability density model.

3 Some Numerical Experiments

The mathematical models have been implemented in Matlab. Some numerical experiments have been performed to validate the developed software technically, and to observe the behaviour of both the LS and the ML approach. One of the experiments consisted of estimating the PDFs of two univariate distributions. The first distribution is uniform in an interval $[0, 1]$. The second distribution behaves like a Gaussian distribution in an interval $[0, 1]$ and is zero elsewhere. Fig. 1 shows the results of the experiments. In [1], we already reported the decrease in quality of the results when there is some mismatch between the estimated PDF and the approximation model. The figure shows that for the LS technique the values of the approximating function may be indeed negative in the neighbourhood of rapid changes. As predicted theoretically, the approximating function obtained from the ML approach continues to behave well.

4 Application in Digital Subtraction Radiography

Digital subtraction radiography (DSR) is a potentially sensitive method for revealing subtle changes in radiolucency between radiographs acquired separated in time. The power of the DSR method stems from its ability to remove so-called structural noise, arising from the invariant structures in the images, leading to a distinctive improvement in the visual acuity of real changes.

The particular application of DSR that we consider here as an illustration of the new method for piecewise linear approximation of a PDF is that of intraoral radiography, and more specifically its application to the detection of approximal caries. Basically, dental caries is a slowly progressing demineralisation of the tooth surface that starts at the surface, and gradually penetrates the

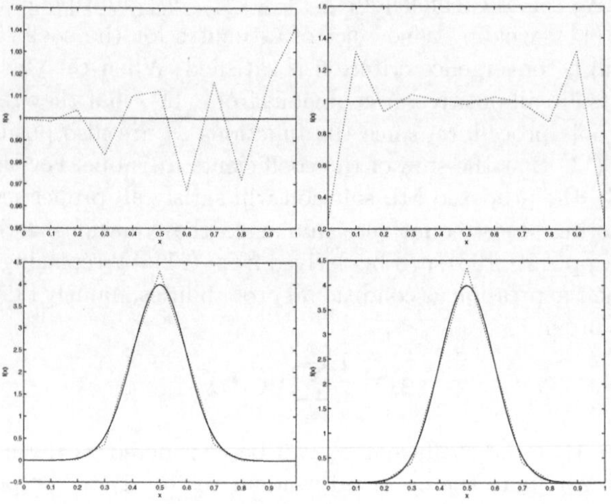

Fig. 1. Graphical representation of the results of the numerical experiments. Solid lines correspond to the real PDF. Top row: results for a uniform distribution, bottom row: results for a (partially) gaussian distribution, left column: results for the LS method, right column: results for the ML method

tooth. Given the treatment ramifications of advanced caries, detection in its early development stage is of prime importance. However, common radiographic examination has been found to yield an insufficient sensitivity for the early detection of approximal caries. Given this observation, DSR is investigated as a more accurate diagnostic method.

Decisive to the success of the DSR method, however, is the ability with which the exposure geometry and the development conditions can be reproduced. Whereas the first requirement can be met by the use of mechanical stabilisation devices or by employing mathematical methods of retrospective geometric correction, changes in the exposure and development conditions necessarily call for a numerical contrast correction procedure. This involves the transformation of the gray value histograms such as to resemble each other as closely as possible.

The standard method of contrast correction used in intra-oral DSR is that, proposed by Ruttiman and co-workers [8], which finds the optimal transformation by equating the cumulative distributions of the respective histograms. Actually, this method has been proposed as a more consistent approach to the problem of contrast correction, compared to an earlier described parametric method [9], based on matching the first and second order moments of the respective distributions.

More recently, a particular interesting method has been suggested by Bidasaria [10], in which the original gray values in the images are randomised in

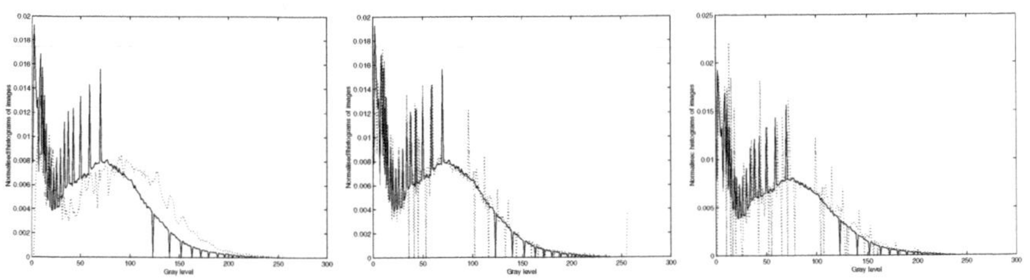

Fig. 2. Plots of desired histogram (solid line) with, (a) Actual starting histogram, (b) Matched histogram from piecewise linear approximation, (c) Matched histogram using the method of Ruttiman

the discretisation intervals, to obtain a piecewise constant approximation of the histograms. Histogram matching then follows immediately.

In our approach, the histogram of image 1, say, is transformed into that of image 2, by first decomposing the histograms into the basis set of triangular functions characterising the PL approximation, prior to using the method of direct histogram specification (DHS) [11] via the uniform distribution. Upon a suitable choice of the points at which the cumulative distribution function (CDF) of image 2 is evaluated, the use of a continuous representation allows to circumvent the explicit inversion of the the transformation to the uniform distribution of histogram 2. Our approach, consisting of a PL approximation, differs fundamentally from that of Bidasaria [10] in which a step approximation of the histogram is proposed. In the preliminary results of this paper, we have shown as in Fig. 2 that our approach yields results comparable to those obtained from the method of Ruttiman et al. [8]. Our approach is a first step towards the use of alternative representations of histograms, as found in [12,13].

5 Discussion and Conclusions

In [1], we introduced the basic concepts and notations for a Piecewise Multilinear (PML) approximation of probability density functions. We showed how to formulate this model, which is basically an interpolation model, as a weighted sum of basis functions, where the weights are the model parameters. We also proposed a solution methodology for estimating the model parameters from a representative learning set of patterns, based on a least squares (LS) fitting criterion.

For a broad class of models that includes the PML model, we show in this paper that the optimal approximation of the PDF by the weighted sum of basis functions, as well as the estimate of this approximation from a learning set of patterns, satisfy the property that their integral over the definition domain equals unity, thus meeting a basic property of a PDF (Theorems 1 and 2). To cope with the problem that another property of PDFs — namely positivity — is not always satisfied by the LS fitting solution, we introduce another approach, which is based on a Maximum Likelihood (ML) criterion. The ML estimate satisfies all properties of a PDF and hence, can be used in applications where these properties are required. Subsequently, we have demonstrated an application of the PL approximation to the problem of contrast correction in DSR.

Acknowledgement

We thank Mandy Runge for her assistance with the preparation of Fig. 2.

References

1. Edgard Nyssen, Luc Van Kempen, and Hichem Sahli. Pattern classification based on a piecewise multi-linear model for the class probability densities. In *Advances in Pattern Recognition — proceedings SSPR2000 and SPR2000*, pages 501–510, 2000. 752, 753, 757, 759
2. Richard O. Duda and Peter E. Hart. *Pattern Classification and Scene Analysis.* John Wiley & Sons, 1973. 752, 753, 757
3. Julius T. Tou and Raphael C. Gonzales. *Pattern Recognition Principles.* Addison Wesley Publishing Company, 1979. 752
4. Robert Schalkoff. *Pattern Recognition — Statistical, Structural and Neural Approaches.* John Wiley & Sons, 1992. 752
5. Fang Sun, Shin'ichiro Omachi, and Hirotomo Aso. An algorithm for estimating mixture distributions of high dimensional vectors and its application to character recognition. In *Proc. 11th Scandinavian Conference on Image Analysis*, pages 267–274, 1999. 752
6. David L. Donoho, Iain M. Johnstone, Gérard Kerkyacharian, and Dominique Picard. Density estimation by wavelet thresholding. *The Annals of Statistics*, 24(2):508–539, 1996. 752
7. A. P. Dempster, N. M. Laird, and D. B. Rubin. Maximum likelihood estimation from incomplete data via the EM algorithm. *J. Royal Statistical Society*, 39:1–38, 1977. 753
8. U. E. Ruttimann, R. L. Webber, and E. Schmidt. A robust digital method for film contrast correction in substraction radiography. *J. Periodont. Res.*, 21:486–495, 1986. 758, 759
9. U. E. Ruttimann, T. Okano, H.-G. Gröndahl, K. Gröndahl, and R. L. Webber. Exposure geometry and film contrast differences as bases for incomplete cancellation of irrelevant structures in dental subtraction radiography. *Proc. SPIE*, 314:372–377, 1981. 758
10. H. B. Bidasaria. A method for almost exact histogram matching for 2 digitized images. *Computer Graphics and Image Processing*, 34, 1986. 758, 759
11. Rafael C. Gonzalez and Paul Wintz. *Digital image processing.* Addison-Wesley Publishing Company, Amsterdam, 1987. 759
12. R. Morandi and P. Constantini. Piecewise monotone quadratic histosplines. *SIAM J. Stat. Comput.*, 10:397–406, 1989. 759
13. J. W. Schmidt, W. Heß, and T. Nordheim. Shape preserving histopolation using rational quadratic splines. *Comput.*, 44:245–258, 1990. 759

Decision Tree Using Class-Dependent Feature Subsets

Kazuaki Aoki and Mineichi Kudo

Division of Systems and Information Engineering, Graduate School of Engineering
Hokkaido University, Kita-13, Nishi-8, Kita-ku, Sapporo 060-8628, Japan
{kazu,mine}@main.eng.hokudai.ac.jp
http://ips9.main.eng.hokudai.ac.jp

Abstract. In pattern recognition, feature selection is an important technique for reducing the measurement cost of features or for improving the performance of classifiers, or both. Removal of features with no discriminative information is effective for improving the precision of estimated parameters of parametric classifiers. Many feature selection algorithms choose a feature subset that is useful for all classes in common. However, the best feature subset for separating one group of classes from another may depend on groups. In this study, we investigate the effectiveness of choosing feature subsets depending on groups of classes (class-dependent features), and propose a classifier system that is built as a decision tree in which nodes have class-dependent feature subsets.

1 Introduction

Feature selection is to find a feature subset that is effective for classification from a given feature set. This technique is effective for both improving the performance of classifiers and reducing the measurement cost of features. Particularly when the scale of the problem is large (in the sense of the number of features or the number of classes, or both), there are some features that have little or no discriminative information. It is well known that such features, called *garbage features*, weaken the performance of classifiers (peaking phenomenon) as long as a finite number of training samples is used for designing the classifiers. Thus, removal of such garbage features should result in an improvement in the performance of such classifiers.

Many techniques for feature selection have been proposed [1,2,3,4,5,6,7]. All of these approaches choose the same feature subset in all classes. However, it seems reasonable to assume that effective feature subsets are different, depending on classes. For instance, in the case of more than two classes, a feature subset that is powerful in discriminating one class from the remaining classes does not always work in discriminating another class from the remaining classes. Thus, when treating many classes, such as in character recognition, selection of feature subsets depending on groups of the classes is effective. We call such a feature subset a *"class-dependent feature subset."*

T. Caelli et al. (Eds.): SSPR&SPR 2002, LNCS 2396, pp. 761–769, 2002.
© Springer-Verlag Berlin Heidelberg 2002

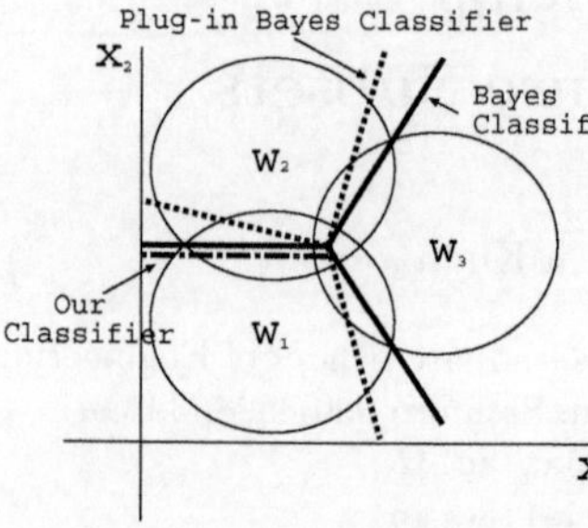

Fig. 1. Example

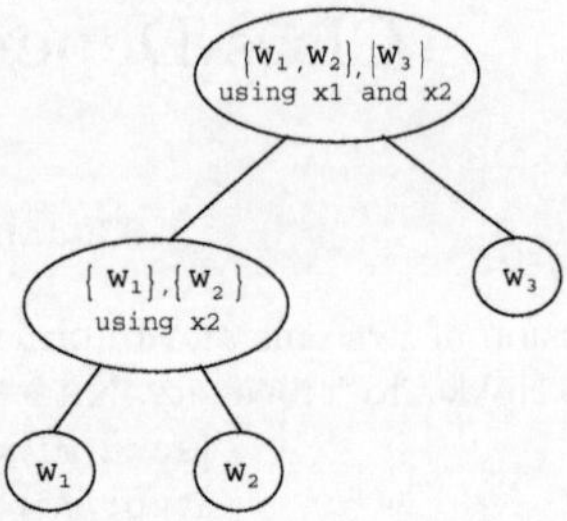

Fig. 2. Decision tree for Fig.1

For instance, in Chinese character recognition, there are more than one thousand characters. A group of similar characters has almost the same values in almost all features but differs in a small number of features, e.g., the number of strokes or whether a short stroke exists or not. Only a small number of features are effective in discrimination of these similar characters. However, these features are not always useful for discrimination between the group of these similar characters and other groups of characters.

Therefore, it is expected that the performance of classifiers can be improved by choosing feature subsets depending on groups of classes. In fact, there have been some studies in which class-dependent features worked well in handwritten character recognition [8,9]. However, theoretical analysis is expected. In this paper, we present a formalization of the usage of *class-dependent feature subsets* and propose a classification system using these subsets.

2 Illustrative Example

Our concept is explained by the example shown in Fig. 1. In Fig. 1, there are three classes according to normal distributions with the same covariance matrix and different means. The Bayes rule therefore becomes linear. As long as a given training sample is finite, we cannot avoid misclassification in the plug-in Bayes classifier estimated from the training sample. However, we can reduce such a misclassification using class-dependent feature subsets. Indeed, in this problem, only feature x_2 has discriminative information between ω_1 and ω_2. Thus, a classifier using only x_2 is expected to perform better than that using x_1 and x_2 in this case. A decision tree designed naturally is shown in Fig. 2.

3 Decision Trees

3.1 Several Types of Decision Tree

We described how class-dependent feature subsets are used to improve classifiers. Then, the next question is what uses of class-dependent feature subsets

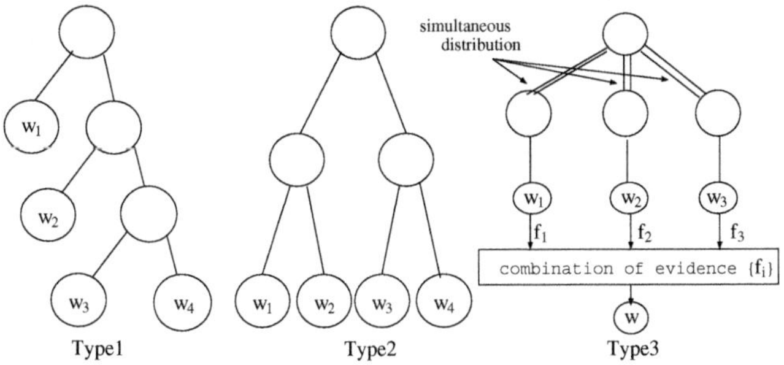

Fig. 3. Three types of decision tree

are possible. Some decision trees are naturally considered (Fig. 3). In all these configurations, the recognition process starts from a whole set of classes at a root down to subsets of classes and reaches a single class at a leaf. Each type has the following aspects.

(*Type* 1) This type of decision tree has the simplest architecture and separates one class from the other classes in each node. The problem shown in Fig. 1 can be solved using this type of decision tree.

(*Type* 2) This is a generalization of type 1. In each node, data are separated into two subsets of classes.

(*Type* 3) The process goes from the root to its children in parallel, and in each child one class is separated from the other classes. Usually, each node outputs a value of the evidence about how well the decision is firm. In the process of gathering evidence, the final decision is made.

Type 3 turns into regular classifiers like a linear or quadratic classifier in multi-class cases when all nodes have the same feature subsets and the evidence is combined by the maximum likelihood method. Another approach called a modular network [8,9] is also included in this type.

In each node of a decision tree, the problem is to classify one group of classes and another group of classes. Here, let two groups of classes be Ω^1 and Ω^2. Then, a node is identified by the following information.

1. (Ω^1, Ω^2): two groups of classes to be classified
2. F: feature subset
3. ϕ: classifier

Thus, an internal node t is denoted as

$$Node^t = \{\Omega_t^1, \Omega_t^2, F_t, \phi_t\}.$$

Our approach is different from conventional decision tree approaches [10,11] in the following two points:

1. In conventional decision tree approaches, each node is split in terms of *impurity* in a feature, whereas in our approach, each node is split in terms of the degree of separation between two groups of classes.
2. In conventional decision tree approaches, classification in each node is simple, and usually only one feature is used. In other words, the performance of classification in each node is not so good. Instead, those approaches complement the simplicity with splitting of data many times. In our approach, we split each node by a class-dependent feature subset. Thus, it is expected that the classification performance is improved in individual small problems.

In this paper, we considered only type 1 and type 2 decision trees.

3.2 Experiments Using Artificial Data 1

We examined the possibility of this approach under the assumption that a decision tree is ideally constructed and ideal class-dependent feature subsets are chosen. The data are shown in Table 1. In Table 1, a feature with 1 means that data are generated according to a normal distribution with average 0.5 and standard deviation 0.1 in the feature, and a feature with 0 means that data are generated according to a normal distribution with average 0 and standard deviation 0.1. There is no correlation between features. For example, it is sufficient to use only features x_1 and x_2 to separate ω_1 and ω_2. As a whole, all features are needed to classify this dataset.

In the ideal tree, at the root node ω_1 is separated from the others by all features $x_1 - x_c$, and then ω_2 by $x_2 - x_c$, and so on (Fig. 4), where c is the number of classes and is also the number of features. In every nodes, a linear classifier was used. Compared with the linear classifier with the full feature set, this tree can estimate the parameters accurately in the deeper nodes.

The results of the experiment are shown in Fig. 5. In the experiment, three, ten or thirty classes and 5, 10, 100 or 1000 training samples per class were used. The number of test samples was fixed at 1000 per class. The average recognition rate of 10 different training datasets is shown. For comparison, the recognition rate by the linear classifier with all features is shown. It was found that the decision tree with class-dependent feature subsets worked better than the linear classifier with all features, when the number of training samples was comparatively small and the number of classes, and also the number of features, was large.

4 Construction of Decision Trees

In this section, we propose an algorithm to construct a decision tree from given data. A decision tree is constructed in a bottom-up way like Huffman coding. The algorithm is as follows.

1. Initialization step: Set $\Omega_i = \{\omega_i\}, (i = 1, 2, \cdots, C)$, $c = C, t = 1$. Attach an unprocessed mark to all Ω_i. These Ω_i correspond to leaves.

Table 1. Artificial data

	features				
class	x_1	x_2	x_3	$\cdots$	x_c
ω_1	1	0	0	$\cdots$	0
ω_2	0	1	0	$\cdots$	0
ω_c	0	0	0	$\cdots$	1

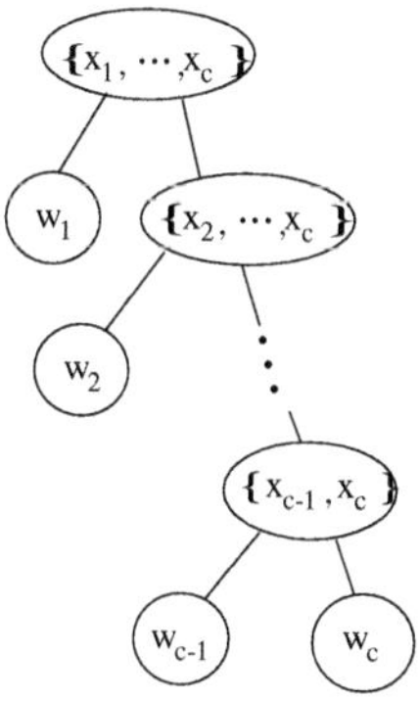

Fig. 4. Decision tree used in this experiments

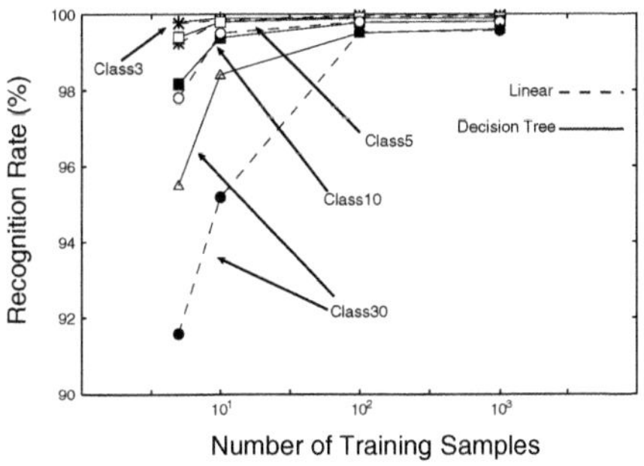

Fig. 5. Result of experiments using data shown in Table.1

2. Calculate the separability S_{ij} of pair (Ω_i, Ω_j) for all unprocessed nodes Ω_i and Ω_j, $(i, j = 1, \cdots, c)$.
3. Choose the pair $(\Omega_{i^*}, \Omega_{j^*})$ with the smallest separability $S_{i^*j^*}$. Let Ω_{i^*} be Ω^1_{c+1} and Ω_{j^*} be Ω^2_{c+1}. Mark Ω_{i^*} and Ω_{j^*} as processed. Select a feature subset F_{c+1} that is effective in discrimination between Ω^1_{c+1} and Ω^2_{c+1}.
4. Construct a classifier ϕ_{c+1} to classify Ω^1_{c+1} and Ω^2_{c+1} with feature subset F_{c+1}. In this step, we have a new node, $Node^{c+1} = \{\Omega^1_{c+1}, \Omega^2_{c+1}, F_{c+1}, \phi_{c+1}\}$.
5. $\Omega_{c+1} = \Omega^1_{c+1} \bigcup \Omega^1_{c+2}$ and $c \leftarrow c + 1$üA$t \leftarrow t + 2$.
6. Repeat steps 2-5 until $t = c$.

In steps 3-5, two nodes are merged into one new node, $(c \leftarrow c + 1)$, and the two merged nodes are marked as being processed, $(t \leftarrow t + 2)$. Finally, a decision tree with $2C - 1$ nodes is constructed.

5 Experiments

We dealt with 4-class and 9-class artificial datasets shown in Fig.6 and real 'mfeat' data from UCI Machine Learning Database [12].

5.1 Artificial Data 2

The separability measure, the classifier, and the feature selecton method we used are as follows:

1. Separability: the recognition rate estimated by the leave-one-out technique with 1-NN (the nearest neighbor) classifier.
2. Feature selection method: an approach based on the structual indices of categories [13] .
3. Classifier: plug-in Bayes linear classifiers

In this experiment, the same type of classifiers were used in all nodes but trained differently in each node.

In this dataset, we compared two decision trees: (1) a decision tree with ideal feature subsets and with an ideal configuration and (2) a decision tree constructed by the proposed algorithm. We dealt with the same type of problems in 4-class and 9-class cases (Fig. 6). The number of training samples was 10 per class, and the number of test samples was 1000 per class. The constructed decision tree is shown in Fig. 7, and the classification boundaries are shown in Fig. 8 and Fig. 9. The figures in parentheses in those figures show the recognition rates for test samples.

The ideal boundary and the boundary by the proposed method are almost comparable. Here, it should be noted that if a single feature is used, the boundary becomes a straight line regardless of the classifier used. We succeeded in improving the performance of the decision tree with the differently trained linear classifiers, compared to that of the single linear classifier using all features.

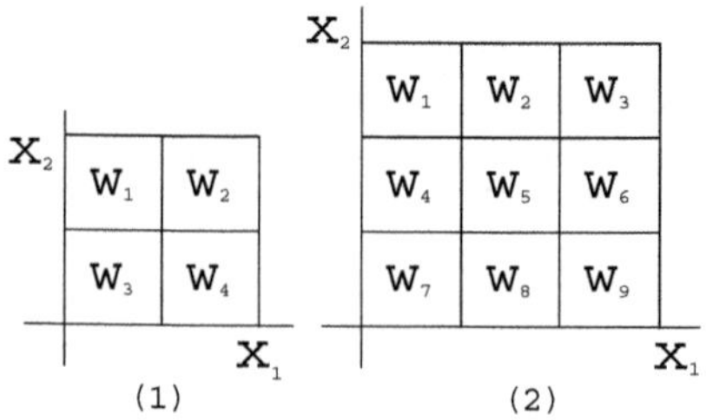

Fig. 6. (1) 4-class and (2) 9-class problem

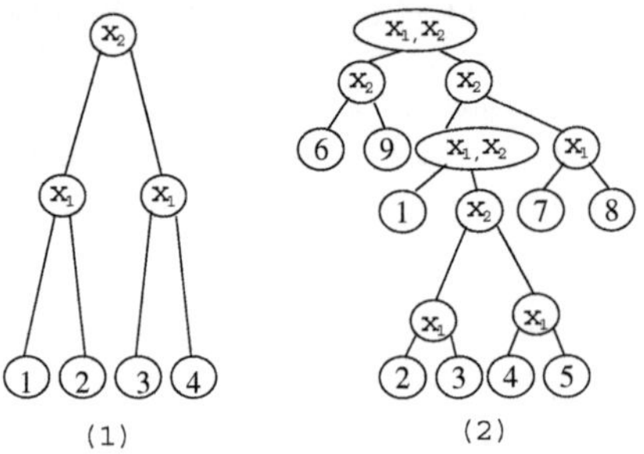

Fig. 7. Constructed decision trees (1) in a 4-class problem and (2) in a 9-class problem

5.2 'mfeat' Data

Next, we examined the mfeat data. This dataset is a handwritten numeric '0'-'9' database. The feature dimension is 76, and the number of training samples is

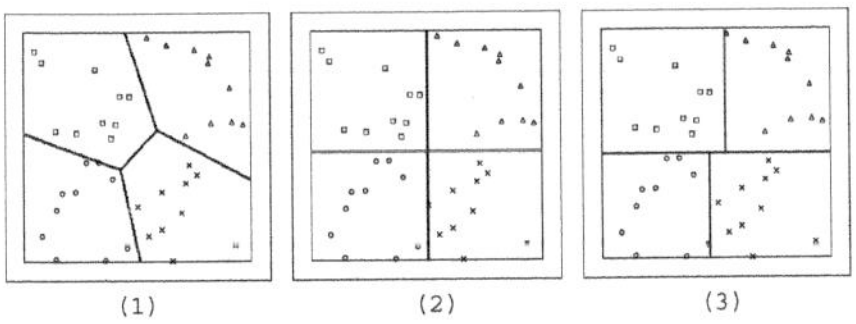

Fig. 8. Constructed boundary in a 4-class problem: (1) Linear (90.4%), (2) Ideal decision tree (97.5%), and (3) The proposed method (92.8%)

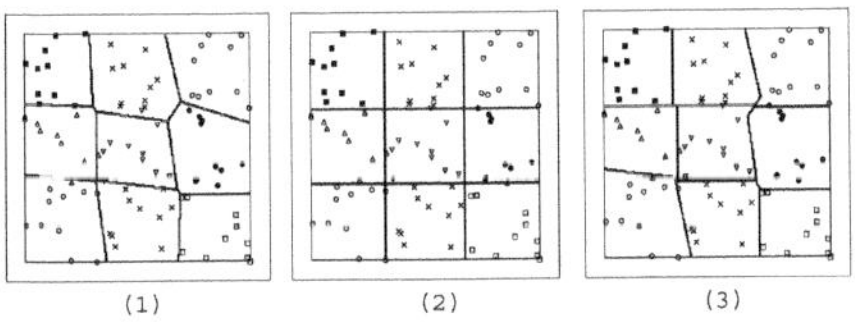

Fig. 9. Constructed boundary in a 9-class problem: (1) Linear (91.2%), (2) Ideal decision tree (97.8%), and (3) The proposed method (93.5%)

200 per class. Those features are Fourier coefficients of a character shape. The top 100 samples of each class were used for training, and the bottom 100 samples were used for testing. The same way as Artificial DAta 2 for the decision tree. The constructed decision tree is shown in Fig. 10. For comparison, we constructed a linear classifier and a 1-NN classifier with all features. The recognition rate of the linear classifier was 80.7%, that of the 1-NN classifier was 82.4%, and that of the proposed method was 83.8%. The 1-NN classifier attained 83.3% when a feature subset common to all classes was chosen.

From the decision tree, we can see that the number of selected features is appropriate depending on the problem. For instance, in node classifying class '0' and '8', the local problem is easy to solve and the number of features is very small. The effectiveness of our approach depends on the number of training samples and the number of classes. Thus, this approach, like other approaches, does not always work well for all kinds of datasets. It is expected that this approach will work well for problems in which the number of classes is large.

6 Discussion

Our approach using class-dependent feature subsets works effectively when the number of classes is large, and it is superior to conventional approaches using a common feature subset. It is expected that it works better when the number of training samples is smaller, because less features take advantage in such a case and individual small problems with only a few classes require less features than the total problem with many classes.

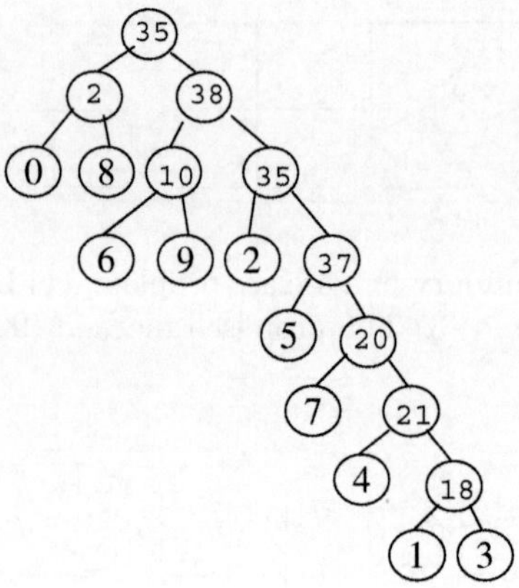

Fig. 10. Decision tree for mfeat (figures in internal nodes are the numbers of features)

Another merit of our approach is that the results are interpretable. In each node, we can show how easy it is to solve a local problem and what set of features is necessary. With such information, it should be possible to improve the performance of the classifier.

7 Conclusion

We have discussed the effectiveness of a class-dependent feature subset, and have presented an algorithm of a classification system as a decision tree. In addition, we can see separability in each node of the decision tree, so we may use this information to improve the decision tree. We will consider the design of the optimum decision tree.

References

1. P. A. Devijver and J. Kittler, *Pattern Recognition: A Statistical Approach.* Prentice-Hall, 1982. 761
2. P. Pudil, J. Novovičová and J. Kittler, Floating Search Methods in Feature Selection. *Pattern Recognition Letters*, **15**(1998), 1119–1125. 761
3. P. Somol, P. Pudil, J. Novovičová and P. Paclík, Adaptive Floating Search Methods in Feature Selection. *Pattern Recognition Letters*, **20**(1999), 1157–1163. 761
4. F. J. Ferri, P. Pudil, M. Hatef and J. Kittler, Comparative Study of Techniques for Large-Scale Feature Selection. *Pattern Recognition in Practice IV*(1994), 403–413. 761
5. M. Kudo and J. Sklansky, A Comparative Evaluation of Medium- and Large-scale Feature Selectors for Pattern Classifiers. *1st International Workshop on Statistical Techniques in Pattern Recognition*(1997), 91–96. 761

6. M. Kudo and J. Sklansky, Classifier-Independent Feature Selection for Two-stage Feature Selection. *Advances in Pattern Recognition*, **1451**(1998), 548–554. 761
7. D. Zongker and A. Jain, Algorithms for Feature Selection: An Evaluation. *13th International Conference on Pattern Recognition*, **2**(1996), 18–22. 761
8. I. S. Oh, J. S. Lee and C. Y. Suen, Analysis of Class Separation and Combination of Class-Dependent Features for Handwriting Recognition. *IEEE Transactions on Pattern Analysis and Machine Intelligence*, **21**(1999), 1089–1094. 762, 763
9. I. S. Oh, J. S. Lee, K. C. Hong and S. M. Choi, Class Expert Approach to Handwritten Numerical Recognition. *Proceedings of IWFHR '96*(1996), 35–40. 762, 763
10. R. O. Duda, P. E. Hart and D. G. Stork, *Pattern Classification: Second Edition.* John Wiley & Sons, 2000. 763
11. L. Breiman, J. H. Friedman, R. A. Olshen and C. J. Stone, *Classification and Regression Trees.* Wadsworth & Brooks / Cole Advanced Books & Software, 1984. 763
12. P. M. Murphy and D. W. Aha, UCI Repsitory of Machine Learning Databases [Machine-readable data repository]. *University of California Irnive, Department of Informaiton and Computations Science*(1996). 765
13. M. Kudo and M. Shimbo, Feature Selection Based on the Structural Indices of Categories. *Pattern Recognition*, **26**(1993), 891–901. 766

Fusion of n-Tuple Based Classifiers for High Performance Handwritten Character Recognition

Konstantinos Sirlantzis[1], Sanaul Hoque[1],
Michael C. Fairhurst[1], and Ahmad Fuad Rezaur Rahman[2]

[1] Department of Electronics, University of Kent
Canterbury, Kent, United Kingdom
{ks30,msh4,mcf}@ukc.ac.uk
[2] BCL Technologies Inc.,
990 Linden Drive, Suite #203, Santa Clara, CA 95050, USA
fuad@bcltechnologies.com

Abstract. In this paper we propose a novel system for handwritten character recognition which exploits the representational power of n-tuple based classifiers while addressing successfully the issues of extensive memory size requirements usually associated with them. To achieve this we develop a scheme based on the ideas of multiple classifier fusion in which the constituent classifiers are simplified versions of the highly successful scanning n-tuple classifier. In order to explore the behaviour and statistical properties of our architecture we perform a series of cross-validation experiments drawn from the field of handwritten character recognition. The paper concludes with a number of comparisons with results on the same data set achieved by a diverse set of classifiers. Our findings clearly demonstrate the significant gains that can be obtained, simultaneously in performance and memory space reduction, by the proposed system.

1 Introduction

Handwritten character recognition is still one of the most challenging problems in pattern classification. Over the years a great number of algorithms have been developed to achieve improved performance. Some of the simplest yet most successful among them are the so-called 'n-tuple' based classifiers. Unfortunately, there is usually a trade-off between high performance and either increased computational load or increased memory requirements. The Scanning n-tuple classifier (SNT) [8] is a typical example of the case where superior recognition rates are attained at the expense of significant storage requirements, especially in applications with samples of realistic size. On the other hand, in recent years there has been a significant shift of interest from the development of powerful but demanding individual classifiers to the development of strategies to fuse the outcomes of a number of relatively simpler classifiers [7]. Such structures have the

T. Caelli et al. (Eds.): SSPR&SPR 2002, LNCS 2396, pp. 770–779, 2002.
© Springer-Verlag Berlin Heidelberg 2002

inherent ability to exploit the diverse recognition mechanisms of the participating classifiers resulting, in most cases, in a scheme demonstrably more successful than the best of its constituent members [12]. In this paper we propose a novel system for handwritten character recognition which exploits the representational power of n-tuple based classifiers while addressing successfully the issues of extensive storage requirements. This is achieved by developing a scheme based on the ideas of multiple classifier fusion in which the constituent classifiers are simplified versions of the highly efficient Scanning n-tuple classifier. Although the individual performances of these classifiers can be comparatively inferior, our findings demonstrate the significant gains that can be obtained, simultaneously in performance and storage space reduction, by the proposed system. In the next sections we initially give a brief description of the components of the system we propose followed by the derivation of the multiple classifier combination rule we adopted which also provides a justification for this choice. To demonstrate the statistical properties of our scheme, we present results obtained over a series of cross-validation experiments drawn from the field of handwritten character recognition. The paper concludes with a number of comparisons with results on the same data set achieved by a diverse set of classifiers reported in the literature, and discussion of our findings.

2 The Proposed Scheme

We start the presentation of the components of our system with the Frequency Weighted Scheme (FWS) which is an n-tuple based classifier reported to demonstrate reasonable levels of performance while requiring comparatively low memory size. We then continue with a description of the original Scanning n-tuple scheme (SNT). Subsequently, the *Bit-Plane Decomposition* technique is described, which is the method we employed to simplify the original feature space used by the SNT and thus reduce the size of the storage space needed by the classifiers constructed on it. Figure 1 gives a schematic representation of the system illustrating the information flow in the parallel combination architecture through the components of the scheme we call 'Layered Sampling of the Bit-Plane Decomposition'.

2.1 Frequency Weighted Scheme (FWS)

In a conventinal n-tuple classifier, the n-tuples are formed by selecting multiple sets of n distinct locations from the pattern space. Each n-tuple thus sees an n-bit feature derived from the pattern. For classification, a pattern is assigned to that class for which the number of matching features found in the training set is maximum. The training process, therefore, requires remembering the occurrences of different features as seen by individual n-tuples. This is usually achieved by setting a 'flag' bit in a large memory array [3].

The FWS is the simplest enhancement of the basic n-tuple classification system. In the basic scheme, both the common and rare feature occurrences are

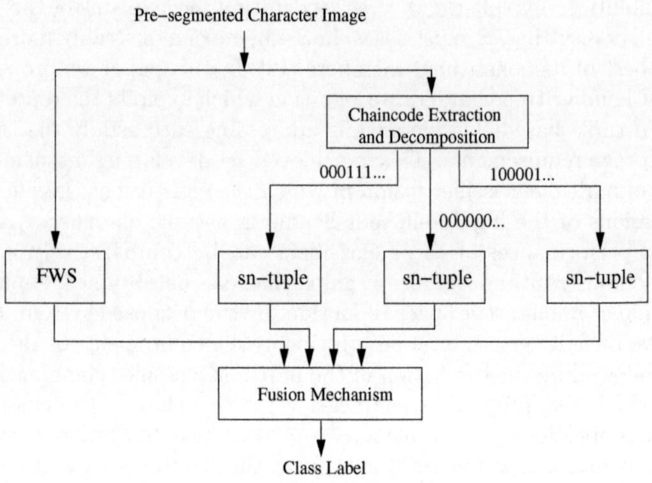

Fig. 1. Schematic of a classification based on Layered Sampling of the Bit-Plane Decomposition

accorded the same discriminatory weight. Thus, the presence of even one rogue pattern in the training set of a class can reduce the discriminatory power of the n-tuple network significantly. As a remedy, in the FWS, instead of setting the flag to record the occurrence of a certain feature in the training set, the relative frequencies are recorded. The frequency counts need to be normalized when different classes have different numbers of training images. The sum of these frequencies corresponding to a particular test image determine its class label.

2.2 Scanning n-Tuple Classifier (SNT)

The Scanning n-tuple (or simply sn-tuple) classifier [8] has been introduced as a statistical-cum-syntactic method for high performance character recognition applications. This is also a variant of the n-tuple classifier except that instead of using the two dimensional raw images directly, the operation is conducted on a one dimensional gray scale representation of the bitmap image. Another difference between the n-tuple and the sn-tuple is that, whereas each n-tuple samples a set of fixed points in the input space, each sn-tuple defines a set of relative offsets between its input points. Each sn-tuple is then scanned over the entire input space.

The one dimensional representation of the binary pattern image is obtained by tracing the contour edges of the image and representing the path by Freeman chain codes [4]. The sn-tuple algorithm is designed to model only one chain code string per pattern and a difficulty arises for images consisting of more than one contour. This is dealt with by mapping a set of strings to a single string

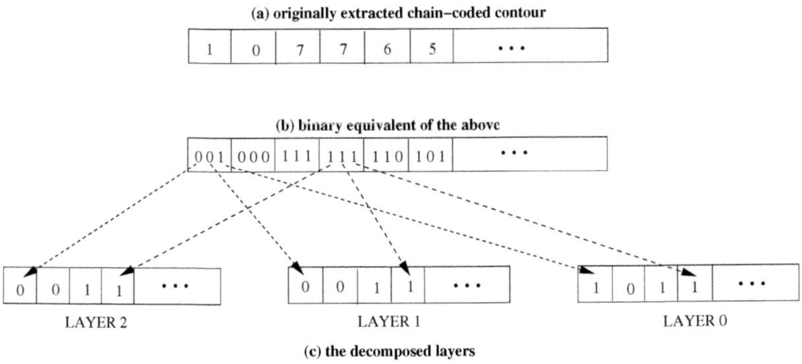

Fig. 2. The Proposed Bit Plane Decomposition Technique

by discarding the positional information (i.e., the start coordinates) and then concatenating the strings together. Besides, the length of a chain coded string is dependent on the character class as well as writing style, degree of slant, image quality etc. Since image classes with short chain codes may be adversely affected, all chains are expanded to a predefined fixed length before training and testing.

2.3 Bit Plane Decomposition

The size of the memory space required by a typical n-tuple based scheme is σ^n units per tuple per class, where σ is the number of distinct values a pixel may have levels and n is the size of the tuples. It can readily be seen that this can become excessively large even with a fairly small number of gray levels. Bit-Plane Decomposition was initially introduced by Schwarz [11] as a means of data compression. However, it can also be used to handle the memory space problem faced by n-tuple based systems [5]. The basic idea is to decompose an image into a collection of binary images ($\sigma = 2$).

For Bit-Plane Decomposition, the individual chain codes extracted from the image are represented in binary. Each is represented by a 3 bit binary code. The chain-coded string is decomposed, subsequently, into 3 layers where layer 'i' is composed of the ith bits of the binary code values. Thus, for example, Layer '0' is formed by collecting all the *least significant bits* of the binary coded string. The decomposition we used here can be seen in Figure 2. The sn-tuple classifiers constructed on each one of the 3 layers extracted from the original chain codes are indicated by SNTL0, SNTL1, and SNTL2 in our Tables (with SNTL0 indicating the one trained on the least significant bit layer).

2.4 Multiple Classifier Decision Fusion

In the present work the four individually trained classifiers used are arranged in a parallel structure to form a multi-classifier recognition system. The choice of this simple architecture to form the ensemble was preferred so that our experimental results could be more easily studied and interpreted.

Although a variety of fusion rules have been devised by researchers [7], the choice of the most appropriate scheme is usually dependent on the degree of diversity of the feature spaces on which the participant classifiers are trained, and the nature of their outputs.

To better understand this, let us consider that a pattern q is to be assigned to one of the m possible classes $\{\omega_1, \ldots, \omega_m\}$ and there are K independent classifiers, in each q is represented by a distinct feature vector $\mathbf{x}_i$, $i = 1, \ldots, K$, each drawn from a corresponding feature space χ_i, $i = 1, \ldots, K$. Let each class ω_k be modelled by the probability density function $P(\mathbf{x}_i|\omega_k)$. Following a Bayesian perspective, each classifier is considered to provide an estimate of the true class posterior probability $P(\omega_j|\mathbf{x}_i)$ given $\mathbf{x}_i$. The idea underlying multiple classifier fusion it to obtain a better estimator by combining the resulting individual estimates [7]. The pattern q should be assigned, consequently, to the class having the highest *posterior* probability. Assuming equal *a priori* probabilities for all the classes, the corresponding decision rule is:

$$\text{assign} \quad \theta \rightarrow \omega_j \quad \text{if}$$

$$P(\omega_j|\mathbf{x}_1, \ldots, \mathbf{x}_K) = \max_{k=1}^{m} P(\omega_k|\mathbf{x}_1, \ldots, \mathbf{x}_K),$$

where θ is the class label of the pattern under consideration q.

Following this line of reasoning in the case where the individual classifiers are sampling identical feature spaces (i.e. $\chi_1 = \ldots = \chi_K$), averaging the estimates will suppress the estimation error as well as the effects of individual classifier overtraining (bias), subsequently reducing the classification error [7]. This gives rise to the well-known ‘sum’ or ‘mean’ rule. Alternatively, if the product of the individual estimates is used in this case it will result in an amplification of the estimation noise. In contrast, the latter combination method will obtain maximal gains from independent pattern representations (i.e. if the classifiers sample independent feature spaces χ_i [1]).

From the preceding discussion it becomes clear that the averaging process is the most beneficial choice of fusion scheme here, since the sn-tuple based classifiers sample the Bit-Plane Decomposition of the same original chain codes. The corresponding ‘mean’ combination rule we used can be formally expressed as follows:

$$\text{assign} \quad \theta \rightarrow \omega_j \quad \text{if}$$

$$K^{-1} \sum_{i=1}^{K} P(\omega_j|\mathbf{x}_i) = \max_{k=1}^{m} \left[K^{-1} \sum_{i=1}^{K} P(\omega_k|\mathbf{x}_i) \right].$$

3 Experimental Results

To observe the behaviour of our system and its statistical properties we employed an inhouse database which consists of 34 classes of pre-segmented characters (numerals 0-9, and upper case letters A-Z, without differentiation between the pairs

Table 1. Mean Error Rates (%) of the components of the proposed system

	Digits (10 classes)	Digits & Letters (34 classes)
FWS	10.00	22.28
SNTL0	23.89	44.92
SNTL1	21.28	40.67
SNTL2	22.45	42.08
SNTL0 + SNTL1 + SNTL2	11.41	23.69

0/O and 1/I). The database corresponds to handwritten characters, every class has 300 samples (10200 characters in total), and the images are provided at a resolution of 16 × 24 pixels. Two recognition problems were constructed. The first included only the numerals (a 10-class task), while the second consisted of all the 34 classes (numerals and letters). For each problem randomly defined disjoint partitions in training and test sets are used to produce the cross-validation estimates of the statistics reported here. The training sets contain 150 samples per class (1500 characters in the digits case and 5100 in the alphanumeric case). The test set size is fixed at 75 samples per class.

We first examine the individual performances of the participants of the multiple classifier system. The upper part of Table 1 shows the recognition error rates of the FWS as well as the sn-tuple classifiers trained on the 3 layers of the Bit-Plane Decomposition for the two task domains. It becomes readily apparent that the performance of the FWS classifier is superior compared to the others. In fact, in the best case the 3 sn-tuple classifiers present more than double the error rate of FWS. Considering the fact that the original sn-tuple algorithm results in error rates of 4.59% and 12.42% for the 10 and 34-class problem respectively, we may safely conclude that significant discriminatory information has been lost by the decomposition. However, observing the significant improvement showed in the lower part of Table 1 representing the 'mean' rule combination of the classifiers trained on the decomposed layers, leads us to hypothesise that they encapsulate information significantly complementary to each other. It should be noted here that despite their relatively poor performance the are very efficient with respect to memory space (they require only memory sizes of the order of 2^n instead of 8^n of the original SNT, n being the tuple size used).

In Table 2 we present the classification error rates achieved by combining the FWS with one or more of the layer-based sn-tuple classifiers. It is easy to observe that additional gains are obtained by these multiple classifier systems, since the highest reduction in error rates achievable, in comparison to the best of the constituent members of the system (i.e. the FWS), are of the order of 75% for the 10-class problem and of the order of 65% for the 34-class task.

Figures 3 and 4 provide a comparative illustration of the achievable gains in recognition accuracy with respect to a diverse set of classification schemes tested on the same database. The figures plot mean values of the performance statistics obtained in our cross-validation experiments. In addition to identifying the best

Table 2. Mean Error Rates (%) of the proposed multiple classifier systems

	Digits (10 classes)	Digits & Letters (34 classes)
FWS + SNTL0	6.59	16.81
FWS + SNTL1	4.51	14.74
FWS + SNTL2	5.63	14.10
FWS + SNTL0 + SNTL1	3.23	11.51
FWS + SNTL0 + SNTL2	3.95	11.34
FWS + SNTL1 + SNTL2	2.40	10.23

performing of the proposed architectures [indicated as Best of Proposed (BoP)], error rates for five other classifiers are shown. The FWS and the conventional sn-tuple scheme (SNT) have been described previously. The remaining classifiers included in the Figures are briefly described below:
Moment-based Pattern Classifier(MPC): This is a statistical classifier which explores the possible cluster formation with respect to a distance measure. This particular implementation used Mahalanobis distance calculated on the nth order mathematical moments derived from the binary image [9].
Multilayer Perceptron (MLP): This is the well-known Artificial Neural Network architecture with 40 hidden units trained using Backpropagation Learning [10].
Moving Window Classifier (MWC): This is again an $n-$tuple based scheme which utilizes the idea of a window scanning the binarised image to provide partial classification indices which are finally combined to obtain the overall classification decision [2, 6]. A 21×13 pixel window with 12-tuples were used for this particular implementation.

Finally, it is worth noting that the schemes proposed in this paper perform favorably even in comparison to a multiple classifier system (denoted by GA in the Figures) optimised by a genetic algorithm, introduced in [12]. The corresponding error rates for the same database achieved by the genetically designed multi-classifier system were 3.40% for the 10-class, and 8.62% for the 34-class tasks.

4 Conclusions

In this paper we have introduced a novel system for high performance handwritten character recognition based on n-tuple classifiers. Our primary idea was to exploit the superior performance characteristics of the Scanning n-tuple (SNT) classifier, while at the same time reducing its excessive requirements for memory space. To this end we proposed the development of a scheme based on multiple classifier systems, which have been shown to achieve increased performance by fusing relatively weaker classifiers. The participants in this system are chosen, then, from a class of appropriately simplified versions of the original SNT algorithm, which have significantly reduced memory size requirements at the expense of considerably higher error rates. A series of cross-validation experiments on a

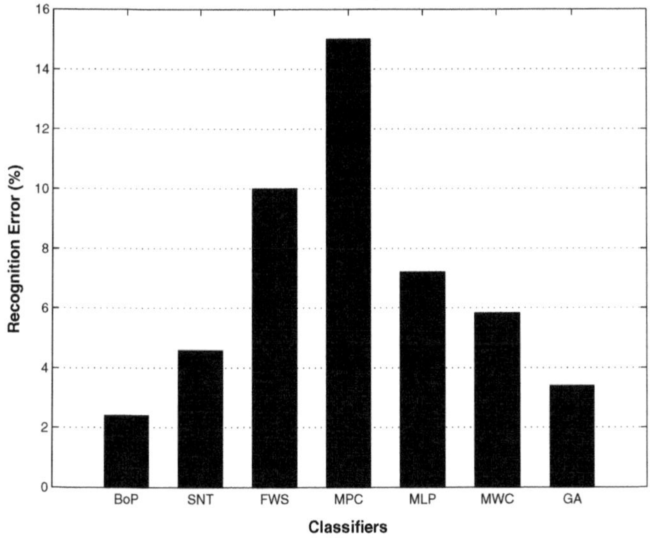

Fig. 3. Comparison of the Error Rates (%) between the Best of the Proposed Schemes (BoP) and Other Classifiers for the 10-class problem

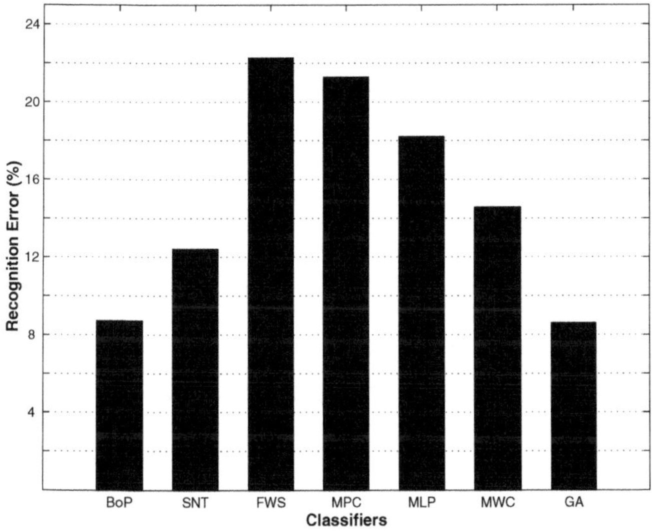

Fig. 4. Comparison of the Error Rates (%) between the Best of the Proposed Schemes (BoP) and Other Classifiers for the 34-class problem

10-class and a 34-class problems from the area of handwritten character recognition serve, finally, to demonstrate the statistical properties of our proposals. Our findings show that although the constituent members of the fusion scheme present poor performance, our system outperforms even established algorithms well-known for their efficiency in the task at hand.

Acknowledgement

The authors gratefully acknowledge the support of the UK Engineering and Physical Sciences Research Council.

References

[1] F. M. Alkoot and J. Kittler. Improving product by moderating k-nn classifiers. In J. Kittler and F. Roli, editors, *Second International Workshop on Multiple Classifier Systems*, pages 429–439. Springer, 2001. 774

[2] M. C. Fairhurst and M. S. Hoque. Moving window classifier: Approach to off-line image recognition. *Electronics Letters*, 36(7):628–630, March 2000. 776

[3] M. C. Fairhurst and T. J. Stonham. A classification system for alpha-numeric characters based on learning network techniques. *Digital Processes*, 2:321–329, 1976. 771

[4] H. Freeman. Computer processing of line-drawing images. *ACM Computing Surveys*, 6(1):57–98, March 1974. 772

[5] M. S. Hoque and M. C. Fairhurst. Face recognition using the moving window classifier. In *Proceedings of 11th British Machine Vision Conference (BMVC2000)*, volume 1, pages 312–321, Bristol, UK, September 2000. 773

[6] M. S. Hoque and M. C. Fairhurst. A moving window classifier for off-line character recognition. In *Proceedings of 7th International Workshop on Frontiers in Handwriting Recognition*, pages 595–600, Amsterdam, The Netherlands, September 2000. 776

[7] J. Kittler, M. Hatef, R. P. W. Duin, and J. Matas. On combining classifiers. *IEEE Transactions on Pattern Analysis and Machine Intelligence*, 20(3):226–239, 1998. 770, 774

[8] S. Lucas and A. Amiri. Recognition of chain-coded handwritten character images with scanning n-tuple method. *Electronic Letters*, 31(24):2088–2089, November 1995. 770, 772

[9] A. F. R. Rahman and M. C. Fairhurst. Machine-printed character recognition revisited: Re-application of recent advances in handwritten character recognition research. *Special Issue on Document Image Processing and Multimedia Environments, Image & Vision Computing*, 16(12-13):819–842, 1998. 776

[10] D. E. Rumelhart, G. E. Hinton, and R. J. Williams. *Learning internal representations by error propagation, in Parallel Distributed Processing*, volume 1, pages 318–362. MIT Press, Cambridge, MA, 1986. D. E. Rumelhart and J. L. McClelland(Eds.). 776

[11] J. W. Schwarz and R. C. Barker. Bit-plane encoding: A technique for source encoding. *IEEE Transaction on Aerospace and Electronic Systems*, 2(4):385–392, 1966. 773

[12] K. Sirlantzis, M. C. Fairhurst, and M. S. Hoque. *Genetic Algorithms for Multiple Classifier System Configuration: A Case Study in Character Recognition*, volume 2096 of *LNCS*, pages 99–108. Springer, 2001. 771, 776

A Biologically Plausible Approach to Cat and Dog Discrimination

Bruce A. Draper, Kyungim Baek, and Jeff Boody

Department of Computer Science, Colorado State University
Fort Collins, CO 80523-1873 U.S.A.
{draper,beak,boody}@cs.colostate.edu

Abstract. The paper describes a computational model of human expert object recognition in terms of pattern recognition algorithms. In particular, we model the process by which people quickly recognize familiar objects seen from familiar viewpoints at both the instance and category level. We propose a sequence of unsupervised pattern recognition algorithms that is consistent with all known biological data. It combines the standard Gabor-filter model of early vision with a novel cluster-based local linear projection model of expert object recognition in the ventral visual stream. This model is shown to be better than standard algorithms at distinguishing between cats and dogs.

1 The Human Visual System

The basic anatomical stages of the human visual process are well known. Images form on the retina, and pass via the optic nerve to the lateral geniculate nucleus (LGN) and superior colliculus (SC), and on to the primary visual cortex (area V1). From here, the human visual system divides into two streams: the dorsal visual pathway and the ventral visual pathway. As described by Milner and Goodale [1], the dorsal stream is responsible for vision in support of immediate physical action. It models the world in egocentric coordinates and has virtually no memory. The ventral stream is responsible for vision in support of cognition. It is responsible for both object recognition and 3D allocentric (i.e. object-centered) object modeling, and maintains a complex visual memory.

Although the early visual process (up to and including V1) is fairly uniform, the dorsal/ventral dichotomy is just one of many divisions that can be drawn in later stages of the human visual system. The dorsal stream, for example, can be further divided into anatomically distinct components for specific egocentric coordinates, e.g. eye-centered, head-centered, and shoulder-centered subsystems ([1], p. 53-55). By analogy, the ventral stream should also be composed of multiple, anatomically distinct systems. This hypothesis is verified by brain imaging studies, which show activity in different anatomical locations depending on whether the subject is viewing, for example, faces or places [2, 3].

It is important, therefore, for claims of biologically plausible object recognition to be specific: *which* object recognition subsystem is being modeled? In this paper, we

T. Caelli et al. (Eds.): SSPR&SPR 2002, LNCS 2396, pp. 779-788, 2002.
© Springer-Verlag Berlin Heidelberg 2002

focus on the recognition of familiar objects from familiar viewpoints at both the category and instance level, a process sometimes called "expert object recognition" [4]. We suggest that expert object recognition is a very fast process in humans, and uses an anatomical pathway from the primary visual cortex (V1) to the fusiform gyrus and the right inferior frontal gyrus. We also suggest that this pathway can be modeled as a sequence of statistical pattern recognition algorithms.

2 The Expert Object Recognition Pathway

The pathway associated with expert object recognition was first identified in fMRI studies of the more limited task of human face recognition [5-7]. In these studies, patients were shown images of human faces while in a scanner. The resulting fMRI images revealed activation not only in the primary visual cortex, but also in the fusiform gyrus. Subsequent PET studies (which imaged a larger portion of the brain) confirmed the activation in the fusiform gyrus, while adding another locus of activity in the right inferior frontal gyrus, an area previously associated through lesion studies with visual memory [8] (see also [2]). Moreover, in both the fMRI and PET studies, the activation was unique to the task of face recognition. Images of places triggered another distinct pathway with activation in the parahippocampal gyrus [2, 3, 8]. This led to speculation that evolution had created a special visual pathway for recognizing faces, and the locus of activation within the fusiform gyrus was dubbed the Fusiform Face Area (FFA; [3]).

More recent evidence suggests, however, that this pathway is used for more than just recognizing human faces. Tong, et al. report that the FFA is activated by animal faces and cartoon faces as well as human faces [9]. Chao et al. report that the FFA is activated by images of full-bodied animals, and animals with obscured faces [10]. Ishai et al. find an area in the fusiform gyrus that responds to chairs [11]. Tarr and Gauthier factored in the past experience of their subjects, and found FFA activation in dog show judges when they view dogs, and in bird experts when they view birds [12]. Most convincing of all, Tarr and Gauthier show that as people become expert at recognizing a class of objects, their recognition mechanism changes. Tarr & Gauthier created a class of cartoon characters called greebles, which in addition to individual identities can be grouped according to gender and family. When novice subjects view greebles, fMRIs show no activity in the FFA. Subjects are then trained to be greeble experts, where the definition of expert is that they can identify a greeble's identity, gender or family with an equal response time. After training, the FFAs of experts become active when they view greebles [12]. It is therefore reasonable to conclude that the FFA is part of a general mechanism for recognizing familiar objects.

3 Properties of Expert Object Recognition

What constitutes expert object recognition? People become expert at recognizing objects when they encounter them often and when instances look alike, as with faces, animals and chairs. Just as important, people become experts at recognizing objects when they have to do so at multiple levels. For example, people recognize faces they have never seen before as being human faces. At the same time, people almost instantly recognize the identity of familiar faces. Gauthier and Tarr use this multiple-level categorization as the defining characteristic of expert object recognition in their greeble studies [13], and it is a critical property of expert object recognition: objects are identified at both the class and instance level.

Expert object recognition is also very fast. While fMRI and PET studies do not give timing information, face recognition in humans can also be detected in ERP studies through a negative N170 signal. This signal occurs, on average, 164 milliseconds post stimulus onset [14]. This implies that the unique stages of expert object recognition – which we equate with the activation of the FFA and right inferior frontal gyrus – must begin within 164 milliseconds of the presentation of the target. Since visual processing begins in V1, this implies that the early stages of visual processing must also be quick. In particular, the early responses of simple and complex cells in V1 mimic Gabor filters and their combinations, and appear within 40 milliseconds of stimulus onset -- quickly enough to serve as input to the FFA. Later responses in the same cells reflect textural and boundary completion properties, but appear as late as 200 milliseconds post onset [15], probably too late to influence the expert recognition process.

Finally, expert object recognition is viewpoint dependent. The response of the FFA, for example, to images of faces presented upside-down is minimal [16]. The FFA responds to faces viewed head-on or in profile, but not to images of the back of the head [9]. In [17], upright and inverted greebles are presented to both novices and expert subjects. Expert subjects only show activation in the FFA for upright greebles (novice subjects have no FFA activation at either orientation).

3.1 Kosslyn's Model of Object Recognition

Brain imaging studies delineate the path within the ventral visual stream for recognizing familiar objects from familiar viewpoints at both the category and instance level. We believe this process accesses visual memory (as opposed to symbolic memory) because of the activation of the right inferior frontal gyrus. We also believe that the input to this process can be modeled in terms of Gabor filters because of the temporal constraints on V1 responses.

Brain imaging studies do not, however, indicate how objects are recognized, or even what the output of the object recognition process might be. For this we turn to a psychological model of object recognition originally proposed by Kosslyn and shown in Figures 1 and 2 [18]. It should be noted that Kosslyn originally proposed this model for visual perception in general, but that we apply it in the more limited context of expert object recognition.

Kosslyn's model makes a strong distinction between the strictly visual system (shown in white in Figure 1) and other mixed-modality systems, including high-level reasoning (shown in gray). Although high-level reasoning can influence vision, particularly in terms of visual attention, vision is distinct from other parts of the brain. The goal of vision is to "see again", in the sense that object recognition retrieves a visual memory that closely matches the current image [19]. Semantics in the form of object labels or other facts are assigned later by an associative memory, which receives data from many systems, including other sensors and high-level reasoning systems.

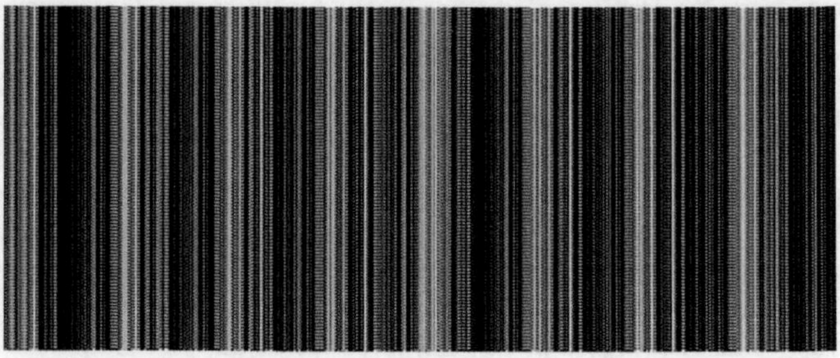

Fig. 1. Our interpretation of Kosslyn's model of the human visual system [18]

Expert object recognition is performed within the ventral stream in a multi-step process, as shown in Figure 2. The visual buffer corresponds roughly to V1. The attention window is a mechanism that selects a small portion of the image for further processing. There is evidence that it can select scales as well as positions [19]. The data selected by the attention window is passed to a preprocessing subsystem, which according to Kosslyn computes non-accidental features and object-specific signal features. For the limited case of expert object recognition, we adopt a simpler model in which the preprocessing system simply computes edge magnitude responses from the complex Gabor filter responses.

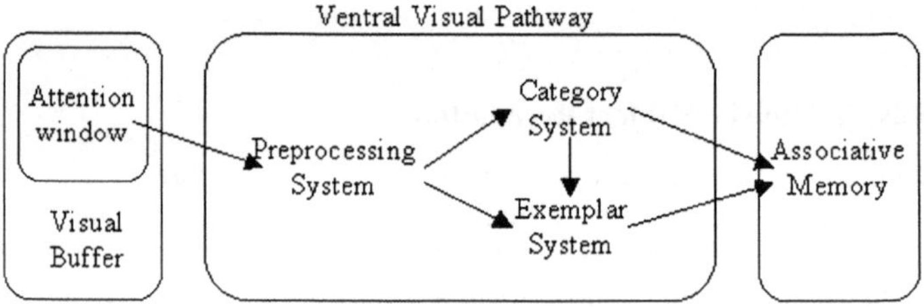

Fig. 2. The expert object recognition within the ventral visual stream

After the attention window, the most significant subsystems are the category subsystem and the exemplar subsystem. As the name implies, the category subsystem assigns images to categories, although the categories are not defined in terms of symbolic object labels. Instead, the category system groups images in memory that look alike (as measured by their V1 responses). New images are then "categorized" in the sense of being assigned to one group or another. As a result, there is no one-to-one mapping between image clusters and object labels. If an object class has many variants (e.g. dogs or chairs) or is commonly seen from many viewpoints, its images may occur in many groups. Alternatively, if two distinct objects look similar to each other, they may fall within a single group, and the category subsystem will not distinguish between them.

The exemplar subsystem matches the current image to a single visual memory. Kosslyn describes visual memories as "compressed images", based in part on the anatomical structure of visual memory and in part on evidence that visual memories can be reconstructed in V1 as mental images [20]. We interpret this as implying that the exemplar subsystem performs subspace matching. The outputs of the category and exemplar subsystems are then passed to the associative memory, which can draw semantic conclusions based on both the best visual match in memory and a cluster of similar images, as well as non-visual inputs.

4 A Computational Model of Expert Object Recognition

We implement Kosslyn's model through the EOR (Expert Object Recognition) system shown in Figure 3. EOR is initially trained on a set of unlabeled training images as shown above the dotted line in Figure 3. The images are filtered through a pyramid of orientation-selective Gabor filters, using the Gabor parameters suggested in [21] for biological systems and the complex cell models suggested in [22], and then responses are combined into edges. We assume that the attention window can consistently select both the position and scale of the target object, and that it can compensate for small in-plane rotations. In effect, the attention window registers images of target objects. We do not know how the attention window works algorithmically, but we finesse the issue by giving the system small, registered images as input.

During training, the categorization system is modeled as an unsupervised clustering algorithm operating on edge data. We currently implement this using K-Means [23]. K-Means is simple and robust, and can be applied to high-dimensional data. Unfortunately, K-Means is also limited to modeling every data cluster as a symmetric Gaussian distribution. (We are experimenting with other clustering algorithms.)

The exemplar subsystem is implemented as a subspace projection and matching system. We have tested three different subspace projection algorithms: principal component analysis (PCA [24]); independent component analysis (ICA [25]), and factor analysis (FA [26]). So far, PCA has proven as effective as other techniques, although FA is useful as a pre-process for suppressing background pixels [27]. ICA can be applied so as to produce either (1) spatially independent basis vectors or (2) statistically independent compressed images ([28], 3.2-3.3). Although some have argued for the benefits of localized basis vectors in biological object recognition [29], we find they perform very poorly in practice [30]. Linear discriminant analysis (LDA [31]) has not been considered, since biological constraints dictate that it be possible to reconstruct an approximation of a source image from its compressed form. The experiments with EOR described below use PCA to model the exemplar subsystem.

At run-time (i.e. during testing) the process is very much simpler, as shown below the dotted line in Figure 3. Test images are Gabor filtered, and the edge responses are compared to the cluster centers learned during training using a nearest neighbor algorithm. Then images are compressed by projecting them into cluster-specific subspaces, and nearest neighbors is applied again, this time to match the compressed images to compressed memories.

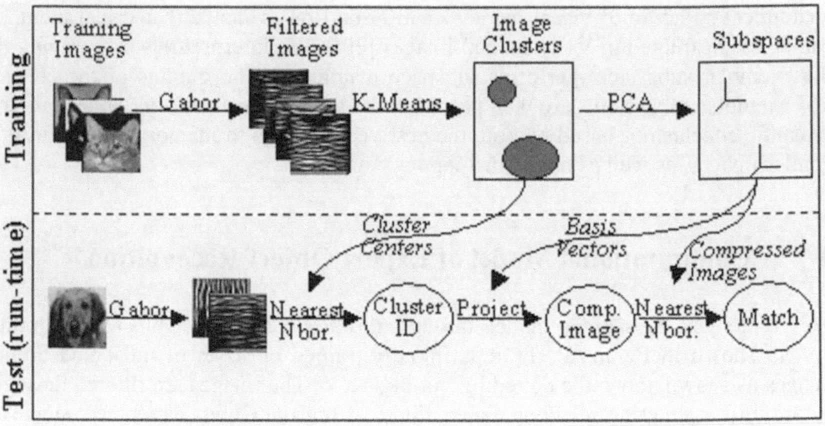

Fig. 3. EOR: Expert Object Recognition System

There are several practical advantages to this biologically inspired design. First, the subspace matching algorithm is local, not global. We compute a unique subspace for every image cluster defined by the category system, and project only the images from that cluster into it. This creates a set of localized subspaces, rather than a single, global subspace as in most PCA-based systems. Localized subspaces have previously been used with face images [32, 33], but never in the context of multiple object classes. The argument for local subspaces is that while global PCA basis vectors are optimal for images drawn from a single, normal distribution, they are not optimal for images drawn from a mixture of Gaussian distributions. In the context of expert object recognition, people are expert are recognizing many types of objects, so the images are drawn from a mixture of distributions.

Another advantage of EOR's design is that the category and exemplar subsystems exploit different properties of the data. The vectors clustered by the category system are combinations of Gabor responses. As a result, they exploit (multi-scale) edge information in images. The exemplar subsystem, on the other hand, projects raw images into the subspace. As a result, the first stage groups according to boundary information, while the second phase includes information about intensities.

5 Experiment

To test EOR, we collected a dataset of 100 images of cat faces and 100 images of dog faces, some of which are shown in Figure 4. (Biological studies clearly show that cat and dog face activate the FFA [9, 10].) No subjects are repeated in the database, which contains images of 200 different animals. The images are 64x64 pixels, and have been registered by hand so that the eyes are in approximately the same position in every image.

The system is trained on 160 images; the remaining 40 images are saved for testing. Test images are then presented to the system, which retrieves the closest matches from memory. If the retrieved image is of the same species (dog or cat) as the test image, the trial is a success, otherwise it is a failure. The system was trained 25 times (using randomly selected sets of 160 training images), yielding a total of 1,000 trials. Overall, EOR succeeded 89.9% of the time, as shown in Table 1.

Fig. 4. Samples of images from Cat and Dog data set

Is this a good result? We compare our results to several standard techniques. The first baseline is global PCA followed by nearest neighbor image retrieval (labeled "PCA" in Table 1). The second is even simpler: we simply correlate the test image to the training images, and retrieve the training image with the highest correlation score. The third baseline correlates the Gabor edge responses of the training and test images. These approaches are labeled "Corr" and "Edge Corr." in Table 1. For completeness, we also clustered the edge responses of the training data, and then labeled test images as cat or dog according to the dominant label in the nearest cluster. To our surprise, this worked better if we gave it only the highest resolution Gabor responses, rather than a pyramid (starting at 32x32) of Gabor responses (see the last two columns in Table 1).

EOR outperformed all five baseline techniques. As shown in Table 1, the performance improvement of EOR over PCA and clustering is statistically significant at the 95% confidence level, according to McNemar's significance test for paired binomial values. This is interesting, since these are the techniques combined inside EOR. The improvement over correlation is significant at only a 90% confidence level, and therefore needs to be verified in other studies.

Table 1. Recognition rates for EOR, PCA, correlation, and K-Means, and McNemar's confidence values for EOR vs. other techniques

	EOR	PCA	Corr	Edge Corr	Cluster (full res)	Cluster (multiscale)
% Correct	89.9%	88.3%	88.6%	89.2%	85.1%	73.7%
$P(H_0)$	--	4.44%	8.27%	9.79%	0.03%	0%

6 Conclusions

People are experts at recognizing objects they see often, even when many instances look alike. Moreover, they recognize familiar objects very quickly, and categorize them at both the class and instance level. Brain imaging studies identify this type of expert object recognition as a specific visual skill, and suggest an anatomical pathway involving the fusiform gyrus and right inferior frontal gyrus. This paper proposes a computational model of this pathway as unsupervised clustering followed by localized subspace projection, and shows that this model outperforms global PCA, correlation, and K-Means clustering on the task of discriminating between cats and dogs.

References

[1] A. D. Milner and M. A. Goodale, *The Visual Brain in Action*. Oxford: Oxford University Press, 1995.

[2] K. Nakamura, R. Kawashima, N. Sata, A. Nakamura, M. Sugiura, T. Kato, K. Hatano, K. Ito, H. Fukuda, T. Schormann, and K. Zilles, "Functional delineation of the human occipito-temporal areas related to face and scene processing: a PET study," *Brain*, vol. 123, pp. 1903-1912, 2000.

[3] K. M. O'Craven and N. Kanwisher, "Mental Imagery of Faces and Places Activates Corresponding Stimulus-Specific Brain Regions," *Journal of Cognitive Neuroscience*, vol. 12, pp. 1013-1023, 2000.

[4] I. Gauthier and M. J. Tarr, "Unraveling mechanisms for expert object recognition: Bridging Brain Activity and Behavior," *Journal of Experimental Psychology: Human Perception and Performance*, vol. in press, 2002.

[5] A. Puce, T. Allison, J. C. Gore, and G. McCarthy, "Face-sensitive regions in human extrastriate cortex studied by functional MRI," *Journal of Neurophysiology*, vol. 74, pp. 1192-1199, 1995.

[6] V. P. Clark, K. Keil, J. M. Maisog, S. Courtney, L. G. Ungeleider, and J. V. Haxby, "Functional Magnetic Resonance Imaging of Human Visual Cortex during Face Matching: A Comparison with Positron Emission Tomography," *NeuroImage*, vol. 4, pp. 1-15, 1996.

[7] N. Kanwisher, M. Chun, J. McDermott, and P. Ledden, "Functional Imaging of Human Visual Recognition," *Cognitive Brain Research*, vol. 5, pp. 55-67, 1996.

[8] E. Maguire, C. D. Frith, and L. Cipolotti, "Distinct Neural Systems for the Encoding and Recognition of Topography and Faces," *NeuroImage*, vol. 13, pp. 743-750, 2001.

[9] F. Tong, K. Nakayama, M. Moscovitch, O. Weinrib, and N. Kanwisher, "Response Properties of the Human Fusiform Face Area," *Cognitive Neuropsychology*, vol. 17, pp. 257-279, 2000.

[10] L. L. Chao, A. Martin, and J. V. Haxby, "Are face-responsive regions selective only for faces?," *NeuroReport*, vol. 10, pp. 2945-2950, 1999.

[11] A. Ishai, L. G. Ungerleider, A. Martin, J. L. Schouten, and J. V. Haxby, "Distributed representation of objects in the human ventral visual pathway," *Science*, vol. 96, pp. 9379-9384, 1999.

[12] M. J. Tarr and I. Gauthier, "FFA: a flexible fusiform area for subordinate-level visual processing automatized by expertise," *Neuroscience*, vol. 3, pp. 764-769, 2000.

[13] I. Gauthier, M. J. Tarr, J. Moylan, A. W. Anderson, P. Skudlarski, and J. C. Gore, "Does Visual Subordinate-level Categorization Engage the Functionally Defined Fusiform Face Area?," *Cognitive Neuropsychology*, vol. 17, pp. 143-163, 2000.

[14] J. W. Tanaka and T. Curran, "A Neural Basis for Expert Object Recognition," *Psychological Science*, vol. 12, pp. 43-47, 2001.

[15] T. S. Lee, D. Mumford, R. Romero, and V. A. F. Lamme, "The role of the primary visual cortex in higher level vision," *Vision Research*, vol. 38, pp. 2429-2454, 1998.

[16] J. V. Haxby, L. G. Ungerleider, V. P. Clark, J. L. Schouten, E. A. Hoffman, and A. Martin, "The Effect of Face Inversion on Activity in Human Neural Systems for Face and Object Recognition," *Neuron*, vol. 22, pp. 189-199, 199.

[17] I. Gauthier, M. J. Tarr, A. W. Anderson, P. Skudlarski, and J. C. Gore, "Behavioral and Neural Changes Following Expertise Training," presented at Annual Meeting of the Psychonomic Society, Philadelphia, PA, 1997.

[18] S. M. Kosslyn, *Image and Brain: The Resolution of the Imagery Debate*. Cambridge, MA: MIT Press, 1994.

[19] S. M. Kosslyn, "Visual Mental Images and Re-Presentations of the World: A Cognitive Neuroscience Approach," presented at Visual and Spatial Reasoning in Design, Cambridge, MA, 1999.

[20] S. M. Kosslyn, A. Pascual-Leone, O. Felician, S. Camposano, J. P. Keenan, W. L. Thompson, G. Ganis, K. E. Sukel, and N. M. Alpert, "The Role of Area 17 in Visual Imagery: Convergent Evidence from PET and rTMS," *Science*, vol. 284, pp. 167-170, 1999.

[21] N. Petkov and P. Kruizinga, "Computational models of visual neurons specialised in the detection of periodic and aperiodic oriented stimuli: bar and grating cells," *Biological cybernetics*, vol. 76, pp. 83-96, 1997.

[22] D. A. Pollen, J. P. Gaska, and L. D. Jacobson, "Physiological Constraints on Models of Visual Cortical Function," in *Models of Brain Functions*, M. Rodney and J. Cotterill, Eds. New York: Cambridge University Press, 1989, pp. 115-135.

[23] R. O. Duda and P. E. Hart, *Pattern Classification and Scene Analysis*. New York: John Wiley and Sons, 1973.

[24] M. Turk and A. Pentland, "Eigenfaces for Recognition," *Journal of Cognitive Neuroscience*, vol. 3, pp. 71-86, 1991.

[25] A. Hyvärinen and E. Oja, "Independent Component Analysis: Algorithms and Applications," *Neural Networks*, vol. 13, pp. 411-430, 2000.

[26] B. G. Tabachnick and L. S. Fidell, *Using Multivariate Statistics*. Boston: Allyn & Bacon, Inc., 2000.

[27] K. Baek and B. A. Draper, "Factor Analysis for Background Suppression," presented at International Conference on Pattern Recognition, Quebec City, 2002.

[28] M. S. Bartlett, *Face Image Analysis by Unsupervised Learning*: Kluwer Academic, 2001.

[29] D. D. Lee and H. S. Seung, "Learning the parts of objects by non-negative matrix factorization," *Nature*, vol. 401, pp. 788-791, 1999.

[30] K. Baek, B. A. Draper, J. R. Beveridge, and K. She, "PCA vs ICA: A comparison on the FERET data set," presented at Joint Conference on Information Sciences, Durham, N.C., 2002.

[31] P. Belhumeur, J. Hespanha, and D. Kriegman, "Eigenfaces vs. Fisherfaces: Recognition Using Class Specific Linear Projection," *IEEE Transaction on Pattern Analysis and Machine Intelligence*, vol. 19, pp. 711-720, 1997.
[32] B. J. Frey, A. Colmenarez, and T. S. Huang, "Mixtures of Local Linear Subspaces for Face Recognition," presented at IEEE Conference on Computer Vision and Pattern Recognition, Santa Barbara, CA, 1998.
[33] N. Kambhatla and T. K. Leen, "Dimension Reduction by Local PCA," *Neural Computation*, vol. 9, pp. 1493-1516, 1997.

Morphologically Unbiased Classifier Combination through Graphical PDF Correlation

David Windridge and Josef Kittler

Centre for Vision, Speech and Signal Processing
Dept. of Electronic & Electrical Engineering, University of Surrey
Guildford, GU2 5XH Surrey, United Kingdom
Telephone: +44 1483 876043
d.windridge@eim.surrey.ac.uk

Abstract. We reinterpret the morphologically unbiased 'tomographic' method of multiple classifier combination developed previously by the authors as a methodology for *graphical PDF correlation*. That is, the original procedure for eliminating what are effectively the back-projection artifacts implicit in any linear feature-space combination regime is shown to be replicable by a piecewise morphology matching process. Implementing this alternative methodology computationally permits a several orders-of-magnitude reduction in the complexity of the problem, such that the method falls within practical feasibility even for very high dimensionality problems, as well as resulting in a more intuitive description of the process in graphical terms.

1 Introduction

Within the field of machine learning there has been a considerable recent interest in the development of *Multiple Classifier Systems* [1-6], which seek to make use of the divergence of classifier design methodologies to limit *a priori* impositions on the morphology applicable to the decision boundary, such that a consistent boost in classification performance is observed.

In establishing a general theoretical framework for such approaches, the authors have determined previously [7-10] that classifier combination in virtually all of its variant forms has an aspect which may be regarded as an approximate attempt at the reconstruction of the combined pattern space by tomographic means, the feature selection process in this scenario constituting an implicit Radon integration along the lines of the physical processes involved in NMR scanning, etc (albeit of a multi-dimensional nature). It was thereby ascertained that an optimal strategy for classifier combination can be achieved by appropriately restructuring the feature-selection algorithm such that a fully-constituted tomographic combination (rather than its approximation) acts in its most appropriate domain: that is, when the combination is comprised of classifiers with *distinct* feature sets.

T. Caelli et al. (Eds.): SSPR&SPR 2002, LNCS 2396, pp. 789–797, 2002.
© Springer-Verlag Berlin Heidelberg 2002

As in medical imaging, this fully constituted tomographic combination necessarily involves the application of a deconvolution procedure to a *back-projected*[1] space, which, in the context of pattern-recognition, we were able to demonstrate amounted to the composite probability density function (PDF) constructed implicitly by the Sum-Rule decision scheme [7]. In conventional implementations of tomography [eg 11], such deconvolution is most usually accomplished via a collective prior filtering of the Radon integrals. This would take the form of a differentiation operator that acts to remove what, in the reconstructive space, would (for the case of perfect angular sample coverage) amount to convolution by an $|1/r|$ function. The very low angular sampling implied by feature-selection, however, means that only the broadest-scale structure of the back-projection artifacts can be removed in this fashion, leaving a deconvolution with angular artifacts that are still overtly and unrepresentatively correlated with the feature axes - precisely the eventuality that we are seeking to eliminate, having tested for the possibility of *actual* correlation at an earlier stage of feature selection.

The most appropriate approach [8] to removing these spurious correlations is therefore that of unfiltered (post-)deconvolution, via an adaptation of a procedure developed for use with astrophysical data: namely, the Högbom deconvolution algorithm [12], which was specifically engineered for the removal of telescopically-correlated morphology. The iterative nature of this technique allows a piece-by-piece removal of systematic artifacts such that, in its unmodified and mathematically ideal form, the procedure can be considered to impose an *a priori* condition of least possible dependence of the recovered morphology on the individual classifiers' feature-axis geometry. Thus, the procedure embodies a distinct methodology for distinguishing between the degenerate solutions that all methods of deconvolution are required to address whenever there exist zeros in the Fourier transform of the entity to be deconvolved.

It is, however, possible to view the degenerate form that the Högbom methodology reduces to in the particular environment of the stochastic domain from an entirely different perspective: that of graphical PDF correlation. In setting out this relation precisely, we shall first seek to describe how the Högbom algorithm is implemented within the Sum-Rule domain for the two-dimensional case[2].

2 Nature of Högbom Deconvolution in the Two-Dimensional Sum-Rule Domain

It was was established in [7-10] that the back-projection artifact implied by the composition (via the sum rule) of two classifiers representing the differing decision spaces; x and y, is the equivalent of a "cross" of infinitesimal width (ie $P_{\text{artefact}}(x, y) = \delta(x) + \delta(y)$, with δ the Dirac delta function). It is consequently this spurious entity (modified appropriately to account for the discrete

[1] For definitions of this, and other tomographic terms, refer to [11]

[2] We shall retain the two-dimensional paradigm for simplicity throughout the paper, the generalisation to higher dimensionalities being self-evident unless otherwise stated (or else see [7-10] for a full description).

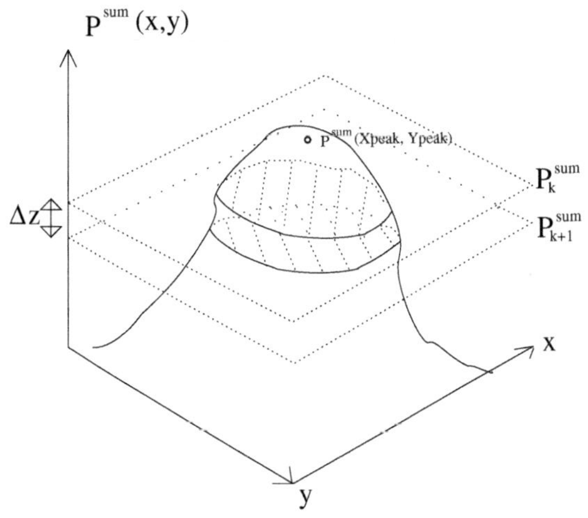

Fig. 1. The composite PDF in the Sum-Rule space

sampling of the PDF inherent in a computational methodology) that we are seeking to remove (deconvolve) from the composite feature-space PDF through recursive Högbom subtraction. In the two-dimensional case this is enacted as follows: a counter value, z, is set at the peak value of the Sum-Rule space, with a recursive scanning cycle then initiated to establish the set of all coordinates within a probability density value $| < \Delta z|$ below this. After registration of these points in a deconvolution matrix (so called because it will ultimately constitute the proposed deconvolution), through the addition of a value Δz to any existing value at the designated coordinates, a set of cross-artifacts centred on the corresponding points of the sum-rule space are then subtracted consecutively. This process is repeated until a subtraction is proposed by the procedure that would yield negative values in the Sum-Rule space, with, hence, a *complete* deconvolution resulting in a residual-free, zero-valued space. (Note that in the application to astronomical data, the procedure must rely instead on a stochastic criterion of completion in the absence of an absolute zero-point, namely the indistinguishability of the histogram of point values from a Poissonion "noise" distribution). The terminal point of the procedure therefore invariably represents (even in the absence of a proper termination) a positive-definite solution in the deconvolution matrix, as demanded by probability theory. This procedure will be more fully quantified at the computational level in the following section, however, we must first address a significant difficulty that arises with this approach:

2.1 Finite Δz Issues

It rapidly becomes apparent in any computational implementation of the Högbom deconvolution algorithm in the tomographic domain that the issue of the necessarily finite setting of the value Δz becomes non-trivial. It is intuitively obvious that the process achieves mathematical ideality only in the asymptotic limit: $\Delta z \to 0$, in which case each iterative stage registers an unambiguous set of discrete points at uniform height. However, the fact that any computational implementation must rely on a finite value of Δz gives rise to complications that have consequences that go far beyond issues of sampling accuracy: selecting different values of Δz for the situation set out above would in fact generate a vastly divergent set of completions at the termination of the procedure. Mitigating this consideration, however, is the fact that these terminal sets *do* represent consistent deconvolutions given the initial data, in the sense that the recovered distributions all revert, if re-convolved by the cross-artifact, to the originally specified Sum-Rule space.

It would perhaps, therefore, seem logical to choose $\Delta z = 0$ as being in some sense a favoured option on (as yet not fully specified) *a priori* grounds. However, any practical implementation must take place within a discretely-sampled computational setting: in proposing a finite Δz procedure that does not experience the above problem (ie, whose solution has no explicit dependence upon the value of Δz), we have to consider more systematically what is taking place during the *simultaneous* subtraction of cross-artifacts implicit in each iteration. As is uniquely the case for tomographic reconstruction of a pattern-space, these subtraction entities share an identity with the form of the axial system (that is to say, constitute a complex of intersecting quadrilaterals of varying dimensionality [8]). We can therefore appreciate that the simultaneity of the subtraction immediately gives rise to an irreconcilable ambiguity: we see that the *overlap* of these entities necessarily gives rise to further intersections at specific points of the pattern space, the artifacts around which *are of the same form as the axial system*, which are hence not in any real sense distinguishable from the original points at which axial artifacts are subtracted.

It is therefore apposite to propose as a modification of the Högbom method (when acting in the expert fusion domain), that these additional points are put forward as candidates for registration *alongside* the originals. It is, in a sense, therefore possible to regard this modification as summing over all possible deconvolution solutions that we earlier encountered at the iterative level. This amounts to applying the most conservative criterion of PDF correlation *within* the terms of the Högbom approach, while maintaining the most presumptive *a priori* condition on the feature correlation in more general terms (which is to say, imposing an assumption of minimal feature dependence on the axes, the alternative having been eliminated at the feature selection stage).

In visualising this alternative approach, it is most useful to focus on the effect that the Högbom algorithm has on the PDFs *constituting* the combination, rather than the Sum-Rule space, as we have hitherto done. The nature of the Högbom iteration is also rendered far more graphically evident from such a perspective:

2.2 PDF-Centred Approach

As we have thus far understood it, then, the commencement of the Högbom procedure consists the determination of the peak position of the Sum-Rule space

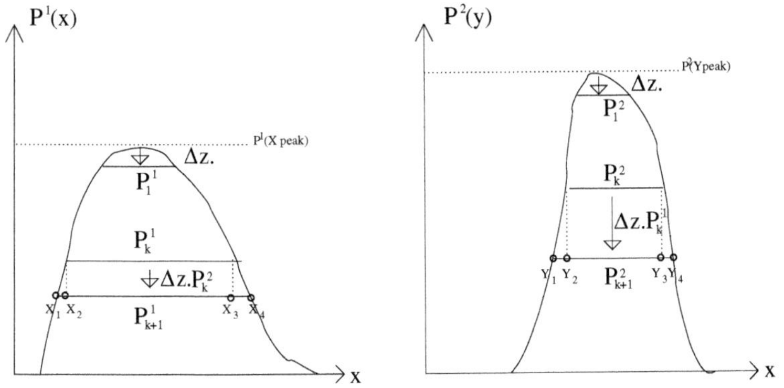

Fig. 2. Requisite subtractions from the two PDFs constituting the combination in the modified methodology: note the presence of P_k^2 in the first diagram's subtraction (and vice-versa)

($P^{sum}(X_{peak}, Y_{peak})$ from fig. 1), and the derivation of the set of points, P_1^{sum}, that lie in the probability density range ($P^{sum}(X_{peak}, Y_{peak}) \rightarrow P^{sum}(X_{peak}, Y_{peak}) - \Delta z$), prior to subtracting a series of cross-artifacts centred on those points. We should now like to associate these points with particular sets of ordinates in the PDF domain such that it is possible to view the 3-dimensional process of fig. 1 within the 2-dimensional format of fig. 2. This would not in general be possible to do in a straightforward fashion if the subtraction entity were of an arbitrary form. However, the fact that the subtraction artifact mirrors the axial system means that it may be equivalently represented as the *independent* summation of 1-dimensional Dirac delta functions (convolved by the sampling element Δx) centred on the appropriate ordinates of the *PDF domain*. The process of subtraction of a *single* artifact in this domain therefore acquires the intuitive aspect of a subtraction of individual delta functions from the appropriate points of the respective classifier PDFs ($\delta(x - x_0)\Delta x$ from $P^1(x)$, and $\delta(y - y_0)\Delta x$ from $P^2(y)$, in our case).

Although this situation readily generalises to the arbitrarily-dimensioned case, it becomes somewhat more complex for multiple subtractions of the type indicated earlier, in that the subtraction of cross-artifacts centred on the additional set of points created by the intersections of the artifacts (arising from the originally detected points) leads to an asymmetry in the corresponding PDF domain subtractions: the particular value to be subtracted from each of the ordinals in a particular PDF turns out to require a proportionality to the subtractions in the remaining PDFs constituting the combination. This is illustrated in fig. 2 for a mid-point of the deconvolution's execution (since we are required to externally impose an infinitesimal subtraction on the *first* iteration of the sequence $k = 1, 2 \ldots$, which cannot, therefore, exhibit this effect explicitly).

A subtraction, then, of the points above P^{sum}_{k+1} (points above P^{sum}_{k} having been assumed to have been removed by previous iterations) leads to a replacement of the ordinal sets: $\{x|P^1(x) = P^1_k\}$ with $\{x|P^1(x) = P^1_{k+1}\}$ and $\{y|P^2(y) = P^2_k\}$ with $\{y|P^2(y) = P^2_{k+1}\}$: that is to say, a reduction of $\Delta z|P^2_k|$ and $\Delta z|P^1_k|$ in $P^1_k(x)$ and $P^2_k(y)$, respectively (with a corresponding registration of Δz in the deconvolution matrix for the coordinate-set $\{x|X_1 \leq x \leq X_2\} \times \{y|Y_1 \leq y \leq Y_2\}$, that is, all combinations of ordinals over this range). Note in particular the transfer of width information from one PDF to the other, giving rise to the mutually morphologically dependent convergence alluded to earlier: we are then now implicitly regarding the PDFs, not as maps $\mathcal{R} \rightarrow \mathcal{R}$, but rather as morphological entities delineating 'areas' in an ordinate-probability space.

The fact that these points lie in bands is critical to the method's economy, and a consequence both of the explicit inclusion of the intersection point-sets (of which more later), but also of the particular nature of this stage-by-stage remapping. For the set of ordinates newly incorporated into the $(k+1)$'th iteration to be consistent with the line defined by the ordinate set arising from the k'th iteration, this involves imposing a transformation:

$$\{P^1_x\} \rightarrow \{P^1_{X_1}\} \quad \forall\,(X_1 < x < X_2) \text{ and } \forall\,(X_3 < x < X_4) \tag{1}$$

$$\{P^2_y\} \rightarrow \{P^2_{Y_1}\} \quad \forall\,(Y_1 < y < Y_2) \text{ and } \forall\,(Y_3 < y < Y_4), \tag{2}$$

at each new stage of the process, such that each new ordinate set is contained within its predecessor. Thus, the algorithmic recursion applies solely now to these ordinal sets (two single-dimensioned entities, rather than to a single Sum-Rule density function of three dimensions). It should also be noted that this approach is equally valid for the more complex case of multiply-peaked PDFs, the extension to the mapping protocol being a matter of straightforward extrapolation.

The other issue which we have yet to approach systematically within this framework arises in relation to multiple subtractions, and concerns the aforementioned ambiguity arising from the cross-correlation between subtractive entities. In fact, it transpires that a quantitative treatment of this effect is rendered significantly more straightforward on consideration within the PDF domain: in removing multiple delta-function elements from the individual density functions, all of the interstitial "overlap" artifacts are implicitly dealt with at the same time. This can be illustrated in the two-dimensional case via an appreciation of the fact that the subtraction of delta-function elements centred on the P^1 ordinals; X_1 and X_2, and the P^2 ordinals; Y_1 and Y_1, would imply a subtraction of cross artifacts centred on; $P^{sum}(X_1, Y_1)$, $P^{sum}(X_2, Y_1)$, $P^{sum}(X_2, Y_2)$ and $P^{sum}(X_1, Y_2)$: that is to say, the complete set of detectable points in the Sum-Rule domain *as well as* their subtraction-artifact overlaps. The only remaining issue to address in relation to the PDF-centred approach to Högbom deconvolution is then the construction of the actual *co*-ordinates for registration in the deconvolution matrix, which, it is readily seen, are just the set of all permutations of detected ordinals within the current iteration.

In this manner, by switching to a PDF-oriented approach, necessitating what is effectively a varying Δz methodology within which the issue of multiple registrations and subtractions is dealt with *automatically*, we have effectively dissolved the distinction between PDF point-detection, artifact-correlation and artifact subtraction, generating a significant speed increase through the fusion of the three space-scanning processes implicit in the tomographic method, as well as a further, arbitrarily large speed increase determined by the implicit fusion of the Δz parameter with the morphology of the PDFs (through the inclusion of cross-sectional magnitude terms *within* each iteration).

We shall determine more precisely the effect that this has on the computational efficiency of the tomographic method as follows:

2.3 Computational Implementation

The first economization attributable to the new approach, arising as a consequence of the implicit identification of the peak-search, peak-correlation and artifact-subtraction procedures, reduces a process of originally $\sim [X]^{2n}[X^{n-1} + X]$ cycles to around X^{n-1} cycles (n being the dimension of the reconstructive space, and X its linear sampling resolution: the square brackets denote a maximum value). This is determined in the following way: within the unmodified Högbom procedure each iterative scan of the Sum-Rule space to obtain a set of points for subtraction carries with it a penalty of X^n cycles. Because Δz is not correlated with the PDF cross-sections as it is in the modified case, the requisite analysis of subtraction-artifact overlapping will require that the additional interstitial points are all individually constructed and registered within the deconvolution matrix. In the worst case scenario, when the ordinates of the detected points cover the entirety of the feature axes, this would amount to an implicit scan over the entire reconstructive space, requiring an additional computation of $[X]^n$ cycles (a scan being effectively the exhaustive cyclic construction of ordinal permutations).

A deconvolution-artifact subtraction at each of these points would then require a further scanning agent to act over the reconstructive space, ostensibly involving a further X^n cycles *per point*. However, it is possible to break the artifacts down into their constituent iterations to obtain a reduction in this. That is, if the set of classifiers constituting the combination have an individual feature-dimensionality given by d_i, then this would represent a required per-point cycle count of magnitude $(X^{d_1} + X^{d_2} + X^{d_3} \ldots)$ in order to perform the subtraction. In execution terms, this represents a *maximum* of $X^{n-1} + X$ cycles (the best case scenario being just nX cycles, or $2X$ in our example). The total cycle count per iteration for the Högbom method is therefore: $X^n[X]^n[X^{n-1} + X]$, where it is understood that this (and all following terms) represent worst-case scenarios.

By contrast, the proposed alternative, in combining the detection, correlation and subtraction procedures, permits a cycle count of only X^n per iteration. This comes about through combining the activity of a detection/subtraction scan that acts over just the constituent PDF feature dimensions (which would in itself now carry only a $[X^{n-1} + X]$ cycle penalty) with a correlation analysis (which would normally constitute an *additional* $[X]^{d_1} + [X]^{d_2} + [X]^{d_3} \ldots = [X]^n$ cycles per point), such that the correlation analysis, in generating every possibly ordinal permutation, now implicitly performs both the detection and subtraction operations in the manner described above.

It is possible, within the proposed alternative to tomographic combination, to further improve on this performance for the particular case of the constituent classifiers constituting point-wise continuous PDFs, through the introduction of a second-order computational economy. We note in fig. 2 that P_k^1 is fully contained within the set P_{k+1}^1, with only the sets $P^1(X_1) \to P^1(X_2)$ and $P^1(Y_1) \to P^1(Y_2)$ then contributing a new behavioural aspect to the $(k+1)$'th iteration (and similarly for $P^2(Y)$). Thus, the newly correlated and registered points in the $(k+1)$'th iteration will all lie <u>inside</u> of the P^{sum} region defined by the coordinate range: $(X_1 \to X_4, Y_1 \to Y_4)$, and <u>outside</u> of the smaller region $(X_2 \to X_3, Y_2 \to Y_3)$. Hence (and this is equally true for multiply-peaked PDFs), it becomes possible to simply discard this region within the correlation analysis (by far the most computationally expensive part of the proposed methodology), leaving only the originally specified artifact subtraction to perform, at a penalty of $[X^{n-1} + X]$ cycles. In algebraic terms this results in a cycle count reduction to:

$$\{[X^{n-1} + X]\} + \{(X + dx)^n - X^n\} \approx \{[X^{n-1} + X]\} + \{n\, dx\, X^{n-1}\} \quad (3)$$

(the later bracketed term in the addition constituting the generalisation of the above reasoning to arbitrary dimensionality, and dx being the sampling element [of similar fractional width to Δz]). This is clearly, then, a very substantial additional saving.

As a final note, it is evident that the number of iterations is itself a key dictator of execution time and, as we have observed, is a quantity that need not necessarily be fixed, a fact from which we have considerably benefited. However, the actual *value* of the number of iterations is governed by PDF morphology, and consequently not straightforwardly enumerable. The original Högbom method, however, does not suffer this limitation, requiring a fixed $(P_{max}^1 + P_{max}^2 + \ldots)/\Delta z$ iterations to execute, and serves as an upper limit for the modified procedure (although in practice we would expect the actual value to be a small fraction of this).

Thus, in the final analysis, the total cycle count for the more efficient methodology can be written:

$$\begin{aligned}&[(P_{max}^1 + P_{max}^2 + \ldots)/\Delta z]\{[X^{n-1} + X]\} + \{ndxX^{n-1}\}\\ &\approx [(P_{max}^1 + P_{max}^2 + \ldots)]\{[X^{n-1} + X]/\Delta z\} + \{nX^{n-1}\}\end{aligned} \quad (4)$$

as opposed to:

$$(P_{max}^1 + P_{max}^2 + \ldots)/\Delta z\{X^n[X]^n[X^{n-1} + X]\} \quad (5)$$

under the original proposal.

3 Conclusion

We have set out to reinterpret the tomographic method of classifier combination within its most natural context, significantly reducing the computation time involved to the extent that the method now poses very little obstacle to practical implementation.

The basis of this efficiency gain is the observation that, viewed in terms of the constituent PDFs, the three chief computational components of the recursive tomographic procedure (the peak-seek, the peak correlation analysis and the subtraction/registration of correlated components) need not actually be performed on an individual basis, reducing an iteration requirement of $X^n[X]^n[X^{n-1}+X]$ computational cycles to a maximum of X^n, with the further possibility of an order of magnitude decrease in this figure for point-wise continuous classifiers. Finally, there are further (if not precisely quantifiable) gains arising from dynamically varying the Δz parameter throughout the procedure.

The authors would like to gratefully acknowledge the support of EPSRC under the terms of research grant number GR/M61320, carried out at the University of Surrey, UK.

References

1. R. A. Jacobs, "Methods for combining experts' probability assessments", Neural Computation, 3, pp. 79-87, 1991
2. J. Kittler, M. Hatef, R. P. W. Duin, and J. Matas, "On combining classifiers", IEEE Transactions on Pattern Analysis and Machine Intelligence, vol. 20, no. 3, 1998, 226-239
3. L. Lam and C. Y. Suen, "Optimal combinations of pattern classifiers", Pattern Recognition Letters, vol. 16, no. 9, 1995, 945-954.
4. A. F. R. Rahman and M C Fairhurst, "An evaluation of multi-expert configurations for the recognition of handwritten numerals", Pattern Recognition Letters, 31, pp. 1255-1273, 1998
5. A. F. R. Rahman and M C Fairhurst, "A new hybrid approach in combining multiple experts to recognise handwritten numerals", Pattern Recognition Letters, 18, pp. 781-790, 1997
6. K. Woods, W. P. Kegelmeyer and K Bowyer, "Combination of multiple classifiers using local accuracy estimates", IEEE Trans. Pattern Analysis and Machine Intelligence, 19, pp. 405-410, 1997
7. D. Windridge, J. Kittler, "An Optimal Strategy for Classifier Combination: Part 1: Multiple Expert Fusion as a Tomographic Process", (PAMI, Submitted)
8. D. Windridge, J. Kittler, "An Optimal Strategy for Classifier Combination: Part 2: General Application of the Tomographic Procedure", (PAMI, Submitted)
9. D. Windridge, J. Kittler, "Classifier Combination as a Tomographic Process", (Multiple Classifier Systems, LNCS. Vol. 2096 , 2001.)
10. D. Windridge, J. Kittler, "A Generalised Solution to the Problem of Multiple Expert Fusion.", (Univ. of Surrey Technical Report: VSSP-TR-5/2000)
11. F. Natterer, Proceedings "State of the Art in Numerical Analysis", York, April1-4, 1996.
12. J. Högbom, "Aperture synthesis with a non-regular distribution of interferometer baselines", Astrophys. J. Suppl. Ser., 15, 417-426, 1974

Classifiers under Continuous Observations

Hitoshi Sakano and Takashi Suenaga

NTT Data Corp.
Kayabacho Tower, 1-21-2, Shinkawa, Chuo-ku
Tokyo, 104-0033, Japan
{sakano,suenaga}@rd.nttdata.co.jp

Abstract. Many researchers have reported that recognition accuracy improves when several images are continuously input into a recognition system. We call this recognition scheme a *continuous observation-based scheme* (CObS). The CObS is not only a useful and robust object recognition technique, it also offers a new direction in statistical pattern classification research. The main problem in statistical pattern recognition for the CObS is how to define the measure of similarity between two distributions. In this paper, we introduce some classifiers for use with continuous observations. We also experimentally demonstrate the effectiveness of continuous observation by comparing various classifiers.

1 Introduction

Our research opens new directions in the field of statistical pattern recognition for highly accurate 3D object recognition.

Continuous observations may improve the robustness of object recognition systems. They enable a model to be extracted that accounts for changes in input when the pose or lighting conditions change and that also reduces the noise in an object image recognition problem (see Fig. 1.). In this paper, we discuss the problem for statistical pattern recognition when continuously input data are assumed. We call the method a *continuous observation-based scheme* (CObS).

It is reasonable to assume continuous observations in experimental studies of object recognition because we ordinarily use video cameras to capture object images. When we are working on image streams, statistical processing of the images in the stream may improve the accuracy of object recognition systems. Statistical processing reduces noise and enables invariant features to be extracted from the images .

We considered the following research issues from the viewpoint of statistical pattern recognition:

1. What type of statistical techniques can be effectively applied to input images?
2. How can we define the similarities between training images or templates and input images?

T. Caelli et al. (Eds.): SSPR&SPR 2002, LNCS 2396, pp. 798–805, 2002.
© Springer-Verlag Berlin Heidelberg 2002

noise reduction
Continuous input samples → feature extraction → Recognition
3D reconstruction (?)

Fig. 1. Continuous observation-based scheme (CObS)

Some procedures have already been proposed under the CObS. Maeda proposed the *mutual subspace method* (MSM) [1], which uses principal component analysis to extract invariant features from input images. Yamaguchi demonstrated the effectiveness of the MSM for facial image recognition experimentally [2]. We proposed the *kernel mutual subspace method* (KMS) [3,4]. The KMS, which is one of the most powerful algorithms for object recognition, is based on kernel principal component analysis (KPCA) [5].

In this paper, we propose some classifiers for use when continuous observations are assumed. We also describe experiments to compare the classifiers experimentally.

In the next section, we describe [a problem in the CObS?] and in section 3 we introduce some classifiers used in the CObS. In section 4, we describe experiments carried out to clarify the properties of the classifiers.

2 Problem Setting

When CObS is assumed, we must consider the two issues described above. Generally, an object recognition algorithm consists of templates $T(\cdot)$ and the calculation of the similarity measure $S(T;\boldsymbol{x}')$ between the template and the input image $\boldsymbol{x}'$. The template is calculated from training data $\boldsymbol{x}_i, i = 1, \ldots, m$, where m is the number of training images. PCA, a multilayer perceptron, sample means, and other statistical techniques are used to calculate the templates T.

Under the CObS, we must define a new measure of similarity between the templates and object images $s(T;T')$ where T' is calculated from features extracted from the input images $\boldsymbol{x}'_i, i = 1, \ldots, m'$, where m' is the number of input images.

We then have to decide what kind of statistical analysis to apply? We can create many classifiers under the CObS.

3 Classifiers in CObS

We describe some of the classifiers used in the CObS. These classifiers are based on CLAFIC [6] (except for the multiple potential function classifier) because the subspace method is regarded as the standard method for object image recognition [7,8].

3.1 Sample Mean

First, we use a sample mean as the simplest statistical technique under CobS. It also has the advantage of reducing noise. When several images have been input into the recognition system, the system uses the sample mean

$$\boldsymbol{\mu}' = \frac{1}{m'} \sum_{i=1}^{m'} \boldsymbol{x}'_i \tag{1}$$

as the input image, where m' is the number of input images. For example, the similarity measure in CLAFIC is defined as

$$S_{CLAFIC}(V_1, \ldots, V_l; x') = \sum_{i=1}^{l} (V_i \cdot \boldsymbol{x})^2, \tag{2}$$

when a single input is assumed, where V_i is the ith eigenvector of a covariance matrix and l is the number of dimensions of the subspace. However, when multiple data input is assumed, the similarity measure is defined as

$$s_{mean}(V_1, \ldots, V_l; \boldsymbol{\mu}') = \sum_{i=1}^{l} (V_i \cdot \boldsymbol{\mu}')^2, \tag{3}$$

using the sample mean.

3.2 Multiple Potential Function Classifier

The definition of a conventional *potential function classifier* (PFC) [9] is

$$S_{PFC}(T; \boldsymbol{x}') = \sum_{i=1}^{m} k(\boldsymbol{x}_i, \boldsymbol{x}'), \tag{4}$$

where $k(\cdot, \cdot)$ is a bell-like nonlinear function called the potential function or kernel, $\boldsymbol{x}'$ is an input image, and m is the number of training samples. The templates of the potential function classifiers are themselves training samples. The classifier is simply extended under the CObS as

$$s_{MPFC}(T; \boldsymbol{x}'_1, \boldsymbol{x}'_2, \ldots, \boldsymbol{x}'_{m'}) = \sum_{j=1}^{m'} \sum_{i=1}^{m} k(\boldsymbol{x}_i, \boldsymbol{x}'_j). \tag{5}$$

In this paper, the extended form of the potential function classifier is called *a multiple potential function classifier* (MPFC) in the CObS.

3.3 Mutual Subspace Method

Yamaguchi demonstrated the effectiveness of the MSM for facial image recognition experimentally [2]. In the MSM, PCA is performed on images obtained by continuous observations and "MSM similarity" is measured. MSM similarity is defined by the angle between two subspaces of the input images and training images. The two subspaces used for the MSM similarity measurement are computed from continuous observation data or preregistered training data.

In the learning phase of the MSM, the [basis'OK?] obtained from the training data is registered as a template. (This is the same as in the conventional subspace method.) In the recognition phase,

1. the PCA basis is calculated from the input images.
2. the following matrix is calculated.

$$Z = (Z_{ij}) = \sum_{k=1}^{M} (\boldsymbol{V}_i \cdot \boldsymbol{V}'_k)(\boldsymbol{V}'_k \cdot \boldsymbol{V}_j), \tag{6}$$

   where $\boldsymbol{V}$ is the basis of the training data subspace and $\boldsymbol{V}'$ is the basis of the subspace obtained from the input data.
3. The maximum eigenvalue of Z, which is the angle between two subspaces, is obtained [10].

The maximum eigenvalue of Z is regarded as the similarity measure $s_{MSM}(V_1, \ldots, V_l; V'_1, \ldots, V'_{l'})$, where l is the number of dimensions of the subspace of the template and l' is the number of extracted features. The MSM therefore consists of continuous observation samples. This ensures the method has robust capability for object image recognition because it is easy to obtain several images by observing the object in the form of a motion image sequence.

3.4 Kernel Mutual Subspace Method

In this section, we describe the *kernel mutual subspace method*. The effectiveness of the method was demonstrated in facial image recognition experiments, which were reported previously [3].

First, we must briefly review kernel principal component analysis (KPCA). KPCA is performed by carrying out singular value decomposition in functional space $\mathcal{F}$ for a given set of data $\boldsymbol{x}_i, i = 1, \ldots, m$ in the n dimensional feature space R^n. We can define the functional space $\mathcal{F}$, which is related to the feature space, by a possibly non-linear map

$$\Psi : \mathcal{R}^N \to \mathcal{F}, \boldsymbol{x} \to \boldsymbol{X}. \tag{7}$$

Note that the functional space $\mathcal{F}$ may have large, possibly infinite dimensionality.

In functional space $\mathcal{F}$, the covariance matrix takes the form

$$\bar{C} = \frac{1}{m} \sum_{j=1}^{m} (\Psi(\boldsymbol{x}_j)\Psi(\boldsymbol{x}_j)^T). \tag{8}$$

The basis of KPCA is given by diagonalizing the matrix. However, the matrix is too large (sometimes infinitely so) to compute. We use the $m \times m$ kernel matrix

$$K_{ij} = \Psi(\boldsymbol{x}_i) \cdot \Psi(\boldsymbol{x}_j), \tag{9}$$

to solve the eigenvalue problem,

$$m\lambda\alpha = \alpha K \tag{10}$$

for non-zero eigenvalues. The α_i denotes the column vector with entries $\alpha_1, \ldots, \alpha_m$.

To extract the principal component, we need to compute the projection onto eigenvectors V in $\mathcal{F}$. Let $\boldsymbol{x}'$ be the input feature vector, with an image $\Psi(\boldsymbol{x}')$ in $\mathcal{F}$. Then

$$V \cdot \Psi(\boldsymbol{x}') = \frac{1}{\lambda} \sum_{i=1}^{m} \alpha_i \Psi(\boldsymbol{x}_i) \cdot \Psi(\boldsymbol{x}') \tag{11}$$

may be called the nonlinear principle components corresponding to Ψ . Since map $\Psi(\cdot)$ makes a very large, possibly infinite, functional space, the cost of computing the dot product is extremely large (or infinite). To avoid this problem, Schölkopf introduced the Mercer kernel, which satisfies

$$k(\boldsymbol{x}, \boldsymbol{y}) = \Psi(\boldsymbol{x}) \cdot \Psi(\boldsymbol{y}). \tag{12}$$

When we use this kernel function, the computation of the dot product $\Psi(\boldsymbol{x}) \cdot \Psi(\boldsymbol{y})$ replaces the computation of function $k(\boldsymbol{x}, \boldsymbol{y})$. That is,

$$V \cdot \Psi(\boldsymbol{x}) = \frac{1}{\lambda} \sum_{i=1}^{m} \alpha_i k(\boldsymbol{x}_i, \boldsymbol{x}). \tag{13}$$

This result shows that we can calculate a projection for the nonlinear principal components in finite time without an explicit form of the nonlinear basis V.

Now we can describe our proposed kernel mutual subspace method (KMS), which combines MSM and KPCA. We first define a similarity measure for the KMS in functional space $\mathcal{F}$. Practical applications demand lower computational costs. Therefore, we must prove that the proposed method takes a finite time to compute the angle in functional space $\mathcal{F}$.

Let W be an eigenvector calculated from continuous images input into the recognition system. Then, we can describe V and W by

$$V = \sum_{i=1}^{m} \alpha_i \Psi(\boldsymbol{x}_i)$$

$$W = \sum_{j=1}^{m'} \alpha_j \Psi(\boldsymbol{x}'_j), \tag{14}$$

where m and m' are the number of samples for training and test images. The similarity measure is then computed by dot product $V \cdot W$:

$$V \cdot W = \sum_{i=1}^{m} \alpha_i \Psi(\boldsymbol{x}_i) \cdot \sum_{j=1}^{m'} \alpha'_j \Psi(\boldsymbol{x}'_j) \tag{15}$$

$$= \sum_{i=1}^{m} \sum_{j=1}^{m'} \alpha_i \alpha'_j \Psi(\boldsymbol{x}_i) \cdot \Psi(\boldsymbol{x}'_j). \tag{16}$$

If we substitute (12), the equations can be written as,

$$V \cdot W = \sum_{i=1}^{m} \sum_{j=1}^{m'} \alpha_i \alpha'_j k(\boldsymbol{x}_i, \boldsymbol{x}'_j). \tag{17}$$

Because the numbers of m and m' are limited, this form shows that this method takes a finite time to compute the dot product of the basis of two subspaces. The method for obtaining the angles between two subspaces can be derived by substituting (17) into (6).

4 Experiment

We compared the classifiers described above in facial image recognition experiments. We employed CLAFIC as a reference recognition method when single image input was assumed because the subspace method is regarded as the standard object recognition method [7,8].

We used 15 individuals from the UMIST data set [11]. The data had a nonlinear structure, as shown in Fig. 2. The facial images were manually segmented and normalized to 15×15 pixel images (225 dimensionality vector). The number of training images was 10 per person, and 599 test images were created from the remaining data. We determined $m' = 5$ in this experiment.

We used a Gaussian radial basis function as a nonlinear kernel function for the KMS and MPFC. The kernel parameter and the number of dimensions of subspaces were selected based on the results of a preliminary experiment. Experiments were performed to compare the effectiveness of using CLAFIC, the sample mean, the MPFC, the MSM, and the KMS. The results are shown in Table 1. These results showed the effectiveness of continuous observations. They also showed that:

- in this case, redundant information was apparently needed to improve the accuracy of the linear methods (sample mean, MSM).
- unexpectedly, the accuracy of the sample mean method was high. This result is inconsistent with Yamaguchi's results [2]. We think the inconsistency was caused by the non-linearity of the distribution. Linear PCA fails when the distribution has a nonlinear structure.
- the MPFC was the least accurate method.

Table 2 lists the properties of the methods.

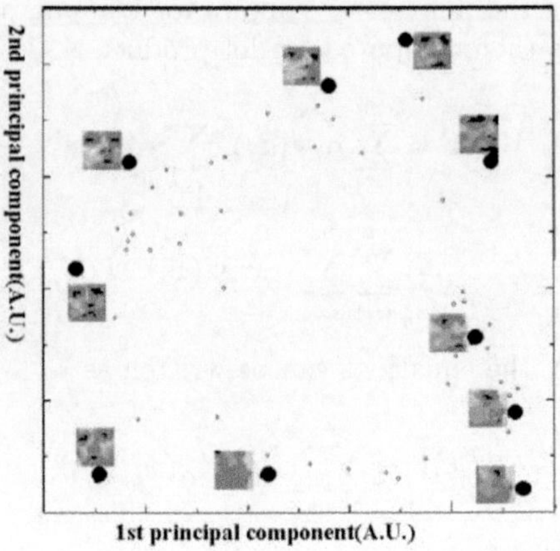

Fig. 2. Scatter plot of facial images in the UMIST data using PCA

Table 1. Recognition rate for each method

method	CLAFIC	Mean	MPFC	MSM	KMS
accuracy(%)	95.7	99.8	94.0	99.6	100.0
# of dim(train.)	5	5	-	6	3
# of dim(input)	-	0	-	2	2

5 Conclusion

We have described a *continuous observation-based system* for object image recognition and some classifiers used in the system. We also clarified the properties of the classifiers in an experimental comparison. The experimental results show the effectiveness of the classifiers in the CObS in terms of recognition accuracy. We believe the CObS offers a new research direction for statistical pattern recognition.

In future work, we will introduce other classifiers under the CObS and clarify their properties.

Acknowledgment

We are grateful to Prof. N.S. Allinson of the University of Manchester Institute of Science and Technology for permitting us to use the UMIST face database.

Table 2. Properties of each classifier

Method	Accuracy	cal. cost.	comments
CLAFIC	low	low	single observation
Sample Mean	high	low	most simple
MPFC	low	high	easy training
MSM	high	mid	redundant expression
KMS	high	high	compact expression

References

1. Ken-ichi Maeda and Sadaichi Watanabe, "Pattern Matching Method with Local Structure", Trans. IEICE(D), Vol. 68-D. No. 3, pp. 345-352(1985) (in Japanese) 799
2. O. Yamaguchi, K. Fukui and K. Maeda, "Face Recognition using Temporal Image Sequence", In Proc. IEEE 4thtl. Conf. on Face and Gesture Recognition, pp. 318-323 (1998) 799, 801, 803
3. H. Sakano, et. al., "Kernel mutual subspace method for robust facial image recognition", in Proc. IEEE Intl. Conf. of Knowledge Engineering System, pp.245-248, (2000) 799, 801
4. H. Sakano, "Kernel Mutual Subspace Method for Object Recognition", Trans. IEICE(D-II), Vol. J84-D-II, No. 8, pp. 1549-1556, (2001) (in Japanese) 799
5. B. Schölkopf, et al., "Nonlinear component analysis as a kernel eigenvalue problem", Neural Computation, Vol. 10, pp. 1299-1319 (1998) 799
6. S. Watanabe and N. Pakvasa, "Subspace method of pattern recognition", Proc. 1st IJCPR, pp. 25- 32 (1973) 799
7. M. Turk and A. Pentland, "Recognition Using Eigenface", Proc. CVPR, pp. 568-591 (1991) 799, 803
8. H. Murase and S. K. Nayer, "Visual learning and recognition of 3-D object from appearance", International Journal of Computer Vision, Vol. 14, pp. 5-24, (1995) 799, 803
9. M. A. Aizerman, et. al., "Theoretical foundations of the potential function method in pattern recognition learning", Automation and Remote Control, Vol. 25, pp.821-837, (1964) 800
10. F. Chatelin, "Veleurs propres de matrices", Masson, Paris (1988) 801
11. D. B. Graham and N. S. Allinson, "Characterizing Virtual Eigensignatures for General Purpose Face Recognition", in H. Wechsler, et al. ed. "Face Recognition From Theory to Applications", Springer Verlag, (1998) 803

Texture Classification Based on Coevolution Approach in Multiwavelet Feature Space

Jing-Wein Wang

Center of General Studies, National Kaohsiung University of Applied Sciences
415 Chien-Kung Road, Kaohsiung 807, Taiwan, R.O.C.
Tel.: 886-7-3814526 Ext. 3350, Fax.: 886-7-5590462
jwwang@cc.kuas.edu.tw

Abstract. To test the effectiveness of multiwavelets in texture classification with respect to scalar Daubechies wavelets, we study the evolutionary-based algorithm to evaluate the classification performance of each subset of selected feature. The approach creates two populations that have interdependent evolutions corresponding to inter and intra distance measure, respectively. With the proposed fitness function composed of the individuals in competition, the evolution of the distinct populations is performed simultaneously through a coevolutionary process and selects frequency channel features of greater discriminatory power. Consistently better performance of the experiments suggests that the multiwavelet transform features may contain more texture information for classification than the scalar wavelet transform features. Classification performance comparisons using a set of twelve Brodatz textured images and wavelet packet decompositions with the novel packet-tree feature selection support this conclusion.

1 Introduction

Multiwavelets have recently attracted a lot of theoretical attention and provided a good indication of a potential impact on signal processing [1]. In this paper, a novel texture classification scheme using Geronimo-Hardin-Massopust (GHM) discrete multiwavelet transform (DMWT) [2] is proposed. The goal is both to extend the experimentation made in [1], and to test the effectiveness of multiwavelets in texture classification with respect to the scalar Daubechies wavelet [3]. An important problem in wavelet texture analysis is that the numbers of features tend to become huge. Inspiration by Siedlecki and Sklansky [4], Wang *et al.* [5] proposed the Max-Max method based on Genetic Algorithms (GAs) [6] to evaluate the classification performance of each subset of selected features. A feature of GAs is that the chromosomes interact only with the fitness function, but not with each other. This method precludes the evolution of collective solutions to problems, which can be very powerful. The approach proposed here an evolutionary framework in which

T. Caelli et al. (Eds.): SSPR&SPR 2002, LNCS 2396, pp. 806-813, 2002.
© Springer-Verlag Berlin Heidelberg 2002

successive generations adaptively develop behavior in accordance with their natural needs.

This paper is organized as follows. A brief review of DMWT is presented in Section 2. The coevolutionary-based feature selection algorithm is proposed in Section 3. The texture classification experimental results are presented in Section 4. We summarize the conclusions in Section 5.

2 Discrete Multiwavelet Transforms

For a multiresolution analysis of multiplicity $r > 1$, (MRA), an orthonormal compact support multiwavelet system consists one multiscaling function vector $\Phi(x) = (\phi_1(x), ..., \phi_r(x))^T$ and one multiwavelet function vector $\Psi(x) = (\psi_1(x), ..., \psi_r(x))^T$. Both Φ and Ψ satisfy the following two-scale relations:

$$\Phi(x) = \sqrt{2} \sum_{k \in \mathbf{Z}} H_k \Phi(2x - k) \tag{1}$$

$$\Psi(x) = \sqrt{2} \sum_{k \in \mathbf{Z}} G_k \Phi(2x - k). \tag{2}$$

Note that multifilters $\{H_k\}$ and $\{G_k\}$ are finite sequences of $r \times r$ matrices for each integer k. Let $V_j, j \in \mathbf{Z}$, be the closure of the linear span of $\phi_{l,j,k} = 2^{j/2}\phi_l(2^j x - k)$, $l = 1, 2, \ldots, r$. By exploiting the properties of the MRA, as in the scalar case, any continuous-time signal $f(x) \in V_0$ can be expanded as

$$\begin{aligned} f(x) &= \sum_{l=1}^{r} \sum_{k \in \mathbf{Z}} c_{l,0,k} \phi_l(x - k) \\ &= \sum_{l=1}^{r} \sum_{k \in \mathbf{Z}} c_{l,J,k} 2^{J/2} \phi_l(2^J x - k) + \sum_{l=1}^{r} \sum_{J \le j < 0} \sum_{k \in \mathbf{Z}} d_{l,j,k} 2^{j/2} \psi_l(2^j x - k) \end{aligned} \tag{3}$$

where

$$\boldsymbol{c}_{j,k} = \left(c_{1,j,k}, ..., c_{r,j,k} \right)^T \tag{4}$$

$$\boldsymbol{d}_{j,k} = \left(d_{1,j,k}, ..., d_{r,j,k} \right)^T \tag{5}$$

and

$$c_{l,j,k} = \int f(x) 2^{j/2} \phi_l(2^j x - k) dx \tag{6}$$

$$d_{l,j,k} = \int f(x) 2^{j/2} \psi_l(2^j x - k) dx. \tag{7}$$

For the two-dimensional discrete multiwavelet transform, a 2-D MRA of multiplicity N for $L^2(R^2)$ can be obtained by using the tensor product of two MRA's of multiplicity N of $L^2(R)$. Fig. 1 shows a textured image of size 512×512 and its one-level decomposition with the GHM multiwavelet transform.

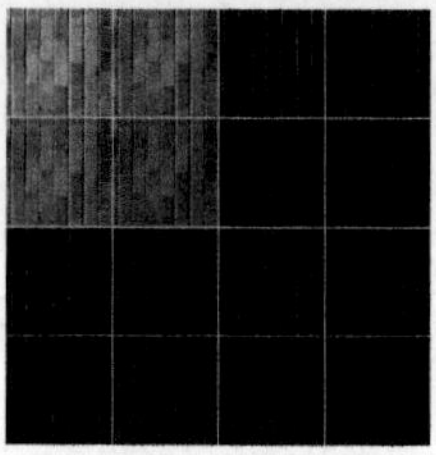

Fig. 1. One-level decomposition of the DMWT for the Brodatz texture D15

3 Feature Selection Using Evolutionary Algorithms

3.1 Initialization

In the proposed method that is derived from the principles of the natural species evolution theory [7] individuals are grouped in populations and thereafter referred to as *inter* population P_b and *intra* population P_w are randomly created. In our case, the two populations have interdependent evolutions (coevolution). The term *inter* reflects the reluctance of this individual for the opposite class. This reluctance is quantified by the mean square distance between pattern points that belong to different classes. An individual of the population P_b, I_x, will compete with each individual of the population kernel K_b which is the collection of individuals with best inter distances. The term *Inter* is formulated as follows:

$$[Inter]_{I_x \in P_b} = \sum_m (I_x \Leftrightarrow I_m) \text{ with } I_m \in K_b,\ m = 1,\ldots, M,$$

$$(I_x \Leftrightarrow I_m) = \begin{cases} D_b^x - D_b^m & \text{if } D_b^x > D_b^m, \\ p & \text{if } D_b^x \le D_b^m, \end{cases} \tag{8}$$

where D_b is the Euclidean distance between classes and p is a penalty. Conversely, the term *Intra* reflects the attraction of this individual for its own class. An individual of the population P_w, I_x, will compete with each individual of the population kernel K_w which is the collection of individuals with best intra distances.

3.2 Evaluation

A character string whose length is equal to the number of features, called the chromosome, as done in GAs represents an individual of the population. With a direct encoding scheme, a bit of one indicates that the feature is used and zero indicates that the feature is not used. The fitness of an individual is derived from her genotype made up of the genes which guide the discrete wavelet decomposition of each textured image, in accordance with our proposed packet-tree representation described in Section 3.4. The genetic representation is used to evolve potential solutions under a set of twelve Brodatz textured images [8]. A best individual of the population kernel K_b will compete with each of the best individuals of the opposite population kernel K_w. The combined results of these competitions directly provide the fitness function. Following the abovementioned, the fitness function Θ is defined as a number composed of two terms:

$$\Theta = (1 - \xi \cdot \delta/\chi) \cdot ([Inter] - [Intra]), \tag{9}$$

where ξ is the weighting constant greater or equal to one, δ is the number of features selected, χ is the number of training samples. The evaluation process of Θ is randomly combined with the *Inter* individual of the population kernel K_b and the *Intra* individual of the population kernel K_w.

3.3 Genetic Operations

According to the roulette wheel selection strategy [6], the combination of populations P_b and P_w individuals with higher fitness value in equation (9) will survive more at the next generation while ones with lower fitness will be removed from the population. The size of each of the population remains constant during evolution. Reproduction step consists of generating new individuals from the combinative individuals previously selected and is performed with crossover and mutation operations. The crossover operator is performed to create new chromosomes and the mutation operation randomly changes a bit of the string. In our method, the combinative individuals selected in the previous step are used to as the parent individuals and then their chromosomes are combined by the following criterion so as to toward the chromosomes of two offspring individuals.

Combinative Crossover Criterion

If the *i*-th gene of the *inter* individual and the *i*-th gene of the *intra* individual are the same, then the *i*-th gene of the offspring individual is set as either individual. Where *i* is the index of a gene in an individual. If not, the *i*-th gene of the offspring individual will be set as either individual at random.

3.4 Feature Selection

After computation of the fitness function for all the combination of the two kernel individuals, a feature selection step is activated for choosing the individuals allowed reproducing at the next generation. The strategy of feature selection involves selecting the best subset A_q,

$$A_q = \{\alpha_u \mid u = 1, \ldots, q; \alpha_u \in B\} \tag{10}$$

from an original feature set B,

$$B = \{\beta_v \mid v = 1, \ldots, Q\},\ Q > q. \tag{11}$$

In other words, the combination of q features from A_q will maximize equation (9) with respect to any other combination of q features taken from Q, respectively. The new feature β_v is chosen as the (λ+1)st feature if it yields

$$\underset{\forall \beta_v}{\text{Max}}\ \underset{\forall \alpha_u}{\text{Max}}\ \Delta[Inter]\ (\alpha_u, \beta_v), \tag{12}$$

where $\alpha_u \in A_\lambda$, $\beta_v \in B - A_\lambda$, and $\Delta[Inter](\alpha_u, \beta_v) = [Inter](\alpha_u, \beta_v) - [Inter](\alpha_u)$. $[Inter](\alpha_u)$ is the evaluation value of equation (8) while the feature α_u is selected and $[Inter](\alpha_u, \beta_v)$ is the evaluation value of equation (8) while the candidate β_v is added to the already selected feature α_u. In a similar way, the feature selection mechanism minimizes intra measure and helps to facilitate classification by removing redundant features that may impede recognition. The proposed schemes consider both the accuracy of classification and the cost of performing classification.

To speed up such a selection process, we consider a packet-tree scheme that is based on fitness value of equation (9) to locate dominant wavelet subbands. The idea is given as below.

1) For a predetermined number of levels, given textured subbands that were decomposed with wavelet packet transforms into.
2) Initially select subbands that can be viewed as the parent and children nodes in a tree at random.
3) After initialization, the subbands at the current level will be selected only if the predecessor at the previous level was selected. Otherwise, we just skip these successors and consider the next subbands.
4) Generate a representative tree of selected features for each texture by averaging the selected feature vectors over all the sample images.
5) Repeat the process for all textures.

4 Teyxture Classification Experiments

4.1 Experiment Design

In this section, we present the experimental results on twelve 512 × 512 images with 256 gray levels found in the Brodatz's album. The reported results for each classification task have the following parameter settings: population size = 20, number of generation = 1000, and the probability of crossover = 0.5. A mutation probability value starts with a value of 0.9 and then varied as a step function of the number of iterations until it reaches a value of 0.01. For the scalar wavelet we randomly chose one hundred 256 × 256 overlapping subimages as training samples for each obtained original textured image and used one thousand samples in the multiwavelet. Due to the curse of dimensionality, this arrangement considers the subband structure resulting from passing a signal through a multifilter bank is different from a scalar wavelet. We tackled the classification problem also with the D4 wavelet transform and the GHM multiwavelet transform. Textural features are given by the extrema number of wavelet coefficients [5], which can be used as a measure of coarseness of the texture at multiple resolutions. Then, texture classifications without and with feature selection were performed using the simplified Mahalanobis distance measure [9] to discriminate textures and to optimize classification by searching for near-optimal feature subsets. The mean and variance of the decomposed subbands are calculated with the leave-one-out algorithm [9] in classification.

The training phase of texture classification composed of the evaluation, selection, and reproduction steps correspond to a generation is repeated until the fitness function does not progress, or when a maximum number of generations is reached. Then, the best string with the highest fitness value in equation (9) at the last generation is preserved for further use in the classification phase. During the classification phase, the unknown texture (one of the twelve Brodatz textures) is matched against the database and the best match is taken as the classification result.

4.2 Experimental Results and Discussions

4.2.1 Texture Classification without Feature Selection

The performance of the classifier was evaluated with three different randomly chosen training and test sets. Algorithms based on the two types of wavelets have been shown to work well in texture discrimination. At the levels 1-3 of Table 1, the percentage of correct classification rate without feature selection improves as the number of levels increases. This observation is expected since there is no problem of curse of dimensionality. However, the classification performance is not monotonically dependent on the feature space dimension, but decreases after having reached a local maximum, as has been shown in Table 1. At the level 4, the inclusion of further feature results in performance degradation. In a similar way, the classification rate is even down to 95.30% at the level 3 of Table 2. Theoretically, multiwavelets should perform even better because scalar wavelets cannot simultaneously posses all of the desire properties, such as orthogonality and symmetry. However, average classification accuracies in the latter have produced performance capable of surpassing the results of the former. This owes to the fact that the number of features should be reduced more to alleviate the serious Hughes phenomenon [9] when the training samples are limited.

4.2.2 Texture Classification with CGA Feature Selection

By making a comparison with the previous results, the classification errors in Tables 3 and 4 mostly decrease when the used features are selectively removed from all the features at the level 4 of Table 1 and level 3 of Table 2, respectively. This decrease is due to the fact that less parameter used in place of the true value of the class conditional probability density functions need to be estimated from the same number of samples. The smaller the number of the parameters that need to be estimated, the less severe the Hughes effect can become. In the meanwhile, we also noticed that the multiwavelet outperforms the scalar wavelet with the packet-tree selection. This is because the extracted features in the former are more discriminative than the latter and, therefore, the selection of a subband for discrimination is not only dependent on the wavelet bases, wavelet decompositions, and decomposed levels but also the fitness function.

Table 1. Classification results (correct rate in %) of the pyramidal decomposition using wavelet transforms without feature selection

Level / Sample set	1	2	3	4
1	95.17	98.42	97.83	97.33
2	95.17	98.92	98.75	97.25
3	94.83	97.75	99.00	96.92
Average	95.00	98.36	98.53	97.17

Table 2. Classification results (correct rate in %) of the pyramidal decomposition using multiwavelet transforms without feature selection

Level / Sample Set	1	2	3
1	95.42	96.17	95.25
2	95.00	96.25	95.91
3	94.42	96.67	94.75
Average	94.95	96.36	95.30

Table 3. Classification results (correct rate in %) using the wavelet packet decomposition with coevolutionary GA feature selection

Sample Set	$\xi = 1$	$\xi = 2$	$\xi = 3$	$\xi = 4$	$\xi = 5$
1	98.47	98.48	98.29	98.49	98.43
2	98.41	98.38	98.62	98.29	98.49
3	98.41	98.43	98.38	98.51	98.49
Average	98.43	98.43	98.43	98.43	98.47

Table 4. Classification results (correct rate in %) using the multiwavelet packet decomposition with coevolutionary GA feature selection

Sample Set	$\xi = 1$	$\xi = 2$	$\xi = 3$	$\xi = 4$	$\xi = 5$
1	98.83	98.80	98.54	98.82	98.58
2	98.85	98.72	98.80	98.71	98.66
3	98.96	98.73	98.90	98.79	98.73
Average	98.88	98.75	98.75	98.77	98.79

5 Conclusions

This paper introduces a promising evolutionary algorithm approach for solving the texture classification problem. Further work is to explore the feasibility of our new methods by incorporating with recent preprocessing techniques of multiwavelets.

Acknowledgements

The author would like to acknowledge the support received from NSC through project number NSC 90-2213-E-151-010.

References

1. Strela, V., Heller, N., Strang, G., Topiwala, P., and Heil, C.: The Application of Multiwavelet Filter Banks to Image Processing. IEEE Trans. Image Process., 8 (1999) 548-563
2. Xia, X. G., Geronimo, J. S., Hardin, D. P., and Suter, B. W.: Design of Prefilters for Discrete Multiwavelet Transforms. IEEE Trans. Signal Process., 44 (1996) 25-35
3. Daubechies, I. (ed.): Ten Lectures on Wavelets. SIAM, Philadelphia, Penn. (1992)
4. Siedlecki, W. and Sklansky, J.: A Note on Genetic Algorithm for Large-Scale Feature Selection. Pattern Recognition Letters, 10 (1989) 335-347
5. Wang, J. W., Chen, C. H., Chien, W. M., and Tsai, C. M.: Texture Classification using Non-Separable Two-Dimensional Wavelets. Pattern Recognition Letters, 19 (1998) 1225-1234
6. Goldberg, D. E. (ed.): Genetic Algorithms in Search, Optimization, and Machine Learning. MA: Addison-Wesley (1989)
7. Bäck, T.: Evolutionary Algorithms in Theory and Practice: Evolution Strategies, Evolutionary Programming, Genetic Algorithms, Oxford University Press, New York (1996)
8. Brodatz, P. (ed.): Textures: A Photographic Album for Artists and Designers. NY: Dover (1966)
9. Devijver, P. A. and Kittler, J. (ed.): Pattern Recognition: A Statistical Approach. Prentice-Hall, Englewood Cliffs, NJ (1982)

Probabilistic Signal Models to Regularise Dynamic Programming Stereo

Georgy Gimel'farb[1]* and Uri Lipowezky[2]

[1] CITR, Department of Computer Science, Tamaki Campus, University of Auckland
Private Bag 92019, Auckland 1, New Zealand
g.gimelfarb@auckland.ac.nz
[2] Tiltan System Engineering Ltd.
35 Hayarkon Street, Beney - Beraq 51204, Israel
Uri@tiltan-se.co.il

Abstract. Ill-posedness of the binocular stereo problem stems from partial occlusions and homogeneous textures of a 3D surface. We consider the symmetric dynamic programming stereo regularised with respect to partial occlusions. The regularisation is based on Markovian models of epipolar profiles and stereo signals that allow for measuring similarity of stereo images with due account of binocular and monocular visibility of the surface points. Experiments show that the probabilistic regularisation yields mostly accurate elevation maps but fails in excessively occluded or shaded areas.

1 Introduction

Computational binocular stereo is an ill-posed problem because the same stereo pair can be produced by very different optical surfaces. The ill-posedness stems from partial occlusions yielding no stereo correspondence and from uniform or repetitive textures resulting in multiple equivalent correspondences. To obtain a unique solution closely approaching visual or photogrammetric reconstruction, the stereo problem has to be regularised.

We consider dynamic programming stereo (DPS) that reconstructs a 3D surface as a collection of independent continuous epipolar profiles. The reconstruction is based on the best correspondence between stereo images, each profile maximising the total similarity of signals (grey values or colours) in the corresponding pixels or of local image features derived from the signals [1,2,3,4,6,7,8,9].

The intensity-based symmetric DPS (SDPS) can be regularised with respect to partial occlusions by modelling the profiles with explicit Markov chains of the 3D points and signals [5]. We compare three models of the epipolar profiles with respect to the overall accuracy of stereo reconstruction. Experiments are conducted with a large-size digitised aerial stereo pair of an urban scene in Fig. 1 having various partially occluded areas and the known ground control.

* This work was supported by the University of Auckland Research Committee grant 3600327/9343.

T. Caelli et al. (Eds.): SSPR&SPR 2002, LNCS 2396, pp. 814–822, 2002.
© Springer-Verlag Berlin Heidelberg 2002

Fig. 1. Left (2329 × 1442) and right (1887 × 1442) stereo images "Town"

2 Markovian Models of an Epipolar Profile

Let x_{left} and x_{right} be discrete x-coordinates of pixels along the corresponding epipolar scan-lines in the left and right image of a stereo pair, respectively. Let x denote a discrete x-coordinate of points of an epipolar profile. Assuming the symmetric epipolar geometry [4], $x = (x_{\text{left}} + x_{\text{right}})/2$. Let $p = x_{\text{left}} - x_{\text{right}}$ be the x-disparity (parallax) of the corresponding pixels. The N-point discrete

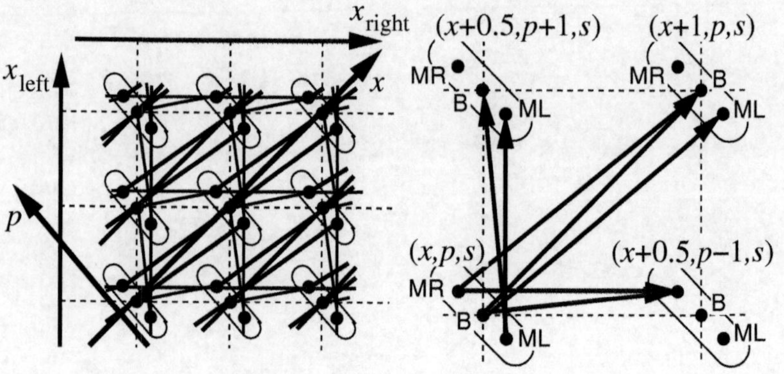

Fig. 2. Fragment of the graph of profile variants and the allowable transitions between the nodes

continuous profile $\mathbf{P} = ((x_i, p_i : i = 0, \ldots, N-1)$ is modelled by a Markov chain of transitions between the successive nodes $v_i(s_i) = (x_i, p_i, s_i)$ of a graph of profile variants (GPV) shown in Fig. 2. Here, $s \in \{\mathrm{B}, \mathrm{ML}, \mathrm{MR}\}$ denotes a visibility state indicating, respectively, the binocularly visible point (BVP) and the monocularly visible point (MVP) observed only in the left or right image. Only the BVP $v(\mathrm{B}) = (x, p, B)$ involves the pair of corresponding pixels with the x-coordinates $x_{\mathrm{left}} = x + p/2$ and $x_{\mathrm{right}} = x - p/2$. The MVPs $v(\mathrm{ML}) = (x, p, \mathrm{ML})$ and $v(\mathrm{MR}) = (x, p, \mathrm{MR})$ are depicted by the single pixels with the x-coordinate x_{left} or x_{right}, respectively.

Transition probabilities for a stationary Markov chain describing the shape of the random profiles $\mathbf{P}$ depend only on the successive visibility states [4]:

$$\Pr(v_{i+1}(s_{i+1})|v_i(s_i)) \equiv \pi(s_{i+1}|s_i) \tag{1}$$

On the assumption that the transitions to the MVPs with $s = \mathrm{ML}$ and $s = \mathrm{MR}$ are equiprobable, the random profile is specified by the two transition probabilities [4]: $\pi_{\mathrm{B|B}}$ and $\pi_{\mathrm{M|M}}$ where M stands for MR or ML. Each profile $\mathbf{P}$ specifies particular stereo correspondence between the left and right images of a stereopair, and similarity between the corresponding pixels is measured with a specific Markovian model of mutual signal adaptation. The adaptation estimates and excludes photometric distortions of the stereo images relative to the surface texture [3,4]. After the images are mutually adapted, the probability of transition to each BVP $v_i(\mathrm{B}) = (x_i, p_i, \mathrm{B})$ depends on the point-wise residual absolute difference Δ_i between the corresponding signals. As follows from the GPV in Fig. 2, the transition probabilities to the BVPs $\Pr(v_i(\mathrm{B}), \Delta_i|v_{i-1}(s_{i-1})$ define uniquely the transition probabilities to the MVPs.

The regularised SDPS relates the point-wise signal similarity for each node $v_i(s_i)$ of the GPV to the log-likelihood ratio

$$l(v_i(s_i), \Delta_i|v_{i-1}(s_{i-1})) = \ln \Pr(v_i(s_i), \Delta_i|v_{i-1}(s_{i-1})) - \ln \Pr_{\mathrm{rand}}(s_i|s_{i-1})$$

of the transition probabilities for the profile specified by a given stereopair and the purely random profile.

Below we experimentally compare three probabilistic models of the signal-depending and random profiles assuming, for simplicity, that $\pi_{\mathrm{B|B}} + \pi_{\mathrm{M|M}} = 1$. The simplest model of the transition probability $\Pr(v_i(\mathrm{B}), \Delta_i | v_{i-1}(s_{i-1})) = F_{\mathrm{B}}(\Delta_i)$ such that $\sum_{\Delta=0}^{\Delta_{\max}} F_{\mathrm{B}}(\Delta) = 1$ is introduced in [5]:

$$F_{\mathrm{B}}(\Delta) \propto \min\{1 - \tau, \max\{\tau, \exp(-\gamma \cdot \Delta)\}\} \tag{2}$$

where γ is a scaling factor and the threshold $\tau > 0$ excludes zero probabilities (in our experiments $\tau = 10^{-10}$).

The adaptation of corresponding signals amplifies the relative number of zero deviations. This can be accounted for by using an additional parameter $\alpha = F_{\mathrm{B}}(0)$ in the transition probability:

$$F_{\mathrm{B}}(\Delta) = \begin{cases} \alpha & \text{if } \Delta = 0 \\ (1-\alpha)\frac{\max\{\tau, \exp(-\gamma \cdot \Delta)\}}{\sum_{\delta=1}^{\Delta_{\max}} \max\{\tau, \exp(-\gamma \cdot \delta)\}} & \text{otherwise} \end{cases} \tag{3}$$

where $\Delta_{\max}$ is the maximum absolute deviation ($\Delta_{\max} = 255$).

The profile models to be compared yield the following likelihood ratios:

- for the transition models in [5]:

$$\begin{aligned} l(v_i(\mathrm{B}), \Delta_i | v_{i-1}(s)) &= \log F_{\mathrm{B}}(\Delta_i) - \log \pi_{\mathrm{B|B}} \\ l(v_i(\mathrm{M}), \Delta_i | v_{i-1}(s)) &= \log(1 - F_{\mathrm{B}}(\Delta_i)) - \log \pi_{\mathrm{M|M}} \end{aligned} \tag{4}$$

- for the conditional Markov model of the profile points depending on the adapted signals:

$$\begin{aligned} l(v_i(\mathrm{B}), \Delta_i | v_{i-1}(s)) &= \log\left(\frac{\pi^{\circ}_{\mathrm{B|B}} \cdot F_{\mathrm{B}}(\Delta_i)}{\pi^{\circ}_{\mathrm{B|B}} \cdot F_{\mathrm{B}}(\Delta_i) + \pi^{\circ}_{\mathrm{M|M}} \cdot P_{\mathrm{M}}}\right) - \log \pi_{\mathrm{B|B}} \\ l(v_i(\mathrm{M}), \Delta_i | v_{i-1}(s)) &= \log\left(\frac{\pi^{\circ}_{\mathrm{M|M}} \cdot P_{\mathrm{M}}}{\pi^{\circ}_{\mathrm{B|B}} \cdot F_{\mathrm{B}}(\Delta_i) + \pi^{\circ}_{\mathrm{M|M}} \cdot P_{\mathrm{M}}}\right) - \log \pi_{\mathrm{M|M}} \end{aligned} \tag{5}$$

  where $\pi^{\circ}_{\mathrm{B|B}}$ and $\pi^{\circ}_{\mathrm{M|M}}$ denote the transition probabilities specifying the actual shape of the profile specified by the stereo images ($\pi^{\circ}_{\mathrm{B|B}} + \pi^{\circ}_{\mathrm{M|M}} = 1$), and P_{M} is the probability of signal deviations for the MVPs ($P_{\mathrm{M}} = 1/\Delta_{\max}$ if the equiprobable deviations are assumed), and
- for the joint Markov models of the profile points and adapted signals:

$$\begin{aligned} l(v_i(\mathrm{B}), \Delta_i | v_{i-1}(s)) &= \log\left(\pi^{\circ}_{\mathrm{B|B}} \cdot F_{\mathrm{B}}(\Delta_i)\right) - \log\left(\pi_{\mathrm{B|B}} \cdot P_{\mathrm{M}}\right) \\ l(v_i(\mathrm{M}), \Delta_i | v_{i-1}(s)) &= \log \pi^{\circ}_{\mathrm{M|M}} - \log \pi_{\mathrm{M|M}} \end{aligned} \tag{6}$$

3 Experimental Results and Conclusions

The original photos of the urban scene in Fig. 1 containing different types of open and partially occluded areas are obtained from the altitude 960 m using

Table 1. Accuracy of the regularised SDPS reconstruction in terms of the cumulative percentage of the control points with the absolute error less than or equal to ε. The notation used: $\bar{\varepsilon}$, σ, and $\varepsilon_{\max}$ are the mean absolute error, standard deviation, and maximum error, respectively, C_{B} is the cross-correlation of the corresponding signals for the reconstructed surface (in the parentheses – after this latter is smoothed by post-processing); ν_{B} is the relative number of the BVPs in the surface

Model (4) – (2) with $\pi_{\mathrm{B|B}} = 0.10$ and $\gamma = 0.1$:
$\bar{\varepsilon} = 3.66$; $\sigma = 6.84$; $\varepsilon_{\max} = 44$; $C_{\mathrm{B}} = 0.944$ (0.856); $\nu_{\mathrm{B}} = 81.8\%$

ε:	0	1	2	3	4	5	10	15	20	25	30	35	40	44
%	18.0	60.1	71.0	78.6	82.2	84.9	90.9	93.0	95.8	97.1	98.2	99.0	99.5	100.0

Model (5) – (3) with $\pi_{\mathrm{B|B}} = 0.10$, $\pi^{\circ}_{\mathrm{B|B}} = 0.90$, $\gamma = 1$, and $\alpha = 0.90$:
$\bar{\varepsilon} = 3.69$; $\sigma = 6.59$; $\varepsilon_{\max} = 48$; $C_{\mathrm{B}} = 0.945$ (0.849); $\nu_{\mathrm{B}} = 81.6\%$

ε:	0	1	2	3	4	5	10	15	20	25	30	35	40	48
%	18.5	58.7	68.4	75.7	80.7	82.8	90.6	94.5	95.8	97.4	98.4	99.0	99.5	100.0

Model (5) – (3) with $\pi_{\mathrm{B|B}} = 0.25$, $\pi^{\circ}_{\mathrm{B|B}} = 0.75$, $\gamma = 1$, and $\alpha = 0.90$:
$\bar{\varepsilon} = 3.72$; $\sigma = 6.73$; $\varepsilon_{\max} = 53$; $C_{\mathrm{B}} = 0.948$ (0.833); $\nu_{\mathrm{B}} = 78.4\%$

ε:	0	1	2	3	4	5	10	15	20	25	30	35	40	53
%	19.3	57.2	66.6	77.0	80.4	83.6	89.0	94.5	96.9	98.2	99.0	99.2	99.2	100.0

Model (6) – (3) with $\pi_{\mathrm{B|B}} = 0.10$, $\pi^{\circ}_{\mathrm{B|B}} = 0.90$, $\gamma = 1$, and $\alpha = 0.90$:
$\bar{\varepsilon} = 2.96$; $\sigma = 5.34$; $\varepsilon_{\max} = 37$; $C_{\mathrm{B}} = 0.950$ (0.861); $\nu_{\mathrm{B}} = 79.6\%$

ε:	0	1	2	3	4	5	10	15	20	25	30	37
%	19.6	63.2	74.2	80.7	84.3	87.7	93.2	94.8	96.6	98.4	99.5	100.0

Model (6) – (3) with $\pi_{\mathrm{B|B}} = 0.25$, $\pi^{\circ}_{\mathrm{B|B}} = 0.75$, $\gamma = 1$, and $\alpha = 0.90$:
$\bar{\varepsilon} = 2.74$; $\sigma = 4.82$; $\varepsilon_{\max} = 36$; $C_{\mathrm{B}} = 0.953$ (0.862); $\nu_{\mathrm{B}} = 77.6\%$

ε:	0	1	2	3	4	5	10	15	20	25	30	36
%	18.8	62.1	74.9	82.2	85.6	88.0	94.3	96.3	97.7	99.0	99.5	100.0

the photogrammetric camera Wild RC30 with the focal length 153.26 mm. For the SDPS reconstruction the digitised images are transformed to the epipolar geometry and scaled down to the resolution of 330 mm per pixel. The left and right images are of the size 2329×1441 and 1887×1441 pixels, respectively. The x-disparity range for these images is $[50, 200]$.

This scene has 383 uniformly distributed ground control points (GCP) found by visual (photogrammetric) processing. Most of them are at the roofs of the buildings because these latter present a real challenge to the reconstruction. These GCPs allow to analyse the accuracy of the regularised SDPS in most complicated conditions and indicate typical reconstruction errors. Table 1 presents results of the reconstruction using the above regularising models with different parameters. The range of successive signal adaptation is $\pm 20\%$ of the grey level difference estimated for the surface texture. To simplify the comparisons, numerical values of the likelihood functions $l_{\mathrm{B}}(\Delta) \equiv l(v_i(\mathrm{B}), \Delta | v_{i-1}(s))$ and

Table 2. Likelihood values for the residual signal deviations

Model (4) – (2) with $\pi_{\mathrm{B|B}} = 0.10$ and $\gamma = 0.1$:

Δ:	0	1	2	3	4	5	6	7	8	9	10	11	12	13
$l_{\mathrm{B}}(\Delta)$:	2.30	2.07	1.84	1.61	1.38	1.15	0.92	0.69	0.46	0.23	0.00	-0.23	-0.46	-0.69
$l_{\mathrm{M}}(\Delta)$:	-22.9	-1.48	-0.89	-0.59	-0.40	-0.27	-0.18	-0.12	-0.07	-0.03	0.00	0.02	0.03	0.03
Δ:	14	15	16	17	18	19	20	30	40	50	100	150	200	255
$l_{\mathrm{B}}(\Delta)$:	-0.92	-1.15	-1.38	-1.61	-1.84	-2.07	-2.30	-4.61	-6.91	-9.21	-20.7	-20.7	-20.7	-20.7
$l_{\mathrm{M}}(\Delta)$:	0.06	0.07	0.08	0.09	0.09	0.09	0.10	0.10	0.11	0.11	0.11	0.11	0.11	0.11

Model (5) – (3) with $\pi_{\mathrm{B|B}} = 0.10$, $\pi^{\circ}_{\mathrm{B|B}} = 0.90$, $\gamma = 1$, and $\alpha = 0.90$

Δ:	0	1	2	3	4	5	6	7	8	9	10	11	12	13
$l_{\mathrm{B}}(\Delta)$:	2.30	2.30	2.28	2.25	2.17	1.98	1.60	0.98	0.16	-0.76	-1.73	-2.72	-3.72	-4.72
$l_{\mathrm{M}}(\Delta)$:	-7.53	-4.88	-3.89	-2.93	-2.00	-1.19	-0.58	-0.20	-0.02	0.06	0.09	0.10	0.10	0.10
Δ:	14	15	16	17	18	19	20	30	40	50	100	150	200	255
$l_{\mathrm{B}}(\Delta)$:	-5.72	-6.72	-7.72	-8.72	-9.72	-10.7	-11.7	-13.7	-13.7	-13.7	-13.7	-13.7	-13.7	-13.7
$l_{\mathrm{M}}(\Delta)$:	0.11	0.11	0.11	0.11	0.11	0.11	0.11	0.11	0.11	0.11	0.11	0.11	0.11	0.11

Model (5) – (3) with $\pi_{\mathrm{B|B}} = 0.25$, $\pi^{\circ}_{\mathrm{B|B}} = 0.75$, $\gamma = 1$, and $\alpha = 0.90$

Δ:	0	1	2	3	4	5	6	7	8	9	10	11	12	13
$l_{\mathrm{B}}(\Delta)$:	1.38	1.37	1.33	1.24	1.04	0.63	-0.01	-0.84	-1.77	-2.75	-3.74	-4.73	-5.73	-6.73
$l_{\mathrm{M}}(\Delta)$:	-6.25	-3.62	-2.65	-1.74	-0.94	-0.35	0.00	0.17	0.24	0.27	0.28	0.29	0.29	0.29
Δ:	14	15	16	17	18	19	20	30	40	50	100	150	200	255
$l_{\mathrm{B}}(\Delta)$:	-7.73	-8.73	-9.73	-10.7	-11.7	-12.7	-13.7	-15.7	-15.7	-15.7	-15.7	-15.7	-15.7	-15.7
$l_{\mathrm{M}}(\Delta)$:	0.29	0.29	0.29	0.29	0.29	0.29	0.29	0.29	0.29	0.29	0.29	0.29	0.29	0.29

Model (6) – (3) with $\pi_{\mathrm{B|B}} = 0.10$, $\pi^{\circ}_{\mathrm{B|B}} = 0.90$, $\gamma = 1$, and $\alpha = 0.90$

Δ:	0	1	2	3	4	5	6	7	8	9	10	11	12	13
$l_{\mathrm{B}}(\Delta)$:	7.64	4.48	3.48	2.48	1.48	-0.02	-1.02	-2.02	-3.02	-4.02	-5.02	-6.02	-7.02	-8.02
$l_{\mathrm{M}}(\Delta)$:	-2.20	-2.20	-2.20	-2.20	-2.20	-2.20	-2.20	-2.20	-2.20	-2.20	-2.20	-2.20	-2.20	-2.20
Δ:	14	15	16	17	18	19	20	30	40	50	100	150	200	255
$l_{\mathrm{B}}(\Delta)$:	-9.02	-10.0	-11.0	-12.0	-13.0	-14.0	-15.0	-16.0	-16.0	-16.0	-16.0	-16.0	-16.0	-16.0
$l_{\mathrm{M}}(\Delta)$:	-2.20	-2.20	-2.20	-2.20	-2.20	-2.20	-2.20	-2.20	-2.20	-2.20	-2.20	-2.20	-2.20	-2.20

Model (6) – (3) with $\pi_{\mathrm{B|B}} = 0.25$, $\pi^{\circ}_{\mathrm{B|B}} = 0.75$, $\gamma = 1$, and $\alpha = 0.90$

Δ:	0	1	2	3	4	5	6	7	8	9	10	11	12	13
$l_{\mathrm{B}}(\Delta)$:	6.54	3.88	2.88	1.88	0.88	-0.12	-1.12	-2.12	-3.12	-4.12	-5.12	-6.12	-7.12	-8.12
$l_{\mathrm{M}}(\Delta)$:	-1.10	-1.10	-1.10	-1.10	-1.10	-1.10	-1.10	-1.10	-1.10	-1.10	-1.10	-1.10	-1.10	-1.10
Δ:	14	15	16	17	18	19	20	30	40	50	100	150	200	255
$l_{\mathrm{B}}(\Delta)$:	-9.12	-10.1	-11.1	-12.1	-13.1	-14.1	-15.1	-17.1	-17.1	-17.1	-17.1	-17.1	-17.1	-17.1
$l_{\mathrm{M}}(\Delta)$:	-1.10	-1.10	-1.10	-1.10	-1.10	-1.10	-1.10	-1.10	-1.10	-1.10	-1.10	-1.10	-1.10	-1.10

$l_{\mathrm{M}}(\Delta) \equiv l(v_i(\mathrm{M}), \Delta | v_{i-1}(s))$ for the model parameters in use are given in Table 2. For the models (5) and (6), $l_{\mathrm{B}}(\Delta) > l_{\mathrm{M}}(\Delta)$ if $\Delta \geq 6 \ldots 8$.

Figure 3 presents the greycoded digital x-parallax map (DPM), or the range image of the reconstructed x-disparities, and the orthoimage formed by fusing the left and right images in Fig. 1 in accord with the DPM. The fusion takes account of the visibility of each point. These results are obtained for the best in Table 1 regularising model (6) using the transition probability (3) with $\gamma = 1$

Fig. 3. Greycoded range image of the reconstructed DPM and the orthoimage 2108 × 1442 of the "Town" (vertical white pointers show positions of the 383 GCPs)

and $\alpha = 0.90$ and for the regularising parameters $\pi_{\mathrm{B|B}} = 0.25$ and $\pi^{\circ}_{\mathrm{B|B}} = 0.75$. In this case the reconstructed DPM has the mean absolute error $\bar{\varepsilon} = 2.74$ with the standard deviation $\sigma = 4.82$. The absolute error for 74.9% of the GCPs is

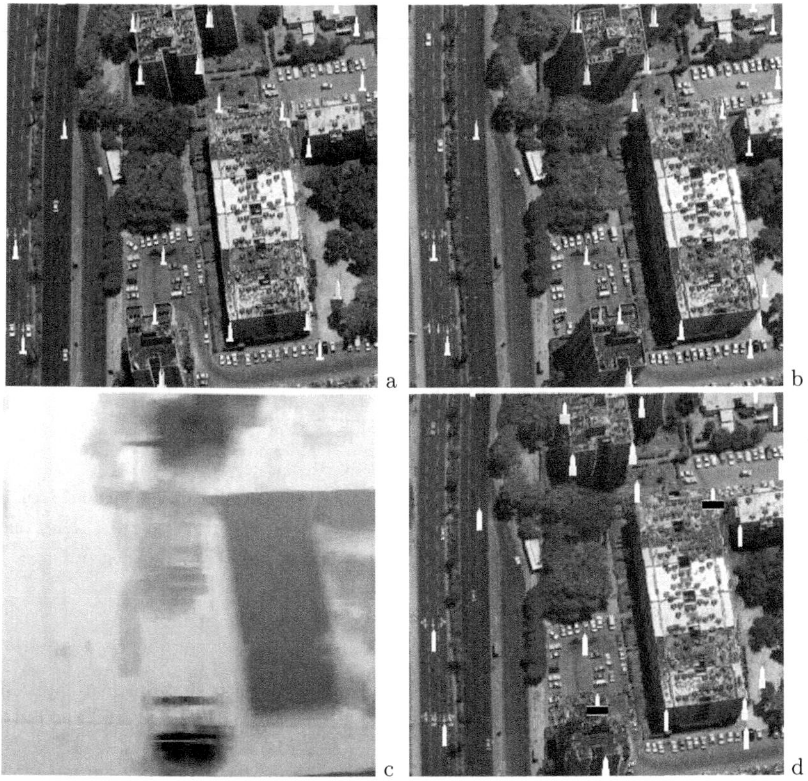

Fig. 4. Reconstruction around the GCP (368, 1277): the fragments 351 × 351 of the left (a) and right (b) images (white and grey pointers indicate the GCPs and reconstruction results, respectively; for a small error, the white pointer is superposed on the grey one), the greycoded DPM (c), and the orthoimage (d) with the GCPs (the larger the bottom rectangle of the pointer, the larger the error)

equal to or less than 2. The maximum error is $\varepsilon_{\max} = 36$ but only 5.7% of the GCPs have the error greater than or equal to 10.

To exclude local y-discontinuities due to homogeneous texture, the reconstructed DPMs are post-processed. The post-processing consists of the in-column median filtering within the moving window 3 × 15 with the subsequent in-line median filtering of the results of the in-column filtering.

A few typical reconstruction errors are shown in Fig. 4. Here, partially occluded areas around the tall buildings are quite large comparing to the adjacent open areas with a high contrast of the texture. In such cases, especially, if the deep shadows create large uniform areas at and around the walls, the SDPS reconstruction fails. But some of these areas are challenging even for visual stereo perception although the corresponding points can be easily found by visual comparisons of the images. But the reconstruction is accurate when the surface texture is not uniform and the occluded parts are not similar do not prevail Large reconstruction errors ($\varepsilon \geq 6$, that is, larger than 4% of the x-disparity range) are encountered in less than 10% of all the GCPs.

The errors are localised to within small areas of the overall scene so that the cross-correlation of the corresponding pixels in the stereo images is relatively high, namely, 0.953 and 0.862 without and with the post-processing, respectively. But the cross-correlation of the corresponding pixels for the GCPs is only 0.654 so that the accurate visual matching does not mean the highest correlation-based similarity of the stereo images.

Our experiments show that the regularised SDPS yields accurate overall reconstruction accuracy with the absolute errors less than 3 pixels (2% of the total x-disparity range) for more than 80–82% of the surface points depending on a proper choice of the regularising signal model. But the regularisation fails in the cases where too large occluded or deeply shaded areas are involved.

Because large local errors do not effect notably the overall similarity of the corresponding points in the stereo images, it is the regularisation rather than stereo matching that plays a crucial role in solving the stereo problem.

References

1. Bensrhair, A., Miché, P., Debrie, R.:Fast and automatic stereo vision matching algorithm based on dynamic programming method. Pattern Recognition Letters **17** (1996) 457–466 814
2. Cox, I. J., Hingorani, S. L., Rao, S. B.: A maximum likelihood stereo algorithm. Computer Vision and Image Understanding **63** (1996) 542–567 814
3. Gimel'farb, G. L.: Intensity-based computer binocular stereo vision: signal models and algorithms. Int. Journal of Imaging Systems and Technology **3** (1991) 189–200 814, 816
4. Gimel'farb, G.: Stereo terrain reconstruction by dynamic programming. In: Jaehne, B., Haussecker, H., Geisser, P. (Eds.): Handbook of Computer Vision and Applications. Academic Press, San Diego, vol. 2 (1999) 505–530 814, 815, 816
5. Gimel'farb, G.: Binocular stereo by maximizing the likelihood ratio relative to a random terrain. In: Klette, R., Peleg, S., Sommer, G. (Eds.): Robot Vision (Lecture Notes in Computer Science **1998**). Springer, Berlin (2001) 201–208 814, 817
6. Li, Z.-N.: Stereo correspondence based on line matching in Hough space using dynamic programming. IEEE Trans. on Systems, Man, and Cybernetics 24 (1994) 144–152 814
7. Rojas, A., Calvo, A., Munõz, J.: A dense disparity map of stereo images. Pattern Recognition Letters **18** (1997) 385–393 814
8. Vergauwen, M., Pollefeys, M., Van Gool, L.: A stereo vision system for support of planetary surface exploration. In: Schiele, B, Sagerer, G. (Eds.): Computer Vision Systems Second Int. Workshop (ICVS 2001), Vancouver, Canada, July 7–8, 2001 (Lecture Notes in Computer Science **2095**). Springer, Berlin (2001) 298–312 814
9. Yip, R. K. K., Ho, W. P.: A multi-level dynamic programming method for stereo line matching. Pattern Recognition Letters **19** (1998) 839–855 814

The Hough Transform without the Accumulators

Atsushi Imiya[1,2], Tetsu Hada[2], and Ken Tatara[2]

[1] National Institute of Informatics
2-1-2 Hitotsubashi, Chiyoda-ku, Tokyo 101-8430
[2] IMIT, Chiba University
1-33 Yayoi-cho, Inage-ku, 263-8522, Chiba, Japan
imiya@media.imit.chiba-u.ac.jp
imiya@nii.ac.jp

Abstract. The least-squares method (LSM) efficiently solves the model-fitting problem, if we assume a model equation. For the fitting to a collection of models, the classification of data is required as pre-processing. The Hough transform, achieves both the classification of sample points and the model fitting concurrently. However, as far as adopting the voting process is concerned, the maintenance of the accumulator during the computation cannot be neglected. We propose a Hough transform without the accumulator expressing the classification of data for the model fitting problems as the permutation of matrices which are defined by data.

1 Introduction

The Hough transform simultaneously detects many conics on a plane. The basic idea of the Hough transform is the classification of sample points and identification of parameters of figures by voting dual figures to the accumulator space [, , ,]. For the achievement of the Hough transform, the maintenance of dual figures in the accumulator space is a fundamental task. For the maintenance of the accumulator space, we are required to prepare large memory areas. Therefore, to derive a simple method for the detection of many conics in a plane, we, in this paper introduce the Hough transform without the accumulator space. The method is based on the property that the classification of sample points is achieved by the permutation of a sequence of sample points and and the matrix representation of a permutation defines a zero-one orthogonal matrix. Furthermore, once the classification of sample points is established, the estimation of parameters of conics is achieved by solving the least-mean-squares criterion. This second process is achieved by the eigenvalue decomposition of the moment matrix constructed sample points.

The eigenvalue decomposition is established by computing orthogonal matrix which diagonalizes the moment matrix. We first define the minimization criterion for the Hough transform. Second, we derive a dynamic system which guarantees the convergence of the Hough transform. Finally, we derive a relaxation method to solve the criterion for the Hough transform without using any accumulator spaces. We apply the method for the detection of conics on a plane.

T. Caelli et al. (Eds.): SSPR&SPR 2002, LNCS 2396, pp. 823–832, 2002.
© Springer-Verlag Berlin Heidelberg 2002

2 Mathematical Aspects of the Hough Transform

2.1 Figures and Dual Figures

Setting $\boldsymbol{x} = (x, y)^\top$ to be a vector in two-dimensional Euclidean space $\mathbf{R}^2$, we define the homogeneous coordinate $\boldsymbol{\xi}$ of a vector $\boldsymbol{x} \in \mathbf{R}^k$ for $k \geq 3$ as $\boldsymbol{\xi} = (\boldsymbol{x}^\top, 1)^\top$. We denote the Euclidean length of vector $\boldsymbol{x}$ in k-dimensional Euclidean space $\mathbf{R}^k$ for $k \geq 1$ as $|\boldsymbol{a}|$. Let S^{n-1} be the unit sphere of $\mathbf{R}^n$ consisting of all points $\boldsymbol{x}$ with distance 1 from the origin. For $n = 1$, $S^0 = [-1, 1]$. Furthermore, the positive half-space is defined by $\mathbf{R}^n_+ = \{\boldsymbol{x}|x_n > 0\}$, for $n \geq 1$. Now, setting $H^{n-1}_+ = S^{n-1} \bigcap \mathbf{R}^n_+$, for $n \geq 1$, the positive unit semi-sphere is defined as $S^{n-1}_+ = S^{n-2}_+ \bigcup H^{n-1}_+$, for $n \geq 1$.

Let $\mathcal{P}^k_2 = \{p_i(x, y)\}^k_{i=1}$ be a set of independent monomials of x and y, and let $\mathcal{A}^k = \{\boldsymbol{a} \,|\, \boldsymbol{a} = \{a_i\}^k_{i=1}\}$ be a set of all n-tuple real numbers, where at least one of a_i is nonzero. Then, setting $P(x, y) = \sum_{i=1}^k a_i \, p_i(x, y)$, a set $\mathcal{C}_2(\mathcal{P}^k_2, \mathcal{A}^k) = \{ (x, y)^\top \,|P(x, y) = 0\}$ defines a family of curves on the plane for $\boldsymbol{a} \in \mathcal{A}^n$ [5]. Here, the suffix 2 of $\mathcal{C}_2(\mathcal{P}^n_2, \mathcal{A}^k)$ indicates a set of algebraic curves of two real arguments. The following theorems are held.

Theorem 1 *An element of $\mathcal{A}^k$ corresponds to a point in the n-dimensional vector space.*

From theorem 1, we define a coefficient vector of $P(x, y)$ as $\boldsymbol{a} = (a_1, a_2, \cdots, a_k)^\top$. For a positive real value λ, $\lambda P(x, y) = 0$ and $-\lambda P(x, y) = 0$ define the same curve. Conversely, once a point $\boldsymbol{a}$ on S^{k-1}_+ is fixed, we obtain a curve of $\mathcal{C}_2(\mathcal{P}^k_2, \mathcal{A}^k)$. This leads to the following theorem.

Theorem 2 *There is one-to-one mapping between $\mathcal{C}_2(\mathcal{P}^k_2, \mathcal{A}^n)$ and S^{k-1}_+.*

If elements of $\mathcal{P}^6_2$ are $p_1(x, y) = 1$, $p_2(x, y) = 2x$, $p_3(x, y) = 2y$, $p_4(x, y) = x^2$, $p_5(x, y) = 2xy$, and $p_6(x, y) = y^2$, we obtain

$$P(x, y) = ax^2 + 2bxy + cy^2 + 2dx + 2ey + f. \tag{1}$$

Then, $\mathcal{C}_2(\mathcal{P}^6_2, \mathcal{A}^6)$ determines the family of conics in the Euclidean plane if at least one of a, b, and c, and f are not zero. Therefore, the Hough transform is a method for the detection of points on S^5_+, which coresspond to conics, from noisy samples on the image in plane. We call a point on S^{k-1}_+ the dual figure of a curve on $\mathbf{R}^2$.

2.2 The Hough Transform as LMS

Let m conics exist on $\mathbf{R}^2$ and sample-points P be separated to clusters of points as $P = \bigcup_{i=1}^m P_i$ such that $P_i = \{\boldsymbol{x}_{ij}\}^{n(i)}_{j=1}$. Assuming that points in each cluster distribute in the neighborhood of conics, the conic fitting problem for a collection of conics is achieved minimizing the functional

$$J(\boldsymbol{a}_1, \cdots, \boldsymbol{a}_m) = \sum_{i=1}^{m} \left\{ \sum_{j=1}^{n(i)} w_i(j)(\boldsymbol{\xi}_{ij}^\top \boldsymbol{a}_i)^2 - \lambda_i(1 - |\boldsymbol{a}_i|^2) \right\}, \tag{2}$$

where $w_i(j) \geq 0$ and $\sum_{i=1}^{m} \sum_{j=1}^{n(i)} w_i(j) = 1$. This minimization problem yields a system of eigenvalue equations

$$\boldsymbol{M}_i \boldsymbol{a}_i = \lambda_i \boldsymbol{a}_i, \text{ s.t. } \boldsymbol{M}_i = \sum_{j=1}^{n(i)} w_i(j) \boldsymbol{\xi}_{ij} \boldsymbol{\xi}_{ij}^\top. \tag{3}$$

for $i = 1, 2, \cdots, m$. Furthermore, the solutions are the eigenvectors which associate to the smallest eigenvalues of each problem. In this paper, we assume that both $w(i)$ and $w_i(j)$ are $1/n$. Furthermore, we assume that in each cluster there exists k points, that, is $k = n(i)$ and $k \times m = n$.

The Hough transform for the conic detection achieves both the classification of sample points and the model fitting concurrently. Therefore, in this section, we formulate the Hough transform as the LSM for the model fitting problem. Setting $\boldsymbol{\Xi} = (\boldsymbol{\xi}_1, \boldsymbol{\xi}_2, \cdots, \boldsymbol{\xi}_n)^\top$, if there is no error in sample points, the parameters of a conic satisfies the equation $\boldsymbol{\Xi}^\top \boldsymbol{a} = \boldsymbol{0}$. For the noisy data, the parameter of a conic is the solution of the functional

$$J(\boldsymbol{a}) = |\boldsymbol{\Xi}^\top \boldsymbol{a}|^2 - \lambda(1 - |\boldsymbol{a}|^2). \tag{4}$$

If there exist m conics, after clustering sample-points, we have the equation, $\boldsymbol{\Xi}_i^\top \boldsymbol{a}_i = \boldsymbol{0}$, for $i = 1, 2, \cdots m$, where $\boldsymbol{\Xi} = (\boldsymbol{\xi}_{i(1)}, \boldsymbol{\xi}_{i(2)}, \cdots, \boldsymbol{\xi}_{i(k)})$. If we do not know any clusters of sample-points, the minimization criterion is expressed as $(\boldsymbol{Q}\bar{\boldsymbol{\xi}})^\top \bar{\boldsymbol{a}} = 0$, where $\boldsymbol{Q}$ is an appropriate permutation matrix and, $\bar{\boldsymbol{\xi}} = (\boldsymbol{\xi}_1^\top, \boldsymbol{\xi}_2^\top, \cdots, \boldsymbol{\xi}_n^\top)^\top$, $\bar{\boldsymbol{a}} = (\bar{\boldsymbol{a}}_1^\top, \bar{\boldsymbol{a}}_2^\top, \cdots, \bar{\boldsymbol{a}}_m^\top)^\top$, and $\bar{\boldsymbol{a}}_i = (\underbrace{\boldsymbol{a}_i^\top, \boldsymbol{a}_i^\top, \cdots, \boldsymbol{a}_i^\top}_{k\text{'s}})^\top$.

Therefore, the parameters minimize the functional

$$J(\boldsymbol{a}, \boldsymbol{Q}) = |(\boldsymbol{Q}\bar{\boldsymbol{\xi}})^\top \bar{\boldsymbol{a}}|^2 - \sum_{i=1}^{m} \lambda_i(1 - |\boldsymbol{a}_i|^2). \tag{5}$$

This property implies that the classification of sample data is equivalent to the permutation of elements of $\boldsymbol{\Xi}$. There exist many possibilities for the selection of a permutation $\boldsymbol{Q}$, if we do not know the estimate of $\{\boldsymbol{a}_i\}_{i=1}^{m}$ [6]. These expressions of the conic fitting problem conclude that the Hough transform achieves both the permutation of data matrix $\boldsymbol{\Xi}$ and the computation of the parameters of conics concurrently.

2.3 Dyanamics of the Hough Transform

We consider the problem to determine a r-dimensional linear subspace in $\mathbf{R}^n$, where $1 < r < n$ which approximate the distribution of data-set $\{\boldsymbol{y}_i\}_{i=1}^{k}$, the mean of which are zero, that is, $\sum_{i=1}^{k} \boldsymbol{y}_i = 0$. The orthogonal projection matrix $\boldsymbol{P}$ such that $rank\boldsymbol{P} = r$ which minimizes the criterion $\varepsilon^2 = tr(\boldsymbol{PM})$, for $\boldsymbol{M} = \frac{1}{k}\sum_{i=1}^{k} \boldsymbol{y}_i\boldsymbol{y}_i^\top$, determines the r-dimensional linear subspace which approximate the distribution of $\{\boldsymbol{y}_i\}_{i=1}^{k}$.

Since matrices $\boldsymbol{P}$ and $\boldsymbol{M}$ are symmetry matrices, there exist orthogonal matrices $\boldsymbol{U}$ and $\boldsymbol{V}$ which diagonalize $\boldsymbol{P}$ and $\boldsymbol{M}$, respectively. If we set $\boldsymbol{P} = \boldsymbol{U\Lambda U}^\top$ and $\boldsymbol{M} = \boldsymbol{VDV}^\top$, where $\boldsymbol{\Lambda}$ is a diagonal matrix whose entries are 0 or 1 such that $tr\boldsymbol{\Lambda} = r$, and $\boldsymbol{D}$ is a diagonal matrix. These decomposition of matrices derive $\varepsilon^2 = tr(\boldsymbol{W\Lambda W}^\top\boldsymbol{D})$, where $\boldsymbol{W} = \boldsymbol{V}^\top\boldsymbol{U}$, since $tr(\boldsymbol{PM}) = tr(\boldsymbol{U\Lambda U}^\top\boldsymbol{VDV}^\top) = tr(\boldsymbol{V}^\top\boldsymbol{U\Lambda U}^\top\boldsymbol{VD})$. Therefore, our problem is mathematically equivalent to the determination of an orthogonal matrix which minimizes the criterion for the detection of conics from unclusterd data.

The gradient flow for $\boldsymbol{W}$

$$\frac{d}{dt}\boldsymbol{W} = -[\boldsymbol{\Lambda}, \boldsymbol{WDW}^\top]\boldsymbol{W}, \tag{6}$$

where $[\boldsymbol{X}, \boldsymbol{Y}] = \boldsymbol{XY} - \boldsymbol{YX}$, is the continuous version of the gradient decent equation to search an orthogonal matrix which minimizes the criterion [7,8] for the detection of conics from unclustered data. Furthermore, setting $\boldsymbol{G} = \boldsymbol{W\Lambda W}^\top$, eq. (6) is equivalent to the double bracket equation,

$$\frac{d}{dt}\boldsymbol{G} = -[\boldsymbol{G}, [\boldsymbol{G}, \boldsymbol{D}]]. \tag{7}$$

If $n = 6$ and $r = 1$, our equation determines a conic on a plane using the homogeneous coordinate system. For the detection of many conics on a plane, the criterion for the minimization is $\varepsilon^2 = |(\boldsymbol{Q}\bar{\boldsymbol{\xi}})^\top\bar{\boldsymbol{a}}|^2 = tr(\boldsymbol{AQMQ}^\top)$, where $\boldsymbol{A} = \bar{\boldsymbol{a}}\bar{\boldsymbol{a}}^\top$, and $\boldsymbol{M} = \bar{\boldsymbol{\xi}}\bar{\boldsymbol{\xi}}^\top$. Let $\boldsymbol{\Theta}$ be an orthogonal matrix which digonalizes $\boldsymbol{A}$ such that $\boldsymbol{A} = \boldsymbol{\Theta}(\boldsymbol{I} \otimes \mathrm{Diag}(1,0,0,0,0,0))\boldsymbol{\Theta}^\top$. Since $\boldsymbol{A}$ is the covariance matrix, matrix $\boldsymbol{A}$ is expressed as the Kroneker product of the identity matrix and $\mathrm{Diag}(1,0,0,0,0,0)$ using an appropriate coordinate system, where m is the number of conics which exist on a plane. The minimization of ϵ^2 is equivalent to minimize $tr(\boldsymbol{QU\Lambda U}^\top\boldsymbol{Q}^\top\boldsymbol{\Theta D\Theta}^\top) = tr(\boldsymbol{W\Lambda W}^\top\boldsymbol{D})$, if we set $\boldsymbol{D} = \boldsymbol{I} \otimes \mathrm{Diag}(1,0,0,0,0,0)$ and $\boldsymbol{W} = \boldsymbol{\Theta}^\top\boldsymbol{QU}$. This expression of the criterion for the detection of many conics implies that the conic detection trough the Hough transform is also achieved by the gradient flow.

The double bracket dynamics has clarified mathematical property of the Hough transform which concurrently detects many conics on a plane from noisy data. This dynamics implies that the permutation which is described as a orthogonal matrix achieves grouping of data for the detection of conics. Furthermore, this dynamics implies that if we can generate a decresing sequence with respect to the energy function $tr(\boldsymbol{AQMQ}^\top)$, we can compute the minimum value of the criterion.

For $tr\boldsymbol{\Lambda} = r$, we have the relation $tr(\boldsymbol{IUPU}^\top) = tr(\boldsymbol{IP}) = tr\boldsymbol{P}$. Therefore, projection matrix $\boldsymbol{P}$ is defined as the solution of a semidefinite programming problem. From the orthogonal decomposition of projection matrix $\boldsymbol{P}$ and moment matrix $\boldsymbol{M}$, we have $tr(\boldsymbol{IA}) = m$ if the number of conics is m. Therefore, the detection of conics is achieved by minimizing $tr(\boldsymbol{Q}^\top \boldsymbol{AQM})$ subject to $tr(\boldsymbol{IA}) = m$.

3 Combinatorial Hough Transform

Vectors$\{\boldsymbol{a}_i^*\}_{i=1}^m$ and matrix $\boldsymbol{Q}^*$, which minimize $J(\boldsymbol{a}_1, \boldsymbol{a}_2, \cdots, \boldsymbol{a}_m, \boldsymbol{Q}) = |\boldsymbol{\Xi}^\top \bar{\boldsymbol{Q}}\boldsymbol{a}|^2$, w.r.t. $|\boldsymbol{a}_i| = 1$, determine conics which fit to sample points $\{\boldsymbol{x}_i\}_{i=1}^m$. It is, however, not so easy to determine all vectors $\boldsymbol{a}_i$, for $i = 1, 2, \cdots, m$ and matrix $\boldsymbol{Q}$, simultaneously. If $\boldsymbol{Q}$ is determined, $\{\boldsymbol{a}_i\}_{i=1}^m$ is computed using semidefinite programming. Furthermore, if the set of parameters $\{\boldsymbol{a}_i\}_{i=1}^m$ is determined, $\boldsymbol{Q}$ is computed using integer programming. Using the relaxation, for $\{\boldsymbol{a}_i\}_{i=1}^n$ and $\boldsymbol{Q}$, we compute the minimum of $J(\boldsymbol{a}_1, \boldsymbol{a}_2, \cdots, \boldsymbol{a}_m, \boldsymbol{Q})$ as follows.

Algorithm 1

1. Set initial estimation of $\boldsymbol{Q}$ as $\boldsymbol{Q}_0$.
2. Compute $\boldsymbol{a}_i^*$ $i = 1, 2, \cdots, m$, which minimize $J(\boldsymbol{a}_1, \boldsymbol{a}_2, \cdots, \boldsymbol{a}_m, \boldsymbol{Q}_0)$.
3. Compute $\boldsymbol{Q}^*$ which minimizes $J(\boldsymbol{a}_1^*, \boldsymbol{a}_2^*, \cdots, \boldsymbol{a}_m^*, \boldsymbol{Q})$ and satisfies the inequality such that $J(\boldsymbol{a}_1^*, \boldsymbol{a}_2^*, \cdots, \boldsymbol{a}_m^*, \boldsymbol{Q}^*) < J(\boldsymbol{a}_1^*, \boldsymbol{a}_2^*, \cdots, \boldsymbol{a}_m^*, \boldsymbol{Q}_0)$.
4. If $\boldsymbol{Q}_0 \neq \boldsymbol{Q}^*$ then $\boldsymbol{Q}_0 := \boldsymbol{Q}^*$ and go to 2 else if $J(\boldsymbol{a}_1^*, \boldsymbol{a}_2^*, \cdots, \boldsymbol{a}_m^*, \boldsymbol{Q}^*) > \varepsilon$, for a small positive constant ε, then go to 1 else accept $\boldsymbol{a}_i^*$, for $i = 1, 2, \cdots, m$ as the solutions.
5. Draw conics $\boldsymbol{\xi}^\top \boldsymbol{a}^* = 0$, for $i = 1, 2, \cdots, m$.

Here, $\varepsilon \leq m\delta$, where δ is introduced in the next section using the discretization method of conics. We call this process the combinatorial Hough transform. Step 2 is achieved by semidefinite programming [9,10] and step 3 is solved by integer programming. In this paper, for step 3, we define a new matrix $\boldsymbol{Q}^*$ computing the distance between sample points and detected figures. If both $\boldsymbol{Q}^*$ and $\boldsymbol{a}_k^*$ are correct ones, the total sum of the distances between points and figures become minimum. For the evaluation of these distances, we compute the minimum distance from each sample point to figures. This process determines a $\boldsymbol{Q}^*$ which satisfies the condition of step 3.

Employing the marge sort, and divide and conquer methods, we divide the array of sample points into subsets. Dividing both intervals $I(x) = \{x| -\frac{L}{2} \leq x \leq \frac{L}{2}\}$ and $I(y) = \{v| -\frac{L}{2} \leq y \leq \frac{L}{2}\}$, into k subintervals equally as $\cup_{i=1}^k I_i(x) = I(x)$ and $\cup_{i=1}^k I_i(y) = I(y)$ such that $I_i(x) \cap I_j(x) = \emptyset$ and $I_i(y) \cap I_j(y) = \emptyset$. From these subintervals, we have k^2 non-overlapping region such as $I_{ij}(x, y) = I_i(x) \times I_j(y)$. In each $I_{ij}(x, y)$, if we recursively divide the region, we have $(k^2)n$ subregions for $n \geq 1$. This decomposition derives k^2-tree description of the image region. Then, the divide and conquer version of our algorithm is described as follows.

Algorithm 2

1. Divide the imaging region to $(k^2)^n$ regions.

2. Apply **Algorithm 1** in each region.
3. Marge the solutions of subregions in the whole region.
4. Set merged data as the initial estimation of the **Algorithm 1** for whole region.

If there exist $O(k^2)$ ellipses which are not mutually overlapping in the imaging region, approximately, there exist one ellipse in each subregion. For the ellipse and circle fitting problem, approximately there also exists a segment of a ellipse or a ellipse in each subregion. In each region, we can solve the usual model fitting problem for a curve which does not contain data classification process by the permutation process to the data array. Therefore, each subproblem which detects a curve in a subregion is solved faster than the original problem which requires the permutation process to all data array. For a practical application, if an ellipse approximately exists in a 300×300 pixel subregion of the 1000×1000 pixel region, we adopt $k = 3$ and $n = 1$ for the partition of regions which yield 9 subregion in the imaging region.

4 Numerical Examples

We estimate the size of the neighborhood on the image plane employing imaging geometry of the pin-hole camera. As usual geometry, we assume that the optical center is at the origin of the world coordinate system, the z-axis is the optical axis and that the imaging plane is at $z = f$. Therefore, a point $(x, y, z)^\top$ in the space is transformed to $(f\frac{x}{z}, f\frac{y}{z})^\top$ on the imaging plane. Here, we select the focal length f, the imaging area $l \times l$, the number of pixels $L \times L$ as 6mm, 512×512mm, and 1024×1024, respectively.

If the neighborhood of each point is the desk whose radius is 5mm on the plane at $z = 1000$mm, the neighborhood of each point on the imaging plane is 0.03mm which is equivalent to 6 pixels if the resolution of CCD is 0.005cm. The length a such that $15\text{mm} \leq a \leq 25\text{mm}$ on the plane at $z = 1000$mm is transformed to $180 \leq a' \leq 300$ pixels.

An ellipse $f(x, y) = 0$, for $f(x, y) = ax^2 + 2bxy + cy^2 + 2dx + 2ey + f$, is expressed as

$$(\boldsymbol{U}(\boldsymbol{x}-\boldsymbol{a}))^\top \begin{pmatrix} l_1^{-1} & 0 \\ 0 & l_2^{-1} \end{pmatrix} (\boldsymbol{U}(\boldsymbol{x}-\boldsymbol{a})) = 0, \ \boldsymbol{U} = \begin{pmatrix} \cos\theta & \sin\theta \\ -\sin\theta & \cos\theta \end{pmatrix}, \ \boldsymbol{a} = \begin{pmatrix} p \\ q \end{pmatrix}, \quad (8)$$

where $\boldsymbol{a}$, l_1, l_2, and θ are the center of an ellipse, the length of the major axis, the length of the minor axis, and the angle between the major axis and the x-axis of the coordinate system. Parameters $k_1 = \frac{\sqrt{(bq+d)^2-4(bq^2+eq+f)}}{2}$ and $k_2 = \frac{\sqrt{(bp+d)^2-4(ap^2+dp+f)}}{2}$ are the half of the lengths of line segments defined by $\{(x, y)^\top | f(x, y) = 0, \ y = q\}$ and $\{(x, y)^\top | f(x, y) = 0, \ x = p\}$, respectively. Setting $\hat{l}_i$, $\hat{k}_i$, for $i = 1, 2$, and $\hat{\boldsymbol{a}}$ to be the parameters of the reconstructed ellipse, if $\hat{\boldsymbol{a}} \cong \boldsymbol{a}$, $\hat{l}_i \cong l_i$ and $\hat{\theta} \cong \theta$, then $\hat{k}_i \cong k_i$. Furthermore, if $\theta = 0$, then $l_i = k_i$ and if $\theta = \frac{\pi}{2}$ $l_1 = k_2$ and $l_2 = k_1$. Therefore, parameters $|\hat{k}_i - k_i|$

for $i = 1, 2$ act as a parameter for the evaluation of the angle between the major axes of two ellipses if two ellipses are almost overlapping. From these geometric properties, in this paper, we evaluate $|k_i - \hat{k}_i|$.

Setting r to be the radius of the neighborhood of a point on the imaging plane, if the reconstructed ellipse $\hat{f}(x, y) = 0$ exists in $\mathbf{B} \oplus \mathbf{D}(r)$ for $\mathbf{B} = \{(x, y)^\top | f(x, y) = 0\}$ and $\mathbf{D}(r) = \{(x, y)^\top | x^2 + y^2 \leq r^2\}$, parameters $\hat{l}_i$ and $\hat{k}_i$, and vector $\hat{\boldsymbol{a}}$ satisfy the relations $|l_i - \hat{l}_i| \leq r$, $|k_i - \hat{k}_i| \leq \sqrt{2}r$, and $|\boldsymbol{a} - \hat{\boldsymbol{a}}| \leq r$. Next, if the neighborhood $\mathbf{D}(r)$ is approximated by 24-neighborhood in $\mathbf{Z}^2$, r is approximated as $\frac{\sqrt{5}}{2}$ pixels.

Next, we define a digital ellipse as a sampling model for the numerical evaluation. Setting $\lambda_1(\beta) \leq \lambda_2(\beta)$ to be the real solutions of $f(x, \beta) = 0$, for $\beta \in \mathbf{Z}$, we define four discrete point sets as

$$\mathbf{O}_1 = \{(\lceil \lambda_1(\beta) \rceil, \beta)^\top | f(x, \beta) = 0\},\ \mathbf{O}_2 = \{(\lfloor \lambda_2(\beta) \rfloor, \beta)^\top | f(x, \beta) = 0\}$$
$$\mathbf{O}_3 = \{(\lfloor \lambda_1(\beta) \rfloor, \beta)^\top | f(x, \beta) = 0\},\ \mathbf{O}_4 = \{(\lceil \lambda_2(\beta) \rceil, \beta)^\top | f(x, \beta) = 0\}.$$

With the same manner, setting $\mu_1(\alpha) \leq \mu_2(\alpha)$ to be the real solutions of $f(\alpha, y) = 0$, for $\alpha \in \mathbf{Z}$, we define four discrete point sets as

$$\mathbf{O}_5 = \{(\alpha, \lceil \mu_1(\alpha) \rceil)^\top | f(\alpha, y) = 0\},\ \mathbf{O}_6 = \{(\alpha, \lfloor \mu_2(\alpha) \rfloor)^\top | f(\alpha, y) = 0\}$$
$$\mathbf{O}_7 = \{(\alpha, \lfloor \mu_1(\alpha) \rfloor)^\top | f(\alpha, y) = 0\},\ \mathbf{O}_8 = \{(\alpha, \lceil \mu_2(\alpha) \rceil)^\top | f(\alpha, y) = 0\}.$$

We adopt $\mathbf{O} = \cup_{i=1} \mathbf{O}_i$ as the discrete ellipse derived from $f(x, y) = 0$. Furthermore, setting $\mathbf{R}_{24}$ to be randomly selected points in $\mathbf{N}_{24} = \{\boldsymbol{x} = (x, y)^\top | x^2 + y^2 \leq 8,\ \boldsymbol{x} \in \mathbf{Z}^2\}$, we adopt $\mathbf{E} = (\mathbf{O} \oplus \mathbf{R}) \setminus \mathbf{O}$ as the collection of discrete sample points from ellipse $f(x, y) =$.

We have evaluated the averages and averages variances of $P = |\boldsymbol{a} - \hat{\boldsymbol{a}}|$, $L_i = |l_i - \hat{l}_i|$, and $K_i = |k_i - \hat{k}_i|$for 10 images in each group. We express the average and average of variance of each parameter as $e_{E(\cdot)}$ and $e_{V(\cdot)}$. For each group, we generated 5 ellipses and the region of interest is separated into 9 regions. In tables, we list the values a and b which determine the density of sample points and signal-to-noise ratio. We set $a = 10, 50, 100$, and $b = 0, 20$ $a/100$ is the ratio of the selected sample points from discrete approximation of each ellipse. Furthermore $b/100$ is the ratio of random noise in the background. Table 1 shows the figures for evaluation and Table 2 shows the computational times for each group. Figures 1 (a) and 1 (b) show a noisy image of ellipses for $a = 10$ and $b = 20$.

If we set $\boldsymbol{\xi} = (x, y, 1)^\top$, which is equivalent to set $a = b = c = 0$ in eq. (1), the method also detects lines. Therefore, we apply the method for the detection of conics and lines which exist in an image. We first detect lines, since for the detection of lines sample points lie on conics affect as background noise during line detection. After detecting lines, we apply the ellipse-detection algorithm. The endpoints of line segment and parts of ellipses are detected back-voting lines and ellipes to the image plane. We extract back-voted lines and parts of ellipses which lie in the union of the neighborhoods of sample points on the imaging plane. Figures 2 (a), and 2 (b) show detected lines and ellipse from an

Table 1. Computational results of fitting of ellipses

Group	1	2	3	4	5	6
a	100	50	10	100	50	10
b	0	0	0	20	20	20
$e_{E(L_1)}$	0.036024	0.047197	0.129199	0.033195	0.039478	0.099252
$e_{E(L_2)}$	0.039044	0.051161	0.100132	0.026181	0.040938	0.112451
$e_{E(K_1)}$	0.047152	0.060962	0.164751	0.046178	0.056796	0.138115
$e_{E(K_2)}$	0.048523	0.052492	0.105435	0.027857	0.039804	0.090192
$e_{E(P)}$	0.048479	0.068985	0.167061	0.029707	0.050235	0.129238
$e_{V(L_1)}$	0.000904	0.001432	0.027780	0.000760	0.001184	0.003777
$e_{V(L_2)}$	0.001621	0.002185	0.005338	0.000500	0.000876	0.007313
$e_{V(K_1)}$	0.004803	0.005625	0.035757	0.001111	0.001723	0.010337
$e_{V(K_2)}$	0.003622	0.003567	0.006834	0.000457	0.001519	0.004738
$e_{V(P)}$	0.002581	0.003924	0.015863	0.015863	0.000705	0.004161

Table 2. Computational times for each group

Group	1	2	3	4	5	6
a	100	50	10	100	50	10
b	0	0	0	20	20	20
time(s)	34.72	21.41	10.71	146.67	97.89	81.94

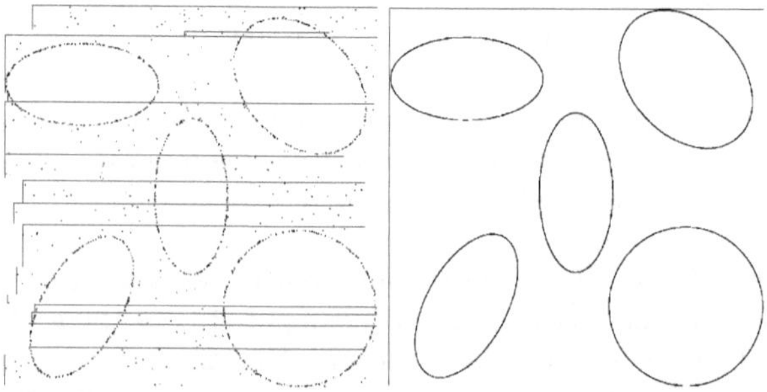

(a) (b)

Fig. 1. (a)Ellipses with background noise, and (b) detected ellipses image with a house and a cylinder, and detected line-segments and elliptic arcs from an image, respectively. These results for synthetic data and a real image show that our method effectively detects lines and conics in an image without using any acumulators.

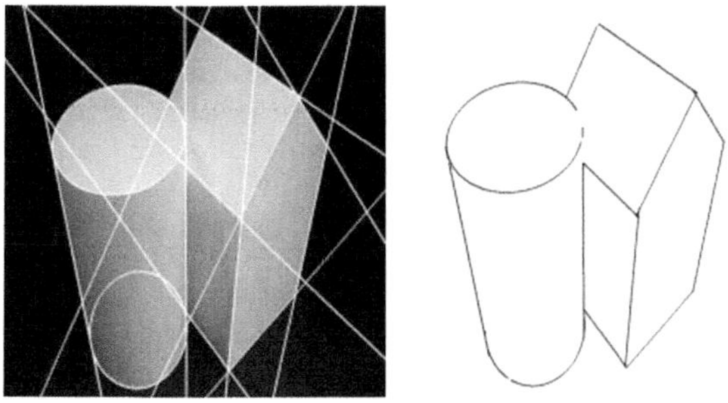

(a) (b)

Fig. 2. (a) Detected lines and ellipes from an image with a house and a cylinder, (b) Detected line-segments and elliptic arcs from an image with a house and a cylinder

5 Conclusions

We introduced the combinatorial Hough transform which is the Hough transform without any accumulators. The combinatorial Hough transform is based on the mathematical property that the grouping of sample points is achieved by the permutation for the sequence of the sample points. We also showed the convergence of the algorithm deriving a dynamic system which achieves the minimization of the criterion for the detection of figures from unclassified sample points.

References

1. Ballard, D. and Brown, Ch. M., *Computer Vision,* Prentice-Hall; New Jersey, 1982. 823
2. Deans, S. R., Hough transform from the Radon transform, IEEE Trans. Pattern Analysis and Machine Intelligence, **PAMI-3**, 185-188, 1981. 823
3. Levers, V. F., Which Hough transform? CVGIP: Image Understanding, **58**, 250-264, 1993. 823

4. Becker, J.-M., Grousson, S., and Guieu, D., Space of circles: its application in image processing, Vision Geometry IX, Proceedings of SPIE, **4117**, 243-250, 2000. 823
5. Cox, D., Little, J., and O'Shea, D., *Ideals, Varieties, and Algorithms: An Introduction to Computational Algebraic Geometry and Commutative Algebra*, Springer-Verlag; New York, 1992. 824
6. Mattavelli, M., Noel, V., and Ammaldi, E., Fast line detection algorithms based on combinatorial optimization, LNCS, **2051**, 410-419, 2001. 825
7. Brockett, R. W., Least square matching problem, *Linear Algebra and its Applications*, **122/123/124**, 1989, 761-777. 826
8. Brockett, R. W., Dynamical system that sort list, diagonalize matrices, and solve linear programming problems, *Linear Algebra and its Applications*, **146**, 1991, 79-91. 826
9. Vandenberghe, L. and Boyd, S., Semdefnite programming, *SIAM Review*, **38**, 49-95, 1996. 827
10. Alizaden, F., Interir point methods in semidefinite programming with application to combinatorial optimization, *SIAM, Journal on Optimization*, **5**, 13-51, 1995. 827

Robust Gray-Level Histogram Gaussian Characterisation*

José Manuel Iñesta and Jorge Calera-Rubio

Universidad de Alicante
Departamento de Lenguajes y Sistemas Informáticos
{inesta,calera}@dlsi.ua.es

Abstract. One of the most utilised criteria for segmenting an image is the gray level values of the pixels in it. The information for identifying similar gray values is usually extracted from the image histogram. We have analysed the problems that may arise when the histogram is automatically characterised in terms of multiple Gaussian distributions and solutions have been proposed for special situations that we have named *degenerated modes*. The convergence of the method is based in the expectation maximisation algorithm and its performance has been tested on images from different application fields like medical imaging, robotic vision and quality control.

1 Introduction

Image segmentation is one of the most challenging problems in computer vision. A lot of work has been devoted to solve this problem [1,2,3,4,5], but it seems still impossible to find a general solution able to deal with all the problems that may arise in a successful and robust way. In this work we are going to focus in the methods that rely on the grey level similarity for selecting the regions of interest in an image. This is one of the most utilised approaches to image segmentation and it is based on the hypothesis that pixels with similar intensities will belong to the same region. This is not true in general but it is in a large number of computer vision applications, specially in indoor scenes and illumination controlled environments.

In such cases the images can present dark objects on a bright background or vice-versa. Then, the image is said to be bi-modal. On the same basis, if n layers of intensity are found in an image then it is called s n-modal image and each layer is called a *mode*. If a meaning in terms of regions of interest can be assigned to some of the layers, then the identification and isolation of these layers can be a good way to segment the image into meaningful parts and the question is, where to look for the information for such an identification.

The image histogram, $h(z)$, is the most valuable information about the grey level distribution in an image and a number of authors have used different algorithms for histogram characterisation in order to extract from it the parameters

* This work has been funded by the Spanish CICYT project TAR; code TIC2000–1703–CO3–02

T. Caelli et al. (Eds.): SSPR&SPR 2002, LNCS 2396, pp. 833–841, 2002.
© Springer-Verlag Berlin Heidelberg 2002

needed for image segmentation [2,6,7,8]. The algorithms dealing with histograms are usually fast because operations are $O(N_z)$, where N_z is the number of grey level intensities. This characterisation is usually performed using parametric functions that permit to describe $h(z)$ in terms of the parameters that characterise those functions. The problem is, therefore, to select the functions and then determine the values of the parameters from the values of the histogram frequencies $h(z_j), j = 0, ..., N_z - 1$.

The majority of the literature works with gaussians as parametric functions $p(z|\omega_i) = G(z|\mu_i, \sigma_i^2)$, that are able to characterise the histograms in terms of their means μ_i and variances σ_i^2, in addition to the a priori probabilities, $P(\omega_i)$ associated to each mode. This way, the histogram is described as a mixture of gaussians

$$h(z) = \sum_i P(\omega_i)\, p(z|\omega_i). \tag{1}$$

and the characterisation can be computed as an expectation maximisation problem using the EM algorithm [9] which converges to the maximum likelihood estimate of the mixture parameters. Also, the Gaussian parameters can be used to calculate thresholds to separate different modes in the histogram [10,11].

The most common case in artificial vision is to have a bimodal image in which $h(z) = p(z|\omega_0)\, P(\omega_0) + p(z|\omega_1)\, P(\omega_1)$, where $p(z|\omega_i)$ is the density probability function of grey level z in mode $i, i = 0, 1$ (two modes: dark and bright), but there are also a number of applications in which the images present more than two models in a natural way, like medical image analysis or industrial inspection environments.

We are going to focus in the convergence of the algorithm in some situations in which degenerated modes may occur. This mode degeneration can be caused by a number of situations: 1) border effects due to sensor effects, one or two modes can be biased to the extremes of the histogram due to the saturation of the sensor; 2) highly-modal analysis, a large number of functions in multi-modal analysis may cause that some of the modes are restricted to very few pixels, not enough to apply a normality criterion on them; and 3) preprocessing stages like compression or normalisation can cause uneven frequency distributions in the histogram that lead the traditional algorithms to fail. We propose a methodology to isolate the degenerated modes that may appear from a background of normality during the convergence and then continue with the method until a successful characterisation of the overall histogram.

2 Histogram Robust Characterisation

2.1 Maximum Likelihood Parameter Estimation

As stated above a possible method to estimate the parameters defining the mixture model is the expectation maximisation algorithm (EM).

Due to the stochastic component of the image histogram $h(z)$, a natural way to deal with it is to consider the histogram as a mixture of Gaussian densities

$h(z) = \sum_i P(\omega_i)p(z|\omega_i)$, and $P(\omega_i)$ is the *a priori* probability of mode i; then, we can consider the histogram characterisation like a parametric unsupervised learning problem [12] where the means μ_i, standard deviations σ_i, *a priori* probability functions $P(\omega_i)$, and *a posteriori* probability functions $P(\omega_i|z)$ of the modes are unknowns, while the number of modes remain constant and equal to n. If we assume normal distributions, the *maximum likelihood* estimators for these quantities cam be computed, using a iterative procedure, through the following equations [12]:

$$\hat{\mu}_i^{[t+1]} = \frac{\sum_{j=0}^{N_z-1} h(z_j)\hat{P}^{[t]}(\omega_i|z_j) \cdot z_j}{\sum_{j=0}^{N_z-1} h(z_j)\hat{P}^{[t]}(\omega_i|z_j)} \tag{2}$$

$$\hat{\sigma}_i^{2\,[t+1]} = \frac{\sum_{j=0}^{N_z-1} h(z_j)\hat{P}^{[t]}(\omega_i|z_j) \cdot (z_j - \hat{\mu}_i^{[t+1]})^2}{\sum_{j=0}^{N_z-1} h(z_j)\hat{P}^{[t]}(\omega_i|z_j)} \tag{3}$$

$$\hat{P}^{[t+1]}(\omega_i) = \frac{1}{N} \sum_{j=0}^{N_z-1} h(z_j)\hat{P}^{[t]}(\omega_i|z_j) \tag{4}$$

$$\hat{P}^{[t+1]}(\omega_i|z_j) = \frac{P(z_j|\omega_i) \cdot \hat{P}^{[t+1]}(\omega_i)}{\sum_{l=0}^{n-1} P(z_j|\omega_l)\hat{P}^{[t+1]}(\omega_l)} \tag{5}$$

where $h(z_j)$ represents the frequencies for each grey level z_j, $j \in [0, N_z]$ and

$$P(z_j|\omega_i) = \int_{z_j - \frac{1}{2}}^{z_j + \frac{1}{2}} p(z|\omega_i)\, dz$$

is the probability of grey level z in mode i taking into account that grey levels are discrete while $p(z|\omega_i)$ are continuous density functions.

Initialisation These equations can be solved starting from some reasonable values. For this, we have used the k-means clustering algorithm [12] in order to found $\hat{\mu}_i^{[0]}$ and the initial data classification to compute the rest of the initial parameters. This algorithm provide approximate initial values for the parameters in a fast way and its use is recommended for this task [12].

2.2 Degenerated Modes

The presence of degenerated modes during the iterative procedure is detected in running time through the condition $\hat{\sigma}_i^{2\,[t]} \leq 0$. This means that in class ω_i, only a grey level (designed by $\bar{z}_i$) remains, or mathematically, that the Gaussian $p(z|\omega_i)$ has converged to a function Dirac delta $\delta(z - \bar{z}_i)$. Therefore, in successive iterations, it should be assumed that, for the ith class, $\hat{\mu}_i = \bar{z}_i$, $\hat{\sigma}_i^2 = 0$, $\hat{P}(\omega_i|z_j) = \delta_{z_j \bar{z}_i}$ and $\hat{P}(z_j|\omega_i) = \delta_{z_j \bar{z}_i}$, where δ_{ab} is now the Kronecker delta, that is equal to 1 if $a = b$ and 0 otherwise.

In order to be sure about the convergence of the method for the whole parameter set during the iterative process, and to prevent the possible variation in the values for $\hat{P}(\omega_j|\bar{z}_i)$ $j \neq i$ due to class overlapping, it is preferable to assume that normality is held in the next iteration by actually taking a Gaussian characterised by $\hat{\sigma}_i^2 = \epsilon$ with $\epsilon > 0$, such that the contributions of the tails of the distribution function $p(z|\omega_i)$ are negligible outside the interval $[\bar{z}_i - \frac{1}{2}, \bar{z}_i + \frac{1}{2}]$ and $\hat{P}(\omega_i)$ is computed with this criterion.

2.3 Convergence

The computational cost associated to the iterative algorithm is very sensitive to the stop conditions imposed to it. According to the discrete and one-dimensional condition of our problem it is not necessary for these stop conditions to be very restrictive. Thus, the algorithm stops at the iteration t when all the following conditions are held:

$$|\hat{\mu}_i^{[t]} - \hat{\mu}_i^{[t-1]}| \leq \frac{1}{2} \tag{6}$$

$$|\hat{\sigma}_i^{2\,[t]} - \hat{\sigma}_i^{2\,[t-1]}| \leq \epsilon^2 \tag{7}$$

$$|\hat{P}^{[t]}(\omega_i) - \hat{P}^{[t-1]}(\omega_i)| \leq \frac{1}{N_z} \tag{8}$$

This set of conditions permits us to be sure that any gray level will not have an appreciable probability of being assigned to an improper mode.

With these conditions the iterative procedure converges very quickly and is very fast for application purposes.

3 Results and Discussion

3.1 Bimodal Case

The ideal situation in computer vision systems is to deal with images that present an histogram with two Gaussian-like hills like that of figure 1. In that case the proposed method performs in a similar way to other based on maximum likelihood estimation [10], entropy maximisation [13], or moment preservation [14,15].

The threshold t is determined as the grey level for which both gaussians cross. That value is the z_j satisfying

$$P(\omega_0)p(t|\omega_0) = P(\omega_1)p(t|\omega_1). \tag{9}$$

This equation can be solved for t and a second degree equation is obtained:

$$at^2 + bt + c = 0 \tag{10}$$

being

$$\begin{aligned} a &= \sigma_0^2 - \sigma_1^2 \\ b &= 2(\mu_0\sigma_1^2 - \mu_1\sigma_0^2) \\ c &= \sigma_0^2\mu_1^2 - \sigma_1^2\mu_0^2 + 2\sigma_0^2\sigma_1^2 \ln \tfrac{\sigma_1 P(\omega_0)}{\sigma_0 P(\omega_1)} \end{aligned} \tag{11}$$

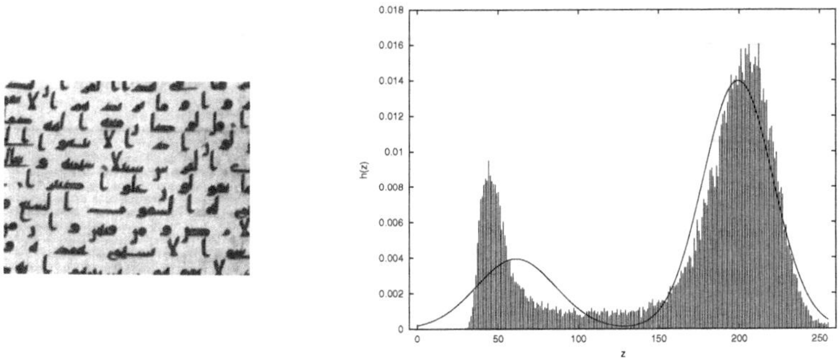

Fig. 1. Bimodal histogram (bars) and the mixture (line)

Table 1. Comparative for threshold calculation in bimodal histograms

METHOD	THRESHOLD
Proposed	130
Maximum likelihood[10]	134
Entropy maximisation[13]	168
Moment preservation[14]	150
Moment preservation[15]	154

From both possible solutions, only one is valid.

See table 1 for a comparative calculation of the threshold that separates both modes for the histogram in figure 1.

The described situation can be found mostly in the laboratory or in industrial environments where the light conditions are perfectly controlled, but this is not always possible.

Even in those controlled environments the sensor can work in not so comfortable situations, providing images that are very bright or very dark, and then one of the modes will be biased (or even *"smashed"*) to the right or to the left zone of the grey level range, and in this situation the assumption of normality is no longer valid. We analyse what happens with this kind of histogram in the next section.

3.2 Degenerated Modes Appear in Histogram Extremes

Due to saturation of sensors, the histogram can present one of its modes displaced to one of the extremes of the histogram. In that case a degenerated mode appears as a spike, usually for the limit value 0 or N_z. An example can be found in Fig. 2. There, a big spike is presented at $z = N_z$ and $h(N_z) = 0.28$, therefore, close to a 30% of the total amount of pixels in the image have a grey level value of 255. This spike is a mode by itself, corresponding to the bright part of the image, but this histogram can not be solved by the traditional methods.

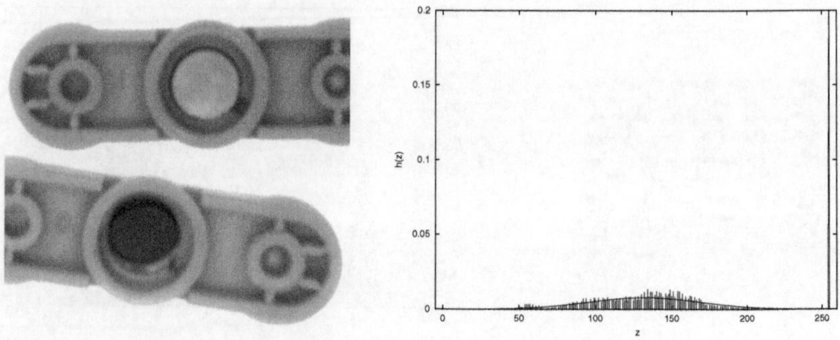

Fig. 2. Degenerated modes present in the histogram due to sensor saturation. Note that the dark mode can be easily explained as a Gaussian but the bright mode appears to the right as a very high spike

Nevertheless, during the convergence of the proposed method, this degenerated mode is detected and isolated from the rest of the histogram, avoiding the rest of the data to converge to the rest of modes that have been requested. In the example of Fig. 2, two modes were requested. The bright mode converged to the degenerated one and the dark mode fitted the remaining data with a Gaussian.

3.3 Multi-modal Case

There are a number of applications in which the images present more than two models in a natural way, for example in medical image analysis (where background, bone tissue and soft tissue appear) or industrial inspection environments (background, objects and shadows). In Fig. 3 an example is shown of a radiographic image that has been characterised using 4 modes. It is observed how the Gaussian modes fit to the data in the histogram.

If the number of modes is large, it is very likely that some of them will contain a small number of pixels, not enough to provide a Gaussian mode and therefore, a degenerated mode will appear. In those cases, the algorithm detects and isolates those modes and converges without any problems.

We have tried to run our method with a high number of modes (up to 50 modes) without problems. When these problems appeared it was due to the initialisation algorithm. When the number of modes is very high, the k-means clustering algorithm can provide modes to which no pixels are found. In this situation the algorithm can not run.

3.4 Processed Images

If the source of the images are not sensors but other low-level stages in a computer imaging system, like image enhancing, restoring, compression, etc. then the image histogram can present a profile that can not be processed with the

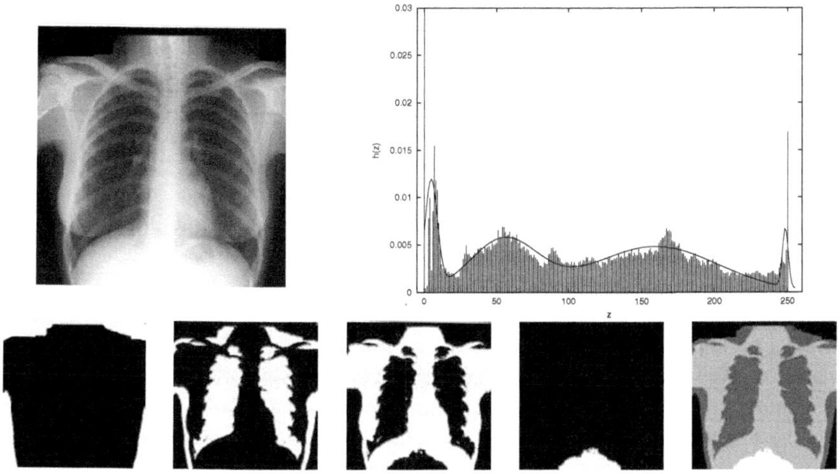

Fig. 3. (top): Multi-modal radiographic image and characterisation using 4 modes. (bottom): Pixels belonging to each mode (in white) and the combined image

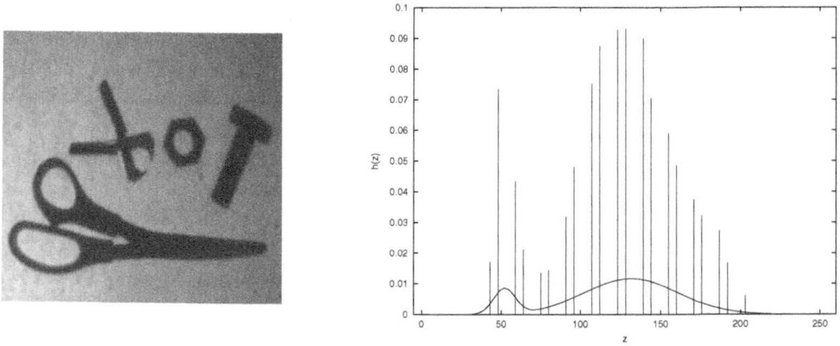

Fig. 4. Histogram of a processed image and its parameterisation

usual methods. See Fig. 4 for an example of the histogram of an image after a compression/decompression process.

Note that both modes (dark and bright, corresponding to tools and background respectively) have been properly characterised and the height of the gaussians is clearly lower than the frequencies of the histogram because there are a lot of zeroes between each pair of non zero histogram values and the gaussians try to fit the total density where they are defined.

This kind of histogram could be also regarded as a collection of degenerated modes, one for each $h(z) \neq 0$. We have tried to run our algorithm in such conditions but the k-means initialisation algorithm is not able to provide a good set of parameters with such kind of data. This is the same situation as in the very high-modal problem described above. On the other hand, finding those values using k-means would be the same as to solve the problem in that case without the aid of our method.

4 Conclusions

We have studied the problems that may arise when an image histogram is automatically characterised in terms of multiple Gaussian distributions. In the general case, specially when the images are very contrasted or a high number of modes are wanted, the frequencies in the histogram do not hold the hypothesis of being described as a Gaussian mixture. We have proposed a method able to detect and separate those *degenerated modes* from the convergence of an expectation maximisation algorithm (EM) with the rest of normal modes present in the histogram.

The algorithm and its performance has been tested successfully on images from different application fields like medical imaging, robotic vision and quality control and in critical situations like sensor saturation, multi-modal analysis and histograms of processed images.

References

1. S. D. Zenzo. Advances in image segmentation. *Image and Vision Computing*, 1(4):196–210, 1983. 833
2. P. K. Sahoo, A. K. C. Wong, and Y. C. Chen. A survey of thresholding techniques. *Computer Vision, Graphics and Image Processing*, 41:233–260, 1988. 833, 834
3. Nikhil R. Pal and Sankar K. Pal. A review on image segmentation techniques. *Pattern Recognition*, 26(9):1277–1294, 1993. 833
4. F. Meyer and S. Beucher. Morphological segmentation. *J. Visual Commun. Image Repres.*, 1(1):21–45, 1990. 833
5. Punam K. Saha and Jayaram K. Udupa. Optimum image thresholding via class uncertainty and region homogeneity. *IEEE Transactions on Pattern Analysis and Machine Intelligence*, 23(7):689–706, july 2001. 833
6. K. Price. Image segmentation: a comment on studies in global and local histogram-guided relaxation algorithms. *IEEE Transactions on Pattern Analysis and Machine Intelligence*, 6:247–249, 1984. 834
7. S. U. Lee, Y. S. Chung, and R. H. Park. A comparative performance study of several global thresholding techniques for segmentation. *Computer Vision, Graphics, and Image Processing*, 52(2):171–190, 1990. 834
8. C. A. Glasbey. An analysis of histogram-based thresholding algorithms. *Computer Vision, Graphics, and Image Processing. Graphical Models and Image Processing*, 55(6):532–537, November 1993. 834
9. D. Titterington, A. Smith, and U. Makov. *Statistical Analysis of Finite Mixture Distributions*. John Wiley and Sons, Chichester, UK, 1985. 834
10. J. Kittler and J. Illingworth. Minimum error thresholding. *Pattern Recognition*, 19(1):41–47, 1986. KITTLER86b. 834, 836, 837
11. N. Papamarkos and B. Gatos. A new approach for multilevel threshold selection. *CVGIP: Graphical Models and Image Processing*, 56(5):357–370, September 1994. 834

12. R. O. Duda, P. E. Hart, and D. G. Stork. *Pattern Classification.* John Wiley and Sons, 2001. 835
13. J. N. Kapur, P. K. Sahoo, and A. K. C. Wong. A new method for gray-level picture thresholding using the entropy of the histogram. *Computer Vision, Graphics and Image Processing*, 29:273–285, 1985. 836, 837
14. N. Otsu. A threshold selection method from gray-level histograms. *IEEE Transactions on System, man and Cybernetics*, 9(1):62–66, 1979. 836, 837
15. W.-H. Tsai. Moment-preserving thresholding: a new approach. *Computer Vision, Graphics and Image Processing*, 29:377–393, 1979. 836, 837

Model-Based Fatique Fractographs Texture Analysis

Michal Haindl[1] and Hynek Lauschmann[2]

[1] Institute of Information Theory and Automation, Academy of Sciences CR
Prague, CZ182 08, Czech Republic
haindl@utia.cas.cz
[2] Faculty of Nuclear Science and Physical Engineering, Czech Technical University
Prague, CZ120 00, Czech Republic
hynek.lauschmannl@fjfi.cvut.cz

Abstract. A novel model-based approach for estimation of the velocity of crack growth from microfractographical images is proposed. These images are represented by a Gaussian Markov random field model and the crack growth rate is modelled by a linear regression model in the Gaussian-Markov parameter space. The method is numerically very efficient because both crack growth rate model parameters as well as the underlying random field model parameters are estimated using fast analytical estimators.

1 Introduction

The quantitative microfractography of fatigue failures is concerned mainly with the investigation of the history of a fatigue crack growth process. Specimens of the material are loaded under service conditions and the crack growth process is recorded. Fracture surfaces images produced by a scanning electron microscope (SEM) are studied to relate image morphological information of the crack surface with the macroscopic crack growth rate (CGR). The crack growth process is reconstituted using integration of CGR along the crack growth direction.

Traditional fractographical methods are based on strictly defined fractographic features measurable in the morphology of a fracture surface. In the case of fatigue analysis, these features are striations [12,13], i.e., fine parallel grooves in the fracture surface. However such methods cannot be used when striations are partially occluded, typically due to corrosion. For such cases, a family of methods is being developed called textural fractography [14]-[22]. The proposed method estimates CGR from textural features derived from a Markovian underlying model.

For the application of the textural method, especially suitable is the mezoscopic dimensional area with SEM magnifications between macro- and microfractography (about $30-500\times$). These magnifications were traditionally seldom used in the past due to the absence of measurable objects in corresponding images (for example see Fig.1). Setting the magnification is limited by several conditions related to individual images, to the whole set of images and to image

T. Caelli et al. (Eds.): SSPR&SPR 2002, LNCS 2396, pp. 842–849, 2002.
© Springer-Verlag Berlin Heidelberg 2002

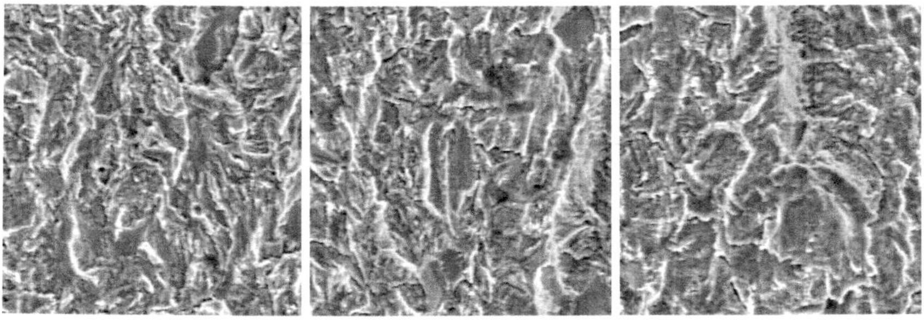

Fig. 1. Cuttings (256 × 256) from small, medium and great CGR

discretization. Images require preprocessing to eliminate lighting variations to get homogeneous image set for subsequent textural analysis. Fractographic information is extracted in the form of integral parameters of the whole image. Such parameters might be different statistical (e.g., correlations, statistical moments, etc.) or textural features.

In the presented method, the fatique image is assumed to be described by a Markov random field (MRF) fitted to this image. Although the MRF models generally suffer with time consuming iterative methods for both parameter estimation as well as for the synthesis, the Gaussian Markov random field (GMRF) model used in this paper belongs to few exceptional Markovian models which avoid the time consuming Markov chain Monte Carlo simulations so typical for the rest of Markovian models family.

Modelling monospectral still images require two dimensional models. Among such possible models the Gaussian Markov random fields are appropriate for fractographical texture modelling not only because they do not suffer with some problems (e.g., sacrificing considerable amount of image information, nonlinear parameter estimation, etc.) of alternative options (see [3,4,5,6,10,11] for details) but they are also easy to synthesize and still flexible enough to imitate a large set of fractographical textures. While the random field based models quite successfully represent high frequencies present in natural fatique textures low frequencies are much more difficult for them. However, the model does not need to generate a realistic fatique texture. For the crack velocity estimation it is sufficient to produce discriminative features.

2 Fatique Images Model

Single monospectral SEM images are assumed to be independently modelled by their dedicated Gaussian Markov random field models (GMRF) as follows. The Markov random field (MRF) is a family of random variables with a joint probability density on the set of all possible realisations Y indexed (in our application) on a finite two-dimensional rectangular ($N \times M$) lattice I, subject to following conditions:

$$p(Y) > 0 \ , \tag{1}$$

and

$$p(Y_r \,|\, Y_s \ \forall s \in I \setminus \{r\}) = p(Y_r \,|\, Y_s \ \forall s \in I_r) \ , \tag{2}$$

where $r = \{r_1, r_2\}$ is the multiindex with the row and column indices, respectively. $I_r \subset I$ is a 2D symmetric contextual support set of the monospectral random field.

If the local conditional density of the MRF model (3) is Gaussian, we obtain the continuous Gaussian Markov random field model (GMRF):

$$p(Y_r \,|\, Y_s \ \forall s \in I_r) = (2\pi\sigma^2)^{-\frac{1}{2}} \exp\left\{-\frac{1}{2\sigma^2}(Y_r - \tilde{\mu}_r)^2\right\} \tag{3}$$

where the mean value is

$$\begin{aligned} E\{Y_r \,|\, Y_s \ \forall s \in I \setminus \{r\}\} &= \tilde{\mu}_r \\ &= \mu_r + \sum_{s \in I_r} a_s (Y_{r-s} - \mu_{r-s}) \end{aligned} \tag{4}$$

and $\sigma, a_s \ \ \forall s \in I_r$ are unknown parameters. The 2D GMRF model can be also expressed [3] as a stationary non-causal correlated noise driven 2D autoregressive process:

$$\tilde{Y}_r = \sum_{s \in I_r} a_s \tilde{Y}_{r-s} + e_r \tag{5}$$

where $\tilde{Y}_r = Y_r - \mu_r$ are centered variables, the noise e_r are random variables with zero mean

$$E\{e_r\} = 0 \ .$$

The e_r noise variables are mutually correlated

$$\begin{aligned} R_e &= E\{e_r e_{r-s}\} \\ &= \begin{cases} \sigma^2 & \text{if } s = (0,0), \\ -\sigma^2 a_s & \text{if } s \ \in I_r, \\ 0 & \text{otherwise.} \end{cases} \end{aligned} \tag{6}$$

Correlation functions have the symmetry property

$$E\{e_r e_{r+s}\} = E\{e_r \, e_{r-s}\}$$

hence the neighbourhood support set and their associated coefficients have to be symmetric, i.e.,

$$s \in I_r \ \Rightarrow \ -s \in I_r$$

and $a_s = a_{-s}$.

2.1 Parameter Estimation

The selection of an appropriate GMRF model support is important to obtain good results in modelling of a given random field. If the contextual neighbourhood is too small it can not capture all details of the random field. Inclusion of the unnecessary neighbours on the other hand add to the computational burden and can potentially degrade the performance of the model as an additional source of noise. We use the hierarchical neighbourhood system I_r, e.g., the first-order neighbourhood is $I_r^1 = \{r-(0,1), r+(0,1), r-(1,0), r+(1,0)\}$,

$$I_r^1 = \begin{matrix} & \otimes & \\ \otimes & r & \otimes \\ & \otimes & \end{matrix} \qquad\qquad I_r^2 = \begin{matrix} \otimes & \otimes & \otimes \\ \otimes & r & \otimes \\ \otimes & \otimes & \otimes \end{matrix} ,$$

etc. An optimal neighbourhood is detected using the correlation method [7] favouring neighbours locations corresponding to large correlations over those with small correlations.

Parameter estimation of a MRF model is complicated by the difficulty associated with computing the normalization constant. Fortunately the GMRF model is an exception where the normalization constant is easy to obtain. However either Bayesian or ML estimate requires iterative minimization of a nonlinear function. Therefore we use the pseudo-likelihood estimator which is computationally simple although not efficient. The pseudo-likelihood estimate for a_s parameters has the form

$$\begin{aligned} \hat{\gamma}^T &= [\hat{a}_s \quad \forall s \in I_r]^T \\ &= \left[\sum_{r \in I} X_r X_r^T\right]^{-1} \sum_{r \in I} X_r \tilde{Y}_r \end{aligned} \tag{7}$$

where

$$X_r = [\tilde{Y}_{r-s} \quad \forall s \in I_r]^T \tag{8}$$

and

$$\hat{\sigma}^2 = \frac{1}{MN} \sum_{r=1}^{MN} (\tilde{Y}_r - \hat{\gamma} X_r)^2 . \tag{9}$$

Alternatively this estimator can be computed recursively [8,9].

2.2 Crack Growth Rate Model

We assume that the crack growth rate $v(i)$ is linearly dependent on GMRF parameters describing corresponding fatique images, i.e.,

$$v(i) = \sum_{j=1}^{\nu} b_j a_{j,i} + \epsilon_i = \theta \gamma_i^T + \epsilon_i , \qquad i = 1, \ldots, n \tag{10}$$

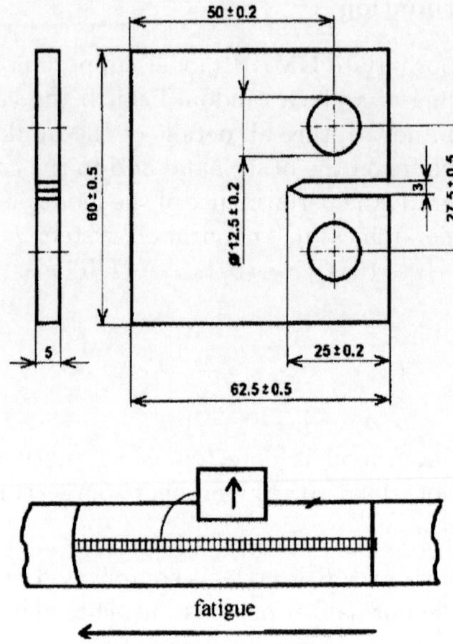

Fig. 2. The fatique test specimen and the location of images in the fatique crack surface

where b_s are unknown parameters, $\theta = [b_1, \ldots, b_\nu]$,
$\nu = \, card\{I_r\} + 1$,
n is the number of fatique images and a_s are RF pseudo-likelihood estimates (7),(9). The growth rate is assumed to have independent Gaussian measurement error ϵ_i with standard deviation $\eta_j \;\; j = 1, \ldots, \nu$. We can assume an overestimated set of equations, i.e., $n \gg \nu$ hence the b_s parameters can be for example estimated using the least square estimator

$$\hat{\theta}^T = (\Gamma^T \Gamma)^{-1} \Gamma^T V \tag{11}$$

where

$$V = [v(1), \ldots, v(n)]^T$$

and

$$\Gamma = [\gamma_1^T, \ldots, \gamma_n^T]^T$$

is a $n \times \nu$ design matrix. Finally the velocity estimator is

$$\hat{v}(i) = \hat{\theta}\, \gamma_i^T \; . \tag{12}$$

The alternative option is a Bayesian estimator for both - unknown parameters from θ as well as for the optimal model selection (i.e., selection of an optimal subset of the variables $a_{j,i}$).

3 Results

The method was applied on data from four fatigue experiments with specimens from stainless steel AISI 304L used in nuclear power plants. The specimen type was CT (Fig.2) with the initial notch length 12.5 mm. Constant cycle loading with parameters $\triangle F = 3400$ N, $R = 0.3$, $f = 1$ Hz occured in water at 20° C. The crack length was measured by COD. Fatigue crack surfaces were documented using SEM with magnification 200×. The sequence of images was located in the middle of the crack surface (Fig.2) and the images were distanced by 0.4 mm. The direction of the crack growth in images is bottom-up. The real area of one image is about 0.6 × 0.45 mm (the images overlap by 0.05 mm). The whole experimental set contains 164 images. Fig.1 shows examples of typical textures - cuttings 256 × 256 pixels from normalized images (size 1200 × 1600 pixels).

The estimation quality was evaluated using the mean absolute error:

$$\zeta = \frac{1}{n}\sum_{i=1}^{n} |v(i) - \hat{v}(i)|$$

and the overall velocity estimation error

$$\varsigma = \frac{100\zeta}{\bar{v}}$$

where $\bar{v}$ is the average velocity in the measurements set. These textures were modelled using the fifth order GMRF model. Although all modelled textures are non stationary and thus violate the GMRF model assumption, the crack rate estimates are fairly accurate. The rate estimates can be further improved if we select a subset of variables γ_i for the model for example by eliminating variables with low correlation with velocity.

4 Conclusions

Our test results of the algorithm on stainless steel fatigue images are encouraging. Some estimated crack rates match true velocities within measurement accuracy. Overall velocity estimation error was $\varsigma = 30\%$ but further improvement is possible if we increase the GMRF model order or introduce a multiresolution MRF model. The proposed method allows quantitative estimation of the crack growth rate while it has still moderate computation complexity. The method does not need any time-consuming numerical optimization like for example some Markov chain Monte Carlo method.

Acknowledgements

This research was supported by the Grant Agency of the Czech Republic under Grants 102/00/0030 and 106/00/1715.

References

1. Bennett, J., Khotanzad, A.: Multispectral random field models for synthesis and analysis of color images. IEEE Trans. on Pattern Analysis and Machine Intelligence **20** (1998) 327–332
2. Bennett, J., Khotanzad, A.: Maximum likelihood estimation methods for multispectral random field image models. IEEE Trans. on Pattern Analysis and Machine Intelligence **21** (1999) 537–543
3. Besag, J.: Spatial interaction and the statistical analysis of lattice systems. Journal of the Royal Statistical Society, Series B **36** (1974) 192–236 843, 844
4. Gagalowicz, A., Ma, S., Tournier-Laserve, C.: Efficient models for color textures. In: Proceedings of Int. Conf. Pattern Recognition, IEEE, Paris, (1986) 412–414 843
5. Haindl, M.: Texture synthesis. CWI Quarterly **4** (1991) 305–331 843
6. Haindl, M.: Texture modelling. In: Proceedings of the World Multiconference on Systemics, Cybernetics and Informatics Int. Inst. of Informatics and Systemics, Orlando, (2000) 634–639 843
7. Haindl, M., Havlíček, V.: Multiresolution colour texture synthesis. In: Proceedings of the 7th International Workshop on Robotics in Alpe-A dria-Danube Region, ASCO Art, Bratislava, (1998) 297–302 845
8. Haindl, M.: Texture Segmentation Using Recursive Markov Random Field Parameter Estimation. In: Proceedings of the 11th Scandinavian Conference on Image Analysis, Pattern Recognition Society of Denmark, Lyngby, (1999) 771-776 845
9. Haindl, M.: Recursive Square-Root Filters. In: Proceedings of the 15th IAPR Int. Conf. on Pattern Recognition II, IEEE Press, Los Alamitos, (2000) 1018-1021 845
10. Kashyap, R. L.: Analysis and Synthesis of Image Patterns by Spatial Interaction Models, In: Progress in Pattern Recognition 1, Elsevier, North-Holland, (1981) 43–50 843
11. Kashyap, R. L., Eom, K.: Estimation in long-memory time series model. J. of Time Series Anal. **9** (1988) 35-41 843
12. Lauschmann, H.: I Computer Aided Fractography: The automatical evaluation of striation parameters. Engineering Mechanics **5** (1998) 377-380 842
13. Lauschmann H.: Textural fractography: estimation of the mean striation spacing and direction. In: Int. Conf. on Stereology and Image Analysis in Materials Science, Polish Society for Stereology, Cracow, (2000) 241-246 842
14. Lauschmann H.: Computer aided fractography. In: Proceedings of international conference Fractography 97, Institute of Materials Research of the Slovak Academy of Sciences, KoÜice, (1997) 181-188 842
15. Lauschmann H., Benes V.: Spatial statistics in material research. In: Industrial statistics, Physica-Verlag, Hiedelberg, (1997) 285-293
16. Cejka V., Benes V.: Computer Aided Fractography: Methods for evaluation of image anisotropy. In: Proceedings Int. Conf. on Stereology, Spatial Statistics and Stochastic Geometry, Union of Czech Mathematicians and Physicists, Prague, (1999) 89-94
17. Lauschmann H.: Computer Aided Fractography: The spectral analysis of fatigue crack images. In: Int. Conf. on Stereology, Spatial Statistics and Stochastic Geometry, Union of Czech Mathematicians and Physicists, Prague, (1999) 171-176
18. Lauschmann H.: Textural analysis of fatigue crack surfaces - Image pre-processing. Acta Polytechnica **40** (2000) 123-129

19. Lauschmann H., Adamek J., Nedbal I. Textural fractography: Spectral analysis of images of fatigue crack surfaces. In: Fractography 2000, Institute of Materials Research of the Slovak Academy of Sciences, KoÜice (2000) 313-320
20. Lauschmann H.: A database-oriented analysis of a fibre process in fractography. Image Analysis and Stereology (Suppl.1) **20** (2001) 379-385
21. Lauschmann H., Racek O.: Textural fractography: Application of Gibbs random fields. In: Proc. 3rd Int. Conf. Materials Structure & Micromechanics of Fracture, University of Technology, Brno, (2001)
22. Lauschmann H., Tùma M., Racek O., Nedbal I.: Textural fractography. Image Analysis and Stereology (Suppl.1) **20** (2001) 842

Hierarchical Multiscale Modeling of Wavelet-Based Correlations

Zohreh Azimifar, Paul Fieguth, and Ed Jernigan*

Department of Systems Design Engineering, University of Waterloo
Waterloo, Ontario, Canada, N2L-3G1

Abstract. This paper presents a multiscale-based analysis of the statistical dependencies between the wavelet coefficients of random fields. In particular, in contrast to common decorrelated-coefficient models, we find that the correlation between wavelet scales can be surprisingly substantial, even across several scales. In this paper we investigate eight possible choices of statistical-interaction models, from trivial models to wavelet-based hierarchical Markov stochastic processes. Finally, the importance of our statistical approach is examined in the context of Bayesian estimation.

1 Introduction

This paper presents a hierarchical multiscale (MS) model to describe the statistical dependencies between the wavelet coefficients as a first-order Markov process. The model is premised on the fact that, regardless of their spatial locations, wavelet coefficients are highly correlated across scales, even those separated by several scales. The virtue of this model is its ability to capture coefficients correlations by concentrating on a very sparse statistical structure. Furthermore, the within-subband coefficients in MS model framework exhibit a clear a Markovian nature.

Our motivation is model-based statistical image processing. That is, we are interested in the statistical manipulation of images, which requires some probabilistic description of the underlying image characteristics. The image pixel interactions in the spatial domain lead to extremely complicated (in particular, highly correlated) statistical structures, which are computationally inconvenient to be used in estimation algorithms. In order to simplify the raw statistics of pixel values, a spatial transformation is considered. The transform is chosen to simplify or nearly decorrelate, as much as possible, the starting statistics, analogous to the preconditioning of complicated linear system problems. The popularity of the wavelet transform (WT) stems from its effectiveness in this task: many operations, such as interpolation, estimation, compression, and denoising are simplified in the wavelet domain, because of its energy compaction and decorrelative properties [1,2].

* The support of the Natural Science & Engineering Research Council of Canada and Communications & Information Technology Ontario are acknowledged.

T. Caelli et al. (Eds.): SSPR&SPR 2002, LNCS 2396, pp. 850–859, 2002.
© Springer-Verlag Berlin Heidelberg 2002

A conspicuously common assumption is that the WT is a perfect whitener, such that all of the wavelet coefficients are independent, and ideally Gaussian. There is, however, a growing recognition that neither of these assumptions are accurate, nor even adequate for many image processing needs. Indeed, significant dependencies still exist between wavelet coefficients. There have been several recent efforts to study the wavelet statistics; mostly marginal models. Each statistical wavelet model focuses on a certain type of dependencies, in which a relatively simple and tractable model is considered. We classify them into the following two categories:

1. Marginal Models:
   (a) Non-Gaussian, i.e., heavy tail distribution [3],
   (b) Mixture of Gaussians [3],
   (c) Generalized Gaussian distribution [2],
   (d) Bessel functions [4].
2. Joint Models:
   Hidden Markov tree models [1].

In virtually all marginal models currently being used in wavelet shrinkage [2], the coefficients are treated individually and are modelled as independent, i.e., only the diagonal elements of wavelet based covariance matrix are considered. This approach, however, is not optimal in a sense that the WT is not a perfect whitening process.

The latter approach, however, examines the joint statistics of coefficients [5]. Normally an assumption is present that the correlation between coefficients does not exceed the parent-child dependencies, e.g. given the state of its parent, a child is decoupled from the entire wavelet tree [1,6].

It is difficult to study both aspects simultaneously: that is, the development of non-Gaussian joint models with non-trivial neighborhood. The study of independent non-Gaussian models has been thorough; the complementary study, the development of Gaussian joint models, is the focus of this paper. The goal, of course, is the ultimate merging of the two fields. However for the purpose of this paper, we are willing to limit ourselves to simplifying marginal assumptions (Gaussianity) which we know to be incorrect, but which allow us to undertake a correspondingly more sophisticated study of joint models.

The main theme of this paper is, then, to concentrate on studying within and across scale statistical dependencies of the wavelet coefficients for a variety of wavelet basis functions and random fields. These correlations are modelled: from complete independent assumption to full dependency between the wavelet coefficients over the entire resolutions. Since correlations are present both within and across scales, we are interested to model them in a wavelet-based MS framework. Finally, the effectiveness of our statistical-based approach is tested through numerical experiments by exploiting Bayesian estimation technique and we show that adding significant dependencies to the wavelet prior model causes dramatic RMSE reductions.

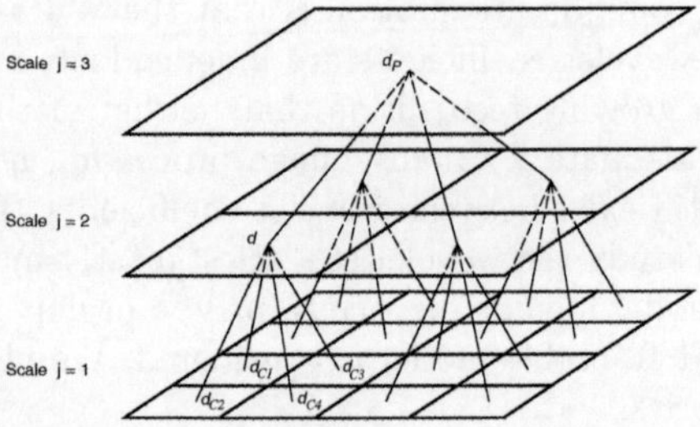

Fig. 1. Illustration of a typical coefficient d along with its parent and children within one wavelet tree subband

2 Discrete Wavelet Transform

The WT of an image f is a process in which the low and high frequency components of f are represented by separate sets of coefficients, namely the approximation $\{a_J\}$ and the detail $\{d_j\}, 1 \leq j \leq J$, with J denoting the coarsest resolution. If, as usual, we define the linear operators H_j and L_j as high- and low-pass filters respectively, then clearly the coefficient vectors may be recursively computed in scale j

$$\begin{aligned} a_j &= L_{j-1}L_{j-1}a_{j-1}, & d_j^h &= H_{j-1}L_{j-1}a_{j-1}, \\ d_j^v &= L_{j-1}H_{j-1}a_{j-1}, & d_j^d &= H_{j-1}H_{j-1}a_{j-1} \end{aligned} \tag{1}$$

with $\{d_j^h, d_j^v, d_j^d\}$ denote the horizontal, vertical, and diagonal subbands of the wavelet decomposition at scale j, respectively. The maximum decomposition level for a discrete image with size $n = N \times N$, would be $J = log_2 N$, with $n/4^j$ detail coefficients in every subband at scale j.

Figure 1 illustrates a natural tree of wavelet subbands. Each wavelet coefficient d is shown as a node with d_p as its parent and $\{d_{ci}\}$ $1 \leq i \leq 4$, as the set of its four children, which represent information about this node at the next finer scale. As the scale j decreases, the children add finer and finer details into the spatial regions occupied by their ancestors [1].

2.1 Basic Notations of Wavelet Image Modeling

In order to perform a precise assessment of correlation between the wavelet coefficients of the finest-scale image $f \sim (0, \Sigma_f)$, we consider a variety of prior models based on Gaussian Markov random field (GMRF) covariance structures. The chosen priors, shown in Figure 2 are the tree-bark and thin-plate models. They are spatially stationary, an assumption for convenience only and is not fundamental to our analysis.

The selected covariance structure Σ_f is transformed into the wavelet domain by computing the 2-D wavelet transform W, containing all translated and dilated

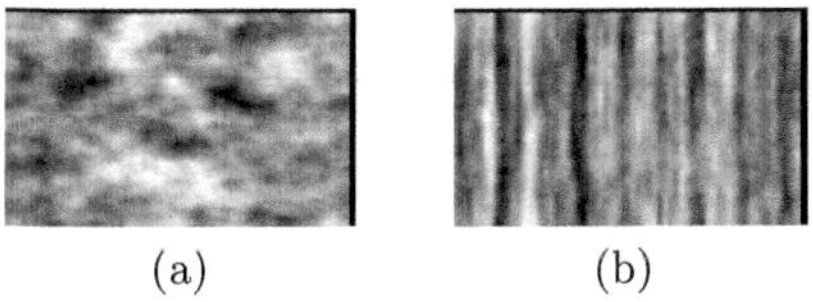

Fig. 2. Two GMRF models used in our simulations: (a) Thin-plate, (b) Tree-bark texture

versions of the selected wavelet basis functions:

$$\Sigma_{Wf} = W \Sigma_f W^T \qquad (2)$$

where we limit our attention to the set of Daubechies basis functions. As more regularity is added to the wavelet function, the within scale decoupling effects increases. Nevertheless, the qualitative structures are similar, and the across-scale correlations are no less significant.

Although in actual data processing we use the covariance matrix, for convenience in understanding the results, the covariance values are normalized, so that the inter-coefficient relationships are measured as correlation coefficients

$$\rho = \frac{E[(d_i - \mu_{d_i})(d_j - \mu_{d_j})]}{\sigma_{d_i}\, \sigma_{d_j}}, \qquad -1 \le \rho \le 1 \qquad (3)$$

where d_i and d_j are two typical wavelet coefficients with mean and standard deviation μ_{d_i}, σ_{d_i} and μ_{d_j}, σ_{d_j}, respectively.

In [7] we defined a recursive method to calculate within and across scale covariances for 1-D signals from the covariance Σ_{a_j,a_j} at the finest scale $j = 1$:

$$\Sigma_{d_{j+1},d_{j+1}} = H_j \Sigma_{a_j,a_j} H_j^T, \qquad \Sigma_{a_{j+1},d_{j+1}} = L_j \Sigma_{a_j,a_j} H_j^T, \qquad etc \qquad (4)$$

Having this tool one can easily assess the extent of *correlation* between the coefficients at the same scale or across different resolutions.

Figure 3 illustrates the correlation structure of the 2-D wavelet coefficients of a 4-level wavelet decomposition. Due to dramatic increase in covariance matrix size, the empirical results are limited to considering the correlation structure of 32×32 images. The main diagonal blocks show the autocorrelation of coefficients located at the same scale and orientation. Due to the column-wise 2-D to 1-D data stacking, large magnitude auto-correlations of the vertical coefficients (labeled as v) tend to concentrate near the main diagonal, whereas those of the horizontal coefficients (h) are distributed on the diagonals 32 pixels apart. The off-diagonal blocks contain those cross-correlations of across orientations or across scales.

It is clear that the within-scale correlations tend to decay very quickly, consistent with the understanding that the WT is decoupling the original signal, while the dependencies across different resolutions remain surprisingly strong,

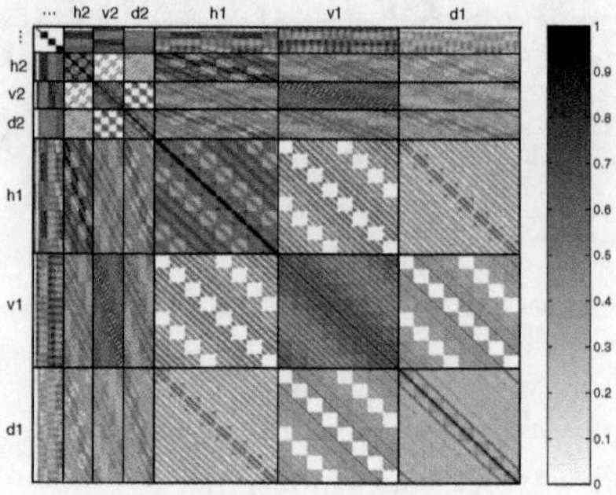

Fig. 3. Scaled values of four-level correlation structure of a thin-plate model decomposed by the Daubechies "db1" wavelet. The main diagonal blocks show autocorrelation of coefficients located at the same scale and orientation, whereas the off-diagonal blocks illustrate cross-correlations across orientations or scales

even for coefficients located several scales apart. This result confirms that although the wavelet coefficients are expected to be decorrelated, there exist cases in which the correlation can be quite significant.

2.2 Numerical Experiments

It is generally infeasible to directly utilize the huge covariance matrix Σ_{Wf} in an estimation process, due to space and time complexities. Our goal is to study the properties of Σ_{Wf} in order to deduce a simple, but still accurate, representation of the underlying correlation model; that is, to construct a new sparse covariance matrix, which contains the most significant information from the prior model. Of course, the study of large covariance matrix is for diagnostic and research purpose; ultimately any practical estimation algorithm will be based on some implicit sparse model of the statistics.

In our experiments the wavelet coefficients are treated in various ways, from complete independence to full dependency among all coefficients over the entire wavelet tree. As is shown in Table 1, eight different cases of adding more features to the new covariance matrix are considered. For each case, except the diagonal case, at least one of the three important neighborhood correlation factors, (*intra-orientation*, *intra-scale*, *and inter-scale*) is considered.

Figure 4 visualizes all eight structures obtained from the original correlation matrix. Note that the standard wavelet-based algorithms, in which the coefficients are treated as statistically independent, only consider diagonal entries of the covariance matrix, shown in Figure 4(a). These structures indicate that adding intra-scale correlations increases the structure's density (Figure 4(f)) much more than the inter-scale dependencies (Figure 4(g)). As is evident a large portion of intra-scale correlation values are very close to zero, which says almost nothing about the correlation structure. This fact suggests devising a hierarchical correlation model which keeps its across-scale strength up to several scales,

Table 1. Eight different ways to obtain a new wavelet-based covariance structure which contains a combination of three important neighborhood correlation factors, namely intra-orientation, intra-scale, and inter-scale

Notation	intra orientation	inter orientation	inter scale
diagonal	0	0	0
interorient	0	1	0
interscale	0	0	1
interorient-interscale	0	1	1
inorient	1	0	0
inorient-interorient	1	1	0
inorient-interscale	1	0	1
full	1	1	1

while reducing the within-scale neighborhood to the very close spatial neighbors, i.e., 3×3 spatially located coefficients.

3 Gaussian Multiscale Modeling

As discussed in Section 2.2, the numerical simulations with covariance structure have revealed the importance of taking into consideration a small within-scale correlation range along with a large extent of across-scale dependencies. In order to meet this requirement, there are two alternatives to consider:

1. Imposing models which describe the long range statistical dependencies, such as the full covariance matrix. Such models, however, lead to estimation algorithms that are highly complex and difficult to implement.
2. Proposing a statistical model which tends to approximate the structural correlations over the entire wavelet tree. The advantage of this approach is the existence of estimation techniques which are fast and very easy to implement [8].

Therefore, first-order MS modeling is used to devise an approximation model of the wavelet coefficient correlations. The MS method [8] models each node on the tree as a stochastic process $X(d)$ and recursively describes the interrelation of parent and child as:

$$X(d) = A_d X(d_p) + B_d \nu_d \tag{5}$$

As seen in Figure 1, d represents each node on the tree with its parent denoted as d_p. Here $\nu_d \sim N(0, I)$ is a white noise process and A_d, B_d are parameters to be determined. At the coarsest resolution (root node), the stochastic process $X(d_J)$ is assumed to obey:

$$E[X(d_J)] = 0, \quad E[X(d_J)X^T(d_J)] = \mathbf{P}_J \tag{6}$$

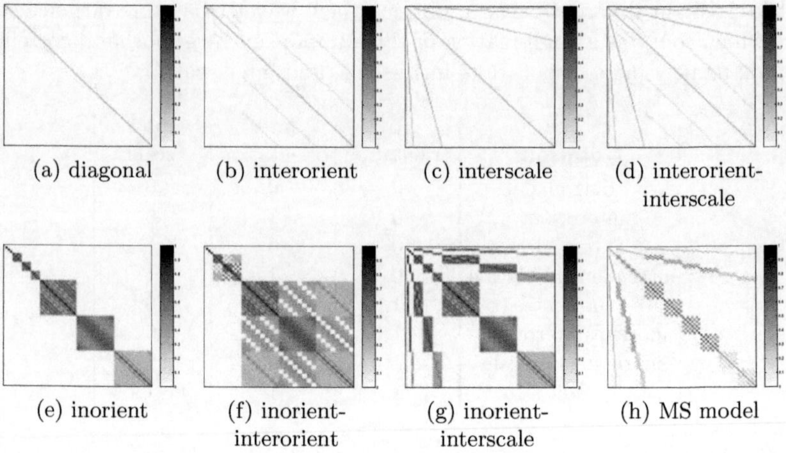

Fig. 4. Various correlation structures achieved from the original covariance matrix Σ_{Wf}. (a-g) Seven structures as in Table 1. As within-scale dependencies are considered (f), the structural density increases dramatically. The across scale correlations (g) add significant information, but have less impact on density increment. (h) The correlation structure presented by a multiscale model

Having the initial conditions defined in (6), one can easily calculate the parameters A and B given in the first-order MS model (5). The cross-correlation of each node d and its parent is computed as [8]:

$$\begin{aligned} & & \mathbf{P}_{d,d_p} &= E[X(d)X^T(d_p)] \\ \mathbf{P}_{d,d_p} = A_d \mathbf{P}_{d_p}^T & \implies & A_d &= (\mathbf{P}_{d_p}^{-1} \mathbf{P}_{d,d_p})^T \\ & & B_d B_d^T &= \mathbf{P}_d - A_d \mathbf{P}_{d_p} A_d^T \end{aligned} \tag{7}$$

Figure 4(h) shows the correlation structure obtained by imposing a first-order stochastic process on the original model (Figure 3). Note that inter-scale correlations, even up to distantly separated scales, are well absorbed by this stochastic model. Also observe that the clear locality of neighborhood dependencies demands within-scale Markovanity. These advantages plus the sparse representation of the MS model makes it an elegant tool to capture wavelet-based hierarchical correlations. The corresponding estimation algorithms can thus be implemented with very low computational effort [8].

The accuracy of the MS model can be increased from first-order, (the state of parent is sufficient for a child to be decoupled from all other nodes) to second-order, (the state of its grand-parent is also needed for a node to be independent from the rest of the tree), etc. Another important issue is the number of coefficients that form a node on the wavelet tree. A particular node may contain only a single wavelet coefficient, or two or more coefficients. To illustrate the tradeoff

Table 2. Summary of computational effort required for the MS model to be imposed on a wavelet binary-tree for a 1-D signal of size N. Each number shows the complexity for a combination of MS order and number of coefficients per node

	MS order 1	2	3	4	...	$(log_2N)^{th}$-order
	1	-	-	-	...	-
Pixels	2	3	-	-	...	-
per node	4	6	7	-	-	-
↓	8	12	14	21	⋮	⋮
	⋮	⋮	⋮	⋮		
	$N/2$	$\Sigma_{i=1}^{2} N/2^i$	$\Sigma_{i=1}^{3} N/2^i$	...	...	$N-1$
Complexity	$O(N)$	...	$\sum_{i=0}^{logN-p}[\sum_{j=i}^{i+p-1} 2^j]^3$		...	$O(N^3)$

between the order (accuracy) of MS model and computational complexity of the estimation process, a wavelet binary-tree for an exponentially distributed 1-D signal of size N is considered. Various ways of MS modeling, from first-order to log_2Nth-order and from single values to vectors of coefficients per node, are examined. Table 2 summarizes time complexity for each MS model. From top left to bottom right the correlation structure becomes more dense while the complexity of even simple estimation algorithms gets harder.

4 Bayesian Wavelet Estimation Approach

A simple estimation algorithm is adopted in this part to evaluate and compare the achieved various statistical structures . To exploit these statistical dependencies we implement a method that estimates the original coefficients by explicit use of wavelet covariance structure. Due to the linearity and orthogonality of the WT, Bayesian Least Square (BLS) method which directly takes into account the covariance structure is:

$$\hat{\mathbf{f}} = \Sigma_f^T(\Sigma_f + R)^{-1}\mathbf{g} \tag{8}$$

The goal is to estimate $\mathbf{f} \sim N(0, \Sigma_f)$ from noisy observation $\mathbf{g}$, where additive noise $\mathbf{v} \sim N(0, R)$ is decorrelated with original data $\mathbf{f}$. Since the BLS is applied in the wavelet domain, it is necessary to substitute (2) into (8). Then the orthogonal wavelet transform of the BLS method is obtained as:

$$\hat{f} = W^{-1}[W\Sigma_f W^T(W\Sigma_f W^T + WRW^T)^{-1}Wg] \tag{9}$$

In order to perform appropriate comparisons and also to emphasize the importance of considering wavelet coefficient correlations – within and across scales,

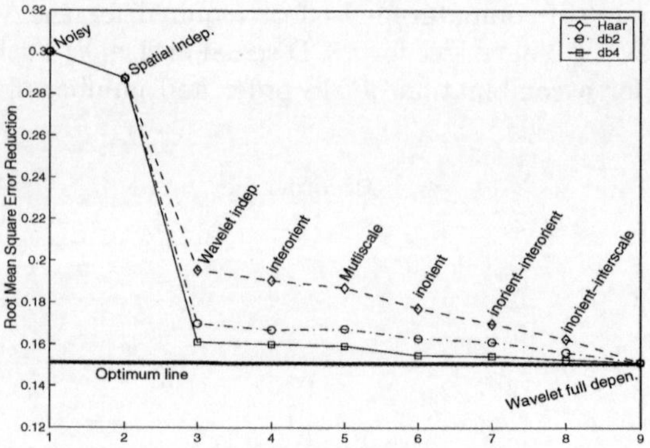

Fig. 5. RMSE measure of noisy observation g and denoised images obtained by BLS method and different covariance structures shown in Figure 4

all structures of Σ_{WF} illustrated in Figure 4 are considered in BLS framework, except those shown in Figure 4(c),(d), due to not being positive definite.

Figure 5 displays the RMSE noise reduction achieved as more correlations are taken into the estimation process. The RMSE performance shows that the more partial correlations (Table 1) are considered, the lower the RMSE. It is extremely important to notice that the rate of RMSE reduction is faster especially if more inter-scale correlations are considered. Larger extent of intra-scale dependencies, however, does not lead to significant RMSE reduction. This fact confirms our earlier discussion of reducing the within-scale neighborhood dependency in our model. As seen in this Figure, the MS-based correlation structure is promising and outperforms the decoupling assumption of the WT, in addition to being a sparse structure of the huge covariance matrix. The MS-based structure with relatively few coefficients vastly reduces the RMSE. Regardless of its well capturing of the across scale dependencies, this model still demands improvements in describing the within scale relations.

5 Conclusions

A multiscale-based analysis of statistical dependencies between the wavelet coefficients was presented. Since correlations are present both within and across scales, wavelet-based hierarchical Markov stochastic processes were proposed and investigated.

The proposed MS model exhibits a sparse locality to the coefficient activities, which results in a dramatic RMSE reduction. The virtue of the model is its ability to capture the most significant statistical information between tree parents and children, however the interrelationship of pixels within a scale is only implicit, and very limited. To complete our development of MS model, we will consider higher local spatial neighboring activities towards a MRF modeling of the wavelet coefficients statistics. The development of MRF methods on hierarchies has some past literature, but is still relatively new and we are willing to extend this work to the proper MRF modeling of statistical dependencies on spatial neighbors.

References

1. Romberg K., Choi H., and Baraniuk R., "Bayesian tree-structured image modeling using wavelet-domain hidden markov models," *IEEE trans. an IP,* vol. 10, pp. 1056-68,2001. 850, 851, 852
2. Portilla J. and Simoncelli E., "Image denoising via adjustment of wavelet coeflicient magnitude correlation," *Proceedings of the 7th ICIP, Cunudu.,* 2000. 850, 851
3. Mumford D. and Huang J., "Statistics of natural images and models," *Proccedings of International Conference an Computer Vision und Pattern Recognition,* 1999. 851
4. Srivastava A., Liu X., and Grenander U., "Analytical models for reduced spectral representations of images," *Proceedings of the 8th ICIP,* 2001. 851
5. E. P. Simoncelli, "Modeling the joint statistics of images in the wavelet domain," *Proceedings of the SPIE 44th Annuul Meeting,* 1999. 851
6. Crouse M. S., Nowak R. D., and Baraniuk R. G., "Wavelet-based statistical signal processing using hidden markov models," *IEEE trans. an Signal Processing,* vol. 46, pp. 886-902,1998. 851
7. Azimifar Z., Fieguth P., and Jemigan E., "Wavelet shrinkage with correlated wavelet coeflicients," *Proceedings of the 8th ICIP,* 2001. 853
8. Chou K., Willsky A., and Benveniste A., "Multiscale recuresive estimation, data fusion, and regularization," *IEEE trans. an Automutic Control,* vol. 39, pp. 468-478, 1994. 855, 856

Author Index

Lecture Notes in Computer Science

For information about Vols. 1–2328
please contact your bookseller or Springer-Verlag

Vol. 2329: P.M.A. Sloot, C.J.K. Tan, J.J. Dongarra, A.G.Hoekstra (Eds.), Computational Science – ICCS 2002. Proceedings, Part I. XLI, 1095 pages. 2002.

Vol. 2330: P.M.A. Sloot, C.J.K. Tan, J.J. Dongarra, A.G. Hoekstra (Eds.), Computational Science – ICCS 2002. Proceedings, Part II. XLI, 1115 pages. 2002.

Vol. 2331: P.M.A. Sloot, C.J.K. Tan, J.J. Dongarra, A.G. Hoekstra (Eds.), Computational Science – ICCS 2002. Proceedings, Part III. XLI, 1227 pages. 2002.

Vol. 2332: L. Knudsen (Ed.), Advances in Cryptology – EUROCRYPT 2002. Proceedings, 2002. XII, 547 pages. 2002.

Vol. 2333: J.-J.Ch. Meyer, M. Tambe (Eds.), Intelligent Agents VIII. Revised Papers, 2001. XI, 461 pages. 2001. (Subseries LNAI).

Vol. 2334: G. Carle, M. Zitterbart (Eds.), Protocols for High Speed Networks. Proceedings, 2002. X, 267 pages. 2002.

Vol. 2335: M. Butler, L. Petre, K. Sere (Eds.), Integrated Formal Methods. Proceedings, 2002. X, 401 pages. 2002.

Vol. 2336: M.-S. Chen, P.S. Yu, B. Liu (Eds.), Advances in Knowledge Discovery and Data Mining. Proceedings, 2002. XIII, 568 pages. 2002. (Subseries LNAI).

Vol. 2337: W.J. Cook, A.S. Schulz (Eds.), Integer Programming and Combinatorial Optimization. Proceedings, 2002. XI, 487 pages. 2002.

Vol. 2338: R. Cohen, B. Spencer (Eds.), Advances in Artificial Intelligence. Proceedings, 2002. XII, 373 pages. 2002. (Subseries LNAI).

Vol. 2340: N. Jonoska, N.C. Seeman (Eds.), DNA Computing. Proceedings, 2001. XI, 392 pages. 2002.

Vol. 2342: I. Horrocks, J. Hendler (Eds.), The Semantic Web – ISCW 2002. Proceedings, 2002. XVI, 476 pages. 2002.

Vol. 2345: E. Gregori, M. Conti, A.T. Campbell, G. Omidyar, M. Zukerman (Eds.), NETWORKING 2002. Proceedings, 2002. XXVI, 1256 pages. 2002.

Vol. 2346: H. Unger, T. Böhme, A. Mikler (Eds.), Innovative Internet Computing Systems. Proceedings, 2002. VIII, 251 pages. 2002.

Vol. 2347: P. De Bra, P. Brusilovsky, R. Conejo (Eds.), Adaptive Hypermedia and Adaptive Web-Based Systems. Proceedings, 2002. XV, 615 pages. 2002.

Vol. 2348: A. Banks Pidduck, J. Mylopoulos, C.C. Woo, M. Tamer Ozsu (Eds.), Advanced Information Systems Engineering. Proceedings, 2002. XIV, 799 pages. 2002.

Vol. 2349: J. Kontio, R. Conradi (Eds.), Software Quality – ECSQ 2002. Proceedings, 2002. XIV, 363 pages. 2002.

Vol. 2350: A. Heyden, G. Sparr, M. Nielsen, P. Johansen (Eds.), Computer Vision – ECCV 2002. Proceedings, Part I. XXVIII, 817 pages. 2002.

Vol. 2351: A. Heyden, G. Sparr, M. Nielsen, P. Johansen (Eds.), Computer Vision – ECCV 2002. Proceedings, Part II. XXVIII, 903 pages. 2002.

Vol. 2352: A. Heyden, G. Sparr, M. Nielsen, P. Johansen (Eds.), Computer Vision – ECCV 2002. Proceedings, Part III. XXVIII, 919 pages. 2002.

Vol. 2353: A. Heyden, G. Sparr, M. Nielsen, P. Johansen (Eds.), Computer Vision – ECCV 2002. Proceedings, Part IV. XXVIII, 841 pages. 2002.

Vol. 2355: M. Matsui (Ed.), Fast Software Encryption. Proceedings, 2001. VIII, 169 pages. 2001.

Vol. 2356: R. Kohavi, B.M. Masand, M. Spiliopoulou, J. Srivastava (Eds.), WEBKDD 2002 – Mining Log Data Across All Customers Touch Points. Proceedings, 2001. XI, 167 pages. 2002. (Subseries LNAI).

Vol. 2358: T. Hendtlass, M. Ali (Eds.), Developments in Applied Artificial Intelligence. Proceedings, 2002 XIII, 833 pages. 2002. (Subseries LNAI).

Vol. 2359: M. Tistarelli, J. Bigun, A.K. Jain (Eds.), Biometric Authentication. Proceedings, 2002. X, 197 pages. 2002.

Vol. 2360: J. Esparza, C. Lakos (Eds.), Application and Theory of Petri Nets 2002. Proceedings, 2002. X, 445 pages. 2002.

Vol. 2361: J. Blieberger, A. Strohmeier (Eds.), Reliable Software Technologies – Ada-Europe 2002. Proceedings, 2002 XIII, 367 pages. 2002.

Vol. 2362: M. Tanabe, P. van den Besselaar, T. Ishida (Eds.), Digital Cities II. Proceedings, 2001. XI, 399 pages. 2002.

Vol. 2363: S.A. Cerri, G. Gouardères, F. Paraguaçu (Eds.), Intelligent Tutoring Systems. Proceedings, 2002. XXVIII, 1016 pages. 2002.

Vol. 2364: F. Roli, J. Kittler (Eds.), Multiple Classifier Systems. Proceedings, 2002. XI, 337 pages. 2002.

Vol. 2365: J. Daemen, V. Rijmen (Eds.), Fast Software Encryption. Proceedings, 2002. XI, 277 pages. 2002.

Vol. 2366: M.-S. Hacid, Z.W. Raś, D.A. Zighed, Y. Kodratoff (Eds.), Foundations of Intelligent Systems. Proceedings, 2002. XII, 614 pages. 2002. (Subseries LNAI).

Vol. 2367: J. Fagerholm, J. Haataja, J. Järvinen, M. Lyly. P. Råback, V. Savolainen (Eds.), Applied Parallel Computing. Proceedings, 2002. XIV, 612 pages. 2002.

Vol. 2368: M. Penttonen, E. Meineche Schmidt (Eds.), Algorithm Theory – SWAT 2002. Proceedings, 2002. XIV, 450 pages. 2002.

Vol. 2369: C. Fieker, D.R. Kohel (Eds.), Algorithmic Number Theory. Proceedings, 2002. IX, 517 pages. 2002.

Vol. 2370: J. Bishop (Ed.), Component Deployment. Proceedings, 2002. XII, 269 pages. 2002.

Vol. 2371: S. Koenig, R.C. Holte (Eds.), Abstraction, Reformulation, and Approximation. Proceedings, 2002. XI, 349 pages. 2002. (Subseries LNAI).

Vol. 2372: A. Pettorossi (Ed.), Logic Based Program Synthesis and Transformation. Proceedings, 2001. VIII, 267 pages. 2002.

Vol. 2373: A. Apostolico, M. Takeda (Eds.), Combinatorial Pattern Matching. Proceedings, 2002. VIII, 289 pages. 2002.

Vol. 2374: B. Magnusson (Ed.), ECOOP 2002 – Object-Oriented Programming. XI, 637 pages. 2002.

Vol. 2375: J. Kivinen, R.H. Sloan (Eds.), Computational Learning Theory. Proceedings, 2002. XI, 397 pages. 2002. (Subseries LNAI).

Vol. 2377: A. Birk, S. Coradeschi, T. Satoshi (Eds.), RoboCup 2001: Robot Soccer World Cup V. XIX, 763 pages. 2002. (Subseries LNAI).

Vol. 2378: S. Tison (Ed.), Rewriting Techniques and Applications. Proceedings, 2002. XI, 387 pages. 2002.

Vol. 2379: G.J. Chastek (Ed.), Software Product Lines. Proceedings, 2002. X, 399 pages. 2002.

Vol. 2380: P. Widmayer, F. Triguero, R. Morales, M. Hennessy, S. Eidenbenz, R. Conejo (Eds.), Automata, Languages and Programming. Proceedings, 2002. XXI, 1069 pages. 2002.

Vol. 2381: U. Egly, C.G. Fermüller (Eds.), Automated Reasoning with Analytic Tableaux and Related Methods. Proceedings, 2002. X, 341 pages. 2002 .(Subseries LNAI).

Vol. 2382: A. Halevy, A. Gal (Eds.), Next Generation Information Technologies and Systems. Proceedings, 2002. VIII, 169 pages. 2002.

Vol. 2383: M.S. Lew, N. Sebe, J.P. Eakins (Eds.), Image and Video Retrieval. Proceedings, 2002. XII, 388 pages. 2002.

Vol. 2384: L. Batten, J. Seberry (Eds.), Information Security and Privacy. Proceedings, 2002. XII, 514 pages. 2002.

Vol. 2385: J. Calmet, B. Benhamou, O. Caprotti, L. Henocque, V. Sorge (Eds.), Artificial Intelligence, Automated Reasoning, and Symbolic Computation. Proceedings, 2002. XI, 343 pages. 2002. (Subseries LNAI).

Vol. 2386: E.A. Boiten, B. Möller (Eds.), Mathematics of Program Construction. Proceedings, 2002. X, 263 pages. 2002.

Vol. 2387: O.H. Ibarra, L. Zhang (Eds.), Computing and Combinatorics. Proceedings, 2002. XIII, 606 pages. 2002.

Vol. 2388: S.-W. Lee, A. Verri (Eds.), Pattern Recognition with Support Vector Machines. Proceedings, 2002. XI, 420 pages. 2002.

Vol. 2389: E. Ranchhod, N.J. Mamede (Eds.), Advances in Natural Language Processing. Proceedings, 2002. XII, 275 pages. 2002. (Subseries LNAI).

Vol. 2391: L.-H. Eriksson, P.A. Lindsay (Eds.), FME 2002: Formal Methods – Getting IT Right. Proceedings, 2002. XI, 625 pages. 2002.

Vol. 2392: A. Voronkov (Ed.), Automated Deduction – CADE-18. Proceedings, 2002. XII, 534 pages. 2002. (Subseries LNAI).

Vol. 2393: U. Priss, D. Corbett, G. Angelova (Eds.), Conceptual Structures: Integration and Interfaces. Proceedings, 2002. XI, 397 pages. 2002. (Subseries LNAI).

Vol. 2396: T. Caelli, A. Amin, R.P.W. Duin, M. Kamel, D. de Ridder (Eds.), Structural, Syntactic, and Statistical Pattern Recognition. Proceedings, 2002. XVI, 863 pages. 2002.

Vol. 2398: K. Miesenberger, J. Klaus, W. Zagler (Eds.), Computers Helping People with Special Needs. Proceedings, 2002. XXII, 794 pages. 2002.

Vol. 2399: H. Hermanns, R. Segala (Eds.), Process Algebra and Probabilistic Methods. Proceedings, 2002. X, 215 pages. 2002.

Vol. 2401: P.J. Stuckey (Ed.), Logic Programming. Proceedings, 2002. XI, 486 pages. 2002.

Vol. 2402: W. Chang (Ed.), Advanced Internet Services and Applications. Proceedings, 2002. XI, 307 pages. 2002.

Vol. 2403: Mark d'Inverno, M. Luck, M. Fisher, C. Preist (Eds.), Foundations and Applications of Multi-Agent Systems. Proceedings, 1996-2000. X, 261 pages. 2002. (Subseries LNAI).

Vol. 2404: E. Brinksma, K.G. Larsen (Eds.), Computer Aided Verification. Proceedings, 2002. XIII, 626 pages. 2002.

Vol. 2405: B. Eaglestone, S. North, A. Poulovassilis (Eds.), Advances in Databases. Proceedings, 2002. XII, 199 pages. 2002.

Vol. 2406: C.A. Peters, M. Braschler, J. Gonzalo, M. Kluck (Eds.), Evaluation of Cross-Language Information Retrieval Systems. Proceedings, 2001. X, 601 pages. 2002.

Vol. 2407: A.C. Kakas, F. Sadri (Eds.), Computational Logic: Logic Programming and Beyond. Part I. XII, 678 pages. 2002. (Subseries LNAI).

Vol. 2408: A.C. Kakas, F. Sadri (Eds.), Computational Logic: Logic Programming and Beyond. Part II. XII, 628 pages. 2002. (Subseries LNAI).

Vol. 2409: D.M. Mount, C. Stein (Eds.), Algorithm Engineering and Experiments. Proceedings, 2002. VIII, 207 pages. 2002.

Vol. 2412: H. Yin, N. Allinson, R. Freeman, J. Keane, S. Hubbard (Eds.), Intelligent Data Engineering and Automated Learning – IDEAL 2002. Proceedings, 2002. XV, 597 pages. 2002.

Vol. 2413: K. Kuwabara, J. Lee (Eds.), Intelligent Agents and Multi-Agent Systems. Proceedings, 2002. X, 221 pages. 2002. (Subseries LNAI).

Vol. 2418: D. Wells, L. Williams (Eds.), Extreme Programming and Agile Methods – XP/Agile Universe 2002. Proceedings, 2002. XII, 292 pages. 2002.

Vol. 2419: X. Meng, J. Su, Y. Wang (Eds.), Advances in Web-Age Information Management. Proceedings, 2002. XV, 446 pages. 2002.

Vol. 2430: T. Elomaa, H. Mannila, H. Toivonen (Eds.), Machine Learning: ECML 2002. Proceedings, 2002. XIII, 532 pages. 2002. (Subseries LNAI).

Vol. 2431: T. Elomaa, H. Mannila, H. Toivonen (Eds.), Principles of Data Mining and Knowledge Discovery. Proceedings, 2002. XIV, 514 pages. 2002. (Subseries LNAI).

GPSR Compliance

The European Union's (EU) General Product Safety Regulation (GPSR) is a set of rules that requires consumer products to be safe and our obligations to ensure this.

If you have any concerns about our products, you can contact us on ProductSafety@springernature.com

In case Publisher is established outside the EU, the EU authorized representative is:

Springer Nature Customer Service Center GmbH
Europaplatz 3
69115 Heidelberg, Germany

Batch number: 10682269

Printed by Printforce, the Netherlands